Methodicum Chimicum Volume 8

Methodicum Chimicum

A Critical Survey of Proven Methods and Their Application in Chemistry, Natural Science, and Medicine

Editor-in-Chief

Friedhelm Korte

Volume Editors

H. Aebi
H. Batzer
E. Baumgartner
K.-H. Buchel
J. Falbe
R. Gompper
M. Goto
U. Hasserodt
H. Machleidt
K. Niedenzu
G. Ohloff
H. Zimmer
F. Zymalkowski

Academic Press · New York · London · San Francisco 1976

Georg Thieme Publishers Stuttgart

Volume 8

Preparation of Transition Metal Derivatives

Edited by

Kurt Niedenzu
Hans Zimmer

Contributions from

D. M. Adams, Leicester/England
C. W. Blewett, Cincinnati, Ohio/USA
K. E. Blick, Prestonsburg, Kentucky/USA
F. Bonati, Milan/Italy
R. J. H. Clark, London/England
G. Davies, Boston, Massachusetts/USA
H. Falk, Vienna/Austria
R. D. Gillard, Cardiff, Wales
B. T. Heaton, Canterbury, Kent/England
M. Hoch, Cincinnati, Ohio/USA
R. D. W. Kemmitt, Leicester/England
R. B. King, Athens, Georgia/USA
L. Kolditz, Berlin/GDR
K. R. Manolov, Plovdiv/Bulgaria
K. S. Mazdiyasni, Wright-Patterson Air Force Base, Ohio/USA
J. A. McCleverty, Sheffield/England
B. J. McCormick, Morgantown, West Virginia/USA
P. R. Mitchell, Basel/Switzerland
F. W. Moore, Ann Arbor, Michigan/USA
G. H. Posner, Baltimore, Maryland/USA
H. Rosenberg, Wright-Patterson Air Force Base, Ohio/USA
E. F. Rothgery, Lexington, Kentucky/USA
K. Schlögl, Vienna/Austria
R. Schmid, Genf/Switzerland
D. G. Tisley, Lafayette, Indiana/USA
G. A. Tsigdinos, Ann Arbor, Michigan/USA
R. A. Walton, Lafayette, Indiana/USA
J. R. Wasson, Lexington Kentucky/USA
H. Zimmer, Cincinnati, Ohio/USA

Academic Press · New York · London · San Francisco 1976

Georg Thieme Publishers Stuttgart

Editors

In this handbook many registered trade marks, trade names, etc. are listed (although specific reference to this fact is not always made in the text), also BIOS and FIAT reports, patents, and methods of manufacture and application. The editors and publishers draw attention to the fact that the legal situation must be carefully checked before commercial use is made of these. Only a selection of industrially produced apparatus and appliances is mentioned. This in no way implies a reflection on the products not included in this volume.

Journal titles are abbreviated chiefly in accordance with the system of Chemical Abstracts.

Printed in Great Britain by William Clowes & Sons, Limited, London, Beccles and Colchester.

Library of Congress Catalog Card Number: 74-21580
ISBN 0-12-460708-x (Academic Press)
ISBN 3-13-480801-3 (Thieme)

Series Preface

The METHODICUM CHIMICUM is a short critical description of chemical methods applied in scientific research and practice. It is particularly aimed at chemists as well as scientists working in associated areas including medicine who make use of chemical methods to solve their 'interrelated' problems.
Considering the present development of science and the necessity for concise and unambiguous information, the series provides a guide to rapid and reliable detection of a suitable method to solve the problem concerned. Thus, particular emphasis is placed on the description of proved procedure whereas a complete and exhaustive compilation of all reported methods and also a detailed description of experimental techniques have been deliberately omitted. Newer methods as well as those which have not yet been reported in review articles are treated more extensively, while conventional methods are dealt with concisely. Biological procedures will be discussed in the analytical volume. The interrelated methods and concepts which are constantly gaining importance will be fully discussed in the third 'Specific Part'.
The METHODICUM CHIMICUM is comprised of three parts. The first 'General Part' consists of Volumes 1, 2 and 3. Volume 1 (Analytical Methods) is concerned with chemical physical, and biological analytical methods including those necessary for the elucidation of structures of compounds.
Volume 2 (Planning of Syntheses) contains a review of fundamentals, principles and models with particular respect to the concepts and applications of theoretical chemistry essential to the practically working scientist.
Volume 3 (Types of Reactions) is designed to illustrate the scope and utility of proved working techniques and syntheses.
The second part (Vols. 4-8), which is particularly devoted to 'Systematic Syntheses', deals with proved methods for syntheses of specific compounds. These procedures are classified according to functional groups linked together in the last step of reaction.
Volume 4 (Syntheses of Skeletons) describes the construction of hydrocarbons and heterocyclic compounds.
Volume 5 describes the formation of C—O-bonds, Volume 6 the formation of C—N-bonds, Volume 7 the syntheses of compounds containing main group elements, and Volume 8 compounds containing transition metal elements.
The third 'Special Part' (Vols. 9-11) is concerned with the chemical aspects connected with the formulation of a question or problem.
Volume 9 deals with nonmetallic synthetic fibers and synthetic materials as well as their additives, Volume 10 with synthetic compounds and Volume 11 with natural products and naturally-occurring compounds.
All Volumes should not contain more than 900 printed pages. They are intended to give the chemist and any person working in fields related to chemistry a suitable answer to his problem. Selected review articles or important original works are cited for the sake of detailed information.
We wish to thank Georg Thieme Verlag, Stuttgart, for making possible the realization of the basic concept of METHODICUM CHIMICUM and for the excellent presentation of the work.
Bonn, September 1974 Friedhelm Korte

Preface to Volume 8

The present volume is a virtually identical English language version of the German edition published in 1973. It selectively presents tested methods for the preparation of derivatives of the transition elements with coverage of the literature through 1970.

Due to the individual characteristics of the elements no uniform style for all chapters has been attempted. However, the arrangement of the periodic table was used in conjunction with particular oxidation states of the central element to serve as a guideline for the presentation of the material. The discussion normally starts with the lowest oxidation state and the hydrogen derivatives to be followed by the halides, chalcogenides, and so on.

Besides those chapters devoted to the preparative chemistry of the individual transition elements, four chapters combine that of significant compound types, i.e., the carbonyls, ferrocenes, other sandwich compounds, and heteropoly ions. Also, the section on optical active cobalt complexes is more general in nature and may benefit the reader in work involving chiral compounds of other metals.

Technical editing including nomenclature and literature references has been the responsibility of the publisher. The editors sincerely regret the extreme delay between submission of the manuscripts and the final completion of the works which was beyond their control or responsibilities.

Kurt Niedenzu
August 1975 Hans Zimmer

Contents

1 Copper

G. H. Posner
Department of Chemistry, Johns Hopkins University,
Baltimore, Maryland 21218, U.S.A.

1.1 Copper: Recovery and Purification

Copper is found in nature in the free state, in sulfides, arsenides, chlorides, and carbonates. It is extracted from the ores by oxidative roasting and smelting, followed by electrodeposition from sulfate solutions.[1,2] Most recently a new continuous flow hydrometallurgical process has been developed to make pure copper from ocean nodules.[3] Copper metal in various forms (*e.g.*, powder, wire, foil, rod) is commercially available. Some of the properties of copper are given in Table 1.[1]

Table 1 Some Properties of Copper

Property (units)	Cu	$Cu^{\oplus}$ or Cu_2O	$Cu^{2\oplus}$ or CuO
Melting point (°C)	1083 ±0·1	—	—
Boiling point (°C)	2595	—	—
Latent heat of fusion (cal/g)	48·9	—	—
Latent heat of vaporization (cal/g)	1150	—	—
Heat of oxidation (cal/g Cu)	—	334	607
Electrochemical equivalent (mg/coulomb)	—	0·6588	0·3294
Electrolytic solution potential *v.* hydrogen (V)	—	−0·470	−0·344
Temperature coefficient of solution potential *v.* hydrogen (mV/°C)	−0·01	—	—

Copper metal is attacked by halogens. It reacts with oxygen and sulfur at high temperatures to form oxides or sulfides, and it dissolves in nitric and sulfuric acids and in ammonia in the presence of oxygen.

Most uses of copper are based on its high (second only to silver) thermal and electrical conductivities. Apparently, copper is to be used now also as the active component of an intrauterine contraceptive.[4]

1.2 Copper(0) Compounds

Very few derivatives of copper(0) are known. Reduction of $K_2[Cu(C{\equiv}CPh)_3]$ with lithium in liquid ammonia followed by precipitation with barium thiocyanate produces the *phenylethynyl copper(0) complex*, $Ba_3[Cu(C{\equiv}CPh)_3]_2$,[5] and zerovalent *dipotassium phthalocyanine cuprate(0)* is formed in a similar way from potassium in liquid ammonia reduction of a copper(II) phthalocyanine.[6] Electrochemical reduction of a copper(II) mixed nitrogen and sulfur complex has been reported to give a complex in which copper is formally copper(0).[7]

[1] *A. Butts*, Copper, The Science and Technology of the Metal, Its Alloys and Compounds. Hafner Publishing Co., New York, 1954.

[2] *W. C. Cooper, I. M. Kolthoff*, in Treatise on Analytical Chemistry, Vol. III/2, pp. 1–42. Wiley Interscience, New York, 1961.

[3] Chem. Eng. News, May 10, 1971, p. 56.

[4] Chem. Eng. News, March 29, 1971, p. 30 and April 12, 1971, p. 29 and March 11, 1974, p. 20.

[5] *R. Nast, P. G. Kirst, G. Beck, J. Gremm*, Chem. Ber. *96*, 3302 (1963).

1.3 Copper(I) Compounds

1.3.1 Hydrides and Hydridoborates

Claims have been made to the preparation of *copper(I) hydride* by reduction of copper(II) salts with hypophosphorous acid[8] and by treatment of copper(I) iodide with lithium tetrahydridoaluminate.[9] Copper(I) hydride (deuteride) of high purity, however, is best prepared by treating copper(I) bromide with diisobutylaluminum hydride (deuteride) in anhydrous pyridine at −50°.[10]

$$CuBr + i\text{-}Bu_2AlH \longrightarrow CuH + i\text{-}Bu_2AlBr \qquad (1)$$

Phosphine and phosphite complexes have been prepared by adding the respective ligands to copper(I) hydride[10,11] and in one case by treating a phosphinecopper(I) halide with sodium trimethoxyhydridoborate which yielded an isolable complex.[12]

Copper(I) hydridoborate complexes are most often prepared by treating a copper salt with the appropriate hydridoborate anion (e.g., $BH_4^{\ominus}$, $B_3H_8^{\ominus}$, $B_9H_{14}^{\ominus}$, $B_{10}H_{13}^{\ominus}$, $B_{10}H_{15}^{\ominus}$, and $B_{11}H_{14}^{\ominus}$).[13,14] Many such complexes have been shown to contain hydrogen bridges between copper and boron.[15]

[6] *G. W. Watt, J. W. Dawes*, J. Inorg. Nucl. Chem. *14*, 32 (1960).

[7] *R. H. Holm, A. L. Balch, A. Davison, A. H. Maki, T. E. Berry*, J. Amer. Chem. Soc. *89*, 2866 (1967).

[8] *W. M. Mueller, J. P. Blackledge, G. C. Libowitz*, Metal Hydrides, pp. 546–550. Academic Press, New York and London, 1968.

[9] *E. Wiberg, W. Henle*, Z. Naturforsch. *7b*, 250 (1952); see also *J. A. Dilts, D. F. Shriver*, J. Amer. Chem. Soc. *90*, 5769 (1968) and *91*, 4088 (1969).

[10] *G. M. Whitesides, J. San Filippo, Jr., E. R. Stedronsky, C. P. Casey*, J. Amer. Chem. Soc. *91*, 6542 (1969).

[11] *J. A. Dilts, D. F. Shriver*, J. Amer. Chem. Soc. *91*, 4088 (1969).

[12] *S. A. Bezman, M. R. Churchill, J. A. Osborn, J. Wormald*, J. Amer. Chem. Soc. *93*, 2063 (1971).

[13] *S. J. Lippard, D. A. Ucko*, Inorg. Chem. *7*, 1051 (1968).

[14] *E. L. Muetterties, G. W. Peet, P. A. Wegner, C. W. Alegranti*, Inorg. Chem. *9*, 2447 (1970).

[15] *B. D. James*, J. Chem. Ed. *48*, 176 (1971).

1.3.2 Halides and Complex Halides

Pure copper(I) fluoride has not been prepared.[16] Copper(I) chloride, bromide, and iodide are commercially available.

Copper(I) carbonyl chloride, $Cu(CO)Cl \cdot H_2O$, is formed by adding carbon monoxide to a solution of copper(I) chloride in hydrochloric acid, and *copper(I) hydroxybromide*, $Cu_2(OH)Br$, is formed by reaction of potassium bromide with copper(I) hydroxide.[17]

The many *halocuprate(I) complexes* containing $CuX_2^{\ominus}$, $CuX_3^{2\ominus}$, and $Cu_2X_3^{\ominus}$ groups (where X=Cl, Br, or I) are easily prepared by treating the copper(I) monohalide with the corresponding hydrohalic acid or with an inorganic or organic halide.[16,18,19,20] Recent X-ray crystallography of one dichlorocuprate shows this anion to be linear.[21]

A complex containing the $Cu_5Cl_{17}^{12\ominus}$ anion has been prepared by treating copper(I) chloride with ammonium chloride, titanium chloride, and hexaminechromium(III) chloride.[22]

Complexes formed between copper(I) halides and oxygen, nitrogen, and carbon ligands are discussed in Sections 1.3.3, 1.3.4, and 1.3.5, respectively.

1.3.3 Oxygen and Group IV Compounds

1.3.3.1 Simple Compounds

Copper(I) oxide is most readily prepared by heating copper in technical-grade nitrogen (1% oxygen) or by treating copper(II) acetate in water with hydrazine hydrate.[23] For control of particle size, copper(I) oxide is prepared by electrolysis of a weakly alkaline solution of sodium chloride using copper electrodes.[23] Recently copper(I) oxide has been prepared from a mixture of sodium potassium tartrate, copper(II) sulfate, and hydrazinium sulfate in alkaline solution.[24]

Unstable *copper(I) methoxide* has been reported from the reaction of methylcopper with methanol.

$$CH_3Cu + CH_3OH \longrightarrow CuOCH_3 + CH_4 \qquad (2)$$

Upon exposure of cuprous methoxide to air, *copper(II) dimethoxide* is formed.[25]

Cuprous carboxylates have not been studied extensively. *Copper(I) formate* is formed when copper(I) oxide is dissolved in formic acid, and *copper(I) acetate* is produced by heating copper(II) acetate *in vacuo* and by reducing the copper(II) complex with hydroxylamine sulfate.[26]

Copper(I) sulfide, selenide, or *telluride* is formed by heating stoichiometric quantities of copper and sulfur, selenium, or tellurium in an evacuated, sealed tube.[23]

$$2Cu + S(Se, Te) \longrightarrow Cu_2S(Se, Te) \qquad (3)$$

Copper dichalcogenides are formed at high temperatures and pressures according to Eqns (4) and (5) (Y=S, Se, Te).[27,28]

$$Cu + 2Y \longrightarrow CuY_2 \qquad (4)$$

$$CuS + S \longrightarrow CuS_2 \qquad (5)$$

Likewise, *mixed dichalcogenides* CuSSe and CuSeTe can be obtained from the elements.[28]

Copper(I) mercaptides, CuSR (R=Et, *n*-Bu, Ph, $PhCH_2$), are most reliably prepared from copper(II) oxide and mercaptans.[29]

$$Cu + S + Se \longrightarrow CuSSe \qquad (6)$$

Copper(I) sulfate is best produced from copper turnings and concentrated sulfuric acid at 200°

---

[16] *R. Colton, J. Canterford*, Halides of the First Row Transition Metals, Chap. 9. Wiley Interscience, London, 1969.

[17] *M. C. Sneed, J. L. Maynard, R. C. Brasted*, Comprehensive Inorganic Chemistry, Vol. II, pp. 1–113. Van Nostrand—Reinhold, Princeton, New Jersey, 1954.

[18] *W. E. Hatfield, R. Whyman, R. L. Carlin*, in Transition Metal Chemistry, Vol. V, pp. 43–179. Dekker, New York, 1969.

[19] *J. R. Clifton, J. T. Yoke*, Inorg. Chem. *5*, 1630 (1966).

[20] *H. D. Caughman, R. C. Taylor*, Inorg. Nucl. Chem. Lett. *6*, 623 (1970).

[21] *M. G. Newton, H. D. Caughman, R. C. Taylor*, Chem. Commun. p. 1227 (1970).

[22] *M. Mori*, Bull. Chem. Soc. Japan *33*, 985 (1960).

[23] *O. Glemser, H. Sauer*, in *G. Bauer*, Handbook of Preparative Inorganic Chemistry, 2nd ed., Vol. II, pp. 1003–1027. Academic Press, New York, 1965.

[24] *J. K. Kochi, A. Bemis*, J. Amer. Chem. Soc. *90*, 4049 (1968); *G. Schlessinger*, Inorganic Preparations, p. 11. Chemical Publ. Co., New York, 1962.

[25] *G. Costa, A. Camus, N. Marsich*, J. Inorg. Nucl. Chem. *27*, 281 (1965).

[26] *C. Oldham*, Progr. Inorg. Chem. *10*, 247 (1968).

[27] *R. A. Munson*, Inorg. Chem. *5*, 1296 (1966).

[28] *T. A. Bither, C. T. Prewitt, J. L. Gillson, P. E. Bierstedt, R. B. Flippen, H. S. Young*, Solid State Commun. *4*, 533 (1966); Chem. Abstr. *66*, 23407q (1967).

[29] *R. Adams, W. Riefschneider, M. D. Nair*, Croatica Chem. Acta *29*, 277 (1957); Chem. Abstr. *53*, 16145d (1959); *T. Saegusa, Y. Ito, T. Shimizu*, J. Org. Chem. *35*, 2979 (1970), for preparation of cuprous mercaptide from copper(I) halide and mercaptan.

Table 2 Copper(I) and Group VI Complexes

Donor ligand (L)			
Structure	Name	Complex	Ref.
$R-COO^{\ominus}$	Carboxylate	$Cu(CO)(CF_3CO_2)$	*a*
$H_3C-CO-CH=C(O^{\ominus})-CH_3$	Acetylacetonate	CuL	*b*
$R-S-R^1$	Sulfide	$[CuL_2]I$	*c*
$R-C(S^{\ominus})=C(S^{\ominus})-R^1$	1,2-Dithiolate	$[C_5H_5N-CuL]^{\ominus}$	*d*
$R-C(=S)-C(=S)-R^1$	1,2-Dithiolene	$[CuL_2]I$	*d*
$(NH_2)_2C=S$	Thiourea	$[CuL_2]Cl$	*e*
$R_2N-C(=S)-S^{\ominus}$	Dithiocarbamate	$[CuL]_4$ $(R_3P)CuL$	*f* *g*
$[(CH_3)_2N]_3P=S$	Tris(dimethylamino)phosphine sulfide	$[CuL_4]ClO_4$	*h*
$R_3P=S$	Phosphine sulfide	$[CuL_3]ClO_4$	*i,j*
$R_2P(=S)-P(=S)R_2$	Diphosphine disulfide	$[CuL_2]ClO_4$	*j*
$R-C(=Se)-C(=Se)-R^1$	1,2-Diselenolene	$[CuL_2]X$	*k*

[a] *A. F. Scott, L. L. Wilkening, B. Rubin,* Inorg. Chem. *8,* 2533 (1969).
[b] *R. Nast, R. Mohr, C. Schultze,* Chem. Ber. *96,* 2127 (1963); *H. D. Gafney, R. L. Lintvedt,* J. Amer. Chem. Soc. *93,* 1623 (1971).
[c] *H. O. House, W. F. Fischer, Jr.,* J. Org. Chem. *34,* 3615 (1969).
[d] *J. A. McCleverty,* Progr. Inorg. Chem. *10,* 49 (1968).
[e] *W. A. Spofford, E. L. Amma,* Chem. Commun. p. 405 (1968).
[f] *R. Hesse,* Arkiv. Kemi. *20,* 481 (1963).
[g] *C. Kowala, J. M. Swan,* Aust. J. Chem. *19,* 555 (1966).
[h] *W. E. Slinkard, D. W. Meek,* Inorg. Chem. *8,* 1811 (1969).
[i] *P. G. Eller, P. W. R. Corfield,* Chem. Commun. p. 105 (1971).
[j] *D. W. Meek, P. Nicpon,* J. Amer. Chem. Soc. *87,* 4951 (1965).
[k] *A. Davison, E. T. Shawl,* Inorg. Chem. *9,* 1820 (1970).

or from copper(I) oxide and neutral, anhydrous dimethyl sulfate.[23]

1.3.3.2 Complexes

Most copper(I) coordination compounds are formed directly by admixture of a copper(I) salt and the desired donor ligand in an appropriate solvent (e.g., an alcohol or water). Table 2 lists some recent examples of typical donor ligands which have been used to form copper(I) complexes.

1.3.4 Nitrogen and Group V Compounds

1.3.4.1 Simple Compounds

Copper(I) nitride, Cu_3N, is formed by passing ammonia over anhydrous copper(II) difluoride[30] at 280°. Unstable *copper(I) amides* have been reported from reaction of lithium or magnesium amides with copper(I) chloride.[31]

$$R_2NM + CuCl \longrightarrow [R_2NCu] \xrightarrow{O_2} R_2N{-}NR_2 \qquad (7)$$
$$M = Li, MgBr$$

Heating stoichiometric amounts of metallic copper and red phosphorus in a sealed tube produces

[30] *R. Juza, H. Hahn,* Z. Anorg. Allg. Chem. *239,* 282 (1938); *R. Juza,* Advan. Inorg. Chem. Radiochem. *9,* 94 (1966).
[31] *T. Kauffmann, J. Albrecht, D. Berger, J. Legler,* Angew. Chem. Int. Ed. Engl. *6,* 633 (1967); *T. Kauffmann, G. Beissner, E. Köppelmann, D. Kuhlmann, A. Schott, H. Schrecken, ibid.,* *7,* 131 (1968).

copper(I) phosphide.[32] Reaction of copper(I) phosphide with additional red phosphorus forms *copper diphosphide.*[32]

$$3Cu + P \longrightarrow Cu_3P \quad (8)$$

$$Cu_3P + 5P \longrightarrow 3CuP_2 \quad (9)$$

Copper(I) dialkyl-phosphides such as $CuPR_2$ or $KCu(PR_2)_2$ are readily prepared from potassium or lithium phosphides and copper(I) halides.[33]

1.3.4.2 Complexes

Most copper(I) coordination compounds are easily formed by mixing a copper(I) salt with a donor ligand in a suitable solvent (*e.g.*, water, alcohol, or acetonitrile). Copper(I) complexes are also prepared by treating copper(II) salts with tertiary phosphines or arsines which reduce copper(II) to copper(I).[18] Recent efforts have been directed primarily at studying the structures and properties of copper complexes containing novel ligands (Table 3).

1.3.5 Carbon Compounds

1.3.5.1 Simple Compounds

The preparation and chemistry of organocopper(I) compounds have recently been reviewed.[34,35] Three types of organocopper(I) species have been extensively studied: RCu, R_2CuLi, and RCu–ligand. The nature of organocopper RCu species depends strongly on the method and the reaction conditions used for its preparation; a good discussion of the difficulties in preparing pure stoichiometric RCu reagents is available.[36]

Two methods, the first general and the second of limited generality, have been used to form RCu reagents:

(a) Metathetical reaction of an organometallic reagent (usually organolithium) with a copper(I) salt [Eqn (10)]:

$$R{-}M + CuX \longrightarrow R{-}Cu + MX \quad (10)$$

(b) Thermal decarboxylation of cuprous carboxylates [Eqn (11)]: [37, 38]

$$RCOOCu \xrightarrow{\text{solvent}} R{-}Cu\ \text{solvent} + CO_2 \quad (11)$$

Pure RCu species have been prepared only *via* method (a) [Eqn (10)], although, in principle, they probably could also be prepared *via* cuprous carboxylate decarboxylation; method (b) has so far not been widely used and may, in practice, be limited to preparation of arylcopper species only.

If RLi is added to RCu then a *lithium diorganocuprate(I)* species represented as R_2CuLi is formed[39,40] [Eqn (12)]. The R_2CuLi species can be formed either without [Eqn (12)] or with [Eqn (13)] a lithium salt.

$$RCu + RLi \longrightarrow R_2CuLi \quad (12)$$

$$2RLi + CuX \longrightarrow R_2CuLi + LiX \quad (13)$$

To solubilize the copper salt used to prepare the organocopper reagent and to stabilize this reagent once formed, a variety of heteroatom ligands have been used to complex or chelate with copper: *amines,*[41,42] *phosphines,*[39,42] *phosphites,*[43] and *sulfides.*[44]

$$RLi + XCu{-}\text{ligand} \longrightarrow [RCuX]Li + \text{ligand} \quad (14)$$

(see [45])

Because of their high reactivity and low thermal stability stoichiometric organocopper reagents are prepared *in situ* and are used immediately. Air and moisture must be rigorously excluded; reactions are run in an atmosphere of argon or prepurified nitrogen, and liquid reagents are best transferred via dry hypodermic syringes and introduced into the reaction mixture through a rubber septum-capped side arm of the reaction flash. Solid reagents should be added through a funnel with its stem extending into the neck of the reaction flask, out of which a rapid and constant flow of inert gas is maintained.

Of the three types of organocopper(I) species just discussed, *lithium diorganocuprates(I)* have received the most attention because of their synthetic utility in conjugate addition[35] and coupling[46] reactions.

---

[32] *H. Haraldsen*, Z. Anorg. All. Chem. *240*, 337 (1939).
[33] *K. Issleib, H. O. Fröhlich*, Chem. Ber. *95*, 375 (1962); *K. Issleib, E. Wenschuh*, Z. Naturforsch. *19b*, 199 (1964).
[34] *G. Bähr, P. Burba, Houben-Wehl*, Methoden der Organischen Chemie, Band XIII/I, pp. 731–777, Thieme Verlag, Stuttgart 1970.
[35] *G. H. Posner*, Org. Reactions *19*, i (1972).
[36] *G. Costa, A. Camus, L. Gatti, N. Marsich*, J. Organometal. Chem. *5*, 568 (1966).
[37] *A. Cairncross, J. R. Roland, R. M. Henderson, W. A. Sheppard*, J. Amer. Chem. Soc. *92*, 3187 (1970).
[38] *T. Cohen, R. A. Schambach*, J. Amer. Chem. Soc. *92*, 3189 (1970).
[39] *H. O. House, W. L. Respass, G. M. Whitesides*, J. Org. Chem. *31*, 3128 (1966).
[40] *H. Gilman, R. G. Jones, L. A. Woods*, J. Org. Chem. *17*, 1630 (1952).
[41] *K. H. Thiele, J. Köhler*, J. Organometal. Chem. *12*, 225 (1968).
[42] *J. B. Siddall, M. Biskup, J. H. Fried*, J. Amer. Chem. Soc. *91*, 1853 (1969).
[43] *H. O. House, W. F. Fischer, Jr.*, J. Org. Chem. *33*, 949 (1968).
[44] *H. O. House, W. F. Fischer, Jr.*, J. Org. Chem. *34*, 3615 (1969).
[45] *N. T. Luong-Thi, H. Rivière*, Tetrahedron Lett. No. *19*, 1579, 1583 (1970).
[46] *G. H. Posner*, Org. Reactions *22*, 253 (1975).

Table 3 Copper(I) Nitrogen and Group V Complexes

Donor ligand (L) Structure	Name	Complex	Ref.
R_2NH	Amine	$[CuL_2]Cl$, $[CuL]Cl$	*a*
$H_2N-CH_2-CH_2-NH_2$	Ethylenediamine	$[CuL(CO)]Cl$	*b*
R_3P	Phosphine	R–CuL $(R-Cu)_2L$, $(R-Cu)_3L$	*c* *d*
$(RO)_3P$	Phosphite	$R-CuL_3$	*e*
(bicyclic structure)	Bicyclic phosphite	$[CuL_4]X$	*f*
$C_6H_4(N{=}CH-C_6H_5)_2$	*o*-Phenylenediamine Schiff base	$[CuL]ClO_4$	*g*
$R-N{=}N-R^1$	Azoalkane	$[CuL]Cl$	*h*
$C_5H_4N-N{=}N-C_5H_4N$	2,2′-Azopyridine	$(CuX)_2L$	*i*
$C_6H_4(NH)(SH)$	*o*-Aminobenzenethiol	CuL	*j*
$F_2P-N(CH_3)_2$	Dimethylaminodifluorophosphine	$[CuL]Cl$	*k*
$H_2N-CH_2-CH_2-CH_2-P(C_6H_5)_2$	3-Aminopropyldiphenylphosphine	$[CuL_2][CuCl_2]$	*l*
$(H_5C_6)_2P-CH_2-CH_2-P(C_6H_5)_2$	1,2-Bis(diphenylphosphino)ethane	$(RCu)_2L_3$	*m*
$C_5H_4N-CH_2-CH_2-As(C_6H_5)_2$	[2-(Diphenylarsinoethyl)]pyridine	$[CuL_2][CuX_2]$	*n*
$C_6H_4[P(S)(C_6H_5)_2][As(C_6H_5)_2]$	Diphenyl(*o*-diphenylarsino)phosphine sulfide	CuL_2ClO_4	*o*
$C_6H_4(CH_2-CH{=}CH_2)[As(CH_3)_2]$	*o*-(2-Propenylphenyl)dimethylarsine	CuL_2I	*p*
$(H_3C)_2As-CH{=}CH-As(CH_3)_2$	*cis*-1,2-Bis(dimethylarsino)ethylene	CuL_2ClO_4	*q*

[a] *J. R. Clifton, J. T. Yoke*, Inorg. Chem. *5*, 1630 (1966); *6*, 1258 (1967).
[b] *G. Rucci, C. Zanzottera, M. P. Lachi, M. Camia*, Chem. Commun. p. 652 (1971).
[c] *H. O. House, W. L. Respess, G. M. Whitesides*, J. Org. Chem. *31*, 3128 (1966).
[d] *G. Costa, A. Camus, N. Marsich, L. Gatti*, J. Organometal. Chem. (Amsterdam) *8*, 339 (1967).
[e] *H. O. House, W. F. Fischer, Jr.*, J. Org. Chem. *33*, 949 (1968).
[f] *J. G. Verkade, T. S. Piper*, Inorg. Chem. *1*, 453 (1962); *R. L. Keiter, J. G. Verkade*, Inorg. Chem. *9*, 404 (1970).
[g] *K. S. Bose, C. C. Patel*, J. Inorg. Nucl. Chem. *32*, 2423 (1970).
[h] *M. Heyman, V. T. Bandurco, J. P. Snyder*, Chem. Commun. 297 (1971).
[i] *D. A. Baldwin, A. B. P. Lever, R. V. Parish*, Inorg. Chem. *8*, 107 (1969).
[j] *L. F. Larkworthy, J. M. Murphy, D. J. Phillips*, Inorg. Chem. *7*, 1436 (1968).
[k] *K. Cohn, R. W. Parry*, Inorg. Chem. *7*, 46 (1968).
[l] *J. W. Corr*, Dissertation Abstr. *27B*, 2638 (1967).
[m] *A. Camus, N. Marsich*, J. Organometal. Chem. *21*, 249 (1970).
[n] *E. Uhlig, M. Maaser*, Z. Anorg. Allg. Chem. *349*, 300 (1967).
[o] *P. Nicpon, D. W. Meek*, Inorg. Chem. *6*, 145 (1967).
[p] *M. A. Bennett, W. R. Kneen, R. S. Nyholm*, Inorg. Chem. *7*, 552 (1968).
[q] *R. D. Feltham, H. G. Metzger, W. Silverthorn*, Inorg. Chem. *7*, 2003 (1968).

Generally ethereal solutions of lithium diorganocuprate(I) reagents are easily prepared under nitrogen or argon at 0° ($R=CH_3$) or below −20° ($R \neq CH_3$) from organolithium compounds and cuprous iodide or bromide in a 2:1 molar ratio. As the second equivalent of organolithium is added, the reaction mixture becomes essentially homogeneous—colorless in the case of $(CH_3)_2CuLi$ and deeply colored in most of the other cases.[47] Formation of the lithium diorganocuprate(I) reagent is rapid, being complete in less than 5 min. at 0°, as shown, for example, by a negative Gilman test with Michler's ketone,[48] or, less accurately, by visually following the dissolution of copper salt. Alternatively, an aliquot quenched at low temperature with benzoyl chloride,[49] for example, might indicate whether organocopper formation is complete (*i.e.*, if any alcoholic product is produced, then organocopper formation is probably incomplete).[50]

Since most of the copper salts used to prepare the organocopper reagents are not hygroscopic, glove-bag or dry-box procedures are unnecessary; highly effective stoichiometric organocopper reagents have been prepared from carefully purified and dried[39] as well as from commercial samples (*e.g.*, Fisher Chemical Co.)[51] of cuprous iodide. Similarly, commercially available organolithium reagents are usually satisfactory, a limitation occasionally being the solvent in which the reagent is prepared; this solvent becomes part of the reaction mixture and may influence the course of conjugate addition.

Stoichiometric organocopper reagents are often highly sensitive to such variables as reaction temperature and solvent.

Temperature variation has been observed to affect both the rate of formation of organocopper reagents and their stability once formed. *Lithium di-n-butylcuprate(I)*, for example, is formed in ether from *n*-butyllithium and cuprous iodide[52] rapidly ($\ll 1$ min.) at 0° but slowly (>10 min.) at −40°. Optimum conditions for formation of such organocopper reagents, therefore, must involve careful control of reaction temperature.

The thermal stability of organocopper reagents depends on the nature of the reagent and on the structure of the organic moiety. *Phenylcopper*, for example, is stable below 80°, but different *complexes of phenylcopper* [*e.g.*, $(C_6H_5Cu)_4C_6H_5Li \cdot [(C_2H_5)_2O]_n$ $(C_6H_5Cu)_2(C_6H_5)_2Mg(THF)_n$ or $C_6H_5CuP(C_6H_5)_3$] have different thermal stabilities.[36,53] Generally the order of stability of organocopper reagents appears to be RCu–ligand > RCu > R_2CuLi.

The structure of R strongly influences the stability and reactivity of organocopper reagents. Thus, whereas *phenylcopper* is stable below 80°, *methylcopper* decomposes at room temperature and *ethylcopper* at −18° in ether.[40,49] Similarly, lithium dimethylcuprate(I) in ether solution under an inert atmosphere is stable for hours at 0°, but the more reactive lithium di-*n*-alkylcuprate(I) and secondary and tertiary organocopper reagents rapidly decompose in ether above −20°. Thus for most effective preparation and use of these less stable, more reactive organocopper reagents the reaction temperature should be carefully controlled. Usually stoichiometric *n*-, *sec*-, and *tert*-organocopper reagents are prepared either below −20° for a sufficient amount of time (usually >5 min.) to allow complete formation of reagent[54] or at 0° for several minutes (rapid reagent formation) followed immediately by cooling to below −20°.[52]

Only two kinds of solvent have been widely used for organocopper preparations and reactions: aromatic hydrocarbons (benzene and toluene) and ethers (tetrahydrofuran and diethyl ether). Stable arylcopper(I) compounds containing 2-(dimethylamino)methyl or 2-methoxymethyl groups have recently been prepared from the corresponding aryllithium and copper(I) bromide.[55]

Cuprous acetylides of various stoichiometries have been prepared from acetylene, ammonium chloride, and copper(I) chloride.[23,56] *Dicopper(I)*

---

[47] *G. M. Whitesides, J. San Filippo, Jr., C. P. Casey, E. J. Panek*, J. Amer. Chem. Soc. *89*, 5302 (1967).

[48] *H. Gilman, F. Schulze*, J. Amer. Chem. Soc. *47*, 2002 (1925).

[49] *H. Gilman, J. M. Straley*, Rec. Trav. Chim. *55*, 821 (1936).

[50] *G. H. Posner, C. E. Whitten*, Tetrahedron Lett. No. *53*, 4647 (1970).

[51] *E. J. Corey, J. A. Katzenellenbogen*, J. Amer. Chem. Soc. *91*, 1851 (1969).

[52] *J. A. Katzenellenbogen*, Ph.D. Thesis, Harvard University, Cambridge, Mass., 1969; Dissertation Abstr. *31*, 1826–B (1970).

[53] *G. Costa, A. Camus, N. Marsich, L. Gatti*, J. Organometal. Chem. *8*, 339 (1967).

[54] *G. M. Whitesides, W. F. Fischer, Jr., J. San Filippo, Jr., R. W. Bashe, H. O. House*, J. Amer. Chem. Soc. *91*, 4871 (1969).

[55] *G. Van Koten, A. J. Leusink, J. G. Noltes*, Chem. Commun. 1107 (1970).

[56] *I. M. Dolgopol'skii, A. L. Klebanskii, Z. F. Dobler*, J. Gen. Chem. USSR Engl. Transl. *33*, 1062, 1064 (1963).

α,ω-diacetylides are similarly prepared.[57] More recently reaction of acetylene and cuprous chloride in hexamethylphosphoric triamide[58] has been used successfully to prepare copper(I) acetylides directly.

1.3.5.2 Complexes

Copper(I) olefin complexes have been extensively reviewed.[59-61] The most satisfactory method for generation of such complexes appears to be bubbling sulfur dioxide into an alcohol solution containing a copper(II) salt and the olefinic ligand; reduction of copper(II) is followed usually by precipitation of the copper(I) complex.[62] Reduction of copper(II) to copper(I) in the presence of an unsaturated ligand has also been achieved by trialkylphosphites[63] and by electrolysis.[64] Alternatively, reaction of a copper(I) salt directly with an olefin in the absence of solvent[65,66] or in acrylonitrile solvent[67] produces a copper(I) olefin complex.

Some recent examples of *copper(I) carbon π-complexes* involve the following ligands: 2-chloro-1,3-butadiene [$(CuCl)_2L$],[68] 2-chloro-2-butene ($NH_3CuL(SO_4)$),[69] acrolein (CuLCl),[65] unsaturated nitriles of the type $RR'C{=}(CN)_2$ (CuLCl),[67] a variety of cyclic dienes (CuLCl, $(CuX)_2L$),[64,70] and cyclic allenes [$(CuCl)_2L$].[71]

Several *pentahapto*cyclopentadienylcopper(I) species have been prepared according to Eqn (15).[72]

$$\tfrac{1}{4}(LCuX)_4 + C_5H_5Tl \longrightarrow C_5H_5CuL + TlX \qquad (15)$$

A study has been made recently of the thermodynamics of copper(I) cycloolefin formation,[73] and the photochemistry of the 1,5-cyclooctadienecopper(I) chloride complex has been examined.[74]

Phosphine complexed copper(I) acetylides have been prepared by treating polymeric copper(I) acetylides with a *tert.* phosphine,[75] and complex cuprate(I) acetylides have been prepared according to the accompanying equations.[76]

$$3KC{\equiv}CR + CuI \longrightarrow K_2[Cu(C{\equiv}CR)_3] + KI \qquad (16)$$

$$KC{\equiv}CR + CuC{\equiv}CR \longrightarrow K[Cu(C{\equiv}CR)_2] \qquad (17)$$

$R{=}CH_3, C_6H_5$

Copper(I) complex cyanides have been recently reviewed.[18,77,78] Mixing copper(I) cyanide with potassium cyanide in an aqueous medium produces $KCu(CN)_2$ and $K_3Cu(CN)_4$,[79] whereas direct oxidation of copper by mercuric cyanide in molten potassium cyanide forms $K_2Cu(CN)_3$ and $K_3Cu(CN)_4$.[80] Nitrogen ($Cu_3L_4(CN)_4$), phosphine, and arsine (CuLCN and CuL_2CN) complexed copper(I) cyanides have also been reported.[81]

1.3.6 Transition Metal Derivatives

Bimetallic complexes of copper(I) with other first-row transition metals are of particular interest because of the metal–metal interac-

---

[57] *F. Sondheimer*, Pure Appl. Chem. *7*, 363 (1963).
[58] *M. Bourgain, J. F. Normant*, Bull. Soc. Chim. France, p. 2477 (1969).
[59] *E. O. Fischer, H. Werner*, Metal π-Complexes, Vol. 1. Elsevier, Amsterdam, 1966; Angew. Chem. Int. Ed. Engl. *2*, 80 (1963).
[60] *M. A. Bennett*, Chem. Rev. *62*, 611 (1962).
[61] *R. G. Guy, B. L. Shaw*, Advan. Inorg. Chem. Radiochem. *4*, 77 (1962).
[62] *H. L. Haight, J. R. Doyle, N. C. Baenziger, G. F. Richards*, Inorg. Chem. *2*, 1301 (1963).
[63] *B. Cook, R. Miller, P. Todd*, J. Organometal, Chem. *19*, 421 (1969).
[64] *S. E. Mahahan*, Inorg. Chem. *5*, 2063 (1966); Inorg. Nucl. Chem. Lett. *3*, 383 (1967).
[65] *S. Kawaguchi, T. Ogura*, Inorg. Chem. *5*, 844 (1966).
[66] *P. J. Hendra, D. B. Powell*, Spectrochim. Acta *18*, 1195 (1962).
[67] *G. N. Schrauzer, S. Eichler*, Chem. Ber. *95*, 260 (1962).
[68] *S. N. Avakyan, R. A. Karapetyan*, J. Gen. Chem. USSR Engl. Transl. *35*, 1681 (1965); Russ. J. Inorg. Chem. Engl. Transl. *10*, 956 (1965).
[69] *R. A. Karapetyan, S. N. Avakyan*, J. Gen. Chem. USSR Engl. Transl. *36*, 766 (1966).
[70] *J. C. Trebellas, J. R. Olechowski, H. B. Jonassen*, Inorg. Chem. *4*, 1818 (1965).
[71] *G. Nagendrappa, G. C. Joshi, D. Devaprabhakara*, J. Organometal. Chem. *27*, 421 (1971).
[72] *F. A. Cotton, T. J. Marks*, J. Amer. Chem. Soc. *92*, 5114 (1970).
[73] *J. M. Harvilchuck, D. A. Aikens, R. C. Murray, Jr.*, Inorg. Chem. *8*, 539 (1969).
[74] *G. M. Whitesides, G. L. Goe, A. C. Cope*, J. Amer. Chem. Soc. *91*, 2608 (1969).
[75] *L. Vallarino*, Chem. Soc. (London), Spec. Publ. *13*, 123 (1959); *A. Sacco, M. Freni*, Gazz. Chim. Ital. *89*, 1800 (1959).
[76] *R. Nast, W. Pfab*, Chem. Ber. *89*, 415 (1956).
[77] *B. M. Chadwick, A. G. Sharpe*, Advan. Inorg. Chem. Radiochem. *8*, 152 (1966).
[78] *M. H. Ford-Smite*, The Chemistry of Complex Cyanides, p. 34. Her Majesty's Stationery Office, London, 1964.
[79] *M. J. Reisfield, L. H. Jones*, J. Mol. Spectros. *18*, 222 (1965).
[80] *J. G. Reinstein, E. Griswold, J. Kleinberg*, Inorg. Chem. *8*, 2499 (1969).
[81] *D. Cooper, R. A. Plane*, Inorg. Chem. *5*, 1677, 2209 (1966).

tions.[82] Pentacarbonylmanganate and tetracarbonylferrate react with triarsinecopper(I) halides according to Eqns (18) and (19).[83]

$$(R_3As)_3CuCl + Mn(CO)_5^{\ominus} \longrightarrow (R_3As)_3CuMn(CO)_5 + Cl^{\ominus} \quad (18)$$

$$2(R_3As)_3CuBr + Fe(CO)_4^{2\ominus} \longrightarrow [(R_3As)_3Cu]_2Fe(CO)_4 + 2Br^{\ominus} \quad (19)$$

1.4 Copper(II) Compounds

1.4.1 Halides, Oxyhalides, and Complex Halides

Copper difluoride, copper difluoride dihydrate, and copper dichloride are commercially available.

Copper dichloride dihydrate is prepared by dissolving copper(II) oxide or carbonate in hydrochloric acid and evaporating the resulting solution, or directly from copper by dissolving the metal in aqua regia and evaporating the solution several times to dryness with hydrochloric acid.[17]

Copper dibromide is synthesized by dissolving copper(II) oxide or carbonate in aqueous hydrobromic acid and evaporating the resulting solution under reduced pressure over sulfuric acid,[23] by treating copper(II) acetate with acetyl bromide,[23] by mixing copper dichloride and boron tribromide,[84] or by heating copper and bromine[85] in a sealed tube at 300°.

Although *copper diiodide* cannot be isolated, several *complexes* can be prepared using appropriate techniques.[16] The complex with bipyridyl, for example, is well characterized and can be formed by mixing bis(bipyridyl)copper(II) perchlorate and sodium iodide.[86]

Copper(II) oxychloride, $CuCl_2 \cdot Cu(OH)_2$, is best made by heating stoichiometric quantities of cupric chloride, calcium carbonate, and water in a bomb[23]; this oxychloride decomposes in the presence of excess water to form one of the varieties of $Cu_2(OH)_3Cl$.[17] Copper(II) oxybromides have been prepared by long heating of copper(II) oxide or hydroxide in a copper dibromide solution.[17]

A wide variety of halocuprates(II) are known. Most of these are prepared from a copper(II) dihalide and the appropriate inorganic or organic halide.

Copper difluoride is fused with barium fluoride to form *barium hexafluorocuprate(II)*[87] and with an alkali metal [Eqn (21)] fluoride to form an *alkali metal tetrafluorocuprate(II)*.[88]

$$CuF_2 + 2BaF_2 \longrightarrow Ba_2CuF_6 \quad (20)$$

$$CuF_2 + 2MF \longrightarrow M_2CuF_4 \quad M = \text{alkali metal} \quad (21)$$

Ammonium tetrafluorocuprate(II) has been prepared by mixing copper difluoride and ammonium fluoride at high pressures and temperatures[89] and also by adding the requisite amount of ammonium fluoride to a solution of copper nitrate in methanol.[90]

Alkali trifluorocuprates(II) are most conveniently prepared by fusing stoichiometric amounts of copper difluoride and alkali metal fluoride.[88]

Ammonium trifluorocuprate(II) can be produced by mixing copper difluoride and ammonium fluoride at high pressures and temperatures,[89] by adding ammonium fluoride to a solution of copper nitrate in methanol,[90] or by mixing ammonium fluoride and copper dibromide in methanol.[91]

Hexamminechromium(III) pentachlorocuprate(II), $[Cr(NH_3)_6][CuCl_5]$, has been prepared by mixing hexamminechromium(III) chloride and copper dichloride in water followed by addition to concentrated hydrochloric acid. X-Ray study confirms the presence of the pentachlorocuprate(II) anion and indicates it to be a regular trigonal bipyramid.[92] *Hexamminecobalt(III) pentachlorocuprate(II)* has been similarly prepared.[93]

Ammonium and *arsonium tetrachlorocuprates(II)*

[82] *M. C. Baird*, Progr. Inorg. Chem. *9*, 1 (1968).

[83] *A. S. Kasenally, R. S. Nyholm, M. H. B. Stiddard*, J. Chem. Soc., p. 5343 (1965); J. Amer. Chem. Soc. *86*, 1884 (1964).

[84] *P. M. Druce, M. F. Lappert, P. N. K. Riley*, Chem. Commun. p. 486 (1967).

[85] *R. R. Hammer, N. W. Gregory*, J. Phys. Chem. *68*, 314 (1964).

[86] *G. A. Barclay, B. F. Haskins, C. H. L. Kennard*, J. Chem. Soc. p. 5691 (1963).

[87] *H. G. Schnering*, Z. Anorg. Allg. Chem. *353*, 1 (1967); *M. Samouel, A. De Kozak*, C.R. Hebd. Séanc. Acad. Sci. Paris, Sér. C, *268*, 1789 (1969).

[88] *O. Schmitz-DuMont, D. Grimm*, Z. Anorg. Allg. Chem. *355*, 280 (1967); *W. Rüdorff, G. Lincke, D. Babel*, *ibid.*, *320*, 150 (1963).

[89] *D. S. Crocket, R. A. Grossman*, Inorg. Chem. *3*, 644 (1964).

[90] *W. G. Bottjer, H. M. Haendler*, Inorg. Chem. *4*, 913 (1965).

[91] *H. M. Haendler, F. A. Johnson, D. S. Crocket*, J. Amer. Chem. Soc. *80*, 2662 (1958).

[92] *K. N. Raymond, D. W. Meek, J. A. Ibers*, Inorg. Chem. *7*, 1111 (1968); *M. Mori*, Bull. Chem. Soc. Jap. *33*, 985 (1960).

[93] *T. V. Long, II, A. W. Herlinger, E. Epstein, I. Bernal*, Inorg. Chem. *9*, 459 (1970).

are usually prepared by mixing an amine or arsine hydrochloride and copper dichloride in water or alcohol.[16] Thus, *acetamidinium tetrachlorocuprate* is formed from stoichiometric amounts of acetamidine hydrochloride and anhydrous copper dichloride in methanol.[94]

$$2CH_3C(NH_2)_2Cl + CuCl_2 \longrightarrow [CH_3C(NH_2)_2]_2CuCl_4 \quad (22)$$

Ammonium trichlorocuprates are produced in a similar fashion, except that reactants are mixed in a 1:1 ratio.[16]

$$R_3NHCl + CuCl_2 \longrightarrow [R_3NH][CuCl_3] \quad (23)$$

Alkali trichlorocuprates have been formed from alkali chloride and excess copper dichloride in hydrochloric acid[95] and directly in the melt from stoichiometric amounts of the constituent chlorides.[96]

The dihydrates of both *tetra-* and *trichlorocuprates* are prepared by crystallization from aqueous solutions containing stoichiometric quantities of the constituent chlorides.[16]

A claim has been made to the preparation of polynuclear copper(II) chloro complexes, $(R_4N)_5Cu_3Cl_{11}$, but a detailed structural determination has not been given.[97]

Recently a *copper(II) complex oxyhalide* has been prepared as a tetramethylammonium salt by mixing copper dichloride with copper(II) oxide at reflux in methanol followed by adding the stoichiometric amount of tetramethylammonium chloride.[98]

Tetra- and *tribromocuprates(II)* are generated similarly as the corresponding chlorocuprates.[16] Reaction of copper dibromide with concentrated hydrobromic acid produces an intensely purple tribromocuprate(II); this reaction has been used as a test for copper(II).[17]

1.4.2 Oxygen and Group VI Compounds

1.4.2.1 Simple Compounds

Copper(II) oxide is commercially available. Treating copper(II) sulfate with aqueous ammonia and sodium hydroxide produces *copper(II) hydroxide*; the literature procedure must be followed carefully to avoid contamination of the hydroxide with sulfate.[99]

$$CuSO_4 + 2NaOH \longrightarrow Cu(OH)_2 + Na_2SO_4 \quad (24)$$

Tri- and *tetrahydroxycuprates(II)* have been reported from copper(II) hydroxide in sodium hydroxide.[100]

Copper(II) dialkoxides ($R = CH_3$, C_3H_5) are prepared by mixing copper(II) chloride or bromide with lithium or sodium alkoxides.[101,102]

$$2LiOR + CuCl_2 \longrightarrow Cu(OR)_2 + 2LiCl \quad (25)$$

The reaction presumably proceeds through the intermediacy of a *copper(I) alkoxide chloride*, Cu(OR)Cl, which for $R = CH_3$ can be isolated before the dimethoxide is formed. Alternatively, $Cu(OCH_3)Cl$ can be prepared by treating the dimethoxide with excess copper(II) chloride in methanol.[102]

Claims have been made to preparation of copper(II) peroxide, but the constitution of the substance(s) obtained is uncertain.[103]

Copper(II) carboxylates[57] have been studied extensively, mainly because of their interesting magnetic properties.[26,104] Synthesis of copper(II) aryl,[105] alkyl-,[106] and alkenylcarboxylates[107] has been achieved in a straightforward manner by mixing the carboxylic acid or its alkali metal salt with a copper(II) carbonate, sulfate, or perchlorate.

Salts of dioxalato-[108] and tetracarboxylato-

---

94 *L. A. Bares, K. Emerson, J. E. Drumheller*, Inorg. Chem. *8*, 131 (1969).

95 *A. W. Schuleter, R. A. Jacobsen, R. E. Rundle*, Inorg. Chem. *5*, 277 (1966).

96 *H. J. Seifert, K. Klatyk*, Z. Anorg. Allg. Chem. *334*, 113 (1964).

97 *O. Piovesana, J. Selbin*, J. Inorg. Nucl. Chem. *32*, 2093 (1970).

98 *J. A. Bertrand, J. A. Kelley*, Inorg. Chem. *8*, 1982 (1969).

99 *R. Fricke, J. Kubach*, Z. Elektrochem. *53*, 76 (1949).

100 *E. A. Buketor, M. Z. Ugorets, K. M. Akhmetov*, Tr. Khim-Met. Inst. Akad. Nauk. Kaz. SSR. *3*, 3 (1967); Chem. Abstr. *69*, 100034g (1968).

101 *T. Saegusa, T. Tsuda, K. Isayama*, J. Org. Chem. *35*, 2976 (1970).

102 *C. H. Brubaker, Jr., M. Wicholas*, J. Inorg. Nucl. Chem. *27*, 59 (1965).

103 *J. A. Connor, E. A. V. Ebsworth*, Advan. Inorg. Chem. Radiochem. *6*, 342 (1964).

104 For reviews of the literature see *M. Kato, H. B. Jonassen, J. C. Fanning*, Chem. Rev. *64*, 99 (1964); *J. Lewis*, Pure Appl. Chem. *10*, 27 (1965); *G. A. Popovich, A. V. Ablov*, Russ. J. Inorg. Chem. Engl. Transl. *9*, 586 (1964).

105 *J. Lewis, Y. C. Lin, L. K. Royston, R. C. Thompson*, J. Chem. Soc. p. 6464 (1965).

106 *W. E. Hatfield, H. M. McGuire, J. S. Paschal, R. Whyman*, J. Chem. Soc. A. p. 1194 (1966); *W. R. May, M. M. Jones*, J. Inorg. Nucl. Chem. *24*, 511 (1962).

107 *B. J. Edmondson, A. P. B. Lever*, Inorg. Chem. *4*, 1608 (1965).

108 *M. A. Viswamitra*, J. Chem. Phys. *37*, 1408 (1962); Z. Kristallogr. Kristallgro metrie, Kristallphys. Kristallchem. *117*, 437 (1962).

cuprate(II)[109] anions have been prepared and their structures determined by X-ray analysis.

$$Cu(OAc)_2 + Ca(OAc)_2 \xrightarrow{H_2O} Ca[Cu(OAc)_4] \cdot 6H_2O \quad (26)$$

Copper(II) sulfide is best prepared by treating a solution of sulfur in carbon disulfide with pure copper powder obtained from copper oxalate.[23] The most common form of copper(II) sulfate is the pentahydrate which is produced both industrially and on a small scale by dissolving copper(II) oxide or carbonate in dilute sulfuric acid and evaporating the resulting solution until the pentahydrate crystals separate.[17]

1.4.2.2 Complexes

Interest in copper coordination compounds relates primarily to their structural, magnetic,[110] and chemical[111] properties; little effort has been devoted to developing new or systematizing current synthetic approaches to these compounds. The vast majority of copper(II) coordination compounds are formed directly by mixing a copper(II) salt with the desired donor ligand in a suitable solvent like ethanol or water.

Selection of an appropriate copper(II) salt is often governed by two rules of thumb: (*a*) if the donor ligand has an acidic proton, copper(II) acetate is usually chosen so that acetic acid is produced as an easily separable by-product;

$$2Z\text{–}H + Cu(OAc)_2 \longrightarrow CuZ_2 + 2HOAc \quad (27)$$

and (*b*) if a crystalline complex is desired, copper(II) salts having bulky anions (*e.g.*, perchlorate) are frequently used to increase chances of crystallinity.

Representative examples of oxygen (Table 4), mixed oxygen and nitrogen (Table 5), and sulphur and selenium (Table 6) donor ligands recently used to form copper(II) complexes are given.

[109] *F. Rüdorff*, Chem. Ber. *21*, 279 (1888); *D. A. Langs*, Dissertation Abstr. *30*, 2589-B (1969).

[110] *B. N. Figgis, D. J. Martin*, Progr. Inorg. Chem. *6*, 37 (1965).

[111] *J. P. Collman*, Transition Metal Chem. *2*, 2 (1966).

Table 4 Copper(II) Oxygen Complexes

Donor ligand (L)			
Structure	Name	Complex	Ref.
$R\text{–}O\text{–}R^1$	Ether	CuL_2X_2, $(CuX_2)_3L_2$	*a*
$H_3C\text{–}CO\text{–}CH{=}C(O^{\ominus})\text{–}CH_3$	Acetylacetonate	CuL_2	*b*
		$CuL_2(C_5H_5N)$	*c*
$H_3C\text{–}S(O)\text{–}CH_2\text{–}CH_2\text{–}S(O)\text{–}CH_3$	2,5-Dithiahexane 2,5-dioxide	$CuL_2(ClO_4)_2$	*d*
R_2SeO	Selenoxide	CuL_2Cl_2, CuL_3Cl_2	*e*
R_3NO	*N*-Oxides	CuL_6X_2, CuL_4X_2, CuL_2X_2, $CuLX_2$	*f*
2,6-dimethylpyridine N-oxide (H_3C, CH_3, N→O)	2,6-Lutidine *N*-oxide	CuL_2Cl_2	*g*
$[(R_2N)_2PO]_2Y$	Pyrophosphoramide (Y=O)	$CuL_3(ClO_4)_2$, $(CuL_2)(CuX_4)$	*h*
	Imidodiphosphoramide ($Y{=}NCH_3$)	$CuL_2(ClO_4)_2$	*i*
	Methylenediphosphonic acid diamide ($Y = CH_2$)	$CuL_2(ClO_4)_2$	*i*
R_3AsO	Arsine oxide	CuL_2X_2, $(CuL_4)(CuCl_3)_2$	*j*

[a] *J. C. Barnes*, J. Inorg. Nucl. Chem. *31*, 95 (1969).

[b] *J. P. Fackler, Jr.*, Progr. Inorg. Chem. *7*, 361 (1966); *W. C. Fernelius, B. E. Bryant*, Inorg. Syn. *5*, 105 (1957).

[c] *W. Partenheimer, R. S. Drago*, Inorg. Chem. *9*, 47 (1970).

[d] *S. K. Madan, C. M. Hull, L. J. Herman*, Inorg. Chem. *7*, 491 (1968).

[e] *R. Paetzold, P. Vordank*, Z. Anorg. Allg. Chem. *347*, 294 (1966).

[f] *W. H. Watson*, Inorg. Chem. *9*, 1879 (1969).

[g] *R. S. Sager, W. H. Watson*, Inorg. Chem. *8*, 308 (1969).

[h] *M. D. Joesten, K. M. Nykerk*, Inorg. Chem. *3*, 548 (1964); *M. D. Joesten, J. F. Forbes*, J. Amer. Chem. Soc. *88*, 5465 (1966); *M. D. Joesten, M. S. Hussain, P. G. Lenhert, J. H. Venable*, J. Amer. Chem. Soc. *90*, 5623 (1968); *M. S. Hussain, M. D. Joesten, P. G. Lenhert*, Inorg. Chem. *9*, 162 (1970).

[i] *K. P. Lannert, M. D. Joesten*, Inorg. Chem. *8*, 1775 (1969).

[j] *G. E. Parris, G. G. Long*, J. Inorg. Nucl. Chem. *32*, 1585 (1970).

Table 5 Copper(II) Mixed Oxygen and Nitrogen Complexes

Donor ligand (L)			
Structure	Name	Complex	Ref.
HO–CH_2–CH_2–NH_2	Ethanolamine	$(CuLX)_2$	*a*
2-(CH=N–R)-phenol (OH ortho)	Salicylaldimines	CuL_2	*b–d*
R–NH–CO–R[1]	Amides	$CuL_6(ClO_4)_2$, $CuL_4(ClO_4)_2$	*e*
H_2N–CH(R)–COO–R[1]	α-Amino acids	CuL_2	*f, g*
	α-Amino acid esters	CuL_2Cl_2	*h*
R–CO–C(R[1])=N–NH_2	Monohydrazones	CuLX, CuL_2X_2	*i*
HO–CH_2–C(=N–R)–NH_2	α-Hydroxyamidines	CuL_2	*j*
H_2N–CO–NH–NH_2	Semicarbazide	CuL_2Cl_2	*k*
$(^{\ominus}OOC–CH_2)_2N–CH_2–CH_2–N(CH_2–COO^{\ominus})_2$	Ethylenediamine tetraacetate	$Cu(LH_2)$	*l*

[a] *G. Douheret*, Bull. Soc. Chim. Fr., pp. 2915, 2921 (1965).
[b] *L. Sacconi*, Coordination Chem. Rev. *1*, 126 (1966); *R. H. Holm, G. W. Everett, Jr., A. Chakravorty*, Progr. Inorg. Chem. *7*, 151 (1966).
[c] *S. Yamada*, Coordination Chem. Rev. *1*, 434 (1966); *G. N. Weinstein, M. J. O'Connor, R. H. Holm*, Inorg. Chem. *9*, 2104 (1970).
[d] *T. S. Kannan, A. Chakravorty*, Inorg. Chem. *9*, 1153 (1970).
[e] *H. Hubacek, B. Stanviv, V. Guttman*, Monatsh. *94*, 1118 (1963); *W. Schneider*, Helv. Chim. Acta *46*, 1842 (1963).
[f] *C. K. Jorgensen*, Inorganic Complexes. Academic Press, New York, 1963.
[g] *A. Nakahara, K. Hamada, I. Miyachi, K. Sakurai*, Bull. Chem. Soc. Jap. *40*, 2826 (1967); *S. E. Livingstone, J. D. Dolan*, Inorg. Chem. *7*, 1447 (1968).
[h] *M. P. Springer, C. Curran*, Inorg. Chem. *2*, 1270 (1963); *S. Yamada, S. Terashima, M. Wagatsuma*, Tetrahedron Lett. p. 1501 (1970).
[i] *B. Chiswell, F. Lions, M. L. Tomlinson*, Inorg. Chem. *3*, 492 (1964).
[j] *R. O. Gould, R. F. Jameson*, J. Chem. Soc. p. 15 (1963),
[k] *J. Iball, C. H. Morgan*, Nature *202*, 689 (1964); J. Chem. Soc. A, p. 52 (1967).
[l] *M. Nardelli, G. F. Gasparri, P. Boldrini, G. G. Battistini*, Acta Crystallogr. *19*, 491 (1965).
C. L. Fox, J. L. Lambert, Inorg. Chem. *8*, 2220 (1969).

Table 6 Copper(II) Sulfur and Selenium Complexes

Donor ligand (L)			
Structure	Name	Complex	Ref.
R–C($S^{\ominus}$)=C($S^{\ominus}$)–R[1]	1,2-Dithiolates	$[(C_6H_5)_4As]_2CuL_2$	*a*
	Arene 1,2-dithiolates	$[R_4N]_2CuL_2$	*b*
$NCS^{\ominus}$	Thiocyanates	$[CuL_4]^{2\ominus}$	*c*
$R_2N–CS–S^{\ominus}$	Dithiocarbamates	CuL_2	*d*
$R_2N–CSe–Se^{\ominus}$	Diselenocarbamates	CuL_2	*e*
$(RO)_2PS–S^{\ominus}$	Dithiophosphates	CuL_2	*f*
$R_2P(=S)–P(=S)R_2$	Diphosphine disulfides	$CuLX_2$	*g*

[a] *A. Davison, D. V. Houre, E. T. Shawl*, Inorg. Chem. *6*, 458 (1967).
[b] *M. J. Baker-Hawkes, E. Billig, H. B. Gray*, J. Amer. Chem. Soc. *88*, 4870 (1966).
[c] *D. Forster, D. M. L. Goodgame*, Inorg. Chem. *4*, 715, 823 (1965).
[d] *M. Delépine*, Bull. Soc. Chim. Fr. p. 5 (1958).
[e] *C. Furlani, E. Cervone, F. D. Camessei*, Inorg. Chem. *7*, 265 (1968).
[f] *A. I. Busev, M. L. Ivanyutin*, Tr. Kom. Anal. Khim. Akad. Nauk SSSR Inst. Geokkhim i Anal. Khim. *11*, 172 (1960); Chem. Abstr. *53*, 5954h (1960).
[g] *M. A. A. Beg, K. S. Hussain*, Chem. Ind. (London), p. 1181 (1966).

1.4.3 Nitrogen and Group V Compounds

1.4.3.1 Simple Compounds

Copper(II) azide is prepared from copper(II) nitrate and lithium or sodium azide.[112]

$$Cu(NO_3)_2 + 2NaN_3 \longrightarrow Cu(N_3)_2 + 2NaNO_3 \quad (28)$$

Copper(II) nitrate is best formed by treating copper metal with either nitrogen tetroxide[113] or pentoxide.[114]

1.4.3.2 Complexes

Preparation of most copper(II) coordination compounds involves no serious problems; usually the formation constant of the complex and its rate of formation are so favourable that simple mixing of a copper(II) salt and a donor ligand by itself or in a suitable solvent causes precipitation of a complex. In general, aromatic amines coordinate with copper more weakly than do aliphatic amines, and especially for unidentate ligands the stoichiometry of the complex often depends on the ratio of reactants used. Some recent (1965–1970) examples of nitrogen (Table 7) and mixed nitrogen and sulfur (Table 8) ligands are given.

1.4.4 Carbon Derivatives

Although diorganocopper(II) species have never been isolated, they have been suggested as reactive intermediates in conjugate addition reactions of organocopper(I) reagents[39,46] and in the reaction of copper(II) chloride with two equivalents of Grignard reagent.[115]

1.4.5 Copper(II) Transition Metal Derivatives

Two *copper(II) complexes with other transition metals* have recently been reported from the interaction of divalent copper with either trisoxalatochromium(III)[116] or a hydroxocyanatotungstate(IV)[117] leading to $CuCr(C_2O_4)_3^{\ominus}$ and $Cu[W(CN)_4(OH)_4]^{2\ominus}$, respectively.

1.5 Copper(III) Compounds

1.5.1 Halides and Complex Halides

Relatively few derivatives of copper(III) are known. *Dibromo-N,N-di-n-butyldithiocarbamatocopper(III)* is prepared by treating *N,N*-di-*n*-butyldithiocarbamatocopper(I) with an equimolar amount of bromine in carbon disulfide,[118] and potassium hexafluorocuprate(III) is prepared by fluorinating a 3:1 mixture of potassium chloride and copper dichloride at 250° in a flow system.[119]

Copper(III) periodates, $K_7CuI_2O_{12}\cdot 7H_2O$ and $Na_7CuI_2O_{12}\cdot 12H_2O$, have been prepared by anodic oxidation of strongly basic copper(II) solution in the presence of excess periodate ion.[120]

1.5.2 Oxygen and Group VI Derivatives

Potassium cuprate(III), $KCuO_2$, has been prepared from finely divided potassium oxide and copper(II) oxide at 400°–500° in dried oxygen at 760 mm Hg.[23] *Sodium cuprate(III)*, although unstable in solution, is formed by mixing copper(II) hydroxide and sodium hypobromite in strong base.[121]

Several dithiolene complexes of copper(II), for example, $[Cu(S_2C_2(CN)_2]^{2\ominus}$, have been oxidized to give copper formally in the III oxidation state.[122] A copper(III) complex with the tellurate anion, $[Cu(HTeO_6)_2]^{7\ominus}$, is reported to be produced when sodium tellurate in sulfuric acid is added to a basic solution of copper(II) chloride and sodium hypochlorite.[123] *Potassium bis(tellurato)cuprate(III)*, $K_9[Cu(TeO_6)_2]$, is prepared by oxidizing copper(II) tellurate with potassium thiosulfite.[124]

---

[112] *M. Straumanis, A. Cirulis*, Z. Anorg. Allg. Chem *251*, 315 (1943).

[113] *G. C. Addison, N. Logan*, Prep. Inorg. Reactions *1*, 154 (1964).

[114] *B. O. Field, C. J. Hardy*, J. Chem. Soc. p. 4428 (1964).

[115] *K. Wada, M. Tamura, J. Kochi*, J. Amer. Chem. Soc. *92*, 6656 (1970).

[116] *J. Aggett*, J. Inorg. Nucl. Chem. *31*, 3319 (1969).

[117] *Kabir-Ud-Din, A. Aziz Khan, M. Aijaz Beg*, J. Inorg. Nucl. Chem. *31*, 3657 (1969).

[118] *P. T. Beurskens, J. A. Cras, J. J. Steggerda*, Inorg. Chem. *7*, 810 (1968).

[119] *W. Klemm, E. Huss*, Z. Anorg. Allg. Chem. *258*, 221 (1949); *A. G. Sharpe*, Advan. Fluorine Chem. Butterworths, London, Vol. I (1960).

[120] *L. Malatesta*, Gazz. Chim. Ital. *71*, 467–474 (1941); Chem. Abstr. *36*, 6929 (1942); *G. Rozovskis, A. Prokopcikas, R. Jankauskas*, Izobret. Prom. Obraztsy. Tovarnye Znaki *44*, 29 (1967); Chem. Abstr. *69*, 11847g (1968).

[121] *J. S. Magee, Jr., R. H. Wood*, Can. J. Chem. *43*, 1234 (1965).

[122] *J. A. McCleverty*, Progr. Inorg. Chem. *10*, 49 (1962); for arguments in favour of considering such species as copper(II)-radical complexes, see *E. I. Stiefel, J. H. Waters, E. Billig, H. B. Gray*, J. Amer. Chem. Soc. *87*, 3016 (1965).

[123] *R. S. Banerjee, S. Basu*, J. Inorg. Nucl. Chem. *27*, 353 (1965); *M. W. Lister*, Can. J. Chem. *31*, 638 (1953).

[124] *M. Mraz, L. Jensovsky, J. Zyka*, Coll. Czech. Chem. Commun. *34*, 512 (1969).

Table 7 Copper(II) Nitrogen Complexes

Donor ligand (L)			
Structure	Name	Complex	Ref.
NH_3 or H_3N	Ammonia	CuL_2X_2, CuL_4X_2 CuL_6X_2	*a* *b*
$R-NH_2$	Amines	$CuL_2(OAc)_2$	*c*
$NO_2^{\ominus}$	Nitrate anion	$CuL_6^{4\ominus}$ $CuL_5^{3\ominus}$ $Cu_3L_{11}^{5\ominus}$	*d* *e* *f*
$OCN^{\ominus}$	Isocyanate anion	$CuL_4^{2\ominus}$	*g*
$SCN^{\ominus}$	Isothiocyanate anion	$CuL_4^{2\ominus}$	*h*
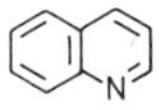	Quinoline	$CuL(acac)_2$	*i*
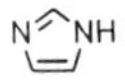	Imidazole	CuL_2 CuL_4Cl_2	*j* *k*
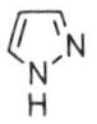	Pyrazole	CuL_4X_2	*l*
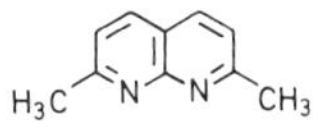	2,7-Dimethyl-1,8-naphthyridine	$CuL_3(ClO_4)_2$	*m*
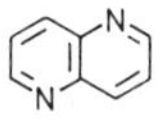	1,5-Diazanaphthalene	$CuLCl_2$	*n*
$R_2N-CH_2-CH_2-NR_2$	Aliphatic diamines	$CuLX_2$	*o*
$(H_2N-CH_2-CH_2-NH-CH_2-)_2CH_2$	1,4,8,11-Tetrazaundecane	$CuLX_2$	*p*
$[(H_3C)_2N-CH_2-CH_2-]_3N$	Tris(2-dimethylaminoethyl)amine	$[CuLX]X$	*q*
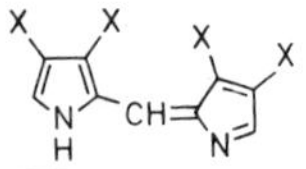	3,3′,4,4′-Tetrahalodipyrromethenes	CuL_2	*r*
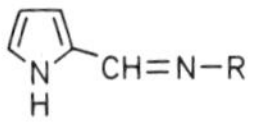	Pyrrole-2-aldimines	CuL_2	*s*
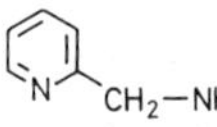	2-Aminomethylpyridine	CuL_2X_2	*t*
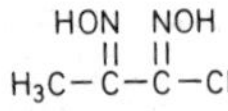 $H_3C-C(=NOH)-C(=NOH)-CH_3$	Dimethylglyoxime	CLX_2	*u*
$H_2N-CO-NH-CO-NH_2$	Biuret	CL_2	*v*
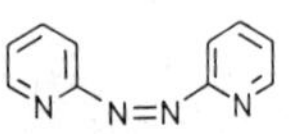	2,2′-Azopyridine	$(CuX_2)_2L$	*w*
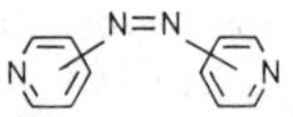	3,3′- and 4,4′-azopyridines	$CuLX_2$	*x*
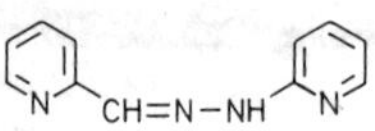	2-Pyridinecarboxaldehyde 2′-pyridylhydrazone	$CuLX_2$	*y*
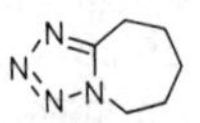	Pentamethylenetetrazole	$CuLX_2$	*z*
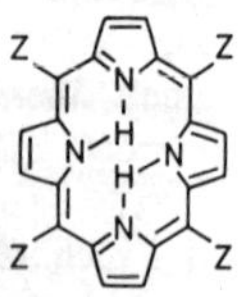	$\alpha,\beta,\gamma,\delta$-Tetra(4-pyridyl)porphine (Z = 4-pyridyl)	CuL	*aa*
	Phthalocyanine	—	*bb, cc*

[a] *B. J. Hathaway, A. A. G. Tomlinson*, Coordination Chem. Rev. *5*, 1 (1970).
[b] *H. Elliott, B. J. Hathaway*, Inorg. Chem. *5*, 885 (1966); *T. Distler, P. A. Vaughan*, Inorg. Chem. *6*, 126 (1967); *A. A. G. Tomlinson, B. J. Hathaway*, J. Chem. Soc. A. 1905 (1968).
[c] *G. Narain*, Can. J. Chem. *44*, 895 (1966).
[d] *H. Elliott, B. J. Hathaway, R. C. Slade*, Inorg. Chem. *5*, 669 (1966).
[e] *A. Garnier*, C.r. Acad. Sci. Ser. B *266*, 930 (1968); *I. Bernal*, Inorg. Chem. *3*, 1465 (1964); *A. Kurtennacker*, Z. Anorg. Allg. Chem. *82*, 207 (1913).
[f] *Z. I. Iorysh, I. M. Rumanova*, Kristallografiya *13*, 513 (1968); Chem. Abstr. *69*, 39514v (1968).
[g] *A. R. Chughtai, R. N. Keller*, J. Inorg. Nucl. Chem. *31*, 633 (1969).
[h] *D. Forster, M. L. Goodgame*, Inorg. Chem. *4*, 823 (1965); *P. Day*, Inorg. Chem. *5*, 1619 (1966).
[i] *P. Jose, S. Ooi, Q. Fernando*, J. Inorg. Nucl. Chem. *31*, 1971 (1969).
[j] *M. Inoue, M. Kishita, M. Kubo*, Bull. Chem. Soc. Jap. *39*, 1352 (1966).
[k] *W. J. Eilbeck, F. Holmes, A. E. Underhill*, J. Chem. Soc. A, p. 757 (1967); *M. Goodgame, L. I. B. Haines*, J. Chem. Soc. A, p. 174 (1966).
[l] *N. A. Daugherty, J. H. Swisher*, Inorg. Chem. *7*, 1651 (1968).
[m] *D. G. Hendricker, R. L. Bonder*, Inorg. Chem. *9*, 273 (1970).
[n] *R. W. Stotz, J. A. Walmsley, F. Walmsley*, Inorg. Chem. *8*, 807 (1969).
[o] *D. W. Meek, S. A. Ehrhardt*, Inorg. Chem. *4*, 584 (1965).
[p] *B. Bosnich, R. D. Gillard, E. D. McKenzie, G. A. Webb*, J. Chem. Soc. A, p. 1331 (1966); *D. C. Weatherburn, E. J. Billo, J. P. Jones, D. W. Margerum*, Inorg. Chem. *9*, 1557 (1970).
[q] *M. Ciampolini, N. Nardi*, Inorg. Chem. *5*, 41 (1966); *P. Paoletti, M. Ciampolini*, Inorg. Chem. *6*, 64 (1967).
[r] *R. J. Motekaitis, A. E. Martell*, Inorg. Chem. *9*, 1832 (1970).
[s] *K. Yeh, R. H. Barker*, Inorg. Chem. *6*, 830 (1967); *R. H. Holm, A. Chakravorty, L. J. Theriot*, Inorg. Chem. *5*, 625 (1966).
[t] *G. J. Sutton*, Aust. J. Chem. *16*, 371 (1963); *S. Utsono, K. Sone*, Bull. Chem. Soc. Jap. *37*, 1038 (1964), *40*, 105 (1967); J. Inorg. Nucl. Chem. *28*, 2647 (1966).
[u] *A. V. Ablov, N. I. Belichuk*, Russ. J. Inorg. Chem. Engl. Transl. *7*, 401 (1962); *K. Falk, E. Ivanova, B. Roos, T. Vänngard*, Inorg. Chem. *9*, 556 (1970).
[v] *A. W. McLellan, G. A. Melson*, J. Chem. Soc. A, p. 137 (1967).
[w] *D. A. Baldwin, A. B. P. Lever, R. V. Parish*, Inorg. Chem. *8*, 107, (1969).
[x] *P. J. Beadle, M. Goldstein, D. M. L. Goodgame, R. Grzeskowiak*, Inorg. Chem. *8*, 1490 (1969).
[y] *F. Lions, I. G. Dance, J. Lewis*, J. Chem. Soc. A, p. 565 (1967).
[z] *D. M. Bowers, A. I. Popov*, Inorg. Chem. *7*, 1594 (1968).
[aa] *E. B. Fleischer*, J. Amer. Chem. Soc. *85*, 1353 (1963); *P. Hambright, E. B. Fleischer*, Inorg. Chem. *9*, 1757 (1970).
[bb] *B. I. Knudsen*, Acta Chem. Scand. *20*, 1344 (1966).
[cc] *A. B. P. Lever*, Advan. Inorg. Chem. Radiochem. *7*, 63 (1965).

Table 8 Copper(II) Mixed Nitrogen and Sulfur Complexes

Donor ligand (L)			
Structure	Name	Complex	Ref.
$H_2N-CH_2-CH_2-SH$	β-Mercaptoethylamine	$CuLX_2$, CuL_2X_2, CuL_4X_2	*a*
$H_2N-NH-CS-NH_2$	Thiosemicarbazide	$CuLX_2$, CUL_2X_2	*b*
$H_2N-NH-CS-NH-NH_2$	Thiocarbohydrazide	CuL_3Cl_2	*c*
$R-NH-CS-NH-NH-CO-NH-R$	*N*-Thiocarbamoyl-*N'*-carbamoyl hydrazines	CuL	*d*
H_5C_2O $CH-CH_3$ $H_2N-CS-NH-N=CH-C=N-NH-CS-NH_2$	2-Keto-3-ethoxybutyraldehyde bis(thiosemicarbazone)	CuL	*e*
N $CH_2-CH_2-S-CH_3$	2-(2-Methylthioethyl)pyridine	$CuLX_2$, CuL_2X_2	*f*
N, SH, N, SH	2,4(1H, 3H)quinazolinodithione	CuL_3	*g*

[a] *P. Spacu, O. Constantinescu, I. Pascaru, M. Brezeanu, F. Zalaru*, Z. Anorg. Allg. Chem. *349*, 107 (1967).
[b] *M. J. Campbell, R. Greskowiak*, J. Chem. Soc. A, p. 396 (1967).
[c] *G. R. Burns*, Inorg. Chem. *7*, 277 (1968).
[d] *A. D. Ahmed, P. K. Mandal*, J. Inorg. Nucl. Chem. *28*, 1633 (1966).
[e] *M. R. Taylor, E. J. Gabe, J. P. Glusker, J. A. Minkin, A. L. Patterson*, J. Amer. Chem. Soc. *88*, 1845 (1966).
[f] *P. S. Chia, S. E. Livingstone, T. N. Lockyer*, Aust. J. Chem. *19*, 1835 (1966); *20*, 239 (1967); *L. F. Lindoy, S. E. Livingstone, T. N. Lockyer*, Aust. J. Chem. *19*, 1391 (1966); *20*, 471 (1967).
[g] *B. Singh, Lakshmi, U. Agarwala*, Inorg. Chem. *8*, 2341 (1969).

1.5.3 Copper(III) Nitrogen Derivatives

Biuretato ($^{\ominus}$HN—CO—NH—CO—NH$^{\ominus}$ = bi) complexes of copper(III) have been prepared by oxidizing biuretatocopper(II) complexes with $S_2O_8^{2\ominus}$ [125] or $IrCl_6^{2\ominus}$.[126]

$$Cu^{2\oplus} + IrCl_6^{2\ominus} + 2biH_2 + 4OH^{\ominus} \longrightarrow [Cu^{III}(bi)_2]^{2\ominus} + IrCl_6^{3\ominus} + 4H_2O \quad (29)$$

1.5.4 Carbon Derivatives

Copper(III) alkyl and aryl species have been postulated as transient intermediates in several reactions promoted by mono- and divalent copper, but have never been isolated or even detected.[127,128] A *carborane complex of copper(III)* has recently been prepared by treating 2,2′-dilithiobiscarborane with copper(II) dichloride and then with tetraethylammonium bromide to form $[(C_2H_5)_4N]Cu[(B_{10}C_2H_{10})_2]_2$ which was shown by X-ray analysis to contain a copper(III)–carbon σ bond.[129]

1.6 Mixed Valence Copper Complexes

Mixed valence copper complexes are of interest in terms of their physical (*e.g.*, electrical and optical) and chemical properties and in terms of bonding theory. A good review of such complexes[130] and a discussion of them as models for biological electron transfer systems are available.[131]

A binuclear mixed valence *copper(0)–copper(I) complex* was detected spectroscopically when a copper(I) alkyl was treated with copper(0),[132] and a stable copper(0)–copper(I) *cluster compound is* formed upon pyrolysis of a copper(I) aryl octamer.[133]

$$Ar_8Cu_8 \longrightarrow Ar_2 + Ar_6Cu_8 \quad (30)$$

Many mixed valence *copper(I)–copper(II) complexes* have been prepared but relatively few have been studied in detail.[130] Tetramminecopper(II) dihalocuprates(I), $Cu(NH_3)_4(CuX_2)_2$ (where X = I, Br, Cl) are prepared from tetramminecopper(II) halides, copper(I) halides and potassium or ammonium halides.[134] Mixing copper(I) and copper(II) perchlorates in acetate-buffered methanol produces a copper(I)–copper(II) acetate complex.[135] A sulfur-bridged copper(I)–copper(II) complex, $Cu_2(NCS)_3(NH_3)_3$, is formed by mixing copper(II) sulfate, ammonium carbonate, ammonia, ammonium thiocyanate,[136] and a tetracyanatotriammine complex of mono- and divalent copper, $Cu_3(CN)_4(NH_3)_3$, has been reported.[137]

1.7 Bibliography

M. C. Sneed, J. L. Margrave, R. C. Brasted, in Comprehensive Inorganic Chemistry, Vol. II, pp. 1–113. Van Nostrand, New York, 1954.

A. Butts (ed.), Copper, the Science and Technology of the Metal, Its Alloys and Compounds. Hafner Publishing, New York, 1954.

Gmelins Handbuch der Anorganischem. Chemie System No. 60—Kupfer. Verlag Chemie, Weinheim.

O. Glemser, H. Sauer, in *G. Brauer*, Handbook of Preparative Inorganic Chemistry, 2nd ed., Vol. II, pp. 1003–1027. Academic Press, New York, 1965.

J. Peisach, P. Aisen, W. E. Blumberg, The Biochemistry of Copper. Academic Press, New York, 1966.

J. Canterford, R. Colton, Halides of the First Row Transition Metals. Wiley, New York, 1968.

W. E. Hatfield, R. Whyman, in *R. L. Carlin*, Transition Metal Chemistry, Vol. V, pp. 47–179. Dekker, New York, 1969. Detailed review on structural aspects of copper complexes, 1962–1967.

J. Powell, Copper, Silver, Gold–Annual Survey Covering the Year 1969, Organometallic Chem. Rev. B *6*, 905 (1970); see also surveys covering previous years.

---

[125] *J. J. Bour, J. J. Steggerda*, Chem. Commun. p. 85 (1967).

[126] *A. Levitzki, M. Anbar*, Chem. Commun. p. 403 (1968).

[127] *J. K. Kochi, A. Bemis, C. L. Jenkins*, J. Amer. Chem. Soc. *90*, 4616 (1968).

[128] *G. H. Posner*, Ph.D. Thesis, Harvard University, Cambridge, Mass., 1968; Dissertation Abstr. *29*, 1613-B (1968).

[129] *D. A. Owen, M. F. Hawthorne*, J. Amer. Chem. Soc. *92*, 3194 (1970).

[130] *M. B. Robin, P. Day*, Advan. Inorg. Chem. Radiochem. *10*, 312–321 (1967).

[131] *R. J. P. Williams*, in *J. Peisach, P. Aisen, W. E. Blumberg*, The Biochemistry of Copper. Academic Press, New York and London, 1966.

[132] *K. Wada, M. Tamura, J. Kochi*, J. Amer. Chem. Soc. *92*, 6650 (1970).

[133] *A. Cairncross, W. M. Sheppard*, J. Amer. Chem. Soc. *93*, 247 (1971).

[134] *J. A. Baglio, H. A. Weakliem, F. Demelio, P. A. Vaughan*, J. Inorg. Nucl. Chem. *32*, 795, 803 (1970); *C. M. Harris*, J. Proc. Roy. Soc. N.S. Wales *82*, 218 (1948); Chem. Abstr. *44*, 6330d (1950).

[135] *C. Sigwart, P. Hemmerich, J. T. Spence*, Inorg. Chem. *7*, 2545 (1968).

[136] *J. Garaj*, Inorg. Chem. *8*, 304 (1969); *J. Garaj, J. Gazo*, Chem. Zvesti *19*, 593 (1965); Chem. Abstr. *63*, 14119g (1965).

[137] *M. Dunaj-Jurczo, M. A. Poraj-Koshic*, Chem. Zvesti *20*, 241 (1966); Chem. Abstr. *67*, 77036g (1967).

2 Silver

E. F. Rothgery
Department of Chemistry, University of Kentucky,
Lexington, Kentucky 40506, U.S.A.

2.1 Recovery and Purification

Owing to its ever increasing cost, silver residues are usually saved and the silver recovered. This can be accomplished by dissolving such residues in *aqua regia*, filtering through glass wool, re-dissolving several times and finally reducing to the metal with formic acid.[1]

2.2 Silver(0)

The complexes in which the oxidation state of silver is formally zero all belong to the (*o*-triars)silver–transition metal carbonyl compounds[2]: (*o*-triars)Ag–$Co(CO)_4$, (*o*-triars)Ag–$Mn(CO)_5$, and [(*o*-triars)Ag]$_2Fe(CO)_4$ [where *o*-triars = *o*-bis(trimethylarsenyl)benzene].

2.3 Silver(I)

The chemistry of silver(I) includes the large majority of compounds. The only silver(II) and (III) compounds are those with anions or ligands capable of stabilizing the oxidation state.

2.3.1 Halides and Halocomplexes

Although it is not actually a silver(I) compound, *silver subfluoride*, Ag_2F, is included here. It can be obtained by the electrolysis of *silver(I) fluoride*.[3] The reaction of silver carbonate with dilute hydrofluoric acid will give a silver(I) fluoride solution.[4]

Refluxing solutions of silver halides with the corresponding cesium halide yields the *halo complexes* Cs_2AgCl_3,[5] $CsAgBr_2$,[6] Cs_2AgBr_3,[6] and Cs_2AgI_3.[5]

The addition of boron trifluoride or phosphorus(V) fluoride to silver(I) fluoride in liquid sulfur dioxide results in the formation of *silver tetrafluoroborate* and *silver hexafluorophosphate*, respectively.[7]

The oxoacids of the halogens form an almost complete series of silver(I) salts. *Silver hypochlorite*, AgClO, is slightly stable in solution and is produced by the action of chlorine on a suspension of silver(I) oxide in water.[8] *Silver hypobromite* and *hypoiodite* decompose very rapidly even in solution. Silver also forms a chlorite, $AgClO_2$, which is an explosive solid.[9] The reaction of chloric acid with silver carbonate results in the formation of *silver chlorate*, $AgClO_3$.[10] *Silver bromate* and *iodate* are only slightly water-soluble and may be obtained by precipitation from aqueous solution. *Silver sulfate* should be employed as starting material since the nitrate will contaminate the product. *Silver periodate*, $AgIO_4$, slowly forms on mixing silver(I) oxide with periodic acid.[11]

2.3.2 Oxygen and Group VI Compounds

2.3.2.1 Binary Group VI Compounds and Salts of Oxoacids

Silver forms binary compounds with all the Group VI nonmetals. The *oxide* is prepared by the addition of an alkalimetal hydroxide to silver nitrate.[12] *Silver sulfide*,[13,16] *selenide*,[14,16] and *telluride*[15,16] are prepared by direct interaction of silver with the respective element. The basic permanganate oxidation of silver(I) oxide has been reported to lead to *silver(I) peroxide*.[17] *Silver thiosulfate*, $Ag_2S_2O_3$, is precipitated from aqueous solution in a metathetical reaction of silver acetate with sodium thiosulfate.[18] In higher concentrations of thiosulfate, soluble species such as $(AgS_2O_3)_2^{2-}$ are formed.[19] *Silver sulfite* is likewise only slightly soluble in water and can be precipitated using sodium sulfite and a soluble silver salt.[20]

---

1 *R. Maxon*, Inorg. Syn. *1*, 2 (1939).
2 *A. Kasenally, R. Nyholm, M. Stiddard*, J. Chem. Soc. (London) 5343 (1965).
3 *A. Hettich*, Z. Anorg. Allgem. Chem. *167*, 67 (1927).
4 *L. Poyer, M. Fielder, H. Harrison, B. Bryant*, Inorg. Synth. *5*, 18 (1957).
5 *C. Brink, C. MacGillavry*, Acta Cryst. *2*, 158 (1949).
6 *K. Lyaliakov*, Dokl. Akad. Nauk SSSR *65*, 171 (1949).
7 *D. Russell, D. Sharp*, J. Chem. Soc. (London), 4689 (1961).
8 *B. Ballard*, Ann. Chim. Phys. *57*, 239 (1834).
9 *G. Levi*, Gazz. Chim. Ital. *53*, 522 (1923).
10 *H. Foote*, J. Amer. Chem. Soc. *27*, 346 (1902).
11 *P. Gyani*, J. Indian Chem. Soc. *27*, 5 (1950).
12 *E. Madsen*, Z. Anorg. Allgem. Chem. *79*, 197 (1913).
13 *O. Hönigschmid, R. Sachtben*, Z. Anorg. Allgem. Chem. *195*, 207 (1931).
14 *O. Hönigschmid, W. Kapfenberger*, Z. Anorg. Allgem. Chem. *212*, 198 (1933).
15 *O. Hönigschmid*, Z. Anorg. Allgem. Chem. *214*, 281 (1933).
16 *B. Tavernier*, Z. Anorg. Allgem. Chem. *343*, 323 (1966).
17 *F. Jirsa*, Z. Anorg. Allgem. Chem. *225*, 302 (1935).
18 *W. Fernelius, K. Detlin*, J. Chem. Educ. *11*, 176 (1934).
19 *H. Brintzinger, W. Eckandt*, Z. Anorg. Allgem. Chem. *231*, 327 (1937).
20 *K. Seubert, M. Elten*, Z. Anorg. Allgem. Chem. *44*, 217 (1894).

Silver forms insoluble salts with the oxoacids of selenium. *Silver selenite*, Ag_2SeO_3, is formed on the reaction of selenous acid with silver carbonate.[16] *Silver selenate*, Ag_2SeO_4, is obtained in the same manner.[21]

The oxoacids of tellurium form three silver salts. *Silver tellurite*, Ag_2TeO_3, precipitates on mixing solutions of silver nitrate and potassium tellurite[16]. *Silver tellurate*, Ag_2TeO_4, and *orthotellurate*, Ag_2TeO_6, are likewise precipitated from solution.[22]

Silver(I) forms salts with practically all the phosphoric acids.[23] Recently, various *silver(I) fluorophosphates*[24] as well as a series of *silver aminophosphates*[25] have been prepared.

The reaction of silver(I) oxide with chromium(III) oxide at elevated temperature results in the formation of *silver chromite*, $AgCrO_2$,[26] and with iron(III) hydroxide in refluxing base *silver ferrate(III)*, $AgFeO_2$ is formed.[27] Metallic silver reacts with cobalt(III) oxyhydroxide in base to yield *silver cobaltite*, $AgCoO_2$,[28] and with Mn_3O_4 in different ratios to form Ag_2MnO_2 and $AgMn_2O_4$.[20] *Silver aluminate*, $AgAlO_2$, results from the high-temperature reaction between sodium aluminate and silver nitrate.[30] *Silver metavanadate*, $AgVO_3$, precipitates on the addition of silver nitrate to ammonium vanadate in solution.[31] Heating stoichiometric amounts of silver nitrate and molybdenum(VI) oxide will yield *silver molybdate*, $AgMoO_4$.[32] When soluble silver salts are added to sodium dihydrogen germanate, *silver gemanate*, Ag_4GeO_4, precipitates from solution.[33]

21 *A. Zachariasen*, Z. Kristallogr. *80*, 402 (1931).
22 *G. Jander*, *F. Kienbaum*, Z. Anorg. Allgem. Chem. *316*, 41 (1962).
23 *J. Laist*, Comprehensive Inorganic Chemistry, Vol. 2, p. 171. Van Nostrand, New York (1954).
24 *U. Schülke*, Z. Anorg. Allgem. Chem. *361*, 225 (1968).
25 *R. Klement*, *W. Gassner*, Z. Anorg. Allgem. Chem. *320*, 235 (1963).
26 *H. Hahn*, *C. deLorent*, Z. Anorg. Allgem. Chem. *290*, 69 (1957).
27 *W. Feitknecht*, *K. Moser*, Z. Anorg. Allgem. Chem. *304*, 181 (1960).
28 *W. Stählin*, *H. Oswald*, Z. Anorg. Allgem. Chem. *367*, 206 (1969).
29 *G. Rienäcker*, *K. Werner*, Z. Anorg. Allgem. Chem. *320*, 141 (1963).
30 *E. Thilo*, *W. Gassner*, Z. Anorg. Allgem. Chem. *345*, 151 (1967).
31 *L. Malprade*, Bull. Soc. Chim. France, 765 (1950).
32 *I. Kulesho*, Zh. Obshch. Khim. *21*, 451 (1951).
33 *G. Krüger*, Z. Anorg. Allgem. Chem. *330*, 267 (1964).

2.3.2.2 Complexes

A number of Group VI complexes with silver salts have been reported. In general, they are obtained by mixing solutions of the ligand with the silver salt. The following compounds have recently been described $(CH_3)_2NPS \cdot AgNO_3$,[34] $(H_2N)_2CS \cdot (AgNO_3)_2$, $OC_4H_8S \cdot AgNO_3$, and $2(CH_3)_2SeO \cdot AgClO_4$.

2.3.3 Group V Derivatives of Silver

2.3.3.1 Binary Compounds and Related Species

Hydrazoic acid precipitates colorless *silver azide*, AgN_3, from a silver nitrate solution.[35] Hydrazine likewise will give silver azide.[35] This material explodes violently at 300°. *Silver nitride*, Ag_3N, has been obtained by crystallization from an ammoniacal solution of silver oxide.[36] It is also violently explosive, detonating above 140° or from shock.

Silver amide, $AgNH_2$, results on the addition of potassium amide to a solution of silver nitrate.[37] Like the other silver–nitrogen compounds it is shock- and temperature-sensitive and is subject to photodecomposition.

When calcium cyanamide is added to a solution of ammoniacal silver(I), the *cyanamide* salt, Ag_2CN_2, precipitates from the solution.[38] *Silver tricyanomethide*, $AgC(CN)_3$, is formed by the reaction between potassium tricyanomethide and silver nitrate.[39]

A novel derivative of sulfamic acid is produced according to reaction (1).[40]

$$3Ag^{\oplus} + 2OH^{\ominus} + SO_3NH_2^{\ominus} \longrightarrow Ag_3NSO_3\text{–}2H_2O \quad (1)$$

The addition of lead nitrate to a mixture of silver nitrate and potassium thiocyanate yields the light-sensitive compound, $KPb[Ag(SCN)_2]_3$.[41] Diphenylphosphine oxide tautomerizes on reaction with silver nitrate to give $(C_6H_5)_2POAg$.[42]

34 *W. Slinkard*, *D. Meck*, Inorg. Chem. *8*, 1811 (1969).
35 *G. Tammann*, *C. Kröger*, Z. Anorg. Allgem. Chem. *169*, 16 (1928).
36 *H. Hahn*, *E. Gilbert*, Z. Anorg. Allgem. Chem. *258*, 77 (1949).
37 *R. Juza*, Z. Anorg. Allgem. Chem. *231*, 121 (1937).
38 *A. Chretien*, *B. Woringer*, C.R. Acad. Sci. *232*, 1114 (1951).
39 *J. Konnert*, *D. Britton*, Inorg. Chem. *5*, 1193 (1966).
40 *R. Paetzold*, *K. Dostal*, *Ruzicka*, Z. Anorg. Allgem. Chem. *348*, 1 (1966).
41 *K. Manolov*, Z. Anorg. Allgem. Chem. *323*, 207 (1963).
42 *L. Quin*, *R. Montgomery*, J. Inorg. Nucl. Chem. *28*, 1750 (1966).

2.3.3.2 Complexes

Quite a number of Group V silver complexes are known. They are generally prepared by mixing solutions of the silver salt and the ligand. For example, the addition of ethylenediamine (en) to a pyridine solution of silver thiocyanate or silver cyanide yields the compounds $(AgSCN)_2 \cdot en$ and $(AgCN)_2 \cdot en$.[43] The reaction of 1,2-ethanedithiocyanate with silver perchlorate in acetone gives an explosive white precipitate, $C_2H_4(SCN)_2 \cdot AgClO_4$.[44] Silver perchlorate forms adducts in benzene with adiponitrile[45] and succinonitrile.[46] A series of complexes of silver salts with substituted pyridines has been prepared in the same manner. The compounds include *bis(2-aminopyridine)silver perchlorate*,[47] *bis(4-methyl-2-cyanopyridine)silver perchlorate*,[48] *bis(cyanopyridine)silver perchlorate*,[49] *2-allylpyridinesilver nitrate*,[50] and silver perchlorate complexes with 2-, 3-, and 4-cyanopyridine oxide.[51]

Other nitrogen heterocycles and substituted heterocycles also form silver complexes. *Bis(imidazole)silver halides*[52] and *nitrate*[53] are formed in ethanol and crystallize on evaporation of the solvent. In similar manner *bis[3-phenyl-1-(2-pyridyl)-1,2-diaza-2-prop-ene]silver perchlorate* can be made.[54] Also known are *bis(5-nitro-1,10-phenanthroline)silver nitrate*,[55] *tetrakis(3,8-dimethyl-5,6-diaza-1,10-phenanthroline)tris(silver nitrate)*,[56] and *1,5-diazanaphthalene bis(silver nitrate)*.[57] Combining ligand and silver perchlorate in methanol gives a precipitate of *bis(2,7-dimethyl-1,8-naphthyridine)silver perchlorate*.[58]

The addition of triphenylphosphinesilver chloride to tin(II) chloride in acetone yields *triphenylphosphinesilver trichlorotin(II)*.[59] Likewise, mixing silver iodide and ligand in a saturated potassium iodide solution gives *iodo(o-allylphenyldiphenylphosphine)silver(I)*.[60] Silver nitrate and pyrazine when combined in warm water give crystals of $N_2C_4H_4 \cdot AgNO_3$[61] on cooling. The unusual ligand containing both phosphorus and nitrogen, dipiperidino(dicyclohexyl)biphosphine, forms the complex $(C_5H_{10}N)_2(C_6H_{11})_2P_2 \cdot (AgI)_2$.[62] Mixing silver nitrate, sodium perchlorate, and 4-methyl-2,6,7-trioxa-1-phosphabicyclo[2.2.2]octane in ethanol precipitates $(CH_3C(CH_2O)_3P)_4 \cdot AgClO_4$.[63] Diphenylarsinomethylpyridine gives a high melting complex with silver nitrate,[64] $C_5H_4NCH_2As(C_6H_5)_2 \cdot AgNO_3$.

2.3.4 Organometallic Compounds

2.3.4.1 Derivatives with Ag–C σ bonds

Alkylsilver compounds are known but are very unstable, decomposing rapidly even at −80° and have never been fully characterized.[65–67] However, the preparation of perfluoroalkylsilver compounds by the reaction of silver fluoride with fluoroalkens has been reported.[68]

$$AgF + >C=C< \longrightarrow F-\overset{|}{\underset{|}{C}}-\overset{|}{\underset{|}{C}}-Ag \quad (2)$$

The following compounds have been isolated in this manner: $(CF_3)_2CFAg$, $(CF_3)_2CClAg$, and $C_3F_3(CF_3)CFAg$.

Bis(phenylsilver)silver nitrate, $(C_6H_5Ag)_2AgNO_3$, has been prepared by the reaction of silver nitrate with triphenylethyltin(IV).[69] This compound is

43 *K. Brodersen, T. Kahlert*, Z. Anorg. Allgem. Chem. *348*, 273 (1966).
44 *R. Frank, H. Droll*, J. Inorg. Nucl. Chem. *32*, 3110 (1970).
45 *D. Barnhart, C. Caughlin, M. Haque*, Inorg. Chem. *8*, 2768 (1969).
46 *M. Kubota, D. Johnson, T. Matsubara*, Inorg. Chem. *5*, 386 (1966).
47 *E. Uhlig, M. Mädler*, Z. Anorg. Allgem. Chem. *338*, 199 (1965).
48 *L. El-Sayed, R. Ragsdale*, J. Inorg. Nucl. Chem. *30*, 651 (1968).
49 *B. Chiswell, F. Lions*, Aust. J. Chem. *22*, 71 (1969).
50 *R. Yingst, B. Douglas*, Inorg. Chem. *3*, 1177 (1964).
51 *L. Nathan, J. Nelson, G. Rich, R. Ragsdale*, Inorg. Chem. *8*, 1495 (1969).
52 *C. Perchard, A. Novak*, J. Chim. Phys. *65*, 1964 (1968).
53 *J. Bauman, J. Wang*, Inorg. Chem. *3*, 368 (1964).
54 *F. Farha, R. Iwamoto*, Inorg. Chem. *4*, 844 (1965).
55 *A. Synmal*, Sci. Cult. *34*, 258 (1968).
56 *J. Porter, J. Murray*, J. Amer. Chem. Soc. *87*, 1628 (1965).
57 *R. Stoltz, J. Walmsley, F. Walmsley*, Inorg. Chem. *8*, 807 (1966).
58 *D. Hendricker, R. Bodner*, Inorg. Chem. *9*, 273 (1970).
59 *J. Dilts, M. Johnson*, Inorg. Chem. *5*, 2079 (1966).
60 *M. Bennett, W. Kneen, R. Nyholm*, Inorg. Chem. *7*, 552 (1968).
61 *J. Schmidt, R. Trimble*, J. Phys. Chem. *66*, 1063 (1962).
62 *W. Seidel*, Z. Anorg. Allgem. Chem. *335*, 316 (1965).
63 *J. Verkade, T. Piper*, Inorg. Chem. *1*, 453 (1962).
64 *E. Uhlig, M. Schäfer*, Z. Anorg. Allgem. Chem. *359*, 178 (1968).
65 *G. Semerano, I. Riccaboni*, Ber. Deut. Chem. Ges. *74*, 1089 (1941).
66 *G. Semerano, I. Riccaboni*, Ber. Deut. Chem. Ges. *74*, 1297 (1941).
67 *C. Brown, R. Johnson*, J. Chem. Soc. 3923 (1960).
68 *W. Miller, R. Burnard*, J. Amer. Chem. Soc. *90*, 7367 (1968).
69 *E. Krause, M. Schmitz*, Ber. Deut. Chem. Ges. *52*, 2159 (1919).

extremely light sensitive and slowly decomposes at room temperature, even in the dark. *Phenylsilver* and several substituted derivatives have been obtained *via* Grignard reactions with silver halides.[70] However, they are unstable and have never been obtained in a pure state. Reaction (3) of silver chloride with lithium perfluorophenyl leads to a stable, moisture-sensitive solid.[71]

$$AgCl + 2LiC_6F_5 \longrightarrow Li[Ag(C_6F_5)_2] + LiCl \qquad (3)$$

The reaction of silver nitrate with tetravinyllead at $-78°$ in ethanol yields a precipitate of *vinylsilver*[72] which decomposes above $-30°$. The most stable compound of the alkenylsilver species is *styrylsilver* which is prepared in the same manner.[73] The reaction of silver(I) fluoride with perfluoro-2-butyne gave the first *perfluorovinylsilver compound*, $CF_3CF{=}C(CF_3)Ag$; this compound is thermally stable.[74]

Silver acetylides have been known for about 100 years. They are prepared by passing the alkyne through ammoniacal silver(I) solutions. The following compounds were obtained in this manner: $AgC{\equiv}CAg$,[75] $CH_3C{\equiv}CAg$,[76] and $C_6H_5C{\equiv}CAg$.[77] The use of neutral silver solutions results in compounds of the type $RC{\equiv}CAg \cdot nAgNO_3$.[75] The insoluble *polymeric alkynylsilver compounds*, $(RC{\equiv}CAg)_n$, react with ligands which are sufficiently strong donators to form adducts which are less associated. Examples of this type of adducts are $C_6H_5C{\equiv}CAg \cdot P(CH_3)_3$,[78] $C_6H_5C{\equiv}CAg \cdot As(C_2H_5)_3$,[78] and $C_6H_5C{\equiv}CAg \cdot NH_2C_3H_7$.[79] Several *complex silver acetylides* of the general type $K[(RC{\equiv}C)_2Ag]$ (where R = H, CH_3, C_6H_5) are known.[80] The complexes are prepared from silver nitrate and excess potassium acetylide in liquid ammonia. They are not explosive, but are very sensitive to light and moisture.

70 *E. Krause, B. Wendt*, Ber. Deut. Chem. Ges. *56*, 2064 (1923).
71 *V. Smith, A. Massey*, J. Organometal. Chem. *23*, C9 (1970).
72 *A. Holliday, R. Pendleburg*, J. Organometal. Chem. 7, 281 (1967).
73 *F. Glockling, D. Kingston*, J. Chem. Soc. 3001 (1959).
74 *W. Miller, R. Snider, R. Hummel*, J. Amer. Chem. Soc. *91*, 6532 (1969).
75 *R. Vestin, A. Somersalo*, Acta Chem. Scand. *3*, 125 (1949).
76 *C. Liebermann*, Ann. Chem. *135*, 268 (1865).
77 *C. Liebermann, G. Damerov*, Ber. Deut. Chem. Ges. *25*, 1096 (1892).
78 *G. Coates, C. Parkin*, J. Inorg. Nucl. Chem. *22*, 59 (1961).
79 *D. Blake, G. Calvin, G. Coates*, Proc. Chem. Soc. 396 (1959).
80 *R. Nast, H. Schindel*, Z. Anorg. Allgem. Chem. *201*, 326 (1963).

2.3.4.2 π-Complexes of Silver Salts

π-Complexes of silver salts are prepared by direct reaction of an organic ligand with the silver salt. This reaction can be accomplished by dissolving the silver salt in the hydrocarbon, passing the hydrocarbon through a silver salt solution, or by mixing solutions of the two components. Many compounds of this type exist only in solution and will not be included in this work; however, many can be isolated as crystalline solids.

Alkenesilver salt complexes. White, needlelike crystals of $C_2H_4 \cdot 2AgNO_3$ have been prepared by passing ethylene into a silver nitrate solution at 0°.[81] Liquid complexes of propene and 1-butene with silver nitrate are known.[82] They contain 1.3 mol of alkene per mol of silver and dissociate unless kept under a partial pressure of the alkene. Ethylene, propene and butenes form a series of complexes with silver tetrafluoroborate,[83] the most stable having the general formula $2C_nH_{2n} \cdot AgBF_4$.

Dienesilver salt complexes. Two crystalline complexes of 1,3-butadiene with anhydrous silver tetrafluoroborate, $C_4H_6 \cdot AgBF_4$ and $3C_4H_6 \cdot 2AgBF_4$, have been isolated.[84] A 3:2 complex of 2-methyl-1,3-butadiene and 1:1 complexes of 1,3- and 1,4-pentadiene with silver tetrafluoroborate have been reported.[84]

Addition of 1,5-hexadiene, 1,7-octadiene, or 1,9-decadiene to solutions of silver perchlorate in anhydrous ethanol/diethyl ether solutions gave the crystalline complexes $3C_6H_{10} \cdot 2AgClO_4$, $C_8H_{14} \cdot AgClO_4$, and $C_{10}H_{18} \cdot AgClO_4$.[85]

Cycloalkenesilver salt complexes. Crystalline complexes of cyclohexene with silver nitrate and perchlorate have been prepared by dissolving the silver salt in the warm alkene and cooling the solution.[86] A crystalline complex of *cis*-cyclooctene has been isolated; it can be crystallized from methanol at 0°.[87] The *cis* and *trans* isomers

81 *K. Tarama, M. Sano, K. Tatsuoka*, Bull. Chem. Soc. Jap. *36*, 1366 (1963).
82 *A. Francis*, J. Amer. Chem. Soc. *73*, 3709 (1951).
83 *H. Quinn, D. Glew*, Can. J. Chem. *40*, 1103 (1962).
84 *H. Quinn*, Can. J. Chem. *45*, 1329 (1967).
85 *G. Bressan, R. Broggi, M. Lachi, A. Segre*, J. Organometal. Chem. *9*, 355 (1967).
86 *A. Comyns, H. Lucas*, J. Amer. Chem. Soc. *79*, 4339 (1957).
87 *W. Jones*, J. Chem. Soc. 1808 (1954).

of both cyclononene and cyclodecene form 2:1 complexes with silver nitrate.[88]

Cyclodienesilver salt complexes. The reaction of cyclopentadiene with silver tetrafluoroborate in diethyl ether gives the complex $C_5H_6 \cdot AgBF_4$ in good yield.[89] The perchlorate complex is sensitive to light and air and is explosive.[90] 1,3-Cyclohexadiene forms a stable adduct in the same manner.[89] The three isomeric cyclooctadienes form complexes with silver nitrate, of different stability.[91] This property has been employed as a means of separating the isomers. A solid silver complex has been prepared by stirring 1,5-cyclononadiene with 1 M aqueous silver nitrate.[92] Silver nitrate complexes of *cis,trans*-1,5- and *cis,cis*-1,6-cyclodecadiene with the formula $C_{10}H_{16} \cdot AgNO_3$ have been obtained as colorless precipitates from the cyclodecadienes and concentrated aqueous silver nitrate.[93]

Cyclopolyenesilver salt complexes. The two isomers of cyclooctatriene form complexes with silver nitrate[87]; the 1,3,6-cyclooctatriene is formed more readily than the 1,3,5-isomer. *cis,cis,cis*-1,4,7-Cyclononatriene reacts readily with excess silver nitrate to form $C_9H_{12} \cdot 3AgNO_3$, which is unusually stable toward light and oxygen and melts at a higher temperature than silver nitrate.[94] Several complexes of 1,5,9-cyclododecatriene are known,[95] the stabilities of which vary with the configuration of the triene.

Cyclooctatetraene forms three complexes with silver nitrate,[96] which on heating revert to the same species. A number of adducts of silver nitrate with various substituted cyclooctatetraenes have likewise been reported.[97] Silver nitrate and tetrafluoroborate form complexes with several tricyclodienes and tricyclotrienes.[98]

Solid bicyclic compounds result from interaction of α- and β-pinene[86] and 2,5-norbornadiene[99,100] with silver salts.

Silver salt–arene complexes are generally less stable than are alkene adducts. Benzene complexes have been reported with silver perchlorate[101] and silver tetrachloroaluminate.[102] In addition, silver perchlorate forms adducts with toluene[103] and the xylenes.[104]

2.3.5 Boron-containing Silver Compounds

Dibromoboro[bis(2-dimethylarsinophenyl)-methylarsine]silver(I), (triars)$AgBBr_2$, is obtained on the reaction of (triars)silver(I) bromide with bis(diphenylphosphino)ethanebis(dibromomborane) cobalt(II).[105]

The two white solid *silver phosphine complexes*, $(C_6H_5)_3PAgB_9H_{12}S$ and $(C_6H_5)_3PAgB_{10}H_{13}$, are produced when caesium salts of the boron anion are added to ethanol–water solutions of silver nitrate and triphenylphosphine.[106] Mixing solutions of silver nitrate, alkali metal salt of 1,2-dicarbanonaborane, and tetraethylammonium bromide results in the formation of $[(C_2H_5)_4N^{\oplus}]$ $[Ag(C_2B_9H_{11})_2]^{3\ominus}$.[107]

2.4 Silver(II)

There are only a few silver(II) compounds known. They fall into three categories: fluorides, oxides, and nitrogen complexes.

2.4.1 Halides

Silver(II) fluoride is a black hygroscopic solid which reacts violently with water. It is a very strong oxidizing and fluorinating agent and is prepared by the fluorination of silver(I) halides.[108,109]

---

[88] *A. Cope, D. McLean, N. Nelson*, J. Amer. Chem. Soc. *77*, 1628 (1955).
[89] *G. Stotzle*, Ph.D. Thesis, Univ. of Munich. Munich, Germany, 1961.
[90] *T. Ulbricht*, Chem. Ind. (London), 1570 (1961).
[91] *W. Jones*, 312 (1954).
[92] *G. Nagendrappa, D. Devaprabhakara*, J. Organometal. Chem. *15*, 225 (1968).
[93] *J. Trebellas, R. Olechowski, H. Jonassen*, Inorg. Chem. *4*, 1818 (1964).
[94] *R. Jackson, W. Streib*, J. Amer. Chem. Soc. *89*, 2539 (1967).
[95] *G. Wilke*, Angew. Chem. *75*, 10 (1963).
[96] *A. Cope, F. Hochstein*, J. Amer. Chem. Soc. *72*, 2515 (1950).
[97] *A. Cope, H. Campbell*, J. Amer. Chem. Soc. *74*, 179 (1952).
[98] *M. Avram, E. Slian, C. Nenitzescu*, Ann. Chem. *636*, 184 (1960).
[99] *H. Quinn*, Can. J. Chem. *46*, 117 (1968).
[100] *J. Traynham, J. Olechowski*, J. Amer. Chem. Soc. *81*, 571 (1959).
[101] *A. Hill*, J. Amer. Chem. Soc. *44*, 1163 (1922).
[102] *R. Turner, E. Amma*, J. Amer. Chem. Soc. *88*, 3243 (1966).
[103] *A. Hill, F. Miller*, J. Amer. Chem. Soc. *47*, 2702 (1925).
[104] *G. Peyronel, G. Belmondi, J. Vezzosi*, J. Inorg. Nucl. Chem. *8*, 577 (1958).
[105] *G. Schmid*, Chem. Ber. *102*, 191 (1969).
[106] *F. Klanberg, E. Muetterties, L. Guggenberger*, Inorg. Chem. *7*, 2273 (1968).
[107] *V. Brattsev, V. Stanko*, Zh. Obshch. Khim. *38*, 2721 (1968).
[108] *M. Ebert, Rodowskas, J. Frazer*, J. Amer. Chem. Soc. *55*, 3056 (1933).
[109] *H. Priest*, Inorg. Syn. *3*, 176 (1950).

2.4.2 Silver(II) Oxides

The basic oxidation of silver(I) oxide with persulfate gives what has been described as *silver(II) oxide.*[110] The material is a black solid which has many uses as an oxidizing agent in preparative organic chemistry. The electrolysis of silver nitrate can lead to several complex silver(II) oxides with formulas such as $Ag_7O_8NO_3$ and $Ag_{14}O_{16}SO_4$.[111]

2.4.3 Silver(II)–Nitrogen Complexes

Bivalent silver can be stabilized by several nitrogen ligands. The general procedure for the preparation of these compounds is to oxidize silver(I) in the presence of the ligand with persulfate[112] or with ozone.[113] Ligands employed include pyridine,[114] dipyridyl,[115] phenanthroline,[116] and succinonitrile[117] (Table 1).

Table 1 Silver(II) Compounds

Compound[a]	References
AgF_2	*b*
AgO	*c*
$Ag_7O_8NO_3$	*d*
$Ag(py)_4(NO_3)_2$	*e*
$Ag(dipy)_2(NO_3)_2$	*f*
$Ag(o\text{-phen})_2(NO_3)_2$	*g*
$Ag(sucntr)_2(NO_3)_2$	*h*

[a] py, pyridine; dipy, dipyridyl; *o*-phen, ortho-phenanthroline; sucntr, succinonitrile.
[b] *H. Priest*, Inorg. Syn. *3*, 176 (1950).
[c] *R. Hammer, J. Kleinberg*, Inorg. Syn. *4*, 176 (1953).
[d] *J. McMillan*, Chem. Rev. *62*, 65 (1962).
[e] *G. Barbieri*, Ber. Deut. Chem. Ges. *60*, 2424 (1927).
[f] *W. Hieber*, Ber. Deut. Chem. Ges. *61*, 2149 (1928).
[g] *J. Selbin, B. Shamburger*, J. Inorg. Nucl. Chem. *24*, 1153 (1962).
[h] *M. Suzuki, T. Schimanouchi*, Bull. Chem. Soc. Japan *41*, 2353 (1968).

2.5 Silver(III) Compounds

Silver(III) compounds are of three types: fluorocomplexes, periodates and tellurates, and nitrogen complexes. All are diamagnetic.

2.5.1 Halides

There are no simple binary silver(III) halides but several *tetrafluoroargentate(III) salts* are known and are prepared according to reaction (4).[118]

$$AgNO_3 + MCl + 2F_2 \xrightarrow{400^\circ} MAgF_4 \quad M = K_1Cs \qquad (4)$$

In a similar manner[119] $CsAgF_4$ and $BaAgF_5$ were obtained. All these compounds are moisture-sensitive and react violently with water.

2.5.2 Oxygen-Containing Compounds

Periodate complexes of silver(III) have been prepared according to reaction (5).[120] In like manner:

$$Ag_2 + 8KOH + 4KIO_4 \xrightarrow{H_2O} 2K_6HAg(IO_6)_2 \cdot 10H_2O \qquad (5)$$

entire series of periodates can be produced, *i.e.*, $K_6HAg_6(IO_6)_2 \cdot 10H_2O$, $K_7Ag(IO_6)_2 \cdot KOH \cdot 8H_2O$, and $KNa_6Ag(IO_6)_2 \cdot NaOH \cdot H_2O$.[121] The same compounds can be obtained by electrolysis of silver(I) solutions in the presence of base and periodates[122] or from the reaction of iodate with silver(II) oxide.[123]

Tellurates of the general formula $M_{9-n}H_nAg(TeO_6)_2 \cdot xH_2O$ result from the persulfate oxidation of silver(I) in the presence of base and tellurium(II) oxide[124] (Table 2).

Table 2 Silver(III) Compounds

Compound	References
$KAgF_4$	*a*
$CsAgF_4$	*b*
$BaAgF_5$	*b*
$K_6HAg(IO_6)_2 \cdot 10H_2O$	*c*
$K_7Ag(IO_6)_2 \cdot KOH \cdot 8H_2O$	*c*
$KNa_6Ag(IO_6)_2 \cdot NaOH \cdot H_2O$	*c*
$M_{9-n}HnAg(TeO_6)_2 \cdot xH_2O$	*d*
$Ag(Ethylenedibiguanide)(NO_3)_3$	*e*

[a] *W. Klemm*, Angew. Chem. *66*, 468 (1954).
[b] *R. Hoppe*, Z. Anorg. Allgem. Chem. *292*, 28 (1957).
[c] *L. Malatesta*, Gazz. Chim. Ital. *71*, 467 (1941).
[d] *B. Brauner, B. Kutzna*, Ber. Deut. Chem. Ges. *43*, 3362 (1907).
[e] *P. Ray*, Nature (London), *151*, 643 (1943).

[110] *R. Hammer, J. Kleinberg*, J. Amer. Chem. Soc. *4*, 12 (1953).
[111] *J. McMillan*, Chem. Rev. *62*, 65 (1962).
[112] *A. Malagusta, T. Labianca*, Gazz. Chim. Ital. *84*, 976 (1956).
[113] *T. Morgan, F. Burnstall*, J. Chem. Soc., 2594 (1930).
[114] *G. Barbieri*, Ber. Deut. Chem. Ges. *60*, 2424 (1927).
[115] *W. Hieber*, Ber. Deut. Chem. Ges. *61*, 2149 (1928).
[116] *J. Selbin, B. Shamburger*, J. Inorg. Nucl. Chem. *24*, 1153 (1962).
[117] *M. Suzuki, T. Schimanouchi*, Bull. Chem. Soc. Jap. *41*, 2353 (1968).
[118] *W. Klemm*, Angew. Chem. *66*, 468 (1954).
[119] *R. Hoppe*, Z. Anorg. Allgem. Chem. *292*, 28 (1957).
[120] *L. Malaprade*, C.R. Acad. Sci. *210*, 504 (1940).
[121] *L. Malatesta*, Gazz. Chim. Ital. *71*, 467 (1941).
[122] *L. Jenovshy, M. Skala*, Z. Anorg. Allgem. Chem. *312*, 26 (1961).
[123] *G. Cohn, G. Atkinson*, Inorg. Chem. *3*, 1741 (1964).
[124] *B. Brauner, B. Kutzna*, Ber. Deut. Chem. Ges. *43*, 3362 (1907).

2.5.3 Nitrogen Complexes

The literature contains two examples of silver(III)–nitrogen complexes. The first reported[125] was *silver(ethylenedibiguanide)*X_3 ($X = SO_4$, NO_3, OH, and ClO_4) obtained on the persulfate oxidation of silver salts in presence of the ligand. Conductance studies have shown it to be a tripositive complex cation.[126] In like manner *bis(biguanidium)silver(III) salts* have been prepared.[127]

[125] *P. Ray*, Nature (London), *151*, 643 (1943).
[126] *P. Ray*, *N. Chakravarty*, J. Indian Chem. Soc. *21*, 47 (1944).

2.6 Bibliography

C. Beverwijk, *G. van der Kerk*, *A. Leusink*, *J. Noltes*, Organosilver Chemistry, Organometal. Chem. Revs. (A), 215 (1970).
J. McMillan, Higher Oxidation States of Silver, Chem. Revs. *62*, 65 (1962).
R. Vestin, *E. Ralf*, Silver Acetylides, Acta Chem. Scand. *3*, 101 (1949).
Organometallic Chemistry Reviews, Section B, Contains yearly reviews of the literature.

[127] *D. Sen*, J. Chem. Soc. (A), 1304 (1969).

3 Gold

E. F. Rothgery
Department of Chemistry, University of Kentucky,
Lexington, Kentucky 40506, U.S.A.

3.1 Recovery and Purification

The high cost of gold renders its recovery from residues mandatory. It can be purified by dissolving it in *aqua regia*. Excess nitric acid is removed by heating, diluting with distilled water, and allowing the solution to stand to precipitate any silver present. Sulfur dioxide is then bubbled through the solution to reduce the gold to the metal form; the solution is boiled to remove chlorine and coagulate the gold precipitate.[1]

Since gold metal costs about the same, weight for weight, as most gold compounds, it has been suggested[2] that it would be much more economical to purchase metallic gold, convert it to a powder by dissolving the metal in *aqua regia* and reducing the resulting $Au^{3\oplus}$ with hydroquinone. This gold powder can be reacted with bromine and potassium bromide to yield pure *potassium tetrabromoaurate(III)*, $K[AuBr_4]$, which is a common starting material in gold chemistry.

3.2 Gold(0)

The reaction of sodium salts of metal carbonyls with triphenylphosphinegold(I) chloride [(Eqn (1)] leads to several compounds which contain gold in the zero oxidation state,[3] *e.g.*,

$(C_6H_5)_3P \cdot Au—Mn(CO)_5$, $(C_6H_5)_3P \cdot Au—Co(CO)_4$, and

$[(C_6H_5)_3P \cdot Au]_2—Fe(CO)_4 \cdot$

$$(C_6H_5)_3P \cdot AuCl + NaM(CO)_x \longrightarrow NaCl(C_6H_5)_3P \cdot Au—M(CO)_x \quad (1)$$

3.3 Gold(I)

Gold(I) compounds are usually not stable in aqueous solution, disproportionating to give gold metal and gold(III) oxide.

3.3.1 Gold(I) Halides

Gold(I) chloride[4] is prepared by heating gold(III) chloride or $H[AuCl_4]$ to 175°. In the same manner *gold(I) bromide* is obtained from tetrabromoauric acid at 100°.[4] Warming gold powder with elemental iodine gives *gold(I) iodide*.[5] Reducing gold(III) chloride with sulfur dioxide in the presence of a high concentration of sodium chloride reportedly gives $NaAuCl_2$.[6] Adding gold powder to silver perchlorate in acetonitrile yields $Au(CH_3CN)_2ClO_4$.[7]

Gold(I) halides form a number of *complexes* with phosphines, isocyanides, and olefin ligands. These compounds are discussed in Sections 3.3.3 and 3.3.4, respectively.

3.3.2 Group VI Derivatives of Gold(I)

The existence of gold(I) oxide and hydroxide is still debated; the structure of the product obtained on the addition of base to gold(I) solutions has not been elucidated. *Cesium oxoaurate(I)*, CsAuO, results on heating CsAu at 400° in oxygen.[8]

The addition of hydrogen sulfide to an acidic solution of potassium dicyanoaurate(I) precipitates *gold(I) sulfide*, Au_2S.[9] *Polysulfides of gold(I)* such as $NH_4[AuS_3]$ form when gold(I) sulfide is added to ammonium sulfide.[10]

Triphenylphosphinegold(I) chloride and $CH_3SSi(CH_3)_3$ react to form $(C_6H_5)_3P \cdot AuSCH_3$.[11] Two mole of trifluoromethyl dithietene add to triphenylphosphinegold(I) chloride in refluxing benzene to produce $(C_6H_5)_3P \cdot AuCl \cdot [(CF_3)_2C_2S_2]_2$.[12] *Sodium bis(thiosulfato)-aurate(I)*, $Na_3[Au(S_2O_3)_2]$, is a very stable compound prepared by the action of an alkali metal thiosulfate on aqueous gold(III) chloride.[13]

Gold(I) telluride, Au_2Te, occurs naturally as a mineral and can be prepared in the laboratory by heating these elements[14] to 400°.

3.3.3 Gold(I) Group V Derivatives

When potassium azide is added to a solution of gold(III) chloride the very explosive *gold(I)*

---

[1] *T. Thorpe, A. Laurie*, J. Chem. Soc. *51*, 565 (1887).
[2] *B. Block*, Inorg. Syn. *4*, 14 (1953).
[3] *C. Coffey, J. Lewis, R. Nyholm*, J. Chem. Soc. 174 (1964).
[4] *W. Biltig, W. Wein*, Z. Anorg. Allgem. Chem. *148*, 192 (1925).
[5] *L. Brüll, F. Griffi*, Ann. Chim. Appl. *28*, 536 (1938).
[6] *M. Diemer*, J. Amer. Chem. Soc. *35*, 552 (1913).
[7] *G. Bergerhoff*, Z. Anorg. Allgem. Chem. *327*, 127 (1964).
[8] *H. Wasel-Nielson, R. Hoppe*, Z. Anorg. Allgem. Chem. *359*, 36 (1968).
[9] *L. Hoffmann, G. Krüse*, Ber. Deut. Chem. Ges. *20*, 2361 (1887).
[10] *K. Hoffmann, F. Hochtlen*, Z. Anorg. Allgem. Chem. *37*, 245 (1904).
[11] *E. Abel, C. Jenkins*, J. Organometal. Chem. *4*, 285 (1968).
[12] *A. Davidson, D. Howe, E. Shawl*, Inorg. Chem. *6*, 458 (1967).
[13] *H. Brown*, J. Amer. Chem. Soc. *49*, 958 (1927).
[14] *B. Brauner*, Monatsh. *10*, 411 (1889).

azide, AuN_3, forms.[15] When ammonia is reacted with gold(I) oxide $Au_3N \cdot NH_3$ results.[16] Dissolving gold in a solution of potassium cyanide and *aqua regia* gives the salt $K[Au(CN)_2]$.[17] Heating this material in a solution of hydrochloric acid produces *gold(I) cyanide*, AuCN.[18] Reacting gold(III) chloride with a saturated solution of phosphine in water gives *gold(I) phosphide* [Eqn (2)].[19]

$$3AuCl_3 + 2PH_3 \xrightarrow{3H_2O} Au_3P + 9HCl + H_3PO_3 \quad (2)$$

A number of *isocyanide complexes* of gold(I) compounds are listed in Table 1.

Table 1. Isocyanide Complexes of Gold(I)

Compound	M.p. (°C)	Reference
$C_6H_5{-}NC \cdot AuCl$	—	*a*
p-$CH_3{-}C_6H_5NC \cdot AuCl$	—	*a*
p-$CH_3{-}C_6H_4NC \cdot AuCN$	244 d	*a*
p-$F{-}C_6H_4NC \cdot AuCl$	247 d	*b*
p-$F{-}C_6H_4NC \cdot AuC_6H_4(p\text{-}F)$	154 d	*b*

[a] *A. Sacco, M. Freni*, Gazz. Chim. Ital. *86*, 195 (1956).
[b] *L. Vaughan, W. Sheppard*, J. Amer. Chem. Soc. *91*, 6161 (1969).

Gold(I) halides form stable adducts with triphenylphosphine. The adduct $(C_6H_5)_3P \cdot AuCl$ is prepared by the addition of freshly distilled triphenylphosphine to an ethanol solution of gold(III) chloride.[20] This compound is used as the starting material for the preparation of the corresponding bromide and iodide compounds which result from a metathesis reaction in acetone.[21] The addition of excess triphenylphosphine to acetone solutions of these adducts will give the bis adducts, $[(C_6H_5)_3P]_2 \cdot AuX$.[21] The addition of diphenyl(*o*-diphenylarsinophenyl)phosphine in hot butanol to sodium tetrachloroaurate(III) gives the white crystalline adduct,[22]

$As(C_6H_5)_2$ / $P(C_6H_5)_2$ · AuCl

The reaction of tetrachloroauric acid with the ligand $P(OCH_2)_3CCH_3$ in ether precipitates the dimer $[ClAuL_2]_2$.[23]

$[(H_3C{-}C(CH_2O)_3P)_2\ AuCl]_2$

3.3.4 Group IV Derivatives

One *alkenylgold(I) compound* has been described. The reaction of sodium cyclopentadienide with an olefin·AuCl complex results in the formation of C_5H_5Au.[24] *Gold(I) acetylide*, Au_2C_2, is highly explosive.[25] Several other acetylides can be prepared as shown in Eqn (3).[25]

$$K{-}C{\equiv}C{-}R + AuI \xrightarrow{NH_3} Au{-}C{\equiv}C{-}R + KI \quad (3)$$

If an excess of potassium acetylide is used, complexes of the type $KAu(C{\equiv}CR)_2$ result.[25] *Alkinylgold(I) derivatives*, $RC{\equiv}CAu$, are formed on the reaction of freshly reduced gold(III) chloride with alkali metal alkynides, the products being slightly soluble polymers. In the presence of donor molecules soluble monomeric species of the type $RC{\equiv}CAu \cdot L$ result[20] (Table 2).

Table 2. Complexes of Phenylethynylgold(I) $C_6H_5C{\equiv}CAu \cdot L$[a]

Ligand (L)	M.p. (°C)
$(C_2H_5)_3P$	83–85
$(C_6H_5)_3P$	163–164
$(C_2H_5)_3As$	67–68
$(C_2H_5)_3Sb$	96–97
Pyridine	80 d
$C_5H_{11}NH_2$	124–125
C_4H_9NC	40–41

[a] From *G. Coates, C. Parkin*, J. Chem. Soc. 3220 (1962).

Sigma-bonded alkyl and aryl derivatives of gold(I) (Table 3) appear to exist only in the presence of stabilizing complexing agents, primarily phosphines. These materials are prepared by reaction of organolithium compounds with phosphine complexes of gold(I) halides[27] as illustrated in Eqn (4).

$$R^1_3P \cdot AuX + LiR^2 \longrightarrow R^1_3P \cdot AuR^2 + LiX \quad (4)$$

[15] *T. Curtius, J. Rissom*, J. Prakt. Chem. *58*, 304 (1898).
[16] *R. Raschig*, Ann. Chem. *235*, 341 (1886).
[17] *F. Chemnetius*, Chem. Zt. *51*, 823 (1927).
[18] *K. Himly*, Ann. Chem. *42*, 157 (1842).
[19] *F. Moser, K. Brukl*, Z. Anorg. Allgem. Chem. *121*, 73 (1922).
[20] *B. Mann, A. Wells, D. Purdie*, J. Chem. Soc. 1828 (1937).
[21] *J. Meyer, A. Allred*, J. Inorg. Nucl. Chem. *30*, 1328 (1968).
[22] *P. Nicpon, D. Meek*, Inorg. Chem. *6*, 145 (1967).
[23] *J. Verkade, T. Piper*, Inorg. Chem. *1*, 453 (1962).
[24] *R. Hüttel, U. Raffay, H. Reinheimer*, Angew. Chem. *79*, 859 (1967).
[25] *R. Nast, U. Kirner*, Z. Anorg. Allgem. Chem. *330*, 311 (1964).
[26] *G. Coates, C. Parkin*, J. Chem. Soc. 3220 (1962).
[27] *G. Calvin, G. Coates, P. Dixon*, Chem. Ind. (London). 1628 (1959).

Table 3. Alkylgold(I) and Arylgold(I) Complexes

Compound	M.p. (°C)	Ref.
$CH_3Au \cdot P(C_2H_5)_3$	62	*a*
$CH_3Au \cdot P(C_6H_5)_3$	175	*a*
$C_2H_5Au \cdot P(C_2H_5)_3$	51	*a*
$C_2H_5Au \cdot P(C_6H_5)_3$	130	*a*
$C_6H_5Au \cdot P(C_2H_5)_3$	168	*a*
$C_6H_5Au \cdot P(C_6H_5)_3$	152	*b*
$F-C_6H_4-Au \cdot P(C_6H_5)_3$	154	*c*
Mesityl-$Au \cdot P(C_6H_5)_3$	100 (dec.)	*a*
$C_5H_5Au \cdot P(C_6H_5)_3$	100 (dec.)	*d*
$C_6F_5Au \cdot P(C_6H_5)_3$	160 (dec.)	*e*

[a] *G. Coates, C. Parkin,* J. Chem. Soc. 3220 (1962).
[b] *F. Glockling, K. Hooton,* J. Chem. Soc. 2658 (1962).
[c] *L. Vaughan, W. Sheppard,* J. Amer. Chem. Soc. *91,* 6161 (1969).
[d] *R. Hüttel, U. Raffay, H. Reinheimer,* Angew. Chem. *79,* 857 (1967).
[e] *R. Nyholm, P. Royo,* Chem. Commun. 421 (1969).

Olefin–gold(I) complexes have been primarily prepared by the reaction of gold(III) chloride or tetrachloroauric acid with an olefin at −20° to −60°.[28, 29] Under carefully controlled conditions compounds with the formula olefin·Au_2X_4 result. They have been shown to be mixtures of gold(I) and gold(III); no gold(II) is present.[30] Table 4 contains a list of olefin–gold(I) complexes.

The reaction of triphenylphosphine gold(I) chloride with triphenylsilyl lithium yields the light- and air-sensitive compound $(C_6H_5)_3P \cdot AuSi(C_6H_5)_3$.[31] In a similar fashion the compound $(C_6H_5)_3P \cdot AuGe(C_6H_5)_3$ has been prepared.[32] The addition of tin(II) chloride dihydrate to tris(triphenylphosphine)gold(I) chloride results in $[(C_6H_5)_3P]_3 \cdot AuSnCl_3$.[33]

3.3.5 Gold(I) Boron Compounds

The reaction of diphenylboron bis(diphenylphosphinoethane)cobalt with triphenylphosphinegold(I) chloride gives the gold boron complex $(C_6H_5)_3P \cdot AuB(C_6H_5)_2$.[34] The reaction of sodium tetrachloroaurate and triphenylphosphine with $CsB_9H_{12}S$ leads to the compound $[(C_6H_5)_3P]_3AuB_9H_{12}S$.[35]

3.4 Gold(II)

Several gold(II) compounds have been mentioned in the literature, but in most cases the actual oxidation state of the gold is questionable. One definitely characterized gold(II) compound is $[(C_2H_5)_4N]^{\oplus}[Au(C_2B_9H_{11})_2]^{\ominus}$ which is obtained on reducing the corresponding gold(III) compound with sodium amalgam[36] (see Section 3.5.4).

A series of compounds AuX_2L_n have been produced where the ligand is bis(diphenylphosphino)acetylene, $n = 1, 3, 4$, and X = halogen, $NCS^{\ominus}$, $BF_4^{\ominus}$, and $PF_6^{\ominus}$.[37]

3.5 Gold(III)

3.5.1 Gold(III) Halides

Heating gold powder with bromine trifluoride has been reported to give AuF_3.[38] The fluorina-

Table 4. Olefin Complexes of Gold(I) Chloride

Ligand (L)	Decomposition temperature (°C)	Ref.
$L \cdot AuCl$		
Cyclopentene	55–60	*a*
Cyclohexene	55	*a*
Cycloheptene	93–98	*a*
cis-Cyclooctene	81–85	*a*
1-Dodecene	23–24	*b*
1-Hexadecene	43–45	*b*
1,5-Cyclooctadiene	110–114	*a*
Norbornadiene	75–78	*a*
Dicyclopentadiene	90–95	*a*
1,4-Hexadiene	50	*b*
$L_nAu_2Cl_4$		
Norbornadiene ($n = 1$)	50	*c*
Norbornadiene ($n = 2$)	62–64	
Norbornadiene ($n = 3$)	78–80	
Dicyclopentadiene ($n = 3$)	80–85	*c*

[a] *R. Hüttel, H. Reinheimer, H. Deitl,* Chem. Ber. *99,* 462 (1966).
[b] *R. Hüttel, H. Reinheimer,* Chem. Ber. *99,* 2778 (1966).
[c] *R. Hüttel, H. Reinheimer, K. Nowak,* Chem. Ber. *101,* 3761 (1968).

[28] *R. Hüttel, H. Dietl,* Angew. Chem. *77,* 456 (1965).
[29] *H. Chalk,* J. Amer. Chem. Soc. *86,* 4733 (1964).
[30] *R. Hüttel, R. Reinheimer, K. Nowak,* Chem. Ber. *101,* 3761 (1968).
[31] *M. Baurd,* J. Inorg. Nucl. Chem. *29,* 367 (1967).
[32] *F. Glockling, M. Wilbey,* J. Chem. Soc. (A), 2168 (1968).
[33] *J. Dilts, M. Johnson,* Inorg. Chem. *5,* 2079 (1966).
[34] *G. Schmid, H. Nöth,* Chem. Ber. *100,* 2899 (1967).
[35] *F. Klansberg, E. Muetterties, L. Guggenheierm,* Inorg. Chem. *7,* 2273 (1968).
[36] *L. Warren, M. Hawthorne,* J. Amer. Chem. Soc. *90,* 4823 (1968).
[37] *A. Carty, A. Efraty,* Inorg. Chem. *8,* 543 (1969).
[38] *L. Asprey, F. Kruse, K. Jack, R. Maitland,* Inorg. Chem. *3,* 602 (1964).

tion of $K[AuCl_4]$ at 400° yields $K[AuF_4]$.[39] *Gold(III) chloride*, $(AuCl_3)_2$, may be formed by direct combination of the elements[40] at 200° or by heating $KAuCl_4$.[41] The *potassium tetrachloroaurate* may be obtained by dissolving gold powder in a solution of chlorine and potassium chloride.[42] In a similar manner the addition of gold to bromine water gives $(AuBr_3)_2$.[43] If potassium bromide is present, $K(AuBr_4)$ results[2]; $HAuBr_4$ is obtained with hydrobromic acid.[2] Gold(III) chloride and potassium iodide will give a precipitate of AuI_3 in water; in the presence of excess potassium iodide $K[AuI_4]$ is formed.[44] When 5-nitro-1,1,10-phenanthroline and tetrachloroauric acid are refluxed in a water–ethanol solution and sodium perchlorate is added $[(C_{10}H_7N_3O_2)AuCl_2]^{\oplus}ClO_4^{\ominus}$ precipitates.[45] If pyridine and sodium carbonate are added to a solution of tetrachloroauric acid the adduct py·$AuCl_3$ is produced.[46] If pyridine is added to tetrabromoauric acid *pyridinium tetrabromoaurate(III)* results,[47] $pyH^{\oplus}AuBr_4^{\ominus}$. When halogen is added to a solution of potassium dicyanoaurate(I), $KAu(CN)_2$, the series of compounds, $KAu(CN)_2Cl_2$,[48] $KAu(CN)_2Br_2$,[48] and $KAu(CN)_2I_2$[49] results. When cesium carbonate and hydrochloric acid are reacted with $(CH_3)_2AuOH$ the salt $Cs(CH_3)_2AuCl_2$ is produced.[50] Cesium can form a series of *hexachloroaurate(III) salts* having the general formulas $Cs_4M(II)(AuCl_6)_2$ and $Cs_2M(I)(AuCl_6)$.[51]
Gold(III) forms a large number of *alkylgold halides* which are discussed in Section 3.5.3.

3.5.2 Gold(III) Group VI Compounds

The action of sodium carbonate on potassium tetrachloroaurate gives *gold(III) hydroxide*, $Au(OH)_3$.[52] On heating the hydroxide to 100° it converts to the *oxide*, Au_2O_3.[53] When hydrogen sulphide is added to a solution of $K[AuCl_4]$ the *gold(III) sulfide*, Au_2S_3, precipitates.[54,55] When gold is heated with alkali metal oxides or hydroxides moisture-sensitive alkali gold oxides with the formulas Na_3AuO_3, $KAuO_2$, $RbAuO_2$, and $CsAuO_2$ are produced.[56]
When tetramethylammonium bromide is added to a solution of tetrachloroauric acid and potassium cyanothiocarbamate a precipitate of $[(CH_3)_4N]^{\oplus}[Au(S_2C_2N_2)_2]^{\ominus}$ forms.[57] In the same manner tetrabutylammonium bromide precipitates $[(C_4H_9)_4N]^{\oplus}[Au(C_6Cl_4S_2)_2]^{\ominus}$ from a solution of $KAuCl_4$ and the ligand tetrachlorobenzene-1,2-dithiol.[58]
Complex sulfites[59] *of gold(III)*, $K_5Au(SO_3)_4 \cdot 5H_2O$ and $Na_5Au(SO_3)_4 \cdot 14H_2O$, form on addition of potassium or sodium sulfite to sodium aurate(III). Gold metal dissolves in 87% selenic acid, the only acid to dissolve gold unaided by additional or complexing agents. Refluxing of the metal and selenic acid in bromobenzene results in the formulation of $Au_2(SeO_4)_3$.[60]
Gold(III) oxide or hydroxide dissolve in nitric acid; cooling such solutions to 0° will precipitate $HAu(NO_3)_4$. The corresponding alkali metal salts may also be obtained.[61] *Gold(III) orthoarsenate* has been prepared as the hydrate, $AuAsO_3 \cdot H_2O$, by precipitation from a solution of potassium orthoarsenate and gold(III) chloride in 50% ethanol.[62] Gold(III) chloride also reacts with potassium carbonate and potassium thiocyanate to give $K[Au(SCN)_4]$.[63]

---

39 *R. Hoppe, W. Klemm*, Z. Anorg. Allgem. Chem. *268*, 364 (1952).
40 *V. Gutmann*, Z. Anorg. Allgem. Chem. *264*, 169 (1951).
41 *H. Topsöe*, Ber. Wien. Acad. *69*, 261 (1874).
42 *W. Billig, W. Wein*, Z. Anorg. Allgem. Chem. *148*, 192 (1925).
43 *A. Buraway, C. Gibson*, J. Chem. Soc. 217 (1935).
44 *B. Johnson*, Phil. Mag. *9*, 266 (1836).
45 *L. Cattalini, A. Doni, A. Ono*, Inorg. Chem. *6*, 280 (1967).
46 *L. Cattalini, M. Tobe*, Inorg. Chem. *5*, 1145 (1966).
47 *G. Reinaker, G. Blumenthal*, Z. Anorg. Allgem. Chem. *328*, 8 (1964).
48 *W. Mason*, Inorg. Chem. *9*, 2688 (1970).
49 *V. Gerdy*, J. Prakt. Chem. *29*, 181 (1843).
50 *W. Scovell, R. Tobias*, Inorg. Chem. *9*, 945 (1970).
51 *A. Ferrari, R. Cecconi, L. Cavalea*, Gazz. Chim. Ital. *73*, 23 (1943).
52 *R. Lyden*, Z. Anorg. Allgem. Chem. *240*, 157 (1939).
53 *H. Schütza, J. Schütza*, Z. Anorg. Allgem. Chem. *245*, 59 (1940).
54 *A. Gutbier, E. Dürrwächter*, Z. Anorg. Allgem. Chem. *121*, 266 (1922).
55 *K. Dubey*, Z. Anorg. Allgem. Chem. *327*, 309 (1965).
56 *R. Hoppe, K. Arenal*, Z. Anorg. Allgem. Chem. *314*, 4 (1962).
57 *F. Cotton, J. McCleverty*, Inorg. Chem. *6*, 229 (1967).
58 *M. Baker-Hawkes, E. Billig, H. Gray*, J. Amer. Chem. Soc. *88*, 4870 (1966).
59 *A. Rosenheim*, Z. Anorg. Allgem. Chem. *59*, 198 (1908).
60 *W. Caldwell, L. Eddy*, J. Amer. Chem. Soc. *71*, 2247 (1949).
61 *R. Schottlander*, Ann. Chem. *217*, 365 (1883).
62 *K. Stavenhagen*, J. Prakt. Chem. *51*, 1 (1895).
63 *R. Cleve*, J. Prakt. Chem. *94*, 14 (1865).

3.5.3 Gold(III) Carbon Compounds

Trimethylgold[64] is formed on reaction of gold(III) bromide with methyllithium in ether at −65°; however, it begins to decompose at −45°. Trimethylgold is stabilized by donors such as nitrogen and especially phosphorus ligands[65] (see Table 5).

Table 5. Trimethylgold Adducts

Compound	M.p. (°C)	Ref.
$(CH_3)_3Au \cdot O(CH_2CH_3)_2$	−40 (dec.)	*a*
$(CH_3)_2Au \cdot$ 2-aminopyridine	—	*a*
$2(CH_3)_3Au \cdot$ ethylenediamine	94 (dec.)	*a*
$(CH_3)_3Au \cdot P(C_6H_5)_3$	120 (dec.)	*b*
$(CH_3)_3Au \cdot P(CH_3)_3$	23	*b*

[a] *H. Gilman, L. Woods*, J. Amer. Chem. Soc. *70*, 550 (1948).
[b] *G. Coates, C. Parkin*, J. Chem. Soc. 421 (1963).

Triarylgold compounds are virtually unknown. Recently, *tris(pentafluorophenyl)gold* and its triphenylphosphine adduct were prepared via the Grignard reaction of gold(III) chloride with pentafluorophenylmagnesium bromide.[66]

Dialkylgold halides are generally formed by the reaction of gold(III) halides with alkylmagnesium halides in ether.[67] *Dialkylgold cyanides* can be obtained from the bromides by reaction with silver cyanide[67] (see Table 6). The reaction of gold(III) chloride with excess potassium cyanide will yield $K[Au(CN)_4]$.[68] When a strong acid is added to this material gold(III) cyanide, $Au(CN)_3$ is liberated.[68]

Table 6. Dialkylgold Halides and Cyanides

Compound	M.p. (°C)	Ref.
$(CH_3)_2AuCl$	71	*a*
$(CH_3)_2AuBr$	69	*a*
$(CH_3)_2AuI$	95	*c*
$(CH_3CH_2)_2AuCl$	48	*b*
$(CH_3CH_2)_2AuBr$	58	*b*
$(CH_3CH_2)_2AuI$	70	*d*
$(CH_3CH_2)_2AuCN$	104	*d*
$(C_3H_7)_2AuCl$	Liquid	*b*
$(C_3H_7)_2AuBr$	Liquid	*b*
$(C_3H_7)_2AuCN$	95	*b*
$(Benzyl)_2AuCl$	70 (dec.)	*b*
$(Benzyl)_2AuBr$	77 (dec.)	*b*
$(Benzyl)_2AuCN$	122 (dec.)	*b*

[a] *H. Gilman, L. Woods*, J. Amer. Chem. Soc. *70*, 550 (1948).
[b] *M. Kharasch, H. Isbell*, J. Amer. Chem. Soc. *53*, 2701 (1931).
[c] *S. Harris, R. Tobias*, Inorg. Chem. *8*, 2259 (1969).
[d] *C. Gibson, J. Simonsen*, J. Chem. Soc. 2531 (1930).

Dimethylgold hydroxide, $[(CH_3)_2AuOH]_4$, is obtained from dimethylgold iodide and sodium hydroxide.[69] In like manner the reaction of sodium trimethylsilanoate, $NaOSi(CH_3)_3$, with dimethylgold bromide gives *trimethylsiloxodimethylgold*,[70] $(CH_3)_2AuOSi(CH_3)_3$. Colorless, crystalline *dimethylgold acetate* was recently prepared from dimethylgold bromide and silver acetate in tetrahydrofuran.[71]

Dialkylgold compounds of dicarboxylic acids have been known for some time. They are prepared by reaction of aqueous solutions of bis(dialkylgold) sulfates with sodium salts of the dicarboxylic acid[72] (Table 7). The reaction of dialkylgold bromides with silver sulfate in acetone yields *bis(dialkylgold) sulfates*.[67] In a similar manner dialkylgold phenylphosphates and phenylarsenates have been prepared.[73]

Table 7. Dialkylgold(III) Compounds Containing Au–O Bonds

Compounds[a]	M.p. (°C)	Ref.
$(CH_3)_2Auacac$	84	*b*
$[(CH_3)_2AuOH]_4$	120 (dec.)	*c*
$[(CH_3)_2AuOSi(CH_3)_3]_2$	40	*d*
$[(CH_3)_2AuO_2CCH_3]_2$	97	*e*
$\{[(C_2H_5)_2Au]_2SO_4\}_2$	97	*f*
$[(C_2H_5)_2Au]_2(COO)_2$	81	*f*
$[(C_2H_5)_2Au]_2C_nH_{2n}(COO)_2$	—	*f*
$[(C_2H_5)_2Au]_3PO_4$	—	*g*
$[(C_2H_5)_2AuO_2P(OC_6H_5)_2]_2$	71	*h*

[a] acac=acetylacetonate.
[b] *F. Brain, C. Gibson*, J. Chem. Soc. 762 (1939).
[c] *C. Gibson, J. Simonsen*, J. Chem. Soc. 2531 (1930).
[d] *H. Schmidbaur, M. Bergfeld*, Inorg. Chem. *5*, 2069 (1966).
[e] *M. Bergfeld, H. Schmidbaur*, Chem. Ber. *102*, 2048 (1969).
[f] *C. Gibson, W. Weller*, J. Chem. Soc. 102 (1941).
[g] *G. Glass, J. Konnert, M. Miles, D. Britton, R. Tobias*, J. Amer. Chem. Soc. *90*, 1131 (1968).
[h] *M. Foss, C. Gibson*, J. Chem. Soc. 3075 (1949).

[64] *H. Gilman, L. Woods*, J. Amer. Chem. Soc. *70*, 550 (1948).
[65] *G. Coates, E. Parkin*, J. Chem. Soc. 421 (1963).
[66] *L. Vaughan, W. Sheppard*, J. Organometal. Chem. *22*, 739 (1970).
[67] *M. Kharasch, H. Isbell*, J. Amer. Chem. Soc. *53*, 2701 (1931).
[68] *C. Hinly*, Ann. Chem. *42*, 337 (1842).
[69] *M. Miles, G. Glass, R. Tobias*, J. Amer. Chem. Soc. *88*, 5738 (1966).
[70] *H. Schmidbaur, M. Bergfeld*, Inorg. Chem. *5*, 2069 (1966).
[71] *M. Bergfeld, H. Schmidbaur*, Chem. Ber. *102*, 2048 (1969).
[72] *C. Gibson, W. Weller*, J. Chem. Soc. 102 (1941).
[73] *M. Foss, C. Gibson*, J. Chem. Soc. 3075 (1949).

Dialkylgold thiolates, R_2AuSR', can be prepared from dialkylgold halides and thiols in the presence of base.[67]

Dimeric diethylgold thiocyanate, $[(C_2H_5)_2AuSCN]_2$, is formed in the reaction of diethylgold bromide with silver thiocyanate.[74] Similar compounds are prepared using substituted thiocyanates.[75]

Alkylgold dihalides of the type $RAuX_2$ are very unstable and only the bromides are known. They are formed on reaction of dialkylgold bromides with bromine in carbon tetrachloride[67,76] (see Eqn 5).

$$R_2AuBr + Br_2 \longrightarrow RAuBr_2 + RBr \tag{5}$$

[74] *W. Gent, C. Gibson,* J. Chem. Soc. 1835 (1949).

[75] *H. Blaauw, R. Nivard, G. van der Kerk,* J. Organometal. Chem. *2*, 236 (1964).

[76] *F. Brain, C. Gibson,* J. Chem. Soc. 762 (1939).

Several *arygold dichlorides* are known.[77] They are produced in small yield by the reaction of benzene with gold(III) chloride.

Diethylgold bromide forms adducts with ammonia,[78] pyridine,[78] and dibenzyl sulfide.[76]

3.5.4 Gold(III) Boron Derivatives

When tetraethylammonium bromide is added to a solution containing gold(III) chloride and the $[B_9C_2H_{11}]^{\ominus}$ ion the compound $[(C_2H_5)_4N]^{\oplus}$-$[Au(B_9C_2H_{11})_2]^{\ominus}$ results.[36]

3.6 Bibliography

B. Armer, H. Schmidbaur, Organogold Chemistry, Angew. Chem. Int. Ed. *9*, 101 (1970).

[77] *M. Karasch, H. Isbell,* J. Amer. Chem. Soc. *53*, 3053 (1931).

[78] *C. Gibson, J. Simonsen,* J. Chem. Soc. 2531 (1930).

4 Zinc

C. W. Blewett
Research Laboratory, Emery Industries, Cincinnati, Ohio, U.S.A.

R. Schmid
Battelle Memorial Institute, Genf, Schweiz

H. Zimmer
Department of Chemistry, University of Cincinnati, Cincinnati, Ohio, U.S.A.

A wide variety of inorganic salts of zinc, cadmium, and mercury are commercially available. However, preparation of the *anhydrous* materials cannot be accomplished by direct heating of hydrated salts, since the latter are basic salts. A general method for the dehydration of hydrates involves their interaction with 2,2′-dimethoxypropane resulting in the removal of water and formation of acetone and methanol.[1]

$$H_3C{-}CO{-}CH_3 + H_2O \qquad (CH_3)_2C(OCH_3)_2$$

The anhydrous salts can also be prepared by refluxing the hydrates with thionyl chloride. In this process hydrogen chloride and sulfur dioxide are formed as volatile materials which are readily removed from the reaction mixture.[2]

$$M_nX_m \cdot H_2O + SOCl_2 \longrightarrow SO_2 + 2HCl + M_nX_m$$

4.1 Inorganic Compounds of Zinc

4.1.1 Zinc Halides

Zinc fluoride, ZnF_2, is formed on interaction of zinc carbonate with hydrofluoric acid. Evaporation of the solution yields a partially dehydrated material which can be transformed into anhydrous zinc fluoride by heating to 800°. On passing dry hydrogen chloride over metallic zinc at 700° zinc chloride is formed. The material sublimes and is deposited in the colder section of the reaction vessel.[3]

$$Zn + 2HCl \longrightarrow ZnCl_2 + H_2$$

At higher reaction temperatures the resultant product is partially contaminated with elemental zinc.

Anhydrous *zinc chloride* is also obtained on interaction of zinc with hydrogen chloride in ether. Heating *in vacuo* is sufficient to remove excess ether and HCl.[4]

Electrolysis of copper(I) chloride in acetonitrile using a zinc anode and working in a nitrogen atmosphere provides (after evaporation of the solvent) a solvatized zinc chloride. Gentle heating of the latter yields the pure zinc chloride.[5] It can be further purified by sublimation in a stream of hydrogen chloride.

Hydrated zinc chloride can be converted to the anhydrous material as described above.[1,2] On heating a concentrated solution of zinc chloride and zinc oxide in water to 150° *zinc hydroxychloride* is formed which slowly precipitates on cooling of the reaction mixture. The crystalline product is washed with acetone and dried *in vacuo* over $CaCl_2$.

$$ZnO + ZnCl_2 + H_2O \longrightarrow 2Zn(OH)Cl$$

Heating a mixture of zinc chloride, ammonium chloride, and water provides for *ammonium tetrachlorozincate*, $(NH_4)_2ZnCl_4$.[4]

Zinc bromide, $ZnBr_2$, is obtained by electrolysis of copper(I) bromide in anhydrous acetonitrile using a zinc anode. After evaporation of the solvent the acetonitrile adduct of the desired salt is obtained and is converted by gentle heating into $ZnBr_2$.[5] This compound is also obtained by anodic dissolution of a zinc electrode in an aqueous solution of hydrobromic acid and bromine. The extremely hygroscopic crystals of $ZnBr_2$ have a m.p. 394°, b.p. 650°.[4]

Zinc iodide, ZnI_2, is formed on interaction of zinc dust with iodine in aqueous solution or by refluxing a zinc–iodine mixture in anhydrous ether. It can also be prepared by the electrolytic method using copper(I) iodide. Zinc iodide is very hygroscopic (m.p. 446°, b.p. 624°).[4]

4.1.2 Zinc–Chalcogen Compounds

Zinc hydroxide exists in several modified forms; ϵ-$Zn(OH)_2$ is formed on boiling zinc oxide with aqueous sodium hydroxide. On standing for several weeks the desired crystalline compound precipitates from the solution and is dried over sulfuric acid. On addition of a stoichiometric amount of ammonia to an aqueous solution of zinc sulfate zinc hydroxide precipitates. The precipitate is washed and dissolved in ammonia; slow evaporation of the solvent yields crystalline zinc hydroxide. The colorless crystals decompose at elevated temperatures.[4]

Zinc sulfide, ZnS, is formed by introducing hydrogen sulfide into an ammonium acetate-buffered solution of zinc sulfate at pH 2–3 or by adding ammonium sulfide to a zinc sulfate solution. The resulting precipitate is washed with 2% acetic acid (saturated with hydrogen sulfide). In order to prepare oxygen-free material all operations have to be performed in the absence of air and the material is dried at 150° in a nitrogen atmosphere. On heating the precipitated zinc sulfide in a stream of

---

[1] *K. Starke*, J. Inorg. Nucl. Chem. *11*, 77 (1959).
[2] *R. Pray*, Inorg. Synth. *5*, 153 (1957).
[3] *W. Kwansik*, in *G. Brauer*, Handbook of Preparative Inorganic Chemistry, Vol. I, p. 27, 242, 2nd Ed. Academic Press, New York and London, 1963.
[4] *F. Wagenknecht*, *R. Juza*, in *G. Brauer*, Handbook of Preparative Inorganic Chemistry, Vol. II, p. 1067, 2nd Ed. Academic Press, New York and London, 1965.
[5] *H. Schmidt*, Z. Anorg. Allgem. Chem. *271*, 305 (1953).

nitrogen to temperatures of about 600°–650° for 8 hr in a porcelain tube, zincblende crystals are obtained. Heating to 1150° for 1 hr provides the wurtzite modification (high-temperature modification, transition temperaturc near 900°) of zinc sulfide.

White, powdery zinc sulfide melts with partial vaporization near 1650°. Under high vacuum (5×10^{-4} mm Hg) the material can be distilled without decomposition.[4,6]

Zinc selenide, ZnSe, is formed by slow addition of an ammonium acetate-buffered solution of zinc sulfate to a hot saturated aqueous solution of hydrogen selenide with simultaneous addition of a dilute (with oxygen-free nitrogen or hydrogen) stream of hydrogen selenide. Also, zinc selenide precipitates from an ammoniacal zinc sulfate solution on treatment with hydrogen selenide. The initial voluminous white suspension turns yellow at pH 6 and precipitates. The product is dried over calcium chloride and phosphorus pentoxide at 120° and is heated to 400° for 4 hr in a stream of hydrogen selenide and hydrogen.[4,6] Another preparation of the compound involves heating a mixture of zinc oxide, zinc sulfide, and selenium or zinc sulfide and selenous acid, respectively, for 15 min. to 800° in a quartz crucible.[4]

$$2ZnO + ZnS + 3Se \longrightarrow 3ZnSe + SO_2$$
$$ZnS + SeO_2 \longrightarrow ZnSe + SO_2$$

Zinc telluride, ZnTe, is formed on heating of zinc filings with a slight excess of powdered tellurium in an evacuated quartz tube to 800°–900° for 24 hr. The brittle product is powdered and excess tellurium is distilled off in a hydrogen atmosphere at 550°.[6] *Tetrabutylammonium bis(cis-1,2-dicyano-1,2-ethylenedithiolate)-zincate*, $[(n\text{-}C_4H_9)_4N]_2[ZnS_4C_4(CN)_4]$, is formed on interaction of zinc chloride with

```
NC          CN
  \        /
    C==C
  /        \
NaS         SNa
```

in ethanol–water (1:1) and subsequent addition of tetrabutylammonium bromide. The orange-red crystals recrystallized from acetone and isobutanol, now become deep red, are washed with isobutanol and N-pentane, and are air-dried (m.p. 143°–144°).[7]

[6] *R. Juza, A. Rabenau, G. Pascher*, Z. Anorg. Allgem. Chem. *285*, 61 (1956).

[7] *A. A. Davison, R. H. Holm*, Inorg. Synth. *10*, 14 (1967).

4.1.3 Derivatives of Zinc with Group V Elements

Zinc amide, $Zn(NH_2)_2$, can be prepared by reaction of diethylzinc (freshly distilled, under vacuum, dissolved in anhydrous ether) with ammonia gas. The reaction is carried out in a nitrogen atmosphere with careful exclusion of moisture.

$$Zn(C_2H_5)_2 + 2NH_3 \longrightarrow Zn(NH_2)_2 + 2C_2H_6$$

After evaporation of the ether the residue is ground in the reaction vessel and treated with ammonia gas for 5 hr at 150° and for 12 hr at room temperature. The resulting product is normally colorless; it slowly decomposes on exposure to air.[4]

Zinc nitride, Zn_3P_2, is formed when zinc dust is heated to 500°–600° in a pyrex tube in a rapid stream ammonia for 40 hours. The grey-black material is obtained about 99% pure and is fairly stable in air.[4]

Zinc phosphide, Zn_3P_2, can be obtained by heating elemental zinc with a slight excess of red phosphorus to 700° in an evacuated quartz tube. Subsequently, the product is heated in a temperature gradient apparatus (850°–760°) where it sublimes to the colder zone. It is maintained at 760° for an additional 24 hours.

Using a ratio of 2 mol phosphorus to 1 mol zinc, the zinc phosphide, ZnP_2, is obtained as a yellow-orange to dark red crystalline product. It can be distilled in a phosphorus atmosphere without decomposition.

Zn_3P_2 is a metallic grey semiconductor. Mixtures with Cd_3P_2 show semiconducting properties with up to 50 mol% Cd_3P_2. Mixed crystal systems of Zn_3P_2 and Cd_3P_2 are obtained by repeated sublimation of mixtures of the two phosphides in a quartz tube at 850°, the sublimate condensing in the cold zone (700°) of the reaction vessel.[8,9]

Zinc arsenide, Zn_3As_2, is formed by passing arsenic vapor with a stream of nitrogen or hydrogen over zinc at 700°. The same compound is obtained on heating equimolar equivalent amounts of zinc and arsenic to 780° in an evacuated sealed tube. The grey material melts with partial sublimation near 1015°.

If the reaction is performed with an excess of arsenic, grey-black $ZnAs_2$ is obtained. This compound melts near 771° with considerable sublimation.[9]

[8] *R. Juza, K. Bär*, Z. Anorg. Allgem. Chem. *283*, 230 (1956).

[9] *M. von Stackelberg, R. Paulus*, Z. Phys. Chem. (B) *28*, 427 (1935).

Zinc thioantimonate, $[Zn_3(SbS_4)_2]$, is obtained as an orange precipitate by the reaction of sodium thioantimonate with zinc chloride or zinc sulfate in aqueous solution. The material is washed with hot water and dried at 80°–100°. Free sulfur can be removed by extraction with carbon disulfide. The required sodium thioantimonate is readily obtained by boiling antimony(III) sulfide with powdered sulfur in 20% sodium hydroxide solution.

4.1.4 Zinc Complexes with Nitrogen Donor Molecules

Dichlorobishydroxylaminezinc (Crismer's salt), $Zn(NH_2OH)_2Cl_2$, is formed on interaction of stoichiometric amounts of hydroxylammonium chloride and zinc oxide in boiling ethanol.[10]

$$2NH_2OH \cdot HCl + ZnO \longrightarrow Zn(NH_2OH)_2Cl_2 + H_2O$$

The compound melts at 155°–158° and explodes near 170°.

Dichlorodi-2-pyridylaminezinc is formed on dropwise addition of a solution of di-2-pyridyl anion in acetone to anhydrous zinc chloride in acetone, accompanied by vigorous stirring. The white precipitate is filtered, washed with acetone and dried at 90°–95°.[11]

$$ZnX_2 + (C_5H_4N)NH(C_5H_4N) \longrightarrow Zn[HN(C_5H_4N)_2]X_2$$

X = Cl, CH_3COO, CN

The corresponding *diacetatodi-2-pyridylaminezinc*, m.p. 254°–259°, is obtained in an analogous procedure utilizing zinc acetate and methanol as solvent.[11]

Dithioisocyanatodipyridinezinc can be obtained from zinc salts by reaction in aqueous solution with pyridine and ammonium thiocyanate.[12] At 0°. The colorless complex is stable up to about 70°.

$$ZnX_2 \cdot nH_2O + 2C_5H_5N + 2NH_4SCN \longrightarrow Zn(C_5H_5N)_2CNCS)_2 + 2NH_4X + nH_2O$$

4.1.5 Zinc Carboxylates, Pseudohalides, and Silicon Derivatives

Anhydrous *zinc acetate* is formed by boiling zinc nitrate with acetic anhydride. The precipitate is washed with acetic anhydride and ether and is dried over KOH or sulfuric acid. The colorless crystals (m.p. 242°), can be sublimed *in vacuo* without decomposition.[13]

Hydrated *zinc diacetylacetonate*, $Zn(C_5H_7O_2)_2 \cdot H_2O$, is formed by the reaction of zinc sulfate with acetylacetone in aqueous sodium hydroxide solution.

$$ZnSO_4 + 2C_5H_8O_2 + 2NaOH \xrightarrow{H_2O} Zn(C_5H_7O_2)_2 \cdot H_2O + Na_2SO_4$$

The product is recrystallized from ethyl acetate (containing some acetylacetone), m.p. 138°–140°. The anhydrous material is obtained by dissolving the hydrate in methanol. On cooling the solution to −50° the methanol adduct precipitates; it loses methanol on heating to 80° for 8 hr (m.p. 127°).[14] *Zinc uranyl acetate* is formed on refluxing uranyl acetate with zinc amalgam in glacial acetic acid. The solution is decanted from the excess amalgam and boiled with excess acetic anhydride.[15]

Zinc cyanide, $Zn(CN)_2$, is formed on addition of aqueous potassium cyanide to an aqueous solution of zinc sulfate. On heating the reaction mixture, zinc cyanide precipitates; it is filtered and washed with hot water and alcohol (or ether) and is dried at 70°–100°. The colorless compound decomposes near 800°. It can also be obtained by precipitation with hydrogen cyanide from a solution of zinc hydroxide in acetic acid.

$$ZnSO_4 + 2KCN \longrightarrow Zn(CN)_2 + K_2SO_4$$

With an equivalent amount of potassium cyanide, zinc cyanide will form *potassium tetracyanozincate*, which crystallizes from the solution upon concentrating the latter.

$$Zn(CN)_2 + 2KCN \longrightarrow K_2[Zn(CN)_4]$$

The same compound is obtained by adding hydrogen cyanide to a suspension of zinc oxide in aqueous potassium carbonate solution. The product (m.p. 538°) can be dried near 105°.

Two mol of zinc oxide and 1 mol of SiO_2 are finely ground and thoroughly mixed for the preparation of *zinc silicate*, Zn_2SiO_4. The mixture is pelletized and heated in a platinum dish

---

[10] *D. E. Walker, D. M. Howell*, Inorg. Synth. *9*, 2 (1967).

[11] *J. Simkin, B. P. Block*, Inorg. Synth. *8*, 10 (1966).

[12] *G. B. Kauffman, R. A. Albers, F. L. Harlan*, Inorg. Synth. *12*, 251 (1970).

[13] *E. Späth*, Monatsh. Chem. *33*, 240 (1912).

[14] *G. Rudolph, M. C. Henry*, Inorg. Synth. *10*, 74 (1967).

[15] *R. C. Paul, J. S. Ghetra, M. S. Bains*, Inorg. Synth. *9*, 42 (1967).

to above 1512° (the melting point of the desired zinc silicate). Rapid heating is required in order to avoid substantial evaporation of zinc oxide.[4] *Zinc hexafluorosilicate*, $ZnSiF_6 \cdot 6H_2O$, can be prepared by dissolving zinc oxide in a slight excess of aqueous hexafluorosilicic acid. On concentrating the solution the desired material precipitates and can be recrystallized from water.[4]

4.1.6 Zinc Ferrate(III)

On heating a mixture of zinc oxide or zinc carbonate with iron(III) hydroxide in a molar ratio of 1:1 to 800°–1000° brown *zinc ferrate(III)* is obtained. The same compound can be prepared by the reaction of sodium zincate, $Na[Zn(OH)_3]$, with iron(III) chloride in hydrochloric acid solution at 60° and subsequent addition of sodium hydroxide. The precipitate is washed and dried in a desiccator over KOH or phosphorus pentoxide.[4]

$$Na[Zn(OH)_3] + 2FeCl_3 + 5NaOH \longrightarrow ZnFe_2O_4 + 6NaCl + 4H_2O$$

Another method for the preparation of zinc ferrate(III) is the precipitation of an oxalate mixture from a solution of iron ion and zinc acetate in boiling acetic acid and oxalic acid. Subsequent decomposition at 1000° yields the desired zinc ferrate(III).[16]

4.2 Organic Compounds of Zinc

Organozinc compounds were among the first organometallic compounds described, the dialkylzincs being prepared by Frankland in 1849.[17] Organozinc compounds react with oxygen and water and, consequently, must be handled in a dry and inert atmosphere. The volatile dialkylzinc compounds—methyl, ethyl, and propyl—are spontaneously flammable, whereas the higher dialkylzinc and other organozinc compounds only fume in air. Organozinc compounds alone or with other transition metal derivatives are widely used as catalysts in olefin polymerization. Compounds which were previously regarded as organozinc compounds—the well-known Reformatsky reagents—have recently been shown to be bromozinc enolates of esters instead.[18,19]

4.2.1 Organozinc Halides

Earlier workers,[20] on the basis of radiochemical experiments with the diethylzinc–zinc bromide–heptane system, had concluded that alkylzinc halides were actually complexes of R_2Zn and ZnX_2 and that the species RZnX did not exist. However, molecular weight,[21,22] i.r.,[23] Raman,[23] and n.m.r.[24,25] studies have demonstrated that alkylzinc halides participate with dialkylzincs and zinc halides in a Schlenk-type equilibrium. In ethers, the equilibrium strongly favours the alkylzinc halide.

The methods which have been used for the preparation of these compounds are:

(1) the redistribution of substituents between zinc halides and dialkylzinc compounds
(2) the reaction of zinc with alkyl halides
(3) the reaction between Grignard reagents or organolithium compounds and zinc halides
(4) the action of acidic hydrocarbons on alkylzinc halides
(5) the reaction of zinc halides with aliphatic diazo compounds

The dismutation reaction

$$R_2Zn + ZnX_2 \longrightarrow 2RZnX$$

is the most general method since the choice of the organic group and the halogen is not limited. Solvents such as toluene, ether, and dioxane have been used,[26,27] but an excess of the dialkylzinc compound has served the same purpose in the case of the lower dialkylzincs since the excess is readily removed by distillation.[28,29] With dioxane, organozinc halides are obtained as dioxane complexes; the dioxane can be removed under high vacuum.

The reaction of zinc with alkyl halides is of value for the synthesis of *alkylzinc iodides* and

16 *D. G. Wickham*, Inorg. Synth. *9*, 152 (1967).
17 *E. Frankland*, Justus Liebigs Ann. Chem. *71*, 171 (1849).
18 *H. E. Zimmermann, M. D. Traxler*, J. Amer. Chem. Soc. *79*, 920 (1957).
19 *W. R. Vaughn, H. P. Knoess*, J. Org. Chem. *35*, 2394 (1970).
20 *A. B. Garrett, A. Sweet, W. L. Marshall, D. Riley, A. Touma*, Record Chem. Progr. *13*, 155 (1952).
21 *M. H. Abraham, P. H. Rolfe*, J. Organometal. Chem. *7*, 35 (1967).
22 *N. I. Sheverdina, I. E. Paleeva, L. V. Abramova, V. S. Yakoleva, K. A. Kocheshkov*, Izv. Akad. Nauk. SSSR, p. 1077 (1967).
23 *D. F. Evans, I. Wharf*, J. Chem. Soc. A, p. 783 (1968).
24 *D. F. Evans, G. V. Fazakerley*, J. Chem. Soc. A, p. 182 (1971).
25 *J. Boersma, J. G. Noltes*, J. Organometal. Chem. *8*, 551 (1967).
26 *N. I. Sheverdina, L. V. Abramova, K. A. Kocheshkov*, Dokl. Akad. Nauk. SSSR *124*, 602 (1959).
27 *N. I. Sheverdina, L. V. Abramova, K. A. Kocheshkov*, Dokl. Akad. Nauk. *134*, 853 (1960).
28 *G. Jander, L. Fischer*, Z. Elektrochemie *62*, 971 (1958).
29 *J. Boersma, J. G. Noltes*, Tetrahedron Lett. 1521 (1966).

bromides. Alkyl iodides react much faster with zinc than the bromides and the rate of reaction is dependent on two additional factors: the activity of the zinc and the polarity of the solvent. Zinc dust can be used, but in many cases a zinc–copper couple (10% copper) is employed to enhance the reaction rate. The alloy prepared by the procedure of Le Goff[30] has been found to be the most active. Thus, in the absence of a solvent, alkylzinc bromides were obtained by the use of this material, whereas all other zinc alloys were found to be unreactive with alkyl bromides.[31] If the reaction of zinc with alkyl halides is carried out in a solvent, the reaction rate increases with increasing polarity of the solvent. Thus, in tetrahydrofuran zinc dust will react with alkyl iodides and propargyl,[32,33] allyl,[33] and benzyl[33] bromides, but not with normal alkyl bromides or benzyl and allyl chloride.[33] In more polar solvents such as dimethylformamide and dimethyl sulfoxide, zinc dust reacts with alkyl iodides exothermically[34] and also with alkyl bromides if a slight amount of iodide is used to initiate the reaction.[35] If the reaction is carried out in more polar solvents the organozinc halides are not normally isolated.

Organozinc halides have also been prepared by the reaction of zinc halides with Grignard reagents, but this method has limited use since the yields of isolatable products are generally low. Although the reaction was reported many years ago by Blaise,[36] only recently have the organozinc halides been isolated from the reaction mixture because of the difficulty of separating them from magnesium halide. By using dioxane, this separation could be achieved and the organozinc halides were obtained as dioxane complexes (the dioxane could be removed by vacuum drying).[37] Ether solutions containing organozinc halides have also been prepared by the reaction of zinc chloride with phenyllithium or 3-thienyllithium, but no compounds were isolated.[38]

Alkynylzinc halides have been synthesized by the reaction of alkylzinc halides with acetylenes,[39] e.g.,

$$\mathrm{EtZnCl + Ph-C{\equiv}C-H \longrightarrow Ph-C{\equiv}C-ZnCl + EtH}$$

The reaction of diazomethane and its derivatives with zinc halides in ether solution leads to *α-halomethylzinc halides*[40] which react with olefins to give cyclopropanes in a methylene transfer process.[41] These types of organozinc compounds can also be prepared by the reaction of zinc with *gem*-dihalo compounds.[42]

Some of the organozinc halides that have been synthesized are given in Table 1.

Organozinc halides are white, crystalline solids; they react vigorously with water, but only slowly with oxygen. Perfluoropropylzinc iodide is exceptional in this respect since in ethanol it is only slowly hydrolyzed owing to its strong association with the solvent.[43] In the solid state ethylzinc iodide has been shown to be a coordination polymer[44]; this structure probably is true for other organozinc compounds. In benzene there is less association, ethylzinc chloride and bromide have been shown to be tetrameric in this solvent.[29] In ethereal solvents halogen bridging is completely eliminated and the organozinc halides are monomeric, though solvated.[21,22]

Although alkylzinc halides do not form complexes with ether, the aryl derivatives do. With dioxane and other strong donors, both classes of organozinc halides form complexes.

4.2.2 Organozinc–Chalcogen Compounds

The general method of preparation of this class of compounds is slow addition of an alcohol, phenol, or mercaptan to a diorganozinc compound at low temperatures. In many cases the resulting organozinc chalcogens are insoluble

---

[30] *E. LeGoff*, J. Org. Chem. *29*, 2048 (1964).
[31] *R. F. Galiulina*, *N. N. Shubanova*, *G. G. Petukhov*, Zh. Obshch. Khim. *36*, 1290 (1966).
[32] *K. H. Thiele*, *D. Gaudig*, Z. Anorg. Allgem. Chem. *365*, 301 (1969).
[33] *M. Gaudemar*, Bull. Soc. Chim. France *974*, (1962).
[34] *L. I. Zakharkin*, *O. Y. Okhlobystin*, Izv. Akad. Nauk. SSSR, *193* (1963).
[35] *R. Joly*, *R. Bucort*, Compt. Rend. *254*, 1655 (1962).
[36] *E. E. Blaise*, Bull. Soc. Chim. France [4] *9*, 1 (1911).
[37] *I. E. Paleeva*, *N. I. Sheverdina*, *L. V. Abramova*, *K. A. Kocheshkov*, Dokl. Akad. Nauk. SSSR *159*, 609 (1964).
[38] *S. Gronowitz*, Ark. Kemi *13*, 533 (1958).
[39] *L. I. Vereshchagin*, *O. G. Yashina*, *T. V. Zarva*, Zh. Org. Khim. *2*, 1895 (1966).
[40] *G. Wittig*, *K. Schwarzenbach*, Angew. Chem. *71*, 652 (1959).
[41] *G. Wittig*, *F. Wingler*, Justus Liebigs Ann. Chem. *656*, 18 (1962).
[42] *H. E. Simmons*, *R. D. Smith*, J. Amer. Chem. Soc. *80*, 5323 (1958).
[43] *W. T. Miller*, *E. Bergmann*, *A. N. Fainberg*, J. Amer. Chem. Soc. *79*, 4159 (1957).
[44] *P. T. Mosley*, *H. M. M. Shearer*, Chem. Commun. 876 (1966).

Table 1. Organozinc Halides

R	X	Method	Yield (%)	Ref.
CH_3	I	2	98	*a*
CH_3	Br	2	100	*a*
C_2H_5	I	1	—	*b*
C_2H_5	I	2	80·6	*c*
C_2H_5	I	3	41·4	*d*
C_2H_5	Br	1	—	*b*
C_2H_5	Br	2	45	*a*
C_2H_5	Cl	1	—	*b*
n-C_3H_7	I	2	82·3	*d*
i-C_3H_7	Br	2	59	*a*
$HC{\equiv}C{-}CH_2$	Br	2	90	*e*
n-C_3F_7	I	2	75	*f*
n-C_3F_7	Br	2	60	*f*
t-C_4H_9	Br	2	20	*g*
C_6H_{11}	Br	2	30	*g*
$(CH_3)_2CH{-}CH_2^{\ominus}$	Br	2	77	*g*
$C_6H_5CH_2^{\ominus}$	Cl	2	80	*h*
C_6H_5	I	1	70	*i*
C_6H_5	I	1	79	*i*
C_6H_5	I	3	40	*d*
C_6H_5	Br	1	70	*i*
C_6H_5	Br	3	31	*d*
2-thienyl	I	1	98	*i*
2-thienyl	Br	1	91	*i*

[a] *R. Joly, R. Bucort*, Compt. Rend. *254*, 1655 (1962).
[b] *J. Boersma, J. G. Noltes*, Tetrahedron Lett. 1521 (1966).
[c] *N. I. Sheverdina, L. V. Abramova, K. A. Kocheshkov*, Dokl. Akad. Nauk. SSSR *124*, 602 (1959).
[d] *I. E. Paleeva, N. I. Sheverdina, L. V. Abramova, K. A. Kocheshkov*, Dokl. Akad. Nauk. SSSR *159*, 609 (1964).
[e] *K. H. Thiele, D. Gaudig*, Z. Anorg. Allgem. Chem. *365*, 301 (1969).
[f] *W. T. Miller, E. Bergmann, A. N. Fainberg*, J. Amer. Chem. Soc. *79*, 4159 (1957).
[g] *R. F. Galiulina, N. N. Shabanova, G. G. Petukhov*, Z. Obshch. Khim. *36*, 1290 (1966).
[h] *L. I. Zakharkin, O. Y. Okhlobystin*, Izv. Akad. Nauk. SSSR 193 (1963).
[i] *N. I. Sheverdina, L. V. Abramova, K. A. Kocheshkov*, Dokl. Akad. Nauk. SSSR *134*, 853 (1960).

at low temperatures thus eliminating the possibility of further reaction.

$$(X = O, S) \quad R_2Zn + R'XH \longrightarrow R{-}Zn{-}XR' + R^1H \qquad (1)$$

From the reaction of water with diethylzinc at 0°C, a compound formulated as *ethylzinc hydroxide* has been obtained.[45]

Reaction of diorganozincs with alcohols or phenols in hexane at −80° yields the *organozinc alkoxides* which are white, crystalline solids.[46]

[45] *R. J. Herold, S. L. Aggarwal, V. Neff*, Can. J. Chem. *41*, 1368 (1963).
[46] *G. E. Coates, D. Ridley*, J. Chem. Soc. (London), 1870 (1965).

X-Ray studies of ethylzinc *t*-butoxide[47] and methylzinc methoxide[48] indicate that they are tetrameric and have a cubic structure in the solid state. In benzene solution the alkoxides are also tetrameric.[46] They occur as dimeric species only if they are substituted by bulky groups. Thus, *t*-butylzinc butoxide has been reported to be dimeric[49] or trimeric.[50] Nuclear magnetic resonance studies[51,52] reveal that some of the organozinc alkoxides—the methanol and ethanol derivatives of dimethyl- and diphenylzinc—are involved in benzene in equilibria with various complexes of their disproportionation products, the diorganozincs and the zinc dialkoxides, i.e.,

$$[MeZnOMe]_4 \rightleftharpoons [Me_2Zn \cdot Zn(OMe)_2] \qquad (2)$$

Phenyl- and methylzinc isopropoxide and *t*-butoxide (which are obtained by reacting the appropriate organozinc compound with the alcohol) are monomeric. On heating the organozinc alkoxides undergo ligand exchange as mentioned above.

The reaction of organozinc compounds with various carbonyl compounds is of some value for the preparation of certain organozinc alkoxides. Thus, reaction of benzophenone with diethylzinc was found to give trimeric Ph_2CHO-$ZnEt$.[53]

With diphenylzinc and benzophenone an addition reaction occurred yielding $PhZnOCPh_3$, which is dimeric in benzene.[53] Ethylzinc diethylamide reacts vigorously with carbon dioxide to give a carbamate derivative [see Eqn (3)].[54] Organozinc alkoxides have also been obtained from the reaction of ethylzinc diphenylamide with aldehydes.[54]

[47] *Y. Matsui, K. Kamiya, M. Nishikawa*, B. Chem. Soc. Jap. *39*, 1828 (1966).
[48] *H. M. M. Shearer, C. B. Spencer*, Chem. Commun. 194 (1966).
[49] *G. E. Coates, P. D. Roberts*, J. Chem. Soc. A, 1233 (1967).
[50] *J. G. Noltes, J. Boersma*, J. Organometal. Chem. *12*, 425 (1968).
[51] *G. Allen, J. M. Bruce, D. W. Farren, F. G. Hutchison*, J. Chem. Soc. B 799 (1966).
[52] *J. M. Bruce, B. C. Cutsforth, D. W. Farren, F. G. Hutchison*, J. Chem. Soc. B, 1020 (1966).
[53] *G. E. Coates, D. Ridley*, J. Chem. Soc. A, p. 1064 (1966).
[54] *J. G. Noltes*, Rec. Trav. Chim. Pays-Bas *84*, 126 (1965).

$$H_5C_2\text{-}Zn\text{-}N(C_2H_5)_2 + CO_2 \longrightarrow H_5C_2\text{-}Zn\text{-}O\text{-}\overset{O}{\overset{\|}{C}}\text{-}N(C_2H_5)_2 \quad (3)$$

$$H_5C_2\text{-}Zn\text{-}N(C_6H_5)_2 + R\text{-}CHO \longrightarrow H_5C_2\text{-}Zn\text{-}O\text{-}\overset{R}{\overset{|}{C}}H\text{-}N(C_6H_5)_2 \quad (4)$$

Although the addition reaction of aldehydes with dialkylzincs has been known for many years as a method for producing secondary alcohols,[55] the intermediate organozinc alkoxides have not been isolated and characterized. Only the lower dialkylzincs undergo the addition reaction.

A number of compounds in which the oxygen is attached to an element other than carbon have been prepared as shown in the following equations.

$$H_5C_2\text{-}ZnI + (C_2H_5O)_2SO_2 \longrightarrow H_5C_2\text{-}Zn\text{-}O\text{-}SO\text{-}OC_2H_5 + C_2H_5I \quad (5)^{56}$$

$$(CH_3)_2Zn + HOSi(CH_3)_3 \longrightarrow H_3C\text{-}Zn\text{-}O\text{-}Si(CH_3) + CH_4 \quad (6)^{57}$$

$$(C_2H_5)_2Zn + R_2N\text{-}OH \longrightarrow H_5C_2\text{-}Zn\text{-}O\text{-}NR_2 + C_2H_6 \quad (7)^{50}$$

$$(CH_3)_2Zn + SO_2 \longrightarrow H_3C\text{-}Zn\text{-}O\text{-}SO\text{-}CH_3 \quad (8)^{58}$$

$$(CH_3)_2Zn + (CH_3)_2P(O)\text{-}OH \longrightarrow H_3C\text{-}Zn\text{-}O\text{-}P(O)(CH_3)_2 + CH_4 \quad (9)^{46}$$

Organozinc mercaptans have been obtained by reaction of organozinc compounds with thiols at low temperatures.[46] The thio derivatives are more highly associated than the alkoxides, the degree of polymerization being dependent on the bulkiness of the organic groups. Thus, the degree of x polymerization of $(MeZnS\text{-}R)_x$ in benzene solution is greater than nine, when $R = n\text{-}C_3H_7$; when $R = t\text{-}C_4H_9$ it is 5 and when $R = i\text{-}C_3H_7$ it is 6.[46] Compounds containing the C–Zn–S system have also been formed by the addition reaction of carbon disulfide with *alkylzinc amides*[54] and *phosphides*.[59]

4.2.3 Organozinc—Group V Compounds

Although it was reported many years ago by Frankland[60] that diethylzinc reacts with diethylamine, it is only recently that any compounds have been isolated and identified from this reaction. The main method of synthesis of organozinc compounds containing the C–Zn–N and C–Zn–P systems consists of the reaction of diorganozincs with molecules having N–H or P–H groups. Consequently, organozinc amides are obtained by the reaction of diorganozinc compounds with secondary amines.[54] Since the amino hydrogen is not very labile and the organozinc amides are not very reactive towards amines, the reaction can be conducted by simply mixing and refluxing the two reactants in a suitable solvent such as benzene. Dimethylzinc behaves somewhat anomalously forming amides only with weakly basic secondary amines, such as diphenylamine. With amines such as dimethyl- and diethylamine only complexes are obtained, no evolution of methane as expected according to Eqn (10) is observed.[61]

$$R_2Zn + HNR'_2 \longrightarrow RZnNR'_2 + RH \quad (10)$$

The *organozinc amides* are normally white crystalline solids, although $Et\text{-}ZnNEt_2$ is a liquid[54]; in benzene they exist as dimers.[46,54] The reaction of diorganozincs with primary amines has hardly been investigated. Reaction of Et_2Zn with *t*-butylamine proceeds in two stages as shown in Eqn (11). The products

$$(H_5C_2)_2Zn + H_2N\text{-}C(CH_3)_3$$
$$H_5C_2\text{-}Zn\text{-}NH\text{-}C(CH_3)_3 \xrightarrow{(C_2H_5)_2Zn} (H_5C_2\text{-}Zn)_2N\text{-}C(CH_3)_3 \quad (11)$$

of both steps have been isolated and identified.[62] A number of other compounds containing N–H groups have been reacted with diorganozincs to give products containing the C–Zn–N system as shown in Eqns (12)–(15).

$$Ph_2C{=}NH + R_2Zn \longrightarrow RZnN{=}CPh_2 + RH \quad (12)^{63}$$
$$(R = CH_3, C_2H_5, C_6H_5)$$

$$R_3P{=}NH + (CH_3)_2Zn \longrightarrow R_3P{=}NZnCH_3 + CH_4 \quad (13)^{64}$$
$$(R = CH_3, C_2H_5)$$

$$Z\text{—}C(S)\text{—}NPhH + (C_2H_5)_2Zn \longrightarrow C_2H_5ZnNPh\text{—}C(S)\text{—}Z + C_2H_6 \quad (14)^{65}$$
$$(Z = H, CH_3, OCH_3, SCH_3, NPh_2)$$

$$Z\text{—}C(O)\text{—}NPhH + (C_2H_5)_2Zn \longrightarrow C_2H_5ZnNPh\text{—}C(O)\text{—}Z + C_2H_6 \quad (15)^{66}$$
$$(Z = H, CH_3, OCH_3, SCH_3, NPh_2)$$

55 *E. Wagner*, Justus Liebigs Ann. Chem. *181*, 261 (1876).
56 U.S. Patent 2942015 (to Ethyl Corp.), *H. E. Petree*.
57 *F. Schindler, H. Schmidbauer, U. Krueger*, Angew. Chem. 77, 865 (1965).
58 *N. A. D. Carey, H. C. Clark*, Can. J. Chem. *46*, 649 (1968).
59 *J. G. Noltes*, Rec. Trav. Chim. Pays-Bas *84*, 782 (1965).
60 *E. Frankland*, Jahresber. Chem. Tech. Reichsan. 419 (1867).
61 *K. H. Thiele, M. Bendull*, Z. Anorg. Allgem. Chem. *379*, 199 (1970).
62 *H. Tani, N. Oguni*, J. Polymer Sci. B, 7, 769 (1969).
63 *I. Pattison, K. Wade*, J. Chem. Soc. A, 57 (1968).
64 *H. Schmidbauer, G. Jonas*, Ber. *101*, 1271 (1968).
65 *J. Boersma, J. G. Noltes*, J. Organometal. Chem. *17*, 1 (1969).
66 *J. Boersma, J. G. Noltes*, J. Organometal. Chem. *16*, 345 (1969).

Organozinc azides have been prepared by the reaction of diorganozinc compounds with chlorine azide.[67] The azides are white, crystalline, presumably polymeric solids.

$$R_2Zn + ClN_3 \longrightarrow RZnN_3 + RCl (R = Me, Et, Ph) \quad (16)$$

Only a few compounds containing the C–Zn–P system are known. *Organozinc phosphides*, $RZnPR^1_2$, have been prepared by the reaction of diorganozinc compounds with secondary phosphines.[59] They are very reactive, ignite on contact with air, and react rapidly with carbon dioxide.

4.2.4 Diorganozinc Compounds

Diorganozinc compounds have been prepared by the following methods:

The reaction of zinc with alkyl halides

$$2Zn + 2RX \longrightarrow 2R\text{–}ZnX \rightleftharpoons R_2Zn + ZnX_2 \quad (17)$$

The reaction of zinc halides with Grignard reagents

$$2R\text{–}MgX + ZnX_2 \longrightarrow R_2Zn + 2MgX_2 \quad (18)$$

The reaction of zinc halides with organolithium compounds

$$ZnX_2 + 2RLi \longrightarrow R_2Zn + 2LiX \quad (19)$$

The reaction of diorganomercury compounds with zinc

$$R_2Hg + Zn \longrightarrow R_2Zn + Hg \quad (20)$$

The reaction of trialkylboron compounds with dimethyl- or diethylzinc

$$2R_3B + 3Me_2Zn \longrightarrow 3R_2Zn + 2Me_3B \quad (21)$$

The reaction of diorganozincs with acidic hydrocarbons

$$R_2^1Zn + R^2H \longrightarrow R^1ZnR^2 + R^1H \xrightarrow{R^2H} R_2^2Zn + R^1H \quad (22)$$

The thermal decarboxylation of zinc carboxylates

$$(R\text{—}\overset{\overset{\displaystyle O}{\|}}{C}\text{—}O)_2Zn \xrightarrow{\Delta} R_2Zn + 2CO_2 \quad (23)$$

The reaction of zinc or zinc halides with organoluminum compounds

$$2R_3Al + ZnX_2 \longrightarrow R_2Zn + R_2AlX \quad (24)$$

and

$$R_3Al + RX + Zn \longrightarrow R_2Zn + R_2AlX \quad (25)$$

The reaction of zinc halides with diazomethane

$$ZnX_2 + 2CH_2N_2 \longrightarrow (XCH_2)_2Zn + 2N_2 \quad (26)$$

The method given in Eqn (17) is most widely used in the preparation of the *dialkylzinc* compounds. Since it is the one originally used by Frankland for the preparation of dialkylzinc compounds, it is frequently referred to as the Frankland synthesis. The reaction proceeds in two steps. The intermediate organozinc halide is not normally isolated. As indicated in Section 4.2.1, the organozinc halide is involved in a Schlenk-type equilibrium with the diorganozinc and the zinc halide. This equilibrium is shifted towards the dialkylzinc by various means, the most common of which is continuous removal of the dialkylzinc by distillation.

The iodides are the most reactive alkyl halides towards zinc, but mixtures of alkyl iodides and bromides are also utilized.[68] Alkyl chlorides react too slowly and are not used. A zinc–copper couple (10% Cu) is normally used in place of zinc since it is more reactive. Although there are a number of methods for preparing these couples, the one made by the procedure of LeGoff[30] has been found to be the most active. Often no solvent is employed, but various ethers have been found suitable;[69] if the product is to be isolated by distillation, the choice of solvent depends on the boiling point of the dialkylzinc to be synthesized. In certain cases, the use of ethers reduces the hazard of working with the more volatile and inflammable dialkylzincs since they form stable complexes with the organozinc compound. Thus, dimethylzinc is isolated as a tetrahydrofuran (THF) complex from the reaction of methyl iodide with a zinc–copper couple in THF. This complex does not spontaneously ignite in air.[70]

The Frankland synthesis is restricted to primary and secondary alkyl groups. Primary dialkylzinc compounds are readily obtained in high yields by this method; secondary dialkylzinc compounds require more attention to experimental details. Good yields of the secondary dialkylzinc species are obtained by slow addition of a mixture of secondary alkyl bromides or iodides to an excess of the zinc–copper couple in an ether solvent with careful control of temperature followed by high vacuum distillation of the dialkylzinc.[71] A number of compounds

[67] *H. Muller, K. Dehnicke*, J. Organometal. Chem. *10*, P1 (1967).

[68] *C. R. Noller*, Org. Syntheses, Coll. Vol. II, 184 (1943).

[69] *L. F. Hatch, G. Sutherland, W. J. Ross*, J. Org. Chem. *14*, 1130 (1949).

[70] *K. H. Thiele*, Z. Anorg. Allgem. Chem. *319*, 183 (1962).

[71] *H. Sarroos, M. Morgana*, J. Amer. Chem. Soc. *66*, 893 (1944).

prepared by the Frankland synthesis are listed in Table 2.

Table 2. Dialkylzinc Compounds from Zinc and Alkyl Halides

Alkyl group	Yield (%)	Ref.
Methyl	90	*a*
Ethyl	89	*b*
n-Propyl	86	*b*
i-Propyl	85	*c*
n-Butyl	79	*b*
s-Butyl	72	*c*
n-Pentyl	58	*d*
i-Pentyl	55	*b*
n-Hexyl	51	*d*
n-Heptyl	41	*d*

[a] *N. K. Hota, C. J. Willis*, J. Organometal. Chem. *9*, 169 (1967).
[b] *C. R. Noller*, Org. Syntheses, Coll. Vol. II., 184 (1943).
[c] *H. Sarroos, M. Morganna*, J. Amer. Chem. Soc. *66*, 893 (1944).
[d] *L. F. Hatch, G. Sutherland, W. J. Ross*, J. Org. Chem. *14*, 1130 (1949).

As is readily apparent, the yield of the dialkylzinc compound decreases as the number of carbon atoms in the alkyl group increases.

The route involving Grignard reagents to diorganozinc compounds has the advantage that it is not limited by the type of the organic group; allylic, vinylic, tertiary, aryl, primary and secondary diorganozinc compounds can be prepared by this method. However, yields are generally low to moderate; diethylzinc and di-*n*-butylzinc have been reported to be synthesized in 21[72] and 57%[73] yield, respectively. *Unsymmetrical diorganozinc compounds*, RZnR', can be prepared by the reaction of organozinc halides with the appropriate Grignard reagents.[74–76] On standing at room temperature the unsymmetrical compounds slowly convert to a mixture of the symmetrical compounds. A representative group of diorganozinc compounds synthesized by the Grignard method is given in Table 3.

Table 3. Diorganozinc Compounds, RZnR', by the Grignard Method

R	R'	Yield (%)	Ref.
Et	Et	21	*a*
n-Pr	*n*-Pr	53	*b*
n-Bu	*n*-Bu	57	*c*
i-Bu	*i*-Bu	62	*b*
t-Bu	*t*-Bu	—	*c*
Vinyl	Vinyl	10	*d*
Allyl	Allyl	30	*e*
n-C_8H_{17}	*n*-C_8H_{17}	—	*f*
n-C_9H_{19}	*n*-C_9H_{19}	—	*f*
C_6H_5	C_6H_5	62	*b*
$C_6H_5CH_2$	$C_6H_5CH_2$	88	*b*
α-C_4H_3S	α-C_4H_3S	34	*b*
C_6F_5	C_6F_5	15	*g*
Et	*n*-Pr	60	*h*
C_6H_5	α-C_4H_3S	88	*h*
$C_6H_5CH_2$	α-C_4H_3S	46	*h*
C_6H_5	C_5H_5	50	*i*

[a] *F. Kusama, D. Koive*, J. Chem. Soc. Japan *72*, 871 (1951).
[b] *N. I. Sheverdina, I. E. Paleeva, N. A. Zaitseva, K. A. Kocheshkov*, Dokl. Akad. Nauk. SSSR *155*, 623 (1964).
[c] *M. H. Abraham*, J. Chem. Soc. 4130 (1960).
[d] *B. Bartocha, H. D. Kaesz, F. G. A. Stone*, Z. Naturforsch. *14b*, 352 (1959).
[e] *K. H. Thiele, P. Zdunneck*, J. Organometal. Chem. *4*, 10 (1965).
[f] *K. H. Thiele, J. Müller*, J. Prakt. Chem. (4) *33*, 229 (1966).
[g] *J. G. Noltes, J. W. G. van den Hurk*, J. Organometal. Chem. *1*, 377 (1964).
[h] *I. E. Paleeva, N. I. Sheverdina, K. A. Kocheshkov*, Dokl. Akad. Nauk. SSSR *157*, 626 (1964).
[i] *W. Strohmeier, H. Landsfeld*, Z. Naturforsch. *15b*, 332. (1960).

The reactions of zinc halides with organolithium compounds is of value primarily for the preparation of *diarylzinc* compounds. One of the disadvantages of this method is the formation of biaryls which in the case of the larger aryl groups are difficult to separate from the diarylzinc compound. The amount of the biaryl formed can be minimized by using crystalline aryllithium compounds prepared from the aryl bromide and *n*-butyllithium[77] rather than the reagent prepared from the aryl bromide and metallic lithium. The zinc halide employed is normally zinc bromide rather than zinc chloride because the bromide is easier to prepare and to maintain in an anhydrous state.[78] Thermally unstable dialkylzinc compounds have been prepared by the organolithium route, *e.g.*,

[72] *F. Kusama, D. Koive*, J. Chem. Soc. Jap. *72*, 871 (1951).
[73] *M. H. Abraham*, J. Chem. Soc. 4130 (1960).
[74] *E. Krause, W. Fromm*, Ber. *59*, 931 (1926).
[75] *N. I. Sheverdina, I. E. Paleeva, K. A. Kocheshkov*, Izv. Akad. Nauk. SSSR 587 (1967).
[76] *M. H. Abraham, J. A. Hill*, J. Organometal. Chem. *7*, 23 (1967).
[77] *N. I. Sheverdina, L. V. Abramova, K. A. Kocheshkov*, Dokl. Akad. Nauk. SSSR *128*, 320 (1959).
[78] *D. Y. Curtin, J. L. Tveten*, J. Org. Chem. *26*, 1764 (1961).

$(Cl_2CH)_2Zn$ from zinc chloride and dichloromethyllithium at $-74°$.[79]

Another method which found major application in the synthesis of diarylzinc compounds is the reaction of diorganomercury compounds with zinc. Although dialkylzinc compounds have been prepared by this route,[80] the Frankland synthesis is more direct and the handling of the toxic dialkylmercury compounds is avoided. The reaction of the readily available diarylmercury compounds with zinc can be conducted without a solvent in the melt,[81] but use of a solvent such as xylene is much more convenient.[82] Since high temperatures, *e.g.*, refluxing xylene, are necessary in order to obtain an appreciable reaction rate, the diarylzinc compound to be synthesized must be stable at those temperatures. An excess of zinc is also normally employed. The yield of the diarylzinc obtained is generally moderate.[82]

The reaction of triorganoboron compounds with dimethyl- or diethylzinc [Eqn (21)] is an excellent method for preparation in high yields of *diallylic derivatives of zinc*, *e.g.*, the diallyl,[83] dimethallyl,[84] and dicrotyl[84] compounds. The trimethylboron (b.p. $-20°$) or triethylboron (b.p. $0°^{1.25}$) produced in the reaction is readily removed by distillation. The allylzinc compounds are very reactive, reacting rapidly with CO_2 at room temperature[84]; saturated dialkylzinc compounds do not show any signs of reaction with CO_2 up to 150°. Other diorganozinc compounds which have been prepared in moderate yield by the organoboron route are the *o*-tolyl, α-naphthyl, and benzyl derivatives.[85] Attempts to prepare long-chain dialkylzinc compounds were unsuccessful because the boron trialkyls react too slowly.[86] This lack of reactivity is unfortunate because extensive purification is necessary to free the less volatile dialkylzinc compounds from halides and ether when they are prepared by any of the other methods.

Acidic hydrocarbons such as acetylenes react with diorganozinc compounds in two stages as shown in Eqn (22). The products of the first step are unsymmetrical diorganozinc compounds, RZnR′, which are unstable and slowly convert to a mixture of the symmetrical compounds. The rate of the reaction is dependent on the acidity of the hydrocarbon and the polarity of the solvent and increases with an increase in these variables. Thus, the reaction of 1 mol of diethylzinc with 2 mol of phenylacetylene has a half-life of 135 min. in THF but only 200 s in dimethyl sulfoxide (DMSO).[87] Less acidic hydrocarbons such as fluorene and phenylbarene will react with diethylzinc in hexamethylphosphoramide.[87] Some of the acetylenic compounds of zinc that have been prepared by this method include $C_6H_5ZnC{\equiv}C{-}C_6H_5$,[88] $(C_6H_5C{\equiv}C)_2Zn$,[88,89] and $(C_6H_{13}C{\equiv}C)_2Zn$.[89] The *dialkynylzinc compounds* are associated and are regarded as linear polymers in which the bridging group is the acetylenic moiety.[89]

Dicyclopentadienylzinc has been prepared in quantitative yield by reacting $Zn[N(SiMe_3)_2]_2$ with excess cyclopentadiene in ether at room temperature for 1 hr.[90]

The thermal decarboxylation of zinc carboxylates is of value only for the preparation of diorganozinc compounds with negative substituents. Thus, *bispentafluorophenylzinc* was prepared in 62% yield by this method.[91]

The preparation of diorganozinc compounds *via* organoaluminum compounds is primarily of value as an industrial production method rather than a laboratory procedure. Reaction of a zinc halide with trialkylaluminum [Eqn (24)] requires 2 mol of trialkylaluminum per mol of dialkylzinc produced.[92] A newer method using the reaction of trialkylaluminum compounds with zinc and alkyl halides [Eqn (25)] requires only 1 mol of trialkylaluminum per mol of dialkylzinc obtained.[93]

---

[79] *G. Kobrich, H. P. Merkle*, Ber. *99*, 1782 (1966).
[80] *E. Frankland, B. F. Duppa*, Justus Liebigs Ann. Chem. *130*, 118 (1864).
[81] *S. Hilpert, G. Gruttner*, Ber. *46*, 1675 (1913).
[82] *A. N. Nesmeyanov, K. A. Kocheshkov, V. I. Potrosov*, Ber. *67*, 1138 (1934).
[83] *K. H. Thiele, P. Zdunneck*, J. Organometal. Chem. *4*, 10 (1965).
[84] *K. H. Thiele, G. Engelhardt, J. Kohler, M. Arnstedt*, J. Organometal. Chem. *9*, 385 (1967).
[85] *K. H. Thiele, J. Kohler*, J. Prakt. Chem. [4] *32*, 54 (1966).
[86] *K. H. Thiele, J. Muller*, J. Prakt. Chem. [4] *33*, 229 (1966).
[87] *O. Y. Okhlobystin, L. I. Zakharkin*, J. Organometal. Chem. *3*, 257 (1965).
[88] *R. Nast, O. Kunzel, R. Muller*, Ber. *95*, 2155 (1962).
[89] *E. A. Jeffrey, T. Mole*, J. Organometal. Chem. *11*, 393 (1968).
[90] *J. Loberth*, J. Organometal. Chem. *19*, 189 (1969).
[91] *P. Satori, M. Weidenbruch*, Ber. *100*, 3016 (1967).
[92] U.S. Patent 3124604 (to Karl Ziegler), *E. Heuther*.
[93] U.S. Patent 3475475 (to Stauffer Chemical), *S. H. Eidt*.

α-Haloalkyl derivatives of zinc have been prepared by the reaction of zinc halides with diazomethane[94] [Eqn (26)].

In general, the dialkylzinc compounds are liquids, whereas the diaryls are solids. The diorganozinc compounds are monomeric with few exceptions (the alkynyl derivatives) and are nonpolar because of the linear structure of the molecules. *Complexes*[95] of various stoichiometries are formed with a variety of donor molecules such as cyclic ethers,[70,96–100] tertiary amines, heterocyclic nitrogen compounds,[96,100–106] and diphosphines.[107,108] By complex formation the sensitivity of organozinc compounds toward oxygen and water is reduced. Ate complexes of the type $Li_x(ZnR_{x+2})$ are formed from organolithium compounds and diorganozincs.[109–112] Similar ate complexes are formed with organocalcium, -strontium, and -barium compound.[113–115] *Organozinc hydride* complexes of the type $MH(ZnR_2)_n$ (where M=Li or Na and n=1 or 2) are also known.[116–118] The 1:2 species $MH(ZnR_2)_2$ presumably contains a Zn–H–Zn bridge.[118]

4.3 Bibliography

G. E. Coates, K. Wade, Organometallic Compounds, Vol. 1, 3rd ed., pp. 121–141. Methuen, London, 1967.

B. J. Wakefield, Adv. Inorg. Chem. Radiochem. *11*, 342 (1968).

N. I. Sheverdina, K. A. Kocheshkov, The Organic Compounds of Zinc and Cadmium. North-Holland, Amsterdam, 1967.

[94] *G. Wittig, K. Schwarzenbach*, Justus Liebigs Ann. Chem. *650*, 1 (1961).
[95] *K. H. Thiele, P. Zdunneck*, Organometal. Chem. Rev. *1*, 331 (1966).
[96] *G. Allen, J. M. Bruce, F. G. Hutchison*, J. Chem. Soc. (London), 5476 (1965).
[97] *I. E. Paleeva, N. I. Sheverdina, K. A. Kocheshkov*, Dokl. Akad. Nauk. SSSR *155*, 623 (1964).
[98] *N. I. Sheverdina, I. E. Paleeva, N. A. Zaitsev, K. A. Kocheshkov*, Dokl. Akad. Nauk. SSSR *157*, 628 (1964).
[99] *K. H. Thiele*, Z. Anorg. Allgem. Chem. *332*, 71 (1962).
[100] *K. H. Thiele, S. Schroeder*, Z. Anorg. Allgem. Chem. *337*, 14 (1965).
[101] *G. E. Coates, S. I. E. Greene*, J. Chem. Soc. (London) 3340 (1962).
[102] *C. R. McCoy, A. L. Aldred*, J. Amer. Chem. Soc. *84*, 912 (1962).
[103] *G. Pajaro, S. Biagini, D. Fiumani*, Angew. Chem. *74*, 901 (1962).
[104] *K. H. Thiele, J. Kohler*, Z. Anorg. Allgem. Chem. *337*, 260 (1965).
[105] *K. H. Thiele*, Z. Anorg. Allgem. Chem. *325*, 156 (1963).
[106] *Y. Takashi*, Bull. Chem. Soc. Japan *40*, 1001 (1967).
[107] *J. G. Noltes, J. W. G. van den Hurk*, J. Organometal. Chem. *1*, 377 (1964).
[108] *J. Ellerman, W. H. Gruber*, Z. Naturforsch. *22b*, 1248 (1967).
[109] *G. Wittig, F. J. Meyer, C. Lange*, Justus Liebigs Ann. Chem. *571*, 165 (1951).
[110] *L. M. Seitz, T. L. Brown*, J. Amer. Chem. Soc. *88*, 4140 (1966).
[111] *S. Toppet, G. Slinckx, G. Smets*, J. Organometal. Chem. *9*, 205 (1967).
[112] *E. Weiss, R. Wolfrum*, Chem. Ber. *101*, 35 (1968).
[113] *H. Gilman, R. N. Meals, G. O'Donnell, L. A. Woods*, J. Amer. Chem. Soc. *65*, 268 (1943).
[114] *H. Gilman, L. A. Woods*, J. Amer. Chem. Soc. *67*, 520 (1943).
[115] *Y. Kawakami, Y. Yasuda, T. Tsurata*, Bull. Soc. Chem. Japan *44*, 1164 (1971).
[116] *G. Wittig, P. Hornberger*, Justus Liebigs Ann. Chem. *577*, 11 (1952).
[117] *P. Kobetz, W. E. Becker*, Inorg. Chem. *2*, 589 (1963).
[118] *G. J. Kubas, D. F. Shriver*, J. Amer. Chem. Soc. *92*, 1949 (1970).

5 Cadmium

C. W. Blewett
Research Laboratory, Emery Industries, Cincinnati, Ohio, U.S.A.

R. Schmid
Battelle Memorial Institute, Genf, Schweiz

H. Zimmer
Department of Chemistry, University of Cincinnati,
Cincinnati, Ohio, U.S.A.

5.1 Inorganic Cadmium Compounds

5.1.1 Cadmium Halides

Cadmium fluoride, CdF_2, is formed on treating cadmium carbonate with 40% hydrofluoric acid in a platinum dish. After evaporation the product is dried at 150° *in vacuo* (m.p. 1049°, b.p. 1748°).[1]

Cadmium chloride is obtained by dissolving metallic cadmium in nitric acid to yield cadmium nitrate. The nitrate is transformed into the chloride by evaporation in hydrochloric acid. The preparation of the anhydrous material is achieved by the refluxing of the salt with thionyl chloride. Cadmium chloride can also be prepared by the reaction of dry cadmium acetate with acetyl chloride in glacial acetic acid using acetic anhydride as solvent.

$$Cd(CH_3COO)_2 + 2CH_3COCl \longrightarrow CdCl_2 + (CH_3CO)_2O$$

The precipitate is collected, washed with benzene, and dried at 100°–120°.[2,3]

Cadmium chloride forms white hygroscopic plates, (m.p. 568°, b.p. 967°). It is somewhat less soluble in water than zinc chloride.

On hydrolysis of slightly dilute aqueous solutions of cadmium chloride (*c* 1 M) with sodium hydroxide at pH 6 *cadmium hydroxychloride*, Cd(OH)Cl, is formed within a few days. The same material is obtained on heating a mixture of cadmium oxide and cadmium chloride solution to 210° for several days in a sealed tube.[3]

Cadmium bromide, $CdBr_2$, can be prepared by directly combining the elements at 450° in a quartz apparatus under a nitrogen atmosphere. The cadmium bromide is distilled off in a stream of bromine. The same compound is obtained from the reaction of anhydrous cadmium acetate with acetyl bromide in glacial acetic acid–acetic acid anhydride solution. On electrolysis of Cu_2Br_2 in acetonitrile using a platinum cathode and a cadmium anode, cadmium bromide can be formed in a nitrogen atmosphere; copper is precipitated at the cathode during this process. After evaporation of the solvent solvated cadmium bromide remains, which converts to $CdBr_2$ on mild heating. The latter method can also be used for the preparation of similar compounds by proper selection of electrolytes. For example, with Cu_2I_2, it is possible to synthesize *cadmium iodide*.[3,4] Cadmium bromide (m.p. 566°, b.p. 963°) is extremely hygroscopic. On refluxing a suspension of cadmium filings and iodine in water this slowly decolorizes with formation of cadmium iodide. This iodide is also formed on evaporation of an aqueous solution of cadmium sulfate and potassium iodide. The residue is extracted with ethanol to yield colorless CdI_2 (m.p. 387°, b.p. 787°).[3]

5.1.2 Cadmium–Chalcogen Derivatives

Cadmium hydroxide, $Cd(OH)_2$, is formed by precipitation of solutions of cadmium nitrate, cadmium acetate, or cadmium iodide with sodium hydroxide. The precipitate is washed with hot water and dried over phosphorus pentoxide at 60°.[3]

Cadmium sulfide, CdS, is formed on passing hydrogen sulfide into a solution of cadmium perchlorate in 0.1–0.3 N perchloric acid. Cadmium sulfide crystals several millimetres in diameter can be prepared by reacting hydrogen sulfide with cadmium vapour near 800°.

The color of cadmium sulfide depends on the constituency of its surface and grain size. It normally varies from lemon-yellow to orange. The cubic modification of the compound is converted into the hexagonal form by heating to 700°–800° in a hydrogen sulfide atmosphere.

Cadmium digallium(III) tetrasulfide, $CdGa_2S_4$, is best prepared by melting cadmium sulfide with gallium(III) sulfide *in vacuo* at temperatures near 700°.[5] In order to generate large crystals iodine is added to the reaction mixture. The volatile iodides migrate through the reaction tube and react in the colder zone (*c* 675°) with the sulfur vapor to regenerate the desired compound. At room temperature the compound is stable towards air and water.

$$CdS + Ga_2S_3 \longrightarrow CdGa_2S_4 \quad (1)$$

$$CdGa_2S_4 + 4I_2 \longrightarrow CdI_2 + 2GaI_3 + 4S \quad (2)$$

Cadmium selenide, CdSe, is formed on addition of a saturated aqueous solution of hydrogen

---

[1] *W. Kwansig*, in *G. Brauer*, Handbook of Preparative Inorganic Chemistry, Vol. I, p. 27, 242, 2nd Ed. Academic Press, New York and London, 1963.

[2] *R. Pray*, Inorg. Synth. *5*, 153 (1957).

[3] *F. Wagenknecht*, *R. Juza*, Handbook of Preparative Inorganic Chemistry, Vol. I, p. 1067, 2nd Ed. Academic Press, New York and London, 1963.

[4] *H. Schmidt*, Z. Anorg. Allgem. Chem. *271*, 305 (1953).

[5] *A. G. Karipides*, *A. V. Cafiero*, Inorg. Synth. *11*, 5 (1968).

selenide to a hot ammonium acetate buffered solution of cadmium sulfate. Simultaneously, a diluted (with nitrogen or hydrogen) stream of hydrogen selenide is passed through the solution. The hydrogen selenide can be gencrated from dilute nitric acid and aluminum selenide.[3]

5.1.3 Cadmium Compounds with Group V Elements

Cadmium amide, $Cd(NH_2)_2$, can be prepared by reacting potassium amide with a slight excess of cadmium thiocyanate in liquid ammonia (excluding air). White cadmium amide precipitates and is filtered off and washed with

$$Cd(SCN)_2 + 2KNH_2 \longrightarrow Cd(NH_2)_2 + 2KSCN$$

liquid ammonia. All operations must be performed under rigorous exclusion of air. Normally, cadmium amide is slightly yellow and readily darkens on exposure to air.[6]

Cadmium nitride, Cd_3N_2, is obtained on pyrolysis of cadmium amide at 180°. The black compound readily decomposes at higher temperatures. Ammonia is removed several times during the process.

$$3Cd(NH_2)_2 \longrightarrow Cd_3N_2 + 4NH_3$$

Cadmium nitride is very sensitive toward moisture; on exposure to air it oxidizes readily.[7]

Cadmium phosphide, Cd_3P_2, is formed on heating metallic cadmium with red phosphorus in an evacuated quartz tube to 400°–680° for several hours. The product is purified by repeated vacuum sublimation near 680°.[8]

Cd_3P_2 and CdP_2 are formed on contact of phosphorus vapors (diluted with nitrogen or hydrogen) with heated (700°) cadmium in an electric furnace. In an excess of phosphorus, CdP_2 was found to be the major product.[8,9]

Cd_3P_2 is a metallic grey solid which reacts with nonoxidizing acids to form phosphine. The compound sublimes near 700°. CdP_2 is a red crystalline material which is stable toward nonoxidizing acids. It can be sublimed without decomposition in an atmosphere of phosphorus vapor.

CdP_4 is formed on heating a cadmium–lead alloy (5 atom% Cd, 95 atom% Pb) with phosphorus to 565°–570° for several days. The black and strongly reflecting crystals decompose into the elements on heating *in vacuo* CdP_4 is nonreactive, but dissolves in boiling aqua regia.[8]

The *cadmium arsenides*, Cd_3As_2 and $CdAs_2$, are formed by passing arsenic vapor in a stream of hydrogen over heated cadmium metal in an electric furnace. Cd_3As_2 (m.p. 721°) is metallic grey. It slowly decomposes with acids to form arsine, AsH_3, but it does not react with water. $CdAs_2$ (m.p. 621°) is black-grey.[9]

5.1.4 Cadmium Carboxylates and Some Pseudohalide Derivatives

Cadmium acetate is formed in an initially quite vigorous reaction by refluxing cadmium nitrate in acetic acid anhydride. The precipitate is collected, washed with acetic acid anhydride and ether, and dried *in vacuo* over KOH or sulfuric acid[3] (m.p. 254°–256°).

Cadmium carbonate, $CdCO_3$, is formed by precipitation with ammonium carbonate from a solution of cadmium chloride. The precipitate is dissolved in excess ammonia and the solution is evaporated over a water bath. Cadmium carbonate is also formed by reacting a hydrochloric acid solution of cadmium chloride with urea in a sealed tube at 200°. The compound decomposes on heating to 320° with generation of carbon dioxide.[3]

Cadmium cyanide, $Cd(CN)_2$, crystallizes on concentrating an aqueous solution of cadmium hydroxide and hydrogen cyanide. The material is dried at 100°. $K_2[Cd(CN)_4]$ is formed on reaction of $Cd(CN)_2$ with a stoichiometric quantity of KCN in aqueous solution. On concentrating the solution the material precipitates and can be collected and dried at 105°.

Cadmium thiocyanate, $Cd(SCN)_2$, is formed by anion exchange [Eqn (3)] between cadmium sulfate and barium thiocyanate in boiling aqueous solution.

$$CdSO_4 + Ba(SCN)_2 \longrightarrow Cd(SCN)_2 + BaSO_4 \qquad (3)$$

Barium sulfate precipitate is filtered off and the cadmium thiocyanate is isolated by evaporation of the filtrate to dryness.[3]

5.1.5 Cadmium Silicate and Cadmium Ferrate(III)

Cadmium silicate, Cd_2SiO_4 (m.p. 1252°), is obtained on heating a stoichiometric mixture

---

[6] *R. Juza, K. Fasold, W. Kuhn,* Z. Anorg. Allgem. Chem. *234*, 86 (1937).

[7] *H. Hahn, R. Juza,* Z. Anorg. Allgem. Chem. *244*, 111 (1940).

[8] *H. Krebs, K. H. Muller, G. Zurn,* Z. Anorg. Allgem. Chem. *285*, 15 (1956).

[9] *M. von Stackelberg, R. Paulus,* Z. Phys. Chem. (B) *28*, 427 (1935).

of CdO and SiO_2 to temperatures higher than 1252°.[3,10]

Cadmium ferrate(III), $CdFe_2O_4$, is formed on heating a mixture of cadmium carbonate or cadmium hydroxide with γ-FeO(OH) to 800°–1000°, respectively, for 1 hr in a platinum dish. The hygroscopic brown product is cooled in a desiccator.

γ-Iron hydroxide is obtained by reaction of iron(II) chloride with hexamethylenetetramine; the resulting $Fe(OH)_2$ is oxidized to γ-FeO(OH) by treatment with an aqueous solution of sodium nitrite for 3 hr at 60°. The product[3,11] is filtered off, washed with warm water, and dried at 60°.

5.2 Organocadmium Derivatives

Organocadmium compounds, in general, are intermediate between organozinc and organomercury compounds in their reactivity towards air and water. Thus, although the dialkyl compounds of cadmium do react with water, they do not ignite in air as the lower dialkyl derivatives of zinc do. Dialkylcadmium compounds are used in organic synthesis for the preparation of ketones from acid chlorides and have gained some commercial importance as catalyst components in Ziegler polymerization of olefins.

5.2.1 Organocadmium Halides

Three methods have been used for the preparation of these compounds:

(a) the redistribution of substituents between cadmium halides and dialkyl derivatives;
(b) the reaction between Grignard reagents and cadmium halides; and
(c) the reaction of metallic cadmium with alkyl halides.

Until recently, the existence of an "RCdX" species was in doubt. It was formerly held that such compounds were weak complexes of R_2Cd and CdX_2. Evidence[12] for this assumption was considered to be the lack of exchange after 21 days between labelled diethylcadmium and cadmium bromide and between diethylcadmium and labelled cadmium bromide. However, results obtained recently by other methods as well as the characterization of specific compounds provides definitive proof for the existence of RCdX species. Thus, Kavanagh and Evans[13] concluded from i.r. and Raman studies on solutions of dimethylcadmium and cadmium halides in tetrahydrofuran that they were in equilibrium with monomeric alkylcadmium halides and that the Schlenk equilibrium was on the side of the alkylcadmium halide. Measurements[14] by n.m.r. of solutions of dimethylcadmium in tetrahydrofuran (THF) saturated with cadmium bromide exhibited proton resonances due to methylcadmium bromide.

The first isolation of *alkylcadmium halides* was reported[15] in 1959. The compounds were prepared by a redistribution reaction in ether and are listed in Table 1. *Arylcadmium halides* have also been obtained by this method.[16] Owing to poor solubility of cadmium halides in ether, the reaction is fairly slow. Thus, reaction of dipropylcadmium with cadmium iodide requires 2 hr at room temperature, while the reaction with the chloride takes 36 hr. Support for the structure of alkylcadmium halides is provided by infra-red studies[17] and molecular weight

Table 1. Organocadmium Halides Prepared *via* Redistribution Reactions

Compound	Yield (%)	Ref.
C_2H_5CdCl	60·6	*a*
C_2H_5CdBr	66·4	*a*
C_2H_5CdI	84·3	*a*
n-C_3H_7CdCl	43·5	*a*
n-C_3H_7CdBr	64·1	*a*
n-C_3H_7CdI	60·3	*a*
n-C_4H_9CdCl	25·3	*a*
n-C_4H_9CdBr	67·5	*a*
n-C_4H_9CdI	60·3	*a*
C_6H_5CdI	80·3	*b*
2-C_4H_3SCdI	80·7	*b*

[a] *N. I. Sheverdina, I. E. Paleeva, E. D. Delinskaya, K. A. Kocheshkov*, Dokl. Akad. Nauk. SSSR *125*, 348 (1959).
[b] *N. I. Sheverdina, I. E. Paleeva, E. D. Delinskaya, K. A. Kocheshkov*, Dokl. Akad. Nauk. SSSR *143*, 1123 (1962).

[10] *W. Biltz, A. Lemke*, Z. Anorg. Allgem. Chem. *203*, 330 (1932).
[11] *R. Fricke, F. Blaschke*, Z. Anorg. Allgem. Chem. *251*, 396 (1943).
[12] *A. B. Garret, A. Sweet, W. L. Marshall, D. Riley, A. Touma*, Record Chem. Progr. *13*, 155 (1952).
[13] *K. Kavanagh, D. F. Evans*, J. Chem. Soc. A, 2890 (1969).
[14] *W. Bremser, M. Winokur, J. D. Roberts*, J. Amer. Chem. Soc. *92*, 1080 (1970).
[15] *N. I. Sheverdina, I. E. Paleeva, E. D. Delinskaya, K. A. Kocheshkov*, Dokl. Akad. Nauk. SSSR *125*, 348 (1959).
[16] *N. I. Sheverdina, I. E. Paleeva, E. D. Delinskaya, K. A. Kocheshkov*, Izv. Akad. Nauk. SSSR *143*, 1123 (1962).
[17] *A. N. Rodionov, I. E. Paleeva, D. N. Shigorin, N. I. Sheverdina, K. A. Kocheshkov*, Izv. Akad. Nauk. SSSR 1031 (1967).

determinations[18] in dimethyl sulphoxide (DMSO) in which they are monomeric.

Organocadmium halides can also be prepared by reaction of equimolar quantities of a Grignard reagent and cadmium halides in ether, but the yields are low since separation of the RCdX species from the magnesium salts formed is difficult. Thus, *n*-butylcadmium bromide was prepared in only 12·5% yield by this method.[19] Additional evidence for the formation of alkylcadmium halide by reaction of the Grignard reagent and a cadmium halide in equimolar ratio and isolation of $RCdX \cdot MgX_2$ has recently been reported.[20]

Organocadmium halides are white, microcrystalline solids which decompose above 100°. Ethylcadmium halides are insoluble in the usual organic solvents, but are soluble in DMSO. Propyl- and higher alkylcadmium halides are soluble in ether; however, no complexes are formed with ether or dioxane[18]; this behaviour is in contrast to that of organozinc halides.[21]

5.2.2 Organocadmium–Chalcogen Compounds

Alkylcadmium alkoxides have been prepared[22] by the reaction of alcohols with dialkylcadmium compounds at low temperatures according to Eqn (4).

$$R_2Cd + R'OH \longrightarrow RCdOR' + RH \quad (4)$$

$R = CH_3$; $R' = CH_3, C_2H_5, i\text{-}C_3H_7, t\text{-}C_4H_9, Ph$

All the compounds with the exception of the *t*-butyl derivative are tetrameric in benzene; they probably have a cubanite structure. The *t*-butyl derivative is dimeric in benzene and hydrolyzes only slowly.

A few organocadmium compounds in which the oxygen is bonded to an element other than carbon are known. Methylcadmium trimethylsilanolate, $MeCdOSiMe_3$ has been synthesized by interaction of dimethylcadmium with trimethylsilanol.[23] This silicon derivative is also tetrameric in solution and has a cubanite structure. A cadmium stannoxide derivative, $EtCdOSnEt_3$, has been prepared by reaction of triethyltin hydroxide with diethylcadmium.[24] It has been reported that *alkylcadmium alkylperoxides*, RCdOOR′, are formed by the reaction of dialkylcadmium with an alkylhydroperoxide, but only in the case of reactions with dimethylcadmium were any compounds isolated.[25]

The thio analogs of the alkoxides, RCdSR′, have been obtained by reaction[22] of dialkylcadmium compounds with thiols in hexane at −78°; MeCdSMe and $MeCdSC_6H_5$ were found to be highly associated in benzene; in this solvent MeCd–S–*i*–Pr is hexameric and MeCd–S–*t*–Bu is tetrameric.

Alkylcadmium thiocyanates have been obtained by the reaction of dialkylcadmium compounds with thiocanogen in benzene.[26] They are relatively thermostable solids. Their formation proceeds according to Eqn (5).

$$R_2Cd + (SCN)_2 \longrightarrow R\text{-}CdSCN + R\text{-}SCN \quad (5)$$

$R = CH_3, C_2H_5$

5.2.3 Organocadmium Compounds with Group V Elements

Only a few compounds containing the C–Cd–N skeleton have been reported. Reaction of dimethylcadmium with trialkylphosphinimines in benzene was found to give *trialkylphosphine-N-methylcadmium imine*.[27]

$$(CH_3)_2Cd + R_3P{=}NH \longrightarrow CH_3CdN{=}PR_3 + CH_4 \quad (6)$$

$R = CH_3, C_2H_5$

These products are tetrameric and were assigned a cubanite structure on the basis of infra-red, n.m.r., and molecular weight data. The compounds are thermally quite stable, decomposing about 100°, and may be purified by sublimation.

Methylcadmium azide, $MeCdN_3$, is obtained as a white, hygroscopic, crystalline solid in good yield by the reaction of dimethylcadmium with chlorine azide in carbon tetrachloride.[28]

A *methylcadmium dimethylsulfoxime* was syn-

---

[18] *I. E. Paleeva, N. I. Sheverdina, K. A. Kocheshkov*, Izv. Akad. Nauk. SSSR 1263 (1967).

[19] *I. E. Paleeva, N. I. Sheverdina, L. V. Abramova, K. A. Kocheshkov*, Dokl. Akad. Nauk. SSSR *159*, 609 (1964).

[20] *J. R. Sanders, E. C. Ashby*, J. Organometal. Chem. *25*, 227 (1970).

[21] *N. I. Sheverdina, L. V. Abramova, K. A. Kocheshkov*, Dokl. Akad. Nauk. SSSR *134*, 853 (1960).

[22] *G. E. Coates, A. Lauder*, J. Chem. Soc. A, 264 (1966).

[23] *F. Schindler, H. Schmidbauer, U. Kruger*, Angew. Chem. *77*, 865 (1965).

[24] *V. N. Pankratova, L. P. Stepovik*, Zh. Obshch. Khim. *38*, 7244 (1968).

[25] *A. G. Davies, J. E. Packer*, J. Chem. Soc. 3164 (1959).

[26] *T. Wizeman, H. Müller, D. Seybold, K. Dehnicke*, J. Organometal. Chem. *20*, 211 (1960).

[27] *H. Schmidbauer, G. Jonas*, Bull. Soc. Chim. France *101*, 1271 (1968).

[28] *K. Dehnicke, J. Strable, D. Seybold, J. Müller*, J. Organometal. Chem. *6*, 2981 (1966).

thesized by the reaction of dimethyl sulfoxime with dimethylcadmium.[29]

$$(H_3C)_2Cd + (H_3C)_2S(=O)=NH \longrightarrow H_3C{-}Cd{-}N{=}S(=O)(CH_3)_2 + CH_4$$

Little is known about its structure.

5.2.4 Diorganocadmium Compounds

The classic routes to diorganocadmium compounds are the reactions of anhydrous cadmium halides with Grignard reagents or organolithium derivatives. In addition to these general methods, a number of other syntheses for specific diorganocadmium compounds have been reported.

A procedure using Grignard reagents has been employed to prepare dialkyl and diaryl compounds in which the substituents are identical or different.[30] The unsymmetrical compounds, RCdR′, are obtained by the action of the Grignard reagent on an alkylcadmium halide[31] (Table 2).

Table 2. Diorganocadmium RCd–R′ Prepared by the Grignard Method

R	R′	Yield (%)	Ref.
CH_3	CH_3	85	*a*
C_2H_5	C_2H_5	90	*a*
C_4H_9	C_4H_9	70	*a*
C_6H_{11}	C_6H_{11}	22	*b*
C_6H_5	C_6H_5	31	*c*
4-Cl–C_6H_4	4-Cl–C_6H_4	32	*c*
2-CH_3–C_6H_4	2-$CH_3C_6H_4$	53	*d*
C_2H_5	C_3H_7	66	*d*
C_2H_5	C_4H_9	85	*d*
C_2H_5	$(CH_3)_2CH$–CH_2–	38	*d*
C_6H_5	2-C_4H_3S	26	*e*

[a] *E. Krause*, Ber. *50*, 1813 (1917).
[b] *G. A. Razuvaev, V. N. Pankratova, A. M. Bobrova*, Z. Obshch. Khim. *38*, 1723 (1968).
[c] *N. I. Sheverdina, I. E. Paleeva, E. D. Delinskaya, K. A. Kocheshkov*, Dokl. Akad. Nauk. SSSR *143*, 1125 (1962).
[d] *K. A. Kocheshkov, N. I. Sheverdina, I. E. Paleeva*, Izv. Akad. Nauk. SSSR 1472 (1963)
[e] *I. E. Paleeva, N. I. Sheverdina, K. A. Kocheshkov*, Dokl. Akad. Nauk. SSSR *157*, 626 (1964).

$$R^1{-}CdX + R^2MgX \longrightarrow R^1{-}Cd{-}R^2 + MgX_2 \qquad (7)$$

The organolithium method is mainly used for the synthesis of *diarylcadmium compounds*. The yields are generally somewhat lower than those obtained via the Grignard method. The separation of the diarylcadmium from biaryl, formed as a by-product, is often a problem. Diphenyl-[32] and dipentafluorophenylcadmium[33] have been prepared in 43 and 51% yields, respectively, by the organolithium route.

The reaction of metallic cadmium with alkyl halides is normally too slow to be of practical value for the preparation of dialkylcadmium compounds. However, it has recently been found that alkyl iodides and reactive bromides such as allyl and benzyl bromide will react with cadmium in hexamethylphosphoramide to give good yields of the *dialkylcadmium derivatives*.[34,35]

Allylic organocadmium compounds have been synthesized in quantitative yield by the reaction of allylic boron derivatives with dimethylcadmium at low temperatures.[36]

$$3(CH_3)_2Cd + 2B(CH_2{-}C(R'){=}CHR^2)_3 \longrightarrow 3Cd(CH_2{-}C(R'){=}CHR^2)_2 + 2(CH_3)_3B \qquad (8)$$

$$R' = H,\ R^2 = H,\ Me;\ R' = Me,\ R^2 = H$$

The trimethylboron produced in the reaction is readily removed by virtue of its low boiling point (b.p. −20°). The allylic compounds of cadmium are extremely reactive, decomposing slightly above 0° and reacting with carbon dioxide at room temperature to give the corresponding carboxylates, e.g., $(CH_2{=}CH{-}CH_2{-}CO_2)_2Cd$.[37]

Divinylcadmium has been prepared by ligand exchange between divinylmercury and dimethylcadmium (unspecified yields).[38] Attempts to prepare divinylcadmium by the Grignard[39]

[29] *H. Schmidbauer, G. Kammel, W. Stadelmann*, J. Organometal. Chem. *15*, P10 (1968).
[30] *I. E. Paleeva, N. I. Sheverdina, E. D. Delinskaya, K. A. Kocheshkov*, Izv. Akad. Nauk. SSSR *143*, 1123 (1962).
[31] *N. I. Sheverdina, I. E. Paleeva, K. A. Kocheshkov*, Izv. Akad. Nauk. SSSR 1472 (1963).
[32] *S. Witlig, F. J. Meyer, G. Lange*, Justus Liebigs Ann. Chem. *571*, 167 (1951).
[33] *M. Schmiesser, M. Weidenbruch*, Chem. Ber. *100*, 2306 (1967).
[34] *J. Chenault, F. Tatibouet*, Compt. Rend. *262*, 499 (1966).
[35] *J. Chenault, F. Tatibouet*, Compt. Rend. *264*, 213 (1967).
[36] *K. H. Thiele, J. Kohler*, J. Organometal. Chem. *7*, 365 (1967).
[37] *K. H. Thiele, J. Köhler, P. Zdunneck*, Z. Chem. *7*, 307 (1967).
[38] *H. D. Visser, L. P. Stodulski, J. P. Oliver*, J. Organometal. Chem. *24*, 563 (1970).
[39] *B. Bartocha, C. M. Douglas, M. Y. Gray*, Z. Naturforsch. *14b*, 809 (1959).

and organolithium[38] routes have been unsuccessful. The reported[40] synthesis of divinylcadmium from divinylmercury and metallic cadmium could not be duplicated.[38]

Dialkynyl or *diarylynyl derivatives of cadmium* have been synthesized by metallation of acetylenes with dialkyl- and diarylcadmium compounds.[41, 42]

$$R_2Cd + 2R^1{-}C{\equiv}CH \longrightarrow (R^1{-}C{\equiv}C)_2Cd + 2RH$$

Thus, $(C_6H_5C{\equiv}C)_2Cd$ was obtained from reaction of $C_6H_5{-}C{\equiv}C{-}H$ with diphenylcadmium[41] or dimethylcadmium[42] in good yield. The rate of metallation reaction has been shown to increase with an increase in the polarity of the solvent employed.[43] The *acetylenic organocadmium derivatives* were found to be associated, rather than monomeric.[42]

In another metallation reaction, *dicyclopentadienylcadmium* was obtained in quantitative yield [Eqn (9)].[44]

$$Cd[N(SiMe_3)_2]_2 + C_5H_6 \xrightarrow[R.T.]{Et_2O} (C_5H_5)_2Cd + 2HN(SiMe_3)_2 \quad (9)$$

The reaction apparently owes its success to the insolubility of the dicyclopentadienylcadmium in the solvent. Dicyclopentadienylcadmium is one of the thermally most stable organocadmium compounds, decomposing near 250°.

The thermal decarboxylation of cadmium carboxylates is of use primarily as a route to perhaloaryl cadmium compounds.[33, 45]

$$(R{-}\overset{O}{\overset{\|}{C}}{-}O)_2Cd \xrightarrow{\nabla} R_2Cd + 2\,CO_2$$

$$R = C_6F_5,\ C_6Cl_5 \quad (10)$$

It should be noted that ordinary cadmium compounds would not be stable at the high temperatures used in the decarboxylation reaction (220° for the fluoro compound and 240° for the chloro derivative); nevertheless, the yields are surprisingly good [85% $(C_6F_5)_2Cd$[33] and 65% $(C_6Cl_5)_2Cd$].[45]

The dialkylcadmium compounds generally are liquids, whereas the *diarylcadmium compounds* are normally solids. The simple dialkyl compounds decompose above 150°. If the alkyl group is a secondary one, the decomposition temperature is below 150°; thus, dicyclohexylcadmium is unstable on standing at room temperature.[46] The diarylcadmium compounds are slightly more stable. Both the dialkylcadmium and diarylcadmium compounds are monomeric, but form complexes with a variety of donor materials such as dioxane and 2,2¹-bipyridyl.[47] Diorganocadmium compounds also react with organolithium compounds to give "ate" complexes in certain solvents.[48] Thus, ethyllithium and diethylcadmium form the solvated complex $LiEt_3Cd \cdot 2THF$. The reaction occurs in tetrahydroferan but not in benzene.

5.3 Bibliography

G. E. Coates, K. Wade, Organometallic Compounds, Vol. I, 3rd Ed., pp. 141–147. Methuen, London, 1967.

B. J. Wakefield, Adv. Inorg. Chem. Radiochem. *11*, 342 (1968).

N. I. Sheverdina, K. A. Kocheshkov, The Organic Compounds of Zinc and Cadmium. North-Holland, Amsterdam, 1967.

---

40 U.S. Patent 3087947 (to Union Carbide), *D. J. Foster, E. Tobler*.

41 *R. Nast, C. Richers*, Z. Anorg. Allgem. Chem. *319*, 320 (1963).

42 *E. A. Jeffrey, T. Mole*, J. Organometal. Chem. *19*, 189 (1968).

43 *O. Y. Okhlabystin, L. I. Zakharin*, Z. Obshch. Khim. *36*, 1734 (1966).

44 *J. Loberth*, J. Organometal. Chem. *19*, 189 (1969).

45 *M. Schmeisser, S. Boke*, Chem. Ber. *103*, 510 (1970).

46 *G. A. Razuvaev, V. N. Pankratova, A. M. Bobrova*, Z. Obshch. Khim. *38*, 1723 (1968).

47 *K. H. Thiele, P. Zdunnek*, Organometal. Chem. Rev. *1*, 331 (1966).

48 *S. Toppet, G. Slinckx, G. Smets*, J. Organometal. Chem. *9*, 205 (1967).

6 Mercury

C. W. Blewett
Emery Industries,
Cincinnati, Ohio, U.S.A.

R. Schmid
Battelle Memorial Institute,
Genf, Switzerland

H. Zimmer
Department of Chemistry, University of Cincinnati
Cincinnati, Ohio, U.S.A.

6.1 Inorganic Mercury Compounds

6.1.1 Metallic Mercury and Amalgams

In order to *purify mercury*, the metal is filtered through a paper with a tiny hole in the bottom. Material remaining in the filter is discarded. Less noble metals which are dissolved in mercury can be removed by shaking with a potassium permanganate solution or nitric acid or by oxidation with air. Subsequently, the mercury is washed with water, dried under a vacuum, and finally distilled under high vacuum.[1]

Amalgams are solutions of metals in mercury. Sometimes they form definite intermetallic compounds such as LiHg, $NaHg_2$, or $CsHg_4$. The formation of an amalgam is at least in principal always possible by direct combination of the elements or by electrolysis of a metal salt solution using a mercury cathode. Also, certain salts may exchange the cation on contact with amalgams. This latter reaction is illustrated by the following example:

$$3NaHg_x + NdCl_3 \longrightarrow 3NaCl + NdHg_x$$

Sodium amalgam is prepared by addition of small pieces of metallic sodium to mercury or by electrolysis of sodium chloride solutions using a mercury cathode. It is also possible to combine mercury with molten sodium which is covered with an inert solvent such as paraffin oil.

Barium and *strontium amalgam* are prepared by electrolysis of the respective salts using a mercury cathode.

Amalgams of *lanthanum*, *cerium*, or *neodymium* are obtained by electrolysis of the respective anhydrous chlorides in absolute alcohol. *Europium*, *samarium*, or *ytterbium* acetate can be electrolyzed in an aqueous solution containing potassium citrate in order to avoid precipitation of the hydrated metal oxides. The efficiency of this synthesis is greatest in the case of europium amalgam, less for that of ytterbium amalgam, and rather small in the case of samarium amalgam.[2,3]

6.1.2 Mercury(I) Compounds

Mercury(I) compounds readily disproportionate according to the following equation.

$$Hg_2^{2\oplus} \rightleftharpoons Hg + Hg^{2\oplus}$$

Mercury(I) fluoride, Hg_2F_2. On addition of potassium hydrogen carbonate to a solution of mercury(I) nitrate in dilute nitric acid mercury(I) carbonate, Hg_2CO_3, precipitates. It is filtered, washed with carbon dioxide-saturated water, and then is treated with small portions of 40% hydrofluoric acid to give yellow mercury(I) fluoride. The reaction mixture is brought to dryness; the remaining solid is pulverized and heated to 120°–150° for several hours. Mercury(I) fluoride (m.p. 570°) is light-sensitive and, consequently, all operations have to be performed in the dark.[1] The *chloride*, *bromide*, and *iodide* of mercury(I) are best prepared by reduction of the corresponding mercury(II) salts with metallic mercury, hydrogen, sulfur dioxide, or organic reducing agents such as glucose. These mercury(I) halides may also be prepared by a reaction between the corresponding alkali halides and a mercury(I) nitrate solution.[4]

Mercury(I) perchlorate, $Hg_2(ClO_4)_2$, is prepared by reduction of mercury(II) perchlorate with $Hg(ClO_4)_2$, mercury.

Mercury(I) nitrate is formed by the interaction of cold dilute nitric acid with an excess of mercury yielding the dihydrate, which can be dehydrated to yield the anhydrous material by simple drying.[5]

Mercury(I) acetate. On addition of an aqueous sodium acetate solution to a solution of mercury(I) nitrate in dilute nitric acid mercury(I) acetate precipitates. After washing with water it is dried in a desiccator over calcium chloride. The salt readily disproportionates on exposure to air or on heating.

Owing to the tendency of the $Hg_2^{2\oplus}$ cation to disproportionate it forms only a small number of stable complexes. However, potentiometric measurements have confirmed the existence of complexes with diphosphate, oxalate, succinate, and tripolyphosphate.[6] Some *complexes* containing Hg(I)–Rh and Hg(I)–Ir bonds have been obtained by reaction of mercury(I) salts with complex rhodium and iridium hydrides.[7]

---

[1] *W. Kwasnik* in *G. Bauer*, Handbook of Preparative Inorganic Chemistry, Vol. I, p. 27, 242, 2nd Ed., Academic Press, New York, London, 1963.

[2] *L. F. Audrieth*, Inorg. Synth. *1*, 5 (1939).

[3] *H. N. McCoy*, Inorg. Synth. *2*, 65 (1946).

[4] *Gmelins* Handbuch der Anorganischen Chemie, Quecksilber Teil 2, Lieferung 2, p. 426, 471, 722, 820, 834; Verlag Chemie, Weinheim/Bergstr. 1967.

[5] *Gmelins* Handbuch der Anorganischen Chemie, Quecksilber Teil B, Lieferung 1, p. 76, 96, Verlag Chemie, Weinheim/Bergstr. 1965.

[6] *T. Yamane*, *N. Davidson*, J. Amer. Chem. Soc. *81*, 4438 (1959).

[7] *R. S. Nyholm*, *K. Vriese*, J. Chem. Soc. 5331, 5337 (1965).

6.1.3 Mercury(II) Compounds

6.1.3.1 Halides

Mercury(II) fluoride, HgF_2, is formed in a horizontally rotating copper cylinder by reacting diehydrous powdered mercury(II) chloride with elemental fluorine. Mercury(II) fluoride is also formed by reacting mercury(II) oxide for 4 hr with a mixture of excess hydrogen fluoride and parts by oxygen at 380°–450°. The colorless powder (m.p. 645°–650°) hydrolyzes readily on exposure to moist air.[1] *Mercury(II) chloride* ("sublimate") is formed when mercury(II) nitrate or mercury(II) sulfate reacts with alkali chlorides. It is also obtained by passing chlorine into solutions of mercury(II) salts or by dissolving mercury(II) oxide in hydrochloric acid.[4]
Mercury(II) bromide, $HgBr_2$. For the preparation of mercury(II) bromide metallic mercury is covered with water and a slight excess of bromine is added dropwise while maintaining a temperature of 50°. On cooling to 0° $HgBr_2$ (m.p. 238°, b.p. 320°), crystallizes; it can be purified by sublimation. Molten mercury(II) bromide is an excellent solvent for many organic and inorganic compounds.[8] *Mercury(II) iodide* can be prepared by mixing aqueous solution of iodine or alkali iodides with mercury(II) nitrate or mercury(II) perchlorate. HgI_2 reacts with an excess of potassium iodide to form K_2HgI_4 and $KHgI_3 \cdot H_2O$, respectively.[4,9] *Mercury(II) perchlorate* is formed when mercury(II) oxide is dissolved in a slight excess of aqueous perchloric acid.[10]
Copper(I) tetraiodomercurate(II), Cu_2HgI_4, and *silver tetraiodomercurate(II)*. Sulfur dioxide is passed into an aqueous solution of HgI_2 and KI to which copper sulfate was added, whereupon Cn_2HgI_4 precipitates; after washing with water it is dried at 100° and recrystallized from hot concentrated hydrochloric acid.[9] Pure Cu_2HgI_4 and Ag_2HgI_4 are formed by melting a stoichiometric mixture of CuI (or AgI) and HgI_2 *in vacuo*. It also can be obtained by triturating such a mixture with a small amount of alcohol. Both complexes are thermotropic compounds.[11,12]

[8] *G. Jander, K. Brodersen*, Z. Anorg. Allgem. Chemie *261*, 262 (1950).
[9] *F. Wagenknecht, R. Juza* in *G. Bauer*, Handbook of Preparative Inorganic Chemistry, Vol. II, p. 1067, 2nd Ed., Academic Press, New York, London, 1965.
[10] *J. H. Espenson, J. P. Birk*, Inorg. Chem. *4*, 527 (1965).
[11] *G. Tammann, G. Veszi*, Z. Anorg. Allgem. Chemie *168*, 46 (1927).

6.1.3.2 Mercury(II)–Chalcogen Compounds

Mercury(II) oxide, HgO, precipitates as a yellow compound on treatment of an aqueous mercury(II) nitrate or chloride solution with cold 5% aqueous sodium hydroxide. The suspension of freshly precipitated mercury(II) oxide can be used immediately for further reactions.[1,13]

Amorphous *black mercury(II) sulfide* is formed when hydrogen sulfide is passed into a solution of mercury(II) salt in 1–2 N hydrochloric acid. The initially white to brown precipitate blackens in the presence of excess hydrogen sulfide. The solution should contain no oxidizing agents.[9]
The stable red modification of mercury(II) sulfide is formed when hydrogen sulfide is passed into a solution of mercury(II) acetate and ammonium thiocyanate in hot glacial acetic acid. The acetic acid which has been generated is then slowly evaporated. (**Caution:** HCN is generated.) The initial black precipitate converts to the red modification. Continuous stirring throughout the evaporation procedure ensures a completely red product. The product should then be dried on filter paper.[14]

Black *mercury(II) selenide* is formed on dropwise addition of dilute mercury(II) chloride solution to a concentrated aqueous solution of hydrogen selenide in the absence of air. The material can be dried in a desiccator over phosphorus pentoxide. When using a concentrated mercury(II) chloride solution in the described process, a yellow adduct is formed having the composition $HgCl_2 \cdot 2HgSe$.
Mercury(II) selenide can also be obtained by heating a stoichiometric amounts of the two elements to 550°–650° in a sealed tube. The material can be purified by sublimation at 600°–650° in a stream of nitrogen.[9] *Mercury(II) tetrafluoroselenate*, $HgSeF_4$, is formed on refluxing SeF_2 and mercury for 30 hours.[15]

6.1.3.3 Mercury(II)–Nitrogen Derivatives

Mercury(II) nitride, Hg_3N_2, is a brown material which is obtained by reacting mercury(II) bromide or iodide with an excess of potassium amide in liquid ammonia.[5]
Mercury(II) amidochloride, $HgNH_2Cl$, is formed on reacting 6 N ammonia solution with mer-

[12] *H. Hahn, G. Frank, W. Klingler*, Z. Anorg. Allgem. Chemie *279*, 271 (1955).
[13] *A. Meuwsen, G. Weiss*, Z. Anorg. Allgem. Chemie *289*, 5 (1957).
[14] *L. C. Newell, R. N. Maxson, M. H. Filson*, Inorg. Synth. *1*, 19 (1939).
[15] *R. D. Peacock*, J. Chem. Soc. 3617 (1953).

cury(II) chloride. The colorless precipitate is carefully dried at 30° excluding light.

$$HgCl_2 + 2NH_3 \longrightarrow HgNH_2Cl + NH_4Cl$$

When working in too dilute solution the desired compound decomposes with decolorization (yellow) and partial formation of $NHgCl \cdot H_2O$. On heating mercury(II) amidochloride volatilizes without melting.[9]

The corresponding *mercury(II) amidobromide* is formed as a light yellow precipitate on reacting 0·1 N aqueous ammonia with a saturated (at 80°) solution of mercury(II) bromide. The precipitate is washed with a small amount of anhydrous methanol and is dried in vacuum.[16]

Mercury(II) diamminedichloride, $Hg(NH_3)_2Cl_2$, is obtained from the interaction of $HgCl_2$ and ammonium chloride in water with 4·5 N ammonia. The mixture is stored for several days with frequent shaking.

The compound precipitates as colorless needles; these are washed with ethanol, and dried in complete darkness over KOH. The same compound is obtained by action of liquid ammonia on mercury(II) chloride. A reversible equilibrium exists with mercury(II) amidochloride which can be shifted towards the amidochloride by addition of potassium amide.

$$Hg(NH_3)_2Cl_2 \rightleftharpoons HgNH_2Cl + NH_4Cl$$

Also, subsequent reaction to Hg_2NCl is possible.

$$2HgNH_2Cl \longrightarrow Hg_2NCl + NH_4Cl$$

$Hg(NH_3)_2Cl_2$ melts and decomposes [m.p. (in ammonia atmosphere) 247°–253°].[9,17]

The corresponding *dibromide*, $Hg(NH_3)_2Br_2$, is formed as a precipitate on reaction of aqueous 0·1 N ammonia solution (which is 1 N with respect to ammonium bromide) with a saturated (80°) solution of mercury(II) bromide. The product is filtered off and dried initially at 60° and then under vacuum over silica. Even small amounts of water decompose the material according to the following equation.

$$Hg(NH_3)_2Br_2 \xrightarrow{[H_2O]} HgNH_2Br + NH_4Br$$

Larger quantities of water react with the compound to yield the adduct $4NHg_2Br \cdot HgBr_2$.[16]

Mercury(II) iminochloride, $Hg_2(NH)Cl_2$. Freshly precipitated mercury(II) oxide reacts with a concentrated solution of $HgCl_2$ in the presence of ammonium chloride to form a yellow-green precipitate, $Hg_2(NH)Cl_2$. The same compound can be prepared by reaction of mercury(II) acetate with $HgCl_2$ and ammonium chloride in anhydrous methanol.[13]

The corresponding *bromide*, $Hg_2(NH)Br_2$, is formed on interaction of a solution of $HgBr_2$ in boiling water with 0·1 N ammonia (containing ammonium bromide). The precipitate is filtered off, washed with cold water, and dried over sodium hydroxide.

$$2HgBr_2 + 3NH_3 \longrightarrow (HgBr)_2NH + 2NH_4Br$$

The same compound can also be obtained from the reaction of mercury(II) oxide with a concentrated solution of $HgBr_2$ in the presence of ammonium bromide. The compound decomposes on exposure to sunlight.[13]

$(HgCl)_2N{-}C_2H_5$ is obtained by reacting HgO and $HgCl_2$ with ethylammonium chloride at 65° for 2 hr in an aqueous solution. If the reaction is carried out at 80° the *adduct* $[Hg_3(NC_2H_5)_2]OH \cdot HgCl_3$ is obtained.[13] $(BrHg)_2N{-}C_2H_5$ is formed similarly using $HgBr_2$ and ethylammonium bromide. The suspension is left standing for two weeks in the dark. Conducting the reaction in hot water or at a reaction temperature of 80°–85°, the adduct $[Hg_3(NC_2H_5)_2]OH \cdot HgBr_3$ is obtained instead.[13]

In similar fashion, $Hg_2(NC_6H_5)Cl_2$ and $Hg_2(NC_6H_5)Br_2$ are formed on reaction of HgO with $HgCl_2$ or $HgBr_2$, respectively, together with the corresponding anilinium salt in aqueous solution at temperatures of 60°–65°.

$$3HgO + HgCl_2 + 2C_6H_5NH_3Cl \longrightarrow 2(ClHg)_2N{-}C_6H_5 + 3H_2O$$

The latter chloride is a yellow-green material which darkens on exposure to sunlight. The material does not decompose in boiling water. The lemon-colored bromide darkens even in diffuse daylight; it is stable in cold water.[13]

$(ClHg)_2N{-}CO{-}NH_2$ and $(BrHg)_2N{-}CO{-}NH_2$ are obtained from the reaction of HgO with an aqueous solution of $HgCl_2(HgBr_2)$ and the corresponding urea hydrohalide at room temperature (shaking for one hour). The colorless to slightly green (bromide: yellow) solids are readily soluble in dilute hydrochloric acid.[13]

$(ClHg)_2NO{-}C_2H_5$ and $(BrHg)_2NO{-}C_2H_5$ are formed by stirring a slurry of HgO with a solution of $HgCl_2(HgBr_2)$ and $(C_2H_5ONH_3)Cl$ $[(C_2H_5ONH_3)Br]$ for one hour.[13]

---

[16] *W. Rüdorf, K. Brodersen*, Z. Anorg. Allgem. Chemie *270*, 145, 159 (1952).

[17] *J. Jander, C. Lafrenz* in Wasserähnliche Lösungsmittel (Chem. Taschenbücher 3), p. 34, Verlag Chemie Weinheim/Bergstr. 1968.

Mercury(II) chloride–monothiourea, $[Hg(SCN_2H_4)Cl]Cl$, can be prepared by briefly heating equimolar amounts of mercury(II) chloride and thiourea in aqueous solution. However, on prolonged heating black mercury(II) sulfide may be formed. The white needles of the thiourea complex can be washed with small amounts of cold water or ethanol and dried under vacuum. Reacting 1 mol of mercury(II) chloride with 2 mol of thiourea yields a *dithiourea adduct*, $[Hg(SCN_2H_4)]Cl_2$ (m.p. 250°), with blackening. In the presence of excess of thiourea the diadduct dissolves with formation of $[Hg(SCN_2H_4)_3]Cl_2$ or $[Hg(SCN_2H_4)_4]Cl_2$, respectively. These latter compounds can be isolated by evaporation of the solution in a vacuum desiccator; both can be recrystallized from water.[18]

Millon's base, $NHg_2OH \cdot H_2O$ and $NHg_2OH \cdot 2H_2O$. On shaking a suspension of yellow HgO in a 12 N ammonia solution for 2 weeks in the dark a material is obtained which after filtering, washing, and drying in vacuum over silica represents yellow $NHg_2OH \cdot H_2O$. If the product is dried at 110°, the brown monohydrate is obtained. Millon's base is also formed on interaction of HgO with liquid ammonia; the compound is light-sensitive.[17,19]

$$2HgO + NH_3 \longrightarrow NHg_2OH \cdot H_2O$$

The *bromide of this base*, NHg_2Br, is formed on interaction of ammonia and $HgBr_2$ in aqueous solution. The resulting precipitate is washed until free of bromide ion; it is dried at 110°.

$$2HgBr_2 + 4NH_3 \longrightarrow NHg_2Br + 3NH_4Br$$

The same compound is formed on heating $HgNH_2Br$ and HgO in a sealed tube; it can be stored under vacuum over silica.[19] On addition of dilute ammonia to an alkaline (NaOH) solution of K_2HgI_4 (*Nessler's reagent*, prepared by interaction of stoichiometric quantities of HgI_2 and KI), the *iodide* of Millon base precipitates. The brown solid is washed with a small amount of methanol. The preparation of NHg_2I can also be accomplished by saturating a solution of HgI_2 in anhydrous methanol with gaseous ammonia. The same material is also obtained by adding potassium iodide to a solution of mercury(II) nitrate in dilute nitric acid and subsequent addition of ammonia.[19]

Dissolving mercury(II) nitrate in dilute nitric acid and adding dilute ammonia solution results in the formation of a black precipitate of Millon's base. After standing for 1 hr the solid is collected. The solid is then either kept in a vacuum at 80° or triturated with 2 N nitric acid for one day, to remove the elemental mercury. It is subsequently dried *in vacuo* to yield a yellowish material. On exposure to moist air a monohydrate is formed readily.[19]

Mercury(II) thionitrosylate, $Hg(NS)_2$, is formed by adding mercury(II) acetate to tetrahydrosulfur nitride, $(SNH)_4$, in anhydrous pyridine. A dark red solution is obtained and on prolonged standing at room temperature a yellow solid $Hg(SN)_2$, precipitates. $Hg_5(NS)_8$ is formed on reacting mercury(II) acetate with $(SNH)_4$ at −10° in anhydrous methanol.[20]

6.1.3.4 Carboxylates and Some Derivatives with Pseudohalides

Yellow mercury(II) oxide dissolves in hot 50% acetic acid to form *mercury(II) acetate*, $Hg(CH_3COO)_2$. On cooling the solution the light-sensitive material (m.p. 178°–180°) precipitates with ice.[9] *Mercury(II) cyanide* precipitates if a solution of mercury(II) oxide in HCN is evaporated. Dissolving equimolar amounts of mercury(II) cyanide and potassium cyanide in water represents a method for preparing *potassium tetracyanomercurate(II)*, $K_2Hg(CN)_4$. The product crystallizes on evaporation of the solution. It can also be prepared by evaporating a solution of mercury(II) cyanide in liquid hydrogen cyanide after addition of the required quantity of potassium cyanide.[9,21]

Mercury(II) thiocyanide, $Hg(SCN)_2$, can be prepared by the reaction of stoichiometric quantities of mercury(II) nitrate and potassium thiocyanate in dilute nitric acid.

$$Hg(NO_3)_2 + 2KSCN \longrightarrow Hg(SCN)_2 + 2KNO_3$$

The resulting precipitate can be recrystallized from hot ethanol or water. The compound decomposes on heating to 165° with a noticeable increase in volume.[9] *Potassium tetrathiocyanatomercurate(II)*, $K_2Hg(SCN)_4$, can be prepared by the reaction of mercury(II)thiocyanate with potassium thiocyanate in hot aqueous solution.

$$Hg(SCN)_2 + 2KSCN \longrightarrow K_2Hg(SCN)_4$$

On cooling the solution mercury(II) sulfide

---

[18] *I. Aucken, R. S. Drago*, Inorg. Synth. *6*, 26 (1960).
[19] *W. Rüdorf, K. Brodersen*, Z. Anorg. Allgem. Chemie *274*, 338 (1953).
[20] *A. Meuwsen, A. Lösel*, Z. Anorg. Allgem. Chemie *271*, 217 (1953).
[21] *G. Jander, B. Grüttner*, Chem. Ber. *81*, 118 (1948).

precipitates and is filtered. Subsequent evaporation of the filtrate yields colorless needles of the desired compound which can be dried over phosphorus pentoxide. The material is soluble in cold water and ethanol, but is insoluble in ether.[9]

6.1.3.5 Complexes with Rhodium and Iridium

Some complexes containing *mercury–rhodium* and *mercury–iridium* bonds have been prepared in maximum yields of about 30%.[7] Compounds such as $L_3X_2Rh{-}HgY$ [$L=(C_6H_5)_2AsCH_3$; X=Cl, Br; Y=F, Cl, I, CN, SCN] are obtained by the reaction of HgY_2 with L_3X_2RhH in methanol or ethanol.

Similar complexes are formed by employing $C_6H_5{-}HgY$ (Y=Cl, Br, I) in hot ethanol or by using a hot suspension of Hg_2Y_2 (Y=F, Cl, Br, I, CH_3COO) instead of the mercury(I) halide.

$[(C_6H_5)_3P]_2(CO)Cl_2Ir{-}HgCl$ is formed by reacting the corresponding iridium hydrides, $[(C_6H_5)P]_2(CO)Cl_2IrH$ or $[(C_6H_5)P]_2(CO)ClIrH_2$, with mercury halides.[7]

6.2 Organomercury Halides

A wide variety of methods has been employed for synthesis of the organomercury halides. Although some of the methods discussed in this section lead first to an organomercury acetate, nitrate, or other salt, nevertheless, they are considered here because these salts are readily transformed into the halides by the action of aqueous potassium or sodium halides. The organomercury halides are always less soluble than the acetates or most of the other salts. They usually precipitate during preparation and are readily isolated. In many cases the primary products are not isolated, but are converted directly to the organomercury halides.

The methods which will be discussed are as follows.

1 The reaction of alkyl halides with metallic mercury

$$RX + Hg \xrightarrow[h\nu]{\Delta} RHgX \quad (1)$$

2 The reaction of Grignard reagents with mercuric halides

$$RMgX + HgX_2 \longrightarrow RHgX + MgX_2 \quad (2)$$

3 The reaction of organolithium compounds with mercuric halides

$$RLi + HgX_2 \longrightarrow RHgX + LiX \quad (3)$$

4 The reaction of diorganomercury compounds with mercuric halides

$$R_2Hg + HgX_2 \longrightarrow 2RHgX \quad (4)$$

5 The reaction of triorganoboranes with mercuric salts

$$R_3B + 3HgX_2 \longrightarrow 3RHgX + BX_3 \quad (5)$$

6 The reaction of aromatic diazonium halides with mercuric halides and copper powder

$$ArN_2X \cdot HgX_2 + 2Cu \longrightarrow ArHgX + 2CuX + N_2 \quad (6)$$

7 The electrophilic substitution of hydrogen by mercury salts (mercuration)

$$RH + HgX_2 \longrightarrow RHgX + HX \quad (7)$$

8 The addition reaction of mercury salts to unsaturated compounds

$$HgX_2 + \;>C{=}C< \;\longrightarrow\; -\underset{X}{\overset{|}{C}}-\underset{HgX}{\overset{|}{C}}- \quad (8)$$

9 The decarboxylation of mercuric carboxylates

$$(R{-}CO_2)_2Hg \longrightarrow RHgO_2CR + 2CO_2 \quad (9)$$

10 The reaction of organodihydroxyboranes with mercuric halides

$$RB(OH)_2 + HgX_2 + H_2O \longrightarrow RHgX + B(OH)_3 + HX \quad (10)$$

11 The reaction of mercuric halides with tetraorganotin compounds or organotin halides

$$R_3SnR^1 + HgX_2 \longrightarrow R_3SnX + R^1HgX \quad (11)$$

$$R_2SnX_2 + HgX_2 \longrightarrow RSnX_3 + RHgX \quad (12)$$

12 The reaction of aliphatic diazo compounds with mercuric halides

$$R{-}CHN_2 + HgX_2 \longrightarrow R{-}\overset{X}{\overset{|}{C}H}{-}Hg{-}X + N_2 \quad (13)$$

13 The reaction of π-allyl complexes with mercury

$$[R^2C(CHR^1)(CH_2)PdX]_2 \;(\pi\text{-allyl, X-bridged}) + Hg\;(\text{excess}) \longrightarrow 2\,R^1{-}CH{=}\overset{R^2}{\overset{|}{C}}{-}CH_2{-}HgX + 2\,Pd \quad (14)$$

A number of other preparative methods have been employed. Some of these have been widely used in the past, but are of no importance today and will not be discussed in detail. These methods include the reaction of mercury halides with sulfinic acids,

$$RSO_2H + HgCl_2 \longrightarrow RHgCl + SO_2 + HCl \quad (15)$$

the synthesis via iodonium compounds,

$$Ar_2ICl + Hg \longrightarrow ArHgCl + ArI \quad (16)$$

and the replacement of the iodooxy group by mercury.

$$ArIO_2 + HgO + ArOH \longrightarrow ArHgOH + ArIO_3 \quad (17)$$

$$ArHgOH \xrightarrow[NaX]{HOAc} ArHgX \quad (18)$$

The organomercury halides are solids, stable in air except for the compounds with tertiary groups attached to the mercury, and can be handled by the normal procedures of organic chemistry. The halogen of the organomercury halides can readily be interchanged *via* the organomercury hydroxide.

$$RHgX \xrightarrow{OH^{\ominus}} RHgOH + X^{\ominus} \quad (19)$$

$$RHgOH \xrightarrow{HX'} RHgX' + HOH \quad (20)$$

6.2.1. The Reaction of Alkyl Halides with Metallic Mercury

Although the reaction between alkyl halides and metallic mercury is one of the oldest methods of preparing organomercury halides, it is of limited significance today. The major reason for the demise of this method is that, with the exception of reactive halides, high temperatures or ultraviolet irradiation or both are necessary to effect the reaction. Such conditions often bring about undesirable changes in the organic moiety and yields of the desired product are consequently low. Another reason is that of the simple alkyl halides, only the alkyl iodides are sufficiently reactive for this reaction to occur. Recently it has been found that anions such as iodide and thiosulphate catalyze this reaction; their presence permits the reaction to be carried out at room temperature without irradiation, but yields are low.[22]

The reaction of alkyl halides with mercury is therefore of value only for the synthesis of organomercury halides in which the organic group is either a reactive group or one that is stable under the fairly harsh conditions necessary to effect the reaction. With both of these classes of halides, excess mercury is used. In the case of active groups, simply shaking the alkyl halide with the mercury at room temperature is frequently sufficient to bring about reaction. Examples of such types of compounds prepared in high yields include *iodomethylmercury iodide*[23] (69%), *allylmercury iodide*,[24] *cinnamylmercury bromide*[25] (82%), and *β-acylvinylmercury iodides*[26] (78%–90%). The stability of *perfluoroalkyl groups* under stringent conditions allows the *trifluoromethyl*[27] and *pentafluoroethylmercury*[28] *iodides* to be obtained in good yields (80% and 88%, respectively) by this method.

6.2.2 The Reaction of Grignard Reagents with Mercuric Halides

High yields of organomercury halides with a wide variety of organic groups, such as primary, secondary, and tertiary alkyl, vinyl, and aryl, can be obtained by this method. Generally excess mercuric halide is used in order to minimize formation of the diorganomercury compound. The solvent usually employed in the reaction by earlier workers was ether; however, because of the insolubility of the mercuric halides in this solvent, problems were encountered. These difficulties were circumvented by continuously extracting the mercuric chloride from a Soxhlet apparatus and by constant high-speed stirring if mercuric bromide was employed. Either technique is also applicable with mercuric iodide. Now the solubility problem can be avoided entirely by using tetrahydrofuran (THF) as a solvent for the reaction. Mercuric halides are sufficiently soluble in tetrahydrofuran and can be added conveniently as such a solution to the Grignard reagent. The halogen of the alkyl halide and that of the mercuric halide preferably should be the same; or, otherwise owing to rapid halogen exchange reactions, mixtures of organomercury halides, RHgX and RHgX',[29] are obtained. Such mixtures are extremely difficult to separate.

Some of the organomercury halides which have been synthesized by the Grignard method are given in Table 1.

6.2.3 The Reaction of Organolithium Compounds with Mercuric Halides

This route to organomercury halides has been used only sparingly because of the large number of other methods available. The reaction is generally conducted in ethereal solvents with a 1:1 molar ratio of organolithium compound to

---

[22] *M. E. Volpin, E. A. Tevdoradze, K. P. Butin*, Zh. Obshch. Khim. *40*, 315 (1970).

[23] *E. P. Blanchard, D. C. Blomstrom, H. E. Simmons*, J. Organometal. Chem. *3*, 97 (1965).

[24] *N. N. Zinin*, Justus Liebigs Ann. Chem. *96*, 363 (1855).

[25] *A. N. Nesmeyanov, O. A. Reutov*, Akad. Nauk. SSSR 655 (1953).

[26] *A. N. Nesmeyanov, M. I. Rybinskaya, T. V. Popova*, Izv. Akad. Nauk. SSSR, 946 (1970).

[27] *H. J. Emeleus, R. N. Hazeldine*, J. Chem. Soc. (London) 2948 (1949).

[28] *J. Banus, H. J. Emeleus, R. N. Hazeldine*, J. Chem. Soc. (London) 3041 (1950).

[29] *J. D'Ans, H. Zimmer, H. Bergmann*, Chem. Ber. *85*, 583 (1952).

Table 1 Selected Organomercury Halides, RHgX, Prepared by the Grignard Method

R	X	Yield (%)	Ref.
CH_3	Cl	—	*a*
CH_3	Br	94	*b*
CH_3	I	—	*a*
C_2H_5	Br	90	*b*
C_3H_7	Br	86	*b*
$(CH_3)_2CH$	Cl	—	*c*
C_4H_9	Br	50	*b*
C_4H_9(sec.)	Br	70	*d*
$(CH_3)_3C$	Cl	—	*e*
C_5H_{11}	Br	50	*b*
$(CH_3)_3C{-}CH_2{-}$	Cl	90	*f*
C_6H_{13}	Br	50	*b*
$C_{16}H_{33}$	Br	90	*b*
C_6H_{11}	Cl	68	*g*
C_6H_{11}	Br	78	*g*
$C_6H_5CH_2$	Cl	84	*h*
2,4,6-$(CH_3)_3C_6H_4CH_2$	Cl	79	*i*
$H_2C{=}CH{-}$	Cl	78	*j*
$H_2C{=}CH{-}$	Br	67	*j*
$H_2C{=}CH{-}$	I	83	*j*
$H_2C{=}C{-}C{-}$ (with $=CH_2$ on the second C)	Cl	31	*k*
C_6H_5	Br	98	*h*
1-naphthyl-	Br	75	*h*
3-F—C_6H_4	Br	82	*l*
4-F—C_6H_4	Br	69	*l*
$H_5C_6{-}O{-}CH{-}$ (with OCH_3 on CH)	Br	80	*m*

[a] *C. S. Marvel, C. G. Gauerke, E. L. Hill,* J. Amer. Chem. Soc. *47*, 3009 (1925).
[b] *K. H. Slotta, K. R. Jacobi,* J. Prakt. Chem. *120*, 249 (1928).
[c] *I. H. Robson, G. F. Wright,* Can. J. Chem. *38*, 21 (1960).
[d] *H. B. Charman, E. D. Hughes, C. Ingold,* J. Chem. Soc. p. 2523 (1959).
[e] *M. M. Kreevoy, R. L. Hansen,* J. Amer. Chem. Soc. *83*, 626 (1961).
[f] *F. C. Whitmore, E. L. Wittle, B. R. Harriman,* J. Amer. Chem. Soc. *61*, 1585 (1939).
[g] *G. Grüttner,* Chem. Ber. *47*, 1651 (1914).
[h] *S. Hilpert, G. Grüttner,* Chem. Ber. *48*, 906 (1915).
[i] *Yu. G. Bundel, V. I. Rozenburg, G. V. Gavrilova, O. A. Reutov,* Izv. Akad. Nauk SSSR, p. 1389 (1969).
[j] *B. Bartocha, F. G. A. Stone,* Z. Naturforsch. *13*, 347 (1958).
[k] *C. A. Aufdermarsh,* J. Org. Chem. *29*, 1994 (1964).
[l] *D. N. Kravtsov, B. A. Kvasov, E. N. Fedin, B. A. Faingor, L. S. Golovchenko,* Izv. Akad. Nauk SSSR, p. 536 (1969).
[m] *M. A. Kazankova, M. A. Belkina, I. F. Lutsenko,* Zh. Obshch. Khim. *37*, 1710 (1967).

mercuric halide. Some examples of organomercury halides prepared by this route are *o-biphenylmercury chloride*,[30] *p-dimethylaminophenylmercury chloride*,[30] *p-trifluoromethylphenylmercury chloride*,[31] *m-biphenylmercury chloride*,[32] *isopropenylmercury bromide*,[33] and *cis-* and *trans-propenylmercury bromide*.[34]

6.2.4 The Reaction of Diorganomercury Compounds with Mercuric Halides

The synthesis of *organomercury halides* by the ligand exchange reaction of mercuric halides with bis(organo)-mercury compounds has been known and applied for many years. Although the reaction is sometimes depicted as an equilibrium reaction, the equilibrium is shifted so far toward the organomercury halide that there is essentially no equilibrium. Nuclear magnetic resonance studies[35] have shown that in the case of dimethylmercury and mercuric halides (iodide, bromide, and chloride) the equilibrium constant is greater than 10^4. Equilibrium constants of the order of 10^6 to 10^8 were calculated for a number of systems from calorimetrically measured enthalpies of reaction.[36,37]

The most common procedure employed for the reaction is refluxing acetone or alcohol solutions of equimolar quantities of the diorganomercury compound and mercuric halide. Many other solvents have been used. The reaction can also be carried out in the melt phase. Yields are virtually quantitative. The rate of the reaction in methanol has been found to increase with the electronegativity of the halogen involved, i.e., $I < Br < Cl$, but in less polar solvents this order is partially reversed.[38]

Other salts of mercury, such as mercuric acetate, also undergo the ligand exchange reaction with bis(alkyl)mercury compounds. The products can then be converted to alkylmercury halides by standard techniques.

Organomercury fluorides have been prepared by the reaction of diorganomercury compounds with mercurous fluoride.[39]

$$(CH_3)_2Hg + Hg_2F_2 \xrightarrow[1\ hr]{50^\circ} \underset{(40\%)}{2H_3CHgF} + Hg \tag{21}$$

[30] *G. Wittig, H. Herwig,* Chem. Ber. *88*, 962 (1955).

[31] *M. T. Maung, J. J. Lagowski,* J. Chem. Soc. (London) 4257 (1963).

[32] *M. Rausch,* J. Inorg. Nucl. Chem. *1*, 414 (1962).

[33] *A. N. Nesmeyanov, A. E. Borisov, N. V. Novikova,* Dokl. Akad. Nauk SSSR *94*, 289 (1954).

[34] *A. N. Nesmeyanov, A. E. Borisov, N. V. Novikova,* Izv. Akad. Nauk. SSSR, 1216 (1959).

[35] *M. D. Rausch, J. R. Van Wazer,* Inorg. Chem. *3*, 761 (1964).

[36] *H. A. Skinner,* Rec. Trav. Chim. Pays-Bas *73*, 991 (1954).

[37] *H. A. Skinner,* Advan. Organometal. Chem. *2*, 49 (1964).

[38] *R. E. Dessy, Y. K. Lee,* J. Amer. Chem. Soc. *82*, 689 (1960).

[39] *D. Breitinger, A. Zober, M. Neubauer,* J. Organometal. Chem. *30*, 649 (1971).

6.2.5 The Reaction of Triorganoboranes with Mercuric Salts

Although it has been known[40,41] for some time that triorganoboranes react with mercuric salts, it is only recently that a method has been devised for the synthesis of organomercury halides.[42,43] The transmetallation reaction is of greatest value in the preparation of organomercury halides in which the organic group is a primary alkyl group. Although triarylboranes do react with mercuric salts, arylboron compounds are generally obtained by reaction of a Grignard reagent with boron halides or esters. Hence, there is no advantage since the organomercury halides can be prepared directly from the Grignard reagent. The reaction of *sec*-trialkylboranes with mercuric salts is very slow and only two of the three alkyl groups are transferred,[44] as shown in Eqn (22).

$$(C_5Hg)_3B + 2Hg(OOCCH_3)_2 \xrightarrow{\Delta,\ THF} C_5HgHg{-}B(OOCCH_3)_2 + 2C_5HgHg{-}Hg{-}OOC{-}CH_3 \quad (22)$$

However, all three groups are transferred when the alkyl group is primary.

The requisite primary alkylboranes are readily prepared by hydroboration of terminal olefins.[45] Reaction of the boranes with mercuric acetate at room temperature in tetrahydrofuran is very rapid and produces a quantitative yield of the alkylmercury acetate.[42] Other ionic type mercuric salts such as the nitrate and trifluoroacetate can be employed; however, mercuric chloride and bromide do not react under the same conditions.[43] Organomercury acetates are easily converted into the desired organomercury halide by treatment with aqueous sodium halide.

$$3R\ CH{-}CH_2 \xrightarrow[THF]{BH_3} (R{-}CH_2{-}CH_2)_3B \xrightarrow[THF,\ room\ temp.]{3Hg(OAc)_2} \underset{(90\text{–}100\%)}{3R{-}CH_2{-}CH_2{-}HgOAc} \quad (23)$$

Organomercury fluorides may be obtained directly by reaction of the borane with mercuric fluoride.[43]

One of the advantages of the method involving organoboranes is that functional groups can be present, which would interfere with the synthesis of such organomercury compounds via Grignard reagents. Thus, 11-chloromercuriundecanoate can be synthesized in 97% yield by this route.[42]

$$H_2C{=}CH{-}(CH_2)_8{-}\overset{O}{\overset{\|}{C}}{-}OCH_3 \xrightarrow{1.\ BH_3/THF;\ 2.\ Hg(O{-}COCH_3)_2/THF;\ 3.\ NaCl/H_2O} Cl{-}Hg(CH_2)_{10}{-}\overset{O}{\overset{\|}{C}}{-}OCH_3 \quad (24)$$

6.2.6 The Reaction of Aromatic Diazonium Halides with Mercuric Halides and Copper Powder

The diazonium salt method developed by Nesmeyanov[46] is an excellent method of preparing arylmercury halides in which the aryl group contains substituents such as nitro and carboxyl, which are incompatible with other methods such as the Grignard and organolithium routes. The Nesmeyanov procedure also avoids the formation of isomers, a problem encountered in the preparation of arylmercury halides by mercuration. Another advantage of the method is that readily available aromatic amines are used as the starting materials.

Normally, the reaction is conducted in two stages. In the first stage a "double salt" from a mercuric halide and the aryldiazonium halide is prepared and isolated. In the second stage, the "double salt" is slowly added to a stirred suspension of copper powder in a solvent at low temperatures (generally $-5°$ and even lower) when aryl groups containing strongly electron-withdrawing groups are present.

$$[ArN_2]^{\oplus}X^{\ominus} \xrightarrow{HgX_2} [ArN_2]^{\oplus}[HgX_3]^{\ominus} \xrightarrow[-5°]{2Cu} ArHgX + N_2 + 2CuX \quad (25)$$

The most commonly used solvent in the second stage is acetone,[47] although other solvents such as alcohol and ethyl acetate have been employed. Comparable yields can be obtained with water as a solvent at 0° except with compounds in which the aryl group is substituted by a powerful electron-withdrawing group.[48]

---

[40] *G. Wittig, G. Keicher, A. Rueckert, P. Roff*, Justus Liebigs Ann. Chem. *563*, 110 (1949).

[41] *J. B. Honeycutt, J. M. Riddle*, J. Amer. Chem. Soc. *82*, 3051 (1960).

[42] *R. C. Larock, H. C. Brown*, J. Amer. Chem. Soc. *92*, 2467 (1970).

[43] *J. J. Tufariello, M. M. Hovey*, J. Amer. Chem. Soc. *92*, 3221 (1970).

[44] *R. C. Lapack, H. C. Brown*, J. Organometal. Chem. *26*, 35 (1971).

[45] *G. Zweifel, H. C. Brown*, Org. Reactions *13*, 1 (1963).

[46] *A. N. Nesmeyanov*, Ber. *62*, 1010, 1018 (1929).

[47] *A. N. Nesmeyanov, N. F. Glushnev, P. F. Epifanskii, A. M. Flegontov*, Ber. *67*, 130 (1934).

[48] *A. N. Nesmeyanov, L. G. Makerova, I. V. Poloryanuk*, Zh. Obshch. Khim. *35*, 681 (1965).

The generally accepted mechanism[49] of the second stage of the synthesis is that the copper reduces (under generation of nitrogen) the diazonium ion to an aryl radical which then attacks metallic mercury formed also by reduction. Indeed, reaction of aryldiazonium halides with metallic mercury has been shown to give moderate yields of arylmercury halides.[50] In this case, mercury serves as the reducing agent. However, a special high-speed stirrer is required. Other reducing agents such as zinc and stannous chloride have been employed, but copper powder gives the best results.

The yields of arylmercury halides obtained by the Nesmeyanov method are generally moderate. Some examples of arylmercury halides prepared by this route are: *phenylmercury chloride*[46] (77%), *β-naphthylmercury chloride*[51] (45%), and *m-nitrophenylmercury chloride*[47] (59%).

6.2.7 Mercuration

The substitution of hydrogen by mercury is primarily of value in the preparation of arylmercury compounds. The major advantage of the mercuration reaction is that a wide variety of substituents on the aromatic ring does not cause interference under the conditions necessary for this synthesis.

It has been firmly established that mercuration proceeds by an electrophilic mechanism involving attack of the aromatic ring by HgX_2, $HgX^{\oplus}$, or $Hg^{2\oplus}$ species.[52] Consequently, one can predict the ease of mercuration and the position of substitution by mercury from the well-known rules for reactions of this type. Literally thousands of aromatic compounds have been mercurated and anomalous results with regard to the orientation of the entering mercury have been observed in only a few instances, in which mercuric acetate was the mercurating agent at high temperatures. The rules of electrophilic aromatic substitution predict that with many compounds, *e.g.*, toluene, a mixture of isomers with the ortho and para isomers predominating will be observed.[53] In many instances, such as for industrial use, a mixture of isomers is just as suitable as the pure compound. If a pure compound is desired and separation of the isomers is not feasible, then the diazonium salt procedure of Nesmeyanov[46] is employed for its preparation.

The mercurating agent employed depends on the reactivity of the aromatic compound with electrophiles in substitution reactions. In general, the less reactive an aromatic species is, the more ionic the mercuration agent has to be. The reason for this pairing is that dimercuration of activated compounds and lengthy reaction times with deactivated compounds is avoided. In many cases, *e.g.*, with benzene[54] and toluene,[55] the aromatic compound is used in excess to avoid dimercuration; however, this is only done if the aromatic species is readily separable from the organomercury compound. The dimercurated compounds, if they are formed, are separated from the monomercurated products on the basis of solubility, since the dimercurated compounds are generally not soluble in the common solvents.

For the mercuration of a given aromatic compound, the reaction variables which must thus be adjusted to the reactivity of that aromatic compound are (1) the nature of the mercurating agent, (2) the solvent, and (3) the temperature. Mercuric chloride, which does not dissociate to any appreciable extent to give the $HgCl^{\oplus}$ species, reacts very slowly and has been used in only a few instances with highly activated molecules. Furfural can be mercurated with an aqueous solution of mercuric chloride at 100°.[56] A mixture of mercuric chloride and sodium acetate in aqueous or aqueous–alcoholic media is a slightly more reactive mercurating agent and is used in the mercuration of other highly activated molecules. *2-Furylmercury chloride* was synthesized by this procedure.[57]

$$\text{Furan} \xrightarrow[\text{Room temperature}]{\text{NaOAc}-\text{HgCl}_2,\ \text{H}_2\text{O}-\text{ECOH}} \text{2-Furyl-HgCl}\ (33\%) \qquad (26)$$

For mercuration of aromatic amines, mercuric acetate in water or water-alcohol and temperatures below 25° are suitable conditions. Such conditions have been employed in the mercuration of aniline.[58] Phenols are slightly less reactive than aromatic amines and consequently mercuric acetate at moderate temperatures is generally used. Phenol can be mercurated by fusion with

[49] *F. B. Makin, W. A. Waters*, J. Chem. Soc. (London) 843 (1938).

[50] *R. E. McClure, E. Lowry*, J. Amer. Chem. Soc. *53*, 319 (1931).

[51] *A. N. Nesmeyanov*, Org. Synth., Coll. Vol. II, 432 (1943).

[52] *W. Kitching*, Organometal. Chem. Rev. A. *3*, 35 (1968).

[53] *H. C. Brown, C. W. McGary*, J. Amer. Chem. Soc. *77*, 2300, 2306, 2310 (1955).

[54] *K. S. McMahon, K. A. Kobe*, Ind. Eng. Chem. *49*, 42 (1957).

[55] *S. Coffey*, J. Chem. Soc. (London) 1029 (1925).

[56] *A. Baroni, G. B. Marini-Bettolo*, Gazz. Chim. Ital. *70*, 670 (1940).

[57] *H. Gilman, G. F. Wright*, J. Amer. Chem. Soc. *55*, 302 (1933).

[58] *S. Mahapatra, J. L. Ghosh, S. S. Cuha Sicar*, J. Indian Chem. Soc. *32*, 613 (1955).

mercuric acetate at 100° to give a mixture of the ortho and para monomercurated products.[59] Mercuric acetate is the most common reagent for only slightly activated or even deactivated aromatic compounds. Acetic acid is frequently used as a solvent and perchloric acid as a catalyst.[60] These conditions have been used for the preparation of *2,4-dimethylphenylmercury acetate*.[61]

$$C_6H_5CH_3 \xrightarrow{Hg(O-COCH_3)_2/\ H_3C-COOH\ /\ HClO_4,\ 25°} 2,4\text{-}(CH_3)_2C_6H_3HgO-COCH_3 \quad (27)$$

For aromatic compounds with deactivating substituents, mercuric perchlorate (formed from mercuric oxide and perchloric acid) in perchloric acid is often used as mercuration agent, as in the case of nitrobenzene.[62]

$$C_6H_5NO_2 \xrightarrow{Hg(ClO_4)_2/\ 70\%\ HClO_4} [m\text{-}O_2NC_6H_4HgClO_4] \xrightarrow{NaCl/H_2O} m\text{-}O_2NC_6H_4HgCl \quad (28)$$

With mercuric acetate alone at high temperatures, nitrobenzene yields *o-nitrophenylmercury acetate*.[63]

Compounds with highly deactivating substituents require the use of stringent conditions such as *mercuric perchlorate* or *mercuric trifluoroacetate* in trifluoroacetic acid. Rate studies[64] have shown that mercuration of aromatic compounds is 7×10^5 times faster with mercuric trifluoroacetate in trifluoroacetic acid than with mercuric acetate in acetic acid. With the former system practically inert compounds such as pentafluorobenzene can be mercurated.[65]

$$C_6F_5H \xrightarrow{Hg(OOC-CF_3)_2\ F_3C-COOH} C_6F_5HgOOC-CF_3 \quad (29)$$

The mercuration reaction is not used to any significant extent as a route to alkylmercury halides. Fairly acidic compounds such as malonic esters are readily mercurated with aqueous mercuric acetate solutions, but the less acidic CH compounds undergo polymercuration. Undesired side reactions of functional groups present in such acids occur under the stringent conditions (strong base) necessary for the mercuration. A general procedure has been developed for the α-monomercuration of ketones with mercuric nitrate. By this method *acetonylmercury iodide* has been obtained.[66]

$$H_3C-\underset{\underset{O}{\|}}{C}-CH_3 \xrightarrow{Hg(NO_3)_2/\ HgO/CaSO_4} [O_3NHg-CH_2-\overset{\overset{O}{\|}}{C}-CH_3] \xrightarrow{KJ/H_2O} JHg-CH_2-\overset{\overset{O}{\|}}{C}-CH_3 \quad (30)$$

6.2.8 Addition of Mercury Salts to Unsaturated Compounds

A widely explored reaction of mercury salts is their addition reaction either by themselves or in conjunction with nucleophilic reagents to a number of unsaturated or acceptor compounds such as olefins, acetylenes, phosphoranes, and carbon monoxide. These reactions are known as methoxymercuration or hydroxymercuration; thus, the names of the participating nucleophile precedes the term mercuration.

With olefins, the mechanism of addition is fairly well established.[67] Interaction of mercury salts with a carbon–carbon double bond leads to an intermediate mercurinium ion, whose fate is dependent on several factors: (*a*) nature of the solvent, (*b*) nucleophilicity of any added reagents, and (*c*) the nature of the substituents at the double bond. The pathways open to the mercurinium ion are: (1) dissociation to the original olefin, (2) addition of the nucleophile Z, (3) elimination of an allylic proton to give an allylmercury compound, and (4) elimination of a vinyl proton to give a vinylmercury compound [Eqn. (31)].

In addition to much indirect evidence[67] for their intermediacy, mercurinium ions have recently been directly observed by nuclear magnetic resonance.[68]

[59] *F. C. Whitmore, E. Middleton*, J. Amer. Chem. Soc. *43*, 619 (1921).
[60] *A. J. Kresge, M. Dubeck, H. C. Brown*, J. Org. Chem. *32*, 745 (1967).
[61] *R. A. Benkeser, D. I. Hoke, R. A. Hickner*, J. Amer. Chem. Soc. *80*, 5294 (1958).
[62] *R. E. Dessy, J. Y. Kim*, J. Amer. Chem. Soc. *82*, 686 (1960).
[63] *O. Dimroth*, Ber. *35*, 2032 (1902).
[64] *H. C. Brown, R. A. Wirlskala*, J. Amer. Chem. Soc. *88*, 1447, 1453, 1456 (1966).
[65] *G. B. Deacon, F. B. Taylor*, Inorg. Nucl. Chem. Lett. *3*, 181 (1967).
[66] *A. A. Morton, H. P. Penner*, J. Amer. Chem. Soc. *73*, 3300 (1951).
[67] *W. Kitching*, Organometal. Chem. Rev. (A) *3*, 61 (1968).
[68] *G. A. Olah, P. R. Clifford*, J. Amer. Chem. Soc. *93*, 2320 (1971).

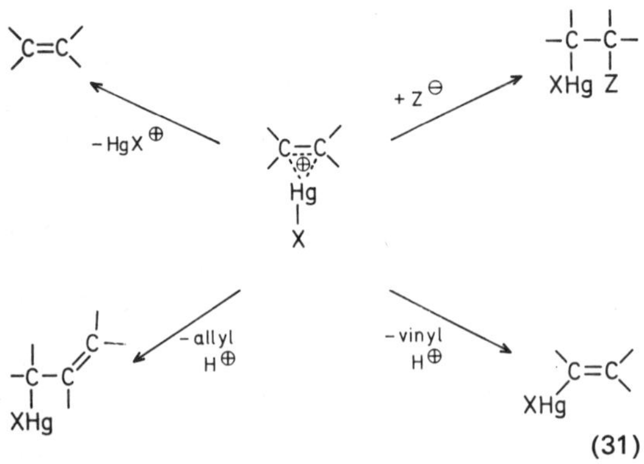

(31)

Cyclohexenyl and norbornylene mercurinium ions, formed from the olefin and mercuric trifluoroacetate in the system fluorosulfonic acid–antimony pentafluoride–sulfur dioxide are sufficiently stable at $-60°$ to be identified by their nmr spectra.
The attack of the nucleophile Z on the mercurinium ion normally occurs at the backside resulting in a *trans* addition. With bicyclic systems where such an approach is not possible, *cis* addition is observed. The orientation of the groups HgX and Z proceeds nearly always according to Markovnikov rules, *i.e.*, Z attacks to yield the most stable intermediate carbonium ion. The attack of Z is subject to steric hindrance; in the reaction of olefins with mercuric acetate in methanol the relative reaction rates of α-, β-, and γ-olefins were found to be 100, 10 and 1, respectively.[69]

The mercuric salt most commonly employed is the acetate because the acetic acid liberated in many of the addition reactions, *e.g.*, hydroxymercuration, does not cause regeneration of the starting olefin as do nitric and sulfuric acids. When mercuric nitrate or mercuric sulfate salts are used, base is frequently added during the course of the addition reaction to prevent acid build up. The mercuric halides are not reactive enough for addition to normal double bonds; only when the double bonds are very reactive can the halides be employed. The addition to strained bicyclic systems represents such a reaction.
A number of nucleophiles readily attack the mercurinium ion and, consequently, an enormous number of alkylmercury halides with various substituents in a position beta to the mercury group can be easily generated. Alcohols and water are the most common nucleophiles, but the use of carboxylates, amines, azides, nitriles, and nitrites have also been reported.

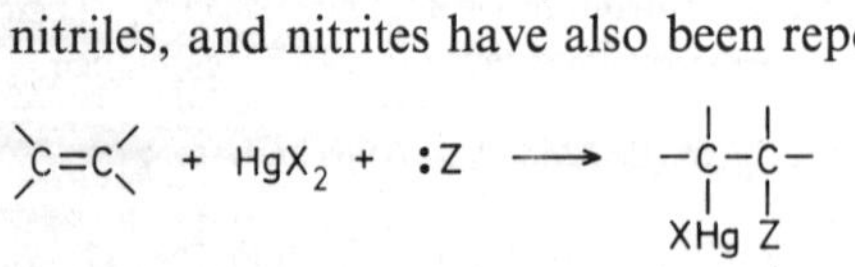

$$\text{>C=C<} + HgX_2 + :Z \longrightarrow \text{XHg–C–C–Z}$$

$$Z = H_2O,\ ROH,\ R{-}C({=}O){-}O^{\ominus},\ R_2NH,\ N_3^{\ominus},\ R{-}C{\equiv}N,\ NO_2^{\ominus},\ ROOH \quad (32)$$

The yields in most of these addition reactions are nearly quantitative (Table 2).

Table 2 Addition Reactions of Mercury Salts with Olefins

Nucleophile	β Substituent	Reaction designation	Ref.
H_2O	OH	Hydroxymercuration	*a*
ROH	OR	Alkoxymercuration	*b*
$R{-}C({=}O){-}O^{\ominus}$	$R{-}C({=}O){-}O$	Carboxymercuration	*c, d*
R_2NH	R_2N	Aminomercuration	*e, f*
$N_3^{\oplus}$	N_3	Azidomercuration	*g*
$R{-}C{\equiv}N$	$NH{-}C({=}O){-}R$[h]	Amidomercuration	*i, j, o*
$NO_2^{\ominus}$	NO_2	Nitromercuration	*k, l*
ROOH	OOR	Peroxymercuration	*m, n*

[a] *H. C. Brown, P. Geoghan,* J. Amer. Chem. Soc. *89*, 1522 (1967).
[b] *J. Chatt,* Chem. Rev. *48*, 1 (1951).
[c] *K. Ichikawa, H. Ouchi,* J. Amer. Chem. Soc. *82*, 3876 (1960).
[d] *G. Slinckx, G. Smets,* Tetrahedron *22*, 3163 (1966).
[e] *R. K. Friedlina, N. S. Kochetkova,* Izv. Akad. Nauk SSSR, p. 128 (1945).
[f] *J. J. Perie, A. Lattes,* Bull. Soc. Chim. France, p. 583 (1970).
[g] *V. I. Sokolov, O. A. Reutov,* Izv. Akad. Nauk SSSR, p. 1632 (1967).
[h] After hydrolysis.
[i] *V. I. Sokolov, O. A. Reutov,* Izv. Akad. Nauk SSSR, p. 222 (1968).
[j] *H. C. Brown, J. T. Kurej,* J. Amer. Chem. Soc. *91*, 5647 (1969).
[k] *G. B. Bachman, M. L. Whitehouse,* J. Org. Chem. *32*, 2303 (1967).
[l] *G. B. Bachman, M. L. Whitehouse,* U.S. Patent 3,428,663.
[m] *D. H. Ballard, A. J. Bloodworth,* J. Chem. Soc. C, p. 945 (1971).
[n] *A. J. Bloodworth, R. J. Bunce,* J. Chem. Soc. C, p. 1453 (1971).
[o] *D. Chow, J. H. Robson, G. F. Wright,* Can. J. Chem. *43*, 312 (1965).

[69] *G. Spengler, H. Frommel, R. Shaff, P. Paul, P. Lonsky,* Brennstoffchem. *37*, 47 (1956).

Loss of a proton from the mercurinium ion is only observed under special conditions. Thus, an allylic proton is generally eliminated only at high temperatures (*c.* 150°C). Allylmercury compounds are postulated[70] as intermediates in the allylic oxidation of olefins by mercuric acetate, a method known as the Treibs reaction.[71] It is conducted by maintaining mercuric acetate and the olefin at ~150°. The allylmercury compound eliminates mercury at these temperatures and the end product is an allyl acetate. The formation of *vinylmercury compounds* is observed only in the presence of extremely poor nucleophiles and in cases where formation of a double bond is favoured by conjugation. Accordingly, 2,2-diphenylvinylmercury chloride was obtained from 1,1-diphenylethylene.[72] If the solvent is changed from acetonitrile to water or alcohol, then normal addition occurs.

$$\mathrm{H_2C{=}C(C_6H_5)_2} \xrightarrow{\mathrm{Hg(NO_3)_2\,/\,H_3C{-}CN}} \mathrm{O_3NHg{-}CH{=}C(C_6H_5)_2} \xrightarrow{\mathrm{NaCl\,/\,H_2O}} \mathrm{ClHg{-}CH{=}C(C_6H_5)_2} \quad (33)$$

The products obtained by the reaction of mercuric salts with acetylenes depend upon the substitution of the triple bond and the nature of the mercury salt. With acetylene itself, mercuric chloride reacts to give *trans*-β-chlorovinylmercury chloride in alcohol[73] or 15% aqueous hydrochloric acid.[74] At high temperatures the *cis* isomer is formed.[75]

$$\mathrm{HC{\equiv}CH} \xrightarrow{+\,\mathrm{HgCl_2};\ \mathrm{HCl(15\%)}\ \mathrm{in\ H_2O}} \begin{cases} \sim 20^\circ: & \mathrm{ClCH{=}CHHgCl}\ (\textit{trans}) \\ \text{reflux}: & \mathrm{ClCH{=}CHHgCl}\ (\textit{cis}) \end{cases} \quad (34)$$

With more ionic mercuric salts, *polymercurated products* are formed by a combination of substitution and addition reactions.[76] Monosubstituted acetylenes also yield polymercurated materials.[77] The reaction of disubstituted acetylenes with mercuric acetate gives a mixture of *cis* and *trans* isomers of the expected adducts.[78,79] Mercuric salts have also been found to add to carbon monoxide[80] and stable phosphoranes.[81,82]

$$\mathrm{R{-}HgX} + \mathrm{CO_2} \longrightarrow \mathrm{R{-}C({=}O){-}O{-}HgX} \quad (35)$$

$$\mathrm{(H_5C_6)_3P{=}CH{-}R} \begin{cases} \xrightarrow{\mathrm{Hg(O{-}COCH_3)_2}} \left[\mathrm{(H_5C_6)_3\overset{\oplus}{P}{-}CH(R){-}HgO{-}COCH_3}\right] \mathrm{H_3C{-}COO^{\ominus}} \\ \xrightarrow{\mathrm{HgCl}} \left[\mathrm{(H_5C_6)_3\overset{\oplus}{P}{-}CH(R){-}HgCl}\right] \mathrm{Cl^{\ominus}} \end{cases} \quad (36)$$

6.2.9 The Decarboxylation of Mercuric Carboxylates

Satisfactory yields of *organomercury carboxylates* can be obtained by the action of peroxides, ultraviolet irradiation, or heat on mercuric carboxylates. One of the advantages of this method is that the mercury salts serving as starting materials can readily be prepared from the carboxylic acid. Another advantage is that a variety of functional groups can be present in the molecule, especially those which are not radical terminators. The decarboxylation method is primarily of value in the preparation of aliphatic organomercury compounds. Its use as a route to aromatic organomercury compounds is very limited, because of the occurrence of

---

[70] *K. B. Wiberg, S. D. Nielsen*, J. Org. Chem. *29*, 3353 (1964).
[71] *W. Treibs, H. Bast*, Justus Liebigs Ann. Chem. *561*, 165 (1949).
[72] *V. I. Sokalov, V. V. Bashilov, O. A. Reutov*, Dokl. Akad. Nauk. SSSR *188*, 127 (1969).
[73] *E. T. Chapman, W. J. Jenkins*, J. Chem. Soc. 847 (1919).
[74] *A. N. Nesmeyanov, R. K. Freidlina*, Dokl. Akad. Nauk. SSSR *24*, 59 (1940).
[75] *R. K. Freidlina, O. V. Nogina*, Izv. Akad. Nauk. SSSR, 99 (1947).
[76] *V. V. Korshak, V. A. Zamyatina*, Izv. Akad. Nauk. SSSR, 111 (1946).
[77] *W. W. Myddleton, A. W. Barrett, J. H. Saeger*, J. Amer. Chem. Soc. *52*, 4405 (1930).
[78] *A. N. Nesmeyanov, A. E. Borisov, V. D. Vil'chevskaya*, Izv. Akad. Nauk. SSSR, 1008 (1954).
[79] *H. Lemaire, H. J. Lucas*, J. Amer. Chem. Soc. *77*, 939 (1955).
[80] *J. Halpern, S. F. A. Kettle*, Chem. & Ind. (London) 668 (1961).
[81] *A. N. Nesmeyanov, V. M. Novikov, O. A. Reutov*, J. Organometal. Chem. *4*, 202 (1965).
[82] *A. N. Nesmeyanov, A. V. Kalinin, O. A. Reutov*, Dokl. Akad. Nauk. SSSR *195*, 98 (1970).

side reactions, the principal one being mercuration of the aromatic ring.

The decarboxylation induced by acyl peroxides is the one most extensively explored and utilized.[83] Although a variety of acyl peroxides can be employed for this

$$(R—COO)_2Hg \xrightarrow{(R^1—CO)_2O_2} R—CO—OHgR + R—CO—O—HgR^1 \quad (37)$$

purpose, yields are highest and separation of a mixture of organomercury compounds is avoided if the acyl groups of the mercury carboxylate and the peroxide are the same (R = R′). The ease of separation will determine whether it is necessary to prepare the peroxide with the same acyl group or whether readily available peroxides, such as benzoyl peroxide, are sufficient. The best solvent for the decarboxylation is the acid from which the mercury salt has been prepared. Benzene[84] and chlorobenzene[85] have been used as solvents, but they should probably be avoided because of the formation of arylmercury compounds by mercuration. Moderate temperatures (~100°) are generally employed.

The yields of organomercury salts obtained by peroxide-induced decarboxylation generally decrease as the chain length of the alkyl group increases unless the acyl groups are the same. Examples of organomercury salts prepared and their yields are: methyl[84] (89%), ethyl[86] (89%), cyclopentyl[87] (54%), and cyclohexyl[87] (70%). As mentioned previously, this method is compatible with the presence of a number of functional groups: an example is given in Eqn (38).[88]

$$\left[H_3CO-\overset{O}{\overset{\|}{C}}-(CH_2)_4-\overset{O}{\overset{\|}{C}}-O\right]_2 Hg \xrightarrow{\left[(H_5C_6)-\overset{O}{\overset{\|}{C}}-O\right]_2} H_3CO-\overset{O}{\overset{\|}{C}}-(CH_2)_4-Hg-O-\overset{O}{\overset{\|}{C}}-R$$

$$\xrightarrow[H_2O]{NaCl} H_3CO-\overset{O}{\overset{\|}{C}}-(CH_2)_4-HgCl \quad (38)$$

In a slight variation of this method, the mercury oxide carboxylate prepared from mercury, hydrogen peroxide, and a carboxylic acid is decomposed by peroxide. The yields of methyl- and ethylmercury salts are 99% and 90%, respectively.[89]

$$Hg + H_2O_2 + R-COOH \longrightarrow (R-\overset{O}{\overset{\|}{C}}-O-Hg)_2O \xrightarrow{\left(R-\overset{O}{\overset{\|}{C}}-O\right)_2} R-Hg-O-\overset{O}{\overset{\|}{C}}-R \quad (39)$$

Mercury carboxylates can also be decarboxylated by the action of u.v. light, but long reaction times and moderate yields preclude its use as a general synthetic method.[87,88,90–92]

Thermal decarboxylation is accomplished at suitably low temperatures to be of use as a preparative route only when electron-withdrawing groups such as halogen are present on the alkyl chain. Examples of the thermal method are the syntheses of *trifluoromethylmercury trifluoroacetate*[93] and *trichloromethylmercury chloride*.[94]

$$Hg(O_2CCF_3)_2 \xrightarrow{300^\circ} CF_3HgO_2CCF_3 \quad (40)$$

$$CCl_3CO_2Na + HgCl_2 \xrightarrow[\Delta]{\text{monoglyme}} \underset{(76\%)}{Cl_3CHgCl} \quad (41)$$

6.2.10 The Reaction of Organohydroxyboranes with Mercuric Halides

Although arylhydroxyboranes readily transfer the aryl group to mercury when treated with aqueous solutions of mercuric chloride, bromide, or acetate at 100° and although yields of the organomercury halide are virtually quantitative,[95–99] this method is of limited value be-

---

[83] *Y. A. Ol'dekop, N. A. Maier*, Izv. Akad. Nauk. SSSR, 1171 (1966).

[84] *G. A. Ruzuvaev, Y. A. Ol'dekop, N. A. Maier*, Zh. Obshch. Khim. *25*, 697 (1955).

[85] *Y. A. Ol'dekop, T. V. Vasilevskaya*, Zh. Obshch. Khim. *28*, 3008 (1958).

[86] *Y. A. Ol'dekop, N. A. Maier*, Zh. Obshch. Khim. *30*, 619 (1960).

[87] *Y. A. Ol'dekop, N. A. Maier, Y. D. But'kov*, Zh. Obshch. Khim. *40*, 641 (1970).

[88] *Y. A. Ol'dekop, N. A. Maier, A. A. Erdman, S. S. Gusev*, Zh. Obshch. Khim. *39*, 1110 (1969).

[89] *Y. A. Ol'dekop, N. A. Maier, Y. A. Dzhomidava*, Zh. Obshch. Khim. *39*, 687 (1969).

[90] *Y. A. Ol'dekop, N. A. Maier*, Zh. Obshch. Khim. *30*, 303 (1960).

[91] *Y. A. Ol'dekop, N. A. Maier*, Zh. Obshch. Khim. *32*, 1441 (1962).

[92] *N. A. Maier, V. I. Gesel'berg, Y. A. Ol'dekop*, Zh. Obshch. Khim. *32*, 2030 (1962).

[93] U.S. Pat. 3 043 859. Erf.: *P. E. Aldrich*.

[94] *T. J. Logan*, J. Org. Chem. *28*, 1129 (1963).

[95] *W. Konig, W. Scharsnbeck*, J. Prakt. Chem. *128*, 153 (1930).

[96] *W. Seaman, J. R. Johnson*, J. Amer. Chem. Soc. *53*, 711 (1931).

[97] *F. R. Bean, J. R. Johnson*, J. Amer. Chem. Soc. *54*, 4415 (1932).

[98] *K. Torsell*, Acta Chem. Scand. *13*, 115 (1959).

[99] *A. N. Nesmeyanov, N. E. Kolabova, Y. K. Kakarov. K. N. Anisemov*, Izv. Akad. Nauk. SSSR, 1992 (1969).

cause the organohydroxyboranes are themselves prepared by the action of Grignard reagents on alkoxyboranes. The alkylhydroxyboranes, which are readily autoxidized[100] do not react at 100°; with mercury(II) chloride; if, however, acetic acid is employed as a solvent, good yields of the alkylmercury halides can be obtained.[98] An interesting application of the hydroxyborane synthesis is the preparation of the geminal dimercury compound, *ethylidenedimercury chloride*.[101]

$$H_3C{-}CH(B[OC(CH_3)_3]_2){-}B[OC(CH_3)_3]_2 \xrightarrow{2\,HgCl_2/15\,OH^{\ominus}} H_3C{-}CH(HgCl){-}HgCl + 2\,B(OH)_3 + 5\,(H_3C)_3C{-}OH + 2\,HCl \quad (42)$$

6.2.11 The Reaction of Mercuric Halides with Organotin Compounds

This method, like the hydroxyborane route, is also of limited value only, because the organotin compounds are generally prepared by the Grignard route. However, the method does have application in the synthesis of organomercury halides with sensitive groups such as *allenylmercury chloride*[102] and *perfluorovinylmercury chloride*,[103] because the structure of the group transferred to mercury is preserved.

$$(CH_3)_3Sn{-}CH{=}C{=}CH_2 + HgCl_2 \xrightarrow{(C_2H_5)_2O,\ 20°} H_2C{=}C{=}CH{-}HgCl + (CH_3)_3SnCl \quad (43)$$

The organotin route has also been employed in the preparation of the vinylmercury compound shown in Eqn (44).[104]

$$(C_3H_7)_3Sn{-}CH{=}CH{-}Sn(C_3H_7)_3 \xrightarrow{2HgCl_2} ClHg{-}CH{=}CH{-}HgCl + (C_3H_7)_3SnCl \quad (44)$$

6.2.12 The Reaction of Aliphatic Diazo Compounds with Mercuric Halides

Diazomethane and its homologues react with mercuric halides to give excellent yields of *α-haloalkyl(aryl)mercury compounds*.[105,106] With a molar ratio of 1:1, the organomercury halide is obtained, whereas use of a 2:1 molar ratio of diazomethane to mercuric halide gives the bis(α-haloalkyl)mercury. Synthesis of these compounds would be difficult by any other method. By this route *chloromethylmercury chloride*[105] and *chlorodiphenylmethylmercury chloride*[105] using diazodiphenylmethylmethane have been prepared.

$$\underset{\text{(excess)}}{CH_2N_2} + HgCl_2 \xrightarrow[-N_2]{(C_2H_5)_2O} \underset{(100\%)}{ClCH_2HgCl} \quad (45)$$

$$(C_6H_5)_2CN_2 + HgCl_2 \xrightarrow[-N_2]{(C_2H_5)_2O} (C_6H_5)_2C(Cl)HgCl \quad (46)$$

6.2.13 The Reaction of π-Allyl Palladium Complexes with Mercury

Allylmercury halides are best prepared by treating π-allyl complexes of palladium with excess mercury.[107] The π complexes are readily obtained by the reaction of the allyl alcohol and the palladous halide. Yields of the allylmercury halides are quantitative. Allylmercury chloride, bromide, and iodide have been prepared by this method, along with a number of other allyl derivatives.

6.3 Organomercury–Chalcogen Compounds

With the exception of the organomercury carboxylates, compounds containing the C–Hg–O system have not been investigated extensively. One of the reasons undoubtedly resides in the instability of some of the compounds even under ambient conditions, a feature quite in contrast to most organomercury compounds.

Organomercury hydroxides are formed by the reaction of organomercury halides or acetates with potassium or sodium hydroxide.

$$RHgX + MOH \longrightarrow RHgOH + MX \quad (47)$$
$$X = Hal,\ M = Na,\ K$$

However, on attempted isolation, the hydroxides may transform into other products depending

[100] *H. R. Snyder, J. A. Kuck, J. R. Johnson*, J. Am. Chem. Soc. *60*, 105 (1938).
[101] *D. S. Matteson, J. G. Shdo*, J. Org. Chem. *29*, 2742 (1964).
[102] *A. Jean, G. Guillerm, M. Lequan*, J. Organometal Chem. *21*, 1 (1970).
[103] *D. Seyferth, R. H. Towe*, Inorg. Chem. *1*, 185 (1962).
[104] *A. N. Nesmeyanov, A. E. Borisov, S. H. Wang*, Izv. Akad. Nauk. SSSR, 1141 (1967).
[105] *L. Hellerman, M. D. Newman*, J. Amer. Chem. Soc. *54*, 2859 (1932).
[106] *R. K. Freidlina, A. N. Nesmeyanov, F. A. Tokareva*, Ber. *69*, 2019 (1936).
[107] *A. N. Nesmeyanov, A. N. Rubezhov, L. A. Leites, S. P. Gubin*, J. Organometal. Chem. *12*, 187 (1968).

on the nature of the organic group. Only two compounds, methylmercury hydroxide and phenylmercury hydroxide, have been studied in detail.

Grdenic and Zado[108] have claimed that all the products obtained by previous workers and reported to be methylmercury hydroxide are actually mixtures of the oxide, $(CH_3Hg)_2O$, and the oxonium salt, $[(CH_3Hg)_3O]^{\oplus}OH^{\ominus}$, in various proportions. The ratio of these two compounds is dependent on the isolation procedure, because the oxonium salt can also be dehydrated to the oxide under the conditions used.

$$3CH_3HgOH \longrightarrow [(CH_3Hg)_3O]^{\oplus}OH^{\ominus} + H_2O \quad (48)$$

$$2[(CH_3Hg)_3O]^{\oplus}OH^{\ominus} \longrightarrow 3(CH_3Hg)_2O + H_2O \quad (49)$$

The hydroxide, if it does exist at all, occurs only in solution and cannot be isolated. The oxonium salt was prepared according to Eq. (50).

$$3(CH_3Hg)_2O + 2HClO_4 \longrightarrow 2[(CH_3Hg)_3O]^{\oplus}ClO_4^{\ominus} \xrightarrow{2KOH} 2[(CH_3Hg)_3O]^{\oplus}OH^{\ominus} \quad (50)$$

Loberth and Weller[109] have recently presented spectroscopic evidence that the compound claimed to be the oxonium salt is actually a hydrate of the oxide, $(CH_3Hg)_2O \cdot xH_2O$. A compound with a melting point identical with the one reported by Grdenic and Zado for their compound was obtained by low-temperature hydrolysis of methyl-mercury (bistrimethylsilyl)amide. Hydrolysis of the amide at room temperature in ether produced the oxide.

$$2CH_3HgN(SiMe_3)_2 + 2H_2O \xrightarrow[-70^\circ]{(C_2H_5)_2O} (CH_3Hg)_2O \cdot xH_2O \quad (51)$$

$$2CH_3HgN(SiMe_3)_2 + H_2O \xrightarrow[25^\circ]{(C_2H_5)_2O} (CH_3Hg)_2O \quad (52)$$

In contrast to methylmercury hydroxide, *phenylmercury hydroxide* is sufficiently stable to be isolated.[110] The latter can be prepared by the reaction of phenylmercury acetate with sodium hydroxide in benzene–water,[110] or better in methanol–water.[111] On heating to 80° the hydroxide dehydrates to phenylmercury oxide in a reversible reaction.

$$C_6H_5HgOAc \xrightarrow{NaOH} C_6H_5HgOH \overset{80^\circ}{\rightleftharpoons} (C_6H_5Hg)_2O + H_2O \quad (53)$$

[108] *D. Grdenic, F. Zado*, J. Chem. Soc. (London) 521 (1962).
[109] *J. Loberth, F. Weller*, J. Organometal. Chem. *32*, 145 (1971).
[110] *A. J. Bloodworth*, J. Organometal. Chem. *23*, 27 (1970).

The organomercury oxides absorb carbon dioxide from the air; *alkylmercury carbonates* can be obtained in satisfactory yields by the reaction of the oxides with CO_2 in methanol.

$$(\text{alkyl-Hg})_2O + CO_2 \xrightarrow{CH_3OH} \text{alkyl-Hg—O—}\overset{\overset{O}{\|}}{C}\text{—O—Hg-alkyl} \quad (54)$$

Phenylmercury oxide was found to react with chloroform to give *phenyl trichloromethylmercury*, an interesting compound which is used for the generation of dichlorocarbene.

$$(C_6H_5\text{—}Hg)_2O + CHCl_3 \longrightarrow C_6H_5\text{—}Hg\text{—}CCl_3 + C_6H_5\text{—}HgOH \quad (55)$$

The most general method for preparing *organomercury alkoxides* and *phenoxides* is the reaction of organomercury halides with alkali metal alkoxides or phenoxides.[112–114]

$$R\text{—}HgX + MOR^1 \longrightarrow R\text{—}Hg\text{—}OR^1 + MX \quad (56)$$

However, it is sometimes difficult to separate the finely divided alkali metal halide from the desired organomercury alkoxide or phenoxide. Another method whereby organomercury alkoxides and phenoxides can be obtained is the reaction of organomercury alkoxides with phenols[114] or alcohols[111] that have higher boiling points than that from which the original alkoxy portion is derived.

$$R\text{—}HgOR^1 + R^2OH \longrightarrow R\text{—}Hg\text{—}OR^2 + R^1OH \quad (57)$$
$$(\text{b.p. of } R^2OH > R^1OH)$$

This transalkoxylation is the best method for synthesizing organomercury alkoxides of the higher alcohols.

Organomercury phenoxides can also be prepared by reaction of the "hydroxides" with phenols in good yields.[115,116]

$$R\text{—}Hg\text{—}OH + ArOH \longrightarrow R\text{—}Hg\text{—}OAr + H_2O \quad (58)$$

Phenylmercury alkoxides are obtained in nearly

[111] *A. J. Bloodworth*, J. Chem. Soc. *C*, 2051 (1970).
[112] *G. A. Razuvaev, S. F. Zhil'tsov, Y. A. Aleksandrov, O. N. Druzhkov*, Zh. Obshch. Khim. *35*, 1152 (1965).
[113] *G. A. Razuvaev, V. I. Shcherbakov, S. F. Zhil'tsov*, Izv. Akad. Nauk. SSSR, 2803 (1968).
[114] *R. Scheffold*, Helv. Chim. Acta *52*, 56 (1969).
[115] *A. N. Nesmeyanov, P. N. Kravtsov*, Izv. Akad. Nauk. SSSR, 431 (1962).
[116] *D. N. Kravtsov, A. P. Zhukov, B. A. Faingor, E. M. Rokhlina, G. K. Semin, A. N. Nesmeyanov*, Izv. Akad. Nauk. SSSR, 1703 (1968).

quantitative yield by the reaction of phenylmercury hydroxide or oxide with dialkyl carbonates.[111]

$$C_6H_5HgOH + (RO)_2Co \longrightarrow C_6H_5HgOR + CO_2 + ROH \quad (59)$$

$$(C_6H_5Hg)_2O + (RO)_2CO \longrightarrow 2C_6H_5HgOR + CO_2 \quad (60)$$

Only a limited number of organomercury alkoxides have been characterized. They hydrolyze very readily[111,112] and are thermally and oxidatively unstable.[112] Molecular weight measurements in benzene of phenylmercury alkoxides indicate that a concentration-dependent equilibrium exists between monomeric and dimeric species. The phenoxides are much more stable than the alkoxides and are hydrolyzed only with great difficulty. Thus, phenylmercury phenoxide may be recrystallized even from water.[117]

The most general route for the preparation of *organomercury carboxylates* is the reaction of organomercury "hydroxides" with carboxylic acids.[118]

$$[R\text{—}HgOH] + R^1\text{—}COOH \longrightarrow R\text{—}Hg\text{—}O\text{—}\overset{O}{\overset{\|}{C}}\text{—}R^1 + H_2O \quad (61)$$

Three other methods which are commonly used are based on equilibrium principles. The methods are:

1 Reaction of an organomercury salt of a volatile acid, such as an acetate with a non-volatile acid [Eqn (62)].[119]

$$R^1HgO\text{—}\overset{O}{\overset{\|}{C}}\text{—}R^2 + R^3COOH \longrightarrow R^1HgO\text{—}\overset{O}{\overset{\|}{C}}\text{—}R^3 + R^2COOH \quad (62)$$

2 Reaction of an organomercury halide or other salt with the sodium salt of a carboxylic acid to form an insoluble organomercury carboxylate [Eqn (63)].[120]

$$R^1HgX + NaO\overset{O}{\overset{\|}{C}}\text{—}R^2 \longrightarrow R^1HgO\overset{O}{\overset{\|}{C}}\text{—}R^2 + NaX \quad (63)$$

3 Reaction of organomercury halides with the silver salt of the carboxylic acid [Eqn (64)].[119]

$$R^1HgX + AgO\text{—}\overset{O}{\overset{\|}{C}}\text{—}R^2 \longrightarrow R^1HgO\text{—}\overset{O}{\overset{\|}{C}}\text{—}R^2 + AgX \quad (64)$$

Organomercury carboxylates are also obtained when mercuric carboxylates such as mercuric acetate are used in addition (oxymercuration) and substitution (mercuration) reactions and will be discussed elsewhere. A few compounds in which the oxygen of the C–Hg–O system is bonded to an atom other than carbon have been synthesized. *Methylmercury trimethylsilanolate* was prepared as indicated in Eqn (65).[121] This compound was found to be monomeric in dilute solution, but tetrameric (cubanite) in the solid state. An organomercury stannoxide, $C_6H_5HgOSnBuCl_2$, was obtained from the reaction of dibutyltin oxide with phenylmercury chloride.[122]

$$2\ H_5C_6\text{—}HgCl + \left((H_9C_4)_2Sn\right)_2O \longrightarrow 2\ H_5C_6\text{—}Hg\text{—}O\text{—}\underset{C_4H_9}{\overset{C_4H_9}{Sn}}\text{—}Cl \quad (65)$$

The first organomercury peroxides to be isolated were prepared according to Eqn (66).[123]

$$R\text{—}HgCl + H_5C_6\text{—}\underset{CH_3}{\overset{CH_3}{C}}\text{—}O\text{—}O\text{—}Na \longrightarrow R\text{—}Hg\text{—}O\text{—}O\text{—}\underset{CH_3}{\overset{CH_3}{C}}\text{—}C_6H_5 \quad (66)$$

In comparison with R–HgO compounds, a large number of organomercury–sulphur compounds are known. This is undoubtedly due to the stability of the Hg–S bond. The behaviour of organomercury thiols, R–Hg–SH is similar to that of the hydroxides; they transform with extreme ease to the sulphide and thus, cannot be isolated.[124]

The *organomercury thiols* and *thiophenolates* are generally prepared by either of two methods: the reaction of organomercury halides, acetates,

---

[117] *M. A. Smalt, C. W. Kreke, E. S. Cook*, J. Biol. Chem. *224*, 999 (1957).

[118] *V. N. Kalinin, L. A. Fedorov, K. G. Gasonov, E. I. Fedin, L. I. Zakhorkin*, Izv. Akad. Nauk. SSSR, 2404 (1970).

[119] *J. E. Connett, A. G. Davies, J. H. S. Green*, J. Chem. Soc. *C*, 106 (1966).

[120] Brit. Pat. 737 753 Ethyl Corp.; C.A. *50*, 15577 (1956).

[121] *H. Schmidbauer, M. Bergfield, F. Schindler*, Z. Anorg. Allgem. Chemie *363*, 73 (1968).

[122] *A. G. Davies, P. G. Harrison*, J. Organometal. Chem. *10*, P31 (1967).

[123] *G. A. Razuvaev, E. I. Fedotova*, Dokl. Akad. Nauk. SSSR *169*, 355 (1966).

[124] *M. Kesler*, Croat. Chem. Acta *36*, 165 (1964).

or other salts with thiols or their alkali metal derivatives,[125] and the reaction of organomercury "hydroxides" with thiols.[126] Both of these reactions are normally run in aqueous-alcoholic media, in which the resulting organomercury-sulfur compounds are insoluble.

$$R{-}HgX + MSR^1 \xrightarrow{C_2H_5OH/H_2O} R{-}Hg{-}S{-}R^1 + MX \quad (67)$$

$$[R{-}HgOH] + R^1SH \xrightarrow{C_2H_5OH/H_2O} R{-}Hg{-}S{-}R + H_2O \quad (68)$$

In addition to the mercaptans, organomercury derivatives of other classes of compounds with SH groups such as thio acids[125] (R–Hg–S–CO–R) and thio amides[127] (R–Hg–S–CO–NHR) have been synthesized by these reactions. The organomercury mercaptans are stable under ambient conditions.

Organomercury sulfides, $(RHg)_2S$, are prepared most satisfactorily by the reaction of organomercury halides or acetates with sodium sulfide.[128,129] Unsymmetrical organomercury sulfides, RHgSHgR′, are obtained by a two stage process[129] as illustrated in Eqn (70).

$$2R{-}HgX + Na_2S \xrightarrow{C_2H_5OH/H_2O} (RHg)_2S + 2NaX \quad (69)$$

$$R{-}HgX + Na_2S \xrightarrow{C_2H_5OH/H_2O} [R{-}HgS]^{\ominus}Na^{\oplus} \xrightarrow[-NaK]{R_2HgX} RHg{-}S{-}Hg{-}R^1 \quad (70)$$

On heating the sulfides decompose to the diorganomercury compounds and mercuric sulfide. This reaction may also occur during storage of these compounds.

Organomercury thiocyanates have been prepared in high yields by the reaction of organomercury acetates with sodium thiocyanate.[130] They may also be obtained by the reaction of diorganomercury compounds with thiocyanogen.[131] Spectroscopic evidence indicates that these compounds are dimeric.

$$RHgOOCCH_3 + NaSCN \longrightarrow (RHgSCN)_2 + NaOOCCH_3 \quad (71)$$

$$R_2Hg + (SCN)_2 \longrightarrow (RHgSCN)_2 + RSCN \quad (72)$$

Organomercury xanthates can be prepared by the reaction of carbon disulfide and alcohols with organomercury "hydroxides" and by the interchange reaction of organomercury acetates with alkali metal xanthates.[125]

$$R{-}Hg{-}OH + CS_2 + R^1OH \longrightarrow R{-}Hg{-}S{-}\overset{S}{\overset{\|}{C}}{-}OR^1 + H_2O \quad (73)$$

$$R{-}Hg{-}O{-}COCH_3 + M{-}S{-}\overset{S}{\overset{\|}{C}}{-}OR \longrightarrow R{-}Hg{-}S{-}\overset{S}{\overset{\|}{C}}{-}OR^1 + MO{-}COCH_3 \quad (74)$$

Organomercury selenocyanates, containing the C–Hg–Se system, have been synthesized in fair yields by the reaction of diorganomercury compounds with either selenium selenocyanate or mercuric selenocyanate.[132]

$$R_2Hg + Se(SeCN)_2 \longrightarrow RHgSeCN + RSeCN + Se \quad (75)$$

$$R_2Hg + Hg(SeCN)_2 \longrightarrow 2RHgSeCN \quad (76)$$

6.4 Organomercury Group V Compounds

A large number of compounds containing the C–Hg–N system are known. They can be prepared by displacement of hydroxide, halogen, acetate, and even alkyl and aryl groups from mercury by certain nitrogen compounds.

Most of the compounds containing the C–Hg–N system have been obtained by the reaction of compounds containing an acidic ⟩N–H group, such as amides, imides, sulfonamides, and weakly basic amines, with organomercury hydroxides. (The organomercury hydroxides are

[125] *G. Spengler, A. Weber*, Brennstoff. Chemie *40*, 55 (1969).

[126] Brit. Pat. 1 121 920, Cosan Chemical Corp.; C.A. *69*, 8159 (1968).

[127] *D. N. Kravtsov, A. N. Nesmeyanov*, Izv. Akad. Nauk. SSSR, 1747 (1967).

[128] *D. Grdenic, B. Markusic*, J. Chem. Soc. 2434 (1958).

[129] *M. Dadic, D. Grdenic*, Croat. Chem. Acta *32*, 39 (1960).

[130] *E. Tobler, D. J. Foster*, Z. Naturforsch. *17B*, 136 (1962).

[131] *K. Dehnicke*, J. Organometal. Chem. *9*, 11 (1967).

[132] *E. E. Aynsley, N. N. Greenwood, M. J. Sprague*, J. Chem. Soc. 2395 (1965).

depicted as RHgOH for convenience; a discussion of their structure is given in Section 6.3.)

$$\rangle NH + RHgOH \longrightarrow RHgN\langle + H_2O \quad (77)$$

Several examples of this reaction are given in the following equations.

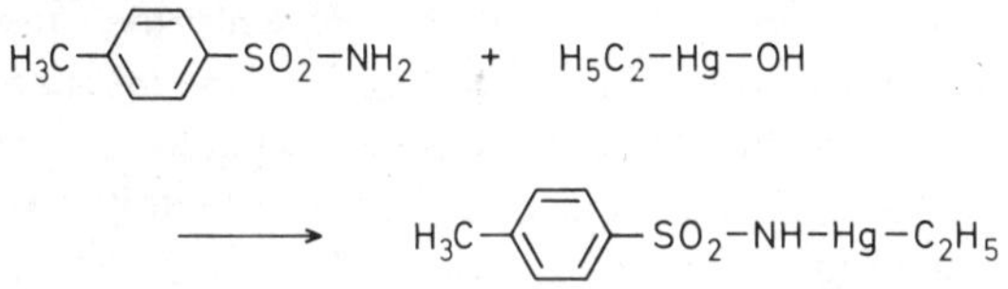

$$H_3C-C_6H_4-SO_2-NH_2 + H_5C_2-Hg-OH \longrightarrow H_3C-C_6H_4-SO_2-NH-Hg-C_2H_5 \quad \text{(Ref. 133)} \quad (78)$$

$$Cl_3C-C(=O)-NH_2 + H_5C_6-Hg-OH \longrightarrow H_5C_6-Hg-NH-C(=O)-CCl_3 \quad \text{(Ref. 134)} \quad (79)$$

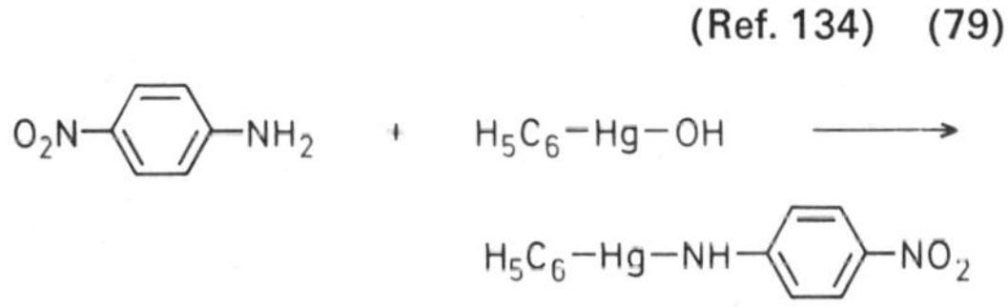

$$O_2N-C_6H_4-NH_2 + H_5C_6-Hg-OH \longrightarrow H_5C_6-Hg-NH-C_6H_4-NO_2 \quad \text{(Ref. 134)} \quad (80)$$

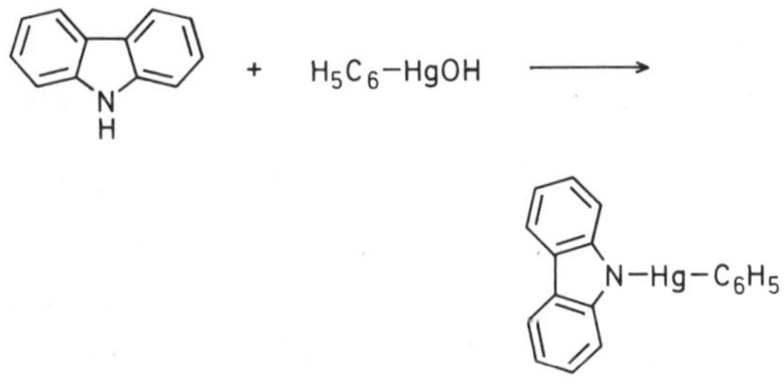

$$\text{carbazole (N-H)} + H_5C_6-HgOH \longrightarrow \text{carbazolyl}-N-Hg-C_6H_5 \quad \text{(Ref. 135)} \quad (81)$$

A study[134] of the reaction of organomercury hydroxides and weakly basic amines indicates that a certain threshold of acidity of the $\rangle$N–H group must be reached before any reaction will occur. Thus, aniline, *p*-bromoaniline, and *p*-chloroaniline do not react with phenylmercury hydroxide, whereas *p*-nitroaniline does, as shown in Eqn (80). The amines which do react with organomercury hydroxides do so quickly and in good yields.

However, not all compounds containing a sufficiently acidic N–H group will react with the hydroxides. Although picramide and *p*-nitroaniline give good yields of the organomercury–nitrogen species, *N*-phenylpicramide and 2,4-dinitroaniline do not react at all with phenylmercury hydroxide. The explanation which has been offered is that steric hindrance of the nitro-group in the ortho position to the amine function prevents attack of the nucleophilic nitrogen on mercury.

If the NH group of a nitrogen compound is very acidic and/or the carbon–mercury bond is weak it might outweigh steric factors. All interaction of diorganomercury compounds with such acidic amines proceeds smoothly and leads to derivatives containing the C–Hg–N system as illustrated by the following examples:

$$Cl_5C_6-SO_2-NH-C(=O)-CH_2-O-C_6Cl_5 + (H_5C_2)_2Hg \longrightarrow H_5C_2-Hg-N(C(=O)-CH_2-O-C_6Cl_5)(SO_2-C_6Cl_5) \quad \text{(Ref. 136)} \quad (82)$$

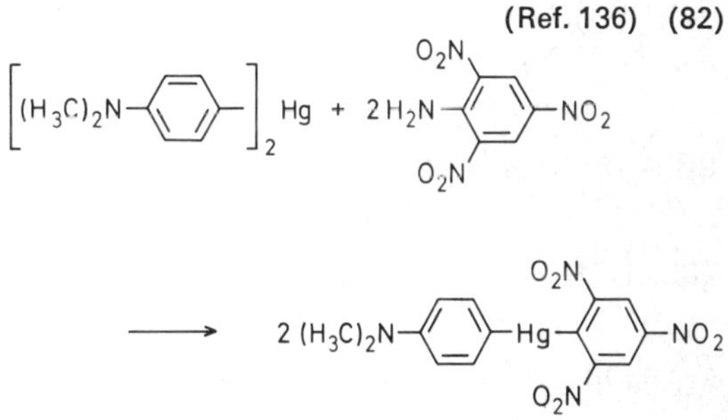

$$[(H_3C)_2N-C_6H_4]_2Hg + 2\,H_2N-C_6H_2(NO_2)_3 \longrightarrow 2\,(H_3C)_2N-C_6H_4-Hg-C_6H_2(NO_2)_3 \quad \text{(Ref. 134)} \quad (83)$$

$$H_5C_6—NH_2 + [(NO_2)_3C]Hg \longrightarrow H_5C_6—NH—Hg—C(NO_2)_3 \; (59\%) \quad \text{(Ref. 137)} \quad (84)$$

The interaction of *N*-halogen compounds with diorganomercury compounds is another method of preparing compounds with the C–Hg–N moiety. Some examples of this method are given in the following equations.

$$(CH_3)_2Hg + (CF_3)_2N—Br \longrightarrow (CF_3)_2N—HgCH \; (70\%) \quad \text{(Ref. 138)} \quad (85)$$

$$[(NO_2)_3]C_2Hg + (CH_3)_2NCl \longrightarrow (CH_3)_2NHgC(NO_2)_3 \; (91\%) \quad \text{(Ref. 139)} \quad (86)$$

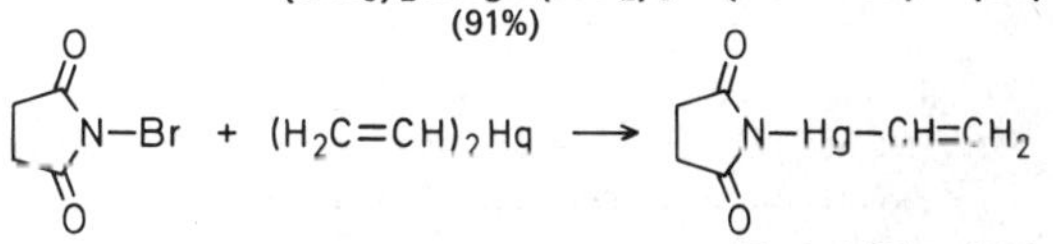

$$\text{succinimide}-N-Br + (H_2C{=}CH)_2Hg \longrightarrow \text{succinimide}-N-Hg-CH{=}CH_2 \quad \text{(Ref. 140)} \quad (87)$$

[133] U.S. Pat. *2*, 452, 595, D. F. Mowery.

[134] *D. N. Kravtsov, A. N. Nesmeyanov*, Izv. Akad. Nauk. SSSR, 1487 (1967).

[135] *D. N. Kravtsov, A. N. Nesmeyanov*, Izv. Akad. Nauk. SSSR, 1741 (1967).

[136] Ger. Patent 1 009 420. Erf.: *W. Konz, R. Sehring*.

[137] *S. S. Novikov, T. I. Godovikova, V. A. Tartakovskii*, Izv. Akad. Nauk. SSSR, 505 (1960).

[138] *R. C. Dobbie, H. J. Emeleus*, J. Chem. Soc. *A*, 367 (1966).

[139] *V. I. Erashko, S. A. Shevelev, A. A. Fainzil'berg*, Izv. Akad. Nauk. SSSR, 2708 (1967).

[140] *E. Tobler, D. J. Foster*, Z. Naturforsch. *17B*, 135 (1962).

Displacement of acetate and halide from mercury is also a useful method for generating compounds containing the C–Hg–N system, as indicated in the following equations:

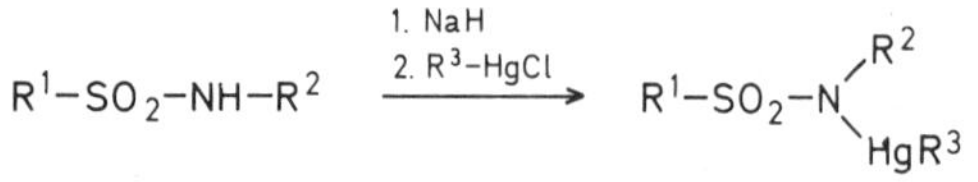

$$R^1{-}SO_2{-}NH{-}R^2 \xrightarrow[\text{2. } R^3{-}HgCl]{\text{1. NaH}} R^1{-}SO_2{-}N(R^2)(HgR^3) \quad \text{(Ref. 141)} \quad (88)$$

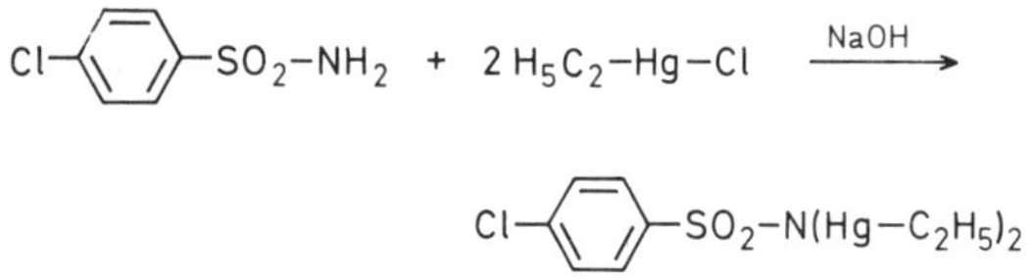

$$Cl{-}C_6H_4{-}SO_2{-}NH_2 + 2\,H_5C_2{-}Hg{-}Cl \xrightarrow{NaOH} Cl{-}C_6H_4{-}SO_2{-}N(Hg{-}C_2H_5)_2 \quad \text{(Ref. 142)} \quad (89)$$

$$CH_3HgBr + LiN(Si(CH_3)_2)_2 \longrightarrow \underset{(85\%)}{CH_3HgN[SiCH_3]_2} \quad \text{(Ref. 109)} \quad (90)$$

A series of complexes of the general formula $[(CH_3Hg)_xNH_y]^{\oplus}ClO_4^{\ominus}$ ($x+y=4$) also containing an N–Hg bond has been prepared by the following reactions[143]:

$$CH_3HgOH + NH_3 + HClO_4 \longrightarrow [NH_3(HgCH_3)]^{\oplus}ClO_4^{\ominus} \quad (91)$$

$$2CH_3HgOH + NH_3 + HClO_4 \longrightarrow [NH_2(HgCH_3)_2]^{\oplus}ClO_4^{\ominus} \quad (92)$$

$$CH_3HgOH + NH_2(HgCH_3)_2ClO_4 \longrightarrow [NH(HgCH_3)_3]^{\oplus}ClO_4^{\ominus} \quad (93)$$

$$2CH_3HgOH + NH_2(HgCH_3)_2ClO_4 \longrightarrow [N(HgCH_3)_4]^{\oplus}ClO_4^{\ominus} \quad (94)$$

Similarly, $(C_6H_5Hg)_2NH_2ClO_4$ was obtained in 36% yield from the reaction of phenylmercury acetate, ammonia and perchloric acid.[144]

Organomercury azides have been obtained in high yields from the reaction of chlorine azide with diorganomercury compounds[145–147]

$$R_2Hg + ClN_3 \longrightarrow RCl + R{-}HgN_3 \quad (95)$$

$R = CH_3$; C_2H_5; $(CH_3)_2CH$; C_4H_9; C_6H_5; cycl. C_3H_5; C_5H_9; C_6H_{11}

Azides have also been prepared by the reaction of organomercury hydroxides with hydrazoic acid.[148,149]

$$R{-}HgOH + NH_3 \longrightarrow R{-}HgN_3 + H_2O \quad (96)$$

$R = CH_3$, C_2H_5, C_6H_5

Benzylmercury azide has been synthesized by the reaction of benzylmercury chloride and sodium azide in methanol.[150] *Pentafluorophenylmercury azide* has been prepared by the reaction of mercuric azide and bis(pentafluorophenyl)mercury in 90% yield.[146]

Organomercury cyanamides are obtained in nearly quantitative yield by the reaction of alkylmercury bis(trimethylsilyl)amine with cyanamide.[109]

$$R{-}Hg{-}NSi(CH_3)_3 + H_2NCN \xrightarrow{(C_2H_5)_2O} (R{-}Hg)_2NCN \quad (97)$$

Only a few compounds containing the C–Hg–P or C–Hg–As systems are known. Complexes of the type $[R^1{-}Hg{-}PR_3]^{\oplus}X^{\ominus}$ and $[R^1{-}Hg{-}AsR_3^2]^{\oplus}X^{\ominus}$ have been isolated as intermediates from the reaction of organomercury halides[151] and perchlorates[152] with trialkylphosphines or arsines. On standing these compounds disproportionate as illustrated in Eqn (98).

$$2R_3^2Y + 2R^1{-}HgX \longrightarrow [2R^1{-}Hg{-}YR_3^2]^{\oplus}X^{\ominus} \longrightarrow R_2Hg + R_3^2Y + HgX_2 \quad (98)$$

$Y = As, P$; $X = Hal, ClO_4, NO_3$

The rate of this disproportionation is dependent on the nature of R^2 and X. As the electron-withdrawing nature of R^2 increases the rate of disproportionation increases. Thus, triphenylphosphine reacts too fast to yield an isolable complex.[152] The complexes in which X is nitrate or perchlorate are more stable than those in which the anion is halide.

Diethyl phenylmercury phosphonate was obtained by the reaction of phenylmercury acetate and triethyl phosphite.[153]

---

141 U.S. Pat. 3 035 073, *R. S. Waritz.*

142 *J. Bialas, E. Eckstein, E. Ejmocki, B. Hetnarski, W. Sobotka, J. Szymaszkiewicz,* Przem. Chem. *40,* 567 (1961); C.A. *57,* 12523g (1962).

143 *D. Breitinger, N. Q. Dao,* J. Organometal. Chem. *15,* P21 (1968).

144 *G. B. Deacon, J. H. S. Green,* J. Chem. Soc. *A,* 1182 (1968).

145 *K. Dehnicke, J. Strahle, D. Seybold, J. Muller,* J. Organometal. Chem. *6,* 298 (1966).

146 *K. Dehnicke, D. Seybold, J. Muller,* J. Organometal. Chem. *11,* 227 (1968).

147 *A. F. Shibada, K. Dehnicke,* J. Organometal. Chem. *26,* 157 (1971).

148 *A. Perret, R. Perrot,* Helv. Chim. Acta *16,* 848 (1933)

149 *J. S. Thayer,* Organometal. Chem. Rev. *1,* 157 (1966).

150 *W. Beck, E. Schuierer, K. Fedl,* Angew. Chem. *78,* 267 (1966).

151 *R. J. Cross, A. Lander, G. E. Coates,* Chem. Ind. 2013 (1962).

152 *G. E. Coates, A. Lauder,* J. Chem. Soc. (London) 1857 (1965).

153 *T. Mukaiyama, I. Kuwajima, Z. Suzuki,* J. Org. Chem. *28,* 2024 (1963).

$$H_5C_6—Hg—OOC—CH_3 + P(OC_2H_5)_3 \longrightarrow H_5C_6—Hg—P(OC_2H_5) + H_3CCOOC_2H_5 \quad (99)$$

In spite of the fact that large numbers of compounds containing the C–Hg–N system have been prepared and that some are even in commercial use,[154] very little is known of their chemistry, especially that of the simpler compounds.

6.5 Diorganomercury Compounds

A substantial number of methods are available for the synthesis of compounds containing the C–Hg–C system. The most important of these are:

1 The reaction of Grignard reagents with mercuric halides:

$$2RMgX + HgX_2 \longrightarrow R_2Hg + 2MgX_2 \quad (100)$$

2 The reaction of organolithium compounds with mercuric halides

$$2RLi + HgX_2 \longrightarrow R_2Hg + 2LiX \quad (101)$$

3 The reaction of amalgams and metallic mercury with organic halides

$$2RX + Na_2Hg \longrightarrow R_2Hg + 2NaX \quad (102)$$

4 The symmetrization of organomercury halides

$$2R—HgX \longrightarrow R_2Hg + HgX_2 \quad (103)$$

5 The mercuration of compounds containing active hydrogen atoms

$$2RH + HgO \longrightarrow R_2Hg + H_2O \quad (104)$$

6 The addition of mercury salts to unsaturated compounds

$$(CF_3)_2C{=}CF_2 + HgF_2 \longrightarrow [(CF_3)_3C]_2Hg \quad (105)$$

7 The thermal decarboxylation of mercuric carboxylates

$$(R—COO)_2Hg \xrightarrow{\Delta} R_2Hg + 2CO_2 \quad (106)$$

8 The reaction of aliphatic diazo compounds with mercuric halides

$$R—CHN_2 + HgX_2 \longrightarrow (R—\underset{|}{\overset{X}{C}}H—)_2Hg \quad (107)$$

A number of other methods exist, but they are of minor importance because the synthesis of the desired compounds can generally be accomplished more readily by the methods listed above. Some examples of these synthetic routes, which will not be discussed further are illustrated by the following equations.

9 The iodonium salt route

$$C_6H_5I^{\oplus}BF_4^{\ominus} + Hg \xrightarrow[CH_3OH]{Fe} (C_6H_5)_2Hg \ (53\%) \quad \text{(Ref. 155)} \quad (108)$$

10 The reaction of organotin compounds with mercuric halides

$$\left(\overset{H_3C}{\underset{H}{}}C{=}C\overset{}{\underset{H}{}}\right)_2 SnBr_2 + HgBr_2 \xrightarrow{-SnBr_4} \left(\overset{H_3C}{\underset{H}{}}C{=}C\overset{}{\underset{H}{}}\right)_2 Hg \ (89\%) \quad \text{(Ref. 156)} \quad (109)$$

11 The reaction of organoboranes with mercuric oxide

$$HgO + (C_2H_5)_3B \xrightarrow[H_2O]{KOH} (C_2H_5)_2Hg \ (95\%) \quad \text{(Ref. 157)} \quad (110)$$

12 The alkylation of mercury salts by organozinc compounds:

$$I—CH_2ZnI + HgI_2 \longrightarrow (ICH_2)_2Hg \ (86\%) \quad \text{(Ref. 158)} \quad (111)$$

13 The alkylation of mercury salts by organoaluminum compounds:

$$(C_2H_5)_3Al + Hg(OOCCH_3)_2 \xrightarrow{CH_3OCH_2CH_2OCH_3} (C_2H_5)_2Hg \ (89\%) \quad \text{(Ref. 159)} \quad (112)$$

The organoaluminum route (Eqn 112) is valuable as an industrial production method for the lower dialkylmercury compounds, but it does not find much use in the laboratory.

In addition to the *symmetrical diorganomercury* compounds, RHgR, the *mixed compounds* $RHgR^1$, are also known. The mixed compounds can be prepared by modifying some of the above methods by employing organomercury halides as the starting material. In some cases it is also possible to synthesize the mixed compounds by reaction of the two symmetrical compounds with each other.

$$R_2Hg + R_2^1Hg \rightleftharpoons 2R—Hg—R^1 \quad (113)$$

The success of any synthesis depends on the stability of the mixed compounds to ligand

[154] *M. S. Whelen*, Metal-Organic Compounds, Advances in Chemistry Series *23*, 84 (1959).

[155] *O. A. Ptitsyna, S. I. Orlov, M. N. Il'ina, O. A. Reutov*, Dokl. Akad. Nauk. SSSR *177*, 862 (1967).

[156] *A. N. Nesmeyanov, A. E. Borisov, N. V. Novikova*, Izv. Akad. Nauk. SSSR, 1216 (1959).

[157] *J. B. Honeycutt, J. M. Riddle*, J. Amer. Chem. Soc. *81*, 2593 (1959).

[158] *D. Seyferth, S. B. Andrews*, J. Organometal. Chem. *30*, 151 (1971).

[159] U.S. Pat. 2 969 381. Erf.: *S. M. Blitzer, T. H. Pearson.*

exchange to the symmetrical compounds. From equilibrium studies[160] at 90°, it was found that there are three different levels of stability. When R and R′ are both saturated alkyl, the equilibrium constant of the forward reaction of Eqn (113) approaches the random value of 4. When R is a saturated alkyl group and R′ is unsaturated or pseudounsaturated (*e.g.*, cyclopropyl), then the equilibrium constant is large (about 100). Finally, if R is a saturated alkyl group and R′ is a very electronegative group such as perfluoroalkyl, then the equilibrium constant is very large. In general, the more electronegative a group is, the more it stabilizes the mixed compound toward ligand exchange. A very highly electronegative group thus behaves as a pseudohalogen.

6.5.1 The Grignard Method

The preparation of diorganomercury compounds *via* Grignard reagents is probably the most widely utilized method. The action of Grignard reagents on mercuric halides involves two stages, the first of which is the formation of the organomercury halide, which is then further alkylated by the Grignard reagent in the second stage. In many cases the organomercury halide represents the starting material of the synthesis. The second stage is considerably slower than the first and, in order to obtain the diorganomercury compound free from organomercury halide, an excess of the Grignard reagent is frequently employed. The yields are also improved if the unreacted magnesium is removed from the Grignard reagent. Earlier workers employed ether as the solvent, but the use of this solvent suffers from the insolubility of the mercuric halides and often the intermediate organomercury halides in it. Some of these difficulties have been alleviated by such techniques as addition of the halide via Soxhlet extraction,[161] removal of the ether by distillation,[162] or replacement of ether by a higher boiling solvent such as xylene.[29] Later workers have avoided the troubles associated with the use of ether and employed tetrahydrofuran (THF) as the solvent; the mercuric halides are soluble in THF. Even the most volatile of diorganomercury compounds, *dimethylmercury*, is separable from THF by fractional distillation.

[160] *G. F. Reynolds, S. R. Daniel*, Inorg. Chem. *6*, 480 (1967).

[161] *H. Gilman, R. E. Brown*, J. Amer. Chem. Soc. *51*, 928 (1929).

[162] *H. Gilman, R. E. Brown*, J. Amer. Chem. Soc. *52*, 3314 (1930).

A large number of symmetrical diorganomercury compounds, with a variety of organic groups, have been prepared by the Grignard method. Some of these are listed in Table 3. The first authentic monomeric mercury heterocycle was prepared by the Grignard route;[163] the mercury atom is part of an eleven-membered ring.

Br, Br (dibromo diaryl ether macrocycle precursor) $\xrightarrow[\text{2) HgBr}_2]{\text{1) Mg, THF}}$ Hg heterocycle (38%) (114)

The Grignard method has also been employed for the synthesis of mixed diorganomercury compounds, $RHgR^1$, by the reaction of Grignard reagents with organomercury halides.

$$R{-}MgX + R^1{-}HgX \longrightarrow R^1{-}Hg{-}R + MgX_2 \quad (115)$$

The reaction is generally conducted at low temperatures (about 0°) in order to avoid ligand exchange of the mixed compounds. Many types of organic groups have been combined at the mercury atom. Examples of the synthesis of some mixed compounds are given in the following equations.

$$C_6H_5CH_2{-}HgCl + C_2H_5MgBr \longrightarrow C_6H_5CH_2{-}HgC_2H_5 \; (70\%) \quad \text{(Ref. 164)} \quad (116)$$

$$CH_3HgI + C_6F_5MgBr \longrightarrow CH_3HgC_6F_5 \; (69\%) \quad \text{(Ref. 165)} \quad (117)$$

$$C_6H_5HgCl + C_6F_5MgBr \longrightarrow C_6H_5HgC_6F_5 \; (40\%) \quad \text{(Ref. 165)} \quad (118)$$

$$H_2C{=}CH{-}HgBr + C_2H_5MgBr \longrightarrow H_2C{=}CH{-}HgC_2H_5 \; (78\%) \quad \text{(Ref. 166)} \quad (119)$$

$$H_2C{=}CH{-}HgBr + CF_2{=}CF{-}MgBr \longrightarrow H_2C{=}CH{-}Hg{-}CF{=}CF_2 \; (35\%) \quad \text{(Ref. 167)} \quad (120)$$

$$C_2H_5HgI + HC{\equiv}C{-}MgBr \longrightarrow C_2H_5Hg{-}C{\equiv}CH \; (73\%) \quad \text{(Ref. 168)} \quad (121)$$

[163] *G. Bohr, F. W. Kupper*, Ber. *100*, 3992 (1967).

[164] *W. E. French, N. Inamoto, G. F. Wright*, Can. J. Chem. *42*, 2228 (1964).

[165] *R. D. Chambers, G. E. Coates, J. G. Livingstone, W. R. K. Musgrave*, J. Chem. Soc. (London) 4367 (1962).

[166] *B. Bartocha, F. G. A. Stone*, Z. Naturforsch. *13B*, 347 (1958).

[167] *R. N. Sterlin, W. K. Li, I. L. Knunyants*, Dokl. Akad. Nauk. SSSR *140*, 137 (1961).

[168] *M. Kraut, L. C. Leitch*, Can. J. Chem. *41*, 549 (1963).

Table 3 Symmetrical Diorganomercury Compounds, R_2Hg, by the Grignard Method

R–Mercury	Yield (%)	Ref.
Dimethyl-	70	*a*
Diethyl-	82	*a*
Dipropyl-	75	*b*
Diisopropyl-	87	*c*
Dibutyl-	80	*a*
Dibut-l-yl-	66	*d*
Di-tert-butyl-	9	*d*
Dipent-2-yl-	80	*e*
Bis-[2-methyl-but-2-yl]-	21	*d*
Bis-[2,2-dimethyl-propyl]-	64	*f*
Dipebtyl-	90	*g*
Dinonyl-	77	*h*
Diundecyl-	73	*h*
Bis-[2-methyl-2-phenylpropyl]-	70	*i*
Dicyclopropyl-	64	*b*
Dicyclohexyl-	86	*j*
Diallyl-	73	*k*
Dibentyl-	73	*l*
Divinyl-	85	*m*
Bis-[trifluor-vinyl]-	50	*n*
Bis-[phenyl-ethinyl]-	84	*o*
Diphenyl-	85	*p*
Bis-[pentafluorphenyl]-	73	*q*
Bis-[pentachlorphenyl]-	50	*r*
Bis-[3-fluor-phenyl]-	71	*s*
Bis-[3-chlor-phenyl]-	68	*k*
Bis-[4-methoxy-phenyl]-	66	*k*
Bis-[4-formyl-phenyl]-	90	*t*
Di-1-naphthyl-	23·5	*l*

[a] *H. Gilman, R. E. Brown,* J. Amer. Chem. Soc. *52*, 3314 (1930).
[b] *S. F. Reynolds, R. E. Dessy, H. H. Jaffé,* J. Org. Chem. *23*, 1217 (1958).
[c] *D. Q. Cowan, H. S. Mosher,* J. Org. Chem. *27*, 1 (1962).
[d] *C. S. Marvel, O. H. Calvery,* J. Amer. Chem. Soc. *45*, 820 (1923).
[e] *S. Murahashi, S. Nozakura, S. Takeuchi,* Bull. Chem. Soc. Jap. *33*, 658 (1960).
[f] *C. W. Blewett,* Ph.D. Thesis, University of Cincinnati, Cincinnati, Ohio, 1968.
[g] *F. D. Hager, C. S. Marvel,* J. Amer. Chem. Soc. *48*, 2689 (1926).
[h] *G. Bahr, G. Meier,* Z. Anorg. Allg. Chem. *294*, 22 (1958).
[i] *R. Criegee, P. Dimroth, R. Schempf,* Chem. Ber. *90,* 1337 (1957).
[j] *W. E. French, N. Inamoto, G. Wright,* Can. J. Chem. *42*, 2228 (1964).
[k] *A. E. Borisov, I. S. Savel'eva, S. R. Seryduk,* Izv. Akad. Nauk. SSSR, p. 924 (1965).
[l] *F. Nerdel, S. Makower,* Naturwissenschaften, *45*, 490 (1958).
[m] *B. Bartocha, F. G. A. Stone,* Z. Naturforsch. *13*, 347 (1958).
[n] *R. N. Sterlin, W. K. Li, I. L. Knunyants,* Izv. Akad. Nauk, SSSR, p. 1506 (1959).
[o] *G. Eglington, W. McCrae,* J. Chem. Soc., p. 2295 (1963).
[p] *J. D'Ans, H. Zimmer, H. Bergmann,* Chem. Ber. *85*, 583 (1952).
[q] *R. D. Chambers, G. E. Coates, J. G. Livingstone, W. K. R. Musgrave,* J. Chem. Soc., p. 4367 (1962).
[r] *F. E. Paulik, S. E. I. Green, R. E. Dessy,* J. Organometal. Chem. *3*, 229 (1965).
[s] *D. N. Kravtsov, B. A. Kvasov, E. N. Fedin, B. A. Faingor, L. N. Golovchenko,* Izv. Akad. Nauk. SSSR, p. 536 (1969).
[t] *G. Drefahl, D. Lorenz,* J. Prakt. Chem. *24*, 106 (1964).

6.5.2 The Organolithium Route

Organolithium compounds have not been used as extensively in the preparation of diorganomercury compounds as the Grignard reagents, even though they are more reactive than the Grignard reagents. The primary use of the organolithium method has been in the synthesis of *aryl-* and *vinylmercury* compounds, as well as in the preparation of reactive diorganomercury compounds. In many cases yields of the desired compounds are poor because of competing Wurtz-type coupling. Some of the symmetrical diorganomercury compounds which have been prepared by this method are given in Table 4. Unsymmetrical diorganomercury compounds can also be prepared by the action of organolithium compounds on organomercury halides; several examples are given in the following equations.

$$C_2H_5HgCl + C_6H_5Li \longrightarrow C_2H_5HgC_6H_5 \quad (42\%) \quad \text{(Ref. 169)} \quad (122)$$

$$H_5C_6{-}C{-}C{-}Li\ (B_{10}H_{10}) + H_5C_6{-}HgCl \longrightarrow H_5C_6{-}C{-}C{-}Hg{-}C_6H_5\ (B_{10}H_{10}) \quad (71\%) \quad \text{(Ref. 170)} \quad (123)$$

Table 4 Symmetrical Diorganomercury Compounds, R_2Hg, by the Organolithium Method

R–Mercury	Yield (%)	Ref.
Bis-[trichlormethyl]-	91	*a*
Bis-[dichlormethyl]-	97	*a*
Bis-[cyclopent-1-enyl]-	85	*b*
Bis-[cyclohex-1-enyl]-	—	*b*
Dibentyl	51	*c*
Di-cis-propenyl	85	*d*
Bis-[2-brom-phenyl-]	04	*e*
Bis-[tetrachlor-pyrid-4-yl]-	12	*f*
Bis-[1,2-dicarbodecaboran-1-yl]-	53	*g*
Bis-[2-phenyl-1,2-dicarbodecaboran-1-yl]-	70	*g*
Bis-[pentamethyl-cyclopentadienyl]-	—	*h*

[a] *G. Köbrich, K. Flory, W. Drischel,* Angew. Chem. *76*, 536 (1964).
[b] *D. J. Foster, E. Tobler,* J. Amer. Chem. Soc. *83*, 851 (1961).
[c] *K. Ziegler,* Chem. Ber. *64*, 445 (1931).
[d] *A. E. Borisov,* Izv. Akad. Nauk. SSSR, p. 1036 (1961).
[e] *C. Tamborski, E. J. Soloski,* J. Organometal. Chem. *10*, 385 (1967).
[f] *J. D. Cook, B. J. Wakefield,* ibid. *13*, 15 (1968).
[g] *L. I. Zakharkin, V. I. Bregadze, O. Y. Okhlobystin,* ibid. *4*, 211 (1965).
[h] *B. Floris, G. Illuminati, G. Ortaggi,* Chem. Commun., p. 492 (1969).

[169] *R. E. Dessy, Y. K. Lee, J. Y. Kim,* J. Amer. Chem. Soc. *83*, 1163 (1961).
[170] *L. I. Zakharkin, V. I. Bregadze, O. Y. Okhlobystin,* J. Organometal. Chem. *6*, 228 (1966).

$$C_6H_5HgCl + C_6H_5CCl_2Li \xrightarrow{-100^\circ} C_6H_5Hg{-}CCl_2{-}C_6H_5 \quad (54\%) \qquad \text{(Ref. 171)} \quad (124)$$

6.5.3 Organic Halides and Metallic Mercury

Only a few organic systems are sufficiently stable to survive the stringent conditions required for production of diorganomercury compounds by reaction of organic halides with metallic mercury. *Bis(pentafluorophenyl)mercury* can be prepared in high yield by reaction of mercury with pentafluorophenyl iodide at high temperatures.[172, 173]

$$2C_6H_5I + 2Hg \xrightarrow[30\text{ hr}]{300^\circ} (C_6F_5)_2Hg + HgI_2 \quad (75\%) \qquad (125)$$

In contrast to the severe conditions required for reaction with metallic mercury, the halides react readily in the presence of various catalysts with a number of amalgams to give good yields of diorganomercury compounds. Sodium amalgam is most widely employed for the synthesis of both dialkyl- and diarylmercury compounds. Lower yields of the dialkylmercury compounds are obtained with a secondary alkyl group because of Wurtz coupling in a side reaction. With active halides, such as benzyl bromide, the Wurtz reaction predominates. Only organic bromides and iodides are sufficiently reactive to be practical starting materials. The most commonly employed catalysts for the reaction are ethyl acetate, methyl acetate, or other carbonyl compounds.[174] The reaction of alkyl halides with sodium amalgam is normally conducted at low temperatures (about 0°), whereas temperatures of about 150° are required in the preparation of the diarylmercury compounds. The disadvantage of the sodium amalgam method for synthesis of dialkylmercury compounds is that a low sodium content (0·2–0·5%) amalgam must be employed in order to minimize occurrence of the Wurtz reaction. Consequently, a large quantity of amalgam is required, *e.g.*, 21 kg of 0·5% amalgam was employed for the preparation of 393 g of diisobutylmercury.[175] For the synthesis of diarylmercury compounds a higher sodium level (3–4%) amalgam can be used. Some of the diorganomercury compounds which have been prepared by the sodium amalgam method are listed in Table 5.

[171] *D. Seyferth, D. C. Mueller*, J. Organometal. Chem. *25*, 293 (1970).
[172] *S. C. Cohen, M. L. N. Reddy, A. G. Massy*, Chem. Commun. 451 (1967).
[173] *J. M. Birchall, R. Hazard, R. N. Hazeldine, A. W. Wakalski*, J. Chem. Soc. *C*, 47 (1967).
[174] *H. F. Lewis, E. Chamberlin*, J. Amer. Chem. Soc. *51*, 291 (1929).
[175] *N. I. Mel'nikov*, Zh. Obshch. Khim. *16*, 2065 (1946).

Table 5 Diorganomercury compounds, R_2Hg, by the Sodium Amalgam Method

R–Mercury	Yield (%)	Ref.
Dimethyl-	65	*a*
Diethyl-	67	*a*
Dipropyl-	69	*b*
Dibutyl-	82	*b*
Diisobutyl-	63	*b*
Bis-[3-methyl-butyl]-	50	*b*
Diphenyl-	47	*c*
Bis-[4-methyl-phenyl]-	50	*d, e*
Bis-[4-dimethylamino-phenyl]-	—	*f*

[a] *W. K. Wilde*, J. Chem. Soc., p. 72 (1949).
[b] *N. I. Mel'nikov*, Zh. Obshch. Khim. *16*, 2065 (1946).
[c] *O. H. Calvery*, J. Amer. Chem. Soc. *48*, 1009 (1926).
[d] *A. Michaelis, I. Rabinerson*, Chem. Ber. *23*, 2344 (1890).
[e] *R. E. Dessy, J. Y. Kim*, J. Amer. Chem. Soc. *82*, 686 (1960).
[f] *A. Schenk, A. Michaelis*, Chem. Ber. *21*, 1501 (1888).

Amalgams of mercury with silver[176] and cadmium[177] have found application in the synthesis of *bis(perfluoroalkyl)mercury* compounds.

$$CF_3I \xrightarrow[80^\circ,\ 10\text{ hr}]{\{Hg,\ Ag\}} (CF_3)_2Hg \quad (68\%) \qquad (126)$$

$$C_2F_5I \xrightarrow[30^\circ,\ 14\text{ days}]{\{Hg,\ Cd\}} (C_2F_5)_2Hg \quad (60\%) \qquad (127)$$

6.5.4 The Symmetrization of Organomercury Halides

The ligand exchange of organomercury halides into diorganomercury compounds and mercury halides is one of the most important preparative methods to obtain symmetric bisorganomercury compounds because there are quite a number of routes to both simple and substituted organomercury halides.

$$RHgX \rightleftharpoons R_2Hg + HgX_2 \qquad (128)$$

As discussed in Section 6.2, Eqn (128) represents an equilibrium which is always greatly on the side of the organomercury halide. Hence, in order to drive the equilibrium to the right, one of the products must be removed. The task of removing the mercuric halide, HgX_2, can be accomplished in two ways: (1) reduction of the mercuric halide to metallic mercury and (2)

[176] *H. J. Emeleus, R. N. Hazeldine*, J. Chem. Soc. (London) 2953 (1949).
[177] *J. Banus, H. J. Emeleus, R. N. Hazeldine*, J. Chem. Soc. (London) 3041 (1950).

complexation of the mercuric halide. A number of reagents, commonly called symmetrizing agents, are known for both ways. Although the action of these agents can be regarded as simply reduction or complexation, they are undoubtedly involved more intimately in the reaction, particularly in the latter case.

Of the symmetrizing agents which function by reduction, the most commonly employed one is alkaline sodium stannite solution, which is readily prepared by the addition of stannous chloride to sodium hydroxide solution.

$$2RHgX + Na_2SnO_2 \xrightarrow[\text{room temp.}]{H_2O} R_2Hg + Hg + 2NaX + H_2SnO_3 \quad (129)$$

The reaction is normally conducted by stirring an aqueous stannite solution and the organomercury halide at room temperature. In addition to normal alkyl and aryl groups, sensitive groups such as allyl,[178] benzyl,[179] and vinyl[180] are tolerated by this reagent. Other organomercury salts, such as the acetates, will also function as the starting material. Several examples of the sodium stannite method are given in the following equations.

$$C_8H_{15}HgBr \xrightarrow[SnCl_2]{NaOH} (C_8H_{15})_2Hg \text{ (63\%)} \quad \text{(Ref. 181)} \quad (130)$$

$$(C_6H_5O)_2CHHgBr \xrightarrow[SnCl_2]{NaOH} [(C_6H_5O)_2CH]_2Hg \text{ (80\%)} \quad \text{(Ref. 182)} \quad (131)$$

$$C_6H_5HgCl \xrightarrow[\text{pyridine}]{NaOH/SnCl_2} (C_6H_5)_2Hg \text{ (70\%)} \quad \text{(Ref. 183)} \quad (132)$$

$$\text{2-thienyl-HgCl} \xrightarrow{NaOH/SnCl_2/\text{pyridine}} (\text{2-thienyl})_2Hg \quad \text{(Ref. 183)} \quad (133)$$

Another frequently employed symmetrizing agent which functions by reduction is hydrazine. This reagent has been used primarily in the preparation of *diarylmercury* compounds.[184, 185]

$$\text{Br-}C_6H_4\text{-Hg-Cl} + N_2H_4 \cdot H_2O \xrightarrow{C_2H_5OH} (\text{Br-}C_6H_4)_2Hg \quad \text{(Ref. 185)} \quad (134)$$

Other reducing agents which are of value as symmetrization reagents include copper, copper/pyridine, magnesium,[186] and chromous ion.[187] Examples of the application of these reagents are given in the following equations.

$$2\ \text{F-}C_6H_4\text{-Hg-Br} \xrightarrow{2\ Cu, \nabla} [\text{F-}C_6H_4]_2Hg \text{ (68\%)} + 2\ CuBr + Hg \quad \text{(Ref. 188)} \quad (135)$$

$$2\ \text{o-}NO_2C_6H_4\text{-Hg-Cl} \xrightarrow{2\ Cu, Py, \nabla} [\text{o-}NO_2C_6H_4]_2Hg + 2\ CuCl \cdot Py + Hg \quad \text{(Ref. 189)} \quad (136)$$

$$2\ H_3C\text{-}C_6H_4\text{-Hg-Br} + Mg \longrightarrow [H_3C\text{-}C_6H_4]_2Hg \text{ (96\%)} + MgBr_2 + Hg \quad \text{(Ref. 186)} \quad (137)$$

$$2C_6H_5CH_2HgClO_4 + 2Cr^{2\oplus} \longrightarrow (C_6H_5CH_2)_2Hg + 2Cr^{3\oplus} + Hg + 2ClO_4^{\ominus} \quad \text{(Ref. 187)} \quad (138)$$

Copper–ammonia has also found extensive use as a symmetrizing agent, particularly in the synthesis of diarylmercury compounds from

[178] *A. Kinman, M. Kleine-Peter*, Bull. Soc. Chim. France, 894 (1957).
[179] *I. L. Maynard*, J. Amer. Chem. Soc. *54*, 2118 (1932).
[180] *A. N. Nesmeyanov, A. E. Borisov, I. S. Savel'eva, E. I. Golubeva*, Izv. Akad. Nauk. SSSR, 1490 (1958).
[181] *A. C. Cope, J. E. Englehart*, J. Amer. Chem. Soc. *90*, 7092 (1958).
[182] *M. M. Kazankova, M. A. Belkina, I. F. Lutsenko*, Zh. Obshch. Khim. *37*, 1710 (1967).
[183] *W. Steinkopf, W. Bielenberg, H. Augenstend-Hensen*, Justus Liebigs Ann. Chem. *430A*, 41 (1923).
[184] *H. Gilmann, M. M. Barnett*, Rec. Trav. Chim. Pays-Bas *55*, 563 (1936).
[185] *D. Spinelli, A. Salvemine*, Ann. Chim. *50*, 1423 (1960).
[186] *F. R. Jensen, J. A. Landgrebe*, J. Amer. Chem. Soc. *82*, 1004 (1960).
[187] *R. J. Ouellette, B. G. van Leuwen*, J. Org. Chem. *30*, 3967 (1965).
[188] *D. N. Kravtsov, B. A. Kvasov, E. N. Fedin, B. A. Baingar, L. S. Golovchenko*, Izv. Akad. Nauk. SSSR, 536 (1969).
[189] *F. Hein, K. Wagler*, Ber. *58*, 1499 (1925).

mercuric halide–diazonium chloride double salts.[190]

$$2ArN_2^{\oplus}[HgCl_3]^{\ominus} + 6Cu + 6NH_3 \longrightarrow Ar_2Hg + 2N_2 + 6CuX \cdot NH_3 + Hg \quad (139)$$

In this method, the action of copper on the double salt produces the arylmercury halide which is not isolated. Addition of ammonia to the reaction mixture then effects the symmetrization of the arylmercury halide. By this method, for example, *bis(β-naphthyl)mercury* was obtained in 48% yield.[191]

Cadmium amalgam is employed as the symmetrizing agent for the perfluoroalkylmercury halides.[176, 192]

$$Cl—Hg—CF_2—CO_2C_2H_5 \xrightarrow[(CH_3)_2CO]{\{Cd, Hg\}} \underset{(78\%)}{Hg(CF_2—CO_2C_2H_5)_2} \quad \text{(Ref. 192)} \quad (140)$$

A number of reagents are available for removal of the mercuric halide by the formation of anionic and cationic mercuric complexes. The most important of reagents which form anionic complexes are: (1) iodide ion which forms the $HgI_4^{2\ominus}$ anion, (2) cyanide ion and (3) thiocyanate ion which give $[Hg(CN)_4]^{2\ominus}$ and $[Hg(CNS)_4]^{2\ominus}$ respectively and (4) the thiosulfate ion resulting in the formation of $Hg(S_2O_3)_2^{2\ominus}$ ion. The symmetrization agents which operate via cationic complex formation include ammonia, amines, and tertiary phosphines.

Iodide ion is one of the more widely used symmetrizing agents because of its lack of reactivity toward many functional groups. *Diaryl-* and *divinylmercury* compounds can be produced in high yields with this reagent. However, alkylmercury chlorides and bromides, with only a few exceptions, are transformed to alkylmercury iodides, which are not symmetrized by iodide.[193] Consequently, this method is not suitable for the preparation of dialkylmercury compounds. Generally, the reaction is conducted with a large excess (about 1000%) of either sodium or potassium iodide to obtain complete conversion of the organomercury halide to the diorganomercury compound. Only with the perhalovinylmercury halide does one molar equivalent suffice.[103] The solvents most frequently employed for this reaction are ethanol and acetone. Other organomercury salts, such as the acetates, may be employed as the starting material. Some representative examples of iodide-induced symmetrization are given in the following equations.

$$C_6F_5Hg—O_2CCF_3 \xrightarrow[C_2H_5OH/H_2O]{NaI} \underset{(85\%)}{(C_6F_5)_2Hg} \quad \text{(Ref. 194)} \quad (141)$$

$$Cl_2C{=}CCl—HgCl \xrightarrow[(CH_3)_2CO]{NaI} \underset{(88\%)}{(Cl_2C{=}CCl)_2} \quad \text{(Ref. 103)} \quad (142)$$

$$\text{2-(HgCl)-4-}NO_2C_6H_3COOC_4H_9 \xrightarrow{KJ / C_2H_5OH} \underset{(95\%)}{[COOC_4H_9(NO_2)C_6H_3]_2Hg} \quad \text{(Ref. 195)} \quad (143)$$

$$\text{2-thienyl}—Hg—Cl \xrightarrow{NaJ / acetone} \underset{(93\%)}{(\text{2-thienyl})_2Hg} \quad \text{(Ref. 196)} \quad (144)$$

Cyanide ion has the distinction of being the first symmetrizing agent discovered; potassium cyanide was employed for the synthesis of dimethyl- and diethylmercury in 1858.[197] A recently found modification which allows the diarylmercury compounds to be readily prepared in high yields consists of passage of the arylmercury halide through a column of sodium cyanide-treated alumina, followed by elution with chloroform.[198] *Diphenylmercury* and *bis-4-phenoxyphenylmercury* were obtained in yields of 96 and 90%, respectively, using this technique. The alkylmercury halides are converted to the cyanides under the same conditions.

$$C_6H_5HgCl \xrightarrow[column]{NaCN/Al_2O_3} \underset{(96\%)}{(C_6H_5)_2Hg} \quad (145)$$

Examples of the use of cyanide under conventional reaction conditions include potassium cyanide for the preparation of *diallylmercury* from allylmercury iodide[199] and sodium cyanide for the preparation of

[190] *A. N. Nesmeyanov, E. I. Kan*, Ber. *62*, 1018 (1929).

[191] *A. N. Nesmeyanov, E. D. Kohn*, Org. Synth., Coll. Vol. II, 381 (1943).

[192] *V. R. Polishchuk, L. S. German, I. L. Knunyants*, Izv. Akad. Nauk. SSSR, 795 (1971).

[193] *F. C. Whitmore, R. J. Sobatzki*, J. Amer. Chem. Soc. *55*, 1128 (1933).

[194] *G. B. Deacon, F. B. Taylor*, Austral. J. Chem. *21*, 2675 (1968).

[195] *F. C. Whitmore, E. B. Middleton*, J. Amer. Chem. Soc. *44*, 1546 (1922).

[196] *W. Sleinkopf*, Justus Liebigs Ann. Chem. *413*, 310 (1917).

[197] *G. B. Buckton*, Justus Liebigs Ann. Chem. *108*, 105 (1858).

[198] *P. V. Roling, S. B. Roling, M. D. Rausch*, Synth. Inorg. Chem. *1*, 97 (1971).

[199] *K. V. Vijayaraghavan*, J. Indien Chem. Soc. *20*, 318 (1943).

bis-β-naphthylmercury and *di-p-tolylmercury* from the chlorides.[200]

Thiocyanate ion, although widely employed in the past, is not used to any extent today. *Bis(perchlorovinyl)-mercury* has been prepared by thiocyanate-induced symmetrization.[103]

$$Cl_2C{=}CCl{-}HgCl \xrightarrow[(CH_3)_2CO]{KSN} \underset{(88\%)}{(Cl_2C{=}CCl{-})_2Hg} \quad (146)$$

Thiosulfate ion is an excellent symmetrization agent for compounds containing reactive functional groups such as amino and hydroxyl. The reaction is conducted in aqueous solution, from which the diorganomercury compound precipitates. Representative examples of the application of thiosulfate for this purpose are given in the following equations.

$$m\text{-}HOC_6H_4{-}Hg{-}Cl \xrightarrow{Na_2S_2O_3/H_2O} (m\text{-}HOC_6H_4)_2Hg \quad \text{(Ref. 201)} \quad (147)$$
(100%)

$$o\text{-}CH_3COOC_6H_4{-}Hg{-}Cl \xrightarrow{Na_2S_2O_3/H_2O} (o\text{-}CH_3COOC_6H_4)_2Hg \quad \text{(Ref. 59)} \quad (148)$$
(80%)

$$H_2N{-}C_6H_4{-}Hg{-}O{-}COCH_3 \xrightarrow{Na_2S_2O_3/H_2O} (O_2N{-}C_6H_4)_2Hg \quad \text{(Ref. 202)} \quad (149)$$
(73%)

$$\text{2-furyl}{-}Hg{-}Cl \xrightarrow{Na_2S_2O_3/H_2O} (\text{2-furyl})_2Hg \quad \text{(Ref. 57)} \quad (150)$$
(95%)

$$\text{3-pyridyl}{-}Hg{-}O{-}COCH_3 \xrightarrow{Na_2S_2O_3/H_2O} (\text{3-pyridyl})_2Hg \quad \text{(Ref. 203)} \quad (151)$$
(85%)

Ammonia is a very mild reagent which is specific for the symmetrization of many organomercury compounds derived from addition of mercuric salts to multiple bonds. These products are generally decomposed by other symmetrizing agents. Ammonia is also the reagent of choice for the symmetrization of organomercury derivatives of carbonyl compounds in which mercury is bonded to the α-carbon atom. However, reaction of organomercury salts with ammonia is not a general method for the preparation of simple dialkyl- and diarylmercury compounds. The reaction procedure normally consists of the bubbling of ammonia through a solution of the organomercury salt in an inert solvent such as chloroform or benzene. Examples of ammonia-induced symmetrization are given in the following equations.

$$H_5C_6{-}CH(COOR){-}Hg{-}Br \xrightarrow{NH_3/CHCl_3} (H_5C_6{-}CH(COOR){-})_2Hg \quad \text{(Ref. 204)} \quad (152)$$

$$H_3C{-}C(=O){-}CH_2{-}Hg{-}Cl \longrightarrow (H_3C{-}C(=O){-}CH_2{-})_2Hg \quad \text{(Ref. 205)} \quad (153)$$

$$Cl(H)C{=}C(H){-}Hg{-}Cl \xrightarrow{NH_3/C_6H_6} (Cl(H)C{=}C(H){-})_2Hg \quad \text{(Ref. 206)} \quad (154)$$

The mercury substituted phosphorus ylid shown in Eqn (155) is formed by the action of ammonia, both as a base and as a symmetrization agent. Reaction of the ylid with aldehydes gives divinylmercury compounds with a variety of substituents on the vinyl group.[82]

$$\left[(H_5C_6)_3\overset{\oplus}{P}{-}CH(COOCH_3){-}Hg{-}O{-}COCH_3\right] H_3C{-}COO^{\ominus} \xrightarrow{NH_3}$$

$$\left[(H_5C_6)_3P{=}C(COOCH_3){-}\right]_2 Hg \xrightarrow[-(H_5C_6)_3PO]{R{-}CHO} \left[R{-}CH{=}C(COOCH_3){-}\right]_2 Hg \quad \text{(Ref. 82)} \quad (155)$$

(156)

Amines have also been employed as symmetrization agents. Both alkyl- and arylmercury halides can be converted in nearly quantitative yield to diorganomercury compounds by treatment with polyethyleneimine (mol. wt. 600) in the presence of water.[207]

---

[200] *R. W. Beattie, F. C. Whitmore*, J. Amer. Chem. Soc. *55*, 1567-(1933).

[201] *A. N. Nesmeyanov, E. M. Toropova*, Zh. Obshch. Khim. *4*, 667 (1934).

[202] *V. P. Chalov*, Zh. Obshch. Khim. *18*, 608 (1948).

[203] *C. D. Hurd, C. J. Morrissey*, J. Amer. Chem. Soc. *77*, 4658 (1955).

[204] *A. N. Nesmeyanov, O. A. Reutov, I. P. Beletskaya, R. E. Mardaleishvili*, Dokl. Akad. Nauk. SSSR *166*, 617 (1957).

[205] *A. N. Nesmeyanov, I. F. Lutsenko, R. M. Khomutov*, Dokl. Akad. Nauk. SSSR *88*, 837 (1953).

[206] *A. N. Nesmeyanov, A. E. Borisov, N. V. Novikova*, Izv. Akad. Nauk. SSSR, 857 (1970).

[207] *R. C. Wade, D. Seyferth*, J. Organometal. Chem. *22*, 265 (1970).

$$2R\text{—}HgX + (\text{—}CH_2\text{—}CH_2\text{—}NH\text{—})_n \xrightarrow[-HgX_2]{H_2O} R_2Hg + (\text{—}CH_2\text{—}CH_2\text{—}NH\text{—})_n \quad (157)$$

Pyridine has also been used in a few instances as symmetrization agent. The sydone derivative shown in Eqn (158) was obtained through pyridine-induced symmetrization.[208]

pyridine, 20° → Hg, (92%) (158)

Tertiary phosphines are effective symmetrizing agents for organomercury salts in which the organic group is either an aryl[153] or a fairly electronegative nonaromatic group such as perchlorovinyl[103] or carbomethoxy.[209]

$$C_6H_5HgOOCCH_3 \xrightarrow[C_6H_6]{(C_4H_9)_3P} (C_6H_5)_2Hg \quad (159)$$

$$Cl_2C{=}CCl\text{—}HgCl \xrightarrow[C_2H_5OH]{(C_6H_5)_3P} (Cl_2C{=}CCl\text{—})_2Hg \quad (160)$$
(88%)

$$H_3CO\text{—}\overset{O}{\overset{\|}{C}}\text{—}Hg\text{—}Cl \xrightarrow{(H_5C_6)_3P/C_6H_6} (H_3CO\text{—}\overset{O}{\overset{\|}{C}}\text{—})_2Hg \quad (161)$$

Phosphines generally are not suitable for the symmetrization of alkylmercury compounds because the intermediate *alkylmercury phosphonium salts*, $[RHgPR_3]^{\oplus}X^{\ominus}$ are too stable;[151,152] the reaction stops at this stage.

6.5.5 Mercuration

The replacement of activated hydrogen by mercury is an excellent preparative route to several classes of diorganomercury compounds. Unsaturated bisalkylmercury compounds are best synthesized by this method. The two procedures which are commonly employed both involve the reaction of acetylenes with mercuric halides or other mercury salts under basic conditions. The first procedure uses potassium or sodium hydroxide in aqueous ethanol as the base.[210] Mercuric iodide, solubilized by potassium iodide as the complex, is the halide generally employed. It is conveniently prepared *in situ* from mercuric bromide or chloride and potassium iodide. Mercury derivatives of a large number of acetylenes have been synthesized by this method in high yields.

$$2R\text{—}C{\equiv}C\text{—}H + 2KOH + K_2[HgI_4] \longrightarrow (R\text{—}C{\equiv}C)_2Hg + 4KI + 2H_2O \quad (162)$$

$R = CH_3$, C_2H_5, $(CH_3)_3C$, $n\text{-}C_5H_{11}$, $n\text{-}C_8H_{17}$, C_6H_5, $C_6H_5CH_2$, $C_6H_5\text{—}CH_2\text{—}CH_2$, $4\text{-}CH_3\text{—}C_6H_4$, $C_6H_{11}\text{—}CH_2$

The second procedure employs an amine in an organic solvent as the base. Several examples resulting from the application of this method are given in the following equations.

$$C_6H_5\text{—}C{\equiv}C\text{—}H + Hg(OOCCH_3)_2 \xrightarrow{C_4H_9NH_2} (C_6H_5\text{—}C{\equiv}C\text{—})_2Hg \quad \text{(Ref. 211)} \quad (163)$$
(98%)

$$C_4HgC{\equiv}C\text{—}H + HgCl_2 \xrightarrow{(C_2H_5)_3N} (H_9C_4\text{—}C{\equiv}C\text{—})_2Hg \quad \text{(Ref. 212)} \quad (164)$$

Both these procedures can be utilized for the preparation of unsymmetrical compounds, $RHgC{\equiv}C\text{–}R'$, by substitution of an organomercury halide for the mercuric halide. Examples of unsymmetrical compounds prepared by this procedure are given in the following equations.

$$2R\text{—}HgCl + HC{\equiv}CH + 2KOH \longrightarrow R\text{—}Hg\text{—}C{\equiv}C\text{—}HgR + 2KCl + 2H_2O \quad \text{(Ref. 213)} \quad (165)$$

$R = CH_3$, C_2H_5, C_3H_7, C_4H_9, $n\text{-}C_5H_{11}$, $n\text{-}C_6H_{13}$, C_6H_5, $C_6H_5CH_2$

$$C_6H_5\text{—}Hg\text{—}C{\equiv}C\text{—}C_6H_5 \xrightarrow[\text{dioxane–}H_2O]{(C_2H_5)} C_6H_5\text{—}Hg\text{—}C{\equiv}C_6H_5 \quad \text{(Ref. 214)} \quad (166)$$
(80%)

Mercury derivatives of a number of aliphatic diazo compounds containing electron-withdrawing groups have been prepared in high yields by reaction of the diazo compounds with mercuric oxide.[215–219] Several examples are shown in the following equations.

[208] *J. M. Tien, I. M. Hunsberger*, J. Amer. Chem. Soc. *83*, 178 (1961).
[209] *F. E. Paulik, R. E. Dessy*, Chem. & Ind. (London) 1650 (1962).
[210] *J. R. Johnson, W. L. McEwen*, J. Amer. Chem. Soc. *48*, 468 (1926).
[211] *G. Eglington, W. McCrae*, J. Chem. Soc. (London) 2295 (1963).
[212] *F. G. Kleiner, W. P. Neumann*, Justus Liebigs Ann. Chem. *716*, 19 (1968).
[213] *R. J. Spohr, R. R. Vogt, J. A. Niewland*, J. Amer. Chem. Soc. *55*, 2465 (1933).
[214] *R. E. Dessy, W. L. Budde, C. Woodruff*, J. Amer. Chem. Soc. *84*, 1172 (1962).
[215] *E. Buchner*, Ber. *28*, 215 (1895).

$$H_5C_6-\overset{O}{\overset{\|}{C}}-CHN_2 + HgO \xrightarrow{\text{pet. ether}} \left(H_5C_6-\overset{O}{\overset{\|}{C}}-CN_2-\right)_2Hg$$
(97%) (Ref. 216) (167)

$$F_3C-CHN_2 \xrightarrow[(C_2H_5)_2O]{HgO} (CF_3-CN_2-)_2Hg$$
(95%) (Ref. 217) (168)

$$(H_5C_6)_2\overset{O}{\overset{\|}{P}}-CHN_2 + HgO \longrightarrow \left[(H_5C_6)_2\overset{O}{\overset{\|}{P}}-CN_2-\right]_2Hg$$
(Ref. 219) (169)

Other CH–acidic compounds such as trinitromethane[220] and fluorodinitromethane[221] can be mercurated with mercuric oxide.

$$HgO + HC(NO_2)_3 \longrightarrow Hg[C(NO_2)_3]_2 \quad (170)$$
(80%)

$$HgO + HCF(NO_2)_2 \longrightarrow Hg[C(NO_2)_2F]_2 \quad (171)$$
(86%)

Mercuration is the method of choice for the synthesis of *dicyclopentadienylmercury*. Two routes to this compound have been developed and are shown in the following equations.

$$HgO + 2C_5H_6 \xrightarrow[85\%]{C_3H_7NH_2;\ 20^\circ} (C_5H_5)_2Hg$$
(Ref. 222) (172)

$$Hg\{N[Si(CH_3)]_2\}_2 + 2C_5H_6 \xrightarrow[100\%]{(C_2H_5)_2O_1\ 20^\circ} (C_5H_5)_2Hg$$
(Ref. 223) (173)

Phenylmercuryhaloalkyl compounds, a class of compounds employed as halocarbene precursors, are readily prepared by mercuration of halogenated hydrocarbons by phenylmercury alkoxides. The alkoxides are formed *in situ* by the reaction of phenylmercury halides with sodium or potassium alkoxides. Several examples of the synthesis of this class of compounds are given in the following equations.

$$H_5C_6-Hg-Cl + F_3C-CHCl_2 \xrightarrow{KOC(CH_3)_3,\ THF,\ 0^\circ} H_5C_6-Hg-CCl_2-CF_3$$
(75%) (Ref. 224) (174)

$$H_5C_6-Hg-Cl + HC(Cl)(Br)(J) \xrightarrow{KOC(CH_3)_3,\ THF/(H_5C_2)_2O} H_5C_6-Hg-C(Cl)(Br)(J)$$
(76%) (Ref. 225) (175)

6.5.6 Addition

Bis(perfluoroalkyl)mercury compounds can be obtained by the reaction of mercuric fluoride with terminal fluoroolefins. Originally, high temperatures and hydrogen fluoride[226] or arsenic trifluoride[227] as the solvent, were employed; it has since been found that the reactions proceed rapidly in dimethylformamide at moderate temperatures.[228, 229] The mercuric fluoride is conveniently prepared *in situ* from mercuric chloride and potassium fluoride. Some examples of the reaction in dimethylformamide, which presumably involves perfluoroalkyl carbanions as intermediates, are given in the following equations.

$$HgCl_2 + KF + (CF_3)_2C{=}CF_2 \xrightarrow{HCON(CH_3)_2} [(CF_3)_3C]_2Hg \quad (176)$$
(66%)

$$HgCl_2 + KF + F_3C-CF{=}CF_2 \xrightarrow{HCON(CH_3)_2} [(CF_3)_2CF]_2Hg \quad (177)$$
(65%)

A number of unique diorganomercury com-

---

[216] *P. Yates, F. Garneau*, Tetrahedron Lett., 71 (1967).
[217] *T. DoMink, O. P. Strausz, H. E. Gunning*, Tetrahedron Lett., 5327 (1968).
[218] *J. Loberth*, J. Organometal. Chem. *27*, 303 (1971).
[219] *M. Regitz, A. Liedhegener, U. Eckstein, M. Martin, W. Anschuetz*, Justus Liebigs Ann. Chem. *748*, 207 (1971).
[220] *S. S. Novikov, T. I. Godovikova, V. A. Tartakovskii*, Izv. Akad. Nauk. SSSR, 505 (1960).
[221] *L. V. Okhlobystina, G. Y. Legin, A. A. Fainzil'berg*, Izv. Akad. Nauk. SSSR, 708 (1969).
[222] *S. Lenzer*, Australian J. Chem. *22*, 1303 (1969).
[223] *J. Loberth*, J. Organometal. Chem. *19*, 189 (1969).
[224] *D. Seyferth, D. C. Mueller*, J. Amer. Chem. Soc. *93*, 3714 (1971).
[225] *D. Seyferth, C. K. Hass, S. P. Happer*, J. Organometal. Chem. *33*, C1 (1971).
[226] *P. E. Aldrich, E. G. Howard, W. J. Linn, W. J. Middleton, W. H. Sharkey*, J. Org. Chem. *28*, 184 (1963).
[227] *C. G. Krespan*, J. Org. Chem. *25*, 105 (1963).
[228] *B. L. Dyatkin, S. R. Sterlin, B. I. Martynov, I. L. Knunyants*, Tetradedron Lett., 1387 (1970).
[229] *B. L. Dyatkin, S. R. Sterlin, B. I. Martynov, I. L. Knunyants*, Tetrahedron Lett. *27*, 2843 (1971).

pounds can be obtained by addition of mercury oxide or mercury salts to ketenes or carbon monoxide, as shown in the following equations.

$$HgO + H_2C{=}C{=}O + CH_3OH \longrightarrow \underset{(97\%)}{Hg(-CH_2-\overset{O}{\overset{\|}{C}}-OCH_3)_2} \quad \text{(Ref. 230)} \quad (178)$$

$$HgF_2 + (F_3C)_2C{=}C{=}O \xrightarrow{HCON(CH_3)_2} \underset{(70\%)}{[(F_3C)_2C(-C(=O)-F)-]_2Hg} \quad \text{(Ref. 231)} \quad (179)$$

$$(H_5C_2)_2NH + CO + Hg(O-COCH_3)_2 \xrightarrow[1\,atm]{RT} \underset{(64\%)}{[(H_5C_2)_2N-\overset{O}{\overset{\|}{C}}-]_2Hg} \quad \text{(Ref. 232)} \quad (180)$$

6.5.7 Decarboxylation

Pyrolysis of mercury carboxylates is a general route to diorganomercury compounds containing electron-withdrawing groups. Both dialkyl- and diarylmercury compounds can be obtained by this method. It has been found that pyridine is essential for the preparation by this method of the unsymmetrical compounds in which one of the organic groups is a typical alkyl or aryl group and the other is a perhaloaryl group.[119] Some examples of the decarboxylation method are shown below.

$$Hg(OOC-C_6Cl_5)_2 \xrightarrow{\Delta,\ pyridine} \underset{(62\%)}{Hg(C_6Cl_5)_2} + 2CO_2 \quad \text{(Ref. 233)} \quad (181)$$

$$H_3C-C_6H_4-HgOOC-C_6Cl_5 \xrightarrow{\Delta,\ Py} \underset{(93\%)}{H_3C-C_6H_4-HgC_6Cl_5} \quad \text{(Ref. 233)} \quad (182)$$

[230] *I. F. Lutsekno, V. L. Foss, N. L. Ivanova*, Dokl. Akad. Nauk. SSSR *141*, 1107 (1961).

[231] *B. L. Dyatkin, L. G. Ahurzkova, B. I. Martynov, E. I. Mysov, S. R. Sterlin, I. L. Knunyants*, J. Organometal. Chem. *31*, C15 (1971).

[232] *U. Schollkopf, F. Gerhart*, Angew. Chem. *78*, 675 (1966).

[233] *G. B. Deacon, P. W. Felder*, J. Chem. Soc. *C*, 2313 (1967).

$$HgCl_2 + 2NaOOC-CBr_3 \xrightarrow{\Delta,\ glyme} \underset{(85\%)}{Hg(CBr_3)_2} + 2CO_2 + 2NaCl \quad \text{(Ref. 234)} \quad (183)$$

$$C_6H_5HgCl + Cl_3CCOONa \xrightarrow{\Delta,\ CH_3OCH_2CH_2OCH_3} \underset{(65\%)}{C_6H_5HgCCl_3} + 2CO_2 + 2NaCl \quad \text{(Ref. 235)} \quad (184)$$

6.5.8 Aliphatic Diazo Compounds

Bis(α-haloalkyl)mercury compounds can be prepared in good yields by reaction of mercuric halides with aliphatic diazo compounds.[105, 236]

$$HgX_2 + CH_2N_2 \longrightarrow (XCH_2)_2Hg \quad (185)$$
$$X = Cl, Br, I$$

$$F_3C-CHN_2 + HgCl_2 \longrightarrow \underset{(46\%)}{(F_3C-CHCl-)_2Hg} \quad \text{(Ref. 237)} \quad (186)$$

Unsymmetrical diorganomercury compounds containing one α-haloalkyl group can be prepared analogously from an organomercury halide, but in many cases they are not stable to symmetrization.

$$HO-CH_2-CH_2-HgBr + CH_2N_2 \longrightarrow HO-CH_2-CH_2-Hg-CH_2-Br \quad \text{(Ref. 238)} \quad (187)$$

The lower dialkylmercury compounds are liquids, whereas the higher dialkyl and diarylmercury compounds are solids. Both classes of compounds are monomeric and do not form any complexes with donor compounds such as amines unless both of the organic groups are very strongly electron withdrawing[239] such as perfluoroalkyl[240] or perfluoroaryl.[241] Primary alkyl and aryl derivatives of mercury are very stable in the presence of air and water. However, compounds in which mercury is bonded to a secondary or tertiary alkyl group or also to various reactive groups such as allyl and cyclopentadienyl groups are unstable because of fairly rapid autoxication. *Dibenzylmercury* is an exception to this generalization as it is quite stable

[234] *R. Robson, I. E. Dickson*, J. Organometal. Chem. *15*, 7 (1968).

[235] *T. J. Logan*, Org. Synth. *46*, 98 (1966).

[236] *G. Wittig, K. Schwarzenbach*, Justus Liebigs Ann. Chem. *656*, 1 (1961).

[237] *B. L. Dyatkin, E. P. Mochalina*, Izv. Akad. Nauk. SSSR, 1225 (1964).

[238] *A. N. Nesmeyanov, R. K. Freidlina, F. A. Tokareva*, Ber. *69*, 2019 (1936).

[239] *K. H. Thiele, P. Zdunneck*, Organometal. Chem. Rev. *1*, 331 (1966).

[240] *H. B. Powell, M. T. Maung, J. J. Lagowski*, J. Chem. Soc. 2848 (1963).

[241] *A. J. Canty, G. B. Deacon*, Australian. J. Chem. *24*, 489 (1971).

under ambient conditions. Consequently, the only commercially available diorganomercury compounds are those in which the organic group is either primary alkyl or aryl.

All organomercury compounds are toxic. Dimethylmercury and methylmercury salts are especially hazardous as far as their role in the environment is concerned. Dimethylmercury and methylmercury salts which have been identified in the environment only partially represent residues of pesticides. However, they are also formed by microorganisms from inorganic mercury compounds. Both organomercury species are secreted only very slowly by living organisms and accumulate therefore in food chains.

Attention should be directed toward the "Minamata disease" which is caused by these organomercury compounds.[242]

6.6 Bibliography

G. E. Coates, K. Wade, Organometallic Compounds, Vol. 1, 3rd Ed., pp. 147–176. Methuen, London (1967).

B. J. Wakefield, Adv. Inorg. Chem. Radiochem. *11*, 342 (1968).

L. G. Makarova, A. N. Nesmeyanov, The Organic Compounds of Mercury. North-Holland Publ., Amsterdam (1967).

[242] *P. Nuorteva*, Naturwissenschaftliche Rundschau *24*, 233 (1971).

7 Scandium

K. E. Blick
Division of Science, Prestonsburg Community College,
Prestonsburg, Kentucky 41654, U.S.A.

7.1 Preparation and Purification of Scandium

Metallic scandium is commercially available. Some common trace impurities found in commercial-grade scandium are listed in Table 1.

Table 1. Trace Impurities of Metallic Scandium

Metals	Impurity level (ppm)
Ca, Fe	100
Si, Mg	<100
Al	50
Yb, Cu	20
Others	<20

A convenient procedure for the preparation of *highly pure scandium* in good yields involves the reduction of scandium(III) fluoride with calcium; the resultant scandium metal is purified by distillation.[1,2] This process is carried out at high temperatures (1500°) and in an inert atmosphere, preferably argon. If higher purity levels are desired, the material may be further purified by repeated vacuum distillation in a metal distillation system[3] (Fig. 1).

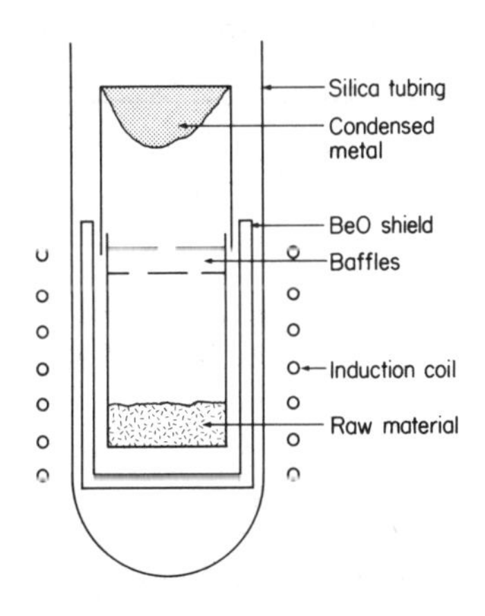

In order to free scandium of accompanying metal impurities, it is convenient to remove the scandium by extraction of its thiocyanate with ether.[4] Pure scandium can also be obtained by employing ion exchange techniques.[5]

[1] *J. Loriers*, Compt. Rend. *234*, 493 (1956).
[2] *F. Petru*, Chem. Listy. *50*, 2025 (1956).
[3] *A. H. Daane* in *E. V. Kleber*, Rare Earth Research, p. 264, Macmillan, New York 1961.
[4] *W. Fisher*, *R. Brock*, Z. Anorg. Allgem. Chemie *249*, 168 (1942).
[5] *R. C. Vickery*, J. Chem. Soc. (London), 245 (1955).

7.2 Binary Interstitial Compounds

Several interstitial compounds of scandium are known. The existence of Sc_4C_3 has been suggested;[6,7] however, more recent studies have shown that the stable carbide of scandium is actually ScC.[8] *Scandium carbide* can be formed by heating scandium oxide with carbon [Eqn (1)].

$$Sc_2O_3 + 5C \longrightarrow 2ScC + 3CO \quad (1)$$

Also, scandium combines with gaseous nitrogen at elevated temperatures to form *scandium nitride*, ScN.[9,10]

Scandium borides, ScB_2 and ScB_{12}, have been prepared by heating a pellet of Sc_2O_3 and boron to temperatures of 1350–1625° *in vacuo*[11] [Eqn (2)].

$$Sc_2O_3 + (2+2n)B \longrightarrow B_2O_3 + 2ScB_n \quad (2)$$

7.3 Scandium Halides

Of the known scandium halides, only the scandium fluoride, ScF_3, and hydrated scandium chloride, $ScCl_3 \cdot 6H_2O$, are generally commercially available. It is noteworthy that all the scandium halides, with the exception of ScF_3, are extremely hygroscopic in the anhydrous state.

Scandium fluoride, ScF_3, is conveniently prepared by treating Sc_2O_3 [or $Sc(OH)_3$] with excess hydrogen fluoride at 700° [Eqn (3)]; the use of

$$Sc_2O_3 + 6HF \longrightarrow 2ScF_3 + 3H_2O \quad (3)$$

platinum or Monel equipment is suggested for this procedure.[12] An alternate method for the preparation of ScF_3 involves the fluorination of Sc_2O_3 with ammonium hydrogen fluoride [Eqn (4)].[3]

$$Sc_2O_3 + 6(NH_4)HF_2 \longrightarrow 2ScF_3 + 6NH_4F + 3H_2O \quad (4)$$

Anhydrous *scandium trichloride*, $ScCl_3$, can be obtained by passing a mixture of chlorine and

[6] *L. Sittig*, Z. Anorg. Allgem. Chemie *144*, 169 (1925).
[7] *P. B. Sarkar*, Ann. Chim. *8*, 207 (1927).
[8] *R. C. Vickery*, *R. Sedlacek*, *A. Ruben*, J. Chem. Soc. (London), 498 (1959).
[9] *F. Newmann*, *D. Kruger*, *K. Kunz*, Z. Anorg. Allgem. Chemie *218*, 379 (1934).
[10] *E. Friederich*, Z. Phys. *31*, 813 (1925).
[11] *M. Pryzbylska*, *A. H. Reddoch*, *G. J. Ritter*, J. Amer. Chem. Soc. *85*, 407 (1963).
[12] Gmelin-Kraut VI, 2, p. 681.

sulfur dichloride over heated Sc_2O_3.[13,14] Alternately, scandium chloride is conveniently prepared by the dehydration of the hydrated salt (*e.g.*, $ScCl_3 \cdot 6H_2O$) with gaseous hydrogen chloride.[15] More recently, a procedure for the preparation of $ScCl_3$ has been reported by treating scandium metal with hydrochloric acid in molybdenum containers.[16,17]

The synthesis of $ScCl_2$ has been suggested;[17] however, subsequent investigations have not confirmed the existence of this species.[18]

Scandium tribromide hexahydrate, $ScBr_3 \cdot 6H_2O$, can be obtained by treating scandium hydroxide (or carbonate) with hydrobromic acid.[19,20] The anhydrous $ScBr_3$ is prepared by heating a mixture of scandium oxide and carbon in a stream of bromine gas[21] [Eqn (5)]. The resultant scandium bromide can be purified by sublimation under vacuum.

$$Sc_2O_3 + 3C + 3Br_2 \longrightarrow 2ScBr_3 + 3CO \qquad (5)$$

NOTE: At sublimation temperatures (900°–1000°), $ScBr_3$ attacks glass [Eqn (6)].

$$4ScBr_3 + SiO_2 \longrightarrow 3SiBr_4 + 2Sc_2O_3 \qquad (6)$$

Compounds such as ScOX(X=F, Cl) can be produced as by-products in the formation of scandium trihalides by halogenation of scandium oxide.[3]

7.4 Oxides and Sulfides

Scandium oxide, Sc_2O_3, is commercially available. The oxide can be obtained by a variety of procedures from thortveitite, a common source of scandium.

For example, Spedding *et al.*[22] have described a procedure in which nearly half of the silicon dioxide of the ore is readily removed as volatile SiF_4; this technique involves heating ground thortveitite with ammonium hydrogen fluoride followed by solvent extraction and ion-exchange techniques to isolate the scandium oxide.

The oxide can also be generated from metallic scandium by heating (500°–1000°) in air.[23]

Scandium sesquisulfide, Sc_2S_3, and the less common *scandium(II) sulfide*, ScS, are prepared by a high temperature reaction (1550°; 2–3 hr) of hydrogen sulfide and scandium sesquioxide, Sc_2O_3, in a high-purity graphite crucible.[24] Similarly, scandium sesquisulfide has been prepared by the interaction of H_2S and $ScCl_3$ at elevated temperatures.[25] Sc_2O_2S and a non-stoichiometric scandium sulfide are apparently formed utilizing this same system.

The weakly basic *scandium hydroxide*, $Sc(OH)_3$, is formed upon treating a solution of a scandium salt with an alkali hydroxide or ammonia solution.[26,27]

7.5 Scandium Salts of Oxoacids

The more common salts of oxoacids of scandium and their most stable hydrates are listed in Table 2.

Table 2. Common Scandium Salts and Their Most Stable Hydrate at Room Temperature

Anhydrous salt	Name	Hydrate
$Sc_2(SO_4)_3$	Scandium sulfate	$Sc_2(SO_4)_3 \cdot 5H_2O$
$Sc_2O(SO_4)_2$	Scandium oxosulfate	—
—	Scandium nitrate	$Sc(NO_3)_3 \cdot 4H_2O$
$Sc_2O(NO_3)_4$	Scandium oxonitrate	—
—	Scandium carbonate	$Sc_2(CO_3)_3 \cdot 6H_2O$
$Sc_2(C_2O_4)_3$(?)	Scandium oxalate	$Sc_2(C_2O_4)_3 \cdot 6H_2O$

Anhydrous *scandium sulfate*, $Sc_2(SO_4)_3$, can be conveniently prepared by evaporation of the

[13] *C. Matignon, F. Bourion*, Compt. Rend. *138*, 631 (1904).
[14] *R. J. Meyer, H. Winter*, Z. Anorg. Allgem. Chemie *67*, 398 (1910).
[15] *J. H. Kleinheksel, H. C. Kremers*, J. Amer. Chem. Soc. *50*, 959 (1928).
[16] *L. F. Druding, J. D. Corbett*, J. Amer. Chem. Soc. *83*, 2462 (1961).
[17] *J. D. Corbett, B. Ramsey*, Inorg. Chem. *4*, 260 (1965).
[18] *O. Polyachenok, G. Novikov*, Zh. Neorg. Khim. *8*, 2819 (1963).
[19] *W. Fisher, R. Gewehr, H. Wingschen*, Zh. Neorg. Khim. *242*, 161 (1939).
[20] *W. Crookes*, Phil. Trans. *209*, 15 (1909).
[21] *O. Honigschmidt*, Z. Elektrochemie *25*, 91 (1919).
[22] *F. H. Spedding, J. E. Powell, A. H. Daane, M. H. Miller, W. H. Adams*, J. Elektrochem. Soc. *105*, 683 (1958).
[23] *P. Pascal*, Nouveau Traite De Chimie Mineral *7*, 777 (1959).
[24] *J. P. Dismukes, J. G. White*, Inorg. Chem. *3*, 1220 (1964).
[25] *W. Klemm, K. Meisel, H. V. von Vogel*, Z. Anorg. Allgem. Chemie *190*, 123 (1930).
[26] *C. R. J. Sterba-Bohm, M. Melichar*, Collec. Czech. Chem. Commun. *7*, 131 (1935).
[27] *H. B. Weiser, W. O. Milligan*, J. Phys. Chem. *42*, 669 (1938).

metal sulfate solution and warming the resultant hydrated salt to 250°.[28–30] Slow crystallization of a neutral sulfate solution yields the *pentahydrate*, $Sc_2(SO_4)_3 \cdot 5H_2O$, which is stable at ambient temperatures.[28]

Scandium nitrate tetrahydrate, $Sc(NO_3)_3 \cdot 4H_2O$, can be prepared by dissolving scandium hydroxide (or oxide) in dilute nitric acid or hot, concentrated nitric acid.[20,31] The *scandium oxonitrate*, $Sc_2O(NO_3)_4$, is formed *via* the thermal decomposition (> 100°) of the normal nitrate, $Sc(NO_3)_3 \cdot 4H_2O$.[7,20]

Scandium carbonate hexahydrate, $Sc_2(CO_3)_3 \cdot 6H_2O$, is precipitated from solution on addition of excess sodium (or ammonium) carbonate to a solution of scandium salt.[20,32–35]

Scandium oxalate as the *hexahydrate*, $Sc_2(C_2O_4)_3 \cdot 6H_2O$, is formed by the addition of oxalic acid to a solution of scandium ions; precipitation of the hexahydrated salt is incomplete, however.[5,32] The formation of the anhydrous oxalate by the careful dehydration of the hydrated salt has been claimed;[29,36] however, subsequent studies indicated that decomposition of the oxalate salt occurs prior to the formation of the anhydrous material.[37]

Scandium perchlorate, $Sc(ClO_4)_3$, can be synthesized by the reaction of stoichiometric amounts of scandium carbonate and vacuum-distilled 70–72% perchloric acid.[38] The hydrated material can be prepared by dissolving Sc_2O_3 in $HClO_4$ followed by recrystallization from perchloric acid.[30]

The *scandium iodate hydrate*, $Sc(IO_3)_3 \cdot 18H_2O$, has been prepared by the metathesis of a soluble scandium salt with ammonium iodate.[40] The *anhydrous iodate* salt, *i.e.*, $Sc(IO_3)_3$, can be generated by gentle heating of the hydrate to 250°.[40,41]

7.6 Complexes of Scandium

Most neutral *complexes of scandium* are simple adducts of scandium halides such as $ScX_3 \cdot nNH_3$. The interaction of alcohol vapors with anhydrous scandium chloride results in the formation of a variety of *alcoholates*, $ScCl_3 \cdot nROH$[42,43] (R = alkyl).

Under anhydrous conditions, the halides react with ammonia to give *ammoniates* of the type $ScX_3 \cdot nNH_3$ (X = Cl, Br; n = 2, 4, or 5).[44] Similarly, scandium chloride reacts with primary and secondary amines to form adducts, *e.g.*, $ScCl_3 \cdot NH_{3-n}R_n$ (n = 1, 2).[45] Aminolysis of scandium halides (Cl, Br) in anhydrous ethanol with $C_2H_4(NH_2)_2$, $(i\text{-}C_3H_7)_2NH$, $N(C_2H_5)_3$, and $(C_2H_5)_2NH$ also results in simple adduct formation.[46] Similar adducts of the scandium halides are formed by reacting the anhydrous halides with pyridine and piperidine, *i.e.*, $ScX_3 \cdot 2py$, $ScX_3 \cdot 4pip$, X = Cl, Br.[47] Interaction of scandium chloride with 1,10-phenanthroline, 2,2′-bipyridine, or 4,4′-bipyridine results in the formation of the following compounds: $ScCl_3 \cdot$3-(1,10-phenanthroline); $ScCl_3 \cdot$2(2,2′-bipyridyl); and $ScCl_3 \cdot$2(4,4′-bipyridyl), respectively.[48]

Several other neutral complexes of scandium are known. For example, *scandium bipyridyl*, $Sc(bipy)_3$, has been prepared as a reduction product of a reaction with dilithium bipyridyl.[49]

Scandium acetylacetonate, $Sc(acac)_3$, can be obtained by the reaction of ammonium acetylacetonate and a scandium chloride solution;[50–52] the resultant scandium acetylacetonate is purified by recrystallization. (*Note:* At reduced pressures

[28] *Z. Trousil*, Collec. Czech. Chem. Commun. *10*, 290 (1938).
[29] *F. Wirth*, Z. Anorg. Allgem. Chemie *87*, 9 (1914).
[30] *F. Wirth*, Z. Anorg. Allgem. Chemie *87*, 9 (1914).
[31] *R. Bock, E. Bock*, Naturwissenschaften *36*, 344 (1949).
[32] *W. Fisher, R. Bock*, Z. Anorg. Allgem. Chemie *249*, 146 (1942).
[33] *W. Crookes*, Chem. News *98*, 274 (1908).
[34] *B. N. Ivanov-Einin, E. A. Ostromouv*, J. Gen. Chem. (USSR) *14*, 772 (1944).
[35] *R. C. Vickery*, J. Chem. Soc. (London), 3113 (1956).
[36] *W. W. Wendlandt*, Anal. Chem. *30*, 58 (1958); *31*, 408 (1959).
[37] *M. Kilpatrick, L. Pokras*, J. Elektrochem. Soc. *100*, 84 (1953).
[38] *L. Pokras*, Thesis, Illinois 1962.
[39] *W. Crookes*, Chem. News *101*, 49 (1910).
[40] *W. Crookes*, Phil. Trans. *210*, 359 (1910).
[41] *W. Crookes*, Chem. News *102*, 73 (1910).
[42] *F. Petru, B. Hajek, F. Jost*, Croat Chem. Acta *29*, 457 (1957).
[43] *E. M. Kirmse*, Z. Chem., 332 (1961).
[44] *Ya. Ya. Kharitonov, V. P. Orlouskii, I. V. Tananaev*, Zh. Neorg. Khim. *8*, 1093 (1963).
[45] *E. M. Kirmse*, Z. Chem. *1*, 334 (1961).
[46] *I. V. Tananaev, V. P. Orlovskii*, Zh. Neorg. Khim. *6*, 1909 (1961).
[47] *F. Petru, F. Jost*, Croat, Chem. Acta *52*, 1645 (1958);
[48] *F. Petru, F. Jost*, Collec. Czech. Chem. Commun. *24*, 2041 (1959).
[49] *B. N. Ivanov-Einin, L. A. Niselson, L. A. Larionova*, Zh. Neorg. Khim. *6*, 334 (1961).
[50] *B. E. Bryant*, Inorg. Synth. *5*, 105 (1957).
[51] *J. K. Marsh*, J. Chem. Soc. (London), 1084 (1941).
[52] *J. G. Stites, C. N. McCarthy, L. L. Quill*, J. Amer. Chem. Soc. *70*, 3142 (1948).

of 8–10 mmHg, the material can be sublimed without decomposition at temperatures of 157°–187°.)

The water-insoluble $Sc(C_9H_6NO)_3 \cdot HC_9H_6NO$ has been prepared by the interaction of a scandium ion solution with 8-quinolinol;[53] the plausibility of a $H[Sc(C_9H_6NO)_4]$ arrangement has also been discussed.[54] Scandium ion reacts with *phenylphosphonic acid* to give an insoluble *complex*, $Sc(C_6H_5PO_3)_3 \cdot 3H_2O$.[55]

Recently, several *scandium complexes with urea* have been prepared by dissolving the appropriate salt, ScX_3 ($X = ClO_4$, NO_3, Cl and Br), in anhydrous methanol, and treating the resultant solution with urea under reflux.[56]

Complexes of the type $(NH_4)_2(ScF_5)$ and $NH_4(ScF_4)$ are synthesized by the addition of ammonium fluoride to scandium fluoride in hydrofluoric acid solution.[57] Scandium ion can be quantitatively precipitated as $[Co(NH_3)_6](ScF_6)$[58] by the addition of a hexamine–cobalt(III) chloride solution to a solution containing the $(ScF_6)^{3\ominus}$ moiety.

[53] *L. Pokras*, *P. M. Bernays*, J. Amer. Chem. Soc. *73*, 7 (1951).

[54] *J. H. Van Tassel*, *W. W. Wendlandt*, *E. Sturm*, J. Amer. Chem. Soc. *83*, 810 (1961).

[55] *J. P. Alimarin*, *V. I. Fadeeva*, *T. N. Petrova*, Zh. Analit. Khim. *16*, 549 (1961).

[56] *F. Kutek*, *B. Dusek*, J. Inorg. Nucl. Chem. *31*, 1543 (1969).

[57] *J. Sterba-Bohm*, Z. Elektrochem. *20*, 289 (1914).

[58] *J. Sterba-Bohm*, Bull. Soc. Chim. France *27*, 185 (1920).

7.7 Organometallic Derivatives

Triethylscandium etherate, $(CH_3CH_2)_3Sc \cdot O(CH_2CH_3)_2$, was prepared by the interaction of scandium chloride and Grignard reagent;[59] however, the existence of this organometallic compound has been questioned in subsequent studies.[60]

Tri(cyclopentadienyl)scandium, $Sc(C_5H_5)_3$, can be prepared in high yields (85%) by reacting scandium trifluoride with molten magnesium cyclopentadienide.[61] Alternatively, the treatment of sodium cyclopentadienide with scandium chloride in tetrahydrofuran solution produces $Sc(C_5H_5)_3$ in reasonably good yields.[62, 63]

7.8 Bibliography

R. C. Vickery, The Chemistry of Yttrium and Scandium, Pergamon Press, Inc., New York, 1960.

E. V. Kleber, Rare Earth Research, Macmillan, New York, 1961.

T. Moeller, *D. F. Martin*, *L. C. Thompson*, *R. Ferrüs*, *G. F. Feistel*, *W. J. Randall*, Chem. Rev. *65*, 1 (1965).

[59] *V. M. Pletus*, Compt. Rend. Acad. Sci. USSR *20*, 27 (1938).

[60] *B. N. Afanasev*, *P. A. Tsyganova*, J. Gen. Chem. (USSR) *18*, 306 (1948).

[61] *A. F. Reid*, *P. C. Wailes*, Inorg. Chem. *5*, 1213 (1966).

[62] *J. M. Birmingham*, *G. Wilkinson*, J. Amer. Chem. Soc. *78*, 42 (1956).

[63] *J. M. Birmingham*, Organomet. Chem. *2*, 377 (1964).

8 Yttrium and Lanthanides

K. S. Mazdiyasni
Air Force Materials Laboratory,
Wright-Patterson Air Force Base, Ohio 45433, U.S.A.

The chemistry of the lanthanides has received much attention during the past few years. In part, new applications for the lanthanides in electronic, laser, magnetic, optical and phosphor materials have prompted this intensive effort. Also, the advances in technology, instruments and data interpretation have contributed much to the understanding of the lanthanides as a whole. It should be noted that yttrium, although not a lanthanide, has considerable chemical similarity with the lanthanides and therefore is included in this chapter.

Due to the lanthanide contraction there is little difference in the atomic size of these elements. Yttrium and the lanthanides have covalent radii ranging from 1·56–1·7 Å and electronegativities of 1·11 to 1·20. Therefore, the reactions they undergo and the chemical properties of the compounds formed are expected to be similar. However, in a discussion of the chemistry of the rare earth elements, the similarity of their behavior and the predominance of the tervalent state are often stressed, thus creating popular confusion and misconceptions. It is informative to state that frequently the divalent and tetravalent states are utilized in effecting the chemical separation of some elements from (naturally occurring) mixtures. Hence, it is apparent that several types of chemical behavior of the rare earth elements must be considered if one is to study representative rare earth materials.

Among the tetravalent states, high stability is associated with the electronic configuration of the $La^{3\oplus}$, $Gd^{3\oplus}$, and $Lu^{3\oplus}$ ions in which the $4f$ levels are empty, half-filled and filled, respectively. The divalent and the tetravalent state of the neighboring rare earth elements appear to arise as a result of approach to, or achievement of, the stable $4f^0$, $4f^7$, and $4f^{14}$ configurations. Among these the divalent states of samarium, europium and ytterbium, and the tetravalent state of cerium are the most important; the corresponding states of thulium, preseodymium and terbium appear to be less stable and are observed only under special conditions. Therefore, within a given class of compounds, it is reasonable to expect that different characteristics will be obtained as the chemical nature of the rare earth element changes.

8.1 Yttrium and Lanthanide Hydrides

The rare earth metals, nearly all of which exist at room temperature in cubic or hexagonal modifications, take up hydrogen at ordinary or elevated temperatures and pressures to form *dihydrides*.

$$M + H_2 \xrightarrow[1\ \text{atm}]{250°\text{–}350°} MH_{2-x} \qquad (1)$$

The (hydrogen deficient) dihydrides are all face-centred cubic, except for those of europium and ytterbium which are orthorhombic. These last two compounds do not react with additional hydrogen at 1 atm pressure but others do and approach the *trihydride* in their composition.[1]

[1] *W. L. Korst, J. C. Wrap*, Inorg. Chem. *5*, 1719 (1966).

$$RH_{2-x} + \tfrac{1}{2}H_2 \xrightarrow[1\text{–}40\ \text{atm}]{250°\text{–}750°} RH_{3-x} \qquad (2)$$

The rare earth dihydrides are increasingly thermally stable with increasing atomic number up to a maximum for Ho or Er hydride. The trihydrides of La, Ce, Pr, and Nd are face-centred cubic, but in the case of Sm and beyond as well as yttrium, a hexagonal close packed phase appears before RH_3 is reached.[1]

Among all the rare earth hydrides only stoichiometric *cerium trihydride*, CeH_3, is readily obtainable. The freshly prepared cerium hydride reacts with nitrogen even at room temperature[2] and when heated to 800°–900°, is completely converted to the *nitride*.[3,4] It is spontaneously inflammable in air and burns to form a mixture of oxide and nitride. However, short treatment with dry CO_2 temporarily makes the surface inactive to an oxidizing atmosphere.[5] Cerium hydride and presumably all other rare earth hydrides dissolve readily in dilute acids with evolution of hydrogen.

$$CeH_3 + 3HCl \longrightarrow CeCl_3 + 3H_2 \qquad (3)$$

Cerium trihydride reacts fairly rapidly with water at 0° liberating hydrogen and forming cerium hydroxide.[5]

$$CeH_3 + 3H_2O \longrightarrow Ce(OH)_3 + 3H_2 \qquad (4)$$

All of the hydrides are ionic pyrophoric, graphitic-appearing materials and are semiconductors.[6] It is assumed that hydrogen is essentially anionic in the dihydride. The dihydrides are ferromagnetic or antiferromagnetic.[7,8]

Certain properties of the rare earth hydrides, such as the thermal stability of the dihydrides and the change in the nature of the chemical bond in the MH_2–MH_3 range, indicate that it should be possible to use these compounds advantageously in special fields of metallurgy, certain types of heterogeneous catalytic synthesis, and in semiconductor materials technology.[6]

8.2 Yttrium and Lanthanide Halides

The lanthanide *trihalides* are prepared by react-

[2] *J. Lipsky*, Z. Electrochem. *15*, 196 (1909).

[3] *G. L. Winkler*, Rept. German Chem. Soc. *24*, 1966 (1891).

[4] *W. Muthman, E. Bauer*, Justus Liebigs Ann. Chem. *331*, 261 (1904).

[5] *M. E. Kost*, Zh. Neorg. Khim. *2*, 2689 (1957).

[6] *V. I. Mikheeva, M. E. Kost*, Russian Chem. Reviews *29*, 28 (1960).

[7] *K. I. Hardcastle, J. C. Warf*, Inorg. Chem. *5*, 1730 (1966).

[8] *J. C. Warf, K. I. Hardcastle*, Inorg. Chem. *5*, 1736 (1966).

ing metal turnings with HCl or I_2.[9,10] Another method commonly used to prepare the trihalides is to heat the metal oxides to ~700° in a stream of chlorine gas saturated with S_2Cl_2 or CCl_4.[11] For purification it is advantageous to vacuum sublime the trihalides in tantalum or molybdenum containers at least once.

The rare earth metals invariably react with their molten trihalides. The electrical behavior of the lanthanide metals dissolved in their molten trichlorides or triiodides are quite different. For example, La at 910° behaves like Ce at 855° showing an accelerated rise in conductivity with increasing metal concentration. The conductivity of Nd + $NdCl_3$ at 855°, on the other hand, rises much more slowly with increasing metal concentration.[12–14]
The metal–metal triiodide systems of the lighter rare earth elements show a considerable variation in the degree of their solution in their respective molten trihalides. In contrast to the respective trichlorides, La, Ce, and Pr form stable, metallic diiodides that are higher melting than the triiodides. However, the stability on melting and the melting points decrease from LaI_2 to PrI_2.[15,16] The properties of the metal-like diiodides are perhaps due to the presence of the "metallic" lattice electrons plus the tripositive metal cation, $M^{3\oplus}e^{\ominus}(I^{\ominus})_2$, while the salt-like rare earth dihalide in the case of $NdCl_2$ is simply $Nd^{2\oplus}(Cl^{\ominus})_2$[17]. Magnetic measurements confirm that the neodymium ion is truly $Nd^{2\oplus}$, but the lanthanum ion, and presumably also the cerium and praseodymium ions, are present as $M^{\oplus 3}$ in their diiodides.[15] The corresponding bromide system of lanthanum and cerium merely dissolves the metal without formation of intermediate solid phases whereas praseodymium forms a praseodymium subbromide.[15]

The lower valent or *divalent halides* of some of the lanthanides such as Sm, Eu and Yb are well known. In recent years, lower halides of other lanthanides, such as Nd,[18] Pr,[14,19,20] and Tm[21] have been obtained by reducing the trihalide with the metal. In addition all of the lanthanides have now been obtained in the divalent state in dilute solutions in a CaF_2 matrix by γ irradiation,[22] fused salt electrolysis,[23] or alkaline earth metal reduction.[24] The *halo complexes* of the lanthanide(III) and hexalolanthanide(III) have also been reported.[25,26] Many lanthanide halo complex species have been reported for fused salt melts ranging, in the case of chlorides, from the simple $MCl_4^{\ominus}$ ion to $M_2Cl_9^{3\ominus}$, and $M_3Cl_{10}^{3\ominus}$ ions. Some examples are shown in Table 1.

Table 1. Anionic Chloro Complexes in Fused Salts

Anion	Metal (M)	Reference
$MCl^{\ominus}$	La	27
$MCl_5^{2\ominus}$	La, Ce, Pr, Nd, Sm	26, 27, 28, 29, 30, 31
$MCl_6^{3\ominus}$	Y, La, Ce, Pr, Nd, Sm Sm and Yb	26, 27, 28, 29, 30, 31, 32, 33
$M_2Cl_7^{3\ominus}$	Sm	26
$M_2Cl_9^{3\ominus}$	Ce, Pr, Nd	27, 32
$M_3Cl_{10}^{3\ominus}$	Y, La, Ce	27, 33

8.3 Binary Lanthanide Derivatives

8.3.1 Oxides

The rare earth oxides are prepared by oxidation of the metal in a stream of oxygen or by reaction

[9] *J. E. Mec, J. D. Corbett*, Inorg. Chem. *4*, 88 (1965).
[10] *J. D. Corbett, L. D. Pollard, J. E. Mee*, Inorg. Chem. *5*, 761 (1966).
[11] *G. I. Novikor, O. G. Polychenok*, Russian J. Inorg. Chem. *8*, 1478 (1963).
[12] *A. S. Dworkin, H. R. Bronstein, M. A. Breding*, Discuss. Faraday Soc. *32*, 188 (1961).
[13] *H. R. Bronstein, A. S. Dworkin, M. A. Breding*, J. Phys. Chem. *66*, 44 (1962).
[14] *A. S. Dworkin, H. R. Bronstein, M. A. Breding*, J. Phys. Chem. *66*, 1201 (1962).
[15] *A. S. Dworkin, R. A. Sallach, H. R. Bronstein, M. A. Breding, J. D. Corbett*, J. Phys. Chem. *67*, 1145 (1963).
[16] *J. D. Corbett, L. F. Druding, W. J. Burkhard, C. B. Lindahl*, Discuss. Faraday Soc. *32*, 79 (1961).
[17] *M. A. Breding*, Molten Salt Chemistry, p. 408, Wiley Intersci. Publ., New York 1964.
[18] *L. F. Druding, J. D. Corbett*, J. Amer. Chem. Soc. *83*, 2462 (1961).
[19] *L. F. Druding, J. D. Corbett, B. N. Ramsey*, Inorg. Chem. *2*, 869 (1963).
[20] *R. A. Sallach, J. D. Corbett*, Inorg. Chem. *2*, 475 (1963).
[21] *L. B. Asprey, F. H. Kruse*, J. Inorg. Nucl. Chem. *13*, 32 (1960).
[22] *D. S. McClure, Z. Kiss*, J. Chem. Phys. *39*, 3251 (1963).
[23] *F. K. Fong*, J. Chem. Phys. *41*, 2291 (1964).
[24] *Z. J. Kiss, P. N. Yocom*, J. Chem. Phys, *41*, 1511 (1964).
[25] *C. Keller and H. Schmutz*, Z. Naturforsch. *19B*, 1080 (1964).
[26] *J. A. Leary*, U.S. Report LA-2661 (1962).
[27] *A. K. Baev, G. I. Novikov*, Zh. Neorg. Khim. *6*, 2610 (1961).
[28] *I. S. Morozov, V. I. Ionov, B. G. Korschunov*, Zh. Neorg. Khim. *4*, 1457 (1959).
[29] *G. I. Novikov, B. G. Polyachenok, S. A. Frid*, Zh. Neorg. Khim. *9*, 472 (1964).
[30] *I. C. Sun, I. S. Morozov*, Zh. Neorg. Khim. *3*, 1914 (1958).
[31] *Sung, Yu-Lin, G. I. Novikov*, Zh. Neorg. Khim. *8*, 700 (1963).
[32] *R. Gut, D. M. Gruen*, J. Inorg. Nucl. Chem. *21*, 259 (1961).
[33] *B. G. Korshunov, D. V. Drobot*, Zh. Neorg. Khim. *9*, 222 (1964).

of the metal halide with stream and CO_2 at relatively high temperatures. Recently, thermal and hydrolytic decomposition of metal alkoxides has also been used to synthesize high purity submicron hydrated oxides.[34, 35]

Up-dated compilations of the enthalpies and entropies of the lanthanide(III) oxides, together with the Gibbs energies derived from them, are given in the literature.[36] Decomposition pressure measurements on the TbO_x system[37] have been supplemented by similar and related studies on the PrO_x system[38] and on other lanthanide-oxygen systems.[39, 40] Extensive and systematic studies of the vaporization process in lanthanide oxide systems have been compiled[41–43] and the rates of vaporization of 17 lanthanide oxide systems[44] and the vaporization of lanthanium, neodymium and yttrium oxides at temperatures between 22° and 2700°K have been reported.[42] The thermodynamic properties of gaseous yttrium monoxide and the atomization energies for the suboxides Y_2O, Y_2O_2, La_2O, and La_2O_2 have also been determined.[45]

8.3.2 Yttrium and Lanthanide Sulfides

The rare earth *sesquisulfide* powders and single crystals are formed by sulfuration of M_2O_3 with H_2S or with CS_2[46] in graphite containers with induction heating at ~1000°–1500°

$$M_2O_3 + 3H_2S \longrightarrow M_2S_3 + 3H_2O \quad (5)$$

$$M_2O_3 + 3CS_2 \longrightarrow 2M_2S_3 + 3CO \quad (6)$$

For Eqn (5), $\Delta F = -50{\cdot}9$ kcal/mol, while $\Delta F = -137$ kcal/mol for Eqn (6). Both H_2S and CS_2 are noxious, toxic, and flammable, and extreme precaution must be taken to prevent leaks during the synthesis of R_2S_3 powder or single crystals. Synthetic graphite, boron nitride, and ceramic materials as containers at high temperatures are not suitable for synthesis of the rare earth sesquisulfides.

Another equally good method for the preparation of the lanthanide sesquisulfides is to react the highest purity metal with sulfur and iodine in a sealed quartz ampoule initially at 400° and finally to raise the temperature to ~600°–1250°. The chemical bonding in these compounds is such that the sulfur atoms are covalently bonded to one another and ionically bonded to the metal.[47, 48]

The X-ray crystallographic analysis indicates that there are only three structure types for the lanthanide sesquisulfides. The A type exists from La through Dy.[49] The D type sesquisulfide structure is found for Dy through Tm, and also for Y. The E type (corundum type) is found only for the two smallest rare earths, Yb and Lu.[49]
The structure type, reaction temperature range, and density of some of the lanthanide sesquisulfides are given in Table 2. The temperature ranges given are not the limits of stability but merely indicate the range over which these types were successfully prepared. An upper limit of 1250° could not be exceeded by the closed system method. Attempts to prepare Eu_2S_3 have been unsuccessful by the above two methods. The highest sulfur to metal ratio obtained has been Eu_3S_4 with a cubic structure.

Type A single crystals of Nd through Ho sesquisulfides with the exception of Eu and E type Yb and Lu sesquisulfides are prepared by 1:1:1 atomic mixtures of metal, sulfur and iodine in the temperature range of 600°–1275°.[48] Only Dy_2S_3 is found to be dimorphic.[48]

8.3.3 Selenides and Tellurides

The selenide and telluride compounds are prepared by reacting selenium or tellurium vapor with rare earth metal shavings at moderate temperatures. The reactions are usually carried out in evacuated, sealed Vycor tubes at ~950°.[50]

---

[34] *K. S. Mazdiyasni, C. T. Lynch, J. S. Smith*, J. Amer. Ceram. Soc. *50*, 532 (1967).
[35] *L. M. Brown, K. S. Mazdiyasni*, J. Amer. Ceram. Soc. *53*, 590 (1970).
[36] *E. F. Westrum jr.*, Advan. Chem. Ser. *71*, 25 (1967).
[37] *E. D. Guth, L. Eyring*, J. Amer. Chem. Soc. *76*, 5242 (1954).
[38] *R. E. Ferguson, E. D. Guth, L. Eyring*, J. Amer. Chem. Soc. *76*, 3890 (1954).
[39] *L. Eyring, H. S. Schuldt, K. Vorres*, U.S. At. Energy Comm. Rept TID-5914 (1959); Nucl. Sci. Abst. *14*, 15629 (1960).
[40] *L. Eyring, K. Vorres, H. S. Schuldt*, U.S. Atom. Energy Comm. Rept TID-5956 (1960); Nucl. Sci. Abst. *14*, 15997 (1960).
[41] *L. L. Ames*, Univ. Microfilms (Ann Arbor, Mich.) Order No. 65–13194; Diss. Abstracts *26*, 3643 (1966).
[42] *P. N. Walsh, H. W. Goldstein D. White*, J. Amer. Ceram. Soc. *43*, 229 (1960).
[43] *D. White*, Air Force Office of Technical Research Report 60–177, Ohio State University Research Foundation, Columbus, Ohio 1960.
[44] *P. K. Smith*, Univ. Microfilms (Ann Arbor, Mich.) Order No. 65–1587; Diss. Abstracts *25*, 5591 (1965).
[45] *D. White, P. N. Walsh, H. W. Goldstein, D. F. Dever*, J. Phys. Chem. *65*, 1404 (1961).
D. White, P. N. Walsh, L. L. Ames, H. W. Goldstein, in Thermodynamics of Nuclear Materials, p. 417. International Atomic Energy Agency, Vienna (1962).
[46] *J. R. Henderson, M. Muramoto, E. Loh*, J. Chem. Phys. 47, 9, 3347 (1967).
[47] *G. V. Samsonov, S. V. Radzikovskaya*, Russian Chem. Rev. *30*, 1, 28 (1961).
[48] *V. I. Marchenko, G. V. Samsonov*, Russian J. Inorg. Chem. *8*, 1061 (1963).
[49] *A. W. Sleight, C. T. Prewitt*, Inorg. Chem. 7, 2282 (1968).
[50] *J. F. Miller, F. J. Reid, R. C. Himes*, J. Elektrochem. Soc. *106*, 12, 1044 (1959).

Table 2. Reaction Temperature Range, Structure Type, and Density for Some of the Lanthanide Sesquisulfides

Compound	Type	Color	Temp. range (°C)	Density (g/cm³) Calc.	Found
La_2S_3	A		600–1250	4·98	4·93
Ce_2S_3	A		600–1250	5·18	5·15
Pr_2S_3	A		600–1250	5·13	5·35
Nd_2S_3	A	Black	600–1250	5·49	5·36
Sm_2S_3	A	Red	600–1250	5·84	5·87
Gd_2S_3	A	Red	600–1250	6·19	6·16
Tb_2S_3	A	Red	600–1200	6·35	
Dy_2S_3	A	Green	600–1100	6·55	6·50
Dy_2S_3	D	Yellow-orange	1250	5·92	5·75
Y_2S_3	D		900–1250	6·07	
Ho_2S_3	D		900–1200	6·21	
Er_2S_3	D		900–1100	6·34	6·09
Tm_2S_3	D		900–1000	6·14	6·08
Yb_2S_3	E	Yellow	700–1250	6·25	
Lu_2S_3	E	Black	900–1100	3·86	3·87

The granular reaction product is melted inductively in tantalum or graphite containers under argon at 1 atm pressure. The estimated melting point temperatures with an accuracy of $\pm 25°$ are 1520° for Er_2Se_3, 1420° for Gd_2Te_3 and 1525° for Y_2Te_3. The melting points for some of the other compounds may be higher, ranging up to about 2000°.[50]

The rare earth monoselenides and monotellurides form a family of refractory materials with interesting and potentially useful electrical properties. These compounds are crystalline face-centred cubic structures and have melting points in the range of 1700°–2100°.[48]

8.3.4 Nitrides

The rare earth nitrides with the exception of Sm and Yb are prepared mostly by first converting a metal ingot or sponge into the hydride and then treating the hydride with ammonia at 1000°.[51] Sm and Yb nitrides are prepared by treating metal filings directly with anhydrous ammonia at 750°–800°.[52] A second method involves the direct reaction of the rare earth with nitrogen in an arc furnace.[52] The former method yields fine powder and the latter produces dense consolidated (NaCl structure) specimens.

The rare earth nitrides, with the exception of SmN, EuN, and YbN which evaporate above 1200°, are stable *in vacuo* or inert atmosphere up to 1500° and decompose to the oxide in moist air liberating ammonia.

8.3.5 Carbides

Yttrium and the lanthanides are generally accepted as forming carbides containing $C_2^{2\ominus}$ ions in their lattices. The general reactions involved in preparing the lowest carbides (silver color) and the highest carbides (gold color) are the reduction of the oxides with carbon black and the direct reaction of metal powder with carbon black *in vacuo* or an inert atmosphere such as helium or dry argon at ~1800°–2000°.[53]

$$2M_2O_3 + 7C \longrightarrow 4MC + 3CO_2 \qquad (7)$$

$$2M_2O_3 + 11C \longrightarrow 4MC_2 + 3CO_2 \qquad (8)$$

$$2M_2O_3 + 15C \longrightarrow 4MC_3 + 3CO_2 \qquad (9)$$

The rare earth carbides may be classified as acetylides although they do not yield pure C_2H_2 on hydrolysis. The yttrium and lanthanide carbides oxidize readily in air even at 20°. In powder form they are readily decomposed by water and dilute acid and alkali solutions. Gd_3C is ferromagnetic at room temperature, while Tb_3C appears to be antiferromagnetic at liquid nitrogen temperature.[54] The yttrium and rare earth dicarbides are thermodynamically the most stable compounds. The melting points of the dicarbides are in the temperature range of 2000°–2300°.[55] The melting point of YC_2 is said to be $2415 \pm 25°$.[56] This value is somewhat higher than the previously reported value of $2300 \pm 50°$.[55] The latter difference may be entirely experimental although the coincidence between the latter temperature and that of the proposed YC_2–C eutectic at 2305° suggests that the earlier measurements might have been made on samples slightly off the stoichiometric composition. The monocarbides, however, are isostructurally hexagonal with considerably lower melting points than the dicarbides. The lanthanide sesquicarbides, M_2C_3, have been reported by a number of investigators.[54] The existence of these carbides, however, has not been established and in some instances has been questioned.

[51] *R. Didchenko, F. P. Gortsema*, J. Phys. Chem. Solid p. 24, 863 (1963).

[52] *N. Sclar*, J. Appl. Phys. *35*, 5, 1534 (1965).

[53] *R. Vickery, R. Sedlacek, A. Ruben*, J. Chem. Soc. 496 (1959).

[54] *F. H. Spedding, K. Gschneider, jr., Danne*, J. Amer. Chem. Soc. *80*, 4498 (1958).

[55] *G. V. Samsonov, T. Ya. Kosolopova, M. D. Lyutaya, G. M. Makarenko*, USAEC Report JPRS-26877, p. 1, Joint Publication Research Service, New York 1964.

[56] *P. Schwarzkopf, R. Kieffer*, Refractory Hard Metals, Macmillan, New York 1953.

8.3.6 Borides

In the laboratory the rare earth borides are almost always prepared by synthesis from the metal or by the reduction of the metal oxide with boron. Most generally, laboratory preparations of pure materials from the elements are accomplished by heating pressed mixtures of powdered metal and boron. Considerable care must be exercised in the preparation of borides in this manner because the reaction is highly exothermic. At temperatures ranging from 900° to 1200° the reaction is initiated and is followed by a very rapid rise in temperature. Frequently the materials are scattered throughout the apparatus and the container is destroyed.
Another way of preparing pure borides in the laboratory is by boron reduction of the oxides, provided that there are no ternary compounds. The reaction proceeds smoothly with the loss of $B_2O_2(g)$ or $B_2O_3(g)$. Boron carbide may be used as the reducing agent but contamination by carbon probably occurs.

Reviews of the preparation and properties of the borides have appeared.[56,57,58,59] and the commercial preparation of metallic borides has been summarized.[57] The diborides of Gd through Lu, including Y, are hexagonal in structure with melting points of approximately 2100°. However, the tri- and tetraborides are tetragonal with the melting point of yttrium tetraboride only reported to be ~2800°. The hexaborides and duodecaborides of the lanthanides are all cubic with a melting point range of 2000°–2500°.

8.4 Yttrium and Lanthanide Alkoxides

The electronegativity of yttrium and the lanthanides places these elements between metals such as aluminum, which form covalent alkoxides, and sodium, which form ionic alkoxides. The degree of ionic character of the M–O bond, which is dependent on the size and electronegativity of the metal atom, is important in determining the character of the alkoxide. This significant factor is made apparent by comparing the electronegativity of the metal atom with that of the oxygen atom which is illustrated in Table 3 by using the Pauling Scale.[60]

Table 3. Ionic Character and Alkoxide Form

Metal	Electronegativity χ[92]	$\chi_o - \chi_m$	Remark
Li	1·0	2·5	Solid alkoxide[a]
Na	0·9	2·6	Solid alkoxide
K	0·8	2·7	Solid alkoxide
Ca	1·0	2·5	Solid alkoxide
Y	1·2	2·3	Mostly solid alkoxide
La	1·1	2·4	Mostly solid alkoxide
Er	1·1	2·4	Mostly solid alkoxide
Dy	1·1	2·4	Mostly solid alkoxide
Yb	1·1	2·4	Mostly solid alkoxide
Lu	1·2	2·3	Mostly solid alkoxide
Th	1·2	2·3	Mostly solid alkoxide
Hf	1·3	2·2	Often liquid alkoxide
Zr	1·4	2·1	Often liquid alkoxide
Ti	1·5	2·0	Often liquid alkoxide
Al	1·5	2·0	Often liquid alkoxide
U	1·7	1·8	Often liquid alkoxide
Si	1·8	1·7	Often liquid alkoxide
Ge	1·8	1·7	Often liquid alkoxide
B	2·0	1·5	Liquid or gaseous alkoxide

[a] At room temperature and pressure.

Disregarding the contribution of the strongly covalent alkyl C–O bond where $\chi_o - \chi_c = 3{\cdot}5 - 2{\cdot}5$, whenever $\chi_o - \chi_m$ is $>2{\cdot}4$ more of the ionic alkoxides are found; where $\chi_o - \chi_m$ is $<2{\cdot}3$, the alkoxides are predominantly covalent in nature. Some metals, notably the rare earths, fall between these differences and only highly branched organic radicals impart sufficient covalent character to the alkoxides. It should be noted, however, that all the alkoxides would be classified as ionic according to the original Pauling concept in which the electronegativity difference of 1·7 and above means the bonds are 50% ionic. Hannay and Smith[61] have placed this difference at 2·1 which is consistent with the covalent character observed in this simple model.
The reactions within the lanthanide series to form alkoxides are all slow and follow no significant trend. The size, shape, and oxygen content of the metal used vary the surface area thus influencing both the reaction rate and % yield. When mixed rare earth (misch-metal) is used, a synergistic effect is found. The reaction rate is increased twofold or more over that for a single-phase rare earth.[62]

[57] *C. F. Powell*, High Temperature Technology, p. 131, John Wiley & Sons, New York 1956.
[58] *C. F. Powell, I. E. Campbell, B. W. Gonser*, Vapor Plating, John Wiley & Sons, New York 1955.
[59] *G. V. Samsonov, L. Y. Markovskii*, Usp. Khim. *25*, 190–273 (1956), Translation RJ-631, Associated Technical Service, Box 271, East Orange, New Jersey.
[60] *L. Pauling*, The Nature of the Chemical Bond, 3rd Ed. pp. 65–105. Cornell University Press, Ithaca, New York, 1960.
[61] *M. B. Hannay, C. P. Smith*, J. Am. Chem. Soc. *68*, 171 (1964).
[62] *L. M. Brown, K. S. Mazdiyasni*, Inorg. Chem. *9*, 2783 (1970).

The study of alkoxides of yttrium and the lanthanides has been limited by expensive starting materials and by the difficulties in preparation and handling. Successful characterization of most of the alkoxides is complicated by their extreme sensitivity to moisture, heat, light, and atmospheric conditions.

8.4.1 Syntheses

The yttrium and lanthanide *tris(isopropoxides)* are prepared by the reaction of metal turnings with excess isopropyl alcohol and a small amount of $HgCl_2$ (10^{-4} mol per mol of metal) as a catalyst.[63]

$$Ln + 3C_3H_7OH \xrightarrow[\text{reflux 24 h}]{HgCl_2} Ln(OC_3H_7)_3 + \tfrac{3}{2}H_2 \quad (10)$$

After filtration, the crude product is purified by recrystallization from hot isopropyl alcohol or vacuum sublimation. Yields of 75% or better are realized with this method. For some of the larger metal ions (lanthanum through neodymium) the reaction rate and % yield are increased by using for the catalyst a mixture of $HgCl_2$ and $Hg(C_2H_3O_2)$ or HgI_2.[62]

The following list is indicative of noted volatility changes with the change in R groups for yttrium alkoxides.

Compounds	Characteristics
$Y(i\text{-}OC_3H_7)_3$	Sublimes at 200°/0·1 mm Hg
$Y(t\text{-}OC_4H_9)_3$	Sublimes at 242°/2 mm Hg
$Y(t\text{-}OC_5H_{11})_3$	m.p. ~225°
$Y(s\text{-}OC_6H_{13})_3$	m.p. ~225°
$Y(t\text{-}OC_7H_{15})_3$	m.p. 224°/partial dec.
$Y(t\text{-}OC_8H_{17})_3$	m.p. 275° dec.
$Y(OC_6H_5)_3 \cdot 4C_6H_{12}$	m.p. ~310° dec.

Table 4 lists the sublimation temperatures and colors associated with various lanthanide metal alkoxides. Experimental observations indicate that when large amounts of $HgCl_2$ are used, the formation of excess free mercury and the amalgamation of the metal are readily observed:[63]

$$Ln(OC_3H_7)_3 + 3HgCl_2 \longrightarrow Ln(OC_3H_5)_3 + 6HCl + 3Hg \quad (11)$$

$$Ln + 4HgCl_2 + 3C_3H_7OH \longrightarrow Ln(OC_3H_5)_3 + 8HCl + 4Hg + \tfrac{1}{2}H_2 \quad (12)$$

In excess $HgCl_2$, using a reflux time of 48 hr or longer, a quantitative reaction to form an alkene is observed. In the case of isopropoxide, the product was identified by i.r. as an isopropene oxide.[63]

Table 4. Physical Properties of Yttrium and Lanthanide Tris(isopropoxides)

Compound	Color	Sublimation point (°C/mm Hg)
$Y(O\text{-}i\text{-}C_3H_7)_3$	Colorless	200/0·10
$Pr(O\text{-}i\text{-}C_3H_7)_3$	Green	175/0·04
$Nd(O\text{-}i\text{-}C_3H_7)_3$	Blue	
$Sm(O\text{-}i\text{-}C_3H_7)_3$	Lt. yellow	180/0·04
$Eu(O\text{-}i\text{-}C_3H_7)_2$	Orange	
$Gd(O\text{-}i\text{-}C_3H_7)_3$	Colorless	200/0·15
$Tb(O\text{-}i\text{-}C_3H_7)_3$	Colorless	190/0·10
$Dy(O\text{-}i\text{-}C_3H_7)_3$	Lt. yellow	190/0·17
$Ho(O\text{-}i\text{-}C_3H_7)_3$	Peach	195/0·18
$Er(O\text{-}i\text{-}C_3H_7)_3$	Pink	195/0·35
$Tm(O\text{-}i\text{-}C_3H_7)_3$	Lt. green	185/0·06
$Yb(O\text{-}i\text{-}C_3H_7)_3$	Colorless	195/0·02
$Lu(O\text{-}i\text{-}C_3H_7)_3$	Colorless	198/0·10

The alkoxides of the Group IVB transition metals have been prepared by the reaction of anhydrous chlorides with alcohols in the presence of a hydrogen chloride acceptor such as ammonia,[64]

$$MCl_4 + 4ROH + 4NH_3 \longrightarrow M(OR)_4 + 4NH_4Cl \quad (13)$$

However, the preparation of yttrium and the lanthanide tris(isopropoxides) by the ammonia method was unsuccessful.[65, 66] This reaction (13) with metal chlorides is extremely slow producing very low yields and products always contaminated with chlorine. Apparently the metal chloride undergoes a small degree of solvolysis in alcoholic solution which in turn forms a stable, highly insoluble ammoniate, $M(NH_3)_yCl_x$, in stead of the alkoxide. This event can be interpreted in terms of yttrium and the lanthanide alkoxides being stronger bases than ammonia.

[63] *K. S. Mazdiyasni, C. T. Lynch, J. S. Smith II,* Inorg. Chem. *5*, 342 (1966).

[64] *D. C. Bradley, W. Wardlaw,* J. Chem. Soc. 208 (1951).

[65] *D. C. Bradley,* Record Chem. Progress (Kresge-Hooker Sci. Lib) *21*, 179 (1960).

[66] *R. N. P. Sinha,* Sci. Cult. (Calcutta) *25* (10) 594 (1960).

The pertinent equilibria are shown in the following equations:

$$Ln(OR)_n \rightleftharpoons Ln(OR)^{\oplus}_{n-1} + RO^{\ominus} \quad (14)$$

$$Ln(OR)^{\oplus}_{n-1} + Cl^{\ominus} \rightleftharpoons Ln(OR)_{n-1}Cl \quad (15)$$

$$ROH \rightleftharpoons RO^{\ominus} + H^{\oplus} \quad (16)$$

$$NH_3 + ROH \rightleftharpoons NH_4^{\oplus} + RO^{\ominus} \quad (17)$$

$$NH_4^{\oplus} + Cl^{\ominus} \rightleftharpoons NH_4Cl \quad (18)$$

Thus the competition between NH_3 and $M(OR)_n$ to donate $RO^{\ominus}$ groups to the solution is dominated by the alkoxides leading to either a chloroalkoxide or a stable ammoniate as contaminants.

It has been found that insoluble crystalline lanthanum *tris(methoxide)* may be obtained by the slow addition of anhydrous lanthanum trichloride in methanol to a solution of lithium methoxide.[67,68]

$$LaCl_3 + 3LiOCH_3 \longrightarrow La(OCH_3)_3 + 3LiCl \quad (19)$$

After filtration, the lanthanum tris(methoxide) is purified by extraction with methanol to a colorless crystalline compound. A similar method was used to prepare yttrium tris(isopropoxide) using THF as a solvent:[63]

$$YCl_3 + 3LiOC_3H_7 \xrightarrow[C_3H_7OH]{THF} Y(OC_3H_7)_3 + LiCl \quad (20)$$

After filtration, the yttrium tris(isopropoxide) is purified by repeated recrystallization or vacuum sublimation.

The method involving the use of sodium isopropoxide to replace the chlorine of some of the lanthanide trichlorides has been successful;[69–71] however, the compounds obtained are not free from sodium and chlorine.[67]

The rare earth metal isopropoxides may be prepared from the reaction of anhydrous metal chloride alcohol solvates and a stoichiometric amount of $NaOC_3H_7$ in benzene.

$$LnCl_3 + 3i\text{-}C_3H_7OH \longrightarrow LnCl_3 \cdot 3i\text{-}C_3H_7OH \quad (21)$$

$$LnCl_3 \cdot 3i\text{-}C_3H_7OH + 3NaOi\text{-}C_3H_7 \xrightarrow[24\,h]{reflux} Ln(Oi\text{-}C_3H_7)_3 + 3i\text{-}C_3H_7OH + 3NaCl \quad (22)$$

[67] *H. Gilman, R. G. Jones, G. Karmas, G. A. Martin,* J. Amer. Chem. Soc. 78, 4285 (1956).

[68] *D. C. Bradley, M. M. Facktor,* Chem. & Ind. (London) 1332 (1958).

[69] *D. C. Bradley, M. A. Saad, W. Wardlaw,* J. Chem. Soc. (London) 1091 (1954); 3488 (1954).

[70] *R. C. Mehratra, J. M. Batwara,* Inorg. Chem. 9, 2505 (1970).

[71] *T. Moeller, D. F. Martin, L. C. Thompson, R. Ferrus, G. R. Feistel, W. J. Randall,* Chem. Rev. 65 (1965).

Repeated experiments have shown that for the separation of easily filterable granular sodium chloride and a yield of greater than 95%, it is essential that sodium isopropoxide in isopropanol and benzene solution be added to the lanthanum trichloride isopropanolate and not in the reverse direction.[70] The isopropoxide can in some cases be purified by recrystallization or sublimation.

Substitution of other R groups for the isopropoxy groups is done by the alcohol interchange technique.[63,70]

$$Ln(Oi\text{-}C_3H_7)_3 + 3ROH \rightleftharpoons Ln(OR)_3 + 3i\text{-}C_3H_7OH \quad (23)$$

where R is C_4H_9, $s\text{-}C_4H_9$, $t\text{-}C_4H_9$, $n\text{-}C_5H_{11}$, $i\text{-}C_5H_{11}$, $s\text{-}C_5H_{11}$, $t\text{-}C_5H_{11}$, etc.

The equilibrium shown in Eqn (23) is an overall reaction and is assumed to include the intermediate stages of mixed alkoxides such as $Ln(OR)_x(OC_3H_7)_{y-x}$ as well. It is best to use a small excess ($\approx 10\%$) of higher boiling ROH; otherwise, the rate of exchange, especially the last stages of the interchange, will be very slow. The isopropyl alcohol liberated during the course of the alcoholysis reaction is fractionated out azeotropically. A wide variety of solvents (benzene, toluene, carbon tetrachloride, cyclohexane, *etc.*) may be used to act as an inert diluent.

The i.r. spectra of the yttrium and lanthanide tris(*iso*-propoxides) are quite similar.[62] The spectra have proved useful in controlling the preparative results for completeness of reaction and identification of products.

8.4.2 Reactions of Lanthanide Tris(isopropoxides)

The lanthanide tris(isopropoxides) have been found to react exothermally with stoichiometric amounts of acetyl chloride and acetyl bromide in benzene with the formation of the corresponding halide derivatives.[70]

$$Ln(Oi\text{-}C_3H_7)_3 + nCH_3COX \longrightarrow Ln(Oi\text{-}C_3H_7)_{3-n}X_n \cdot mCH_3COOC_3H_7 + (n-m)CH_3COOi\text{-}C_3H_7 \quad (24)$$

where $n = 1$, 2 or 3, and $X = $ Cl or Br.

The *halide alkoxides* show increasing tendency to add molecules of isopropyl acetate as the substitution proceeds.

Several new derivatives of lanthanide tris(isopropoxides) with stoichiometric quantities of various glycols[70] have been reported and the following three types of derivatives have been isolated.

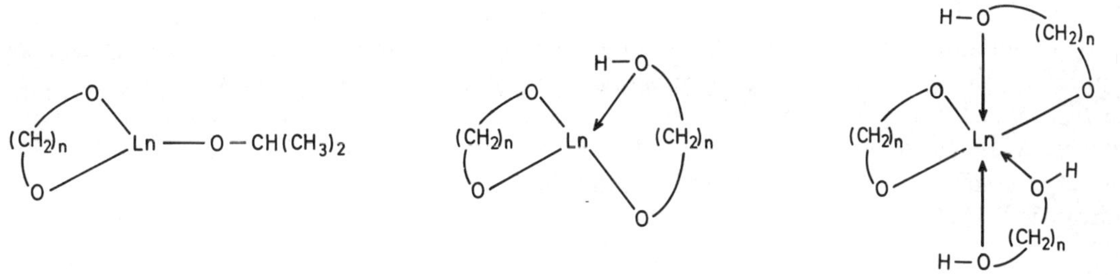

Most of these derivatives are insoluble in organic solvents. However, the pinacol and hexylene glycol derivatives probably give soluble products.

The reaction of lanthanide tris(isopropoxides) and various β-diketones and β-ketoesters have been found to yield corresponding derivatives according to the general reaction[70]

$$Ln(O\text{-}i\text{-}C_3H_7)_3 + RCOCH_2COR' \longrightarrow Ln(O\text{-}i\text{-}C_3H_7)_{3-x}(RCOCHCOR')_x + i\text{-}C_3H_7OH \quad (25)$$

where x is 1, 2, or 3 and R is CH_3, C_6H_5, OCH_3, OC_2H_5 and R' is CH_3 or C_6H_5.

Tris(isopropoxy) methylphosphonato complexes of yttrium and the lanthanides [Ln(IMP)$_3$] are prepared by the reaction of (diisopropyl)methylphosphonate (DIMP) with the trichloride at temperatures between 40°–200° with precipitation of [Ln(IMP)$_3$] and evolution of a mixture of isopropyl chloride, hydrogen chloride and propane.

Organophosphorus compounds have been extensively utilized in effecting the chemical separation of rare earth metal ions.[71] Complexes of the type $Ln[H(DBP_2)_3]_3$[72] and $Ln(DBP)_3$[73,74] have been isolated from the reaction of dibutyl phosphate (DBP) with trivalent lanthanide salts. Moreover, a number of complexes of the general type LnL_3Cl_3 (where L = triphenylphosphide oxide or hexamethylphosphoramide) have been reported.[75,76]

8.5 Carboxylates

The lanthanide carboxylates can be synthesized by the following method.[70]

$$LnX_3 + 3NaCOOR \longrightarrow Ln(COOR)_3 + 3NaX \quad (26)$$

$R^{2\ominus}C_{11}H_{23}$, $C_{15}H_{31}$, $C_{17}H_{35}$; X=Cl, NO_3.

The tris(carboxylates) are readily precipitated in aqueous solution and can be further purified by recrystallization from benzene.

It has been reported that the carboxylates of the lanthanides are stable up to 300° and slowly decompose at higher temperatures to their corresponding oxides.

$$ROOC\text{—}Ln(O\overset{\overset{\displaystyle C}{\|}}{C}\text{—}R)_3 \xrightarrow{>300^\circ} \tfrac{1}{2}Ln_2O_3 + \tfrac{3}{2}CO_2 + \tfrac{3}{2}\,R_2C{=}O \quad (27)$$

8.6 ß-Diketonates

Much research has been concerned with the adducts of acetylacetonate,[77–82] benzoyl acetonate and dibenzoylmethanate,[79,81,83,84] dipivalololy-methanate,[83,85] and fluorinated β-diketonate.[78,82,85,86]. The coordination numbers, seven, eight, and nine for their complexity are either suggested or demonstrated.

---

[72] *E. E. Kriss, Z. A. Sheka*, Dokl. Akad. Nauk. SSSR *138*, 846 (1961).

[73] *J. D. Smith*, J. Inorg. Nucl. Chem. *9*, 150 (1959).

[74] *W. H. Baldwin, C. E. Higgins*, J. Inorg. Nucl. Chem. *17*, 334 (1961).

[75] *D. R. Cousins, F. A. Hart*, J. Inorg. Nucl. Chem. *30*, 3009 (1968).

[76] *J. T. Donoghue, D. A. Peters*, J. Inorg. Nucl. Chem. *31*, 467 (1969).

[77] *G. W. Pope, J. F. Steinbach, W. F. Wagner*, J. Inorg. Nucl. Chem. *20*, 304 (1961).

[78] *L. R. Melby, N. J. Rare, E. Abramson, J. C. Caris*, J. Amer. Chem. Soc. *86*, 5117 (1964).

[79] *G. Jantsch, E. Meyer*, Ber. *53B*, 1577 (1920).

[80] *L. I. Kononenko, E. V. Melenteva, R. A. Vitkun, N. S. Poluenktov*, Ukr. Khim. Zh. *31*, 1031 (1965).

[81] *E. Butter, K. Kreher*, Z. Naturforsch. *20a*, 408 (1965).

[82] *C. Brecher, H. Samelson, Lempicki*, J. Chem. Phys. *42*, 1081 (1965).

[83] *F. Halverson, J. S. Brinen, J. R. Leto*, J. Chem. Phys. *41*, 157 (1964).

[84] *D. G. Karraker*, Inorg. Chem. *6*, 1863 (1967).

[85] *J. E. Schwarberg, D. R. Greer, R. E. Sievers, K. J. Eisentrant*, Inorg. Chem. *6*, 1933 (1967).

[86] *T. Sekine, D. Dyrssen*, J. Inorg. Nucl. Chem. *29*, 1481 (1967).

[87] *T. Moeller, H. E. Kremers*, Chem. Rev. *37*, 114 (1945).

The yttrium and lanthanide *acetylacetonates* are prepared by the addition of an aqueous solution of metal chloride, pH 5, to aqueous ammonium acetylacetonate. The pH value of the reacting mixture is maintained at a value below the pH of precipitation of the corresponding rare earth hydroxide.[87] The mixture is stirred thoroughly to insure the conversion of any basic acetylacetonate to the normal compound. The crystalline precipitate metal acetylacetonate trihydrate is filtered and air dried. The corresponding monohydrate is obtained by stirring the trihydrate in a desiccator over magnesium perchlorate. The physical properties of yttrium and some of the rare earth acetylacetonates (acac), prepared by the above method are listed in Table 5.

Yttrium and lanthanide *tris(trifluoroacetylacetonates)*, (tfacac), can be prepared by the addition of an aqueous solution of the ammonium salt of trifluoroacetylacetonate, NH_4(tfacac), to an aqueous solution of yttrium and lanthanide chlorides. As soon as the pH of the solution mixture reaches 7, the M(tfacac)$_3$·2H_2O chelate dihydrate precipitates.[88] The product is washed with water and recrystallized from benzene. The yttrium and lanthanide trifluoroacetylacetonate dihydrated chelates are highly soluble in methanol, they do not exhibit sharp melting points. Some physical properties of these chelates are listed in Table 5.

The hexafluoroacetylacetonate of the lanthanides, are prepared by shaking an aqueous solution of the metal chloride with a diethyl ether solution of ammonium hexafluoroacetylacetonate, NH_4(hfacac).[89] The chelate is extracted into the ether phase. The ether solution of the chelate is dried with sodium sulfate and then the ether is evaporated. Except for the chelates of the La, Ce, Pr, Ho, Er, and Lu which are purified by vacuum sublimation at a temperature of 180° and 0·05 mm Hg, the metal chelates of the lanthanides are recrystallized from a 9:1 solution of water-methanol. The anhydrous chelates may be obtained by drying the hydrated compounds

[88] *R. D. Young, W. C. Fernelius*, Inorg. Synth. 2 (1948).

[89] *F. Halversen, J. S. Brinen, J. R. Leto*, J. Chem. Phys. *40*, 2790 (1964).

Table 5. Physical Properties of Yttrium and Lanthanide Metal β-Diketonates

	Metal Chelate Ligand									
	(acac)$_3$		(tfacac)$_3$		(hfacac)$_3$		[H(thd)]$_3$		(fod)	
	m.p. (°)	Sub. Temp. Zone	m.p. (°)	Sub. Temp. Zone	m.p. (°)	Sub. Temp. Zone	m.p. (°)	Sub. Temp. Zone	m.p. (°)	Color
Y	141–142	n.v.	146–148	n.v.	142–144	80–40	145–155	92–40	162–167	White
La	150–152	Slightly volatile	142–144	123–93	144–146	116–82	239–245	132–92	215–230	White
Ce	148–150	n.v.	127–129	65–49	125–126	114–60	276–278	118–88	—	Yellow orange
Pr			136–138	125–97	148–151	162–50	222–224	124–90	218–223	Light Green
Nd			147–149	127–94	141–142	114–72	218–219	120–91	210–215	Lavender
Sm	143–145	n.v.	147–149	134–97	144–145	130–50	200–201	116–66	206–210	Cream white
Eu			144–146	126–92	176–177	104–60	190–191	116–65	205–212	Light yellow
Gd			145–147	118–94	170–173	94–50	183–184	138–66	203–213	White
Tb			154–156	132–100	170–172	97–60	150–152	97–61	190–196	Yellow
Dy			152–154	118–76	185–188	80–46	182–183	94–57	180–188	White
Ho			154–156	123–104	214–215	92–50	178–180	92–50	172–178	Peach
Er	126–128	n.v.	148–150	n.v.	194–198	87–40	179–180	93–53	158–164	Pink
Tm			156–158	n.v.	194–196	88–60	170–173	86–44	140–146	White
Yb			140–142	n.v.	177–178	92–74	165–167	87–48	125–132	White
Lu			188–190	n.v.	222–223	80–46	173–174	92–50	118–125	White

over P_2O_5 at 0·02 mm Hg for 72 hr or longer. The anhydrous chelates are generally darker in color than the hydrated compounds and exhibit great absorption affinity for atmospheric moisture. A summary of the physical properties is given in Table 5.

The metal *dipivialoylmethanates*, $M[H(thd)]_3$, prepared by mixing the metal chloride with a 1:1 ethanol-water solution and an alcoholic solution of the sodium salt of H(thd). The crystalline metal chelate precipitates immediately.[90] The crystalline solid is sublimed in the temperature range of 144°–200° at 0·1 mm Hg and finally recrystallized from hexane. The physical properties of $M[H(thd)]_3$ are listed in Table 5.

Yttrium and lanthanide *tris(1,1,1,2,2,3,3-heptafluoro-7,7-dimethyl-4,6-octandione) monohydrates*, $M(fod)_3 \cdot H_2O$, are prepared by dissolving the hydrated metal nitrate in absolute methanol. Sodium hydroxide is added to the solution to bring the pH to approximately 5. In a separate container, H(fod) is dissolved in absolute methanol and treated with NaOH until neutral to litmus paper. The two mixtures are added together with stirring and finally the entire solution is added dropwise to water. The resultant solid complex monohydrate precipitate is isolated by filtration and recrystallization from methylene chloride at $<0°$.[91] The anhydrous product is obtained by vacuum drying over P_2O_5. Some physical properties are shown in Table 5.

8.7 Yttrium and Lanthanide Diphosphoramides

Complexes of yttrium and rare earth ions with octamethyldiphosphoramide, (OMPA) (OMPA is extremely toxic and must be handled with care) $[(CH_3)_2N]_2PO{-}O{-}PO[N(CH_3)_2]_2$, are as stable as those of the alkaline earth ions. The yttrium and lanthanide OMPA are prepared[92] by dissolving hydrated metal chloride, $Ln(Cl)_3 \cdot xH_2O$, in methanol. A stoichiometric amount of $AgClO_4 \cdot H_2O$ is added to precipitate $Cl^{\ominus}$ as AgCl. The filtrate is treated with 2,2-dimethoxypropane and OMPA is added.

[90] *K. J. Eisentraunt, R. E. Sievers*, J. Amer. Chem. Soc. *87*, 5254 (1965).

[91] *R. E. Sievers, K. J. Eisentraut, C. S. Springer jr., D. W. Meek*, Advan. Chem. Ser. *71*, 141 (1967).

[92] *M. D. Johnsten, R. A. Jacob*, in Advances in Chemistry Series 71, *13*, American Chem. Soc. Washington, D.C. (1967).

8.8 Phosphates and Vanadates

Addition of dihydrogen phosphate to the yttrium and rare earth salts at pH of 4–5 results in a highly gelatinous precipitate of the phosphate accompanied by a marked increase in the acidity of the solution. This reaction could be represented by the following equation

$$M^{3\oplus} + H_2PO_4^{\ominus} \longrightarrow MPO_4 + 2H^{\oplus} \quad (28)$$

Since phosphoric and polyphosphoric acids contain only one strongly acidic hydrogen atom per phosphorus atom in the acid molecule, the hydrogen displacement reaction may be extended to the formation of the respective yttrium and rare earth orthophosphates, diphosphates and triphosphates[93] in accordance with the following equations.

$$2M^{3\oplus} + 3H_2PO_4^{\ominus} \longrightarrow \underset{\text{orthophosphate}}{M_2(HPO_4)_3} + 3H^{\oplus} \quad (29)$$

$$4M^{3\oplus} + 3H_2P_2O_7^{2\ominus} \longrightarrow \underset{\text{diphosphate}}{M_4(P_2O_7)_3} + 6H^{\oplus} \quad (30)$$

$$5M^{3\oplus} + 3H_2P_3O_{10}^{2\ominus} \longrightarrow \underset{\text{triphosphate}}{M_5(P_3O_{10})_3} + 6H^{\oplus} \quad (31)$$

The possible usefulness of the phosphate and polyphosphate under controlled pH are to effect separation of the rare earth elements from each other and of the rare earth elements from accompanying species. The use of the triphosphate as a complexing agent for heavy metal ions has been suggested.[94,95]

The fluorescent inorganic yttrium and lanthanide compounds derived from the orthovanadates of the general formula RVO_4 are prepared by balanced cation substitution. Two series of compounds are described[96] and the general method for the preparation of orthovanadates is as follows:

Vanadium pentoxide or ammonium metavanadate and oxides or carbonates of europium, rare earth metals, magnesium, alkaline earths or alkaline elements are thoroughly mixed and first fired in a platinum container at 300°–400° in ambient atmosphere. The powder is then remixed and refired at a final temperature of 900°–1100° for several hours.

[93] *A. G. Buyers, E. Giesbrecht, L. F. Audrieth*, J. Inorg. Nucl. Chem. *5*, 133 (1957).

[94] *E. Giesbrecht, L. F. Audrieth*, J. Inorg. Nucl. Chem. *6*, 308 (1958).

[95] *E. Giesbrecht*, J. Inorg. Nucl. Chem. *15*, 265 (1960).

[96] *A. Bril, W. L. Wanmaker, J. Broos*, J. Chem. Phys. *43*, 311 (1965).

8.9 Organometallic Derivatives

There is no significant information on the alkyl and aryl derivatives of yttrium and the lanthanides found in the literature. Triethyl-yttrium and scandium etherate, $(C_2H_5)_3MO(C_2H_5)_2$, b.p. 220°, have been reported;[97] however, later their existence was questioned.[98] Attempts to prepare lanthanide alkyls from the reaction of metal chloride with alkylmagnesium halides have been unsuccessful.[99] It seems that alkyl and aryl derivatives of yttrium and the rare earth elements are either very difficult to prepare and isolate or perhaps non-existent.

Dicyclopentadienylyttrium chloride is prepared in 10%–20% yields from the reaction of anhydrous yttrium trichloride with cyclopentadienylsodium with tetrahydrofuran as the solvent at ambient temperature.[100]

$$YCl_3 + 2C_5H_5Na \xrightarrow[20\ hr]{THF} YCl(C_5H_5)_2 + 2NaCl \qquad (32)$$

The NaCl is filtered and the THF removed by vacuum. The crude product is sublimed at 250° in a stream of purified hydrogen or argon gas. The pure product is stable in inert gas indefinitely but reacts rapidly in air to produce cyclopentadiene and yttrium chlorohydroxide.
Yttrium and lanthanide *tris(cyclopentadienyls)* can be prepared as outlined above by proper adjustments of the stoichiometry of the reactants[101]

$$MCl_3 + 3C_5H_5Na \xrightarrow{THF} M(C_5H_5)_3 + 3NaCl \qquad (33)$$

The rare earth tricyclopentadienyl compounds are all crystalline solids, thermally stable to approximately 400°, and sublime above 220° at 10^{-4} mm Hg. The compounds decompose in water giving cyclopentadiene and the metal hydroxides. They are insoluble in hydrocarbon solvents but dissolve readily in THF, glycol and dimethylether. The tricyclopentadienylyttrium and lanthanides react instantaneously and quantitatively with ferrous chloride in THF to give ferrocene.

[97] *V. M. Pletts*, C. R. Acad. Sci. USSR *20*, 27 (1938).
[98] *F. A. Carlton*, Chem. Rev. *55*, 551 (1955).
[99] *B. N. Anfanasyev, P. A. Tayganova*, J. Gen. Chem. USSR *21*, 485 (1951).
[100] *J. M. Birmingham, G. Wilkinson*, J. Amer. Chem. Soc. *76*, 6210 (1954);
J. M. Birmingham, G. Wilkinson, J. Amer. Chem. Soc. *78*, 42 (1956).
[101] *Y. Okamoto, W. Brenner, J. C. Gaswamic*, Rare Earth Research II 3rd Conference, p. 99, 1963.

8.10 Bibliography

A. N. Nesmeyanov, U. A. Priselkov, V. V. Karelin, Thermodynamics of Nuclear Materials, p. 667, International Atomic Energy Agency, Vienna 1962.
E. F. Westrum jr., Adv. Chem. Series *71*, 33 (1967).
T. R. P. Gibb jr., Progr. Inorg. Chem. *3*, 355 (1962);
T. R. P. Gibb jr., J. Inorg. Nucl. Chem. *24*, 349 (1962);
T. R. P. Gibb jr., Nonstoichiometric Compounds, American Chem. Soc., Washington, D.C., 99–110 (1963).
G. G. Libowitz, J. Nucl. Mater. *2*, 1 (1960);
G. G. Libowitz, Nonstoichiometric Compounds, Amer. Chem. Soc., 74–86 (1963);
G. G. Libowitz, The Solid-State Chemistry of Binary Metal Hydrides, *W. A. Benjamin*, New York 1965.
J. D. Corbett, L. F. Druding, W. J. Burkhard, C. B. Lindahl, Discuss. Faraday Soc. *32*, 79 (1961).
M. A. Breding in *M. Blander*, Molten Salt Chemistry, p. 408, Wiley Intersci. Publ. New York 1964.
G. V. Samsonov, T. Ya. Kosolopova, M. D. Lyutaya, G. M. Makarenko, USAEC Report JPRS-26877, p. 1, Joint Publication Research Service, New York 1964.
P. Schwarzkopf, R. Kieffer, Refractory Hard Metals, MacMillan, New York 1953.
C. F. Powell in *I. F. Campbell*, High Temperature Technology, p. 131, Wiley, Intersci. Publ., New York 1956.
G. V. Samsonov, L. Y. Markovskii, Usp. Khim. *25*, 190 (1956), Translation RJ-631 Associated Technical Service, Box 271, East Orange, New Jersey.
E. F. Westrum jr. in *L. Eyring*, Progress in the Science and Technology of the Rare Earths, Vol. II, Pergamon Press, London.
D. White, P. N. Walsh, L. L. Ames, H. W. Goldstein in Thermodynamics of Nuclear Materials, p. 417, International Atomic Energy Agency, Vienna 1962.
D. C. Bradley, Advan. Chem. Ser. *23*, 10 (1959).
L. Pauling, The Nature of the Chemical Bond, 3rd Ed. p. 65, Cornell University Press, Ithaca, New York 1960.
D. C. Bradley, Record Chem. Progress (Kresge-Hooker Sci. Lib.) *21*, 179 (1960).
R. C. Mehrotra, J. M. Batwara, S. N. Misra, J. N. Misra, U. D. Tripathi, Proceedings of the 6th Rare Earth Conference, May 3–5, p. 392, Oakridge, Tenn. 1967.
T. Moeller, D. F. Martin, L. C. Thompson, R. Ferrus, G. R. Feistel, W. J. Ranadll, Chem. Rev. 65 (1965).
R. D. Young, W. C. Fernelius, Inorg. Synth. 2 (1948).

9 Actinides

K. S. Mazdiyasni
Air Force Materials Laboratory,
Wright-Patterson Air Force Base, Ohio 45433, U.S.A.

The actinides are often referred to as "a second family of rare earth metals". Although it is true that the chemical behavior of many of the actinides is similar to that of the lanthanides, such a description is oversimplified. The first four elements in the actinide series (atomic number 89 to 92) have chemistries very much like those of the transition elements, lanthanum, hafnium, tantalum, and tungsten, respectively. Elements 93 and 95, neptunium and plutonium, are generally classified as inner transition elements, that is, elements with partially filled 5*f* subshells. However, the chemistries of neptunium and plutonium are sufficiently different from those of the presumed "congeners", prometheum and samarium, so that such a classification is not completely correct.

9.1 Hydrides

The *actinide–hydrogen system* up to MH_2 is similar to a number of metal–hydrogen systems in which solid hydride phases are formed. There is generally a significant solubility of hydrogen in the metal, a solid immiscibility region, and a range of composition for the hydride phase. The solid solubility of the hydride in the metal usually increases with temperature and the lower hydrogen concentration limit of the hydride decreases with increasing temperature.

It has been shown that the parabolic reaction rate law is applicable to the formation of some of the *actinide hydrides*[1] and that the reaction rate constant is a function of temperature and pressure. For example, thorium reacts with hydrogen in the temperature range of 500°–800° and pressure range of 1–24 mm Hg to form the compounds ThH_2 and Th_4H_{15}, the metal acquiring a coherent ThH_2 surface layer. It appears that the composition of *thorium hydride* approaches $ThH_{2.00}$ at lower temperature and higher pressure.[2,3] The solubility of ThH_2 in thorium has been reported[4] to increase rapidly with temperature from 1 atom % at 300° to approximately 26 atm % at 800°. The hydrogen content of the hydride phase in equilibrium with thorium decreases from $ThH_{1.96}$ at 800°. Therefore, it is reasonable to assume that the thorium–thorium hydride immiscibility gap will close at about 1200°, if not interrupted by a lowering of the phase transition in thorium. However, at room temperature the equilibrium phases are thorium metal with only about 2×10^{-3} atm % hydrogen in solution and ThH_2 with a composition very close to $ThH_{2.00}$.
Another actinide hydride is UH_3 which has an electrical conductivity of the same order of magnitude as that of uranium metal[5] and is ferromagnetic below about 173°K.[6,7] The *uranium trihydride* is prepared by reacting high-purity uranium metal with hydrogen at 200° and cooling the reaction slowly to room temperature in the presence of hydrogen at 1 atm. pressure.[8] Although the above method of preparation of uranium trihydride is known to produce only β-phase material, a crystalline cubic phase with $a_0 = 6.644$ Å, α-UH_3 also forms in varying fractional amounts when the reaction is carried out below 200°.

9.2 Halides and Derivatives

The *actinide halides* show similar chemical behavior to the lanthanide halides. However, experimental difficulties with the trivalent actinides up to plutonium are considerable because of the ready oxidation of this state. Some correlations exist with the actinides in studies of the lanthanide tetrafluorides and fluoro complexes. For other compounds of tetravalent actinides, protactinium shows almost as many similarities as differences between thorium and the uranium–americium set. Thus, investigating the complex-forming properties of the actinide halides is most important. In the penta- and hexavalent states, the elements from uranium to americium show considerable similarities. Protactinium(V) behaves in much the same way as these elements in the pentavalent state, except in water where its hydrolytic behavior is more like that of niobium and tantalum.

The *halo complexes* of the trivalent actinides and lanthanides as mentioned above show considerable similarities. The *tetrafluoroactinides(II)* have been reported,[9,10] all having hexagonal symmetry. Some *hexahaloactinides* have also been prepared. The *triphenylphosphonium hexachloroamericium(III) salt* has been isolated from

---

[1] *D. T. Peterson, D. G. Westlake*, J. Phys. Chem. *63*, 1514 (1959).

[2] *M. W. Mallett, I. E. Campbell*, J. Amer. Chem. Soc. *73*, 4850 (1951).

[3] *D. T. Peterson, J. Rexer*, J. Less Common Metals *4*, 92 (1967).

[4] *D. T. Peterson, D. G. Westlake*, Trans. Met. Soc. AIME *215*, 444 (1959).

[5] *F. H. Spedding, A. S. Newton, J. C. Warf, O. Johnson, R. W. Nottorf, I. B. Johns, A. H. Daane*, Nucleonics *4*, 4 (1949).

[6] *W. Jrzebiatowski, A. Sliwa, B. Stalinski*, Roczniki Chem. *26*, 110 (1952); b *28*, 12 (1954);

[7] *M. K. Wilkinson, C. G. Shull, R. E. Rundle*, Phys. Rev. *99*, 627 (1955).

[8] *H. E. Flatow, D. W. Osborne*, Phys. Rev. *164*, 755 (1967).

[9] *C. Keller, H. Schmutz*, Z. Naturforsch. *19B*, 1080 (1964).

[10] *J. A. Leary*, U.S. Report LA-2661 (1962).

nonaqueous solvents[11]; however, the aqueous alkali chloride–$AmCl_3$ system is more complicated, the species $CsAmCl_4 \cdot 4H_2O$ and $Cs_2NaAmCl_6$ being isolated from aqueous solution and Cs_3AmCl_6 from ethanolic hydrochloric acid solutions of the components. This last species may, however, be $Cs_8Am_3Cl_{17}$. The plutonium compound, $Cs_3PuCl_6 \cdot 2H_2O$, has been isolated from aqueous hydrochloric acid.[12] Many more *actinide halo complex species* have been reported for fused salt melts.

9.2.1 Tetravalent Halides

In the tetravalent state, the actinide elements from protactinium to curium[13] form a variety of fluoro complexes. Also, a layer structure has been found for ThI_4.[14] *Protactinium tetrachloride* is prepared[15,16] either by reduction of the pentachloride with hydrogen at 800°[15] or, less satisfactorily, by direct chlorination of the dioxide, PaO_2, with carbon tetrachloride at 500°. It is said that the hydrogenation is more easily carried out at 400° using stoichiometric quantities of the reagents for the reaction,

$$2PaCl_5 + H_2 \longrightarrow 2PaCl_4 + 2HCl \quad (1)$$

in a sealed Pyrex vessel.[17] The tetrabromide is prepared similarly in virtually quantitative yield, but the analogous reaction with the pentaiodide is less satisfactory; only 50% conversion is observed at about 520°. This can probably be attributed to the known instability of hydrogen iodide ($\Delta H = 31$ kcal) and the fact that with the reaction carried out in a sealed tube, the hydrogen iodide is not removed from the reaction zone as it is generated.

Although such reactions can be carried out safely using about 50 mg of pentahalide in a 10 ml vessel, the combination of radiation hazards associated with protactinium-231 and the dangers inherent in sealing glass vessels containing hydrogen renders the method unsafe with larger amounts. In addition, the use of a continuous flow of hydrogen over the pentahalide is not practical owing to the fairly high volatility of the pentahalide. Vigorous drying of the reaction gas is also necessary owing to the rapid hydrolysis of $PaCl_5$.

On a much larger scale, the *protactinium(IV) halides* are prepared by reduction of the pentahalide with aluminum metal at 400°–500°.[17]

$$3PaX_5 + Al \longrightarrow 3PaX_4 + AlX_3 \quad (2)$$

Crystals of yellow-green *tetrachloride*, bright red *tetrabromide*, or black *tetraiodide* are deposited in a region of the vessel at 350° and the aluminum halide in a cooler region. Protactinium halides are sublimed slowly above 500° *in vacuo* (10^{-4} mm). At this temperature the tetraiodide invariably reacts with borosilicate glass leaving a pink residue of the nonvolatile oxodiiodide.

Protactinium tetrahalides, like their actinide analogs, are moisture-sensitive and hydrolysis, followed by oxidation, occurs fairly rapidly in the atmosphere. Apart from the tetrafluoride, the tetrahalides all dissolve readily in aqueous mineral acids (excepting hydrofluoric acid) to form solutions which are quite stable in the absence of dissolved oxygen.

Protactinium halides, like their thorium and uranium analogs,[18,19] react with anhydrous, oxygen-free acetonitrile to form[20] the sparingly soluble complex, $MX_4 \cdot CH_3CN$ (X=Cl, Br, or I).

Protoactinium oxodihalide is prepared by the reaction of the tetrahalide and dioxide at 600°; however, the reaction does not go to completion even in the presence of excess tetrahalide. An alternative method is to react the tetrahalide at 150°–200° with antimony trioxide.[17]

$$3PaX_4 + Sb_2O_3 \longrightarrow 3PaOX_2 + 2SbX_3 \quad (3)$$
$$X = Cl, Br$$

Protactinium tetrafluoride is prepared by the reaction of Pa_2O_5 with hydrogen and hydrogen fluoride in a nickel reactor heated to 350°–500°.[21] The product at 500° is red-brown PaF_4, which sinters. At 400° the product is powdery PaF_4 with a typical X-ray pattern, isostructural with UF_4. Protactinium tetrafluoride is an excellent oxygen scavenger and readily oxidizes to Pa_2OF_8. The alkali fluoride complexes of tetra-

[11] *J. L. Ryan*, Advan. Chem. Ser. *71*, 331 (1967).
[12] *R. E. Stevens*, J. Inorg. Nucl. Chem. *27*, 1873 (1965).
[13] *R. A. Penneman, J. K. Keenan, L. B. Asprey*, Advan. Chem. Ser. *71*, 248 (1967).
[14] *A. Zalkin, J. D. Forester, D. H. Templeton*, Inorg. Chem. *3*, 639 (1964).
[15] *P. A. Sellers, S. Fried, R. E. Elson, W. H. Zachariasen*, J. Amer. Chem. Soc. *76*, 5935 (1954).
[16] *R. E. Elson, S. Fried, P. A. Sellers, W. H. Zachariasen*, J. Amer. Chem. Soc. *72*, 5791 (1950).
[17] *D. Brown, P. J. Jones*, J. Chem. Soc. (A) 719 (1967).
[18] *K. W. Bagnall, D. Brown, P. J. Jones, J. G. H. du-Preez*, J. Chem. Soc. 350 (1965).
[19] *K. W. Bagnall, D. Brown, P. J. Jones*, J. Chem. Soc. (A) 1763 (1966).
[20] *D. Brown, P. J. Jones*, Chem. Commun. 279 (1966).
[21] *L. B. Asprey, F. H. Kruse, R. A. Penneman*, Inorg. Chem. *6*, 544 (1967).

valent protactinium are formed when alkali (Li, Na, K, Rb) fluorides are reacted with PaF_4 at 400°–450° in the presence of hydrogen.[21] The compound $LiFPaF_4$ is tetragonal and $7MF \cdot 6PaF_4$ (M = Na, K and Rb) is rhombohedral in crystal structure.

Hexachloro complexes of all the actinides(IV) from thorium to plutonium are now known and only $[N(C_2H_5)_4]ThCl_6$ is reported as dimorphic. All the analogous hexabromo complexes are also known since the neptunium(IV) and plutonium(IV) compounds have been isolated from ethanolic hydrobromic acid.[22]

Plutonium tetrachloride is unknown; however, it complexes with oxygen donor ligands, such as amides, can be prepared by treating Cs_2PuCl_6 with a solution of the ligand in a nonaqueous solvent.[23] Some of the *complexes* formed by actinide tetrachlorides *with amides* are as follows: $MCl_4 \cdot 6H_3C–CO–NH_2$ and $2MCl_4 \cdot 5H_3C–CO–N(CH_3)_2$, (M = U, Np, Pu),[23] $MCl_4 \cdot 4H_3C–CO–N(CH_3)_2$, and $MCl_4 \cdot 4H_3C–CO–N(CH_3)_2$, (M = U or Th).[24]

9.2.2 Mixed Complexes

The *Th, U,* and *Np tetranitrate complexes with* N,N-*dimethylacetamide* are probably nitrate-bridge dimers analogous to the UCl_4 complex, but the thorium and uranium *tetrathiocyanate* complexes with *N,N*-dimethylacetamide are 1:4 monomers and are eight-coordinated like the *octaisothiocyanate* complex.[25]

Hydrated $NpOF_3$ has been prepared by the action of hydrogen fluoride on Np_2O_5,[23,26] but NpF_5 itself has not been prepared. Salts of the fluoro complex ions, $MF_6^{\ominus}$, $MF_7^{2\ominus}$, and $MF_8^{3\ominus}$, for pentavalent actinides (Pa, U, Np, and Pu) have been reported.[27–41]

9.2.3 Uranium Compounds

Uranium tetrachloride is prepared by the vapor-phase reaction of strong halogenating reagents, such as carbon tetrachloride, phosphorus pentachloride, thionyl chloride, phosgene, or sulfur monochloride, with uranium dioxide.

$$UO_2 + CCl_4 \xrightarrow{475°} UCl_4 + CO_2 \quad (4)$$

Another method involves the reaction of carbon tetrachloride with uranium trioxide at 250° under pressure. Yet another method, but more convenient than the above two methods, is the reaction of hexachloropropene, $Cl_2C{=}CCl–CCl_3$ (b.p. 210°), with uranium trioxide under reflux and atmospheric pressure. Uranium tetrachloride is a dark green solid (m.p. 590°, b.p. 792°).

Complexes of uranyl chloride with a variety of oxygen donor ligands are reported, namely, with phosphine oxides,[42] pyridine *N*-oxide,[43] *N,N*-dimethyl formamide,[44] acetamide,[45] and with *N,N,N′,N′*-tetramethyldicarboxylic acid amides, the last being mainly polymeric compounds. Similar *complexes of uranyl bromide*[45] and *iodide*[44] are also reported. *Uranium tetrafluoride* is a green, nonvolatile, crystalline solid which can

---

[22] *J. L. Ryan, C. K. Jorgensen,* Mol. Phys. *7,* 17 (1963).
[23] *K. W. Bagnall, A. M. Deane, T. L. Markin, P. S. Robinson, M. A. A. Stewart,* J. Chem. Soc. (London) 1611 (1961).
[24] *K. W. Bagnall, D. Brown, P. J. Jones, P. S. Robinson,* J. Chem. Soc. 2531 (1964).
[25] *V. P. Markov, E. N. Traggeim,* Zh. Neorg. Khim. *6,* 2316 (1961).
[26] *K. W. Bagnall, D. Brown, J. G. H. duPreez,* J. Chem. Soc. (London) 5217 (1965).
[27] *L. B. Asprey, F. H. Kruse, R. A. Penneman,* J. Amer. Chem. Soc. *87,* 3518 (1965).
[28] *L. B. Asprey, R. A. Penneman,* Science, *145,* 924 (1964).
[29] *D. Brown, J. F. Easey,* Nature *205,* 589 (1965).
[30] *D. Brown, J. F. Easey,* J. Chem. Soc. 254 (1966).
[31] *M. N. Bukhsh, J. Flegenheimer, F. M. Hall, A. G. Maddock, C. Miranda, J. de Ferreira,* Inorg. Nucl. Chem. *28,* 421 (1966).
[32] *O. L. Keller, A. Chetham Strode,* Proc. Colloque Phys-Chim Protactinium 119 (1965).
[33] *L. B. Asprey, R. A. Penneman,* Inorg. Chem. *3,* 727 (1964).
[34] *J. R. Geichman, E. A. Smith, P. R. Ogle,* Inorg. Chem. *2,* 1012 (1963).
[35] *J. R. Geichman, E. A. Smith, S. S. Trond, P. R. Ogle,* Inorg. Chem. *1,* 661 (1962).
[36] *R. A. Penneman, L. B. Asprey, G. D. Sturgeon,* J. Chem. Soc. (London) *84,* 4608 (1962).
[37] *R. A. Penneman, F. H. Kruse, R. S. George, J. S. Coleman,* Inorg. Chem. *3,* 309 (1964).
[38] *R. A. Penneman, G. D. Sturgeon, L. B. Asprey,* Inorg. Chem. *3,* 126 (1964).
[39] *W. Rüdorff, H. Leutner,* Justus Liebigs Ann. Chem. *632,* 1 (1960).
[40] *G. D. Sturgeon, R. A. Penneman, F. H. Kruse, L. B. Asprey,* Inorg. Chem. *4,* 748 (1965).
[41] *L. B. Asprey, T. K. Keenan, R. A. Penneman, G. D. Sturgeon,* Inorg. Nucl. Chem. Lett. *2,* 19 (1966).
[42] *P. Gans, B. C. Smith,* J. Chem. Soc. (London) 4172 (1964).
[43] *P. V. Balakrishnan, S. K. Patil, H. V. Venkatasetty,* J. Inorg. Nucl. Chem. *28,* 537 (1966).
[44] *M. Lamisse, R. Heimburger, R. Rohmer,* Compt. Rend. *258,* 2078 (1964).
[45] *K. W. Bagnall, D. Brown, P. J. Jones,* J. Chem. Soc. 741 (1966).

be prepared by high-temperature hydrofluorination of uranium dioxide.

$$UO_2 + 4HF \xrightarrow{550^\circ} UF_4 + 2H_2O \quad (5)$$

Uranium tetrafluoride is an important chemical in nuclear technology serving as the raw material for the preparation of uranium hexafluoride. Other fluorinating agents can be used in its preparation. The reaction of Freon 114 ($CClF_2$-$CClF_2$) with uranium trioxide at 600°–700° results in good conversion to uranium tetrafluoride. Another method for the large-scale production of uranium tetrafluoride starts with uranyl peroxide, $UO_4 \cdot 2H_2O$, which is converted to uranium trioxide by thermal decomposition at 250°. This oxide is then reduced with hydrogen to UO_2 at 600° and fluorinated with ammonium hydrogen fluoride, NH_4HF_2. The reaction mixture is then heated at 150° and the product decomposed at 350°–500° to the final product, uranium tetrafluoride.

Uranium hexafluoride is commonly prepared by the direct fluorination of uranium tetrafluoride at 250°.

$$UF_4 + F_2 \longrightarrow UF_6 \quad (6)$$

Without the use of elemental fluorine, the hexafluoride is prepared by the following method.

$$2UF_4 + O_2 \xrightarrow{800^\circ} UF_6 + UO_2F_2 \quad (7)$$

In the above method the yield is low and recycling of uranyl fluoride is required.

Uranium hexafluoride is moisture-sensitive and very reactive. It should be handled with care and stored in copper, nickel, or aluminum containers. A mild reduction of uranium hexafluoride with hydrogen bromide results in *α-uranium pentafluoride*.[46]

$$2UF_6 + 2HBr \xrightarrow{65^\circ} 2UF_5 + 2HF + Br_2 \quad (8)$$

It has been reported that fluorine oxidation of UF_4 in liquid anhydrous HF proceeds readily to UF_5 and then goes only slowly to UF_6.[47]

Uranium tetrabromide is prepared by passing a nitrogen stream containing bromine over electrolytic uranium metal at 650°. The product is further purified by sublimation *in vacuo* at 600°.[48]

9.2.4 Thorium Compounds

Thorium tri- and tetrafluorides are light blue and grey in color, respectively, and are prepared by the reaction of thorium hydride with HF and hydrogen at 350°–500°.[49]

$$2ThH_2 + 7HF \xrightarrow{H_2} ThF_3 + ThF_4 \quad (9)$$

Thorium tetrachloride reacts with methyl, ethyl, or isopropyl alcohol to give metal alcoholates with the general formula $ThCl_4 \cdot 4ROH$, in which thorium shows a covalency of 8. The alcoholate is a monomeric substance in solution. With regard to the reactions of tertiary alcohols or other alcohols containing strongly electron-releasing alkyl groups, it is suggested[55] that secondary reactions take place involving the alcohol and hydrogen chloride liberated in the first step,

$$R^tOH + ThCl_4 \longrightarrow (R^tO)ThCl_3 + HCl \quad (10)$$

resulting in the formation of alkyl chloride and water.

9.3 Oxides

The *actinide oxides* are prepared by the reaction of the pure metal with oxygen at elevated temperatures. The most stable actinide oxides in air at room temperature have fluorite-type structures.

Thorium dioxide is one of the most thermally stable oxides known, but it forms a slightly oxygen-deficient congruently vaporizing solid $ThO_{1.998}$ at a temperature of about 2500°. ThO_2 is a fluorite-type dioxide with $a = 5{\cdot}999$ Å.

Protactinium dioxide is known to have a fluorite-type structure. No evidence exists for a PaO_{2-x} phase; however, there should be no reason that oxygen-deficient protactinium oxide could not be readily made. Very limited studies of nonfluorite-type higher oxides (i.e., Pa_2O_5) have been reported.

Among the actinide *oxides*, those of *uranium* are the most thoroughly examined revealing it to be one of the most complex binary systems known. The phase diagram in the $UO_{2\pm x}$ region has been reported[50,51] and neutron diffraction studies[52] have been described.

In general, gain or loss of oxygen in the uranium–oxygen system is compensated for by a change

[46] *A. S. Wolf, W. E. Hobbs, K. E. Rapp*, Inorg. Chem. *4*, 756 (1965).

[47] *R. A. Penneman, T. K. Keenan, L. B. Asprey*, Advan. Chem. Ser. *71*, 248 (1967).

[48] *J. R. Clifton, D. M. Gruen, A. Ron*, J. Chem. Phys. *51*, 224 (1969).

[49] *P. Ehrlich, G. Kaupa*, Z. Anorg. Allgem. Chemie *333*, 214 (1964).

[50] *L. E. J. Roberts*, Advan. Chem. Ser. *39*, 66 (1963).

[51] *A. E. Martin, R. K. Edwards*, J. Phys. Chem. *69*, 1788 (1965).

[52] *B. T. M. Willis*, I.A.E.A. Symp. Vienna 1964.

in the oxidation state of the uranium atom. In the uranium–oxygen system, it is possible to vary oxygen concentration as much as 15% as a function of temperature without marked effect in the crystal structure of the oxide. Table 1 summarizes in a condensed but simplified form some aspects of the phase relationships in the uranium–oxygen system. However, much emphasis is placed on the uranium oxides of "composition" UO_2, U_3O_8, UO_3, and $UO_4 \cdot 2H_2O$, which are discussed in some detail below. Table 2 shows some important physical properties of the oxides.

Uranium dioxide is prepared by the reduction of UO_3 or U_3O_8.

$$UO_3 + H_2 \xrightarrow{700^\circ} UO_2 + H_2O \quad (11)$$

$$UO_3 + CO \xrightarrow{350^\circ} UO_2 + CO_2 \quad (12)$$

$$U_3O_8 \underset{+O_2,\ 700^\circ}{\overset{+CO,\ 750^\circ}{\rightleftharpoons}} 3UO_2 + 2CO_2 \quad (13)$$

The color of uranium dioxide varies from brown (less reactive) to black (pyrophoric). Very finely divided uranium dioxide prepared, for example, by the thermal decomposition of uranyl oxalate is frequently pyrophoric. Ignition at 700° in air forms U_3O_8. High-purity, submicron uranium oxide may also be formed by hydrolytic decomposition of uranium alkoxides, $U(OR)_4$, to $U(OH)_4$ which is dehydrated and calcined at 600°–700° to the oxide powder.

Triuranium octoxide is that oxide of uranium which results from the ignition of any uranium oxide and, indeed, of almost all uranium compounds in air. At 700°, the oxide obtained has a composition very close to $UO_{2.667}$ or U_3O_8. The oxide can be olive-green to black-green to black in color. The higher the temperature of ignition the blacker the color, although even the blackest U_3O_8 samples show a green streak. Since U_3O_8 is the stable uranium oxide on ignition in air, it occupies an important place in uranium technology with many uranium compounds being prepared from U_3O_8 and ultimately ending up as U_3O_8 again.

Uranium trioxide can be obtained by the thermal decomposition of uranyl nitrate hexahydrate or ammonium diuranate. These two compounds are the usual final products from the ore extraction and purification procedure.

$$UO_2(NO_3)_2 \cdot 6H_2O \xrightarrow{350^\circ} UO_3 + N_2O_4 + 6H_2O + \tfrac{1}{2}O_2 \quad (14)$$

Table 1 Phase Relations in the Uranium-Oxygen System

at 200°C		at 1000°C	
Composition	Crystal type	Composition	Crystal type
$UO_{2.00}$–$UO_{2.25}$	Fluorite	$UO_{2.00}$–$UO_{2.25}$	Fluorite+β-UO_2–($UO_{2.25}$)
$UO_{2.25}$–$UO_{2.3}$	Fluorite+(β-UO_2) changes to tetragonal	$UO_{2.25}$–$UO_{2.50}$	βUO_2+orthorhombic
$UO_{2.3}$–$UO_{2.6}$	Fluorite+tetragonal	$UO_{2.50}$–$UO_{2.62}$	Orthorhombic
$UO_{2.60}$–UO_3	Orthorhombic (U_3O_8 structure)	$UO_{2.62}$–UO_3	Thermally unstable

Table 2 Physical Properties of Uranium Oxides

	UO_2	U_3O_8	UO_3	$UO_4 . 2H_2O$
M.p., °C	3000 ± 200	1450 (dec.)	450 (dec.)	100 (dec.)
Thermochemical data				
ΔH°, kg–cal./mole	259·2	853·5	291·6	436
ΔF°, kg–cal./mole	246·6	804	273·1	—
S°, cal./(mole)(°C)	18·63	66	23·6	—
Crystallographic data Cryst. form	Face-centered cubic	Orthorhombic	Hexagonal	Orthorhombic
Lattice Parameters, A.				
a_1	5·468	6·721	3·971	8·74
a_2	—	3·988	—	6·50
a_3	—	4·149	4·168	4·21
No. of molecules per cell	4	2/3	1	2
X-ray density, g/cu. cm	10·97	8·39	8·34	5·15

$$(NH_4)_2U_2O_7 \xrightarrow{350^\circ} 2UO_3 + H_2O + 2NH_3 \quad (15)$$

Since uranium trioxide may undergo thermal decomposition at temperatures above 450°, it is usually desirable to carry out its preparation at temperatures below 400° in order to avoid formation of U_3O_8. It requires very careful control of the temperature to prepare a trioxide which, on the one hand, does not contain nitrogen because of incomplete ignition and, on the other hand, is free of U_3O_8.

Uranium trioxide ranges in color from light yellow to orange to brick red. It exists in an amorphous form and at least four crystalline modifications. The amorphous form may be prepared by the ignition of uranyl peroxide, $UO_4 \cdot 2H_2O$, or uranyl oxalate at 380°–400°. Different phases of uranium oxides are prepared by the oxidation of U_3O_8 with oxygen at 450°–750° and oxygen pressures of 30–150 atm. The most thermally stable U_3O_8 also results from the decomposition of uranyl nitrate at temperatures below 400°. Further oxidation of U_3O_8 with dinitrogen tetroxide or atomic oxygen at 250°–350° results in a brick-red colored oxide. Amorphous uranium trioxide begins to decompose at 400° in air, between 520° and 600° an oxide of composition $UO_{2.90}$ forms, and further heating at 600° results in the formation of U_3O_8.

Uranium Peroxide. Although anhydrous UO_4 does not exist, the *dihydrate*, $UO_4 \cdot 2H_2O$, is readily prepared by the addition of hydrogen peroxide to aqueous solutions of uranyl salts. This hydrate is one of the least soluble compounds of uranium known and can be precipitated from an acid solution. When it is dehydrated, it loses water and oxygen simultaneously until the lower oxides UO_3 or U_3O_8, depending on temperature, result. Finely divided and reactive uranium trioxide can be made by the careful ignition of $UO_4 \cdot 2H_2O$.

Neptunium Compounds. The fluorite-type NpO_2 is the stable oxide formed in air when neptunium oxo salts are decomposed. There is some evidence that A-type Np_2O_3 is formed from a nonstoichiometric NpO_{2-x} phase.[53] Irreversible decomposition of Np_3O_8 has been investigated,[54] but no phases intermediate to Np_3O_8 and NpO_2 have been observed.

The most obvious feature is that the actinide dioxides are more easily reduced from ThO_2 to CmO_2 showing an approach to the behavior of the lanthanides.

[53] *R. J. Ackermann, R. L. Faircloth, E. G. Rauk, R. J. Thorn*, Inorg. Nucl. Chem. *28*, 111 (1966).

[54] *L. E. J. Roberts, A. J. Walter*, Rept. German Chem. Soc. 3624 (1963).

Actinide Alkoxides. The *actinide alkoxides* may be prepared by any one of the four following reaction schemes.[55, 56, 57]

$$MCl_4 + NaOR \longrightarrow M(OR)_4 \xrightarrow{O_2} M(OR)_5 \quad (16)$$

$$MCl_4 + ROH + NH_3 \longrightarrow M(OR)_4 \xrightarrow{O_2} M(OR)_5 \quad (17)$$

$$MCl_4 + NaOR \longrightarrow M(OR)_4 \xrightarrow{Br_2} M(OR)_4Br \xrightarrow{NaOR} M(OR)_5 \quad (18)$$

$$MCl_5 + ROH + NH_3 \longrightarrow M(OR)_5 \quad (19)$$

M = Th or U

A fifth and most generally satisfactory method involves an exchange of alkoxide groups; *e.g.*, the readily available uranium(V) ethoxide is mixed with another alcohol, and after heating for a short time, the alcohol is removed by distillation *in vacuo*.

$$U(OC_2H_5)_5 + 5ROH \longrightarrow U(OR)_5 + 5C_2H_5OH \quad (20)$$

The *uranium(V) 2,2,2-trifluoroethoxide* has also been prepared by reaction (17). The product is a green-brown crystalline solid (b.p. 130° 0·008 mm Hg). Also, uranium alkoxides can be prepared by the reaction of uranium(IV) diethylamide with alcohol[58] [Eqn (21)].

$$U[N(C_2H_5)_2]_4 + 4ROH \longrightarrow U(OR)_4 + 4(C_2H_5)_2NH \quad (21)$$

9.4 Sulfides and Sulfates

Uranium sulfide is prepared by the reaction of uranium metal or uranium hydride with sulfur or H_2S at 400°–600°.[59] Since the β-US_2 formed is pyrophoric, it must be annealed at 900°. Direct degradation of β-US_2 to US is not a feasible preparative route; however, β-US_2 may be reduced to US by heating with an equivalent amount of UH_3.

[55] *D. C. Bradley, M. A. Saad, W. Wardlaw*, J. Chem. Soc. (London) 2002 (1954).

[56] *D. C. Bradley*, Record Chem. Prog. (Kresge Hooker Sci. Lib.) *21*, 179 (1960).

[57] *R. G. Jones, E. Bindschadler, G. Karmas, G. A. Martin jr., J. R. Thirtle, F. A. Yeoman, H. Gilman*, Amer. Chem. Soc. *78*, 4289 (1956).

[58] *R. G. Jones, G. Karmas, G. A. Martin, H. Gilman*, J. Amer. Chem. Soc. *78*, 4285 (1956).

[59] *M. Allbut, R. M. Dell*, J. Nucl. Mater. North-Holland Publishing, Amsterdam *24*, 1 (1967).

Weighed amounts of UH_3 and US_2 are intimately mixed in the atmosphere of an argon dry box, pelleted, and heated at 1400° under vacuum. This results in a sintered mixture of US, U_2S_3, U_3S_5, and free uranium which can be handled in air without ignition. The mixture is then homogenized to pure US by heating at 1800° for 1 hr *in vacuo* in a tantalum container. Similar procedures may also be used for the preparation of ThS with only slight modifications, but cannot be used for the preparation of PuS since the reaction of PuH_{3-x} and H_2S is apparently too sluggish even at 600°.

Plutonium sulfide is prepared by the reaction of plutonium metal and sulfur in an evacuated, sealed silica tube heated at 400°–500° for 2 to 3 days. This is followed by homogenization with the addition of PuH_2 in stoichiometric amounts to form PuS at 1600° to 1700° *in vacuo.*[59]

Uranium sulfate, $U(SO_4)_2$, forms a tetra- and octahydrate on precipitation from water. The *octahydrate* is precipitated from weakly acid solution between 0° and 75°, whereas the *tetrahydrate* is formed from strongly acid solution in this temperature range. The stability ranges of the various uranium(IV) sulfate hydrates have been given as:

$$U(SO_4)_2\cdot 8H_2O \xrightarrow{68^\circ} U(SO_4)_2\cdot 4H_2O \xrightarrow{100^\circ} U(SO_4)_2\cdot 2H_2O \xrightarrow{200^\circ} U(SO_4)_2\cdot \tfrac{1}{2}H_2O \xrightarrow{800^\circ} U(SO_4)_2 \qquad (22)$$

Uranium(IV) sulfate is more soluble in cold water than hot water and undergoes hydrolysis to form the sparingly soluble basic sulfate, $UOSO_4\cdot 2H_2O$. Uranyl sulfate exists usually as the trihydrate, $UO_2SO_4\cdot 3H_2O$.

9.5 Nitrides and Group V Compounds

The preparation of *uranium nitride*, UN, is performed in the following ways:

(a) By passing nitrogen or ammonia over uranium metal.[60,61]

(b) By vacuum decomposition or hydrogen reduction of a higher nitride.[60,62,63]

(c) By the reaction of a higher nitride with uranium under high nitrogen pressure.[60]

(d) By nitriding a stoichiometric mixture of UO_2 and carbon.[64]

(e) By the reaction sintering of stoichiometric mixtures of U_2N_3 and uranium (hot pressing at 1000° followed by annealing at 1100°).[65]

(f) By the reaction of UF_4 with ammonia at 890°–920° and then vacuum reduction (10^{-4} torr) at 1000°.

It has been reported that the reaction with ammonia yields the higher nitrides in a finer form than the reaction with nitrogen.[62] In the preparation of uranium nitride, a two-stage process is used. The higher nitride is prepared in a stainless steel reaction vessel and then loaded, under helium atmosphere, into a graphite container for decomposition under vacuum at 1650°–1700° to the mononitride. However, the use of ammonia has the advantage that the high reactivity of the monatomic form of nitrogen produced by the dissociation of ammonia accounts for a slightly lower temperature of formation than that necessary when N_2 gas is used. Very fine particulate uranium nitride can be prepared by first hydriding uranium chips and subsequently dehydriding under an inert atmosphere. The reaction rate of hydrogen with uranium in the form of wire reaches a maximum at 225°,[66] at higher temperatures the rate decreases. The hydride formed at 225° is the β phase.

Sesquinitride can be formed by the reaction of the metal *uranium* or hydride with nitrogen under pressure and temperatures up to 800°[61] for extended periods of time. The *mononitride* is formed by vacuum decomposition of higher nitrides above 1050°. Uranium mononitride decomposes to uranium metal at 1800° and higher.

Intermediate products are obtained during nitriding of uranium metal when the reaction is carried out in

[60] *R. Rundle, N. Baenziger, A. Wilson, R. McDonald,* J. Amer. Chem. Soc. *70*, 99 (1948).

[61] *K. M. Taylor, C. A. Lenie, P. E. Doherty, L. N. Harley, J. J. Keaty,* R & D Division Carborundum Co., Niagara Falls, Third Quarterly Report, Microcard ORO 248, Sept. 1st to Nov. 30th (1959).

[62] *P. Chiatti,* J. Amer. Chem. Soc. *35*, 123 (1952).

[63] *S. A. J. Sambell, J. Williams,* UKAFA (Harwell) Report AERE M/R-2854.

[64] *K. M. Taylor, C. A. Lenie, P. E. Doherty, C. H. McMurty,* Fourth Quarterly Report, Microcard TID-6591 (1960).

[65] *A. Accary, R. Caillat,* Recent Advances in Reaction Sintering, Amer. Ceram. Soc. Symposium, Toronto 1961.

[66] *P. E. Evans, J. J. Davies,* J. Nucl. Mater. North Holland Publishing Co., Amsterdam *10*, 43 (1963).

an insufficient protective atmosphere. The uranium nitride is light gray in color; the structure of U_2N_3 is body-centered cubic and that of UN is face-centered cubic. The hardness of UN is in the range of 815–880 kg/mm^2 and that of the higher nitride in the range of 1400–1910 kg/mm^2.

9.5.1 Uranium Amides

Uranium(IV) diethylamide is prepared by the reaction of lithium diethylamide with uranium(IV) chloride using ether as a solvent. The product is an emerald green liquid isolated by distillation *in vacuo*. An identical is isolated when lithium diethylamide is reacted with uranium(V) chloride.
The reaction of uranium(V) diethylamide with ethyl mercaptan in ether solution gives *uranium (IV) ethyl mercaptide* as a light green solid which is purified by washing with ether.[58] On alcoholysis [Eqn (24)] uranium alkoxides are formed.

$$U[N(C_2H_5)]_4 + 4RSH \longrightarrow U(SR)_4 + 4(C_2H_5)_2NH \quad (23)$$

$$U[N(C_2H_5)_2]_4 + 4ROH \longrightarrow U(OR)_4 + 4(C_2H_5)_2NH \quad (24)$$

The *thorium* and *uranium dibenzoylmethane complexes* have also been reported.[67,68]
Uranium phosphide is prepared by the reaction of metal powder with phosphine at low temperature, followed by homogenization *in vacuo* at 1400°.[59,69,70] The above reaction, although very satisfactory for the preparation of uranium phosphide, involves some hazard and extreme care must be exercised. The sequence of operations is as follows:

(a) Clean uranium pellets are loaded into the reaction vessel and hydrided at 250°–275° for several hours.
(b) The reaction temperature is raised to 600° while phosphine diluted with argon free of air is admitted into the reaction vessel.
(c) Condensation of white phosphorus in the outlet tube indicates completion of the reaction. While argon gas is allowed to flow, the operation is stopped and the temperature is raised to 900° to decompose any UP_2.
(d) The U_3P_4 product is degraded up UP by raising the temperature to 1400° under vacuum.

[67] *W. Forsling*, Acta Chem. Scand. *3*, 1133 (1949).
[68] *C. Wiedenheft*, Inorg. Chem. *8*, 1174 (1969).
[69] *Y. Baskin, P. D. Shalek*, J. Inorg. Nucl. Chem. *26*, 1679 (1964).
[70] *Y. Baskin*, J. Amer. Ceram. Soc. *49*, 541 (1966).

9.6 Carbides

Thorium carbide is prepared either by melting the unpowdered elements together or by heating the powdered elements at a temperature below the melting point. Owing to the rapid oxidation of the metal, the arc melting technique should be used if the highest purity thorium carbide is to be produced. It is essential to recover and store the resulting carbide in an inert atmosphere.
The metal oxide, ThO_2, will react with carbon at 1800°–1900° to give ThC or ThC_2 of uncertain purity.[71] It has been reported that 99% of the ThO_2 can be converted to the carbide when heated at 2130° for 30 min.[72,73]

ThC is cubic (NaCl type) with a wide composition range. ThC_2 has a monoclinic structure at room temperature, converting to tetragonal, and finally to a cubic structure before melting. Compositions below ThC_2 are metallic gray, whereas freshly fractured surfaces of ThC_2 have a very pale metallic yellow color which darkens with time. Very pure thorium metal is nonreactive at ambient temperature, but the carbides, especially ThC, are so reactive to water vapor that they should be handled either in an inert atmosphere or under a dry liquid. Both carbides can be dissolved by H_2O, 1:1 HCl, H_2SO_4, HNO_3, 25% tartaric acid, and 5% NaOH. ThC_2 forms only a thin coat of ThO_2 after 24 hr at 300° in pure O_2, while at 500° it is fully oxidized.[74] In moist air it reacts ten times faster than UC_2, being completely converted to the oxide after approximately 10 hr at 30°. The carbon-deficient material is somewhat less reactive.

Uranium carbide is prepared in very high purity by arc melting the elements together. If a graphite electrode is used, the melt tends to pick up additional carbon, and for this reason, a tungsten electrode is preferred. Uranium also will react with hydrocarbon gases to produce UC at low temperatures and UC_2 when the temperature is raised. Methane will give mainly UC below

[71] *G. V. Samsonov, T. Ya Kosalapova, V. N. Paderno*, Zh. Prikl. Khim. *33*, 1661 (1960).
[72] *D. E. Scaife, A. W. Wylie*, Australian At. Energy Symp. Proc., Sydney, 172 (1958), Melbourne Univ. Press, Victoria, Australia.
[73] *C. H. Prescott jr., W. B. Hincke*, J. Amer. Chem. Soc. *49*, 2744 (1927).
[74] *N. H. Brett, D. Law, D. T. Lwey*, J. Inorg. Nucl. Chem. *13*, 44 (1960).

650°, and UC_2 above 900°.[75] Propane and butane will react much more rapidly than methane, but the conditions are more critical to prevent deposition of free carbon. From these gases, UC is formed below 750°.

From an economic point of view the reaction between the uranium oxide and carbon is very attractive. However, unless considerable effort is made in high-vacuum purification or the product is arc melted, impure material is certain to result. Uranium carbides are gray, pyrophoric when powdered, and metallic looking when seen as arc-melted specimens.

Plutonium dicarbide is prepared by either arc melting the elements or PuO_2 and carbon. Both techniques are equally satisfactory for making the carbides. The reaction between the elements is rapid and leads to oxygen-free, dense carbide specimens which are relatively stable in an ambient atmosphere. However, compositions near Pu_2C_3 tend to shatter when cooled rapidly or when remelting is attempted.[76,77]

Reaction between the oxide and carbon *in vacuo* at temperatures as high as 1750° can give a product which is free of the oxide phase, but composition control is difficult.[78] In addition, Pu_2C_3 is always present when PuC is prepared, even after the carbon content is varied over a wide range of concentration. The reaction in H_2 at 1600° always leaves some oxide phase.[77] It is reported that the above reaction becomes appreciable at 1100° and is complete after the temperature has reached 1600°. Even when the oxide phase is absent, considerable oxygen contamination can remain through the formation of PuC_xO_y.[79,80,81]

[75] *H. S. Kailsh*, 2nd Uranium Carbide Meeting, BMI, Columbus, Ohio, TID *7589*, 59 (1960).

[76] *O. L. Kruger*, J. Nucl. Mater. *7*, 142 (1962).

[77] *M. Palfreyman, L. E. Russell* in *F. Benesovsky*, Powder Metallurgy in the Nuclear Age, Plansee Proceedings, Springer, Vienna, 417 (1962).

[78] *A. E. Ogard, W. C. Pritchard, R. M. Douglass, J. A. Leary* in *F. Benesovsky*, Powder Metallurgy in the Nuclear Age, Plansee Proceedings, Springer, Vienna, Wien, 364 (1962), J. Inorg. Nucl. Chem. *24*, 29 (1962).

[79] *F. Anselin, G. Dean, R. Lorenzelli, R. Pascord* in *L. E. Russell*, Carbides in Nuclear Energy, Vol. I: Physical and Chemical Properties/Phase Diagram 113, Macmillan, New York 1964.

[80] *R. E. Skavdahl*, Trans. Amer. Nucl. Soc. *6*, 393 (1963).

[81] *R. Ainsley, D. C. Wood, R. G. Souden* in *L. E. Russell*, Carbides in Nuclear Energy, Vol. II, Preparation Fabrication/Irradiation Behavior 540, Macmillan, New York 1964.

Another method is heating the plutonium hydride with carbon. In general, the product is porous, impure, and easily attacked by water vapor. The powdered carbides are very reactive to water vapor, to a greater extent than UC, and pyrophoric in some cases. The Pu_2C_3 is more reactive than PuC and a dry box atmosphere of 1000 ppm O_2+H_2O produces excessive surface corrosion within a few minutes.

All compositions in the Pu–C system appear to be metallic gray. PuC has a NaCl-type crystal structure; there has been no crystal structure assigned to Pu_3C_2. Pu_2C_3 is also cubic, isotypic with U_2C_3. The PuC_2 apparently exists at high temperatures and is isotypic with tetragonal α-UC_2. PuC oxidizes slowly when heated to 200°–300° in air and burns brightly in oxygen at 400°. However, it is relatively stable at room temperature for prolonged periods of time. Plutonium carbide is not attacked by cold water, but effervesces steadily in hot water. Cold, dilute HCl or H_2SO_4 causes hydrolysis, although HNO_3 shows only a slight action. It should be noted that Pu_2C_3 is somewhat more stable towards acids than PuC. Also, PuC_2 is much less stable in moist air than the other carbides, but more oxidation-resistant.

9.7 Bibliography

F. H. Spedding, A. S. Newton, J. C. Warf, O. Johnson, R. W. Nottorf, I. B. Johns, A. H. Daane, Nucleonic *4*, 4 (1949).

M. K. Wilkinson, C. G. Shull, R. E. Rundle, Phys. Rev. *99*, 627 (1955).

J. L. Ryan, Advan. Chem. Ser. *71*, 331 (1967).

L. B. Asprey, F. H. Kruse, R. A. Penneman, J. Amer. Chem. Soc. *87*, 3518 (1965).

G. D. Sturgeon, R. A. Penneman, F. H. Kruse, L. B. Asprey, Inorg. Chem. *4*, 748 (1965).

L. E. J. Roberts, Advan. Chem. Ser. *39*, 66 (1963).

G. V. Samsonov, T. Ya. Kosalapova, V. N. Paderno, Zh. Prikl. Khim. *33*, 1661 (1960).

D. E. Scaife, A. W. Wylie, Australian At. Energy Symp. Proc., Sydney, 172, Melbourne Univ. Press, Victoria, Australia 1958.

H. S. Kailsh, 2nd Uranium Carbide Meeting, BMI, Columbus, Ohio, TID 7589, 59 (1960).

M. Palfreyman, L. E. Russell in *F. Benesovsky*, Powder Metallurgy in the Nuclear Age, Plansee Proceedings Springer Verlag Wien, 417 (1962).

F. Anselin, G. Dean, R. Lorenzelli, R. Pascord in *L. E. Russell*, Carbides in Nuclear Energy, Vol. I: Physical and Chemical Properties/Phase Diagram 113, Macmillan, New York 1964.

R. Ainsley, D. C. Wood, R. G. Souden in *R. E. Russell*, Carbides in Nuclear Energy, Vol. II, Preparation Fabrication/Irradiation Behaviour, 540, Macmillan, New York 1964.

A. Accary, R. Caillat, Recent Advances in Reaction Sintering, Amer. Ceram. Soc. Symposium, Toronto 1961.

M. Allbutt, R. M. Dell, J. Nucl. Mater. North-Holland Publishing, Amsterdam 24, 1 (1967).

10 Titanium

K. S. Mazdiyasni
in cooperation with
B. J. Schaper, L. M. Brown, and J. Gwinn
Air Force Materials Laboratory,
Wright-Patterson Air Force Base, Ohio 45433, U.S.A.

M. Hoch
Department of Materials Science and Metallurgical Engineering,
University of Cincinnati, Cincinnati, Ohio 45221, U.S.A.

Although titanium may be considered a relatively high-priced metal, it is, nevertheless, the most desirable material for many aerospace systems. The high strength-to-weight ratio and excellent corrosion resistance make titanium and its alloys unique for applications in high-speed aircraft and various space vehicles. The demand for titanium compounds is considerably greater than that for the metal itself.

Since titanium has a high chemical affinity for other elements, production of the metal is a difficult process. The most economic and efficient process for manufacturing *titanium metal* is the magnesium process (*Kroll*).

$$TiCl_4(g) + 2Mg(l) \longrightarrow Ti(s) + 2MgCl_2(l) \quad (1)$$
$$\Delta F^\circ = -129,200 + 45T \text{ [cal/mol]}$$

In some plants liquid sodium and $TiCl_4$ are fed simultaneously into an argon-filled steel reactor heated to about 650°. After the reaction begins, the temperature is allowed to climb to about 900°.

$$TicL_4(g) + 4Na(l) \longrightarrow Ti(s) + 4NaCl(l) \quad (2)$$
$$\Delta F^\circ = -226,200 + 65T \text{ [cal/mol]}$$

Another method is the direct reduction of TiO_2 with calcium or calcium hydride at about 1000°. The product can be crushed and then leached with a dilute acid to remove the calcium oxide. The oxygen content of this material (between 0·2 and 0·4%) is undesirably high. Pure metallic titanium has also been produced by the decomposition of TiI_4 on tungsten wire heated to 1300°–1500°.

10.1 Titanium Hydrides

There has long been speculation about the possible existence of TiH_4 but this compound has never been prepared. Of the known *hydrides of titanium* nonstoichiometric TIH_{2-x} is a crystalline powder having a cubic structure with a=4·461 Å. *Titanium dihydride* is prepared by reacting the spongy metal with prepurified hydrogen gas at ~400° in a quartz reaction tube.[1] However, experimental results indicate that in order to obtain the highest hydrogen content the hydrogenation reaction should be carried out in a stainless steel (USS 32) reaction vessel.[1] The resultant composition obtained as the mean value of many different runs is $TiH_{1.998 \pm 0.006}$. Titanium dihydride is reported to be a light gray powder, metallic in appearance, d=3·72 g/cm³ and stable in an ambient atmosphere.

Titanium hydride is used as a precursor for the preparation of titanium borides, nitrides and silicides. It is used for forming glass- or ceramic-to-metal seals and for producing a titanium coating on copper or copper-plated metals. It acts as a catalyst in the hydrogenation of certain unsaturated organic compounds.

It is known that $TiCl_4$ can be reduced with H_2 at high pressures and temperatures,[2-4] and also by electric discharge[5] to form *titanium subhalides*. Titanium chlorohydride (brown color) has been prepared by allowing $TiCl_4$–H_2 mixtures to react for an extended period of time at 90 mmHg pressure.

$$TiCl_4 + H_2 \longrightarrow HTiCl_3(g) + HCl \quad (3)$$

It has been further suggested that at lower pressures $TiCl_3$ may first be converted into TiH_4 and thus then subsequently into a higher chlorohydride according to the following reactions.

$$TiCl_3H + H_2 \longrightarrow HCl + TiCl_2H \quad (4)$$

$$TiCl_2H + H_2 \longrightarrow TiCl_2H_2, \text{ etc.} \quad (5)$$

At high pressures where the concentrations of hydrogen are small as compared with $TiCl_3(g)$ or $TiCl_2H(g)$ it is not surprising that crystalline subhalides form rather than a chlorohydride or TiH_4. Under these conditions,

$$TiCl_3(g) \longrightarrow TiCl_3(s)$$
and
$$TiCl_2(g) \longrightarrow TiCl_2(s)$$

are more probable steps than those leading to hydride formation.

10.2 Titanium Halides

10.2.1 Titanium Fluorides

Titanium tetrafluoride, TiF_4, is a colorless solid having a density of 2·80 g/cm³ with a sublimation point of 284°. It is hygroscopic, soluble in water, and in aqueous hydrofluoric acid gives the $TiF_6^{2\ominus}$ ion. It is best prepared by the reaction of titanium tetrachloride with anhydrous HF or by

[1] *Oka, Akira*, Bull. Chem. Soc. Japan *40*, 2284 (1967).

[2] *K. Funaki, K. Uchemusa*, J. Chem. Soc. Japan 2nd Chem. Sec. *59*, 14 (1956).

[3] *N. C. Baenzinger, R. E. Rundle*, Acta Cryst. *I*, 274 (1948).

[4] *W. E. Reid jr.*, J. Electrochem. Soc. *104*, 21 (1957).

[5] *V. Gutmann, H. Nowotny, G. Ofner*, Z. Anorg. Allgem. Chemie *278*, 78 (1955).

the reaction of the metal with fluorine at 250°. Exposure to moist air, especially at elevated temperatures, leads to the oxofluoride, $TiOF_2$.[6] *Titanium trifluoride* is a bluish crystalline solid with a density of 2·98 g/cm^3. It is insoluble in water but somewhat soluble in dilute acids and bases. A crude product may be obtained by passing HF over heated titanium hydride or by reacting HF with $TiCl_3$. Purification may be achieved by vacuum sublimation at 900°–950°.

10.2.2 Titanium Chlorides

Titanium tetrachloride is the most important of the titanium halides. It results in one of the two principle methods of attacking an ore of titanium and is the starting point for the preparation of titanium metal and the organic compounds of titanium. Large quantities are being used in the production of pigmentary titanium dioxide by the "chloride route" and in the utilization for smoke screens for military purposes. Titanium tetrachloride has a wide variety of applications in organic synthesis such as low-pressure polymerization of ethylene, propylene, and other olefins and as a catalyst in the Friedel–Crafts reaction. A most comprehensive patent review has been made[7] covering the application of titanium tetrachloride (and also of titanium trichloride, dichloride, and other titanium halides) as a catalyst for hydrocarbon polymerization.

Titanium tetrachloride is prepared by the chlorination of the mineral rutile in the presence of carbon at 800°–1000°. The reactions involved are as follows.

$$TiO_2 + 2Cl_2 + C \longrightarrow TiCl_4 + CO_2 \qquad (6)$$

At 1100°K
$\Delta H = -52{\cdot}51$ kcal/mol
$\Delta G = -66{\cdot}93$ kcal/mol
At 1300°K
$\Delta H = -53$ kcal/mol
$\Delta G = -69{\cdot}51$ kcal/mol

$$TiO_2 + 2Cl_2 + 2C \longrightarrow TiCl_4 + 2CO \qquad (7)$$

At 1100°K
$\Delta H = -11{\cdot}97$ kcal/mol
$\Delta G = -72{\cdot}19$ kcal/mol
At 1300°K
$\Delta H = -12{\cdot}97$ kcal/mol
$\Delta G = -73{\cdot}06$ kcal/mol

The preferred raw material is the mineral rutile which contains 95% or more TiO_2.
A process for the chlorination of titanium slags in chloride melts has been developed[8,9] and has been described in detail.[10]

Titanium tetrachloride is very susceptible to hydrolysis, it causes burns, and fumes strongly when in contact with moist air. The fumes are dense, noxious and consist of finely divided oxochlorides. Titanium tetrachloride is miscible with many common solvents such as hydrocarbons, carbon tetrachloride, and chlorinated hydrocarbons but reacts with solvents containing hydroxyl, carboxyl or diketone groups; the latter react as enols to form substitution products with the elimination of HCl.[11,12] Water forms clear mixtures with titanium tetrachloride at room temperature yielding an acid from the hydrolysis reaction. A series of oxochlorides and ultimately hydrated titanium dioxide may be obtained from these solutions according to the conditions of temperature, acidity, and concentration.
Reduction of titanium tetrachloride by sodium, calcium, or magnesium yields the metal. Magnesium or sodium reduction is used for the commercial preparation of titanium metal.
Titanium tetrachloride is reduced by hydrogen. The product of the reaction at 700° is a dull purple *titanium trichloride* which may be collected as a crystalline solid at a temperature above the boiling point of the tetrachloride. Hydrogen and aluminum are both used as reducing agents in the commercial production of titanium trichloride from the tetrachloride.

There are four known modifications of *titanium trichloride*. The α-$TiCl_3$ was the first modification to be discovered. Reduction of titanium tetrachloride by hydrogen yields this product in practically pure form. β-$TiCl_3$ is obtained by the reaction of $TiCl_4$ with $Al(C_2H_5)_3$ in heptane at 80°, while γ-$TiCl_3$ is formed by the reduction of $TiCl_4$ with organoaluminum compounds at 150°–200°, in which case a small amount of aluminum chloride is present as a major contaminant. δ-$TiCl_3$ is made by dry grinding of either the α- or γ-form. A method by which the different phases of $TiCl_3$ are distinguished from one another has been reported.[13] All modifications must be handled and stored in a dry oxygen-free atmosphere.

The application of titanium trichloride in the preparation of polymerization catalysts is reported in a large

---

[6] *K. S. Vorres, F. B. Dutton*, J. Amer. Chem. Soc. *77*, 2019 (1955).
[7] *J. Barksdale*, Titanium, Its Occurrence, Chemistry and Technology, 2nd Ed., Ronald Press Co., New York 1966.
[8] *E. C. Perkins*, U.S. Bur. Mines, Rept. Invest., 6317 (1963).
[9] *A. Z. Berzukladnikor*, J. Appl. Chem. USSR *40*, 25 (1967).
[10] *D. J. Jones*, The Production of Titanium Chloride, R. H. Chandler, London 1969.
[11] *P. Pascal*, Nouveau Traité de Chimie Minérale, Tome IX, Masson et Cie, Paris 1963.
[12] *A. Slawisch*, Chem.-Ztg. *92*, 311 (1968).
[13] *G. Natta, Pa. Corradini, G. Allegra*, J. Polymer. Sci. *51*, 399 (1961).

number of patents. Anhydrous titanium trichloride dissolves in water to give a violet solution from which, after saturation with hydrogen chloride, a stable *crystalline hydrate*, $TiCl_3 \cdot 6H_2O$, may be crystallized.

Titanium dichloride, $TiCl_2$, is formed by reduction of $TiCl_4$ with titanium or sodium metal at 900°–1000°.

$$Ti + 2TiCl_4 \longrightarrow 2TiCl_3 + TiCl_2 \quad (8)$$

In this process the $TiCl_3$ is condensed on the reactor wall whereas the $TiCl_2$ solidifies in the receiver.[14] Titanium dichloride is a black powder with a hexagonal structure having a melting point of $1035 \pm 10°$ and boiling point of $1500 \pm 40°$. $TiCl_2$ disproportionates to the metal and the tetrachloride at ~1600° in a hydrogen atmosphere.

$$2TiCl_2 \longrightarrow TiCl_4 + Ti \quad (9)$$

Powdered $TiCl_2$ is spontaneously combustible in air and is a strong reducing agent. It dissolves in water with the generation of hydrogen and formation of a solution containing trivalent titanium. Thermodynamic data have been reviewed and reported.[9,10,15]

10.2.3 Titanium Bromides and Iodides

Titanium tetrabromide or *tetraiodide* may be prepared by direct synthesis from titanium metal and bromine or iodine and by the reaction of HBr or HI with titanium tetrachloride.

Titanium tetrabromide is a lemon-yellow crystalline solid with a cubic structure and a melting point of 38°–39° and a boiling point of 233°. *Titanium tetraiodide*[16,17] is a reddish-brown octahedral crystalline material with a melting point of 150° and a boiling point of 377°. The *titanium tri-* and *dibromides* and *tri-* and *diiodides* may be prepared by methods similar to those described for either tri- and dichlorides or tri- and difluorides.[16]

10.3 Other Binary Titanium Derivatives

10.3.1 Oxides

Titanium dioxide occurs naturally in three crystalline forms, anatase, brookite, and rutile. These materials are formed from relatively pure TiO_2 with residual concentrations of impurities such as iron which gives them dark color.

Brookite is rare, is of little commercial importance, and is orthorhombic in crystal structure. The other two forms are tetragonal but are not isomorphous. The mineral *rutile* is an important raw material for the production of $TiCl_4$ and then titanium metal, titanium dioxide pigment, and organic derivatives of titanium. A considerable amount of this material is also used in welding rod coatings.

The phase diagrams of the titanium–oxygen system are known.[17,18] Metallic titanium can dissolve oxygen up to a composition of approximately $TiO_{0.42}$. This phase, a solid solution of oxygen in α-titanium, retains the hexagonal structure of the γ-form of the metal. The effect on the parameter *a* of the unit cell is small, that on *c* is greater; however, both parameters increase as the oxygen content increases. As the oxygen concentration increases, the transition temperature to the β-form of the metal increases.

Titanium dioxide is prepared by the hydrolytic decomposition of prepurified $TiCl_4$ in the presence of ammonia. The crude product ("*titanic acid*") is then washed and dried at 110° and finally calcined at 800°. Preparation of *ultra-high purity* TiO_2 has been reported.[19] The method used involves precipitation of $TiO_2 \cdot xH_2O$ from $TiCl_4$, conversion of the hydrated oxide to the double oxalate, $(NH_4)_2TiO(C_2O_4)_2$, recrystallization of the latter from methanol, and subsequent calcination. Alternatively, titanium dioxide may be prepared by the hydrolytic decomposition of titanium alkoxides to $Ti(OH)_4$ which, after filtration and drying at 110° is calcined at 500°–700°.

TiO_2 is thermally stable and quite resistant to chemical attack. Hydrogen, carbon monoxide, and carbon at high temperatures reduce TiO_2 to a lower oxide or mixtures of the carbide and lower oxides. Oxygen-deficient dark blue TiO_{2-x} is formed when TiO_2 is heated *in vacuo*.

The *lower titanium oxides*, TiO, Ti_2O_3, and Ti_3O_5, may be prepared by the reduction of TiO_2 (rutile or anatase) with Ti at ~1600° *in vacuo* or by the reduction of TiO_2 by hydrogen and carbon monoxide at ~1200°–1750°.[20,21]

---

[14] *V. V. Radiyakin, B. N. Kushkin, M. P. Volynskaya*, Inst. Titana *2*, 171 (1968).

[15] Thermomechanical Tables, The Dow Chemical Co., Midland, Mich., 945 (1965).

[16] *L. R. Blair, H. H. Becham, W. K. Nelson*, Titanium Compounds, Inorganic, in ETC *20*, 381 (1969).

[17] *E. S. Bumps, H. D. Kessler, M. Hansen*, Trans. Amer. Soc. Metals *45*, 1008 (1953).

[18] *A. D. McQuillan*, Proc. Roy. Soc. London *204A*, 248 (1950).

[19] *W. Piekarczyk*, Internl. Symp. Reinstoffe Wiss. Tech. Tagungsberichte *1*, 213 (1966); C.A. *70*, 21376 (1969).

[20] *V. Dufek, V. Brozek*, Chem. Prumysl. *17*, 185 (1967); C.A. *67*, 34513 (1967).

[21] *H. J. Washaki, N. F. H. Bright, J. F. Rowland*, J. Less-Common Metals *17*, 99 (1969).

TiO oxidizes at 150°–200° to give Ti_2O_3, whereas Ti_3O_5 is formed in a temperature range of 250°–350°. Thin surface films of oxide play an important role in the resistance of titanium metal to corrosion. The lower oxides of titanium are used as semiconductors.

10.3.2 Sulfides and Selenides

The preparation and properties of *titanium sulfides* have been described.[11,22] *Titanium trisulfide* is a black crystalline powder, mono-clinic in structure, with a density of 3·21 g/cm³ obtained by heating spongy titanium with excess sulfur in a sealed tube or by the reaction of $TiCl_4$ with dry hydrogen sulfide.

$$TiCl_4 + 3H_2S \longrightarrow TiS_3 + 4HCL + H_2 \quad (10)$$

Titanium sulfide, TiS_2, may be prepared by heating spongy titanium with sulfur in a sealed tube, by reacting $TiCl_4$ with H_2S, or by pyrolyzing titanium trisulfide at 550°. The lower sulfides such as Ti_2S_3, TiS, and Ti_2S may be obtained by the reduction of TiS_2 with hydrogen or titanium metal or *in vacuo* at ~1000°.

The *titanium selenide compounds* include TiSe and $TiSe_2$. The structure is found to be a NiAs-like type for TiSe and a $Cd(OH)_2$-like type for $TiSe_2$. Titanium selenides are prepared by alloying the elements at 800°–1000° in alumina containers sealed inside an evacuated silica tube.[23]

10.3.3 Nitrides and Phosphide

Preparation of stoichiometric titanium nitrides is very difficult. Usually the final product contains a considerable amount of the metal or metal oxide.[24,25]

Titanium nitride may be prepared by the direct reaction of titanium and nitrogen, and by the reaction of titanium tetrachloride with nitrogen and hydrogen. The latter reaction occurs readily at the surface of a heated tungsten wire. The nitride is also obtained when titanium dioxide is reduced with carbon in a nitrogen atmosphere. Another method is the vacuum evaporation synthesis technique of titanium metal thin films in nitrogen, which results in maximum compositions of $TiN_{1.0}$ at a substrate temperature of 320°.[26]

Titanium nitride is a light yellow crystalline powder of cubic structure, $a = 4{\cdot}245$ Å, $d = 5{\cdot}213$ g/cm³, which after pressing and sintering, can be polished to a golden-yellow mirror. Titanium nitride melts at 2950°, but there is some evidence of decomposition at this temperature. The specific resistivity of a pure evaporated titanium thin film is $1{\cdot}2 \times 10^{-4}$ Ω cm which increases gradually with the nitrogen concentration and reaches a final value of $2{\cdot}8 \times 10^{-4}$ Ω cm corresponding to $TiN_{1.0}$ through a maximum value of $5{\cdot}0 \times 10^{-4}$ Ω cm at the composition $N/Ti = 0{\cdot}5$. Titanium nitride is a superconductor in the temperature range of 1·2°–1·6°K. A thin layer deposited on silica, however, shows semiconductivity.[27]

Titanium nitride is not attacked by acid, except by boiling aqua regia. It is decomposed by boiling alkalis with the evolution of ammonia. It is stable *in vacuo*, and in the presence of hydrogen and carbon monoxide at high temperature. Titanium nitride, however, is rapidly oxidized at 1200° by oxygen, nitric oxide, and carbon dioxide.[28]

The most widely used method for preparing *titanium phosphide* is to heat titanium metal with red phosphorus in a sealed tube at 550°–850°. The reaction of phosphine with $TiCl_4$ or with spongy titanium has also been used.[29]

Titanium monophosphide is a gray metallic powder, having a hexagonal structure, and a density of 4·27 g/cm³. The powder burns in air and is insoluble in acid and alkali media.

10.3.4 Carbides and Silicides

Titanium carbide is prepared industrially by the reduction of TiO_2 with carbon. The oxide reduction begins at 935°;[30] between 1000°–1500° it proceeds through the following steps.[31]

$$TiO_2 \longrightarrow Ti_3O_5 \longrightarrow Ti_2O_3 \longrightarrow TiO \longrightarrow \begin{matrix} TiC \\ \text{or Ti} \end{matrix}$$

Titanium carbide is obtained when $TiCl_4$ gas is brought into a reaction chamber with CH_4 after bubbling it through liquid $TiCl_4$ at 1050°.[32]

---

[22] *Y. Jeannin*, Ann. Chim. (Paris) *7*, 57 (1962).

[23] *F. Gronvald, F. J. Langmyer*, Acta Chem. Scand. *15*, 1949 (1961).

[24] *E. E. Vainshtein, V. I. Cherkov*, Dan SSSR *157*, 300 (1964).

[25] *G. Hass*, Vacuum *2*, 331 (1952).

[26] *A. Itoh*, Proc. of the Fourth Internat. Vacuum Congress, 536 (1968).

[27] *P. Pascal*, Nouveau Traite de Chimie Minerale, Tome IX, Masson et Cie, Paris 1963.

[28] *G. H. J. Neville*, Kirk-Othmer, Encyclopedia of Chemical Technology, 2nd Ed. Vol. 20, pp. 381–424 (1970).

[29] *P. O. Snell*, Acta Chem. Scand. *27*, 1773 (1967).

[30] *V. P. Elyutin, P. F. Merkulova, Yu. A. Pavlav*, Proizv. i Obrabotka Stali i Splavov *38*, 79 (1958).

[31] *G. V. Samsonov*, Zh. Ukr. Khim, *23*, 287 (1956)

[32] Japanese Patent 634 (1957), Erf.: *Y. Jmai, I. Ishizaki, N. Yano*.

$$TiCl_4 + CH_4 \longrightarrow TiC + 4HCl \quad (11)$$

Titanium carbide has also been prepared according to the following reaction.[33]

$$CaC_2 + 2TiCl_4 + 3H_2 \xrightarrow{800^\circ} 2TiC + CaCl_2 + 6HCl \quad (12)$$

Washing with water easily frees the TiC of $CaCl_2$ and CaC_2. Another method is to react TiS_2 and carbon at ~2000° for a short time.[34] Another method commonly described in the literature is the reaction of titanium hydride and carbon *in vacuo* at 900°–1200°.[35]

Titanium carbide is a colorless very hard metallic compound d=4·91 g/cm³, with a face-centred cubic structure, a=4·285 Å. It is not attacked by hot or cold HCl, H_3PO_4, CH_3COOH, or H_2SO_4. However, HNO_3+HCl, HNO_3, and H_2SO_4+HNO_3 cause complete dissolution at room temperature; $HClO_3$ causes dissolution only when heated. TiC is considered structurally stable in static H_2 up to 2400°. However, the action of H_2 is complicated by the rapid change in carbon and titanium activity with composition. TiC reacts with oxygen at ~450°, the oxidation producing a nonadherent anatase-TiO_2 layer which is nonprotecting. Also, complete nitriding will occur between 1000°–1300° in 1 atm of N_2. At lower pressures or higher temperatures, the carbide will form a solution with variable nitrogen content depending on the conditions.

The *titanium silicides* include Ti_2Si, TiSi, and $TiSi_2$ which in general are prepared by heating titanium and silicon powders together in appropriate containers. Titanium hydride may be used instead of titanium metal and provides its own protective atmosphere. Other methods of preparation include the electrolysis of a fused melt of alkali metal fluorosilicates containing TiO_2 with an iron cathode, and decomposition of $TiCl_4$ on silicon powder.

Titanium disilicide has a slightly distorted orthorhombic structure, silvery gray color and a melting point of 1540°. Titanium disilicide oxidizes slowly in air at 700°–800° and is attacked violently by fused alkali. It is resistant to mineral acids except for hydrofluoric acid.

The *Pentatitanium trisilicide*, Ti_5Si_3, is prepared by the direct reaction of TiO_2 and Ca_2Si initially at 900° and finally in a temperature range of 1200°–1300°.

$$4Ca_2Si + 5TiO_2 \longrightarrow Ti_5Si_3 + 6CaO + 2CaO \cdot SiO_2 \quad (13)$$

CaO and calcium silicate are dissolved in acetic acid and a pure product is obtained.

10.3.5 Borides

The methods generally used for the preparation of *titanium diboride* include the following:

(a) Synthesis from the elements or from boron and the metal hydride at ~1700°.[36,37]
(b) Electrolysis of molten oxides and salts.[38,39]
(c) Reduction with carbon of a mixture of metal dioxide (or the corresponding hydroxide and elemental boron B_2O_3, or boric acid.[40–42]
(d) Solid phase reactions which involve B_4C.[43–50]
(e) Synthesis from the metal and boron trihalide by a vapor phase deposition process[51,52] according to reaction (14).

$$TiCl_4 + 2BCl_3 + 5H_2 \longrightarrow TiB_2 + 10HCl \quad (14)$$

Starting with pure substances and with careful handling (avoiding the introduction of impurities), a pure product may be obtained by method (a). However, the high temperatures necessary for synthesis from the elements, make this method inconvenient, while synthesis from the gas phase (method e) is very expensive. Products prepared by the electrolysis procedure (method b) always contain large amounts of impurities. Methods (c) and (d) lead to products containing various amounts of carbon as major impurities.

---

[33] *D. Schuler*, Dissertation, Techn. Hochschule, Zurich 1952.
[34] *K. Ogawa, Y. Bando*, Fundai Oyobi Funmatsuakin *6*, 160 (1959).
[35] *G. Ebersbach, R. Mey, G. Ulbrich*, Technik *24*, 7, 460 (1969).

[36] *A. Polty, H. Margolin, J. Nielsen*, Trans. Amer. Soc. Metals *46*, 312 (1954).
[37] *V. A. Epel'baum, M. A. Gurevich*, Zh. Fis. Khim. *32*, 2275 (1958).
[38] *F. Glaser*, Powder Met. Bull. *6*, 51 (1951).
[39] *S. Sinderband, P. Schwarzkopf*, Powder Met. Bull. *5*, 42 (1950).
[40] *T. Stoda, S. Yamaguchi, S. Kitahara*, Sci. Papers Inst. Phys. Chem. Res. (Tokyo) *53*, 68 (1959).
[41] U.S. Patent No. 293247 (1961).
[42] *H. Bluementhal*, Powder Met. Bull. *7*, 79 (1956).
[43] *R. Meyer, H. Pastor*, Bull. Soc. Fr. Ceram. *66*, 59 (1965).
[44] *R. Kieffer, F. Benesovsky, E. Honak*, Z. Anorg. Allgem. Chemie *268*, 191 (1952).
[45] *T. Kubo, T. Hanazawa*, Kogyo Kagaku Zasshi *63*, 1144 (1960).
[46] *V. F. Funke, S. I. Uydkovskii, G. V. Samsonov*, Zh. Prikl. Khim. *33*, 831 (1960).
[47] *G. A. Meerson, G. V. Samsonov*, Zh. Prikl. Khim. *27*, 1130 (1954).
[48] *C. Baruch, T. Evans*, J. Metals *7*, 909 (1959).
[49] *V. F. Funke, S. I. Jukovskii, Poroshkovaya*, Met. Akad. Nauk. Ukr. SSR 3 (4) 49 (1963).
[50] *Yu. B. Paderno, J. J. Serebryakova, G. V. Samsonov*, Metal. *11*, 48 (1955).
[51] *R. E. Gannon, R. C. Folweiler, T. Vasilas*, J. Amer. Ceram. Soc. *46*, 496 (1963).
[52] *P. Preshev*, Mitt. Inst. Allgem. Anorg. Chem. (Sofia) *4*, 53 (1966).

There are four *intermediate phases* in the titanium boron system, Ti_2B, TiB, TiB_2, and Ti_2B_5. Ti_2B is believed to disproportionate at temperatures both below 1800° and above 2200°. In both cases a titanium-rich liquid is formed and the other phase is TiB, at low temperature, and TiB_2. TiB_2 melts congruently at ~2800° but the behavior of Ti_2B_5 is not known. TiB_2, with a hexagonal structure, is not soluble in acids except for $HClO_4$, but will dissolve in alkaline media. At high temperatures, TiB_2 is fairly resistant to oxygen attack.

10.4 Titanium Acids and Salts

10.4.1 Meta- and Orthotitanic Acid

Titanium hydroxide, $Ti(OH)_4$, and *hydrated oxide*, $TiO_2 \cdot 2H_2O$, are called *orthotitanic acid.* The orthotitanic acids are hydrogels with considerable adsorbent properties. On boiling the suspensions, the hydrogels become more granular and less adsorbent, and their compositions approach $TiO_2 \cdot H_2O$ or $TiO(OH)_2$, *metatitanic acid.* In the presence of acid electrolytes, colloidal titanium hydrogel may be obtained. The coagulation of such a solution is an important factor in the production of titanium dioxide pigments by the sulfate route.

The use of titanium hydroxide as an adsorbent in the extraction of uranium from sea water has been studied.[53] Although economically feasible, this method is not likely to be of importance while adequate sources of uranium mineral exist.

10.4.2 Peroxo Complexes of Titanium

The appearance of a yellow color on addition of H_2O_2 to a solution of $TiCl_4$ has been known for a long time.[54] However, it seems to have been impossible to isolate the products in the form of well-defined crystalline compounds. Amorphous *perosulfates*, $Ti(O_2)SO_4 \cdot 3H_2O$[55] and $K_2SO_4 \cdot Ti(O_2)SO_4 \cdot 3H_2O$[56] have been obtained from such yellow solutions by precipitation with ethanol. Corresponding amorphous *peroxotitanium oxalates, malonates,* and *maleinates* have also been described.[57] Moderately acidic solutions (pH > 2) slowly deposit amorphous precipitates of *peroxotitanium hydrate*, $TiO_3(H_2)_x$, which can be obtained at a faster rate by adding a base. The existence of one peroxo group, $O_2^{2\ominus}$, per one titanium atom has been established by analysis but there are disagreements among the investigators as to the number of water molecules (x), *i.e.*, $[Ti(O_2)(OH)x]$ $x=1$[58–60] or 2.[61,62] Addition of ammonium fluoride and NH_3 precipitates a yellow solid $(NH_4)_3Ti(O_2)F_5$[63] which is the only crystalline compound of this type obtained so far. In all these preparations the ratio of peroxo groups to titanium is 1:1. Compounds with higher ratios can be obtained only from strongly alkaline solutions.[62–65] Recently, the peroxo complexes of titanium have been studied at higher acidities.[66,67] Chelates of mononuclear $TiO_2^{2\oplus}$, as well as of dinuclear $Ti_2O_5^{2\oplus}$, are produced depending on the acidity of the mixtures and the dentation of the chelating ligand.

10.4.3 Titanium Polysulfide and Polyselenide

Titanium–sulfur and titanium–selenium bonds are formed[68] by the reaction of $(C_5H_5)_2TiCl_2$ with polysulfide or polyselenide in accordance with the reactions (15) and (16).

$$(C_5H_5)_2TiCl_2 + (NH_4)_2S_5 \longrightarrow (C_5H_5)_2TiS_5 + 2NH_4Cl \quad (15)$$

$$(C_5H_5)_2TiCl_2 + (NH_4)_2Se_5 \longrightarrow (C_5H_5)_2TiSe_5 + 2NH_4Cl \quad (16)$$

10.4.4 Titanium Sulfates

Titanium sulfate, $Ti(SO_4)_2$, is prepared by reacting titanium tetrachloride with sulfur trioxide.

---

[53] *R. V. Davies*, Nature *203*, 1110 (1964).
[54] *G. Schöm*, Z. Anal. Chem. *9*, 41 (1870).
[55] *C. C. Patel, M. Mohan*, Nature *186*, 803 (1960).
[56] *R. Schwarz*, Z. Anorg. Allgem. Chemie *210*, 303 (1933).
[57] *G. V. Jere, C. C. Patel*, Nature *194*, 471 (1962).
[58] *F. Fitcher, A. Goldach*, Helv. Chim. Acta *13*, 1200 (1930).
[59] *G. V. Jere, C. C. Patel*, Z. Anorg. Allgem. Chemie *319*, 175 (1962).
[60] *F. Riveng*, Bull. Soc. Chim. France *512*, 283 (1945).
[61] *R. Schwarz, H. Giese*, Z. Anorg. Allgem. Chemie *176*, 209 (1928).
[62] *M. Schenck*, Helv. Chim. Acta *19*, 625 (1936).
[63] *A. C. R. Piccini*, Acad. Sci. *97*, 1064 (1883); *A. C. R. Piccini*, Gass Chim. Ital. *17*, 479 (1887); *A. C. R. Piccini*, Z. Anorg. Allgem. Chemie *10*, 438 (1895).
[64] *M. Mori, N. Shibata, E. Kyono, S. Ito*, Bull. Chem. Soc. Japan *29*, 904 (1956).
[65] *P. Melikov, L. Pissarevski*, Ber. *31*, 953 (1898).
[66] *J. Mühlebach, K. Muller, G. Schwarzenbach*, Inorg. Chem. *9*, 2381 (1970).
[67] *D. Schwarzenbach*, Inorg. Chem. *9*, 2391 (1970).
[68] *H. Köpf, B. Block, M. Schmidt*, Chem. Ber. *101*, 272 (1968).

$$TiCl_4 + 6SO_3 \longrightarrow Ti(SO_4)_2 + 2S_2O_5Cl_2 \quad (17)$$

Titanium sulfate is very sensitive to moisture and is used in large quantities in the production of TiO_2 pigments. When heated, it becomes *titanyl sulfate*, $TiOSO_4$, and finally titanium oxide. A stable *dihydrate*, $TiOSO_4 \cdot 2H_2O$, may be prepared by heating a solution containing about 10% TiO_2 and 40–50% sulfuric acid until crystallization occurs. Commercial titanyl sulfate is obtained by treating $TiO_2 \cdot xH_2O$ with concentrated sulfuric acid.

10.4.5 Titanium Phosphate

Titanium(III) phosphate may be prepared by adding sodium phosphate to either a titanium(III) sulfate or chloride solution, then adding alkali to decrease the acidity until precipitation occurs. *Titanium diphosphate*, TiP_2O_7, m.p. 1490°, is a white crystalline powder in the cubic system and has a density of 3·106 g/cm³. Titanium diphosphate is prepared by heating a mixture of TiO_2 and P_2O_5 at 900°. It is used as an ultraviolet reflecting pigment. *Hydrated titanium phosphate* is a gelatinous material of variable composition and when properly prepared, it is highly insoluble in acids. Its insolubility in acid has led to some commercial interest for separating titanium from iron in ilmenite sulfate solution. It has also been used in the dyeing and leather tanning industries. Titanium phosphate has also been used as an ion exchange material and as a catalyst.

10.5 Titanium Alkoxides

The chemistry of *titanium alkoxides* has been a subject of extensive research[69–72] because of their industrial importance as precursor materials for various applications.[73] Among these are the preparation of ultrahigh purity submicron oxide powders,[74–76] the vapor deposition of oxide thin films,[77, 78] and the preparation of single and mixed phase ferroelectric materials.[79–82] The methods described here are most useful for laboratory-scale preparation of titanium alkoxides and with minor changes in the preparative procedures may be well suited for industrial production.

Most metal alkoxides are extremely sensitive to moisture, heat, and light. Therefore, the main emphasis must be directed toward the utilization of very clean glass or stainless steel reaction vessels in order to obtain a high purity product in ~90% theoretical yield. The specific experimental conditions for particular preparations have been adequately described in the literature.

10.5.1 General Preparations

Titanium tetrakis(methoxide), *-ethoxide* and numerous titanium normal alkoxides can be prepared by the reaction of sodium or lithium alkoxides with anhydrous titanium tetrachloride.[83, 84]

$$4MOR + TiCl_4 \xrightarrow{ROH} Ti(OR)_4 + 4MCl_4 \quad (18)$$

M = Li, Na

R = Ch_3, C_2H_5, C_3H_7, $(F_3C)_2CH$–

The titanium ethoxide is distilled *in vacuo*. The sodium or lithium alkoxide method, although satisfactory for the synthesis of the lower molecular weight straight-chain alkoxides, is not suited for the preparation of many other alkoxides.[84, 85] However, it has been successfully employed for the synthesis of *1,1,1,3,3,3,-hexafluoroisopropoxide* of titanium.[86]

The most economic and simple method for the large-scale production of alkoxides involves the addition of commercial grade anhydrous metal halide to a mixture of 10% anhydrous alcohol in

69 *D. C. Bradley*, Progr. Inorg. Chem. *2*, 303 (1960).
70 *D. C. Bradley* in *Jolly*, Preparative Inorganic Reaction, Vol. II, p. 169, Wiley Intersci. Publ. New York 1962.
71 *R. C. Mehrotra*, Inorg. Chim. Acta, Rev. *1*, 99 (1967).
72 *L. M. Brown, K. S. Mazdiyasni*, Anal. Chem. *41*, (10) 1243 (1969).
73 *R. H. Stanley*, Kirk-Othmer Encyclopedia of Chemical Technology 2nd Ed., Vol. 20, pp. 475–503 (1970).
74 *K. S. Mazdiyasni, C. T. Lynch, J. S. Smith*, J. Amer. Ceram. Soc. *50*, 532 (1967).
75 *K. S. Mazdiyasni*, Proceedings of the VI International Symposium on Reactivity of Solids, Eds. *J. W. Mitchell, R. C. DeVries, R. W. Roberts, P. J. Cannon*, Wiley, Intersci. Publ. p. 115–125, Aug. 25–30, 1968.
76 *K. S. Mazdiyasni, L. M. Brown*, J. Amer. Ceram. Soc. *53* [*1*] 43–45 (1970).
77 *K. S. Mazdiyasni, C. T. Lynch* in *P. Popper*, Special Ceramics p. 115, Academic Press, New York 1965.
78 *J. A. Aboaf*, J. Elektrochem. Soc. Solid State Science *114* (9) 948–52 (1967).
79 *J. H. Harwood*, Chem. Process Eng. *48*, 100 (1967).
80 *K. S. Mazdiyasni, R. T. Dolloff, J. S. Smith*, J. Amer. Ceram. Soc. *52*, 523 (1969).
81 *J. S. Smith II, R. T. Dolloff, K. S. Mazdiyasni*, J. Amer. Ceram. Soc. *53*, 91 (1970).
82 *K. S. Mazdiyasni, C. T. Lynch, J. S. Smith II*, J. Amer. Ceram. Soc. *49*, (5) 286 (1966).
83 *F. Bischoff, H. Adkins*, J. Amer. Ceram. Soc. *46*, 256 (1954).
84 *D. C. Bradley, R. Gaze, W. Wardlaw*, J. Chem. Soc. (London) 721 (1955);
D. C. Bradley, W. Wardlaw, Nature *165*, 75 (1950);
D. C. Bradley, W. Wardlaw, J. Chem. Soc. (London) 208 (1951).
85 U.S. Patent *2*, 187, 721 (1940); Erf.: *J. Nelles;* Brit. Patent 512, 452 (1939).
86 *K. S. Mazdiyasni, B. J. Schaper, L. M. Brown*, Inorg. Chem. *10*, 889 (1971).

a diluent (benzene or toluene) in the presence of anhydrous ammonia.[86–88]

$$TiCl_4 + 4ROH + 4NH_3 \xrightarrow[25°]{C_6H_6} Ti(OR)_4 + 4NH_4Cl \quad (19)$$

$R = i\text{-}C_3H_7$

The titanium *tetrakis(isopropoxide)* is readily purified by fractional distillation. Because the removal of NH_4Cl by filtration is always cumbersome and time-consuming, the ammonia method may be carried out in the presence of an amide or nitrile. In this particular method the metal alkoxide separates out as the upper layer, while the ammonium chloride remains in solution in the amide or nitrile in the lower layer, thus the filtration step is eliminated.[88, 89]

An improved method of preparation of titanium alkoxides based on a recent report for the preparation of Zr and Hf analogs has been suggested.[90] This procedure consists of treating the anhydrous metal halide saturated in benzene with HCl with various esters, such as ethyl formate, ethyl acetate, diethyl oxalate, *n*-propyl acetate, isopropyl acetate, *n*-butyl acetate, and isobutyl acetate, in the presence of anhydrous ammonia. The alcohol formed in the reaction is itself consumed in the formation of metal alkoxides.

$$TiCl_4 + 4HCl + 4Ch_3COOR + 4NH_3 \longrightarrow Ti(OR)_4 + 4CH_3COCl + 4NH_3Cl \quad (20)$$

A valuable method for converting one alkoxide to another is an ester exchange reaction.[91] The method is particularly suited for the preparation of the *tertiary butoxide* from the isopropoxide and *t*-butyl acetate. The reaction is as follows:

$$Ti[O\text{-}CH(CH_3)_2]_4 + 4CH_3COOC(CH_3)_3 \longrightarrow Ti[O\text{-}C(CH_3)_3]_4 + 4CH_3COOCH(CH_3)_2 \quad (21)$$

Since there is a large difference between the boiling points of the esters, the fractionation is simple and quite rapid. Another advantage appears to be the lower rate of oxidation of the esters as compared to that of the alcohol.

Substitution of other *branched R groups* with the lower straight chain alcohols has been carried out in an alcohol interchange reaction as shown in Eqn (22).

$$Ti(OR)_4 + 4R^2OH \underset{C_6H_6}{\rightleftharpoons} Ti(OR^2)_4 + 4ROH \quad (22)$$

The distillation of an azeotrope drives the reaction to completion. The interchange becomes slow in the final stages with highly branched alcohols, probably owing to steric hindrance. The rise in temperature to the higher boiling alcohol, however, reliably signals the end of the reaction. In this method of preparation, speed, reversibility and absence of side reactions play major roles. It should be further emphasized that the alcoholysis reactions are carried out in the presence of benzene for the following reasons:

(1) Only a slight excess of higher alcohol is required for completing the reaction.
(2) The ethanol (or isopropanol) produced is continuously removed azeotropically with benzene, thus driving the reaction to completion; the alcohol which is fractionated out azeotropically can be determined by a simple oxidimetric method and the progress of the reaction can be followed.
(3) Side reactions, such as the formation of alkyl chloride and water, which lower the yield considerably in the case of branched alcohols are avoided.
(4) When a higher boiling alcohol is used, the reaction is carried out at a lower refluxing temperature owing to the presence of benzene and, thus, side reactions are minimized.
(5) The oxidizing properties of the metals are lowered when they are linked covalently in accordance to the following scheme.[92]

$$M(OR)_4 + R'OH \longrightarrow [R'(H)O \rightarrow M(OR)_{4-1}(O\text{-}R)] \longrightarrow (R'O)M(OR)_{4-1} + ROH \quad (23)$$

A number of *mixed alkoxides*[92] have been prepared by the following method:

$$M(OR^1)_4 + nR^2OH \longrightarrow M(OR^1)_{4-n}(OR^2)_n + nR^1OH \quad (24)$$

$R^1 = t\text{-sehyl}$, $R^2 = DH_3, C_2H_5$

$n = 1, 2$

[87] *H. Meerwein*, J. Bersin. Ann. *455*, 23 (1927). *H. Meerwein*, J. Bersin. Ann. *476*, 113 (1929).

[88] *D. C. Bradley, R. C. Mehrotra, W. Wardlaw*, J. Chem. Soc. (London) 1634 (1953).

[89] U.S. Patent *2*, 654, 770 (1953); *2*, 655, 523 (1953); Erf.: *D. F. Herman.*

[90] *S. K. Anand, J. J. Singh, R. K. Multani, B. D. Jain*, Israel J. Chem. *7*, 171 (1969).

[91] *R. C. Mehrotra*, J. Amer. Chem. Soc. *76*, 2266 (1954).

[92] *R. C. Mehrotra*, Inorg. Chim. Acta Rev. *1*, 99–112 (1967).

Another example is the preparation of mixed titanium alkoxides[93] as outlined in Eqn (25).

$$TiCl(OR^1)_3 + R^2OH + R^3{}_3N \longrightarrow Ti(OR^2)(OR^1)_3 + R^3{}_3NHCl \quad (25)$$

It is possible to prepare mixed alkoxides of titanium containing up to three different alkoxide groups by the following methods.[94]

$$TiCl_2(OR^1)_2 + R^2OH + C_5H_5N \longrightarrow TiCl(OR^2)(OR^1)_2 + C_5H_5NHCl \quad (26)$$

A very convenient method for the preparation of primary and secondary titanium alkoxides is the Meerwein–Ponndorf redox reaction involving a metal alkoxide and a ketone or aldehyde.[95,96]

$$M(OCHR_2)_x + xR_2'CO \rightleftharpoons M(OCHR_2')_x + xR_2CO \quad (27)$$

This reaction has been utilized effectively in the preparation of *2,2,2,-trichloroethoxides* of titanium by the addition of isopropoxides to chloral.

$$M[O{-}CH(CH_3)_2]_4 + 4CCl_3CHO \longrightarrow M(O{-}CH_2{-}CCl_3)_4 + 4(CH_3)_2CO \quad (28)$$

This reaction has its limitations owing to the occurrence of side reactions, especially with ionic or more electropositive alkoxides and aldehydes.

10.5.2 Titanium Alkoxyhalides

The *titanium haloalkoxides* have been well reviewed.[97,98] This type of compound has been recognized for over a century; in almost all cases they were prepared by mixing the hydrogen halides with the titanium alkoxides in an inert solvent.

$$Ti(OR)_4 + nHX \xrightarrow{X = Cl,\ Br} TiX_n(OR)_{4-n} + nROH \quad (29)$$

The action of excess hydrogen halide results in the formation of essentially the same product as that obtained by the reaction of $TiCl_4$ and an alcohol.[99]

$$Ti(OR)_4 + 2HX \longrightarrow TiX_2(OR)_2 \cdot ROH + ROH \quad (30)$$

$$TiX_4 + 3ROH \longrightarrow TiX_2(OR)_2 \cdot ROH + 2HX \quad (31)$$

$R = C_2H_5$, $CH(CH_3)_2$, $X = Cl, Br$

Also, titanium alkoxides react with acetyl chloride to give titanium monochloride alkoxides.[99]

$$Ti(OR)_4 + CH_3COCl \longrightarrow TiCl(OR)_3 + CH_3CO_2R \quad (32)$$

10.5.3 Acetylacetone Derivatives of Titanium

The reaction of titanium alkoxides with acetylacetone (acac) yields two types of acetylacetone derivatives of titanium, $Ti(OR)_3$ (acac) and $Ti(OR)_2$ $(acac)_2$.[100,101] Titanium dichloro diacetylacetone, $TiCl_2$ $(acac)_2$, may also be prepared by the reaction of titanium dialkoxide diacetylacetone and hydrogen chloride.[102]

$$Ti(OR)_2(acac)_2 + 2HCl \longrightarrow TiCl_2(acac)_2 + 2ROH \quad (33)$$

Another route by which the same compound can be prepared in quantitative yield is the reaction of titanium dialkoxychloride and acetylacetone.[102]

$$TiCl_2(OR)_2 + 2H(acac) \longrightarrow TiCl_2(acac)_2 + 2ROH \quad (34)$$

10.6 Titanium Alkylamides

Tetrakis(diethylamino) titanium is obtained from the reaction of the metal halide with lithium diethylamide.[103]

$$TiCl_4 + 4LiN(C_2H_5)_2 \longrightarrow Ti[N(C_2H_5)_2]_4 + 4LiCl \quad (35)$$

Furthermore, the tetrakis(dialkylamino) derivative may be prepared by a transamination reaction in which the metal derivative of a lower secondary amine is treated with a higher secondary amine.[104]

$$Ti(NR_2)_4 + 4R_2'NH \longrightarrow Ti(NR_2')_4 + 4R_2NH \quad (36)$$

By treating titanium dialkylamino derivatives with an alcohol, metal alkoxides are formed.[104]

$$Ti(NR_2)_4 + 4ROH \longrightarrow Ti(OR)_4 + 4R_2NH \quad (37)$$

---

[93] *J. C. Ghash, B. N. Ghash-Mezumdar, A. K. Base*, J. Indian. Chem. Soc. *31*, 683 (1955).
[94] *A. N. Nesmeyanov, O. V. Nogina*, Izv. Akad. Nauk. SSSR; Otdel Khim. Nauk. 41 (1954).
[95] *H. Meerwein, R. Schmidt*, Justus Liebigs Ann. Chem. *444*, 221 (1925).
[96] *H. Meerwein, B. von Back, B. W. Kirschnick, W. Lenz, A. Migge*, J. Prakt. Chem. *1472*, 113 (1937).
[97] *R. C. Mehrotra*, J. Indian Chem. Soc. *50*, 731 (1953) *52*, 759 (1955).
[98] *N. M. Cullinane, S. J. Chard, G. F. Price, B. B. Millward*, J. Appl. Chem. *2*, 250 (1952).
[99] *D. C. Bradley, I. M. Thomas*, J. Chem. Soc. (London), 3857 (1960).
[100] *D. C. Bradley*, in *A. Cotton*, Progress in Inorganic Chemistry, Vol. II, Interscience Publishers, New York 1960.
[101] *D. M. Puri, K. C. Pande, R. C. Mehrotra*, J. Less Common Metals *4*, 393 (1962).
[102] *D. M. Puri, K. C. Pande, R. C. Mehrotra*, J. Less Common Metals *4*, 481 (1962).
[103] *A. Slawisch*, Chemiker-Ztg. *9*, 311 (1968).
[104] *R. Feld, P. L. Cowe*, The Organic Chemistry of Titanium, p. 114 Butterworth & Co., London 1960.

10.7 Titanium Alkyls and Aryls

Titanium alkyl derivatives, R_nTiX_{4-n}, are known in which $n=1$, 2, 3, or 4 and where R—methyl, ethyl, butyl, etc., and X is a halogen or alkoxy group. Of the complete series, *tetramethyltitanium*,[105] *trimethyltitanium chloride*, *dimethyltitanium dichloride*[106] and *methyltitanium trichloride*[107] only the trimethyltitanium halide is unknown; however, the *trimethyltitanium iodide*[108] has been identified.

All alkyltitanium halide syntheses are conducted *via* the interaction or organometallic compounds of elements in Group I, II, or III with titanium halides at low temperature (0° to −80°) in an inert solvent such as diethyl ether and in the absence of oxygen. A completely different behavior is observed when the medium of the reaction is a nonpolar solvent such as hexane.[109] At temperatures ranging from 0° to −20° addition of $TiCl_4$ to a suspension of methylmagnesium chloride in hexane results in the formation of CH_3TiCl_3 independent of reaction time and molar ratio of the Grignard reagent.

$$CH_3MgCl + TiCl_4 \xrightarrow{\text{hexane}} Ch_3TiCl_3 + MgCl_2 \quad (38)$$

Titanium tetrachloride and allylmagnesium chloride react in benzene at 0° as follows:

$$H_2C{=}CH{-}CH_2{-}MgCl + TiCl_4 \xrightarrow[0^\circ]{C_6H_6} H_2C{=}CH{-}CH_2{-}TiCl_3 + MgCl_2 \quad (39)$$

Only two types of *titanium aryls*, $ArTiX_{4-n}$, (X=alkoxy) are known, $ArTiX_3$ and Ar_4Ti. *Phenyltitanium tris(isopropoxide)* is prepared by the reaction of phenyllithium with titanium tris-(isopropoxy) chloride in ether.[110]

$$[(CH_3)_2CH{-}O]_3TiCl + H_5C_6{-}Li \longrightarrow H_5C_6{-}Li[O{-}CH(CH_3)_2]_3 + LiCl \quad (40)$$

The product is a colorless crystalline solid (m.p. 88°–90°) stable at 10° in the dark for extended periods of time.

Other known compounds having $n=1$ in the general formula Ar_nTiX_{4-n} (X=alkoxy) are *p*-tolyl- and *p*-methoxyphenyltitanium tris(isopropoxide),[111] and 1-naphthyltitanium tris(butoxide).[112] All three may be prepared by the action of organolithium or organomagnesium compounds on the appropriate titanium halide. *Tetraphenyltitanium*, Ar_4Ti, is obtained by reacting phenyllithium with $TiCl_4$ in diethyl ether at −70°.[113]

$$TiCl_4 + H_54C_6{-}Li \longrightarrow (C_6H_5)_4Ti + 4LiCl \quad (41)$$

The compound is orange-yellow, crystalline, and decomposes above −20°.

$$(C_6H_5)_4Ti \longrightarrow (C_6H_5)_2Ti + H_5C_6{-}C_6H_5 \quad (42)$$

10.8 Cyclopentadienyl Derivatives of Titanium

Cyclopentadienyltitanium compounds are the most stable organotitanium derivatives of titanium which so far have been prepared. The general method of synthesis of the simpler compounds involves the reaction of a metal (sodium, lithium, or magnesium) derivative of cyclopentadiene with titanium halides, alkoxides, *etc.* usually in diethyl ether and under a dry inert atmosphere. *Bis(cyclopentadienyl)titanium dihalides* are prepared by the reaction of titanium tetrachloride with sodium[114] or lithium[115] cyclopentadienide or the cyclopentadienyl Grignard compound[114] with subsequent displacement of chlorine by the desired halogen. A more direct and simplified route for preparation of the biscyclopentadienyl)-titanium dihalide and *cyclopentadienyltitanium trihalide* is to react titanium tetrahalide with magnesium cyclopentadienide in an inert atmosphere.[116] Generally a 1:2 mole ratio of $Mg(C_5H_5)_2$ to TiX_4 seems to favor the formation of the substituted compounds, whereas a 1:1 mol ratio favors the formation of the disubstituted compound. Because the rate and order of addition also effect the nature of the product, mixtures often result. Attempts to prepare $C_5H_5TiCl_3$ from $TiCl_4$ and NaC_5H_5 in both xylene and tetrahydrofurane (THF) have been less successful.[116]

---

105 Brit. Pat. 858, 541 (1961); Farbweke Hoechst.

106 *C. Beerman, H. Bestian*, Angew. Chem. *71*, 618 (1959).

107 Ger. Pat. *1*, 089, 283, (1960); Erf.: *C. Berman*, Farbwerke Hoechst.

108 *H. J. Berthold, C. Groh*, Angew. Chem. Intern. Ed. Engl. *2*, 398 (1963).

109 *K. H. Thiele and K. Jacob*, Z. Anorg. Allgem. Chem. *356*, 190 (1968).

110 *D. C. Bradley, H. Holloway*, Can. J. Chem. *40*, 62, 1176 (1962).

111 U.S. Pat. 2, *886*, 579 (1959); Erf.: *D. F. Herman*.

112 Brit. Pat. 779, 490 (1957).

113 *V. N. Latjaeva*, J. Organometal Chem. *2*, 388 (1964).

114 *G. Wilkinson, J. M. Birmingham*, J. Amer. Chem. Soc. *76*, 4281 (1954).

115 *L. Summers, R. H. Uloth, A. Holmes*, J. Amer. Chem. Soc. *77*, 3604 (1955).

116 *C. L. Sloan, W. A. Barber*, J. Amer. Chem. Soc. *81*, 1364 (1959).

Several new derivatives of titanium(IV) containing the $Ti(C_5H_5)_2$ moiety have been reported in the literature. The *cis*-1,2-ethylenedithiolate derivative, $(C_5H_5)_2TiS_2C_2H_2$, has been prepared[117,118] from disodium ethylenedithiolate with $(C_5H_5)_2TiCl_2$.

$$(C_5H_5)_2TiCl_2 + NaS-CH=CH-SNa \longrightarrow (H_5C_5)_2Ti(S-CH=CH-S) + 2\,NaCl \quad (43)$$

Organic compounds containing Ti–O–S, Ti–O–P and Ti–O–B bond systems are known and apparently are only of theoretical interest.[119,120] Titanium(IV) derivatives containing titanium–oxygen bonds are prepared from $(C_5H_5)_2TiCl_2$ by reaction with phenol or substituted phenols in the presence of sodium amide in benzene.[121]

$$(C_5H_5)TiCl_2 + 2ROH + 2NaNH_2 \longrightarrow (C_5H_5)_2Ti(OR)_2 + 2NH_3 + 2NaCl \quad (44)$$

The reactions of bis(cyclopentadienyl)titanium dichloride with silver perfluorocarboxylates result in *dicarboxylatotitanium derivatives.*[122] Aqueous solutions containing the acetylacetonato complex cation $[(C_5H_5)_2Ti(acac)]^{\oplus}$ are also obtained[123] by stirring a suspension of $(C_5H_5)_2TiCl_2$ in water with acetylacetone.

Reactions have been carried out successfully with cyclopentadienyltitanium trichloride and pentamethylcyclopentadienyltitanium trichloride with ethanol[124] and with thiophenol in the presence of triethylamine.

$$(C_5H_5)TiCl_3 + 3C_2H_5OH + 3N(C_2H_5)_3 \longrightarrow (C_5H_5)Ti(OC_2H_5)_3 + 2N(C_2H_5)_3\cdot HCl \quad (45)$$

$$(C_5H_5)TiCl_3 + 3C_6H_5SH + 3N(C_2H_5)_3 \longrightarrow (C_5H_5)Ti(SC_6H_5)_3 + 3N(C_2H_5)_3\cdot HCl \quad (46)$$

Cyclopentadienyl tris(thiophenolato)titanium-(IV) is unstable and decomposes at ambient temperature even under an inert gas atmosphere.

[117] *R. B. King, C. A. Eggers*, Inorg. Chem. *7*, 340 (1968).
[118] *H. Köpf*, Z. Naturforsch. *23b*, 1531 (1968).
[119] *K. Andrä*, J. Organometal Chem. *11*, 567 (1968).
[120] *R. B. King, R. N. Kapoor*, J. Organometal Chem. *15*, 457 (1968).
[121] *G. Doyle, R. S. Tobias*, Inorg. Chem. 7, 2484 (1968).
[122] Japanese Patent No. 16349/63.
[123] *R. Field*, J. Chem. Soc. (London), 3963 (1964).
[124] *A. N. Nesmeyanov, O. V. Nogina, V. A. Dubovitsky*, Izv. Akad. Nauk. SSSR, Ser. Khim, 527 (1968).

10.9 Bimetallic Derivatives of Titanium

Bimetallic derivatives of titanium are prepared by the reaction of the hexacarbonyls of chromium, molybdenum, or tungsten with tris(dimethylamino)cyclopentadienyltitanium.[125]

$$M(CO)_6 + (C_5H_5)Ti[N(CH_3)_2]_3 \longrightarrow 3CO + (C_5H_5)Ti[N(CH_3)_2]_3M(CO)_3 \quad (47)$$

The formation of titanium–silicon bonds was achieved[126] by the reaction of $TiCl_4$ with $(C_6H_5)N_3SiK$. The resulting compound of formula $Ti[Si(C_5H_5)_3]_4$ is isolated from the reaction mixture as a colorless solid which is slowly hydrolyzed by aqueous alkali. Also, the red-brown $(C_5H_5)_2Ti[Si(C_6H_5)_3]_2$ is prepared from a similar reaction with $(C_5H_5)_2TiCl_2$.

The preparation of *(cyclopentadienyl)allyl titanium(III)* compounds, have been reported in detail.[127] The preparation involves the reaction of the proper Grignard reagent with $Ti(C_5H_5)_2Cl_2$ in the presence of dienes.

In the case of 1,3-pentadiene the reaction leading to the allyl derivatives may occur either by addition of the diene to intermediate hydride or by intermolecular hydrogen transfer within an intermediate alkylolefin complex.

Dicyclopentadienylpropylchlorotitanium(IV) can be reduced with alkali metals in the presence of excess BCl_3 to obtain a *dicyclopentadienyltitanium-tetrachloroborate* complex.[128] Reduction of $(C_5H_5)_2TiCl_2$ with zinc in toluene leads to the preparation of a compound having a Ti–Zn bond.[129]

A dark blue solid, $Ti(C_5H_5)_2B_3H_8$, (m.p. 117°–120°), a titanium(III) derivative containing the $B_3H_8^{\ominus}$ group, is prepared by the reaction of $(C_5H_5)_2TiCl_2$ with B_3H_8.[130]

Adducts of $[(C_5H_5)_2TiCl]_2$ with bidentate organic nitrogen bases (NN) such as 1,10-phenanthroline, 2,2′-dipyridyl, *o*-phenylenediamine, and α-picolylamine, have been reported.[131]

[125] *D. C. Bradley, A. S. Kasenally*, Chem. Commun., 1430 (1968).
[126] *E. Hengge, H. Zimmerman*, Angew. Chem. *80*, 153 (1968).
[127] *H. A. Martin, F. Jellinek*, J. Organometal Chem. *12*, 149 (1968).
[128] *G. Henrici Olive, S. Olive*, Agnew. Chem. *80*, 796 (1968).
[129] *J. J. Salzmann*, Helv. Chim Acta *51*, 526 (1968).
[130] *F. Klanberg, E. L. Muetterties, L. T. Guggenberger*, Inorg. Chem. *7*, 2272 (1968).
[131] *R. S. P. Coutts, P. C. Wailer*, Australian J. Chem. *21*, 2199 (1968).

These adducts have the general formula $[(C_6H_5)_2TiCl]_2(NN)_3$, *i.e.*, they contain 1·5-bidentate nitrogen groups per metal atom.

The *dicyclopentadienyltitanium(III) cation*, $[Ti(C_5H_5)_2]^{\oplus}$ has been precipitated from its aqueous solutions as carbonato or sulfato complexes, $[Ti(C_5H_5)_2]_2SO_4$ and $[Ti(C_5H_5)_2]_2CO_3$.[132] The carbonato complex has a pale blue color and the sulfato has a pale green color.

Oxygen-free aqueous solutions containing the $[TiC_5H_5)_2]^{\oplus}$ cation have been used to prepare *dithiocarbamates* $[(C_6H_5)TiS_2CNR_2]$.[133] The compounds of formula $(C_5H_5)_2TiS_2NR_2$ are suggested to contain tetracoordinated titanium. It is also said that the magnetic moment at room temperature corresponds to one unpaired electron per titanium. *Thioformate*, *thioacetate*, *thiopropionate*, *thiostearate*, and *thiobenzoate* derivatives of *dicyclopentadienyltitanium(III)* with the same magnetic properties have been reported.[134] However, the problem concerning the structure of dicyclopentadienyltitanium is still unanswered.

10.10 Bibliography

J. Barksdale, Titanium, Its Occurrence, Chemistry and Technology, 2nd Ed., Ronald Press Co., New York 1966.

D. J. Jones, The Production of Titanium Chloride, R. H. Chandler, London 1969.

P. Pascal, Nouveau Traite de Chimie Minerale, Tome IX, Masson et Cie, Paris 1963.

L. R. Blair, H. H. Becham, W. K. Nelson, Titanium Compounds, Inorganic, in ETC. *20*, 381 (1969).

A. D. McQuillan, Proc. Roy. Soc., London *204A*, 248 (1950).

G. H. J. Neville, Kirk-Othmer, Encyclopedia of Chemical Technology, 2nd Ed. *20*, 381 (1970).

D. C. Bradley in *Jolly*, Preparative Inorganic Reactions, Vol. II, p. 169, Wiley Intersci. Publ., New York 1962.

R. C. Mehrotra, Inorg. Chim. Acta Rev. *1*, 99 (1967).

D. C. Bradley in *A. Cotton*, Progress in Inorganic Chemistry, Vol. II, Interscience Publishers, New York, London 1960.

---

[132] *R. S. P. Coutts, P. C. Wailer*, Australian J. Chem. *21*, 1181 (1968).

[133] *R. S. P. Coutts, P. C. Wailer, J. V. Kingston*, Chem. Commun., 1170 (1968).

[134] *R. S. P. Coutts, P. C. Wailer*, Australian J. Chem. *21*, 373 (1968).

11 Zirconium and Hafnium

K. S. Mazdiyasni
Air Force Materials Laboratory,
Wright-Patterson Air Force Base, Ohio 45433, U.S.A.

Both zirconium and hafnium are hard ductile metals with a silver gray color and are much like stainless steel in appearance. Zirconium and hafnium metals in bulk form are nonreactive at room temperature, but they oxidize severely in air at ~700° and ~1000°, respectively. The finely divided or spongy zirconium or hafnium metals are easily ignited in air by a spark or blow. Zirconium and hafnium react readily with halogens, sulfur, and hydrogen at temperatures of about 250°–400°, with the formation of halides, sulfides, and hydrides, respectively. The hydride formation, however, is reversible at elevated temperatures *in vacuo*. Some of the physical properties of zirconium and hafnium are listed in Table 1.

Table 1 Physical Properties

Property	Zirconium	Hafnium
Density (g/cm³)	6·586	13·09 ± 0·01
M.p. (°C)	1830 ± 40	2222 ± 30
B.p. (°C)	> 2900	3100
α–β transformation (°C)	865	1777
Thermal expansion linear coefficient (°K)	$5{\cdot}5 \times 10^{-6}$	$5{\cdot}9 \times 10^{-6}$
Crystal structure		
α	Hexagonal close-packed	Hexagonal close-packed
β	Body-centered cubic	Body-centered cubic
Thermal conductivity [cal/(sec)(cm³)(°C/cm)]	0·035 at 125°	0·05333 at 50° 0·0494 at 400°
Specific heat [cal/(g)(°K)] at 298·15°K	0·067 ± 0·001	6·15
Electrical resistivity [Ω-cm]	45×10^{-6}	$35{\cdot}1 \times 10^{-6}$

11.1 Hydrides

Zirconium hydride, ZrH_x, and *hafnium hydride*, HfH_x, are readily prepared by the reaction of the corresponding metals with prepurified hydrogen gas at 400°.[1] The maximum hydrogen content attainable for 99·7% pure zirconium metal is $x = 1{\cdot}999 \pm 001$, and for hafnium $x = 2{\cdot}016$. The reported value larger than the stoichiometric value may be due to impurities in the metal.

11.2 Halides

11.2.1 Tetrahalides and Chlorides

The *zirconium and hafnium tetrahalides* are prepared by the reaction of the metals or metal hydrides with hydrogen halides and hydrogen at ~700°–900°. The anhydrous zirconium and hafnium *di-* and *trichlorides* are prepared by two simultaneous step reductions of the metal tetrachloride[2] (vapor) as indicated by the following equations.

$$M + MCl_4 \xrightarrow[1\ \text{atm}]{700} 2MCl_2 \qquad (1)$$

$$2MCl_2 + 2MCl_4 \xrightarrow[1\ \text{atm}]{400} 4MCl_3 \qquad (2)$$

$$3MCl_4 + M \xrightarrow[1\ \text{atm}]{} 4MCl_3 \qquad (3)$$

However, the direct preparation of the metal dichloride as shown in Eqn (1) does not always yield pure MCl_2 because of the difficulty in preventing the formation of MCl_3 during the cooling cycle of the reaction vessel. Zirconium and hafnium dichloride may be prepared by reducing MCl_3 with pure powdered metals [Eqn (4)].

$$M + 2MCl_3 \xrightarrow{675°} 3MCl_2 \qquad (4)$$

The zirconium and hafnium dichlorides are black powders and relatively more stable in air than the tetra or trichlorides. The dichlorides are not readily soluble in water. The solution after long standing produces a grayish precipitate.

11.2.2 Fluorides

The *zirconium* and *hafnium trifluorides* are prepared by treating the metal hydrides with stoichiometric amounts of hydrogen fluoride and hydrogen at 750° in a copper vessel.[3] The critical reaction temperature is reported to be within 20° of the given temperature. At higher temperatures the metal tetrafluoride is formed and at lower temperatures the reaction is incomplete. Another method involves the reduction of $(NH_4)_2MF_6$ with hydrogen.

The zirconium and hafnium trifluorides are bluish gray in color, insoluble in hot water, and relatively stable in air, but they decompose at 300°. At 850°–1000° they disproportionate to the corresponding metal and metal tetrafluoride.

11.2.3 Bromides

The *zirconium* and *hafnium tribromides* are prepared by reduction of the metal tetrabromide with pure powdered metal in a bomb.

$$MBr_4 + M \xrightarrow{300°\text{–}450°} MBr_3 \qquad (5)$$

[1] *Oka, Akira,* Bull. Chem. Soc., Japan *40*, 2284 (1967).

[2] *B. Swaroop, S. N. Flengas,* Can. J. Chem. *43*, 215 (1965).

[3] *P. Ehrich, F. Ploger, E. Koch,* Z. Anorg. Allgem. Chemie *333*, 209 (1964).

An alternative method is to react the metal tetrabromide with fine aluminum powder in a closed vessel.

$$3MBr_4 + Al \xrightarrow{470^\circ} 3MBr_3 + AlBr_3 \qquad (6)$$

Both zirconium and hafnium tribromides are fine crystalline materials, dark blue or black in color, and stable in a dry oxygen-free atmosphere. Traces of moisture decompose the metal tribromide to a red-brown and light gray material. The zirconium and hafnium tribromides are insoluble in THF, dioxane, chloroform, or pyridine.

11.2.4 Iodides

The *zirconium* and *hafnium* triiodides may be made by the reduction of liquid metal tetrahalides (under pressure) with metal at 510°[4]

$$M + 3MI_4 \xrightarrow{510^\circ} 4MI_3 \qquad (7)$$

The disproportionation of the triiodide at 360–390 is used to prepare the *diiodide* according to Eqn (8).

$$2MI_3 \xrightarrow{360^\circ-390^\circ} MI_2 + MI_4 \qquad (8)$$

The X-ray diffraction powder patterns for the metal tri- and diiodides are very similar to each other. This is perhaps due to the fact that the structure of the metal iodides is largely determined by the hexagonal arrangement of the iodide units.

11.3 Binary Zirconium and Hafnium Derivatives

11.3.1 Oxides

The *zirconium* and *hafnium oxides* are found together in the mineral baddeleyite. Purification processes have been described.[5] Usually the hydroxides or hydrous oxides of zirconium and hafnium are converted to the tetrachlorides or phosphates and separated by such techniques as ion-exchange, fractional crystallization, or fractional distillation.[6] A convenient recent separation method uses differential solvent extraction with fluorosubstituted acetylacetone. Using these fractional or differential methods, zirconium compounds are readily available with low hafnium contents or vice versa.

The zirconium and hafnium oxides are nearly identical in their chemical properties. Differences in physical properties are of great significance, however, and increased interest has focused on some of these differences in the last few years.[7,8] Both zirconium and hafnium oxides have distorted fluorite crystal structures which are intermediate to those in common rutile and fluorite structures. The phase relationships of the zirconium–oxygen system and hafnium–oxygen system are found in the literature.[9–12]

Zirconium and *hafnium dioxides* are prepared by hydrolytic decomposition of the metal tetrachloride in the presence of ammonia. The crude product is washed and dried at 100°–110° and finally calcined at 800°–1000°. Preparation of ultrahigh-purity ZrO_2 and HfO_2 has been described.[13–17] The method used involves precipitation from metal alkoxides $M(OR)_4$ as $MO_2 \cdot xH_2O$ and subsequent calcination at 500°–600°.

ZrO_2 and HfO_2 are thermally stable and quite resistant to chemical attack. Carbon monoxide and carbon at high temperature reduce ZrO_2 and HfO_2 to lower oxides or mixtures of the carbide and lower oxides. Oxygen-deficient gray or black ZrO_{2-x} and HfO_{2-x} are formed when zirconium oxides or hafnium oxides are heated *in vacuo*.[18] The lower oxides or suboxides of zirconium and hafnium may be prepared by the reduction of the metal oxides with metal at 1400°–1800°.

11.3.2 Carbides

The *zirconium* and *hafnium carbides* are prepared by the reduction of the metal oxides with

---

[4] *F. R. Sale, R. A. J. Shelton*, J. Less Common Metals *9*, 60 (1965).

[5] *D. R. Martin, P. J. Pizzolato* in *C. A. Hempel*, Rare Metals Handbook, 2nd Ed. Reinhold Publishing Co., New York 1961.

[6] *E. S. Gould*, Inorganic Reactions and Structure, Holt, New York 1955.

[7] *R. C. Garvie* in *A. M. Alper*, High Temperature Oxides, Part II, p. 117, Academic Press, New York 1970.

[8] *C. T. Lynch* in High Temperature Oxides, Part II, p. 193, Academic Press, New York 1970.

[9] *R. Ruh, H. J. Garrett*, J. Amer. Ceram. Soc. *50*, 257 (1967).

[10] *R. Ruh, J. J. Rockett*, J. Amer. Ceram. Soc. *53*, 360 (1970).

[11] *R. F. Domagala, R. Ruh*, Amer. Soc. Metals, Trans. Quart. Chem. Soc. *58*, 164 (1955).

[12] *E. Rudy, P. Stecher*, J. Less Common Metals *5*, 78 (1963).

[13] *K. S. Mazdiyasni, C. T. Lynch, J. S. Smith*, J. Amer. Ceram. Soc. *48*, 372 (1965).

[14] *K. S. Mazdiyasni, C. T. Lynch, J. S. Smith*, J. Amer. Ceram. Soc. *50*, 532 (1967).

[15] *K. S. Mazdiyasni*, Proceedings of the VI International Symposium on Reactivity of Solids, Edited by *J. W. Mitchel, R. C. De Vries, R. W. Roberts, P. Cannon*, p. 115, August 25–30, 1968.

[16] *K. S. Mazdiyasni, L. M. Brown*, J. Amer. Ceram. Soc. 53, 43 (1970).

[17] *L. M. Brown, K. S. Mazdiyasni*, J. Amer. Ceram. Soc. 53, 590 (1970).

[18] *P. Kofstad, D. Ruzicka*, J. Elektrochem. Soc. *110*, 181 (1963).

graphite in the temperature range of 1600°–2500°. The reduction of the metal oxide proceeds in three steps with the formation of M_2O_3, then MO, and finally the carbide.[19]

$$ZrO_2 + 2{\cdot}63C \longrightarrow ZrC_{0\cdot71}O_{0\cdot08} + 1{\cdot}92CO \quad (9)$$

$$HfO_2 + 2{\cdot}9C \longrightarrow HfC_{0\cdot95}O_{0\cdot05} + 1{\cdot}95CO \quad (10)$$

Another method involves the reaction of the metal halide with methane in a hydrogen atmosphere.[20]

$$MCl_4 + CH_4 + H_2 \longrightarrow MC + 4HCl + H_2 \quad (11)$$

Yet another method is to react the powdered element either by arc melting or powder metallurgy.

The *hafnium carbide* is very difficult to obtain in an oxygen-free state. Experience has shown that of all the carbides HfC is the most difficult to purify. Only by melting or heating in a good vacuum above ~2500° can the oxygen content be reduced to acceptable levels. The reaction between hafnium hydride and carbon goes much more quickly and also can produce an oxygen-free product if precautions are taken to purify the resulting carbide at high temperature.[21]

The phase relationships and physical, chemical, and thermomechanical properties of both zirconium and hafnium diborides have been reviewed[22] in great detail.

11.3.3 Borides

The methods used for the preparation of *zirconium* and *hafnium diborides* can be divided into the following groups:

(a) Synthesis from elements.[23–28]
(b) Electrolysis of molten oxides and salts.[29–31]
(c) Reduction with carbon of a mixture of metal dioxide (or corresponding hydroxide) and B_2O_3 (or boric acid).[32–36]
(d) Solid phase reactions which involve B_4C.[37–44]
(e) Synthesis from the metal and boron halides by a vapor deposition process according to the following reaction.[45–47]

$$MX_4 + 2BX_3 + 5H_2 \longrightarrow MB_2 + 10HX \quad (12)$$

Starting from pure elements and exercising great care (avoiding the introduction of impurities), pure products may be obtained by methods (a) and (e). The high temperatures necessary for synthesis from the elements, however, make these methods inconvenient, whereas synthesis from the gas phase is very expensive. Methods (c) and (d) lead to products containing smaller or larger amounts of carbon. Method (b) has been employed for pound-scale production of some borides; however, the product prepared by this method contains large amounts of impurities.

The chemical properties, method of fabrication, and practical applications of refractory borides have been reported.[48,49]

---

[19] *G. A. Meerson, G. V. Samsonov*, J. Appl. Chem. USSR *25*, 823 (1952).
[20] *I. E. Campbell, C. F. Powell, D. H. Nowicki, B. W. Gonser*, J. Elektrochem. Soc. *96*, 318 (1949).
[21] *H. Nowotny, R. Kieffer, F. Benesovsky, E. Brukl*, Monatsh. Chem. *90*, 86 (1959).
[22] *E. K. Storms* in *J. L. Margrave*, The Refractory Carbides, Vol. II, Academic Press, New York 1967.
[23] *A. Moissan*, Compt. Rend. *70*, 290 (1895); *A. Moissan*, Ann. Chim. Phys. *7*, 229 (1896).
[24] *E. Wderkind*, Ber. *46*, 1198 (1913).
[25] *P. Earlich*, Z. Anorg. Allgem. Chemie *1*, 259 (1949).
[26] *E. Decker, J. Kasper*, Acta Cryst. *7*, 77 (1954).
[27] *A. Polty, H. Margolin, J. Nielsen*, Trans. Amer. Soc. Metals *46*, 312 (1954).
[28] *V. A. Epel'baum, M. A. Gurevich*, Zh. Fiz. Khim. *32* 2275 (1958).
[29] *L. Andrieux*, Ann. Chim. Phys. *12*, 42 (1929); *L. Andrieux*, J. Four. Elec. *57*, 54 (1948).
[30] *F. Glaser*, Powder Met. Bull. No. 6, 51 (1951).
[31] *P. McKenna*, Ind. Eng. Chem. *28*, 767 (1936).
[32] *G. A. Kudintseva, B. M. Tsarev, V. A. Epel'baum*, Proc. Conf. on Boron, Its Chemistry and Its Compounds, Moscow 1955, p. 106, Gaskhimizdat, Moscow, 1958.
[33] *T. Atoda, S. Yamaguchi, S. Kitahara*, Sci. Paper Inst., Phys. Chem. Res (Tokyo) *53*, 68 (1959).
[34] U.S. Patent 2, *973*, 247 (1961).
[35] *H. Blumenthal*, Powder Met. Bull., No. 7, 79 (1956).
[36] *R. Kieffer, F. Benesovsky, E. Honak*, Z. Anorg. Allgem. Chemie, *268*, 191 (1952).
[37] *R. Meyer, H. Pastor*, Bull. Soc. Fr. Ceram. *66*, 59 (1965).
[38] *J. Kubo, T. Hanazawa*, Kogyo Kagaku Zasshi *63*, 1144 (1960).
[39] *V. F. Funke, S. I. Uvdkovskii, G. V. Samsonov*, Zh. Prikl. Khim. *33*, 831 (1960).
[40] *G. A. Meerson, G. V. Samsonov*, Zh. Prikl. Khim. *27*, 1135 (1954).
[41] *C. Baruch, T. J. Evans*, Metals 7, 909 (1955).
[42] *V. F. Funke, S. I. Judkovskii*, Porashkovzya Met. Akad. Nauk. Ukr. SSR 3 (4) 49 (1963).
[43] *Yu, B. Paderno, T. J. Serebryakova, G. V. Samsonov*, Tsvetn. Metal *11*, 48 (1959).
[44] *K. Moers*, Z. Anorg. Allgem. Chemie *198*, 243 (1931).
[45] *R. E. Gannon, R. C. Falweiler, T. V. Vasilos*, J. Amer. Ceram. Soc. *46*, 496 (1963).
[46] *P. Peshev*, Mitt. Inst. Allgem. Anorg. Chem. (Sofia) *4*, 53 (1966).
[47] *G. Bliznakov, P. Peshev*, J. Less Common Metals *7*, 441 (1964).
[48] *R. Thompson, A. A. R. Wood*, Chemical Engineer *3*, 51 (1963).
[49] *P. Peshev, G. Bliznakov*, J. Less Common Metals *14*, 23 (1968).

11.4 Zirconium and Hafnium Alkoxides

Metal alkoxides, $M(OR)_4$ (M = Zr or Hf), have been the subject of extensive research in recent years[50–53] owing to their industrial importance as precursor materials for several applications. Among these are the preparation of ultrahigh-purity submicron refractory oxide powders,[14–16] the vapor deposition of thin oxide films,[54,55] and the preparation of single and mixed phase ferroelectric and piezoelectric materials.[56–58] Zyttrite (Zyttrite®, Cherrybrooke Co., Fairborn, Ohio) which is an yttria or rare earth stabilized zirconium dioxide ceramic exhibiting excellent microstructural, thermomechanical, and thermophysical properties at elevated temperature (~2400°), also resulted from the processing of the alkoxy-derived mixed oxides.[14,17]

The following methods are most appropriate for laboratory-scale preparation of Zr and Hf alkoxides and (with minor change in the preparative procedures) may be suited for industrial production. Another method of preparing alkoxides is given in Eqn (29).

Most metal alkoxides are extremely sensitive to moisture, heat, and light. Therefore, the main emphasis must be directed to the utilization of very clean glass or stainless steel reaction vessels equipped with condenser and take-off in order to obtain a high purity, 90% theoretical yield product. The specific experimental conditions for particular preparations have been adequately described in the literature.

11.4.1 Ammonia Method

The most economic and simple method for a large-scale production of alkoxides involves the addition of commercial-grade anhydrous metal halide to a mixture of 10% anhydrous alcohol in the diluent (benzene or toluene) in the presence of anhydrous ammonia.[58–62]

$$MCl_4 + 4ROH + 4NH_3 \xrightarrow[25^\circ]{} M(OR)_4 + 4NH_4Cl \quad (13)$$

M = Zr, Hf
R = i-C_3H_7

The metal *tetrakis(isopropoxides)* are readily purified by recrystallization in hot isopropyl alcohol. The removal of NH_4Cl by filtration is always cumbersome and time consuming, the ammonia method may be carried out in the presence of an amide or nitrile. In this particular variation the metal alkoxide separates out as the upper layer, while the ammonium chloride remains in the amide or nitrile solution in the lower layer and the filtration step is eliminated.[63,64]

An improved preparation for Zr and Hf alkoxides has been reported recently.[65] This method consists of treating the anhydrous metal halide in benzene with various esters such as ethyl formate, ethyl acetate, diethyl oxalate, *n*-propyl acetate, isopropyl acetate, *n*-butyl acetate, and isobutyl acetate in the presence of anhydrous ammonia. The alcohol formed in the exothermic reaction is itself consumed in the formation of metal alkoxides.

$$MCl_4 + 4CH_3COOR \xrightarrow[\text{Exothermic}]{+4NH_3/5^\circ} M(OR)_4 + 4CH_3COOH \quad (14)$$

11.4.2 Heavier Alkoxides

The reaction generally employed to synthesize heavier alkoxides of Zr and Hf is an ester exchange reaction.[66] The method is particularly suited for the preparation of the tertiary butoxide from the isopropoxide. The reaction is shown in Eqn (15).

$$M(OCH(CH_3)_2)_4 + 4CH_3COOC(CH_3)_3 \xrightarrow{\Delta} M(OC(CH_3)_3 + 4CH_3COOCH(CH_3)_2 \quad (15)$$

[50] *D. C. Bradley*, Progr. Inorg. Chem. *2*, 303 (1960).
[51] *D. C. Bradley* in *Jolly*, Preparative Inorganic Reactions, Vol. II, p. 169, Wiley Intersci. Publ., New York 1962.
[52] *R. C. Mehrotra*, Inorg. Chim. Acta Rev. *1*, 99 (1967).
[53] *L. M. Brown, K. S. Mazdiyasni*, Anal. Chem. *41*, (10) 1243 (1969).
[54] *K. S. Mazdiyasni, C. T. Lynch* in *P. Popper*, Special Ceramics 1964, p. 115, Academic Press, New York 1965.
[55] *J. A. Aboaf*, J. Elektrochem. Soc. *114* (9) 948 (1967).
[56] *J. H. Harwood*, Chem. Process, Eng. *48* (16) 100 (1967).
[57] *K. S. Mazdiyasni, R. T. Dolloff, J. S. Smith*, J. Amer. Ceram. Soc. *52*, 523 (1969).
[58] *J. S. Smith, R. T. Dolloff, K. S. Mazdiyasni*, J. Amer. Ceram. Soc. *53*, 91 (1970).
[59] *H. J. Meerwein, J. Bersin*, Ann. *476*, 113 (1929).
[60] U.S. Patent 2, 187, 721 (1940); Brit. Patent 512, 452 (1939); Erf.: *J. Nelles*.
[61] *D. C. Bradley, W. Wardlaw*, Nature *165*, 75 (1950).
[62] *D. C. Bradley, W. Wardlaw*, J. Chem. Soc. (London), 208 (1951).
[63] *D. C. Bradley, R. C. Mehrotra, W. Wardlaw*, J. Chem. Soc. (London), 1634 (1953).
[64] U.S. Patent 2, *654*, 770 (1953); U.S. Patent 2, *655*, 523 (1953); Erf.: *D. F. Herman*.
[65] *S. K. Anand, J. J. Singh, R. K. Multani, B. D. Jain*, Israel J. Chem. *7*, 171 (1969).
[66] *R. C. Mehrotra*, J. Amer. Chem. Soc. *76*, 2266 (1954).

Since there is a large boiling point difference between the esters, the fractionation is simple and quite rapid. Another advantage appears to be the lower rate of oxidation of the esters compared to that of the alcohols.

11.4.3 Alcoholysis Reactions

Substitution of other branched R groups with the lower straight-chain alcohols has been achieved by an alcohol interchange reaction as illustrated in Eqn (16).

$$M(OR)_4 + 4R^1OH \underset{C_6H_6}{\overset{\Delta}{\rightleftharpoons}} M(OR^1)_4 + 4ROH \quad (16)$$

The distillation of an azeotrope drives the reaction to completion. This interchange is slow in the final stages with highly branched alcohols probably because of steric hindrance. The rise in temperature to the higher boiling alcohol signals the end of the reaction. In this method of preparation, speed, reversibility, and absence of side reactions are of major significance.

It should be emphasized that the alcoholysis reactions are carried out in the presence of benzene for the following reasons.

(1) Only a slight excess of higher alcohol is required for completing the reaction.
(2) The ethanol (or isopropanol) produced is removed continuously, pushing the reaction to completion. The amount of alcohol which is fractionated out azeotropically can be estimated by a simple oxidimetric method and the progress of the reaction can be followed.
(3) Side reactions such as the formation of alkyl chloride and water are avoided, these lower the yield considerably in the case of branched alcohols.
(4) When a higher boiling alcohol is used, the reaction is carried out at a lower refluxing temperature owing to the presence of benzene and, thus, side reactions are minimized.
(5) The oxidizing properties of the metals are lowered when they are linked covalently in accordance to the following scheme (Eqn 17).

$$M(OR^1)_4 + R^2OH \longrightarrow \begin{array}{c} R^2 \\ \diagdown \\ O-M(OR^1)_3 \\ \diagup \quad | \\ H \quad O-R^1 \end{array} \longrightarrow (R^2O)M(OR^1)_3 + R^1OH \quad (17)$$

11.4.4 Mixed Alkoxides

A number of *mixed Hf* and *Zr alkoxides*[52] have been prepared by the method shown in Eqn (18).

$$M(OR)_4 + nR'OH \longrightarrow M(OR)_{4-n}(OR')_n + nROH \quad (18)$$

R = tertiary group,
$R' = CH_3, C_2H_5$.
$n = 1$

11.4.5 Meerwein–Ponndorf Reaction

A very convenient method for the preparation of *primary and secondary Hf and Zr alkoxides* is the Meerwin–Ponndorf redox reaction involving a metal alkoxide and a ketone or aldehyde[67] [Eqn (19)].

$$M(OCHR_2^1)_x + xR_2^2CO \rightleftharpoons M(OCHR_2^2)_x + xR_2^1CO \quad (19)$$

This reaction has been utilized effectively in the preparation of *zirconium and hafnium 2,2,2-trichloroethoxides* by the addition of isopropoxides to chloral.

$$M[OCH(CH_3)_2]_4 + 4Cl_3CCHO \longrightarrow M(OCH_2CCl_3)_4 + 4(CH_3)_2CO \quad (20)$$

This reaction has its limitations owing to the occurrence of side reactions, especially with ionic or more electropositive alkoxides and aldehydes. However, it should be noted that the mechanism is not well understood.

11.4.6 Zirconium and Hafnium Alkoxyhalides

The *zirconium* and *hafnium alkoxyhalides* have been well reviewed.[68,69] This type of compound has been recognized for over a century and in almost all cases they are prepared by mixing hydrogen halides with the metal alkoxides in an inert solvent.

$$M(OR)_4 + nHX \longrightarrow MX_n(OR)_{4-n} + nROH \quad (21)$$
$X = Cl, Br$

The action of excess hydrogen halide results in the formation of essentially the same product as is obtained by the reaction of MX_4 and an alcohol.[70]

$$M(OR)_4 + 2HX \longrightarrow MX_2(OR)_2 \cdot ROH + ROH \quad (22)$$

$$MX_4 + 3ROH \longrightarrow MX_2(OR)_2\ ROH + 2HX \quad (23)$$

Also metal alkoxides react with acetyl chloride to give metal monochloride alkoxides.[70]

$$M(OR)_4 + CH_3COCl \longrightarrow MCl(OR)_3 + CH_3CO_2R \quad (24)$$

11.4.7 Acetylacetone Derivatives

The reaction of metal alkoxides with acetylacetone (acac) yields two types of derivatives,

[67] *H. Meerwein, R. Schmidt*, Justus Liebigs. Ann. Chem. *444*, 221 (1925).
[68] *R. C. Mehrotra*, J. Indian Chem. Soc. *50*, 731 (1953); *52*, 759 (1955).
[69] *N. M. Cullinane, S. J. Chard, G. F. Price, B. B. Millward*, J. Appl. Chem. *2*, 250 (1952).
[70] *D. C. Bradley, I. M. Thomas*, J. Chem. Soc. (London), 3857 (1960).

$M(OR)_3(acac)$ and $M(OR)_2(acac)_2$.[71,72] Also metal dichlorodiacetylacetone, $MCl_2(acac)_2$, may be prepared by the reaction of the metal dialkoxide diacetylacetone and hydrogen chloride.[73]

$$M(OR)_2(acac)_2 + 2HCl \longrightarrow MCl_2(acac)_2 + 2ROH \quad (25)$$

Another route by which the same compound can be prepared in quantitative yield is the reaction of the metal dialkoxychloride and acetylacetone[73] as outlined in Eqn (26).

$$MCl_2(OR)_2 + 2H(acac) \longrightarrow MCl_2(acac)_2 + 2ROH \quad (26)$$

11.4.8 Zirconium and Hafnium Alkylamide

The tetrakis(diethylamino) derivative is prepared by reaction of the metal halide with lithium diethylamide.[74]

$$MCl_4 + 4LiNOO(C_2H_5)_2 \longrightarrow M[NOO(C_2H_5)_2]_4 + 4LiCl \quad (27)$$

Furthermore, the tetrakis(dialkylamino) derivative can be prepared by aminolysis in which the metal derivative of a lower secondary amine is treated with a higher boiling secondary amine.[75]

$$M(NR^1_2)_4 + 4R_2{}^2NH \longrightarrow M(NR_2{}^2)_4 + 4R_2{}^1NH \quad (28)$$

By treating the metal dialkylamino derivatives with an alcohol, metal alkoxides are formed.[75]

$$M(NR_2)_4 + 4ROH \longrightarrow M(OR)_4 + 4R_2NH \quad (29)$$

11.4.9 Zirconium and Hafnium Fluoroalkoxides

Zirconium and *hafnium fluorosubstituted isopropoxides* are prepared by the reaction of the metal tetrachloride with sodium hexafluoroisopropoxide.[76] For example, 1,1,1,3,3,3-hexafluoroisopropoxide of zirconium or hafnium is prepared in the following manner.[77]

$$MCl_4 + 4NaOCH(CF_3)_2 \longrightarrow M[OCH(CF_3)_2]_4 + 4NaCl \quad (30)$$

The product is purified by sublimation.

Zirconium and *hafnium tetrakis(hexafluoroacetylacetonates)* $Zr(hfa)_4$ and $Hf(hfa)_4$ are prepared by the reaction of the metal halides and Hhfa in carbon tetrachloride.[78] The reaction is exothermic with the evolution of hydrogen chloride. The crude reaction product (yield >95%) for both zirconium and hafnium is a white crystalline material and is further purified by vacuum sublimation at room temperature and 0·03–0·02 mm Hg.

11.5 Cyclopentadienyl Derivatives

Biscyclopentadienyl derivatives of zirconium and hafnium are prepared by the reaction of cyclopentadienylmagnesium chloride or bromide with the metal chloride or bromide, respectively, in benzene–ether solution. A more convenient alternative procedure (generally giving higher yields) is the reaction of cyclopentadienylsodium with the metal halide in either tetrahydrofuran or 1,2-dimethyloxyethane solution.[79]

Dicyclopentadienyldiallyl derivatives of *zirconium* have been described.[80] The cream-colored $Zr(C_5H_5)_2(C_3H_5)_2$ is obtained by the reaction of dicyclopentadienyl zirconium dichloride with allylmagnesium chloride [Eqn (31)].

$$(C_5H_5)_2ZrCl_2 + 2H_2C{=}CH_2{-}CH{-}MgCl \longrightarrow 2MgCl_2 + (C_5H_5)_2Zr(CH_2{-}CH{=}CH_2)_2 \quad (31)$$

The zirconium derivative is sensitive to air and light at −18° but can be stored for long periods in an inert atmosphere.

Zirconium–sulfur bonds in dicyclopentadienyl derivatives of zirconium dichloride and $Na_2S_2C_2H_2$ react in chloroform[81,82] [Eqn (32)].

$$(C_5H_5)_2ZrCl_2 + NaS{-}CH{=}CH{-}SNa \longrightarrow (H_5C_5)_2Zr(S{-}CH{=}CH{-}S) + 2\,NaCl \quad (32)$$

The ethylene dithiolate derivative is a very unstable compound which is sensitive to air and solvolysis, but is stable in oxygen-free dry solvents such as benzene, chloroform and carbon disulfide.

[71] *D. C. Bradley* in *A. Cotton*, Progress in Inorganic Chemistry, Vol. II, Interscience Publishers, New York 1960.

[72] *D. M. Puri, K. C. Pande, R. C. Mehrotra*, J. Less Common Metals *4*, 393 (1962).

[73] *D. M. Puri, K. C. Pande, R. C. Mehrotra*, J. Less Common Metals *4*, 481 (1962).

[74] *A. Slawisch*, Chemiker-Ztg. *9*, 311 (1968).

[75] *R. Feld, P. L. Cowe*, The Organic Chemistry of Titanium, p. 114, Butterworth Co., London 1960.

[76] *H. J. Koetzsch*, Chem. Ber. *99*, 1143 (1966).

[77] *K. S. Mazdiyasni, B. J. Schaper, L. M. Branon*, Inorg. Chem. *10*, 889 (1971).

[78] *S. C. Chattoraj, C. T. Lynch, K. S. Mazdiyasni*, Inorg. Chem. *7*, 2501 (1968).

[79] *G. Wilkinson, J. M. Birmingham*, J. Amer. Chem. Soc. *76*, 4281 (1954).

[80] *H. A. Martin, P. J. Lemaire, F. J. Jellinck*, J. Organometal Chem. *14*, 149 (1968).

[81] *R. B. King, C. A. Eggers*, Inorg. Chem. *7*, 340, (1968).

[82] *H. Kopf*, J. Organometal Chem. *14*, 353 (1968).

Dicyclopentadienyl zirconium dichloride is used[82] as a starting material for the preparation of compounds containing the zirconium-sulfur bond [Eqn (33)].

$$(C_5H_5)_2ZrCl_2 + 2C_6H_5SH \xrightarrow{N(C_2H_5)_3} (C_5H_5)_2Zr(SC_6H_5)_2 + 2HCl \quad (33)$$

Derivatives with zirconium as part of a five-membered ring are also prepared by the reaction of dicyclopentadienyl zirconium dichloride with benzene 1,2-dithiol [Eqn (34)] and disodium malenitrile dithiolate [Eqn (35)].

$$(C_5H_5)_2ZrCl_2 + C_6H_4(SH)_2 \rightarrow (C_5H_5)_2Zr(S_2C_6H_4) \quad (34)$$

$$(C_5H_5)_2ZrCl_2 + (HS)_2C_2(CN)_2 \rightarrow (C_5H_5)_2Zr(S_2C_2(CN)_2) \quad (35)$$

The preparation of dicyclopentadienyl(*o*-hydrox-quinolato) hafnium chloride and the corresponding bromo derivatives has been described.[83]

11.6 Bibliography

G. A. Kudintseva, B. M. Tsarev, V. A. Epel'baum, Proc. Conf. Boron, Its Chemistry, and Its Compounds, Moscow, 1955, 106 pp., Gaskhimizdat, Moscow, 1958.

D. C. Bradley, Preparative Inorganic Reactions (W. L., Jolly, ed.), Vol. 2, pp. 169–186. Wiley, New York, 1962.

R. C. Mehrotra, Inorg. Chim. Acta *1*, 99 (1967).

E. K. Storms, The Refractory Carbides (J. L. Margrave, ed.), Vol. 2. Academic Press, New York, 1967.

K. S. Mazdiyasni, Proc. VI Int. Symp. Reactivity of Solids (J. W. Mitchel, R. C. DeVries, R. W. Roberts, and P. Cannon, eds.), pp. 115–125.

C. T. Lynch, in High Temperature Oxides (A. M. Alper, ed.), Part II, p. 193. Academic Press, New York, 1970.

[83] *E. M. Brainina, M. K. Minacheva, B. V. Lokshin,* Izv. Akad. Nauk. SSSR, Ser. Khim. 817 (1968).

12 Vanadium

R. J. H. Clark
Christopher Ingold Laboratories,
University College, London W.C.1, England

12.1 Hydride

In the presence of hydrogen at 1 atm pressure and at 300° metallic vanadium will absorb hydrogen over a period of 5 hr to form a hydride with the final composition $VH_{0.94}$ (48·4 atm % of hydrogen).[1]

Vanadium hydride is a gray metallic material which increases in brittleness with increasing hydrogen content. X-ray diffraction work has shown it to be a single phase with a tetragonal unit cell.

12.2 Halides

The halides of vanadium are important starting materials for the preparations of complexes. They vary in their composition from the six-coordinate polymeric and strongly reducing bivalent halides to the monomeric and oxidizing pentafluoride.

12.2.1 Halides of Vanadium(II)

Vanadium difluoride is prepared by reduction of the trifluoride at 1150° with a mixture of hydrogen fluoride and hydrogen.[2] It crystallizes as blue needles which possess the rutile structure.

Vanadium dichloride can be prepared by heating the trichloride for about 40 hr at 400°–675° in a current of hydrogen[3] or by disproportionation of the trichloride at about 800° in a stream of nitrogen.[4]

Vanadium dibromide can in a similar way be prepared by reduction of the tribromide with hydrogen. However, *vanadium diiodide* is best prepared[5] by thermal decomposition of the triiodide above 280°. Alternatively, the diiodide can be prepared[6,7] by direct reaction *in vacuo* of a stoichiometric mixture of the elements in a short quartz tube at 160°–170°.

The dichloride, dibromide, and diiodide are colored, hygroscopic, and antiferromagnetic solids, which are strongly reducing and possess the cadmium iodide structure. Other key properties of the dihalides are given in Table 1.

12.2.2 Halides of Vanadium(III)

Vanadium trifluoride can be prepared by passing anhydrous hydrogen fluoride in a stream of nitrogen over the trichloride or tribromide at

[1] *W. Rostoker*, The Metallurgy of Vanadium, John Wiley & Sons, New York 1958.

[2] *J. W. Stout, W. O. J. Boo*, J. Appl. Phys. *37*, 966 (1966).

[3] *W. Klemm, E. Hoschek*, Z. Anorg. Allgem. Chemie *226*, 359 (1936); *R. C. Young, M. E. Smith*, Inorg. Synth. *4*, 126 (1953).

[4] *P. Ehrlich, H. J. Seifert*, Z. Anorg. Allgem. Chemie *301*, 282 (1959).

[5] *A. Morette*, Compt. Rend. *207*, 1218 (1938).

[6] *W. Klemm, L. Grimm*, Z. Anorg. Allgem. Chemie *249*, 198 (1942).

[7] *D. Juza, D. Giegling, H. Schäfer*, Z. Anorg. Allgem. Chemie *366*, 121 (1969).

Table 1 Properties of Halides of Vanadium[a]

Halides	Color	M.p. (°C)	B.p. (°C)	D_4^{20}(pyk) (g/ml)	$\Delta H^\circ_{f,298}$ (kcal/mole)	S°_{298} (cal/deg/mole)	μ (BM)
VF_2	Blue	—	—	3·96	—	—	3·36 (300°K)
VCl_2	Green	*ca.* 910	Subl.p.	3·09	−110	23·2	2·41 (300°K)
VBr_2	Brown-orange	*ca.* 800	Subl.p.	4·52	−83	29	2·80 (300°K)
VI_2	Red-bronze	750–800	Subl.p.	5·0	−63	32	3·21 (288°K)
VF_3	Green	1406	—	3·36	−319	—	2·55 (293°K)
VCl_3	Red-violet	425	Disprop.	2·82	−134	31·3	2·74 (293°K)
VBr_3	Gray-brown	400	Disprop.	4·20	−118	40	2·67 (293°K)
VI_3	Brown-black	280	Dec.	4·2	−67	44	—
VF_4	Green	100	Subl.p. (dec.)	3·15	−321	—	1·68 (293°K)
VCl_4	Dark brown	−25·7	152	1·820	−136·2	58	1·61 (300°K)
VBr_4	Magenta	−23	Dec.	—	−108·5	*ca.* 80	—
VF_5	White	19·5	48·3	>2·5	−352	—	Diamagnetic

[a] *R. J. H. Clark, in* Comprehensive Inorganic Chemistry (J. C. Bailar, H. J. Emeléus, R. S. Nyholm, and A. F. Trotman-Dickenson, eds.). Pergamon, Oxford, 1973; see also Ref. 37.

Pyk = pyknometrically determined

about 600° in a platinum crucible.[8] It can also be prepared[9] by heating the dichloride in gaseous hydrogen fluoride at 700°, by heating the hydride with hydrogen fluoride at 400°, by disproportionation[10] of the tetrafluoride at 100°–120° and, together with other fluorides, by direct fluorination[9] of the metal at 200°. A further method involves the thermal decomposition of ammonium hexafluorovanadate(III) the latter being prepared by fusion of ammonium hydrogen fluoride with vanadium(III) oxide.

Vanadium trichloride is most simply prepared[11] by thermal decomposition of the tetrachloride on reflux for 2 days at 150°. Any tetrachloride which still remains at this stage may be distilled away at 200° in a stream of hydrogen. An alternative procedure[12] is to heat vanadium(III) oxide with thionyl chloride in a bomb for 24 hr at 200° [Eq. (1)].

$$V_2O_3 + 3SOCl_2 \longrightarrow 2VCl_3 + 3SO_2 \quad (1)$$

The bomb is cooled to 0° and opened to allow the sulfur dioxide to escape. The trichloride is then washed with carbon disulfide and dried *in vacuo* at 80°.

A further method involves refluxing vanadium(V) oxide with sulfur monochloride for 8 hr.[13] The excess of the latter, which contains dissolved sulfur, is then decanted off and the trichloride dried as described above.

$$2V_2O_5 + 6S_2Cl_2 \longrightarrow 4VCl_3 + 5SO_2 + 7S \quad (2)$$

The trichloride can also be prepared[14] by refluxing the powdered metal with iodine monochloride. After cooling the mixture, the iodine is removed in carbon tetrachloride and the trichloride dried as before.

Vanadium tribromide can be prepared by the action of dry bromine on the following substances: vanadium nitride at red heat, the carbide at 500°–600°, the metal in a sealed evacuated tube at 400°, ferrovanadium at red heat (the bromine being carried in a stream of dry carbon dioxide), or an intimate mixture of vanadium(V) oxide and carbon.[15] Alternative procedures which do not involve the use of bromine itself include (*a*) treatment[14] of the powdered metal with iodine monobromide and (*b*) the action[16] of carbon tetrabromide on vanadium(V) oxide in a sealed ampoule for 10–12 hr at 350°.

Early procedures for the synthesis of *vanadium triiodide* have given only a poorly characterized product. The most convenient method for the production of the pure, crystalline triiodide makes use of the reaction between vanadium metal and iodine in a sealed tube under a temperature gradient such that the vanadium metal is held at about 550° and the iodine at about 250°. Under these conditions, the triiodide is transported as it is formed to the region of the tube at about 350°, where it is deposited as large micaceous black platelets.[17]

The trihalides are all highly colored, paramagnetic, polymeric crystalline solids in which the coordination number of the metal atom is six (Table 1). The trichloride, tribromide, and triiodide have the hexagonal bismuth triiodide structure. The trihalides are insoluble in nonpolar solvents, but react with coordinating solvents with complex formation (see Section 12.13.3).

Some care must be taken on heating certain of the trihalides. Whereas the trifluoride will sublime unchanged, both the trichloride and the tribromide sublime with partial disproportionation (to $VX_2 + VX_4$) and the triiodide sublimes with some decomposition (to $VI_2 + I_2$).

12.2.3 Halides of Vanadium(IV)

Vanadium tetrafluoride is prepared[18] by treating the tetrachloride with anhydrous HF at −28°. The reaction is best carried out[10] in an inert solvent such as trichlorofluoromethane which remains liquid at −78°. The tetrafluoride may also be prepared by direct fluorination of the metal. Both tri- and pentafluorides are also produced during this reaction, but a clean separation of the different fluorides is possible owing to their different volatilities.

Vanadium tetrachloride can be prepared by the action of dry chlorine on the metal[19] at 250°–

[8] *A. Morette*, Compt. Rend. *207*, 1218 (1938).
[9] *H. J. Emeléus, V. Gutmann*, J. Chem. Soc. 2979 (1949).
[10] *R. G. Cavell, H. C. Clark*, J. Chem. Soc. 2692 (1962).
[11] *F. Ephraim, E. Ammann*, Helv. Chim. Acta *16*, 1273 (1933).
[12] *H. Hecht, G. Jander, H. Schlapmann*, Z. Anorg. Allgem. Chemie *254*, 255 (1947); *R. B. Johannesen*, Inorg. Synth. *6*, 119 (1960).
[13] *H. Funk, W. Weiss*, Z. Anorg. Allgem. Chemie *295*, 327 (1958).
[14] *V. Gutmann*, Monatsh. Chem. *81*, 1155 (1950).
[15] *G. Brauer*, Handbook of Preparative Inorganic Chemistry, 2nd Ed., Academic Press, New York 1965.
[16] *S. A. Shchukarev, T. A. Tolmacheva, V. M. Tsintsius*, Russian J. Inorg. Chem. (Engl. Transl.) *7*, 777 (1962).
[17] *K. O. Berry, R. R. Smardzewski, R. E. McCarley*, Inorg. Chem. *8*, 1994 (1969).
[18] *O. Ruff, H. Lickfett*, Ber. *44*, 2539 (1911).
[19] *P. Gross, C. Hayman*, Trans. Faraday Soc. *60*, 45 (1964).

300°, on the carbide, nitride, or silicide of vanadium, or on ferrovanadium. Alternatively, it may be prepared by disproportionation of the trichloride at 425° or above,[20] and collected by vacuum distillation. Further procedures[21] involve the action of various chlorinating agents (e.g., $SOCl_2$, SO_2Cl_2, or $COCl_2$) on the metal or on vanadium(V) oxide at about 350°.

Vanadium tetrachloride is purified by fractional distillation in an atmosphere of chlorine, and freed from the latter by repeated freezing and pumping *in vacuo*.

Vanadium tetrabromide was first isolated in 1964 by a procedure which involved[22] disproportionation of the tribromide at 325°. The tetrabromide vapor was trapped in a condenser held at −78°. The compound is, however, not easy to work with and begins to decompose at around −23°. Vanadium tetraiodide has never been isolated, but evidence for its presence in the vapor phase has been obtained through a study of the transpiration of vanadium diiodide in iodine vapor.[17]

The tetrahalides are all very sensitive to moisture, and rigorous precautions must be taken during all preparative operations, in order to avoid oxovanadium(IV) impurities. The tetrafluoride is a six-coordinate polymeric material which contains fluorine bridges; on the other hand, the tetrachloride and tetrabromide are four-coordinate monomeric species. All the tetrahalides are powerful Lewis acids, and readily coordinate to a wide range of ligands (Section 12.13.4).

Vanadium tetrachloride is barely stable thermodynamically, and even when held in a Pyrex tube over a pressure of chlorine it slowly decomposes into vanadium trichloride and chlorine. Whether this decomposition is catalytically or photochemically induced is not known, but it clearly imposes a limitation on both the storage and the transportation of the compound. The tetrahalides are all strongly colored (Table 1) and paramagnetic to the extent of one unpaired electron.

Mixed vanadium(IV) halides. Mixed chloride bromides, VBr_xCl_{4-x} ($x = 1, 2$, or 3), are believed to be formed in the vapor phase by passing bromine over heated vanadium trichloride, but they have not been isolated. Halogen redistribution reactions are presumed to take place instantaneously by analogy with the corresponding titanium(IV) system.[23]

[20] *O. Ruff, H. Lickfett*, Ber. *44*, 506 (1911).
[21] *E. Chauvenet*, Compt. Rend. *152*, 87 (1911).
[22] *R. E. McCarley, J. W. Roddy*, Inorg. Chem. *3*, 54 (1964).
[23] *R. J. H. Clark, C. J. Willis*, Inorg. Chem. *10*, 1118 (1971).

12.2.4 Halides of Vanadium(V)

The only established pentahalide of *vanadium* is the *pentafluoride*. This compound may be prepared[9] by passing gaseous fluorine diluted with nitrogen (or bromine trifluoride) over the powdered metal contained in a nickel boat at 300°. The product is collected in a rigorously dried, all-glass apparatus at −78° and purified by trap-to-trap distillation from dry sodium fluoride. It is essential in this preparation to eliminate not only moisture but also all traces of vacuum grease and hydrogen fluoride from the apparatus.

The pentafluoride is a colorless polymeric solid which melts to a pale yellow viscous liquid which is undoubtedly associated in the liquid state and consists of a chain polymer with *cis*-bridging fluorine atoms in the solid state. However, in the vapor state the molecule is monomeric, both electron diffraction and vibrational spectroscopic work indicating that it possesses the trigonal-bipyramidal structure.

Vanadium pentafluoride is believed to attack glass at room temperature and to be a powerful oxidizing and fluorinating agent.

A report of the preparation of vanadium pentachloride[24] has been shown to be spurious.[25]

12.3 Oxohalides

The oxohalides of vanadium, in contrast to the oxovanadium ions (see Section 12.5), have been little studied.

12.3.1 Oxohalides of Vanadium(III)

Only two *oxohalides of vanadium(III)* have been characterized, the oxochloride and oxobromide. *Vanadium oxochloride* can be prepared[26] by a chemical transport reaction between vanadium(III) oxide and vanadium trichloride by use of a temperature gradient of 620°–720°.

$$V_2O_3 + VCl_3 \longrightarrow 3VOCl \qquad (3)$$

It may also be prepared[27] by heating the oxodichloride in a stream of nitrogen. The oxochloride is a brown, slightly paramagnetic material.

Vanadium oxobromide is prepared similarly, either by thermal decomposition of the oxodi-

[24] *A. Slawisch, A. Jannopulos*, Z. Anorg. Allgem. Chemie *374*, 101 (1970).
[25] *I. M. Griffiths, D. Nicholls*, Chem. Commun. *713*, (1970).
[26] *H. Schäfer, F. Wartenpfuhl*, J. Less-Common Metals *3*, 29 (1961).
[27] *P. Ehrlich, H. J. Seifert*, Z. Anorg. Allgem. Chemie *301*, 282 (1959).

bromide at about 360° *in vacuo* or by a sealed tube reaction between vanadium tribromide and arsenic(III) oxide at 400° for 6 days.

Both oxohalides belong to the orthorhombic FeOCl structural type; other key properties are given in Table 2.

12.3.2 Oxohalides of Vanadium(IV)

Vanadium oxodifluoride is prepared by heating the oxodibromide in a stream of anhydrous hydrogen fluoride initially at 150°–200°, and then at 600°–700° for 6 hr.

Vanadium oxodichloride was originally prepared by reduction of the oxotrichloride with zinc powder at 400° in a sealed tube. However, a more efficient method is to heat vanadium(V) oxide, the trichloride, and the oxotrichloride in a sealed tube. The latter is maintained in a temperature gradient, the hotter end at 600°. Under these conditions, the oxodichloride is transported from the reaction zone over a period of 4–5 days according to Eqn (4).[13,15]

$$V_2O_5 + 3VCl_3 + VOCl_3 \longrightarrow 6VOCl_2 \quad (4)$$

Vanadium oxodibromide can be prepared either by thermal decomposition of the oxotribromide at 180° or by the action of bromine vapor on a mixture of vanadium(V) oxide and sulfur held in a sealed tube. If the hot end of the tube is maintained at 500°–600° and the cold end at 260°, the oxodibromide sublimes to the cool end.

The oxodiiodide has not, apparently, yet been prepared anhydrous.

Few physical properties and none of the structures of the oxodihalides have yet been established; however, there is no doubt that they are all polymeric.

12.3.3 Oxohalides of Vanadium(V)

Two types of *oxohalide of vanadium(V)* are known, VOX_3 and VO_2X. Of these, the former are the better established.

Vanadium oxotrifluoride is prepared by treating vanadium trifluoride, held at red heat in a platinum crucible, with oxygen[18] or by the action of fluorine[28] or bromine trifluoride[29] on vanadium(V) oxide.

Vanadium oxotrichloride is prepared[18] by the action of dry chlorine on vanadium(III) oxide or vanadium(V) oxide at 600°–800°, although temperatures of 300°–500° are sufficient in the presence of charcoal or sulfur. It may also be prepared by treatment of vanadium(V) oxide with hydrogen chloride, acetyl chloride, or thionyl chloride.[12]

$$V_2O_5 + 3SOCl_2 \longrightarrow 2VOCl_3 + 3SO_2 \quad (5)$$

In this case, the reaction mixture is held under reflux for 6–8 hr on a water bath under rigorously anhydrous conditions; the method yields pure vanadium oxotrichloride provided no excess thionyl chloride is used.

Another method[12] involves heating a mixture of vanadium(V) oxide and aluminum trichloride,

[28] *H. M. Haendler, S. F. Bartram, R. S. Becker, W. J. Bernard, S. W. Bukata*, J. Amer. Chem. Soc. 76, 2177 (1954).

[29] *H. J. Emeléus, A. A. Woolf*, J. Chem. Soc. (London) 164 (1950).

Table 2 Properties of Oxohalides of Vanadium[a]

Oxohalide	Color	M.p. (°C)	B.p. (°C)	D_4^{20} (pyk) (g/ml)	$\Delta H^\circ_{f,298}$ (kcal/mole)	S°_{298} (cal/deg/mole)
$VOCl$	Brown	620	Dec.	3·44	−144	18
$VOBr$	Violet	*ca.* 480	Dec.	4·00	—	—
VOF_2	Yellow	—	—	3·396	—	—
$VOCl_2$	Green	*ca.* 300	Disprop.	2·88	−165	28·5
$VOBr_2$	Yellow-brown	320	Dec.	—	—	—
VOF_3	Pale yellow	110	Subl.	2·459	—	—
$VOCl_3$	Yellow	−78·9	127·2	1·830	−177	—
$VOBr_3$	Deep red	−59	180 (dec.)	2·993	—	—
VO_2F	Brown	>300	—	—	—	—
VO_2Cl	Orange	—	180 (dec.)	2·29	−183	—

[a] *R. J. H. Clark, in* Comprehensive Inorganic Chemistry (J. C. Bailar, H. J. Emeléus, R. S. Nyholm, A. F. Trotman-Dickenson, eds.). Pergamon, Oxford, 1973; see also Ref. 37.

Pyk=pyknometrically determined

the required oxotrichloride being formed and distilled out of the reaction mixture at about 400°. The oxotrichloride is extremely sensitive to moisture; even very brief exposure to the atmosphere turns its color to orange or red.

A further method involves direct reaction between the trichloride and oxygen.[13] The liquid oxotrichloride is best purified by fractional distillation in the presence of sodium.

Vanadium oxotribromide can be prepared[30] by passing bromine vapor over vanadium(III) oxide or over a mixture of vanadium(V) oxide and carbon, temperatures of about 600° being required to bring about reaction. It is purified by distillation under reduced pressure. The oxotriiodide is unknown.

Vanadium oxotrifluoride attacks glass, especially when warm. All the oxotrihalides are hygroscopic, but especially the oxotrichloride and oxotribromide which react violently with moisture (rapid and profound intensification of color takes place in the presence of moisture). The oxotrihalides are all monomeric (at least in the vapor phase) with pseudotetrahedral symmetry.

The *dioxohalides of vanadium(V)* have only recently been established. The *dioxofluoride* is prepared by fluorination of the *dioxochloride*.[31] This, in turn, is prepared[32] by passing a stream of Cl_2O gas diluted with oxygen into vanadium(V) oxotrichloride at room temperature, by passing ozone into boiling vanadium oxotrichloride, or by reacting arsenic(III) oxide and vanadium oxotrichloride as follows.

$$VOCl_3 + Cl_2O \longrightarrow VO_2Cl + 2Cl_2 \quad (6)$$

$$VOCl_3 + O_3 \longrightarrow VO_2Cl + Cl_2 + O_2 \quad (7)$$

$$3VOCl_3 + As_2O_3 \longrightarrow 3VO_2Cl + 2AsCl_3 \quad (8)$$

The dioxochloride is an orange, very hygroscopic crystalline solid, which dissolves without apparent decomposition in solvents such as tetrahydrofuran or acetoacetic acid. It is probably polymeric in the solid state.

12.4 Oxides

A large number of different phases have been reported for the vanadium–oxygen system, but not all of these appear to be distinct. An incomplete phase diagram has been published by Rostocker[1] and a more complete one by Stringer,[33] who has also reviewed the subject critically.

12.4.1 Primary Vanadium–Oxygen Solid Solution (α Phase)

Vanadium-oxygen phases. Oxygen enters the body-centred cubic lattice of vanadium metal interstitially up to the solubility limit of 3·2 atom % (1 wt. %).[33] In so doing, it brings about a slight increase in the lattice parameter from 3·025 to 3·049 Å, as well as a marked hardening of the metal from 120 to 450 VHN (Vicker's hardness). At least two other vanadium-rich phases are known, with stability ranges $V_{0\cdot18}$ to $VO_{0\cdot33}$ (β phase) and $VO_{0\cdot53}$ (γ phase). The Vickers hardness increases uniformly over the two-phase $\alpha+\beta$ region to a maximum of 1100 VHN at 15 atom % (5 wt. %) oxygen.

12.4.2 Vanadium(II) Oxide (δ Phase)

Vanadium(II) oxide is formed as a gray metallic powder by reduction of higher oxides with a number of different reducing agents. A typical procedure involves heating a stoichiometric mixture of vanadium(III) oxide with vanadium metal powder [contained in an aluminum(III) oxide crucible] inside a small evacuated quartz tube. The optimum reaction conditions range from 1200°–1600°, reaction times of about 24 hr and 1 hr respectively, being required at these limiting temperatures.[15]

This oxide possesses the sodium chloride structure over a wide range of composition, variously quoted as $VO_{0\cdot80}$–$VO_{1\cdot30}$ or $VO_{0\cdot85}$–$VO_{1\cdot24}$. The X-ray density of stoichiometric vanadium(II) oxide is some 15% higher than the pyknometric density (5·55 g/cm^3), and, thus, the lattice is highly defective at the stoichiometric composition. At the vanadium-rich phase limit the cation lattice contains about 13% vacancies. Other key properties of the (stoichiometric) oxides are given in Table 3.

12.4.3 Vanadium(III) Oxide

Vanadium(III) oxide is prepared by reduction of vanadium(V) oxide with very pure hydrogen in two steps: (1) for 2 hr at 600° (i.e., below the melting point of V_2O_5) and (2) for a further 6 hr at 900°–1000°.

$$V_2O_5 + 2H_2 \longrightarrow V_2O_3 + 2H_2O \quad (9)$$

Its preparation in the form of single crystals has

---

[30] *F. Gonzâléz Nunez, E. Figueroa,* Compt. Rend. *206*, 437 (1938).

[31] *J. Weidlein, K. Dehnicke,* Z. Anorg. Allgem. Chemie *348*, 278 (1966).

[32] *K. Dehnicke,* Chem. Ber. *97*, 3354 (1964).

[33] *J. Stringer,* J. Less-Common Metals *8*, 1 (1965).

Table 3 Properties of the Principal Vanadium Oxides

Property and unit		V_2O_5	V_2O_4	V_2O_3	VO
Color		Orange	Blue	Gray-black	Gray
Melting point	(°C)	658[a]	1637[b]	1967[b]	950[c] disprop. vac.
Density (pyk)	(g/ml)	3·352[d] (22·6°)	4·260[d] (21·4°)	4·843[d] (22·0°)	5·55[e] (25·0°)
Heat of formation	$\Delta H^\circ_{f,298\cdot15}$ (kcal/mole)	−370·6[f]	−341·2[f]	−291·3[f]	−103·2[f]
Absolute entropy	$S^\circ_{298\cdot15}$ (cal/deg/mole)	31·3[d]	24·5[d]	23·5[d]	10·2[g]
Magnetic susceptibility (corrected for diamagnetism)	$10^6\chi^1$ (c.g.s.u.)	66[h] (293°K)	99[i] (293°K)	940[h] (293°K)	3376[f] (288°K)
Magnetic moment	(BM)	0	0·41[i]	1·49[h]	2·80[j]
Structure		Distorted tetragonal pyramid[k]	α Form distorted rutile[l]	Corundum[m]	Sodium chloride[n]
Space group		$Pmnm\text{–}D_{2h}^{13}$	$P2_1/c\text{–}C_{2h}^{5}$	$R\bar{3}c\text{–}D_{3d}^{6}$	$Fm3m\text{–}O_h^{5}$
No. molecules/unit cell		2	4	2	4
V–O distances (Å)		1·585–2·02[k]	1·76–2·05[l]	1·96–2·06[m]	2·05[n]

[a] *T. Carnelley*, J. Chem. Soc. *33*, 273 (1878).
[b] *E. Friederich, L. Sittig*, Z. Anorg. Allgem. Chem. *145*, 127 (1925).
[c] *H. Hartmann, W. Mässing*, Z. Anorg. Allgem. Chem. *266*, 98 (1951).
[d] *C. T. Anderson*, J. Amer. Chem. Soc. *58*, 564 (1936).
[e] *M. Frandson*, J. Amer. Chem. Soc. *74*, 5046 (1952).
[f] *A. D. Mah, K. K. Kelley*, U.S. Bureau Mines Rept. Invest., p. 5858 (1961).
[g] *N. P. Allen, O. Kubaschewski, O. von Goldbeck*, J. Electrochem. Soc. *98*, 417 (1951).
[h] *W. Klemm, E. Hoschek*, Z. Anorg. Allgem. Chem. *226*, 359 (1936).
[i] *W. Rüdorff, G. Walter, J. Stadler*, Z. Anorg. Allgem. Chem. *297*, 1 (1958).
[j] *E. Wedekind, C. Horst*, Chem. Ber. *45*, 262 (1912).
[k] *H. G. Bachmann, F. R. Ahmed, W. H. Barnes*, Z. Kristallogr. Kristallgeometrie, Kristallphys., Kristallchem. *115*, 110 (1961).
[l] *A. Magnéli, G. Andersson*, Acta Chem. Scand. *9*, 1378 (1955); *G. Andersson*, Acta Chem. Scand. *10*, 623 (1956).
[m] *G. Andersson*, Acta Chem. Scand. *8*, 1599 (1954); *R. E. Newnham, Y. M. de Haan*, Z. Kristallogr. Kristallgeometrie, Kristallphys., Kristallchem. *117*, 235 (1962).
[n] *N. Schönberg*, Acta Chem. Scand. *8*, 221 (1954).

also been described.[34] Many other reducing agents (*e.g.*, carbon or carbon monoxide) are capable of reducing V_2O_5 to V_2O_3. Like many of the vanadium oxides, vanadium(III) oxide is antiferromagnetic.

12.4.4 The Homologous Series V_nO_{2n-1}

By heating appropriate mixtures of the components V_2O_5, V_2O_3, and vanadium metal for 2–20 days at 650°–1000°, it is possible to prepare different members of a series of oxides with the general formula V_nO_{2n-1} (n=4–8).[33] These oxide phases have only very narrow regions of homogeneity. Their structures appear to be based on that of rutile but with periodic defects.

12.4.5 Vanadium(IV) Oxide

Vanadium(IV) oxide is prepared by heating an intimate mixture of V_2O_5 and V_2O_3 in stoichiometric proportions for some 40°–60 hr at 750°–800° in a small evacuated quartz tube. Alternatively, reduction of the pentoxide may be effected using other reducing agents such as carbon, carbon monoxide, sulfur dioxide, or oxalic acid. In the latter case, the V_2O_5 is fused with an excess of oxalic acid until a greenish blue, completely water-soluble oxovanadium(IV) oxalate is obtained. The latter is then calcined to the dioxide in the absence of air. Vanadium(IV) oxide is a deep blue crystalline material with a distorted rutile structure; it is antiferromagnetic.

12.4.6 Vanadium(V) Oxide

Vanadium(V) oxide is the final product of the oxidation of vanadium metal, lower vanadium oxides, ammonium vanadate, or sulfides or nitrides of vanadium. It is made in purest form by heating ammonium metavanadate to 500°–550°.

$$2NH_4VO_3 \longrightarrow V_2O_5 + 2NH_3 + H_2O \quad (10)$$

[34] *H. Hahn, C. de Laurent*, Angew. Chem. *68*, 523 (1956).

Any traces of nitrogen contaminating the product can be virtually eliminated by heating the latter for 18 hr at 530°–570° in a moist stream of oxygen. This oxide can also be prepared by burning the finely powdered metal in an excess of oxygen.

Vanadium(V) oxide is a reddish orange solid with a relatively low melting point (658°). It is slightly soluble in water, and also dissolves in both acids and bases. It possesses moderate oxidizing properties. The aqueous solutions readily become colloidal.
The pentoxide loses oxygen reversibly at 700°–1125°, a phenomenon which presumably accounts for its catalytic properties. Indeed, it is an important catalyst for the conversion of sulfur dioxide to sulfur trioxide, the sulfonation of aromatic hydrocarbons, the reduction of olefins, the oxidation of hydriodic acid by bromic acid or by hydrogen peroxide, of sugar by nitric acid, of alcohol by air, of stannous salts by nitric or chloric acids, of cyclic organic compounds by hydrogen peroxide, of naphthalene and its substitution products by air, and the reduction of aromatic hydrocarbons by hydrogen.[35]

12.5 Oxovanadium Species

12.5.1 Oxovanadium(IV) Species

Many oxo–metal species have been characterized, the most stable of which are the $VO_2^{\oplus}$ and $VO^{2\oplus}$ ions. Indeed, the latter is considered to be the most stable diatomic ion known.[36] It forms a wide variety of stable complexes, which may be cationic, neutral, or anionic.
Most *oxovanadium(IV) complexes* are of the following types:

$[VOL_5]^{n\pm}$ e.g. $R_3[VOF_5]$

$[VOL_4]^{n\pm}$ e.g. $R_2[VOF_4]$

$[VOL_xL^1_{5-x}]^{n\pm}$ e.g. $VOCl_2[(CH_3)_2SO]_3$

$[VOL_xL^1_{4-x}]^{n\pm}$ e.g. $VOCl_2[C_5H_5N]_2$

The $VO^{2\oplus}$ entity bonds most effectively to the more electronegative atoms, *e.g.*, F, Cl, O, or N, although bonds to S and P are also known. The fluoro complexes are the most stable. There are a very large number of oxovanadium(IV) compounds, and only typical preparations can be described here (for the oxohalides see Section 12.3).
The hydrated sulfate is probably the most common starting material for the preparation of oxovanadium(IV) derivatives, and is available commercially in at least two forms which differ only in their extent of hydration. Of these two forms, the blue one is water-soluble, the gray-green is not. The former is prepared by bubbling sulfur dioxide through a solution of vanadium(V) oxide in sulfuric acid; deep blue crystals of the required compound can be extracted from the resulting deep blue solution. Many oxovanadium(IV) adducts have been prepared by treating solutions of the sulfate with cationic, neutral, or anionic ligands.[36,37]
Ligands with acidic protons react with the oxovanadium(IV) ion in a different way; the procedure is best illustrated in the case of *vanadium(IV) oxo-bis-(acetylacetonate) i.e.* [*bis(2,4-pentanedione)oxovanadium(IV)*]. The procedure of treating vanadium(V) oxide with acetylacetone relies on the facts that acetylacetone (*a*) can reduce vanadium(V) to vanadium(IV), (*b*) reacts as an acid with the pentoxide to give the desired product, and (*c*) can behave as a solvent for the product.[38]

$$2V_2O_5 + 9C_5H_8O_2 \longrightarrow 4VO(C_5H_7O_2)_2 + (CH_3CO)_2CO + 5H_2O \quad (11)$$

Oxovanadium(IV) derivatives are typically five-coordinate (tetragonal-pyramidal) or six-coordinate (octahedral), but in all cases, the unique V–O bond is very short and dominates the structure (its typical length is about 1·6 Å).[37] There is particular interest in the study of the electronic spectra [*e.g.*, $VO(H_2O)_5^{2\oplus}$ ion],[39] magnetism, and ESR spectra of these complexes.[37] The most characteristic feature of all oxovanadium(IV) complexes is their very strong infrared-active band at $985 \pm 50\ cm^{-1}$ associated with the stretching vibration of the unique V–O bond.
Most *oxovanadium(IV) complexes* which have the tetragonal-pyramidal structure will attach a sixth ligand in the position *trans* to the unique V–O bond. Several studies of the heats, entropies, and free energies of coordination of these sixth ligands have been made.[40]

12.5.2 Oxovanadium(V) Species

Vanadium(V) forms two oxo species, typified by

[35] *N. V. Sidgwick*, The Chemical Elements and Their Compounds, Oxford University Press, Oxford 1950.
[36] *J. Selbin*, Chem. Rev. *65*, 153 (1965); *J. Selbin*, Coord. Chem. Rev. *1*, 293 (1966).
[37] *R. J. H. Clark*, The Chemistry of Titanium and Vanadium, Elsevier Publ. Co., Amsterdam 1968.
[38] *B. E. Bryant*, *W. C. Fernelius*, Inorg. Synth. *5*, 115 (1957).
[39] *C. J. Ballhausen*, *H. B. Gray*, Inorg. Chem. *1*, 111 (1962).
[40] *R. L. Carlin*, *F. A. Walker*, J. Amer. Chem. Soc. *87*, 2128 (1965).

the $VO^{3\oplus}$ and $VO_2^{\oplus}$ ions. These groupings have already been mentioned in respect of the oxohalides of vanadium(IV) (Section 12.3). These entities do not occur as free ions but are found in complexes of the types $VOCl_3 \cdot L$ and $VOCl_3 \cdot 2L$ (L = monodentate ligand). Such complexes are readily prepared by treating vanadium oxotrichloride, in an inert solvent and with rigorous exclusion of moisture, with the ligand (typically an oxygen or nitrogen donor ligand), which results in the immediate precipitation of the required complex.[41,42]

These complexes, which are highly colored, appear to be five- or six-coordinate monomers.

Vanadium oxotrichloride also reacts with ligands containing replaceable hydrogen atoms to produce substitution products of the types $VO(OMe)_3$, $VOCl_2(OMe)$, $VOCl_2(OEt)$, $VOCl(OEt)_2$, etc.[41]

The second structural type of oxovanadium(V) species is not well characterized, but may occur in such complexes as $VO_2(NO_3)$, VO_2F, and $VO_2(SbF_6)$, and also (as a *cis*-VO_2 unit) in the complexes $K_3[VO_2$ oxalate] and $K_3[VO_2F_4]$.[43]

12.6 Peroxides

In the preparation of peroxo complexes of vanadium a number of guidelines have been established.[44] These are that in aqueous solutions (a) the number of peroxo groups per vanadium atom increases with the alkalinity of the solution, (b) increase in the acidity increases the degree of polymerization and decreases the number of peroxo groups per vanadium atom, and (c) increase in the concentration of hydrogen peroxide decreases the degree of polymerization. Although many different kinds of peroxovanadium complexes have been reported, few have been well characterized. The best known are the tetraperoxovanadates K_3MO_8 (M = V, Cr, Nb, or Ta), all of which are isomorphous and possess the dodecahedral stereochemistry (D_{2d} symmetry).[45] The vanadium compound is prepared[46] by dropping a solution of the orthovanadate in hydrogen peroxide into a cold 50:50 mixture of alcohol and water.

12.7 Vanadates

In aqueous solution, vanadates on acidification undergo a series of hydrolysis–polymerization reactions. The reactions are complex and difficult to formulate correctly, and the situation is not helped by the fact that the attainment of equilibrium is slow. Several discussions of the species present in solution at closely defined conditions of pH and vanadium concentration have appeared.[47–49]

The most common vanadate is *ammonium metavanadate*, NH_4VO_3. This is readily prepared[50] by dissolving vanadium(V) oxide in sodium carbonate solution and then precipitating the metavanadate NH_4VO_3 from this solution with ammonium chloride. It is a colorless crystalline salt, which may be yellowish on account of traces of vanadium(V) oxide as impurity. It liberates ammonia above 50°, and is readily converted to vanadium(V) oxide on ignition. Alkali metal metavanadates can be prepared by related procedures.[51]

If a vanadate solution is acidified beyond pH 6·5, it turns bright orange and at appropriate concentrations will yield[52] crystalline salts of the *decavanadate ions* $[H_2V_{10}O_{28}]^{4\ominus}$, $[HV_{10}O_{28}]^{5\ominus}$, and $[V_{10}O_{28}]^{6\ominus}$ in various degrees of hydration, *e.g.*,

$$10[VO_3]^{\ominus} + 4H^{\oplus} \rightleftharpoons [V_{10}O_{28}]^{6\ominus} + 2H_2O \quad (12)$$

However, if sufficient acid is added to make the net anionic charge per metal atom exactly 0·6 and the solution then held at 60°, the light orange vanadate $M^I[V_3O_8]$ (M^I = monovalent cation) is precipitated. If a solution containing slightly less acid is rapidly evaporated at 40°, dark red

---

[41] *H. Funk, W. Weiss, M. Zeising*, Z. Anorg. Allgem. Chemie *296*, 36 (1958).
[42] *H. L. Krauss, G. Gnatz*, Chem. Ber. *95*, 1023 (1962).
[43] *W. P. Griffiths, T. D. Wickens*, J. Chem. Soc. (A), 400 (1968).
[44] *J. A. Connor, E. A. V. Ebsworth*, Adv. Inorg. Chem. Radiochem. *6*, 279 (1964).
[45] *J. D. Swalen, J. A. Ibers*, J. Chem. Phys. *37*, 17 (1962).
[46] *W. P. Griffith*, J. Chem. Soc. 5345 (1963); *W. P. Griffith*, J. Chem. Soc. 5248 (1964); *J. E. Fergusson, C. J. Wilkins, J. F. Young*, J. Chem. Soc. 2136 (1962).
[47] *L. G. Sillén*, Quart. Rev. *13*, 146 (1959).
[48] *K. F. Jahr*, Angew. Chem. Intern. Ed. Engl. *5*, 689 (1966).
[49] *M. T. Pope, B. W. Dale*, Quart. Rev. *22*, 527 (1968).
[50] *R. H. Baker, H. Zimmermann, R. N. Maxson*, Inorg. Synth. *3*, 117 (1950).
[51] *F. Holtzberg, A. Reisman, M. Berry, M. Berkenblit*, J. Amer. Chem. Soc. *78*, 1536 (1956).
[52] *H. T. Evans*, Inorg. Chem. *5*, 967 (1966).

crystals of the *pentavanadate* $M^I_3[V_5O_{14}]$ are formed.

The structures of the vanadates are very diverse, and contain examples of chain, layer, and discrete polymeric species, all of which involve oxygen bridging.

12.8 Sulfides and Selenides

A large number of chalcogenide phases have been reported, but some (*e.g.*, VS_5) do not seem to be well established. Much less information is available on the selenides and tellurides than on the sulfides; structural aspects of these compounds have been reviewed.[53]

All *vanadium sulfides* can be prepared by heating an intimate mixture of the elements as fine powders. The latter are placed in sintered clay crucibles inside quartz tubes which are then evacuated, sealed, and heated for many days *in vacuo* at 1000°–1300°. Contact between vanadium metal and quartz should be avoided.

Other procedures can be used to prepare particular sulfide phases, as indicated below.

Vanadium(II) sulfide can be prepared by thermal decomposition of vanadium(III) sulfide at 1000° in a stream of hydrogen; the product after 20 hr has the composition $VS_{1.02}$. Vanadium(II) sulfide is brown-black, and possesses the nickel arsenide structure.

Vanadium(III) sulfide is prepared by passing a moderately fast stream of hydrogen sulfide over a thin layer (~0·5 g) of vanadium(III) oxide for about 10 hr at 750°.

$$V_2O_3 + 3H_2S \longrightarrow V_2S_3 + 3H_2O \quad (13)$$

Procedures for the preparation of single crystals of this sulfide have also been established.[34] It is a black, paramagnetic material which is homogeneous between $VS_{1.17}$ and $VS_{1.53}$. It is resistant to dilute acids and insoluble in sodium hydroxide.

Vanadium polysulfide, VS_4, is prepared by heating vanadium(III) sulfide with an excess of sulfur for 15 hr at 400° in a sealed tube. The mixture is then annealed at 90° for 12 hr in order to convert the excess of sulfur into the soluble α form, which is then removed by exhaustive extraction in a Soxhlet apparatus. The polysulfide which remains behind is a black powder, which is resistant to nonoxidizing acids but is dissolved immediately by potassium hydroxide solution. Above 500° it decomposes to vanadium(III) sulfide and sulfur. It occurs naturally as the ore patronite.

Vanadium(II) selenide is prepared by thermal degradation of the tervalent selenide at 1000°–1100°. The latter is prepared according to Eqn (14). The reaction, which is carried out in a

$$V_2O_3 + 3H_2 + 3Se \longrightarrow V_2Se_3 + 3H_2O \quad (14)$$

quartz tube at 600°–900°, yields a product whose precise composition varies markedly (between $VSe_{1.9}$ and $VSe_{1.4}$) with the exact experimental conditions.[54]

12.9 Nitrogen Compounds, Phosphides, Arsenides, and Antimonides

Vanadium nitride, VN, is prepared in purest form by way of the following reactions.[15]

$$V + \tfrac{1}{2}N_2 \longrightarrow VN \quad (15)$$

$$V + NH_3 \longrightarrow VN + \tfrac{3}{2}H_2 \quad (16)$$

The powdered metal is contained in an alumina or molybdenum boat and reaction temperatures in the range 900°–1300° are required (depending on the particle size of the metal). Alternate methods involve (*a*) heating vanadium(III) oxide (in intimate mixture with carbon) with nitrogen at 1250° and (*b*) thermal decomposition[55] of the complex $(NH_4)_3VF_6$ at 600°. The nitride is a grayish violet powder (m.p. 2050°), which has the sodium chloride structure, and a range of homogeneity extending from $VN_{0.71}$ to $VN_{1.00}$. It liberates ammonia in the presence of alkalis and is also attacked by boiling nitric acid.

A second nitride, with a range of homogeneity extending from $VN_{0.37}$ to $VN_{0.50}$, can be prepared in a similar way by heating stoichiometric quantities of the nitride VN and vanadium metal to 1100°–1400°.

These nitrides are much less stable than those of Group IV elements, and this has ruled out their possible use as ceramic materials.[56]

Vanadium phosphides are prepared by direct reaction between the elements on their being heated in a quartz tube at high temperatures

[53] *F. Hulliger*, Structure and Bonding *4*, 83 (1968).

[54] *E. Hoschek, W. Klemm*, Z. Anorg. Allgem. Chemie *242*, 49 (1939).

[55] *H. Funk, H. Böhland*, Z. Anorg. Allgem. Chemie *334*, 155 (1964).

[56] *H. J. Goldschmidt*, Interstitial Alloys, Butterworths & Co., London 1967.

in vacuo for about 2 days.[57] The metal (contained in an alumina crucible) is held at 700°–1000°, while the phosphorus is held at 480°–550°. The phosphides are grayish black materials which are metallic in appearance. They are attacked by hot nitric acid, aqua regia, and concentrated sulfuric acid.

Little is known about other compounds of vanadium with elements of Group V of the periodic chart.[53]

Vanadium tetrachloride reacts with chlorine azide to produce[58] the explosive *vanadium tetrachloride azide*, $VCl_4 \cdot N_3$. This material splits off nitrogen in various solvents to give the volatile nitride chloride, $Cl_3V{=}N{-}Cl$. The latter, which reacts with both Lewis acids (*e.g.*, $SbCl_5$) and bases (*e.g.*, pyridine), has a crystal structure consisting of Cl_3VNCl units linked together as dimers.

12.10 Carbides, Silicides, and Borides

The *vanadium carbides*, VC and V_2C, are prepared in purest form by direct reaction of the elements in high vacuum at about 1300° for 1 day. The reactant mixture should be held in a graphite crucible. Alternate methods include reduction of vanadium oxides with carbon and hydrogen at high temperatures (1700°–2100°).

The carbide, VC, has the sodium chloride structure and a wide range of homogeneity. Both carbides are dark, very hard, chemically resistant substances, nitric being the only mineral acid to attack them in the cold.

The *vanadium silicides*, V_3Si, V_5Si_3, VSi_2, and *vanadium borides*, V_3B_2, VB, V_3B_4, and VB_2, have similar physical and chemical properties to those of the carbides, but are not important compounds.[56]

12.11 Alkoxides

The only established alkoxides of vanadium(III) are *vanadium(III) methoxide* $V(OCH_3)_3$, and *vanadium(III) ethoxide* $V(OC_2H_5)_3$. These can be prepared[59] under argon by treating the anhydrous vanadium trichloride with the appropriate lithium alkoxide in dry alkanol. They are green, nonvolatile solids which are undoubtedly polymeric.

The *vanadium(IV) tetraalkoxides* can be prepared by alcoholysis of the diethylamino compound [Eqns (17) and (18)].

$$VCl_4 + 4LiN(C_2H_5)_2 \longrightarrow V[N(C_2H_5)_2]_4 + 4LiCl \quad (17)$$

$$V[N(C_2H_5)_2]_4 + 4ROH \longrightarrow V(OR)_4 + 4(C_2H_5)_2NH \quad (18)$$

The primary alkoxides, $V(OR)_4$ ($R{=}CH_3$ or C_2H_5), are brown solids, the *n*-pentyl analog is a green solid, but most compounds of this type are dark brown liquids at room temperature. The tetramethoxide is probably a tetramer in the solid state.

The secondary alkoxides are dark green solids or liquids which are probably monomeric in benzene. The analogous tertiary alkoxides are also dark green liquids with the exception of the butoxide, $V(O\text{-}t\text{-}Bu)_4$, which is a blue liquid.

The preparations and properties of *alkoxide halides of vanadium(IV)* have been reviewed by Masthoff and co-workers,[60] together with information on *alkoxides of vanadium(V)* of the types $VO(OR)_3$ and $VO(OSiR_3)_3$, etc.[61]

12.12 Organometallic Compounds

The key organometallic compounds of this type, the *vanadium hexacarbonyl*, $V(CO)_6$, and *dibenzene vanadium(O)*, together with their derivatives, are discussed in Chapters 29 and 31, respectively. It remains here to draw attention only to the *σ-bonded derivatives of vanadium*, of which few have yet been characterized.

Treatment of the complex $(\pi\text{-}C_5H_5)_2VCl$ with a stoichiometric quantity of phenyllithium in dimethoxyethane at −50° results in the formation[62,63] of the *σ-bonded vanadium complex* $(\pi\text{-}C_5H_5)_2V(\sigma\text{-}C_6H_5)$ in solution; it can be isolated from petroleum ether at −80° as black

---

[57] *M. Zumbusch, W. Biltz*, Z. Anorg. Allgem. Chemie *249*, 1 (1942).

[58] *J. Strähle, K. Dehnicke*, Z. Anorg. Allgem. Chemie *338*, 287 (1965).

[59] *D. C. Bradley, M. L. Mehta*, Can. J. Chem. *40*, 1710 (1962).

[60] *R. Masthoff, H. Köhler, H. Böhland, F. Schmeil*, Z Chem. *5*, 122 (1965).

[61] *F. Schindler, H. Schmidbauer*, Angew. Chem. Intern. Ed. Engl. *6*, 683 (1967).

[62] *H.-J. de L. Meijer, M. J. Janssen, G. J. M. van der Kerk*, Rec. Trav. Chim. Pays-Bas *80*, 831, (1961); *H. J. de L. Meijer, M. J. Janssen, G. J. M. van der Kerk*, Chem. Ind. (London) 119 (1960).

[63] *H. J. de L. Meijer, G. J. M. van der Kerk*, Rec. Trav. Chim. Pays-Bas *84*, 1420 (1965).

crystals (m.p. 92°). The same compound can also be obtained by treatment of the complex $(\pi\text{-}C_5H_5)_2VCl_2$ with phenyllithium. σ-Bonded derivatives of vanadium(III) of the type $(\pi\text{-}C_5H_5)_2VR$ (where $R = C_6H_5$, $p\text{-}CH_3C_6H_4$, $p\text{-}(CH_3)_2NC_6H_4$, $C_6H_5CH_2$, or CH_3) melt in the range 90°–111°, are air-sensitive, and have magnetic amounts of ~2·7 BM at 280°K.

Organic halides react with the complex $(\pi\text{-}C_5H_5)_2V$ in two distinct ways as illustrated by Eqns (19) and (20).

$$(\pi\text{-}C_5H_5)_2V + RX(\text{excess}) \longrightarrow (\pi\text{-}C_5H_5)_2VX + \tfrac{1}{2}RR \quad (19)$$

$$(\pi\text{-}C_5H_5)_2V(\text{excess}) + RX \longrightarrow (\pi\text{-}C_5H_5)_2VR + (\pi\text{-}C_5H_5)_2VX \quad (20)$$

$R = CH_2C_6H_5, C_2H_5, CH_3$

It seems likely that the initial step in these reactions is the oxidative addition of the organic halide to form complexes of the type $(\pi\text{-}C_5H_5)_2VRX$, although the latter have not yet been isolated.

The purple *lithium vanadium complex* $Li_4[V(C_6H_5)_6]\cdot 3\cdot 5H_2O$ is prepared by treatment of an ether solution of phenyllithium with the complex $VCl_3\cdot 3(THF)$ (THF = tetrahydrofuran), the former being in tenfold excess.[64] The resulting complex anion apparently contains six phenyl groups σ-bonded to the vanadium atom, the latter being in the bivalent state (μ = 3·85 BM). The complex loses ether at 50° and biphenyl at 80°, and is readily decomposed by acids or water.

12.13 Vanadium Complexes

The complexes of vanadium vary widely in their stability, from those of vanadium(V) which are generally oxidizing, to those of vanadium in the bivalent and lower oxidation states which are strongly reducing. A summary of the oxidation states and stereochemistries of vanadium complexes is given in Table 4.

Table 4 Stereochemistries of Vanadium Compounds[a]

Oxidation State	Coordination number	Stereochemistry	Example[b]
1⊖	6	Octahedral	$[V(bipy)_3]^{\ominus}$, $[V(phen)_3]^{\ominus}$, $V(CO)_6^{\ominus}$
0	6	Octahedral	$V(bipy)_3$, $V(phen)_3$, $V(terpy)_2$, $V(CO)_6$, $V[C_2H_4(PMe_2)_2]_3$
1 (d^4)	6	Octahedral	$[V(bipy)_3]^{\oplus}$
	7	Octahedral + 1	$(Ph_3P)AuV(CO)_6$
2 (d^3)	6	Octahedral	VCl_2, $V(H_2O)_6^{2\oplus}$, $V(CN)_6^{4\ominus}$
3 (d^2)	6	Octahedral	$[V(C_2O_4)_3]^{3\ominus}$, $[VCl_3\cdot 3C_4H_8O]$, VF_3, $[V(CH_3OH)_4Cl_2]Cl$, $V[OC(NH_2)_2]_6^{3\oplus}$
	6	Trigonal prism	$V[S_2C_2(C_6H_5)_2]_3$[c]
	5	Trigonal-bipyramidal	$VCl_3\cdot 2NMe_3$
	4	Tetrahedral	$VCl_4^{\ominus}$, $VBr_4^{\ominus}$
4 (d^1)	8	Dodecahedral	$VCl_4\cdot 2(diars)$
	7	"4:3" Coordination	$VCl_4\cdot triars$
	6	Octahedral	$VCl_6^{2\ominus}$, VO_2, VCl_4bipy
	5	Tetragonal-pyramidal	$VO(acac)_2$, $[VO(NCS)_4]^{2\ominus}$
	5	Trigonal-bipyramidal	$VOCl_2\cdot 2NMe_3$
	4	Tetrahedral	VCl_4, $V(OSiPh_3)_4$
5 (d^0)	8	Dodecahedral	$V(O_2)_4^{3\ominus}$
	6	Octahedral	$VF_6^{\ominus}$
	5	Trigonal-bipyramidal	VF_5
	4	Tetrahedral	$VOCl_3$, $VO_4^{3\ominus}$

[a] *R. J. H. Clark*, The Chemistry of Titanium and Vanadium Elsevier, Amsterdam, 1968.

[b] $Ph = C_6H_5$, bipy = 2,2′-bipyridyl, Hacac = acetylacetone, diars = *o*-phenylenebisdimethylarsine, terpy = 2, 2′, 2″-terpyridyl, phen = 1. 10-phenanthroline, triars = *bis*-(*o*-dimethylarsinophenyl) methylarsine

[c] Oxidation state of the metal is not well-defined.

12.13.1 Electron-Rich Systems Involving Formally Vanadium (−I), (0), (I), and (II)

The lowest oxidation states of vanadium are most readily established with bidentate ligands such as 2,2′-bipyridyl, 1,10-phenathroline, and 1,2-bis(dimethylphosphino)ethane. In the first case, reduction of an alcoholic or aqueous solution of the complex $[V(bipy)_3]I_2$ with metallic magnesium or zinc yields[65] the zerovalent *vanadium bipyridyl complex* $[V(bipy)_3]$. Further reduction of the latter with lithium aluminum hydride in tetrahydrofuran (THF) yields the complex $Li[V(bipy)_3]\cdot 4THF$. Such compounds are all extremely air-sensitive. The colors of the various complexes in tetrahydrofuran and the magnetic moments of the solid complexes (at 293°K) are sensitive monitors of the formal oxidation states of the vanadium atom, namely $[V(bipy)_3]^{2\oplus}$, green, 3·75 BM (t_{2g}^3 configuration); $[V(bipy)_3]^{\oplus}$, red, 2·80 BM (t_{2g}^4); $[V(bipy)_3]$, blue, 1·80 BM (t_{2g}^5); and $[V(bipy)_3]^{\ominus}$, red 0 BM (t_{2g}^6).

Many of the synthetic procedures used for the preparation of low-valent bipyridyl complexes of

[64] *E. Kurras*, Montatsber. Deut. Akad. Wiss. Berlin. *2*, 109 (1960).

[65] *S. Herzog*, Z. Anorg. Allgem. Chemie *294*, 155 (1958).

vanadium fail in the case of *vanadium 1,10-phenanthroline complexes*. However, the readily prepared, dark blue complex $[V(phen)_3]I_2$ can be reduced with dilithium naphthalide or dilithium benzophenone in tetrahydrofuran [Eqn (21)].[66]

$$[V(phen)_3]I_2 + Li_2(C_6H_4)_2CO \longrightarrow [V(phen)_3] + (C_6H_5)_2CO + 2LiI \quad (21)$$

The neutral complex $[V(phen)_3]$ thus prepared consists of dark green crystals ($\mu = 1{\cdot}90$ BM) which are soluble in tetrahydrofuran, benzene, and pyridine to give reddish brown solutions. Two equivalents of iodine as expected are required to oxidize the zerovalent complex back to the original bivalent vanadium complex. Unlike the analogous bipyridyl complex, $[V(phen)_3]$ is not sublimable in high vacuum.

Further reduction[66] of the complex $[V(phen)_3]$ with the same reducing agent results in the formation of the corresponding complex ion of vanadium(-I) [Eqn (22)].

$$[V(phen)_3] + Li_2(C_6H_4)_2CO \xrightarrow{THF} Li[V(phen)_3]\cdot 3{\cdot}5(THF) + LiC_6H_4\text{—}CO\text{—}C_6H_5 \quad (22)$$

The required complex crystallizes as black rods from the dark blue tetrahydrofuran solution; these are extremely air-sensitive.

2,2′,2″-terpyridyl, like bipyridyl and phenanthroline, is also capable of forming electron-rich compounds of vanadium.[67]

Addition of an alcoholic solution of terpyridyl to an aqueous solution of vanadium(II) sulfate yields a deep green solution, from which deep green crystals of the *terpyridyl vanadium complex* $[V(terpy)_2]I_2$ can be precipitated with potassium iodide. If a solution of this complex in dimethylformamide is shaken with magnesium powder, it turns wine red. Black crystals of the neutral complex $[V(terpy)_2]$ can be isolated; the latter decomposes at $100°/10^{-3}$ mm Hg. Other reducing agents such as lithium aluminum hydride are also capable of reducing $[V(terpy)_2]I_2$ to $[V(terpy)_2]$. The complex ions are all *pseudo*-octahedral.

Tertiary phosphines, owing to their capacity to form π bonds in addition to σ bonds, are well capable of stabilizing low oxidation states of metals. In this connection, vanadium trichloride in tetrahydrofuran and in the presence of 1,2-bis(dimethylphosphino)ethane can be reduced[68] with sodium naphthalenide to the dark brown zerovalent complex $\{V[(CH_3)_2PCH_2CH_2P(CH_3)_2]_3\}$. This complex is thermally stable, but is rapidly oxidized in air.

Other electron-rich systems which have recently been studied include[69,70] the series of tris-bidentate complexes of the type

$$[V(S_6C_6R_6)]^n \qquad \left(\left[\begin{matrix} R\ \ R \\ -S-C=C-S- \end{matrix}\right]_3 V\right)^n$$

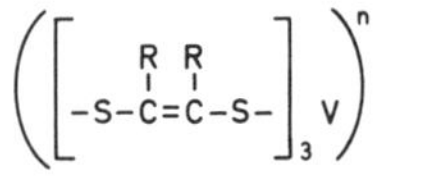

$R = CF_3$, $n = -2$ or -1;
$R = C_6H_5$, $n = -2$, -1, or 0;
$R = CN$, $n = -2$ or -1
and

$$[V(S_2C_6H_3X)_3]^{2\ominus}$$

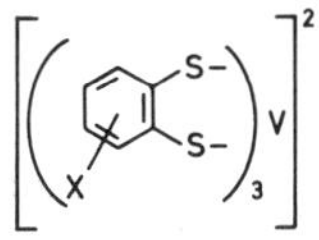

$X = H$ or CH_3.

The oxidation state of the vanadium atom in these complexes is not well defined.

12.13.2 Vanadium(II)

The best known complex is the violet *vanadium(II) sulfate*, $VSO_4\cdot 6H_2O$, which can be prepared by the sodium amalgam or zinc electrolytic reduction of a solution of vanadium(V) oxide in sulfuric acid.[71] Perhaps the simplest procedure for its preparation is the electrolytic reduction of oxovanadium(IV) sulfate ($VOSO_4\cdot 3H_2O$) followed by precipitation of the required complex with ethanol from the resulting vanadium(II) solution. The salt contains the $V(H_2O)_6^{2\oplus}$ ion.

The *ammonium, potassium, and rubidium vanadium sulfates* $M_2V(SO_4)_2\cdot 6H_2O$ (Tutton's salts) can likewise be prepared by the electrolytic reduction of a sulfuric acid solution of vanadium(V) oxide and the alkali metal sulfate. These amethyst-colored salts have the schönite structure and, hence, contain the $V(H_2O)_6^{2\oplus}$ ion. They are more stable to oxidation than the simple sulfate, especially in the absence of acid.

---

[66] *S. Herzog, U. Grimm*, Z. Chem. *4*, 32 (1964); *S. Herzog, U. Grimm*, Z. Chem. *6*, 380 (1966).

[67] *S. Herzog, H. Aul*, Z. Chem. *6*, 343 (1966).

[68] *J. Chatt, H. R. Watson*, J. Chem. Soc. (London) 2545 (1962).

[69] *J. H. Waters, R. Williams, H. B. Gray, G. N. Schrauzer, H. W. Finck*, J. Amer. Chem. Soc. *86*, 4198 (1964).

[70] *A. Davidson, N. Edelstein, R. H. Holm. A. H. Maki*, J. Amer. Chem. Soc. *86*, 2799 (1964); *A. Davidson, N. Edelstein, R. H. Holm, A. H. Maki*, Inorg. Chem. *3*, 814 (1964); *A. Davidson, N. Edelstein, R. H. Holm, A. H. Maki*, Inorg. Chem. *4*, 55 (1965).

[71] *L. F. Larkworthy, J. M. Murphy, K. C. Patel, D. J. Phillips*, J. Chem. Soc. (A) 2936 (1968).

Vanadium(II) hexacyanide, $K_4V(CN)_6 \cdot 3H_2O$, is prepared by reduction of vanadium(III) acetate with potassium amalgam in the presence of excess potassium cyanide.[72] It is a yellow-brown complex which is isostructural with the well-known iron analog, $K_4Fe(CN)_6 \cdot 3H_2O$.

Most studies involving the preparations of complexes of vanadium(II) involve treatment of the above sulfate complexes with solutions of the appropriate ligand with the rigorous exclusion of air. The most detailed of these have established two general types of complexes, namely $VCl_2 \cdot 4L$ and $VCl_2 \cdot 2L$.[73,74] The former type is found where L = H_2O, CH_3CN, C_5H_5N, and HCN; most of these complexes are blue with magnetic moments of ~3·89 BM, and all are considered to be octahedral monomers. The second type is known where L = H_2O, CH_3OH, C_4H_8O, $C_4H_8O_2$, C_5H_5N, and $HCON(CH_3)_2$; these are usually green, have magnetic moments of ~3·2 BM, and are considered to be chlorine-bridged polymers.

Many complexes of the type $[VL_6]^{2\oplus}$ have been prepared in association with the $[V(CO)_6]^{\ominus}$ ion (*e.g.*, where L = aniline, acetonitrile, or benzonitrile).[75] Such reactions take place directly between vanadium hexacarbonyl and the ligand in benzene, and rely on the fact that the hexacarbonyl, in the presence of Lewis bases, readily disproportionates [Eqn (23)] (see also Chapter 29).

$$3V^0 \longrightarrow V^{II} + 2V^{-I} \qquad (23)$$

12.13.3 Vanadium(III)

Addition complexes of vanadium(III) are usually prepared by reaction between the vanadium trihalide and the ligand, all operations being carried out on a vacuum line in order to avoid oxygen and moisture. Complexes with the following stoichiometries are known with monodentate ligands (L) (see Table 5): $[VL_6]^{3\oplus}(X^{\ominus})_3$, $[VL_4X_2]^{\oplus}X^{\ominus}$, $[VL_3X_3]$, $R^{\oplus}[VX_4L_2]^{\ominus}$, $(R^{\oplus})_2[VX_5L]^{2\ominus}$, and $(R^{\oplus})_3[VL_6]^{3\ominus}$ (R = monovalent cation, X = halide ion). In addition, many analogous complexes are known in which bidentate and multidentate ligands are involved.

[72] *J. M. Baker, B. Bleaney*, Proc. Phys. Soc. *65A*, 952 (1952).

[73] *H. J. Seifert, B. Gerstenberg*, Z. Anorg. Allgem. Chemie *315*, 56 (1962).

[74] *H. J. Seifert, T. Auel*, Z. Anorg. Allgem. Chemie *360*, 50 (1968); *H. J. Seifert, T. Auel*, J. Inorg. Nucl. Chem. *30*, 2081 (1968).

[75] *W. Hieber, J. Peterhans, E. Winter*, Chem. Ber. *94*, 2572 (1961).

Table 5 Typical Complexes of Vanadium(III) with Monodentate Ligands

Compound	Color	μ(BM)	Temperature °K	Ref.
$(NH_4)V(SO_4)_2 \cdot 12H_2O$	Blue-violet	2·80	300	*a*
$\{V[CO(NH_2)_2]_6\}Br_3$	Green	2·75	298	*b*
$[VCl_3 \cdot 3CH_3CN]$	Green	2·79	300	*c*
$[VCl_3 \cdot 3C_4H_8O]$	Orange	2·80	300	*d*
$[VCl_3 \cdot 2NMe_3]$	Purple-red	2·72	298	*e*
$Et_4N[VCl_4 \cdot 2CH_3CN]$	Yellow	2·81	300	*c*
$K_3[VF_6]$	Green	2·79	300	*f*
$(C_5H_6N)_3[VCl_6]$	Purple-pink	2·71	300	*g*
$(C_5H_6N)_3[VBr_6]$	Orange-brown	2·81	300	*g*

[a] *B. N. Figgis, J. Lewis, F. E. Mabbs*, J. Chem. Soc., p. 2480 (1960).

[b] *R. L. Carlin, E. G. Terezakis*, J. Chem. Phys. *47*, 4901 (1967).

[c] *R. J. H. Clark, R. S. Nyholm, D. E. Scaife*, J. Chem. Soc. A, p. 1296 (1966).

[d] *R. J. Kern*, J. Inorg. Nucl. Chem. *24*, 1105 (1962).

[e] *G. W. A. Fowles, C. M. Pleass*, J. Chem. Soc., p. 1674 (1957).

[f] *R. S. Nyholm, A. G. Sharpe*, J. Chem. Soc., p. 3579 (1952).

[g] *G. W. A. Fowles, B. J. Russ*, J. Chem. Soc. A, p. 517 (1967).

Typical of complex ions of the first type is the *vanadium(III) hexahydrate* ion, which is known from crystallographic work to be present as discrete units in the crystalline alum $CsV(SO_4)_2 \cdot 12H_2O$.[76] The ammonium alum is prepared by treating ammonium metavanadate, NH_4VO_3, in sulfuric acid with sulfur dioxide until the clear dark blue solution of the oxovanadium(IV) ion is produced, and then the latter is electrolytically reduced to vanadium(III). The reduced solution is allowed to stand in a closed vessel for ~3 days whereupon the alum crystallizes as red or blue crystals (depending on the concentration of sulfuric acid in the mother liquor). The alum effloresces slowly in air with loss of water and with oxidation. At 40°–50° the alum melts in its water of crystallization. The alums of other cations are prepared in a similar way. The $V(H_2O)_6^{3\oplus}$ ion is the principal species present in strong perchloric acid solutions of vanadium(III).

Another complex ion of this type is the *vanadium(III) hexaurea* ion which can be readily crystallized as the iodide or perchlorate.[77]

Complex ions of the type $[VL_4X_2]^{\oplus}$ are readily formed in solution by treating vanadium trichloride with alcohols such as methanol,

[76] *H. Hartmann, H. L. Schläfer*, Z. Naturforsch. *6A*, 754 (1961).

[77] *G. Barbieri*, Atti Reale Accad. Naz. Lincei, Rend. *24*, 435 (1915).

ethanol, *n*-propanol, isopropanol, *n*-butanol, isobutanol, and sec-butanol.[78] The complexes crystallizing from these solutions have either $[VL_4X_2]X$ or $[VL_3X_3]$ as the structural formula. Many *neutral complexes of vanadium(III)* may be formed by reflux of the appropriate trihalide either with the ligand itself or with a solution of the ligand.[79,80] These have the formula $[VL_3X_3]$ (*e.g.*, where L=ethanol, tetrahydrofuran, pyridine, or acetonitrile) and are six-coordinate octahedral monomers, or the formula $[VL_3X_2]$ (*e.g.*, where L=trimethylamine, triethylphosphine, or quinuclidine) and are five-coordinate trigonal-bipyramidal monomers with D_{3h} symmetry.[80–83]

Anionic Complexes of Vanadium(III). For example,[80] if a solution of the green complex $[VCl_3 \cdot 3CH_3CN]$ in acetonitrile is treated with a large organic halide, creamy yellow complexes of of the type $R^{\oplus}[VCl_4(CH_3CN)_2]^{\ominus}$ are precipitated (where R=$(C_2H_5)_4N$, $(C_6H_5)_3CH_3As$, or $(C_6H_5)_4As$). These behave as 1:1 electrolytes in acetonitrile, and involve an octahedrally coordinated anion. Ions of the types $[VCl_3Br \cdot 2CH_3CN]^{\ominus}$ and $[VBr_4 \cdot 2CH_3CN]^{\ominus}$ can be prepared similarly. Moreover, the coordinated acetonitrile in these anions can be replaced by ligands having stronger donor properties (*e.g.*, 2,2′-bipyridyl, 1,10-phenanthroline, or pyridine) to produce the new ions $[VCl_4 \cdot bipy]^{\ominus}$, $[VCl_4 \cdot phen]^{\ominus}$, and $[VCl_4 \cdot 2(py)]^{\ominus}$.

Several examples of ions of the type $[VL_6]^{3\ominus}$ are known. The $VCl_6{}^{3\ominus}$ and $VBr_6{}^{3\ominus}$ ions, known only as pyridinium salts,[84] are prepared by mixing the acetonitrile adducts $[VX_3 \cdot 3CH_3CN]$ with pyridinium hydrohalide in chloroform at ratios of greater than 1:3. The salt $K_3[V(NCS)_6]$ is prepared by treating a solution of vanadium(V) oxide in sulfuric acid first with sulfur dioxide to produce vanadium(IV) (the excess of sulfur dioxide being removed by boiling) and then by reduction of the resulting solution to vanadium(III) by electrolytic means. A stoichiometric quantity of potassium thiocyanate is then added. After removal of potassium sulfate by filtration, the required complex may be isolated as brownish red crystals.[85] An alternative procedure, which yields the tetraethylammonium salt, involves heating a mixture of potassium thiocyanate, tetraethylammonium bromide, and anhydrous vanadium trichloride under reflux in anhydrous ethanol[86] Related procedures[87] are available for the preparation of piperidinium salts of this anion, but similar attempts to prepare the pyridinium salt have led instead to the salt $pyH[V(NCS)_4(py)_2]$, analogous to Reinecke's salt of chromium(III) (py=pyridine).

The only *tetrahedral complex anions of vanadium(III)* are formed by thermal decomposition of, for example, the yellow salt $(C_6H_5)_4As[VCl_4(CH_3CN)_2]$. The overall reaction scheme is given in Eqn (24).[80,88]

$$VCl_3 + CH_3CN \xrightarrow{\nabla} [VCl_3(CH_3CN)_3] \xrightarrow{(C_6H_5)_4AsCl} [(C_6H_5)_4As]^{\oplus}[VCl_4(CH_3CN)_2]^{\ominus} \xrightarrow{80^\circ} [(C_6H_5)_4As]^{\oplus}[VCl_4]^{\ominus} \quad (24)$$

The blue product is isomorphous with the salt $(C_6H_5)As[FeCl_4]$, which is known from X-ray work to contain a tetrahedral anion. Similarly, thermal decomposition of the complex $[(C_2H_5)_4N]^{\oplus}[VBr_4 \cdot 2CH_3CN]^{\ominus}$ yields the complex $Et_4N[VBr_4]$, which likewise contains a tetrahedral anion.

A further anionic species of interest is the $[V_2Cl_9]^{3\ominus}$ ion, which can be prepared[88,89] as the tetraethylammonium salt by shaking a mixture of vanadium trichloride with tetraethylammonium chloride in thionyl chloride for 24 hr. The resulting solution is filtered, and the required deep red complex is precipitated with ether. The magnetic properties and electronic spectra of this anion indicate that there is significant interaction between the unpaired electrons on each paramagnetic centre.[88,90]

---

[78] *A. T. Casey, R. J. H. Clark*, Inorg. Chem. *8*, 1216 (1969).

[79] *R. J. H. Clark, J. Lewis, D. J. Machin, R. S. Nyholm*, J. Chem. Soc. (London), 379 (1963).

[80] *R. J. H. Clark, R. S. Nyholm, D. E. Scaife*, J. Chem. Soc. 1296 (1966).

[81] *M. W. Duckworth, G. W. A. Fowles, P. T. Greene*, J. Chem. Soc. (London) 1592 (1967).

[82] *K. Isslieb, G. Bohn*, Z. Anorg. Allgem. Chem. *301*, 188 (1959).

[83] *R. J. H. Clark, G. Natile*, Inorg. Chim. Acta *4*, 533 (1970).

[84] *B. J. Russ, G. W. A. Fowles*, J. Chem. Soc. (A) 517 (1967).

[85] *O. Schmitz-Dumont, G. Broja*, Z. Anorg. Allgem. Chemie *255*, 299 (1948).

[86] *R. J. H. Clark, A. D. J. Goodwin*, Spectrochim. Acta *26A*, 323 (1970).

[87] *H. Böhland, P. Malitzke*, Z. Anorg. Allgem. Chemie *350*, 70 (1967).

[88] *A. T. Casey, R. J. H. Clark*, Inorg. Chem. *7*, 1598 (1968).

[89] *D. M. Adams, J. Chatt, J. M. Davidson, J. Gerratt*, J. Chem. Soc. 2189 (1963).

[90] *R. Saillant, R. A. D. Wentworth*, Inorg. Chem. *7*, 1606 (1968).

Complexes of vanadium(III) with bidentate ligands (B) are prepared in the same way as those with monodentate ligands and, hence, further comment is not required. Neutral complexes of the type $[VB_3]$ (*e.g.*, where HB = benzoylacetone or dibenzoylmethane) can be prepared by addition of the ligand to an ethanolic solution of vanadium trichloride, followed by neutralization of the solution with 3 mol of piperidine. Related procedures have given a large number of complexes of this type with tris-β-diketones and tris-β-ketoamines.[91]

12.13.4 Vanadium(IV)

Addition compounds of vanadium tetrafluoride are often best made by treatment of the pentafluoride with excess ligand. For instance,[92] the pentafluoride reacts with ammonia to form an unidentified orange-brown solid, which then loses ammonia, followed by ammonium fluoride to yield a buff complex $VF_4 \cdot NH_3$. Likewise pyridine reacts with the pentafluoride to yield the vanadium(IV) complex $VF_4 \cdot py$. Both of these complexes are insoluble in common organic solvents and are presumed to be fluorine-bridged polymers. *Vanadium(IV) complexes* can also be prepared by direct reaction between vanadium tetrafluoride and the ligand.

Vanadium tetrachloride forms a large number of addition complexes with monodentate (L) and bidentate ligands (B) of the types $VCl_4 \cdot 2L$ and $VCl_4 \cdot B$; these are presumed to be six-coordinate octahedral complexes, although in few cases this has been definitely established owing to their low solubilities and ease of hydrolysis. The syntheses of such addition compounds are carried out in an inert solvent such as carbon tetrachloride, benzene, or cyclohexane. Addition complexes of the type $VCl_4 \cdot 2L$ are formed[93,94] (where L = acetonitrile, pyridine, phosphorus oxotrichloride, aldehydes, pyrazine, 2,6-dimethylpyrazine, benzophenone, tetrahydrofuran, tetrahydropyran, substituted pyridine *N*-oxides and many aromatic amines). In general, these complexes are reddish brown and have magnetic moments, close to 1·7 BM. With some ligands, such as pyridine, dialkyl sulfides, alkyl cyanides, or tetrahydrofuran, if present in excess, the vanadium is reduced to the tervalent state and complexes of the types $VCl_3 \cdot 3L$ or $VCl_3 \cdot 2L$ result. With typical bidentate ligands,[95] such as 2,2′-bipyridyl, 1,10-phenanthroline, 1,2-dimethoxyethane, or *o*-phenylenebisdiethylarsine, vanadium tetrachloride forms brown complexes of the expected type $VCl_4 \cdot B$.

Many of the above complexes of vanadium tetrachloride hydrolyze to discrete complexes of the types $VOCl_2 \cdot 2L$ or $VOCl_2 \cdot B$;[95] these are typical oxovanadium(IV) species, and are generally green.

On treatment with a ligand which possesses replaceable hydrogen ions, vanadium tetrachloride rapidly loses two molecules of hydrogen chloride. For example,[93] acetylacetone and benzoylacetone react with vanadium tetrachloride in an inert solvent to form the compounds $[VCl_2(C_5H_7O_2)_2]$ and $[VCl_2(C_{10}H_9O_2)_2]$.

Complete replacement of the chloride in vanadium tetrachloride can be brought about by treatment with lithium dialkylamides,[96,97] to give, for example, the dark green liquid $V(NR_2)_4$ ($R = C_2H_5$). These compounds react vigorously with alcohols to give the tetraalkoxides $V(OR)_4$ (Section 12.11).

Finally, the preparations of the *hexahalo anions of vanadium(IV)* are considered. The compounds M_2VF_6 (M = K, Rb, or Cs) are best prepared by heating the tetravalent salts M_2VF_5 with 1% fluorine gas (diluted with carbon dioxide) at about 500° for 10–15 min.[98] The products are then cooled rapidly to −78° in order to prevent the formation of vanadium(V) species. These salts are pinkish yellow, are stable in dry air, and have magnetic moments as expected of about 1·7 BM. The hexachloro salts M_2VCl_6 (M = pyridinium, quinolinium, or isoquinolinium) can be prepared by a reaction involving the fission of the oxovanadium bond of oxotetrachlorovanadates by thionyl chloride.[99]

$$M_2VOCl_4 + SOCl_2 \longrightarrow M_2VCl_6 + SO_2 \qquad (25)$$

[91] *F. Rohrscheid, R. E. Ernst, R. H. Holm*, Inorg. Chem. *6*, 1315, 1607 (1967);
F. Rohrscheid, R. E. Ernst, R. H. Holm, J. Amer. Chem. Soc. *89*, 6472 (1967).

[92] *R. G. Cavell, H. C. Clark*, J. Chem. Soc. (London) 2692 (1962).

[93] *H. Funk, G. Mohaupt, A. Paul*, Z. Anorg. Allgem. Chemie *302*, 199 (1959).

[94] *M. W. Duckworth, G. W. A. Fowles, R. A. Hoodless*, J. Chem. Soc. (London) 5665 (1963).

[95] *R. J. H. Clark*, J. Chem. Soc. 1377 (1963);
R. J. H. Clark, J. Chem. Soc. 5699 (1965).

[96] *D. C. Bradley, M. L. Mehta*, Can. J. Chem. *40*, 1183 (1962).

[97] *I. M. Thomas*, Can. J. Chem. *39*, 1386 (1961).

[98] *W. Liebe, E. Weise, W. Klemm*, Z. Anorg. Allgem. Chemie *311*, 281 (1961).

[99] *P. A. Kilty, D. A. Nicholls*, J. Chem. Soc. (London) 4915 (1965).

The cesium, potassium, and ammonium salts of the $VCl_6^{2\ominus}$ ion can be prepared by a similar procedure using nitromethane as solvent. The diethyl- and triethylammonium salts are formed by direct reaction between the appropriate alkylammonium halide and vanadium tetrachloride in chloroform. The *hexachlorovanadates(IV)* are extremely sensitive to moisture, but are otherwise stable up to at least 50°. Typical complexes of vanadium(IV) are listed in Table 6.

Table 6 Typical Complexes of Vanadium(IV)[a]

Compound	Color	μ(BM) (20°C)	M.p. (°C)	Ref.
$VF_4 \cdot NH_3$	Buff	1·83	250 (dec.)	*b*
$VF_4 \cdot py$	Gray-pink	1·79	150 (dec.)	*b*
$VCl_4 \cdot bipy$	Brown	1·77	—	*c*
$VCl_2(acac)_2$	Dark blue	—	160–165	*d*
$VCl_4(EtS{-}CH_2{-})_2$	Black	1·69	130 (dec.)	*e*
$VCl_4(diars)_2$	Orange	1·74	134 (dec.)	*f*
K_2VF_6	Pink-yellow	1·73	—	*g*
Cs_2VCl_6	Dark red	1·75	—	*h*

[a] py = Pyridine, bipy = 2,2′-bipyridyl, diars = *o*-phenylenebisdimethylarsine.

[b] *R. G. Cavell, H. C. Clark*, J. Chem. Soc., p. 2692 (1962).

[c] *R. J. H. Clark*, J. Chem. Soc., p. 1377 (1963).

[d] *H. Funk, H. Mohaupt, A. Paul*, Z. Anorg. Allgem. Chem. *302*, 199 (1959).

[e] *R. J. H. Clark, W. Errington*, Inorg. Chem. *5*, 650 (1966).

[f] *R. J. H. Clark, J. Lewis, R. S. Nyholm*, J. Chem. Soc., p. 2460 (1962).

[g] *W. Liebe, E. Weise, W. Klemm*, Z. Anorg. Allgem. Chem. *311*, 281 (1961).

[h] *P. A. Kilty, D. Nicholls*, J. Chem. Soc., p. 4915 (1965).

12.13.5 Vanadium(V)

The best defined complexes of vanadium(V) are the *hexafluorovanadates(V)* M^IVF_6 (M = K or Ag) and $Ba(VF_6)_2$. These can be prepared by the action of bromine trifluoride on a mixture of the appropriate chloride and vanadium trichloride.[100] The salts are colorless (with the exception of the orange-red silver salt) and diamagnetic. They are very readily hydrolyzed. There is no tendency for the $VF_6^{\ominus}$ ion to attach further fluoride ions to become seven- or eight-coordinate (*cf.* $NbF_7^{2\ominus}$ and $TaF_8^{3\ominus}$ ions).

A nitrosyl derivative, $NO^{\oplus}[VF_6]^{\ominus}$, is also known; this may be prepared by the action of excess of NOCl and bromine trifluoride on vanadium(V) oxide.[101]

12.14 Bibliography

P. Pascal, Nouveau Traité de Chimie Minérale. Masson, Paris, 1958.

G. Brauer, Handbook of Preparative Inorganic Chemistry, 2nd ed. Academic Press, New York and London, 1965.

J. Selbin, Chem. Rev. *65*, 153 (1965).

R. J. H. Clark, The Chemistry of Titanium and Vanadium. Elsevier, Amsterdam, 1968.

J. C. Bailar, H. J. Emeléus, R. S. Nyholm, A. F. Trotman-Dickenson (Editors), Comprehensive Inorganic Chemistry. Pergamon, Oxford, 1973.

[100] *H. J. Emeléus, V. Gutmann*, J. Chem. Soc. (London) 2979 (1949).

[101] *A. G. Sharpe, A. A. Woolf*, J. Chem. Soc. (London) 798 (1951).

13 Niobium and Tantalum

Lothar Kolditz
Sektion Chemie der Humboldt-Universität Berlin,
German Democratic Republic

There are three principal methods for preparing *metallic niobium* and *tantalum*: (*a*) thermal decomposition of oxides or halides; (*b*) reduction of halides or halide complexes with alkali metals or hydrogen; and (*c*) electrolysis of halide complexes in molten state.

Thermal decomposition of oxides is employed if the material is a conductor and can be brought quickly to higher temperatures by inductive electric heating. Niobium(IV) oxide (which is prepared from the pentoxide by heating with carbon) satisfies this condition. Application of alternating current results in elimination of oxygen and formation of the metal.[1] Thermal reduction with aluminum produces a niobium–aluminum alloy which only slowly releases aluminum on melting *in vacuo*; complete removal of aluminum cannot be accomplished by this procedure.

Thermal decomposition of halides is a useful method for producing very pure metal. Niobium is formed by decomposing NbI_3 at a wire heated to 1300°–1600° in an evacuated quartz vessel which has a wall temperature of 400°–600°.[2] Niobium and tantalum have also been prepared from the pentachlorides in an analogous manner.[3]

Reduction with alkali metal is used to prepare niobium or tantalum utilizing the complex K_2MF_7 (M = Nb, Ta). The latter reacts with sodium on heating according to Eqn (1).[4]

$$K_2MF_7 + 5Na \longrightarrow M + 5NaF + 2KF \qquad (1)$$

The crude product usually contains some oxide, but oxygen is readily eliminated on heating.[5] Reduction with hydrogen is possible with the pentachlorides. For example, $NbCl_5$ is reduced by hydrogen at temperatures between 1000° and 1300°.[6] Electrolysis of a melt is an excellent method to prepare the metals. Previous electrolysis experiments in aqueous solution did not succeed. However, a cryolite melt containing oxides of Nb and Ta can be used for the separation of Nb and Ta, since niobium is deposited at low current density (5 Å/dm²).[7] Electrolysis can also be performed in molten alkali fluoride or chloride utilizing heptafluoroniobate[8,9] or -tantalate.[10–12] For the preparation of niobium add niobium carbide to fluoroniobate–NaCl–KCl melts.[13] Solution of pure tantalum oxide or in combination with fluorotantalate in melts is also possible.[14,15] Electrolytic refining of tantalum has been achieved by use of tantalum anodes in NaCl or KCl melts which contained less than 2% $TaCl_5$ (by weight).[16]

13.1 Halides Having Oxidation States 4+ and 5+

13.1.1 Pure and Mixed Halides

The best methods for obtaining *Nb(V)* or *Ta(V) fluoride* are the fluorination of the powdered metal with elemental fluorine[17] or ClF_3[18] or exchange reactions of the chlorides with AsF_3 at room temperature.[19,20]

Niobium or *tantalum(V) halides* can be prepared by reaction of the corresponding oxides with halogenating agents such as $SOCl_2$,[21] CCl_4,[22]

[1] *W. v. Bolton*, Z. Elektrochemie *1113*, 145 (1907).
[2] *D. M. Tschishikov, A. M. Grinko*, Zh. Neorg. Khim. *4*, 982 (1959).
[3] *W. G. Burgers, J. C. M. Basart*, Z. Anorg. Allgem. Chemie *216*, 223 (1934).
[4] *O. P. Kolchin, M. A. Vol'dman*, Izv. Akad. Nauk SSSR, Neorgan Mater. *3*, (6) 1099 (1967).
[5] *W. v. Bolton*, Z. Elektrochemie *11*, 45 (1905).
[6] *L. A. Nisselson, Ja. M. Polyakov, A. N. Kresstovnikov*, Zh. Prikl. Khim. (Leningrad) *37*, 669 (1964).
[7] *R. Monier, Ph. Grandjean, J. Zahler*, Helv. Chim. Acta *46*, 2966 (1963).
[8] *M. Sakawa, T. Kuroda*, Denki Kagaku Oyobi Kogyo Butsuri Kagaku *37*, 99 (1969); C.A. *71*, 8985m (1969).
[9] *Cl. Decroly, A. Mukhtar, R. Winand*, J. Elektrochem. Soc. *115*, 905 (1968).
[10] *I. D. Efross, M. F. Lantratov*, Zh. Prikl. Khim. *36*, 2659 (1963).
[11] *S. Senderoff, G. W. Mellors, W. J. Reinhart*, J. Elektrochem. Soc. *112*, 840 (1965).
[12] *J. J. Rameau, M. J. Barbier*, C.R. Acad. Sci. Paris, Ser. *C 269*, 670 (1969).
[13] *M. Sakawa, T. Kuroda*, Denki Kagaku Oyobi Kogyo Butsuri Kagaku *37*, 69 (1969); C.A. *70*, 100299t (1969).
[14] *T. Tuchi, O. Kenji*, Sci. Repts. Res. Inst. Tohoku Univ. Ser. *A13*, 456 (1961); Ser. *A14*, 42 (1962).
[15] *V. M. Amosov*, Izv. Vyssh. Ucheb. Zaved, Tsvet. Met. *8*, 110 (1965).
[16] *I. Nakagawa, T. Kirihara*, Kogyo Kagaku Zasshi *68*, 1854 (1965); C.A. 13 208g (1966).
[17] *B. Frlec*, Vestn. Slov. Kem. Drus. *16*, 47 (1969); C.A. *72*, 85 705t (1970).
[18] *N. Ss. Nikolayev, Yu. A. Busslayev, A. A. Opalovski*, Zh. Neorg. Khim. *3*, 1731 (1958).
[19] *L. Kolditz, G. Furcht*, Z. Anorg. Allgem. Chemie *312*, 11 (1961).
[20] *L. Kolditz, Ch. Kürschner, U. Calov*, Z. Anorg. Allgem. Chemie *329*, 172 (1964).
[21] *H. Funk, W. Weiß*, Z. Anorg. Allgem. Chemie *295*, 327 (1958).
[22] *H. Schäfer, C. Pietruck*, Z. Anorg. Allgem. Chemie *264*, 2 (1951).

or CBr_4.[23] By circulating boiling $SOCl_2$ crude products are formed which contain $NbCl_5 \cdot SCl_4$ or $TaCl_5 \cdot SCl_4$. The latter can be refined by sublimation *in vacuo*. A suitable method for separating $NbCl_5$ and $TaCl_5$ involves fractional distillation in the absence of oxide chloride.[24] At normal pressure $NbCl_5$ boils at 248·3° and $TaCl_5$ at 234°. If the oxides are reacted with CBr_4 at 200° in a sealed evacuated tube, only $TaBr_5$ is produced, whereas niobium pentoxide yields $NbOBr_3$.

No oxide halides are formed if the metals react with bromine or iodine at temperatures above 250°; hence, this reaction yields *pure bromides* and *iodides*.[25]

Halogen exchange reactions are useful for the preparation of *Nb(V)* and *Ta(V) iodide*;[26] suitable starting materials are the corresponding chlorides which react *in vacuo* with AlI_3 at 98°–160° or SiI_4 at 170°–246°. It should be noted that, in general, the iodine content of these reaction products is somewhat lower than expected.

The general method for the preparation of *Nb(IV)* and *Ta(IV) halides* involves the reduction of the corresponding pentahalides with Nb or Ta metal, although Al, Si, or H_2 can be employed for the same purpose. NbI_4 is best prepared by thermal decomposition of the pentaiodide. References to reaction conditions and pertinent literature are included in Table 1.

Mixed halides can generally be formed by melting halide mixtures; however, this process results in mixtures which are not readily separable as was shown for the chloride fluoride species.[26a] Even readily formed compounds such as $(NbCl_4F)$ and $(TaCl_4F)$ cannot be obtained in a pure form by this procedure. The best method for obtaining these mixed halides is Cl–F exchange performed on coordination compounds $[PCl_4][MCl_6]$ (M = Nb or Ta) using AsF_3.

$$4[PCl_4][MCl_6] + 8AsF_3 \rightarrow (MCl_4F)_4 + 4PF_5 + 8AsCl_3 \quad (2)$$

This reaction can be carried out with $AsCl_3$ as a solvent at room temperature.[19,20]

Table 1 Niobium(IV) and Tantalum(IV) Halides

Compound	Reaction temperature (°C)	Reducing agent	Ref.
NbF_4	250°–350°	Si	*a*
	250°–350°	Nb	*a, b*
$NbCl_4$	400°	Nb, H_2, Al	*c*
$NbBr_4$	350°–410°	Nb	*d*
NbI_4	300°	Thermal decomposition of NbI_5	*e*
$TaCl_4$	620° (deposition at 300°)	Ta	*f*
$TaBr_4$	630° (deposition	Ta, Al	*g*
TaI_4	at 400°)		

[a] *F. P. Gortsema, R. Didchenko*, Inorg. Chem. *4*, 182 (1965).
[b] *H. Schäfer, F. G. von Schnering, K.-J. Niehues, H.-G. Nieder-Varenholtz*, J. Less Common Metals *9*, 95 (1965).
[c] *H. Schäfer, C. Pietruck*, Z. Anorg. Allgem. Chem. *266*, 151 (1951).
[d] *R. E. McCarley, A. T. Bruce*, Inorg. Chem. *2*, 540 (1963).
[e] *J. D. Corbett, P. W. Seabaugh*, J. Inorg. Nucl. Chem. *6*, 207 (1958).
[f] *H. Schäfer, H. Scholz, R. Gerken*, Z. Anorg. Allgem. Chem. *381*, 154 (1964).
[g] *R. E. McCarley, J. C. Boatman*, Inorg. Chem. *2*, 547 (1963).

The mixed halides $NbBr_3I_2$, $TaBr_4I$, and $TaBrI_4$[27] have not yet been shown to be pure compounds. They are thought to be formed from Nb_2O_5 or Ta_2O_5 in a melt with $AlBr_3$ or AlI_3 at 230°. $NbCl_5$ and $NbBr_5$ combine in a complete series of solid solutions.[28]

13.1.2 Halogen Complexes

Fluoro complexes of Nb(V) and *Ta(V)* can be oxidized electrochemically in salt melts (KF–NaF–LiF eutectic) to the 6+ state. The resultant compounds are stable at temperatures between 700° and 850° in the melt; at lower temperatures they decompose with elimination of fluorine. Graphite anodes and copper cathodes have been used. The composition of the complexes is not known.[29] Fluoro complexes having an oxidation state of 5+ can generally be prepared from the corresponding oxides and hydrofluoric acid in the presence of fluorides such as KF or NaF,[30,31] heptafluoro complexes are formed preferentially by this procedure. In KF LiF melts these compounds predominate.[32] Deposi-

[23] *M. Chaigneau*, Compt. Rend. *248*, 3173 (1959).
[24] *L. A. Nisselson*, Zh. Neorg. Khim. *3*, 2603 (1958).
[25] *R. F. Rolsten*, J. Amer. Chem. Soc. *79*, 5409 (1957); *R. F. Rolsten*, J. Phys. Chem. *62*, 126 (1958); *R. F. Rolsten*, J. Amer. Chem. Soc. *80*, 2952 (1958).
[26] *L. A. Nisselson, I. V. Petrusevich*, Zh. Neorg. Khim. *5*, 249 (1960).
[26a] *L. Kolditz, U. Calov*, Z. Anorg. Allgem. Chem. *376*, 1 (1970).

[27] *M. Chaigneau*, Compt. Rend. *247*, 300 (1958).
[28] *R. Dorschner, J. Dehand*, Bull. Soc. Chim. France 2056 (1967).
[29] *G. W. Mellors, S. Senderoff*, J. Elektrochem. Soc. *112*, 642 (1965).
[30] *M. C. Marignac*, Ann. Chim. (Paris) *8*, 5 (1866).
[31] *C. W. Balke, E. F. Schmith*, J. Amer. Chem. Soc. *30* 1637 (1908).
[32] *J. St. Fordyce, R. L. Baum*, J. Chem. Phys. *44*, 1166 (1966).

tion of pure complexes requires special concentrations of HF and cations.[33,34] A very pure *heptafluorotantalate* has been prepared by an anion exchange. *Hexafluoro-* and *octafluoro* complexes can be isolated as well as heptafluoro compounds, as by Eberts and Pink.[35] Formation constants have been given[36] for $TaF_9^{4\ominus}$ complexes. Hexafluoro complexes are obtained from pentafluorides and XeF_2,[37] SeF_4,[38,39] and $SeOF_2$[40] in 1:1 molar ratio, whereas $SeF_4 \cdot 2NbF_5$ must be formulated $[SeF_3]^{\oplus}[F_5NbFNbF_5]^{\ominus}$.[41] XeF_2 also forms a 1:2 compound.[37] In BrF_3 hexafluoro complexes can be prepared from the metals or oxides in the presence of KF.[42] A neutralization occurs in the BrF_3 system according to Eqn (3).

$$[BrF_2]^{\oplus}[NbF_6]^{\ominus} + K^{\oplus}[BrF_4]^{\ominus} \longrightarrow K[NbF_6] + 2BrF_3 \quad (3)$$

There are numerous reports on the systems of fluoro complexes with KCl and KF[43] and the pentachlorides with KCl and $MgCl_2$;[44,45] NaCl, KCl and $AlCl_3$;[46-48] NaCl and $FeCl_3$;[49] KCl and $ZrCl_4$,[50] because halo complexes are useful for the electrolytical preparation of the metals in salt melts (see Section 13.1.1). In fluoro complex systems the compounds cited have been observed as well as $KCl \cdot K_2[TaF_7]$.

In the *chloro systems* only hexacomplexes are formed with Nb and Ta, higher coordination numbers have not been observed. Experiments for direct preparation of octachloro complexes have failed[51] and in all binary systems of the pentachlorides with LiCl,[52,53] NaCl, KCl,[54-55] TlCl,[56] and CuCl[52] only hexachloro complexes could be detected.

In general, the preparation of *chloro or bromo complexes of Nb(V) and Ta(V)* proceeds best in nonaqueous solvents such as CH_3CN, $CHCl_3$, $AsCl_3$, $SOCl_2$, and ICl according to Eqn (4).

$$M = Nb, Ta; \quad X = Cl, Br; \quad MX_5 + X^{\ominus} \longrightarrow MX_6^{\ominus} \quad (4)$$

Whereas *hexabromo* complexes are stable materials,[57] the corresponding hexaiodo complexes have not yet been obtained. The same preparative procedure is used for the synthesis of mixed halo complexes $[N(C_2H_4)_4][MX_5Y]$ (X=Cl, Br; Y=Cl, Br, I).[58] All mixed complexes $[MF_nCl_{6-n}]^{\ominus}$ (n=1–5) have been observed in a solution of MCl_5 in HF (molar ratio 1:3).[59] $NbCl_5$ and $TaCl_5$ form 1:1 compounds with NOCl[60] and CNCl.[61] $TaCl_5$ and $SnCl_2$ interact to yield $SnCl_2 \cdot 2TaCl_5$[62]; $NbCl_5$ and

33 *M. Sakawa, T. Kuroda*, Denki Kagaku *36*, 146 (1968); C.A. *69*, 24075d (1968).
34 *G. S. Savchenko, I. V. Tananayev*, Zh. Prikl. Khim. *19*, 1093 (1946);
G. S. Savchenko, I. V. Tananayev, Zh. Prikl. Khim. *20*, 385 (1947).
35 *R. E. Eberts, F. X. Pink*, J. Inorg. Nucl. Chem. *30*, 457 (1968).
36 *L. P. Varga, H. Freund*, J. Phys. Chem. *66*, 21 (1962).
37 *J. H. Holloway, J. G. Knowles*, J. Chem. Soc. A 756 (1969).
38 *R. J. Gillespie, A. Whitla*, Can. J. Chem. *48*, 657 (1970).
39 *A. J. Edwards, G. R. Jones*, J. Chem. Soc. A 1891 (1970).
40 *A. J. Edwards, G. R. Jones*, J. Chem. Soc. A 2858 (1969).
41 *A. J. Edwards, G. R. Jones*, J. Chem. Soc. A 1491 (1970).
42 *V. Gutmann, H. J. Emeléus*, J. Chem. Soc. (London) 1046 (1950).
43 *Bin-ssin Zui, N. P. Lushnaya, W. I. Konstantinov*, Zh. Neorg. Khim. *8*, 389 (1963).
44 *N. D. Chikanov, A. P. Palkin*, Zh. Prikl. Khim. *37*, 1830 (1964).
45 *N. D. Chikanov, V. T. Ilginova*, Zh. Neorg. Khim. *11*, 1455 (1966);
N. D. Chikanov, V. T. Ilginova, Zh. Neorg. Khim. *11*, 2822 (1966).
46 *I. Ss. Morosov, W. A. Krochin*, Zh. Neorg. Khim. *7*, 2400 (1962).
47 *N. D. Chikanov, A. A. Ivanova*, Zh. Neorg. Khim. *12*, 2219 (1967).
48 *N. D. Chikanov, A. A. Dubyanskaya*, Zh. Neorg. Khim. *12*, 3187 (1967).
49 *I. Ss. Morosov, A. T. Ssimonich*, Zh. Neorg. Khim. *6*, 937 (1961).
50 *I. Ss. Morosov, N. P. Lipatova, A. T. Ssimonich*, Zh. Neorg. Khim. *8*, 172 (1963).
51 *D. Brown*, Proc. Int. Conf. Coord. Chem. 8th. 437 Vienna 1964.
52 *D. J. Toptigin*, Zh. Neorg. Khim. *8*, 1187 (1963).
53 *K. Huber, E. Jost*, Congr. intern. chimpur et. appl. 16. Paris 1957, published 1958.
54 *N. D. Chikanov, A. P. Palkin*, Zh. Neorg. Khim. *10*, 1259 (1965).
A. P. Palkin, N. D. Chikanov, Zh. Neorg. Khim. *7*, 1370, 2388, 2394 (1962).
55 *O. R. Gavrilov, L. A. Nisselson*, Zh. Neorg. Khim. *11*, 209 (1966).
56 *N. D. Chikanov, L. A. Kazberova*, Zh. Neorg. Khim. *12*, 2509 (1967).
57 *C. Furlani, E. Zinato*, Z. Anorg. Allgem. Chemie *351*, 210 (1967).
58 *G. A. Ozin, G. W. A. Fowles, D. J. Tichmarsh, R. A. Walton*, J. Chem. Soc. (London) A 642 (1969).
59 *Yu. Y. Buslaev, G. Ilin, V. Bainova, M. N. Krutkina*, Dokl. Akad. Nauk SSSR *196*, 374 (1971).
60 *J. A. MacCordick, R. Rohmer*, Compt. Rend. *263*, 1369 (1966).
61 *J. A. MacCordick*, Compt. Rend. *266*, 1296 (1968).
62 *I. Ss. Morosov, Tschi-Fa Li*, Zh. Neorg. Khim. *8*, 2733 (1963).

$GaCl_3$ form $GaCl_3 \cdot NbCl_5$.[63] PCl_5 and $NbCl_5$ or $TaCl_5$ form $[PCl_4][NbCl_6]$ and $[PCl_4][TaCl_6]$ in $POCl_3$ or $AsCl_3$ as solvents.[64] These compounds crystallize from $AsCl_3$ coordinated with one molecule of solvate.

Of *Nb(IV)* and *Ta(IV) halo complexes* only the hexachloro derivatives are known. They are formed from tetrachlorides and alkali chlorides[65–68] and the Cs salts were found to be thermally most stable.[69] Cs, Rb, and NH_4 salts can be separated from a HCl-saturated aqueous solution of Nb(IV) by addition of the corresponding chlorides.[70] Another preparative method involves heating mixtures of $CsMCl_6$ (M=Nb, Ta) and CsCl in the presence of red phosphorus as reducing agent.[71]

13.1.3 Complexes with Other Ligands

Nb(V) and *Ta(V) halides* exist in oligomer units linked by halogen bridges. In mixed halides fluorine bridging is preferred to chlorine bridging.[26a, 72] Donor molecules can cleave the bridging units to form complexes. In the case of mixed halides, pure halide complexes are obtained by mutual halogen exchange.[26a] In principle, the same holds true for Nb(IV) and Ta(IV) compounds, although mixed halides are not yet known in this case. Table 2 contains a selection of derivatives having an oxidation state of 5+.

Cyclopentadienyl compounds of Nb and *Ta* are

[63] *P. J. Fedorov, L. P. Chagleeva*, Zh. Neorg. Khim. *12*, 818 (1967).

[64] *R. Gut, G. Schwarzenbach*, Helv. Chim. Acta XLII, 2156 (1959).

[65] *B. G. Korshunov, W. W. Ssafonov*, Zh. Neorg. Khim. *6*, 733 (1961).

[66] *W. W. Ssafonov, B. G. Korshunov, S. N. Schevzova*, Zh. Neorg. Khim. *7*, 1979 (1962);
W. W. Ssavonov, B. G. Korshunov, S. N. Schevzova,

[67] *L. G. Schadrova*, Zh. Neorg. Khim. *9*, 1406 (1964);
W. W. Ssafonov, B. G. Korshunov, S. N. Schevzova,

[68] *L. G. Schadrova*, Zh. Neorg. Khim. *10*, 669 (1965);
Je. K. Smirnova, I. V. Vassilkova, Vestnik Leningradskogo Univ. *30*, Ser. Fiz. Khim. *2*, 161 (1965);
Je. K. Smirnova, I. V. Vassilkova, Zh. Neorg. Khim. *12*, 566 (1967).

[69] *W. W. Ssafonov, B. G. Korshunov*, Izv. Akad. Nauk SSSR, Neorg. an. Mater. *1*, 604 (1965).

[70] *I. Ss. Morosov, N. P. Lipatova*, Zh. Neorg. Khim. *11*, 1018 (1966).

[71] *S. M. Horner, F. M. Gollier jr., S. Y. Tyree jr.*, J. Less-Common Metals *13*, 85 (1967).

[72] *H. Preis*, Z. Anorg. Allgem. Chemie *346*, 272 (1966);
H. Preis, Z. Anorg. Allgem. Chemie *362*, 13 (1968).

Table 2 Niobium and Tantalum Pentahalide Complexes

Compound	Remarks	Ref.
$MF_5 \cdot R_2O$	$R = CH_3, C_2H_5$	*a*
$MF_5 \cdot R_2S$	$R = CH_3, C_2H_5$	*a*
$MF_5 \cdot 2(CH_3)_2SO$	—	*b*
$MF_5 \cdot 2(py)$	$py = C_6H_5N$	*c*
$MF_5 \cdot 2NH_3$	—	*d*
$MF_5 \cdot 2C_2H_5NH_2$	—	*e*
$MF_5 \cdot A$	$A = (C_2H_5)_2NH, (C_2H_5)_3N$	*e*
$MF_5 \cdot CH_3CN$	—	*f*
$MCl_5 \cdot R_2O$	$R = CH_3, C_2H_5, n\text{-}C_3H_7$	*g*
$MCl_5 \cdot POCl_3$	Coordination via oxygen	*h, i*
$MCl_5 \cdot R_3PO$	$R = C_6H_5, C_6H_5 \cdot CH_2$	*j*
$MCl_5 \cdot dipy$	dipy = 2,2′-bipyridyl	*k, l*
$MCl_5 \cdot CH_3CN$	—	*f*
$MCl_5 \cdot 3POCl_3 \cdot TiCl_4$	—	*m*
$MCl_5 \cdot S_4N_4$	—	*n*
$MX_5 \cdot R_2S$	X = Cl, Br, I; $R = CH_3, C_2H_5, n\text{-}C_3H_7$	*g, o*
$MX_5 \cdot 2(CH_2)_4S$	$(CH_2)_4S$ = tetrahydrothiophene	*o*
$MX_5 \cdot C_4H_8O$	X = Cl, Br; C_4H_8O = dioxane	*p*
$MX_5 \cdot R_3EO$	X = Cl, Br; $R = C_6H_5$; E = P, As	*q*
$MX_5 \cdot py$	X = Cl, Br	*r*
$MX_5 \cdot diars$	X = Cl, Br; diars = *o*-phenylenebis(dimethylarsine) (coordination number 7)	*s*

[a] *F. Fairbrother, K. H. Grundy, A. Thompson*, J. Chem. Soc. p. 765 (1965).

[b] *F. Fairbrother, K. H. Grundy, A. Thompson*, J. Less Common Metals *10*, 38 (1966).

[c] *H. C. Clark, H. J. Emeléus*, J. Chem. Soc. p. 190 (1958).

[d] *R. G. Cavell, H. C. Clark*, J. Inorg. Nucl. Chem. *17*, 257 (1961).

[e] *Yu. A. Buslaev, M. A. Glushkova, M. M. Er Ershova, V. A. Bochkarova*, Zh. Neorg. Khim. *13*, 63 (1968).

[f] *L. Kolditz, U. Calov*, Z. Anorg. Allgem. Chem. *376*, 1 (1970).

[g] *D. B. Copley, F. Fairbrother, A. Thompson*, J. Chem. Soc. pp. 271, 315 (1964).

[h] *I. A. Sheka, B. A. Woitowich, L. A. Nisselson*, Zh. Neorg. Khim. *4*, 1803 (1959).

[i] *C. I. Bränden*, Acta Chem. Scand. *16*, 1806 (1962).

[j] *D. Brown, J. F. Easey, J. G. H. du Preez*, J. Chem. Soc. A p. 258 (1966).

[k] *C. Djordjevitsch, V. Katowitsch*, Proc. 9th Int. Conf. Coord. Chem. St. Moritz-Bad, p. 464 (1966).

[l] *G. W. A. Fowles, D. J. Tidmarsh, R. A. Walton*, J. Chem. Soc. A, p. 1546 (1969).

[m] *B. A. Woitovich, S. Ss. Barabanova*, Dokl. Akad. Nauk Ukr. SSSR, p. 1068 (1963).

[n] *P. J. Ashley, E. G. Torrible*, Can. J. Chem. *47*, 2587 (1969).

[o] *F. Fairbrother, J. F. Nixon*, J. Chem. Soc., p. 150 (1962).

[p] *K. Feenan, G. W. A. Fowles*, J. Chem. Soc., p. 2449 (1965).

[q] *D. B. Copley, F. Fairbrother, A. Thompson*, J. Less Common Metals *8*, 256 (1965).

[r] *R. E. McCarley, B. G. Hughes, J. C. Boatman, B. A. Torp*, Advan. Chem. *37*, 243 (1963).

[s] *R. J. H. Clark, D. L. Kepert, R. S. Nyholm*, J. Chem. Soc., p. 2877 (1965).

known, for example, $(C_5H_5)_2MBr_3$.[73] They are prepared by reacting the pentabromides with Grignard reagent. It should be noted that these materials contain M–C bonds. Complex formation of Nb(V) and Ta(V) derivatives with compounds such as tributyl phosphate in aqueous solution is used for their extraction and separation.[74]

Nb(IV) and Ta(IV) halides also form numerous coordination compounds with donor molecules either by direct reaction of the tetrahalides with the ligands or by reaction of the pentachlorides with an excess of the ligands which causes reduction. Thus, this procedure presents an additional method for the preparation of tetrahalides (see Section 13.1.1). Some of these compounds are compiled in Table 3.

Table 3 Niobium and Tantalum Tetrahalide Complexes

Compound	Starting material	Remarks	Ref.
$MX_4 \cdot 2(py)$	MX_4	X = Cl, Br, I	*a, b*
$MX_4 \cdot 2(py)$	MX_5	X = Cl, Br	*c*
$MX_4 \cdot 2(pi)$	MX_5	X = Cl, Br; pi = 2-methyl-pyridin	*c*
$MX_4 \cdot dipy$	MX_5	X = Cl, Br; dipy = 2,2′-bipyridyl	*c*
$MX_4 \cdot diars$	MX_5	X = Cl, Br, I; diars = *o*-phenylenebis(dimethylarsine) (coordination number 8)	*d*
$MCl_4 \cdot phen$	MCl_5	phen = 1,10-phenanthroline	*c*

[a] *R. E. McCarley, J. C. Boatman*, Inorg. Chem. *2*, 547 (1963).
[b] *R. E. McCarley, B. A. Torp*, Inorg. Chem. *2*, 540 (1963).
[c] *M. Allbutt, K. Feenan, G. W. A. Fowles*, J. Less Common Metals *6*, 299 (1964).
[d] *R. J. H. Clark, D. L. Kepert, J. Lewis, R. S. Nyholm* J. Chem. Soc., p. 2865 (1965).

13.2 Halides Having Oxidation States Less Than 4+

The pentahalides do not exist as monomeric molecules under normal conditions, but are oligomerized by halogen bridges. This behavior can be related to the electrophilic character of the MX_5 units which is demonstrated by the readiness of the pentahalides to form coordination compounds with donor molecules. Even halides having an oxidation state of 4+ contain metal–metal bonds,[75] illustrating the electronically unsaturated character of these materials.

Halides having oxidation states less than 4+ show a remarkable tendency to form metal–metal bonds in cluster units. For this reason the monomolecular formulation of these compounds, which is not even quite correct for halides having a higher oxidation state (but can be used in those cases) cannot be justified here and can cause misinterpretations.

Several methods for the preparation of *niobium trifluoride* have been reported; subsequently it has been recognized that NbF_3 does not exist. All substances with the alleged composition NbF_3 contain oxygen and NbF_3 does not seem to be stable. A defined fluorine compound of Nb in lower oxidation species has the composition $NbF_{2.5}$; it is formed from Nb and NbF_5 in a transport reaction when the temperature gradient is 900°–400°. The compound should be formulated as $[Nb_6F_{12}]F_3$, with the structural unit of the Nb atoms being an octahedron.[76]

Transport reactions of the type cited are a general method for the preparation of the lower halides.

Also, the other halides having lower oxidation states can only be prepared as clusters, in which $[M_6X_{12}]^{n\oplus}$ units ($n = 2$, 3, or 4) are predominant. Anions in these compounds include halides, hydroxides, or sulfates, though other anions might possibly exist in solution. In the solid state the lattice structure is the limiting factor. The compounds of type Nb_3X_8 [which can be obtained in a transport reaction from Nb and NbX_5 (X = Br, I) at temperatures above 500°] contain Nb_3 groups as basic units.[77] The compound Nb_6I_{11} containing a $Nb_6I_8^{3\oplus}$ cluster is formed[78] by reduction or thermal decomposition of Nb_3I_8. This compound absorbs hydrogen above 300° to yield HNb_6I_{11} in which hydrogen occupies the centre of the Nb_6 octahedron.[79]

[73] *G. Wilkinson, J. M. Birmingham*, J. Amer. Chem. Soc. *76*, 4281 (1954).
[74] *M. Bonnet, R. Guillaumont*, Radiochim. Acta 12 (2), 98 (1969).
[75] *H. Schäfer, H.-G. v. Schnering*, Angew. Chem. *76*, 833 (1964).
[76] *H. Schäfer, H. G. v. Schnering, K.-J. Niehues, H.-G. Nieder-Varenholtz*, J. Less-Common Metals *9*, 95 (1965).
[77] *A. Simon, H.-G. v. Schnering*, J. Less-Common Metals *11*, 31 (1966).
[78] *A. Simon, H.-G. v. Schnering, H. Schäfer*, Z. Anorg. Allgem. Chemie *355*, 295 (1967).
[79] *A. Simon*, Z. Anorg. Allgem. Chemie *355*, 311 (1967).

Simple formulations such as MX_2 or MX_3 are obviously to be considered only as an approximation. It is characteristic that frequently broad homogeneous ranges have been found, *e.g.*, for $NbCl_3$ from $NbCl_{2.67}$ to $NcCl_{3.1}$.[80] Table 4 lists some compounds and the literature references

Table 4 Cluster Compounds of Nb and Ta

Compound	Remarks	Ref.
$[Nb_6F_{12}]F_3$	—	*a*
$[Nb_6Cl_{12}]Cl_2$	—	*b*
$[Nb_6Cl_{12}]Cl_2 \cdot 7H_2O$	—	*c*
$[Nb_6Cl_{12}]Cl_2 \cdot 8H_2O$	—	*d*
$[Nb_6Cl_{12}]Cl_2 \cdot L_4$	$L = (CH_3)_2SO$ and other organic compounds	*e*
Nb_3X_8	X = Cl, Br, I	*f, g*
$[Nb_6I_8]I_3$	—	*h*
$[HNb_6I_8]I_3$	—	*i*
$[Et_4N]_xNb_6Cl_{18}$	$Et = C_2H_5$, $x = 2–4$	*j, k*
$[Ph_4As]_2Nb_6Cl_{16}(OH)(H_2O)$	$Ph = C_6H_5$	*j*
$[Ph_4As]_2Nb_6Cl_{17}X$	X = Oh, H_2O	*j*
$H_x[Nb_6Cl_{12}]Cl_{3+x} \cdot L$	$L = H_2O$, C_2H_5OH	*l*
ANb_4X_{11}	A = Cs, Rb; X = Cl, Br; planar Nb_4 clusters	*m*
$K_4[Nb_6Cl_{12}]Cl_6$	—	*n*
$[PyH]_2[M_6X_{12}]Cl_6$	$Py = C_6H_5N$; X = Cl, Br	*o*
$[PyH]_2[Ta_6Cl_{12}]Br_6$	—	*o*
$[Ta_6Cl_{12}]Cl_3$	—	*p*
$[Ta_6Cl_{12}]Cl_4 \cdot 2H_2O$	—	*q*
$Ta_6Cl_{14} \cdot 8H_2O$	—	*r*
$TaBr_{2.5} = Ta_6Br_{15}$	—	*s*
$[Ta_6Br_{12}]Br_3 \cdot 8H_2O$	—	*q*
$[Ta_6Br_{12}]Br_4 \cdot 7H_2O$	—	*q*
$[Ta_6Br_{12}]^{n\oplus}$	$n = 2–4$	*t*
$[Ta_6X_{12}][SO_4]_2$	X = Cl, Br	*u*
$[Ta_6X_{12}]X_2$	X = Br, I	*p*
$[Ta_6X_{12}]X_n$	X = Br,I $n = 2,3,5$	*v*
$[Ta_6I_{12}]I_2$	—	*w*
$[M_6Cl_{12}]Cl_3 \cdot 7H_2O$	M = Nb, Ta	*l*
$[M_6Br_{12}]Br_2 \cdot 8H_2O$	M = Nb, Ta	*x*
$[M_6X_{12}]^{2\oplus}$	M = Nb, Ta; X = Cl, Br	*k*
$[M_6Br_{12}]^{n\oplus}$	$n = 3,4$	*y*
$[M_6I_{12}]T_3$	—	*z*

[a] *H. Schäfer, H. G. von Schnering, K.-J. Niehues, H.-G. Nieder-Varenholtz*, J. Less Common Metals *9*, 95 (1965).
[b] *A. Simon, H.-G. von Schnering, H. Wöhrle, H. Schäfer*, Z. Anorg. Allgem. Chem. *339*, 155 (1965).
[c] *H. S. Harned, C. Pauling, R. B. Corey*, J. Amer. Chem. Soc. *82*, 4815 (1960).
[d] *P. B. Fleming, L. A. Mueller, R. E. McCarley*, Inorg. Chem. *6*, 1 (1967).
[e] *R. A. Field, D. L. Kepert*, J. Less Common Metals *13*, 378 (1967).
[f] *A. Simon, H.-G. von Schnering*, J. Less Common Metals *11*, 31 (1966).
[g] *P. W. Scabaugh, J. D. Corbett*, Inorg. Chem. *4*, 176 (1965).
[h] *A. Simon, H.-G. von Schnering, H. Schäfer*, Z. Anorg. Allgem. Chem. *355*, 295 (1967).
[i] *A. Simon*, Z. Anorg. Allgem. Chem. *355*, 311 (1967).
[j] *P. B. Fleming, Th. A. Dougherty, R. E. McCarley*, J. Amer. Chem. Soc. *89*, 159 (1967).
[k] *R. A. Mackay, R. F. Schneider*, Inorg. Chem. *6*, 549 (1967).
[l] *B. Spreckelmeyer, H. Schäfer*, J. Less Common Metals *13*, 122 (1967).
[m] *A. Broll, A. Simon, H.-G. von Schnering, H. Schäfer*, Z. Anorg Allgem. Chem. *367*, 1 (1969).
[n] *A. Simon, H.-G. von Schnering, H. Schäfer*, Z. Anorg. Allgem. Chem. *361*, 235 (1968).
[o] *B. Spreckelmeyer*, Z. Anorg. Allgem. Chem. *365*, 18 (1969).
[p] *Ph. J. Kühn, R. E. McCarley*, Inorg. Chem. *4*, 1482 (1965).
[q] *B. Spreckelmeyer*, Z. Anorg. Allgem. Chem. *358*, 147 (1968).
[r] *H. Schäfer, D. Bauer*, Z. Anorg. Allgem. Chem. *340*, 62 (1965).
[s] *H. Schäfer, R. Gerken, H. Scholz*, Z. Anorg. Allgem. Chem. *355*, 96 (1965).
[t] *B. Spreckelmeyer, H. Schäfer*, J. Less Common Metals *13*, 127 (1967).
[u] *R. E. McCarley, B. G. Hughes, F. A. Cotton, R. Zimmerman*, Inorg. Chem. *4*, 1491 (1965).
[v] *R. E. McCarley, J. C. Boatman*, Inorg. Chem. *4*, 1486 (1965).
[w] *D. Bauer, H.-G. von Schnering, H. Schäfer*, J. Less Common Metals *8*, 388 (1965).
[x] *H. Schäfer, B. Spreckelmeyer*, J. Less Common Metals *11*, 73 (1966).
[y] *B. Spreckelmeyer*, Z. Anorg. Allgem. Chem. *365*, 225 (1969).
[z] *D. Bauer, H. Schäfer*, J. Less Common Metals *14*, 476 (1968).

for their preparations. Papers which report only reduction of Nb or Ta halides without investigation of the cluster character were not considered. There are also numerous studies on systems of lower Nb or Ta halides in salt melts in which compounds such as A_2NbF_4, A_3NbF_6,[81] A_2NbCl_5,[82] and A_2TaCl_5[83] (A = Na, K, Rb, or Cs) are formulated.

[80] *H. Schäfer, K.-D. Dohmann*, Z. Anorg. Allgem. Chemie *300*, 1 (1959).

[81] *L. E. Ivanovskii, M. T. Krasilnikov*, Tr. Inst. Elektrokhim. Akad. Nauk SSSR, Ural. Filial *10*, 61 (1967); *69*, 113 071s (1968).

[82] *W. W. Ssafonov, B. G. Korshunov, T. N. Zimina*, Zh. Neorg. Khim. *11*, 906 (1966).

[83] *W. W. Ssafonov, B. G. Korshunov, S. N. Schevzova, S. I. Bakum*, Zh. Neorg. Khim. *9*, 1687 (1964).

13.3 Oxygen Compounds

13.3.1 Oxides

Nb(V) and *Ta(V) oxides* are obtained on hydrolysis of the corresponding halides; however, small amounts of halogen, particularly fluorine, remain in the products. The hydrolysis is accomplished with aqueous NH_3; oxide hydroxide hydrates are formed first. They can be converted to the oxides by heating. Alkoxy compounds also can be used as starting materials for the hydrolysis. The preparation of fairly pure materials requires working in dilute solutions.[84] The hydrolysis of chlorides is accomplished by boiling with dilute $[NH_4]_2SO_4$ solution.[85]

Simple *alkali niobates* and *tantalates* A_3MO_4 and AMO_3 (A=alkali) are known and various polymeric niobates and tantalates exist which are not yet fully characterized. In the system $NaNbO_3$–Nb_2O_5 the compounds $NaNb_3O_8$ and $NaNb_{13}O_{33}$[86] are formed at temperatures between 1100° and 1200°. Investigation of KNO_3 systems also gave compounds of complex composition such as $7K_2O \cdot 6Nb_2O_5 \cdot 27H_2O$,[87] which have not been identified as uniform species. *Zirconium niobates* and *tantalates* having the composition $(ZrO)_2Nb_2O_7$ and $(ZrO)_2Ta_2O_7$ have been prepared by precipitation reaction.[88]

Oxalic or tartaric acid forms coordination compounds with *Nb(V)* or *Ta(V)* oxides. The insoluble *tantalum hydroxy oxalate* $Ta(C_2O_4)(OH)_3$ is obtained from $Ta(OH)_5$ and oxalic acid solution in a solid state reaction. The solubility of this compound is enhanced in acid medium by formation of cationic complexes such as $[Ta(C_2O_4)(OH)_2]^{\oplus}$. At pH > 3 and with an excess of oxalic acid the anionic complexes $[Ta(C_2O_4)_2(OH)_2]^{\ominus}$ and $[Ta(C_2O_4)_2(OH)_3]^{2\ominus}$ are obtained.[89] The compounds $A_3NbO(C_2O_4)_3 \cdot 2H_2O$ (A=K, Na, NH_4) crystallize from solutions of Nb(V) hydroxide in oxalic acid on addition of oxalates. The composition of the corresponding acid is presumed to be $H_3NbO(C_2O_4)_3 \cdot 7{\cdot}5H_2O$.[90,91] On reaction with tartaric acid the complexes $A_4[Nb_4(C_4H_4O_6)_3O_9 \cdot (H_2O)_{12}] \cdot nH_2O$ (n=2 for A=H; n=1 for A=Na, K; n=0 for A=NH_4) have been obtained.[92] Lower oxides of Nb and Ta apparently have complicated structures. NbO_2 is the alleged product from the reaction pentoxide with carbon.[1] It is also formed on oxidation of metallic Nb.[93] Seven phases have been verified between the limiting compositions $NbO_{2{\cdot}00}$ and $NbO_{2{\cdot}50}$.[94] In the case of Ta analogies to *wolfram bronzes* have been noted.[95]

13.3.2 Oxide Halides and Coordination Compounds

There are repeated reports of products which are oxide-trifluorides, although their identity was not established rigorously. Oxide trifluorides are formed during hydrolysis of the pentafluorides or by pyrolysis of coordination compounds of pentafluorides with SO_3.[96] Reaction of SiO_2 with TaF_5 yields *tantalum oxide trifluoride.*[97] Niobium and tantalum dioxide fluorides are more readily obtained than the oxide trifluorides. They can be prepared by evaporation of solutions of the pentoxides in hydrofluoric acid.[98]

A number of oxygen-rich compounds such as Nb_3O_7F[99] have been prepared by reaction of these MO_2F derivatives with oxides. The cited compound was also prepared by thermal decomposition of NbO_2F.[100] Thermolysis of

---

[84] *A. I. Vaisenberg, T. F. Zhitkova, G. V. Lilyanova, K. I. Marvina, A. F. Mindel,* Nauch.T r., Gos, Nauch-Issled. Prockt. Inst. Redkometal. Prom. *20.* 93 (1968); C.A. *71,* 18 330h (1969).

[85] *Y. Saeki, R. Matsuzaki,* Kogoyo Kagaku Zasshi *67,* 874 (1964); C.A. *61,* 10 306b (1964).

[86] *S. Andersson,* Acta Chem. Scand. *21,* 1777 (1967).

[87] *A. Reismann, F. Holtzberg, M. Berkenblit,* J. Amer. Chem. Soc. *81,* 1292 (1959).

[88] *V. I. Spitzin, L. N. Komissarova, S. A. Vladimirova, Yu. P. Simanov, N. N. Tyutyueva,* Dokl. Akad. Nauk SSSR *131,* 857 (1960).

[89] *A. K. Babko, V. V. Likachina, B. I. Nabivanez,* Zh. Neorg. Khim. *10,* 865 (1965).

[90] *L. G. Vlasov, A. W. Lapizkij, M. A. Slimow, W. Strishkov,* Zh. Neorg. Khim. *7,* 2534 (1962).

[91] *T. F. Limar, T. G. Slatinskaya, O. P. Ssikora,* Zh. Anorg. Khim. *9,* 2381 (1964).

[92] *N. R. Srinivasan,* Proc. Indian Acad. Sci. *A36,* 185 (1952).

[93] *J. Stringer,* Acta Met. *17,* 1227 (1969).

[94] *H. Schäfer, D. Bergmann, R. Gruehn,* Z. Anorg. Allgem. Chemie *365,* 31 (1969).

[95] *F. Galasso, L. Katz, R. Ward,* J. Amer. Chem. Soc. *81,* 5898 (1959).

[96] *H. C. Clark, H. J. Emeléus,* J. Chem. Soc. (London) 765 (1965).

[97] *H. Schäfer, D. Bauer, W. Beckmann, R. Gerken, G. Nieder-Varenholtz, K. J. Niehues, H. Scholz,* Naturwissenschaften *51,* 241 (1964).

[98] *L. K. Frevel, H. W. Rinn,* Acta Cryst. *9,* 626 (1956).

[99] *S. Andersson, A. Aström,* Acta Chem. Scand. *18,* 2233 (1964).

[100] *S. Andersson, A. Aström,* Acta Chem. Scand. *19,* 2136 (1965).

TaO_2F gives Ta_3O_7F in analogous manner.[101] Other substances include $Nb_5O_{13}F$[99] and $Nb_{31}O_{77}F$ as well as $Nb_{34}O_{84}F_2$,[102] which belong to a series of structurally related compounds of the general formula $M_{3n+1}X_{8n-2}$ (X=O, F). Combinations of these phases allegedly give other materials such as $Nb_{65}O_{161}F_3$ ($=Nb_{31}O_{77}F+Nb_{34}O_{84}F_2$)[103] and combinations with oxide phases are also possible, *e.g.*, $Nb_{59}O_{147}F$ ($=Nb_{28}O_{70}+Nb_{31}O_{77}F$).[103] The generally accepted ways for preparing *oxofluoroniobates or -tantalates* include melting or heating of alkali fluorides with the oxides, heating of MO_2F and niobates or tantalates, hydrolyzing fluorine complexes, or reacting pentafluorides with oxides in hydrofluoric acid. Table 5 depicts the most important compounds of this type.

On heating Nb(V) oxide in the presence of AF (A=Li, Na, K) compounds of the type $ANbO_2F$[104] are obtained.

Table 5 Fluoro-oxoniobates and tantalates (M = Nb, Ta)

Compound	Remarks	Ref.
A_0MOF_0	A = Na, K, NH_4	*a, b*
A_2NbOF_5	A = Li, K, NH_4	*b, c, d, e*
$[NH_4]_3NbOF_6$	—	*e*
H_2MOF_5	—	*f, g*
K_2MO_2F	—	*b, h*
$A_3MO_2F_4$	A = alkali	*h, i*
K_2NbO_3F	—	*b, j*
$KNbO_2F_2$	—	*k*
KM_2O_5F	Analog to wolfram bronzes, K_xWO_3	*l*
$[NH_4]_3Nb_2O_2F_9$	—	*e*
$K_xNbO_{2+x}F_{1-x}$	—	*m*
$A_xTaO_{2+x}F_{1-x}$	A = K, Tl	*n, o*
$ANb_6O_{15}X$	A = Li, Na; X = F, OH	*p, q, r*
$CdNaNb_2O_6F$	Pyrochlore structure	*s*
$CdTa_2O_5F_2$	Perovskite structure	*r*
$Cd_2NbB^{IV}O_6F$	B^{IV} = Ti, Ge, Sn; pyrochlore structure	*t*

[a] *O. L. Keller, Jr., A. Chetham-Strode, Jr.*, Inorg. Chem. *5*, 367 (1966).
[b] *N. K. Voskresenskaya, G. P. Budova*, Dokl. Akad. Nauk SSSR *170*, 329 (1966).
[c] *Yu. I. Koltzov*, Izv. Akad. Nauk SSSR, Neorg. Mater. *1*, 907 (1965).
[d] *J. Galy, S. Andersson, J. Portier*, Acta Chem. Scand. *23*, 2949 (1949).
[e] *Yu. A. Buslaev, E. G. Ilin, V. D. Kopanev*, Dokl. Akad Nauk SSSR *196*, 829 (1971).
[f] *N. S. Nikolaev, Yu. A. Buslaev*, Zh. Neorg. Khim. *4*, 205 (1959);
Yu. A. Buslaev, N. S. Nikolaev, Zh. Neorg. Khim. *4*, 465 (1959).
[g] *L. M. Ferris*, J. Chem. Eng. Data *11*, 343 (1966).
[h] *Bin-ssin Zui, N. P. Luzhnaya, V. I. Konstantinov*, Zh. Neorg. Khim. *8*, 396 (1963).
[i] *W. Rüdorff*, Z. Anorg. Allgem. Chem. *364*, *69* (1969).
[j] *G. P. Budova, N. K. Voskresenskaya*, Dokl. Akad. Nauk SSSR *177*, 1077 (1967).
[k] *A. A. Lastokhkina, I. A. Sheka, L. A. Malinko*, Izv. Akad. Nauk SSSR, Neorg. Mater. *5*, 757 (1969).
[l] *A. Magneli, St. Nord*, Acta Chem. Scand. *19*, 1510 (1965).
[m] *R. de Pape, G. Gauthier, P. Hagenmüller*, C.R. Acad. Sci. Ser. C *266*, 803 (1968).
[n] *R. de Pape, J. L. Fourquet, G. Gauthier*, C.R. Acad. Sci. Ser. C *269*, 1298 (1969).
[o] *J. L. Fourquet, G. Ory, G. Gauthier, R. de Pape*, C.R. Acad. Sci. Ser. C *271* 773 (1970).
[p] *M. Lundberg, St. Andersson*, Acta Chem. Scand. *19*, 1376 (1965).
[q] *St. Andersson*, Acta Chem. Scand. *19*, 2285 (1965).
[r] *L. Jahnberg*, Acta Chem. Scand. *21*, 615 (1967).
[s] *E. Aleshin, R. Roy*, J. Amer. Ceram. Soc. *45*, 18 (1962).
[t] *D. Laguitton, J. Lucas*, C.R. Acad. Sci. Ser. C *269*, 105 (1969).

Oxide trichlorides have been studied in more detail and are better characterized than the corresponding fluorides. They are prepared by transport reaction of pentoxides or pentachlorides[105,106] and by reaction of the pentachlorides with oxygen at temperatures of about 500°.[107–109] *Niobium oxide trichloride* is formed by chlorination of the oxides with CCl_4 above 600°[110] and by reaction of Cl_2 on pyrochlore, $Ca_2Nb_2O_7$, in the presence of carbon, near 500°.[111] Coordination compounds of $NbCl_5$ with oxygen-containing ligands such as ether,[112]

[101] *L. Jahnberg*, Acta Chem. Scand. *21*, 615 (1967).
[102] *S. Andersson*, Acta Chem. Scand. *19*, 1401 (1965).
[103] *R. Gruehn*, Naturwissenschaften *54*, 645 (1967).
[104] *W. Rüdorf, D. Krug*, Z. Anorg. Allgem. Chemie *329*, 211 (1964).
[105] *H. Schäfer, F. Kahlenberg*, Z. Anorg. Allgem. Chemie *305*, 327 (1960).
[106] *H. Schäfer, E. Sibbing*, Z. Anorg. Allgem. Chemie *305*, 341 (1960).
[107] *A. N. Ketov, I. M. Kolesov*, Izv. Vyssh. Ucheb. Zaved. Tsvet. Met. *9*, 77 (1966).
[108] *Y. Saeki, T. Matsushima*, Denki Kagaku Oyobi Kogyo Butsuri *32*, 667 (1964); C.A. *62*, 6133f (1965).
[109] *F. Fairbrother, A. H. Cowley, N. Scott*, J. Less-Common Metals *1*, 206 (1959).
[110] *Ss. A. Shchukarev, K. Smirnova, T. S. Shemyakina, E. N. Ryabov*, Zh. Neorg. Khim. *7*, 1216 (1962).
[111] *S. L. Stefanyuk, I. S. Morosov*, Zh. Prikl. Khim. *37*, *1665* (1964).
[112] *A. H. Cowley, F. Fairbrother, N. Scott*, J. Chem. Soc. 3133 (1958).

sulfoxides, or phosphine oxides also produce $NbOCl_3$ when heated. This observation again illustrates that this material forms readily. This type of reaction can be used also for the preparation of *tantalum oxide trichloride*[112] though it does not proceed quite as favorably as in the case of $NbOCl_3$. A noteworthy way of preparing $TaOCl_3$ involves the reaction of $TaCl_5$ with Cl_2O.[113]

Sublimation of $NbOCl_3$ in a temperature gradient of 350° to room temperature results in the formation of the oxide chlorides Nb_3O_7Cl and NbO_2Cl[114] as well as $NbOCl_3$; hydrolysis of $NbOCl_3$ at temperatures between 350° and 550° produces $Nb_4O_9Cl_2$.[114a] A compound with niobium in oxidation state 4+ is $NbOCl_2$; this material can be prepared by reduction of $NbOCl_3$[70] with $SnCl_2$. Methods for the preparation of oxide bromides are analogous to those described for the oxide trichlorides: heating the adduct $MBr_5 \cdot O(C_2H_5)_2$, reacting oxygen with pentabromides,[109] and brominating oxides with CBr_4.[23] Niobium oxide trichloride is the product of a transport reaction of Nb, I_2, and Nb_2O_5. Thermal decomposition of NbOI produces NbO_2I.[115] The substances MO_2Cl, MO_2Br, and Ta_3O_7Cl are obtained from the metals, the oxides, and halogens or halides in transport reactions.[116]

Oxide halides are polymeric and contain predominantly oxygen bridges as shown for $NbOCl_3$.[117] Donor molecules give adducts by bridge cleavage.

$NbOCl_3$ react with *N,N*-diethylaniline or substituted quinones to form 1:1 adducts[118]; $NbOX_3 \cdot dipy$ (X=Cl, Br; dipy=bipyridyl) and $NbOBr_3 \cdot o\text{-phen}$ (*o*-phen=*o*-phenanthroline[119,120] are also known.

Pentachlorides and pentabromides exchange halogen with oxygen in ether solution or on reaction with sulfoxides to give adducts such as $NbOCl_3 \cdot 2R_2O$ ($R=CH_3$, C_2H_5), $Ta_2O_6Cl_3 \cdot 3(CH_3)_2SO$, $Ta_3O_4X_7 \cdot 5(CH_3)_2SO$ (X=Cl, Br), $Nb_4O_6Br_8 \cdot (CH_3)_2SO$, $Ta_2O_5Br \cdot 3(CH_3)_2SO$,[121] $Nb_2OCl_8 \cdot 2(dipy)$ and $Ta_4OCl_{18} \cdot 4(dipy)$.[120]

Oxo-chloroniobates are formed by melting alkali chlorides with $NbOCl_3$. The following compounds have been characterized: $ANbOCl_4$, A_2NbOCl_5, and A_3NbOCl_6 (A=Na, K, Rb, Cs).[122–126] Compounds of the type A_2NbOCl_5 ($A=NH_4$, K, Rb, Cs) precipitate from hydrochloric acid solutions of $NbCl_5$ or $NbOCl_3$ on addition of the corresponding chlorides.[127] Tetra- and pentachlorooxoniobates have also been observed in systems containing $NbOCl_3$, alkali chlorides, and $AlCl_3$.[128,129] Penta- and hexachlorooxoniobates of aryldiazonium cations can be prepared.[130] Owing to the thermal instability of $TaOCl_3$ no studies of systems of the latter with alkali chlorides have been carried out. The preparation of $A[TaOCl_4]$ (A=K, Rb, Cs, NH_4) is accomplished by two methods [Eqns (5) and (6)].

$$3A[TaCl_6] + Sb_2O_3 \longrightarrow 3A[TaOCl_4] + 2SbCl_3 \quad (5)$$

$$3ACl + 3TaCl_5 + Sb_2O_3 \longrightarrow 3A[TaOCl_4] + 2SbCl_3 \quad (6)$$

Reactions occur above 180° in a stream of chlorine.[131] Reduction of A_2NbOCl_5 (A=Rb, Cs) with $SnCl_2$ results in the formation of A_2NbOCl_4 with niobium in oxidation state 4+;

---

[113] *K. Dehnicke*, Naturwissenschaften *52*, 58 (1965).
[114] *K. Huber, I. Baunok*, Chimia Swiz. *15*, 365 (1961).
[114a] *St. Andersson*, Acta Chem. Scand. *19*, 2285 (1965).
[115] *H. Schäfer, R. Gerken*, Z. Anorg. Allgem. Chemie *317*, 105 (1962).
[116] *H. Schäfer, L. Zylka*, Z. Anorg. Allgem. Chemie *338*, 309 (1965).
[117] *D. E. Sands, Z. Zalkin, R. E. Elson*, Acta Cryst. *12*, 21 (1959).
[118] *A. V. Leshchenko, V. T. Panyushkin, A. D. Garnovskii, O. A. Osipov*, Zh. Neorg. Khim. *11*, 2156 (1966).
[119] *N. Bruicevic, C. Djordjevic*, J. Less-Common Metals *13*, 470 (1967).
[120] *C. Djordjevitsch, V. Katowitsch*, Proc. 9th intern. Conf. Coord. Chem. p. 464, St. Moritz, 1966.
[121] *D. B. Copley, F. Fairbrother, K. H. Grundy, A. Thompson*, J. Less-Common Metals *6*, 407 (1964).
[122] *S. A. Shchukarve, Je. K. Smirnova, T. S. Shchemiyakina*, Zh. Neorg. Khim. *7*, 2217 (1962).
[123] *S. A. Shchukarev, Je. K. Smirnova, I. W. Wassilkova*, Izv. Leningradskowo Univ. *18*, Nr. 16 Ser. Fys. Khim. Nr. 3, 132 (1963).
[124] *Je. K. Smirnova, I. W. Wassilkova*, Nr. 4, 164 (1964).
[125] *I. S. Morosov, W. A. Krochin*, Zh. Neorg. Khim. *8*, 2376 (1963).
[126] *O. R. Gavrilov, L. A. Nisselson*, Zh. Neorg. Khim. *11*, 1941 (1966).
[127] *N. P. Lipatova, I. S. Morosov*, Zh. Neorg. Khim. *10*, 429 (1965).
[128] *W. A. Krochin, I. S. Morosov*, Zh. Neorg. Khim. *8*, 2727 (1963).
[129] *O. R. Gavrilov, L. A. Nisselson, A. A. Shirokov*, Zh. Prikl. Khim. *40*, 762 (1967).
[130] *W. M. Cherkassov, I. F. Wladimirtsev*, Zh. Obshch. Khim. *30* (92), 2235 (1960).
[131] *I. S. Morosov, A. I. Morosov*, Zh. Neorg. Khim. *11*, 335 (1966).

reaction temperatures are between 450° to 500°.[70]

13.3.3 Peroxo Derivatives

Niobates and tantalates react with H_2O_2 to form peroxo complexes which have been investigated spectroscopically. Peroxo orthometallates having the composition A_3MO_8 (A = alkali) have been isolated.[31,132–135] They react with hydrofluoric acids to yield peroxofluoro complexes.[136,137] *Peroxoniobates* with the anions $[NbO_2(O_2)]^{\ominus}$ and $[Nb(O_2)_3]^{\ominus}$ are known, they result from the reaction of polyniobates with H_2O_2.[135] Dissolving peroxoniobates in 12 N HCl yields pentachloro compounds of the type $A_2[Nb(O_2)Cl_5]$ (A = K, Rb, Cs, NH_4).[138] *Peroxotantalates* do not readily dissolve in 12 N HCl only an uncharacterized chlorooxotantalate has been obtained from a $TaCl_5$ solution in hydrochloric acid. The material was dissolved in 12 N HCl and treated with H_2O_2 to give compounds of the type $A_2[Ta(O_2)Cl_5]$ (A = Rb, Cs, NH_4).[139]

13.3.4 Alkoxy Compounds, Alkoxy Halides, and Related Materials

Pentaalkoxy compounds of Nb and Ta having the general formula $M(OR)_5$ have been obtained by dissolving the pentachlorides in the corresponding alcohol and adding the solution to ammoniacal benzene. NH_4Cl precipitates and the alkoxy compounds can be isolated by evaporating the solvent.[139a] Reaction of dialkylamides, $M(NR_2)_5$, with alcohols also produces *niobium* or *tantalum penta alkoxides* $M(OR)_5$.[140] $Nb(OR)_5$ adds $NaOC_2H_5$ to form the salt $Na[Nb(OC_2H_5)_6]$.[141] Exchange of alkoxy groups of $M(OR)_5$ by other alcohols or esters results in the formation of a number of compounds containing different alkoxy groups.[142,143] Reaction with trialkyl silanols or silyl acetates gives the corresponding siloxy derivates $(RO)_nM(OSiR)_{5-n}$ (n = 1–4),[144,145] and β-ketoesters react in the enol form with $M(OR)_5$ to yield mixed enolates.[146] Hydrolysis of alkoxy compounds produces polymeric oxide alkoxides $MO_x(OR)_{5-2x}$ with M having the coordination number six.[147] Niobium or tantalum alkoxyhalides $MX_n(OR)_{5-n}$ (X = F, Cl, Br) can be prepared either by interaction of stoichiometric amounts of alkoxides with pentahalides in CH_3CN, CCl_4, etc.[141,148–150] or by halogenation of the alkoxides, $M(OR)_5$, with acetyl halides.[151,152] Reaction of pentahalides with alcohol or $SO(OR)_2$ is also possible.[141,148,150,153–155] Halogen exchange of alkoxychlorides with alkali fluorides produces the corresponding fluorine derivatives.[156] Halogen alkoxides dimerize in nonpolar solvents such as benzene, but polar solvents cause transfer of alkoxy groups or halide with formation of cations and anions, *e.g.*,

$$2NbCl_4OC_2H_5 \rightleftharpoons [NbCl_3OC_2H_5]^{\oplus} + [NbCl_5OC_2H_5]^{\ominus} \quad (7)$$

Addition of alkali or ammonium halides results

---

132 *E. H. Reisenfeld, H. E. Wohler, W. Kutsch*, Ber. 38, 1885 (1905).
133 *J. E. Guerchais, R. Rohmer*, Compt. Rend. 1135 (1964).
134 *C. W. Balke*, J. Amer. Chem. Soc. *27*, 1140 (1905).
135 *B. Spinner*, Rev. Chim. Miner. *6*, 319 (1969).
136 *A. Piccini*, Z. Anorg. Allgem. Chemie *2*, 21 (1892).
137 *R. D. Hall, E. F. Smith*, Proc. Am. Phil. Soc. *44*, 171 (1905); Chemical News *92*, 252 (1905).
138 *J. E. Guerchais, B. Spinner, R. Rohmer*, Bull. Soc. Chim. France 55 (1965).
139 *J. Dehand, J. E. Guerchais, R. Rohmer*, Bull. Soc. Chim. France 346 (1966).
139a *D. E. Bradley, B. N. Charkravarti, W. Wardley*, J. Chem. Soc. (London) 2383 (1956); *J. Fuchs, K.-F. Jahr, G. Heller*, Chem. Ber. *96*, 2472 (1963).
140 *I. M. Thomas*, Can. J. Chem. *39*, 1386 (1961).
141 *L. Kolditz, M. Schönherr*, Omagiu Acad. Prof. R. Ripan, p. 321, Bukarest (1966).
142 *P. N. Kapoor, R. C. Mehrotra*, J. Less-Common Metals *10*, 66 (1966).
143 *R. C. Mehrotra, R. N. Kapoor, S. Prakash, P. N. Kapoor*, Australian J. Chem. *19*, 2079 (1966).
144 *D. C. Bradley, I. M. Thomas*, J. Chem. Soc. 3404 (1959); Chem. & Ind. (London) *17*, 1231 (1958).
145 *R. C. Mehrotra, P. N. Kapoor*, J. Indian Chem. Soc. *44*, 345 (1967).
146 *R. C. Mehrotra, P. N. Kapoor*, J. Less-Common Metals *7*, 453 (1964).
147 *D. C. Bradley, H. Holloway*, Can. J. Chem. *40*, 62 (1962).
148 *L. Kolditz, M. Schönherr*, Z. Chem. *5*, 349 (1965).
149 *R. Gut, H. Busar, E. Schmid*, Helv. Chim. Acta *48*, 878 (1965).
150 *M. Schönherr, L. Kolditz*, Z. Chem. *10*, 72 (1970).
151 *R. C. Mehrotra, P. N. Kapoor*, J. Less-Common Metals, *10*, 348 (1966).
152 *S. Prakash, P. N. Kapoor, R. N. Kapoor*, J. Prakt. Chem. *36*, 24 (1967).
153 *H. Köppel, M. Schönherr, L. Kolditz*, 1. Chem. *11*, 28 (1971).
154 *A. M. Golub, A. M. Ssytsch*, Zh. Neorg. Khim. *10*, 889 (1965).
155 *D. C. Bradley, R. K. Multani, W. Wardlaw*, J. Chem. Soc. (London) 4647 (1958).
156 *Yu. A. Buslaev, Yu. V. Kokunov, S. M. Kremer, V. A. Shcherbakov*, Izv. Akad. Nauk SSSR, Neorg. Mat. *3*, *8*, 1424, 1505 (1967).

in the formation of the corresponding *haloalkoxyniobates* or *haloalkoxytantalates*. The molecular compounds undergo the same reaction. Of the *chloroalkoxyniobates* $[NbCl_{6-n}(OR)_n]^{\ominus}$ compositions with $n=1$ or 2 and $R=CH_3$ or C_2H_5 are known.[141,148,150] On warming of solutions or addition of pyridinium chloride salts with anions the $NbOCl_5^{2\ominus}$ and $NbOCl_4^{\ominus}$ are formed by splitting off alkyl chlorides.[57,141,148,150] $NbOCl_4^{\ominus}$ is dimeric and probably contains double oxygen bridges. The tendency to condensation is also noted for fluoroalkoxy compounds. Salts of the compositions $CsNbF_2(OR)_2O$, $Cs_2Nb_2F_5(OR)O_3$, $Cs_3Nb_3F_9(OR)O_4$, $Cs_3Nb_3F_{10}O_4$, $Cs_4Ta_4F_{10}(OR)_4O_5$, $Cs_3Ta_2F_7(OR)O_2$, and $CsTaOF_4$ have been described.[157,158] Heating $NbCl_2(OC_2H_5)_3$ to 170° will cause condensation and formation of the compound $Nb_3O_5Cl(OC_2H_5)_4$.[148] In the presence of ligands such as dipyridyl the adducts $NbOCl_2(OR)dipy$, $Nb_4O_6(OH)_2Cl_4(OR)_2(dipy)_2$ and $Ta_6O_2(OH)_{18}Cl_6(OR)_2(dipy)_2$ (with $R=C_2H_5$, n-C_3H_7) are formed in alcoholic solutions of the pentachlorides; $Nb_2OCl_8(dipy)_2$ and $Ta_4OCl_{18}(dipy)_4$ precipitate from ether solutions of the pentachlorides.[120,158] $NbOBr_3$ reacts with acetylacetone to form $NbOBr(acac)_2$, a polymeric compound containing Nb–O–Nb bridges; the latter are solvolyzed with an excess of acetylacetone to yield $NbBr_2(acac)(OR)_2$.

Complexes containing niobium in oxidation state 4+ have been prepared by electrolytic reduction of $NbCl_5$ in HCl-saturated alcoholic solution. The resultant composition is $[BH]_2Nb(OR)Cl_5$ (B = methylamine, pyridine or quinoline).[159] Another compound of niobium (oxidation state 4+ is $Nb_2Cl_5(OC_2H_5)_3(dipy)_2$, which is obtained from an alcoholic solution of Nb(IV) chloride.[158]

Alkoxyhalides show an increased tendency to form solvates with higher halogen content.[160] $NbCl_4OC_2H_5$ precipitates from acetonitrile as a 1:1 adduct. Attempts to obtain the unsolvated compound led to its total destruction. A solvate-free material can be prepared simply by using a nonpolar solvent such as CCl_4. On the other hand, $NbCl(OC_2H_5)_4$ crystallizes from acetonitrile without a solvate molecule.[141] The same is true for tantalum compounds where materials such as $Ta(OR)_nBr_{5-n}$ ($n=1$, 2) form 1:1 adducts with acetyl acetate, but in the case of $n=3$ or 4 solvate-free compounds are obtained.[152]

13.4 Chalcogen Compounds

Different *chalcogenides of niobium* are described, *e.g.*, Nb_3S_4, NbS, NbS_2, $NbSe_2$, and Nb_2Se, which can all be prepared from the elements.[161,162] It appears that phases with various compositions exist and the same holds true for combinations with tellurium and for some tantalum chalcogenides which occur in steel. TaS_3 and other TaS_x molecules have been detected in the gas phase[163], and thio and seleno complexes such as Cu_3MX_4 and Tl_3MX_4 (X = S, Se) have been prepared.[164]

Niobium chalcogenide halides of the composition $NbCh_2Cl_2$, $NbCh_2Br_2$, and $NbCh_2I_2$ (Ch = S, Se) are known. They are formed on reaction of Nb with halogens in the presence of sulfur or selenium and can be transported in a temperature gradient.[165] In these compounds niobium has a formal oxidation state of 4+; they contain Nb–Nb bonds and the chalcogens are present in the form of $Ch_2^{2\ominus}$ groups.[166] Thermal decomposition of NbS_2Cl_2 and $NbCh_2Br_2$ proceeds as in Eqn (8).

$$3NbCh_2X_2 \longrightarrow NbS_2 + S_2 + NbChX_3 \quad (8)$$

In case of NbS_2I_2, reaction (9) has been established.

$$NbS_2I_2 \longrightarrow NbS_2 + I_2 \quad (9)$$

$TaSCl_3$ and $TaSeBr_3$ are also known.

157 *Yu. A. Buslaev, Yu. V. Kokunov*, Izv. Akad. Nauk SSSR, Neorg. Mat. *4*, 537 (1968); Izv. Sibirsk. Otd. Akad. Nauk SSSR, Ser. Khim. Nauk *1*, 50 (1968).

158 *Yu. A. Buslaev, Yu. V. Kokunov, C. Djordjevic, V. Katovic*, Chem. Commun. *8*, 224 (1966).

159 *R. A. D. Wentworth, C. H. Brubaker jr.*, Inorg. Chem. *2*, 554 (1963).

160 *D. C. Bradley*, Progress in Inorganic Chemistry, Vol. II, p. 334, Interscience Publishers, New York, London 1960.

161 *D. Houdouin*, Compt. Rend. *C268*, 1943 (1969).

162 *H. Schäfer, F. Wehmeier, M. Trenkel*, J. Less-Common Metals *16*, 290 (1968).

163 *Je. O. Shurakovskij, V. V. Gorskij*, Dopovidi Akad. Nauk Ukr. SSR 1428 (1966).

164 *W. P. F. A. M. Omlov, F. Jellinek*, Rec. Trav. Chim. Pays-Bas *88*, 1205 (1969).

165 *H. Schäfer, W. Beckmann*, Z. Anorg. Allgem. Chemie *347*, 225 (1966).

166 *H.-G. v. Schnering, W. Beckmann*, Z. Anorg. Allgem. Chemie *347*, 231 (1966).

13.5 Nitrogen Derivatives

The composition of *niobium* and *tantalum nitrides* is reported to be NbN and TaN;[163, 167] they are polymeric and on thermal decomposition form nitride chlorides. An exact characterization has not yet been reported.

Heating $NbCl_5$ with NH_4Cl to temperatures between 250° and 260° produces a nitride chloride which almost has the composition $NbNCl_2$.[167] It reacts with fluorine (diluted with argon) to form $NbNF_2 \cdot 0{\cdot}5HF$[168] with elimination of chlorine. The HF in the compound seems to indicate that the starting material contains hydrogen.

Reaction of $NbCl_5$ with NH_4Cl in nitrobenzene at 180° gives a material having the composition $Nb_2N_3Cl_7H_6$; the latter can be decomposed to $NbNCl_2$ by heating to 230°. Removal of NH_4Cl by washing the hydrogen-containing product with water results in the formation of $Nb_3N\text{-}Cl_2(OH)_{10}$. In the presence of PCl_5 a product having the formula $P_5NbN_{11}Cl_9H_2$ has been obtained from which $P_5NbN_8Cl_4(OH)_2$ was isolated after washing with water. Since for niobium nitride chlorides occurrence of Nb–N–Nb– chains have been established P–N–Nb–N chains are supposedly present in the phosphorus derivative.[169]

Tantalum ammonium derivatives. On heating $[NH_4][TaCl_6]$ to 325°, $TaNCl_2$ is formed, during this reaction the not fully characterized $Ta(NH_3)_2Cl_3$ is observed as well as HCl and $TaCl_5$. Heating to 400° affords Ta_2N_3Cl that is hydrolyzed to TaON, a compound stable towards mineral acids and alkali; it is dissolved only by a mixture of HF and H_2SO_4. TaON is stable up to 900° in an inert atmosphere[170]; it has also been prepared by a transport reaction in the presence of NH_4Cl.[171] Reaction of $TaBr_5$ with NH_4Cl yields TaNBr which is decomposed above 450° with the elimination of bromine.[172, 173]

Niobium and tantalum nitrides or oxide nitrides are used in electronics. These thin films may contain more complicated compositions than formulated, but it has not yet been demonstrated that these materials also exist in compact form. *Amides of niobium and tantalum* can be prepared from the halides by reaction with ammonia or amines. Dialkyl amines, $M(NR_2)_5$ (which result from halides and $LiNR_2$), are important as starting materials for the preparation of alkoxy compounds (see Section 13.3.4).

13.6 Binary Compounds Containing B, C, P, As, or Sb

The synthesis from the elements in the presence of iodine is the general method for the preparation of these compounds. The compounds can be transported in a temperature gradient and can be obtained in well-crystallized form. Compounds such as NbAs, $NbAs_2$, Nb_3Sb, Nb_5Sb_4, and $NbSb_2$ are known. In all NbX_n systems (X = S, Se, Te) the X-poor compounds move to the hot temperature zone in transport reactions, X-rich compounds deposit in the colder zone.[174] System investigations have substantiated the following compositions: Nb_3As, Nb_7As_4, $Nb_5\text{-}As_3$, $NbAs_{0.75}$, Ta_2As, and Ta_5As_4.[175]

A further possibility for preparing simple binary compounds such as the tantalum borides is a magnesiothermic reaction. MgO formed in this reaction is extracted with hydrochloric or phosphoric acid.[176]

It has not yet been shown that borides and

[167] *I. V. Tananaev, G. B. Seifer, E. A. Ionova*, Dokl. Akad. Nauk SSSR *127*, 584 (1959).

[168] *Yu. A. Buslaev, S. M. Sinitrsyna, V. A. Bochkareva, M. A. Polikarpova*, Izv. Akad. Nauk SSSR, Neorg. Mater. *4*, 453 (1968).

[169] *Yu. A. Buslaev, M. A. Gluskova, Ss. M. Ssinitzyna, M. M. Jerschova, M. A. Polikarpova*, Izv. Akad. Nauk SSSR, Neorg. Mater. *1*, 498 (1965).

[170] *Yu. A. Buslaev, M. A. Gluskova, M. M. Ershova, E. M. Sustrovich*, Izv. Akad. Nauk SSSR, Neorg. Mater. *2*, 2120 (1966).

[171] *Yu. A. Buslaev, G. M. Safronov, V. I. Pakhomov, M. A. Glushkova, V. P. Repko, M. M. Ershova, A. N. Zhukov, I. A. Zhdanova*, Izv. Akad. Nauk SSSR, Neorg. Mater. *5*, 45 (1969).

[172] *M. A. Glushkova, M. M. Ershova, Yu. A. Buslaev* Izv. Akad. Nauk SSSR, Neorg. Mater. *3*, 2259.

[173] *D. C. Bradley, I. M. Thomas*, J. Chem. Soc. 3857 (1960); Proc. Chem. Soc. 225 (1959).

[174] *S. Rundqvist, B. Carlsson, Cu. Pontchour*, Acta Chim. Scand. *23*, 2188 (1969).

[175] *H. Schäfer, W. Fuhr*, J. Less-Common Metals *8*, 375 (1965).

[176] *L. Ya. Markovskii, N. V. Vekshina, E. T. Berzuk, G. E. Shakareva, T. K. Voevodskaya*, Porosh. Met. *9*, 13 (1969).

carbides commonly formulated as Nb_3B_2, NbB_2, TaB, Ta_3B_4, TaB_2, NbC, and TaC[177,178] truly have these compositions. There are claims to postulatory compositions such as $NbB_{1.875}$ or $TaB_{1.919}$.[179] Possibly, cluster formulations similar to the halides (see Section 13.2) have to be considered. Carbides are distinguished by their high melting points (2050° for NbC and 3909° for TaC).

[177] *Yu. B. Kuzma, V. S. Telegus, D. A. Kovalyk*, Porosh. Met. *9*, 79 (1969).

[178] *Y. Murata, E. D. Whitney*, Am. Ceram. Soc. Bull. *48*, 698 (1969).

[179] *G. K. Johnson, E. Greenberg, J. L. Margrave, W. N. Hubband*, J. Chem. Eng. Data *12*, 597 (1967).

14 Chromium

J. R. Wasson
Department of Chemistry, University of Kentucky,
Lexington, Kentucky 40506, U.S.A.

14.1 Elemental Chromium

Chromium (m.p. 1890°) exhibits formal oxidation states in its compounds ranging from 2− in the $Cr(CO)_5^{2\ominus}$ anion to 6+ in the chromate, $CrO_4^{2\ominus}$, ion. Chromium occurs in nature in the ore, chromite, $FeCr_2O_4$, which is used directly for the production of ferrochromium alloys. Chromium can replace aluminum in aluminum minerals giving rise to gem stones, *e.g.*, emeralds and rubies. Chromium can be electroplated from chromic acid–chromium(III) sulfate solutions at high current densities with simultaneous hydrogen evolution. Colloidal chromium can be obtained by electrical dispersion under isobutyl alcohol or by treatment of very finely divided metal alternatively with dilute acid and alkali. The metal dissolves fairly readily in nonoxidizing mineral acids, but not in concentrated or dilute cold aqua regia or nitric acid. These latter reagents passivate the metal. The metal is available commercially in high purity in a number of forms. Many thousands of chromium compounds have been characterized and these are collected in standard compilations.[1–3] The reactions of chromium compounds have been the subject of several reviews.[4,5] The vibrational spectra of chromium complexes has been extensively reviewed.[6,7]

14.2 Simple Binary Compounds

Chromium borides with compositions CrB, CrB_2, and Cr_3B_2 may be obtained by reaction of the elements at elevated temperatures or by electrolysis of fused borates.[8] The structures of chromium borides have been reviewed.[9] The *nitride*, CrN, and *silicides* are formed by reaction of the elements at high temperatures.[1]

Chromium forms a number of *carbides*. Powder compacts containing three parts chromium(III) oxide and one part carbon black, intimately mixed and pressed, are heated to 1600° in a carbon tube furnace under hydrogen to produce[10] Cr_4C_2. This temperature must be maintained fairly closely since lower carbides form under the same conditions at lower temperatures. Below 1300° Cr_7C_3 is the major product. Thermal decomposition of chromium carbonyl[11] produces a layered mixture of chromium and Cr_3C_2. Chromium carbides can also be prepared by arc heating of the elemental powders but metal losses by evaporation are considerable.

Cr_3C_2 has been used in conjunction with other carbides in carbide cutting tools low in tungsten or tungsten-free. Large amounts of this carbide (~75%) used with a nickel or copper–nickel binder free of cobalt form a material which is acid- and wear-resistant and is used for valve parts in chemical industry.[12]

The reaction of sulfur with chromium at elevated temperatures gives rise[13] to a number of *sulfides*, *e.g.*, CrS, Cr_7S_8, Cr_5S_6, Cr_3S_4, and Cr_2S_3. With *selenium* the phases CrSe, Cr_7Se_8, Cr_3Se_4, and Cr_2Se_3 have been characterized.[14–17] With *tellurium* the phases CrTe, Cr_7Te_8, Cr_5Te_6, Cr_3Te_4, Cr_2Te_3, and Cr_5Te_8 have been obtained.[18–20] The chalcogenides Cr_5S_8, Cr_5Se_8, and $Cr_5S_{4.8}Se_{3.2}$ are prepared[21] by reaction of the elements at elevated temperatures at pressures of 65–89 kbars.

The solubility of hydrogen in chromium metal has been extensively studied.[22] A number of *hydrides*, *e.g.*, CrH, CrH_2, and CrH_3 have been

---

[1] *Gmelins* Handbuch der Anorganischen Chemie, No. 52, Chrom, Verlag Chemie, Weinheim/Bergstr. Jahr.

[2] *Sidgwick*, The Chemical Elements and Their Compounds, Clarendon Press, Oxford 1950.

[3] Pascal's Nouveau Traité de Chimie Minérale, Masson, Paris.

[4] *J. P. Candin, K. A. Taylor, D. T. Thompson*, Reactions of Transition-Metal Complexes, Elsevier Publ. Co., New York 1968.

[5] *R. D. Archer*, Coord. Chem. Rev. *4*, 243 (1969).

[6] *D. M. Adams*, Metal-Ligand and Related Vibrations, St. Martins Press, New York 1968.

[7] *K. Nakamoto*, Infrared Spectra of Inorganic and Co-ordination Compounds, 2nd. Ed., John Wiley & Sons, New York 1970.

[8] *J. L. Andrieux*, Ann. Chim. 10 *12*, 463 (1929); *J. L. Andrieux, P. Blum*, Compt. Rend. *229*, 210 (1949).

[9] *A. F. Wells*, Structural Inorganic Chemistry, 3rd. Ed., p. 822, 1023, Clarendon Pres, Oxford 1962.

[10] *P. Schwarzkopf, R. Kieffer*, Refractory Hard Metals, p. 124, Macmillan, New York 1953.

[11] *B. B. Owen, R. T. Webber*, Trans. Met. Soc. AIME *175*, 693 (1948).

[12] *R. Kieffer, F. Benesovsky* in Encyclopedia of Chemical Technology, 2nd. Ed. Vol. IV, p. 77, John Wiley & Sons, New York 1964.

[13] *F. Jellinck*, Acta Cryst. *10*, 620 (1957).

[14] *I. M. Corliss, N. Elliot, J. M. Hastings, R. L. Sass*, Phys. Rev. *122*, 1402 (1961).

[15] *M. Chevreton, M. Murat, C. Eyraud, E. F. Bertaut*, J. Phys. (Paris) *24*, 443 (1963).

[16] *V. Ivanova, D. Abdinov, G. Aliev*, Phys. Status Soldi *24*, K 145 (1967).

[17] *F. H. Wehmeier, E. T. Keve, S. C. Abrahams*, Inorg. Chem. *9*, 2125 (1970).

[18] *M. Chevreton, E. F. Bertaut, F. Jellinek*, Acta Cryst. *16*, 431 (1963).

[19] *A. F. Andresen*, Acta Chem. Scand. *17*, 1335 (1963); *A. F. Andresen*, Acta Chem. Scand. *24*, 3495 (1970).

[20] *M. Cherveton, M. Murat, E. F. Bertaut*, Colloq. Int. Cent. Nat. Rech. Sci. (Paris), No. *157*, 49 (1967).

[21] *A. W. Sleight, T. A. Bither*, Inorg. Chem. *8*, 566 (1969).

[22] *W. M. Mueller, J. P. Blackledge, G. G. Libowitz*, Metal Hydrides, p. 623, Academic Press, New York.

reported. Gaseous[23] CrH and several non-stoichiometric[22] hydrides have also been characterized. Numerous *carbonyl hydrides*,[22] *e.g.*, salts[24] of $HCr_2(CO)_{10}^{\ominus}$, have been prepared and mixed complex cyclopentadienyl-carbonyl and related hydrides[25,26] have been reported. The complex hydrides of chromium have been reviewed.[27]

14.3 Chromium(−II, −I, 0, I) Compounds

A limited number of compounds with these oxidation states are known. By far, the largest number of compounds contain chromium(0) — the carbonyls and substituted carbonyl compounds comprise most of this class. The carbonyls are discussed in Chapter 29 of this volume. Chromium(−II) compounds apparently also belong to this class.

Heating the hexacarbonyl of chromium with o-phenylenebisdimethylarsine (diars) *in vacuo* gives[28] first Cr(diars) $(CO)_4$.

Further heating at elevated temperature yields $[Cr(diars)_2(CO)_2]$. The products are monomeric nonelectrolytes. These and complexes with related ligands serve as useful starting materials for the preparation of *low-valent chromium compounds*.

The compounds alleged to contain chromium(I) are frequently those for which the electronic configurations of the ligands are not well-defined, *e.g.*, *2,2'-bipyridyl derivatives* (bipy). The blue complex $[Cr(bipy)_3](ClO_4)$ is prepared by the action of magnesium on the chromium(II) tris-complex in the absence of air. It is insoluble in water, but is soluble in coordinating solvents. Black *tris(2,2'-bipyridine)chromium(0)* can be prepared[29] by reacting chromium(II) acetate hydrate with 2,2'-bipyridine. The reaction involves disproportionation and chromium(II) salts are also obtained. The product can also be obtained by the reduction of the chromium(II) chelate with sodium in tetrahydrofuran. The compound gives red solutions in benzene, tetrahydrofuran, pyridine, and dimethylformamide. It ignites in air with oxidation to give chromium(III) oxide. Tris(2,2'-bipyridine)chromium(III) perchlorate, $[Cr(bipy)_3](ClO_4)_3$, is reduced[30] at the dropping mercury electrode in acetonitrile with tetraethylammonium perchlorate as supporting electrolyte in consecutive one-electron steps to yield $Cr(bipy)_3^{\ominus}$, the salts of which may be isolated.

Garnet-red *hexaphenylisonitrilochromium(0)*, $Cr(CNC_6H_5)_6$ (m.p. 178·5°), results[31] from the disproportionation of chromous acetate in phenyl isonitrile. The diamagnetic compound is stable in air. It is soluble in cold benzene and chloroform and is readily soluble in most hot solvents. Isonitrile complexes have recently been reviewed.[32]

The reaction[33,34] of anhydrous chromium(III) chloride with sodium cyclopentadienide in tetrahydrofuran in 1:2 molar ratio leads to a green solution. By reacting nitric oxide with this solution greenish yellow *cyclopentadienyldinitrosylchromium chloride* may be prepared. The nitrate salt, prepared by treating $[Cr(C_5H_5)(NO)_2]Cl$ with aqueous silver nitrate, reacts with excess sodium thiocyanate to yield the yellow-green thiocyanate which may be extracted into chloroform. Cyanide, nitrite, fluoride, bromide, and iodide salts are similarly prepared.

Bright green *potassium pentacyanonitrosylchromate(I)* is obtained by treating[35] chromium trioxide in cold saturated potassium hydroxide with hydroxylamine hydrochloride. It is stable in the solid state and in aqueous solution.

Deep red-brown crystals of *pentaaquonitrosylchromium sulfate* are obtained[36] by concen-

[23] *B. Kleman, B. Liljequist*, Ark. Fys. *9*, 345 (1955).
[24] *H. Behrens, W. Haag*, Z. Naturforsch. B *14*, 600 (1959).
[25] *E. O. Fischer, W. Hafner, H. O. Stahl*, Z. Anorg. Allgem. Chemie *282*, 47 (1955).
[26] *E. O. Fischer, W. Hafner*, Z. Naturforsch. *10*, 140 (1955).
[27] *A. P. Ginsberg*, Transition Metal Chem. *1*, 111 (1965).
[28] *H. L. Nigam, R. S. Nyholm, M. H. B. Stiddard*, J. Chem. Soc. (London) 1803 (1960).
[29] *S. Herzog, K. C. Renner, W. Schon*, Z. Naturforsch. *12b*, 809 (1957).
[30] *Y. Sato, N. Tanaka*, Bull. Chem. Soc., Japan *42*, 1021 (1969).
[31] *G. Brauer*, Handbook of Preparative Inorganic Chemistry, Vol. II, p. 1363, Academic Press, New York 1965.
[32] *L. Malatesta, F. Bonati*, Isocyanide Complexes of Metals: John Wiley & Sons, New York 1969
[33] *T. S. Piper, G. Wilkinson*, J. Inorg. Nucl. Chem. 2, 38 (1956).
[34] *O. L. Carter, A. T. McPhall, G. A. Sim*, Chem. Commun. 49 (1966).
[35] *W. P. Griffith, J. Lewis, G. Wilkinson*, J. Chem. Soc. (London) 872 (1959).
[36] *M. Ardon, J. I. Herman*, J. Chem. Soc. (London) 507 (1962).

tration of solutions of the complex prepared by oxidation of aqueous chromous salts with nitrate ion or nitric oxide. It is stable for months in a dry atmosphere, but it slowly decomposes in aqueous solution.

Yellow *pentamminenitrosylchromium(I) dichloride*, $[Cr(NO)(NH_3)_5]Cl_2$ is obtained[37,38] by treating a suspension of chromous chloride in liquid ammonia with nitric oxide. It is stable to air, but is slowly hydrolyzed in aqueous solution. It can be converted to the cyanide complex by reaction with potassium cyanide. Pentamethanol and pentaethanol complexes, *i.e.*, $[Cr(NO)(CH_3OH)_5]Cl_2$ and $[Cr(NO)(C_2H_5OH)_5]Cl_2$, are prepared by passing nitric oxide into alcoholic solutions of chromous chloride. Electrolytic reduction of an aqueous solution of potassium pentacyanonitrosylchromate(I) yields the blue salt *potassium pentacyanonitrosylchromate(0)*, $K_4[Cr(NO)(CN)_5]$, which is very sensitive to oxidation in air.

Nitric oxide reacts[39] smoothly with aqueous solutions of chromous salts to form $CrNO^{2\oplus}$ species along with a variety of other materials. In the presence of *o*-phenylenebis(dimethylarsine) (diars), addition of sodium perchlorate yields yellow crystals of *chloronitrosyl[o-phenylenebis(dimethylarsine)]chromium perchlorate*, $[CrClNO(diars)_2]ClO_4$. Treatment of a methanolic solution of this product with sodium dithionite in methanol yields bright orange *chloronitrosylbis[o-phenylenebis(dimethylarsine)]chromium*, $[CrClNO(diars)_2]$.

14.4 Chromium(II) Compounds

Blue-green *chromium difluoride* (m.p. 894°) is best prepared[40] by reducing chromium trifluoride with a stoichiometric amount of chromium metal in an evacuated nickel bomb at 1000° or by the thermal decomposition of ammonium hexafluorochromate(IV) at 1100° in a flow system. Pale blue *sodium tetrafluorochromate(II)*, Na_2CrF_4, results[41,42] from the interaction of stoichiometric quantities of sodium fluoride and chromium dichloride or chromium(II) sulfamate in aqueous solution. It is extremely hygroscopic. Blue alkali metal *trifluorochromates(II)* such as $CsCrF_3$ result[42–45] from the interaction of chromium(II) acetate and an alkali hydrogen difluoride in an inert solvent or chromium(II) acetate and alkali fluoride in hydrofluoric acid. Pure compounds are prepared with great difficulty.

Colorless, hygroscopic *chromium dichloride* is commonly prepared[46–50] by the hydrogen reduction of chromium trichloride at about 500°. A mixture of hydrogen and hydrogen chloride is used to prevent reduction to the metal. It can also be obtained[51] by reduction of chromium trichloride with chromium metal in a bomb at about 900°. It is readily soluble in water and forms a blue solution; it reacts[52–54] with pyridine in ethanol to yield green $CrCl_2(pyridine)_2$. Chromium dichloride reacts[55,56] with a wide variety of phosphines to yield 1:2 adducts which decompose to 1:1 polymeric compounds. Phosphine and arsine oxides form[55,57,58] stable 1:2 complexes.

Chromium dichloride tetrahydrate is prepared[59–61]

---

[37] *W. P. Griffith*, J. Chem. Soc., 3286 (1963).
[38] *H. Kobayashi, I. Tsujikawa, M. Mori, Y. Yamamoto*, Bull. Chem. Soc. Japan *42*, 709 (1969).
[39] *R. D. Feltham, W. Silverthron, G. Mcpherson*, Inorg. Chem. *8*, 344 (1969).
[40] *B. J. Sturm*, Inorg. Chem. *1*, 665 (1962).
[41] *A. J. Deyrup*, Inorg. Chem. *3*, 1645 (1964).
[42] *A. Earnshaw, L. F. Larkworthy, K. S. Patel*, J. Chem. Soc. A, 363 (1966).
[43] *A. J. Edwards, R. D. Peacock*, J. Chem. Soc. (London).
[44] *K. Knox*, Acta Cryst. *14*, 583 (1961).
[45] *J. C. Cousseins, A. Dekozak*, Compt. Rend. *263C*, 1533 (1966).
[46] *G. Brauer*, Handbook of Preparative Inorganic Chemistry, Academic Press, New York 1963.
[47] *C. T. Anderson*, J. Amer. Chem. Soc. *59*, 488 (1937).
[48] *A. Burg*, Inorg. Synth. *3*, 150 (1950).
[49] *J. W. Tracey, N. W. Gregory, E. C. Lingafelter*, Acta Cryst. *14*, 927 (1961).
[50] *J. W. Stout, R. C. Chisholm*, J. Chem. Phys. *36*, 979 (1962).
[51] *J. D. Corbett, R. J. Clark, R. F. Mundy*, J. Inorg. Nucl. Chem. *25*, 1287 (1963).
[52] *N. S. Gill, R. S. Nyholm, G. A. Barclay, T. I. Christie, P. J. Pauling*, J. Inorg. Nucl. Chem. *18*, 88 (1961).
[53] *D. G. Holah, J. P. Fackler*, Inorg. Chem. *4*, 1112 (1965).
[54] *H. Lux, L. Eberle, D. Sarre*, Chem. Ber. *97*, 503 (1964).
[55] *K. Issleib, A. Tzschach, H. O. Frohlich*, Z. Anorg. Allgem. Chemie *298*, 164 (1958).
[56] *K. Issleib, H. O. Frohlich*, Z. Anorg. Allgem. Chemie *298*, 84 (1958).
[57] *L. F. Larkworthy, D. J. Phillips*, J. Inorg. Nucl. Chem. *29*, 2101 (1967).
[58] *D. E. Scaife*, Australian J. Chem. *20*, 845 (1967).
[59] *J. P. Fackler, D. G. Holah*, Inorg. Chem. *4*, 954 (1965).
[60] *A. Earnshaw, L. F. Larkworthy, K. S. Patel*, J. Chem. Soc. (London) 3267 (1965).
[61] *A. Earnshaw, L. F. Larkworthy*, Proc. Chem. Soc. 281 (1963).

by dissolving chromium metal in dilute hydrochloric acid; evaporation of the solution at 80° or precipitation with acetone gives the blue hygroscopic product. The only compound obtained[62,63] by the fusion of sodium chloride and chromium dichloride is *sodium pentachlorochromate(II)*, Na_3CrCl_5. Other alkali chlorides react in the melt with chromium dichloride to yield *tetrachlorochromates(II)*.[64,65] *Trichlorochromates, e.g.*, $RbCrCl_3$, are prepared similarly. For both types of compounds stoichiometric quantities of materials must be employed.

Adducts of the type $CrCl_2L_2$ (L = acetonitrile)[66] have been reported. A number of *tetrachloro-*, *trichloro*, and *tribromochromates(II)* can be obtained[67] by adding an alkali acetate to a solution of chromium(II) in acetyl halide–acetic acid mixture. Pentachlorochromates(III) can be prepared similarly using a chromium(III) solution.

Colorless *chromium dibromide* (m.p. 842°) is obtained[68,69] by passing hydrogen bromide over chromium metal at about 750° in a flow system. It dissolves in oxygen-free water to a yield a blue solution and reacts[70] with pyridine to give $CrBr_2(C_5H_5N)_2$. Two isomers of the bis adduct with triphenylphosphine oxide[58] have been characterized and 1:2 adducts with triphenylarsine oxide[57] and aliphatic[56] phosphines have been obtained. *Chromium dibromide hexahydrate* is produced[70,71] by dissolving chromium metal in 40% hydrobromic acid and evaporating *in vacuo* or precipitating the dark blue complex with acetone.

Red-brown *chromium diiodide* may be obtained[51,72] by interaction of stoichiometric amounts of the elements in a sealed tube at 550°–700°. Vacuum sublimation of the product at 700° decomposes any triiodide impurity into the diiodide. 1:2 Adducts with triphenylarsine oxide[58] and pyridine[70] are known. *Chromium diiodide hydrate* results[53,60,66,70] when chromium metal dissolves in hydriodic acid, from which the blue hydrate may be obtained.

Chromium(II) halides react[32,73] with isonitriles, *e.g.*, phenyl isonitrile, to yield orange-red chlorides or olive-brown bromides of the type $[Cr(CNR)_4X_2]$ (X = Cl, Br) which are stable in air. They can be heated in water without decomposing and are soluble in chloroform and methylene dichloride. Light blue-green $Cr(PMT)_2Cl_2$ (PMT = pentamethylenetetrazole) (m.p. 160° dec.) is prepared[74] by the reaction of hydrated chromium(II) chloride with PMT in 50% dimethoxypropane–methanol mixture. The compound decomposes rapidly in moist air.

Breaking a flake of spectroscopically pure chromium metal into deoxygenated hydrochloric acid affords a way of preparing[75] highly pure chromous solutions. A variety of hydrated salts can be similarly prepared.[60,76] Chromous solutions can react[41] with solutions of calcium chloride, calcium carbonate, and boric acid to form air-stable compounds of the type $CrCl\text{-}OH \cdot H_3BO_3$ and with amines to form amine complexes.[60,76,77] Potassium tricyanomethide, $KC(CN)_3$, reacts[78] with chromous solution to yield pale blue *chromium(II) tricyanomethide*, $Cr[C(CN)_3]_2 \cdot \frac{1}{8}H_2O$.

Tris(1,10-phenanthroline)- and *tris(2,2'-bipyridine) chromium(II)* chelates as perchlorate salts[79–81] as

---

62 *H. J. Seifert, K. Klatyk*, Z. Anorg. Allgem. Chemie *334*, 113 (1964).
63 *J. C. Shiloff*, J. Phys. Chem. *64*, 1566 (1960).
64 *H. J. Seifert, K. Klatyk*, Z. Anorg. Allgem. Chemie *334*, 113 (1964);
H. J. Seifert, K. Klatyk, Naturwissenschaften *49*, 539 (1962).
65 *R. Gut, R. Gnehur*, Chimia *16*, 289 (1962).
66 *D. G. Holah, J. P. Fackler*, Inorg. Synth. *10*, 26 (1967).
67 *H. D. Hardt, G. Streit*, Naturwissenschaften *55*, 443 (1968).
68 *R. C. Schoonmaker, A. H. Friedman, R. F. Porter*, J. Chem. Phys. *31*, 1586 (1959).
69 *R. J. Sime, N. W. Gregory*, J. Amer. Chem. Soc. *82*, 800 (1960).
70 *D. G. Holah, J. P. Fackler*, Inorg. Chem. *4*, 954, 1112 (1965);
D. G. Holah, J. P. Fackler, Inorg. Synth. *10*, 26 (1967).
71 *H. Lux, G. Illmann*, Chem. Ber. *91*, 2143 (1958).
72 *F. Hein, J. Wintner-Holder*, Z. Anorg. Allgem. Chemie *202*, 81 (1931).
73 *G. Brauer*, Handbook of Preparative Inorganic Chemistry, Vol. II, p. 1364, Academic Press, New York, 1965.
74 *D. M. Bowers, A. J. Popov*, Inorg. Chem. *7*, 1594 (1968).
75 *J. M. Crabtree*, J. Chem. Soc. (London) 4647 (1964).
76 *A. Earnshaw, L. F. Larkworthy, K. S. Patel*, J. Chem. Soc. 2276 (1969).
77 *M. Ciampolini*, Chem. Commun. 47 (1966).
78 *G. L. Silver*, Inorg. Nucl. Chem. Lett. *4*, 533 (1968).
79 *G. Brauer*, Handbook of Preparative Inorganic Chemistry, Vol. II, p. 1361, Academic Press, New York 1965.
80 *G. N. Lamar, G. R. Van Hecke*, J. Amer. Chem. Soc. *91*, 3442 (1969).
81 *P. M. Lutz, G. J. Long, W. A. Baker jr.*, Inorg. Chem. *8*, 2529 (1969).

well as methyl-substituted *o*-phenanthroline[82] tris chelates are prepared by reaction of chromous chloride with a methanolic solution of the ligand, followed by addition of aqueous sodium perchlorate–perchloric acid solution. The complexes readily oxidize in moist air. The bipyridine complex is readily reduced to the chromium(I) chelate by magnesium powder. All reactions and handling of the chelates must be performed under a nitrogen atmosphere using degassed solvents. The 2,2′-bipyridine compounds of chromium have been reviewed.[83]

Tris(ethylenediamine)chromium(III) perchlorate is reduced[83a] at the dropping mercury electrode in acetonitride with tetraethylammonium perchlorate as supporting electrolyte to yield the chromium(II) complex.

Divalent *chromium mesoporphyrin IX dimethyl ester* is prepared[84,85] by a facile metal-insertion reaction using chromium carbonyl and the porphyrin in high boiling solvents, *e.g.*, decalin. The related phthalocyanines of chromium have been reviewed.[86]

Oxidation of the compound $[Cr(diars)_2(CO)_2]$ (diars = *o*-phenylenebisdimethylarsine) at room temperature with bromine or iodine yields[87] $[Cr(CO)_2(diars)_2X]X$, (X = Br, I). 1-Triars [methylbis(dimethylarsino-3-propyl)arsine] and *v*-triars [tris-1,1,1-(dimethylarsinomethyl)ethane] react[88] with chromium hexacarbonyl to produce *tricarbonyltriarsine complexes*. The compounds react with elemental iodine at −70° to yield compounds of the type $[Cr(CO)_3(triars)I]I$ which can be isolated as red to brown tetraphenylborate salts.

Diamagnetic compounds of the type $[Cr(XR_{1,2,\text{or }3})_3NO]$ (where X = N, O or Si and NO is formally considered to be NO) result[89] from the interaction of nitric oxide with the parent chromium(III) compounds. The compounds are soluble in organic solvents and can be sublimed *in vacuo*.

Dark red crystals of *dimeric chromium(II) acetate monohydrate* result[90] when chromous chloride is treated with sodium acetate. The dry compound is stable in air for several hours and is stable indefinitely in a nitrogen atmosphere. The yellow-green oxalate, $CrC_2O_4 \cdot 2H_2O$, is obtained from a metathetical reaction of sodium oxalate with chromium(II) sulfate. The preparation of a number of *n*-alkanoate analogus of chromous acetate have been described.[91]

Chromium(II) sulfide is obtained[92] by heating a stoichiometric mixture of the elements in a quartz tube at 1000°.

The magnetic properties of chromium(II) complexes have been reviewed.[93]

14.5 Chromium(III) Compounds

Reviews have appeared concerning the optical activity of *chromium(III) complexes*,[94] the use in the stabilization of unusual counterions,[95] linkage isomerism,[96] five-coordinate complexes,[97] electronic spectra of tetragonal complexes,[98–100] and photochemistry.[101,102] The separation of isomeric chromium(III) compounds and reaction products has been reviewed.[103] The species formed on dissolving chromium(III) salts in non-

---

[82] *G. N. Lamar, G. R. Van Hecke*, Inorg. Chem. *9*, 1546 (1970).

[83] *W. R. McWhinnie, J. D. Miller*, Adv. Inorg. Chem. Radiochem. *12*, 135 (1969).

[83a] *Y. Sato, N. Tanaka*, Bull. Chem. Soc. Japan *42*, 1091 (1969).

[84] *M. Tsutsui, M. Ichikuwa, F. Vohwinkel, K. Suzuki*, J. Amer. Chem. Soc. *88*, 854 (1966).

[85] *M. Tsutsui, R. A. Velapddi, K. Suzuki, F. Vohwinkel, M. Ichikawa, T. Koyano*, J. Amer. Chem. Soc. *91*, 6262 (1969).

[86] *A. B. P. Lever*, Adv. Inorg. Chem. Radiochem. *7*, 28 (1965).

[87] *J. Lewis, R. S. Nyholm, C. S. Pande, S. S. Sandhu, M. H. B. Stiddard*, J. Chem. Soc. (London) 3009 (1964).

[88] *R. S. Nyholm, M. R. Snow, M. H. B. Stiddard*, J. Chem. Soc. (London) 6570 (1965).

[89] *D. C. Bradley, C. W. Newing*, Chem. Commun. *219* (1970).

[90] *G. Brauer*, Handbook of Preparative Inorganic Chemistry, Vol. II, p. 1368, Academic Press, New York 1965.

[91] *S. Herzog, W. Kalies*, Z. Anorg. Allgem. Chemie *329*, 83 (1964).

[92] *G. Brauer*, Handbook of Preparative Inorganic Chemistry, Vol. II, p. 1346, Academic Press, New York 1965.

[93] *G. A. Webb*, Coord. Chem. Rev. *4*, 107 (1969).

[94] *S. Kirschner, N. Ahmad, K. Magnell*, Coord. Chem. Rev. *3*, 201 (1968);
S. Kirschner, Coord. Chem. Rev. *2*, 461 (1967).

[95] *F. Basolo*, Coord. Chem. Rev. *3*, 213 (1968).

[96] *J. L. Burmeister*, Coord. Chem. Rev. *3*, 225 (1968).

[97] *C. Furlani*, Coord. Chem. Rev. *3*, 141 (1968).

[98] *A. B. P. Lever*, Coord. Chem. Rev. *3*, 119 (1968).

[99] *J. R. Perumareddi*, Coord, Chem. Rev. *4*, 73 (1969).

[100] *L. S. Forster*, Trans. Metal. Chem. *5*, 1 (1969).

[101] *A. W. Adamson*, Coord. Chem. Rev. *3*, 169 (1968).

[102] *V. Balzani, V. Carassiti*, Photochemistry of Coordination Compounds, Academic Press, New York 1970.

[103] *L. F. Druding, G. B. Kauffman*, Coord. Chem. Rev. *3*, 409 (1968).

aqueous solvents have been discussed.[104] The magnetic properties of dimeric chromium(III) compounds have been reviewed,[105] as has the aqueous solution chemistry[106] of chromium(III).

14.5.1 Halogen Compounds and Derivatives

Green *chromium trifluoride* can be produced[107] by the reaction of chromium metal with anhydrous hydrogen fluoride in a bomb at 300°; at 900° chromium difluoride is obtained. Chromium trifluoride forms complex fluorides and adducts with selenium tetrafluoride[108] and bromine trifluoride.[109] Green *chromium trifluoride trihydrate* is prepared[110–112] by reducing chromium trioxide in 10% hydrofluoric acid with ethanol in the presence of glucose.

A compound having the composition $CrF_{2.5}$ can be prepared[40,113,114] by reacting equimolar amounts of chromium difluoride and trifluoride in a sealed tube at 800°–900°. Crystallographic studies have shown that $CrF_{2.5}$ contains two types of CrF_6 octahedra.

Hexafluorochromates(III), *e.g.*, K_3CrF_6, can be prepared[40,115,116] by the fusion of chromium salts with ammonium hydrogen fluoride. The hydrazinium salt is obtained[117] from the reaction of hydrazinium fluoride and hydrofluoric acid with chromium metal.

The solubility of sodium hexafluorochromate(III) in the eutectic mixture of sodium fluoride and sodium tetrafluoroborate used in molten salt breeder reactors has been reported[118] as a function of temperature.

Pentafluorochromates(III), *e.g.*, Cs_2CrF_5, can be isolated[119] from saturated aqueous solutions of alkali fluorides and chromium trifluoride. They can be dehydrated[120] by heating *in vacuo*.

Violet anhydrous *chromium trichloride* (m.p. 1150°) is prepared[51,121,122] by the chlorination of chromium metal in a flow system at 960°–1000°. It is readily purified by sublimation in a stream of chlorine at 850° or in a vacuum at slightly higher temperatures.

It forms[56,123–132] *adducts* with a wide variety of donor molecules such as alkyl nitriles, ethers, amines, alcohols, phosphines, *etc.* Amines[133–140] form adducts of the general type $CrCl_3 \cdot nL$ (where n may be 2 or 3); in some instances, chloride coordinated to chromium may be displaced. Trimethylamine forms the five-coordinate

[104] *V. Gutmann*, Coord. Chem. Rev. *2*, 239 (1967).
[105] *P. W. Ball*, Coord. Chem. Rev. *4*, 361 (1969).
[106] *J. E. Earley, R. D. Cannon*, Trans. Metal Chem. *1*, 33 (1965).
[107] *E. L. Muetterties, J. E. Castle*, J. Inorg. Nucl. Chem. *18*, 148 (1961).
[108] *H. C. Clark, Y. N. Sadana*, Can. J. Chem. *42*, 50 (1964).
[109] *H. C. Clark, Y. N. Sadana*, Can. J. Chem. *42*, 702 (1964).
[110] *S. T. Talipov, V. E. Antipov*, Dokl. Akad. Nauk. SSSR *12*, 27 (1949).
[111] *J. Maak, P. Eckerlin, A. Rabenau*, Naturwissenschaften *48*, 218 (1964).
[112] *H. L. Schlafer, H. Gausmann, H. V. Zander*, Inorg. Chem. *6*, 1528 (1967).
[113] *H. Steinfink, J. H. Burns*, Acta Cryst. *17*, 823 (1964).
[114] *W. P. Osmond*, Proc. Phys. Soc. *87*, 767 (1966).
[115] *H. Bode, E. Voss*, Z. Anorg. Allgem. Chemie *290*, 1 (1957).
[116] *R. D. Peacock*, J. Chem. Soc. 4684 (1957).
[117] *J. Slivnivk, J. Pezdiic, B. Sedej*, Monatsh. Chem. *98*, 204 (1967).
[118] *C. J. Barton*, J. Inorg. Nucl. Chem. *33*, 1948 (1971).
[119] *R. S. Nyholm, A. G. Sharpe*, J. Chem. Soc. 3579 (1959).
[120] *R. D. Peacock*, Progr. Inorg. Chem. *2* (1960).
[121] *N. Wooster*, Z. Kristallogr. *74*, 363 (1930).
[122] *B. Morosin, A. Narath*, J. Chem. Phys. *40*, 1958 (1964).
[123] *B. J. Hathaway, D. G. Holah*, J. Chem. Soc. 537 (1965).
[124] *R. J. H. Clark, M. L. Greenfield, R. S. Nyholm*, J. Chem. Soc. A, 1254 (1966).
[125] *B. J. Kern*, J. Inorg. Nucl. Chem. *25*, 5 (1963), *24*, 1105 (1962).
[126] *M. J. Frazer, W. Gerrard, R. Twaits*, J. Inorg. Nucl. Chem. *25*, 637 (1963).
[127] *G. W. A. Fowles, P. T. Greene, T. E. Lester*, J. Inorg. Nucl. Chem. *29*, 2365 (1967).
[128] *F. A. Cotton, R. Francis*, J. Amer. Chem. Soc. *82*, 2986 (1960).
[129] *F. A. Cotton, R. Francis, W. D. Horrocks*, J. Phys. Chem. *64*, 1534 (1960).
[130] *W. E. Bull, S. K. Madan, J. E. Willis*, Inorg. Chem. *2*, 303 (1963).
[131] *H. Herwig, H. Zeiss*, J. Org. Chem. *23*, 1404 (1958).
[132] *K. Issleib, G. Doll*, Z. Anorg. Allgem. Chemie *305*, 1 (1960)
[133] *J. C. Taft, M. M. Jones*, J. Amer. Chem. Soc. *82*, 4196 (1960).
[134] *G. Morgan, F. H. Burstall*, J. Chem. Soc. 1649 (1937).
[135] *R. J. H. Clark, M. L. Greenfield*, J. Chem. Soc. *409* (1967).
[136] *B. Bosnich, R. D. Gillard, E. D. McKenzie, G. A. Webb*, J. Chem. Soc. (A) 1331 (1966).
[137] *C. L. Robinson, R. C. White*, Inorg. Chem. *1*, 281 (1962).
[138] *T. B. Hackson, J. O. Edwards*, Inorg. Chem. *1*, 398 (1962).
[139] *H. L. Schlafer, O. Kling*, Z. Anorg. Allgem. Chemie *302*, 1 (1959);
H. L. Schlafer, O. Kling, Z. Anorg. Allgem. Chemie *309*, 245 (1961);
H. L. Schlafer, O. Kling, Z. Anorg. Allgem. Chemie *313*, 187 (1961).
[140] *M. Linhard, M. Weigel*, Z. Anorg. Allgem. Chemie *271*, 101, 115 (1952).

adduct, $CrCl_3[N(CH_3)_3]_2$.[141–143] Complexes with bidentate nitrogen heterocyclic ligands, e.g., *o*-phenanthroline and bipyridine are well known.[144,145] Chromium trichloride reacts[146] with thiourea or substituted thiourea to form adducts of the type CrL_3Cl_3. Three isomeric forms of chromium chloride hexahydrate are known[146]: dark green $[CrCl_2(H_2O)_4]Cl \cdot 2H_2O$, violet $[Cr(H_2O)_6]Cl_3$, and pale green $[CrCl(H_2O)_5]Cl \cdot H_2O$. The dark green compound is available commercially and is the starting material for the preparation of the other isomers.

Dark red *alkali hexachlorochromates(III)* are obtained[147–150] from the reaction of chromium trichloride with the molten alkali chloride. Tris(1,2-propanediamine)metal(III) hexachlorochromates(III) can be prepared[151] by reaction of the complex metal chloride and chromium chloride hexahydrate in hydrochloric acid.

Green $Cr(OCH_3)Cl_2 \cdot 2CH_3OH$ is prepared[152] by adding an equimolar amount of sodium methoxide in methanol to a methanolic solution of chromium trichloride. One molecule of methanol may be replaced by one molecule of dioxane, acetone, or acetonitrile. The dimethanolate loses one molecule of methanol at 100°.

Green *chromium oxide chloride* is obtained[153] from the reaction of chromium(III) oxide with excess chromium trichloride over a thermal gradient of 1040–840° in a silica tube. It is stable to hot water, alkali, acids, and air at room temperature. It decomposes *in vacuo* on heating to yield chromium(III) chloride and chromium(III) oxide. When heated in air, it decomposes to chromium(III) oxide and chlorine.

141 *M. W. Duckworth, G. W. A. Gowles, P. T. Green*, J. Chem. Soc. (A), 1592 (1967).
142 *G. W. A. Fowles, P. T. Greene, J. S. Wood*, Chem. Commun. 971 (1967).
143 *G. W. A. Fowles, P. T. Greene*, Chem. Commun. 784 (1966).
144 *J. A. Broomhead, F. P. Dwyer*, Australian J. Chem. *14*, 250 (1961).
145 *W. W. Brandt, F. P. Dwyer, E. A. Byarfas*, Chem. Rev. *54*, 960 (1954).
146 *E. Cervone, P. Cancellieri, C. Furlani*, J. Inorg. Nucl. Chem. *30*, 2431 (1968).
147 *A. I. Efimov, B. Z. Titerimov*, Russ. J. Inorg. Chem. *8*, 1042 (1963).
148 *B. G. Korshunov, B. Y. Raskin*, J. Inorg. Chem. *7*, 584 (1962).
149 *S. A. Shchukarev, I. V. Vasil'kova, A. I. Efimov, B. Z. Titirimov*, J. Inorg. Chem. *11*, 268 (1966).
150 *C. M. Cook*, J. Inorg. Nucl. Chem. *25*, 123 (1963).
151 *W. E. Hatfield, R. C. Fay, C. E. Pfluger, T. S. Piper*, J. Amer. Chem. Soc. *85*, 265 (1963).
152 *L. Dubicki, G. A. Kakos, G. Winter*, Australian J. Chem. *21*, 1461 (1968).
153 *H. Schafer, F. Wartenpfuhl*, Z. Anorg. Allgem. Chemie *308*, 282 (1961).

The interaction of chromium dioxide dichloride with a slight excess of phosphorus pentachloride in a sealed tube at 140° yields[154] a blue compound having the composition $PCrCl_8$. It has been suggested that it contains the tetrahedral tetrachlorochromate(III) anion; however, a polymetric octachedral structure has not been eliminated.

Black *chromium tribromide* is best prepared[46,155] by bromination of the metal at 750° in a flow system. At high temperatures *in vacuo* it decomposes to the dibromide and bromide. Its chemistry[123,133,156] is very close to that of chromium trichloride. *Chromium oxide bromide* results[157] from the reaction of chromium tribromide and arsenic trioxide at 380° in a sealed tube. Black *chromium triiodide* is obtained[158,159] by the reaction of stoichiometric quantities of the elements at 200°–225° in a sealed tube. At higher temperatures or if an excess of chromium metal is used, the product is contaminated with the diiodide. The compound is thermally unstable.

$Cr(CO)_4$(diars) reacts[87] with bromine and iodine to yield complexes with the composition $Cr(diars)X_3$ (X = Br, I). These are believed to be halogen-bridged octahedrally coordinated dimers. The bromide readily hydrates to give $[Cr(diars)(H_2O)Br_3]^0$. The reaction of $Cr(diars)_2(CO)_2$ with iodine in boiling carbon tetrachloride yields the complex $[Cr(diars)_2I_2]^{\oplus}I^{\ominus}$. Green complexes of the type $[Cr(diars)_2X_2][Cr(diars)X_4]$ (X = Cl) result[160,161] on treating chromium trichloride hexahydrate in concentrated hydrochloric acid with diars in acetone. Modification of the procedure, *i.e.*, using concentrated perchloric acid instead of hydrochloric acid can be employed to produce complexes of the type $[Cr(diars)_2X_2]ClO_4$. Displacement with lithium halide and hydrohalic acid can be used to prepare the bromo and iodo analogs.

154 *D. J. Machin, D. F. C. Morris, E. L. Short*, J. Chem. Soc. 4658 (1964).
155 *R. J. Sime, N. W. Gregory*, J. Amer. Chem. Soc. *82*, 93 (1960).
156 *I. V. Howell, L. M. Venanyi, D. C. Goodall*, J. Chem. Soc. (A) 395 (1967).
157 *M. Danot, J. Rouxel*, Compt. Rend. *262C*, 1879 (1966).
158 *J. F. Dillon, C. E. Olson*, J. Appl. Phys. *36*, 1259 (1965).
159 *N. W. Gregory, L. L. Handy*, Inorg. Synth. *5*, 128 (1957).
160 *R. S. Nyholm, G. J. Sutton*, J. Chem. Soc. 560 (1958).
161 *R. D. Feltham, W. Silverthorn*, Inorg. Chem. *7*, 1154 (1968).

Dark blue *alkali μ-trichlorohexachlorodichromates(III)*, e.g., $Rb_3Cr_2Cl_9$, are obtained[147,150,162–164] from the reaction of stoichiometric quantities of alkali chloride and chromium trichloride in the melt. Quaternary ammonium salts can be prepared[165] by refluxing the chloride salt and chromium trichloride in thionyl chloride. The tetraethylammonium salt is also obtained[166] by the reaction of tetraethylammonium chloride and chromium trichloride hexahydrate in acetonitrile solution.

14.5.2 Oxygen Donor Compounds

Green *chromium(III) oxide*, Cr_2O_3, formed by burning the metal in oxygen or roasting the hydrated oxide ("hydroxide"), is commercially available in a number of forms. If the oxide is heated too strongly, it becomes inert to both acids and bases. It dissolves readily in acids to give hexaaquo salts and in concentrated alkali to form chromites. Chromium(III) oxide can be fused with other metal oxides to form *mixed oxide systems*.

Black salts containing chromium in oxidation states 3+ and 6+, of composition $M^ICr_3O_8$ (M^I = Li, K, Cs), are prepared[167] by reacting the alkali dichromate with chromium(VI) oxide in a sealed tube. The compounds decompose on heating in air.

Hexaaquochromium(III) perchlorate is prepared[168,169] by reducing reagent-grade chromium trioxide with hydrogen peroxide in 1 M perchloric acid. It can also be obtained by dissolving chromium(III) hydroxide in concentrated perchloric acid.

Pale green anhydrous *chromium(III) nitrate* results[170] from the reaction of chromium hexacarbonyl with dinitrogen pentoxide in carbon tetrachloride. It is extremely deliquescent. It dissolves in dry ethyl acetate, acetonitrile, and dimethylsulfoxide without apparent decomposition and is insoluble in benzene, carbon tetrachloride, and chloroform.

A number of *dimethyl sulfoxide* (DMSO) *complexes* of chromium(III) have been described.[128] $CrCl_3(DMSO)_3$ is prepared by adding chromium metal to DMSO kept saturated with hydrogen chloride. The resulting green solution is filtered to remove excess metal and taken to dryness *in vacuo* at 10°, affording a red lilac powder. *Hexakis(dimethylsulfoxide)chromium(III) perchlorate* is prepared by adding DMSO to a concentrated solution of chromium(III) perchlorate. Emerald green needles are obtained by removing most of the solvent under vacuum at about 100°. Mixed complexes have been examined in aqueous solution.[169]

Hexakis complexes of the type $Cr(L)_6(ClO_4)_3$ (L = tetramethylene sulfoxide) may generally be prepared[171] by addition of a slight excess of the ligand to the hydrated perchlorate in methanol. For moisture-sensitive ligands or complexes, 2,2-dimethoxypropane[172] or triethyl orthoformate may be added to the reaction mixture to remove water.

Blue-white to green pyridine *N*-oxide and substituted pyridine *N*-oxides form[173,174] hexakis complexes on reaction of the ligands with methanolic solutions of hydrated chromium(III) perchlorate. Bright green *bis(perchlorato)tetrakis(tri-n-butylphosphine oxide)chromium(III) perchlorate* (m.p. 131°–132°) results[175] from the interaction of the phosphine oxide and chromium(III) perchlorate in triethyl orthoformate solution. Dark green *potassium trisoxalatochromate(III)*, $K_3[Cr(C_2O_4)_3]\cdot 3H_2O$, is obtained[176] by the reduction of potassium dichromate with a mixture of oxalic acid and potassium oxalate.

Green *hexaureachromium(III) chloride trihydrate*, a useful starting material for the prepara-

---

[162] *I. V. Vasil'kova, A. I. Efimov, B. Z. Pitirimov*, Russian J. Inorg. Chem. *9*, 493 (1964).
[163] *G. J. Wessel, D. J. W. Ijdo*, Acta Cryst. *10*, 466 (1957).
[164] *R. Saillant, R. A. D. Wentworth*, Inorg. Chem. *7*, 1606 (1968).
[165] *D. M. Adams, J. Chatt, J. M. Davidson, J. Gerratt*, J. Chem. Soc. 2189 (1963).
[166] *A. Earnshaw, J. Lewis*, J. Chem. Soc. 396 (1961).
[167] *K. A. Wilhelmi*, Ark. Kemi *26*, 131, 141 (1966).
[168] *R. A. Plane, J. A. Laswick*, J. Amer. Chem. Soc. *81*, 3564 (1959).
[169] *K. R. Ashley, R. E. Hamm, R. H. Magnusson*, Inorg. Chem. *6*, 413 (1967).
[170] *C. C. Addison, D. J. Chapman*, J. Chem. Soc. 539 (1964).
[171] *D. W. Meek, R. S. Drago, T. S. Piper*, Inorg. Chem. *1*, 285 (1962).
[172] *K. Starke*, J. Inorg. Nucl. Chem. *11*, 77 (1959).
[173] *R. Whyman, W. E. Hatfield, J. S. Paschal*, Inorg. Chim. Acta *1*, 113 (1967).
[174] *R. G. Garvey, J. N. Nelson, R. O. Radsdale*, Coord. Chem. Rev. *3*, 375 (1968).
[175] *N. M. Karayannis, C. M. Mikulski, L. L. Pytlewski, M. M. Labes*, Inorg. Chem. *9*, 582 (1970).
[176] *G. Brauer*, Handbook of Preparative Inorganic Chemistry, Vol. II, p. 1372, Academic Press, New York 1965.

tion of chromium(III) complexes, results[177] from the interaction of slightly more than a stoichiometric quantity of urea with an aqueous solution of commercial chromium trichloride hexahydrate. The crystals are readily soluble in water, but insoluble in absolute alcohol.

Violet crystals of *tris(acetylacetonato)chromium(III)* (m.p. 216°) can be obtained[178] by refluxing ethanolic chromium(III) nitrate hexahydrate with acetylacetone(2,4-pentanedione). The compound is soluble in alcohol, chloroform, and benzene. It is insoluble in water and petroleum ether. Other β-diketonate complexes can be prepared similarly.[179] *Tris(3-cyanomethyl-2,4-pentanedionato)chromium(III)*, $Cr[CH_3C(0)C(CH_2CN)C(O)CH_3]_3$, is prepared[180] by the aminomethylation of tris(acetylacetonato)chromium(III) followed by quaternization of the amine and subsequent displacement of trimethyl amine with alkali metal cyanide. Tris(acetylacetonato)chromium(III) undergoes a wide variety of substitution reactions.[181–185]

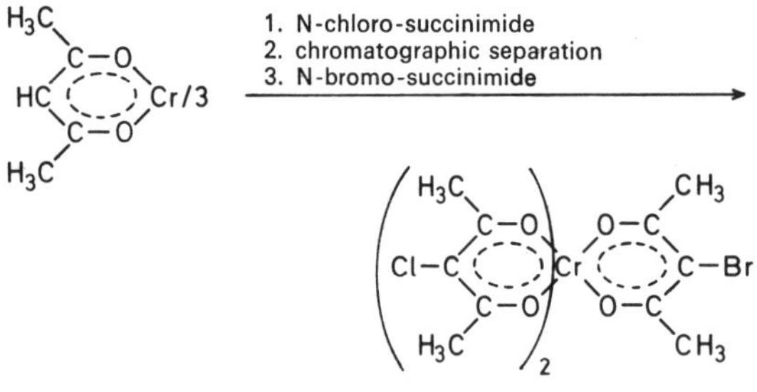

$(H_3C)_2(C_2O_2)HC$ Cr/3 → (1. N-chloro-succinimide; 2. chromatographic separation; 3. N-bromo-succinimide) → $[(H_3C)_2(C_2O_2)C\text{-}Cl]Cr[(CH_3)_2(C_2O_2)C\text{-}Br]$ /2

(1)

These and other reactions of ligands coordinated to chromium(III) have been reviewed.[181–185]

A number of simple as well as polymeric acetates have been reported.[186, 187]

14.5.3 Sulfur and Selenium Derivatives

Blue to purple chromium(III) complexes with bidentate sulfur ligands are usually prepared by adding the ligand to an alcoholic solution of chromium trichloride trihydrate or chromium(III) perchlorate hydrate. Several of these complexes are listed in Table 1. Purple *tetra(n-butyl)-ammoniumoxobis(toluene-3,4-dithiolato)-chromium(III)* is prepared[188] by the reaction of hydrated chromium chloride in ethanol with toluene-3,4-dithiol and potassium ethoxide.

Table 1. Chromium(III) Complexes with Sulfur and Selenium Ligands

Complex	Ref.	Ligand
$^{\ominus}S-C(=O)-C(=O)-S^{\ominus}$	*a*	Dithiooxalate
$(RO)_2P(=S)-S^{\ominus}$	*b, p, c*	Dithiophosphate
$R_2N-C(=S)-S^{\ominus}$	*c, p*	Dithiocarbamate
$R_2N-C(=S)-NR_2$ and CrL_3Cl_3	*d*	Thiourea or substituted thiourea
$NC-C(S^{\ominus})=C(S^{\ominus})-CN$	*e*	Malonitriledithiolate
$CrO(tdt)_2$	*f*	tdt = Toluene-3,4-dithiolate
$RO-C(=S)-S^{\ominus}$	*g, p, q, r*	Xanthate
$R-C(=S)-S^{\ominus}$ $R = C_6H_5, C_6H_5CH_2$	*h*	Dithiobenzoate, dithophenylacetate
$(C_2H_5)_2N-C(=Se)-Se^{\ominus}$ R = ethyl	*i*	Diselenocarbamate

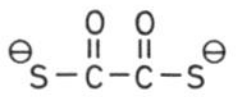

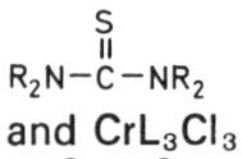

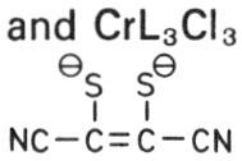

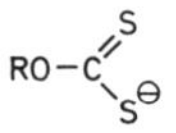

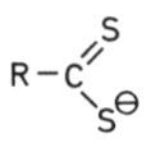

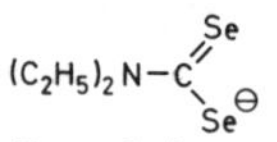

[177] *G. Brauer*, Handbook of Preparative Inorganic Chemistry, Vol. II, p. 1359, Academic Press, New York 1965.

[178] *G. Brauer*, Handbook of Preparative Inorganic Chemistry, Vol. II, p. 1384, Academic Press, New York 1965.

[179] *A. M. Fatta, R. L. Lintvedt*, Inorg. Chem. *10*, 478 (1971).

[180] *K. B. Takvorian, R. H. Barker*, Inorg. Synth. *12*, 85 (1970).

[181] *J. P. Collman, M. Yamada*, J. Org. Chem. *28*, 3017 (1963).

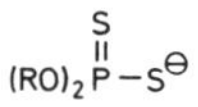

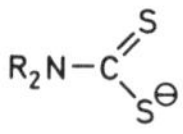

[182] *W. L. Young*, Ph. D. Dissertation, University of North Carolina, 1964.

[183] *R. H. Barker, J. P. Collman, R. L. Marshall*, J. Org. Chem. *29*, 3216 (1964).

[184] *J. P. Collman*, Transition Metal Chemistry, *2*, 1 (1966).

[185] *B. Bock, K. Flatau, H. Junge, M. Kuhr, H. Musso*, Angew. Chem. Internat. Ed. Engl. *10*, 225 (1971).

[186] *G. Brauer*, Handbook of Preparative Inorganic Chemistry, Vol. II, pp. 1359–1384, Academic Press, New York 1965.

[187] *S. C. Chang, G. A. Jeffrey*, Acta Cryst. *B26*, 673 (1970).

[188] *E. I. Steifel, R. Eisenberg, R. C. Rosenberg, H. B. Gray*, J. Amer. Chem. Soc. *88*, 2956 (1966).

Complex	Ref.	Ligand
$(C_2H_5O)_2P(=Se)Se^{\ominus}$ R = ethyl	*j*, *s*	Diselenophosphate
$Cr(SR)_3$ R = CH_3, C_2H_5, *i*-C_3H_7, C_6H_5, $CH_2C_6H_5$	*k*	Mercaptides
$M^I[CrS_2]$ M^I = $Li^{\oplus}$, $Na^{\oplus}$, $K^{\oplus}$, $Rb^{\oplus}$	*l*, *u*	Thiochromite
$H_5C_6-C(=O)S^{\ominus}$	*m*	Monothiobenzoate
$(NC)_2C=C(S^{\ominus})_2$	*n*	
$X_2P(=S)S^{\ominus}$ X = CF_3, F, CH_3, C_6H_5	*t*	Dithiophosphinates
$R_2N-C(=S)-S-S-C(=S)-NR_2$	*u*	ts – Tetraalkylthiuramdisulfide

[a] *F. P. Dwyer, A. M. Sargeson*, J. Amer. Chem. Soc. *81*, 2335 (1959); *R. L. Carlin, F. Canziani*, J. Chem. Phys. *40*, 371 (1964).

[b] *J. R. Wasson, S. J. Wasson, G. M. Woltermann*, Inorg. Chem. *9*, 1576 (1970); *C. K. Jørgensen*, J. Inorg. Nucl. Chem. *24*, 1571 (1962); *D. E. Goldberg, W. C. Fernelius, M. Shamma*, Inorg. Syn. *6*, 142 (1960); *A. A. G. Tomlinson*, J. Chem. Soc. A, p. 1509 (1971).

[c] *C. K. Jørgensen*, J. Inorg. Nucl. Chem. *24*, 1571 (1962).

[d] *P. Askalani, R. A. Bailey*, Can. J. Chem. *47*, 2275 (1969); *E. Cervone, P. Cancellieri C. Furlani*, J. Inorg. Nucl. Chem. *30*, 2431 (1968).

[e] *E. I. Stiefel, L. E. Bennett, Z. Dori, T. H. Crawford, C. Sims, H. B. Gray*, Inorg. Chem. *9*, 281 (1970); *A. Davison, W. Edelstein, R. H. Holm, A. H. Maki*, J. Amer. Chem. Soc. *86*, 2799 (1964).

[f] *E. I. Stiefel, R. Eisenberg, R. C. Rosenberg, H. B. Gray*, J. Amer. Chem. Soc. *88*, 2956 (1966).

[g] *G. W. Watt, B. J. McCormick*, Spectrochim. Acta *21*, 753 (1965).

[h] *C. Furlani, L. Luciani*, Inorg. Chem. *7*, 1586 (1968); *M. Bonanico, G. Dessy*, Ric. Sci. *38*, 1106 (1968).

[i] *C. Furlani, E. Cervone, F. D. Camassei*, Inorg. Chem. *7*, 265 (1968).

[j] *A. A. G. Tomlinson*, J. Chem. Soc. A, p. 1409 (1971); *R. C. Kudchadker, R. Zingaro, A. Irgolic*, Can. J. Chem. *46*, 1415 (1968).

[k] *D. A. Brown, W. K. Glass, B. Kumar*, J. Chem. Soc. A, p. 1510 (1969); Chem. Commun., p. 736 (1967).

[l] *J. G. White, H. L. Puich*, Inorg. Chem. *9*, 2581 (1970); *W. Rudorff, K. Stegmann*, Z. Anorg. Allg. Chem. *251*, 376 (1943); *W. Rudorff, W. R. Ruston, A. Scherhaufer*, Acta Crystallagr. *1*, 196 (1948); *P. F. Bongers, C. F. van Brugen, J. Koopstra et al.*, J. Phys. Chem. Solids *29*, 977 (1968).

[m] *C. Furlani, M. L. Luciani, R. Candori*, J. Inorg. Nucl. Chem. *30*, 2121 (1968).

[n] *J. P. Fackler, Jr., D. Coucouvanis*, J. Amer. Chem. Soc. *88*, 3913 (1966); *D. Coucouvanis*, Progr. Inorg. Chem. *11*, 233 (1970).

[o] *G. Contreras, H. Cortes*, J. Inorg. Nucl. Chem., *33*, 1337 (1971).

[p] *F. Galsbøl, C. E. Schaffer*, Inorg. Syn. *10*, 42 (1967).

[q] *G. W. Watt, B. J. McCormick*, J. Inorg. Nucl. Chem. *27*, 898 (1965).

[r] *G. Brauer* (ed.), Handbook of Preparative Inorganic Chemistry, Vol. 2, p. 1383. Academic Press, New York, 1965.

[s] *C. K. Jørgensen*, Mol. Phys. *5*, 485 (1962).

[t] *R. G. Cavell, W. Byers, E. D. Day*, Inorg. Chem. *10*, 2710 (1971).

[u] *G. Brauer*, Ref. *r*, p. 1394.

Tris(dithiooxalato)chromate(III) salts, *e.g.*, $KCa[Cr(C_2O_2S_2)_3]$, are obtained[189,190] by adding potassium dithiooxalate to hexakis(urea)-chromium(III) chloride. Care must be employed in the preparation of these salts to minimize impurities.

Malonitriledithiolate complexes of chromium-(III) are readily prepared under conditions similar to that employed for toluene-3,4-dithiol compounds, but a nitrogen atmosphere must be used since oxidation to complexes having chromium in oxidation states formally greater than three are formed. This behavior is characteristic for ligands of the class[191–196]

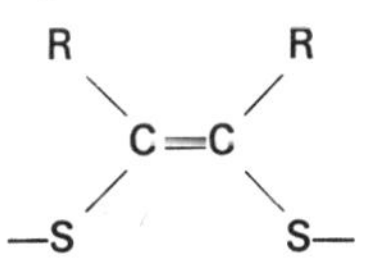

R = CF_3, C_6H_5, CN

which give rise to trigonal prismatic complexes, *e.g.*, $Cr[S_2C_2(C_6H_5)_2]_3$.[191–195,197,198] The spec-

[189] *F. P. Dwyer, A. M. Sargeson*, J. Amer. Chem. Soc. *81*, 2335 (1959).

[190] *R. L. Carlin, F. Canziani*, J. Chem. Phys. *40*, 371 (1964).

[191] *E. I. Stiefel, L. E. Bennett, Z. Dori, T. H. Crawford, C. Sims, H. B. Gray*, Inorg. Chem. *9*, 281 (1970).

[192] *J. A. McCleverty*, Progr. Inorg. Chem. *10*, 49 (1968).

[193] *R. Eisenberg*, Progr. Inorg. Chem. *12*, 295 (1970).

[194] *J. Locke, J. A. McCleverty*, Inorg. Chem. *5*, 1157 (1966).

[195] *G. N. Schrauzer*, Trans. Metal Chem. *4*, 299 (1968).

[196] *R. Huisman, R. DeJonge, C. Haas, F. Jellinek*, J. Solid State Chem. *3*, 56 (1971).

[197] *G. N. Schrauzer, H. W. Finck, V. Mayweg*, Angew. Chem. *76*, 715 (1964).

[198] *J. H. Waters, R. Williams, H. B. Gray, G. N. Schrauzer, H. W. Finck*, J. Amer. Chem. Soc. *86*, 4199 (1964).

tral properties of sulfur complexes have been reviewed.[199]

Green nonvolatile *chromium(III) mercaptides*, *e.g.*, $Cr(SCH_3)_3$, are prepared[200] by photolytic oxidative decarbonylation of tricarbonylbenzene-chromium or chromium hexacarbonyl in dimethyl disulfide, reaction of chromium(III) chloride with excess dimethyl disulfide, and reaction of sodium methyl mercaptide with anhydrous chromium chloride in excess dimethyl disulfide. The photolytic method is the one preferred.

Black *lithium thiochromite*, $LiCrS_2$, is prepared[201] by heating dried stoichiometric mixtures of lithium carbonate and chromium(III) oxide at 800° in an argon gas flow saturated with carbon disulfide. Compounds of the type MCr_2S_4[202–205] (M = Ni, Co) have been prepared by heating the elements at elevated temperatures. Selenides of the type ACr_2Se_4 (A = Fe, Co, Ni) are prepared[205] similarly.

Black *chromium(III) sulfide*, Cr_2S_3, cannot be precipitated from aqueous solution owing to its instability toward hydrolysis to the hydrated oxide. It can be prepared[206] by treating chromium(III) chloride with hydrogen sulfide at red heat or directly from the elements. It is quite stable toward nonoxidizing acids.

14.5.4 Nitrogen Donor Complexes

Nitrogen donor ligand complexes are the most extensively examined compounds of chromium(III). The compounds are prepared by a variety of methods. Many of these compounds have played important roles in the development of inorganic stereochemistry. The ammines of chromium(III) have been the subject of an extensive recent review[207] and no attempt is made to consider ammine compounds in detail. The preparation of ammine complexes is extensively documented.[208–211]

Yellow *hexamminechromium(III) chloride* may be obtained[212] by oxidation of an ammoniacal solution of a chromous salt which contains ammonium chloride or by the reaction of anhydrous chromium(III) chloride with liquid ammonia using sodium amide as a catalyst. The nitrate salt may be prepared by treating the chloride with concentrated nitric acid. Hexamminechromium(III) salts are sensitive to light in solution or in the solid state and decompose slowly in solution and more rapidly on boiling, depositing chromium hydroxide. Heating with concentrated hydrochloric acid yields red *chloropentamminechromium(III) chloride*, $[Cr(NH_3)_5Cl]Cl_2$.

Orange-yellow *tris(ethylenediamine)chromium(III) sulfate* may be prepared[213] from the reaction of ethylenediamine (en) with anhydrous chromium(III) sulfate (prepared by heating the hydrated material for 2–3 days at 110°). Chloride and thiocyanate salts are obtained metathetically using hydrochloric acid and ammonium thiocyanate, respectively. The chloride and thiocyanate salts are readily converted by heating to *cis* and *trans* salts of the type $[Cr(en)_2X_2]X$. The resolution of optical isomers of the tris(ethylenediamine)chromium cation, $Cr(en)_3^{3\oplus}$ can be accomplished[214] *via* formation of the complex salt Li $[(+)\text{D-}Cr(en)_3][(+)\text{-tartrate}]_2 \cdot H_2O$. *Trisdiaminechromium(III)* salts can be conveniently[215] prepared by heating the anhydrous diamine with commercial chromium(III) chloride hydrate in methanol using granulated zinc as a catalyst. Large cations like the tris-

---

[199] *C. K. Jorgensen*, Inorg. Chim. Acta Revs. *2*, 65 (1968).

[200] *D. A. Brown, W. K. Glass, B. Kumar*, J. Chem. Soc. (A), 1510 (1969).

[201] *J. G. White, H. L. Puich*, Inorg. Chem. *9*, 2581 (1970).

[202] *R. J. Bouchard, W. T. Robinson, A. Wold*, Inorg. Chem. *5*, 977 (1966).

[203] *R. J. Bouchard, A. Wold*, J. Phys. Chem. Solids *27*, 591 (1966).

[204] *S. L. Holt, R. Bouchard, A. Wold*, J. Phys. Chem. Solids *27*, 755 (1966).

[205] *B. L. Morris, R. Russo, A. Wold*, J. Phys. Chem. Solids *31*, 635 (1970).

[206] *G. Brauer*, Handbook of Preparative Inorganic Chemistry, Vol. II, p. 1346, Academic Press, New York 1965.

[207] *C. S. Garner, D. A. House*, Trans. Metal Chem. *6*, 59 (1970).

[208] *R. Abegg*, Handbuch der Anorganischen Chemie, Vol. IV/1 Section E, Elements of Group VI, S. Hirzel Verlag, Leipzig 1921.

[209] *M. M. J. Sutherland*, in *J. N. Friend*, Textbook of Inorganic Chemistry, Vol. X, p. 74, C. Griffin and Co., London 1928; Gmelins Handbuch der Anorganischen Chemie 52C, 8th Ed., Verlag Chemie, Weinheim/Bergstr. 1965.

[210] *J. W. Mellor*, A Comprehensive Treatise on Inorganic and Theoretical Chemistry, Vol. XI, p. 400, Longmans Green and Co., London 1931.

[211] *P. Pascal*, Nouveau Traite de Chimie Minerale, Vol. XIV, Masson and Co., Paris 1959.

[212] *G. Brauer*, Handbook of Preparative Inorganic Chemistry, Vol. II, p. 1351, Academic Press, New York 1965.

[213] Ibid; p. 1354.

[214] *F. Galsbøl*, Inorg. Synth. *12*, 269 (1970).

[215] *R. D. Gillard, P. R. Mitchell*, J. Chem. Soc. A, 2129 (1968).

(ethylenediamine)chromium (III) and hexamminechromium(III) cations can be employed for the stabilization of complex anions including those not found in appreciable concentrations in solution, *e.g.*, $CuCl_5^{3\ominus}$.[216] Orange-red crystals of *hexammine-chromium(III) pentacyanonickelate(III) dihydrate*, $[Cr(NH_3)_6]Ni(CN)_5 \cdot 2H_2O$, are obtained[217] by adding hexamminechromium(III) chloride to an aqueous solution of tetracyanonickelate(II) and potassium cyanide. Thermal deamination of $Cr(en)_3Cl_3$ in a sodium chloride matrix can be used to prepare[218] *cis*-$[Cr(en)_2Cl_2]Cl$, which can be converted to the perchlorate salt, *cis*-$[Cr(en)_2Cl_2]ClO_4$ by recrystallization from cold water addition of perchloric acid or sodium perchlorate. The thermal matrix method has been extensively developed[219] by Wendlandt and co-workers.

Substitution reactions can be used to obtain a wide variety of ammine derivatives, *e.g. aquopentamminechromium(III) nitrate* is obtained[220] by dissolving chloropentamminechromium(III) chloride in hot dilute nitric acid.

The dark red, air- and water-stable σ-bonded organochromium(III) species, *cis-bis(2-methoxy phenyl)bis(2,2′-bipyridyl)chromium(III) iodide monohydrate* is readily synthesized[221] from the reaction of 2-methoxyphenylmagnesium bromide with dibromobis(tetrahydrofuran)chromium(II), 2,2′-bipyridine, and aqueous potassium iodide in the presence of air. The phenyl derivative has also been prepared.[222] Orange-yellow *lithium hexaphenylchromate(III)*, $Li_3Cr(C_6H_5)_6 \cdot 2{\cdot}5(C_2H_5)_2O$ is obtained[223, 224] by treating anhydrous chromium trichloride with phenyllithium in diethyl ether. The crystals are soluble in diethyl ether, benzene, and tetrahydrofuran. The compound is sensitive to air and moisture and is completely hydrolyzed by water.

A variety of *trigonal amide complexes* of the type $Cr(NR_2)_3$, e.g., $R = C_2H_5$, are prepared[225–228] by the reaction of the lithium amide with anhydrous chromium(III) chloride in tetrahydrofuran. The crystal structures of $M[N(SiMe_3)_2]_3$ (M = Cr, Fe) have shown[229] that these trigonally coordinated metal complexes have a configuration corresponding to the $M(NSi_2)_3$ unit with planar $MNSi_2$.

1:1 Complexes with a large number of naphtholazopyrazolone dyes can be prepared[230] by the reaction of chromium trichloride hexahydrate or (diethylenetriamine)$Cr(CO)_3$ with the dyes in the presence of other ligands.

Chromium(II) phthalocyanine, CrPc, can be prepared[231, 232] by the reaction of phthalonitrile with chromium hexacarbonyl in high boiling solvents. The dark purple crystals oxidize readily in air to form CrPcOH. The hydroxy complex undergoes a variety of reactions, *e.g.*, with methylphenylphosphinic acid to yield green $CrPc[O_2P(CH_3)(C_6H_5)]$. Chromium(III) phthalocyanide hydroxide results[233] from the interaction of chromium(III) acetate with phthalonitrile.

A number of compounds such as *tetraphenylporphinechromium(III) ethoxide* (ethanol adduct), tetraphenylporphinechromium(III) methoxide (methanol adduct), and tetraphenylporphinechromium(III) hydroxide (hydrate) can be obtained[234] from the reaction of chromium hexacarbonyl with tetraphenylporphine. The phthalocyanines of chromium have been reviewed.[234a]

---

[216] *M. Mori, Y. Saito, T. Watanabe*, Bull. Chem. Soc. Japan *34*, 295 (1961).

[217] *W. E. Hatfield, R. Whyman, R. C. Fay, K. N. Raymond, F. Basolo*, Inorg. Synth. *11*, 51 (1958).

[218] *W. W. Wendlandt, C. H. Stembridge*, J. Inorg. Nucl. Chem. *27*, 575 (1965).

[219] *W. W. Wendlandt, L. K. Sveum*, J. Inorg. Nucl. Chem. *29*, 975 (1967).

[220] *W. K. Wilmarth, H. Graaf, S. T. Gustin*, J. Amer. Chem. Soc. *78*, 2683 (1956).

[221] *J. J. Daly, F. Sanz, R. P. A. Sneeden, H. H. Zeiss*, Chem. Commun. 243 (1971).

[222] *H. Muller*, Z. Chem. *9*, 311 (1969).

[223] *F. Hein, R. Weiss*, Z. Anorg. Allgem. Chemie *295*, 145 (1958).

[224] *G. Brauer*, Handbook of Preparative Inorganic Chemistry, Vol. II, p. 1375, Academic Press, New York 1965.

[225] *J. S. Basi, D. C. Bradley, M. H. Chisholm*, J. Chem. Soc. (A) 1433 (1971).

[226] *E. C. Alyea, J. S. Basi, D. C. Bradley, M. H. Chisholm*, Chem. Commun. 495 (1968).

[227] *D. C. Bradley, M. B. Hursthouse, C. W. Newing*, Chem. Commun. 411 (1971).

[228] *J. C. W. Chien, W. Kruse*, Inorg. Chem. *9*, 2615 (1970).

[229] *D. C. Bradley, M. B. Hursthouse, D. F. Rodesiler*, Chem. Commun. 14 (1969).

[230] *M. Idelson, I. R. Karady, B. H. Mark, D. O. Rickter, V. H. Hooper*, Inorg. Chem. *6*, 450 (1967).

[231] *J. A. Elvidge, A. B. P. Lever*, J. Chem. Soc. 1257 (1961).

[232] *E. G. Meloni, L. R. Ocone, B. P. Block*, Inorg. Chem. *6*, 424 (1967).

[233] *J. A. Elvidge, A. B. P. Lever*, Proc. Chem. Soc. 123 (1959).

[234] *E. B. Fleischer, T. S. Srivastava*, Inorg. Chim. Acta *5*, 151 (1971).

[234a] *A. B. P. Lever*, Adv. Inorg. Chem. Radiochem. *7*, 27 (1965).

A number of *six-coordinate chromium(III) complexes* with *N,N'*-ethylenebis(salicylideneiminate (salen) (**1**) have been described.[235, 236] [Cr(salen)-$(OH_2)_2$]Cl results from the reaction of chromium(III) chloride hexahydrate with *N,N'*-

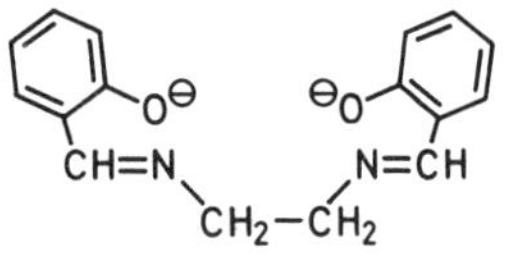

$O^{\ominus}$ $^{\ominus}O$
CH=N N=CH
CH_2–CH_2

ethylenebis(salicylaldimine) in ethylene glycol–water–methanol at 120°. The complex is also prepared by refluxing a suspension of anhydrous chromium trichloride and *N,N'*-ethylenebis-(salicylaldimine) in tetrahydrofuran. Complexes such as [Cr(salen)pyCl] can be prepared by metathetical reactions in methanol and ethanol near 50°.

Tris(salicylaldiminato)chromium(III) chelates may be obtained[237, 238] from the reaction of chromium(III) acetate hydrate, the salicylaldehyde, sodium carbonate, and arylamine in ethylene glycol. The crystals can be recrystallized from ethanol.

The *tetraethylammonium salt of hexathiocyanatochromium(III)* is obtained[239] by refluxing a stoichiometric mixture of tetraethylammonium bromide, chromium trichloride hexahydrate, and potassium thiocyanate in acetone for 2 hours. The compound is isolated by filtering the hot solution and cooling the filtrate. Dark red *potassium hexathiocyanatochromate(III)*, K_3-$[Cr(NCS)_6]$, results from heating an aqueous mixture of chrome alum, $KCr(SO_4)_2 \cdot 12H_2O$, with potassium thiocyanate for 2 hours. The tetrahydrate of the corresponding ammonium salt may be similarly prepared using ammonium thiocyanate. A considerable number of complexes of the type $K[Cr(NCS)_4(amine)_2]$ can be obtained from the reaction of potassium hexathiocyanatochromate(III) with amines. Potassium hexathiocyanatochromate(III) precipitates a number of transition metal and heavy metal, *e.g.*, $Ag^{\oplus}$ and $Hg^{2\oplus}$, ions from aqueous solution. These are polymeric materials[240] which are similar to the related hexacyanochromate(III) salts.

Tetramethylammonium hexaselenocyanatochromate(III) results[241] on addition of anhydrous chromium(III) chloride to a boiling solution of potassium selenocyanate in ethanol. The purple complex precipitates on addition of tetramethylammonium chloride. It rapidly decomposes under the influence of water and oxygen or carbon dioxide and water. It is soluble in acetone and dimethyl sulfoxide, slightly soluble in ethanol, and nearly insoluble in water, benzene, and ether. Dry oxygen and dry carbon dioxide do not harm the complex. Orange *selenocyanatopentamminechromium(III) bromide* is prepared[242] by the reaction of the aquopentammine complex with sodium selenocyanate in dimethyl sulfoxide.

14.5.5 Cyanide Derivatives

The *cyano complexes* of chromium(II) and chromium(III) are not markedly stable, the lower oxidation state complexes being very sensitive to oxidation in air, while the tetravalent compounds are prone to hydrolysis in aqueous solution and are also light-sensitive. The salts of *hexacyanochromate(II)* ion can be oxidized by air or hydrogen peroxide to *hexacyanochromates(III)*, Bright yellow crystals of potassium hexacyanochromate(III) are readily obtained[243] from the reaction of chromium(III) acetate with potassium cyanide. The aquation of hexacyanochromate(III) ion has been extensively examined and a number of cyanoaquo complexes have been isolated.[244]

A polymeric light brown complex with the composition $Fe_3[Cr(CN)_6]_2$ is prepared[245] by slow addition at 0° of a solution of potassium hexacyanochromate(III) to a solution containing excess ferrous sulfate. A complex of composition $Cr_4[Fe(CN)_6]_3$ is obtained by oxidation

---

[235] *S. Yamada, K. Iwasaki*, Inorg. Chim. Acta *5*, 3 (1971).

[236] *P. Coggon, A. T. McPhial, F. E. Mabbs, A. Richards, A. S. Thornely*, J. Chem. Soc. (A) 3296 (1970).

[237] *S. Yamada, K. Iwasaki*, Bull. Chem. Soc. Japan *41*, 1972 (1968).

[238] *A. P. Gardner, B. M. Gatehouse, J. V. B. White*, Chem. Commun. 694 (1968).

[239] *A. Sabatini, I. Bertini*, Inorg. Chem. *4*, 959 (1965); *G. Brauer*, Handbook of Preparative Inorganic Chemistry, Vol. II, p. 1374, Academic Press, New York 1965.

[240] *J. R. Wasson, C. Trapp*, J. Inorg. Nucl. Chem. *30*, 2437 (1968).

[241] *K. Michelson*, Acta Chem. Scand. *17*, 1811 (1963).

[242] *N. V. Duffy, F. G. Kosel*, Inorg. Nucl. Chem. Lett. *5*, 519 (1969).

[243] *G. Brauer*, Handbook of Preparative Inorganic Chemistry, Vol. II, p. 1373, Academic Press, New York 1965.

[244] *D. K. Wakefield, W. B. Schaap*, Inorg. Chem. *8*, 512 (1969).

[245] *D. B. Brown, D. F. Shriver, H. L. Schwartz*, Inorg. Chem. *7*, 77 (1968).

in air of the chromium(II) salt similarly prepared to the iron(II) salt. These compounds undergo linkage isomerization, *e.g.*, $Fe^{3\oplus}—N{\equiv}C—Cr^{3\oplus}$ and $Fe^{2\oplus}—C{\equiv}N—Cr^{3\oplus}$, under a variety of conditions.

The reduction of potassium hexacyanochromate(III) by potassium in liquid ammonia results in the formation of a green precipitate with the composition $K_6Cr(CN)_6$, which has properties consistent with its formulation as a chromium(0) complex. It must be handled in an inert atmosphere at all times. It is oxidized by water to the chromium(III) starting material.

The photochemistry of cyanochromium(III) complexes has been extensively examined,[246] and a review of cyanide complexes is available.[247]

14.5.6 Dimeric and Polymeric Compounds

Numerous dimeric and polymeric chromium(III) compounds are known. Usually, the formation of polymeric compounds results in the coordination number six being satisfied by the sharing of an edge or a face of an octahedron. Only a representative sampling of polymeric compounds are cited in this chapter. The magnetic properties and bonding in binuclear chromium(III) compounds have been reviewed.[248, 249]

Turquoise-green *chromium(III) orthophosphate*, $(CrPO_4)_x$, results[250] from the hydrazine reduction of chromium(VI) oxide in the presence of orthophosphoric acid. The hydrated material initially obtained is dehydrated at 800° *in vacuo* for 2 hours.

Bright blue-green *chromium(III) hydroxide*, α-$Cr(OH)_3 \cdot nH_2O$, results[251] on treating hexaaquochromium(III) chloride with dilute aqueous ammonia. It yields blue salts of the type $[Cr(H_2O)_6]X_3$ with dilute acids. The β form of chromium(III) hydroxide is similarly prepared only starting with dichlorotetraaquochromium(III) chloride. The β form is insoluble in acetic acid and reacts with dilute acids to yield salts of the type $[Cr(H_2O)_4X_2]X$. An impure, but catalytically active form of the hydroxide, results from the reduction of chromium(IV) oxide with ethanol.

Black *chromium nitride*, CrN, results[252] from the reaction of nitrogen gas with electrolytic chromium at 900° or the interaction of anhydrous chromium chloride with ammonia gas at elevated temperatures. It is insoluble in acids and alkalis.

The interaction of anhydrous chromium trichloride with neutral phosphate and phosphonate esters at elevated temperatures leads[253, 254] to the formation of green crystalline *tris(dialkoxyphosphato)*- and *tris(alkoxyalkylphosphonato)chromium(III) polymers.* Green poly[di-μ-(diphenylphosphinato)]aquohydroxychromium(III) is produced[255] by reactions (2) and (3).

$$CrCl_2 + 2KO—\overset{\overset{\displaystyle O}{\|}}{P}(C_6H_5)_2 + H_2O \xrightarrow{CH_3OH} \{Cr[O_2P(C_6H_5)_2]_2H_2O\} + 2KCl \quad (2)$$

$$4\{Cr[O_2P(C_6H_5)_2]H_2O\} \xrightarrow[\text{tetrahydrofuran–water}]{\text{air}} \{Cr(H_2O)OH[O_2P(C_6H_5)_2]\}_x \quad (3)$$

The polymer is readily soluble in chloroform, benzene, and tetrahydrofuran, but is insoluble in water and diethyl ether. It does not melt before decomposing.

Phosphine complexes are obtained[256] on heating anhydrous chromium trichloride with tertiary phosphines in benzene or toluene at 60°. Green complexes of the type $[CrCl_3(PR_3)_2]$ can be isolated when $R = CH_3$, C_2H_5, or n-C_4H_9. These complexes lose 1 mol of trialkylphosphine on gentle warming in 1:1 carbon tetrachloride–methylene dichloride to form insoluble purple complexes of the type $[CrCl_3PR_3]_n$, which are probably polymeric with halogen bridging. Treatment of the green complexes with tetraphenylphosphine chloride results in green salts of

---

[246] *V. Balzani, V. Carassiti*, Photochemistry of Coordination Compounds, Academic Press, New York 1970.

[247] *B. M. Chadwick, A. G. Sharpe*, Adv. Inorg. Chem. Radiochem. *8* (1965).

[248] *B. Jezowska-Trzebiatowska, W. Wojciechowski*, Trans. Met. Chem. *6*, 1 (1970).

[249] *G. F. Kokoszka, G. Gordon*, Trans. Met. Chem. *5*, 181 (1969).

[250] *G. Brauer*, Handbook of Preparative Inorganic Chemistry, Vol. II, p. 1364, Academic Press, New York 1965.

[251] *G. Brauer*, Handbook of Preparative Inorganic Chemistry, Vol. II, pp. 1345–1346, Academic Press, New York 1965.

[252] *G. Brauer*, Handbook of Preparative Inorganic Chemistry, Vol. II, p. 1347, Academic Press, New York 1965.

[253] *V. Gutmann, G. Beer*, Inorg. Chim. Acta *3*. 87 (1969).

[254] *C. M. Mikulski, N. M. Karayannis, M. J. Strocko, L. L. Pytlewski, M. M. Laebs*, Inorg. Chem. *9*, 2053 (1970).

[255] *K. D. Maguire*, Inorg. Synth. *12*, 258 (1970).

[256] *M. A. Bennett, R. J. H. Clark, A. D. J. Goodwin*, J. Chem. Soc. (A), 541 (1970).

the type $[(C_6H_5)_4P]^{\oplus}[CrCl_4(PR_3)_2]^{\ominus}$ which are unstable in solution.
Red-brown *di-μ-hydroxobis[bis(2,2′-bipyridine)-chromium(III)] perchlorate* dihydrate results[257] from the neutralization of a perchloric acid solution of the appropriate quantities of hydrated chromium(III) perchlorate and 2,2′-bipyridine with lithium hydroxide. The corresponding 1,10-phenanthroline complex is prepared similarly. Adjustment of reaction conditions results in the formation of oxo- and oxohydroxo-bridged species. Oxidation of aqueous solutions of chromous salts made basic with ammonium hydroxide gives rise[220,258] to a variety of dimeric *chromium(III) ammine complexes*, *e.g.*, $[(NH_3)_5Cr(OH)Cr(NH_3)_5]Cl_5$, which are frequently in equilibrium with each other. Slight variations of the reaction conditions, *e.g.*, temperature, alkalinity, and concentrations of reactants, can cause variation of products obtained. Chromatographic methods are quite useful for the separation of the resulting products and isomers. Dimeric amine complexes are listed in the standard reference works. It is noted that many of these complexes bear reinvestigation.
Conventional methods for the synthesis of tris(ethylenediamine)chromium(III) salts, when employed with isobutylenediamine (ibn) either give[259] no reaction or yield bis(ibn) complexes. Dimeric oxo- and dihydroxo-bridged complexes are also obtained.
Pink *di-μ-hydroxybis[di(glycine)chromium(III)] monohydrate* results[260,261] on treatment of an aqueous solution of chromium(III) chloride hydrate and glycine with aqueous sodium hydroxide. *Di-μ-hydroxybis[di(phenylalanine)-chromium(III)]* is prepared[262] by refluxing phenylalanine with an aqueous solution of tris(ethylenediamine)chromium(III) chloride. The chelates of other amino acids have been similarly[263] prepared. The preparation and structure of the red trisglycinate monohydrate have been reported.[264]
Hexamminechromium(III) nitrate reacts[265] with potassium amide in liquid ammonia to give the bright red *amide*.

$$[Cr(NH_3)_6](NO_3)_3 + 3KNH_2 \longrightarrow 3KNO_3 + 6NH_3 + Cr(NH_2)_3 \quad (4)$$

With an excess of potassium amide a potassium salt of a polymeric anion is obtained.

$$[Cr(NH_2)_3]_n + KNH_2 \longrightarrow [K[Cr(NH_2)_4]]_n \quad (5)$$

Low-temperature treatment of polymeric chromium(III) amides with ammonium salts results in dissolution of the amides. It has been suggested that the residues remaining after removal of ammonia from these solutions contain high polymers such as $\{[Cr(NH_3)_3(NH_2)Br]Br\}_n$.
Polynuclear compounds are useful starting materials for the synthesis of monomeric complexes.[266]

14.6 Chromium(IV) Compounds

Green *chromium tetrafluoride* is obtained[108,267] by the fluorination of chromium metal at about 350° in a flow system. Impurities are removed by vacuum sublimation. It turns brown on the slightest contact with moisture. It can be sublimed *in vacuo* at temperatures greater than 100°. Refluxing with selenium tetrafluoride produces pink $CrF_2 \cdot SeF_4$ and buff-colored $CrF_3 \cdot SeF_4$, which can be readily separated.
Pink *alkali hexafluorochromates(IV)*, *e.g.*, K_2CrF_6, are prepared[268,269] by fluorinating the appropriate mixture of alkali chloride and chromium trichloride at about 150° in a flow system. *Alkali pentafluorochromates(IV)*, *e.g.*, $CsCrF_5$, result[108,267] from the fluorination of stoichiometric amounts of chromium tetrafluoride and alkali chlorides in bromine trifluoride, residual bromine trifluoride being removed at 160° *in vacuo*. *Chromium tetra-t-but-*

[257] *J. Josephsen, C. E. Schaffer*, Acta Chem. Scand. *24*, 2929 (1970).
[258] *G. Brauer*, Handbook of Preparative Inorganic Chemistry, Vol. II, pp. 1359–1361, Academic Press, New York 1965.
[259] *K. G. Poulsen, C. S. Garner*, J. Amer. Chem. Soc. *81*, 2615 (1959).
[260] *A. Earnshaw, J. Lewis*, J. Chem. Soc. 396 (1961).
[261] *G. Brauer*, Handbook of Preparative Inorganic Chemistry, Vol. II, p. 1382, Academic Press, New York 1965.
[262] *P. Vieles, N. Israily*, Bull. Soc. Chim. France 139 (1967).
[263] *J. A. Weyh, R. E. Hamm*, Inorg. Chem. *7*, 2431 (1968).
[264] *R. F. Bryan, P. T. Greene, P. F. Stokely, E. W. Wilson jr.*, Inorg. Chem. *10*, 1468 (1971).
[265] *O. Schmitz-Dumont, J. Pilzecker, H. F. Pepenbrink*, Z. Anorg. Allgem. Chemie *248*, 175 (1941).
[266] *D. W. Hoppenjans, J. B. Hunt, M. J. Dechant*, Chem. Commun. 510 (1968);
D. W. Hoppenjans, J. B. Hunt, Inorg. Chem. *8*, 505 (1969).
[267] *H. von Wartenburg*, Z. Anorg. Allgem. Chemie *247*, 135 (1941).
[268] *D. H. Brown, K. R. Dixon, R. D. W. Kemmitt, D. W. A. Sharp*, J. Chem. Soc. 1559 (1965).
[269] *E. Huss, W. Klenum*, Z. Anorg. Allgem. Chemie *262*, 25 (1950).

oxide (m.p. 36°–37°) has been prepared by the following methods.

(a) The action of di-*t*-butyl peroxide on dibenzene chromium.[270]
(b) Alcoholysis of $Cr[N(C_2H_5)_2]_4$ with *t*-butanol.[271]
(c) Oxidation of $Cr(O\text{-}t\text{-}Bu)_3$ in the presence of *t*-butanol by Br_2, $Pb(OA_c)_4$, O_2, and other oxidizing reagents.[271]
(d) From the reaction of $CrCl_3 \cdot 3THF$ with NaO-*t*-Bu and cuprous chloride[272] (THF = tetrahydrofuran).
(e) The addition of cuprous chloride to a suspension of $LiCr(O\text{-}t\text{-}Bu)_4$ in refluxing THF.[226]

Recently, a number of Cr(IV) *tertiary alkoxides* and the *triethylsilyloxide* have been prepared by method (b).[273] $Cr[OSi(C_2H_5)_3]_4$ is a viscous royal blue liquid which distills with decomposition. *Mixed tertiary alkoxides* of Cr(IV) can be obtained from *slow* exchange reactions with tertiary alcohols.

Attempts to synthesize Cr(IV) compounds by exchange reactions with chelating ligands were reported to lead to redox reactions and the formation of tris-Cr(III) chelates. Chromium(IV) tertiary alkoxides are volatile, monomeric, paramagnetic blue compounds which are less readily hydrolyzed than the corresponding titanium(IV) compounds. Chromium(IV) *t*-butoxide reacts[271] with triethylsilanol to yield tetrakistriethylsiloxychromium(IV) and *t*-butyl alcohol. With primary or secondary aliphatic alcohols the appropriate chromium(III) alkoxide and aldehyde or ketone and *t*-butyl alcohol are obtained. With acetylacetone the chromium is reduced, giving tris(acetylacetonato)chromium(III), acetone, and *t*-butyl alcohol. Chromium(IV) *t*-butoxide is also obtained by the oxidation of chromium(III) butoxide [obtained by treating trisdialkylaminochromium(III) compounds with *t*-butanol] in an excess of *t*-butyl alcohol. Instead of molecular oxygen, oxidizing agents such as bromine, lead tetraacetate, or di-*t*-butyl peroxide may be used.

Chromium(III) trisdiethylamide is prepared[274] by the action of anhydrous chromium chloride on lithium diethylamide in tetrahydrofuran followed by evaporation of the solvent *in vacuo* and extraction of the residue with pentane. Evaporation of the pentane extract followed by high-vacuum disproportionation gives the green liquid *chromium(IV) tetrakisdiethylamide*, $Cr(NEt_2)_4$, in high yield based on reaction (6).

$$2Cr(NEt_2)_3 \longrightarrow Cr(NEt_2)_4 + Cr(NEt_2)_2 \qquad (6)$$

The reaction is general and can be used to prepare numerous chromium(IV) dialkylamides.
Violet *diperoxychromates(IV)* with the general formula $M^I[CrO(O_2)_2OH]$ are prepared[275,276] by the reaction of hydrogen peroxide with neutral or slightly acidic solutions of chromium (VI). Light brown *diperoxotriamminechromium*(VI), $[Cr(O_2)_2(NH_3)_3]$, is prepared[277,278] by the hydrogen peroxide reduction of chromium(VI) oxide in aqueous ammonia. It reacts with potassium cyanide to yield red *potassium diperoxotricyanochromate(IV)*.
Emerald green *barium orthochromate*, Ba_2CrO_4, results[279,280] from reaction (7).

$$BaCrO_4 + Cr_2O_3 + 5Ba(OH)_2 \longrightarrow 3Ba_2CrO_4 + 5H_2O \qquad (7)$$

A large excess of base can lead to the formation of black-green tribarum chromate(IV), Ba_3CrO_5. The complexes *diperoxoammineethylenediaminechromium(IV) monohydrate*, $[Cr(en)(NH_3)(O_2)_2] \cdot H_2O$, *diperoxoaquopropylenediaminechromium(IV) dihydrate*, $[Cr(pn)(OH_2)(O_2)] \cdot 2H_2O$, and *diperoxoaquoisobutylenediaminechromium(IV) monohydrate*, $[Cr(ibn)(OH_2)(O_2)_2] \cdot H_2O$ (en = ethylenediamine, pn = 1,2-diaminopropane, and ibn = isobutylenediamine) are prepared[281] by the reaction of mixtures of chromium trioxide and hydrogen peroxide with the diamines or mixtures of ethylenediamine and ammonia. A large number of chromium(III)–amine complexes can be prepared from these complexes.

---

[270] *M. Hagihara, H. Yamasaki*, J. Amer. Chem. Soc. *81*, 3160 (1959).
[271] *J. S. Basi, D. C. Bradley*, Proc. Chem. Soc. *305* (1963).
[272] *H. L. Krauss, G. Munster*, Z. Anorg. Allgem. Chemie *352*, 24 (1967).
[273] *E. C. Alyea, J. S. Basi, D. C. Bradley, M. H. Chisholm*, J. Chem. Soc. (A), 772 (1971).
[274] *J. S. Basi, D. C. Bradley, M. H. Chisholm*, J. Chem. Soc. (A), 1433 (1971).
[275] *W. P. Griffith*, J. Chem. Soc. 3948 (1962).
[276] *J. A. Connar, E. A. V. Ebsworth*, Adv. Inorg. Chem. Radiochem. *6*, 280 (1964).
[277] *R. Stomberg*, Ark. Kemi *22*, 49 (1964).
[278] *G. Brauer*, Handbook of Preparative Inorganic Chemistry, Vol. II, p. 1392, Academic Press, New York 1965.
[279] *G. Brauer*, Handbook of Preparative Inorganic Chemistry, Vol. II, p. 1393, Academic Press, New York 1965.
[280] *K. A. Wilkelmi, O. Jenssen*, Acta Chem. Scand. *15*, 1415 (1961).
[281] *D. A. House, R. G. Hughes, C. S. Garner*, Inorg. Chem. *6*, 1077 (1971).

Dark red complexes of the type $[Cr(S_2C_2R_2)_3]$ ($R=C_6H_5$) are prepared[282] by the reaction of the "thioester solution" obtained by heating benzoin and tetraphosphorus decasulfide in xylene with an aqueous solution of chromium trichloride hexahydrate which has been reduced with hydrazine. A number of complexes in this category have been reported.[195]

14.7 Chromium(V) Compounds

Red *chromium pentafluoride* results[283, 284] from the direct fluorination of chromium. It is a powerful oxidant and fluorinating agent.

Attempts to prepare pure chromium oxide trifluoride[108, 285] have been without success. The compound is always contaminated with starting materials and/or unidentified products. Purple *oxotetrafluorochromates(V)*, *e.g.* $Ag[CrOF_4]$, are prepared[119, 285] by refluxing the corresponding dichromate with bromine trifluoride and subsequently removing the solvent. The compounds are moisture-sensitive.

Very dark red *chromium oxide trichloride*, $CrOCl_3$, is obtained[286–288] by the interaction of chromium oxide dichloride and boron trichloride or thionyl chloride, and chromium trioxide with thionyl chloride or sulfuryl chloride. It is purified by vacuum sublimation at room temperature. It is light-sensitive and disproportionates slowly to chromium dioxide dichloride and chromium(III) compounds.

Organic ammonium salts of the *oxotetrachlorochromate(V)* ion are obtained[289] by reduction of chromium trioxide with hydrogen chloride gas, followed by addition of the stoichiometric quantity of organic base in acetic acid. These salts can also be prepared[287] by reaction of the same materials in thionyl chloride.

Oxopentachlorochromates(V) may be prepared[287] by reacting chromium trioxide or chromium dioxide dichloride with an organic base in thionyl chloride. Alkali metal salts may be obtained[290] by dissolving chromium trioxide in anhydrous acetic acid saturated with hydrogen chloride and adding to this a solution of the alkali chloride in hydrochloric acid which is also saturated with hydrogen chloride.

Red-brown crystals of *potassium tetraperoxochromate(V)*, K_3CrO_8, result[291] from the interaction of chromium(VI) oxide, potassium hydroxide, and hydrogen peroxide. The compound may be stored for months without decomposition. It is soluble in cold water and insoluble in ethanol or ether. Substitution of ammonium chloride for potassium hydroxide in the reaction yields violet-black unstable crystals of *ammonium pentaperoxodichromate*, $(NH_4)_2Cr_2O_{12}\cdot 2H_2O$. The compound explodes at 50°. Solutions of potassium tetraperoxychromate(V) rapidly evolve oxygen on addition to hydrogen peroxide according to reaction (8). This equation has been

$$4CrO_8^{3\ominus} + 8H^{\oplus} \longrightarrow 4HCrO_6^{\ominus} + 3O_2 + 2H_2O \quad (8)$$

used[292] to account for the solution behavior of tetraperoxychromates(V).

Black-green *barium chromate(V)*, $Ba_3(CrO_4)_2$, results[293] from reaction (9). It is completely

$$2BaCrO_4 + BaCO_3 \xrightarrow{1000^\circ} Ba_3(CrO_4)_2 + CO_2 + \tfrac{1}{2}O_2 \quad (9)$$

soluble in dilute acids with disproportionation to chromium(III) and chromium(VI).

Blue-green *calcium chromate(V) hydroxoapatite*, $Ca_5(CrO_4)_3OH$, and the blue-black chromate(V), $Ca_3(CrO_4)_2$, are prepared[294] by thermal decomposition of calcium chromate–calcium oxide mixtures in nitrogen–water vapor and nitrogen atmospheres, respectively. Extraction with glycol–methanol mixtures removes excess calcium oxide. Alkali chromates(V) are obtained by similar procedures.

---

282 *G. N. Schrauzer, V. P. Mayweg*, J. Amer. Chem. Soc. *88*, 3235 (1966).
283 *A. J. Edwards*, Proc. Chem. Soc. *205* (1963).
284 *T. A. O'Donnell, D. F. Stewart*, Inorg. Chem. *5*, 1434 (1966).
285 *A. G. Sharpe, A. A. Woolf*, J. Chem. Soc. 798 (1951).
286 *H. L. Krauss, G. Muenster*, Z. Naturforsch, *17b*. 344 (1962).
287 *H. O. Krauss, M. Leder, G. Muenster*, Chem. Ber. *96*, 3008 (1963).
288 *R. B. Johannesen, H. L. Krauss*, Chem. Ber. *97*, 2094 (1964).
289 *H. Kon*, J. Inorg. Nucl. Chem. *25*, 933 (1963).
290 *D. Brown*, J. Chem. Soc. 4944 (1964).
291 *G. Brauer*, Handbook of Preparative Inorganic Chemistry, Vol. II, p. 1391, Academic Press, New York 1965.
292 *D. Quane, B. Bartlett*, J. Chem. Phys. *53*, 4404 (1970).
293 *G. Brauer*, Handbook of Preparative Inorganic Chemistry, Vol. II, p. 1394, Academic Press, New York 1963.
294 *R. Scholder, H. Schwarz*, Z. Anorg. Allgem. Chemie *326*, 11 (1963).

14.8 Chromium(VI) Compounds

Lemon yellow *chromium hexafluoride* is prepared[295] by heating chromium metal with fluorine at 200 atm to 400° in a bomb. Small amounts of manganese are added to increase the yield. The compound decomposes above −100° in the absence of a large pressure of fluorine to yield fluorine and the pentafluoride.

Dark red *chromium oxide tetrafluoride*[283] (m.p. 55°) is a by-product of the reaction of chromium metal with fluorine in a heated glass flow system. The compound is readily hydrolyzed.

The best method[296] for the preparation of dark red *chromium dioxide difluoride* (chromyl fluoride), CrO_2F_2 (m.p. 31·6°), involves the reaction of chromium trioxide and anhydrous hydrogen fluoride at room temperature. The compound can be recrystallized from liquid hydrogen fluoride. It attacks glass and silica and quickly polymerizes to a white solid on exposure to sunlight. It undergoes halogen exchange[297] with chromium dioxide dichloride yielding CrO_2FCl.

Red *trioxofluorochromate(VI)* salts such as $KCrO_3F$ are obtained[298,299] by reacting the appropriate dichromate with 40% hydrofluoric acid. They are readily hydrolyzed.

Orange *potassium trioxochlorochromate(VI)* precipitates[31,300–302] on cooling from a solution in which potassium dichromate and hot concentrated hydrochloric acid have been reacted. The anion hydrolyzes in water, but is stable in dilute hydrochloric acid.

Dark red *chromium dioxide dichloride* (chromyl chloride), CrO_2Cl_2 (m.p. −96·5°, b.p. 117°), can be prepared[303] by the reaction of chromium trioxide or dichromate with thionyl chloride. The preparation, purification, and properties of the compound have been reviewed.[304]

It is readily hydrolyzed and is sensitive to light and elevated temperatures. Thermal decomposition[305] of the compound produces a number of materials (depending on the temperature of pyrolysis), including the oxide, Cr_5O_9, and ultimately Cr_2O_3. At room temperature it reacts with peroxydisulfuryldifluoride[306] to yield $CrO_2(SO_3)_2$. With pyridine[307] mono and bis adducts have been prepared. Chromyl chloride forms[308] a wide variety of adducts. The reactions are simple additions to solutions of chromyl chloride in carbon tetrachloride under a nitrogen atmosphere.

Very dark red *chromium dioxide dibromide* (chromyl bromide), CrO_2Br_2, is formed[309] by the reaction of chromium trioxide and hydrogen bromide in the presence of phosphorus pentoxide. It is thermally unstable even below room temperature.

Dark red *chromium(VI) oxide nitrate* (chromyl nitrate), $CrO_2(NO_3)_2$ (b.p. 63°–65° at 0·7 mm Hg pressure), is prepared[310,311] by the interaction of dinitrogen pentoxide with chromium(VI) oxide. It is soluble in carbon tetrachloride and reacts with water to form chromic and nitric acids. It is corrosive to most metallic surfaces except aluminum and attacks wood, paper, and rubber in the same manner as fuming nitric acid does. It cannot be stored for a long time, but is relatively stable in a sealed ampoule in the absence of light and moisture. It can be purified by vacuum distillation over lead dioxide.

Red *chromium(VI) dioxide perchlorate* (chromyl perchlorate), $CrO_2(ClO_4)_2$ (m.p. −1°, b.p. 175°), is produced[311] by the interaction of chromium(VI) oxide with dichlorine trioxide. It dissolves in carbon tetrachloride. It may be stored for extended periods of time in the dark at

---

295 *O. Glemser, H. Roesky, K. H. Hellberg*, Angew. Chem. Intern. Ed. Engl. *2*, 266 (1963).

296 *A. Engelbrecht, A. V. Grosse*, J. Amer. Chem. Soc. *74*, 5262 (1952).

297 *G. D. Flesch, H. J. Svec*, J. Amer. Chem. Soc. *80*, 3189 (1958);
G. D. Flesch, H. J. Svec, J. Amer. Chem. Soc. *81*, 1787 (1959).

298 *H. Stammreich, O. Sala, D. Bassi*, Spektrochim Acta *10*, 593 (1963).

299 *R. Bogvad, A. H. Nielsen*, Acta Cryst. *4*, 77 (1951).

300 *H. Stammreich, O. Sala, K. Kawai*, Spectrochim. Acta, *17*, 226 (1961).

301 *L. Helmholz, W. R. Foster*, J. Amer. Chem. Soc. *72*, 4971 (1950).

302 *G. P. Haight, D. C. Richardson, N. C. Coburn*, Inorg. Chem. *3*, 1777 (1964).

303 *J. H. Freeman, C. E. C. Richards*, J. Inorg. Nucl. Chem. *7*, 287 (1958).

304 *W. H. Hartford, M. Darrin*, Chem. Rev. *58*, 1 (1958).

305 *S. Z. Markarov, A. A. Vakhrushev*, Izv. Akad. Nauk. SSSR, Otdee. Khim. Nauk. 1731 (1960).

306 *M. Lustig, G. H. Cady*, Inorg. Chem. *1*, 714 (1962).

307 *J. Bernard, M. Camelot*, Compt. Rend. *258*, 5881 (1964).

308 *R. C. Makhija, R. A. Stairs*, Can. J. Chem. *47*, 2293 (1969).

309 *H. L. Krauss, K. Stark*, Z. Naturforsch. *17b*, 1 (1962).

310 *A. D. Harris, J. C. Terbellas, H. B. Honassen*, Inorg. Synth. *9*, 83 (1967).

311 *G. Brauer*, Handbook of Preparative Inorganic Chemistry, Vol. II, p. 1386, Academic Press, New York 1963.

the temperature of dry ice. It is a powerful oxidant and frequently explodes at 80°.

Chromates ($CrO_4^{2\ominus}$) and *dichromates* ($Cr_2O_7^{2\ominus}$) have been extensively studied and many such salts are commercially available. The salts are readily prepared[167] from commercial chromium trioxide, e.g., lithium carbonate reacts with the appropriate amount of hot aqueous chromium(VI) oxide to yield crystals of *lithium dichromate*, $Li_2Cr_2O_7 \cdot 2H_2O$. Details of similar preparations are available.[312]

Blue *peroxochromic acid*, $CrO_5 \cdot (H_2O)_x$, is formed in aqueous solution by treating chromium(VI) oxide with hydrogen peroxide.[313,314] Salts with organic bases, *e.g.*, pyridine 1,10-phenanthroline, aniline, quinoline, and 2,2′-bipyridyl are used to stabilize the acid. The pyridine adduct is prepared by addition of pyridine to a solution of peroxychromic acid. The principal equilibrium involved in the formation of "perchromic acid" in aqueous solution is[315]

$$HCrO_4^{\ominus} + 2H_2O_2 + H^{\oplus} \rightleftharpoons CrO_5 + 3H_2O \qquad (10)$$

The reaction of dichromates with acidified hydrogen peroxide can also be employed[316] to produce solutions of "perchromic acid". A wide variety of adducts can be obtained by addition of the appropriate ligands. Yellow to dark red crystals of the *pyridine* (py) *diadduct of chromium(VI) oxide*, $CrO_3 \cdot 2py$, are obtained [311] by mixing chromium(VI) oxide with excess pyridine at ice–salt temperatures. The complex is soluble in pyridine, but insoluble in benzene and diethyl ether. It decomposes slowly at 100° and hydrolyzes instantly in water. It is stable indefinitely in the dark and is stored in sealed containers at room temperature. Two modifications of blue *oxodiperoxido-2,2′-bipyridinechromium(VI)* have been characterized.[317]

---

[312] *K. A. Wilhelmi*, Ark. Kemi *26*, 157 (1966).
[313] *K. A. Hofmann, H. Hiendlmaier*, Ber. *38*, 3066 (1905).
[314] *W. P. Griffith*, J. Chem. Soc. 3948 (1962).
[315] *D. F. Evans*, J. Chem. Soc. 4013 (1957).
[316] *R. Armstrong, N. A. Gibson*, Australian J. Chem. *21*, 897 (1968).
[317] *R. Stomberg, I. B. Ainalem*, Acta Chem. Scand. *221*, 1439 (1968).

15 Molybdenum

F. W. Moore
Research Laboratory, Climax Molybdenum Company,
Ann Arbor, Michigan 48105, U.S.A.

15.1 Molybdenum Metal and Molybdenum(0) Compounds

Pure *molybdenum metal* is available commercially in a variety of forms. Molybdenum powder is prepared by first reducing sublimed molybdenum trioxide[1,2] or purified ammonium molybdate[3] to non-volatile molybdenum dioxide with hydrogen at 500°, followed by reduction to the metal at temperatures above 800°. It can also be prepared by reducing molybdenum pentachloride with hydrogen at temperatures above 700°.[4] The powder is converted to useful forms by powder metallurgy or vacuum arc-casting. Reduction of molybdenum oxides with carbon or metals such as aluminum yields an impure product.[2]
Molybdenum carbonyl compounds are discussed in Chapter 29, and molybdenum cyclopentadienyl-type complexes are considered in Chapter 31. Molybdenum does not form a definite hydride, but the finely divided metal adsorbs a small amount of hydrogen.[5–7]

The zerovalent compounds *molybdenum dinitrosonium dihalide*, $MoX_2(NO_2)$, where X = Cl or Br, are obtained in high yield by passing the nitrosyl halide through a dichloromethane solution of the hexacarbonyl.[8] The dichloride readily forms *complexes* of the type $Mo(NO)_2Cl_2 \cdot 2L$ with amines, arsines, and phosphines on direct reaction with the ligand. The reaction of 2,2′,2″-tripyridyl with the hexacarbonyl yields $Mo(tripy)_2$[9] and the reaction of dilithium 2,2′-dipyridyl with $[Mo(bipy)_3]Cl_3$ in tetrahydrofuran yields $Mo(bipy)_3$.[10] The trifluorophosphine complex $Mo(PF_3)_6$ is isolated from the reaction of phosphorus trifluoride with dibenzene molybdenum.[9]

[1] *G. Brauer*, Handbook of Preparative Inorganic Chemistry, 2nd Ed., Vol. II, p. 1401, Academic Press, New York, 1965. (Engl. Transl.)

[2] Gmelin's Handbuch der Anorganischen Chemie, Molybdän, 8th Ed., p. 32, Verlag Chemie, Weinheim/Bergstr. 1935.

[3] *G. Lorang, V. G. Kinh, J. Langeron*, C.R. Acad. Sci., Paris *271C*, 284 (1970).

[4] *Y. Saeki, R. Matsuzaki, T. Matnishima*, Denki Kagaku Oyobi Kogyo Butsuri Kagaku 35, 46 (1967). C.A. *67*, 92993 (1967).

[5] *V. I. Mikheeva*, Hydrides of Transition Metals, Translated by United States Atomic Energy Commission, A.E.C. tr 5224, p. 98 (1962).

[6] *D. T. Hurd*, An Introduction to the Chemistry of Hydrides, John Wiley & Sons, New York, 1952; *K. M. MacKay*, Hydrogen Compounds of the Metallic Elements, E. and F. N. Spon, London 1966.

[7] *M. L. Hill*, J. Metals *12*, 725 (1960).

[8] *F. A. Cotton, B. F. G. Johnson*, Inorg. Chem. *3*, 1609 (1964); *B. E. G. Johnson, K. B. Al-Obadi*, Inorg. Synth. *12*, 264 (1970).

[9] *H. Behrens, U. Anders*, Z. Naturforsch. *196*, 767 (1964).

[10] *S. Herzog, I. Schneider*, Z. Chem. *2*, 24 (1962).

15.2 Intermetallic Compounds

The discussion of intermetallic compounds is limited to borides, carbides, nitrides, aluminides, silicides, phosphides, germanides, and arsenides.

The following borides have been prepared[11–14] by heating molybdenum and boron *in vacuo* or hydrogen at temperatures of 1200 to 1800°; Mo_2B,[15,16] MoB,[15] MoB_2,[17] Mo_2B_5,[15,16] and MoB_4.[18] The boride Mo_2B_5 has also been prepared by magnesiothermic reduction of the oxides[19] and by the reaction of molybdenum with hydrogen and boron trichloride at 900°.[20]
Molybdenum reacts with carbon to form two *carbides:* Mo_2C and MoC.[21,22] The first compound is prepared[23] by heating a stoichiometric mixture of pure carbon and Mo in a hydrogen atmosphere at 1400° to 1500°. The monocarbide is difficult to prepare as a pure phase, but it can be prepared[24] from pressed mixtures of

[11] *M. Hansen, K. Anderko*, Constitution of Binary Alloys, 2nd Ed., p. 253, McGraw-Hill Book Co., New York 1958.

[12] *R. P. Elliot*, Constitution of Binary Alloys, First Supplement, p. 125, McGraw-Hill Book Co., New York 1965.

[13] *F. A. Shunk*, Constitution of Binary Alloys, Second Supplement, p. 90, McGraw-Hill Book Co., New York 1969.

[14] *B. Aronsson, T. Lundstrom, S. Rundqvist*, Borides, Silicides and Phosphides, Methuen, London 1965.

[15] *P. W. Gilles, B. D. Pollock*, Trans. Met. Soc. AIME *197*, 1537 (1953).

[16] *K. I. Portnoi*, Izv. Akad. Nauk SSSR, Metal 4, 175 (1967), C.A. *68*, 63132 (1968).

[17] *E. Rudy, F. Benesonsky, L. Toth*, Z. Metallk. *54*, 345 (1963).

[18] *A. Chretein, J. Helgorsky*, C.R. Acad. Sci., Paris *252*, 742 (1961).

[19] *L. Ya. Markovskii*, Porosh. Met. *9* (5), 13–18 (1969); C.A. *71*, 62615 (1969).

[20] *V. N. Korev, A. F. Netserov, J. P. Glozkova*, Fiz. Metal. Metalloved. *6* (1), 86 (1963); C.A. *59*, 10781 (1963).

[21] *M. Hansen, K. Anderko*, Constitution of Binary Alloys, 2nd Edn, Ref. 11, p. 370, Ref. 12, p. 220.

[22] Refractory Molybdenum Carbides and Nitrides, Climax Molybdenum Company, New York, 1956.

[23] *R. Schwarzkopf, R. Kieffer*, Refractory Hard Metals, Macmillan Co., New York, 1953.

[24] *A. E. Kovolskii, S. V. Semeovskaya*, Kristollografiya *4*, 923–4 (1959).

molybdenum and lampblack at 1750° under a pressure of 600 kg/cm^2.

Two *nitrides*, Mo_2N and MoN, are formed when molybdenum is heated with nitrogen or ammonia.[22,25] Heating the metal with nitrogen or ammonia usually gives mixtures of these nitrides, but pure Mo_2N can be obtained by carrying out the reaction at 1100° to 1400° under a pressure of 300 atm.[26] Reportedly, it can also be prepared by passing ammonia over finely divided molybdenum at 400°–700° and similarly the mononitride is prepared by passing ammonia over the metal at 700° for 120 hr.[23] A new nitride, $Mo_{16}N_7$, has been prepared by heating the metal and calcium nitride at 750° in a nitrogen atmosphere.[27]

At least four *aluminides* are definitely known to exist: $Al_{12}Mo$, Al_5Mo, Al_3Mo, and Mo_3Al.[28,29] They are generally prepared by fusion of the elements in an inert atmosphere.

Molybdenum forms three well-defined *silicides:* $MoSi_2$, Mo_5Si_3, and Mo_3Si.[14,30,31] The disilicide is prepared from the finely divided elements at 1040° in an inert atmosphere.[30,32] The other two silicides are also prepared directly by heating a stoichiometric mixture of the elements.[32–34]

Germanium reacts with molybdenum at high temperatures in an inert atmosphere to form several compounds: $MoGe_2$, $Mo_{13}Ge_{23}$, and Mo_3Ge.[35–37]

At least four *phosphides* of molybdenum have been isolated: MoP_2, MoP, Mo_4P_3, and Mo_3P.[14,38] These compounds have been prepared by heating the elements in evacuated, sealed tubes at very high temperatures.[39] The monophosphide can be prepared by heating a mixture of Ca_3P_2 and molybdenum[40] and Mo_3P has been prepared by electrolysis of molybdic oxide in a sodium phosphate melt.[41]

Arsenic forms several *arsenides* when the elements are heated at 700° to 1100°: $MoAs_2$, Mo_4As_5, MoAs, and Mo_5As_4.[42,43] The diarsenide is prepared by heating the elements at 570° in a sealed tube and distilling off the excess arsenic,[44] and the preparation of Mo_5As_4 from the elements has recently been reported.[45]

15.3 Molybdenum(II) Compounds

Molybdenum(II) oxide, MoO,[46] and fluoride, MoF_2,[47] have not been isolated to date.

---

[25] *M. Hansen, K. Anderko*, Constitution of Binary Alloys, 2nd Ed., Ref. 11, p. 966, Ref. 12, p. 622; *F. A. Shunk*, Constitution of Binary Alloys, 2nd Suppl., p. 515, McGraw-Hill Book Co., New York 1969.

[26] *P. Ettmayer*, Monatsh. Chem. *101*, 127 (1970).

[27] *R. Karam, R. Ward*, Inorg. Chem. *9*, 1385 (1970).

[28] *M. Hansen, K. Anderko*, Constitution of Binary Alloys, 2nd Ed., Ref. 11, p. 114, Ref. 12, p. 45; *F. A. Shunk*, Constitution of Binary Alloys, 2nd Suppl., p. 29, McGraw-Hill Book Co., New York 1969.

[29] *M. Pötzschke, K. Schubert*, Z. Metallk. *53*, 548 (1962).

[30] Refractory Metal Silicides, Climax Molybdenum Co., New York, 1963;

[31] *M. Hansen, K. Anderko*, Constitution of Binary Alloys, 2nd Ed., Ref. 11, p. 973, Ref. 12, p. 633; *F. A. Shunk*, Constitution of Binary Alloys, 2nd Suppl., p. 524, McGraw-Hill Book Co., New York 1969.

[32] *C. H. Dauben, A. W. Searey, D. H. Templecon, L. Brewer*, J. Amer. Ceram. Soc. *33*, 291 (1950).

[33] *D. H. Templeton, C. H. Dauben*, Acta Cryst. *3*, 261 (1950).

[34] *B. Aronsson*, Arkiv. Kemi *16*, 379 (1960).

[35] *M. Hansen, K. Anderko*, Constitution of Binary Alloys, 2nd Ed., Ref. 11, p. 709, Ref. 10, p. 482; *F. A. Shunk*, Constitution of Binary Alloys, 2nd Suppl., p. 388, McGraw-Hill Book Co., New York 1969.

[36] *O. I. Popova, R. S. Biryukova*, Porosh. Met. *10* (3), 89–93 (1970); *73*, 41303 (1970).

[37] *R. S. Biryukova, O. I. Popova, V. N. Bondarey*, Russian J. Inorg. Chem. *15* (8), 1186 (1970).

[38] *M. Hansen, K. Anderko*, Constitution of Binary Alloys, 2nd Ed., Ref. 17, p. 970, Ref. 12, p. 625; *F. A. Shunk*, Constitution of Binary Alloys, 2nd Suppl., p. 519, McGraw-Hill Book Co., New York 1969.

[39] *S. Rundqvist, T. Lundstrom*, Acta Chem. Scand. *17*, 37 (1963).

[40] *R. L. Ripley*, J. Less-Common Metals *4*, 496 (1962).

[41] *R. D. Blougher, J. K. Hulm, P. N. Yocum*, J. Phys. Chem. Solids *26* (12), 2037 (1965).

[42] *F. A. Shunk*, Constitution of Binary Alloys, 2nd Suppl., p. 56, McGraw-Hill Book Co., New York 1969.

[43] *H. Baller, H. Nowotny*, Monatsh. Chem. *95*, 1272 (1964).

[44] *E. Heinerth, W. Biltz*, Z. Anorg. Allgem. Chemie *198*, 171 (1931).

[45] *B. Reiss, H. Wagini*, Z. Naturforsch. *21a* (11), 2008 (1966).

[46] *L. Kihlborg*, Acta Chem. Scand. *28*, 1571 (1964).

[47] *H. J. Emeléus, V. Gutmann*, J. Chem. Soc. (London) 2115 (1950).

The polynuclear halides of Mo(II) will be discussed in order of the increasing atomic number of the halide in the cation. A monomeric chloride complex, $MoCl_2 \cdot (diars)_2$ (where diars is *o*-phenylenedimethylarsine), has been prepared by reaction of the ligand with an ethanolic hydrogen chloride solution of hexachloromolybdate(III) ion.[48]

Molybdenum(II) chloride, Mo_6Cl_{12}, is prepared[49,50] by passing $MoCl_5$ vapor over the metal at 600°–700° or by the thermal decomposition of molybdenum trichloride in an inert gas stream.[51] It has also been prepared[52] by passing a stream of phosgene over the metal at 610°. The crude product is purified by treatment with hydrochloric acid and thermal decomposition of the acid $[H_3O]_2[Mo_6Cl_8] \cdot Cl_6 \cdot 6H_2O$ that is formed at 200° *in vacuo*. A second form of the dichloride identified as β-$MoCl_2$ has been isolated by treating molybdenum(II) acetate with hydrogen chloride at 250° to 350°.[53] The chloride reacts directly with excess ligand to form complexes of type $[(Mo_6Cl_8)Cl_4 \cdot 2L]$ with a variety of ligands such as pyridine, trimethylamine, triphenylphosphine oxide, triphenylarsine oxide,[54] triphenylphosphine,[55] dimethylformamide,[55,56] dimethyl sulfoxide,[55,56] and nitriles.[57] Several compounds have been prepared that contain the $[Mo_6Cl_8]^{4\oplus}$ group as a central cation by replacement of some of the chlorines of Mo_6Cl_{12}. The *mixed halides* $[Mo_6Cl_8]Br_4$ and $[Mo_6Cl_8]I_4$ are prepared by dissolving Mo_6Cl_{12} in the appropriate hydrohalic acid and removing the excess acid by heating under vacuum.[58] Alkoxides of the type $Na_2[(Mo_6Cl_8)(OR)_6]$ are formed by reacting the chloride with the corresponding sodium alkoxide.[59] *Polynuclear chloromolybdates* of the type $(Mo_3Cl_{13})^{7\ominus}$, $(Mo_3Cl_{12})^{6\ominus}$, $(Mo_3Cl_{11})^{5\ominus}$, $(Mo_2Cl_9)^{5\ominus}$, and $(Mo_2Cl_8)^{4\ominus}$ have been prepared by addition of alkali metal chlorides to a solution of molybdenum(II) acetate in concentrated hydrochloric acid.[60–62]

Molybdenum(II) bromide is prepared[63] by passing a bromine–nitrogen mixture over the metal at 600° to 700° or by fusion at 650° to 700° of a mixture of molybdenum(II) chloride with lithium bromide.[59,64] Like the corresponding chloride, molybdenum(II) bromide forms *mixed halides* of the type $(Mo_6Br_8)Cl_4$ and $(Mo_6Br_8)I_4$ on treatment with the appropriate hydrohalic acid.[59] A polynuclear salt, $Cs_2[(Mo_6Br_8)Br_6]$ has been prepared by reacting molybdenum dibromide with cesium bromide in iodine monobromide at 50°.[65] The bromide forms a pyridine complex, $Mo_6Br_{12} \cdot 2C_5H_5N$, on direct reaction with the ligand.[58]

Molybdenum(II) iodide is formed by the thermal decomposition of the triiodide or by the reaction[66,67] of molybdenum with iodine at temperatures above 400°. The iodide is also prepared by the reaction of molybdic oxide with aluminum iodide[68] at 230° and by fusion of the dichloride with lithium iodide.[59]

---

[48] *J. Lewis, R. S. Nyholm, P. W. Smith*, J. Chem. Soc. (London) 2592 (1963).

[49] *P. Nannelli, B. P. Block*, Inorg. Synth. *12*, 172 (1970).

[50] *W. L. Jolly*, The Synthesis and Characterization of Inorganic Compounds, p. 456, Prentice Hall, New York 1970, Englewood Cliffs.

[51] *D. E. Couch, A. Brenner*, J. Res. Nat. Bur. Stan. *63A*, 185 (1959).

[52] *K. Lindner, E. Haller, H. Helwig*, Z. Anorg. Allgem. Chemie *130*, 210 (1923).

[53] *G. B. Allison, I. R. Anderson, J. C. Sheldon*, Australian J. Chem. *22*, 1091 (1969).

[54] *J. C. Sheldon*, J. Chem. Soc. (London) 4183 (1963), and 1287 (1964).

[55] *J. C. Fergusson, B. H. Robinson, C. J. Wilkins*, J. Chem. Soc. (London) 486A (1967).

[56] *F. A. Cotton, N. F. Curtis*, Inorg. Chem. *4*, 241 (1965).

[57] *W. M. Carmichael, D. A. Edwards*, J. Inorg. Nucl. Chem. *20*, 1535 (1967).

[58] *J. Sheldon*, J. Chem. Soc. (London) 410 (1962); 1007 (1960).

[59] *R. Nannelli, P. Block*, Inorg. Chem. *7*, 2423 (1968).

[60] *I. R. Anderson, J. C. Sheldon*, Australian J. Chem. *18*, 271 (1965).

[61] *G. B. Allison, I. R. Anderson, J. C. Sheldon*, Australian J. Chem. *20*, 869 (1967);
G. B. Allison, I. R. Anderson, J. C. Sheldon, Australian J. Chem. *22*, 1097 (1969).

[62] *M. J. Bennett, J. V. Brencic, F. A. Cotton*, Inorg. Chem. *8*, 1061 (1969);
J. V. Brencic, F. A. Cotton, Inorg. Chem. *8*, 2698 (1969);
J. V. Brencic, F. A. Cotton, Inorg. Chem. *9*, 346, 351 (1970).

[63] *K. Lindner, H. Helwig*, Z. Anorg. Allgem. Chemie *142*, 180 (1925).

[64] *P. Nannelli, B. P. Block*, Inorg. Synth. *12*, 176 (1970).

[65] *A. A. Opalovskii, P. P. Smailov*, Dokl. Akad. Nauk. SSSR *174* (5), 1109–10 (1967); C.A. *67*, 70173 (1967).

[66] *F. Klanberg, H. W. Kohlschütter*, Z. Naturforsch. *15b*, 616 (1960).

[67] *J. Lewis, D. J. Machin, R. S. Nyholm, P. Pauling, P. W. Smith*, Chem. & Ind. (London) 259 (1960).

[68] *M. Chaigneau*, Bull. Soc. Chim. France 886 (1957).

A molybdenum(II) cyanide complex $K_4[Mo(CN)_6]$, can be prepared by reducing potassium octacyanomolybdate(IV) with hydrogen[69] at 390°.

Molybdenum(II) carboxylates, $Mo(RCO_2)_2$, are formed when molybdenum hexacarbonyl is refluxed with a variety of mono- and dicarboxylic acids.[70] Conditions for preparing the diacetate were optimized by using diglyme as a cosolvent.[60]

15.4 Molybdenum(III) Compounds

15.4.1 Halides, Oxohalides, and Halide- and Pseudohalide Salts

Molybdenum(III) fluoride, MoF_3, can be prepared[71] by treating the trichloride or bromide with a stream of hydrogen fluoride at 600°. Reduction of molybdenum pentafluoride with the metal at 400° or antimony trifluoride at 200° also yields a pure product.[72]

Molybdenum(III) chloride of high purity is readily prepared[73,74] by reduction of molybdenum pentachloride with stannous chloride at 300°. It can also be prepared on a small scale by the reaction[75] of red phosphorous with molybdenum pentachloride in a sealed tube at 200°. Reductions with molybdenum metal[76] or hydrogen[47] are difficult to carry out and usually yield an impure product, but small samples of crystalline $MoCl_3$ of high purity have been obtained using the metal as reductant followed by vapor transport.[77]

Molybdenum(III) bromide is best prepared[78] by passing bromine over finely divided molybdenum at 600°. It is claimed that it can also be prepared directly from the elements at room temperature if ether is used as a solvent.[79] It can also be prepared from molybdenum pentachloride and boron tribromide at room temperature.[80]

Molybdenum triiodide is obtained by reacting the elements in a sealed tube at 300° and subliming off the excess iodine.[67] A carbon disulfide solution of molybdenum pentachloride is reduced by gaseous hydrogen iodide to yield triiodide.[66] It is also readily prepared from molybdenum hexacarbonyl and iodine in a sealed tube[81] at 105°.

Potassium hexafluoromolybdate(III), K_3MoF_6, is prepared[82] by heating potassium fluoride and molybdenum trifluoride in a sealed platinum tube at 750°. It has also been obtained by the fusion of *potassium hexachloromolybdate(III)* with potassium hydrogen fluoride.[83] The existence of the earlier described[84] complexes of the type $KMoF_4 \cdot H_2O$ have not been confirmed in the modern literature.

Salts containing the *hexachloromolybdate(III)* anion, $[MoCl_6]^{3\ominus}$, and *aquopentachloromolybdate(III)* anion, $[MoCl_5 \cdot H_2O]^{2\ominus}$, are prepared by electrolytic reduction of hydrochloric acid solutions of the molybdate or molybdic oxide to the trivalent state followed by addition of the alkali metal chloride to precipitate the salt.[85–89]

It should be noted that in the preparation of these salts that less than about 1·5 mol of the alkali chloride

[69] *J. S. Yoo, E. Griswold, J. Kleinberg*, Inorg. Chem. *4*, 365 (1965).

[70] *A. T. Stephenson, E. Bannister, G. Wilkinson*, J. Chem. Soc. (London) 2538 (1965).

[71] *H. J. Emeléus, V. Gutmann*, J. Chem. Soc. (London) 2979 (1949).

[72] *D. E. LaValle, R. M. Steele, M. K. Williamson, H. L. Yakel*, J. Amer. Chem. Soc. *82*, 2433 (1950).

[73] *A. K. Mallock*, Inorg. Synth. *12*, 178 (1970); *A. K. Mallock*, Inorg. Nucl Chem. Lett. *3*, 441 (1967).

[74] *R. Saillant, R. A. D. Wentworth*, Inorg. Chem. *8*, 1226 (1969).

[75] *S. M. Horner, F. N. Collier, S. Y. Tyree*, J. Less-Common Metals *13*, 85 (1967).

[76] *T. T. Campbell*, J. Elektrochem. Soc. *106*, 1196 (1959).

[77] *H. Schäfer and co-workers*, Z. Anorg. Allgem. Chemie *353*, 281 (1967).

[78] *P. J. H. Cornell, R. E. McCarley, R. D. Hogue*, Inorg. Synth. *10*, 50 (1967).

[79] *J. R. M. Fernandez, A. B. Duran*, Anales Real Soc. Espan. Fis. Quim (Madrid) *55B*, 823 (1959).

[80] *F. M. Druce, M. F. Lappert, P. N. K. Riley*, Chem. Commun. 486 (1967).

[81] *C. Djordjevic, R. S. Nyholm, C. S. Pande, M. H. B. Stiddard*, J. Chem. Soc. *A*, 16 (1966).

[82] *L. M. Toth, G. D. Branten, G. P. Smith*, Inorg. Chem. *8*, 2694 (1969).

[83] *R. D. Peacock*, Progress in Inorganic Chemistry, p. 205, Interscience Publishers, New York, 1960.

[84] *A. Rosenheim, H. J. Braun*, Z. Anorg. Allgem. Chemie *46*, 320 (1905).

[85] *S. Senderoff, A. Brenner*, J. Elektrochem. Soc. *101*, 28 (1954).

[86] *H. Hartmann, H. J. Schmidt*, Z. Physik. Chem. (Frankfurt am Main) *11*, 234 (1957).

[87] *L. F. Lindoy, S. E. Livingstone, T. N. Lockyer*, Australian J. Chem. *18*, 1549 (1965).

[88] *G. Brauer*, Handbook of Preparative Inorganic Chemistry, 2nd Ed., vol. II, p. 1408, Academic Press, New York, 1965. (English Translation).

[89] *W. G. Palmer*, Practical Inorganic Chemistry, p. 413, Cambridge University Press, Cambridge 1954.

per molybdenum should be used to prevent contamination of the product.[87] If the molybdenum concentration of the molybdenum(III) solution is high then it is possible to isolate dimeric salts[90,91] of the type $[Mo_2Cl_9]^{3\ominus}$. These salts are also obtained by evaporating the electrolytically produced molybdenum(III) chloride solution to dryness and reacting the solid with the dry alkali halide by grinding or in a saturated hydrochloric acid solution.[91–93] The cesium and rubidium salts have also been prepared[74] by heating the alkali chloride with molybdenum trichloride in a sealed tube at 800°. The potassium salt could not be prepared by this method.

The corresponding *hexabromomolybdate(III)* salts, $[MoBr_6]^{3\ominus}$, can be prepared by the electrolytic reduction of molybdenum(VI) species in hydrobromic acid and precipitated with the alkali bromide.[87,94] The dimeric salts of the type $[Mo_2Br_9]^{3\ominus}$ have been obtained from the reaction of rubidium hexabromomolybdate(III) with refluxing ammonia[95] and by heating caesium bromide with molybdenum tribromide[74] at 770°. The *hexaiodomolybdate(III)* salt could not be isolated from a solution of potassium hexabromomolybdate and hydriodic acid.[96]

The molybdenum(III) cyanide salt, $K_4[Mo(CN)_7]\cdot 2H_2O$, is obtained when a solution of potassium hexachloromolybdate is treated with potassium cyanide in a nitrogen atmosphere.[97,98] Another molybdenum(III) complex, $K_2[Mo(CN)_5]$, is formed when potassium octacyanomolybdate(IV) is treated with molten potassium cyanide.[99] A hexacoordinated complex, $K_3[Mo(CN)_6]$, is obtained when $K_4[Mo(CN)_6]$ is repeatedly extracted with methanol in air.[69]

The reaction of potassium cyanate with potassium hexachloromolybdate(III) in molten dimethyl sulfone yields a complex containing $[Mo(OCN)_6]^{3\ominus}$ group.[100] Complexes containing the anion, $[Mo(NCS)_6]^{3\ominus}$ can be prepared from solutions containing Mo(III) and thiocyanate ion with potassium, ammonium, and pyridinium as cations.[97,101,102] Complexes of the type $R_3PH[Mo(SCN)_4(PR_3)_3]$ are formed by reacting $K_3[Mo(SCN)_6]$ with phosphines.[103]

The *oxofluoride*, $MoOF\cdot 4H_2O$ is reported to result from the reaction of the corresponding chloride with ammonium fluoride solution.[104] The brown *oxochloride* is precipitated by the addition of acetone to a concentrated molybdenum(III) chloride solution obtained by electrolytic reduction of a solution of molybdic oxide in hydrochloric acid. A green isomer is also reported to form if the concentrated molybdenum(III) chloride solution is further electrolyzed and then treated with acetone.[105] Attempts to repeat this work resulted in isolation of a buff solid in less than 20% yield which was analyzed as $MoOCl\cdot 3H_2O$ and the green isomer could not be obtained.[106] The *oxobromide* is obtained in a manner similar to that of the oxochloride from an electrolytically reduced solution of molybdic oxide in hydrobromic acid.[107]

15.4.2 Chalcogenides

Molybdenum(III) oxide, MO_2O_3, has been isolated as an amorphous black powder by reducing molybdic oxide in liquid ammonia with potassium metal and by dehydrating the oxide hydrate.[108] The *oxide hydrate* or hydroxide $[Mo_2O_3\cdot 3H_2O$ or $Mo(OH)_3]$ is prepared by addition of water to the oxide[108] or obtained as a coating on the cathode when a neutral solution of ammonium molybdate is electrolytically

[90] *J. Lewis, R. S. Nyholm, P. W. Smith*, J. Chem. Soc. *A*, 57 (1969).
[91] *I. E. Grey, P. W. Smith*, Australian J. Chem. *22*, 121 (1969).
[92] *I. E. Grey, P. W. Smith*, Australian J. Chem. *22*, 1627 (1969).
[93] *P. W. Smith, A. G. Wedd*, J. Chem. Soc. *A*, 2447 (1970).
[94] *W. A. Wardlaw, A. J. Harding*, J. Chem. Soc. (London) 1592 (1926).
[95] *A. J. Edwards, R. D. Peacock, A. Said*, J. Chem. Soc. (London) 4643 (1962).
[96] *C. Furloni, O. Piovesana*, Mol. Phys. *9*, 341 (1965).
[97] *J. Lewis, R. S. Nyholm, P. W. Smith*, J. Chem. Soc. (London) 4590 (1961).
[98] *P. C. H. Mitchell, R. J. P. Williams*, J. Chem. Soc. (London) 4570 (1962).
[99] *W. L. Magnusson, E. Griswold, J. Kleinberg*, Inorg. Chem. *3*, 88 (1964).
[100] *R. A. Bailey, S. G. Kozok*, J. Inorg. Nucl. Chem. *31*. 689 (1969).
[101] *W. R. Bucknall, S. R. Carter, W. Wardlaw*, J. Chem, Soc. (London) 512 (1927).
[102] *H. H. Schmidtke, D. Garthoff*, Helv. Chem. Acta *50* (6), 1631 (1967).
[103] *K. Isslieb, B. Bierman*, Z. Anorg. Allgem. Chemie *347*, 39 (1966).
[104] *W. Wardlaw, R. L. Wormell*, J. Chem. Soc. (London) 1089 (1928).
[105] *W. Wardlaw, R. L. Wormell*, J. Chem. Soc. *125*, 2370 (1924); 130, 133 (1927).
[106] *G. A. Tsigdinos*, Unpubl. Climax Molybdenum Co., Research Laboratory, Ann Arbor, Michigan 1966.
[107] *W. Wardlaw, R. L. Wormell*, J. Chem. Soc. (London) 1092 (1927).
[108] *G. W. Watt, D. D. Davies*, J. Amer. Chem. Soc. *70*, 3751 (1948).

reduced.[108,109] It can also be obtained by the addition of an ammonium hydroxide solution to a molybdenum(III) solution formed by electrolytic reduction of a solution of molybdic oxide in sulfuric acid.[110,111]

The *sesquisulfide*, Mo_2S_3, has been prepared by heating the elements[112] and by the thermal decomposition of molybdenum disulfide.[113] By modifying an older procedure,[114] the pure sulfide can be obtained by heating molybdenum disulfide in an inert atmosphere at 1650° and treating the product with cold, dilute aqua regia to remove impurities.[115]

Molybdenum sesquiselenide, Mo_2Se_3,[116] and *molybdenum telluride*, Mo_2Te_3,[117] are prepared from the elements by heating at high temperatures in an inert atmosphere.[118] These compounds have recently been shown to have the stoichiometry Mo_3X_4 rather than Mo_2X_3.[118,119]

15.4.3 Coordination Compounds

Molybdenum trichloride is reported to react with pyridine in a sealed tube at 270° to form $MoCl_3 \cdot 3C_5H_5N$.[120] Although this complex is reported to form readily at reflux,[121] molybdenum(III) chloride was found to react only slightly with refluxing pyridine after several days.[122] This tris(pyridine) complex is readily prepared in high yield by refluxing potassium hexachloromolybdate(III) with pyridine for 3 hr.[123,124] This procedure can be used with K_3MoCl_6 or $(NH_4)_2[MoCl_5(H_2O)]$ to form other tris complexes with ligands such as thiourea,[124] urea,[125] dimethylformamide,[126] and alkyl nitriles.[127] Attempts to prepare the tris(2,2′-bipyridyl) complex using the trichloride[128] or K_3MoCl_6[129,130] yielded $[MoCl_2 \cdot 2bipy][MoCl_4 \cdot bipy]$ and the use of *o*-phenanthroline as ligand gave similar results.

The tribromide reacts with refluxing pyridine to form the tris-complex, $MoBr_3 \cdot 3$pyridine.[121,131,132] Like the corresponding chloride, the bromide forms a complex of the type $[MoBr_2(bipy)_2][MoBr_4(bipy)]$ on reaction with 2,2′-dipyridyl.[129]

The tris–pyridine adduct, $MoI_3 \cdot 3$pyridine, is readily formed on brief reflux of the iodide with the ligand.[81] Unlike the corresponding chloride and bromide, the iodide reacts with 2,2′-bipyridyl to form a complex of the type $[Mo(bipy)_3]I_3$.[129]

The *molybdenum(III) acetylacetonate*, $MoO(acac) \cdot 2{\cdot}5H_2O$, is obtained by heating acetylacetone and $MoOCl \cdot 2H_2O$ in ethanol.[133] The *tris(acetylacetonate)*, $Mo(acac)_3$, is prepared by reacting an aqueous solution of acetylacetone with potassium hexachloromolybdate(III)[134,135] or ammonium aquopentachloro-

[109] *E. T. Wherry, E. F. Smith*, J. Amer. Chem. Soc. *29*, 806 (1907).
[110] *W. Wardlaw, W. H. Parker*, J. Chem. Soc. (London) *127*, 1312 (1925).
[111] *H. M. Spittle, W. Wardlaw*, J. Chem. Soc. (London) 794 (1929).
[112] *J. R. Stubbles, F. D. Richardson*, Trans. Faraday Soc. *56*, 1460 (1960).
[113] *C. L. McCabe*, J. Metals *7*, 61 (1955).
[114] *M. Guichard*, C.R. Acad. Sci., Paris *130*, 137 (1900);
M. Guichard, Ann. Chim. (Paris) *32*, 557 (1901).
[115] Unpubl. Climax Molybdenum Company, Research Laboratory, Ann Arbor, Michigan.
[116] *E. Wenderhorst*, Z. Anorg. Allgem. Chemie *173*, 270 (1928).
[117] *A. Morette*, C.R. Acad. Sci., Paris *215*, 86 (1942);
A. Morette, Ann. Chim. (Paris) *19*, 130 (1944).
[118] *A. A. Opalovskii, V. E. Federov*, Dokl. Akad. Nauk SSSR *163* (5), 1163–4 (1965).
[119] *C. J. Hallada*, Unpubl. Climax Molybdenum Company Research Laboratory, Ann Arbor, Michigan, 1966.
[120] *A. Rosenheim, G. Abel, R. Lewy*, Z. Anorg. Allgem. Chemie *197*, 189 (1931).
[121] *D. A. Edwards, G. W. A. Fowles*, J. Less-Common Metals *4*, 512 (1962).
[122] *A. K. Mallock, G. A. Tsigdinos*, Abstracts of Papers, 155th National Meeting of American Chemical Society, M129 (1968).
[123] *H. B. Jonassen, L. J. Bailin*, Inorg. Synth. *7*, 140 (1963).
[124] *O. Piovesena, C. Furlani*, Atti Accad. Naz. Lincei, Classe Sci. Fis., Mat. Natur., Rend. *41*, 324 (1966).
[125] *V. I. Spitsyn, I. D. Kolli, T. Wen-hsia*, Russian J. Inorg. Chem. *9*, 54 (1964).
[126] *T. Karmorita, S. Miki, S. Yamada*, Bull. Chem. Soc. Japan *38*, 123 (1965).
[127] *P. W. Smith, A. G. Wedd*, J. Chem. Soc. *A*, 231 (1966); *A*, 1377 (1968).
[128] *P. A. Marzilli, D. A. Buckingham*, Australian J. Chem. *19*, 2259 (1966).
[129] *W. M. Carmichael, D. A. Edwards, R. W. Walton*, J. Chem. Soc. *A*, 97 (1966).
[130] *D. W. DuBois, R. T. Iwamoto, J. Kleinberg*, Inorg. Chem. *8*, 815 (1969).
[131] *G. Brauer*, Handbook of Preparative Inorganic Chemistry, 2nd Ed., Vol. II, p. 1408, Academic Press, New York, 1965. (English Translation).
[132] *E. A. Allen, K. Feenan, G. W. A. Fowles*, J. Chem. Soc. (London) 1636 (1965).
[133] *G. T. Morgan, R. A. S. Castell*, J. Chem. Soc. (London) 3252 (1928).
[134] *M. L. Larson, F. W. Moore*, Inorg. Chem. *1*, 852 (1962);
M. L. Larson, F. W. Moore, Inorg. Synth. *8*, 153 (1966).
[135] *K. Christ, H. L. Schafer*, Angew. Chem. Intern. Ed. Engl. *2*, 97 (1963).

molybdate(III)[136] at about 60° under an inert atmosphere. This compound can also be prepared by refluxing molybdenum hexacarbonyl with acetylacetone.[134, 137] Tris complexes have been obtained from other β-diketones using these techniques.[136, 137] The tris(8-hydroxyquinolinato) complex was isolated by refluxing a methanol solution of the ligand with $K_2[MoCl_5(H_2O)]$.[138] The reaction of an ethanolic solution of lithium hexachloromolybdate(III) with various thiol-containing ligands produced a series of dimeric molybdenum(III) complexes which could be formulated as $Mo_2(ligand)_3Cl_3$, $Mo_2(ligand)_4Cl_2(H_2O)_3$, and $Mo_2(ligand)_2Cl_4(H_2O)_3$ depending on the ligand used.[87] Similar complexes containing bromine rather than chlorine were also isolated. The *tris(pheny mercaptide)*, $Mo(SC_6H_5)_3$, is obtained from the reaction of the hexacarbonyl with diphenyldisulfide.[139]

A *molybdenum(III) oxalate*, $Mo_2O(C_2O_4)_2(H_2O)_6$, can be prepared by dissolving molybdenum(III) hydroxide in hot aqueous oxalic acid and treating the concentrated solution with acetone to cause precipitation.[110] Longer heating with oxalic acid yields a second oxalate formulated as $Mo_4O_3(C_2O_4)_3 \cdot 12H_2O$.

15.5 Molybdenum(IV) Compounds

15.5.1 Halides, Oxochloride, and Halide- and Pseudohalide Salts

Molybdenum tetrafluoride can be prepared by heating molybdenum disulfide with sulfur tetrafluoride at 350° in a bomb.[140] It is also obtained[141, 142] as a green residue by thermally decomposing at 170° the compound formed on reacting molybdenum hexacarbonyl with fluorine at −70°.

Molybdenum tetrachloride can be prepared by refluxing molybdenum pentachloride with tetrachloroethylene[143, 144] or benzene.[145] It has also been prepared with greater difficulty by reacting molybdenum pentachloride with the trichloride.[77] The chlorination of molybdenum dioxide with carbon tetrachloride or hexachlorobutadiene gives impure materials.[146–148]

Molybdenum tetrabromide is formed in high purity by the reaction of molybdenum tribromide with liquid bromine[149] at 55°. It has also been prepared by treating molybdenum hexacarbonyl with liquid bromine[150–152] and by brominating molybdenum tribromide at 400°.[153] Molybdenum tetraiodide could not be prepared from the elements[67] and has not been isolated to date.

Reaction of excess sodium iodide with molybdenum hexafluoride in liquid sulfur dioxide yields *sodium hexafluoromolybdate(IV)*, Na_2MoF_6.[154, 155]

Salts of *hexachloromolybdate(IV)* are obtained by reacting molybdenum pentachloride with an alkali metal chloride in liquid sulfur dioxide,[156]

---

[136] *J. H. Balthis*, J. Inorg. Nucl. Chem. *24*, 1016 (1962).
[137] *T. G. Dunne, F. A. Cotton*, Inorg. Chem. *2*, 263 (1963).
[138] *Q. W. Choi, J. S. Oh, K. W. Lee, W. Lee*, Daehan Hwahek Hwoejec *12* (4), 146 (1968); C.A. *71*, 27048 (1969).
[139] *R. N. Jowitt, P. C. H. Mitchell*, Inorg. Nucl. Chem. Lett. *4*, 39 (1968).
[140] *A. L. Oppegard, W. C. Smith, E. Muetterties, V. A. Engelhardt*, J. Amer. Chem. Soc. *83*, 2835 (1960).
[141] *R. D. Peacock*, Proc. Chem. Soc. 59 (1957); *A. J. Edward, R. D. Peacock, R. W. H. Small*, J. Chem. Soc. (London) 4486 (1962).
[142] *J. Lewis, R. Whyman*, J. Chem. Soc. (London) 6027 (1965).
[143] *T. M. Brown, E. L. McCann III*, Inorg. Chem. *7*, 1227 (1968); *T. M. Brown, E. L. McCann III*, Inorg. Synth. *12*, 181 (1970).
[144] *D. L. Kepert, R. Mandyczewsky*, Inorg. Chem. *7*, 2091 (1968).
[145] *M. L. Larson, F. W. Moore*, Inorg. Chem. *3*, 285 (1964).
[146] *S. A. Shchukarev et al.*, Russian J. Inorg. Chem. *5*, 802 (1960).
[147] *S. M. Horner, S. Y. Tyree jr.*, Inorg. Chem. *1*, 947 (1962).
[148] *G. B. Allison, J. C. Sheldon*, Inorg. Chem. *6*, 1493 (1967).
[149] *P. J. H. Cornell, T. E. McCarley, D. Hogue*, Inorg. Synth. *10*, 49 (1967).
[150] *W. Hieber, E. Romberg*, Z. Anorg. Allgem. Chemie *221*, 321 (1935).
[151] *S. A. Shchukarev, I. V. Vasil'kove, N. D. Zatiseva*, Vestn. Leningr. Univ. *16* (22), Ser. Fiz. i. Khim. No. 4, 127 (1961); C.A. *56*, 9510 (1962).
[152] *A. A. Opalovskii, P. P. Samoilov*, Russian J. Inorg. Chem. *13*, 196 (1968).
[153] *W. Klemm, H. Steinberg*, Z. Anorg. Allgem. Chemie *227*, 193 (1936).
[154] *A. J. Edwards, R. D. Peacock*, Chem. & Ind. (London) 1441 (1960).
[155] *D. H. Brown, K. R. Dixon, R. D. W. Kemmit, D. W. A. Sharp*, J. Chem. Soc. (London) 1559 (1965).
[156] *E. A. Allen, B. J. Brisdon, D. A. Edwards, G. W. A. Fowles, R. G. Williams*, J. Chem. Soc. (London) 4649 (1963).

and iodine monochloride,[157,158] or from the melt.[157,159] The pyridinium salt has been prepared by reacting pyridinium hydrochloride with molybdenum tetrachloride bis(propylcyanide) in chloroform.[156,160] The cesium and rubidium salts of *hexabromomolybdate(IV)* were prepared by heating molybdenum tribromide, the alkali metal bromide, and iodine monobromide[152,157] at 300°. The corresponding iodine-containing salts could not be prepared.[157] Several complexes of Mo(IV) containing cyanide as ligand have been isolated. *Potassium octacyanomolybdate(IV)*, $K_4[Mo(CN)_8] \cdot 2H_2O$, can be prepared by treating a crude molybdenum(V) thiocyanate complex[161–164] or a hydrazine-reduced molybdenum(VI) solution[153,165] with potassium cyanide. It can also be obtained by aerial oxidation of the molybdenum(III) compound, $K_4[Mo(CN)_7]$[166,167] or the reaction of a solution of potassium cyanide with potassium hexachloromolybdate(III) in the presence of air.[168] Attempts to repeat the synthesis[167] of $K[Mo(CN)_5]$ have been unsuccessful.[48,169] *Potassium dioxotetracyanomolybdate(IV)*, $K_4[MoO_2(CN)_4] \cdot 6H_2O$ can be prepared by modifying[170] an older procedure[168] and by photolysis of solutions containing the $[Mo(CN)_8]^{4\ominus}$ ion.[171] It can also be isolated by treating the molybdenum(V) hydroxide, $MoO(OH)_3$, with potassium cyanide.[172] The monohydroxy compound, $K_3[MoO(OH)(CN)_4] \cdot H_2O$, has been prepared by modification[170] of a previous preparation[168] and by treatment of a solution of $K_4[MoO_2(CN)_4]$ with an acidic ion exchange resin at pH 11.[171] The use of more ion exchange resin to obtain pH 7 yields the dihydroxy complex, $K_2[Mo(OH)_2(CN)_4]$.[171] The trihydroxy complex, $K_4[Mo(CN)_4(OH)_3]$, and the tetrahydroxy complex, $K_4[Mo(CN)_4(OH)_4]$ are reportedly formed when the molybdenum(V) hydroxide $MoO(OH)_3$ is treated with a solution of potassium cyanide and evaporated.[168,173] The black compound $Mo(OH)_2(CN)_2$ is isolated by treating a solution of $K_4[MoO_2(CN)_4]$ with 6 M hydrochloric acid.[171,172] A sulfur-bridged cyano complex, $K_6[(CN)_6MoSMo(CN)_6]$, is obtained by passing hydrogen sulfide into a potassium molybdate solution containing a high concentration of potassium cyanide.[174] The pyridinium salt of hexacoordinated thiocyanate complex, $[pyH]_2[Mo(NCS)_6]$, is obtained by mixing a suspension of $(pyH)_0$-$[Mo(CNS)_6]$ with a concentrated solution of potassium ferricyanide.[175]

The brown *oxochloride*, $MoOCl_2$, is obtained by heating together molybdic oxide and molybdenum trichloride in a thermal gradient.[176,177] It can also be prepared by reacting an excess of molybdenum oxotetrachloride with aluminum powder in a sealed tube[178] at 130°.

15.5.2 Oxide, Hydroxides, Molybdites, Sulfide, Selenide and Telluride

Molybdenum dioxide can be prepared by reducing molybdic oxide or ammonium molybdate with the metal[179–181] at 1000°, with hydro-

[157] *A. J. Edwards, R. D. Peacock, A. Said*, J. Chem. Soc. (London) 4643 (1962).
[158] *T. L. Brown, W. G. McDugle, Jr, L. G. Kent*, J. Amer. Chem. Soc. *92*, 3645 (1970).
[159] *A. J. Efimov, A. I. Efimov, L. P. Belorukova, A. M. Ryndina, B. Z. Pitrimov*, Russian J. Inorg. Chem. *8*, 605, 1042 (1963).
[160] *S. M. Horner, S. Y. Tyree*, Inorg. Chem. *2*, 568 (1963).
[161] *N. H. Furman, C. O. Miller*, Inorg. Synth. *3*, 160 (1950).
[162] *W. G. Palmer*, Practical Inorganic Chemistry, p. 410, Cambridge University Press, Cambridge 1954.
[163] *G. Brauer*, Handbook Preparative Inorganic Chemistry, 2nd Ed., Vol. II, p. 1416, Academic Press, New York 1965. (English Translation.)
[164] *G. G. Schlessinger*, Inorganic Laboratory Preparations, p. 91, Chemical Publ. Co., New York, 1962.
[165] *J. van de Poel, H. M. Neumann*, Inorg. Synth. *11*, 53 (1968).
[166] *R. C. Young*, J. Amer. Chem. Soc. *54*, 1402 (1932).
[167] *M. C. Steele*, Australian J. Chem. *10*, 404 (1957).
[168] *W. R. Bucknall, W. Wardlaw*, J. Chem. Soc. (London) 2981 (1927).
[169] *P. C. H. Mitchell*, Coord. Chem. Rev. *1*, 315 (1966).
[170] *S. J. Lippard, B. J. Russ*, Inorg. Chem. *6*, 1943 (1967).
[171] *J. van de Poel, H. N. Neumann*, Inorg. Chem. *7*, 2086 (1968).
[172] *W. F. Jakob, C. Michaelewicz*, Roczniki Chem. *12*, 576 (1932).
[173] *W. P. Griffiths, J. Lewis, G. Wilkinson*, J. Chem. Soc. (London) 872 (1959).
[174] *A. Miller, P. Christophliemk*, Angew. Chem. Internal. Ed. Engl. *8*, 753 (1969).
[175] *G. A. Barbieri*, Atti Accad. Lincei *12*, 55 (1930).
[176] *H. Schafer, J. Tillack*, J. Less-Common Metals *6*, 152 (1964).
[177] *P. C. Crouch, G. W. A. Fowles, I. B. Tomkins, R. A. Walton*, J. Chem. Soc. *A*, 2412 (1969).
[178] *M. Mercer*, J. Chem. Soc. *A*, 2019 (1969).
[179] *G. Brauer*, Handbook of Preparative Inorganic Chemistry, 2nd Ed., Vol. II, p. 1409, Academic Press, New York 1965. (English Translation.)
[180] *W. Rüdorff, H. Kornelson*, Rev. Chem. Miner. *6*, 153 (1969).
[181] *A. Magnéli, G. Andersson, B. Blomberg, L. Kihlborg*, Anal. Chem. *24*, 1998 (1952).

gen[179,182] at 500°, with a hydrogen–water mixture[183] at 500°, or with ammonia gas[184] at 460°. Highly crystalline material can be obtained by the electrolytic reduction of a melt of the composition $K_2O \cdot 5MoO_3$[185] or Na_2MoO_3–MoO_3.[186]

A *molybdenum(IV) hydroxide* formulated as $MoO(OH)_2$ is reported to precipitate out as a brown solid when a solution containing Mo(IV) is treated with a sodium hydroxide solution.[187] The Mo(IV) solution was obtained by heating a hydrochloric acid solution of Mo(VI) and Mo(III) at 80°. The *polynuclear hydroxide*, $Mo_5O_5(OH)_{10}$, is prepared by reduction of molybdic oxide with zinc and concentrated hydrochloric acid in an inert atmosphere.[188,189]

A *lithium molybdite*, $Li_2Mo_2O_5$, is formed when molybdenum dioxide and lithium oxide are heated[190] at 800°, whereas sodium molybdite, $Na_2Mo_3O_6$, is formed by heating the dioxide with sodium oxide at 800° or by hydrogen reduction of $3MoO_3 \cdot 3Na_2O$.[191] Molybdites of the composition $AMoO_3$ (where A is barium, calcium, magnesium, or strontium) are prepared by reacting the alkaline earth oxide with molybdenum dioxide or by hydrogen reduction of the normal molybdate.[192–194] The reaction of magnesium oxide with the dioxide at 1100° was found to yield $Mg_2Mo_3O_8$ rather than $MgMoO_3$ and other molybdites of the type $A_2Mo_3O_8$ were obtained from the oxides with the following divalent metals: zinc, cadmium, iron, cobalt, nickel, and managanese.[195]

Molybdenum disulfide of high purity is commercially available. It can be prepared from the elements at around 1000° in an inert atmosphere.[2] A rhombohedral form has been prepared by heating molybdic oxide and sulfur in a potassium carbonate melt.[196]

Molybdenum diselenide is prepared from the elements[116,197,198] at 700°. The ditelluride is similarly prepared.[117,198–200]

15.5.3 Coordination Compounds

Molybdenum tetrafluoride forms *complexes* of the type $MoF_4 \cdot 2L$ with ligands such as pyridine, trimethylamine, dimethyl sulfoxide, and dimethylformamide.[201]

Molybdenum tetrachloride forms a *dipyridine* adduct, $MoCl_4 \cdot 2C_5H_5N$, by reacting a benzene suspension of the chloride with pyridine,[145,202] and a monoadduct with 2,2′-bipyridyl.[129,202] This latter complex can also be prepared by reducing molybdenum pentachloride with excess ligand[129,203] or by reaction of the ligand with molybdenum tetrachloride bis(propylcyanide).[132] Bis adducts are formed by the direct reaction of the chloride with acetonitrile, ethyl cyanide, and benzonitrile.[202] The *bis(nitrile) complexes* can be prepared by reducing molybdenum pentachloride with excess nitrile or by direct reaction with the tetrachloride.[204] Displacement of the nitrile from the bis(propylcyanide) adduct yields new bis adducts with tetrahydrofuran, pentamethylene oxide, triphenylarsine, triphenylphosphine, triphenylphosphine oxide, and pyrazine as ligands.[132] This reaction was

---

182 *L. Kihlborg*, Acta Chem. Scand. *13*, 954 (1959).
183 *F. Zado*, J. Inorg. Nucl. Chem. *25*, 1115 (1963).
184 *S. Zador, C. B. Alcock*, J. Chem. Thermodynamics *2*, 9 (1969).
185 *B. C. Brandt, A. C. Skapski*, Acta Chem. Scand. *21*, 661 (1967).
186 *D. S. Perloff, A. Wold*, J. Phys. Chem. Solids, Suppl. No. 1, 361 (1967).
187 *P. Souchay, M. Cadiot, B. Vicossat*, Bull. Soc. Chim. France 892 (1970).
188 *G. Brauer*, Handbook of Preparative Inorganic Chemistry, 2nd Ed., Vol. II, p. 1411, Academic Press, New York 1965. (English Translation).
189 *O. Glemser, G. Lutz*, Z. Anorg. Allgem. Chemie *264*, 17 (1951).
190 *P. H. Hubert*, C.R. Acad. Sci., Paris *259*, 2238 (1964).
191 *P. H. Hubert*, C.R. Acad. Sci., Paris *262C*, 1189 (1966).
192 *R. Scholder, W. Klemm*, Angew. Chem. *66*, 461 (1964)
193 *R. Scholder, L. Brixner*, Z. Naturforsch. *106*, 178 (1955).
194 *J. Debucca, A. Wold, L. H. Brixner*, Inorg. Synth. *10*, 1 (1968).
195 *W. McCarroll, R. Ward, L. Katz*, J. Amer. Chem. Soc. *78*, 2909 (1956); *79*, 5410 (1957).
196 *R. E. Bell, R. E. Herfert*, J. Amer. Chem. Soc. *79*, 3351 (1957).
197 *Yu. M. Ukrainskii, O. V. Novoselova*, Proc. Acad. Sci. SSSR, Chem. Sect. *139*, 828 (1961).
198 *L. H. Brixner*, J. Inorg. Nucl. Chem. *24*, 257 (1962).
199 *O. Knop, R. D. McDonald*, Can. J. Chem. *39*, 897 (1961).
200 *D. Puotinen, E. E. Newnham*, Acta Cryst. *14*, 691 (1961).
201 *E. L. Muetterties*, J. Amer. Chem. Soc. *82*, 1082, 6249 (1960).
202 *T. M. Brown, D. K. Pings, L. R. Lieto, S. J. Delong*, Inorg. Chem. *5*, 1965 (1966).
203 *C. G. Hull, M. H. B. Stiddard*, J. Chem. Soc. *A*, 1933 (1966).
204 *E. A. Allen, B. J. Brisdon, G. W. A. Fowles*, J. Chem. Soc. 4531 (1964).

studied extensively with a variety of alkaryl- and triarylphosphines and arsines.[205] Attempts to repeat the synthesis[147] of the tetrakistriphenylarsine oxide adduct, $MoCl_4[(C_6H_5)_3AsO]_4$ were unsuccessful and apparently a stable complex cannot be formed with this ligand.[148]

Molybdenum tetrabromide, on direct reaction with ligand, forms bis adducts with methyl and phenyl cyanide and a monoadduct with 2,2′-bipyridyl.[202] This latter complex can also be prepared by heating the ligand with rubidium hexabromomolybdate(III) in the presence of benzene.[202]

The reaction of the tetrachloride with acetylacetone yields the red-purple solid formulated as $MoCl_2(C_5H_7O_2)_2$.[145]

The oxidation in air of solutions of molybdenum(III) oxalates yields a *molybdenum(IV) oxalate* formulated as $MoO(C_2O_4)(H_2O)_3$[110]; salts of the type $M_2[Mo_3O_4(C_2O_4)_3(H_2O)_5]$, where M is ammonium, potassium, or pyridinium, can be isolated by precipitation with alcohol.[110, 111, 206, 207]

Complexes of the type $MoOCl_2L_3$, where L is a tertiary arsine or phosphine, are formed when the ligand and $MoCl_4 \cdot 2C_2H_5CN$ are boiled together in ethanol.[205, 208]

An *oxophthalocyanine complex*, $MoO(C_{32}H_{16}N_8)$, is formed in low yield by reacting molybdenum dioxodichloride or molybdenum tetrachloride with phthalonitrile.[209]

An *oxodithiophosphate* complex, $MoO[(i\text{-}C_3H_2O)_2PS_2]$, was isolated on heating a hydrochloric acid solution of potassium hexachloromolybdate(III) with the potassium salt of the ligand.[87] Other complexes of this type where the ligand is diethyldithiocarbamate, diethyl- or diphenyldithiophosphate, can be prepared by reducing the corresponding dimeric molybdenum(V) complex with zinc, thiophenol, or sodium dithionite.[210] Refluxing molybdenum hexacarbonyl with tetramethylthiuram disulfide in benzene yields a red solid formulated as the tetrakis(diethyldithiocarbamate), $Mo[(C_2H_5)_2NCS_2]_4$ complex.[139]

Molybdenum forms a series of tris complexes of the type $Mo(S_2C_2R_2)_3$ with a number of 1,2-dithiols of the type R—C(SH)=C(SH)—R, where R is CF_3, CN, C_6H_5, etc., that are formally in the 4+ oxidation state.[211, 212] The complex where R is CF_3 is prepared by refluxing molybdenum hexacarbonyl with bis(trifluoromethyl)-dithietene in an appropriate solvent.[213, 214] The analogous selenium complex has recently been prepared by a similar reaction.[215] The dimercaptoethylene complex with R=H is prepared by reacting sodium molybdate with the sodium salt of the ligand[216] and the *cis*-(stilbenedithiolato) complexes with $R = C_6H_5$ are similarly prepared by reacting sodium molybdate with the reaction product of benzoin with phosphorus pentasulfide.[216] The bis complex, $Mo[S_2C_2(C_6H_5)_2]_2$, was isolated from the reaction of a dioxane solution of ammonium paramolybdate[217] or molybdenum pentachloride,[218] with the benzoin-P_2S_5 reaction mixture. Reaction of the benzoin–P_2S_5 reaction mixture with molybdenum pentachloride forms the binuclear sulfur-bridged complex, $Mo_2S_2[S_2C_2C_6H_5]_4$, which can also be obtained on reacting the dicarbonyl complex $(CO)_2Mo[S_2C_2(C_6H_5)_2]_2$ with sodium hydrosulfide.[219] The methyl-containing complex, $Mo[S_2C_2(CH_3)_2]_3$, is formed by reacting ammonium paramolybdate with the product of the acetoin–P_2S_5 reaction.[216] The complex containing CN as the substituent has been isolated only as an anion, $\{Mo[S_2C_2(CN)_2]_3\}^{2\ominus}$, by reaction of the sodium salt of the ligand with molybdenum pentachloride.[220, 221] Aromatic compounds with *o*-dithiol groups

[205] *A. V. Butcher, J. Chatt*, J. Chem. Soc. 2652 (1970).
[206] *G. M. French, J. H. Garside*, J. Chem. Soc. 2006 (1962).
[207] *E. Wendling*, Bull. Soc. Chim. France 437 (1965).
[208] *L. K. Atkinson, A. H. Mawby, D. C. Smith*, Chem. Commun. 1399 (1970).
[209] *H. A. O. Hill, M. M. Norget*, J. Chem. Soc. *A*, 1476 (1966).
[210] *R. N. Jowitt, P. C. H. Mitchell*, J. Chem. Soc. *A*, 2632 (1969).
[211] *A. Davidson, N. Edelstein, R. H. Holm, A. H. Maki*, J. Amer. Chem. Soc. *86*, 2799 (1964).
[212] *G. N. Schrauzer* in *R. L. Carlin*, Transition Metal Chemistry, Vol. IV, p. 299, Marcel Dekker, New York 1968.
[213] *R. B. King*, Inorg. Chem. *2*, 641 (1963).
[214] *A. Davidsen, R. H. Holm*, Inorg. Synth. *10*, 8 (1967).
[215] *A. Davidson, E. T. Shawl*, Inorg. Chem. *9*, 1822 (1970).
[216] *G. N. Schrauzer, V. P. Mayweg*, J. Amer. Chem. Soc. *88*, 3235 (1966).
[217] *J. A. McCleverty, J. Locke, B. Ratcliff, E. J. Wharton*, Inorg. Chim. Acta *3*, 283 (1969).
[218] *E. I. Steifel, R. Eisenberg, R. C. Rosenberg, H. B. Gray*, J. Amer. Chem. Soc. *88*, 2956 (1966).
[219] *G. N. Schrauzer, V. P. Mayweg, W. Henrich*, J. Amer. Chem. Soc. *88*, 5174 (1966).
[220] *J. A. McCleverty, J. Locke, E. J. Wharton*, J. Chem. Soc. *A*, 816 (1968).
[221] *E. I. Steifel et al.*, Inorg. Chem. *9*, 281 (1971).

form similar compounds and the tris complexes of *o*-benzenedithiol and 3,4-toluenedithiol were prepared by refluxing a carbon tetrachloride solution of molybdenum pentachloride with the ligand.[218,222]

15.6 Molybdenum(V) Compounds

15.6.1 Halides, Oxohalides, and Halide- and Pseudohalide Salts

Molybdenum pentafluoride is readily prepared by reducing molybdenum hexafluoride with an excess of phosphorus trifluoride at room temperature.[223] Molybdenum metal[224] or the hexacarbonyl can also be utilized as a reductant of the hexafluoride.[141] It can also be prepared by the thermal decomposition of the intermediate product formed on fluorinating the hexacarbonyl at $-70°$ and the pentafluoride is obtained as the volatile product.[141] The reaction of arsenic trifluoride[225] or liquid hydrogen fluoride[226] at 150° with molybdenum pentachloride yields the *chlorofluoride*, $MoCl_2F_3$. The chlorofluoride, $Mo_2Cl_3F_6$, is formed by the reaction of the pentachloride with the hexafluoride and by the reaction of the hexafluoride with phosphorus trichloride, arsenic trichloride, and other chlorides.[227] The bromofluoride, $MoBrF_4$, is isolated from the reaction of the hexafluoride with the tetrabromide.[224]

Molybdenum pentachloride is usually prepared by chlorination of the metal[228-231] at about 400°. The metal should be pretreated with hydrogen at elevated temperatures to prevent formation of oxochloride impurities. The chloride can also be prepared by chlorination of molybdic oxide with refluxing thionyl chloride[71,232] or with carbon tetrachloride under pressure.[233,234] The impure chloride can be freed of oxochloride impurities by fractional vacuum sublimation[229,235] or by treatment with refluxing thionyl chloride.[232]

Molybdenum oxotrifluoride has not been isolated. *Molybdenum oxotrichloride*, $MoOCl_3$, is readily prepared by the thermal decomposition of the oxotetrachloride in a nitrogen stream[236] at 120°. The pure product is also obtained by reducing the oxotetrachloride with refluxing chlorobenzene,[235,237] dry hydrogen,[178] or aluminum powder.[178] The reaction of molybdenum pentachloride with liquid sulfur dioxide also yields the pure product.[238] It can be prepared by heating a mixture of pentachloride with dimolybdenum pentoxide at 310° in a sealed ampoule[239] or with antimony trioxide.[177] The reaction of the pentachloride with molybdic oxide or molybdenum dioxodichloride is complicated by the formation of a by-product, *molybdenum oxotetrachloride*.[240]

Molybdenum dioxomonochloride, MoO_2Cl, can be prepared by vacuum drying of a solution of molybdenum pentachloride in ethanol.[241] It can also be isolated by reducing the dioxodichloride with aluminum powder[178] or stannous chloride.[242] Vacuum sublimation of the dioxodichloride from the reaction mixture formed by heating molybdic oxide with the oxotrichloride leaves the pure dioxomonochloride.[242]

---

[222] *E. I. Steifel, H. B. Gray*, J. Amer. Chem. Soc. *87*, 4012 (1965).
[223] *T. A. O'Donnell, D. F. Stewart*, J. Inorg. Nucl. Chem. *24*, 309 (1962).
[224] *M. Mercer, T. A. Ouelette, C. T. Ratcliff, D. W. A. Sharp*, J. Chem. Soc. *A*, 2532 (1969).
[225] *S. Skramonsky*, Z. Chem. *6*, 431 (1966).
[226] *A. A. Opalovskii, K. A. Khaldoyanidi*, Russian J. Inorg. Chem. *13*, 310 (1968).
[227] *T. A. O'Donnell, D. F. Stewart*, Inorg. Chem. *5*, 1434 (1966).
[228] *I. M. Pearson, C. S. Garner*, J. Phys. Chem. *65*, 690 (1961).
[229] *R. Colton, I. B. Tomkins*, Australian J. Chem. *17*, 496 (1964).
[230] *G. Brauer*, Handbook of Preparative Inorganic Chemistry, 2nd Ed., Vol. II, p. 1405, Academic Press, New York 1965. (English Translation.)
[231] *A. J. Leffler, R. Penque*, Inorg. Synth. *12*, 187 (1970).
[232] *H. J. Seifert, H. P. Quak*, Angew. Chem. *73*, 621 (1961).
[233] *K. Knox*, J. Amer. Chem. Soc. *79*, 3358 (1957); *K. Knox*, Inorg. Synth. *7*, 173 (1963).
[234] *S. A. Shchukarev, I. V. Vasil'sova, B. N. Sharupin*, Zh. Obshch. Khim. *26*, 2093 (1956).
[235] *M. L. Larson, F. W. Moore*, Inorg. Chem. *5*, 801 (1966).
[236] *R. Colton, I. B. Tomkins*, Australian J. Chem. *18*, 447 (1965).
[237] *M. L. Larson, F. W. Moore*, Inorg. Synth. *12*, 190 (1970).
[238] *D. A. Edwards*, J. Inorg. Nucl. Chem. *25*, 1198 (1963).
[239] *B. M. Ninsha, B. G. Korshunova*, Russian J. Inorg. Chem. *14*, 888 (1969).
[240] *I. A. Glukhov, S. S. Eliseev*, J. Inorg. Nucl. Chem. *7*, 40 (1962).
[241] *R. Colton, I. B. Tomkins*, Australian J. Chem. *21*, 1975 (1968).
[242] *S. S. Eliseev, I. A. Glukhov, N. V. Gaidenko*, Russian J. Inorg. Chem. *15*, 1158 (1970).

The *triochloride*, $MoSCl_3$, was obtained by the reaction of the pentachloride with antimony trisulfide.[243]

Molybdenum oxotribromide is prepared by reacting molybdenum dioxodibromide with phosphorus pentabromide in refluxing carbon tetrachloride.[238]

Salts of *hexafluoromolybdate(V)*, $MMoF_6$, can be prepared by reacting an alkali metal iodide with the hexafluoride in liquid sulfur dioxide.[244,245] They can also be prepared by fluorination of the carbonyl in the presence of the metal fluoride or iodide with iodine pentafluoride[246] or sulfur tetrafluoride.[247] When sulfur tetrafluoride is used, molybdenum dioxide or disulfide can be used as the starting material.[244,247] The nitrosyl salts are obtained by reacting the metal with nitrosyl fluoride in liquid hydrogen fluoride or the hexafluoride with nitric oxide.[248,249] Potassium *octafluoromolybdate(V)* is prepared by reacting potassium iodide with the carbonyl in iodine pentafluoride[246,250] or from the hexafluoride and potassium fluoride in liquid sulfur dioxide.[246]

Potassium *oxopentafluoromolybdate(V)*, K_2MoOF_5, has been prepared by fusing potassium hexafluoromolybdate(V) with potassium hydrogen fluoride in a carbon dioxide atmosphere.[245]

Quarternary ammonium salts of *hexachloromolybdate(V)* have been obtained by reaction of ammonium chloride with molybdenum pentachloride in methylene chloride in a sealed tube.[251,252]

Several salts of *oxopentachloromolybdate(V)* and *oxotetrachloromolybdate(V)* have been isolated. The oxopentachloromolybdate salts are prepared by saturating a hydrochloric acid solution of the metal chloride and molybdenum(V) with hydrogen chloride.[156] The molybdenum(V) may be generated electrolytically,[253,254] by solution of molybdenum pentachloride,[156] or by reduction of a molybdate solution with hydriodic acid or other reducing agents.[89,255,256] The free acid, H_2MoOCl_5, has reportedly been obtained by cooling a hydrochloric acid solution containing molybdenum(V).[257] Salts of cations derived from 2,2′-bipyridyl, *o*-phenanthroline, ethylenediamine, and 8-hydroxyquinoline have been isolated.[258,259] Salts containing the $[MoOCl_4]^{\ominus}$ group are obtained when sulfur dioxide is used as the solvent in place of hydrochloric acid[156] or with large cations such as tetrabutylammonium, when the normal procedure is used.[260] The interesting dimeric salt, $Cs_4[Mo_2O_3Cl_8]$, is formed by addition of cesium chloride to a partially neutralized solution of the pentachloride in water.[261] The dimeric dihydroxy salt, $[(CH_3)_4N]_2[Mo_2O_4Cl_4(OH)_2]$, is isolated by the addition of acetone to a molybdenum(V) solution in 4 M hydrochloric acid.[262] The analogous bromine-containing salts, $[MoOBr_5]^{2\ominus}$ and $[MoOBr_4]^{\ominus}$, are prepared like the chloro salts except that hydrobromic

---

[243] *D. Britnell, G. W. A. Fowles, R. Mandyczewsky*, Chem. Commun. 608 (1970).

[244] *R. D. Kemmitt, D. R. Russel, D. W. A. Sharp*, J. Chem. Soc. (London) 4408 (1963).

[245] *G. B. Hargreaves, R. D. Peacock*, J. Chem. Soc. (London) 4212 (1957).

[246] *G. B. Hargreaves, R. D. Peacock*, J. Chem. Soc. (London) 4390 (1958).

[247] *R. D. Kemmitt, D. W. A. Sharp*, J. Chem. Soc. (London) 2496 (1961).

[248] *J. R. Geichman, E. A. Smith, R. R. Ogle*, Inorg. Chem. *1*, 631 (1962);
J. R. Geichman, E. A. Smith, R. R. Ogle, Inorg. Chem. *2*, 1012 (1963).

[249] *F. Seel, W. Birnkraut, D. Werner*, Chem. Ber. *95*, 1264 (1962).

[250] *G. B. Hargreaves, R. D. Peacock*, J. Chem. Soc. (London) 2170 (1958).

[251] *B. J. Brisdon, R. S. Walton*, J. Inorg. Nucl. Chem. *27*, 1101 (1965).

[252] *B. J. Brisdon et al.*, J. Chem. Soc. *A*, 1825 (1967).

[253] *G. Brauer*, Handbook of Preparative Inorganic Chemistry, 2nd Edn., Vol. II, p. 1413, Academic Press, New York 1965. (English Translation.)

[254] *R. G. James, W. Wardlaw*, J. Chem. Soc. (London) 2726 (1928).

[255] *J. P. Simon, P. Souchay*, Bull. Soc. Chim. France 1402 (1956).

[256] *E. Wendling, R. Rohmer, R. Weiss*, C.R. Acad. Sci., Paris *256*, 117 (1963).

[257] *H. K. Saha, M. K. Haldar*, J. Indian Chem. Soc. *44*, 741 (1967).

[258] *H. K. Saha, M. K. Haldar*, J. Indian Chem. Soc. *44*, 231 (1967);
H. K. Saha, M. K. Haldar, Z. Anorg. Allgem. Chemie *377*, 221 (1970);
H. K. Saha, M. K. Haldar, Z. Anorg. Allgem. Chemie *380*, 97 (1971).

[259] *R. Heimberger, R. Rohmer*, Bull. Soc. Chim. France 2556 (1963).

[260] *H. L. Schaefer, K. Christ*, Z. Anorg. Allgem. Chemie *349*, 289 (1967).

[261] *R. Colton, C. C. Rose*, Australian J. Chem. *21*, 883 (1968).

[262] *R. M. Wing, K. P. Callahan*, Inorg. Chem. *8*, 871 (1969).

acid is used as the reaction media.[156, 263–265] The ammonium salt, $(NH_4)_2MoOBr_5$, has been prepared by heating a solution of ammonium paramolybdate in hydrobromic acid followed by evaporation to dryness.[266] Salts containing the $[MoOBr_4]^{\ominus}$ group are formed with large cations such as quinolinium and tetraphenylarsonium.[264, 266, 267]

The *molybdenum(V) octacyanide* salt, $K_3[Mo(CN)_8]$, is prepared by oxidizing the corresponding octacyanomolybdate(IV) salt with ceric ion or permanganate.[98, 168, 268] The tetrabutylammonium salt was also prepared by oxidation of the molybdenum(IV) salt.[269] The unstable acid, $H_3[Mo(CN)_8]$, was isolated by grinding the silver salt with dilute hydrochloric acid and evaporating the filtrate.[168] The tetrahydroxy complex, $K_3[Mo(OH)_4(CN)_4]$ was reportedly prepared by treating an ethanolic suspension of the molybdenum(V) hydroxide, $MoO(OH)_3$, with potassium cyanide.[173] The poorly characterized compound, $H[MoO_2(CN)_2(H_2O)]_2$, has been prepared by oxidation of potassium octacyanomolybdate(IV) with nitric acid.[168, 270]

The reaction of a solution of $MoOCl_5^{2\ominus}$ with thiocyanate ion produces thiocyanate complexes that can be precipitated by addition of large cations such as pyridinium. Thus, the earlier preparation[271] of $(C_5H_6N)_4[MoO_2(SCN)_3]_2$ has been successfully repeated.[89, 161] With trimethyl- or tetramethylammonium as the cation, salts of the type $A_4[Mo_2O_3(NCS)_8]$ can be precipitated with large cations.[271]

15.6.2 Chalcogenides

Dimolybdenum pentoxide, Mo_2O_5, has recently been isolated by heating the molybdenum(V) hydroxide, $MoO(OH)_3$, in a nitrogen stream[239] at 460°. *Molybdenum oxotrihydroxide*, $MoO(OH)_3$, is formed by adding ammonium hydroxide or carbonate to a molybdenum(V) solution.[89, 272]

Although the preparation of dimolybdenum pentasulphide, Mo_2S_5, has been reported in the earlier literature,[273] a pure product could not be obtained in attempts to prepare this compound.[119]

A molybdenum(V) selenide was reportedly prepared by treating a reduced ammonium molybdate solution with hydrogen selenide,[116] but no attempt has apparently been made to duplicate this preparation.

15.6.3 Coordination Compounds

Molybdenum pentachloride forms *complexes* of the type $MoCl_5 \cdot L$ on reaction with ligands such as phosphorus oxotrichloride,[274] phosphorus pentachloride,[275] trimethylamine,[276] triethylamine, and 2,4,6-trimethylpyridine.[277] The bis adduct of trichloroacetonitrile, $MoCl_5 \cdot 2Cl_3CCN$, can be prepared directly from the pentachloride and ligand or by the oxidation of molybdenum tetrachloride with excess ligand.[278]

Molybdenum oxotrichloride reacts directly to form *bis adducts* of the type $MoOCl_3 \cdot 2L$ with ligands such as triphenylphosphine,[279] alkyl nitriles,[279] pyridine,[235] dimethyl sulfoxide,[235] 4,4′-bipyridyl,[280] pyrazine,[280] and tertiary alkyl amines.[281] Monoadducts are formed with ligands such as 2,2′-bipyridyl,[279] 1,10-phenanthroline,[279] pyrazine,[280] and dinitriles.[280] Complexes with dimethyl sulfoxide, triphenylphosphine oxide, and triphenylarsine oxide were prepared by addition of the ligand to an

[263] *P. R. Gray*, J. Inorg. Nucl. Chem. *12*, 304 (1960).
[264] *H. K. Saha, A. K. Bannerjee*, J. Indian Chem. Soc. *45*, 660 (1968).
[265] *J. H. Garside*, J. Chem. Soc. (London) 6634 (1965).
[266] *J. Allen, H. M. Neumann*, Inorg. Chem. *3*, 1612 (1964).
[267] *J. G. Scane*, Acta Crystallogr. *23*, 85 (1967).
[268] *H. W. Willard, R. C. Thielke*, J. Amer. Chem. Soc. *57*, 2609 (1935).
[269] *B. J. Corden, J. A. Cunningham, R. Eisenberg*, Inorg. Chem. *9*, 356 (1970).
[270] *G. A. Barbieri*, Atti Accad. Naz. Lincei, Classe Sci., Fis., Mat. Natur. Rend. *3*, 375 (1931).
[271] *R. G. James, W. Wardlaw*, J. Chem. Soc. (London) 2731 (1928).
[272] *W. F. Jakob, E. Turkiewicz*, Rocziniki Chem. *11*, 569 (1931);
W. F. Jakob, E. Turkiewicz, Rocziniki Chem. *31*, 681 (1957).
[273] *F. Mawrow, M. Nokolow*, Z. Anorg. Allgem. Chemie *95*, 188 (1916).
[274] *V. Gutmann*, Z. Anorg. Allgem. Chemie *269*, 279 (1952).
[275] *W. L. Groenveld*, Rec. Trav. Chim. Pays-Bas *71*, 1152 (1952).
[276] *D. A. Edwards, G. W. A. Fowles*, J. Chem. Soc. (London) 24 (1961).
[277] *T. M. Brown, B. Ruble*, Inorg. Chem. *6*, 1335 (1967).
[278] *G. W. A. Fowles, D. A. Rice, N. Rolfe, R. A. Walton*, Chem. Commun. 459 (1970).
[279] *D. A. Edwards*, J. Inorg. Nucl. Chem. *27*, 303 (1963).
[280] *W. M. Carmichael, D. A. Edwards*, J. Inorg. Nucl. Chem. *32*, 1199 (1970).
[281] *A. A. Kuznetsova, L. F. Goryachova, Y. A. Buslaev*, Izv. Akad. Nauk SSSR, Ser. Khim. 509 (1970).

ethanol solution of the pentachloride,[282] and the monoadduct of octamethyldiphosphoramide was similarly prepared.[283] The monoadducts with 2,2′-bipyridyl and *o*-phenanthroline are formed by refluxing the corresponding organic ammonium oxopentachloromolybdate(V) salt in ethanol or butyronitrile.[284] The bipyridyl complex is also obtained by addition of the ligand to a moist carbon tetrachloride solution of the pentachloride.[285] The reaction of a carbon tetrachloride solution of molybdenum oxotetrachloride with pyridine yields the bis(pyridine) adduct; with benzophenone, a monoadduct is formed, and with acetylacetone the neutral acetylacetone adduct, $MoOCl_3 \cdot C_5H_8O_2$, is formed.[235] The bis(triphenylphosphine) adduct is formed in high yield by boiling a solution of the ligand and ammonium molybdate in ethanol–hydrochloric acid.[286] The bis(triphenylphosphine oxide) adduct is prepared by heating excess ligand with molybdenum trichloride or potassium hexachloromolybdate(III).[287]

The direct reaction of molybdenum oxotribromide with 2,2′-bipyridyl or *o*-phenanthroline forms monoadducts and with acetonitrile or triphenylarsine oxide bis adducts are formed.[287] The bis adduct of triphenylphosphine oxide is formed by heating an excess of the ligand with molybdenum tribromide.[287] The monobipyridyl adduct is formed when an ethanol–methylene chloride solution of $Mo(CO)_4 \cdot$bipy is treated with bromine.[201] A similar treatment with bromine of solutions containing the hexacarbonyl and the ligand yields the bis adduct of triphenylphosphine oxide and the monoadduct of 1,2-bis(diphenylphosphine)ethane (diphos).[142]

The *pentavalent acetylacetonate*, $MoO(OH)(C_5H_7O_2)_2 \cdot 4H_2O$, is formed when acetylacetone is heated with diammonium oxopentachloromolybdate(V).[288, 289] When the reaction is carried out in aqueous solution the dimeric complex, $Mo_2O_3(C_5H_7O_2)_4$ is obtained.[290] This complex is formed by the careful oxidation in air of molybdenum tris(acetylacetonate)[134] or by heating molybdenyl bis(acetylacetonate) with excess acetylacetone[291] at 210°. The dioxomonoacetylacetonate, $MoO_2(C_5H_7O_2)$, is reportedly formed by oxidation in air of the molybdenum(III) complex, $MoO(C_5H_7O_2)$.[133] The reaction of molybdenum oxotrichloride with acetylacetone in refluxing benzene yields a complex formulated as $MoOCl(C_5H_7O_2)_2$.[235] The reaction of acetonitrile solutions of tetraethylammonium salts of $[MoOCl_5]^{2\ominus}$ or $[MoOBr_4(H_2O)]^{\ominus}$ with various β-diketones yields salts[292] of the type $[MoOX_3(\beta\text{-diketonate})]^{\ominus}$.

Salts containing the grouping $[Mo_2O_4(C_2O_4)_4(H_2O)_2]^{2\ominus}$ are formed by precipitation from solutions containing oxalic acid and molybdenum(V) generated either from oxopentachloromolybdate(V)[293–296] or by hydrazine reduction of molybdates.[297] Mixed complexes were formed by reacting salts of the oxalate with ligands such as pyridine.[298]

Molybdenum pentachloride reacts with monocarboxylic acids in refluxing carbon tetrachloride to form compounds of the type $(RCO_2)_2MoCl_3$.[299] Ethylenediaminetetraacetic acid (EDTA) reacts with molybdenum(V) solution formed by electrolytic reduction to form the acid, $H_2[Mo_2O_4(EDTA)]$,[300] and several salts are formed by reacting the barium salt with metal sulfates.[301] Reaction of a molybdenum(V) solution with 1,2-dipropylenediamine-

[282] *S. M. Horner, S. Y. Tyree*, Inorg. Chem. *1*, 122 (1962).
[283] *M. D. Joestin*, Inorg. Chem. *6*, 1598 (1967).
[284] *H. K. Saha, M. C. Haldar*, J. Indian Chem. Soc. *45*, 88 (1968).
[285] *P. C. H. Mitchell*, J. Inorg. Nucl. Chem. *25*, 963 (1963).
[286] *H. Gehrke jr., L. Bowden*, Inorg. Nucl. Chem. Lett. *5*, 151 (1969).
[287] *W. M. Carmichael, D. A. Edwards*, J. Inorg. Nucl. Chem. *30*, 2641 (1968).
[288] *A. Rosenheim, C. Nernst*, Z. Anorg. Allgem. Chemie *209*, 216 (1933).
[289] *P. C. H. Mitchell*, J. Inorg. Nucl. Chem. *25*, 963 (1963).
[290] *M. L. Larson, F. W. Moore*, Inorg. Chem. *2*, 881 (1963).
[291] *H. Gehrke, J. Veal*, Inorg. Chem. Acta *3*, 623 (1969).
[292] *H. E. Pence, J. Selbin*, Inorg. Chem. *8*, 353 (1969).
[293] *R. G. James, W. Wardlaw*, J. Chem. Soc. (London) 2145 (1927).
[294] *H. M. Spittle, W. Wardlaw*, J. Chem. Soc. (London) 2742 (1928).
[295] *P. C. H. Mitchell*, J. Inorg. Nucl. Chem. *26*, 1967 (1964).
[296] *F. A. Cotton, S. M. Morehouse*, Inorg. Chem. *4*, 1377 (1965).
[297] *G. Caviechi*, Gazz. Chim. Ital. *83*, 402 (1963).
[298] *P. C. H. Mitchell*, J. Chem. Soc. *A*, 146 (1969).
[299] *M. L. Larson*, J. Amer. Chem. Soc. *82*, 1223 (1960).
[300] *L. Y. Haynes, D. T. Sawyer*, Inorg. Chem. *6*, 2146 (1969).
[301] *D. Hruskova, J. Podlahova, J. Podlaha*, Collect. Czech. Chem. Commun. *35*, 2738 (1970).

tetraacetic acid yields a similar complex.[300,302] Analogous dimeric complexes are formed with histidine,[303] glutathione,[304] and cysteine.[305,306]
A monohydroxy 8-hydroxyquinolate(oxine), MoO(OH)(oxine)$_2$, is formed from a molybdenum(V) solution and the ligand.[307] Reaction of excess ligand with a solution of the molybdenum(V) oxalate forms the purple dimeric complex, Mo_2O_3(oxine)$_4$.[298,308] This solid is also precipitated at pH 4·5 from molybdenum(V) solutions.[309] With ammonium oxopentachloromolybdate(V) in an ethanol solution the ligand yields MoOCl(oxine)$_2$.[307]
The reaction of tetraphenylporphine ($TPPH_2$) with the hexacarbonyl in decalin yields MoO(OH)(TPP), which can be converted to MoOCl(TPP) on treatment with concentrated hydrochloric acid.[310]
Molybdenum pentachloride reacts with primary amines to form complexes of the type $MoCl_2(NHR)_3$ and with secondary amines to form complexes of the type $MoCl_3(NR_2)_2$.[311] With 2,2′-bipyridyl as donor an ethanolic solution of the pentachloride yields a precipitate of $Mo_2O_3Cl_4$(bipy)$_2$.[284]
The phenoxide, $MoCl_2(C_6H_5O)_3$, is formed by reaction of the pentachloride with phenol in carbon disulfide[307,312] and analogous compounds are obtained with *m*- and *p*-cresol.[312]
Reaction of methanol with the pentachloride at low temperatures yields $MoCl_3(OCH_3)_2 \cdot 3CH_3OH$[313] and vacuum evaporation of alcoholic solutions of the pentachloride forms alkoxides of the type $MoCl_3(OR)_2$.[314] Addition of pyridine and other amines to such solutions in alcohol yields salts of the type pyH[$MoCl_4(OCH_3)_2$],[313,315] and compounds of the type [$(C_2H_5)_4N$][$MoCl_4(OR)_2$] are formed from hydrogen chloride-saturated alcohol solutions of the pentachloride.[316]
Molybdenum(V) solutions react with ligands such as alkyl xanthates,[317–319] dialkyldithiocarbamates,[318–321], and dialkyldithiophosphates[210,319,322] to form dimeric complexes of the type $Mo_2O_3L_4$. The ethylxanthate complex reacts with hydrogen sulfide to form polymeric[$MoS_2(S_2COC_2H_5)$]$_n$.[319]

15.7 Nonstoichiometric Oxides between Valency (V) and (VI)

The heating of molybdic oxide with the metal or the dioxide results in formation of a series of nonstoichiometric oxides with a formal valency between (V) and (VI).[323] These oxides are prepared by heating the calculated amount of molybdic oxide with the metal or dioxide in sealed, evacuated quartz tubes which may be lined with gold or platinum foil, and the conditions for preparation of these oxides are summarized in the table on p. 197.

Some of these oxides, particularly the metastable θ- and *K*-oxides, are difficult to prepare as pure phases and require long heating at constant temperature.

Structurally related to the above oxides are the nonstoichiometric hydrous oxides, the so-

---

[302] *R. M. Wing, K. P. Callahan*, Inorg. Chem. *8*, 2303 (1969).
[303] *L. R. Melby*, Inorg. Chem. *8*, 1539 (1969).
[304] *T. J. Huang, G. P. Haight*, J. Amer. Chem. Soc. *93*, 611 (1971).
[305] *T. J. Huang, G. P. Haight*, J. Amer. Chem. Soc. *92*, 2336 (1970).
[306] *A. Kay, P. C. H. Mitchell*, Nature *219*, 167 (1968).
[307] *P. C. H. Mitchell, R. J. P. Williams*, J. Chem. Soc. (London) 4570 (1962).
[308] *H. Berge, H. L. Kreutzmann*, Z. Anal. Chem. *210*, 81 (1965).
[309] *C. B. Riola, F. Guerrieri*, Ann. Chim. (Rome) *57* (7), 873 (1967).
[310] *T. S. Srivastava, E. E. Fleischer*, J. Amer. Chem. Soc. *92*, 5518 (1970).
[311] *D. A. Edwards, G. W. A. Fowles*, J. Chem. Soc. *24* (1967).
[312] *A. Rosenheim, C. Nernst*, Z. Anorg. Allgem. Chemie *209*, 214 (1933).
[313] *H. Funk, F. Schmid, H. Scholz*, Z. Anorg. Allgem. Chemie *310*, 86 (1961).
[314] *D. C. Bradley, R. K. Multani, W. Wardlaw*, J. Chem. Soc. (London) 4647 (1958).
[315] *D. A. McClung, L. R. Dalton, C. H. Brubaker*, Inorg. Chem. *5*, 1985 (1966).
[316] *D. P. Rillema, C. H. Brubaker*, Inorg. Chem. *8*, 1645 (1969).
[317] *L. Malatesta*, Gazz. Chim. Ital. *69*, 408 (1939).
[318] *F. W. Moore, M. L. Larson*, *6*, 998 (1967).
[319] *R. N. Jowitt, P. C. H. Mitchell*, Chem. Commun. 605 (1966);
R. N. Jowitt, P. C. H. Mitchell, J. Chem. Soc. *A*, 1702 (1970).
[320] *L. Malatesta*, Gazz. Chim. Ital. *69*, 752 (1939).
[321] *F. W. Moore, R. E. Rice*, Inorg. Chem. *7*, 2510 (1968).
[322] *G. Spengler, A. Weber*, Chem. Ber. *92*, 2163 (1959).
[323] *L. Kihlborg*, in *R. Ward*, Nonstoichiometric Compounds, p. 37, Advances in Chemistry No. 39, American Chemical Society, Washington 1963.

Compound	Ref.	Crystal structure	Temperature of formation
Mo_4O_{11}, η-oxide	323–325	Monoclinic	500° to 615°
Mo_4O_{11}, γ-oxide	323–329	Orthorhombic	615° to 800°
Mo_5O_{14}, θ-oxide	324, 330	Tetragonal	500° to 540°
Mo_8O_{23}, β-oxide	324, 326–328	Monoclinic	650° to 780°
Mo_9O_{26}, β'-oxide	324–328	Monoclinic	760° to 780°
$Mo_{18}O_{52}$, ζ-oxide	324, 325, 327	Triclinic	600° to 750°
$Mo_{17}O_{47}$, κ-oxide	324, 331	Orthorhombic	500° to 560°

called "molybdenum blues". The compound $Mo_4O_{10}(OH)_2$ is obtained by oxidation of lower valent molybdenum blues or by reduction of molybdic oxide with the metal–water, stannous chloride, or zinc–dilute hydrochloric acid.[332–334] Reduction of molybdic oxide with zinc and concentrated hydrochloric acid yields the compound $Mo_2O_4(OH)_2$, and the reduction of molybdic oxide dihydrate with the metal and water at 100° yields $Mo_8O_{15}(OH)_{16}$.[332,334] The compound $Mo_2O_4(OH)_2$ can also be prepared by the oxidation in air of $Mo_5O_8(OH)_8$ which itself is formed by boiling $Mo_5O_5(OH)_{10}$ with potassium hydroxide.[189,335]

[324] *L. Kihlborg*, Acta Chem. Scand. *13*, 954 (1954).
[325] *E. Rode*, *G. Lysanova*, Proc. Acad. Sci. SSSR *145*, 614 (1962).
[326] *O. Glemser*, *G. Lutz*, Z. Anorg. Allgem. Chemie *263*, 2 (1950).
[327] *B. Phillips*, *L. Chang*, Trans. Met. Soc. AIME *233*, 1433 (1965).
[328] *J. Bousquet*, *A. Guillon*, Bull. Soc. Chim. France 2250 (1966).
[329] *G. Brauer*, Handbook of Preparative Inorganic Chemistry, 2nd Ed., Vol. II, p. 1410, Academic Press, New York, 1965. (English Translation.)
[330] *L. Kihlborg*, Ark. Kemi *21*, 427 (1963).
[331] *L. Kihlborg*, Acta Chem. Scand. *14*, 1612 (1960).
[332] *O. Glemser*, *G. Lutz*, Z. Anorg. Allgem. Chemie *264*, 173 (1951).
[333] *G. Brauer*, Handbook of Preparative Inorganic Chemistry, 2nd Ed., Vol. II, p. 1411, Academic Press, New York 1965. (English Translation.)
[334] *E. Ya. Rode*, *G. V. Lysanova*, Proc. Acad. Sci. SSSR, Chem. Sect. *145*, 629 (1962).
[335] *O. Glemser*, *G. Lutz*, *G. Meyer*, Z. Anorg. Allgem. Chemie *285*, 178 (1956).

15.8 Molybdenum(VI) Compounds

15.8.1 Halides, Oxohalides, and Halide Salts

Molybdenum hexafluoride is prepared by passing fluorine over the metal[71,336,337] at 150°–300° or by reaction with the carbonyl[141] at 50°. The metal is also completely fluorinated by chlorine trifluoride[338] and bromine trifluoride.[71] It can also be prepared by fluorination of molybdic oxide with sulfur tetrafluoride,[140] bromine trifluoride, or iodine pentafluoride.[339] The reaction of the dibromide with hydrogen fluoride above 550° also yields the hexafluoride.[71,340] as does the reaction of the disulfide with nitrogen trifluoride.[341]

It has been reported that molybdenum hexachloride can be isolated by fractionation of the reaction mixture from molybdic oxide and thionyl chloride,[342] but this preparation could not be confirmed.[239,343]

Molybdenum oxotetrafluoride can be obtained by passing a mixture of oxygen and fluorine over the metal or by treating the corresponding oxochloride with hydrogen fluoride.[344] The dioxodifluoride is prepared by treating the corresponding chloride with hydrogen fluoride or fluorinating molybdic oxide with selenium tetrafluoride or iodine pentafluoride.[345] It is also reportedly formed by the controlled hydrolysis of the hexafluoride.[346]

Molybdenum oxotetrachloride is readily prepared by oxidizing the molten pentachloride with dry oxygen[347] or by the reaction of thionyl

[336] *T. A. O. Donnell*, *D. F. Stewart*, Inorg. Chem. *5*, 1434 (1966).
[337] *E. J. Barber*, *G. H. Cady*, J. Phys. Chem. *60*, 505 (1956).
[338] *N. C. Nikolaev*, *Y. A. Buslaev*, *A. A. Opalovskii*, Zh. Neorg. Khim. *3*, 173 (1958).
[339] *N. C. Nikolaev*, *V. F. Sukhoverkhov*, Dokl. Akad. Nauk SSSR *136*, 621 (1961).
[340] *H. J. Emeléus*, *V. Gutmann*, J. Chem. Soc. (London) 2115 (1950).
[341] *O. Glemser*, *J. Wegner*, *R. Mews*, Chem. Ber. *100*, 2474 (1967).
[342] *M. Mercer*, Chem. Commun. 119 (1966).
[343] *J. H. Canterford*, *R. Cotton*, Halides of the Second and Third Row Transition Metals, p. 256, John Wiley & Sons, New York 1968.
[344] *G. H. Cady*, *G. B. Hargreaves*, J. Chem. Soc. (London) 1568 (1961).
[345] *N. Bartlett*, *P. L. Robinson*, J. Chem. Soc. (London) 3549 (1961).
[346] *N. S. Nikolaev*, *S. S. Vlasov*, *Yu. A. Buslaev*, *A. A. Opalovskii*, Izv. Sibirsk. Otd. Akad. Nauk SSSR (10) 47–56 (1960); C.A. *55*, 12014 (1961).
[347] *A. K. Mallock*, Inorg. Synth. *10*, 54 (1967).

chloride with molybdic oxide[229, 348] or molybdenum dioxodichloride.[236] It can also be prepared by the reaction of sulfuryl chloride with the metal in a sealed tube[349] or by heating a mixture of the pentachloride with the dioxodichloride, but by-product oxotrichloride must be separated from the reaction mixture.[240] The mixed oxohalide, $MoOCl_2Br_2$, is reportedly formed by heating molybdic oxide with bromochlorodifluoromethane.[350]

The best method for preparing the *dioxodichloride* is the chlorination of the dioxide with chlorine.[351–354] It has also been isolated by passing a mixture of chlorine and oxygen over the metal.[236, 355] The hydrate, $MoO_2Cl_2 \cdot H_2O$, is prepared by passage of hydrogen chloride over hot molybdic oxide.[353, 356] The thiochloride, MoS_2Cl_2, is formed by reacting sulfur dichloride with molybdenum metal[357] or by passing hydrogen sulfide into a benzene suspension of molybdenum oxotetrachloride.[358] *Molybdenum dioxodibromide* is prepared by passing a mixture of bromine and oxygen over the metal[236] at 300° or by heating the dioxide with bromine.[354]

The preparation of the *oxotetrabromide* and -iodide has been briefly described, but little is known about these compounds.[359]

The *nitrodichloride*, $MoNCl_3$, is prepared by decomposing the intermediate azide complex formed by the reaction of the pentachloride with chlorine azide.[360]

Salts of *octafluoromolybdate(VI)* can be prepared by reacting an alkali metal fluoride with the hexafluoride in the presence of trace moisture,[361] bromine trifluoride,[362] or iodine pentafluoride.[246, 363] The sodium salt can also be prepared by passing molybdenum hexafluoride over anhydrous sodium fluoride.[364] The corresponding *heptafluoromolybdate(VI)* salts can also be prepared from the alkali metal fluoride and molybdenum hexafluoride in iodine pentafluoride,[246, 363] and the sodium salt has been prepared by passing the hexafluoride vapor over activated sodium fluoride.[364] The *nitrosyl-* and *nitrylheptafluoride* complexes are isolated by reacting the hexafluoride with the appropriate chloride.[248]

Rubidium and cesium *oxopentafluoromolybdate(VI)* are prepared by reacting the moist alkali metal fluoride and the hexafluoride in iodine pentafluoride, arsenic trifluoride, or sulfur dioxide as solvents.[246, 363] Salts containing the $[MoO_2F_4]^{2\ominus}$ group are prepared by fusion of the alkali metal fluoride with the corresponding molybdate or by action of hydrofluoric acid on molybdic oxide[365–367] in the presence of an alkali metal fluoride. Salts containing the $[MoO_4F]^{3\ominus}$, $[MoO_3F_3]^{3\ominus}$, and $[MoO_3F]^{\ominus}$ groups can also be isolated from the reaction of an alkali metal fluoride and hydrofluoric acid with molybdic oxide.[363–370] Addition of hydrogen peroxide to such systems allows the isolation of *peroxyfluoride salts*[370] of the type $[MoO(O_2)F_4]^{2\ominus}$ and $[MoO(O_2)_2F_2]$. Salts containing the $[MoO_2Cl_4]^{2\ominus}$ group are prepared by saturating a hydrochloric acid

---

[348] *R. Cotton, I. B. Tomkins, P. W. Wilson*, Australian J. Chem. *17*, 496 (1964).
[349] *D. A. Edwards, A. A. Woolf*, J. Chem. Soc. *A*, 91 (1966).
[350] *M. Chaigneau, M. Chastagnier*, C.R. Acad. Sci., Paris *271*, 1249 (1970).
[351] *R. L. Graham, L. G. Hepler*, J. Phys. Chem. *63*, 723 (1959).
[352] *F. Zado*, J. Inorg. Nucl. Chem. *25*, 1115 (1963).
[353] *H. M. Neumann, N. C. Cook*, J. Amer. Chem. Soc. *79*, 3026 (1957).
[354] *H. Oppermann*, Z. Anorg. Allgem. Chemie *379*, 262 (1970).
[355] *I. R. Beattie, K. M. S. Livingston, D. J. Reynolds, G. A. Ozin*, J. Chem. Soc. *A*, 1210 (1970).
[356] *H. L. Kraus, W. Huber*, Chem. Ber. *94*, 2864 (1961).
[357] *J. P. Rannou, M. Sergent*, C.R. Acad. Sci., Paris *265C*, 734 (1967).
[358] *K. M. Sharma, S. K. Anand, P. K. Multani, B. D. Jain*, Chem. & Ind. (London) 1556 (1969).
[359] *I. N. Marov, Yu. N. Dubrov, V. K. Belyaeva, A. N. Ermakov*, Dokl. Akad. Nauk SSSR *171*, 2, 385–8 (1966); C.A. *66*, 70570 (1967).
[360] *K. Dehnicke, J. Strahle*, Z. Anorg. Allgem. Chemie *339*, 171 (1965).

---

[361] *H. C. Clark, H. J. Emeléus*, J. Chem. Soc. (London) 4778 (1957).
[362] *B. Cox, D. W. A. Sharp, A. G. Sharpe*, J. Chem. Soc. 1242 (1956).
[363] *G. B. Hargreaves, R. D. Peacock*, J. Chem. Soc. (London) 2170 (1958).
[364] *S. Katz*, Inorg. Chem. *3*, 1598 (1964); *S. Katz*, Inorg. Chem. *5*, 666 (1966).
[365] *O. Schmitz-Dumont, P. Opgenhoff*, Z. Anorg. Allgem. Chemie *275*, 21 (1954).
[366] *Y. A. Busloev, R. L. Davidovich*, Russian J. Inorg. Chem. *10*, 1014 (1965).
[367] *R. Weiss, D. Grandjean, B. Metz*, C.R. Acad. Sci., Paris *260*, 3969 (1965).
[368] *O. Schmitz-Dumont, P. Opgenhoff*, Z. Anorg. Allgem. Chemie *267*, 277 and *268*, 57 (1952).
[369] *A. A. Opalovskii, S. S. Batsanov, Z. M. Kuznetsova, M. N. Nesterenko*, Izv. Sibirsk. Otd. Akad. Nauk SSSR, Ser. Khim. Nauk *1*, 15–9 (1968); C.A. *79*, 70467 (1968).
[370] *W. P. Griffith*, J. Chem. Soc. 5248 (1964).

solution of the molybdate with hydrogen chloride.[353, 371, 372] Like the corresponding fluoride system, addition of hydrogen peroxide to the system yields peroxy salts[371, 373] of the type $[MoO(O_2)Cl_4]^{2\ominus}$.

15.8.2 Chalcogenides

Molybdic oxide is readily available in high purity and easily purified by sublimation in an oxygen stream.[326, 374] The *dihydrate* of molybdic oxide, $MoO_3 \cdot 2H_2O$, is obtained as a yellow, crystalline deposit when solutions of ammonium paramolybdate or sodium molybdate in nitric acid are allowed to stand.[375–377] Despite claims of high yields in short periods of time, it was found that a 40% yield of pure material was the best that could be obtained under a variety of conditions and that a solution of sodium molybdate in 2·5 N nitric acid left to stand 1 month gave the best results.[378] The yellow *monohydrate* is obtained by heating the dihydrate[376, 379] at 100° or by dehydration under vacuum at room temperature.[380] The white monohydrate is prepared by heating the dihydrate with dilute acid[376, 379] or water[381] at about 50°. It can also be prepared by heating a solution of sodium molybdate in hydrochloric[382] or nitric acid.[383]

Molybdenum trisulfide has reportedly been obtained in pure form by hydrolysis of a thiomolybdate solution with ethyl acetate or by heating ammonium tetrathiomolybdate in a hydrogen sulfide stream[384] at 200°. It has been reported that attempts to precipitate the trisulphide from acidified tetrathiomolybdate solutions always results in hydrated materials that from analysis are high in sulfur.[385]

Thiomolybdates are formed by passage of hydrogen sulfide through alkaline molybdate solutions. The treatment of a cold, ammoniacal solution of ammonium paramolybdate with hydrogen sulfide yields ammonium dithiomolybdate, $(NH_4)_2Mo_2S_2$,[386–388] and other salts are similarly prepared.[386] Addition of cesium or quaternary ammonium salts to cold solutions of the *dithiomolybdate* yields[389] salts of *trithiomolybdate*, $[MoOS_3]^{2\ominus}$. Tetrathiomolybdates, $[MoS_4]^{2\ominus}$, are formed by passage of hydrogen sulfide through a molybdate solution and salts of ammonium ion and potassium are formed in this manner.[386–388, 390–391] Reaction of a soluble *tetramolybdate* with various metal ions allows isolation of salts of lead,[392] cobalt, nickel, copper, and zinc,[393] which analyze as tetrathiomolybdates, but which have been shown to be mixtures of the sulfides. A cuprous ammonium salt, $CuNH_4MoS_4$, is prepared by passing hydrogen sulfide into an ammoniacal ammonium molybdate solution containing cupric sulfate.[394]

The selenium analogues of the above salts can be prepared by treating molybdate solutions with hydrogen selenide and the triseleno, $[MoOSe_3]^{2\ominus}$, and tetraseleno salts, $[MoSe_4]^{2\ominus}$; the mixed

---

[371] *W. P. Griffith, T. D. Wickins*, J. Chem. Soc. *A*, 397 (1968).
[372] *E. Wendling, R. Rohmer*, Bull. Soc. Chim. France 8 (1967).
[373] *E. Wendling, R. Rohmer, R. Weiss*, Rev. Chem. Miner. *1*, 225 (1964).
[374] *L. Kihlborg, A. Magneli*, Acta Chem. Scand. *9*, 471 (1955).
[375] *G. Brauer*, Handbook of Preparative Inorganic Chemistry, 2nd Edn., Vol. II, p. 1412, Academic Press, New York 1965. (English Translation.)
[376] *M. L. Freedman*, J. Amer. Chem. Soc. *81*, 3834 (1959);
M. L. Freedman, J. Chem. Eng. Data *8*, 113 (1963).
[377] *M. Murgier, Y. Doucet*, C.R. Acad. Sci., Paris *208*, 1585 (1939).
[378] *F. W. Moore*, Unpubl. Climax Molybdenum Co., Research Laboratory, Ann Arbor, Mich.
[379] *L. M. Ferris*, J. Chem. Eng. Data 6, 600 (1961).
[380] *S. Maricic, J. A. S. Smith*, J. Chem. Soc. 886 (1958).
[381] *A. Rosenheim, I. Davidson*, Z. Anorg. Allgem. Chemie *37*, 316 (1903).
[382] *P. Cannon*, J. Inorg. Nucl. Chem. *11*, 124 (1959).
[383] *V. Auger*, C.R. Acad. Sci., Paris *206*, 913 (1938);
V. Auger, C.R. Acad. Sci., Paris *207*, 164 (1938).
[384] *E. Hayek, U. Pallasser*, Monatsh. Chem. *99*, 2126 (1968).
[385] *E. Ya. Rode, B. A. Lebedev*, Russian J. Inorg. Chem. 6, 608 (1961).
[386] *A. Muller, E. Diemann, E. J. Baran*, Z. Anorg. Allgem. Chemie *375*, 87 (1970).
[387] *F. W. Moore, M. L. Larson*, Inorg. Chem. *6*, 998 (1967).
[388] *J. Bernard, G. Triodot*, Bull. Soc. Chim. France 810 (1961).
[389] *M. J. F. Leroy, F. Kaufmann, R. Charlionet, R. Rohmer*, C.R. Acad. Sci., Paris *263C*, 60 (1966).
[390] *A. Muller, E. Diemann, U. Heidborn*, Z. Anorg. Allgem. Chemie *371*, 136 (1969);
A. Muller, E. Diemann, U. Heidborn, Z. Anorg. Allgem. Chemie *376*, 120 (1970).
[391] *E. Diemann, A. Muller*, Spectrochim. Acta *26A*, 215 (1970).
[392] *R. S. Saxena, M. C. Jain*, Indian J. Chem. *6* (4), 224 (1968).
[393] *G. M. Clark, W. P. Doyle*, J. Inorg. Nucl. Chem. *28*, 381 (1966).
[394] *W. P. Binnie, M. J. Redman, W. J. Mallio*, Inorg. Chem. *9*, 1449 (1970).

salts, Cs_2MoOS_2Se and $Cs_2MoOSSe_2$, have been isolated.[395]

15.8.3 Coordination Compounds

Adducts of *molybdenum dioxodifluoroide* with dimethyl sulfoxide, 2,2′-dipyridyl, *o*-phenanthroline, and triphenylphosphine oxide have been prepared by treating a hydrofluoric acid solution of molybdic oxide with the ligand followed by evaporation.[396]

Several adducts of *molybdenum dioxodichloride* have been prepared by direct reaction of the oxochloride with the ligand. Bis adducts have been isolated with ligands such as pyridine,[396] dialkyl amides,[253, 397, 398] arsine oxides,[282, 399–400] phosphine oxides,[282, 400] amine oxides,[282, 399] dimethyl sulfoxide,[282] acid anhydrides,[356] nitriles,[356, 401] and trimethylamine.[401] Bidentate ligands like bipyridyl, *o*-phenanthroline, cyclic ethers,[356, 401] and diketones form monoadducts.[356] The acid anhydride complexes can be prepared by the reaction of the acid chloride with lead molybdate. The reaction of a hydrochloric acid solution of molybdic oxide with excess ligand reportedly forms dioxodichloride complexes with alkyl amides, dimethyl sulfoxide, and phosphine oxides,[396] and acetic acid solutions of the pentachloride form quinone complexes.[402]

Adducts of the *dioxodibromide* have been prepared from the ligand and the oxobromide with arsine oxides, phosphine oxides,[400] and aliphatic amines.[403] Quinone complexes were isolated by reacting the ligand with a methanolic solution of the metal in bromine,[402] and phosphine oxide[142] and bipyridyl[201] complexes were obtained by oxidation of the ligand carbonyl complexes in the presence of bromine.

Molybdenum *nitridotrichloride* forms tris(pyridine)[360] and mono(triphenylphosphine oxide)[404] complexes on reaction with the ligand.

Molybdic oxide forms a monopyridine adduct when heated with pyridine at 175° for 4 days[397] and a monoadduct with diethylenetriamine on refluxing the ligand and oxide with water.[405]

Molybdenyl bis(acetylacetonate), $MoO_2(C_5H_7O_2)_2$, is prepared in high yield by acidification with dilute nitric acid of an ammonium paramolybdate–acetylacetone solution[291, 406] or by addition of water to a solution of the dioxodichloride in acetylacetone.[235] It is less efficiently prepared by refluxing molybdic oxide with the ligand.[133, 407] With other β-diketones low yields of the molybdenyl complex are obtained[133, 320] by this method. The corresponding 8-hydroxyquinolinate, $MoO_2(C_9H_6NO)_2$, is prepared by acidification of solutions containing the ligand and molybdate ion,[408, 409] and the thio analog is similarly prepared.[410] The acidification technique can also be used to prepare molybdenyl complexes of α-benzoinoxime,[411] aromatic hydroxamic acids,[412–413] and dithiocarbamates.[318–321] The polymeric molybdenylbis(diphenylphosphinate) can be prepared by heating diphenylphosphinic acid with molybdenylbis(acetylacetonate).[414]

The *oxalate complex*, $H_2[MoO_3(C_2O_4)]\cdot 2H_2O$ is formed when molybdic oxide is heated with an oxalic acid solution,[415, 416] and several salts

---

395 *A. Muller, E. Diemann*, Z. Anorg. Allgem. Chemie, *373*, 57 (1970); Chem. Ber. *102*, 945 3277 (1969).

396 *R. Kergoat, J. E. Guerchais*, Bull. Soc. Chim. France 2932 (1970).

397 *J. Bernard, M. Camelot*, C.R. Acad. Sci., Paris 2630 (1966).

398 *C. Ringel, H. A. Lehman*, Z. Anorg. Allgem. Chemie *353*, 158 (1967).

399 *F. Choplin, G. Kaufman, R. Rohmer*, C.R. Acad. Sci., Paris *2690*, 333 (1969).

400 *A. Bartecki, D. Dembicka*, Roczniki Chem. *39*, 1783 (1965).

401 *W. M. Carmichael, D. A. Edwards, G. W. A. Fowles, P. R. Marshall*, Inorg. Chem. Acta *1*, 93 (1967).

402 *P. J. Growly, H. M. Haendler*, Inorg. Chem. Acta *1*, 904 (1962).

403 *K. P. Srivastava*, J. Inst. Chem. Calcutta *1968* (41, no. 5), 204–9 (1968); C.A. *72*, 38384 (1970).

404 *W. Kolitsch, K. Dehnicke*, Z. Naturforsch. *256*, 1080 (1970).

405 *W. F. Marzluff*, Inorg. Chem. *3*, 397 (1964).

406 *M. M. Jones*, J. Amer. Chem. Soc. *81*, 3188 (1959).

407 *W. C. Fernelius, K. Terada, B. E. Bryant*, Inorg. Synth. *6*, 147 (1960).

408 *A. I. Vogel*, A. Textbook of Quantitative Inorganic Analysis p. 508, 540 3rd Ed., John Wiley & Sons, New York 1961.

409 *H. M. Stevens*, Anal. Chim. Acta *14*, 126 (1956)

410 *J. Bankovskis, E. Svarles, A. Ievins*, Zh. Analit. Khim. *14*, 331 (1959).

411 *H. J. Hoenes, K. G. Stone*, Talanta *4*, 250 (1960).

412 *A. N. Bantysh, D. A. Kryazev, O. V. Levina*, Russian J. Inorg. Chem. *9*, 1157 (1964).

413 *R. L. Datta, B. Chatterjii*, J. Indian Chem. Soc. *44*, 780 (1967);
R. L. Datta, B. Chatterjii, J. Indian Chem. Soc. *46*, 268 (1969).

414 *J. J. Pitts, M. A. Robinson, S. I. Trotz*, J. Inorg. Nucl. Chem. *31*, 3685 (1969).

415 *A. Rosenheim*, Z. Anorg. Chem. *11*, 225 (1896);
A. Rosenheim, Ber. *26*, 1191 (1893).

416 *A. A. Vorontsova*, Russian J. Inorg. Chem. *5*, 1373 (1960).

of this complex have been isolated.[417–419] Salts containing the dimeric group $[Mo_2O_5(C_2O_4)_2(H_2O)_2]^{2\ominus}$ can also be obtained by crystallization of solutions of molybdic oxide in metal oxalates[418] or by oxidation of a molybdenum(V) oxalate.[420] The reaction of oxalic acid and hydrogen peroxide with metal molybdates yields *peroxyoxalates*[421] of the type $[MoO(O_2)_2C_2O_4]^{2\ominus}$, and $[MoO_2(O_2)C_2O_4]^{2\ominus}$. Addition of heavy metal salts, such as barium, to a solution of molybdate and malic acid yields salts of the type $Ba[MoO_3 \cdot C_4H_4O_5]$[422] and citric acid gives similar complexes.[423] Similarly, heavy metal salts of the type $[MoO_4(C_8H_8O_3)_2]^{2\ominus}$ and $[MoO_4(C_2H_4O_2)_2]^{2\ominus}$ have been obtained from mandelic and glycolic acids, respectively.[424] Salts containing the $MoO_2^{2\oplus}$ group are isolated when large organic cations are added to solutions of molybdates and α-thiolcarboxylic acids.[425] Also, amino acids such as glycine form complexes of the type $Mo_2O_7H_3$(ligand) from acidic molybdate solutions.[426] The free acid and salts of the ion $[(MoO_3)_2C_{10}H_{12}O_8N_2]^{4\ominus}$ are formed when molybdate solutions react with ethylenediaminetetraacetic acid.[427–429] Similar complexes are also formed with *N*-methyliminodiacetic[430] and iminodiacetic acid.[431]

[417] *A. Rosenheim, A. Bertheim*, Z. Anorg. Chem. *34*, 436 (1903).
[418] *E. Wendling, J. Delavillandre*, Bull. Soc. Chim. France 866 (1968).
[419] *J. Gopalakrishnen, B. Viswanathan, V. Srinivason*, J. Inorg. Nucl. Chem. *32*, 2565 (1970).
[420] *F. A. Cotton, S. M. Moorehouse, J. S. Wood*, Inorg. Chem. *3*, 397 (1964).
[421] *W. P. Griffith, T. D. Wickins*, J. Chem. Soc. *A*, 590 (1967).
[422] *S. Prasad, L. P. Pandey*, J. Indian Chem. Soc. *42*, 783 (1965).
[423] *S. Prasad, L. P. Pandey*, J. Proc. Inst. Chemists (India) *37*, 207 (1965).
[424] *D. Brown*, J. Chem. Soc. (London) 4732 (1961).
[425] *A. T. Busev, G. P. Rudzit*, Zh. Analit. Khim. *18*, 840 (1963).
[426] *H. Eguchi, T. Takeuchi, A. Ouchi, A. Furuhashi*, Bull. Chem. Soc. (Japan) *42*, 3585 (1969).
[427] *M. Naarova, J. Podhabova, J. Podlaha*, Collect. Czech. Chem. Commun. *33*, 1991 (1968).
[428] *R. J. Kula*, Anal. Chem. *38*, 1581 (1966).
[429] *S. I. Chan, R. J. Kula, D. T. Sawyer*, J. Amer. Chem. Soc. *86*, 377 (1964);
D. T. Sawyer, J. M. McKinnie, J. Amer. Chem. Soc. *82*, 4191 (1960).
[430] *R. J. Kula*, Anal. Chem. *38*, 1382 (1966).
[431] *R. J. Kula*, Anal. Chem. *39*, 1171 (1967).

15.9 Simple, Isopoly-, and Peroxy-molybdates

This section will deal only with simple, isopoly- and peroxymolybdates. The thiomolybdates were discussed in Section 15.8.2 and the heteropoly compounds are discussed in Chapter 32. The molybdates are taken up in accordance with the position of the cation in the periodic table.

The *simple molybdates* of the alkali metals are readily prepared by fusion of the alkali metal oxide or carbonate with molybdic oxide or by reacting molybdic oxide with a solution of the hydroxide or carbonate.[432–434] *Polymolybdates* of lithium and sodium with ratios of MoO_3 to metal oxide of 2:1, 3:1, and 4:1 are prepared by heating the simple molybdate with molybdic oxide.[432, 435]

Similar polymolybdates are reported for potassium and rubidium, but there appears to be disagreement about which compounds are obtained.[435–437] Cesium apparently forms several polymolybdates, but there is disagreement about whether the dimolybdate exists.[436, 437]

Ammonium molybdate, $(NH_4)_2MoO_4$, is obtained upon crystallization[438] or addition of ethanol[432, 439] to solutions of molybdic oxide in excess ammonia, but this compound readily loses ammonia to form higher molybdates. *Ammonium dimolybdate* and *paramolybdate*, $(NH_4)_6Mo_7O_{24} \cdot 4H_2O$, are commercially available and these compounds along with the tri- and octamolybdates can be isolated from solutions of molybdic oxide in aqueous ammonia depending on crystallization conditions.[440] The paramolybdate is prepared by evaporation of

[432] Gmelin's Handbuch der Anorganishcen Chemie, Molybdän 8th Ed., p. 210, 214, 229, 281, 285, Verlag Chemie Weineheim/Bergstr. 1935.
[433] *B. M. Gatehouse, P. Leverett*, J. Chem. Soc. 849 (1969);
B. M. Gatehouse, P. Leverett, Chem. Commun. 374 (1967).
[434] *J. Bye*, Ann. Chim. (Paris) *20*, 463 (1945).
[435] *F. Haermann*, Z. Anorg. Allgem. Chemie *177*, 150 (1929).
[436] *V. I. Spitsyn, I. M. Kuleshov*, J. Gen. Chem. (USSR) *21*, 1493–1502 (1951); C.A. *46*, 9006 (1952).
[437] *R. Salmon, P. Caillet*, Bull. Soc. Chim. France 1569 (1969).
[438] *Z. G. Karov, F. M. Perel'man*, Russian J. Inorg. Chem. *5*, 343 (1960).
[439] *H. Guiter*, C.R. Acad. Sci., Paris *220*, 146 (1945).
[440] *K. Funaki, T. Segawa*, J. Elektrochem. Soc. Japan *18*, 152 (1950).

excess ammonia from an ammonium molybdate solution,[441] and the octamolybdate has been obtained by addition of hydrated molybdic oxide to a paramolybdate solution.[442]

The normal *cupric molybdate*, $CuMoO_4$, is prepared by heating an equimolar mixture of the oxides[443–446] at 600°–700° and another cupric molybdate, Cu_2MoO_5, is formed when 2 mol of cupric oxide is heated with 1 mol of molybdic oxide.[443,445] A third molybdate, $Cu_3Mo_2O_9$, is formed by heating the correct mixture of the oxides or by fusing a mixture of $CuMoO_4$ and Cu_2MoO_5.[444,445,447] The *cuprous molybdate*, $Cu_6Mo_4O_{15}$, is obtained by heating a mixture of cuprous oxide and molybdic oxide at 700° in an inert atmosphere.[443] *Silver molybdate*, Ag_2MoO_4, is prepared by precipitation from aqueous solution[447,448] or by fusing silver nitrate with molybdic oxide.[449] Fusion of silver molybdate with molybdic oxide yields the di- and tetramolybdates.[447]

Beryllium molybdate is formed by heating the oxides[450] or by reaction of a beryllium hydroxide solution with molybdic oxide.[451] Magnesium molybdate is similarly prepared by heating the oxides[450] or by reacting a sodium molybdate solution with magnesium chloride.[452,453] The molybdates of calcium,[454] strontium, and barium[455] are readily prepared by heating the oxide or carbonate with molybdic oxide[450,456] or by reacting a soluble molybdate with a solution of a salt of the metal.[448,457] Strontium and barium form molybdates of the type A_3MoO_6 on reaction of molybdic oxide with 3 mol of the metal oxide.[458,459]

Zinc molybdate is readily prepared by heating the oxides[446,460] or by reacting a solution of sodium molybdate with a zinc chloride solution.[452] Similarly, *cadmium molybdate* is readily prepared from the oxides[447,360,461] at 600°, or isolated as a precipitate from the reaction of a soluble molybdate with cadmium nitrate or chloride.[462,463] *Mercurous molybdate*, Hg_2MoO_4, is formed as a precipitate when mercurous ion reacts with molybdate ion,[448,464] and mercuric molybdate is precipitated by treatment of dilute mercuric nitrate solution with sodium molybdate.[465]

Pure *aluminum molybdate* cannot be prepared by aqueous precipitation, but it is obtained by heating the oxides at 700° for several days.[443,466,467] *Indium(III)* and *thallium(III) molybdates* have been isolated by heating the oxides[466,468,469] at 600°–900°. Thallium(I)

---

[441] *G. Brauer*, Handbook of Preparative Inorganic Chemistry, 2nd Ed., Vol. II, p. 1711, Academic Press, New York 1965. (English Translation.)

[442] *J. Aveston, E. W. Anacker, J. S. Johnson*, Inorg. Chem. *3*, 735 (1964).

[443] *W. P. Doyle, G. McGuire, G. M. Clark*, J. Inorg. Nucl. Chem. *28*, 1185 (1966).

[444] *R. Kohlmuller, J. P. Faurie*, C.R. Acad. Sci., Paris *264C*, 1751 (1967).

[445] *K. Nassau, J. W. Shiever*, J. Amer. Ceram. Soc. *52*, 36 (1969).

[446] *A. N. Zelikman*, Russian J. Inorg. Chem. *1*, 2778 (1956).

[447] *P. Kohlmuller, J. P. Faurie*, Bull. Soc. Chim. France 4379 (1968).

[448] *T. Dupuis, C. Duval*, Anal. Chim. Acta *4*, 173 (1950).

[449] *I. M. Kuleshov*, Zh. Obshch. Khim. *21*, 406 (1951).

[450] *G. Tamman, F. Westerhold*, Z. Anorg. Allgem. Chemie *149*, 35 (1925).

[451] Gmelin's Handbuch der Anorganischen Chemie, Molybdän, 8th Ed., p. 287, Verlag Chemie, Weineheim/Bergstr., 1935.

[452] *A. W. Sleight, B. L. Chamberland*, Inorg. Chem. *7*, 1672 (1968).

[453] *J. E. Ricci, W. L. Linke*, J. Amer. Chem. Soc. *73*, 3603 (1957).

[454] *A. N. Zelikman, L. V. Belyaerskoya*, J. Appl. Chem. USSR *27*, 1091 (1958).

[455] *O. A. Ustinov, G. P. Novoselov, M. A. Andrianov, N. T. Chebotarev*, Russian J. Inorg. Chem. *15*, 1320 (1970).

[456] *A. N. Zelikman*, Russian J. Inorg. Chem. *1*, 2778 (1970).

[457] *V. I. Spitsyn, I. A. Sovich*, J. Gen. Chem. USSR *22*, 1323 (1952).

[458] *R. Sabatier, M. Wathle, J. P. Besse, G. Baud*, C.R. Acad. Sci., Paris *271C*, 368 (1970).

[459] *L. Chang, M. G. Seroger, B. Phillips*, J. Amer. Ceram. Soc. *49*, 338 (1966).

[460] *W. P. Doyle, F. Forbes*, J. Inorg. Nucl. Chem. *27*, 1271 (1965).

[461] *L. N. Alymova, V. E. Karapetyan, A. M. Moroz, G. A. Moskovkina*, Izv. Akad. Nauk SSSR, Neorg. Mater. *4* (7), 1199 (1968); C.A. *69*, 90653 (1968).

[462] *A. N. Zobnina, I. P. Kislyakov*, Izv. Akad. Nauk SSSR, Neorg. Mater. *2* (3), 511 (1966).

[463] *L. H. Brixner*, Elektrochem. Tech. *6*, 88 (1968).

[464] *C. M. Gupta, R. S. Saxena*, J. Inorg. Nucl. Chem. *14*, 297 (1960).

[465] *C. M. Cupta, R. S. Saxena*, J. Indian Chem. Soc. *35*, 617 (1958).

[466] *K. Nassau, H. J. Levinstein, G. M. Lociano*, J. Phys. Chem. Solids *26*, 1805 (1965).

[467] *U. K. Trunov, V. V. Lutsenko, L. M. Kovba, Izv. Vyssh. 10*, 375 Ucheb. Zavedenii, Khim. Tekhnol. *10* (4), 375 (1967); C.A. *68*, 73212 (1968).

[468] *V. K. Trunov, L. M. Kovba*, Vestn. Mosk. Univ., Khim. *22* (1), 114 (1967); C.A. *66*, 109 180 (1967).

[469] *P. V. Klevtsov*, Izv. Akad. Nauk SSSR, Neorg. Mater. *4*, 160 (1968).

molybdate is formed as a precipitate from the reaction of thallous ion with molybdate ion.[470]

Molybdic oxide does not react with any of the oxides of titanium.[443] *Zirconium(IV) molybdate*, $Zr(MoO_4)_2$,[460,471] and *hafnium molybdate*, $Hf(MoO_4)_2$,[443] are easily prepared from the oxides[472] at about 700°.

Neither silicon dioxide,[460] germanium dioxide,[473] nor stannic oxide[460,474] reacts with molybdic oxide at elevated temperatures. The normal lead compound, $PbMoO_4$, is readily formed from the oxides[456,460] or by precipitation from aqueous solution.[448,475,476] The other lead molybdate, Pb_2MoO_5, is formed by heating 2 mol of lead monoxide with 1 mol of molybdic oxide.[460,477]

The compound V_2MoO_8[443,478,479] is formed when an equimolar mixture of molybdic oxide and vanadium pentoxide is heated at 600° and the compound $VOMoO_4$ is formed by heating vanadium dioxide with molybdic oxide.[443,480] Niobium pentoxide reacts with molybdic oxide to form $Nb_2O_5 \cdot 3MoO_3$,[481] but there is some doubt about the existence of $4Nb_2O_5 \cdot MoO_3$ and $Nb_2O_5 \cdot 2MoO_3$ which had been reported earlier.[482,483] The compound $4Ta_2O_5 \cdot MoO_3$ has also been reported.[483]

Bismuth trioxide reacts with molybdic oxide to form a variety of compounds, and there is disagreement about which compounds actually exist. It appears that Bi_6MoO_{12}, Bi_2MoO_6, $Bi_2Mo_2O_9$, and $Bi_2(MoO_4)_3$ can definitely be prepared by heating the appropriate mixture of the oxides.[484–488] The compound Bi_4MoO_9 has been prepared by an aqueous precipitation procedure.[489]

Chromium(III) molybdate, $Cr_2(MoO_4)_3$, is obtained by heating chromium sesquioxide with molybdic oxide at 800° in a sealed tube.[433,466,490] The reaction of molybdic oxide with tungstic oxide at high temperatures forms $WMoO_6$ and perhaps $2WO_3 \cdot MoO_3$ and $WO_3 \cdot 2MoO_3$.[491,492] Manganese(II) molybdate, $MnMoO_4$, is prepared by heating any manganese oxide with molybdic oxide[443] or by reacting manganous chloride with sodium molybdate in aqueous solution.[452] *Ferric molybdate* can be prepared by the reaction of the calculated amount of ferric oxide with molybdic oxide[445,490,493,494] at 600° or by precipitation from a solution of ferric chloride/sodium molybdate.[495] Ferrous molybdate can be obtained by reacting molybdic oxide with ferrous oxide or a mixture of ferric oxide and iron in an inert atmosphere[456,493,496] or by precipitation from an aqueous solution containing sodium molybdate and ferrouschloride.[496,497] *Cobalt(II) molybdate* is prepared by heating molybdic oxide with a cobalt oxide[443,498,499] or

[470] *L. Moser, W. Reif*, Monatsh. Chem. *52*, 348 (1929).
[471] *G. P. Novoselov, O. A. Ustinov*, Russian J. Inorg. Chem. *13*, 634 (1968);
G. P. Novoselov, O. A. Ustinov, Russian J. Inorg. Chem. *15*, 301 (1970).
[472] *W. Freundlich, J. Thoret*, C.R. Acad. Sci. Paris *265C*, 96 (1967).
[473] *G. M. Schwab, J. Gerloch*, Z. Physik. Chem. (Frankfurt am Main) *56*, 121 (1967).
[474] *J. Buiten*, J. Catalysis *10*, 188 (1968).
[475] *H. V. Weiss, M. G. Lai*, Talanta *8*, 72 (1961).
[476] *M. V. Moklosoev, Z. E. Bamutra, V. I. Krivobok*, Zh. Prikl. Khim. *40* (12), 2621 (1967).
[477] *I. Belyaev, N. Smolyaminov*, Russian J. Inorg. Chem. *12*, 1703 (1967).
[478] *H. A. Eick, L. Kihlborg*, Acta Chem. Scand. *20*, 1658 (1966).
[479] *G. Tridot, J. Tudo, G. Leman-Delcour, M. Nolf*, C.R. Acad. Sci., Paris *260*, 3410 (1965).
[480] *H. A. Eick, L. Kihlborg*, Acta Chem. Scand. *20*, 722 (1966).
[481] *E. J. Felten*, J. Less-Common Metals *9*, 206 (1965).
[482] *V. K. Trunov, L. M. Kovba, E. I. Strotkina*, Dokl. Akad. Nauk SSSR *153* (5), 1085 (1963).
[483] *L. M. Kovba, V. K. Trunov*, Vestn. Mosk. Univ. Khim. *19* (6), 32 (1964).
[484] *I. Belyaev, N. Smolyaminov*, Russian J. Inorg. Chem. *7*, 579 (1962).
[485] *A. Bleijenberg, B. C. Lippens, G. Schuit*, J. Catalysis *4*, 581 (1965).
[486] *L. Ya. Erman, G. I. Galy'perina*, Russian J. Inorg. Chem. *11*, 122 (1966);
L. Ya. Erman, G. I. Galy'perina, Russian J. Inorg. Chem. *13*, 487 (1968).
[487] *L. Ya. Erman, E. L. Galy'perin*, Russian J. Inorg. Chem. *15*, 441 (1970).
[488] *A. Ianik*, Roczniki Chem. *41*, 1399 (1967).
[489] *G. Gattow*, Z. Anorg. Allgem. Chemie *298*, (1959).
[490] *V. K. Trunov, L. M. Kovba*, Izv. Akad. Nauk SSSR, Neorg. Mater. *2*, 151 (1961).
[491] *L. E. Semikina, V. E. Limonov*, Russian J. Inorg. Chem. *13*, 1006 (1968).
[492] *A. Magneli, B. Blomberg, L. Kihlborg, G. Sundkvist*, Acta Chem. Scand. *1*, 1382 (1955).
[493] *W. Jager, A. Rahmel, K. Becker*, Arch. Eisenhuettenw. *30*, 435 (1958).
[494] *A. Marcu, S. Serban, C. Fagarasanu, E. Belea, A. Faclieru*, Rev. Chim. (Bucharest) *21* (7), 405 (1970).
[495] *A. Kuznetsova, L. M. Skrebkova, L. M. Kefeli, L. M. Plyasova*, Izv. Sibirsk. Otd. Akad. Nauk SSSR, Ser. Khim. Nauk (11), 61 (1966); C.A. *65*, 8310 (1966).
[496] *A. W. Sleight, B. L. Chamberland, J. F. Weiher*, Inorg. Chem. *7*, 1093 (1968).
[497] *K. Shapiro, Yu. Yurkevich*, Zh. Prikl. Khim. *36*, 2584 (1962).
[498] *J. Lipsch, G. Schuit*, J. Catalysis *15*, 163 (1969).
[499] *G. W. Smith*, Acta Cryst. *15*, 1504 (1962).

by aqueous precipitation.[452] Nickel(II) molybdate can be similarly prepared by heating the oxides[443] or by an aqueous precipitation method.[452]

The normal trivalent *lanthanide molybdates*[466,500] have been prepared by ignition of the oxides at temperatures from 700° to 1100°, and lanthanum and yttrium also form compounds of the type A_2MoO_6 and $A_2Mo_4O_{15}$.[501–503] Lanthanum[504,505] and cerium(III)[506] molybdates can also be precipitated from aqueous solution.

Thorium dioxide reacts to form $Th(MoO_4)_2$ at 700° with molybdic oxide.[460,472,507] Several studies have been made of the preparation of UMo_2O_8 from uranium dioxide.[483,507,508] Neptunium(IV) oxide forms $NpMo_2O_8$ and americium(III) forms $Am_2(MoO_4)_3$ on reaction with molybdic oxide.[509]

Tetraperoxymolybdates of the type $[Mo(O_2)_4]^{2\ominus}$ are formed by the interaction of soluble molybdates and hydrogen peroxide in alkaline solution.[510–513] Calcium and strontium salts containing the triperoxy group, $[MoO(O_2)_3]^{2\ominus}$, have also been obtained.[513] Diperoxymolybdates, $[MoO_2(O_2)_2]^{2\ominus}$, are formed by reacting hydrogen peroxide with molybdates in acid solution,[510,514,515] and monoperoxy salts, $[MoO_3(O_2)]^{2\ominus}$, have also been isolated.[513] More *complex peroxymolybdates*, such as $[Mo_7O_{22}(O_2)_2]^{6\ominus}$, have been obtained from neutral selutions using low ratios of peroxide to molybdate.[516]

15.10 Bibliography

J. W. Mellor, A Comprehensive Treatise on Inorganic and Theoretical Chemistry, Vol. 11. Longmans, Green, New York, 1931.

D. H. Killeffer, A. Linz, Molybdenum Compounds. Wiley (Interscience) New York, 1952.

L. Northcott, Molybdenum. Academic Press, New York, 1956.

P. Pascal, Nouveau Traite de Chimie Minerale, Vol. 14. Masson, Paris, 1959.

Molybdenum Metal. Climax Molybdenum Co., New York, 1960.

M. C. Manzone, J. Z. Briggs, Less Common Alloys of Molybdenum. Climax Molybdenum Co., New York, 1962.

Properties of Simple Molybdates. Climax Molybdenum Co., New York, 1962.

A. I. Busev, Analytical Chemistry of Molybdenum. Davey, New York, 1964.

J. A. Connor, E. Ebsworth, Advances in Inorganic Chemistry and Radiochemistry, (H. J. Emeléus and A. G. Sharpe, eds.), Vol. 6. Academic Press, New York, 1964, p. 280.

G. W. A. Fowles in *W. L. Jolly* (ed.), Preparative Inorganic Reactions, Vol. 1, p. 121. Wiley (Interscience), New York, 1964.

M. L. Larson, Organic Complexes of Molybdenum. Climax Molybdenum Co., New York, 1964. Annual Supplements.

J. Selbin, Angew. Chem. Int. Ed. Engl. *5*, 712 (1966); J. Chem. Educ. *41*, 86 (1964).

P. C. H. Mitchell, Quart. Rev. (London) *20*, 103 (1966).

B. M. Chadwick, A. G. Sharpe, in Advances in Inorganic Chemistry and Radiochemistry (H. J. Emeléus and A. G. Sharpe, eds.), Vol. 8, p. 84. Academic Press, New York, 1966.

V. Gutmann (ed.), Halogen Chemistry, Vol. 3. Academic Press, New York, 1967.

R. Puschel, E. Lassner, in Chelates in Analytical Chemistry (H. A. Flascha and A. J. Barnard, Jr., eds.), Vol. 1. Dekker, New York, 1967.

J. A. McCleverty, in Progress in Inorganic Chemistry (F. A. Cotton, ed.), Vol. 10, p. 50. Wiley (Interscience), New York, 1968.

J. T. Spence, Coordination Chem. Rev. *4*, 475 (1969).

G. A. Tsigdinos, C. J. Hallada, Isopoly Compounds of Molybdenum, Tungsten, and Vanadium. Climax Molybdenum Co., New York, 1969.

Properties of Molybdic Oxide. Climax Molybdenum Co., New York, 1969.

W. P. Griffith, Coordination Chem. Rev. *5*, 459 (1970).

500 *E. Ya. Rode, G. V. Lysanova, V. G. Kuznetsov, L. Z. Gakhman*, Russian J. Inorg. Chem. *13*, 678 (1968).

501 *J. P. Fousnier, J. Fousnier, R. Kohlmuller*, Bull. Soc. Chim. France 4277 (1970).

502 *E. I. Get'man, M. V. Mokhosoev*, Izv. Akad. Nauk SSSR, Neorg. Mater. *4* (9), 1554 (1968); C.A. *70*, 23643 (1969);
E. I. Get'man, M. V. Mokhosoev, Izv. Akad. Nauk SSSR, Neorg. Mater. *4* (10), 1743–8 (1968); C.A. *70*, 31931 (1969).

503 *A. N. Pokrovskii, V. K. Rybakov, V. K. Trunov*, Russian J. Inorg. Chem. *14*, 1233 (1969).

504 *R. S. Saxena, M. L. Mittal*, Z. Anorg. Allgem. Chemie *324*, 208 (1963).

505 *N. K. Davidenlo, G. A. Komoshko, K. B. Yatsimirskii*, Russian J. Inorg. Chem. *13*, 58 (1968).

506 *R. S. Saxena, M. L. Mittal*, J. Elektroanal. Chem. Interfacial Elektrochem. *5*, 287 (1963).

507 *V. K. Trunov, O. A. Efremova, L. M. Kovba*, Radiokhimiya *8* (6), 717 (1966); C.A. *66*, 69687 (1967).

508 *P. Paillert*, C.R. Acad. Sci., Paris *265C*, 85 (1967).

509 *W. Freudlich, M. Pager*, C.R. Acad. Sci., Paris *269C*, 392 (1969).

510 *G. Brauer*, Handbook of Preparative Inorganic Chemistry, 2nd Ed., Vol. II, p. 1414, Academic Press, New York 1965. (English Translation.)

511 *K. Gleu*, Z. Anorg. Allgem. Chemie *204*, 67 (1932).

512 *A. Menez, F. Petillon, J. E. Guerchais*, C.R. Acad. Sci., Paris *269C*, 1104 (1969).

513 *V. A. Shcherbinin, G. A. Bogdanov*, Russian J. Inorg. Chem. *4*, 112 (1959).

514 *W. P. Griffith*, J. Chem. Soc. (London) 5345 (1963).

515 *F. Chauveau, P. Souchay, G. Tridot*, Bull. Soc. Chim. France 819 (1960).

516 *R. Stromberg, L. Trysberg*, Acta Chem. Scand. *23*, 314 (1969).

16 Tungsten

D. G. Tisley and R. A. Walton
Department of Chemistry, Purdue University,
Lafayette, Indiana 47907, U.S.A.

16.1 Tungsten Metal

In the 1930s it was thought that tungsten metal existed in two allotropic forms; α-tungsten prepared from tungsten trioxide by hydrogen reduction at 750°[1] and β-tungsten prepared by electrolysis (below 700°) of fused mixtures of tungsten trioxide with alkali metal phosphates[2] or alkali metal tungstates.[3] Since then the β-phase has been prepared by the low-temperature (~550°) hydrogen reduction of tungsten trioxide. It has been shown to contain oxygen and to be of probable ideal composition W_3O,[4,5] although in practice its composition is $WO_{0\cdot005-0\cdot1}$.[6,7] It has been suggested that β-tungsten may be regarded as a metal with a disturbed structure, rather than as a definite oxide or an allotropic form of tungsten. Also, heteroatoms other than oxygen may stabilize the structure.[8]

Tungsten metal is readily available commercially with a tungsten content of at least 99·9%. It may be freed from traces of dissolved oxygen and other gases by heating in hydrogen at 1000°.[9] Extremely pure tungsten, with impurity levels of less than 10 ppm, can be obtained by the reaction of tungsten hexacarbonyl with hydrogen and steam or with hydrogen and carbon dioxide at temperatures in excess of 1000°,[10] or by the decomposition of tungsten hexachloride on heated tungsten wires *in vacuo* at 1500°–2000°.[11] Further purification is then achieved by zone melting in an electron beam furnace.[11]

[1] *C. Hampel*, Rare Metals Handbook, 2nd Ed. p. 582, Reinhold Publishing Co., London 1961.
[2] *H. Hartmann, F. Ebert, O. Bretschneider*, Z. Anorg. Allgem. Chemie *198*, 116 (1931).
[3] *W. Burgers, J. van Liempt*, Rec. Trav. Chim. Pays-Bas, *50*, 1050 (1931).
[4] *G. Hägg, N. Schonberg*, Acta Cryst. *7*, 351 (1954).
[5] *M. Charlton*, Nature *174*, 703 (1954).
[6] *T. Millner*, Zh. Neorg. Khim. *3*, 946 (1958); Engl. *3* (4), 170.
[7] *Do Quang Kim, N. Wallet, F. Marion*, C.R. Acad. Sci. Paris *261*, 2667 (1965).
[8] *J. Neugebauer, A. Hegedüs, T. Millner*, Z. Anorg. Allgem. Chemie *293*, 241 (1958).
[9] *R. McCarley, T. Brown*, Inorg. Chem. *3*, 1232 (1964).
[10] *J. Ward, A. Coon, J. Oxley*, Appl. Fundam. Thermodyn. Met. Processes. Proc. Conf. Thermodyn. Props. Mater. Univ. Pittsburgh (1964), 197. Publ. 1967; C.A. *68*, 61800m (1968).
[11] *G. Guenzler, G. Weise, H. Opperman, H. Liebscher, K. Hempel, M. Saradshow, R. Guenther*, Hermsdorfer Tech. Mitt. *7*, 561 (1967); C. A. *68*, 42434d (1968).

16.2 Binary Refractory Compounds of Tungsten with Elements of Groups V, IV and III

The binary refractory compounds of tungsten are a group of high-melting, hard substances which have metallic character; their strength at elevated temperatures makes them extremely useful for high temperature processes.

16.2.1 Nitrides

The nitrides, which are not as technologically important as the other binary refractories, are generally found as contaminants in the preparation of carbides.
There is reported to be no reaction between tungsten and nitrogen below at least 900°;[12] however, during the course of investigating[13,14] the behavior of tungsten wires at 2500° under low pressures of nitrogen, small amounts of a brown solid of stoichiometry WN_2 were found.

The tungsten nitrides are generally prepared by passing ammonia over heated tungsten; the reaction is reported[12] to start at 140°, but temperatures well above this are required for complete reaction. Heating this system at 700°–800° for up to 2 days produces W_2N.[15–17]

A structural modification of W_2N was thought to exist,[18] but this has since been identified as an oxide nitride.[19] By lowering the reaction temperature from 800° to 300° over a 4-week period, Schonberg[17] obtained a product with 41·6 atom% N; it was suggested on the basis of X-ray data that this contained a phase WN, isomorphous with WC. Since then a polymorph of WN has been claimed[20] to be formed by the reaction of ammonia with β-tungsten at 700°.
More recently there have been investigations of several

[12] *P. Lafitte, P. Grandadam*, C.R. Acad. Sci. Paris *200*, 1039 (1935).
[13] *I. Langmuir*, J. Amer. Chem. Soc. *35*, 931 (1913).
[14] *C. Smithells, H. Rooksby*, J. Chem. Soc. 1182 (1927).
[15] *G. Hägg*, Z. Physik. Chem. *B7*, 356 (1930).
[16] *M. Mathis*, Bull. Soc. Chim. France 443 (1951).
[17] *N. Schonberg*, Acta Chem. Scand. *8*, 204 (1954).
[18] *R. Kiessling, Y. Liu*, J. Metals *3*, 639 (1951).
[19] *R. Kiessling, L. Peterson*, Acta Met. *2*, 675 (1954).
[20] *J. Neugebauer, A. Hegedüs, T. Millner*, Z. Anorg. Allgem. Chemie *302*, 50 (1959).

intermediate nitride phases using electron diffraction techniques, but precise experimental details are lacking.[21]

16.2.2 Phosphides

The two well-established phosphides, WP and WP_2, may be prepared by heating stoichiometric mixtures of tungsten metal powder with red phosphorus in evacuated sealed tubes, at temperatures between 600° and 1000° for several days.[22,23] It is also possible to prepare WP_2 by passing phosphine over tungsten at 800°.[22] Hulliger reports[24] a low-temperature form of WP_2, designated α-WP_2, formed by reacting tungsten and phosphorus below 700°. A lower phosphide, W_3P, is formed[25] by electrolysis (below 900°) of a fused sodium metaphosphate–tungsten(VI) oxide–sodium chloride bath; the phase is also detected in phosphide samples heated in an argon arc furnace.[26] The detection of this phase reopens the question of the subphosphides, W_2P and W_4P, prepared by Hartmann and Orban,[27] which were dismissed as mixtures of tungsten and WP by Faller and Biltz.[28]

16.2.3 Arsenides

Again there are two well-established phases, W_2As_3 and WAs_2.[29] Both may be prepared by heating mixtures of the metal powders in sealed, evacuated quartz tubes at temperatures of up to 950°.[30,31] W_2As_3 is also obtained from the thermal decomposition of WAs_2 above 600°.[30]

A phase W_2As (obtained in the presence of excess tungsten metal) has been reported,[32] but not confirmed. The phase W_4As_5[32a] has subsequently been postulated[33] to be identical with W_2As_3.

16.2.4 Carbides

The carbides are the most utilized of the tungsten refractories, over 60% of the tungsten produced being used as carbides. Only two stoichiometries are established, W_2C, first prepared by Moissan[34] by melting tungsten with carbon, and WC, isolated by Williams[35] from an iron–tungsten(VI) oxide–carbon melt.

The most common method of preparing WC is by heating to 1400° stoichiometric mixtures of tungsten and carbon in a hydrogen atmosphere or under vacuum.[36] It may also be obtained from tungsten(VI) oxide by direct reaction with carbon at 1100° in a hydrogen atmosphere in the presence of hydrogen halides, the most effective one being HI.[36] A hydrogen–methane mixture at 1000° will also yield WC from tungsten oxygen compounds,[37] and it is reported to be the main reaction product of heating tungsten hexacarbonyl to 1000°.[38] This latter reaction also can give rise to oxide carbides[39] when performed at lower temperatures.

The carbide W_2C is prepared by heating tungsten or WC with stoichiometric amounts of carbon under hydrogen at 1600°.[40] It may be obtained pure by heating to above 2000°.[41,42] Both carbides have recently been prepared by the reaction of hydrogen cyanide with tungsten metal powder,[43] WC being formed above 700° and W_2C above 2200°.

Polymorphic modifications of both carbides have been detected. Thus, α-W_2C is converted to β-W_2C at 2400°[44,45]; the β form is easily reconverted to α-W_2C.

---

[21] *V. Khitrova*, Kristallografiya *8*, 873 (1963).
V. Khitrova, Soviet Physics-Cryst. *8*, 701 (1964).
[22] *N. Schonberg*, Acta. Chem. Scand. *8*, 226 (1954).
[23] *S. Rundqvist, T. Lundström*, Acta Chem. Scand. *17*, 37 (1963).
[24] *F. Hulliger*, Nature *204*, 775 (1964).
[25] *P. Yocom*, Dissertation Abstr. *18*, 1974 (1958).
[26] *S. Rundqvist*, Nature *211*, 847 (1966).
[27] *H. Hartmann, J. Orban*, Z. Anorg. Allgem. Chemie *226*, 257 (1936).
[28] *F. Faller, W. Biltz*, Z. Anorg. Allgem. Chemie *248*, 209 (1941).
[29] *H. Haraldsen*, Angew. Chem. *78*, 64 (1966); *H. Haraldsen*, Angew. Chem. Intern. Ed. Engl. *5*, 48 (1966).
[30] *P. Jensen, A. Kjekshus, T. Skansen*, Acta Chem. Scand. *20*, 403 (1966).
[31] *J. Taylor, L. Calvert, M. Hunt*, Can. J. Chem. *43*, 3045 (1965).
[32] *H. Boller, H. Nowotny*, Monatsh. Chem. *85*, 1272 (1964).
[32a] *H. Moissan*, C.R. Acad. Sci. Paris *123*, 15 (1896).
[33] *P. Jensen, A. Kjekshus, T. Skansen*, Acta Chem. Scand. *19*, 1499 (1965).
[34] *H. Moissan*, C.R. Acad. Sci. Paris *116*, 1225 (1893).
[35] *P. Williams*, C.R. Acad. Sci. Paris *126*, 1722 (1898).
[36] *T. Ya Kosolapova*, Carbides. Plenum Press, New York, London 1971.
[37] *A. Newkirk, I. Aliferis*, J. Amer. Chem. Soc. *79*, 4629 (1957).
[38] *D. Hurd, H. McEntee, P. Brisbin*, Ind. Eng. Chem. *44*, 2432 (1952).
[39] *H. Lux, A. Ignatowicz*, Chem. Ber. *101*, 809 (1968).
[40] *K. Becker, R. Holbling*, Angew. Chem. *40*, 512 (1927).
[41] *H. Goldschmidt, J. Brand*, J. Less-Common Metals *5*, 181 (1963).
[42] *K. Yvon, H. Nowotny, F. Benesovsky*, Monatsh. Chem. *99*, 726 (1968).
[43] *M. Caillet, Y. Lagarde, J. Besson*, C.R. Acad. Sci. Paris Ser. C *270*, 1867 (1970).
[44] *K. Becker*, Z. Physik *51*, 481 (1928).
[45] *F. Skaupy*, Z. Elektrochem. *33*, 487 (1927).

A modification of WC, β-WC, has recently been found[46] and is claimed to be stable above 2525°.

16.2.5 Silicides

The silicides are softer than the other refractories, but they exhibit a high resistance to corrosive and oxidizing agents. Only two phases are known, WSi_2 and W_5Si_3; the latter was originally thought to be W_3Si_2, but a structure determination by Aronsson[47,48] finally decided the stoichiometry, although some later papers still use the W_3Si_2 formulation.

Both silicides are prepared by mixing stoichiometric amounts of tungsten metal and silicon and sintering at temperatures up to 1900° in an argon atmosphere.[49] The reaction has also been performed at lower temperatures[50] (1200° for WSi_2 and 1400° for W_5Si_3), and it is claimed that there are significant losses of silicon above 1200° and 1450° for WSi_2 and W_5Si_3, respectively.[50]

16.2.6 Borides

There has been much interest in the borides recently because of their resistance to corrosion and oxidation at high temperatures. Since Moissan's original work,[51] Kiessling[52] and Brewer[53] have prepared tungsten borides of compositions W_2B, WB, and W_2B_5 by sintering stoichiometric mixtures of tungsten and boron at 1200°–1300° *in vacuo* or a partial pressure of argon. The lower two borides have narrow ranges of homogeneity, but W_2B_5 has a defect structure and is only of ideal composition W_2B_5, actual compositions being in the range $WB_{2\cdot00-2\cdot15}$. The WB phase is reported[54,55] to exist in two forms, the tetragonal WB being converted reversibly at 1900° to an orthorhombic form. Further methods for the production of borides have been developed recently; these include the reaction of tungsten(IV) oxide with boron[56] to form borides and boron monoxide, BO, and the reaction between tungsten metal and boron nitride above 1300°.[57] A higher boride of composition WB_x $(4<x<5)$[58–61] has been prepared by heating tungsten with a large excess of boron to 1400°–1500°[58] or in an electric arc.[59] This phase was originally thought to dissociate above 1600°, but has since been found to be stable. A further phase, WB_{12}, has been claimed,[62] but present confirmation is lacking.

[46] *R. Sara*, J. Amer. Ceram. Soc. *48*, 251 (1965).
[47] *B. Aronsson*, Acta Chem. Scand. *9*, 1107 (1955).
[48] *B. Aronsson*, Acta Chem. Scand. *9*, 137 (1955).
[49] *L. Brewer, A. Searcy, D. Templeton, C. Dauben*, J. Amer. Ceram. Soc. *33*, 291 (1950).
[50] *F. Josien*, Rev. Chim. Miner. *1*, 91 (1964).
[51] *H. Moissan*, C.R. Acad. Sci. Paris *123*, 15 (1896).
[52] *P. Kiessling*, Acta Chem. Scand. *1*, 893 (1947).
[53] *L. Brewer, D. Sawyer, D. Templeton, C. Dauben*, J. Amer. Ceram. Soc. *34*, 173 (1951).
[54] *B. Post, F. Glaser*, J. Chem. Phys. *20*, 1050 (1952).
[55] *T. Lundström*, Ark. Kemi. *30*, 115 (1969).
[56] *P. Peshev, G. Blizanakov, L. Leyarovska*, J. Less-Common Metals *13*, 241 (1967).

16.3 Tungsten(0) and (I)

In this section we include the nitric oxide complexes bisnitronium(dihalide)tungsten(0), $W(NO)_2X_2$ (X=Cl or Br), and their derivatives, in which the nitric oxide molecule may *formally* be considered to be present as the nitrosonium ion $NO^{\oplus}$.[63] Also included are the organometallic compounds of tungsten(0) and (I) which do not contain carbonyl groups and are not "sandwich"-type compounds.

16.3.1 Compounds with Group VII

The reaction of nitrosyl chloride or bromide with tungsten hexacarbonyl[63,64] yields *bisnitrosonium(dihalide)tungsten(0)*, $W(NO)_2X_2$ (X=Cl or Br). Unlike the molybdenum analog, reaction with potassium iodide does not form the corresponding iodide compound.

These compounds will react with ligand molecules[63,64] (e.g., L=triphenylphosphine) in benzene solution to form *octahedral complexes* of stoichiometry $W(NO)_2X_2L_2$, in which the halide ions are postulated[65] to be *trans* to one another and the nitric oxide and ligand molecules (L) mutually *cis* to one another. Isomers with all like atoms mutually *cis* or with the ligand molecule *trans* to one another can be made[65,66] by passing nitric oxide into benzene suspensions of dihalo(triscarbonyl)bis(ligand)tungsten(II), $WX_2(CO)_3L_2$.[66] The resulting mixtures may be

[57] *F. Baehren, F. Thümmler, D. Vollath*, J. Less-Common Metals *18*, 295 (1969).
[58] *A. Chretien, J. Helgorsky*, C.R. Acad. Sci. Paris *252*, 742 (1961).
[59] *P. Romans, M. Krug*, Acta Cryst. *20*, 313 (1966).
[60] *H. Nowotny, H. Haschke, F. Benesovsky*, Monatsh. Chem. *98*, 547 (1967).
[61] *G. Samsonov*, Dokl. Akad. Nauk. SSSR *113*, 1299 (1957); Engl. 417.
[62] *E. Rudy, F. Benesovsky, L. Toth*, Z. Metallk. *54*, 345 (1963).
[63] *F. Cotton, B. Johnson*, Inorg. Chem. *3*, 1609 (1964).
[64] *B. Johnson*, J. Chem. Soc. *A*, 475 (1967).
[65] *M. Anker, R. Colton, I. Tomkins*, Australian J. Chem. *21*, 1149 (1968).
[66] *M. Anker, R. Colton, I. Tomkins*, Australian J. Chem. *20*, 9 (1967).

separated by chromatography. When *bisnitrosonium(dibromo)tungsten(0)*, $W(NO)_2Br_2$, is reacted with the dimethyldithiocarbamate anion, elimination of halide occurs and the product *bisnitrosoniumbis (dimethyldithiocarbamate) tungsten(0)*, $W(NO)_2(S_2CNMe_2)_2$, results.[67] Salts of the type $[Ph_4P]_2[W(NO)_2Cl_2Br_2]$, derived from $W(NO)_2Cl_2$, can be formed by reaction with tetraphenylarsonium bromide[68]; these salts are air- and moisture-stable, and have been reacted with a variety of bidentate, dianionic sulfur ligands, (SS), to form complexes of stoichiometry $[Ph_4P][W(NO)_2(SS)_2]$.[68]

16.3.2 Compounds with Group V

The compound *trisbipyridyltungsten(0)*, $W(Bipy)_3$, can be obtained from the reaction of tungsten(VI) chloride with 2,2′-bipyridyl and dilithium bipyridyl in diethyl ether.[69] It will react irreversibly with phenyl isonitrile to form *hexakis(phenylisonitrile)tungsten(0)*, $W(PhNC)_6$.[70,71] With 2,2′,2″-terpyridyl, the compound *bis(terpyridyl)tungsten(0)* may be formed by heating tungsten hexacarbonyl with terpyridyl at 260°.[72] The only example of a tungsten(0)[73] compound with a phosphorus ligand is *tris(1,2-bisdimethylphosphinoethane)-tungsten(0)*, $W(DMPE)_3$, formed by the lithium aluminum hydride reduction of tungsten(VI) chloride in tetrahydrofuran, with subsequent addition of ligand.

16.3.3 Compounds with π-Bond Donors

The reaction of hexafluorobut-2-yne with tris(acetonitrile)tricarbonyltungsten(0), $W(CO)_3(MeCN)_3$, at 100° gives *acetonitriletris (hexafluorobut-2-yne)tungsten* $W[(CF_3)_2C_2]_3$-MeCN.[74] Reaction of the above carbonyl derivative with methyl vinyl ketone in boiling hexane affords *tris(methyvinylketone)tungsten(0)*, $W[MeCOCHCH_2]_3$,[75] in which the ligand molecules are bonded from the C=O and C=C bonds via a π-type interaction. Finally, reacting tetraethylammonium halopentacarbonyltungsten(0), $[Et_4N][W(CO)_5X]$ (X=Cl, Br, or I), with *p*-benzoquinone in tetrahydrofuran forms the complexes *halotris(p-benzoquinone)tungsten(I)*, $WX(quinone)_3$, in which the tungsten-halogen bond is retained.[76]

[67] *B. Johnson, K. Al-Obaidi, J. McCleverty*, J. Chem. Soc. *A*, 1668 (1969).

[68] *N. Connelly, J. Locke, J. McCleverty, D. Phipps, B. Ratcliffe*, Inorg. Chem. *9*, 278 (1970).

[69] *S. Herzog, E. Kubetschek*, Z. Naturforsch. *18B*, 162 (1963).

[70] *S. Herzog, E. Gutsche*, Z. Chem. *3*, 393 (1963).

[71] *L. Malatesta, A. Sacco*, Ann. Chim. (Rome) *43*, 622 (1953).

[72] *H. Behrens, U. Anders*, Z. Naturforsch. *19B*, 767 (1964).

[73] *J. Chatt, H. Watson*, J. Chem. Soc. 2545 (1962).

[74] *R. King, A. Fronzaglia*, Chem. Commun. 547 (1965).

[75] *R. King, A. Fronzaglia*, Chem. Commun. 274 (1966).

[76] *F. Calderazzo, R. Henzl*, J. Organomet. Chem. *10*, 483 (1967).

16.4 Tungsten(II) and (III)

16.4.1 Hexanuclear Halide Clusters

The dihalides of tungsten are, like the dihalides of molybdenum, hexanuclear, and contain the $[W_6X_8]^{4\oplus}$ unit[77] (X=halogen). This cluster may be pictured as a cube with halogen atoms at each corner and tungsten atoms at the centre of each face. The remaining halogens then bond, one to each tungsten, with some intercluster bridging to maintain overall stoichiometry. The situation becomes more complex with the trihalides and intermediate oxidation state compounds with the formation of $[W_6X_{12}]^{n\oplus}$ clusters[78] and polyhalide bridges.[79,80]

16.4.1.1 Halides

Tungsten(II) chloride may be prepared by the hydrogen, phosphorus, or aluminum reduction of tungsten(VI) chloride. However, the hydrogen and phosphorus reactions give impure products and the aluminum method is reported to be inefficient.[81] The most satisfactory method is by the disproportionation of tungsten(IV) chloride (see Section 16.5.1) at 450° 500°.[9] Recrystalliz ing as the chloro acid, $[H_3O]_2[W_6Cl_8]Cl_6 \cdot nH_2O$, from a hydrochloric acid–ethanol mixture and subsequent heating at 325° gives pure tungsten(II) chloride.

The structure of tungsten(II) chloride is that of the $[W_6Cl_8]^{4\oplus}$ cluster with two terminal and four bridging chlorides, which in Schäfer's notation is $[W_6Cl_8]Cl_2Cl_{4/2}$. When tungsten(II) chloride is reacted with an excess of liquid chlorine at 100°, the products are tungsten(VI) chloride and black tungsten(III) chloride. The hexa-

[77] *H. Schäfer, H. Schnering, J. Tillack, F. Kuhnen, H. Woehrle, H. Baumann*, Z. Anorg. Allgem. Chemie *353*, 281 (1967).

[78] *R. Siepmann, H. Schnering, H. Schäfer*, Angew. Chem. *79*, 650 (1967); Engl. *6*, 637.

[79] *R. Siepmann, H. Schnering*, Z. Anorg. Allgem. Chemie *357*, 289 (1968).

[80] *H. Schäfer, H. Schnering, R. Siepmann, A. Simon, D. Giegling, D. Bauer, B. Spreckelmeyer*, J. Less-Common Metals *10*, 154 (1966).

[81] *J. Fergusson*, Halogen Chem. *3*, 227 (1967).

nuclear cluster is retained in this latter halide, but it is now the $[W_6Cl_{12}]^{6\oplus}$ unit, similar in structure to the $[Nb_6Cl_{12}]^{2\oplus}$ moiety, and the halide is formulated as $[W_6Cl_{12}]Cl_6$.[78]

Tungsten(II) bromide is most readily prepared from the disproportionation of tungsten(IV) bromide (see Section 16.5.1) above 450°[9] or by the reduction of tungsten(V) bromide with aluminum in a thermal gradient.[82] Recrystallization from a hydrobromic acid–ethanol mixture yields the hydrate, $[W_6Br_8]Br_4 \cdot 2H_2O$, which on heating gives pure tungsten(II) bromide. Tungsten(II) bromide is isostructural with tungsten(II) chloride[77] and is best represented as $[W_6Br_8]$-$Br_2Br_{4/2}$.

The reaction of tungsten(II) bromide with excess bromine appears to be more complex than the analogous chloride reaction, and compounds with the stoichiometries W_6Br_{14}, W_6Br_{16}, and W_6Br_{18} have been obtained.[83–86] *W_6Br_{14}* is obtained by heating W_6Br_{16} or W_6Br_{18} *in vacuo* at 200° or 110°, respectively, but it is not formed directly from the reaction of tungsten(II) bromide with bromine.[85,86] It is formulated as $[W_6Br_8]Br_6$, *i.e.*, six terminal bromine atoms.[87] W_6Br_{16} is formed by heating tungsten(II) bromide with elemental bromine at 130°–140° *in vacuo*.[85,86] The structure has been determined,[79] and is found to consist of $[W_6Br_8]^{6\oplus}$ units linked by two linear $Br_4^{2\ominus}$ groups and to have four terminal bromine atoms around the cluster; it may be represented as $[W_6Br_8]Br_4[Br_4]_{2/2}$. There are two different clusters with the formulation *W_6Br_{18}*. One, which is formed by the action of bromine on tungsten(II) bromide at temperatures between 80° and 100°,[85,86] exists in two crystalline modifications (α and β forms), and the other is formed [together with tungsten(VI) bromide] by heating tungsten(II) bromide with bromine at 230° under 10^{-4} Torr pressure.[87] The low temperature form is believed to contain the $[W_6Br_8]^{6\oplus}$ cluster, and is formulated as $[W_6Br_8]Br_2[Br_4]_{4/2}$.[87] The high temperature form is isostructural[87] with tungsten(III) chloride and it is thus formulated as $[W_6Br_{12}]Br_6$.

Tungsten(II) iodide may be prepared from the chloride by halogen exchange in a molten mixture of potassium and lithium iodides at 540° for 15 hr.[82] An alternative method is to take the tungsten(III) iodide (see below) and heat it at 600° under nitrogen.[88] The diiodide thus obtained is isostructural with the dichloride and dibromide.

Recently, a compound formulated as W_6I_{15} has been prepared by the reaction of iodine with tungsten(II) iodide at 350° *in vacuo*.[88] It is postulated to have a structure represented by $[W_6I_8]I_4[I_3]_{2/2}$.[88]

A compound with the stoichiometry WI_3, which is *not* isomorphous with MoI_3, has been prepared by the action of iodine on tungsten hexacarbonyl at 120°.[89] It is not certain that this *tungsten(III) iodide* contains a hexanuclear cluster, but it does provide a useful route to the diiodide.

16.4.1.2 Mixed Halide Clusters

Mixed tungsten(II) halides of the type $[W_6X_8]Y_4$ (where $X \neq Y$) are expected from a consideration of the structure of the dihalides and by analogy with the molybdenum systems. Recently,[82] several of these have been prepared by dissolving tungsten(II) halides in solutions of halogen acids, HY, and ethanol, collecting the halo acids, $[H_3O]_2[W_6X_8]Y_6 \cdot nH_2O$, which are formed, and heating them at 325°. By this means it has been possible to obtain $[W_6Cl_8]Br_4$, $[W_6Cl_8]I_4$, and $[W_6Br_8]Cl_4$. The mixed cluster $[W_6Br_8]F_4$ has also been obtained, by the action of silver perchlorate, followed by HF, on $[W_6Br_8]Br_4$.[82]

16.4.1.3 Hexahalo Salts of $[W_6X_8]^{n\oplus}$

These salts may be readily prepared by dissolving the dihalides in hot concentrated halogen acid, with a little ethanol to aid dissolution and adding cesium or tetraalkylammonium halides.[82] Compounds of general formula $M_2[W_6X_8]X_6$ (where M = Cs or R_4N and X = Cl, Br, or I) are obtained. Mixed products are also possible.

16.4.1.4 Neutral Adducts

Hydrated derivatives, $[W_6X_8]Y_4 \cdot 2H_2O$ (X, Y = Cl, Br, or I), may be obtained by heating the corresponding halo acids at 200° *in vacuo* rather than at 325° which gives the anhydrous halides. Derivatives of simple monodentate ligands, L, e.g., acetonitrile, may be obtained simply by stirring tungsten(II) halides with the ligand; the resulting compounds are formulated as $[W_6X_8]$-$X_4 \cdot 2L$.[82] No work has yet been reported involv-

[82] *R. Hogue, R. McCarley*, Inorg. Chem. *9*, 1354 (1970).
[83] *H. Schäfer, R. Siepmann*, J. Less-Common Metals *11*, 76 (1966).
[84] *R. McCarley, T. Brown*, J. Amer. Chem. Soc. *84*, 3216 (1962).
[85] *H. Schäfer, R. Siepmann*, Naturwissenschaften *52*, 344 (1965).
[86] *H. Schäfer, R. Siepmann*, Z. Anorg. Allgem. Chemie *357*, 273 (1968).
[87] *U. Lange, H. Schäfer*, J. Less-Common Metals *21*, 472 (1970).
[88] *H. Schulz, R. Siepmann, H. Schäfer*, J. Less-Common Metals *22*, 136 (1970).
[89] *C. Djordjevic, R. Nyholm, C. Pande, M. Stiddard*, J. Chem. Soc. *A*, 16 (1966).

ing the reaction of bidentate ligands with the dihalides.

16.4.2 Further Halogen Compounds

The best known halogen compounds of tungsten(III) are the *tri-μ-halobis[trichlorotungstate(III)] anions* $W_2X_9^{3\ominus}$ (X=Cl[90,91] or Br,[92,93]), prepared by the reduction (generally using tin) of tungsten(VI) in hydrochloric or hydrobromic acids. The *chloride*, $K_3W_2Cl_9$, was reported[94] to react with refluxing pyridine or aniline under nitrogen to produce the dimers $W_2Cl_6\cdot3$-(ligand); however, a reinvestigation has led to the product being reformulated as $W_2Cl_6\cdot4$-(pyridine), a formulation confirmed by a recent structural investigation.[95] In addition, the parent chloride, $W_2Cl_9^{3\ominus}$, will undergo a one-electron oxidation[96] when treated with elemental halogen in dichloromethane according to Eqn (1).

$$2W_2Cl_9^{3\ominus}+X_2 \longrightarrow 2W_2Cl_9^{2\ominus}+2X^{\ominus} \qquad (1)$$
X = Cl, Br, or I

The anion has been separated as the violet tetra-*n*-butylammonium salt. Finally, the compounds *tetrachlorobis(alkoxide)bisalcoholditungsten(III)* may be obtained[97] by refluxing $W_2Cl_9^{3\ominus}$ with alcohols, ROH (R=methyl, ethyl, or *n*-propyl).

Other tungsten(III) compounds of importance are *trichlorobis(pyridine)tungsten(III)*, $WCl_3\cdot2$-(pyr), obtained[98] from the reflux of tetrachlorobis(acetonitrile)tungsten(IV), $WCl_4\cdot2MeCN$ (see 16.5.3.2) with pyridine, and two derivatives of 1,2-bisdiphenylphosphinoethane, (DPPE). *Trichloro(diphenylphosphinoethane)tungsten(III)*, $WCl_3(DPPE)$, which is postulated to be dimeric, is obtained from the reaction of $WCl_4\cdot2MeCN$ with DPPE in acetonitrile, and *trichlorobis(diphenylphosphinoethane)tungsten(III)*, $WCl_3\cdot2$-(DPPE), which is apparently seven-coordinate, is formed when $WCl_4\cdot2$(propionitrile) is reacted with molten DPPE at 180°.[99]

Finally, mention must be made of the monomeric tungsten(II) compound *diiodobis(o-phenylenebisdimethylarsine)tungsten(II)*, $WI_2(diars)_2$, which is made[89] by the reaction of tungsten(III) iodide with an excess of molten diarsine at 165°. It is of importance since it completes the series of compounds of stoichiometry $MI_2(diars)_2$ of the second and third row transition metals in Groups VI, VII, and VIII.

16.4.3 Cyanotungstates

The hydrogen reduction of potassium octacyanotungstate, $K_4W(CN)_8$, at 390°[100] causes the evolution of hydrogen cyanide and the formation of the tungsten(II) cyano compound, *tetrapotassium hexacyanotungsten(II)*, $K_4W(CN)_6$. Extraction of this compound with methanol in dry air leaves a brown product which appears to be the tungsten(III) cyano compound, *tripotassium hexacyanotungsten(III)*, $K_3W(CN)_6$.

16.5 Tungsten(IV)

16.5.1 Halides

Tungsten(IV) fluoride may be prepared by reducing tungsten hexafluoride with benzene in a nickel bomb at 110° for from 3 to 9 days.[101] It is stable up to 800° *in vacuo*, but decomposes into its constituent elements above this temperature.

Tungsten(IV) chloride can be prepared by the reduction of tungsten(VI) chloride with a variety of reducing agents including hydrogen,[102] red or white phosphorus,[103] and aluminum metal.[9] The reaction with aluminum, which is the most efficient and gives a purer product than other methods, is carried out in a sealed, evacuated tube in a temperature gradient.[9] Tungsten(IV) chloride disproportionates above 400° to tungsten(II) chloride and tungsten(V) chloride.[9]

Tungsten(IV) bromide is formed when tungsten(V) bromide is reduced using tungsten[84] or

---

[90] *E. Heintz*, Inorg. Synth. *7*, 142 (1963).
[91] *R. Saillant, J. Hayden, R. Wentworth*, Inorg. Chem. *6*, 1497 (1967).
[92] *R. Young*, J. Amer. Chem. Soc. *54*, 4515 (1932).
[93] *J. Hayden, R. Wentworth*, J. Amer. Chem. Soc. *90* 5291 (1968).
[94] *H. Jonassen, S. Cantor, A. Tarsey*, J. Amer. Chem. Soc. *78*, 271 (1956).
[95] *R. Jackson, W. Streib*, Inorg. Chem. *10*, 1760 (1971).
[96] *R. Saillant, R. Wentworth*, J. Amer. Chem. Soc. *91*, 2174 (1969).
[97] *P. Clark, R. Wentworth*, Inorg. Chem. *8*, 1223 (1969).
[98] *D. Blight, D. Kepert*, J. Chem. Soc. *A*, 534 (1968).
[99] *P. Boorman, N. Greenwood, M. Hildon*, J. Chem. Soc. *A*, 2466 (1968).
[100] *J. Yoo, E. Griswold, J. Kleinberg*, Inorg. Chem. *4*, 365 (1965).
[101] *H. Priest, W. Schumb*, J. Amer. Chem. Soc. *70*, 3378 (1948).
[102] *W. Biltz, G. Fendius*, Z. Anorg. Allgem. Chemie *172*, 385 (1928).
[103] *G. Novikov, N. Andreeva, O. Polyachenok*, Zh. Neorg. Khim. *6*, 1990 (1961); Engl. 1019.

aluminum[9] metal. Once again, the aluminum reduction is the more efficient and is carried out in a similar fashion to that for the production of the chloride. Tungsten(IV) bromide is also reported[81] to be formed by heating tungsten(V) bromide at 180° *in vacuo*. In a similar fashion to the chloride, the bromide disproportionates above 400° to tungsten(II) bromide and tungsten(V) bromide.[9]

Tungsten(IV) iodide is prepared by the reaction between tungsten(VI) oxide and aluminum triiodide at 230° according to Eqn (2)[104] or by

$$2AlI_3 + WO_3 \longrightarrow WI_4 + I_2 + Al_2O_3 \qquad (2)$$

the reaction of tungsten(VI) chloride with hydrogen iodide at 100°.[105] Tungsten(IV) iodide decomposes readily to give tungsten(II) iodide and iodine.[106]

16.5.2 Tungsten(IV) Oxide Halides

Tungsten(IV) oxide fluoride, WOF_2, was reported by Priest and Schumb[101] to be formed by the reaction of tungsten(IV) oxide with anhydrous hydrogen fluoride at 500°–650°. More recently, its existence has been refuted by Moldavskii *et al.*[107]

Tungsten(IV) oxide chloride, $WOCl_2$, may be prepared from tungsten(V) oxide trichloride, $WOCl_3$ (see 16.6.2), by heating it at temperatures up to 400° *in vacuo*.[108,109] It may also be prepared by heating tungsten(VI) oxide tetrachloride, $WOCl_4$ (see 16.8.3), with stannous chloride at 240° for 3 to 4 hr.[110] In addition, the reaction of a 1:1:1 molar mixture of tungsten metal, tungsten(VI) oxide, and tungsten(VI) chloride (see 16.8.2) in a temperature gradient yields the oxide dichloride.[111]

Tungsten(IV) oxide bromide, $WOBr_2$, has recently been prepared[112] by heating a 2:1:3 molar mixture of tungsten metal, tungsten(VI) oxide, and bromine in a temperature gradient. The experimental conditions have to be accurately controlled because it is possible to obtain an impure compound of empirical composition $W_2O_3Br_3$.[112]

16.5.3 Neutral Complexes of Tungsten(IV) Halides

The oxide halides of tungsten(IV) have been found to be largely unreactive,[109] and will not be considered further.

16.5.3.1 Complexes with Group VI Ligands

It has been found that by reacting tungsten(IV) chloride with methanol at temperatures up to 25°, a green compound of stoichiometry $W_2Cl_4(OMe)_4(MeOH)_2$, *tetrachlorotetramethoxybismethanolditungsten(IV)*, may be obtained.[113] Analogous products may be obtained using ethanol, *n*-propanol, or isopropanol. Also, the two ethanol molecules in the ethanol adduct may be replaced by pyridine. If tetrachlorotetraethoxybisethanolditungsten(IV) is reacted with a solution of potassium ethoxide in ethanol the complex *bischlorohexa(ethoxy)bisethanolditungsten(IV)*, $W_2Cl_2(OEt)_6(EtOH)_2$, is produced.[113] Finally, the reaction of tungsten(V) chloride (see 16.6.1) with methanol and pyridine[114] will produce *bischlorobismethoxybispyridinetungsten(IV)*, $WCl_2(OMe)_2 \cdot 2(pyr)$.

16.5.3.2 Complexes with Group V Ligands

The reaction of tungsten(IV) halides with a variety of nitrogen donors, *e.g.*, alkyl nitriles and heterocyclic tertiary amines, gives compounds of stoichiometry $WX_4 \cdot 2L$[9] (X=halogen). The nitriles are often used as intermediates in the formation of other complexes, *e.g.*, $WCl_4 \cdot 2PPh_3$, the nitrile molecules being relatively easily exchanged.[115] Compounds with the same stoichiometry can also be obtained from the reduction of tungsten(V) and (VI) halides with the same types of ligand.[9,115] Alternative routes to compounds of the same general type are: (*a*) *via* the hexahalotungstate(IV) salts, $WX_6^{2\ominus}$, with pyridine[98,116] and 1,2-bisdiphenylphosphinoethane [DPPE],[99] forming $WCl_4 \cdot 2(pyr)$ and

[104] *M. Chaigneau*, Bull. Soc. Chim. France 886 (1957).
[105] *Gmelin*, Handbuch der Anorganischen Chemie; Wolfram. Verlag Chemie, Berlin 1935.
[106] *L. Brewer, L. Bromley, P. Gilles, N. Lefgren* in *L. Quill*, The Chemistry and Metallurgy of Miscellaneous Materials. p. 276. McGraw-Hill, New York 1950.
[107] *D. Moldavskii, V. Temchenko, N. Podpadaeva*, Zh. Obshch. Khim. *38*, 2125 (1968); Engl. 2026.
[108] *M. Mercer*, J. Chem. Soc. *A*, 2019 (1969).
[109] *P. Crouch, G. Fowles, I. Tomkins, R. Walton*, J. Chem. Soc. *A*, 2412 (1969).
[110] *S. Eliseev, I. Glukhov, N. Gaidaenko*, Zh. Neorg. Khim. *14*, 627 (1969); Engl. 328.
[111] *J. Tillack, R. Kaiser, G. Fisher, P. Eckerlin*, J. Less-Common Metals *20*, 171 (1970).
[112] *J. Tillack, R. Kaiser*, Angew. Chem. *81*, 149 (1969); Engl. *8*, 142.
[113] *W. Reagan, C. Brubaker*, Inorg. Chem. *9*, 827 (1970).
[114] *H. Funk, H. Naumann*, Z. Anorg. Allgem. Chemie *343*, 294 (1966).
[115] *E. Allen, B. Brisdon, G. Fowles*, J. Chem. Soc. *4531* (1964).
[116] *C. Kennedy, R. Peacock*, J. Chem. Soc. 3392 (1963).

WCl_4(DPPE), respectively; (*b*) *via* the tetracarbonylbis(ligand)tungsten(0) compounds, $W(CO)_4L_2$, by their reaction with chlorine or bromine to form, *e.g.*, tetrabromo(bipyridyl)-tungsten(IV),[117] WBr_4(bipy), or tetrachlorobis-(dimethylphenylphosphine)tungsten(IV), $WCl_4 \cdot (PMe_2Ph)_2$.[118] The latter compound will react reversibly with a further mole of dimethyl-phenylphosphine to give the seven-coordinate compound *tetrachlorotris(dimethylphenylphosphine)tungsten(IV)*, $WCl_4(PMe_2Ph)_3$.

A slightly different compound from those above is formed by the prolonged reaction of tungsten(VI) chloride with secondary amines; this gives aminolyzed products of stoichiometry $WCl_3(NR_2) \cdot 2NHR_2$[119] (where R = alkyl group).

16.5.4 Anionic Halide Complexes

Hexahalo salts of tungsten(IV) of the type M_2WX_6 (M = cation, X = Cl, Br) may be prepared by the reaction of the corresponding tungsten(VI) halide with metal iodides[116] in a sealed tube at 130°. By this means salts with M = K, Rb, Cs, Tl, and Ba have been prepared; surprisingly the sodium salt is isolated in an impure state. Dickinson *et al.*[120] found that by heating the hexachlorotungstate(V) salts (see 16.6.4) at 250°–300° *in vacuo*, they were able to produce the hexachlorotungstate(IV) salt and tungsten(VI) chloride; by grinding an equimolar mixture of the two, the hexachloro-tungstate(V) salts could be regenerated. Hexa-chlorotungsten(IV) salts with dialkylammonium cations may be formed by the reaction of certain secondary amines with tungsten(VI) chloride at −34° for 2 hr[119]; further reaction produces the $WCl_3(NR_2) \cdot 2NHR_2$ mentioned earlier (see 16.5.3).

Reduction of a hydrochloric acid solution of potassium tungstate with tin,[121,122] using a slightly different procedure than that adopted for the preparation of tri-μ-chlorobis[trichloro-tungstate(III)] (see Section 16.4.5), produces a green anionic complex which was earlier thought[123] to be of composition $K_2W(OH)Cl_5$. This has since been reformulated[122] as *tetra*-potassium *μ-oxobis[pentachlorotungsten(IV)]*, $K_4[W_2OCl_{10}]$, a reformulation which has been confirmed.[124] This compound has been postulated to have a bent W–O–W bond and to be a mixed valence state tungsten(III)–tungsten(V) complex.[124]

16.5.5 Compounds of Tungsten(IV) with Group VI

16.5.5.1 Oxygen

The simple, binary *tungsten(IV) oxide*, WO_2, may be prepared by the hydrogen or ammonia reduction of tungsten(VI) oxide[125] or by heating a mixture of tungsten metal powder and tungsten(VI) oxide in a sealed, evacuated tube at 950° for 40 hr.[126]

Other oxygen-containing compounds include *tetrakis(8-quinolinato)tungsten(IV)*, $W[C_9H_6NO]_4$, which was postulated to be eight-coordinate.[127] This has now been confirmed by the recent X-ray crystal structure of the analogous 5-bromo-8-quinolinato compound, $W[C_9H_5NOBr]_4$.[128] There are also some incompletely characterized compounds obtained from the prolonged reflux of tungsten hexacarbonyl with acetic acid,[129] which are proposed to be trinuclear.[129]

16.5.5.2 Sulfur

Binary tungsten(IV) sulfide is the only sulfide formed directly from the elements.[130] It may be prepared by heating sulfur with either tungsten metal, tungsten(IV) oxide, or tungsten(VI) oxide at 900° under nitrogen,[131] by heating tungsten(VI) sulfide above 170°,[130,132] or by the action of hydrogen sulfide on either tungsten metal[133] or tungsten(VI) oxide.[134] A rhombohedral form may be obtained by heating 1:2 mixtures of

117 *C. Hull, M. Stiddard*, J. Chem. Soc. *A*, 1633 (1966).
118 *J. Moss, B. Shaw*, J. Chem. Soc. *A*, 595 (1970).
119 *B. Brisdon, G. Fowles, B. Osborne*, J. Chem. Soc. *1330* (1962).
120 *R. Dickinson, S. Feil, F. Collier, W. Horner, S. Horner, S. Tyree*, Inorg. Chem. *3*, 1600 (1964).
121 *O. Olsson*, Ber. *46*, 566 (1913).
122 *R. Colton, G. Rose*, Australian J. Chem. *21*, 883 (1968).
123 *E. König*, Inorg. Chem. *2*, 1238 (1963).
124 *E. König*, Inorg. Chem. *8*, 1278 (1969).
125 *F. Zado*, J. Inorg. Nucl. Chem. *25*, 1115 (1963).
126 *A. Magneli*, Ark. Kemi. Mineralogi Och Geologi *24(2)* 1 (1947).
127 *R. Archer, W. Bonds*, J. Amer. Chem. Soc. *89*, 2236 (1967).
128 *W. Bonds, R. Archer, W. Hamilton*, Inorg. Chem. *10*, 1764 (1971).
129 *T. Stephenson, D. Whittaker*, Inorg. Nucl. Chem. Lett. *5*, 569 (1969).
130 *F. Jellinek*, Ark. Kemi. *20*, 447 (1963).
131 *O. Glemser, H. Sauer, P. König*, Z. Anorg. Allgem. Chemie *257*, 241 (1948).
132 *J. Wildervanck, F. Jellinek*, Z. Anorg. Allgem. Chemie *328*, 309 (1964).
133 *F. Le Boete, C. Mathiron, S. Toesca, D. Delafosse, J. Colson*, Bull. Soc. Chim. France 3869 (1969).
134 *F. Le Boete, J. Colson*, C.R. Acad. Sci. Paris Ser. C *268*, 2142 (1969).

tungsten and sulfur at 1800° under pressures ~45 kbars.[135]

Tetrakis [N,N - diethyldithiocarbamate] tungsten - (IV), W(dtc)$_4$, is reported[136] to be produced from the reaction of tetrachlorobisacetonitriletungsten(IV) with sodium diethyldithiocarbamate in acetonitrile; it is believed to be eight-coordinate.

Tungsten(IV) forms two major dithiolene compounds, *bis(tetraphenylarsonium)tris(maleonitriledithiolato)tungstate(IV)*, $(Ph_4As)_2W(mnt)_3$, formed by the reaction in ethanol of tungsten(V) chloride, disodium maleonitriledithiolate, and tetraphenylarsonium chloride,[137] and *bis(tetraethylammonium)tris(tetrachlorobenzene-1,2-dithiolato)tungstate(IV)*, $(Et_4N)_2W(tcbt)_3$, prepared similarly from tungsten(VI) dioxide dichloride, WO_2Cl_2, tetrachlorobenzene-1,2-dithiol, and tetraethylammonium bromide.[138] No firm conclusions as to the structures of these compounds have yet been reached.

16.5.5.3 Selenium and Tellurium

The only well-known compound of selenium and tungsten(IV) is the binary *selenide*, WSe_2. This is prepared by heating stoichiometric mixtures of tungsten metal and selenium in evacuated, sealed ampoules at 600°–700°.[139, 140]

The only *telluride* of tungsten is the tungsten(IV) telluride, WTe_2. It is prepared in an identical fashion to the selenide.[139, 141]

16.5.6 Octacyanotungstate(IV) anion

The octacyanotungstate(IV) anion, $W(CN)_8{}^{4\ominus}$, has been of interest for a number of years because of its unusual coordination number and structure[142–144] and its interesting photochemical behavior.[145–147] The anion is prepared by the reaction between potassium cyanide and tripotassium tri-μ-chlorobis[trichlorotungstate(III)], $K_3W_2Cl_9$,[148–151] or from the reduction of sodium oxalatotungstate with tin in the presence of potassium cyanide.[152]

An interesting reaction of the octacyanotungstate(IV) anion is that of its silver salt with alkyl halides to form neutral *tetracyanotetraisonitriletungsten(IV)* compounds, $W(CN)_4(CNR)_4$ (where R = ethyl, 1- or 2-propyl, *t*-butyl, and benzyl).[153]

16.6 Tungsten(V)

16.6.1 Halides

Tungsten(V) fluoride, isomorphous with molybdenum(V) fluoride,[154] has recently been prepared by the reaction of tungsten(VI) fluoride, at approximately 10 Torr pressure, with tungsten wires heated to 500°–700°.[155, 156] It disproportionates to tungsten(IV) and tungsten(VI) fluorides below 50°.[156]

Tungsten(V) chloride may be prepared by the reduction of tungsten(VI) chloride in a hydrogen stream at 320°–350°,[102] but this reaction is slow and gives impure products. Reduction with red phosphorus[103] at 250°–300° will also produce the pentachloride, but contamination with phosphorus is sometimes a problem. Reduction with aluminum at 300° is reported[156a] to give a pure product and is the most satisfactory reduction method. By boiling tungsten(VI) chloride under nitrogen and repeatedly distilling the de-

---

135 *M. Silverman*, Inorg. Chem. *6*, 1063 (1967).

136 *J. Smith, T. Brown*, Inorg. Nucl. Chem. Lett. *6*, 441 (1970).
P. Heckley, D. Holah, Inorg. Nucl. Chem. Lett. *6*, 865 (1970).

137 *E. Stiefel, L. Bennett, Z. Dori, T. Crawford, C. Simo, H. Gray*, Inorg. Chem. *9*, 281 (1970).

138 *E. Wharton, J. McCleverty*, J. Chem. Soc. *A*, 2258 (1969).

139 *L. Brixner*, J. Inorg. Nucl. Chem. *24*, 257 (1962).

140 *R. Kershaw, M. Vlasse, A. Wold*, Inorg. Chem. *6*, 1599 (1967).

141 *O. Knop, H. Haraldsen*, Can. J. Chem. *34*, 1142 (1956).

142 *J. Hoard, H. Nordsieck*, J. Amer. Chem. Soc. *61*, 2853 (1939).

143 *S. Basson, L. Bok, J. Leipoldt*, Acta Cryst. *B26*, 1209 (1970).

144 *R. Parish*, Spectrochim. Acta *22*, 1191 (1966).

145 *V. Balzani, V. Carassiti*, Photochemistry of Coordination Compounds, p. 123. Academic Press, London, New York 1970.

146 *S. Lippard, H. Nozaki, B. Russ*, Chem. Commun. 118 (1967).

147 *S. Lippard*, Progr. Inorg. Chem. *8*, 109 (1967).

148 *O. Olsson*, Z. Anorg. Allgem. Chemie *88*, 49 (1914).

149 *H. Baadsgaard, W. Treadwell*, Helv. Chim. Acta *38*, 1669 (1955).

150 *E. Goodenow, C. Garner*, J. Amer. Chem. Soc. *77*, 5268 (1955).

151 *K. Mikhalevich, V. Litvinchuk*, Zh. Neorg. Khim. *3*, 1846 (1958); Engl. *3* (*8*) 177.

152 *A. Kosinska, Z. Wilczewska-Stasicka*, Roczniki Chem. *31*, 1029 (1957).

153 *H. Latka*, Z. Anorg. Allgem. Chemie *353*, 243 (1967).

154 *A. Edwards*, J. Chem. Soc. *A*, 909 (1969).

155 *J. Schröder, F. Grewe*, Angew. Chem. *80*, 118 (1968); Engl. *7*, 132.

156 *J. Schröder, F. Grewe*, Chem. Ber. *103*, 1536 (1970).

156a *J. Moss, B. Shaw*, Chem. Commun 632 (1968).

composition products, it is possible to obtain the pentachloride,[157] which is also obtained from the disproportionation of tungsten(IV) chloride (see 16.5.1.2). A method which is reported to give at least a 90% yield of product[158] is to stir tungsten(VI) chloride in tetrachloroethylene at 100° for 24 hr in the presence of strong light.

Tungsten(V) bromide may be prepared by reacting tungsten metal powder with bromine vapor diluted with nitrogen at 1000°.[115,159] This halide is also formed[160] by the reaction of tungsten metal with bromine in a sealed tube at 600°–800°, but this latter method does not seem to be as satisfactory.

16.6.2 Tungsten(V) Oxide, Sulfide, and Nitride Halides

For oxidation state (V) we have the possibility of oxide halides of stoichiometries WOX_3 and WO_2X. Several of each class are known.

The *tungsten(V) oxide trichloride*, $WOCl_3$, may be prepared by the reduction with aluminum of tungsten(VI) oxide tetrachloride, $WOCl_4$, (see 16.8.3) in a sealed, evacuated tube at 100°–140°.[161,162] It is also formed by the reaction of tungsten(V) chloride with antimony trioxide in a sealed, evacuated tube at 100°.[109]

The *tungsten(V) dioxide chloride*, WO_2Cl, has recently been reported to be formed by the reaction of tungsten(VI) dioxide dichloride, WO_2Cl_2, (see 16.8.3) with stannous chloride at 250° in a sealed tube.[163]

The only known *tungsten(V)* sulfide chloride is the *sulfide trichloride*, $WSCl_3$, which may be prepared by the reaction of tungsten(V) chloride with antimony trisulfide at 150° in an evacuated, sealed tube.[164]

Recently, the *tungsten(V) nitride dichloride*, $WNCl_2$, has been prepared by the reaction of tungsten(V) chloride with ammonium chloride.[164a]

Tungsten(V) oxide tribromide may be prepared in an analogous fashion to the oxide trichloride by reduction of the tungsten(VI) oxide tetrabromide (see 16.8.3) with aluminum powder at 190°[161,162] or by the reaction of tungsten(V) bromide with antimony trioxide at 150°.[109] It may also be prepared by the reaction of a 2:1:4·5 molar mixture of tungsten:tungsten (VI) oxide:bromine in a temperature gradient.[165]

The only known *tungsten(V)* oxide iodide is the tungsten(V) *dioxide iodide*, WO_2I, prepared by heating a 1:2:1·5 molar mixture of tungsten metal:tungsten(VI) oxide:iodine in a temperature gradient.[166]

16.6.3 Neutral Complexes of Tungsten(V) Halides and Oxide Halides

There are not a large number of neutral complexes of tungsten(V) halides, since most ligands cause reduction to tungsten(IV) species.[115] The complexes of the oxide trihalides are generally of the type, $WOX_3 \cdot 2L$ (X = halide, L = monodentate or $\frac{1}{2}$ bidentate ligand), although a few 1:1 adducts WOX_3L, are known.

16.6.3.1 Complexes with Group VI Ligands

A number of alkoxide complexes of tungsten(V) are known. They are prepared by reacting tungsten(V) chloride[113,167] and bromide[114] with alcohols and sodium alkoxides at temperatures below 0°; they include *tetrachlorohexa(alkoxy)ditungsten(V)*, $W_2Cl_4(OR)_6$ and *bischlorooctaalkoxyditungsten(V)*, $W_2Cl_2(OR)_8$. Other complexes of tungsten(V) halides with oxygen ligands such as phenol and a variety of carboxylic acids and their derivatives have been prepared.[159,168]

16.6.3.2 Complexes with Group V Ligands

The reaction of tungsten(V) chloride or bromide with 2,4,6-trimethylpyridine or benzonitrile in dichloromethane solution produces complexes of stoichiometry $WX_5 \cdot 2L$ (where X = Cl or Br

---

157 *R. Colton, I. Tomkins*, Australian J. Chem. *19*, 759 (1966).
158 *T. Brown, E. McCann*, Inorg. Chem. *7*, 1227 (1968).
159 *H. Funk, H. Schauer*, Z. Anorg. Allgem. Chemie *306*, 203 (1960).
160 *S. Shchukarev, G. Kokovin*, Zh. Neorg. Khim. *9*, 1309 (1964); Eng., 715.
161 *G. Fowles, J. Frost*, Chem. Commun 252 (1966).
162 *P. Crouch, G. Fowles, J. Frost, P. Marshall, R. Walton*, J. Chem. Soc. *A*, 1061 (1968).
163 *S. Eliseev, I. Glukhov, N. Gaidaenko*, Zh. Neorg. Khim. *15*, 2243 (1970); Engl. 1158.
164 *D. Britnell, G. Fowles, R. Mandyczewsky*, Chem. Commun 608 (1970).
164a *A. Kuznetsova, Y. Buslaev, L. Goryachova, Y. Podzolko*, Izv. Akad. Nauk. SSSR. Ser. Khim. 463 (1970); Engl. 416.
165 *J. Tillack, R. Kaiser*, Angew Chem. *80*, 286 (1968). Engl. *7*, 294.
166 *J. Tillack*, Z. Anorg. Allgem. Chemie *357*, 11 (1968).
167 *D. Rillema, W. Reagan, C. Brubaker*, Inorg. Chem. *8*, 587 (1969).
168 *H. Funk, H. Hoppe*, Z. Chem. *8*, 31 (1968).

and L=monodentate ligand).[169] Similar products may be obtained by reacting tungsten(V) chloride with ligands in carbon tetrachloride or benzene[170,171]; the ligands in this case include 1,2-bisdiphenylphosphinoethane, 2,2′-bipyridyl, 1,10-phenanthroline, and pyridine. A typical reaction product is *pentachloro(bipyridyl)tungsten(V)*, $WCl_5 \cdot bipy$.[171] These complexes are thought to be ionic, $[WX_4L_2]X$, rather than seven-coordinate neutral species. Funk and Hoppe[168] isolated products of the type $WX_5 \cdot MeCN$ and $WX_5 \cdot PhCN$ from reactions similar to those mentioned above; the source of this apparent discrepancy with the results of Brown and Ruble[169] is still uncertain, although the neutral 1:1 adducts may be intermediates in the formation of the ionic derivatives $[WX_4L_2]X$.

The oxide trihalides form adducts of stoichiometry $WOX_3 \cdot 2L$, *e.g.*, $WOCl_3 \cdot 2MeCN$, by direct reaction with liquid ligands such as acetonitrile.[172] In certain cases only one monodentate ligand molecule will coordinate by direct reaction, *e.g.*, pyridine affords *oxotrichloro(pyridine)-tungsten(V)*, $WOCl_3 \cdot pyr$.[172] It has been suggested[172] that these 1:1 adducts are formed because the polymeric tungsten(V) oxide chloride is not completely disrupted and the strong W–O–W bridges, present in the parent oxide trihalides, are retained in these derivatives.

16.6.4 Anionic Halide Complexes

Hexahalo complexes of tungsten(V) of the type MWX_6 (where M=alkali metal or alkylammonium cation and X=F, Cl, Br) are known. The fluorides are prepared by the action of lithium iodide on tungsten(VI) fluoride in liquid sulfur dioxide.[173,174] The *octafluorotungstate(V) anion*, $WF_8^{3\ominus}$, has also been reported.[175] The chlorides can be prepared with alkali metal cations by reacting alkali metal iodides with equimolar quantities of tungsten(VI) chloride at 80°–130°,[120] and also by grinding an equimolar mixture of the hexahalotungsten(IV) salt with tungsten(VI) chloride.[120] Salts with alkylammonium cations may be prepared by the reaction of tungsten(VI) chloride with the alkylammonium chloride in thionyl chloride[176] or by the reaction of equimolar quantities of tungsten(V) chloride and alkylammonium chloride in chloroform.[177] The corresponding *hexabromotungstate(V) anion* is produced by the reaction of equimolar quantities of tungsten(V) bromide with alkylammonium bromides (e.g., Et_4NBr) in chloroform.[178]

The *tungsten(V) oxopentahalo* and *tetrahalo anions*, $WOX_5^{2\ominus}$ and $WOX_4^{\ominus}$ (X=Br, Cl), are well-established compounds. The former are prepared by treating a solution of tungsten(V) halide in hydrochloric acid with alkali or alkylammonium halides[179,180]; large cations, *e.g.*, quinolinium cation, favor the precipitation of the oxotetrahalo salts.[179,180]

The *haloalkoxotungstate(V) anions*, *e.g.*, tetrachlorobisethoxotungstate(V), $[WCl_4(OEt)_2]^{\ominus}$, have been prepared by treating solutions of tungsten(V) chloride in methanol or ethanol with tetraalkylammonium salts.[113,114,167,181]

The salt, *potassium trichlorotristhiocyanato-tungstate(V)*, $K[W(NCS)_3Cl_3]$, has recently been prepared[182] by the reaction of potassium thiocyanate with tungsten(V) chloride in acetone.

16.6.5 Compounds of Tungsten(V) with Groups VI, V, and IV Ligands

Pentaalkoxides of the type $W(OR)_5$ have been prepared, *e.g.*, *tungsten(V) ethoxide*,[113] from the reaction of tungsten(V) chloride in ethanol with sodium ethoxide. *Tungsten(V) phenoxide* has been made by the reduction of tungsten(VI) phenoxide.[183]

Thiocyanate salts of the type $[RH]_2WO[NCS]_5$, $[RH]_4W_2O_3[NCS]_8$, $[RH]_2WO_2[NCS]_3$, and

---

[169] *T. Brown, B. Ruble*, Inorg. Chem. *6*, 1335 (1967).
[170] *P. Boorman, N. Greenwood, M. Hildon, R. Parish*, Inorg. Nucl. Chem. Lett. *2*, 377 (1966).
[171] *P. Boorman, N. Greenwood, M. Hildon, R. Parish*, J. Chem. Soc. *A*, 2002 (1968).
[172] *P. Crouch, G. Fowles, P. Marshall, R. Walton*, J. Chem. Soc. *A*, 1634 (1968).
[173] *G. Hargreaves, R. Peacock*, J. Chem. Soc. 4212 (1957).
[174] *R. Kemmitt, D. Russell, D. Sharp*, J. Chem. Soc. 4408 (1963).
[175] *G. Hargreaves, R. Peacock*, J. Chem. Soc. 2170 (1958).
[176] *D. Adams, J. Chatt, J. Davidson, J. Gerratt*, J. Chem. Soc. 2189 (1963).
[177] *B. Brisdon, R. Walton*, J. Inorg. Nucl. Chem. *27*, 1101 (1965).
[178] *B. Brisdon, R. Walton*, J. Chem. Soc. 2274 (1965).
[179] *E. Allen, B. Brisdon, D. Edwards, G. Fowles, R. Williams*, J. Chem. Soc. 4649 (1963).
[180] *D. Brown*, J. Chem. Soc. 4944 (1964).
[181] *D. Rillema, C. Brubaker*, Inorg. Chem. *8*, 1645 (1969).
[182] *N. Ul'ko, M. Parubocha*, Visn. Kiiv. Univ. Ser. Chim. *10*, 18 (1969); C. A. *73*, 105 040y (1970).
[183] *H. Funk, H. Matschiner, H. Naumann*, Z. Anorg. Allgem. Chemie *340*, 75 (1965).

$[RH]W(OH)_3(NCS)_3$ (where RH = alkylammonium cation) have been isolated from aqueous thiocyanate solutions of tungsten(V) oxide tetrachloride by addition of the appropriate alkylammonium halides.[184,185]

Finally, the *octacyanotungstate(V) anion*, $W(CN)_8^{3\ominus}$, which cannot be obtained directly from the reaction of tungsten compounds with cyanide, can be prepared by the oxidation of the quadrivalent complex with potassium permanganate[186] or ceric sulfate.[187]

16.7 Non-stoichiometric Oxides

16.7.1 Oxides

It has been found that when tungsten(VI) oxide is heated strongly *in vacuo*, oxygen is lost to give a defect lattice.[188,189] This process continues to a limiting composition of $WO_{2\cdot98}$, beyond which several new phases start to appear. These substoichiometric phases may be prepared by two main methods. The first consists of heating tungsten(VI) oxide under vacuum at temperatures between 1050° and 1225°[189]; this is the simplest method for the preparation of $WO_{2\cdot96}$. The second consists of heating stoichiometric mixtures of tungsten metal powder and tungsten(VI) oxide at 1250°–1300° in an argon atmosphere.[189] Using this second method it is possible to obtain oxides of completely general composition.

Phases which have been definitely identified by X-ray or combined X-ray/density measurements are given in the following table

Composition	Unit cell composition	Ref.
α-$WO_{2\cdot96}$	$W_{50}O_{148}$	189
β-$WO_{2\cdot96}$[a]	$W_{25}O_{74}$	189
$WO_{2\cdot95}$	$W_{40}O_{118}$	190
$WO_{2\cdot90}$	$W_{20}O_{58}$	191
$WO_{2\cdot72}$	$W_{18}O_{49}$	192

[a] β-$WO_{2\cdot96}$ is the high temperature modification (> 1250°), reversibly convertible to the α form.

16.7.2 Bronzes

The term "bronze" is applied to a ternary metal oxide of general formula $M'_xM''_yO_z$ (where M″ is a transition metal, M''_yO_z is its highest binary oxide, M′ is some other metal, and x lies between 0 and 1).[193] Thus, the tungsten bronzes are of general formula M_xWO_3 (with M = alkali metal, Ba, Pb, Tl, Cu, Ag, or a rare earth metal).

The tungsten bronzes were first prepared by Wöhler,[194] who reduced a sodium tungstate melt with hydrogen; tin, iron, zinc and phosphorus have also been used as reducing agents.[195] There are three basic methods used for bronze preparation[193]: (*a*) *vapor phase reaction:* This is of a general type [Eqn (3)]. The metal M must be

$$xM(g) + WO_3(g) \longrightarrow M_xWO_3(s) \quad (3)$$

appreciably volatile at high temperatures, but easily handled at room temperature. An example of this type of reaction is the production of crystals of the thallium tungsten bronzes[196] at 900°. An advantage of this reaction is the high purity of the product formed. (*b*) *Electrolytic reduction:* In this type of reaction, molten mixtures of a metal tungstate and tungsten(VI) oxide are electrolysed with platinum or tungsten electrodes. Crystals of bronze form at the cathode and oxygen is liberated at the anode. This is the most successful method for the growth of large single crystals[196,197]; however, optimum conditions are difficult to find. (*c*) *Solid state reactions:* This is the most generally applicable type of reaction.[190,195] The reactants are ground and heated *in vacuo*, and react in the following fashion, e.g., for sodium bronzes

$$3xNa_2WO_4 + (6-4x)WO_3 + xW \xrightarrow{850^\circ} 6Na_xWO_3 \quad (4)$$

A variation of this method is the reaction of tungsten(VI) oxide with metal halides, using either tungsten(IV) oxide[198] or tungsten metal[199] as reducing agents, e.g.,

$$xBaCl_2 + xWO_2 + WO_3 \longrightarrow Ba_xWO_3 + xWO_2Cl_2 \quad \text{Ref. 198} \quad (5)$$

In addition, some success has recently been

[184] *H. Funk, H. Boehland*, Z. Anorg. Allgem. Chemie *318*, 169 (1962).
[185] *H. Boehland, E. Niemann*, Z. Anorg. Allgem. Chemie *336*, 225 (1965).
[186] *H. Willard, R. Thielcke*, J. Amer. Chem. Soc. *57*, 2609 (1935).
[187] *L. Bok, J. Leipoldt, S. Basson*, Acta Cryst. *B26*, 684 (1970).
[188] *O. Glemser, H. Sauer*, Z. Anorg. Allgem. Chemie *252*, 144 (1943).
[189] *E. Gebert, R. Ackermann*, Inorg. Chem. *5*, 136 (1966).
[190] *P. Gado, A. Magneli*, Acta Chem. Scand. *19*, 1514 (1965).
[191] *A. Magneli*, Ark. Kemi. *1*, 513 (1949).
[192] *A. Magneli*, Ark. Kemi. *1*, 223 (1949).
[193] *P. Dickens, M. Whittingham*, Quart. Rev. *22*, 30 (1968).
[194] *F. Wöhler*, Ann. Physik. *2*, 350 (1824).
[195] *E. Banks, A. Wold*, Prep. Inorg. React. *4*, 237 (1968).
[196] *M. Sienko*, J. Amer. Chem. Soc. *81*, 5556 (1959).
[197] *C. Collins, W. Ostertag*, J. Amer. Chem. Soc. *88*, 3171 (1966).
[198] *L. Conroy, T. Yokokawa*, Inorg. Chem. *4*, 994 (1965).
[199] *L. Conroy, G. Podolsky*, Inorg. Chem. *7*, 614 (1968).

achieved in the preparation of bronzes by reacting tungsten(VI) oxide with alkali metal azides.[200] Also, from tungstate–tungsten mixtures under high pressure (~65 kbars), it is possible to prepare bronzes in unusual polymorphic modifications.[201]

Purification of the products from electrolytic or solid state reactions can generally be achieved by leaching out impurities with hot water and hot dilute acids, but some decomposition may occur if dilute alkalis are used. Any silica impurities can be removed with 48% hydrofluoric acid.[196,199]

Very recently, bronzelike compounds have been prepared, in which some of the oxygen in the bronze phase is replaced by fluorine.[202] This is achieved by reacting an equimolar mixture of sodium fluoride and tungsten(IV) oxide with tungsten(VI) oxide according to Eqn (6).

$$x\mathrm{NaF} + x\mathrm{WO_2} + (1-x)\mathrm{WO_3} \xrightarrow{800^\circ} \mathrm{Na}_x\mathrm{WO}_{3-x}\mathrm{F}_x \quad (6)$$

In addition, bronzes with mixed cations have been prepared, e.g., $Na_xBa_yWO_3$.[198]

One very useful technique for the determination of the composition of bronzes, especially sodium bronzes, is based on the fact that the lattice constants are a function of x. For the sodium bronzes the lattice constant $a = 0{\cdot}0819x + 3{\cdot}7846$,[203] and so by measuring this lattice constant, the composition can be determined.

16.8 Tungsten(VI)

16.8.1 Hydrides

Hexahydridotris(dimethylphenylphosphine)tungsten(VI), $WH_6(PMe_2Ph)_3$, has recently been prepared by the treatment of tetrachlorobis(dimethylphenylphosphine)tungsten(IV), $WCl_4(PMe_2Ph)_2$, with sodium hydridoborate[156a] or sodium amalgam.[204]

16.8.2 Halides

Tungsten(VI) fluoride may be prepared by direct fluorination of tungsten metal at temperatures of 150°–300°.[205] Other fluorinating agents which have been used include chlorine trifluoride,[206,207] chlorine monofluoride,[208] nitrogen trifluoride,[209] and sulfur hexafluoride.[210] The tungsten(VI) fluoride obtained may be purified by low temperature distillation, and stored over sodium fluoride.[211]

Fluoride chlorides of general composition WF_nCl_{6-n} (n=1–5) have been isolated. *Tungsten(VI) pentafluoride chloride*, WF_5Cl, is prepared in the reaction between titanium(IV) chloride, $TiCl_4$, and tungsten(VI) fluoride, WF_6[212,213]; small amounts of *tungsten(VI) difluoride tetrachloride*, WF_2Cl_4, and *tungsten(VI) fluoride pentachloride*, $WFCl_5$, are also obtained from this reaction.[213] Controlled fluorination of tungsten(VI) chloride produces *tungsten(VI) pentafluoride chloride* in 75% yield, and there is also evidence for the formation of unstable *tungsten(VI) tetrafluoride dichloride*, WF_4Cl_2.[213,214] Finally, the reaction of tungsten(VI) fluoride with boron trichloride is reported[205] to produce small amounts of *tungsten(VI) trifluoride trichloride*, WF_3Cl_3.

Tungsten(VI) chloride, WCl_6, is prepared by the chlorination of tungsten metal at 600°.[215,216] It can also be prepared by the action of carbon tetrachloride on tungsten(VI) oxide[217,218] or sulfide[219] in a bomb at 400° or by refluxing a suspension of tungsten(VI) oxide in hexachloropropene for 4 hr.[220]

---

200 *B. Chamberland*, Inorg. Chem. *8*, 1183 (1969).
201 *T. Bither, J. Gilson, H. Young*, Inorg. Chem. *5*, 1559 (1966).
202 *J. Doumerc, M. Pouchard*, C.R. Acad. Sci. Paris Ser. C *270*, 547 (1970).
203 *B. Brown, E. Banks*, J. Amer. Chem. Soc. *76*, 963 (1954).
204 *B. Bell, J. Chatt, G. Leigh*, Chem. Commun. 842 (1970).
205 *T. O'Donnell, D. Stewart*, Inorg. Chem. *5*, 1434 (1966).
206 *N. Nikolaev, Y. Buslaev, A. Opalovskii*, Zh. Neorg. Khim. *3*, 1731 (1958).
207 *N. Nikolaev, Y. Busalev, A. Opalovskii* Russian J. Inorg. Chem. (Engl.) *3*(*8*), 14.
208 *J. Pitts, A. Jache*, Inorg. Chem. *7*, 1661 (1968).
209 *O. Glemser, J. Wegener, R. Mews*, Chem. Ber. *100*, 2474 (1967).
210 *R. Johnson, B. Siegel*, J. Inorg. Nucl. Chem. *31*, 955 (1969).
211 *A. Noble, J. Winfield*, J. Chem. Soc. *A*, 501 (1970).
212 *B. Cohen, A. Edwards, M. Mercer, R. Peacock*, Chem. Commun. 322 (1965).
213 *G. Frazer, M. Mercer, R. Peacock*, J. Chem. Soc. *A*, 1091 (1967).
214 *G. Frazer, C. Gibbs, R. Peacock*, J. Chem. Soc. *A*, 1708 (1970).
215 *J. Canterford, R. Colton*, Halides of the Second and Third Row Transition Metals. Wiley, New York 1968.
216 *M. Lietzke, M. Holt*, Inorg. Synth. *3*, 163 (1950).
217 *K. Knox, S. Tyree, R. Srivastava, V. Norman, J. Bassett, J. Holloway*, J. Amer. Chem. Soc. *79*, 3358 (1957).
218 *E. Epperson, H. Frye*, Inorg. Nucl. Chem. Lett. *2*, 223 (1966).
219 *A. Bardawil, F. Collier, S. Tyree*, Inorg. Chem. *3*, 149 (1964).
220 *W. Porterfield, S. Tyree*, Inorg. Synth. *9*, 133 (1967).

Tungsten(VI) bromide can be obtained by the action of bromine liquid[221] or vapor[222] on gently heated tungsten metal. A method giving a purer product is that involving the reaction of tungsten hexacarbonyl, $W(CO)_6$, with bromine at or below 0°.[160,221] Tungsten(VI) bromide can also be prepared, almost quantitatively, by the reaction of tungsten(VI) chloride with boron tribromide at or below room temperature.[223]

16.8.3 Tungsten(VI) Oxide, Sulfide, and Nitride Halides

Two types of oxide and sulfide halides are possible, *i.e.*, the tungsten(VI) oxide/sulfide tetrahalides, WYX_4, and the tungsten(VI) dioxide/disulfide dihalides, WY_2X_2.

Tungsten(VI) oxide tetrafluoride, WOF_4, has been prepared by the action of anhydrous hydrogen fluoride on the corresponding oxide tetrachloride, $WOCl_4$,[224] or by the reaction of a 3:1 fluorine:oxygen mixture with tungsten metal.[225] Alternatively, the reaction between a 2:1 molar mixturc of tungstcn(VI) fluoridc:tungsten(VI) oxide[226] can be used to prepare this oxide fluoride.

Tungsten(VI) dioxide difluoride, WO_2F_2, has been reported to be formed from the partial hydrolysis of tungsten(VI) oxide tetrafluoride.[224]

Tungsten(VI) oxide tetrachloride, $WOCl_4$, is readily prepared by a variety of methods:

(a) by refluxing tungsten(VI) oxide with thionyl chloride[227];
(b) by the action of carbon tetrachloride on tungsten(VI) oxide in a bomb at 250°[218];
(c) by the reaction between sulfuryl chloride and tungsten metal at 300°[228];
(d) by refluxing tungsten(VI) oxide with octachlorocyclopentene[229];
(e) by the reaction of sulfur dioxide with tungsten(VI) chloride[230]; and
(f) by the reaction of a 2:1 molar mixture of tungsten(VI) chloride:tungsten(VI) oxide at 150°.[221]

Tungsten(VI) dioxide dichloride, WO_2Cl_2, can be prepared by the partial hydrolysis of tungsten(VI) oxide tetrachloride[231] or by the reaction of a 2:1 molar mixture of tungsten(VI) oxide:tungsten(VI) chloride in a sealed evacuated tube at 150°.[221,232]

Tungsten(VI) sulfide tetrachloride, $WSCl_4$, is obtained by the reaction of tungsten(V) or (VI) chloride with elemental sulfur[164,233] or from the reaction between tungsten(VI) chloride and antimony trisulfide.[164]

Tungsten(VI) disulfide dichloride, WS_2Cl_2, has recently been prepared from the reaction of tungsten(VI) oxide tetrachloride, $WOCl_4$, with hydrogen sulfide in benzene.[234]

Finally, *tungsten(VI) nitride trichloride*, $WNCl_3$, is conveniently prepared from the reaction between tungsten(VI) chloride and chlorine azide, ClN_3.[235]

Bromination of tripotassium tri-μ-chlorobis[trichlorotungstate(III)], $K_3W_2Cl_9$, at 450° produces *tungsten(VI) oxide trichloride bromide*, $WOCl_3Br$,[236] not tungsten(IV) trichloride bromide as originally reported.[237]

Tungsten(VI) oxide dichloride dibromide, $WOCl_2Br_2$, has recently been prepared by halogen exchange between phosphorus pentabromide and tungsten(VI) oxide tetrachloride.[238]

Tungsten(VI) oxide tetrabromide, $WOBr_4$, may be prepared by one of the following methods:

(a) by heating tungsten(VI) oxide and carbon tetrabromide in a bomb at 440°[239,240];
(b) by the reaction of tungsten(V) bromide with liquid sulfur dioxide[illegible];

[221] *P. Crouch, G. Fowles, R. Walton*, J. Inorg. Nucl. Chem. *32*, 329 (1970).
[222] *H. Schafer, E. Smith*, J. Amer. Chem. Soc. *18*, 1098 (1896).
[223] *P. Druce, M. Lappert, P. Riley*, Chem. Commun. 486 (1967).
[224] *O. Ruff, F. Eisner, W. Heller*, Z. Anorg. Allgem. Chemie *52*, 256 (1907).
[225] *G. Cady, G. Hargreaves*, J. Chem. Soc. 1568 (1961).
[226] *F. Tebbe, E. Muetterties*, Inorg. Chem. *7*, 172 (1968).
[227] *R. Colton, I. Tomkins*, Australian J. Chem. *18*, 447 (1965).
[228] *D. Edwards, A. Woolf*, J. Chem. Soc. *A*, 91 (1966).
[229] *S. Feil, S. Tyree, F. Collier*, Inorg. Synth. *9*, 123 (1967).
[230] *G. Fowles, J. Frost*, J. Chem. Soc. *A*, 1631 (1966).
[231] *I. Beattie, K. Livingston, D. Reynolds, G. Ozin*, J. Chem. Soc. *A*, 1210 (1970).
[232] *H. Funk, F. Modry*, Z. Chem. *7*, 27 (1967).
[233] *N. Fortunatov, N. Timischenko*, Ukr. Khim. Zh. *35*, 1207 (1969).
[234] *K. Sharma, S. Anand, R. Multani, B. Jain*, Chem. & Ind. (London) *43*, 1556 (1969).
[235] *K. Dehnicke, J. Straehle*, Z. Anorg. Allgem. Chemie *339*, 171 (1965).
[236] *P. Boorman, N. Greenwood, H. Whitfield*, J. Chem. Soc. *A*, 2256 (1968).
[237] *R. Young, R. Laudise*, J. Amer. Chem. Soc. *78*, 4861 (1956).
[238] *D. Kepert, R. Mandyczewsky*, J. Chem. Soc. *A*, 2990 (1969).
[239] *S. Shchukarev, G. Kokovin*, Zh. Neorg. Khim. *9*, 1565 (1964).
[240] *S. Shchukarev, G. Kokovin*, Russian J. Inorg. Chem. (Engl.) p. 849.

(c) by heating tungsten(VI) oxide and boron tribromide at 130°[241];

(d) by heating tungsten metal with an oxygen–bromine mixture at 350°[242]; and

(e) by heating a 1:2 molar mixture of tungsten(VI) oxide:tungsten(VI) bromide at 150°.[221]

Tungsten(VI) dioxide dibromide, WO_2Br_2, can be prepared from the reaction of an oxygen–bromine mixture with tungsten metal,[242] but the mixture must be richer in oxygen than for method (d) above. It is also possible that this oxide bromide could be obtained from the reaction of a 2:1 molar mixture of tungsten(VI) oxide:tungsten(VI) bromide at elevated temperatures.[221]

Tungsten(VI) sulfide tetrabromide, $WSBr_4$, can be made by the reaction between tungsten(VI) bromide and antimony trisulfide,[164] using a procedure similar to that for the analogous sulfide tetrachloride.

Tungsten(VI) dioxide diiodide, WO_2I_2, is prepared by the reaction of a 1:2:3 molar mixture of tungsten metal:tungsten(VI) oxide:iodine in a sealed, evacuated tube in a temperature gradient.[243]

Tungsten(VI) oxide tetraiodide is claimed to have been detected in the vapor phase by reaction of lithium iodide with tungsten(VI) oxide,[244] but has not been isolated in the solid state.

16.8.4 Neutral Complexes of Tungsten(VI) Halides and Oxide Halides

16.8.4.1 Complexes with Group VI Ligands

Noble and Winfield[245] have prepared a series of mixed *fluoride–alkoxide tungsten(VI) complexes* of the type $WF_{6-n}(OR)_n$ ($n=1$–4, $R=CH_3$; $n=1$ or 2, $R=C_6H_5$) by the reaction of tungsten(VI) fluoride with alkylalkoxysilanes, *e.g.*, $Me_2Si(OMe)_2$, at 20° and below. The fluoride–alkoxides, $WF_5(OR)$ (R = Me, Et, Ph) can also be prepared by the reaction of tungsten(VI) fluoride with dialkyl sulfites, $(RO)_2SO$, R = alkyl group.[211] Related chlorides,[246] *e.g.*, *dichlorotetraphenoxytungsten(VI)*, $WCl_2(OPh)_4$, and bromides,[183] *e.g.*, *dibromotetra(dibromophenoxy)tungsten(VI)*, $WBr_2(OC_6H_3Br_2)$, have also been prepared. Unstable *hexafluorobis(diethylsulfide)tungsten(VI)*, $WF_6 \cdot 2Et_2S$, and the analogous *selenide*, are formed from the reaction of diethyl sulfide or selenide with tungsten(VI) fluoride at 20°.[247]

Complexes of the oxide halides are generally six-coordinate. Reaction of tungsten(VI) oxide tetrachloride or bromide with ethers affords 1:1 adducts, *e.g.*, $WOCl_4 \cdot C_4H_8O$[248]; however, it is thought that dialkyl sulphides cause reduction.[248] Six-coordinate species are also obtained from the dioxide dichloride, WO_2Cl_2, *e.g.*, reaction with dimethyl sulfoxide (DMSO) gives *dioxodichlorobis (dimethylsulfoxide) tungsten (VI)*, $WO_2Cl_2 \cdot 2DMSO$.[249] Oxide halide complexes have also been obtained indirectly. For instance, the reaction of tungsten(VI) fluoride with dimethyl ether, Me_2O, yields *oxotetrafluoro(dimethylether)tungsten(VI)*, $WOF_4 \cdot Me_2O$,[247] and the chlorination of triscarbonyltris(triphenylphosphine)tungsten(0) yields *dioxodichlorobis(triphenylphosphineoxide)tungsten(VI)*, $WO_2Cl_2 \cdot 2Ph_3PO$.[250]

16.8.4.2 Complexes with Group V Ligands

Tungsten(VI) fluoride forms a number of addition compounds with Group V ligands; these include *hexafluoro(trimethylamine)tungsten(VI)*, $WF_6 \cdot NMe_3$,[226] *hexafluoro(triphenylphosphine)-tungsten(VI)*, $WF_6 \cdot PPh_3$,[226] and *hexafluoro-mono/bis/tris(pyridine)tungsten(VI)*, $WF_6 \cdot pyr_n$ ($n=1$–3).[226, 251]

Tungsten(VI) chloride undergoes a series of reactions with ammonia.[252] Initially, ammoniates, $WCl_6 \cdot xNH_3$ are formed, but then ammonolysis occurs to form species such as *tetrachlorobis(amido)tungsten(VI)*, $WCl_4(NH_2)_2$. A similar series of reactions occurs between tungsten(VI) chloride and amines; simple adducts, *e.g.*, *hexachloro(trimethylamine)tungsten(VI)*, $WCl_6 \cdot NMe_3$, and amide complexes, *e.g.*, *dichlorotetra(alkylamido)tungsten(VI)*, $WCl_2 \cdot (NHR)_4$, both being formed.[119]

A further addition product of tungsten(VI)

---

[241] *M. Lappert, B. Prokai*, J. Chem. Soc. *A*, 129 (1967).
[242] *R. Colton, I. Tomkins*, Australian J. Chem. *21*, 1975 (1968).
[243] *J. Tillack, P. Eckerlin*, Angew. Chem. *78*, 451 (1966); Engl. *5*, 421.
[244] *B. Ward, F. Stafford*, Inorg. Chem. *7*, 2569 (1968).
[245] *A. Noble, J. Winfield*, J. Chem. Soc. *A*, 2574 (1970).
[246] *H. Funk, W. Baumann*, Z. Anorg. Allgem. Chemie *231*, 264 (1937).
[247] *A. Noble, J. Winfield*, Inorg. Nucl. Chem. Lett. *4*, 339 (1968).
[248] *G. Fowles, J. Frost*, J. Chem. Soc. *A*, 671 (1967).
[249] *B. Brisdon*, Inorg. Chem. *6*, 1791 (1967).
[250] *J. Lewis, R. Whyman*, J. Chem. Soc. 6027 (1965).
[251] *H. Clark, H. Emeléus*, J. Chem. Soc. 4778 (1957).
[252] *G. Fowles, B. Osborne*, J. Chem. Soc. 2275 (1959).

chloride is *hexachlorobis(trichloroacetonitrile)-tungsten(VI)*, $WCl_6 \cdot 2CCl_3CN$, formed from the reaction between tungsten(VI) chloride or tungsten(V) chloride and trichloroacetonitrile.[253] This reaction is in contrast to that with acetonitrile where reduction to tungsten(IV) occurs.[115]
Compounds formed between tungsten(VI) oxide halides and Group V ligands are six-coordinate and similar to those found with Group VI ligands. For example, tungsten(VI) oxide tetrachloride, $WOCl_4$, forms *oxotetrachloro(acetonitrile)tungsten(VI)*, $WOCl_4 \cdot MeCN$, when reacted with acetonitrile,[248] and tungsten(VI) dioxide dichloride forms *dioxodichlorobis(acetonitrile)-tungsten(VI)*, $WO_2Cl_2 \cdot 2MeCN$, when reacted with the same ligand.[249,250]

16.8.5 Anionic Halide Complexes

The *heptafluorotungstate(VI)* and *octafluorotungstate(VI) anions*, $WF_7^{\ominus}$ and $WF_8^{2\ominus}$, are the only binary anionic complexes of tungsten(VI). They can be prepared by the reaction of alkali metal fluorides with tungsten(VI) fluoride[175,254,255] or by the reaction of nitrosyl or nitryl fluoride with tungsten(VI) fluoride.[256,257]
Many oxofluoro anions of tungsten(VI) have been reported and they include the following: *tetraoxofluorotungstate(VI)*, $WO_4F^{3\ominus}$, prepared from the trioxotrifluorotungstate(VI) anion, $WO_3F_3^{3\ominus}$, by melting in air[258]; *trioxotrifluorotungstate(VI)*, $WO_3F_3^{3\ominus}$, prepared by the solid state reaction of tungsten(VI) oxide with alkali metal fluorides[258]; *trioxodifluorotungstate(VI)*, $WO_3F_2^{2\ominus}$, prepared from solutions of $WO_3F_3^{3\ominus}$ by addition of suitable cations[259]; *dioxotetrafluorotungstate(VI)*, $WO_2F_4^{2\ominus}$, prepared either from the reaction of the tungstate anion, $WO_4^{2\ominus}$, with hydrofluoric acid[260] or by the reaction of tungsten(VI) oxide with alkali metal fluorides in aqueous solution[261]; the *trioxofluorotungstate(VI)* anion, $WO_3F^{\ominus}$, can also be obtained from this latter reaction.[261] Finally, the *oxopentafluorotungstate(VI)* anion, $WOF_5^{\ominus}$, is generated by the reaction of a 1:1 molar mixture of tungsten(VI) oxide tetrafluoride:alkylammonium fluoride in dichloromethane[226] or by the reaction of tungsten(VI) oxide with nitrosyl fluoride.[262]
The *oxopentahalo anions*, $WOX_5^{\ominus}$ (X=Cl or Br), are formed by the reaction of the tungsten(VI) oxide tetrahalide with the appropriate alkylammonium halides in chloroform.[230] The cesium salts have also been prepared.[230] The *dioxotetrachlorotungstate(VI)* anion, $WO_2Cl_4^{2\ominus}$, has been prepared by the reaction of a 1:2 molar mixture of tungsten(VI) chloride:alkali metal chloride in hydrochloric acid,[263] or by the reaction of tungsten(VI) oxide with hydrochloric acid.[264] The *trioxochlorotungstate(VI)* anion, $WO_3Cl^{\ominus}$, is reported to be formed by the reaction of a 1:1 molar mixture of tungsten(VI) oxide and alkali metal chloride at 600°.[265]
Finally, the *oxoperoxotetrachlorotungstate(VI)* anion, $W(O_2)OCl_4^{2\ominus}$, can be obtained from the reaction of the tungstate(VI) anion, $WO_4^{2\ominus}$, with hydrogen peroxide and hydrochloric acid at −10°.[266]

16.8.6 Compounds of Tungsten(VI) with Group VI Elements

16.8.6.1 Binary Oxide, Sulfide, and Selenide

Tungsten(VI) oxide, WO_3, can be prepared by heating most tungsten compounds in oxygen or air; it is thought to be polymorphic.[267]
Tungsten(VI) sulfide, WS_3, can be prepared by treating solutions containing the tetrathiotungstate(VI) anion, $WS_4^{2\ominus}$, with hydrochloric acid[268,269] or preferably by heating ammonium

253 *G. Fowles, D. Rice, N. Rolfe, R. Walton*, Chem. Commun. 459 (1970).
254 *B. Cox, D. Sharp, A. Sharpe*, J. Chem. Soc. 1242 (1956).
255 *S. Katz*, Inorg. Chem. *5*, 666 (1966).
256 *J. Geichmann, E. Smith, P. Ogle*, Inorg. Chem. *2*, 1012 (1963).
257 *N. Bartlett, S. Beaton, K. Jha*, Chem. Commun. 168 (1966).
258 *O. Schmitz-Dumont, I. Bruns, I. Heckmann*, Z. Anorg. Allgem. Chemie *271*, 347 (1953).
259 *O. Schmitz-Dumont, P. Opgenhoff*, Z. Anorg. Allgem. Chemie *275*, 21 (1953).
260 *G. Jander, B. Fiedler*, Z. Anorg. Allgem. Chemie *308*, 155 (1961).
261 *Y. Buslaev, R. Davidovich*, Zh. Neorg. Khim. *10*, 1862 (1965); Engl., 1014.
262 *O. Glemser, J. Wegener, R. Mews*, Chem. Ber. *100*, 2474 (1967).
263 *F. Petillon, M. Youinou, J. Guerchais*, Bull. Soc. Chim. France 2375 (1968).
264 *A. Majumdar, R. Bhattacharyya*, J. Inorg. Nucl. Chem. *29*, 2359 (1967).
265 *J. Prigent, P. Caillet*, C.R. Acad. Sci. Paris *256*, 2184 (1963).
266 *M. Youinou, J. Guerchais*, Bull. Soc. Chim. France 40 (1968).
267 *L. Chang, B. Phillips*, J. Amer. Ceram. Soc. *52*, 527 (1969).
268 *P. Ehrlich*, Z. Anorg. Allgem. Chemie *257*, 247 (1948).
269 *R. Voorhoeve, H. Wolters*, Z. Anorg. Allgem. Chemie *376*, 165 (1970).

tetrathiotungstate(VI), $(NH_4)_2WS_4$, above 120°.[269,270]

Tungsten(VI) selenide, WSe_3, is prepared by the action of sulfuric acid on solutions of tetraselenotungstate(VI) anion, $WSe_4^{2\ominus}$.[131,271]

16.8.6.2 Tungstates and Substituted Tungstates

The *tungstate ion*, $WO_4^{2\ominus}$, is generated by dissolving tungsten(VI) oxide in alkali. It can also be prepared in the solid phase by fusing alkali metal carbonates with tungsten(VI) oxide.

Substituted tungstates, WO_xY_{4-x} (x=0–3, Y=S or Se), are also known. The *thiotungstates* were originally prepared by Corleis[272] by passing hydrogen sulfide through solutions of potassium and ammonium tungstates. They have since been made with a variety of cations.[273–277] The analogous selenides can be made by similar methods using hydrogen selenide instead of hydrogen sulfide.[278,279] Recently, mixed *oxothioselenotungstates(VI)*, *e.g.*, $WOS_2Se^{2\ominus}$, have been prepared.[279–281]

16.8.6.3 Isopolytungstates

The subject of the polymeric isopolytungstates, formed on acidification of solutions of normal tungstates, has been reviewed by Kepert,[282] but there is still controversy over the exact nature of these species.[283–285]

The two main species are *paratungstate*, $W_{12}O_{42}^{12\ominus}$ or $H_2W_{12}O_{42}^{10\ominus}$, formed by the treatment of a tungstate ($WO_4^{2\ominus}$) solution with 7/6 mol of hydrochloric acid per mol tungstate[285] and *metatungstate*, $H_2W_{12}O_{40}^{6\ominus}$, formed by treating a tungstate solution with 9/6 mol of hydrochloric acid per mol tungstate.[282,285] It is reported that the tetraphenylarsonium cation is specific for metatungstate, forming $(Ph_4As)_6(H_2W_{12}O_{40})$.[286]

16.8.6.4 Peroxide Complexes

The action of hydrogen peroxide on slightly acid solutions of potassium tungstate produces the *μ-oxotetraperoxodioxobisaquotungstate(VI)* anion, $W_2(O_2)_4O_3 \cdot 2H_2O^{2\ominus}$.[287–289] The action of hydrogen peroxide on tungsten(VI) oxide in the presence of ligands produces monomeric *oxobisperoxobisligandtungsten(VI)*, $WO(O_2)_2 \cdot 2L$ (L=monodentate or ½ bidentate ligand).[290,291] Finally, the reaction of hydrogen peroxide with tungstate(VI) solutions in the presence of oxalic acid produces the *oxobisperoxo(oxalato)tungstate(VI) anion*, $[WO(O_2)_2C_2O_4]^{2\ominus}$.[292,293]

16.8.6.5 Dithiolenes

Several dithiolene compounds of tungsten(VI) are formed by refluxing tungsten hexacarbonyl or tungsten(VI) chloride with dithiolato ligands.[294–296] Compounds prepared in this way include *tris[bis(trifluoromethyl)-1,2-dithiolato]tungsten(VI)*, $W[S_2C_2(CF_3)_2]_3$,[294] *tris(toluene-3,4-dithiolato)tungsten(VI)*,[295,296] and *tris(bisphenyl-1,2-dithiolato)tungsten(VI)*.[295] They are thought to be isostructural with the analogous rhenium compounds.

16.8.6.6 Further Compounds with Oxygen Donors

Refluxing tungsten hexacarbonyl[297] or tungsten(VI) dioxide dichloride[298] with acetylacetone

---

270 *E. Rode, B. Lebeder*, Zh. Neorg. Khim. *9*, 2068 (1964); Engl., 1118.
271 *L. Moser, K. Atynski*, Monatsh. Chem. *45*, 241 (1924).
272 *E. Corleis*, Justus Liebigs Ann. Chem. *232*, 244 (1886).
273 *M. Leroy, G. Kaufmann, R. Charlionet, R. Rohmer*, C.R. Acad. Sci. Paris Ser. C. *263*, 601 (1966).
274 *A. Muller, E. Diemann*, Z. Naturforsch. *23B*, 1607 (1968).
275 *A. Muller, E. Diemann, H. Schulze*, Z. Anorg. Allgem. Chemie *376*, 120 (1970).
276 *A. Muller, E. Diemann, U. Heidborn*, Z. Anorg. Allgem. Chemie *376*, 125 (1970).
277 *M. Leroy, G. Kaufmann*, C.R. Acad. Sci. Paris Ser. C. *265*, 1322 (1967).
278 *A. Muller, E. Diemann*, Naturwissenschaften *55*, 650 (1968).
279 *A. Muller, E. Diemann*, Chem. Ber. *102*, 945 (1969).
280 *E. Diemann, A. Muller*, Inorg. Nucl. Chem. Lett. 5, 339 (1969).
281 *A. Muller, E. Diemann*, Z. Anorg. Allgem. Chemie *373*, 57 (1970).
282 *D. Kepert*, Progr. Inorg. Chem. *4*, 199 (1962).
283 *W. Griffith, P. Lesniak*, J. Chem. Soc. *A*, 1066 (1969).
284 *W. Lipscomb*, Inorg. Chem. *4*, 132 (1965).
285 *J. Aveston*, Inorg. Chem. *3*, 981 (1964).
286 *R. Ripan, C. Calu*, Talanta *14*, 887 (1967).
287 *K. Jahr, E. Lother*, Ber. *71B*, 894, 903, 1127 (1938).
288 *F. Einstein, B. Penfold*, Acta Cryst. *17*, 1127 (1964).
289 *W. Griffith*, J. Chem. Soc. 5345 (1963).
290 *H. Mimoun, I. Seree de Roch, L. Sajus*, Bull. Soc. Chim. France 1481 (1969).
291 *R. Kergoat, J. Guerchais*, C.R. Acad. Sci. Paris Ser. C. *268*, 2304 (1969).
292 *W. Griffith, T. Wickins*, J. Chem. Soc. *A*, 590 (1967).
293 *M. Sljukic, N. Vuletic, B. Matkovic, B. Kojic-Prodic*, Croat Chim. Acta *42*, 499 (1970).
294 *A. Davidson, N. Edelstein, R. Holm, A. Maki*, J. Amer. Chem. Soc. *86*, 2799 (1964).
295 *E. Stiefel, R. Eisenberg, R. Rosenberg, H. Gray*, J. Amer. Chem. Soc. *88*, 2956 (1966).
296 *E. Stiefel, H. Gray*, J. Amer. Chem. Soc. *87*, 4012 (1965).
297 *J. Goan, C. Huether, H. Podall*, Inorg. Chem. *2*, 1078 (1963).
298 *A. Nidolovskii*, Croat Chim. Acta *40*, 143 (1968).

(HAA) in benzene produces *bisoxo(hydroxo)-acetylacetonatotungsten(VI)*, $WO_2(OH)AA$,[297] or *bisoxo(bisacetylacetonato)tungsten(VI)*, $WO_2(AA)_2$,[298] respectively. The former compound is postulated to be dimeric.[297]

Reaction of tungsten(VI) oxide tetrachloride or tungsten(VI) chloride with phenol produces the stable *tungsten(VI) hexaphenoxide*[246,299]; similar products are also obtained with substituted phenols.[299,300]

16.8.7 Compounds of Tungsten(VI) with Group V

The only important compounds are *hexakis(isothiocyanato)tungsten(VI)*, $W(NCS)_6$, prepared by the reaction of ammonium thiocyanate with tungsten(VI) chloride in acetone,[301,302] and *hexakis(dimethylamido)tungsten(VI)*, $W(NMe_2)_6$ obtained from the action of lithium dimethylamide on tungsten(VI) chloride.[303]

16.9 Bibliography

K. Li and *C. Wang*, Tungsten. Van Nostrand-Reinhold, Princeton, New Jersey, 1955.

R. Parish, Advan. Inorg. Chem. Radiochem. *9*, 315 (1966).

J. Fergusson, Halogen Chem. *3*, 227 (1967).

[299] *P. Mortimer*, *M. Strong*, Australian J. Chem. *18*, 1579 (1965).

[300] *H. Funk*, *G. Mohaupt*, Z. Anorg. Allgem. Chemie *315*, 204 (1962).

[301] *H. Funk*, *H. Bohland*, Z. Anorg. Allgem. Chemie *324*, 168 (1963).

[302] *N. Ul'ko*, *R. Savchenko*, Zh. Neorg. Khim. *12*, 328 (1967); Engl., 169.

[303] *D. Bradley*, *M. Chisholm*, *C. Heath*, *M. Hursthouse*, Chem. Commun. 1261 (1969).

17 Manganese

Geoffrey Davies
Department of Chemistry, Northeastern University,
Boston, Massachusetts 02115, U.S.A.

The chemistry of manganese is extensive.[1] However, although all oxidation states from 0 to VII are exhibited in compounds of manganese, those having oxidation states II, IV, and VII are by far the most common. Manganese has a particular affinity for bound oxygen atoms, and the known compounds containing manganese in oxidation states IV through VII are virtually all of this class. In recent years our knowledge of manganese in oxidation state III has increased considerably.

17.1 Recovery and Purification of the Element

Manganese, which ranks ninth of the metals found in igneous rocks, is never found free, but is recovered from ores containing the nonstoichiometric "MnO_2" (pyrolusite), hydrous oxides, or the carbonate.[1] The ores are roasted in air to form Mn_3O_4, which is then reduced with aluminum. An increasing amount of high-purity *metallic manganese* is being produced by electroreduction of *manganous salts*, particularly the sulfate, in the presence of a large excess of $(NH_4)_2SO_4$, at a mercury cathode.[1,2] The mercury can be removed by distillation at 250°. Manganese exists as α,[3] β,[4] γ and δ allotropes. Both the α and β modifications of manganese can be readily obtained at room temperature. Electrolysis produces the γ allotrope, which can be converted to the α modification by heating to 150°.[1] Slightly impure samples of the element will displace H_2 from cold water and acids, and are reported[5] to dissolve in acetylacetone. The metal readily combines with B,[6] C,[7] N,[8] O, F, Cl, S, Se, Te, *etc.*, at elevated temperatures.

Manganese(VII) compounds are the purest starting materials for the synthesis of manganese derivatives. Simple salts can be freed from other transition metal salt impurities (*e.g.*, Fe, Co, Ni) by converting them to $MnSO_4$ and heating to redness, which leaves this salt unchanged and converts the impurities to insoluble oxides.

17.2 Manganese(0)

There are few known derivatives of *manganese(0)*. A yellow paramagnetic compound of approximate formula $K_5Mn(CN)_6K_6Mn(CN)_6 \cdot 2NH_3$ separates when $K_3Mn(CN)_6$ is reduced by K in liquid NH_3.[9] The curious compound *cis*-$K_4[Mn(NO)_2(CN)_2]_2$ is produced in the reaction between $Mn(NO)_3CO$ and KCN in 1:2 molar ratio in liquid NH_3.[10] This compound, which evidently contains an Mn–Mn bond, is converted to the *trans*-isomer on heating in a sealed tube with ethanol.[10] Black needlelike crystals of $Mn^0(bipy)_3$ (bipy = bipyridine) can be prepared by reducing $Mn^{II}(bipy)_3Br_2$ with excess Li_2bipy in tetrahydrofuran.[11] Reduction of *manganese(II) phthalocyanine* (Pc) with dilithiobenzophenone in tetrahydrofuran precipitates the dark green compound $Li_2[MnPc] \cdot 6(THF)$, which is soluble in acetonitrile (violet), and darkens in air, liberating THF.[12]

17.3 Manganese(I)

Hydrated $Mn^{\oplus}$ ions are only observed in solution as unstable intermediates.[13] However, alkali metal derivatives of $Mn(CN)_6^{5\ominus}$ can be prepared by reduction of an alkaline solution of

---

[1] *A. H. Sully*, Manganese, Academic Press, New York 1955.
[2] *A. Lange, A. Shirinskikh, S. Bukhman*, Izv. Akad. Nauk. SSSR. Ser. Khim. *20*, 68 (1970).
[3] *A. Bradley, J. Thewlis*, Proc. Roy. Soc. *115*, 456 (1927).
[4] *H. Moser, E. Raub, E. Vincke*, Z. Anorg. Chem. *210*, 67 (1933);
T. Yamada, J. Phys. Soc. Japan *28*, 1499, 1503 (1970).
[5] *O. Kammori, S. Kmitaka, T. Kenichi, K. Arakawa*, Bunseki Kagaku *15*, 651 (1966); C. A. *65*, 11407g (1966).
[6] *S. Andersson, J. Carlsson*, Acta Chem. Scand. *24*, 1971 (1970).
[7] *W. A. Frad*, Adv. Inorg. Chem. Radiochem. *11*, 153 (1968);
R. Fruchart, Ann. Chim. Paris *4*, 143 (1969).
[8] *R. Juza*, Adv. Inorg. Chem., Radiochem. *9*, 81 (1966);
G. Mamporiya, R. Agladze, Zh. Neorg. Khim. *12*, 2541 (1967).
[9] *A. Davidson, J. Kleinberg*, J. Phys. Chem. *57*, 571 (1953);
V. Christensen, J. Kleinberg, A. Davidson, J. Amer. Chem. Soc. *75*, 2495 (1953).
[10] *H. Behrens, E. Lindner, H. Schindler*, Z. Anorg. Allgem. Chemie *365*, 119 (1965).
[11] *S. Herzog, M. Schmidt*, Z. Chem. *2*, 24 (1962);
S. Herzog, M. Schmidt, Z. Chem. *3*, 392 (1963);
Y. Sato, N. Tanaka, Bull. Chem. Soc. Japan *41*, 2064 (1968);
Y. Kaizu, T. Yazaki, H. Torii, H. Kobayashi, Bull. Chem. Soc. Japan *43*, 2068 (1970).
[12] *R. Taube, H. Munke*, Angew. Chem., Intern. 1 Ed. Engl. *2*, 477 (1963).
[13] *D. Brown, F. Dainton*, Trans. Faraday Soc. *62*, 1139 (1966);
M. Fiti, Rev. Roumaine-Chim. *15*, 77 (1970).

$M_2{}^IMn(CN)_6$ with aluminum powder[14] or amalgams,[15] by electrolytic reduction,[16] or by adding KCN solution to finely divided manganese.[17] The colorless potassium salt $K_5Mn(CN)_6$, which is much less soluble than the sodium analog, forms yellow solutions which are readily oxidized in air, and will reduce water according to Eqn (1).

$$2Mn^{I} + 2H_2O \longrightarrow 2Mn^{II} + H_2 + 2OH^{\ominus} \quad (1)$$

A permanganate-colored solution of the complex anion $Mn(CN)_5NO^{3\ominus}$ is obtained by treating $Mn(OAc)_2$ solution with KCN in an NO atmosphere. The solid salt $K_3Mn(CN)_5NO$ can be prepared by reducing a mixture of $K_3Mn(CN)_6$ and KCN with hydroxylamine.[18] The interesting compounds formed in reaction (2) between anhydrous MnI_2 and alkyl or aryl isocyanides in an organic solvent have been known for some years.[19]

$$2MnI_2 + 12RNC \longrightarrow \underset{\mathbf{1}}{Mn(CNR)_6I} + \underset{\mathbf{2}}{Mn(CNR)I_3} \quad (2)$$

Compound **2** is an $I_3{}^{\ominus}$ complex, and can be reduced to **1** by treatment of the mixture with aqueous thiosulfate. **1** and **2** can be separated by addition of ethanol (**1** dissolves). The corresponding hydroxide is a strong base which forms a hydrogencarbonate salt. This series of diamagnetic solids is indefinitely stable and isoelectronic with the corresponding Cr and Fe^{II} complexes.[20] The nitrate can be prepared by ion exchange.[21] Reduction of *manganese(II) phthalocyanine* with one equivalent of dilithium benzophenone in dry tetrahydrofuran under argon gives the light green salt $Li[MnPc]\cdot 6(THF)$.[12]

17.4 Manganese(II)

Manganese(II) forms a large number of salts and complexes. Two features may be said to characterize the complex chemistry of this oxidation state, thereby determining the strategy to be employed in synthesizing a given compound. One is the general lability of the high-spin d^5 configuration, which is common to the vast majority of manganese(II) compounds and often simplifies interconversion. For this reason the emphasis in this section is on the synthesis of basic building blocks. The other feature is a preference for bound oxygen atoms which, in some cases, gives rise to marked instability of manganese compounds in the presence of oxygen.

17.4.1 Cyanides and Complex Cyanides

The addition of a stoichiometric amount of potassium cyanide to a solution of a manganese(II) salt causes the precipitation of impure $Mn(CN)_2$,[22] which can also be obtained by reduction of $K_3Mn(CN)_5NO$ in H_2 at 330°.[23] The initial rose color becomes green and a green precipitate, which could be $KMn(CN)_3$ or, most likely,[24] $K_2Mn[Mn(CN)_6]$, is formed. Yellow solutions of the complex ion $Mn(CN)_6{}^{4\ominus}$ are obtained by adding KCN so that $[CN^{\ominus}] > 1{\cdot}5$ M,[25] and these are slowly oxidized in air to $Mn(CN)_6{}^{3\ominus}$ and insoluble oxides. The deep blue solid $K_4Mn(CN)_6$ (which is less soluble than the Na salt) is low-spin,[26] as are solutions of the anion in, *e.g.*, *t*-butyl phosphate.[27] Aqueous solutions of this complex are said to contain

---

[14] *G. Brauer*, Handbuch der Präparativen Anorganischen Chemie, Ferdinand Enke Verlag Stuttgart 1954.

[15] *D. Clauss, A. Lissner*, Z. Anorg. Allgem. Chemie *297*, 300 (1958);
W. Treadwell, W. Raths, Helv. Chim. Acta *35*, 2275 (1952);
W. Manchot, H. Gall, Ber. *61*, 1135 (1928).

[16] *A. Lange, A. Shrinskikh, S. Bukhman*, Izv. Akad. Nauk SSSR, Ser. Khim. *20*, 68 (1970);
G. Grube, W. Brause, Ber. *60*, 2273 (1927);
J. Mendez, F. Conde, Ann. Chim. *64*, 65 (1968); C. A. *68*, 119 146 (1968).

[17] *W. Treadwell, W. Raths*, Helv. Chim. Acta *35*, 2259 (1952).

[18] *F. Cotton, R. Monchamp, R. Henry, R. Young*, J. Inorg. Nucl. Chem. *10*, 28 (1959);
W. Hieber, R. Nast, E. Proeschel, Z. Anorg. Allgem. Chemie *256*, 159 (1948).

[19] *A. Sacco, L. Naldini*, Gazz. Chim. Ital. *86*, 207 (1956);
A. Sacco, L. Naldini, Rec. Trav. Chim. *75*, 646 (1956).

[20] *L. Malatesta*, Progr. Inorg. Chem. *1*, 283 (1959); *L. Malatesta, F. Bonati*, Isocyanide Complexes of Metals, p. 84, Wiley, London 1969.

[21a] *D. Matheson, R. Bailey*, J. Amer. Chem. Soc. *91*, 1975 (1969);

[21b] *D. Matheson, R. Bailey*, J. Amer. Chem. Soc. *89*, 6389 (1967).

[22] *A. Qureshi, A. Sharpe*, J. Inorg. Nucl. Chem. *30*, 2269 (1968);
A. Descamps, Ann. Chim. (Paris) *24*, 178 (1881);
N. Goldenberg, Trans. Faraday Soc. *36*, 847 (1940);
E. Rotlevi, D. Eaton, Can. J. Chem. *48*, 1073 (1970).

[23] *D. Banks, J. Kleinberg*, Inorg. Chem. *6*, 1849 (1967).

[24] *B. Chadwick, A. Sharpe*, Adv. Inorg. Chem. Radiochem. *8*, 83 (1966).

[25] *J. Lower, W. Fernelius*, Inorg. Synth. *2*, 214 (1946);
B. Figgis, Trans. Faraday Soc. *57*, 204 (1961).

[26] *S. Freed, W. Kasper*, J. Amer. Chem. Soc. *52*, 1012 (1930).

[27] *A. Vashman, T. Vereschagina, I. Pionin*, Zh. Strukt. Khim. *11*, 433 (1970).

more than one species,[28] and it may be that $[Mn(CN)_5OH_2]^{3\ominus}$ is present in alkaline cyanide solutions. The parent acid $H_4Mn(CN)_6$ can be obtained from $PbMn(CN)_6$ by treatment with H_2SO_4 or H_2S.

Treatment of a *manganese(II) acetate* solution with KCN in a nitric oxide atmosphere produces blue-violet crystals of $K_3Mn(CN)_5NO \cdot 2H_2O$.[29] Oxidation of the corresponding *manganese(I) compounds* (Section 17.3) produces the same species. The compounds are photochemically active[30] and aqueous solutions are unstable.[29]

The low-spin manganese(II) analogs of the interesting *manganese(I) isocyanide complexes* (Section 17.3) can be prepared by oxidation of the latter with nitric acid in glacial acetic acid (nitrate salt), or bromine, or by electrolysis in alcohol.[20] Treatment of a manganese(I) isocyanide complex with manganese(VII) in perchloric acid generally gives the perchlorate salt. Anions which have been used to obtain solid compounds include $(HgI_4)^{2\ominus}$, picrate, and $(CdBr_4)^{2\ominus}$ (the latter, particularly useful for the deeply colored aryl- and vinylisocyanide complexes, may give precipitates contaminated with $CdBr_2$)[21a] and hexafluorophosphates (R = alkyl). The alkyl homologs are generally yellow or pink. The manganese(II) complexes are reduced to manganese(I) in basic solution or in the presence of free isocyanide.[20] The hexafluorophosphate complexes apparently revert to the manganese(I) state if recrystallization is attempted.[21a, 21b]

Manganese(II) forms an insoluble complex with $Co(CN)_6{}^{3\ominus}$,[31] and titration of $Fe(CN)_6{}^{4\ominus}$ with a manganese(III) solution results in a brown precipitate after darkening of an initial yellow deposit.[32] The parent acid $H_2MnFe(CN)_6 \cdot 5{\cdot}5H_2O$ has also been prepared.[33]

[28] *W. Kemula, S. Siekierski, K. Sierski*, Rocz. Chem. *29*, 966 (1955).

[29] *A. Blanchard, F. Magnusson*, J. Amer. Chem. Soc. *63*, 2236 (1941);
H. Gray, C. Ballhausen, J. Chem. Phys. *36*, 1151 (1962).

[30] *J. Wictor, J. Senkowski, J. Czaja, D. Rudowska*, Roczniki Chem. *43*, 253 (1969).

[31] *P. Tribunescu, M. Cristea*, Rev. Roumaine-Chim. *14*, 463 (1969);
A. Ludi, M. Guedel, M. Ruegg, Inorg. Chem. *9*, 2224 (1970).

[32] *W. Malik, R. Agarwal, S. Verma*, J. Indian. Chem. Soc. *43*, 501 (1966);
V. Kuznetsov, Z. Popova, G. Seifer, Zh. Neorg. Khim. *15*, 2077 (1970).

[33] *K. Petrov, I. Tananaev, V. Pervyk, A. Korol'kov*, Izv. Akad. Nauk SSSR, Neorg. Mater. *2*, 495 (1966).

17.4.2 Halides and Halo Complexes

Pink MnF_2, which is only slightly soluble in water, forming an unstable tetrahydrate, can be obtained as a powder from the reactions between $MnCO_3$ and HF,[34] Mn and F_2 in a bomb,[35] $MnCl_2$ and HF,[36] and NH_4MnF_3 and CO or CO_2[37] at elevated temperatures. Methods for the preparation of single crystals are available.[38] Alkali metal trifluorides of the type M^IMnF_3 (M^I = Rb,[39] Na,[40] K,[41] $NH_4{}^{\oplus}$, and Tl[42]) are well known. The *thallium(I) salt* is prepared by evaporating a mixture of MnF_2 and TlF in the ratio 1:1·6,[43] and the *ammonium salt* precipitates on addition of excess NH_4HF_2 solution to a solution of $MnCl_2$.

The higher complex fluorides, $M^I_{x-2}MnF_x$ (x = 3–6), can be made by fusing the calculated stoichiometric mixture of MnF_2, and M^IF or $M^{II}F_2$.[43] An example of this technique is the preparation of Pb_2MnF_6 as depicted in reaction (3).

$$2PbF_2 + MnF_2 \xrightarrow[15\ \text{hr}]{580^\circ} Pb_2MnF_6 \qquad (3)$$

This particular reaction is catalyzed by Pt.[44] $K_3Mn_2F_7$ may be prepared by fusion of the appropriate stoichiometric mixture of MnF_2 and KHF_2.[45]

The pink *dichloride*, $MnCl_2$, has a wide and interesting chemistry. The crystalline solid may

[34] *A. Kurtenacker, W. Finger, F. Hey*, Z. Anorg. Allgem. Chimie, *211*, 83 (1933).

[35] *E. Muetterties, J. Castle*, J. Inorg. Nucl. Chem. *18*, 148 (1961).

[36] *R. Fowler, H. Anderson, J. Hamilton, W. Benford, A. Spadetti, S. Bitterlich, I. Litant*, Ind. Eng. Chem. *39*, 343 (1947)

[37] *P. Nuka* Z. Anorg. Allgem. Chemie *180*, 235 (1929).

[38] *N. Mikhailov, S. Petrov*, Kristallografiya *11*, 443 (1966);
K. Nassau, J. Appl. Phys. (Paris) *32*, 1820 (1961);
M. Griffel, J. Stout, J. Amer. Chem. Soc. *72*, 4351 (1950);
H. Griffel, J. Stout, J. Chem. Phys. *18*, 1455 (1950);
J. Stout, J. Chem. Phys. *18*, 1455 (1950);
M. Bizette, B. Tsai, C. R. Acad. Sci. *238*, 1575 (1954).

[39] *A. Mehra, P. Venkateswarlu*, J. Chem. Phys. *47*, 2334 (1967).

[40] *F. Pompa, F. Siciliano*, Ric. Sci. *39*, 21 (1969).

[41] *H. Battes, F. Kneubuehl*, Phys. Lett. *A30*, 98 (1969).

[42] *T. Strivistava*, Curr. Sci. *38*, 538 (1969).

[43] *I. N. Belyaev, O. Y. Revina*, Russ. J. Inorg. Chem. *11*, 772, 1041 (1966).

[44] *M. Samouel*, C. R. Acad. Sci., Paris *268*, 409 (1969).

[45] *A. Chretien, J. Cousseins*, C. R. Acad. Sci., Paris *259*, 4696 (1964).

be obtained by treating MnO_2[46] or $Mn(OAc)_2$ with acetyl chloride in benzene[47] or by refluxing MnX_2 (X=NO_3, CO_3, HCOO, OAc) with $SOCl_2$.[48] This compound gives a green solution in ether and forms a series of hydrates[49] (anhydrous above 200°) and alcoholates.[50] Although the compound $MnCl_2 \cdot 6NH_3$ seems to be well-characterized,[51] luminescent[52] adducts with less than six ammonia molecules appear to be non-stoichiometric (e.g., $MnCl_2 \cdot 0{\cdot}82NH_3$, $MnCl_2 \cdot 1{\cdot}83NH_3$).[53] $MnCl_2$ forms a large number of molecular adducts with other basic ligands (see below and Section 17.4.4).

Complex chlorides can be prepared by direct interaction in the melt[54] or by mixing stoichiometric quantities of the constituent salts in water.[55] Tetraalkyl cations of Group V elements generally give precipitates on mixing the constituent chlorides in acetone or methanol.[56] Pyridinium salts can be precipitated from concentrated HCl.[57] The stability of the higher chlorides generally decreases as the alkalimetal cation increases in size.[54,58]

Manganese(II) bromide, $MnBr_2$, is a pink, hygroscopic solid obtained by treating $Mn(OAc)_2$ with AcOBr in benzene.[47] Tetra*bromomanganates(II)* are obtained using similar methods to those described above for the corresponding chlorides.[59] MnI_2, which is also pink and hygroscopic, is prepared by reaction between the elements in ether[60] or *via* the reaction between MnO_2 or MnS_2 and aluminum iodide in a sealed tube at 230°.[61] The synthesis of *tetraiodomanganates(II)* closely follows that of the chlorides and bromides.[59]

The manganese(II) halides, in particular $MnCl_2$, have the general property of forming molecular adducts or solvates on mixing with basic molecules and chelates. These complexes provide an interesting variety of stereochemical and ligand environments which is well documented.[62,63] The complexes are usually prepared by mixing the constituents in a suitable solvent such as tetrahydrofuran (THF) or ethanol or by refluxing the manganese halide or complex halide with the ligand. In most cases it is advisable to exclude moisture (e.g., in THF by refluxing the sodium-dried solvent with LiH or drying the alcohol with

---

[46] *A. Chretien, G. Oechsel*, C. R. Acad. Sci. Paris *206*, 254 (1938);
J. Reed, B. Hopkins, L. Audrieth, Inorg. Synth. *1*, 29 (1939).

[47] *G. Watt, P. Gentile, E. Helvenston*, J. Amer. Chem. Soc. *77*, 2752 (1955).

[48] *D. Khristov, S. Karaivanov, V. Kolushki*, Godishnik Sofiskyia Univ. Khim. Fak. *55*, 49 (1960).

[49] *Z. El Saffar*, J. Chem. Phys., *52*, 4097 (1970);
A. Lodzinska, F. Golinska, Roczniki Chem. *43*, 1929 (1969).

[50] *O. Zvjagintzev, A. Tschenkali*, J. Gen. Chem. Russ *11*, 791 (1941).

[51] *G. Fowles*, Prog. Inorg. Chem. *6*, 1 (1964);
E. Weitz, Angew. Chem. *37*, 391 (1924);
E. Weitz, F. Miller, Ber. *58*, 363 (1925).

[52] *H. de la Garanderie*, Phys. Chem. Scintill. Munich 1965, Int. Symp. Lumin. 264 (1965); C. A. *67*, 121 257f (1966).

[53] *F. Remy*, Rev. Chim. Minerale *2*, 693 (1965); *5*, 935 (1968);
B. Wayland, W. Rice, Inorg. Chem. *6*, 2270 (1967).

[54] *H. Seifert, F. Koknat*, Z. Anorg. Allgem. Chemie *341*, 269 (1965);
B. Markov, R. Chernov, Ukr. Khim. Zh. *24*, 139 (1958).

[55] *H. Benrath*, Z. Anorg. Allgem. Chemie *220*, 145 (1934);
A. Greenberg, G. Walden, J. Chem. Phys. *8*, 645 (1940);
S. Jensen, Acta Chem. Scand. *18*, 2085 (1964); *22*, 641, 647 (1968);
J. Mellor, A Comprehensive Treatise on Inorganic and Theoretical Chemistry, Vol. XII, p. 363, Longmans, Green and Co., London 1932.

[56] *P. Paoletti, A. Vacca*, Trans. Faraday Soc. *60*, 50 (1964);
A. Blake, F. Cotton, Inorg. Chem. *3*, 5 (1964);
P. Pauling, Inorg. Chem. *5*, 1498 (1966);
C. Furlani, A. Furlani, J. Inorg. Nucl. Chem. *19*, 51 (1962);
T. Melia, R. Merrifield, J. Chem. Soc. *A*, 1166 (1970).

[57] *H. de la Garanderie*, Compt. Rend. *255*, 2585 (1962).

[58] *M. Kestigian, W. Croft*, Mater. Res. Bull. *4*, 877 (1969);
P. Ehrlich, F. Koknat, H. Seifert, Z. Anorg. Allgem. Chemie *341*, 281 (1965);
A. Le Pailler-Malecot, C. R. Acad. Sci., Paris *261*, 3346 (1965);
G. Butterworth, J. Woollam, Phys. Lett. *A29*, 259 (1969);
S. Jensen, Acta Chem. Scand. *21*, 889 (1967);
S. Jensen, Acta Chem. Scand. *22*, 641 (1968);
N. Gill, F. Taylor, Inorg. Synth. *9*, 136 (1967).

[59] *L. Naldini, A. Sacco*, Gazz. Chim. Ital. *89*, 2258 (1959);
N. Gill, R. Nyholm, J. Chem. Soc. 3997 (1959);
F. Cotton, D. Goodgame, M. Goodgame, J. Amer. Chem. Soc. *84*, 167 (1962).

[60] *J. Grey*, J. Amer. Chem. Soc. *68*, 605 (1946).

[61] *M. Chaigneau*, Bull. Soc. Chim. France 886 (1957);
M. Chaigneau, M. Chastagnier, Bull. Soc. Chim. France 1192 (1958).

[62] *R. Colton, J. H. Canterford*, Halides of the First Row Transition Metals p. 223, 225, Wiley-Intersci. Publ., London 1969.

[63] *I. Lindqvist*, Inorganic Adduct Molecules of Oxo-Compounds, Springer Verlag, Berlin 1963.

2,2-dimethoxypropane).[64] Pyridine forms a polymeric[65] adduct with $MnCl_2$, but the methylpyridines form discrete complexes of the type MnL_2Cl_2.[66]

An example of the distinction between a molecular adduct and a solvate is provided by the compound having the empirical formula $MnI_2(CH_3CN)_3$. Conductance and diffuse reflectance data indicate[67] that the correct formulation should, in fact, be $Mn(CH_3CN)_6^{2\oplus}MnI_4^{2\ominus}$, *i.e.*, this complex is a solvate. A large number of solvates of this type as well as those in which $BF_4^{\ominus}$, tetrachlorometallate, and other anions are employed as precipitators have been described, and some examples are listed in Table 1. Most of these solvates are yellow or brown.

Table 1 Some Solvates of Manganese(II), $Mn(S)_nX_y$

S	n	X[a]	Ref.
Nitromethane	6	$FeCl_4^{\ominus}(2)$	*e*
Dimethyl sulfoxide	6	$MnCl_4^{2\ominus}(1)$	*f*
Diphenyl sulfoxide	6	$ClO_4^{\ominus}(2)$	*g*
Acetonitrile	6·5	$SnCl_6^{2\ominus}(1)$	*h*
Acetonitrile	6	$ClO_4^{\ominus}$, $BF_4^{\ominus}(2)$	*i*
ROH[b]	1–4		*j*
Ethanol	6	$ClO_4^{\ominus}(2)$	*k*
Dimethylmethyl-phosphonate	6	$CrO_4^{2\ominus}(1)$	*l*
2-Pyridone	6	$BF_4^{\ominus}(2)$	*m*
Acetone	6	$FeCl_4^{\ominus}$, $InCl_4^{\ominus}(2)$	*n*
RNH_2	6	$CrO_4^{2\ominus}(1)$	*o*
U[c]	6	$Cl^{\ominus}$, $Br^{\ominus}$, $NO_3^{\ominus}$, $ClO_4^{\ominus}(2)$	*p*
Acetic acid	6	$Cl^{\ominus}$, $Br^{\ominus}(2)$	*q*
M[d]	6	$MX_4^{\ominus}(2)$	*r*

[a] The value of y is shown in parentheses.
[b] C_1–C_4 primary alcohols.
[c] U = *N-n*-butylurea, *N*-methylthiourea, or *N,N'*-dimethylthiourea.
[d] M = methylformate, ethyl acetate, or diethylmalonate.
[e] *W. Driessen, W. Groeneveld*, Rec. Trav. Chim. Pays-Bas *88*, 620 (1969).
[f] *R. Drago, D. Meek*, J. Phys. Chem. *65*, 1446 (1961); and refs. therein.
[g] *P. van Leeuwen, W. Groeneveld*, Rec. Trav. Chim. Pays-Bas *85*, 1173 (1966).
[h] *J. Reedijk, W. Groenveld*, Rec. Trav. Chim. Pays-Bas *86*, 1103 (1967).
[i] *J. Reedijk, J. Vervelde, W. Groeneveld*, Rec. Trav. Chim. Pays-Bas *89*, 42 (1970); see also *J. Reedijk, W. Groeneveld* ibid. *87*, 513 (1968).
[j] *R. Osthoff, R. West*, J. Amer. Chem. Soc. *76*, 4732 (1954); *J. Druce*, J. Chem. Soc., p. 1407 (1937); *J. Partington, A. Whynes*, ibid., p. 1952 (1948).
[k] *P. van Leeuwen*, Rec. Trav. Chem. Pays-Bas *86*, 247 (1967).
[l] *N. Karayannis, C. Owens*, J. Inorg. Nucl. Chem. *31*, 2767 (1969).
[m] *J. Reedijk*, Rec. Trav. Chim. Pays-Bas *88*, 1139 (1969).
[n] *W. Driessen, W. Groeneveld*, Rec. Trav. Chim. Pays-Bas *88*, 977 (1969).
[o] *G. Narain, P. Shukla*, J. Indian Chem. Soc. *43*, 694 (1966).
[p] *M. Pakinam*, Diss. Abstr. *30B*, 1563 (1969).
[q] *P. van Leeuwen, W. Groeneveld*, Rec. Trav. Chim. Pays-Bas *87*, 86 (1968).
[r] *W. Driessen, W. Groeneveld, F. van der Wey*, Rec. Trav. Chim. Pays-Bas *89*, 353 (1970); *J. Reedijk*, ibid. *89*, 605 (1970).

A complication due to atmospheric oxidation of the ligand has been encountered,[68] for example, in the synthesis of triphenylphosphine adducts; a color change from colorless to yellow during the synthesis was attributed to complex formation with the phosphine oxide. The possibility of atmospheric oxidation, though in general not a serious problem, should nevertheless be systematically checked in the synthesis of manganese(II) compounds.

17.4.3 Oxygen and Sulfur Derivatives

The preference for bound oxygen atoms which characterizes manganese chemistry in general is exemplified by the tendency of some manganese(II) compounds to undergo aerobic oxidation, especially on heating.

Addition of sodium hydroxide to the solution of a manganese(II) salt generally results in the precipitation of a gelatinous, colorless precipitate of $Mn(OH)_2$. In the presence of oxygen the solid darkens and is eventually oxidized to $Mn_2O_3 \cdot xH_2O$ *via* a complicated oxidative mechanism.[69] In the absence of oxygen the solid may be redissolved in excess alkali, and is believed to

[64] *N. Karrayannis, C. Paleos, L. Pytlewski, M. Labes*, Inorg. Chem. *8*, 2559 (1969).
[65] *R. Zannetti, R. Serra*, Gazz. Chim. Ital. *90*, 328 (1960).
[66] *G. Beech, C. Mortimer, E. Tyler*, J. Chem. Soc. *A*, 1111 (1967).
[67] *B. Hathaway, D. Holah*, J. Chem. Soc. 2400 (1964).
[68] *D. Negoui*, An. Univ. Bucuresti, Ser. Stiint. Natur. *14*, 145 (1965); C. A. *67*, 60349t (1967).
[69] *R. Fox, D. Swinehart, A. Garrett*, J. Amer. Chem. Soc. *63*, 1779 (1941);
B. Dzanashvili, E. Bogoyavlenskii, K. Purtseladze, Soobshch. Akad. Nauk Gruz. SSR, *43*, 361 (1966); C. A. *65*, 16462h (1966).

exist as $Mn(OH)_4^{2\ominus}$ anions in 5–12 M KOH.[70] The solid salt $Na_2Mn(OH)_4$ has been obtained by heating $Mn(OH)_2$ with NaOH in the absence of oxygen for 15 hr at 50–100°.[71] The insoluble oxide MnO may be obtained by heating the carbonate under N_2 or H_2[72] [Eqn (4)] or the oxalate [Eqn (5)] in air or by reducing the higher oxides with hydrazine.

$$MnCO_3 \longrightarrow MnO + CO_2 \quad (4)$$

$$MnC_2O_4 \longrightarrow MnO + CO_2 + CO \quad (5)$$

The addition of an alkaline or ammonium sulfide solution to a solution of $Mn(OAc)_2$ gives a salmon-colored precipitate of hydrous MnS.[73] Solutions of $MnCl_2$ give a more impure product. The precipitate is soluble in acid and darkens in the presence of oxygen, producing sulfate anions. Anhydrous greenish black MnS can be obtained by heating the hydrous precipitate in H_2 at 320° or by direct reaction between the elements, which produces the α modification. The conditions for interconversion of the three known allotropes[74] are illustrated in the following scheme.

$$\begin{array}{ccccc} & & MnS_2 & & \\ & & \downarrow 260° & & \\ \beta\text{-}MnS & \xrightarrow{200°} & \alpha\text{-}MnS_2 & \xleftarrow{300°} & \gamma\text{-}MnS \end{array}$$

Manganese(II) sulfides are reportedly[75] oxidized to $Mn_3^{IV}S_2O_8$ by oxygen in the temperature range 25°–400°. *Manganese(II) selenide*, MnSe, formed by direct interaction between the elements, is also known in three crystalline modifications.[76] The carbonate, $MnCO_3$, is a potentially useful synthetic starting material, despite its oxidative instability. The solid, which is colorless when pure, is best prepared[77] by mixing solutions of $MnSO_4$, NH_4HCO_3, and NH_4OH in the final concentration ratio 1:1·4:4·5. The precipitate invariably contains $Mn(OH)_2$ and higher oxides if Na_2CO_3 is used as the precipitant. Solutions of this compound darken on standing in air, and the solid is converted to MnO on heating [Eqn (4)]. Commercial samples of $MnCO_3$ are usually buff-colored owing to the presence of oxide impurities. The preparation of manganese(II) perchlorate and nitrate salts by treatment of the carbonate with the respective acids is accompanied by the production of variable amounts of insoluble oxides. The salt $(NH_4)_2Mn(SO_4)_2 \cdot 6H_2O$ is a useful source of pure manganese(II) compounds. It is prepared[78] by reduction of a slight excess of MnO_2 (Section 17.6.3) with a boiling solution of oxalic acid. The salt precipitates on addition of solid ammonium sulfate to the hot filtrate. Fusion at red heat produces the pure salt $MnSO_4 \cdot 4H_2O$, which becomes a pentahydrate on standing in air. Besides being a good synthetic starting material, $MnSO_4$ and its hydrates have an interesting chemistry of their own. For example, they form a complex $MnLSO_4 \cdot 2H_2O$ (L = *1-dimethylamino-2-butyne*) on mixing with the ligand in aqueous solution.[79] Urea also forms an adduct having the formula $MnSO_4 \cdot (urea)_4$ [a similar adduct is formed with $Mn(NO_3)_2$]. The urea is readily replaced by ammonia.[80]

Woolf[81] has described the preparation of fluorosulfates by heating solid salts under anhydrous conditions in HSO_3F. The manganese(II) salt $Mn(SO_3F)_2$ is best prepared from $Mn(OAc)_2$ ($MnSO_4$ reacts much more slowly). Acetic acid is not a suitable reaction medium despite the fact that the product salt is insoluble. The reduction of $Mn(OAc)_3$ by HSO_3F gives the same product,

[70] *A. Kozawa, T. Kalnoki-Kis, J. Yeager*, J. Elektrochem. Soc. *113*, 405 (1966);
A. Kozawa, J. Electrochem. Soc. Japan., Overseas Ed. *36*, 196 (1968);
J. Brenet, Z. Pavlovic, R. Popovic, J. Chim. Phys. Physiochim. Biol. *64*, 719 (1967).

[71] *R. Scholder, F. Schwochow*, Angew. Chem. Intern. Ed. Engl. *5*, 1047 (1966).

[72] *A. Hegedus, K. Martin*, Microchim. Acta. 833 (1966).

[73] *G. Jander, H. Schmidt*, Wiener Chem. Zeit, *46*, 49 (1943);
H. Weiser, W. Milligan, J. Phys. Chem. *35*, 2330 (1931);
H. Komura, J. Phys. Soc. Jap. *26*, 1446 (1969).

[74] *S. Furuseth, A. Kjekshus*, Acta. Chem. Scand. *19*, 1405 (1965);
Y. Fujino, C. Sugiura, S. Kiyono, Tech. Ref. Tohoku Univ. *34*, 301 (1969); C. A. *72*, 105 370k (1970).

[75] *V. Kazakov, S. Perbeneva*, Zh. Neorg. Khim. *12*, 1417 (1967).

[76] *A. Baroni*, Z. Kristallogr. *99*, 336 (1938).

[77] *J. Marcy, F. Matthes*, Chem. Tech. (Leipzig) *21*, 627 (1962).

[78] *W. Palmer*, Experimental Inorganic Chemistry, p. 475, Cambridge 1965.

[79] *S. Avakan, K. Voshkaryan*, Zh. Obschch, Khim. *39*, 1098 (1969).

[80] *G. Lazevka, I. Girei, Y. Zonov*, Vest. Akad. Nauk Bel. SSR, Ser. Khim. Nauk *2*, 50 (1968); C. A. *70*, 44184b (1969);
R. Issa, M. El Shazly, M. Iksander, Z. Anorg. Allgem. Chemie *354*, 90 (1967).

[81] *A. Woolf*, J. Chem. Soc. *A*, 355 (1967).

which is subject to hydrolysis in aqueous solution.[81] Analogous compounds derived from amidosulfonic acid have been described.[82] Methods for the preparation of $Mn(OMe)_2$ include the reaction between $MnCl_2$ and NaOMe in methanol[83] and the general reaction[84] given in Eqn (6).

$$2LiOR + MnX_2 \xrightarrow{MeOH} Mn(OR)_2 + 2LiX \quad (6)$$
$$(X = Cl, Br)$$

The solids are pale pink, and the spectra are very similar to that of the hexaquo ion. Of the dihydroxy alcohols, *perfluoropinacol*, $[(CF_3)_2COH]_2$, reacts in a similar fashion in 1:1 MeOH–water to produce a salt of the type K_2MnL_2.[85]

The neutral phenoxides of manganese(II) with a wide range of mono-, di-, and trihydroxy phenols have been synthesized by reaction between $MnCl_2$ and the phenol or from an aqueous mixture of $MnCl_2$ and sodium phenolate.[86] The compounds are black and, unfortunately, very insoluble in most common solvents.

$Mn(NO_3)_2$ hydrates are a good source of "MnO_2" (Section 17.6.3). The *complex nitrites*, particularly $Mn(NO_2)_4{}^{2\ominus}$ and $Mn(NO_2)_5{}^{3\ominus}$ prepared by Goodgame and Hitchman,[87] are of considerable structural interest.

Among the phosphato complexes of manganese(II) is the salt $NH_4MnPO_4 \cdot H_2O$, which is prepared[88] by heating MnO_2 with H_2SO_4 until effervescence ceases, followed by reduction with sodium bisulfate [precipitation on addition of a solution containing NH_4Cl and $(NH_4)_2HPO_4$] or by heating the manganese(II)–EDTA complex (see below) with a mixture of H_2O_2 and $(NH_4)_2HPO_4$ for 2 hr.[89] The complex is insoluble in excess ammonia and is smoothly converted on heating to the diphosphate $Mn_2P_2O_7$.[90] A *lithium manganese(II) orthophosphate salt* has also been described.[91] One of the more insoluble manganese(II) salts is $Mn_3(PO_4)_2$, precipitated as a colorless amorphous complex on treating a solution of most manganese(II) salts with excess Na_2HPO_4.[92]

Examples of other salts and their preparation include $MnFe_2O_4$, prepared by heating the constituent oxides or oxalates in air,[93] the mixed chromate $K_2(MnCrO_4)_2 \cdot 2H_2O$,[94] and the molybdate $MnMoO_4$ prepared[95] again by heating constituent oxides. The colorless borate $MnH_4(BO_3)_2$, obtained by heating a manganese(II) salt with borax at 100°, is also stoichiometric, but the solid formed[96] on treating an aqueous suspension of MnO_2 with SO_2 is probably a mixture of MnS_2O_6 and $MnSO_3$. The *heterpolyvanadates*, $K_4Mn[V_{10}O_{28}] \cdot xH_2O$ ($x = 10$, 12), can be prepared by mixing the constituents in slightly acid solution.[97] The *uranylvanadates*, $UO_2MnVO_4 \cdot xH_2O$, form on heating the constituents in a sealed tube at 180°.[98]

Carboxylate salts of manganese(II) provide a good source of useful synthetic starting materials. Like the simple halides (Section 17.4.2), they are usually quite soluble in organic solvents such as alcohols and carboxylic acids.

Manganese(II) formate dihydrate,[99] Mn-$(HCOO)_2 \cdot 2H_2O$, is conveniently prepared by neutralization of the carbonate with formic acid. The anhydrous salt forms an adduct with

82 *N. Perakis, T. Karantassis*, C. R. Acad. Sci. Paris *240*, 1407 (1955).

83 *B. Kandelaky, T. Setaschwilli, I. Tewberidze*, Kolloid Z. *73*, 47 (1935);
J. Druce, J. Chem. Soc. 1407 (1937).

84 *R. Adams, E. Bishop, R. Martin, G. Winter*, Australian J. Chem. *19*, 207 (1966),
A. Arnoul, C. Malarde, C. R. Acad. Sci., Paris *262*, 1076, 1372 (1966).

85 *M. Allan, C. Willis*, J. Amer. Chem. Soc. *90*, 5343 (1968).

86 *S. Prasad, K. Kacker*, J. Indian. Chem. Soc. *35*, 890 (1958).

87 *M. Goodgame, M. Hitchman*, Inorg. Chem. *6*, 813 (1967);
M. Goodgame, M. Hitchman, J. Chem. Soc. *A*, 612 (1967).

88 *W. Palmer*, Experimental Inorganic Chemistry, p. 476, Cambridge 1965.

89 *G. Buzagh-Gere, L. Erdey*, Talanta *16*, 14234 (1969).

90 *R. Atkinson, C. Stager*, Can. J. Phys. *47*, 1557 (1969).

91 *P. Elliston, J. Creer, G. Troup*, J. Phys. Chem. Solids *30*, 1335 (1969).

92 *J. Stephens, C. Calvo*, Can. J. Chem. *47*, 2215 (1969).

93 *R. Walden*, Phys. Rev. *99*, 1727 (1955);
D. Wickham, Inorg. Synth. *9*, 154 (1967).

94 *G. Guillem, L. Cot, C. Avinens, A. Norbert*, C. R. Acad. Sci., Paris *c270*, 1870 (1970).

95 *W. Doyle, G. McGuire, G. Clark*, J. Inorg. Nucl. Chem. *28*, 1185 (1966).

96 *J. Meyer, W. Schramm*, Z. Anorg. Chem. *132*, 226 (1924).

97 *C. Flynn jr., M. Pope*, J. Amer. Chem. Soc. *92*, 85 (1970);
V. Bulygina, I. Bezinikov, V. Zolotavin, Zh. Neorg. Khim. *15*, 435 (1970).

98 *F. Cesbron*, Bull. Soc. Fr. Mineral Crystallogr. *93*, 320 (1970).

99 *T. Clarke, J. Thomas*, J. Chem. Soc. *A*, 2227, 2230 (1969).

ammonia.[100] The *acetate tetrahydrate*,[101] $Mn(OAc)_2 \cdot 4H_2O$, can be dehydrated with acetic acid anhydride and is weakly ferromagnetic. Other examples of salts obtained by neutralization are the lactate[101] and the kojate.[102]

An interesting insoluble blue *hydrazine carboxylate salt*, $Mn(N_2H_3COO)_2 \cdot 2H_2O$, has recently been described.[103a,b] The preparation of this compound involves the reaction of CO_2 with an aqueous mixture of hydrazine hydrate and $MnSO_4$. Dehydration is readily carried out at 105°–125°.

The oxalate salt, $Mn(C_2O_4) \cdot 2H_2O$, is only very slightly soluble in water.[104] The allotrope obtained depends on the method of preparation.[105] The salt $K_2Mn(C_2O_4)_2 \cdot 2H_2O$ is obtained by treating a boiling, saturated solution of $K_2C_2O_4$ with the manganese(II) monooxalate.[106]

Manganese(II) complexes with EDTA (ethylenediaminetetraacetic acid) and its derivatives are useful precursors for the production of the corresponding manganese(III) complexes.[107] They are also interesting from a structural viewpoint, as seven-coordination has been demonstrated, for example, in $Mn[Mn(OH_2)HY]_2 \cdot H_2O$ (H_4Y= ethylenediaminetetraacetic acid).[108] The salt $Na_2[Mn(EDTA)] \cdot 2H_2O$ is precipitated[109] on addition of ethanol to a boiling, neutralized mixture of $MnCl_2$ and $Na_2H_2EDTA \cdot 2H_2O$. Unlike the corresponding copper(II) complex, the crystal form of the salt does not depend on the method of precipitation,[110] *i.e.*, the same product is obtained on cooling a supersaturated solution or by slowly evaporating the reaction mixture. Complexes of this ligand form water-soluble complexes of the type $H_4MnY_2 \cdot 3N_2H_4 \cdot 8H_2O$ on long standing in the presence of a 25% excess of hydrazine[103b] (*cf.* the insoluble hydrazinecarboxylates above).

Neutralization of $MnCO_3$ with the acid in Eqn (7)[108] produces the acid salt $Mn_3(HEDTA)_2 \cdot 10H_2O$, which is readily recrystallized from water because of its relative insolubility.

$$3MnCO_3 + 2H_2Y \longrightarrow Mn_3(HY)_2 + 3H_2O + 3CO_2 \quad (7)$$

The *bisacetylacetonate complex* of manganese(II) is a useful synthetic material. Preparative routes involve the reaction of MnO or $MnCO_3$ with the ligand acid,[111] but the best procedure is that employing $MnCl_2$ in the presence of sodium acetate,[112] which yields the air-sensitive yellow dihydrate $Mn(acac)_2 \cdot 2H_2O$[113] [oxidation product is $Mn(acac)_3$ (see Section 17.5.3)]. The anhydrous, tan compound, which is air-stable, is obtained by heating to 100° at a few millimetres pressure (dehydrating agents are unsatisfactory).[112] Although soluble in pyridine and propanol–water, the complex is only slightly soluble in methanol, ethanol, and acetone. Like MnX_2 (Section 17.4.2), this compound forms adducts with a variety of molecules. Examples are the *ammoniate* $Mn(acac)_2 \cdot 2NH_3$ and the *pyridine adduct*,[114] reported to be initially five-coordinate in benzene solution.[115] Other adducts are formed with γ-picoline, 3,5-lutidine, quinoline, and isoquinoline.[110] The complex $Mn[PtX(acac)_2]_2$ (X=Cl, Br) described by Lewis and

[100] *M. Bernard, F. Busnot*, Bull. Soc. Chim. France, *9*, 3061 (1969).

[101] *K. Osaki, Y. Nakai, T. Watanabe*, J. Phys. Soc. (Japan) *18*, 919 (1963);
S. Bhatnagar, M. Nevgi, R. Sharma, Phil. Mag. *22*, 409 (1936).

[102] *J. Wiley, G. Tyson, J. Steller*, J. Amer. Chem. Soc. *64*, 963 (1942).

[103a] *A. Braibanti, G. Bigliardi, A. Tiripiccio*, Nature *211*, 1174 (1966).

[103b] *P. Goroshvili, M. Chkoniya, D. Akhobadze, T. Chkoniya*, Trudy. Gruz. Politekh. Inst. *1*, 7 (1969); C. A. *72*, 139 140a (1970).

[104] *M. Smith, B. Topley*, Proc. Roy. Soc. *A134*, 224 (1931);
B. Topley, M. Smith, J. Chem. Soc. 321 (1935).

[105] *R. Deyrieux, A. Penelux*, Bull. Chim. Soc. France 2160 (1970);
J. Lagier, C. A. *71*, 85 668 (1969).

[106] *H. Matthies, H. Schultz*, Naturwissenschaften *55*, 342 (1968);
S. Smyshlyaev, L. Voltko, A. Taraseuko, C. A. *72*, 127 035 (1970).

[107] *Y. Yoshina, A. Ouchi, Y. Tsunoda, M. Kojima*, Can. J. Chem. *40*, 775 (1962).

[108] *S. Richards, B. Pederson, J. Silverton, J. Hoard*, Inorg. Chem. *3*, 27 (1964);
J. Hoard, G. Smith, M. Lind in *S. Kirschner*, Advances in the Chemistry of Co-ordination Compounds, p. 296, MacMillan Co., New York 1961.

[109] *D. Sawyer, P. Paulsen*, J. Amer. Chem. Soc. *81*, 816 (1959).

[110] *C. Foxx, J. Lambert*, Inorg. Chem. *8*, 2220 (1969).

[111] *B. Emmert, H. Gsottschneider, H. Stanger*, Ber. *69*, 1319 (1936);
K. Nakamoto, A. Martell, J. Chem. Phys. *32*, 588 (1960).

[112] *R. Charles*, Inorg. Synth. *6*, 164 (1960).

[113] *H. Montgomery, E. Lingafetter*, Acta Cryst. *24B*, 1127 (1968);
S. Onuma, S. Shibata, Bull. Chem. Soc. Japan *43*, 2395 (1970).

[114] *A. Das. D. Rao*, Curr. Sci. *39*, 60 (1970).

[115] *D. Graddon, G. Mockler*, Australian J. Chem. *17*, 1119 (1964).

Oldham[116] evidently contains only one type of Mn–O bond.

Compounds of manganese(II) with ligands related to acetylacetone have been described. If a mixture of HX and H_2S (X=Cl, Br) is passed through an ethanolic mixture of $MnCl_2$ and acetylacetone, the solids $Mn(C_5H_7S_2)_2X_4$ are obtained.[117] Complexes with benzoylacetone, acetoacetic ester, acetonedicarboxylic acid ester, and salicylaldehyde are all obtained by the same general methods used to prepare the $Mn(acac)_2$ compounds. They are all yellow or pale pink and are air-sensitive especially when moist. Methyl-substituted acetylacetonates are more air-stable and in methanol are further stabilized by the presence of excess ligand.[118] The ammonium salt $NH_4Mn(FTA)_3$ [FTA= 4,4,4 trifluoro-1-(2-furyl)-1,3-butadione] is precipitated on mixing solutions of $MnCl_2$ or $Mn(NO_3)_2$ and an excess of the ligand in 95% aqueous ethanol.[119] Gerlach and Holm[120] have described the synthesis of an extensive series of ML_2 complexes of the ligand **3**.

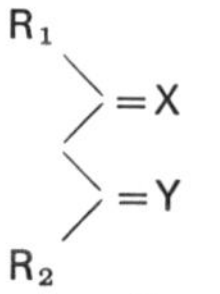

R₁ =X =Y R₂

3

X=Y=O
X=S, Y=O

The method of preparation is similar to that described by Holm *et al.*[121] for preparation of *β-ketoamine complexes* in *t*-butyl alcohol. Thus, the starting material for air-sensitive *bis(dipivaloylmethanido)manganese(II)* (X=Y=O) is $(Et_4N)_2MnBr_2Cl_2$. However, an attempted synthesis by this technique (with X=S, Y=O) was unsuccessful.[120] Of related interest are the manganese(II) complexes with ω-nitroacetophenone (**4**), prepared[122] by reaction of the ligand with

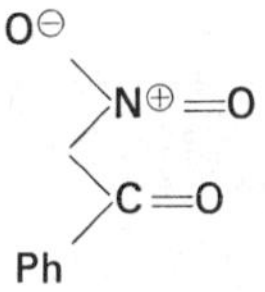

O⊖ N⊕=O C=O Ph

4

metal ethanolates (see above) or with $Mn(OAc)_2$ in acetone. Both the air-stable anhydrous compound and its disolvates with H_2O, EtOH, and pyridine have been reported.[122]

Karayannis and co-workers[123] have recently described the preparation of a five-coordinated complex of the type $Mn(TMP)_4OClO_3^{2\oplus}$ by reaction between $Mn(ClO_4)_2$ and trimethylphosphate (TMP) in triethylorthoformate solution. In a related study, Goodgame *et al.*[124] have proposed that the analogous phosphine and arsine oxide complexes are also five-coordinate.

The yellow *manganese(II) diethylthiocarbamate* prepared by mixing a manganese(II) salt and $Na(Et_2NCS_2)$ in water is pyrophoric.[125] The methyldithiocacodylate salt is obtained by treating a manganese(II) salt with NaS_2AsMe_2 or by passing H_2S through a solution of the cacodylate in methanol containing HCl.[126] Complexes are also formed with *cis*-1,2-dimercaptoethylene and 4,5-dimercapto-*o*-xylene in pyridine solution[127] (the complex is precipitated as $[AsPh_4]$-$[Mn(L)_2py]$).

Thiocyanate and selenocyanate complexes are discussed in Section 17.4.4 below.

17.4.4 Nitrogen and Group V Complexes

Complexes with bound N, P, and As atoms are not common in manganese(II) chemistry, and are nearly always found as chelated molecular species.

As noted above (Section 17.4.2) the manganese halides are particularly noted for their ability to form molecular adducts with N-donating ligands. Some of these species, of particular interest in view of their five-coordination,[128] are listed in

[116] *J. Lewis, C. Oldham*, J. Chem. Soc. *A*, 1456 (1966).
[117] *A. Furuhashi, K. Watanuki, A. Ouchi*, Bull. Chem. Soc. Japan *41*, 110 (1968);
G. Heath, R. Martin, I. Stewart, Australian J. Chem. *22*, 83 (1969).
[118] *A. Schugarman*, Diss. Abstracts *29B*, 925 (1968).
[119] *W. McSharry, M. Cefola*, J. Inorg. Nucl. Chem. *31*, 2777 (1969).
[120] *D. Gerlach, R. Holm*, Inorg. Chem. *8*, 2292 (1969).
[121] *R. Holm, F. Bohrscheid, G. Everett*, Inorg. Synth. *11*, 72 (1968).
[122] *M. Bonamico, I. Collamati, C. Ercolani, G. Dessy, D. Machin*, Chem. Commun. 654 (1967).
[123] *N. Karayannis, E. Bradshaw, L. Pytlewski, M. Labes*, J. Inorg. Nucl. Chem. *32*, 1079 (1970).
[124] *D. Goodgame, M. Goodgame, P. Haywood*, J. Chem. Soc. *A*, 1352 (1970);
W. Rieff, W. Baker, Inorg. Chem. *9*, 570 (1970).
[125] *J. Fackler, D. Holah*, Inorg. Nucl. Chem. Lett. *2*, 251 (1967).
[126] *A. Casey, N. Ham, D. Mackey, R. Martin*, Australian J. Chem. *23*, 1117 (1970).
[127] *E. Hoyer, W. Dietzsh, H. Mueller, W. Schroth*, Z. Chem. *7*, 354 (1967).
[128] *M. Ciampolini, N. Nardi, G. Speroni*, Coord. Chem. Rev. *1*, 223 (1966);
M. Ciampolini, Str. and Bonding *6*, 52 (1969).

Table 2 Complexes of Manganese(II) Halides with Nitrogen Donor Ligands

Ligand	Complex	Reactants	Solvent	Remarks	Ref.
NH_3	$Mn(NH_3)_6Br_2$	$MnBr_2 + NH_3$	Liq. NH_3	White	*i*
en[a]	$Mn(en)_3Br_2$	$MnBr_2 \cdot H_2O$ + en	—	Gray-white	*i*
en	$Mn(en)_3X$	MnX + en	—	$X = S_2O_3$, Se_2O_3	*j*
dien[b]	$Mn(dien)_2Br_2$	$MnBr_2 \cdot H_2O$ + dien	—	White; very insoluble	*i, k*
trien[c]	$Mn(tren)_2Br_2$	$MnBr_2 \cdot H_2O$ + tren	—	White	*i*
trien	$Mn_2L_3Br_4$	$MnBr_2 \cdot H_2O$ + tren	Ethanol	White	*i*
Me_4tren[d]	$MnLX_2$	$MnX_2 + L$	Hot butanol	Yellow; X = Cl, Br, I	*l*
Me_6tren[e]	$MnLBr_2$	$MnBr_2 + L$	Butanol	Yellow; no reaction with $MnCl_2$	*m*
Hexamine[f]	MnL_nX_2	$MnX_2 \cdot nH_2O + L$	Ethanol	$n = 1$, X = Cl; $n = 2$, X = Br; $n = 3$, X = I	*n*
Sal-Me[g]	MnL_2	$Mn(OAc)_2$ + salicylaldehyde + $MeNH_2$	Ethanol	Yellow	*o*
PQ[h]	MnL_nX_2	MnX + L	Ethanol	$n = 2$, X = Cl, Br, I, SCN; $n = 4$, X = I, SCN	*p*

[a] en = ethylenediamine.
[b] dien = diethylenetriamine.
[c] trien = triethylenetetramine.
[d] Me_4tren = bis(2-dimethylaminoethyl)amine.
[e] Me_6tren = tris(2-methylaminoethyl)amine.
[f] hexamine = hexamethylenetetramine.
[g] Sal-Me = *N*-methylsalicylaldimine.
[h] PQ = pyrrole, quinoline, or isoquinoline.
[i] *G. Watt, B. Manhas*, J. Inorg. Nucl. Chem. *28*, 1945 (1966).
[j] *V. Varand, N. Podberezskaya, V. Shill'man, V. Bakakin, E. Ruchkin*, Izv. Sib. Otd. Akad. Nauk SSSR. Ser. Khim. Nauk *44* Izv. (1967); Chem. Abstr. *69. 15472d* (1969).
[k] *C. Jorgensen*, Inorg. Chim. Acta *3*, 313 (1969).
[l] *M. Ciampolini, G. Speroni*, Inorg. Chem. *5*, 45 (1966).
[m] *M. Ciampolini, N. Nardi*, Inorg. Chem. *5*, 1150 (1966); *M. di Vaira, P. Orioli*, Acta Crystallogr. *B24*, 1269 (1968).
[n] *J. Allan, D. Brown, M. Lappin*, J. Inorg. Nucl. Chem. *32*, 2287 (1970).
[o] *L. Sacconi, M. Ciampolini, G. Speroni*, J. Amer. Chem. Soc. *87*, 3102 (1965); *P. Orioli, M. di Vaira, L. Sacconi*, Chem. Commun. *103*, (1965).
[p] *K. Dash, D. Rao*, Indian J. Chem. *5*, 333 (1967).

Table 2. The syntheses are relatively straightforward provided oxygen and moisture are excluded. Jorgensen[129] recommends the use of hydrazine to prevent oxidation of the *tris-(diethylenetriamine) complex.*

By contrast, the series of manganese(II) complexes, with 1,10-phenanthroline and 2,2′-bipyridyl, and their derivatives are easily prepared without special precautions by mixing a manganese(II) salt with the ligand in ethanol and they are air-stable indefinitely.[130] Broomhead and Dwyer[131] recommend *N,N*-dimethylformamide (DMF) as a suitable reaction medium for the monocomplexes. Of particular note are the *2-carboxy-1,10-phenanthroline complexes* described by Goodwin and Sylva[132] and the *monoterpyridyl complex* prepared by Judge *et al.*[133] Complex pseudohalide species $M^{I}_{x-2}MnX_x$ (X = NCS,[134] N_3,[135] and NCSe[136]) are prepared

[129] *C. Jorgensen*, Inorg. Chim. Acta *3*, 313 (1969).

[130] *P. Pfeiffer, B. Werdelman*, Z. Anorg. Allgem. Chemie *261*, 197 (1950);
F. Burstall, R. Nyholm, J. Chem. Soc. 3570 (1952);
J. Novak, H. Ahrend, Talanta *11*, 898 (1964);
T. Glikman, M. Podlinyaeva, Ukr. Khim. Zh. *21*, 211 (1955);
R. Dowsing, J. Gibson, M. Goodgame, J. Chem. Soc. *A* 1133 (1970);
A. Sychev, Y. Tiginyarii, Y. Gromovoi, Zh. Fiz. Khim. *42*, 2081 (1968).

[131] *J. Broomhead, F. Dwyer*, Australian J. Chem. *14*, 250 (1961);
H. Preston, C. Kennard, R. Plowman, J. Inorg. Nucl. Chem. *30*, 1463 (1968).

[132] *H. Goodwin, R. Sylva*, Australian J. Chem. *20*, 217 (1967).

[133] *J. Judge, W. Reiff, G. Intile, P. Ballway, W. Baker*, J. Inorg. Nucl. Chem. *29*, 1711 (1967).

[134] *C. Flint, M. Goodgame*, J. Chem. Soc. *A*, 442 (1970);
H. Schmidtke, D. Garthoff, Z. Naturforsch, *24b*, 126 (1969).

[135] *W. Beck, W. Fehlhammer, P. Poellmann, E. Schuierer, F. Feldi*, Chem. Ber. *100*, 2335 (1967).

by the same general procedures as outlined in Section 17.4.2.

A long-standing interest in the role of manganese in photosynthesis[137] has perhaps stimulated the preparation of *macrocyclic N-bonded manganese(II) complexes*. Metal troponiminates of type **5** (R = Et, Ph) are prepared[138]

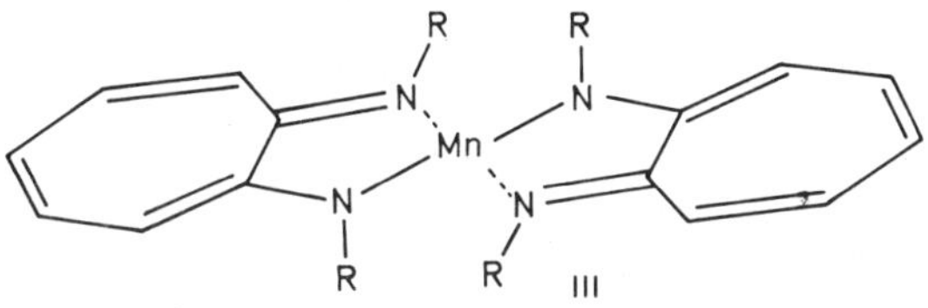

5

by reaction of $MnCl_2$ with the *N*-lithioaminotroponimines in the absence of oxygen. The complexes are oxidized both in solution and in the solid state to the corresponding manganese(III) species. However, only a mono complex MLX_2 (X = Cl, Br, I, NCS, NCSe) is formed with 2,3,5,6-tetrakis(6-methyl-2-pyridyl)pyrazine (**6**) on mixing the ligand with a manganese(II)

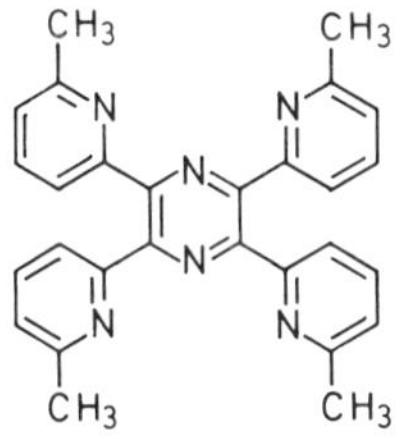

6

salt in hot ethanol, presumably because of the steric requirements of the ligand.[139] Pyrromethane chelates (**7**) of divalent transition metal ions are prepared under very mild conditions by exchange of the metal for $Ca^{2\oplus}$ in THF solution.[140]

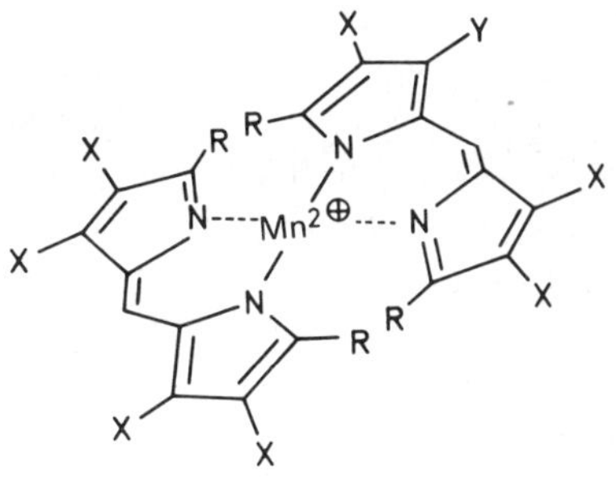

7

The preparation of seven coordinated macrocyclic complexes reported by Martin *et al.*[141] is evidence of further progress in this area. Manganese(II) phthalimide slowly forms a series of black complexes $Mn(C_8H_4O_2N)_2A_4$ with primary and secondary amines in acetone.[142]

Manganese(II) phthalocyanine is prepared by mixing $Mn(OAc)_2$ with phthalonitrile in undecyl alcohol[143] or by heating $MnCl_2$ with the ligand under a stream of HCl in the presence of NaOAc.[144] The complex reversibly forms an oxygen adduct in pyridine solution,[143] and is oxidized to the manganese(III) species by formic acid.[145] The two black-green crystaline modifications[146] can be distinguished by reaction with nitric oxide (only the β forms reacts).[147] The water-soluble *4,4′,4″,4‴-tetrasulfophthalocyaninatomanganese(II) complex*[148] has an interesting photochemistry.[149]

The few complexes containing bound phosphorus atoms include the phosphide $Mn(PH_2)_2\cdot$

---

[136] *J. Burmeister, L. Williams*, Inorg. Chem. *5*, 1113 (1966);
F. Pruchnik, S. Wajda, Roczniki Chem. *44*, 933 (1970);
P. Stancheva, V. Skopenko, G. Tsintsadze, Ukr. Khim. Zh. *35*, 166 (1969).

[137] *M. Calvin*, Rev. Pure Appl. Chem. *15*, 1 (1965);
M. Atassi, Biochem. J. *103*, 29 (1967);
T. Stein, R. Plane, J. Amer. Chem. Soc. *91*, 607 (1969);
L. Boucher, H. Garber, Inorg. Chem. *9*, 2644 (1970).

[138] *D. Eaton, W. McClennan, J. Weiher*, Inorg. Chem. 7, 2040 (1968).

[139] *H. Goodwin, F. Lions*, J. Amer. Chem. Soc. *81*, 6415 (1959).

[140] *R. Motekaitis, A. Martell*, Inorg. Chem. *9*, 1832 (1970).

[141] *R. Martin, A. van Heuvelen, H. Hamilton*, Inorg. Nucl. Chem. Lett. *6*, 445 (1970).

[142] *G. Narain, P. Shukla*, Australian J. Chem. *20*, 227 (1967).

[143] *H. Przywarska-Boniecka* in *B. Jezowska-Trzebiatowska*, Theory and Structure of Inorganic Complexes, p. 651, Oxford, 1964;
H. Przywarska-Boniecka, Roczniki Chem. *39*, 1377 (1965).

[144] *B. Berezin, G. Sennikova*, Izv. Vyssh. Ucheb. Zaved Khim., Khim. Technol. *10*, 563 (1967);
W. Klemm, H. Senff, J. Prakt. Chem. *154*, 73 (1939).

[145] *H. Walter*, Monatsber, Deut. Akad. Wiss. Berlin *11*, 873 (1969); C. A. *73*, 60 584 (1970);
W. Hanke, Z. Anorg. Allgem. Chemie *355*, 160 (1967).

[146] *C. Barraclough, R. Martin, S. Mitra, R. Sherwood*, J. Chem. Phys. *53*, 1638 (1970).

[147] *C. Ercolani, C. Neri, G. Sartori*, J. Chem. Soc. *A*, 2123 (1968).

[148] *J. Weber, D. Busch*, Inorg. Chem. *4*, 469 (1965).

[149] *L. Zavgorodnyaya, T. Glikman*, Zh. Obshch. Khim. *39*, 1443 (1969).

$3NH_3$, prepared[150] by reaction of $(NH_4)_2$-$Mn(SCN)_4$ with KPH_2, and the phosphino salt $K_2Mn(PH_2)_4 \cdot 2NH_3$, obtained[150] from a similar reaction with $MnCl_2 \cdot 6NH_3$. The reaction medium is liquid ammonia in both cases.[150]
Colorless (X=Cl, Br) or pale yellow (X=I) complexes of the type $Mn(pbd)X_2$ (X=Cl, Br, I; pbd=*o*-phenylenebisdimethylarsine) are prepared from MnX_2 in dry dioxane (no reaction occurs in ethanol, methanol, water, or acetone).[151] The complexes are stable to oxidation under normal conditions (diarsine rapidly reduces manganese(III) complexes).

17.5 Manganese(III)

There are three main routes to manganese(III) species. These involve oxidation of manganese(II) or reduction of manganese(IV) or (VII). Trivalent manganese is oxidizing, and disproportionation readily occurs under weakly complexing conditions at low acidity [Eqn (8)].

$$2Mn(III) \rightleftharpoons Mn(II) + Mn(IV) \quad (8)$$

Few solid complexes have been completely characterized, but studies of solutions containing manganese(III) reveal a strong tendency toward hydrolysis which should be borne in mind when establishing the constitution of manganese(III) complexes.

17.5.1 Cyanides and Complex Cyanides

$K_3Mn(CN)_6$ can be prepared by oxidation in air of $K_4Mn(CN)_6$ (Section 17.4.1) or by treatment of a solution of $K_2MnF_5 \cdot H_2O$[152] with excess KCN at room temperature. The brick red solid is air-stable, but solutions are slowly hydrolyzed even in the dark,[153] producing $(CN)_2 \cdot H_3Mn$-$(CN)_6$ which is evidently a strong tribasic acid.[154]
Saha[155] has prepared the pink salt $K_3Mn(CN)_5$-OH by treating $Mn(OAc)_2$ with concentrated KCN solution under aerobic conditions and adding ethanol to the resulting filtrate. This salt is particularly important, as it illustrates the tendency of the manganese(III) state towards hydrolysis. The yellow salts $M^I_2Mn(CN)_5NO$ (M^I=Ag, K, also a Zn salt) are obtained by oxidation of the corresponding manganese(II) salt with concentrated nitric acid.[156] Other *complex cyanides* reported include $K_3Mn_2(CN)_9 \cdot 4KOH$[157] and $1{\cdot}2K_2Mn(CN)_6 \cdot 0{\cdot}7K_3Mn(CN)_6 \cdot 0{\cdot}25H_2O$.[158] By analogy, the first of these compounds may be a mixture of *cyano–hydroxo complexes*, perhaps $K_3Mn(CN)_4(OH)_2 \cdot K_3Mn(CN)_5OH \cdot KOH$.

17.5.2 Halides and Complex Halides

The purple solid MnF_3 is obtained by reaction between fluorine and MnX_2[159] or the oxides[160] (Section 17.5.3) or by treatment of $Mn(IO_3)_2$ with BrF_3.[161] Dissolution of Mn_2O_3 in aqueous HF yields ruby red crystals of the dihydrate. MnF_3 yields MnF_2 and F_2 on heating.
Manganese(III) trichloride is probably present in the solution obtained by dissolving MnO_2 in cold, concentrated HCl. The unstable black solid is prepared by reaction between $Mn(OAc)_3$ and HCl at $-100°$[162] or by treating a suspension of MnO_2 with HCl in ethanol at $-63°$ (precipitation with CCl_4 or ligroin).[163] $MnCl_3$ forms characteristic green solutions in ethanol, acetyl chloride, and other organic solvents. Stable 1:3 adducts with ammonia have been reported.[164]
Dark brown to violet salts of the type M^IMnF_4 (M^I=Li, K) can be prepared by reduction of the pentafluoromanganates(IV) (Section 17.6.2) with hydrogen.[165] The red or dark red *pentafluoro-*

[150] *G. Ueker, O. Schmitz-Dumont*, Z. Anorg. Allgem. Chemie *371*, 318 (1969).
[151] *R. Nyholm, G. Sutton*, J. Chem. Soc. 564 (1958).
[152] *W. Palmer*, Experimental Inorganic Chemistry, p. 479, Cambridge 1965.
[153] *A. MacDiarmid, N. Hall*, J. Amer. Chem. Soc. *75*, 5204 (1953);
R. Schwartz, K. Tede, Ber. *60*, 69 (1927).
[154] *J. Brigando*, Bull. Chim. Soc. France *24*, 503 (1957).
[155] *H. Saha*, Sci. Cult. (Calcutta) *27*, 582 (1961); C. A. *57*, 1833a (1962);
J. Meyer, Z. Anorg. Allgem. Chimie *81*, 385 (1913);
I. Chawla, M. Frank, J. Inorg. Nucl. Chem. *32*, 555 (1970).
[156] *F. Cotton, R. Monchamp, R. Henry, R. Young*, J. Inorg. Nucl. Chem. *10*, 28 (1959).
[157] *N. Goldenberg*, Trans. Faraday Soc. *36*, 847 (1940).
[158] *A. McCarthy*, J. Chem. Soc. *A*, 1379 (1970).
[159] *H. Emeléus, G. Hunt*, J. Chem. Soc. *A*, 396 (1964).
[160] *E. Aynsley, R. Peacock, P. Robinson*, J. Chem. Soc. *A*, 1622 (1950).
[161] *A. Sharp, A. Woolf*, J. Chem. Soc. p. 798 (1951); *M. Hepworth, K. Jack*, *Acta* Cryst. *10*, 345 (1957).
[162] *A. Chretien, G. Varga*, Bull. Soc. Chim. France *3*, 2385 (1936).
[163] *J. Krepelka, J. Kubis*, Collect. Czech. Chem. Commun. 7, 105 (1935).
[164] *H. Funk, H. Kreis*, Z. Anorg. Allgem. Chemie *349*, 45 (1967).
[165] *R. Hoppe, W. Liebe, W. Dahne*, Z. Anorg. Allgem. Chemie *307*, 276 (1961);
R. Hoppe, W. Dahne, W. Klemm, Ann. Chim. (Paris) *658*, 1 (1962).

manganates(III) are obtained by dissolution of Mn_2O_3 in 20–40% HF[152] (precipitation, *e.g.*, with KHF_2), by electrolytic oxidation of manganese(II) in HF[166] or by reduction of manganese(VII) with manganese(II) in the presence of HF and KHF_2.[60] Hydrated salts of NH_4,[167,168] K,[152] Cs,[169] and Ag have been reported. The possibility that the ammonium salt may be, in fact, a hydroxofluoro complex has been considered.[167,168]

K_3MnF_6 has been obtained by adding KHF_2 to the brown supernatant solution obtained in the preparation of K_2MnF_6[170] (Section 17.6.2) or by fusing $KMnF_5 \cdot H_2O$ (above) with KHF_2 in an inert atmosphere.[171] The excess KHF_2 was extracted with formamide.[171]

Pentachloromanganates(III) are generally prepared by reduction of MnO_2. For example, the bis(tetraethylammonium) salt is obtained by treating a suspension of MnO_2 in ether with acetyl chloride and precipitating the salt by addition of Et_4NCl dissolved in ethanol.[172] *Anhydrous phenanthrolinium, dipyridylium*,[173] $K^{\oplus}$, and $NH_4^{\oplus}$ salts can be prepared by this method (the ammonium salt is a monohydrate). Complex cobalt(III) cations have been successfully used to isolate *hexachloromanganate(III)* salts.[174] Manganese(III) acetate forms adducts with 1–4 mol of acetyl chloride in acetic acid; addition of potassium acetate precipitates K_2MnCl_6.[175]

17.5.3 Oxygen and Sulfur Derivatives

Methods for the preparation of aqueous solutions containing manganese(III) and their properties have recently been summarized.[176] Electrolysis under controlled conditions is a useful general technique. Ciavatta and Grimaldi[177] have shown that anodic oxidation of manganese(II) at a gold electrode in 3 M perchloric acid produces a solution containing monomeric Mn(III)*aq* ions. Anodic oxidation at platinum produces an oxide layer with spectacular color changes on the electrode, particularly at low acidities and low manganese(II) concentrations.[178] Ozonolysis of acidic solutions of manganese(II) is another promising synthetic route.[179] These solutions in strong acid are of little use for the preparation of solid complexes because of their tendency to oxidize potential ligands.[176]

Mn(OH)O occurs in nature as manganite.[1] The compound can be prepared by drying air-oxidized $Mn(OH)_2$ (Section 17.4.3) at 100°, by passing Cl_2 through an aqueous suspension of $MnCO_3$, or by electrolytic reduction of an alkaline suspension of MnO_2 (Section 17.6.3).[180] The solid redissolves in concentrated alkaline solution. Oxidation in air of MnO_2 at 550°–900° produces Mn_2O_3.[1] The α and β allotropes of this compound have different magnetic moments.[181] Scholder and Protzer[182] have recently reported that heating Mn_2O_3 with anhydrous alkali metal hydroxides produces salts of the type M^IMnO_2 (M^I=Na) and $M^IMn_4O_7$ (M^I=K, Rb). $NaMnO_2$ can also be prepared by heating $MnCO_3$ with Na_2CO_3 in air.[182]

Solutions of manganese(III) in strong sulfuric acid have been known for some years.[183] They are easily prepared by electrolysis[176] or by dissolving MnO_2 in hot sulfuric acid. A dark green powder, believed to be $Mn_2(SO_4)_3$, is obtained by treating freshly precipitated MnO_2

---

[166] *B. Cox, A. Sharp*, J. Chem. Soc. 1798 (1954).
[167] *D. Sears, J. Hoard*, J. Chem. Phys. *50*, 1066 (1969).
[168] *R. Dingle*, Inorg. Chem. *4*, 1287 (1965).
[169] *I. Ryss, B. Vitokhnovskaya*, Zh. Neorg. Khim. *3*, 1185 (1958).
[170] *R. Peacock*, J. Chem. Soc. 4684 (1957).
[171] *R. Peacock, D. Sharp*, J. Chem. Soc. 2762 (1959).
[172] *T. Davis, J. Fackler, M. Weeks*, Inorg. Chem. *7*, 1994 (1968).
[173] *H. Goodwin, R. Sylva*, Australian J. Chem. *18*, 1743 (1965);
H. Goodwin, R. Sylva, Australian J. Chem. *20*, 627 (1967).
[174] *W. Hatfield, R. Fay, C. Pfluger, T. Piper*, J. Amer. Chem. Soc. *85*, 265 (1963);
W. Hatfield, R. Whyman, R. Fay, K. Raymond, F. Basolo, Inorg. Synth. *11*, 48 (1968);
D. Adams, D. Morris, J. Chem. Soc. *A*, 694 (1958).
[175] *H. Hardt, M. Fleischer*, Z. Anorg. Allgem. Chemie *357*, 113 (1968).
[176] *G. Davies*, Coord. Chem. Rev. *4*, 199 (1969).
[177] *L. Ciavatta, M. Grimaldi*, J. Inorg. Nucl. Chem. *31*, 3071 (1969).
[178] *G. Atkinson, G. Brydon*, Anal. Chim. Acta *46*, 309 (1969);
G. Atkinson, G. Brydon, Anal. Chim. Acta *51*, 539 (1970);
J. Reynaud, Bull. Soc. Chim. France 4353 (1968).
[179] *T. Senzaki, I. Ikehata*, Kogyo Yosui *116*, 46 (1968); C. A. *69*, 100 209 (1968).
[180] *A. Kozawa, J. Yeager*, J. Elektrochem. Soc. *115*, 1003 (1968).
[181] *T. Moore, M. Ellis, P. Selwood*, J. Amer. Chem. Soc. *72*, 3860 (1950);
R. Norrestam, Acta Chem. Scand. *21*, 2871 (1967).
[182] *R. Scholder, U. Protzer*, Z. Anorg. Allgem. Chemie *369*, 313 (1969).
[183] *A. Ubbelohde*, J. Chem. Soc. (London) 1605 (1935).

with concentrated H_2SO_4.[184] Strongly acidic solutions of this compound deposit solid material on dilution. A solid sulfate hydrate is precipitated during electrolysis of manganese(II) in 7–8 M sulfuric acid[185]; this may be the acid sulfate $HMn(SO_4)_2 \cdot 2H_2O$ reported by Houlton and Tartar.[186] The Na, K, and Rb salts of the coral red alum $M^IMn(SO_4)_2 \cdot 12H_2O$ are only stable at low temperature, and are easily dehydrated.[187] The ammonium salt of the corresponding aluminum alum, $NH_4AlMn(SO_4)_2 \cdot 12H_2O$[188] evidently has at least two crystalline forms.[172]

Treatment of NH_4MnPO_4 (Section 17.4.3) or $MnCl_2$ with nitric acid in phosphoric acid produces the rather insoluble gray-green phosphate hydrate, $MnPO_4 \cdot H_2O$.[189] The solid is stable to dilute acids, but is decomposed by alkali, producing hydrated oxides. $MnPO_4$ is reduced to $Mn_2P_2O_7$ on heating (*cf.* NH_4MnPO_4 above). Solutions of diphosphate complexes have been widely used as oxidimetric reagents.[176,190]

The cinnamon brown *triacetate* $Mn(OAc)_3 \cdot 2H_2O$ is the simplest isolable carboxylate complex of manganese(III). This useful synthetic agent is prepared by treating $Mn(OAc)_2$ with MnO_2 or Cl_2 in hot glacial acetic acid.[191] The black anhydrous compound is best prepared by heating crystalline $Mn(NO_3)_2 \cdot 2H_2O$ with acetic anhydride. This compound is hydrolyzed by water, but it can be recrystallized from acetic acid and is soluble in ethanol and pyridine and to some extent in chloroform. The *dicarboxylates* are much more photosensitive, especially when moist. The red-violet trisoxalato complex $K_3Mn(C_2O_4)_3 \cdot 3H_2O$ is prepared either via Eqn (9) [this reaction is achieved best by a preliminary reduction to manganese(II) then oxidation with the stoichiometric amount of manganese(VII) in the presence of K_2CO_3][192] or by treatment of moist MnO_2 with KHC_2O_4 at 0°, precipitating by addition of ethanol.[193] The bisoxalato complex is more soluble than the tris

$$5H_2C_2O_4 + KMnO_4 + K_2CO_3 = K_3Mn(C_2O_4)_3 + 5H_2O + 5CO_2 \quad (9)$$

complex. Treatment of dry $K_3Mn(C_2O_4)_3$ with an aqueous suspension of MnO_2[192,193] or mixing solutions containing stoichiometric quantities of $KMnO_4$ and oxalic acid yields pale green or yellow crystals of salt $KMn(C_2O_4)_2(H_2O)_2 \cdot xH_2O$.[192]

The bismalonato complex is obtained by treating freshly prepared $Mn_2O_3 \cdot xH_2O$ with malonic acid (HMal) [gives dark green $HMn(Mal)_2(H_2O) \cdot 4H_2O$[193]] and by adding the appropriate alkali malonate (the Cs salt is the least soluble). The tris complex $K_3Mn(Mal)_3 \cdot 3H_2O$ is prepared by addition of the bis complex to a concentrated solution of potassium malonate, and can be dehydrated to the monohydrate.[193] Procedures other than the last generally yield the less soluble bis complex even in anhydrous ethanol.[194] Salicylato complexes are prepared by similar procedures.[195]

Ethylenediaminetetraaceticacid and its derivatives form an interesting series of manganese(III) complexes. Reduction of suspended MnO_2, either by excess ligand[196] or with ethanol,[197] or oxidation of the manganese(II) complex (Section 17.4.3) with MnO_2[196] are good synthetic routes; some difficulty may be encountered in initiating precipitation of these complexes using the procedures described.[197] Solutions decompose slow-

[184] *U. Wannagat, E. Horn, G. Valk, F. Hoefler,* Z. Anorg. Allgem. Chemie *340*, 181 (1965);
L. Domange, Bull. Soc. Chim. France *6*, 1452 (1939).

[185] *R. Agladze, N. Kharabadze*, Elektrochim. Margantsa Akad. Nauk Gruzin. SSSR *1*, 297 (1957).

[186] *H. Houlton, H. Tartar*, J. Amer. Chem. Soc. *60*, 549 (1938);
S. Gorbatscher, E. Schpitalski, J. Gen. Chem. Russ. *10*, 1961 (1940).

[187] *H. Bommer*, Z. Anorg. Allgem. Chemie *246*, 275 (1941);
H. Hartmann, H. Schläfer, Z. Naturforsch. *69b*, 760 (1951);
O. Christensen, Z. Anorg. Allgem. Chemie *27*, 321 (1901).

[188] *J. Zernicke*, Rec. Trav. Chim. Pays-Bas *71*, 965 (1952).

[189] *W. Palmer*, Experimental Inorganic Chemistry, p. 480, Cambridge 1965;
J. Lower, W. Fernelius, Inorg. Synth. *2*, 213 (1946).

[190] *A. Berka, J. Vulterin, J. Zyka*, Newer Redox Titrants, Chap. 2. Pergamon Press, Oxford 1965.

[191] *W. der Haas, B. Schultze*, Physica *6*, 481 (1939).

[192] *G. Cartledge, W. Ericks*, J. Amer. Chem. Soc. *58*, 2061 (1936).

[193] *J. Bullock, M. Patel, J. Salmon*, J. Inorg. Nucl. Chem. *31*, 415 (1969).

[194] *G. Cartledge, P. Nichols*, J. Amer. Chem. Soc. *62*, 3057 (1940).

[195] *G. Barbieri*, Ber. *60*, 2421 (1927);
S. Makarov, F. Glikina, Russian J. Inorg. Chem. *5*, 1080 (1960).

[196] *Y. Yoshino, A. Ouchi, Y. Tsunoda, M. Kojima*, Can. J. Chem. *40*, 775 (1962);
T. Takeuchi, Y. Tsunoda, Nippon Kagaku Zasshi *88*, 172 (1967): C. A. *66*, 101 204 (1967).

[197] *R. Hamm, M. Suwyn*, Inorg. Chem. *6*, 139 (1967).

ly due to oxidation of the ligand.[198,199] An addition compound with azide, $K_2Mn(EDTA)N_3 \cdot H_2O$, has been prepared;[200] in addition to other evidence for the presence of one inner-sphere water molecule, both the EDTA and CyDTA complexes have well-defined and similar acid dissociation constants[196,197]

Acetylacetone (acacH) and its derivatives also provide a useful series of bis- and trismanganese(III) complexes. The tris complex is best prepared by the oxidation of manganese(II) by manganese(VII) in the presence of excess ligand[199] or by oxidation in air of the manganese(II) compound (Section 17.4.3) under the same conditions. The black compound, which appears green when powdered, is soluble in most organic solvents. Salts of the type $Mn(acac)_2(H_2O)_2X \cdot nH_2O$ ($X = ClO_4$, $n = 2{\cdot}5$; $X = Cl$, $n = 2$) are prepared by treating the tris complex with $HClO_4$ or HCl at pH 2.[201] The insoluble trispicolinate and trisoxinate complexes are easily prepared from the trisacetylacetonate in ethanol solution.[202] Fackler and coworkers[172,203] have prepared the tris(hexafluoroacetylacetonate) by ligand exchange with the corresponding trifluoroacetylacetonato complex, the 3-cyanoacetylacetonate derivative by ligand exchange with $Mn(DPM)_3$ or $Mn(DIBM)_3$ (HDPM = dipivaloylmethane, HDIBM = diisobutyrylmethane), and the interesting dark green tris(tropolonato)manganese(III) complex by the reaction between $Mn(OAc)_3$ (above) and tropolone in toluene.[172] The biguanidobenzoylacetone adduct complex is prepared by displacement from the tris(benzoylacetone) species in ethanol under reflux conditions.[204]

Nyholm and Turco[205] developed a method for the preparation of tris(dipyridyl 1,1′-dioxide)-manganese(III) complexes involving the oxidation of a solution of $MnSO_4$ by $S_2O_8^{2\ominus}$ in the presence of the ligand at 75°. This produces the salt $Mn(dipyO_2)_3(S_2O_8)_{1{\cdot}5} \cdot 4H_2O$. The perchlorate, $Mn(dipyO_2)_3 \cdot (ClO_4)_3 \cdot 3H_2O$, was obtained when the reaction was run in the presence of excess perchlorate. This general procedure[205] has recently been used to prepare the analogous light-sensitive bis complex with terpyridyl 1,1′,1″-tri-*N*-oxide.[206] Bromine was used to oxidize $Mn(OAc)_2$ in the preparation of the tris(picolinic acid–oxide) species.[207]

The preparation of solvates with dimethylsulfoxide and *N,N*-dimethylformamide involves treatment of $Mn(OAc)_3 \cdot 2H_2O$ with the solvents and precipitation with $Mn(ClO_4)_2$–$HClO_4$ in cold ethanol or ether.[208] The solvates $Mn(S)_6(ClO_4)_3$ are violet and i.r. evidence suggests bonding through oxygen atoms in both cases.[208]

Manganese(III) complexes with dithiocarbamates ($L = R_2NCS_2$; R = Et, *n*-Pr, *n*-Bu, *i*-Bu) are rapidly formed on mixing solutions of $MnSO_4$ and the ligand.[209,210] They are non-electrolytes, MnL_3, and can be extracted with chloroform.[210] Similarly, solutions of manganese(II) are rapidly oxidized in air in the presence of sulfosalicyclic acid[211] and the series of complexes obtained[212] with N-substituted salicylaldimines are generally non-electrolytes.[213]

17.5.4 Nitrogen and Group V Complexes

Compounds with ligands donating nitrogen atoms have so far been restricted to derivatives of ethylenediamine,[214] 2,2′-dipyridyl, 2′,2″-terpyridyl, and 1,10-phenanthroline.[214,215] Methods usually involve reduction of manganese(VII)

---

[198] *K. Schroeder, R. Hamm*, Inorg. Chem. *3*, 391 (1964);
N. Tanaka, T. Shurakashi, Nippon Kagaku Zasshi *90*, 57 (1969); C. A. *70*, 81 371 (1969).

[199] *R. Charles*, Inorg. Synth. *6*, 164 (1960).

[200] *M. Suwyn, R. Hamm*, Inorg. Chem. *6*, 2150 (1967).

[201] *G. Cartledge*, J. Amer. Chem. Soc. *74*, 6015 (1952);
G. Cartledge, J. Amer. Chem. Soc. *73*, 4416 (1951).

[202] *M. Ray, J. Adhya, D. Biswas, S. Poddar*, Australian J. Chem. *19*, 1737 (1966).

[203] *J. Fackler, T. Weeks, I. Chawla*, Inorg. Chem. *4*, 130 (1965).

[204] *M. Ray, J. Adhya, N. Poddar, S. Poddar*, Australian J. Chem. *21*, 801 (1968)

[205] *R. Nyholm, A. Turco*, J. Chem. Soc. (London) 1121 (1962).

[206] *W. Rieff, W. Baker*, Inorg. Chem. *9*, 570 (1970).

[207] *A. Lever, J. Lewis, R. Nyholm*, J. Chem. Soc. (London) 5262 (1962).

[208] *C. Prabhakaran, C. Patel*, J. Inorg. Nucl. Chem. *30*, 867 (1968).

[209] *C. Prabhakaran, C. Patel*, Indian J. Chem. *7*, 1257 (1969);
L. Cambi, L. Szego, Ber. *64*, 2591 (1931).

[210] *R. Golding, P. Healy, P. Newman, E. Sinn, W. Tennant, A. White*, J. Chem. Phys. *52*, 3105 (1970).

[211] *A. Parukilar, P. Subbaraman*, Indian J. Chem. *8*, 266 (1970).

[212] *B. Sharma, C. Patel*, Indian J. Chem. *8*, 747 (1970);
C. Prabhakaran, C. Patel, J. Inorg. Nucl. Chem. *33*, 3316 (1969).

[213] *A. van den Bergen, K. Murray, M. O'Connor, B. West*, Australian J. Chem. *22*, 39 (1969).

[214] *J. Summers*, Dissertation, Univ. Florida 1968;
J. Summers, Diss. Abstracts *30B*, 109 (1969).

(*e.g.*, with HCl in the preparation of $MnCl_3$-terpy)[215] or use of manganese(III) oxides [*e.g.*, for $MnF_3(H_2O)$phen].[215]

Phthalocyanatomanganates(III) are generally prepared by oxidation of the corresponding manganese(II) compounds (Section 17.4.4). Fenkart and Brubaker[216] have recently presented evidence which strongly suggests that the sodium salt of the complex with 4,4′,4″,4‴-tetrasulfophthalocyanine (SPc) is $Na_4Mn(SPc(OH)(H_2O)$ rather than $Na_3Mn(SPc)\cdot 2H_2O$ as previously proposed.[148] At pH 9 the dimeric species $(H_2O)(SPc)Mn–O–Mn(SPc)(H_2O)^{9\ominus}$, which is formed at a measurable rate,[216] is formally analogous to the pyridine adduct py(Pc)Mn–O–Mn(Pc)py·2py investigated by Vogt and coworkers.[217] The dimers are readily dissociated on acidification.[216]

We referred earlier (Section 17.4.4) to the interesting chemistry of the related porphyrin complexes. Boucher and Garber[218] have recently summarized methods for the synthesis of these compounds.

The *manganese(III) hematoporphyrin IX* compound, prepared[219] by alkaline hydrolysis of the corresponding complex with the dimethyl ester,[220] also analyzes as a monohydrate–monohydroxide [see the complex $Na_4Mn(SPc)(H_2O)$-OH], although analysis is, of course, weak evidence in the absence of other information. Like the phthalocyanine, this molecule also behaves as a weak acid,[219] losing both protons from the axial water molecules at pH > 12. Dimer formation was also detected in the manganese(II) complex (prepared by reduction with $S_2O_3^{2\ominus}$) at pH > 7.[219] Displacement of the manganese(II) occurs at low pH.

17.6 Manganese(IV)

The chemistry of manganese(IV) is largely restricted to oxides and complexes with complicated ligands which donate oxygen atoms.

17.6.1 Cyanides

Addition of excess KCN to cold alkaline $KMnO_4$ solution which was originally thought[221] to give a *manganese(IV) cyanide* species probably results in the precipitation of $K_3Mn(CN)_6$ contaminated with KOH,[157] but oxidation of $K_3Mn(CN)_6$ with NOCl in sulfolane or *N,N*-dimethylformamide results in the formation of the canary yellow manganese(IV) compound $K_2Mn(CN_6)$,[222] which evidently decomposes on addition of deaerated water.

17.6.2 Halides and Complex Halides

MnF_4 is the blue, very hygroscopic solid obtained by treatment of manganese with fluorine over a nickel-based fluidized bed at 600°–700°,[223, 224] by treatment of MnF_3 with fluorine at 550°,[165] or by the reaction between fluorine and Li_2MnF_6 at 550°.[165] The compound prepared via the first route can be freed from excess F_2 and HF *in vacuo* at −60°.[224] This compound is a very strong oxidizing and fluorinating agent, and slowly reverts to brown-red MnF_3 on standing at 0°.

Tetrachloromanganese(IV) has proved rather more elusive. A green, neutral solid is precipitated[163] when chloroform is added to an ethereal suspension of MnO_2, which has been treated with anhydrous HCl at −70°. Lisov[225] claims to have prepared $MnCl_4$ electrolytically in HCl solution.

Pentafluoromanganates(IV) can be prepared by dissolution of $KMnO_4$ in BrF_3 ($KMnF_5$)[161] or treatment of a stoichiometric mixture of the alkali metal fluoride and MnF_2 with fluorine at 350° (e.g., $LiMnF_5$).[165] Another possible route is via the reaction between an alkali metal fluoride and $Mn(IO_3)_2$ in acid solution. No satisfactory preparation of the pentachloromanganates(IV) has been reported.

Hexafluoromanganates(IV) are among the best characterized manganese(IV) compounds. K_2MnF_6 is obtained as small golden yellow,

---

[215] *H. Goodwin, R. Sylva*, Australian J. Chem. *18*, 1743 (1965);
H. Goodwin, R. Sylva, Australian J. Chem. *20*, 629 (1967).

[216] *C. Fenkart, C. Brubaker*, J. Inorg. Nucl. Chem. *30*, 3245 (1968).

[217] *L. Vogt, A. Zalkin, D. Templeton*, Science *151*, 569 (1966);
L. Vogt, A. Zalkin, D. Templeton, Inorg. Chem. *6*, 1725 (1966).

[218] *L. Boucher, H. Garber*, Inorg. Chem. *9*, 2644 (1970).

[219] *D. Davis, J. Montalvo*, Anal. Chem. *41*, 1195 (1969).

[220] *P. Loach, M. Calvin*, Biochemistry *2*, 361 (1963).

[221] *A. Yakamich, Compt. rend. 190*, 681 (1930).

[222] *J. Fowler, J. Kleinberg*, Inorg. Chem. *9*, 1005 (1970).

[223] *H. Roesky, O. Glemser*, Angew. Chem. *75*, 920 (1963).

[224] *H. Roesky, O. Glemser, K. Hellberg*, Chem. Ber. *98*, 2046 (1965).

[225] *V. Lisov*, Ukr. Khim. Zh. *33*, 849 (1967).

hygroscopic crystals by reducing $KMnO_4$ with ether in the presence of KHF_2 and HF[226] or with HF at 0°.[227] Salts of most alkali metals,[165,168,228] strontium,[229] and barium[229] have been reported. The *hexachloromanganates(IV)* are more unstable than the corresponding fluorides and liberate chlorine even in dry air at room temperature. Reduction of manganese(VII) with either fuming HCl at 0°[227] (precipitate on addition of KCl) or acetyl chloride in boiling glacial acetic acid (precipitate on saturation of the red-brown solution wth HCl)[175] yields the potassium salt K_2MnCl_6.

17.6.3 Oxygen Derivatives

The two main routes to the important nonstoichiometric compound "MnO_2" (hereafter called manganese dioxide) are *via* oxidation of manganese(II) or reduction of manganese(VII). Oxidation of manganese(II) salts can be carried out electrochemically,[230] by ozonolysis,[231] by roasting a solid salt [*e.g.*, $Mn(NO_3)_2$] in air,[232] or with $S_2O_8^{2\ominus}$ [catalyzed by Ag(I)] or alkaline hypobromite. Reduction of manganese(VII) with most reducing agents in alkaline solution yields colloidal or gelatinous manganese dioxide.[233] The properties of the aggregate obtained and its activity, for example, in catalytic applications, depend critically on the details of the preparative route, age, and subsequent treatment of the solid. The nonstoichiometry which is exhibited by the different materials can be due to occluded water or other impurities or to the presence of mixed oxidation states of manganese. The best methods of preparation[231,232,234] give MnO_2 of about 98% stoichiometric purity. MnO_2 is precipitated slowly from neutral manganese(VII) solutions, especially in the presence of phosphate anions.[235] The yellow–brown species produced, believed to be $H_2MnO_4^{2\ominus}$, evidently obeys Beer's Law.[235] The compound dissolves in fuming sulfuric acid to give a blue solution, which can also be obtained by reducing $KMnO_4$ with fuming sulfuric acid. This color has been attributed[236] to $Mn(SO_4)_2$, which separates as black crystals on cooling the mixture below 0°.

There is some evidence to suggest that the disproportionation equilibrium [Eqn (8)] is shifted to the right in concentrated sulfuric acid solution,[237] presumably through the relative stability of sulfatomanganese(IV) species.

Periodatomanganese(IV) salts of sodium and potassium have been prepared by Lister and coworkers.[238,239] Reaction (10) was used[238] to

$$MnSO_4 + 2Na_2H_3IO_6 \xrightarrow{0{\cdot}4\ \text{M}\ H_2SO_4} NaMnIO_6 + NaIO_3 + Na_2SO_4 + 3H_2O \quad (10)$$

prepare the sodium salt and $KMnIO_6$ was prepared from KIO_4 at a slightly higher acid concentration (1–1·5 M H_2SO_4). These salts are virtually insoluble in water and are stable up to 230°. By contrast, if the oxidation of $MnCl_2$ by NaOCl is carried out in the presence of nitric acid and an *excess* of $Na_2H_3IO_6$, then the much more soluble salts $Na_7H_4Mn(IO_6)_3 \cdot 17H_2O$ and $K_7H_4Mn(IO_6)_3 \cdot 8H_2O$ (from KIO_3 and KOCl) are obtained by precipitation with ethanol.[239] The potasium salt can also be prepared by

---

[226] *W. Palmer*, Experimental Inorganic Chemistry, p. 484, Cambridge 1965.

[227] *P. Moews*, Inorg. Chem. *5*, 5 (1966).

[228] *D. Novotny*, *G. Sturgeon*, Inorg. Nucl. Chem. Lett. *6*, 455 (1970).

[229] *R. Hoppe*, *K. Blinne*, Z. Anorg. Allgem. Chemie *291*, 269 (1957).

[230] *L. Young*, Anodic Oxide Films, p. 274, Academic Press London, 1961;
J. Y. Welsh, Electrochem. Technol. *5*, 504 (1967).

[231] *J. Belew*, *T. Chwang*, Chem. & Ind. (London) 1958 (1967);
V. Lunenok-Burmakina, *T. Kusakovskaya*, *A. Miroshnichenko*, Isotopenpraxis *3*, 146 (1967); C. A. *70*, 102 648 (1969);
J. Marcy, *F. Mattes*, Chem. Tech. (Berlin) *19*, 430 (1960).

[232] *W. Latimer*, Oxidation Potentials, p. 238, Prentice Hall, Englewood Cliffs, 1952;
A. Hegedus, Acta Chem. Sci., Hung. *46*, 311 (1953).

[233] *G. Sterr*, *A. Schmier*, Z. Anorg. Allgem. Chemie *368*, 225 (1969).

[234] *R. Giovanoli*, *E. Staehli*, *W. Feitknecht*, Chimia *23*, 264 (1969);
R. Giovanoli, *E. Staehli*, *W. Feitknecht*, Helv. Chim. Acta *53*, 453 (1970);
R. King, *F. Stone*, Inorg. Synth. *7*, 194 (1963).

[235] *B. Jezowska-Trzebiatowska*, *J. Kalencinski*, Bull. Acad. Polon Sci., Ser. Sci., Chim. *9*, 791 (1961).

[200] *H. Mishra*, *M. Symons*, Proc. Chem. Soc. *23*, (1962);
H. Mishra, *M. Symons*, J. Chem. Soc. 4490 (1963);
F. Lankshear, Z. Anorg. Allgem. Chemie *82*, 97 (1913).

[237] *R. Selim*, *J. Lingane*, Anal. Chim. Acta *21*, 536 (1959);
A. Fenton, *N. Furmann*, Anal. Chem. *32*, 748 (1960);
T. Kemp, *W. Waters*, J. Chem. Soc. 339 (1964).

[238] *I. Riemer*, *M. Lister*, Can. J. Chem. *39*, 2431 (1961).

[239] *M. Lister*, *Y. Yoshino*, Can. J. Chem. *38*, 1291 (1960).

cation exchange from the less soluble sodium species, but the barium salt is much more difficult to purify because of its insolubility. These triperiodatomanganese(IV) complexes decompose to manganese(VII) on standing. Similar pertelluratomanganese(IV) complexes have been reported.[240] Goldenberg[157] has prepared $K_2Mn(IO_3)_6$ by oxidizing iodic acid with freshly prepared MnO_2 in the presence of KIO_3. The oxalate complex $K_2Mn(Ox)_2(OH)_2 \cdot 2H_2O$ reported by Grey[158] may have a Mn–O–Mn bond in view of its low magnetic moment.

The first *heteropolyvanadates* of manganese(IV) have recently been prepared by Flynn and coworkers[97,241] by oxidation of manganese(II) species with peroxydisulfate. This work follows the earlier preparation of the heteropolyniobates by the oxidation of ethylenediaminetetraacetatomanganese(II) with H_2O_2 in the presence of $Na_7HNb_6O_{19} \cdot 15H_2O$.[242] The complexes of manganese(IV) of this type which Flynn[97,241,242] has reported are shown in Table 3.

Another notable advance is the reported preparation[243] of the gray nitrilotriacetato (NTA) complex (**8**) obtained by treating $MnCl_2$ or

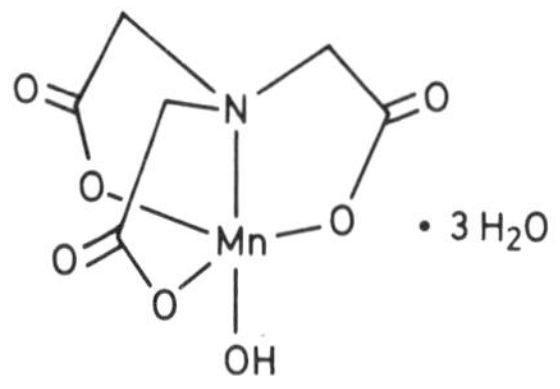

8

$Mn(OAc)_2$ with freshly prepared hydrous oxides of manganese(IV) in the presence of NTA at room temperature, or by heating the ligand–salt mixture in water or *N,N*-dimethylformamide. Possible stabilization of manganese(IV) through O–N donor atoms is also illustrated by the phenanthroline and dipyridyl complexes[215] and by the complex $[Mn(bipy)_2O](S_2O_8)_{1.5} \cdot 3H_2O$ prepared by oxidation of $MnSO_4$ with ammonium persulfate in the presence of 2,2′-bipyridyl.[244]

Table 3 Heteropolyvanadates and Heteropolyniobates of Manganese(IV)

Compound	Ref.
$K_7MnV_{13}O_{38} \cdot 18H_2O$	*a*
$Mn_7MnV_{13}O_{38} \cdot 24H_2O$	*a*
$(NH_4)_7MnV_{13}O_{38} \cdot 5H_2O$	*a*
$Cs_6NaMnV_{13}O_{38} \cdot 8H_2O$	*b*
$K_5MnV_{11}O_{32} \cdot (10–12)H_2O$	*b*
$(NH_4)_{0.5}H_{0.5}MnV_{11}O_{32} \cdot 12H_2O$	*b*
$Cs_{4.5}H_{0.5}MnV_{11}O_{32} \cdot 7H_2O$	*b*
$K_5HMn_3V_{12}O_{39} \cdot 10H_2O$	*b*
$(NH_4)_5HMn_3O_{12}O_{39} \cdot (14–15)H_2O$	*b*
$Na_{12}MnNb_{12}O_{38} \cdot (48–50)H_2O$	*c*
$K_8Na_4MnNb_{12}O_{38} \cdot 21H_2O$	*c*

[a] *C. Flynn, Jr., M. Pope*, J. Amer. Chem. Soc. *92*, 85 (1970).
[b] *C. Flynn, Jr., M. Pope*, Inorg. Chem. *9*, 2009 (1970).
[c] *C. Flynn, Jr., G. Stuckey*, Inorg. Chem. *8*, 332 (1969).

17.7 Manganese(V)

The reduction of manganate(VI) (Section 17.8) with $S_2O_3^{2\ominus}$, $SO_3^{2\ominus}$ or phenol in strong (5 M) alkali produces the sky-blue manganese(V) anion.[245,246] A blue color attributable to manganese(V) is observed in the careful electrolytic reduction of $MnO_4^{2\ominus}$ even in 0·1 M alkaline solution, but this disproportionates [Eqn (11)] within a few minutes. Increasing the hydroxide concentration to 2 M is not sufficient to prevent the disproportionation.

$$2Mn^{V} \rightleftharpoons Mn^{IV} + Mn^{VI} \quad (11)$$

Solid salts of the type $M_3^{I}MnO_4$ (M^{I} = Li, Na, K, Rb, Cs),[183,247] $M_3^{II}(MnO_4)_2$ (M^{II} = Ba),[248] and $M_5^{II}(MnO_4)_3(OH)$ (M^{II} = Sr, Ba)[248] have been prepared. The main routes for their preparation are as follows:[249]

[240] *M. Lister, Y. Yoshino*, Can. J. Chem. *40*, 1490 (1962);
Y. Yoshino, T. Takeuchi, H. Kinoshita, Nippon Kagaku Zasshi *86*, 978 (1965); C. A. *64*, 10738g (1965);
L. Jensovsky, Omagiu Raluca Ripan 293 (1966).
[241] *C. Flynn jr., M. Pope*, Inorg. *9*, 2009 (1970).
[242] *C. Flynn jr., G. Stuckey*, Inorg. Chem. *8*, 332 (1969).
[243] *M. Voronkov, S. Mikhailova*, Khim. Geterotskl. Soedin 795 (1969); C. A. *72*, 106 676b (1970).
[244] *R. Nyholm, A. Turco*, Chem. & Ind. (London) 74 (1960).
[245] *J. Pode, W. Waters*, J. Chem. Soc. (London) 717 (1956).
[246] *R. Scholder, W. Waterstradt*, Z. Anorg. Allgem. Chemie *277*, 172 (1954).
[247] *J. Guerchais, M. Leroy, R. Rohmer*, C. R. Acad. Sci., Paris *261*, 3628 (1965).
[248] *W. Klemm, C. Breudel, G. Whermeyer*, Chem. Ber. *93*, 1506 (1960).
[249] *H. Lux*, Z. Naturforsch. *1*, 281 (1947);
W. Klemm, Intern. Symp. Reactivity Solids Gotteburg *1*, 173 (1952); C. A. *48*, 12 602 (1954);
A. Carrington, M. Symons, J. Chem. Soc. 3373 (1956);
J. Kleinberg, J. Chem. Educ. *33*, 73 (1956).

(1) reduction of alkaline $KMnO_4$ with KI;
(2) fusing a mixture of KOH, $KMnO_4$, and KNO_3;
(3) fusing MnO_2 with KOH in the presence of oxygen,[183] $NaNO_2$, or $RbO_{1.9}$;[183]
(4) mixing $KMnO_4$ and KOH in the melt or in solution;
(5) heating a manganese(VI) salt with alkali; and
(6) heating a mixture of K_2CO_3 and $MnCO_3$.[183]

The mixed manganate(V) $NaBaMnO_4$, which is reported to be stable in air, was obtained by prolonged reaction between $BaCO_3$, MnO_2, and NaOH in the melt.[250] Anhydrous *alkali manganates(V)* are reported to be stable to 1000°,[251] but even cold strongly alkaline solutions slowly precipitate hydrated manganese oxides after standing for a few days. The recent preparation[252] of $MnOCl_3$ must be regarded as a major advance in the chemistry of manganese(V). The compound is obtained by reduction of $KMnO_4$ with HSO_3Cl or a mixture of HSO_3Cl and sucrose in $CHCl_3$. The excess HSO_3Cl is extracted with 100% H_2SO_4 and the product is further purified by vacuum distillation. Solutions in pretreated[252] CCl_4 are reasonably stable at room temperature, but eventually become brown owing to reduction to $MnCl_3$. The compound undergoes vigorous hydrolysis, but produces $MnO_4^{3\ominus}$ anions on dissolution in 5 M KOH solution.

17.8 Manganese(VI)

Like manganese(V), the stable compounds of manganese(VI) which have been obtained are all oxyanion species. Heating alkaline $KMnO_4$ with MnO_2 or ferromanganese[253] or reducing with hydrogen[254] produces a green color from which the green manganese(VI) species can be extracted with carbonate-free alkali.[255] An example of these preparative methods is the production of the K, Rb, and Cs salts by heating the respective peroxides with MnO_2 at 420°.[255]

[250] *J. Barraud, R. Olazcuaga, G. LeFlem*, C. R. Acad. Sci., Paris *C270*, 1175 (1970).
[251] *H. Peters, K. Radeka, L. Till*, Z. Anorg. Allgem. Chemie *346*, 1 (1966).
[252] *T. Briggs*, J. Inorg. Nucl. Chem. *30*, 2866 (1968).
[253] *R. Agladze, V. Kvesclava*, Elektrokhim. Margantsa *4*, 31 (1969).
[254] *A. Webster, J. Halpern*, Trans. Faraday Soc. *53*, 51 (1957);
E. Wilke, H. Kuhn, 1. Physik. Chem. (Leipzig) *76*, 601 (1911).
[255] *G. Puquenoy*, C. R. Acad. Sci., Paris *C268*, 828 (1969); *R. Nyholm, P. Williams*, Inorg. Synth. *11*, 57 (1968).

Teal blue $BaMnO_4$ is only sparingly soluble[256] and can be prepared by reduction of alkaline manganese(VII) in the presence of barium[245] or by addition of $KMnO_4$ to boiling $Ba(OH)_2$ solution. Solutions containing manganate(VI) anions disproportionate slowly even in 1 M alkaline solution at room temperature.[257] These

$$3MnO_4^{2\ominus} + 2H_2O \longrightarrow 2MnO_4^{\ominus} + MnO_2 + 4OH^{\ominus} \quad (12)$$

solutions are rapidly oxidized to manganese(VII) by sodium bismuthate, hypochlorite, or periodate.[258] K_2MnO_3 is obtained on heating K_2MnO_4 above 500°.

The red solid "MnO_3" which is reportedly[236] obtained by dropping a solution of $KMnO_4$ in cold, concentrated sulfuric acid onto anhydrous $NaCO_3$ may, in fact, be a mixture of Mn_2O_7 (Section 17.9) and lower manganese oxides.

The general properties and structure of the alkali metal manganates(VI) are well-documented.[249, 259]

A potentially useful synthetic manganese(VI) compound is MnO_2Cl_2,[252] although it is less stable than MnO_3Cl (Section 17.9) and $MnOCl_3$ (Section 17.7). It is obtained by reduction of MnO_3Cl with excess SO_2 and is separated by vacuum distillation.[252] Its decomposition products include $MnOCl_3$. Dilute solutions in CCl_4 are amber-colored.[252]

17.9 Manganese(VII)

The chemistry of the manganese oxidation state (VII) is entirely restricted to oxide and oxide–halide complexes.

The dark green complex MnO_3F (m.p., −78°; b.p., 60°) is obtained by treating $KMnO_4$ with anhydrous HF[252, 260] or IF_5.[261, 262] However,

[256] *H. McDonald*, Diss. Abstracts *21*, 454 (1960).
[257] *F. Duke*, J. Phys. Chem. *56*, 882 (1952).
[258] *R. Stewart* in *K. Wiberg*, Oxidation in Organic Chemistry, Vol. 5A, p. 6, Academic Press, New York, 1965.
[259] *A. Carrington, D. Schoenland, M. Symons*, J. Chem. Soc. 659 (1957);
K. Jensen, W. Klemm, Z. Anorg. Allgem. Chemie *237*, 47 (1938);
G. Palenik, Inorg. Chem. *6*, 507 (1967).
[260] *A. Englebrecht, A. Grosse*, J. Amer. Chem. Soc. *76*, 2042 (1954);
K. Wiechart, Z. Anorg. Allgem. Chemie *261*, 310 (1950).
[261] *E. Aynsley*, J. Chem. Soc. (London) 2425 (1958).
[262] *P. Aymonino, H. Schultze, A. Mueller*, Z. Naturforsch, *B24*, 1508 (1969).

the most satisfactory method is via HSO_3F[252] (the same is also true for MnO_3Cl, see below). The product must be purified by vacuum distillation in each case. This compound is *violently explosive* at ordinary temperatures producing a mixture of MnF_2, MnO_2, and oxygen.

The preparation of the *explosively unstable* black compound MnO_3Cl follows the same general pattern as that used for MnO_3F. The passage of anhydrous HCl through a solution of $KMnO_4$ in concentrated H_2SO_4 gives a condensate at $-50°$.[263] Briggs[252] has recently described the controlled reduction of Mn_2O_7 (see below) by chlorosulfuric acid at low temperatures. The unused Mn_2O_7 and HSO_3F can be collected at $-30°$ and the product at $-68°$. The product is greatly stabilized by dissolution in pretreated[44] CCl_4 or $CFCl_3$.

The most important compound of manganese(VII) is $KMnO_4$. Acid solutions of manganese(II) salts are readily oxidized to manganese(VII) by solid sodium bismuthate, PbO_2, AgO,[264] KIO_4, or ozone.[265] Manganese(VI) may be oxidized electrolytically to $MnO_4^{\ominus}$ at a smooth platinum anode in alkaline solution. The solid salts $NaMnO_4 \cdot 3H_2O$, $Ba(MnO_4)_2$, and $AgMnO_4$ have all been reported.[266] $Ba(MnO_4)_2$ is a useful precursor for the parent manganese(VII) acid, $HMnO_4$, and is obtained as a solid when a boiling solution of $Ba(NO_3)$ and $Ba(OH)_2$ is treated with $KMnO_4$. The solid is purified by treatment with CO_2 and steam. Another route to $Ba(MnO_4)_2$ is via treatment of $AgMnO_4$ [prepared as red solid on addition of silver(I) to a solution of $MnO_4^{\ominus}$] with $BaCl_2$.

$CsMnO_4$ is the most insoluble permanganate salt. Solid manganese(VII) complexes with $(C_6H_5)_4AsX$[267] (X=Cl, ClO_4, complex soluble in organic solvent) and $Fe(phen)_3^{2\oplus}$ have been reported.[268] The decomposition of $MnO_4^{\ominus}$ is catalyzed by solid hydrous oxides in neutral and alkaline[269,270] solution.

Interesting color changes are observed on addition of sulfuric acid to aqueous $MnO_4^{\ominus}$. In 60% H_2SO_4 the brown color observed is presumably due to a mixture of $MnO_4^{\ominus}$ with protonated manganese(VII) species and manganese(IV) complexes.[271] The characteristic green color of manganese(VII) in very concentrated sulfuric acid is usually attributed to the presence of "permanganic acid," $H_xMnO_4^{(x-1)\oplus}$, although some authors favor $MnO_3^{\oplus}$ or $O_3MnO \cdot SO_3H$ species.[272,273] The parent acid, $HMnO_4$, is obtained by:

(i) treatment of $Ba(MnO_4)_2$ with a slight excess of H_2SO_4;
(ii) oxidation of manganese(II) with PbO_2 in acid solution; or
(iii) passage of a solution of a permanganate salt down a cation exchange column in the $H^{\oplus}$ form.

The strongly oxidizing violet crystals form a dihydrate[274] in moist air and are unstable to heat and sunlight.

Dimanganese(VII) heptoxide, Mn_2O_7, a useful manganese(VII) intermediate (see above), is obtained as a dark oil on addition of powdered $KMnO_4$ to concentrated H_2SO_4.[275] The oil freezes to dark green crystals in liquid air, and distils at 0°/15 mm.[276] This compound is *explosive* above 10°, but dissolves without decomposition in acetic anhydride. The compound can be extracted from these concentrated acid solutions with liquid freons, $SOCl_2$, or CCl_4.[252]

---

[263] *D. Michel, A. Doiwa*, Naturwissenschaften *53*, 129 (1966).

[264] *M. Tanaka*, Bull. Chem. Soc. Japan *26*, 299 (1953); *M. Tanaka*, Bull. Chem. Soc. Japan *27*, 10 (1954).

[265] *H. Willard, L. Merritt*, Ind. Eng. Chem. Anal. Ed. *14*, 486 (1942).

[266] *F. Miller, C. Williams*, Anal. Chem. *24*, 1253 (1952); *F. Miller, G. Carlson, F. Bentley, W. Jones*, Spectrochim. Acta *16*, 135 (1960); *J. Teltow*, Z. Physik. Chem. (Leipzig), *B40*, 397 (1938); *J. Teltow*, Z. Physik. Chem. (Leipzig) *B43*, 198 (1939).

[267] *J. Matuszek, T. Sugihara*, Anal. Chem. *33*, 35 (1961).

[268] *G. Smith, F. Richter*, Phenanthroline and Substituted Phenanthroline Indicators, G. F. Smith, Chemical Company, Columbus, Ohio 1944.

[269] *H. Morse, A. Hopkins, M. Walker*, Amer. Chem. J. *18*, 401 (1896).

[270] *K. Loft, M. Symons, P. Trevalion*, Proc. Chem. Soc. 357 (1960).

[271] *R. Stewart, M. Mocek*, Can. J. Chem. *41*, 1160 (1963); *N. Bailey, A. Carrington, K. Lott, M. Symons*, J. Chem. Soc. (London) 290 (1960).

[272] *H. Mishra, M. Symons*, J. Chem. Soc. (London) 4411 (1952); *D. Royer*, J. Inorg. Nucl. Chem. *17*, 159 (1961).

[273] *R. Stewart*, in *K. Wiberg*, Oxidation in Organic Chemistry, Vol. 5A, p. 6. Academic Press, New York 1965.

[274] *N. Frigerio*, J. Amer. Chem. Soc. *91*, 6200 (1969).

[275] *A. Simon, F. Feher*, Z. Elektrochem. *38*, 137 (1932).

[276] *J. Strickland, G. Spicer*, Anal. Chim. Acta *3*, 543 (1949).

17.10 Bibliography

L. Malatesta, Progr. Inorg. Chem. *1*, 283 (1959).
R. Peacock, Progr. Inorg. Chem. *2*, 193 (1960).
R. Clark, Rec. Chem. Progr. *26*, 269 (1965).
P. Braterman, J. Chem. Soc. *A*, 1471 (1966).
S. Yamada, Coord. Chem. Rev. *1*, 415 (1966).
R. Dingle, Acta Chem. Scand. *20*, 33 (1966).
B. Chadwick, *A. Sharp*, Adv. Inorg. Chem. Radiochem. *8*, 83 (1966).
N. Hush, *M. Hobbs*, Progr. Inorg. Chem. *10*, 259 (1968).
T. Zordan, *L. Hepler*, Chem. Rev. *68*, 737 (1968).
T. Davis, *J. Fackler*, *M. Weeks*, Inorg. Chem. *7*, 1994 (1968).
J. Summers, Thesis, University of Florida, 1968; *J. Summers*, Diss. Abstracts *30*, 109 (1969).
L. Boucher in *S. Kirschner*, Proceedings of the *J.C. Bailar jr.*, Symposium, p. 126, Plenum Press, New York 1969.
M. Ciampolini, *N. Nardi*, *G. Speroni*, Coord Chem. *Rev. 1*, 223 (1966);
M. Ciampolini, Structure and Bonding *6*, 52 (1969).
R. Colton, *J. Canterford*, Halides of the First Row Transition Metals, Wiley Interscience, London 1969.
G. Davies, Coord. Chem. Rev. *4*, 199 (1969).

18 Technetium

John R. Wasson
Department of Chemistry, University of Kentucky,
Lexington, Kentucky 40506, U.S.A.

18.1 Elemental Technetium

Technetium was first prepared in 1937 by bombarding molybdenum with deuterons in a cyclotron[1] *via* the reaction Mo(d,n)Tc. The main source of relatively large amounts of technetium is its recovery from a mixture of the fission products of uranium. One of the longest-lived isotopes of technetium, ^{99}Tc, is produced in approximately 6·2% yield from the fission of ^{235}U produced by thermal neutrons. Sixteen isotopes of technetium with mass numbers from 92 to 107 are known. The majority of technetium isotopes are formed by the irradiation of molybdenum targets with deuterons or protons, or molybdenum or ruthenium targets with neutrons. The short-lived isotopes of greatest practical importance are the isomers ^{95m}Tc (60 days) ^{97m}Tc (90·5 days) and ^{99m}Tc (6·0 hours). The main isotope, ^{99}Tc ($2{\cdot}12 \times 10^5$ years) is best obtained[2] from fission of ^{235}U. It undergoes radioactive decay with emission of a 0·292 meV beta particle. Due to their similar chemistry rhenium compounds are frequently used as carriers for technetium isotopes.

Technetium is a silver-gray metal melting at $2200 \pm 50°$ and boiling at approximately 4700° and has a density of 11·478 g/cm³. Its paramagnetic susceptibility[3] per gram is 270×10^{-6} c.g.s. at 25° and 290×10^{-6} c.g.s. at 507°. In the region 2200–9000 Å over 2000 lines in the optical spectrum have been determined. Several of the lines are useful for the identification[4] of technetium with a sensitivity of 10^{-7} g.

The metal can be obtained by the hydrogen reduction of the heptasulfide[5] at 1100° and by the hydrogen reduction of ammonium pertechnetate, NH_4TcO_4 (see Section 18.7).[6] In the latter method it is best to reduce the ammonium pertechnetate to technetium dioxide at about 200° followed by reduction to the metal at about 500–600°. The metal can also be prepared[7] by heating ammonium hexachlorotechnetate(IV) to red heat in a nitrogen atmosphere. Technetium can also be electroplated[7] from a solution of ammonium pertechnetate in 1 M sulfuric acid.

The chemistry of rhenium and technetium are very similar and a knowledge of rhenium chemistry (*cf.* Chapter 19) can be used to suggest and guide work with technetium. Additional references[8–12] are listed for special techniques employed in handling technetium compounds and discussions of their properties.

18.2 Technetium(−I, 0, I) Compounds

Compounds with technetium in the −1, 0, +1 oxidation states are mainly those with π-bonding ligands such as cyanide ion, carbon monoxide, etc; +1 oxidation state will be of primary concern here.

Besides the carbonyl halides prepared from the oxidation of technetium carbonyl, $Tc_2(CO)_{10}$, with halogens,[13] cyano complexes provide the best example of univalent technetium. When potassium pertechnetate, $KTcO_4$, is reduced[14] with potassium amalgam in the presence of excess potassium cyanide or potassium tetracyanotrihydroxotechnetate(IV), $K_3[Tc(OH)_3(CN)_4]$, an olive-green solution containing *hexacyanotechnetate(I)* ions is formed and potassium, $K_5Tc(CN)_6$, and thallium(I), $Tl_5Tc(CN)_6$, salts may be obtained. The potassium salt does not contain water of crystallization. The brick-red thallium salt is soluble in ammoniacal hydrogen peroxide and dilute acids. All the compounds of univalent technetium are stable in the solid state in dry air but are very sensitive to oxidation in solution. The oxidation product contains the $Tc(OH)_3(CN)_4^{3\ominus}$ anion.

---

[1] *C. Perrier, E. Segre,* J. Chem. Phys. *5,* 712 (1937); *C. Perrier, E. Serge,* Nature *140,* 193 (1937).
[2] *S. Katcoff,* Nucleonics *16,* 78 (1958).
[3] *C. W. Nelson, G. E. Boyd, W. T. Smith,* J. Amer. Chem. Soc. *76,* 348 (1954).
[4] *W. F. Meggers,* Spectrochim. Acta *4,* 317 (1951); *G. E. Boyd, Q. V. Larson,* J. Phys. Chem. *60,* 707 (1956).
[5] *S. Fried,* J. Amer. Chem. Soc. *70,* 442 (1948).
[6] *J. W. Cobble, C. M. Nelson, G. W. Parker, W. T. Smith jr.,* J. Amer. Chem. Soc. *74,* 1852 (1952).
[7] *J. D. Eakins, D. G. Humphries,* J. Inorg. Nucl. Chem. *25,* 737 (1963).
[8] *R. Colton,* The Chemistry of Rhenium and Technetium, John Wiley & Sons, New York 1965.
[9] *R. D. Peacock,* The Chemistry of Technetium and Rhenium, Elsevier Publ. Co., London 1965.
[10] *J. H. Canterford, R. Colton,* Halides of the Second and Third Row Transition Metals, John Wiley & Sons, New York 1968.
[11] *K. V. Kotegov, O. N. Pavlov, V. P. Shvedov,* Adv. Inorg. Chem. Radiochem. *11,* 1 (1968).
[12] *R. Colton, R. D. Peacock,* Quart. Rev. Chem. Soc. *16,* 299 (1962).
[13] *J. C. Hileman, D. K. Huegins, H. D. Kaesz,* Inorg. Chem. *1,* 933 (1962); *M. A. El-Sayed, H. D. Kaesz,* Inorg. Chem. *2,* 158 (1963).
[14] *K. Schwochau, W. Herr,* Z. Anorg. Allgem. Chemie *318,* 198 (1962); *K. Schwochau,* Ber. Kernforschungsanlage, Jülich *68,* 101 (1962).

18.3 Technetium(II) and -(III) Compounds

The reduction of potassium hexachlorotechnetate(IV) with *o*-phenylene-bis-dimethylarsine, diars in alcohol results[15] in *trivalent technetium complex* with the composition $[Tc(diars)_2X_2]X$ (X=Cl, Br, I). The bromide and iodide salts are prepared by metathetical reactions. *Complexes of divalent technetium*, $Tc(diars)_2X_2$, are obtained when the corresponding compounds of trivalent technetium are reduced with sulfur dioxide or boiling alcohol.

The ammonium salt of the $Tc_2Cl_8^{3\ominus}$ anion, formally containing technetium in the +2 and +3 oxidation states, is prepared by the reduction[16] of ammonium hexachlorotechnetate(IV) with zinc in hydrochloric acid. Apparently, the compound cannot be obtained[17] without some ammonium hexachlorotechnetate(IV) impurity.

18.4 Technetium(IV) Compounds

Red *technetium tetrachloride*, which is quite stable, is best prepared[18] by chlorination of the metal. It is readily purified by sublimation at about 300° in a gentle stream of chlorine or nitrogen. It dissolves in hydrochloric acid to yield yellow solutions containing the hexachlorotechnetate(IV) anion. It is not instantly hydrolyzed in water or alkaline solution.

Pink *potassium hexafluorotechnetate(IV)* can be obtained by the reaction of the hexachlorotechnetate(IV) salt[19] or the hexabromotechnetate(IV) salt[20] with potassium hydrogen fluoride melts. The complex is resistant to hydrolysis and may be purified by recrystallization from water. The complex must be heated in concentrated alkaline solution to bring about hydrolysis.[20] Solutions of the free acid and other salts can be prepared by ion-exchange procedures.

Potassium hexachlorotechnetate(IV) can be prepared by reduction of potassium pertechnetate in hydrochloric acid containing some potassium iodide.[21] The salt is unstable in 1 M hydrochloric acid, being oxidized eventually to pertechnetate ion. With this procedure a red binuclear complex of composition $K_4[Tc_2OCl_{10}]$ can be obtained,[22] which can react further with hydrochloric acid to yield the hexachlorotechnetate salt.

Potassium hexabromotechnetate(IV) is prepared[21] by treating the corresponding hexachlorotechnetate(IV) with successive amounts of hydrobromic acid and repeated evaporation to expel the hydrogen chloride formed in the exchange reaction. The very dark red potassium salt is readily hydrolyzed but may be recrystallized from concentrated hydrobromic acid.

Potassium hexaiodotechnetate(IV) is obtained[21] by digesting the corresponding hexachloro- or hexabromotechnetate(IV) salts with successive amounts of hydriodic acid. The black crystals undergo hydrolysis readily.

Technetium dioxide dihydrate can be prepared by reducing pertechnetate solutions with zinc and hydrochloric acid. Dehydrating the hydrate *in vacuo* at 300° yields[3] anhydrous technetium dioxide. The dioxide can be sublimed at about 1000° *in vacuo* without decomposition. It can be reduced to the metal by hydrogen at 500° and heating in oxygen yields the heptoxide.

Amorphous *technetium disulfide* is prepared by the thermal decomposition of the heptasulfide. A crystalline disulfide can be obtained[18] by heating the heptasulfide with sulfur at 1000° for 24 hours.

Technetium carbide, TcC, can be prepared[23] by carburizing technetium metal with hydrogen–benzene mixtures or by heating the metal with a large excess of graphite at temperatures of 700°–1100°. A carbide of unspecified composition was obtained[24] by vacuum inductive heating of technetium metal to its melting point in a graphite crucible. The carbide thus prepared is a superconductor at liquid helium temperatures.

---

[15] *J. E. Fergusson, R. S. Nyholm*, Nature *183*, 1039 (1959);
J. E. Fergusson, R. S. Nyholm, Chem. & Ind. (London) 347 (1960).

[16] *J. D. Eakins, D. G. Humphries, C. E. Mellish*, J. Chem. Soc. (London) 6012 (1963);
F. A. Cotton, W. K. Bratton, J. Amer. Chem. Soc. *87*, 921 (1965).

[17] *W. K. Bratton, F. A. Cotton*, Inorg. Chem. *9*, 789 (1970).

[18] *R. Colton*, Nature *193*, 872 (1962).

[19] *D. E. Lavalle, R. M. Steele, W. T. Smith*, J. Inorg. Nucl. Chem. *28*, 260 (1966).

[20] *K. Schwochau, W. Herr*, Angew. Chem. Intern. Ed. Engl. *2*, 97 (1963).

[21] *J. Dalziel, N. S. Gill, R. S. Nyholm, R. D. Peacock*, J Chem. Soc. (London) 4012 (1958).

[22] *G. E. Boyd*, J. Chem. Educ. *36*, 3 (1959).

[23] *W. Trzebiatowski, J. Rudzinski*, Z. Chem. *2*, 588 (1962).

[24] *A. L. Giorgi, E. G. Szklarz*, J. Less-Common Metals *11*, 455 (1966).

By dissolving hydrated technetium dioxide or potassium hexachlorotechnetate in an alkaline solution of potassium cyanide yellow solutions can be formed from which a dark brown thallium salt of the composition $Tl_3[TcO(OH)(CN)_4]$ or $Tl_3[Tc(OH)_3(CN)_4]$ can be precipitated.[25] The complexes are readily decomposed by acids and are oxidized to pertechnetate by various oxidizing agents. Addition of a potassium cyanide solution in methanol to potassium hexaiodotechnetate(IV) yields dark red *potassium hexacyanotechnetate(IV)* which is readily hydrolyzed.

18.5 Technetium(V) Compounds

Yellow *technetium pentafluoride*, m.p. 50°, is a by-product of the fluorination of technetium metal.[26] The compound is readily hydrolyzed and decomposes in glass containers at about 60°.

Yellow crystalline salts of *sodium* and *potassium hexafluorotechnetates(V)* are obtained[26,27] by the reaction of technetium hexafluoride with alkali metal chlorides in iodine pentafluoride solution.

Technetium oxide trichloride is one of the products of the reaction[28] between chlorine and technetium dioxide. It sublimes near 500° *in vacuo*.

Technetium oxide tribromide results[29] from the reaction of technetium dioxide with bromine at about 350°. It is rather involatile and thermally stable. It is instantly hydrolyzed by water and alkali.

The brown *complex* $[Tc(diars)_2Cl_4]Cl$ (diars = *o*-phenylene-bis-dimethylarsine) is obtained[30] by oxidizing the orange technetium(III) complex, $[Tc(diars)_2Cl_2]Cl$ with gaseous chlorine. The oxidation state for technetium has been confirmed by potentiometric titration with standard titanium trichloride solution.

18.6 Technetium(VI) Compounds

Technetium hexafluoride is prepared[31] by treating the metal with an excess of fluorine in a nickel reactor for two hours at 400°.

The volatile product is purified by fractional sublimation. It is a golden-yellow solid at room temperature melting to a yellow liquid at 37·4°. The vapor is colorless and monomeric.

Blue *technetium oxide tetrafluoride* has been reported[32] to be a minor product of the fluorination of technetium metal in a flow system, the necessary oxygen presumably originating from oxide films on the metal.

Technetium readily reacts[18] with chlorine at moderate temperatures in a flow system. Two products form, volatile green *technetium hexachloride* (m.p. 25°) and technetium tetrachloride. The hexachloride is extremely unstable, decomposing to the tetrachloride even on slight warming. It is possible to separate it from the tetrachloride by distillation at room temperature in a rapid stream of nitrogen and collecting it in a tube cooled in ice.

A blue compound, considered to be *technetium oxide tetrachloride*, subliming at 80°–90° is obtained[28] from the reaction of chlorine with technetium dioxide.

18.7 Technetium(VII) Compounds

An unusual *technetium hydride*, K_2TcH_9, can be obtained by reduction of pertechnetate salts with potassium ethylenediamine/ethanol mixtures.[33]

Technetium trioxide fluoride = pertechnyl fluoride, TcO_3F, is isolated[34] from the reaction of fluorine with heated technetium dioxide.

It is a yellow crystalline compound melting at 18·3° to a yellow liquid and boiling near 100°. It is stable at room

[25] *K. Schwochau, W. Herr.* Z. Anorg. Allgem. Chemie *318*, 198 (1962).

[26] *A. J. Edwards, D. Hugill, R. D. Peacock*, Nature *200*, 672 (1963).

[27] *D. Hugill, R. D. Peacock*, J. Chem. Soc. *A*, 1339 (1966).

[28] *C. M. Nelson, G. E. Boyd, W. T. Smith*, J. Amer. Chem. Soc. *76*, 348 (1954);
G. E. Boyd, J. Chem. Educ. *36*, 7 (1959).

[29] *J. H. Canterford, R. Golton*, Halides of the Second and Third Row Transition Metals, p. 289, John Wiley & Sons, New York 1968.

[30] *J. E. Fergusson, R. S. Nyholm*, Chem. Ind. (Düsseldorf) *13*, 347 (1960).

[31] *H. Selig, C. H. Chernick, J. G. Malm*, J. Inorg. Nucl. Chem. *19*, 377 (1961);
H. Selig, J. G. Malm, J. Inorg. Nucl. Chem. *24*, 641 (1962).

[32] *D. Hugill, R. D. Peacock*, J. Chem. Soc. *A*, 1339 (1966);
A. J. Edwards, D. Hugill, R. D. Peacock, Nature *200*, 672 (1963);
A. J. Edwards, G. R. Jones, R. J. C. Sills, J. Chem. Soc. *A*, 2521 (1970).

[33] *A. P. Ginsberg*, Inorg. Chem. *3*, 567 (1964).

[34] *H. Selig, J. G. Malm*, J. Inorg. Nucl. Chem. *25*, 349 (1963).

temperature in nickel or Monel metal containers but it attacks Pyrex glass. It hydrolyzes to pertechnetic and hydrofluoric acid.

Technetium trioxide chloride = pertechnyl chloride can be obtained[35] by heating technetium tetrachloride gently in a stream of oxygen. Heating colorless liquid pertechnyl chloride gently gives rise to a red solid thought to be technetium trioxide. Heating the red compound in chlorine reproduces pertechnyl chloride.

The combustion of technetium in excess oxygen at 500° yields[36] *technetium heptoxide*, Tc_2O_7 (m.p. 119·5°, b.p. 310·5°) exclusively. The compound can be readily purified by repeated sublimations. At room temperature the heptoxide is a light-yellow crystalline substance soluble in water and dioxane. The crystals absorb water readily dissolving into a red liquid. Dissolution of the heptoxide in water yields a colorless solution which, when slowly evaporated over concentrated sulfuric acid, yields dark red hygroscopic crystals. The composition of these crystals indicate that they are anhydrous pertechnetic acid.

When technetium heptoxide is dissolved in water *pertechnetic acid*, $HTcO_4$, is formed. A large number of salts have been isolated. *Tetraphenylarsonium pertechnetate* and *nitron pertechnetate* are practically insoluble and are used in gravimetric analysis. Considering the stability and ease of obtaining pertechnetate salts, they are useful starting materials for the preparation of technetium complexes, lower oxidation states being achieved using suitable reducing agents.

Technetium heptasulfide is precipitated[37] from a solution of a pertechnetate salt in 2–4 M hydrochloric acid with hydrogen sulfide; free sulphur is removed using carbon disulfide. The heptasulfide is reduced[5] to technetium metal by hydrogen at 1100°.

---

[35] *J. H. Canterford, R. Colton*, Halides of the Second and Third Row Transition Metals, p. 278, John Wiley & Sons, New York 1968.

[36] *G. E. Boyd, J. W. Cobble, C. M. Nelson, W. T. Smith jr.*, J. Amer. Chem. Soc. *74*, 556 (1952);
W. T. Smith jr., L. E. Line jr., W. A. Bell, J. Amer. Chem. Soc. *74*, 4964 (1952).

[37] *C. L. Rulfs, W. W. Meinke*, J. Amer. Chem. Soc. *74*, 235 (1952).

19 Rhenium

John R. Wasson
Department of Chemistry, University of Kentucky,
Lexington, Kentucky 40506, U.S.A.

19.1 Metallic Rhenium

Dark gray to black *powdered rhenium* metal (m.p., 3180°; b.p., 5630°) is produced[1–3] by hydrogen reduction of potassium or ammonium perrhenate or rhenium sulfides.

The metal dissolves in oxidizing acids to yield solutions of perrhenic acid, but does not dissolve in hydrofluoric or hydrochloric acid. The metal combines with all the halogens except iodine and dissolves in hydrogen peroxide. The powder is oxidized when heated in air to 300° or higher. *Colloidal rhenium* is formed[4] by the hydrazine or formaldehyde reduction of potassium hexachlororhenate(IV) in the presence of gum arabic. Metallic rhenium can also be obtained[5–8] by electrolysis of sulfuric acid solutions of perrhenate salts. References 9–17 provide extended discussions of the metallurgy of rhenium and the properties of rhenium compounds.

19.2 Intermetallic Compounds and Hydrides

A well-defined carbide of rhenium does not exist. Hughes[18] found a eutectic temperature on the rhenium-rich side of the system at 2480° and 16·9% carbon. X-Ray diffraction patterns of quenched rhenium–carbon alloys suggest the existence of a dicarbide or monocarbide of rhenium.[19,20]

Although *rhenium heptaselenide* is well established, tellurides of rhenium are less well defined. A number of *nonstoichiometric rhenium tellurides*, dark gray powders which become lighter as the tellurium content increases, can be obtained[21] by treating metallic rhenium or ammonium perrhenate in a tube furnace with a stream of hydrogen telluride. The tellurides are stable in air. They are completely decomposed by nitric acid, but only partly decompose in concentrated hydrochloric or sulfuric acids.

Simple hydrides of rhenium are not established. However, numerous *complex hydrides*, *e.g.*, $[HRe(CO)_5]$[22,23] and $[(\pi\text{-}C_5H_5)_2ReH]$,[24,25] and hydride cations, *e.g.*, $\{[(\pi\text{-}C_5H_5)_2ReH_2]^{\oplus}\}$,[26] are known. A rhenium(VII) complex hydride, K_2ReH_9, has been characterized (see Section 19.9). Some reactions used to produce hydrides are the following:

$$[(C_2H_5)_4N]_2ReHg + CO + H_3C\text{—}CH(OH)\text{—}CH_3 \longrightarrow [(C_2H_5)_4N][Re_2(CO)_6H_3] \quad \text{(Ref. 26a)}$$

$$ReH_4[P(C_6H_5)_3]_2 + H_3C\text{—}C(=O)\text{—}CH_2\text{—}C(=O)\text{—}CH_3 \longrightarrow ReH_2(acac)[P(C_6H_5)_3]_3 \quad \text{(Ref. 26b)}$$

acac = acetylacetonate anion

---

[1] *K. B. Lebedev*, The Chemistry of Rhenium, Butterworth & Co., London 1962.
[2] *R. Colton*, Nature *194*, 374 (1962).
[3] *H. Brauer*, Handbook of Preparative Inorganic Chemistry, Vol. II, p. 1476, Academic Press, New York 1965.
[4] *G. Zenghelis, K. Stathis*, C.R. *209*, 797 (1939).
[5] US Pat. 2.616.840, Erf.; *R. Levi.*
[6] Brit. Pat. 661.153 (1951), Phillips Electrical.
[7] Dutch Pat. 72.568 (1963), Erf.: *N. V. Phillips.*
[8] German Pat. 626.322 (1936), Erf.: *G. Fink, P. Deren.*
[9] *R. Colton*, The Chemistry of Rhenium and Technetium, John Wiley & Sons, New York 1965.
[10] *R. D. Peacock*, The Chemistry of Technetium and Rhenium, Elsevier Publ. Co., London 1965.
[11] *J. H. Canterford, R. Colton*, Halides of the Second and Third Row Transition Metals, John Wiley & Sons, New York 1968.
[12] *R. Colton, R. D. Peacock*, Quart. Rev., Chem. Soc. Publ. Co. *16*, 299 (1962).
[13] *B. W. Ronser*, Rhenium, Elsevier Publ. Co., New York 1962.
[14] *S. Tribalat*, Rhenium et Technetium, Cauthiers–Villars, Paris 1957.
[15] *W. M. Mueller, J. B. Blackledge, George G. Libowitz*, Metal Hydrides, Academic Press, New York 1968.
[16] *Gmelins*, Handbuch der Anorganischen Chemie, No. 70 (Rhenium), Berlin 1941.
[17] *J. C. F. Druce*, Rhenium, Cambridge University Press, Cambridge 1948.
[18] *J. E. Hughes*, J. Less-Common Metals *1*, 377 (1959).
[19] *M. R. Nadler, C. P. Kampter*, J. Phys. Chem. *64*, 1468 (1960).
[20] *A. I. Evstyukhin, Yu. G. Godin, S. A. Kokhtev, I. I. Suchkov*, Met. i Metalloved. Chistykh Metal., Sb. Nauk. Rabot *4*, 149 (1963).
[21] *W. A. Obolonchik, A. A. Yanaki*, Tr. Vses. Soveshch. Probl. Reniya *2*, 59 (1968); C.A. 131004e (1971).
[22] *W. Hieber, G. Braun*, Z. Naturforsch. *B* 14, 132 (1959).
[23] *W. Hieber, L. Schuster*, Z. Anorg. Allgem. Chemie *285*, 205 (1956).
[24] *G. Wilkinson, J. M. Birmingham*, J. Amer. Chem. Soc. *77*, 3421 (1955).
[25] *M. L. H. Green, L. Pratt, G. Wilkinson*, J. Chem, Soc. (London) 3916 (1958).
[26] *E. O. Fischer, Y. Hristidn*, Z. Naturforsch. B *15*, 135 (1960).
[26a] *A. P. Ginsberg, H. J. Hawkes*, J. Amer. Chem. Soc. *90*, 5930 (1968).
[26b] *M. Freni, D. Giusto, P. Romiti, E. Zucca*, J. Inorg. Nucl. Chem. *31*, 3211 (1969); *M. Freni, P. Romiti, D. Giusto*, J. Inorg. Nucl. Chem. *32*, 145 (1970).

$$ReOCl_3[P(C_6H_5)_3]_2 + LiAlH_4 \xrightarrow{THF} \{ReH_7[P(C_6H_5)_3]_2\} \qquad \text{(Ref. 26c)}$$

Ginsberg[27] has published a useful review of complex transition metal hydrides which should be consulted for a discussion of earlier work and related compounds.

19.3 Rhenium(−I, 0, I) Compounds

The best example of a rhenium(−I) compound is the *pentacarbonylrhenium anion* (see Chapter 29), as is the dimeric carbonyl, $Re_2(CO)_{10}$, for zerovalent rhenium. *Substituted zerovalent rhenium carbonyl* compounds, e.g., $\{Re(CO)_4[P(C_6H_5)_3]\}_2$, are also known.[28] A number of rhenium(I) compounds have been established. The reaction[29] of *cis*-$Re(CO)_4Br[P(CH_3)_2C_6H_5]$ with hydrazine leads to the complex $Re(CO)_3(NH_2)N_2[P(CH_3)_2C_6H_5]$ which contains a σ-bonded NH_2 group and molecular nitrogen. Other *molecular nitrogen complexes* are prepared by related reactions.[30–32] The complex *trans*-$\{ReCl(N_2)[P(CH_3)_2C_6H_5]_4\}$ reacts[33] with $CrCl_3(THF)_3$ (THF = tetrahydrofuran) in dichloromethane to give the dinuclear dinitrogen complex $\{[C_6H_5P(CH_3)_2]_4ClReN_2CrCl_3(THF)_2\}$, which probably contains[33] a bridging nitrogen molecule. The compound is monomeric and only weakly conducting in solution. It hydrolyzes and oxidizes rapidly in the atmosphere.
Volatile *nitratopentacarbonylrhenium(I)* results[34] from the reaction of decacarbonyldirhenium and liquid dinitrogen tetroxide. The syntheses and reactions of numerous *carbonylrhenium(I)* compounds have recently received considerable attention.[35–39] In many instances straightforward substitution and oxidation reactions of the parent zerovalent carbonyl are employed.
Rhenium(I) iodide is prepared[40,41] by heating rhenium tetraiodide to constant weight in a stream of nitrogen in the presence of a small amount of iodine at 200° or by thermally decomposing ammonium hexaiodorhenate(IV) at 440° *in vacuo*. It combines with iodine in a sealed tube at 200° to give the triiodide.
Black *hydrated oxides*, $Re_2O \cdot 2H_2O$ and $ReO \cdot H_2O$, have been claimed[42] to result from the reduction of perrhenic acid with zinc or cadmium in hydrochloric acid under an atmosphere of carbon dioxide.

19.4 Rhenium(II) Compounds

Rhenium(II) iodide is prepared[40] by heating the triiodide for 6 hr in a sealed tube at 350° in an atmosphere of carbon dioxide. Iodine is removed by extraction with carbon tetrachloride.
The reaction of rhenium heptaoxide with hydrazine hydrochloride in ethanol in the presence of triphenylphosphine (Ph_3P) yields[43] brick red "$ReCl_2(PPh_2)_2$." The red-brown iodine analog can be obtained by treating the chloro complex with sodium iodide. However, later work has demonstrated[44] that these are nitridorhenium(V) complexes.
A *complex* with the composition ReI_2(bipyridine) has been claimed[45] to result from the interaction of bipyridine with crude rhenium(IV) iodide in acetone. This and similar

[26c] *J. Chatt, R. S. Coffey*, J. Chem. Soc. *A*, 1963 (1969).
[27] *A. P. Ginsberg*, Trans. Metal. Chem. *1*, 111 (1965).
[28] *M. Freni, D. Giusto, P. Romiti*, J. Inorg. Nucl. Chem. *29*, 761 (1967).
[29] *J. T. Moelwyn-Huches, A. W. B. Garner*, Chem. Commun. 1309 (1969).
[30] *J. Chatt, J. R. Dilworth, G. J. Leigh*, Chem. Commun. 687 (1969).
[31] *J. Chatt, J. R. Dilworth, G. J. Leigh, R. L. Richards*, Chem. Commun. 955 (1970).
[32] *B. R. Davis, J. A. Ibers*, Inorg. Chem. *10*, 578 (1971).
[33] *J. Chatt, R. C. Fay, R. L. Richards*, J. Chem. Soc. *A*, 702 (1971).
[34] *C. C. Addison, R. Davis, N. Logan*, J. Chem. Soc. (London) 3333 (1970).
[35] *L. V. Anterrante, G. V. Nelson*, Inorg. Chem. *7*, 2059 (1968).
[36] *H. Behrens, E. Lindner, P. Passler*, Z. Anorg. Allgem. Chemie *361*, 125 (1968).
[37] *F. Zingales, A. Trovati, F. Cariati, P. Uguagliati*, Inorg. Chem. *10*, 507 (1971).
[38] *F. Zingales, A. Grovati, P. Uguagliati*, Inorg. Chem. *10*, 510 (1971).
[39] *R. H. Angelici, G. C. Faber*, Inorg. Chem. *10*, 514 (1971).
[40] *J. E. Fergusson, B. H. Robinson, W. R. Roper*, J. Chem. Soc. 2113 (1962).
[41] *V. G. Tronev, R. A. Dovlyatshina*, Azerb. Khim. Zh. No. 4, 116 (1965).
[42] *R. C. Young, J. W. Irvine jr.*, J. Amer. Chem. Soc. *59*, 2648 (1937).
[43] *J. Chatt, G. A. Rowe*, J. Chem. Soc. (London) 4019 (1962).
[44] *J. Chatt, J. D. Garforth, N. P. Johnson, C. A. Rowe*, J. Chem. Soc. 1012 (1964).
[45] *C. Furlani, G. Ciullo*, J. Inorg. Nucl. Chem. *27*, 1167 (1965).

complexes are suggested to be dimeric and to be mixtures of rhenium(I) and rhenium(III).
σ-Phenylenebisdimethylarsine (diar) forms complexes of the type $[Re(diar)_2X_2]$ (X=Cl, Br). These are discussed in Section 19.5. Related complexes with tri- and quadridentate arsines which are similarly prepared have been reported.[46,47]

19.5 Rhenium(III) Compounds

Red *rhenium trichloride* is best prepared[48,49] by thermal decomposition of rhenium pentachloride in nitrogen.

It has been shown to exist as trinuclear clusters as a solid, in solution, and in the vapor phase. It is reduced to the metal by hydrogen at 250°–350° and reacts with oxygen to give rhenium trioxide chloride and rhenium oxide tetrachloride. Hydrolysis yields the *hydrated rhenium(III) oxide*. It reacts with a number of ligands (L), *e.g.*, pyridine and triphenylphosphine, to form[50] complexes of the type, $L_3Re_3Cl_9$. Interaction of trimeric rhenium trichloride with carboxylic acids in the absence of air yields[51] orange, crystalline *carboxylates*, *e.g.*, $Re_2(C_2H_3O_2)_2Cl_2$. Compounds of this type can also be obtained by refluxing $ReOCl_3$-$[P(C_6H_5)_3]_2$ with carboxylic acids or by reacting dirhenium(III) octachlorides with carboxylic acids.[52]

Red-brown *rhenium tribromide* results from the thermal decomposition of the pentabromide[53] or silver hexabromorhenate(IV).[54] *Mixed halide clusters* $Re_3(Cl,Br)_9$ have been prepared[55] by interaction of rhenium trichloride with the tribromide or subliming rhenium trichloride in an atmosphere of hydrogen bromide.
Black *rhenium triiodide* is obtained[56] by reducing perrhenic acid directly to the triiodide by either the action of concentrated hydriodic acid and ethanol at elevated temperatures or by thermal decomposition[57] of rhenium tetraiodide in an atmosphere of iodine. It evolves iodine slowly *in vacuo*.
Dimeric dark blue or dark green *octachloro-* and *octabromodirhenate(III) salts*, $Re_2X_8^{2\ominus}$ (X=Cl, Br) can be prepared by a variety of methods[52,58–60] and have been reviewed.[11] Reduction of potassium perrhenate in hydrochloric acid with hypophosphorus acid or at 209° with molecular hydrogen at a pressure of 50 atm or reaction of hydrochloric acid with dimeric chlorocarboxylate rhenium(III) complexes[51,60] yields octachlorodirhenate(IV) salts. Sodium thiocyanate reacts with octachlorodirhenates(IV) to yield either $[Re_2(NCS)_8]^{2\ominus}$ or $[Re(NCS)_6]^{2\ominus}$ depending on the conditions.[61] Octahalodirhenate(III) salts can be oxidized[62] by chlorine and bromine to produce salts containing $Re_2X_9^{\ominus}$ anions (X=halogen). Octahalodirhenate(IV) salts react[63,64] with a number of carboxylic acids and sulfur ligands. In most instances the dimer unit is maintained.
A complex containing both Re(II) and Re(III), $Re_2Cl_5(DTH)_2$ (DTH=dithiahexane), is formed[65] by the reaction of tetra-*n*-butylammonium octachlorodirhenate(III) with dithiahexane.
Black *hydrated rhenium sesquioxide*, $Re_2O_3 \cdot xH_2O$, has been claimed[66] to result from the hydrolysis of rhenium trichloride by alkali in the absence of air.

---

[46] *R. J. Mawby, L. M. Venanzi*, J. Chem. Soc. 4447 (1962).
[47] *W. J. Kirkham*, J. Chem. Soc. (London) 550 (1965).
[48] *L. C. Hurd, E. Brimm*, Inorg. Synth. *1*, 182 (1939).
[49] *H. Gehrke jr., D. Bue*, Inorg. Synth. *12*, 193 (1970).
[50] *F. A. Cotton, S. J. Lippard, J. T. Mague*, Inorg. Chem. *4*, 508 (1965).
[51] *F. I. M. Taha, G. Wilkinson*, J. Chem. Soc. (London) 5406 (1964).
[52] *F. A. Cotton, N. F. Curtis, B. F. G. Johnson, W. R. Robinson*, Inorg. Chem. *4*, 326 (1965).
[53] *R. Colton*, J. Chem. Soc. (London) 2078 (1962).
[54] *J. P. King, J. W. Cobble*, J. Amer. Chem. Soc. *82*, 2111 (1960).
[55] *H. Rinke, M. Klein, H. Schafer*, J. Less-Common Metals *12*, 497 (1967).
[56] *L. Malatesta*, Inorg. Synth. *7*, 185 (1963).
[57] *G. W. Watt, R. J. Thompson*, Inorg. Synth. *7*, 187 (1963).
[58] *F. A. Cotton, C. B. Harris*, Inorg. Chem. *4*, 330 (1965).
[59] *V. G. Tronev, S. M. Bondin*, Khim. Redkikh. Elementov. Akad. Nauk SSR, Inst. Obshch. i Neorgan. Khim, *1*, 40 (1954).
[60] *F. A. Cotton, C. Oldham, W. R. Robinson*, Inorg. Chem. *5*, 1798 (1966).
[61] *F. A. Cotton, W. R. Robinson, R. A. Walton, R. Whyman*, Inorg. Chem. *6*, 929 (1967);
F. A. Cotton, W. R. Robinson, R. A. Walton, Inorg. Chem. *6*, 1257 (1967).
[62] *F. Bonati, F. A. Cotton*, Inorg. Chem. *6*, 135 (1967).
[63] *F. A. Cotton, C. Oldham, R. A. Walton*, Inorg. Chem. *6*, 214 (1967).
[64] *M. J. Bennett, W. K. Bratton, F. A. Cotton, W. R. Robinson*, Inorg. Chem. *7*, 1570 (1968).
[65] *M. J. Bennett, F. A. Cotton, R. A. Walton*, J. Amer. Chem. Soc. *88*, 3865 (1966);
M. J. Bennett, F. A. Cotton, R. A. Walton, Proc. Roy. Soc. Ser. *A 303*, 175 (1968).
[66] *W. Geilmann, F. W. Wrigge, W. Biltz*, Z. Anorg. Allgem. Chemie *214*, 239 (1950).

o-Phenylenebisdimethylarsine (diar) forms[67–69] both rhenium(III) and rhenium(II) *complexes.* The tervalent bromo and chloro complexes are prepared by refluxing perrhenic acid, hypophosphorus acid, the arsine, and the appropriate hydrohalic acid in ethanol. Addition of sodium perchlorate yields sparingly soluble complexes of the type $[Re(diar)_2X_2]ClO_4$ (X=Cl, Br). The iodide is obtained by a metathetical reaction of the bromide with sodium iodide. Neutral rhenium(II) compounds of the type $[Re(diar)_2X_2]$ result when the tervalent complexes are reduced with sodium stannite. In the presence of excess ligand, diar, the rhenium(III) compounds can be oxidized with bromine or chlorine to yield rhenium(V) complexes of the type $[Re(diar)_2X_4]ClO_4$.

The reaction of $\{ReO(OC_2H_5)Cl_2[P(C_6H_5)_3]_2\}$ with β-diketones yields[70] complexes of the type $\{ReCl_2(acac)[P(C_6H_5)_3]_2\}$ and $[ReCl(acac)_2P(C_6H_5)_3]$ (acac=acetylacetonate anion).

Blue $K_3[Re(OH)_3(CN)_3]$ results[71] from the tetrahydridoborate reduction of $K_4[ReO_2(CN)_4]$.

The cluster compounds of rhenium(III) have been the subject of a number of reviews.[11, 72, 73]

19.6 Rhenium(IV) Compounds

Pale blue *rhenium tetrafluoride* results[74] from the thermal decomposition of rhenium pentafluoride at about 150°. It can be sublimed without decomposition at about 300° *in vacuo*; it hydrolyzes to rhenium dioxide and hydrofluoric acid.

Black *rhenium tetrachloride* is prepared[75] by the reaction of thionyl chloride with hydrated rhenium dioxide. It can also be obtained by heating[76] rhenium pentachloride in a mixture of carbon tetrachloride and tetrachloroethylene. It can be readily prepared[77–79] by heating rhenium trichloride and pentachloride in a sealed tube at about 300°. The reactions and properties of rhenium tetrachloride have been extensively investigated.[77–79] The properties of the compound are consistent with those of a trimeric species.

Rhenium tetrabromide can be prepared[80] by the reduction of perrhenic acid with hydrobromic acid or by dissolution of rhenium dioxide in hydrobromic acid. It can also be obtained by treating rhenium tetraiodide with successive portions of hydrobromic acid, with evaporation to a small volume between each addition.

Unstable *rhenium tetraiodide* is prepared[81] by the reduction of perrhenic acid with hydriodic acid at room temperature. It loses iodine slowly *in vacuo* at room temperature and rapidly on heating. It is very hygroscopic and dissolves in water to yield a brown solution which rapidly precipitates rhenium dioxide.

Potassium hexafluororhenate(IV) is prepared[82] by treating the corresponding hexabromorhenate(IV) with hydrogen fluoride at 450°. Other hexafluororhenate(IV) salts can be obtained[83] by neutralizing the free acid, prepared from the potassium salt by ion exchange. Hexafluororhenate(IV) salts are generally resistant to hydrolysis. The silver and copper salts decompose if their solutions are evaporated to dryness. Thermal decomposition of ammonium hexafluororhenate(IV) in an argon atmosphere or *in vacuo* at 300° has been reported[84] to yield a black solid with the composition ReNF.

---

[67] *N. F. Curtis, J. E. Fergusson, R. S. Nyholm,* Chem. & Ind. (London) 625 (1958).

[68] *J. E. Fergusson, R. S. Nyholm,* Chem. & Ind. (London) 1555 (1958).

[69] *A. Earnshaw, B. N. Figgis, J. Lewis, R. D. Peacock,* J. Chem. Soc. (London) 3132 (1961).

[70] *D. E. Grove, N. P. Johnson, C. J. L. Lock, G. Wilkinson,* J. Chem. Soc. (London) 490 (1965).

[71] *P. H. L. Walter, J. Kleinberg, E. Griswold,* Inorg. Chem. *1*, 10 (1962).

[72] *J. E. Fergusson,* Coord. Chem. Rev. *1*, 459 (1966).

[73] *B. R. Penfold,* Perspectives in Structural Chemistry 2, 71 (1968).

[74] *G. B. Hargreaves, R. D. Peacock,* J. Chem. Soc. (London) 1099 (1960).

[75] *D. Brown, R. Colton,* Nature *198*, 1300 (1963).

[76] *A. Brignole, F. A. Cotton,* Chem. Commun. 706 (1971).

[77] *J. H. Canterford, R. Colton,* Halides of the Second and Third Row Transition Metals, p. 314, John Wiley & Sons, New York 1968.

[78] *F. A. Cotton, W. R. Robinson, R. A. Walton,* Inorg. Chem. *6*, 223 (1967).

[79] *J. R. Anderson, J. C. Sheldon,* Inorg. Chem. *7*, 2602 (1968).

[80] *R. Colton, G. Wilkinson,* Chem. & Ind. (London) 1314 (1959).

[81] *R. D. Peacock, A. J. E. Welch, L. F. Wilson,* J. Chem. Soc. (London) 2901 (1958).

[82] *E. Weise,* Z. Anorg. Allgem. Chemie *283*, 337 (1956).

[83] *N. S. Nikolaev, E. G. Ippolitov,* Dokl. Akad. Nauk SSSR *140*, 129 (1961).

[84] *D. E. Lavalle, R. M. Steele, W. T. Smith,* J. Inorg. Nucl. Chem. *28*, 260 (1966).

Hexachlororhenates(IV) are usually prepared[85–91] by reduction of perrhenates in hydrochloric acid with any one of several reducing agents or by dissolving rhenium dioxide in hydrochloric acid solution. Hexachlororhenates can also be prepared[92] by reaction of perrhenates with carbon tetrachloride in a bomb at 400°. Tetraalkylammonium hexachlororhenates(IV) can be prepared[93] by the reaction of alkylammonium chlorides and rhenium oxide tetrachloride in thionyl chloride solution. The complex *potassium μ-oxodecachlorodirhenate(IV)*, $K_4[Re_2OCl_{10}]$, can be obtained[89,94–96] from the reduction of potassium perrhenate by potassium iodide in hydrochloric acid. $K_2[Re(OH)Cl_5]$ and $K_4[Re_2OCl_{10}]$ are in equilibrium with each other in solution. The reaction of $ReOCl_3[P(C_6H_5)_3]_2$ with hydrogen chloride in acetone results[97] in the formation of 1,1-dimethyl-3-oxobutyltriphenylphosphonium hexachlororhenate(IV). Reduction of potassium hexachlororhenate(IV) with alkali metals in liquid ammonia leads[98] to mixtures of hydrides and nitrides of rhenium.

Hexabromorhenic(IV) acid can be prepared[80] by reducing perrhenic acid with hydrobromic acid or by dissolving[91] rhenium dioxide in hydrobromic acid. The potassium salt can be obtained by addition of potassium bromide to solutions of hexabromorhenic(IV) acid. Dark red *potassium hexabromorhenate(IV)* hydrolyzes to yield rhenium dioxide quantitatively.

Dark red to black *potassium hexaiodorhenate(IV)* can be obtained by procedures similar to those used to prepare the chloro and bromo complexes. It is hydrolyzed rapidly, but gives somewhat stable solutions in hydriodic acid, dry methanol, and acetone. The salt reacts with sulfuric acid to yield a solution of pentaiodorhenic(IV) acid from which *tetraethylammonium hydroxopentaiodorhenate(IV)*, $[(C_2H_5)_4N]_2[Re(OH)I_5]$, can be precipitated.[99,100]

Anhydrous rhenium dioxide results from the reduction of the heptaoxide with rhenium metal[101,102] or hydrogen[75,103] at 300°. The dihydrate can be prepared by the hydrolysis of rhenium pentachloride or hexahalogenorhenates(IV) or by the action of various reducing agents on solutions of perrhenates. Fusion with alkali in the absence of air yields *rhenites*, $ReO_3^{2\ominus}$. The dioxide is insoluble in water and aqueous alkali, but it dissolves in oxidizing acids and alkaline hydrogen peroxide solutions to give perrhenates. It dissolves in hydrohalic acids to produce hexahalogenorhenates(IV).

Dark black *rhenium disulfide* can be prepared[104,105] by thermal decomposition of the heptasulfide *in vacuo* or by direct combination of the elements. *Rhenium diselenide* is obtained from the thermal decomposition of the heptaselenide *in vacuo* at about 320°. Rhenium disulfide can be chlorinated[106,107] at elevated temperatures to yield rhenium pentachloride. Control of the chlorination can be employed to produce $ReSCl_2$ and $Re_2S_3Cl_4$.

Bis(tetraphenylphosphonium)trisdicyanoethylene-1,2-dithiolatorhenium, $[Ph_4P]_2[Re(S_2C_2(CN_2)]_3$, is prepared[108] by adding warm aqueous K_2ReCl_6 to methanolic disodium *cis*-dicyano-

---

[85] *E. Enk*, Ber. *64*, 791 (1931).
[86] *G. W. Watt, R. J. Thompson*, Inorg. Synth. *7*, 189 (1963).
[87] *L. C. Hurd, V. A. Reinders*, Inorg. Synth. *1*, 178 (1939).
[88] *F. Krauss, H. Steinfeld*, Ber. *64*, 2552 (1931).
[89] *F. Krauss, H. Dahlmann*, Ber. *65*, 877 (1932).
[90] *C. L. Rulfs, R. J. Meyer*, J. Amer. Chem. Soc. (London) *77*, 4505 (1955).
[91] *G. K. Schweitzer, D. L. Wilhelm*, J. Inorg. Nucl. Chem. *3*, 1 (1956).
[92] *W. W. Horner, F. N. Collier, S. Y. Tyree*, Inorg. Chem. *3*, 1388 (1964).
[93] *K. W. Bagnall, D. Brown, R. Colton*, J. Chem. Soc. (London) 3017 (1964).
[94] *B. Jezowska-Trzebiatowska*, Trav. Soc. Sci., Lettres Wroclaw, Ser. B *39*, 5 (1953).
[95] *B. Jezowska-Trzebiatowska, S. Wajda*, Bull. Acad. Polon. Sci., Classe III *2*, 249 (1954).
[96] *J. C. Morrow*, Acta Cryst. *15*, 851 (1962).
[97] *H. Gehrke jr., G. Eastland, M. Leitheiser*, J. Inorg. Nucl. Chem. *32*, 867 (1970).
[98] *C. L. Ottinger, I. E. McFall, C. W. Keenan*, Inorg. Chem. *3*, 1321 (1964).
[99] *W. Biltz, F. W. Wrigge, E. Prange, G. Lange*, Z. Anorg. Allgem. Chemie *234*, 142 (1933).
[100] *R. Colton*, Australian J. Chem. *18*, 435 (1965).
[101] *W. Biltz*, Z. Anorg. Allgem. Chemie *214*, 225 (1933).
[102] *I. Noddack, W. Noddack*, Z. Anorg. Allgem. Chemie *181*, 32 (1929).
[103] *H. V. A. Briscoe, P. L. Robinson, E. M. Stoddart*, J. Chem. Soc. (London) 666 (1931).
[104] *H. V. A. Briscoe, P. L. Robinson, E. M. Stoddart*, J. Chem. Soc. (London) 1439 (1931).
[105] *W. Biltz, F. Weibke*, Z. Anorg. Allgem. Chemie *203*, 3 (1931).
[106] *V. G. Tronev, G. A. Bekhtle, S. B. Davidyants*, Trudy Akad. Nauk Tadzhik SSR *84*, 105 (1958).
[107] *I. A. Glukhov, S. B. Davidyants, M. A. Yanusov, N. A. Elmanova*, Zh. Neorghan. Khim. *6*, 1264 (1961).
[108] *N. G. Connelly, C. J. Jones, J. A. McCleverty*, J. Chem. Soc. *A* 712 (1971).

ethylenedithiolate. The green product is recrystallized from aqueous acetone (m.p., 269°–270°). Oxidation of $[Ph_4P]_2[ReS_2C_2(CN)_2]_3$ with iodine in ethanol yields solutions from which brown *bis(dicyanoethylenedithiolate)oxorhenate*, $[ReO(S_2C_2(CN)_2]_2^{\ominus}$ salts can be precipitated with bulky cations.

The reaction of potassium hexachlororhenate(IV) with potassium thiocyanate at 225° results[109] in the formation of the *hexathiocyanatorhenate(IV)*, $K_2Re(SCN)_6$. A number of these salts can be prepared from the potassium salt by metathetical reactions. *Hexacyanates* of rhenium(IV), *e.g.*, $[(C_6H_5)_4As]_2[Re(OCN)_6]$, can be prepared[110] by a similar reaction sequence in molten dimethyl sulfone.

A *nitrosyl complex*, $Ag_3[Re(CN)_7NO]$, is claimed[111] to precipitate from a nitric acid solution of potassium octacyanorhenate(V) on addition of silver nitrate.

The bonding in dimeric rhenium complexes[112] and the spectra of rhenium(IV) in cubic crystal fields[113] have been the subject of recent reviews.

19.7 Rhenium(V) Compounds

Green *rhenium pentafluoride* is prepared[74] by the reaction of rhenium hexafluoride with a mixture of tungsten hexafluoride and hexacarbonyl. Above 140° it disproportionates to rhenium hexafluoride and rhenium tetrafluoride. Pyrex glass is slowly attacked[74] at 250° by rhenium oxide tetrafluoride to yield black, nonvolatile rhenium oxide trifluoride. It is very hydroscopic and gives a blue aqueous solution.

Deep brown to black *rhenium pentachloride* (m.p., 220°) is prepared[114–116] by the chlorination of rhenium powder (previously heated in hydrogen to reduce oxide films) at 500°–700°. Rhenium hexachloride impurity is removed by distillation. It reacts with oxygen to yield oxide chlorides and gives potassium hexachlororhenate(IV) when heated[117] with potassium chloride. Reactions of rhenium pentachloride with either sulfur dioxide or arsenic(III) oxide leads[118] to mixtures of rhenium(VI) oxide tetrachloride and rhenium trichloride. With anhydrous tin(II) chloride the pentachloride is quantitatively reduced to the trichloride. The pentachloride is reduced by 2,2′-bipyridine (bipy) to give $ReCl_4(bipy)$, and ammonolysis occurs with liquid ammonia to produce polymeric $[ReCl_3(NH_2)_2{\cdot}2NH_3]_n$. The *oxopentabromorhenate(V)* ion results from the interaction of the pentachloride with 48% hydrobromic acid. The reaction[119] of the pentachloride with triphenylphosphine (Ph_3P) in acetone, acetonitrile, and benzene depends on the "age" of the rhenium pentachloride, the solvent, and the amount of water present in the solvent. The reaction in acetone leads to the following products:

$ReOCl_3[P(C_6H_5)_3]_2$, $ReCl_4[P(C_6H_5)_3]_2$,
$[CH_3COCH_2C(CH_3)_2P(C_6H_5)_3]ReCl_5(PPh_3)_2$,
$[CH_3COCH_2C(CH_3)_2P(C_6H_5)_3]_2Re_2Cl_9$, and
$[CH_3COCH_2C(CH_3)_2P(C_6H_5)_3]_2Re_2Cl_8$.

The products isolated in benzene are

$\{ReCl_3[P(C_6H_5)_3]_{2x}$ and $[(C_6H_5)_3PH]_2Re_2Cl_9$,

while in acetonitrile only $ReCl_4(PPh_3)_2$ is isolated. A 1:1 phosphorus trichloride–rhenium pentachloride adduct[120] is obtained by reacting rhenium with phosphorus(V) chloride at 600°. *Adducts of rhenium oxide trichloride*, $ReOCl_3$, of the type $ReOCl_3(PR_3)_2$ are obtained[43, 72, 121, 122] by the interaction of phosphine with alkali perrhenate salt in a mixture of concentrated hydrochloric acid and ethanol.

---

[109] *R. A. Bailey, S. L. Kozak*, Inorg. Chem. *6*, 419 (1967).
[110] *R. A. Bailey, S. L. Kozak*, J. Inorg. Nucl. Chem. *31*, 689 (1969).
[111] *R. Colton, R. D. Peacock, G. Wilkinson*, J. Chem. Soc. 1374 (1960).
[112] *G. Jezowska-Trzebiatowska, W. Wojciechowaki*, Trans. Metal Chem. *6*, 1 (1970).
[113] *P. B. Dorain*, Trans. Metal. Chem. *4*, 1 (1968).
[114] *W. Geilmann, F. W. Wrigge, W. Biltz*, Z. Anorg. Allgem. Chemie *214*, 248 (1933);
R. Colton, Nature *194*, 374 (1962);
[115] *L. C. Hurd, E. Brimm*, Inorg. Synth. *1*, 180 (1939).
[116] *E. R. Epperson, S. M. Horner, K. Knox, S. Y. Trya jr.*, Inorg. Synth. *7*, 163 (1963).
[117] *H. V. A. Briscoe, P. L. Robinson, C. M. Stoddart*, J. Chem. Soc. (London) 2263 (1931).
[118] *D. A. Edwards, R. T. Ward*, J. Chem. Soc. *A* 1617 (1970).
[119] *H. Gehrke jr., G. Eastland*, Inorg. Chem. *9*, 2722 (1970).
[120] *P. Machmer*, Inorg. Nucl. Chem. Lett. *4*, 91 (1968).
[121] *J. Chatt, G. A. Rowe*, Chem. & Ind. (London) 92 (1962);
J. Chatt, G. A. Rowe, J. Chem. Soc. 4019 (1962);
J. E. Fergusson, Coord. Chem. Rev. *1*, 459 (1966).
[122] *N. P. Johnson, C. J. L. Lock, G. Wilkinson*, J. Chem. Soc. (London) 1054 (1964);
N. P. Johnson, C. J. L. Lock, G. Wilkinson, Inorg. Synth. *9*, 145 (1967).

Oxotrichlorobis(triphenylphosphine)rhenium(V) reacts with 2,2'-bipyridyl (bipy) to yield $ReOCl_3$(bipy).[123] *Oxotrichlororhenium(V)* has been obtained[124] in α and β forms by the reaction of rhenium pentachloride with rhenium dioxide and the photochemical decomposition of rhenium oxide tetrachloride using 350 nm light. At temperatures above 78° *in vacuo* and on dissolution in CCl_4 or $TiCl_4$, both forms of $ReOCl_3$ decompose readily to yield $ReOCl_4$, Re_3Cl_9, and ReO_3Cl. The reaction[125] of perrhenates with 1,1,1-tris(diphenylphosphinomethyl)ethane (TDPME) in the presence of hydrochloric and hypophosphorus acids yields $ReOCl_3$·TPDME. A number of TPDME complexes containing the ReO grouping and also those of rhenium(III) halides have been described.[125]

Dark blue *rhenium pentabromide* is prepared[53] by passing bromine vapor in a stream of nitrogen over rhenium at 650°. It melts slightly above room temperature and decomposes on heating to give the tribromide. Colorless *potassium hexafluororhenate(V)*, $KReF_6$, results[126,127] from the reduction of rhenium hexafluoride with potassium iodide.

Hexafluororhenates(V) attack glass at 300°; they are sensitive to moisture and darken on exposure to air. They disproportionate on hydrolysis.

Red *cesium oxopentachlororhenate(V)* can be precipitated[100] from a solution of rhenium pentachloride in concentrated hydrochloric acid by addition of cesium chloride. *Oxotetrahalorhenates(V)*, *e.g.*, $Cs[ReOBr_4]$ and $(C_6H_5)_4As[ReOCl_4]$ have been prepared by the oxidation of trimeric rhenium(III) bromide[52,128,129] and by the reduction of perrhenates with hydrohalic acids.[129] *Hydrates* of the *oxotetrahalorhenates(V)*, e.g., $[ReOCl_4(H_2O)]^{\ominus}$, can also be obtained[129] from the hydrohalic acid reduction of perrhenates. The structures of $[(C_6H_5)_4As][ReOBr_4(CH_3CN)]$ and $[(C_2H_5)_4N][ReOBr_4(H_2O)]$ have been established.[52,128] Neutral complexes of the type $ReOX_3L_2$ [X= halide, L=$(C_6H_5)_3P$ and $(C_6H_5)_3As$] are readily obtained from the oxotetrahalorhenates(V). Oxotetrahalorhenates(V) react with liquid ammonia and amines to form complexes of the type[130,131] $[ReO_2(NH_3)_4]Cl$ and $[ReO_2(en)_2]Cl$ (en = ethylenediamine). Complexes of this type can also be prepared by the oxidation[132] of amine solutions of hexahalorhenates(IV). The complex $[ReO_2(en)_2]Cl$ undergoes successive protonation:

$$\textit{trans}\text{-}[ReO_2(en)_2]Cl \xrightarrow{H^{\oplus}} [ReO(OH)(en)_2]^{2\oplus} \xrightarrow{2H^{\oplus}} [ReO(OH_2)(en)_2]^{3\oplus}$$

brown-green → purple → blue

and salts of the intermediate hydroxo complex can be isolated.[132] The *trans*-dioxo structure of *dioxotetrapyridinerhenium(V)* chloride has been confirmed.[133] Additional reactions of the previously mentioned ethylenediamine complex have been described.[134] Rhenium(V) complexes have been isolated by the following reactions:

$$[Re(en)_2O_2]Cl + HCl + NaNO_2 + CsCl \longrightarrow Cs_2[Re(NO)Cl_5] \quad \text{(Ref. 134a)}$$

en = ethylenediamine

$$ReCl_5 + KOCN + (C_6H_5)_4AsCl \xrightarrow{(CH_3)_2SO} (C_6H_5)_4As[Re(OCN)_6] \quad \text{(Ref. 134b)}$$

$$K_2ReI_6 + KCN \xrightarrow{CH_3OH} K_3[Re(CN)_8] \quad \text{(Refs. 134c, g)}$$

$$NH_4ReO_4 + HCl + SnCl_2 + tu \longrightarrow ReO(tu)_4Cl_3 \cdot H_2O \quad \text{(Ref. 134d)}$$

tu = thiourea

[123] *F. A. Cotton, R. A. Walton*, Inorg. Chem. 5, 1802 (1966).
[124] *P. W. Frais, D. J. L. Lock, A. Guest*, Chem. Commun. 75 (1971).
[125] *R. Davis, J. E. Fergusson*, Inorg. Chim. Acta 4, 16 (1970).
[126] *R. D. Peacock*, J. Chem. Soc. (London) 467 (1957).
[127] *R. D. Peacock, D. W. A. Sharp*, J. Chem. Soc. (London) 2762 (1959).
[128] *F. A. Cotton, S. J. Lippard*, Inorg. Chem. 4, 1621 (1965); 5, 9, 416 (1966).
[129] *J. H. Beard, J. Casey, R. K. Murmann*, Inorg. Chem 4, 797 (1965).
[130] *J. R. Wasson*, M.A. Thesis, University of Missouri, Columbia, 1966.
[131] *J. R. Wasson, R. K. Murmann*, unpubl.
[132] *R. K. Murmann*, Inorg. Synth. 8, 173 (1966).
[133] *C. Calvo, N. Krishnamachari, C. J. L. Lock*, J. Cryst. Mol. Structure 1, 161 (1972).
[134] *T. S. Knodashova, M. A. Porai-Koshits, G. K. Babeshkina, R. S. Gainullina, V. S. Sergienko, V. M. Stepanovich*, Zh. Strukt. Khim. 11, 783 (1970).
[134a] *J. A. Casey, R. K. Murmann*, J. Amer. Chem. Soc, 92, 78 (1970).
[134b] *R. A. Bailey, S. L. Kozak*, J. Inorg. Nucl. Chem. 31, 689 (1969).
[134c] *R. Colton, R. D. Peacock, G. Wilkinson*, J. Chem. Soc. (London) 1374 (1960).
[134d] *L. Morpurgo*, Inorg. Chim. Acta 2, 169 (1968).
[134g] *C. J. L. Lock, G. Wilkinson*, J. Chem. Soc. (London) 2281 (1964).

$$[RO_2(en)_2]Cl + KCN \xrightarrow{CH_3OH\ reflux} K_3[ReO(CN)_4] \quad \text{(Ref. 134e)}$$

$$RBr_3 + HBr + O_2 + (C_2H_5)_4NBr \longrightarrow (C_2H_5)_4N[ReBr_4O(H_2O)] \quad \text{(Ref. 134f)}$$

Numerous *oxopentachlororhenates(V)* have been isolated[135] by reduction of potassium perrhenate in hydrochloric acid by hydrogen iodide, followed by addition of the appropriate chloride. Salts of large organic cations, *e.g.*, 2,2′-bipyridinium, are converted[136] to complexes of the type $ReOCl_3 \cdot L$ when heated in hydrochloric acid solution.

A substantial number of compounds having rhenium–nitrogen multiple bonds have been characterized[43, 44, 137–139]; some of these are given in Table 1. These compounds are of three types: $ReX_3(NAr)(PR_3)_2$, $ReNX_2(PR_3)_3$, and $ReNX_2(PR_3)_2$ (X = halogen, Ar = aryl, R = alkyl or aryl). A cyanide complex, $K_2[ReN(CN)_4] \cdot H_2O$, is prepared[137] from the hydrazine reduction of perrhenic acid in the presence of potassium cyanide. Complexes of the type $ReNX_2(PR_3)_2$ are generally prepared[138] by the hydrazine dihydrochloride reduction of perrhenic acid in the presence of the phosphine. $ReNX_2\{P(C_2H_5)_2C_6H_5\}_2$ compounds in cold dry benzene react[139] with a variety of acceptor molecules to form adducts of the type $[P(C_2H_5)_2C_6H_5]X_2Re{\equiv}N{\rightarrow}L$ (X = Cl, L = BF_3, BCl_3, BBr_3, $PtCl_2[P(C_2H_5)_3]$; X = Br, L = BCl_3, BBr_3).

From spectroscopic and chemical similarities it was concluded that the nitrogen atom in $[P(C_2H_5)_2C_6H_5]X_2Re{\equiv}N$ complexes is very similar to that in organic nitriles. Complexes of the type $ReX_3NR(PR_3')$ can be synthesized by the reduction of perrhenic acid or $ReOCl_3(PPh_3)_2$ with the appropriate hydrazine, followed by phosphine exchange.

Table 1. Compounds Containing Rhenium–Nitrogen Multiple Bonds

Compound	Ref.
$K_2[ReN(CN)_4 \cdot H_2O]$	*a*
$ReNCl_2[P(C_2H_5)_2C_6H_5]_3$	*b*
$ReNCl_2[P(C_6H_5)_3]_2$	*c*
$ReCl_3(NC_6H_4OCH_3)[P(C_2H_5)_2C_6H_5]_2$	*d*
$ReCl_3(NC_5H_4COCH_3)[P(C_2H_5)_2C_6H_5]_2$	*d*
$ReCl_3(NCH_3)[(C_6H_5)_2P(C_2H_5)]_2$	*e*
$K_3[ReN(CN)_5]$	*f*
$Cs_2[ReN(CN)_4]$	*f*

[a] *W. O. Davies, N. P. Johnson, P. Johnson, A. J. Graham*, Chem. Commun., p. 736 (1969).
[b] *P. W. R. Corfield, R. J. Doedens, J. A. Ibers*, Inorg. Chem. *6*, 197 (1967).
[c] *R. J. Doedens, J. A. Ibers*, Inorg. Chem. *6*, 204 (1967).
[d] *D. Bright, J. A. Ibers*, Inorg. Chem. *7*, 1099 (1968).
[e] *D. Bright, J. A. Ibers*, Inorg. Chem. *8*, 703 (1969).
[f] *N. P. Johnson*, J. Chem. Soc. *A*, p. 1843 (1969).

19.8 Rhenium(VI) Compounds

Yellow *rhenium hexafluoride* (m.p., 18·5°) is best prepared[140, 141] by heating the reaction mixture obtained from the fluorination of rhenium metal with an excess of rhenium metal.

ReF_6 readily undergoes a disproportionation reaction in water to yield rhenium 7+ and 4+ species. It does not react with oxygen at 500°. Heptafluoride impurity is readily detected by i.r. spectroscopy.[142] The hexafluoride reacts with nitric oxide and nitrosyl fluoride to yield $NOReF_6$ and $(NO_2)_2ReF_8$, respectively.[143] Unstable red *rhenium pentafluoride chloride* is one of the products[144] of the reaction of fluorine with rhenium pentachloride at 30°.

Blue *rhenium oxide tetrafluoride*, $ReOF_4$ (m.p., 108°; b.p., 171°), results[74] from the reaction of rhenium hexafluoride and metal carbonyls. It attacks glass above 250° and is rapidly hydrolyzed.

Dichroic (red-brown by transmitted light, dark green by reflected light) *rhenium hexachloride*

[134e] *J. H. Beard, C. Calhoun, J. Casey, R. K. Murmann.* J. Amer. Chem. Soc. *90*, 3389 (1968).
[134f] *F. A. Cotton, S. J. Lippard*, Inorg. Chem. *4*, 1621 (1965).
[135] *B. N. Ivanov-Emin, K. C. Dipak, A. I. Ezhov*, Russ. J. Inorg. Chem. *11*, 733 (1966).
[136] *K. C. Dipak, B. N. Ivanov-Emin*, Russ. J. Inorg. Chem. *11*, 736 (1966).
[137] *C. J. L. Lock, G. Wilkinson*, J. Chem. Soc. (London) 2281 (1964).
[138] *J. Chatt, C. D. Falk, G. J. Leigh, R. J. Paske*, J. Chem. Soc. *A* 2288 (1969).
[139] *J. Chatt, B. T. Heaton*, J. Chem. Soc. *A* 705 (1971).
[140] *J. G. Malm, H. Selig*, J. Inorg. Nucl. Chem. *20*, 189 (1961).
[141] *H. Brauer*, Handbook of Preparative Inorganic Chemistry, Vol. I, p. 264, Academic Press, New York 1963.
[142] *H. H. Claassen, J. G. Malm, H. Selig*, J. Chem. Phys. *36*, 2890 (1962).
[143] *N. Bartlett, S. P. Beaton, N. K. Jha*, Chem. Commun. 168, (1966).
[144] *R. D. Peacock, D. F. Stewart*, J. Inorg. Nucl. Chem. *3*, 255 (1967).

(m.p., 250°) is best prepared[145] by absorbing ammonium perrhenate solution into porous brick material, drying, then reducing the salt to the metal with hydrogen at 600°, and finally treating with a 1:1 mixture of chlorine and nitrogen at 650°. It is thermally stable in both nitrogen and chlorine and it does not decompose to the tetrachloride at 300° but hydrolyzes readily.

Brown *rhenium oxide tetrachloride* (m.p., 32°; b.p., 224°) is obtained[146] from the reaction of a rhenium chloride with rhenium heptoxide or the action[147] of oxygen on rhenium trichloride. It dissolves in cold concentrated hydrochloric acid to give a brown solution from which it is claimed[146] that *potassium oxohexachlororhenate(VI)*, K_2ReOCl_6, can be isolated. It has been reported[148] that rhenium oxide tetrachloride reacts with cold dry ammonia gas to yield $ReO(NH_2)_2Cl_2$.

Blue *rhenium oxide tetrabromide* is prepared[53] by heating rhenium metal in a bromine–oxygen stream or by reacting rhenium heptoxide and bromine vapor, rhenium tetrabromide and oxygen, or anhydrous rhenium dioxide and bromine vapor. On heating, it is decomposed to rhenium oxide tribromide.

Eight-coordinate fluoro complexes, e.g., K_2ReF_8, result[83,149,150] from the interaction of rhenium hexafluoride and potassium fluoride. Rubidium and cesium salts have also been described. *Octafluororhenates(VI)* react[83,149,150] with rhenium hexafluoride to yield *heptafluororhenates(VI), e.g.*, $CsReF_7$. Octafluororhenates(VI) slowly turn blue in air or water yielding blue *oxopentafluororhenates(VI), e.g.*, $KReOF_5$.

It has been claimed[93,146] that red potassium *oxohexachlororhenate(VI)*, K_2ReOCl_6, results when rhenium oxide tetrachloride is dissolved in concentrated hydrochloric acid and potassium chloride is added. *Oxopentachlororhenates(VI), e.g.*, $KReOCl_5$, are prepared[151] by the reaction of the chlorides of large organic cations with rhenium oxide tetrachloride in chloroform solution. The complexes dissolve in nitromethane. Diethyldithiocarbamate, thiourea, and thiosemicarbazide complexes of oxorhenium(VI) obtained[152] by reaction of the appropriate ligands with rhenium oxide tetrachloride have been characterized in solution, but not in the solid state.

Red *rhenium trioxide*, ReO_3, is obtained[153,154] in high yield by the pyrolysis of the dioxane adduct of rhenium heptoxide, $Re_2O_7 \cdot 3C_4H_8O$. It dissolves in oxidizing acids and alkaline hydrogen peroxide solutions to give perrhenates. When fused with alkali in the absence of air, it forms *rhenates(VI)*, M_2ReO_4. *In vacuo* above 300° it decomposes to give the heptaoxide and the dioxide.

Black-green *tris(benzenedithiolato)rhenium(VI)* results[155–158] from the reaction of benzene-1,2-dithiol with rhenium pentachloride in carbon tetrachloride in the presence of air. Similar complexes of *cis*-1,2-diphenylethene-1,2-dithiol and toluene-3,4-dithiol are prepared similarly.

Black hexamminecobalt(III) *octacyanorhenate(VI)*, $[Co(NH_3)_6]_2[Re(CN)_8]_3$, precipitates[137] from a solution of potassium octacyanorhenate(V) in oxygenated cold dilute hydrochloric acid on addition of hexamminecobalt(III) chloride. The solid slowly dissolves in water to yield a purple solution.

19.9 Rhenium(VII) Compounds

Yellow *rhenium heptafluoride* (triple point, 48·3°; b.p., 73·7°) is prepared[140,159] by the fluorination of rhenium metal at 300°–400°.

---

[145] *J. H. Canterford, R. Colton*, Halides of the Second and Third Row Transition Metals, p. 284, John Wiley & Sons, New York 1968.

[146] *A. Bruckl, K. Ziegler*, Ber. *65*, 916 (1923).

[147] *O. W. Kolling*, Trans. Kansas Acad. Sci. *56*, 378 (1953).

[148] *A. Bruckl, E. Plettinger*, Ber. *65*, 971 (1933).

[149] *E. G. Ippolitov*, Russ. J. Inorg. Chem. *7*, 485 (1962).

[150] *P. A. Koz'min*, Zh. Strukt. Khim. *5*, 70 (1964).

[151] *B. J. Brisdon, D. A. Edwards*, Chem. Commun. 278 (1966);
B. J. Bridson, D. A. Edwards, Inorg. Chem. *7*, 1898 (1968).

[152] *N. S. Garif'yanov*, Izv. Akad. Nauk SSSR, Ser. Khim. 1902 (1968).

[153] *A. D. Melaven, J. N. Fowle, W. Brickell, C. F. Hiskey*, Inorg. Synth. *3*, 187 (1950).

[154] *H. Nechamkin, A. N. Kurtz, C. F. Hiskey*, J. Amer. Chem. Soc. *73*, 2829 (1951).

[155] *E. I. Stiefel, R. Eisenberg, R. C. Rosenberg, H. B. Gray*, J. Amer. Chem. Soc. *88*, 2956 (1966);
J. Amer. Chem. Soc. *88*, 2874 (1966).

[156] *E. I. Stiefel, H. B. Gray*, J. Amer. Chem. Soc. *87*, 4012 (1965).

[157] *R. Eisenberg, J. A. Ibers*, J. Amer. Chem. Soc. *87*, 3776 (1965).

[158] *E. I. Stiefel, L. E. Bennett, Z. Dori, T. H. Crawford, C. Simo, H. B. Gray*, Inorg. Chem. *9*, 281 (1970).

[159] *J. G. Malm, H. Selig, S. Fried*, J. Amer. Chem. Soc. *82*, 1510 (1960).

It is always contaminated with the hexafluoride. It is thermally stable and can be stored indefinitely in dry Pyrex glass. It is instantly hydrolyzed to perrhenic acid and hydrofluoric acid and is reduced by the metal to the hexafluoride.

Cream-colored *rhenium oxide pentafluoride*, $ReOF_5$ (m.p., 34·5°; b.p., 55°) is prepared[160] concomitantly with rhenium dioxide trifluoride in the ratio 1:10 by the reaction of fluorine with potassium perrhenate at about 100°.

It is instantly hydrolyzed to a mixture of perrhenic and hydrofluoric acids. It does not etch glass in the absence of moisture. Pale yellow *rhenium dioxide trifluoride*, ReO_2F_3 (m.p., 90°; b.p., 185·4°), can be obtained[160] in crystalline form by slow vacuum sublimation. It does not attack glass up to 300°. It is readily hydrolyzed to a mixture of perrhenic and hydrofluoric acids.

Deep yellow *rhenium trioxide fluoride* (perrhenyl fluoride), ReO_3F (m.p., 147°; b.p. (with decomposition), 164°), is best prepared[161] by the action of iodine pentafluoride on finely divided potassium perrhenate.

Slow sublimation *in vacuo* yields crystalline material. It does not attack glass, but is readily hydrolyzed to a mixture of perrhenic and hydrofluoric acids. It fumes in moist air. It can also be prepared[162] from the reaction of perrhenyl chloride with anhydrous hydrogen fluoride.

Colorless *rhenium trioxide chloride* (perrhenyl chloride), ReO_3Cl (m.p., 4·5°; b.p., 130°) is best prepared[163] by the chlorination of rhenium trioxide at 160°–190°. It is stable in dry air, but is rapidly hydrolyzed to a mixture of perrhenic and hydrochloric acids. It is decomposed by mercury, silver, and stopcock grease to yield rhenium trioxide.

Colorless *rhenium trioxide bromide*, ReO_3Br, is prepared by the bromination of potassium perrhenate,[80] rhenium trioxide,[164] or rhenium heptoxide.[165] It is readily decomposed thermally to rhenium trioxide.[80]

Potassium dioxotetrafluororhenate(VII), $KReO_2F_4$, is obtained[166] by dissolving potassium perrhenate in bromine trifluoride. Other alkali metal salts may be similarly prepared; they all hydrolyze rapidly.

Pale yellow *rhenium heptoxide*, Re_2O_7 (m.p., 300°; b.p., 360°), results[167,168] from treating rhenium metal at 150° with excess oxygen. It is very hygroscopic and dissolves in water to yield perrhenic acid. It reacts with dry hydrogen sulfide to give rhenium heptasulphide.

Perrhenic acid, $HReO_4$, prepared by dissolving rhenium heptoxide in water is stable and exhibits little oxidizing power. It can be reduced by hydrazine and various amalgams. Numerous perrhenate salts have been described. The *perrhenate salts* of nitron and the tetraphenylarsonium cation are used for the gravimetric determination of rhenium. Perrhenate ion coordinated to cobalt(III) in pentammineperrhenatocobalt is obtained[169] via vacuum dehydration of aquopentamminecobalt(III) perrhenate. Colorless trimethylsilyl perrhenate, $(CH_3)_3SiOReO_3$, is prepared[170] by either of the following reactions.

$$Re_2O_7 + (CH_3)_3SiOSi(CH_3)_3 \longrightarrow 2(CH_3)_2SiOReO_3$$

$$AgReO_4 + (CH_3)_3SiCl \longrightarrow AgCl + (CH_3)_3SiOReO_3$$

The product is very sensitive to hydrolysis and must be handled in a dry atmosphere. The action of hydrogen sulfide on potassium perrhenate produces[171–173] *potassium thioperrhenate*, $KReO_3S$, which decomposes on standing to yield potassium perrhenate and *tetrathioperrhenate*.

Black *rhenium heptasulfide*, Re_2S_7, is prepared[167,168] by passing hydrogen sulfide into a solution of potassium perrhenate. It is readily converted to perrhenic acid by alkaline hydrogen peroxide or nitric acid.

Rhenium heptaselenide, Re_2Se_7, results[104] from passing hydrogen selenide into a solution of

[160] *E. E. Aynsley, R. D. Peacock, P. L. Robinson*, J. Chem. Soc. (London) 1622 (1950).
[161] *E. E. Aynsley, M. L. Hair*, J. Chem. Soc. (London) 3747 (1958).
[162] *A. Engelbrecht, A. V. Grosse*, J. Amer. Chem. Soc. *76*, 2042 (1954).
[163] *C. J. Wolf, A. F. Clifford, W. H. Johnston*, J. Amer. Chem. Soc. *79*, 4257 (1957).
[164] *E. Amble, S. L. Miller, A. L. Schawlaw, C. H. Townes*, J. Chem. Phys. *20*, 192 (1952).
[165] *A. Bruckl, K. Ziegler*, Monatsh. Chem. *63*, 329 (1933).
[166] *R. D. Peacock*, J. Chem. Soc. (London) 601 (1955).
[167] *A. D. Melaven, J. N. Fowle, W. Bricknell, C. F. Hiskey*, Inorg. Synth. *3*, 188 (1950).
[168] *W. T. Smith jr., L. E. Line jr., W. A. Bell*, J. Amer. Chem. Soc. *74*, 4965 (1954).
[169] *E. Lenz, R. K. Murmann*, Inorg. Chem. *7*, 1880 (1968);
E. Lenz, R. K. Murmann, Inorg. Synth. *12*, 214 (1970).
[170] *M. Schmidt, H. Schmidbaur*, Inorg. Synth. *9*, 149 (1967).
[171] *W. Feit*, Z. Anorg.All gem. Chemie *199*, 262 (1931).
[172] *B. Krebs, A. Muller, H. Beyer*, Z. Anorg. Allgem. Chemie *362*, 44 (1968).
[173] *A. Muller, E. Diemann, V. V. Krishna Rao*, Chem. Ber. *103*, 2961 (1970).

potassium perrhenate. Pale yellow *rhenium trioxide nitrate* (perrhenyl nitrate) is prepared[174] by the action of dinitrogen pentoxide on rhenium trioxide chloride. The compound decom-

$$ReO_3Cl + N_2O_5 \longrightarrow ReO_3NO_3 + NO_2Cl$$

poses at 70°–75° yielding nitrogen dioxide and rhenium heptoxide. Hydrolysis yields only a mixture of nitric and perrhenic acids.

A number of complex oxides containing rhenium have been characterized, *e.g.*, the perovskites,[175] A_2ReMO_6 (A = Ca, M = Co; A = Ba, M = Cr). Oxides of the type $A_4^{II}Re_2^{VIII}M^{II}O_{12}$ (A = Ba; M = Mg, Ca, Co, Zn, Cd) are obtained[176] by heating the appropriate metal oxides in air with rhenium metal.

Pale yellow *potassium nitridorhenate(VII)*, K_2ReO_3N, is obtained[177] from the following reaction.

$$Re_2O_7 + 3KNH_3 \xrightarrow{\text{liquid } NH_3} K_2ReO_3N + KReO_4 + 2NH_3$$

Addition of concentrated aqueous solutions of perrhenates to potassium in ethylenediamine results[178,179] in the production of white crystalline "*rhenide*" salts. Reduction of aqueous alkaline perrhenate solutions by sodium amalgam also yields hydride species. The salt K_2ReH_9 has been isolated and thoroughly characterized.

---

174 *C. C. Addison, R. Davis, N. Logan,* J. Chem. Soc. *A* 1449 (1967).

175 *A. W. Sleight, J. M. Longo, R. Ward,* Inorg. Chem. *1,* 245 (1962).

176 *J. M. Longo, L. Katz, R. Ward,* Inorg. Chem. *4,* 235 (1965).

177 *A. F. Clifford, R. R. Olson,* Inorg. Synth. *6,* 167 (1960).

178 *K. Knox, A. P. Ginsberg,* Inorg. Chem. *3,* 555 (1964).

179 *S. C. Abrahams, A. P. Ginsberg, K. Knox,* Inorg. Chem. *3,* 528 (1964).

20 Iron

Kalojan R. Manolov
Department of Inorganic Chemistry,
Institute of Food Industry,
Plovdiv, Bulgaria

20.1 Iron(0) Compounds

Iron carbonyls are the most stable zerovalent iron compounds (see Chapter 29). They are starting material for the preparation of iron(0) compounds by replacing one or several molecules of carbon monoxide as exemplified by the following reaction (1).

$$Fe(CO)_5 + nL \longrightarrow Fe(L)_n(CO)_{5-n} + nCO \quad (1)$$

The reaction proceeds at room temperature or on moderate heating in organic solvents under inert atmosphere. The number (n) of the exchanged molecules increases with the temperature but does not exceed 2 (Table 1).

Table 1. Synthesis of Mixed Iron(0) Complexes According to Reaction (1)

Donor molecule (L)	n	Ref.
NH_3	2	a, b
ZR_3, $P(OR)_3$; Z = P, As, or Sb; R = alkyl, aryl	1, 2	c, d
$P[Z(CH_3)_3]_3$; Z = S, Ge, Sn	in THF 1, irradiation	e
P_4O_6	equimol. amounts 1 100°	f
cyclic phosphines $(PR)_4$, R = methyl, ethyl, phenyl; $(PR)_4[Fe(CO)_3]_2$		g
conjugated dienes = L/2	2	h
perfluorobutadiene, $F_2C{=}CF-CF{=}CF_2$	1	i
1,2-benzocycloheptatriene = L/2	2 120°	j
R−NC; R = alkyl, aryl	1, 2	k
$H_2C{=}CH-CN$	1	l
pyridine	1 dimer	m
1,10-phenanthroline = L/2	2 dimer	n
bis(dimethylarsino)-*o*-carborane = L/2 in hexane	2 low temp.	o
cyclohexadierones = L/2	2	p

[a] *W. Hieber, F. Sonnekalb, E. Becker*, Ber. *63*, 973 (1930).
[b] *W. Hieber, R. Werner*, Chem. Ber. *90*, 116 (1957),
[c] *W. Hieber, W. Freyer*, Chem. Ber. *91*, 1230 (1958).
[d] *A. Sacco, M. Fremi*, Ann. Chim. (Roma) *48*, 218 (1958).
[e] *H. Schumann, O. Stelzer*, J. Organometal. Chem. *13* (2), 25 (1968).
[f] U.S. Patent 3414390 (1968), Monsanto Co., Erf.: *J. G. Riess, J. R. Van Wazer*; CA *70*, 49123b (1949).
[g] *H. G. Ang, B. O. West*, Australian J. Chem. *20* (6), 1133 (1967).
[h] *G. Wilkinson*, J. Chem. Soc. (London) 3753 (1959).
[i] *R. L. Hunt, D. M. Roundhill, G. Wilkinson*, J. Chem. Soc. A (6) 982 (1967).
[j] *D. J. Bertelli, J. M. Viebrock*, Inorg. Chem. *7* (6), 1240 (1968).
[k] *W. Hieber, D. von Pigenot*, Chem. Ber. *89*, 193 (1956); *W. Hieber, D. von Pigenot*, Chem. Ber. *89*, 610, 616 (1956).
[l] *Kettle, Orgel*, Chem. & Ind. (London) 49 (1950).
[m] *W. Hieber, G. Brendel*, Z. Anorg. Allgem. Chem. *289*, 324 (1957).
[n] *W. Hieber, J. G. Floss*, Chem. Ber. *90*, 1617 (1957).
[o] *R. Zaborowski, K. Cohn*, Inorg. Chem. *8* (3), 678 (1969).
[p] *H. Alper, J. T. Edward*, J. Organometal. Chem. *16* (2), 342 (1969).

The interaction between pentacarbonyliron and donor molecules can lead to dimers (ref. *m*, *n* in Table 1) or binuclear complexes. For example, the alkylthionitrosokomplex $[Fe_2(CO)_6RNS]$, R = *t*-butyl, was prepared in hexane at ambient temperature from pentacarbonyliron and the appropriate butylsulfide $(RN)_2S$.[1]
The displacement reaction is accelerated by u.v.-irradiation of the reaction mixture (ref. *e*, Table 1). The irradiation may result in splitting off carbon monoxide and formation of tetracoordinated complexes. Hence, tributylphosphine (R_3P) and pentacarbonyliron when irradiated in tetrahydrofuran at room temperature produce *tributylphosphinetricarbonyliron(0)* $[(R_3P)(CO)_3Fe]$.[2] The reaction of eneacarbonyldiiron proceeds in similar fashion (Eqn 2).

$$Fe_2(CO)_9 + 2L \longrightarrow 2Fe(L)(CO)_4 + CO \quad (2)$$

The reaction is performed under anhydrous conditions (ether solution) at room temperature and in nitrogen atmosphere with L = spiro[2,4]hepta-4,6-diene.[3] Donor molecules such as conjugated dienes react according to the scheme (3).

$$Fe_2(CO)_9 + 2L \longrightarrow 2(L)Fe(CO)_3 + 3CO \quad (3)$$

The latter reaction proceeds at room temperature in organic solvents (benzene, hexane) with L = conjugated diene,[4] methylenecyclopropane,[5] or under reflux with the donor molecule. The latter procedure was utilized with *o*-bis(bromomethyl)benzene affording quinodimethyneiron,*a*.[6] A compound with a similar constitution, $(L)[Fe(CO)_3]_2$, L = 1,3,5,7-tetramethylcyclooctatetraene, was prepared by boiling the reactants in octane.[7]

Complete displacement of carbon monoxide from eneacarbonyldiiron is achieved on treat-

[1] *S. Otsuka, T. Yoshida, A. Nakamura*, Inorg. Chem. *7* (9), 1833 (1968).
[2] *H. Schumann, O. Stelzer, U. Niederreuter*, J. Organometal. Chem. *16* (3), 64 (1969).
[3] *C. H. De Puy, V. M. Kobal, D. H. Gibson*, J. Organometal. Chem. *13* (1), 266 (1968).
[4] *C. H. De Puy, R. N. Greene, T. E. Schroer*, Chem. Commun. (20), 1225 (1968).
[5] *R. Noyori, T. Nishimura, H. Takaya*, Chem. Commun. (3), 89 (1969).
[6] *W. R. Roth, J. D. Meier*, Tetrahedron Lett. (22), 2053 (1967).
[7] *F. A. Cotton, A. Musco*, J. Amer. Chem. Soc. *90* (6), 1444 (1968).

ment with aciloyl-1,2,3-triazole or 3,5-dimethylpyrazole in tetrahydrofuran and argon atmosphere. The resultant complexes are of the type $(L)_2Fe$ or $(L)_2Fe(CO)$. L=4-butyroyl- or 4-benzoyl-1,2,3-triazole or 3,5-dimethylpyrazole.[8] On utilization of dodecacarbonyltriiron the resultant product is a trinuclear mixed complex of the zero-valent iron. The displacement reaction proceeds stepwise and depends on the temperature and the concentration of the ligand. At low ligand concentrations and moderate temperature only one molecule carbon monoxide is replaced; increased ligand concentration and higher temperature lead to displacement of two molecules carbon monoxide.

For example, refluxing dodecacarbonyltriiron and trimethoxyphosphine in cyclohexane yields complexes of the type $Fe_3(CO)_{12-n}(L)_n$: L=trimethoxyphosphine, $n=1,2,3$.[9] The interaction with dibenzyltetrasulfide in benzene at 50° affords the disulfide complex $[Fe_3(CO)_9S_2]$.[10] Sulfido, selenido, and tellurido complexes of the type $[Fe_3(CO)_9Z_2]$, Z=S, Se, or Te, undergo further displacement under conversion into complexes of the type $[F_3(CO)_8(L)Z_2]$ and $[Fe_3(CO)_7(L)_2Z_2]$, $L=R_3X$ or $(RO)_3X$, R=butyl or phenyl, X=P or As.[11] On refluxing dodecacarbonyltriiron with tris(*p*-fluorophenyl)phosphine in nitrogen atmosphere *tris(p-fluorophenyl)phosphine tetracarbonyliron*, $Fe(CO)_4(PR_3)$, is obtained.[12]

Tetranitrosyliron(0), $Fe(NO)_4$, is prepared by treatment of pentacarbonyliron with nitrogen monoxide at high pressure and temperatures below 45°.[13,14] *Dinitrosyldicarbonyliron(0)* can be obtained at 85° according to reaction (4),[15]

$$Fe_3(CO)_{12} + 2NO \longrightarrow Fe(CO)_2(NO)_2 + 2Fe(CO)_5 \quad (4)$$

or reaction (5)[16]

$$Fe(CO)_5 + 2NO \longrightarrow Fe(CO)_2(NO)_2 + 3CO \quad (5)$$

Phosphines, phosphites, or triphenylarsines react with dinitrosyldicarbonyliron(0) to give *complexes* of the type $Fe(NO)_2(CO)(R_3Z)$, Z=P or As.[17]

The complex $Fe(H_2O)_6[Co(CO)_4]_2$ has been prepared from powdered iron, cobalt dichloride, sodium sulfide, sodium thiosulfate and carbon monoxide at room temperature and 15 atm.[18] A modification of this method involves reduction of iron(III) chloride with phenyllithium in ether to yield *pentakis(phenyllithium)iron(0)*, $Fe(LiC_6H_5)_5$.[19]

Dinitrosyliron halides of the type $[(NO)_2Fe(L)]X$ can be reduced with zinc powder (in the presence of zinc dibromide) to yield *dinitrosyliron complexes*, $(NO_2Fe(L)$, L=*o*-aminobenzenethiol.[20] Bis(diphenylphosphine)ethyleneiron, $Fe(Ph_2PCH_2CH_2PPh_2)_2C_2H_4$, Ph=phenyl, is obtained by reduction of an ethereal suspension of iron(III) acetylacetonate at 0° with ethoxyethylaluminum in the presence of ethylenebis(diphenylphosphine).[21] The reduction of iron(III) acctylacctonate with triisobutylaluminum in the presence of triethylphosphine under nitrogen atmosphere yields the complex $(Et_3P)_4FeN_2$.[22]

Pentacoordinated iron(0) compounds can coordinate with one additional donor molecule. Thus pentacarbonyliron reacts with ammonia, amine, or pyridine to give complexes of the type $Fe(CO)_5(L)$, $L=NH_3$, RNH_2, C_5H_5N.[23] *Tricarbonylphosphine complexes* $Fe(CO)_3(Ph_3Z)_2$, Z=P, As, or Sb; Ph=phenyl, form adducts with one or two molecules mercuric halides, $Fe(CO)_3(Ph_3Z)_2 \cdot (HgX_2)_n$, $n=1,2$.[24]

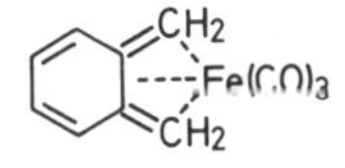

[8] *A. N. Nesmeyanov, M. I. Rybinskaya, N. S. Kochetkova, V. N. Babin, G. B. Shul'pin*, Dokl. Akad. Nauk SSSR *181* (6), 1397 (1968).

[9] *P. J. Pollick, A. Wojcicki*, J. Organometal. Chem. *14* (2), 469 (1968).

[10] *R. A. Krause, C. R. Ruggles*, Inorg. Nucl. Chem. Lett. *4* (10), 555 (1968).

[11] *G. Cetini, P. L. Stanghellini, R. Rossetti, O. Gambino*, J. Organometal. Chem. *15* (2), 373 (1968).

[12] *F. T. Delbeke, G. P. Van der Kelen, Z. Eckhaut*, J. Organometal. Chem. *16* (3), 512 (1969).

[13] *W. Manchot, H. Gall*, Justus Liebigs Ann. Chem. *470*, 271 (1929).

[14] *W. Manchot, E. Enk*, Justus Leibigs Ann. Chem. *470*, 275 (1929).

[15] *W. Hieber, J. S. Anderson*, Z. Anorg. Allgem. Chemie *208*, 238 (1932).

[16] *J. Lewis, R. J. Irwing, G. Wilkinson*, J. Inorg. Nucl. Chem. *7*, 32 (1958).

[17] *D. E. Morris, F. Basolo*, J. Amer. Chem. Soc. *90* (10), 2531, 2536 (1968).

[18] *G. P. Chiusoli, G. Momdelli*, Chim. Ind. (Milan) *49*, (8), 857 (1967).

[19] *B. Sarry*, Angew. Chem. *79* (11), 537 (1967).

[20] *W. Hieber, K. Kaiser*, Z. Anorg. Allgem. Chemie *358* (5–6), 271 (1968).

[21] *G. Hata, H. Kondo, A. Miyake*, J. Amer. Chem. Soc. *90* (9), 2278 (1968).

[22] *C. H. Campbell, A. R. Dias, M. L. H. Green, T. Saito, M. G. Swanwick*, J. Organometal. Chem. *14* (2), 394 (1968).

[23] *W. Hieber, F. Mühlbauer*, Ber. *65*, 1083 (1932).

[24] *D. M. Adams, D. J. Cook, R. D. W. Kemmitt*, J. Chem. Soc. *A* (5), 1067 (1968).

20.2 Iron(I) Compounds

Iron(I) compounds are sensitive to air. They must be prepared in an inert or reducing atmosphere. The oldest method leading to iron(I) nitrosyl complexes is the one of Pavel[25] where a solution of $FeSO_4$ is added to a hot solution of alkali metal sulfide and alkali metal nitrite to afford the salt $M[Fe_4(NO)_7S_3] \cdot nH_2O$, M=Na, $n=2$, M=K, $n=1$. The reduction of iron(II) salts with nitrogen monoxide appears to be more convenient. This reaction proceeds in the presence of donor molecules and yields various iron(I) complexes. For example, treatment of a solution of iron(II) sulfate containing alkali metal thiosulfate, with nitrogen monoxide at room temperature produces *alkali metal dinitrosylthiosulfatoferrate(I)* $M[Fe(NO)_2S_2O_3]$, M= NH_4, K, Na, Rb.[26] In similar fashion the reaction with tetracarbonyliron(II) iodide $[Fe(CO)_4]I_2$, yields *dinitrosyliron(I) iodide* $[Fe(NO)_2]I$[27] and reduction of iron(II) thioethylate, $Fe(SC_2H_5)_2$, gives dimeric *dinitrosyliron(I) thioethylate* $[(NO)_2FeSC_2H_5]_2$.[28]

The reduction of iron(II) halides (X=I or Br)[27] proceeds according to Eqn (6),

$$2FeX_2 + 4NO \longrightarrow 2Fe(NO)_2X + X_2 \qquad (6)$$

but the chloride must be reduced with powdered iron at 70° and in the presence of nitrogen monoxide, as illustrated in reaction (7).

$$FeCl_2 + Fe + 6NO \longrightarrow 2Fe(NO)_3Cl \qquad (7)$$

Dinitrosylsulfidoferrates(I), $M[Fe(NO)_2S]$, M=H, Na, K, NH_4, C_2H_5, C_6H_5, are prepared by passing nitrogen monoxide through a solution containing the appropriate reactants. The scheme (3) illustrates the reaction pattern.[28,29]

$$2FeCl_2 + 4NO + 3Na_2S \longrightarrow 2NaFe(NO)_2S + 4NaCl + s \qquad (8)$$

Stronger reductors than nitrogen monoxide are useful in selected cases. Thus an aqueous solution of sodium pentacyanoaquoferrate(II) can be reduced with sodium hyponitrite. After dilution with methanol the solution affords *sodium pentacyanonitrosylferrate(I) eneahydrate*, $Na_4[Fe(CN)_5NO] \cdot 9H_2O$.[30,31]

The second route for the preparation of iron(I) compounds is based on iron(0) compounds as starting materials. For example, tetranitrosyliron(0) and potassium thiosulfate interact to form *potassium dinitrosylthiosulfatoferrate(I) monohydrate*, $K[Fe(NO)_2S_2O_3] \cdot H_2O$[32] and iodine interacts with *dinitrosyldicarbonyliron(0)*[33] according to reaction (9).

$$2Fe(CO)_2(NO)_2 + I_2 \longrightarrow 2Fe(NO)_2I + 4CO \qquad (9)$$

Reaction of dodecacarbonyltriiron with iodine yields *dimeric tetracarbonyliron(I) iodide* $[Fe(CO)_4I]_2$.[34] Pentacarbonyliron and iodine interact to give tetracarbonyliron(II) iodide which can be reduced with hydrogen to *dicarbonyliron(I) iodide*, $Fe(CO)_2I$.[35] The interaction of zerovalent iron compounds with nitrosylhalides also yields iron(I) derivatives.[36] The latter reaction proceeds in acetonitrile according to scheme (10).

$$Fe(CO)_3(Ph_3P)_2 + NOX \longrightarrow (CO)_2(NO)Fe(Ph_3P)_2X + CO \qquad (10)$$

Dinitrogen tetraoxide in benzene acts as nitrosyl nitrate, its interaction with bis(triphenylphosphine)tricarbonyliron produces the complex *bis(triphenylphosphine)dicarbonylnitrosyliron(I)* nitrate, $(CO)_2(NO)Fe(Ph_3P)_2NO_3$.[36]

Many iron(I) compounds are prepared from appropriate iron(I) compounds.

(a) Metathetical reactions. For example, the ammonium salt $NH_4[Fe_4(NO)_7S_3] \cdot H_2O$ has been prepared by treatment of the sodium salt with ammonium carbonate.[25] The hydroxylammonium and hydrazinium salts of the same anion were obtained from the reaction of the sodium salt, hydroxylammonium (hydrazinium) chloride, and

---

[25] *O. Pavel*, Ber. *12*, 1410 (1879);
O. Pavel, Ber. *15*, 2600 (1882).

[26] *K. A. Hofmann, O. F. Wiede*, Z. Anorg. Allgem. Chemie *8*, 320 (1895);
K. A. Hofmann, O. F. Wiede, Z. Anorg. Allgem. Chemie *9*, 298 (1895).

[27] *W. Hieber, R. Nast*, Z. Anorg. Allgem. Chemie *244*, 23 (1940).

[28] *W. Manchot, H. Gall*, Ber. *60*, 2318 (1927).

[29] *R. V. G. Ewens*, Nature *161*, 530 (1948).

[30] *A. Ungarelli*, Gazz. Chim. Ital. *55*, 118 (1925).

[31] *P. Ray, P. V. Sarkar*, J. Chem. Soc. *119*, 392 (1921).

[32] *W. Manchot, E. Enk*, Justus Liebigs Ann. Chem. *470*, 283 (1929).

[33] *W. Hieber, J. S. Anderson*, Z. Anorg. Allgem. Chemie *211*, 132 (1933).

[34] *F. A. Cotton, B. F. G. Johnson*, Inorg. Chem. *6* (11), 2113 (1967).

[35] *W. Hieber, H. Legally*, Z. Anorg. Allgem. Chemie *245*, 295, 305 (1940).

[36] *G. R. Crooks, B. F. G. Johnson*, J. Chem. Soc. *A* (6), 1239 (1968).

sodium hydroxide.[37] The halide ion in dicarbonylnitrosylbis(triphenylphosphine)-iron(I) chloride or bromide can be replaced by the anions $PF_6^{\ominus}$, $BF_4^{\ominus}$,

(b) By thermal decomposition. On heating of an appropriate material ligands can be split off. The only one simple salt of the monovalent iron, the iodide FeI, was obtained in this manner. *Dicarbonyliron(I) iodide*, $Fe(CO)_2I$, on heating in a stream of carbon dioxide converts to *iron(I) iodide*, FeI.[35]

(c) By additional coordination with donor molecules. For example, dimeric iron(I) halides coordinate with one or two donor molecules and thus are converted into complexes of the type $[(NO)_2Fe(L)_n]X$, L = donor molecule containing nitrogen or sulfur, $n=1$, or 2.[38]

(d) By ligand displacement. The carbon monoxide molecules of monovalent iron complexes are replaceable; this reaction is favored by u.v. irradiation, heating, or refluxing the solution of complex and ligand. Thus a cyclohexane solution of dimeric dimethylphosphidotricarbonyliron(I) $[(CH_3)_2PFe(CO)_3]_2$ was refluxed and irradiated with a tungsten lamp to afford products containing phosphine or diphosphine molecules instead of the carbon monoxide.[39]

(e) *Potassium dinitrosylsulfidoferrate(I)*, $K[Fe(NO)_2S]$, has been prepared from $K[Fe_4(NO)_7S_3]$ by action of dilute potassium hydroxide according to reaction (11).[25]

$$2K[Fe_4(NO)_7S_3] + 4KOH \longrightarrow 6K[Fe(NO)_2S] + Fe_2O_3 + N_2O + H_2O \quad (11)$$

20.3 Iron(II) Compounds

20.3.1 Hydrides and Hydridocomplexes

Both fused and solid iron dissolves hydrogen and under formation of nonstoichiometric compounds which are considered as solid solutions. The *hydride* FeH_2 can be prepared according to reaction (12).

$$FeCl_2 + 2Mg(C_6H_5)Br + 2H_2 \longrightarrow FeH_2 + 2C_6H_6 + 2MgBrCl \quad (12)$$

[37] *I. Bellucci, C. Cecchetti*, Atti Accad. Naz. Lincei, Classe Sci. Fis., Mat. Natur., Rend. *15*, 470 (1906).

[38] *W. Hieber, K. Keiser*, Z. Anorg. Allgem. Chemie *358* (5–6), 271 (1968).

[39] Brit. P. 1 096 404 (1967), Imperial Chem. Ind., Erf.: *D. T. Thompson;* C.A. *68*, 95985r (1968).

The process is performed in anhydrous ether by bubbling hydrogen through the solution. The oil-like product is impure and cannot be purified.[40,41,42]

Iron(II) tetrahydridoborate, $Fe(BH)_{42}$, is formed in ether solution[43] according to reaction (13).

$$FeCl_2 + 2LiBH_4 \longrightarrow Fe(BH_4)_2 + 2LiCl \quad (13)$$

Iron(II) chloride reacts with sodium hydridoborate in ethanol solution containing donor molecules of the type $PRPh_2$ (R = ethyl, butyl, Ph = phenyl) and hydrogen or argon atmosphere to form the *hydridocomplex* $FeH_2(PRPh_2)_3$.[44] Zerovalent iron complexes are oxidized to Fe(II) hydridocomplexes by treatment with hydrogen or hydrogen chloride according to the following reactions (14) and (15).

$$Fe(L)_2C_2H_4 + H_2 \longrightarrow H_2Fe(L)_2 + C_2H_4 \quad (14)$$

$$Fe(L)_2C_2H_4 + HCl \longrightarrow HFe(L)_2Cl + C_2H_4 \quad (15)$$

L = ethylenebis(diphenylphosphine)

The reactions proceed in benzene or xylene at atm. pressure in argon atmosphere and at 60–70°; the reactants of reaction (15) must be present in equimolar amounts.[45]

Zerovalent iron complexes convert to hydridocomplexes by hydrogen transfer from organic ligand to iron according to scheme (16).

$$Fe(HL)_n \longrightarrow HFe(HL)_{n-1} \quad (16)$$

The reaction is enhanced by u.v. irradiation or by heating. Thus a toluene solution of the complex $Fe(HL)_2C_2H_4$, L = ethylenebis(diphenylphosphine), affords $HFe(C_6H_4PPhCH_2CH_2PPh_2)(HL)$ under evolution of ethylene upon heating to 70° or u.v. irradiation.[45] The same starting complex reacts with hydrogen chloride at −15° in toluene or with deuterium chloride at −30° in ether gives $HFeCl(HL)_2$ or $DFeCl(HL)_2$, respectively.[45]

On treatment of iron pentacarbonyl with

[40] *W. Schlenk, Th. Weischselfelder*, Ber. *56*, 2230 (1923).

[41] *T. Weischselfelder, B. Thiede*, Justus Liebigs Ann. Chem. *447*, 64 (1926).

[42] *B. Sarry*, Naturwissenschaften *41*, 115 (1954); *B. Sarry*, Z. Anorg. Allgem. Chemie *280*, 65 (1955); *B. Sarry*, Z. Anorg. Allgem. Chemie *288*, 41, 48 (1956).

[43] *J. Aubry, G. Monnier*, Bull. Soc. Chim., France *4*, 482 (1955).

[44] *A. Sacco, M. Aresta*, Chem. Commun. (20), 1223 (1968).

[45] *G. Hata, H. Kondo, A. Miyake*, J. Amer. Chem. Soc. *90* (9), 2278 (1968).

alkali in solution the corresponding *alkali metal tetracarbonylferrate* is obtained. The latter reacts with acids to yield *dihydrogen tetracarbonylferrate*, $H_2[Fe(CO)_4]$.[46,47] The compound acts as a dibasic acid as indicated by reaction (17).

$$H_2[Fe(CO)_4] + 2L \longrightarrow (HL)_2[Fe(CO)_4] \qquad (17)$$

L = organic base such as amines, pyridine, 2L = *o*-phenanthroline.[48]

20.3.2 Halides and Halocomplexes

Iron(II) chloride is commercial product. The *fluoride*, FeF_2, has been prepared from the chloride by treatment with hydrogen fluoride at room temperature[49] or by reduction of iron(III) fluoride with powdered iron according to (18).[50]

$$2FeF_3 + Fe \longrightarrow 3FeF_2 \qquad (18)$$

Heating of powdered iron with alkali metal bromide or iodide proceeds according to reaction (19).[51]

$$Fe + 2NaX \longrightarrow FeX_2 + 2Na, \; X = Br \text{ or } I \qquad (19)$$

The *bromide*, $FeBr_2$, has been prepared by thermal decomposition of Fe_3Br_8 at 400–500°.[52] Tetracarbonyliron diiodide, $Fe(CO)_4I_2$, decomposes thermally to yield *iron(II) iodide*.[53]
The iodide can be prepared from iron sulfide and iodine according to reaction (20).[54]

$$FeS + I_2 \longrightarrow FeI_2 + S \qquad (20)$$

The halides of bivalent iron form double salts with various metal halides; the resultant compounds crystallize as hydrates from their concentrated aqueous solutions. Thus *lithium iron(II) chloride*, $Li[FeCl_3(H_2O)_3]$ was prepared by freezing of the solution of the simple salts in carbon dioxide atmosphere.[55] A mixture of the chlorides of cadmium and iron gives the salt $2CdCl_2 \cdot FeCl_2 \cdot 12H_2O$ which was considered as hexachloroiron(II) complex $[Cd(H_2O)_6]_2[FeCl_6]$; it is stable in solid state only.[56,57] The iodides and bromides can be prepared in similar fashion. Some double salts crystallize as anhydrous compounds, for example $HgBr_2 \cdot FeBr_2$ or $2HgI_2 \cdot FeI_2$.[58]
Several double fluorides have been prepared by neutralization of a HF-acidified solution of the metal fluoride. Thus neutralization of a solution of iron(II) fluoride and hydrogen fluoride with ammonia yields *ammonium iron(II) fluoride* $NH_4F \cdot FeF_2 \cdot 2H_2O$.[59] The aluminum salt, $Al[FeF_5 \cdot H_2O]$, was obtained from aluminum fluoride, hydrogen fluoride, and iron(II) carbonate.[60]
Some investigators prefer to prepare iron(II) chloride (from powdered iron and hydrogen chloride), add the second chloride and evaporate the solution (cesium-iron(II) chlorides, $n CsCl \cdot FeCl_2 \cdot 2H_2O$, $n = 1$ or 2).[61]
The fusion of alkali metal chlorides with iron(II) chloride leads to various solid phases $n MCl \cdot m FeCl_2$ which are considered as double salts.[62,63]
The tetrachlorocomplex, L_2FeCl_4, L = dithioacetylacetonium ion, crystallizes from ethanol solution.[64]

20.3.3 Binary Compounds with the Elements of the Oxygen Group

Iron forms binary compounds with the elements of the oxygen group of the two types FeZ and FeZ_2, Z = S, Se, or Te. Compounds of the type FeZ are prepared:

(a) by heating of a mixture containing powdered iron and sulfur, selenium (*in vacuo* or in an inert atmosphere) or tellurium (in an inert atmosphere).[65,66]

---

46 *W. Hieber, F. Leutert*, Naturwissenschaften *19*, 360 (1931);
W. Hieber, F. Leutert, Z. Anorg. Allgem. Chemie *204* 145 (1932).
47 *W. Hieber, H. Schulten*, Z. Anorg. Allgem. Chemie *232*, 29 (1937).
48 *W. Hieber, H. Vetter*, Z. Anorg. Allgem. Chemie *212*, 145 (1933).
49 *C. Poulenc*, C. R. Acad. Sci., Paris *115*, 943 (1892).
50 *O. Ruff, E. Ascher*, Z. Anorg. Allgem. Chemie *183*, 196 (1929).
51 *L. Hackspill, R. Grandadam*, C. R. Acad. Sci., Paris *180*, 68 (1925).
52 *P. Höfer*, Kali *21*, 223 (1927).
53 *W. Hieber, G. Bader*, Ber. *61*, 1718 (1928).
54 *E. Filhol, J. Melliès*, Ann. Chim. Phys. *22*, 58 (1871).
55 *A. Chassevant*, C. R. Acad. Sci., Paris *115*, 114 (1892).
56 *T. von Hauer*, Sitzungsberichte Wien. Akad. *17*, 351 (1855).
57 *H. Basset, R. N. C. Strain*, J. Chem. Soc. 1795 (1952).
58 *A. Duboin*, C. R. Acad. Sci., Paris *145*, 713 (1907).
59 *R. Wagner*, Ber. *19*, 897 (1886).
60 *R. F. Weinland, O. Köppen*, Z. Anorg. Allgem. Chemie *22*, 266 (1900).
61 *E. Wilke-Dörfurt, G. Heyne*, Ber. *45*, 1013 (1912).
62 *F. W. Clendinnen*, J. Chem. Soc. *121*, 801 (1922).
63 *N. P. Luzhnaya*, Izv. Akad. Nauk SSSR 27 (1949).
64 *R. Mason, E. D. McKenzie, G. B. Robertson, G. A. Rusholme*, Chem. Commun. (24), 1673 (1968).
65 *W. P. Jorissen, C. Groeneveld*, Recl. Trav. Chim. Pay-Bas *46*, 47 (1927).
66 *G. Tammann, K. Schaarwächter*, Z. Anorg. Allgem. Chemie *167*, 401 (1927).

(b) by treating powdered iron or iron oxide with vapors of the element or with hydrogen sulfide (selenide) at elevated temperatures (S),[67,68] (Se)[69,70], (Te);[71]
(c) by precipitation of iron(II) salts with H_2Z or M_2Z, Z = S, Se, or Te, M = NH_4, Na, K.

Compounds of the type FeZ_2 are natural products, however, various methods for their preparation have been published (Table 2).

Table 2. Preparation of Iron(II) Chalcogenides of the Type FeZ_2, Z = S, Se, Te

	Starting materials	Conditions	Ref.
FeS_2	Fe or FeS and S	450°–500°, N_2 or CO_2 atm.	*a*
	Fe, FeO or Fe_2O_3 and H_2S	200°–400°	*b*
	Fe_2O_3, S and NH_4Cl	heating until NH_4Cl is removed	*c*
	$FeCl_3$ and P_2S_5		*d*
	$FeCl_3$ and Na_2S_5		*e*
	FeS (freshly pptd.) and S	In solution, 100°	*f*
	SeS_2O_3 and $Na_2S_2O_3$	Sealed tube, 100°	*g*
$FeSe_2$	$FeCl_3$ and H_2Se	500°	*h*
	Fe and Se	Fusion, *in vacuo*	*i*
	FeO, Se and $Na_2C_2O_4$	265°–312°, 100 atm., hydrotherm.	*j*
$FeTe_2$	Fe and Te	Fusion *in vacuo*	*i*
	FeO, Te and $Na_2C_2O_4$ or NaO_2CH		*k*

[a] *C. Rammelsberg*, Pogg. Ann. *121*, 337 (1864); *C. Rammelsberg*, Pogg. Ann. 365, 369.
[b] *Doelter*, Z. Krystall. *11*, 29 (1886).
[c] *E. Weinschenk*, Z. Krystall. *17*, 486 (1890).
[d] *E. Glatzel*, Ber. *23*, 57 (1890).
[e] *N. D. Costeanu*, Ann. Chim. Phys. *2* (9), 189 (1914).
[f] *V. Rodt*, Z. Anorg. Allgem. Chem. 188 *102*, 130 (1918).
[g] *E. T. Allen, J. L. Crenshaw, J. Jonston, E. S. Larsen*, Z. Anorg. Allgem. Chem. *76*, 201 (1912).
[h] *H. Fonzes-Diacon*, C. R. Acad. Sci. Paris *130*, 1710 (1900).
[i] *S. Tengner*, Z. Anorg. Allgem. Chem. *239*, 126 (1938).
[j] *L. Cambi, M. Elli*, Chim. Ind. (Milan) *50* (1), 94 (1968).
[k] *L. Cambi, M. Elli*, Chim. Ind. (Milan) *50* (8), 869 (1968).

[67] *F. K. Lotgering*, Phillips Res. Rep. *11*, 190 (1956).
[68] *E. T. Allen, J. L. Crenshaw, J. Jonston, E. S. Larsen*, Z. Anorg. Allgem. Chemie *76*, 201 (1912).
[69] *W. Hempel, M. G. Weber*, Z. Anorg. Allgem. Chemie *77*, 48 (1912).
[70] *J. B. Peel, P. L. Robinson*, Proc. Univ. Durham *8*, 153 (1933).
[71] *L. Moser, K. Ertl*, Z. Anorg. Allgem. Chemie *118*, 269 (1921).

20.3.4 Binary Compounds with the Elements of the Nitrogen Group

The elements of the nitrogen group form non-stoichiometric compounds with iron, which are considered to be solid solutions or compounds of the intermetallic type.

The phase diagram of the system iron–nitrogen shows various nitrides but three of them can be prepared in relatively pure state by treating iron oxides (Fe_2O_3 or Fe_3O_4) with a mixture of hydrogen and ammonia. The constitution of the nitrides depends on the temperature and the ratio hydrogen/ammonia.

Pure ammonia reacts with iron, iron oxide, or iron chloride at 450–500° to give *diiron nitride* Fe_2N.[72,73] A mixture of hydrogen and an excess of ammonia reacts with iron oxide at 500–800° to form *triiron nitride* Fe_3N[74]; if the gas mixture is enriched with hydrogen the conversion at 300–600° affords *tetrairon nitride*, Fe_4N.[75] The reaction can also proceed in a tube furnace at 750° and atm pressure.[76] If powdered iron is treated with ammonia at 350–500° a mixture of the nitrides is formed.[77]
The mixed *lithium iron nitride*, Li_3FeN_2, was prepared from lithium nitride and iron nitrides (Fe_2N or Fe_4N) at 580°. Iron carbonyl can also be used as iron source.[78]

The phase diagram of the system iron–phosphorus illustrated the possible formation of various phosphides.[79] Four of them can be prepared in relatively pure state by heating proper amounts of powdered iron and red phosphorus. Arsenic and antimonium behave similarly; details are presented in Table 3.

A mixture of iron phosphides, Fe_2P and FeP, or arsenides, FeAs or $FeAs_2$, is formed on electrolysis of fused phosphates or arsenates.[80]
Mixed phosphides, arsenides or antimonides were found as stable phases in various ternary systems. For example, the phases LiFeP and

[72] *G. Hägg*, Nature *121*, 826 (1928); *122*, 314 (1928); *G. Hägg*, Z. Phys. Chem. *8*, 455 (1930); *12*, 33 (1931).
[73] *R. Bridelle*, Ann. Chim. *10*, 824 (1955).
[74] *E. Lehrer*, Z. Elektrochem. *37*, 460 (1930).
[75] Brit. P. 1 091 116 (1967), Imperial Chem. Ind., Erf.: *O. Downing;* CA *68*, 15239n (1968).
[76] Brit. P. 1 135 812 (1968). Imperial Chem. Ind., Erf.: *O. Downing;* CA *70*, 39388c (1969).
[77] Brit. P. 1 142 228 (1969), Imperial Chem. Ind., Erf.: *O. Downing, N. Harris;* CA *70*, 89268j (1969).
[78] *M. Fromont*, Rev. Chim. Miner. *4* (2), 447 (1967).
[79] *E. Wachtel, G. Urbain, E. Ubelacker*, C. R. Acad. Sci., Paris *257* (17), 2470 (1963).
[80] *J. L. Andrieux*, Rev. Met. *45*, 49 (1948).

Table 3. Preparation of Iron Phosphides, Arsenides, and Antimonides

	Starting materials	Conditions	Ref.
Fe_3P	Fe, P calcd. amts.	*in vacuo,* at red heat	*a*
	Fe, copper phosphide	fusion, upper layer	*b*
Fe_2P	Fe, P calcd. amts.	fusion under nitrogen	*a, b*
	Fe, copper phosphide	fusion, HNO_3 treatment	*b, c*
FeP	Fe_2P, P-vapors	at red heat	*a, b*
FeP_2	Fe-phosphides, P-excess	900°–1000°, inert atm.	*a*
Fe_2As	Fe, As calcd. amts.	*in vacuo,* 440°, fusion	*d–f*
FeAs	Fe, As calcd. amts.	*in vacuo,* heating	*g*
	$FeAs_2$, H_2	680°	*g*
$FeAs_2$	Fe, As calcd. amts.	*in vacuo,* 700°	*g*
	Fe-arsenides, As vapors	430°–618°	*h*
FeSb	Fe, Sb calcd. amts.	fusion	*i*
$FeSb_2$	Fe, Sb calcd. amts.	fusion	*i*

[a] *W. Biltz, W. Franke, K. Meisel, R. Juza,* Z. Anorg. Allgem. Chem. *218*, 346 (1934).
[b] *A. Le Chatelier, S. Wologdine,* C. R. Acad. Sci., Paris, *149*, 709 (1909).
[c] *J. B. Friauf,* Trans. Amer. Soc. Steal Treat. *17*, 499 (1930).
[d] *G. Tammann, K. Schaarwächter,* Z. Anorg. Allgem. Chem. *167*, 403 (1927).
[e] *R. D. Heyding, L. D. Calvert,* Can. J. Chem. *35*, 449 (1957); *R. D. Heyding, L. D. Calvert,* Can. J. Chem. *38*, 313 (1960).
[f] *L. Hollan,* Ann. Chim. (Paris) *1* (11), 437 (1966).
[g] *S. Hilpert, T. Dieckmann,* Ber. *44*, 2378 (1911).
[h] *A. Beutel, F. Lorenz,* Zh. Min. 367 (1915); *A. Beutel, F. Lorenz,* Zh. Min. 10 (1916).
[i] *N. S. Kurnakow, N. S. Konstantinow,* Z. Anorg. Allgem. Chem. *58*, 1 (1908).

LiFeAs were formed at temperatures below 800°.[81]

In some cases the preparation of iron complexes in nitrogen atmosphere leads to nitrogen complexes. Thus the hydridocomplex $FeH_2(L)_3$, L=ethyl or butyldiphenylphosphine, coordinates with one molecule nitrogen and converts into $N_2FeH_2(L)_3$. The reaction proceeds in solid state or in solution.[82] Some zerovalent iron complexes coordinate with an additional nitrogen molecule under formation of the corresponding nitrogen complexes (see Section 20.1).

20.3.5 Iron(II) Salts

The preparation of most iron(II) salts can be achieved by the conventional methods.

(a) By dissolution of powdered iron in the appropriate acid. This method can be applied to non-oxidizing acids at room temperature or slightly above and an inert atmosphere is needed in some cases. The following compounds were prepared in this manner: *perchlorate* $Fe(ClO_4)_2$ (diluted perchloric acid, <40%),[83] *selenite* $FeSeO_3$ and *selenate* $FeSeO_4{\cdot}xH_2O$,[84] *azide* $Fe(N_3)_2$,[85] *tetrafluoroborate* $Fe(BF_5)_2$,[86] and *formiate* $Fe(HCO_2)_2{\cdot}2H_2O$.[87]

(b) By dissolving iron(II) salts (carbonate, sulfide, etc) in the dilute acid *perchlorate* (from FeS and $HClO_4$ under CO_2 atmosphere),[88] bromate $Fe(BrO_3)_2$ (from $FeCO_3$ and $HBrO_3$),[89] *disulfate* FeS_2O_7 (from $FeSO_4$ and conc. H_2SO_4),[90] *sulfite* $FeSO_3$ (from FeS freshly pptd. and SO_2 in water suspension),[91] *tetrathionate* FeS_4O_6 (from $FeCl_3$ and SO_2),[92] *dithionate* $FeS_2O_6{\cdot}7H_2O$[93] according to reaction (21)

$$Fe_2(SO_4)_3 + 3SO_2 + 2H_2O \longrightarrow 2FeS_2O_6 + 2H_2SO_4 \quad (21)$$

or in water suspension (from $Fe(OH)_3$ and SO_2).[91]

(c) By ion exchange between iron(II) sulfate or chloride and a barium or silver salt. The precipitate of barium sulfate or silver chloride is filtered off and the filtrate is evaporated and left to crystallize. This method is useful for the preparation of iron(II) salts of acids acting as strong oxidizing agents, for example: *nitrate* $Fe(NO_3)_2$ (from $FeCl_2$ and $AgNO_3$),[93] *chlorate* $Fe(ClO_3)_2$,[94] *perchlorate* $Fe(ClO_4)_2$[95] (from $FeSO_4$ and $Ba(ClO_3)_2$ respectively $Ba(ClO_4)_2$), *thio-*

[81] *R. Juza, K. Langer,* Z. Anorg. Allgem. Chemie *361*, (1–2), 58 (1968).
[82] *A. Sacco, M. Aresta,* Chem. Commun. (20), 1223 (1968).
[83] *H. E. Roscoe,* Proc. Roy. Soc. *11*, 502 (1860).
[84] *J. S. Muspratt,* Quart. Rev., Chem. Soc. *2*, 63 (1850).
[85] *Th. Curtius, J. Rissom,* J. Prakt. Chem. *58*, 298 (1898).
[86] *H. Funk, F. Binder,* Z. Anorg. Allgem. Chemie. *155*, 331 (1926).
[87] *M. A. Scheurer-Kestner,* Bull. Soc. Chim., France *5*, 345 (1863).
[88] *F. Lindstrand,* Z. Anorg. Allgem. Chemie *230*, 187 (1936).
[89] *C. Rammelsberg,* J. Prakt. Chem. *25*, 226 (1842).
[90] *T. Bolas,* Justus Liebigs Ann. Chem. *172*, 106 (1874).
[91] *A. Gelis, J. M. Fordos,* J. Pharm. Chim. *4*, 333 (1843).
[92] *U. Antony, E. Manasse,* Gazz. Chim. Ital. *29*, 483 (1899).
[93] *A. Piccini, F. M. Zuco,* Atti Accad. Naz. Lincei, Classe Sci. Fis. Mat. Natur., Rend. *1*, 15 (1885).
[94] *A. Wachter,* J. Prakt. Chem. *30*, 326 (1843).
[95] *M. Serullas,* Ann. Chim. Phys. *46*, 305 (1830).

sulfate $FeS_2O_3 \cdot 5H_2O$ (from alkali earth thiosulfate),[96] *perfluorocarboxylates* $(C_5H_5)Fe(CO)_2$-(OOCR), R=CF_3, C_2F_5, C_3F_7, (from silver carboxylate and cyclopentadienyldicarbonyliron halide).[97]

(d) By precipitation. For example, *tellurite* $Fe(TeO_3)_2 \cdot 4H_2O$ (from Na_2TeO_3),[98] *thiocarbonate* $FeCS_3$ (from Na_2CS_3),[99] and many other well known insoluble salts.

(e) By evaporation. Several salts containing complex anions are obtained by evaporation of a solution containing both iron and a second metal salt: *hexafluorotitanate* $Fe(TiF_6) \cdot 6H_2O$,[100] *hexahalostannate* $Fe(SnX_6) \cdot 6H_2O$, X=Cl or Br, (from FeX_2 and SnX_4 in the presence of SnX_2 to prevent oxidation of iron(II)).[101]

(f) By heating of the reactants at high temperature. This method is useful for the preparation of salts from iron oxide, FeO, and other metal oxides. For example, *titanite* $FeTi_2O_4$ (from Ti_2O_3 at 1000°), *orthotitanate* Fe_2TiO_4, *metatitanate* $FeTiO_3$,[102] *vanadate(IV)* $FeVO_3$ (under 850°),[103] etc.

Some thiosalts have also been prepared at high temperature. For example, *thiophosphate* $Fe_3(PS_4)_2$ (from FeS and PsS_5),[104] *thiohypophosphate* $Fe_2P_2S_6$[105] and *thipyrophosphate* $Fe_2P_2S_7$[106] (from calculated amounts of the elements in a sealed tube at 500°) have been prepared in this manner. The *selenophosphite* Fe_2PSe_3 was also obtained from the elements.[107]

20.3.6 Chelates

Many chelating molecules act as acids and form chelates in the same way as acids form salts. In order to avoid oxidation of iron(II) the preparation must be performed in an inert atmosphere. A convenient method involves treatment of powdered iron with acetic acid in the presence of the chelating molecule or its derivative. For example, *iron(II) benzenesulfonate* $Fe(C_6H_5SO_3)_2 \cdot 6H_2O$ was prepared from benzenesulphonylchloride, iron and acetic acid[108,109] and *iron difluoredithiophosphate*, $Fe(SPSF_2)_2$ was obtained in similar fashion.[110]

Many chelates are insoluble and precipitate. The formation of the chelate is favoured by addition of an alkaline solution to remove hydrogen ion. The alkali metal salt of the chelating acid acts better than the free acid. This procedure was used for the preparation of *chelates with various oximes* (dimethyldioxime, α-benzyldioxime,[111] α-anisyldioxime, furyldioxime,[112] methylethyldioxime.[113]

The chelate precipitates on careful neutralization with sodium diselenocarbamate is formed only after alkalizing with sodium hydroxide.[114] The precipitation procedure requires an inert atmosphere in order to prevent the oxidation of iron(II) to iron(III). For example, the chelate with *o*-aminobenzenethiol was precipitated in an inert atmosphere.[115]

Relatively soluble chelates crystallize from concentrated solution of the reactants. Thus, *iron(II) arylsulfonates* were obtained from concentrated aqueous solution of iron(II) salt and sodium sulfonate.[116]

Iron(II) chloride and the sodium salt of cyclooctatetraene Na_2L (from sodium and cyclooctatetraene in ether) suspended in benzene yield *cyclooctatetraeneiron(II)*, FeL, at 70°.[117] The synthesis of iron(II) chelates *in situ* with the chelating molecule is a convenient method for the preparation of chelates with thioderivatives of ketones or aldehydes. Chelates of the

---

[96] *A. V. Alba, R. U. Lacal*, Rev. Acad. Cienc. Exact. Fis. Quim. Zaragoza *9*, 97 (1954).
[97] *R. B. King, R. N. Kapoor*, J. Organometal. Chem. *15*, (2), 457 (1968).
[98] *J. J. Berzelius*, Pogg. Ann. *32*, 595 (1834).
[99] *J. J. Berzelius*, Pogg. Ann. *6*, 455 (1826).
[100] *R. Weber*, Pogg. Ann. *120*, 287 (1863).
[101] *B. Rayman, K. Preis*, Justus Liebigs Ann. Chem. *223*, 332 (1884).
[102] *F. Halla*, Z. Anorg. Allgem. Chemie *184*, 426 (1929).
[103] *J. C. Bernier, P. Poix*, Ann. Chim. (Paris) *2* (2), 81 (1967).
[104] *E. Glatzel*, Z. Anorg. Allgem. Chemie *4*, 186 (1893).
[105] *C. Friedel*, Bull. Soc. Chim., France *11*, 1057 (1894).
[106] *L. Ferrand*, C. R. Acad. Sci., Paris *122*, 621 (1896).
[107] *C. Friedel, C. Chabrie*, Bull. Soc. Chim. *13*, 164 (1895).
[108] *A. Seyewetz, L. Poizat*, Bull. Soc. Chim., France *9*, 251 (1911).
[109] *H. E. Armstrong, E. H. Rodd*, Proc. Roy. Soc. *90*, 469 (1914).
[110] *F. N. Tebbe, H. W. Roesky, W. C. Rode, E. L. Muetterties*, J. Amer. Chem. Soc. *90* (13), 3578 (1968).
[111] *M. A. Whiteley*, J. Chem. Soc. *83*, 24 (1903).
[112] *L. Tschugaeff*, Ber. *41*, 1678 (1908).
[113] *L. Tschugaeff*, Z. Anorg. Allgem. Chemie *46*, 146 (1905).
[114] *K. A. Jensen, V. Krishnan*, Acta Chem. Scand. *21* (10), 2904 (1967).
[115] *L. F. Larkworthy, J. M. Murphy, D. J. Phillips*, Inorg. Chem. *7* (7), 1436 (1968).
[116] *C. W. Dudley, C. Oldham*, Inorg. Chim. Acta *2* (2), 199 (1968).
[117] Brit. P. 1 128 128 (1968), Studienges. Kohle: CA *70*, 96963k (1969).

type $Fe(L)_2X_4$, HL = thioketone or thioaldehyde, X = Br,[118] HL = β-dithioketone, X = Cl,[119] have been prepared in an alcohol solution of iron(II) salt and the appropriate aldehyde or ketone, by treatment with H_2S, HX, and X_2.

Some bidentate ligands form polymers of the chelating agent bridged by iron atoms. Such polymers have been prepared from iron(II) acetate on heating with *p,p'*-biphnylylenebis-(methylglyoxime) in ethanol[120] or reflux in methanol with pyridine-2,6-dihydroxamic acid.[121]

20.3.7 Complexes

The formation of iron(II) complexes is generally accomplished by conventional methods. Special attention must be paid to the removal of the air since iron(II) compounds are easily oxidized to iron(III).

Insoluble complexes are prepared by mixing a solution of iron(II) salt with a solution of the ligand under inert atmosphere (hydrogen, nitrogen, argon). For example, a methanol solution of iron(II) halides reacts with ethylenediamine to afford the complexes $Fe(L)_nX_2$, X = Cl, Br or I, L = ethylenediamine, $n = 3$.[122] Complexes of the same type can be prepared (in water or alcohol) with various ligands such as α,α'-dipyridyl,[123] methylamine,[124] phenylhydrazine,[125] dimethylsulfoxide ($X = BF_4$, $n = 6$) in acetone,[126] pyrazole (X = Cl, $n = 4$),[127] tetrakis(3-dimethylarsinopropyl)-*o*-phenylenediarsine ($n = 1$, $X = ClO_4$).[128]

The stability of the complex increases if the anion X constitutes an organic radical or has chelating properties. For example, on reflux of iron(II) phthalimidate in ethanol with organic bases (α,α'-bipyridine,*o*-phenanthroline) complexes of the type $Fe(L)_2X_2$ are obtained.[129] A complex of the same type, L = pyridine, X = 1,2-di(hydroxylamino)butylketone, has been prepared by recrystallization from an alcoholic solution containing pyridine.[130]

More *soluble complexes* are prepared by evaporation to crystallization ($Fe(L)_nX_2$, L = urea, $n = 6$, X = Cl or NO_3[131]; $L = N_2H_5Cl$, $n = 3$, X = Cl[132]), or by precipitating dilution with alcohol ($K_2Pb[Fe(NO_2)_6]$,[133] $3Na_2S_2O_3 \cdot FeS_2O_3 \cdot 8H_2O$).[134]

Solvents acting as donors form complexes of iron(II) salt by recrystallization from such solvents. The resultant complex can be considered as solvate. In this manner complexes with liquid ammonia, $[Fe(NH_3)_5(H_2O)]F_2$,[135] methanol $FeCl_2 \cdot 4CH_3OH$,[136] $FeSO_4 \cdot 1\frac{1}{2}CH_3OH$,[137] pyridine, quinoline, and aniline have been prepared.[138,139,125]

A convenient method for the preparation of some iron(II) complexes involves the treatment of a solid salt with an appropriate gaseous ligand. This method has been used particularly for the preparation of *aminocomplexes* $[Fe(NH_3)_6]Cl_2$,[140] $[Fe(NH_3)_6]SO_4$[141,142] and the carbonylcomplexes $Fe(CO)_4X_2$, X = monovalent anion.[143,144] The *nitrosylcomplex* $[Fe(NO)(NH_3)_5]Cl_2$ was obtained from an ammonia solution of iron dichloride by reaction with nitrogen monoxide in hydrogen atmosphere.[145]

---

118 *A. Furuhashi, T. Takeuchi, O. Toshio*, Bull. Chem. Soc., Jap. *41* (9), 2049 (1968).
119 *A. Ouchi, M. Nakatani, Y. Takahashi*, Bull. Chem. Soc., Jap. *41* (9), 2044 (1968).
120 *V. V. Korshak, M. S. Mirkamilova, N. I. Bekasova*, Vysokomolekul Soedin., Ser. B *9* (10), 748 (1967).
121 *A. P. Terent'ev, I. G. Il'ina, E. G. Rukhadze*, Vysokomolekul Soedin., Ser. B *9* (10), 788 (1967).
122 *R. E. Breuil*, C. R. Acad. Sci., Paris *196*, 2009 (1933);
R. E. Breuil, C. R. Acad. Sci., Paris *199*, 298 (1934).
123 *F. Blau*, Ber. *21*, 1077 (1888).
124 *W. Biltz, G. F. Hüttig*, Z. Anorg. Allgem. Chemie *109*, 90 (1920).
125 *H. Grossmann, F. Hünseler*, Z. Anorg. Allgem. Chemie *46*, 392 (1905).
126 *F. Kutek*, Collect. Czech. Chem. Commun. *33* (6), 1930 (1968).
127 *N. A. Daugherty, J. H. Swisher*, Inorg. Chem. *7* (8), 1651 (1968).
128 *G. A. Barclay, C. M. Harris, J. V. Kingston*, Chem. Commun. (16), 965 (1968).
129 *P. P. Shukla*, Indian J. Chem. *5* (11), 583 (1967).
130 *A. V. Ablov, V. N. Zubarev*, Zh. Neorg. Khim, *13* (12), 3235 (1968).
131 *G. A. Barbieri*, Atti Accad. Naz. Lincei, Classe Sci. Fis., Mat. Natur., Rend. *22*, 868 (1913).
132 *Tn. Curtius, F. Schrader*, J. Prakt. Chem. *50*, 341 (1894).
133 *C. Przibilla*, Z. Anorg. Allgem. Chemie *15*, 438 (1897).
134 *H. Euler*, Ber. *37*, 1704 (1904).
135 *W. Biltz, E. Rahlfs*, Z. Anorg. Allgem. Chemie *166*, 371 (1927).
136 *A. Benrath*, J. Prakt. Chem. *72*, 220 (1905).
137 *J. O. Gibson, J. L. Driscoll*, J. Chem. Soc. 1443 (1929).
138 *G. Spacu*, Ann. Univ. Jassy *8*, 30 (1914–15).
139 *F. Hünseler*, Z. Anorg. Allgem. Chemie *46*, 370 (1905).
140 *W. Hieber, G. Bader*, Ber. *61*, 1717 (1928).
141 *F. Ephraim*, Z. Phys. Chem. *83*, 209 (1913).
142 *E. Weitz, H. Müller*, Ber. *58*, 363 (1925).
143 *W. Hieber*, Z. Elektrochem. *43*, 390 (1937).
144 *W. Hieber, H. Legally*, Z. Anorg. Allgem. Chemie *245*, 295 (1940).
145 *H. Müller*, Dissertation, Halle (1926).

Ligand displacement reactions are performed as follows.

A solid iron(II) complex and a gaseous ligand are interacted to prepare aminocomplexes $[Fe(NH_3)_6]X_2$, X=Br, I (from $Fe(CO)_4X_2$),[140] X=SCN (from $Fe(py)_4(SCN)_2$ py=pyridine),[146] and some carbonyl complexes such as $K_3[Fe(CN)_5CO]\cdot 3\frac{1}{2}H_2O$ obtained at 130–135° in a sealed tube according to reaction (22).[147]

$$K_4[Fe(CN)_6] + CO + 2H_2O \longrightarrow K_3[Fe(CN)_5CO] + NH_3 + HCOOK \quad (22)$$

The nitrosyl complex $[Fe(NO)(NH_3)_5](NO_3)_2$ has been prepared from $[Fe(NH_3)_6](NO_3)_6$ and NO[142] and FeL_nX_2 complexes (L=pyridine, n=6, X=I, from $Fe(CO)_4I_2$[148] and L=2,2′-dipyridyl, n=2, X=SCN, from $Fe(py)_4(SCN)_2$ in pyridine under nitrogen)[149] were obtained in similar fashion.

Anion displacements are also useful for the preparation of some species. Thus the *hexammino-iron(II) nitrite* $[Fe(NH_3)_6](NO_2)_2$ was obtained from the corresponding acetate.[142] The *tetracarbonyl iodide* is formed according to the reaction (23).[150]

$$Fe(CO)_4H_2 + I_2 \longrightarrow Fe(CO)_4I_2 + H_2 \quad (23)$$

Complex cyanides react with Lewis acids according to the scheme (24).[151]

$$Fe(L)_2(CN)_2 + 2BX_3 \longrightarrow Fe(L)_2(CNBX_3)_2,\ X = H, CH_3, F, Cl, Br \quad (24)$$

Iron(II) complexes can be prepared from low- and high-valent iron compounds as follows:

(a) From zerovalent iron compounds according to reactions (25),[152–154] (26)[155] and (27).[156]

$$Fe(CO)_5 + X_2 \longrightarrow Fe(CO)_4X_2 + CO \quad X = \text{halogen} \quad (25)$$

$$Fe(CO)_5 + RX \longrightarrow RFe(CO)_4X + CO \quad X = Cl;\ R = \text{trifluoromethylsulfonyl} \quad (26)$$

$$Fe_3(CO)_{12} + 3Br \longrightarrow Fe_3(CO)_9Br_6 + 3CO \quad (27)$$

The zerovalent iron complexes $Fe(CO)_5$ or $L_2Fe(CO)_3$ interact with germanium or tin tetrachloride to give complexes with Fe–Ge or Fe–Sn bonds of the type $(CO)_4Fe(ZX_3)X$, Z=Ge, Sn, X=Br, I, or $(CO)_4Fe(ZX_3)_2$, Z=Ge, Sn, X=Cl.[157]

(b) From monovalent iron compounds according to reaction (28).[158]

$$[Fe(CO)_4I]_2 + I_2 \longrightarrow 2Fe(CO)_2I_2 + 4CO \quad (28)$$

(c) From iron(III) complexes by ligand displacement reactions with simultaneous reduction. For example, the sodium pentacyanonitrosylferrate(III) was treated (in concentrated solution) with ammonia, hydrazine, or sodium hydroxide to yield the corresponding iron(II) complex $Na_3[Fe(CN)_5Z]\cdot xH_2O$, Z=$NH_3$, x=x,[159] Z=N_2H_4, x=0,[160] and Z=H_2O, x=7.[159] Similarly was obtained $Na_4[Fe(CN)_5AsC_2]\cdot 10H_2O$ (from $Na_2[Fe(CN)_5NO)]$ and As_2O_3).[161,162]

(d) From organoiron compounds. π-allyltricarbonyliron iodide when treated with dimethylsulfoxide in pentane at 20° in argon atmosphere affords pentakis(dimethylsulfoxide)iron(II) iodide $[(CH_3)_2SO]_5$–FeI_2.[163]

Some iron(II) complexes are used as clathrating agents. Iron dichloride forms *clathrates* of the type $Fe(NH_3)_2Ni(CN)_4\cdot 2Z$, Z=benzene or aniline[164] in ammoniacal solution and inert atmosphere and in the presence of potassium tetracyanonickelate, $K_2Ni(CN)_4$, and ammonium chloride.

146 *G. Spacu*, Ann. Univ. Jassy *8*, 180 (1915).
147 *J. A. Muller*, C. R. Acad. Sci., Paris *104*, 992 (1887).
148 *H. Hock, H. Stuhlmann*, Ber. *62*, 431 (1929).
149 *A. T. Cassey*, Australian J. Chem. *21*, (9), 2291 (1968).
150 *W. Hieber, H. Vetter*, Z. Anorg. Allgem. Chemie *212*, 145 (1933).
151 *D. F. Shriver, A. Luntz, J. J. Rupp*, Proc. Int. Conf. Coord. Chem. 8th Vienna 320 (1964).
152 *W. Hieber, G. Bader*, Ber. *61* B, 1717 (1928); *W. Hieber, G. Bader*, Chem. Ber. *89*, 193 (1956); *W. Hieber, G. Bader*, Chem. Ber. *89*, 610 (1956); *W. Hieber, G. Bader*, Chem. Ber. Z. Anorg. Allgem. Chemie *190*, 193 (1930).
153 *W. Hieber, A. Wirsching*, Z. Anorg. Allgem. Chemie *245*, 35 (1940).
154 *H. Stuhlmann*, Ber. *62*, 433 (1929).
155 *E. Lindner, H. Weber, G. Vitzthum*, J. Organometal. Chem. *13* (2), 431 (1968).
156 *W. Hieber*, Z. Anorg. Allgem. Chemie *201*, 329 (1931).
157 *R. Kummer, W. A. G. Graham*, Inorg. Chem. *7* (6), 1208 (1968).
158 *F. A. Cotton, B. F. G. Johnson*, Inorg. Chem. *6* (11), 2113 (1967).
159 *K. A. Hofmann*, Justus Liebigs Ann. Chem. *312*, 12 (1900).
160 *E. Biesalski, O. Hauser*, Z. Anorg. Allgem. Chemie *74*, 384 (1912).
161 *F. Hölzl, K. Rokitansky*, Monatsh. Chem. *56*, 90 (1930).
162 *K. A. Hofmann*, Z. Anorg. Allgem. Chemie *12*, 146 (1896).
163 *A. N. Nesmeyanov, I. I. Kritskaya, G. M. Babakhina*, Izv. Akad. Nauk SSSR, Ser. Khim. (1), 192 (1968).
164 *T. Nakano, T. Miyoshi, T. Iwamoto, Y. Sasaki*, Bull. Chem. Soc. Japan *40* (5), 1297 (1967).

Low coordinated complexes can be prepared by thermal decomposition of highly coordinated complexes. For example, the diaminocomplexes $Fe(NH_3)_2X_2$ have been prepared from the hexaminocomplexes $Fe(NH_3)_6X_2$, X = Br or I[165] and the hexaminochloride was also obtained from $FeCl_2 \cdot 10NH_3$.[135]
In some cases decomposition can be accomplished in solution. Thus tris(dipyridyl)iron(II) thiocyanate decomposes (in boiling chloroform containing 1% H_2O) to *bis(dipyridyl)iron(II) thiocyanate.*[149]

20.3.8 Organoiron Compounds

Organoiron compounds are air and water sensitive and must be handled in an inert atmosphere under anhydrous conditions. Pure organoiron compounds are very unstable; however, if the iron atom is also coordinated with donor molecules comparatively stable organoiron complexes are obtained. The classical method for the preparation of metalorganic compounds, *i.e.*, reaction with Grignard reagent, can be applied to cyclopentadienylcarbonyliron halides as illustrated in Eqn (29).

$$C_5H_5Fe(CO)_2X + RMgBr \longrightarrow C_5H_5Fe(CO)_2R + MgXBr \quad (29)$$

R = R'C:C (R' = butyl, phenyl) X = Cl, in tetrahydrofuran,[166] R = phenyl, X = Br.[167, 168]
Grignard reagents of some benzylamines or esters produce organoiron compounds of the type *a* where Z = N or O.[169]

Organolithium compounds react with iron halides according to Eqn (30).

$$FeX_2 + 2LiR \longrightarrow Fe(R)_2 + 2LiX \quad (30)$$

LiR = 2-lithiumtrichlorothiophene[170]

Organoiron halides obtained according to the scheme (31) in ether solution[171] are more stable than the diorganoiron derivatives.

$$FeI_2 + C_2H_5ZnI \longrightarrow C_5H_5FeI + ZnI_2 \quad (31)$$

Iron carbonyls react with organohalides according to reactions (32) and (33).

$$Fe(CO)_5 + RX \longrightarrow RFe(CO)_3X + 2CO \quad (32)$$

R = allyl or substituted allyl, X = Cl or Br, in nitrogen atmosphere at 20° in pentane.[172] RX = substituted cyclohexadienes,[173] R = C_2F_5, X = I,[174] R = heptamethylbenzenonium, $C_6(CH_3))_7$, X = $AlCl_4$, in sealed tube.[175]

$$Fe_2(CO)_9 + 2RX \longrightarrow 2RFe(CO)_3X + 3CO, \text{ R = 2-methallyl}^{176} \quad (33)$$

Iron carbonyls, carbonylnitrosyliron, and other carbonyl derivatives react with sodium hydroxide or amalgam to form carbonylferrates of the type $NaFe(CO)_4$, $Na[Fe(CO)_2C_5H_5]$, $Na[Fe(NO)(CO)_3]$ etc. (see Chapter 36). These compounds react with organohalides to afford the corresponding organoiron compound according to Eqn (34).

$$Na[Fe(CO)_2C_5H_5] + RX \longrightarrow RFe(CO)_2C_5H_5 + NaX \quad (34)$$

Some resultant organoiron compounds and their starting materials are listed in Table 4.
Some compounds have been prepared *in situ* with the carbonylate ion. Thus dicyclopentadienyltetracarbonyldiiron $[C_5H_5Fe(CO)_2]_2$ is reduced with sodium amalgam at −60° to room temperature in tetrahydrofuran and in the presence of *p*-fluorophenylenediiodide, $(p\text{-}FC_6H_4)_2I_2$, to give the complex $(p\text{-}FC_6H_4)Fe(CO)_2(C_5H_5)$ (ref. *f*, Table 4).
Hydrogentetracarbonylferrate, $HFe(CO)_4$, reacts with butadiene according to Eqn (35).[177]

$$HFe(CO)_4 + CH_2{=}CHCH{=}CH_2 \longrightarrow CH_2{=}CHCH(CH_3)Fe(CO)_4 \quad (35)$$

[165] *W. Biltz, G. F. Hüttig,* Z. Anorg. Allgem. Chemie *109*, 93 (1920). Der Hexammin-Komplex erhielt man aus Eisen(II)-chlorid-Hexamin[170].
[166] *M. L. H. Green, T. Mole,* J. Organometal. Chem. *12* (2), 404 (1968).
[167] *B. F. Hallam, P. L. Pauson,* J. Chem. Soc. 3030 (1956).
[168] *T. S. Piper, G. Wilkinson,* Naturwissenschaften *43*, 15 (1956);
T. S. Piper, G. Wilkinson, J. Inorg. Nucl. Chem. *3*, 104 (1956).
[169] *F. W. Kuepper,* J. Organometal. Chem. *13* (1), 219 (1968).
[170] *M. D. Rausch, T. R. Criswell, A. K. Ignatovicz,* J. Organometal. Chem. *13* (2), 419 (1968).
[171] *A. Job, R. Reich,* C. R. Acad. Sci., Paris *174*, 1358 (1922).
[172] US P. 3 338 936 (1967). Hercules Inc., Erf.: *R. F. Heck;* CA *68*, 49788f (1968).
[173] *A. J. Burch, P. E. Cross, J. Lewis, D. A. White, S. B. Wild,* J. Chem. Soc. *A 2*, 332 (1968).
[174] *P. Pascal,* Nouveau Traité Chim. Minér. *17* (1), 920 (1967).
[175] *V. A. Koptyug, R. N. Berezina, V. G. Shubin,* Tetrahedron Lett. *6*, 673 (1968).
[176] *K. Ehrlich, G. F. Emerson,* Chem. Commun. *2*, 59 (1969).
[177] *H. Reihlen, A. Gruhl, G. v. Heßling, O. Pfrengle,* Justus Liebigs Ann. Chem. *482*, 161 (1930).

Table 4. Preparation of Organoiron Compounds from Iron Carbonylates and Related Products

Starting material		Organoiron derivative	Ref.
$Na[Fe(CO)_2C_5H_5]$	$R-Fe(CO)_2C_5H_5$		
	$R{:}H_2C{=}CH-CH_2-$	allyl-cyclopentadienyl-dicarbonyl-iron	*c*
	$HCC{\equiv}-CH_2-$	cyclopentadienyl-dicarbonyl-propargyl-iron	*b*
	$F_5C_6-CH_2-$	cyclopentadienyl-dicarbonyl-(2,3,4,5,6-pentafluor-benzyl)-eisen	*a*
	$H_2C{=}CH-$	cyclopentadienyl-dicarbonyl-vinyl-iron	*c*
	$H_2C{=}C{=}CH-$	allenyl-cyclopentadienyl-dicarbonyl-iron	*d*
	C_6H_5	cyclopentadicarbonyl-phenyl-iron	*a*
	$4-F-C_6H_4$	cyclopentadienyl-dicarbonyl-(4-fluorphenyl)-iron	*f*
	$4-CH_3-C_6H_5$	cyclopentadienyl-dicarbonyl-(4-methyl-phenyl)-iron	*c*
	$4-CH_3O-C_6H_5$	cyclopentadienyl-dicarbonyl-(4-methoxy-phenyl)-iron	*c*
	$Si(CH_3)_2Si(CH_3)_3$	cyclopentadienyl-dicarbonyl-(pentamenthyl-disilyl)-iron	*e*
$Na_2[Fe(CO_3)_4]$	$(C_nF_{2n+1}CO)-Fe(CO)_4$		*g*
$Fe(CO)_5(H_2C{=}CH-CN)$	$K[(NC-CH_2-CH_2)Fe(CO)_3]$		*h*
$Na[Fe(NO)(CO)_3]$	$Fe(NO)(CO)_2(CO\ alkyl)$		*i*
$Na_2[C_2H_5Fe(NO)_2]$	$[H_3C-Fe(NO)(C_2H_5)]_2$		*j*

[a] *M. I. Bruce*, J. Organometal Chem. *10* (3), 495 (1967).
[b] *J. L. Roustan, P. Cadiot*, C. R. Acad. Sci., Paris, Ser. C *268* (8), 734 (1969).
[c] *M. I. Bruce, C. H. Davis*, J. Chem. Soc. *A 7*, 1077 (1969).
[d] *M. D. Johnson, C. Mayle*, Chem. Commun. *5*, 192 (1969).
[e] *R. B. King, K. H. Pannell*, Z. Naturforsch. B *24* (2), 262 (1969).
[f] *A. N. Nesmeyanov, Yu. A. Chapovskii, I. V. Polovyanyuk, L. G. Makarova*, Izv. Akad. Nauk SSSR, Ser. Khim. *7*, 1628 (1968).
[g] *P. Pascal*, Nouveau Traite Chim. Miner. *17* (1), 920 (1967).
[h] Fr. P. 1 453 988 (1966), Soc. Usines Chim. Phone-Poulenc, Erf.: *P. Chabardes, P. Gandilhon, Ch. Grard, M. Thiers*; CA *67*, 32784h (1967).
[i] *F. M. Chaudhari, Y. R. Knox, P. L. Pauson*, J. Chem. Soc. C *21*, 2255 (1967).
[j] *H. Brunner, H. Wachsmann*, J. Organometal. Chem. *15* (2), 409 (1968).

The halogen atom of complex cyclopentadienyliron halides can be substituted by organic radicals. For example, cyclopentadienyldicarbonyliron iodide, $C_5H_5Fe(CO)_2I$, reacts with diphenylmercury, Ph_2Hg, in dry benzene and under u.v. irradiation to produce *phenylcyclopentadienyldicarbonyliron*, $C_6H_5Fe(CO)_2C_5H_5$.[178]

The cyclopentadienyl radicals of bis(cyclopentadienyl) iron can be replaced by heterocyclic radicals, if bis(cyclopentadienyl)iron is heated to 250° in a sealed tube with a 10-fold excess of pyrazole, imidazole, or 1,2,4-triazole; *bisazolyliron derivatives*, FeL_2, result.[179]

Dimeric iron(I) compounds react under cleavage of the Fe–Fe bond and formation of organoiron compounds. Thus when dimeric cyclopentadienyldicarbonyliron, $[Fe(C_5H_5)(CO)_2]_2$, is heated with trifluorosilane SiF_3H the compound $(F_3Si)Fe(CO)_2(C_5H_5)$[180] is formed.

Organoirion carbonyl complexes undergo displacement reactions according to scheme (36).

$$RFe(CO)_2(C_5H_5) + Z \longrightarrow RFe(CO)Z(C_5H_5) + CO \quad (36)$$

Z = phosphines

The reaction proceeds in benzene at 25–80° under u.v. irradiation.[181] Refluxing in benzene with phosphines leads also to displacement of one molecule of carbon monoxide.[182]

[178] *A. N. Nesmeyanov, I. V. Polovyanyuk*, Zh. Obshch. Khim. *37* (9), 2015 (1967).
[179] *F. Seel, V. Sperber*, Angew. Chem., Intern. Ed. Engl. *7* (1), 70 (1968).
[180] *R. R. Schrieke, B. O. West*, Inorg. Nucl. Chem. Let *5* (3), 141 (1969).
[181] *A. N. Nesmeyanov, Yu. A. Chapovskii, I. V. Polovyanyuk, L. G. Makarova*, Izv. Akad. Nauk SSSR, Ser. Khim. (7), 1628 (1968).
[182] *R. J. Haines, A. L. Du Preez*, Inorg. Chem. *8* (7), 1459 (1969).

Irradiation of some zerovalent iron complexes (containing mobile hydrogen) causes hydrogen transfer from the ligand to the iron atom according to scheme (37).

$$Fe(HL)_2Z \xrightarrow[u.v.]{} HFeL(HL)+Z \quad (37)$$

HL=ethylenebis(diphenylphosphine), Z= C_2H_4.[183]

σ-Alkyneiron compounds $(C_5H_5)(CO)_2FeC{\equiv}CCH_3$ have been prepared by acid-catalyzed isomerization of σ-allenyliron compounds $(C_5H_5)(CO)_2FeCH{=}C{=}CH_2$ (ref. *d*, Table 4).

20.4 Iron(III) Compounds

20.4.1 Hydrides

An iron(III) hydride cannot be prepared in pure state; reaction of anhydrous iron(III) chloride in ether with phenylmagnesium bromide and hydrogen affords a black oil[184,185] according to Eqn (38).

$$FeCl_3+3C_6H_5MgBr+3H_2 \longrightarrow FeH_3+3C_6H_6+3MgBrCl \quad (38)$$

The compound contains magnesium halides and organic moieties,[186,187] (see Section 20.3.1).

20.4.2 Halides and Halocomplexes

The *fluoride*, FeF_3 is obtained from $FeCl_3$ and NH_4F or from Fe_2O_3 and HF[188] or by reaction of Fe or $FeCl_3$ with fluorine.[189]

The *chloride* is a commercial product, the bromide is unstable, the iodide has not been isolated.

Mixed bromide chloride, $FeBrCl_2$, has been prepared by heating of a mixture of iron dichloride and bromine in a sealed tube for two days.[190]

The *iodide chloride*, $FeICl_2$, is formed on reaction of an alcoholic solution of iodine with iron dichloride.[191]

Only the fluorides form stable complexes. *Fluoroferrates(III)* have been prepared by the following procedures.

(a) By dissolving the fluorides in aqueous hydrofluoric acid and subsequent evaporation and crystallization: $BaFeF_5$, $Ba_3[FeF_6]_2$,[192] Na_3FeF_6,[193] $K_2[FeF_5H_2O]$,[194] Cs_3FeF_6,[195] and $Ag_2[FeF_5H_2O]\cdot 2H_2O$.[194]

(b) By dissolving the hydroxides, oxides, or carbonates in hydrofluoric acid followed by evaporation and crystallization: $Li_3[FeF_6]$ (from $FeCl_3$ and Li_2CO_3),[196] $KFeF_4$ and K_3FeF_6 (from FeF_3 and K_2CO_3),[194] $(NH_4)_3FeF_6$ (from $Fe(OH)_3$ and NH_4F),[194] NH_4FeF_4 (from $Fe(OH)_3$ and NH_3),[194] $Sr[FeF_5H_2O]$ (from $FeCl_3$ and $SrCO_3$),[197] $Ba[FeF_5H_2O]$ (from $Fe(OH)_3$ and $BaCl_2$),[198] $Zn[FeF_5H_2O]$ (from $Fe(OH)_3$ and $ZnCO_3$).[199]

The *chlorocomplexes* are stable in solid state. They crystallize from a solution containing the appropriate chlorides and concentrated hydrochloric acid, for example $Be[FeCl_5H_2O]$ and $Mg[FeCl_5H_2O]$.[200,201]

20.4.3 Iron(III) Salts and Alcoholates

The preparation of iron(III) salts follows the conventional methods, *i.e.* dissolving of powdered iron, iron(III) oxide or hydroxide in the appropriate acid: *perchlorate* $Fe(ClO_4)_3$ (from Fe and $HClO_4$),[202] *orthoperiodate* $FeH_2IO_6\cdot$

[183] *G. Hata, H. Kondo, A. Miyake*, J. Amer. Chem. Soc. *90* (9), 2278 (1968).
[184] *R. C. Ray, R. B. Sahai*, J. Indian Chem. Soc. *23*, 67 (1946).
[185] *B. Sarry*, Naturwissenschaften *41*, 115 (1954);
B. Sarry, Z. Anorg. Allgem. Chemie *280*, 65 (1955);
B. Sarry, Z. Anorg. Allgem. Chemie *288*, 41 (1956);
B. Sarry, Z. Anorg. Allgem. Chemie *288*, 48 (1956).
[186] *J. Aubry, G. Monnier*, Bull. Soc. Chim., France *4*, 482 (1955).
[187] *K. Shinoki, Y. Takegami, T. Ueno, T. Sakata*, Kogyo Kagaku Zasshi *67*, 316 (1964).
[188] *C. Poulenc*, C. R. Acad. Sci., Paris *115*, 943 (1892).
[189] *O. Ruff, E. Ascher*, Z. Angew. Chem. *41*, 739 (1928).
[190] *C. Lenormand*, C. R. Acad. Sci., Paris *116*, 820 (1893).
[191] *K. Seubert, A. Dorrer*, Z. Anorg. Allgem. Chemie *5*, 411 (1894).
[192] *J. Ravez, J. Viollet, R. de Pape, P. Hagenmuller*, Bull. Soc. Chim., France (4), 1325 (1967).
[193] *R. Wagner*, Ber. *19*, 897 (1886).
[194] *R. Weinland, I. Lang, H. Fikentscher*, Z. Anorg. Allgem. Chemie *150* 47 (1925).
[195] *W. Minder*, Z. Kryst. *96*, 15 (1937).
[196] *A. H. Nielsen*, Z. Anorg. Allgem. Chemie *224*, 84 (1935).
[197] *A. H. Nielsen*, Z. Anorg. Allgem. Chemie *226*, 222 (1936).
[198] *A. H. Nielsen*, Z. Anorg. Allgem. Chemie *227*, 423 (1936).
[199] *R. F. Weinland, O. Köppen*, Z. Anorg. Allgem. Chemie *22*, 226 (1900).
[200] *G. Neumann*, Justus Liebigs Ann. Chem. 244, 329 (1888).
[201] *H. Remy, H. J. Rothe*, J. Prakt. Chem. *114*, 137 (1926).
[202] *A. M. Leko, V. Canic*, Bull. Soc. Chim., Belgrade *14*, 249 (1949).

$4H_2O$ (from $FeCl_3$, I_2, and HNO_3),[203] *sulfaminate*, $Fe(HN_2SO_3)$ (from $Fe(OH)_3$) and *sulfamic* acid),[204] *formiate* $[Fe_3(HCOO)_6(OH)_2]$-$HCOO \cdot 4H_2O$ (from $Fe(OH)_3$ and HCOOH).[205] Ion exchange reactions [including precipitation of insoluble iron(III) salts (*thiocarbonate*, $Fe_2(CS_3)_3$,[206] *phthalate, benzoate, cinnamate etc.*)][207] or precipitation of alkali metal salts nitrite, $Fe(NO_2)_3$ from $FeCl_3$ and KNO_2,[208] *tris(trimethoxysiloxy)iron(III)*, $Fe[OSi(CH_3)_3]_3$ from $FeCl_3$ and $NaOSi(CH_3)_3$ in ether solution)[209] should also be mentioned.

Some mixed salts include $FeSO_4Cl \cdot 6H_2O$ (from $FeSO_4$ and Cl_2 in water).[210]

Iron(III) oxide reacts with metal oxides to form ferrites. They are prepared by heating a mixture of the appropriate oxides, or a mixture of unstable compounds (hydroxides, carbonates, organic salts) which decompose to the oxides during the calcination process. Several *ferrites* have been prepared hydrothermally, *i.e.*, by coprecipitation of their hydroxides and boiling of the precipitate. The preparation of some ferrites is presented in Table 5.

The ferrites of bivalent metals, MFe_2O_4, are of the spinelle type, with the exception of the lead and alkali earth ferrites, which are hexagonal.

Alkali metal *selenoferrates(III)* $MFeSe_2$, M = K, Rb, have been prepared from alkali carbonate, selenium and powdered iron at 800°;[211] the thioferrates $MFeS_2$, M = K, Rb or Cs, were prepared from alkali metal sulfide, sulfur and powdered iron.[212]

Alcoholates can be obtained from iron(III) chloride and sodium alcoholate in alcohol–ether solution according to reaction (39).

$$FeCl_3 + 3NaOR \longrightarrow Fe(OR)_3 + 3NaCl \qquad (39)$$

$R = C_2H_5$[213]

[203] *P. C. Raychoudhury*, J. Indian Chem. Soc. *18*, 576 (1941).

[204] *E. Divers, T. Haga*, J. Chem. Soc. *69*, 1634 (1896).

[205] *G. A. Barbieri*, Atti Accad. Naz. Lincei, Classe Sci. Fis., Mat. Natur., Rend. *25*, 725 (1916).

[206] *J. J. Berzelius*, Pogg. Ann. *6*, 455 (1826); *Kang*, Kunstseide *7*, 278 (1925).

[207] *R. F. Weinland, F. Paschen*, Z. Anorg. Allgem. Chemie *92*, 81 (1915).

[208] *L. Pesci*, Gazz. Chim. Ital. *18*, 183 (1888).

[209] US P. 3 373 178 (1968), Wasag-Chem., Erf.: *M. Schmidt, H. Schmidbaur;* CA *68*, 1065 12n (1968).

[210] *B. Walter*, Chem. Ztg. *45*, 842 (1921).

[211] *W. Bronger*, Naturwissenschaften *53* (20), 525 (1966).

[212] *W. Bronger*, Z. Anorg. Allgem. Chemie *358* (5–6), 225 (1968).

[213] *A. Hantzsch. C. H. Desch*, Justus Liebigs Ann. Chem. *323*, 12 (1902).

Table 5. Preparation of Ferrites

Ferrite	Starting materials	Conditions	Ref.
$LiFeO_2$	LiOH, Fe_2O_3	elec. furnace, 700°	*a*
$CuFeO_2$	$NaFeO_2$, Cu^+	ion exchange	*b*
$NiFe_2O_4$	NiO, Fe_2O_3	sealed tube	*c*
MFe_2O_4	MC_2O_4 (M = Mg, Mn, Ni, Co, Zn), $Fe_2(C_2O_4)_3$	1100°–1200°	*d*
$NiFe_2O_4$	$Ni(OH)_2$, $Fe(OH)_3$	copptn., heating	*e*
$CuFe_2O_4$	$Cu(OH)_2$, $Fe(OH)_3$	copptn., 800° (*f*); 400°, 5–20 atm (*g*)	*f, g*
$BiFeO_3$	Bi_2O_3, Fe_2O_3	800°, 750°	*h, i*
$ZFeO_3$	Z_2O_3 (Z = In or Tl), Fe_2O_3	heating, high press.	*j*
$Fe_4(TiO_4)_3$	Fe_2O_3, TiO_2	1000°, stoichiom. amts.	*k*
(Mn, Cu)Fe_2O_4	$MnCO_3$, CuO, Fe_2O_3	1000°–1265°	*l*

[a] *S. A. Kutolin, A. I. Vulikh*, Metody Poluch. Kim. Reaktivov Prep. *16*, 59 (1967).

[b] *W. Gessner*, Wiss. Z. Hochsch. Architek. Bauw., Weimar *15* (2), 185 (1968).

[c] *P. Kleinert, D. Schmidt, E. Glauche*, Z. Chem. *7* (1), 32 (1967).

[d] *D. G. Wickham*, Inorg. Synth. *9*, 152 (1967).

[e] *Sh. Saito, T. Takei*, Funtai Oyobi Funmatsuyakin *13* (2), 60 (1966).

[f] *A. Krause, Q. Binkowna*, Monatsh. Chem. *96* (4), 1183 (1965).

[g] *T. P. Adamovich, V. V. Sviridov, A. D. Lobanok, Fiz. Svoistva Ferritov*, Inst. Fiz. Tverd. Tela Poluprov. Akad. Nauk Beloruss, SSR, 95 (1967).

[h] *E. I. Speranskaya, V. M. Skorilov*, Izv. Akad. Nauk SSSR, Neorg. Mater. *3* (2), 341 (1967).

[i] *G. D. Achenbach, W. J. James, G. Gerson*, J. Amer. Chem. Soc. *50* (8), 437 (1967).

[j] *R. D. Shannon*, Inorg. Chem. *6* (8), 1474 (1967).

[k] *T. König, O. v. d. Pfordten*, Ber. *22*, 1493 (1889).

[l] *E. Pollert, A. Novak*, Silikaty *11* (3), 279 (1967).

The *thioalcoholates* are prepared in similar fashion.

Halomethylates of the type $FeCl(OCH_3)_2$, $Fe_4X_6(OCH_3)_6 \cdot 4CH_3OH$, and $Fe_4X_3(OCH_3)_9$, X = Cl or Br, are also known.[214] Isopropyoxy groups of iron(III) isopropylate, $Fe(OPr)_3$, can be replaced by trimethylsiloxy groups by refluxing with trimethylacetoxysilane, R_3SiOAc, R = CH_3. The products are of the type $R_3SiOFe(OPr)_2$, $(R_3SiO)_2FeOPr$, and $(R_3SiO)_3Fe$.[215]

20.4.4 Iron(III) Chelates

Trivalent iron chelates are colored and soluble compounds which exist in solution as cations $[FeL]^{2\oplus}$, $[FeL_2]^{\oplus}$, neutral molecules FeL_3, or anions $[FeL_6]^{3\ominus}$. Several compounds can be obtained in solid state.

The methods for the preparation of iron(III) chelates are similar to the methods for the preparation of salts. Strongly acidic chelating agents dissolve powdered iron to the corresponding chelate. The chelate with vitamin C was prepared in this way and its water solution was diluted with acetone to precipitate $C_6H_7O_6FeOH \cdot 2H_2O$.[216]

Iron carbonyls react with the chelating agent under decomposition and the resultant iron powder dissolves to give the corresponding chelate. For example, iron carbonyl and acetylacetone react under u.v. irradiation to form iron(III) acetylacetonate.[217]

Freshly precipitated iron(III) hydroxide dissolves in a solution of the chelating agent; chelates with acetylacetone,[218] benzenesulfonic acid[219] etc. have been prepared by this procedure.

Insoluble chelates can be precipitated from water or in organic solutions. The chelates are of the type FeL_3, HL = α-nitroso-β-naphtole,[220] *o*-nitrophenol,[221] 8-hydroxyquinoline and its derivatives,[222] quinaldinic acid,[223] *N*-acyl-*N*-phenylhydroxylamines.[224]

Alkaline media favor the formation of chelates. Good yields are obtained if the precipitation is carried out with an alkali metal salt of the chelating agent. Thus salicilic acid, H_2L (neutralized with sodium carbonate) gives with iron(III) salts chelates of the type $NaFeL_2$ or Na_3FeL_3[225, 226] and chelates with alizarine are of the same type.[227] Eugenol and vanillal (HL) form chelates of the type $Na[FeL_4] \cdot \frac{1}{2}H_2O$,[228] hydroxamic acid—$FeL_2Cl$,[229] and cinnamohydroxamic acid—$FeL_2OH \cdot H_2O$.[230]

The alkalization can be achieved with organic bases such as triethylamine and chelate formation proceeds according to scheme (40).

$$FeCl_3 + 3HL + 3Et_3N \longrightarrow FeL_3 + 3[Et_3NH]Cl \quad (40)$$

HL = salicylideneimines[231]

Soluble chelates can be isolated as solids by extraction and evaporation. For example, *iron(III) sulfosalicylate* was obtained by extraction with ether.[226] Many anionic iron(III) chelates $[FeL_2]^{\ominus}$, crystallize after treatment with tetraalkylammonium halide, e.g., $(R_4N)[FeL_2]$, R = butyl, H_2L = tetrafluorobenzene-*o*-dithiol.[232]

Some Schiff' bases (*N*-(2-hydroxyphenyl)salicylaldimine) form chelates of the type FeLX, X = Cl or Br.[233]

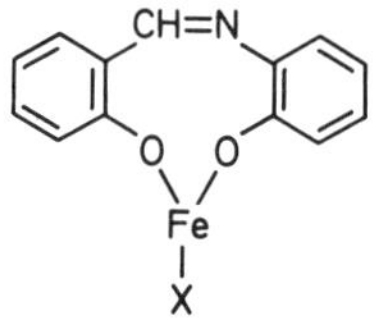

CH=N
O O
Fe
X

20.4.5 Iron(III) Complexes

Trivalent iron forms numerous complexes with

[214] *G. A. Kakos, G. Winter*, Australian J. Chem. *22* (1), 97 (1969).
[215] *P. P. Sharma, R. C. Mehrotra*, Indian J. Chem. *5* (9), 456 (1967).
[216] *W. Diemair, R. Zacharias*, Z. Anal. Chem. *130*, 333 (1950).
[217] *H. Reihlen, A. Gruhl, G. v. Hessling*, Justus Liebigs Ann. Chem. *472*, 275 (1929).
[218] *G. Urbain, A. Debierne*, C. R. Acad. Sci., Paris *129*, 302 (1899).
[219] *J. V. Dubsky*, J. Prakt. Chem. *90*, 81 (1914).
[220] *M. Ilinski, G. von Knorre*, Ber. *18*, 699 (1885).
[221] *O. Baudisch. N. Karzeff*, Ber. *45*, 1164 (1912).
[222] *S. Ishimaru*, J. Chem. Soc., Japan *55*, 201 (1934).
[223] *P. R. Ray, M. K. Bose*, Z. Anal. Chem. *95*, 400 (1933).
[224] *N. N. Ghosh, G. Siddhanta*, J. Indian Chem. Soc. *45* (11), 1049 (1968).
[225] *F. Zetzsche, G. Viel, G. Lilljiqvist, A. Loosli*, Justus Liebigs Ann. Chem. *435*, 233 (1924).
[226] *R. F. Weinland, A. Herz*, Justus Liebigs Ann. Chem. *400*, 223 (1913).
[227] *R. F. Weinland, K. Binder*, Ber. *45*, 1113 (1912); *R. F. Weinland, K. Binder*, Ber. *46*, 874 (1913).
[228] *R. F. Weinland, H. Neff*, Arch. Phar. (Weinheim, Ger.) *252*, 600 (1914).
[229] Swiss P. 440 314 (1967), Seigfried A.G., Erf.: *E. Bayer;* CA *68*, 80060m (1968).
[230] *V. Springer, I. Benedikovic*, Chem. Zvesti *22* (10), 797 (1968).
[231] *K. S. Murray, A. v. d. Bergen, M. J. O'Connor, N. Rehak, B. O. West*, Inorg. Nucl. Chem. Lett. *4* (2), 87 (1968).
[232] *A. Callaghan, A. J. Layton, R. S. Nyholm*, J. Chem. Soc. D. *8*, 399 (1969).
[233] *K. S. Murray, B. O. West*, Inorg. Nucl. Chem. Lett. *4* (8), 439 (1968).

organic and inorganic molecules; most of them are stable in solution only. The preparation of iron(III) complexes follows the conventional methods, *i.e.*, treating a solution of iron(III) salt with a solution of the ligand or *vice versa.* If the complex is less soluble it precipitates immediately; in many cases, however, the solution must be evaporated until the crystallization begins. This simple method has been used for the preparation of complexes of the type $Fe(L)_nX_3$ (Table 6).

Complexes or solvates can be obtained by recrystallization of iron(III) salts in donor solvents. Thus iron(III) chloride crystallizes with one molecule ether,[234] or two molecules ethanol.[235] Aliphatic alcohols form adducts of the type $FeCl_3 \cdot 2ROH$ (C_1–C_4) or $FeCl_3 \cdot ROH$ (C_5–C_7) (in benzene suspension).[236] Liquid ammonia reacts with iron(III) chloride at $-78°$ to form hexamminoiron(III) chloride, $[Fe(NH_3)_6]Cl_3$.[237, 238]

Some complexes have been prepared by treating a solid iron(III) salt with a gaseous ligand. This procedure was used for the preparation of hexamminoiron(III) bromide $[Fe(NH_3)_6]Br_3$,[239] tetramminodiaquosulfate $[Fe(NH_3)_4(H_2O)_2]_2(SO_4)_3$ (from $Fe_2(SO_4)_3 \cdot 9H_2O$ and NH_3),[240] the nitrosylchloride complexes $FeCl_3NOCl$ or $NO[FeCl_4]$ (from $FeCl_3$ and NOCl)[241] and the nitrogen dioxide complex $[Fe(NO_2)_2]Cl_3$ (from $FeCl_3$ and NO_2).[242]

Some iron(III) complexes can be prepared by oxidation of the corresponding iron(II) complex: $Na_2[Fe(CN)_5NH_3]$ from $Na_3[Fe(CN)_5NH_3]$ by oxidation with nitric acid or alkali hypobromite,[243] $Na_2[Fe(CN)_5H_2O]$ from $Na_3[Fe(CN)_5H_2O]$ by oxidation with bromine, nitric acid, or potassium permanganate.[243]

Soluble complexes can be obtained in solid state by extraction and evaporation, *e.g.*, complexes with pyridine, phosphine oxides, nitrogen oxides and sulfoxides.[244]

Complexes containing unsaturated hydrocarbons can be polymerized and are thus converted into the polymer complexes. Thus vinylphosphine oxide complexes polymerize at 150°.[245]

Table 6. Preparation of Iron Complexes $Fe(L)_nX_3$

L	n	X	Ref.
aniline	1	Cl	a
benzidine	3	Cl, $2H_2O$	b
pyridine	3, 4	Cl, SCN	c, d
dimethylselenoxide	6	ClO_4	e
7—oxo—7H— < benzo-benzanthrone	1	Cl, Br	f
2,3-dimethyl-1-phenyl-2,5-dihydro-pyrazol-5(on)-a antipyrine	3, 6	Cl, SCN, ClO_4	g, h, i
furfurylaldehyde	1	$Cl \cdot CH_3OH$	j
N,N-dialkyldithiocarbamate	2	halides	k
1,2-bis(methylsulfonyl)ethane	1	Cl in acetone-1,2-dichlorethane	l

[a] *J. V. Dubsky, E. Wagenhofer,* Z. Anorg. Allgem. Chem. *130,* 112 (1936).
[b] *G. Spacu,* Bull. Soc. Stiinte Cluj *2,* 195 (1924–25).
[c] *R. F. Weinland, A. Kissling,* Z. Anorg. Allgem. Chem. *120,* 218 (1922).
[d] *A. Hantzsch,* Z. Anorg. Allgem. Chem. *166,* 239 (1927).
[e] *R. Paetzold, G. Bochmann,* Z. Chem. *8* (8), 308 (1968).
[f] *R. Ch. Paul, R. Parkash, S. S. Sandhu,* Z. Anorg. Allgem. Chem. 352 (5–6), 322 (1967).
[g] *R. Weinland, O. Schmidt,* Arch. Pharm. *261,* 10 (1923).
[h] *J. V. Dubsky, E. Krametz, J. Trtilek,* Collect. Czechosl. Chem. Commun. *7,* 311 (1935).
[i] *A. Ravi, J. Gopalakrishnan, C. C. Patel,* Indian J. Chem. *5* (8), 356 (1967).
[j] *W. C. Gangloff, W. E. Henderson,* J. Amer. Chem. Soc. *39,* 1425 (1917).
[k] *H. H. Wickman, A. M. Trozzolo,* Inorg. Chem. *7* (1), 63 (1968).
[l] *J. G. H. Du Preez, W. J. A. Steyn, A. J. Basson,* J. S. Afr. Chem. Inst. *21* (1), 8 (1968).

20.5 Iron(IV) Compounds

Derivatives of tetravalent iron are prepared by oxidation of iron(III) compounds at high temperature in alkaline medium. For example,

[234] *A. Forster, C. Cooper, G. Yarrow,* J. Chem. Soc. 810 (1917).
[235] *E. Lloyd, C. B. Brown, D. G. R. Bonnell, W. J. Jones,* J. Chem. Soc. 664 (1928).
[236] *M. I. Usanovich, E. Kh. Ablanova,* Izv. Akad. Nauk Kaz. SSR, Ser. Khim. *18* (1), 26 (1968).
[237] *W. Hüttig,* Z. Anorg. Allgem. Chemie *114,* 167 (1920).
[238] *W. Biltz, E. Birk,* Z. Anorg. Allgem. Chemie *134,* 121 (1924).
[239] *F. Ephraim, S. Millimann,* Ber. *50,* 529 (1917).
[240] *G. Spacu,* Ann. Univ. Jassy *8,* 180 (1915).
[241] *W. J. van Heteren,* Z. Anorg. Allgem. Chemie *22,* 278 (1900).
[242] *V. Thomas,* Bull. Soc. Chim., France *15,* 1090 (1896).
[243] *K. A. Hofmann,* Justus Liebigs Ann. Chem. *312,* 12 (1900).
[244] *G. Stephan, H. Specker,* Naturwissenschaften *55* (9), 443 (1968).
[245] US P. 3 422 079 (1969), Union Carbide Corp., Erf.: *F. J. Welch, H. J. Paxton jr.;* CA *70,* 58396 (1969).

barium or strontium hexahydroxyferrate(III) $M_3[Fe(OH)_6]_2$, were converted into the corresponding ferrate(IV), $MFeO_3$, by heating to 300° (Ba)–500° (Sr)[246] according to reaction (41).

$$Ba_3[Fe(OH)_6]_2 + \frac{1}{2}O_2 \longrightarrow 2BaFeO_3 + Ba(OH)_2 + 5H_2O \quad (41)$$

Pure *orthoferrate(IV)*, Ba_2FeO_4, has been prepared at 800°–900° in the presence of an excess of barium hydroxide[246] according to the following scheme (42).

$$Ba_3[Fe(OH)_6]_2 + Ba(OH)_2 + \frac{1}{2}O_2 \longrightarrow 2Ba_2FeO_4 + 7H_2O \quad (42)$$

The reaction proceeds quantitatively with the corresponding strontium compounds.[247]
Another way for the preparation of iron(IV) compounds involves the thermal decomposition of ferrates(VI). Thus barium *metaferrate(IV)* has been prepared from barium ferrate(VI) by heating to 90°–100° in 0·5N KOH.[248,249] Potassium and lithium metaferrates(IV), M_2FeO_3, M = K, Li,[250,251] were obtained in similar fashion.
Silver metaferrate(IV), Ag_2FeO_3, is best prepared from potassium ferrate(VI) and silver nitrate; the resultant silver ferrate(VI) decomposes immediately according to scheme (43).

$$2Ag_2FeO_4 \longrightarrow 2Ag_2FeO_3 + O_2 \quad (43)$$

20.6 Iron(V) Compounds

Iron(V) compounds are very unstable. They exist only in solid state. The *alkali metal ferrates(V)* M_3FeO_4, M = K, Na, have been obtained from stoichiometric amounts of iron(III) oxide and the appropriate alkali metal hydroxide by heating to 450° in an oxygen atmosphere.[252,253] Potassium peroxide, K_2O_4, oxidizes iron(II) oxide at 450°[254] [reaction (44)].

$$2FeO + 3K_2O_4 \longrightarrow 2K_3FeO_4 + 3O_2 \quad (44)$$

20.7 Iron(VI) Compounds

Sodium and potassium *ferrate(VI)*, M_2FeO_4, M = Na, K, have been prepared by electrolysis of 40% NaOH (KOH) utilizing an iron anode and platinum cathode.[255] The oxidation can also be carried out with alkali metal hypochlorite. The salt M_2FeO_4 crystallizes on standing. A mixture of iron(III) oxide and potassium or sodium hydroxide affords M_2FeO_4 on heating to 450° in an oxygen atmosphere.[256] Lithium, rubidium and cesium ferrate(VI) are prepared by double decomposition between alkali earth metal ferrate(VI) and alkali metal sulfate, carbonate or oxalate.[257]
The ferrates(VI), $MFeO_4$, M = bivalent metal, can be prepared in similar fashion. Barium ferrate, $BaFeO_4$, was obtained by heating of a mixture of barium peroxide, BaO_2, and iron(III) oxide.[258]
The ferrates of many bivalent metals, $MFeO_4$, M = Ca, Ba, Sr, Mg, Mn, Co, Ni, Zn, Pb, Cu, and silver ferrate(VI), Ag_2FeO_4, are less soluble than potassium ferrate(VI). They can be precipitated from concentrated solution[259,260] according to the following scheme (45).

$$M(NO_3)_2 + K_2FeO_4 \longrightarrow MFeO_4 + 2KNO_3 \quad (45)$$

Another convenient method for the preparation of iron(VI) compounds resides in treatment of a suspension of freshly precipitated iron(III) hydroxide with chlorine, bromine, sodium hypochlorite (hypobromite), or hydrogen peroxide in the presence of EDTA.[261]
Stabilized ferrates(VI) have been prepared by

---

[246] *R. Scholder*, Angew. Chem. *65*, 240 (1953); *R. Scholder*, Angew. Chem. *66*, 461 (1954).
[247] *R. Scholder, H. v. Bunsen, W. Zeiss*, Z. Anorg. Allgem. Chemie *283*, 330 (1956).
[248] *R. Scholder, F. Kindervater, W. Zeiss*, Z. Anorg. Allgem. Chemie *283*, 338 (1956).
[249] *I. N. Belyaev, I. G. Koharovtseva*, Zh. Neorg. Khim. *13* (4), 932 (1968).
[250] *M. Wronska*, Bull. Acad. Polon. Sci., Ser. Sci. Khim., Geol., Geograph. 7, 137 (1959).
[251] *W. Kerler, W. Neuwirth, E. Fluck, P. Kuhn, B. Zimmermann*, Z. Phys. *173*, 321 (1963); *173*, 200 (1963).
[252] *K. Wahl, W. Klemm, G. Wehrmeyer*, Z. Anorg. Allgem. Chemie *285*, 322 (1956).
[253] *R. Scholder*, Angew. Chem. *66*, 461 (1954).
[254] *W. Klemm*, Angew. Chem. *66*, 468 (1954).
[255] *J. Tousek*, Collect. Czech. Chem. Commun. *27*, 914 (1962).
[256] *K. Wahl, W. Klemm, G. Wehrmeyer*, Z. Anorg. Allgem. Chemie *285*, 322 (1956).
[257] *R. Scholder, H. v. Bunsen, F. Kindervater, W. Zeiss*, Z. Anorg. Allgem. Chemie *282*, 268 (1955).
[258] *J. A. Hedwall, N. v. Zweigbegk*, Z. Anorg. Allgem. Chemie *108*, 135 (1919).
[259] *L. Losana*, Gazz. Chim. Ital. *55*, 468 (1925).
[260] *W. Heister*, Techn. Mitt. Krupp *12*, 161 (1954).
[261] *G. L. Kochanny, A. Timnick*, J. Amer. Chem. Soc. *83*, 2777 (1961).

fusion of a mixture of the appropriate ferrite and sodium carbonate in an oxygen atmosphere[262] at 1150°.

[262] *Ya. E. Vil'nyanskii, O. I. Pudovkina*, Zh. Prikl. Khim. *22*, 683 (1949).

20.8 Bibliography

R. B. King, Organometal. Chem. Rev. Sect. B *4* (1), 81 (1968).
E. Koenig, Coord. Chem. Rev. *3* (4), 471 (1968).

21 Ruthenium

Jon A. McCleverty
Chemistry Department, The University,
Sheffield S3 7HF, England

21.1 Ruthenium: Occurrence, Purification, Analysis, and General Chemistry

21.1.1 Occurrence

Ruthenium is a rare element, almost as rare as osmium, its abundance in the earth's crust being as low as 0.0004 ppm. The main sources of the element are laurite (mainly RuS_2 with some OsS_2), osmiridium, and platinum ore concentrates.

21.1.2 Purification

Platinum-bearing concentrates are extracted with aqua regia, and the insoluble portion is heated with an oxidizing agent such as sodium peroxide, although laurite and osmiridium may be directly oxidized with peroxide. The resulting mixture is extracted with water, the soluble fraction then containing $[RuO_4]^{2\ominus}$. Addition of ethanol causes precipitation of hydrated RuO_2, any remaining osmium being retained in solution as $[OsO_4]^{2\ominus}$. The hydrated RuO_2, which may still be contaminated by traces of osmium, is dissolved in HCl, evaporated down, and boiled with HNO_3 to remove the volatile OsO_4. The residue is then treated with chlorine to give the volatile RuO_4. In the event that the osmium content of the mineral is low, the ethanol-precipitation step can be omitted and the extract from the peroxide oxidation can be treated directly with chlorine. The ruthenium tetroxide is condensed into HCl in the presence of ammonium salts, giving aquated $[NH_4]_3[RuCl_6]$, which, when isolated as a crystalline solid and heated in an inert atmosphere, gives the metal. For osmium-rich ores, a final treatment with HNO_3 may be necessary to eliminate traces of osmium as OsO_4 and the ruthenium-containing product is then $H_2[Ru(NO)Cl_5] \cdot nH_2O$, which, on heating in an inert atmosphere, affords the metal.

21.1.3 Analysis

Ruthenium can be estimated colorimetrically as a thiourea complex,[1] although osmium and palladium interfere. If osmium is present in less than a 10-fold excess, the ruthenium may be determined colorimetrically (5–15 μg) as a 2,4-diphenylthiosemicarbazide complex.[2] In low concentrations (0·002–0·1 μg) ruthenium may be estimated[3] by its catalytic effect as RuO_4 on the cerium(IV)–arsenic(III) reaction. Gravimetric methods of analysis involve formation of RuS_2,[3] or the thionalide complex, followed by reduction to the metal.[4]

21.1.4 General Chemistry

Ruthenium metal is more resistant to oxidation by oxygen than osmium, but less so than iridium or rhodium. A fast reaction does not occur below 600°, and there is some evidence that a protective coating of RuO_2 is formed. The metal is insoluble in all acids, including aqua regia, although if $KClO_3$ is added to the latter, oxidation of the metal occurs rapidly. The metal is attacked by chlorine, and by fluorine at 300°, and halides of the metal in various oxidation states are formed. It does not react directly with sulfur, but will react with arsenic, silicon, phosphorus, and boron at high temperatures. It is dissolved by molten alkalis and molten oxidizing agents (*e.g.*, peroxides).

Because of its high cost, ruthenium has only a few industrial uses, although it has some potential as a catalyst in certain homogeneous reactions (*e.g.*, hydrogenation of terminal olefins[5,6]) when present as complexes of Ru(II) and Ru(III).

Ruthenium and osmium are very similar to each other in the wide range of oxidation states represented in their compounds (VIII→0). The "normal" oxidation state for the element is (III), and the least common oxidation states are (VII) and (I). Ruthenium(VIII) is less stable than osmium(VIII) and Ru(VII) is represented only by $[RuO_4]^{\ominus}$; few complexes of Ru(VI) are known. Ruthenium(V) is represented only by $[RuF_5]_4$ and salts of $[RuF_6]^{\ominus}$; salts of $[Ru^{IV}Cl_6]^{2\ominus}$ are very easily reduced to $[Ru^{III}Cl_6]^{3\ominus}$, much more so than the corresponding osmium salts. The lower oxidation states (II), (I), and (0) are stabilized particularly by trialkyl- or triarylphosphines, arsines, stibines, isonitriles, carbon monoxide, and olefins.

Two particular aspects of the chemistry of ruthenium deserve special mention. These are (*a*) the ability to form more nitrosyl complexes than any other metal (see Section 21.5.4; here coordinated NO is regarded as a neutral ligand) and (*b*) the tendency to form poly-

---

[1] *G. H. Ayres, F. Young*, J. Anal. Chem. *22*, 1277 (1950).

[2] *W. Gerlmann, R. Neeb*, Z. Anal. Chem. *152*, 96 (1956).

[3] *I. M. Kolthoff, P. J. Elving*, Treatise on Analytical Chemistry, 8, Interscience Publishers New York, 1973.

[4] *F. E. Beamish*, Talanta *12*, 789 (1965);
F. E. Beamish, Talanta *13*, 773 (1966);
F. E. Beamish, Analytical Chemistry of the Noble Metals, Pergamon Press, Oxford 1966;
W. J. Rogers, F. E. Beamish, D. S. Russell, Ind. Eng. Chem. *12*, 561 (1940).

[5] *B. R. James*, Inorg. Chim. Acta *4*, 73 (1970);

[6] *J. Halpern, B. R. James*, Can. J. Chem. *44*, 671 (1966);
J. Halpern, J. F. Harrod, B. R. James, J. Amer. Chem. Soc. *88*, 5150 (1966);
J. Halpern, J. F. Harrod, B. R. James, J. Amer. Chem. Soc. *83*, 753 (1961).

nuclear complexes with $O^{2\ominus}$, $OH^{\ominus}$, or $N^{3\ominus}$ bridging groups, especially in the chemistry of ruthenium(IV).

21.2 Ruthenium(0)

Reaction of cycloocta-1,3,5-triene (C_8H_{10}) and bicyclo[4.2.0]octa-2,4-diene (C_7H_8) with anhydrous $RuCl_3$ and isopropylmagnesium bromide in ether afforded[7] $[Ru(C_8H_{10})(C_7H_8)]$. A related complex containing cycloheptatriene (C_7H_8) and cycloocta-1,5-diene (C_8H_{12}) could be prepared similarly[7] or by reacting the polymeric halide, $[Ru(C_8H_{12})Cl_2]_n$, with C_7H_8 and isopropylmagnesium bromide in ether. All these reactions required u.v. light. Reduction of $[Ru(\pi\text{-}C_6H_6)_2]^{2\oplus}$ with alkali metal amalgams in polar solvents afforded[8] $[Ru(\pi\text{-}C_6H_6)_2]$, and the related compound containing hexamethylbenzene was synthesized in the same way. These compounds are stereochemically nonrigid. Reduction of $[Ru(\pi\text{-}C_6H_6)_2]^{2\oplus}$ with $LiAlH_4$, however, afforded[9] two compounds, $[Ru(\pi\text{-}C_6H_7)_2]$ a bis-π-cyclohexadienyl complex of Ru(II), and $[Ru(\pi\text{-}C_6H_6)(C_6H_8)]$, in which the C_6H_8 is cyclohexa-1,3-diene.

21.3 Ruthenium(I)

The most common compounds of the element in this oxidation state are stabilized by CO, and are discussed in Chapter 29.

21.4 Ruthenium(II)

21.4.1 Halo Complexes

$RuCl_2$, as such, does not apparently exist, but aqueous solutions containing Ru(II) have been obtained by the reduction of $RuCl_3$ or chloro-aquo Ru(III) and Ru(IV) species, either electrolytically or using zinc or aluminum, zinc amalgam, $Ti^{3\oplus}$ or $Cr^{2\oplus}$, or hydrogen catalyzed by platinum black.[10, 11] The ruthenium complex present in these reduced solutions has been identified as the chlorocluster, $[Ru_5Cl_{12}]^{2\ominus}$, and has been isolated as the salts of the $[M(bipyr)_3]^{2\oplus}$ (M = Fe or Ru, bipyr = 2,2'-bipyridyl) or *o*-phenylenemethylenebis (triphenylphosphonium) cations.[11] The bromides and iodides are presumably similar. The *cluster anion* forms blue aqueous solutions which apparently catalyze homogeneously the hydrogenation of terminal olefins by molecular hydrogen.[5, 6] Salts of $[Ru_5Cl_{12}]^{2\ominus}$ are a useful source of complexes of Ru(II) and Ru(III) (see Scheme 1).[12] Addition of pyridinium hydrochloride to solutions of $[Ru_5Cl_{12}]^{2\ominus}$ in HCl afforded the green pyridinium salt of the cluster anion $[Ru_4Cl_{12}]^{4\ominus}$.

21.4.2 Aquo Species and Complexes Containing O- and S-Donor Atoms

Electrolytic reduction of $[Ru(H_2O)_5Cl]^{2\oplus}$ or $[Ru(H_2O)_4Cl_2]^{\oplus}$ in aqueous solution in the presence of $BF_4^{\ominus}$ ions afforded[13] $[Ru(H_2O)_6]^{2\oplus}$.

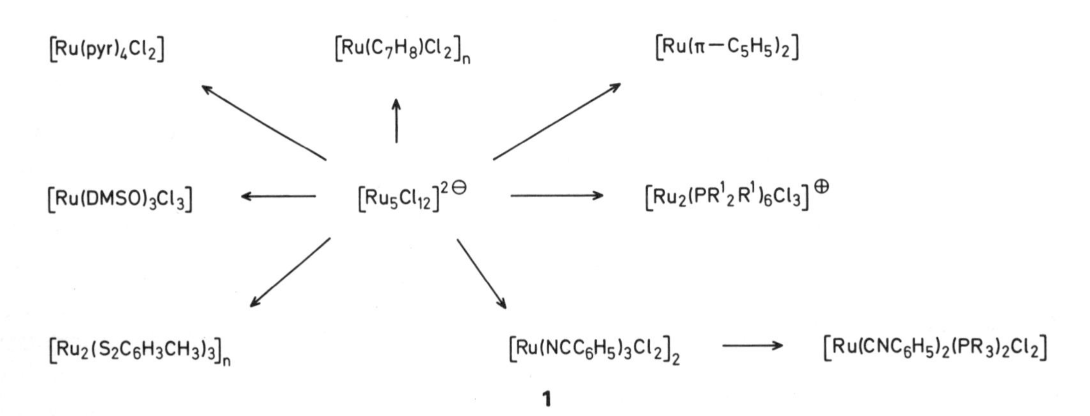

1

[7] *J. Muller, E. O. Fischer*, J. Organometal. Chem. *5*, 275 (1966); *R. R. Schrock, J. Lewis*, J. Amer. Chem. Soc. *95*, 4102 (1973).
[8] *E. O. Fischer, Ch. Elschenbroich*, Chem. Ber. *103*, 162 (1970).
[9] *D. Jones, L. Pratt, G. Wilkinson*, J. Chem. Soc. (London) 4458 (1962).
[10] *G. A. Rechnitz*, Inorg. Chem. *1*, 953 (1962).
[11] *D. Rose, G. Wilkinson*, J. Chem. Soc. *A*, 1791 (1970).
[12] *J. D. Gilbert, D. Rose, G. Wilkinson*, J. Chem. Soc. *A*, 2765 (1970).
[13] *D. K. Atwood, T. de Vries*, J. Amer. Chem. Soc. *84*, 2659 (1962);
E. E. Mercer, R. R. Buckley, Inorg. Chem. *4*, 1692 (1965).

This hexaquo species could be separated from other chloroaquo Ru(II) complexes by cation exchange techniques.

Acidification or oxidation of ruthenium(II) ammine complexes containing the $SO_3^{2\ominus}$ or $HSO_3^{\ominus}$ ions afforded[14] sulfur dioxide addition complexes, e.g., $[Ru(NH_3)_4(SO_2)X]X$ (X=Cl or Br).

Reaction of $RuCl_3$ in ethanol with dithiocarbamate ions, $(S_2CNR_2)^{\ominus}$ (R=Me, Et, or benzyl), in the presence of CO afforded[15] $[Ru(CO)_2(S_2CNR_2)_2]$.

21.4.3 Ammine Complexes, and Species Containing N-Donor Atoms

Salts of $[Ru(NH_3)_6]^{2\oplus}$ can be prepared by the reduction of $RuCl_3$ with zinc dust in ammonia, in the presence of NH_4X (X=desired counteranion).[16] The *pentammine*, $[Ru(NH_3)_5]^{2\oplus}$ or $[Ru(NH_3)_5(H_2O)]^{2\oplus}$, is also formed when ruthenium(III) pentammino species are reduced by $Cr^{2\oplus}$.[17] The *hexammines* are orange salts and they are air-sensitive. The related ethylenediamine (en) complexes, $[Ru(en)_3]^{2\oplus}$, were prepared similarly.[18] Trisbipyridyl (bipyr) and *o*-phenanthroline (phen) compounds, $[Ru(bipyr)_3]^{2\oplus}$ and $[Ru(phen)_3]^{2\oplus}$, were prepared[19,20] by the reaction of the ligands with $RuCl_3$ at 250°. Oxidation of *d*- or *l*-$[Ru(phen)_3]^{2\oplus}$ with $Ce^{4\oplus}$ gave the corresponding blue, optically active $[Ru(phen)_3]^{3\oplus}$ which could be reduced back to the Ru(II) complex without loss of optical configuration.[19,21]

21.4.4 Nitrogen Complexes

The *pentamminonitrogen ruthenium(II)* complex, $[Ru(NH_3)_5(N_2)]^{2\oplus}$, is extraordinarily stable, and can be prepared by a number of routes. Thus, displacement[22] of water from $[Ru(NH_3)_5(H_2O)]^{2\oplus}$, reaction of Ru(III) or Ru(IV) chloro or ammino complexes[23] with hydrazine,[23a] reduction[24] with zinc dust and ammonia or treatment of $[Ru(NH_3)_5L]X_3$ (L=NH_3 or H_2O) with azide ion under slightly acidic conditions[25] all lead to the formation of salts of $[Ru(NH_3)_5(N_2)]^{2\oplus}$. Even in the synthesis of $[Ru(NH_3)_6]^{2\oplus}$ by reduction of $RuCl_3$ with zinc dust in aqueous ammonia (see Section 21.4.3), 10% yields of the pentamminonitrogen complex were obtained.[16,24]

The azidonitrogen complex, *cis*-$[Ru(en)_2(N_3)(N_2)][PF_6]$, was prepared by the route shown below.[25]

$$cis\text{-}[Ru(en)_2Cl_2]Cl \cdot H_2O + Ag^{\oplus}\{p\text{-}MeC_6H_4SO_3\}^{\ominus} \longrightarrow [Ru(en)_2(H_2O)_2]^{3\oplus}$$

$$cis\text{-}[Ru(en)_2(H_2O)_2]^{3\oplus} + N_3^{\ominus} + PF_6^{\ominus} \longrightarrow cis\text{-}[Ru(en)_2(N_3)_2][PF_6]$$

$$cis\text{-}[Ru(en)_2(N_3)_2][PF_6] \xrightarrow{65°,\ 25\ min} cis\text{-}[Ru(en)_2(N_3)(N_2)][PF_6]$$

Reaction of this azidonitrogen complex with nitrous acid and $BPh_4^{\ominus}$ in the cold afforded *cis*-$[Ru(en)_2(N_2)_2][BPh_4]_2$, *cis*-$[Ru(en)_2(H_2O)(N_2)][BPh_4]_2$, and N_2O. The *bisnitrogen complex* was very unstable. The water in $[Ru(NH_3)_5(H_2O)]^{2\oplus}$ can be readily displaced by Lewis bases, in particular, $[Ru(NH_3)_5(N_2)]^{2\oplus}$ giving $[(NH_3)_5Ru{-}N{=}N{-}Ru(NH_3)_5]^{4\oplus}$.[26] The water can also be displaced,[27] under pressure, by nitrous oxide, giving $[Ru(NH_3)_5(N_2O)]^{2\oplus}$.

---

[14] *K. Gleu, W. Breuel, K. Rhem*, Z. Anorg. Allgem. Chemie *235*, 201, 211 (1938);
L. H. Vogt, J. L. Katz, S. E. Wiberley, Inorg. Chem. *4*, 1157 (1965).

[15] *J. V. Kingston, G. Wilkinson*, J. Inorg. Nucl. Chem. *28*, 2709 (1967).

[16] *F. M. Lever, A. R. Powell*, J. Chem. Soc. *A*, 1477 (1969).

[17] *J. F. Endicott, H. Taube*, Inorg. Chem. *4*, 437 (1965).

[18] *A. D. Allen, C. V. Senoff*, Can. J. Chem. *43*, 888 (1965).

[19] *F. H. Burstall*, J. Chem. Soc. 173 (1936);
C. F. Liu, N. C. Liu, J. C. Bailar, Inorg. Chem. *3*, 1197 (1964).

[20] *F. P. Dwyer, E. C. Gyarfas*, Austral. J. Chem. *16*, 544 (1963).

[21] *F. P. Dwyer, E. C. Gyarfas*, Nature *163*, 918 (1949).

[22] *D. E. Harrison, H. Taube*, J. Amer. Chem. Soc. *89*, 5706 (1967);
I. J. Itkovitch, J. A. Page, Can. J. Chem. *46*, 2743 (1968).

[23] *A. D. Allen, F. H. Bottomley, R. O. Harris, V. P. Reinsalu, C. V. Senoff*, J. Amer. Chem. Soc. *89*, 5595 (1967).

[23a] Complexes prepared using hydrazine are likely to be contaminated with hydrazine and this particular synthetic route is not recommended.

[24] *J. Chatt, A. B. Nikolsky, R. L. Richards, J. R. Sanders, J. E. Fergusson, J. L. Love*, J. Chem. Soc. *A*, 1479 (1970).

[25] *L. A. P. Kane-Maguire, P. S. Sheridan, F. Basolo, R. G. Pearson*, J. Amer. Chem. Soc. *92*, 5865 (1970).

[26] D. E. Harrison, E. Weissburger, *H. Taube*, Science, *159*, 320 (1968).

[27] *A. A. Diamantis, G. J. Sparrow*, Chem. Commun. 819 (1970);
J. N. Armor, H. Taube, J. Amer. Chem. Soc. *91*, 6874 (1969).

Treatment of $[Ru(PPh_3)_3HCl]$ with triethylaluminum and nitrogen gas in ether afforded[28] $[Ru(PPh_3)_3(N_2)H_2]$. The nitrogen could be reversibly displaced by hydrogen, forming $[Ru(PPh_3)_3H_4]$, but treatment of the nitrogen complex with PPh_3 resulted in the irreversible formation of $[Ru(PPh_3)_4H_2]$. Reaction of $[Ru(PPh_3)_3(N_2)H_2]$ with NH_3 gave $[Ru(PPh_3)_3(NH_3)H_2]$, and when a mixture of the starting material and ammoniated product was allowed to stand together in tetrahydrofuran (THF), orange crystals of $[Ru_4(PPh_3)_5(NH_3)_3]$ were precipitated.

Reaction of $[Ru(depe)_2HCl]$ (depe = $Et_2PCH_2CH_2PEt_2$) with $NaBPh_4$ and nitrogen gas gave $[Ru(depe)_2(N_2)H][BPh_4]$.[29] Reaction of $[Ru(NO)(diars)_2Cl]Cl_2$ with hydrazine afforded $[Ru(diars)_2(N_3)Cl]$, which, on treatment with $NOPF_6$ in methanol, afforded[30] *trans*-$[Ru(NO)(diars)_2Cl][PF_6]_2$ and *trans*-$[Ru(diars)_2(N_2)Cl][PF_6]$.

21.4.5 Phosphine, Arsine, and Stibine Complexes

A large number of ruthenium(II) complexes containing tertiary alkyl- and arylphosphines, arsines, and stibines, of the type $[Ru(LR_3)_4X_2]$ (X = Cl or Br; L = P or As),[31,32] $[Ru(LR_3)_3X_2]$[32] (X = Cl or Br; L = P or Sb), and $[Ru_2(PR_3)_6Cl_3]Cl$[33] are known. These are prepared by the reaction of $RuCl_3$ with PR_3, in the presence of $X^{\ominus}$, with, or without, solvent (usually an alcohol). An excess of PPh_3 reacts with $RuCl_3$ giving $[Ru(PPh_3)_3Cl_2]$,[32] which is the course of a variety of ruthenium(II) phosphine complexes

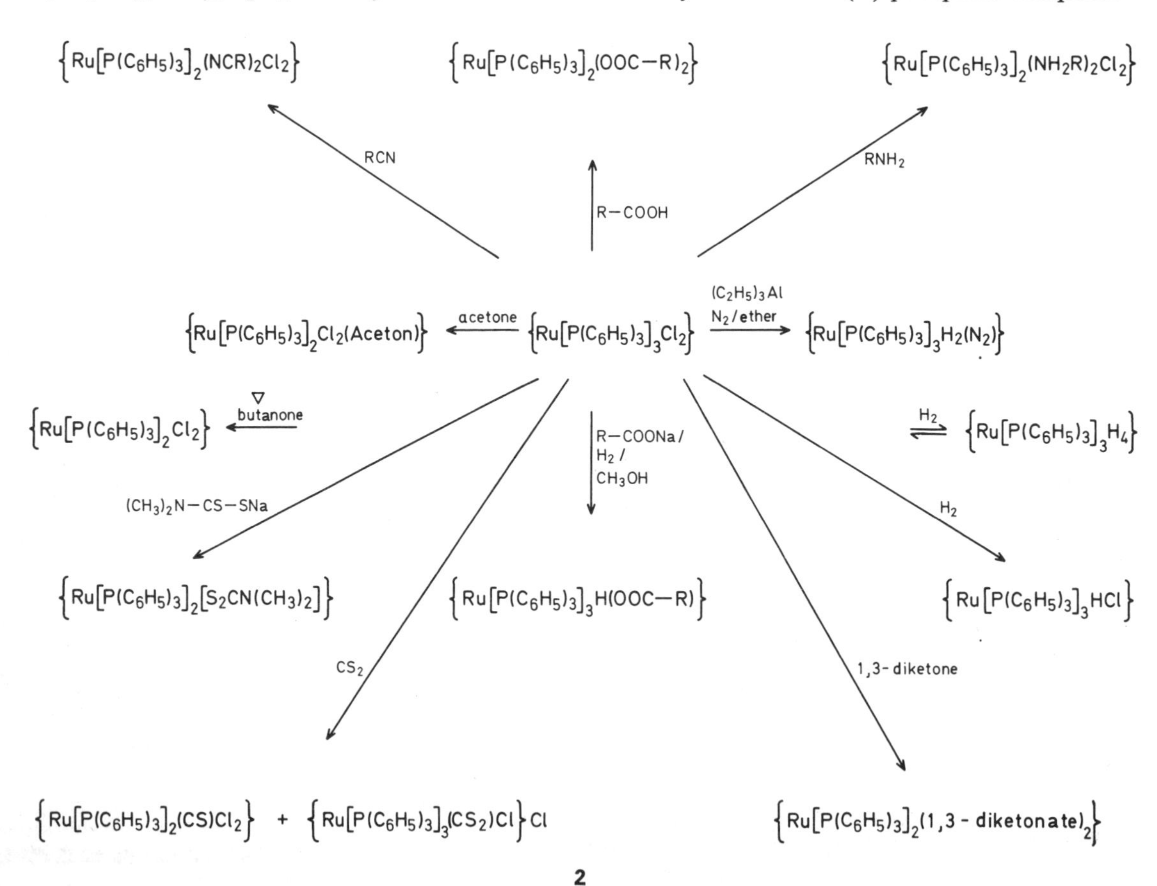

2

[28] *W. H. Knoth*, J. Amer. Chem. Soc. *90*, 7172 (1968); *T. Ito, S. Kitazume, A. Yamamoto, S. Ikeda*, J. Amer. Chem. Soc. *92*, 3011 (1970); *W. H. Knoth*, Inorg. Synth. *15*, 31 (1974).

[29] *G. M. Bancroft, M. J. Mays, B. E. Prater, F. P. Stefanini*, J. Chem. Soc. *A*, 2146 (1970).

[30] *P. G. Douglas, R. D. Feltham, H. G. Metzger*, Chem. Commun. 889 (1970).

[31] *J. Chatt, B. L. Shaw, A. E. Field*, J. Chem. Soc. 3466 (1964); *R. G. Hayter*, Inorg. Chem. *3*, 301 (1964).

[32] *T. A. Stephenson, G. Wilkinson*, J. Inorg. Nucl. Chem. *28*, 945 (1966).

[33] *J. Chatt, R. G. Hayter*, J. Chem. Soc. (London) 896 (1961); *M. S. Lupin, B. L. Shaw*, J. Chem. Soc. *A*, 741 (1968).

(Scheme 2). The dichloride reacts[34] with hydrogen giving $[Ru(PPh_3)_3HCl]$, which catalyzes the homogeneous hydrogenation of terminal olefins and acetylenes and, in the absence of olefin, will catalyze H/D exchange.
The complexes[33,35] of the type

cis- or *trans*-$[RuA_2X_2]$
X = Cl, Br
$A = R_2P—CH_2—CH_2—PR_2$ $R = CH_3, C_2H_5, C_6H_5$ 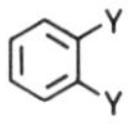
(o-phenylene with two Y substituents)
$Y = PR_2$,
$As(CH_3)_2$ (only *trans*)

were obtained in the same way. These diphos and diars complexes reacted with $LiAlH_4$ or alkly lithium (LiR) giving *trans*-$[RuA_2HX]$,[36] *cis*- and *trans*-$[RuA_2H_2]$,[36] *trans*-$[RuA_2RX]$,[37,38] *trans*-$[RuA_2R_2]$[37] (Table 1).
Reduction of *cis*- or *trans*-$[RuA_2Cl_2]$ with arene anions in tetrahydrofuran[38] led probably to a tautomeric equilibrium.

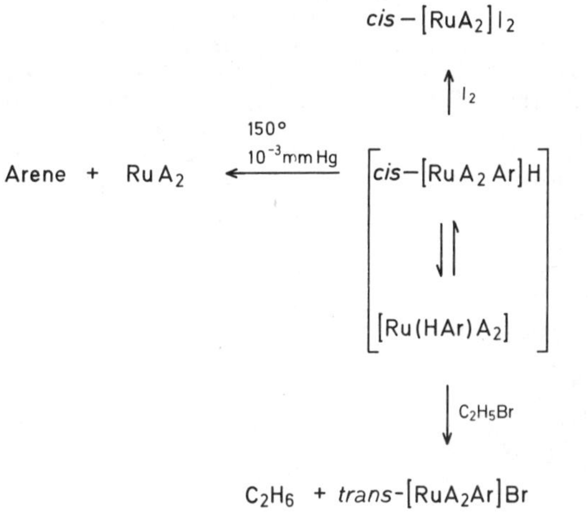

[01] *P. S. Hallman, B. R. McGarvey, G. Wilkinson*, J. Chem. Soc. *A*, 3143 (1968).
[35] *R. S. Nyholm, G. J. Sutton*, J. Chem. Soc. 567 (1958).
[36] *J. Chatt, R. G. Hayter*, J. Chem. Soc. 2605 (1961).
[37] *J. Chatt, R. G. Hayter*, J. Chem. Soc. 6017 (1963).
[38] *J. Chatt, J. M. Davidson*, J. Chem. Soc. 843 (1965).

cis-$[RuA_2]I_2$

↑ I_2

Arene + RuA_2 ← (150°, 10^{-3} mm Hg) [*cis*-$[RuA_2Ar]H$ ⇌ $[Ru(HAr)A_2]$]

↓ C_2H_5Br

C_2H_6 + *trans*-$[RuA_2Ar]Br$

$A = R_2P—CH_2—CH_2—PR_2$ $(R = CH_3, C_2H_5, C_6H_5)$
ArH = benzene, naphthalene, anthrocene

21.4.6 Cyanide Complexes and Olefinic, Allylic and Stannous Compounds

Reaction of KCN with K_2RuO_4 or treatment of a boiling aqueous solution of $RuCl_3$ with KCN afforded $K_4[Ru(CN)_0]$.[39] Both *cis*- and *trans*-$[Ru(CNR)_4X_2]$ (X = Cl, Br or I; R = Me, Et, p-MeC_6H_4) were prepared by reaction of

[39] *J. L. Howe*, J. Amer. Chem. Soc. *18*, 981 (1896).

Table 1 Tertiary Phosphine, Arsine, and Stibine Complexes of Ruthenium(II)[a]

Complex	Substituents on phosphine, arsine, stibine or complex	Synthetic method	X	Ref.
$[Ru_2(PR_2R')_6Cl_3]X$	R = Me, Et, R' = Ph R = Ph, R' = Me, Et	$RuCl_3 \cdot nH_2O$ with PR_2R' in methanol/water, or, better, in 2-methoxyethanol; X by metathesis in ethanol	Cl, ClO_4, BPh_4, SCN, picrate	*b*
trans-$Ru(diphos)_2X_2]$	diphos = $R_2PCH_2CH_2PR_2$ (R = Me, Et, Ph), $Ph_2PCH_2PPh_2$	$RuCl_3$ in aqueous ethanol with slight excess of diphos; X by methathesis in THF or ethanol	Cl, Br, SCN, OCOMe, CN	*b*
cis-$[Ru(diphos)_2X_2]$	diphos = $R_2PCH_2CH_2PR_2$ (R = Me, Et, Ph), $Ph_2PCH_2PPh_2$, o-$C_6H_4(PEt_2)_2$	$[Ru_2(PR_2R')_6Cl_3]Cl$ with diphos in absence of solvent; X by metathesis in water or aqueous ethanol	Cl, Br, I, SCN, OCOMe, CN	*b*
cis-$[Ru(diars)_2X_2]$	diars = o-$C_6H_4(AsMe_2)_2$	$RuCl_3$ and diars in aqueous ethanol or $[Ru_2(PR_2R')_6Cl_3]Cl$ and diars in absence of solvent; X by metathesis in ethanol	Cl, Br, I, SCN	*b, c*
cis-$[Ru(diphos)_2RX]$	diphos = $R_2'PCH_2CH_2PR_2'$ (R' = Me, Et, Ph), $Ph_2PCH_2PPh_2$, R = Me, Et, n-Pr	*trans*-$[Ru(diphos)_2X_2]$ with $(AlR_3)_2$ in the absence of solvent or $Ru(diphos)_2HR$ with HX in benzene; X by metathesis in acetone	Cl, Br, I, SCN	*d*
trans-$[Ru(diphos)_2RX]$	diphos = $R_2'PCH_2CH_2PR_2'$ (R' = Me, Et, Ph), $Ph_2PCH_2PPh_2$, R = Me, Et, n-Pr	*cis*-$[Ru(diphos)_2X_2]$ with $(AlR_3)_2$ in the absence of solvent; metathesis in acetone	Cl, Br, I, SCN	*d*

continued

Complex	Substituents on phosphine, arsine, stibine or complex	Synthetic method	X	Ref.
trans-[Ru(diphos)$_2$ArX]	diphos = $R_2PCH_2CH_2PR_2$ (R = Me, Et, Ph), $Ph_2PCH_2PPh_2$, Ar = Ph, *p*-MeC_6H_4	*cis*- or *trans*-[Ru(diphos)$_2$-Cl_2] with LiAr in ether–benzene; X by metathesis in acetone	Cl, Br, I, SCN	*d*
cis-[Ru(diphos)$_2R_2$]	diphos = $R_2PCH_2CH_2PR_2$ (R = Me, Et, Ph), $Ph_2PCH_2PPh_2$	*cis*-[Ru(diphos)$_2Cl_2$] or *cis*-[Ru(diphos)$_2$MeCl] with LiR in ether–benzene	—	*d*
[Ru(diphos)$_2$RH]	diphos = $R_2'PCH_2CH_2PR_2'$ (R′ = Me, Et, Ph), $Ph_2PCH_2PPh_2$, R = Me, Et, *n*-Pr, Ph, *p*-MeC_6H_4	[Ru(diphos)$_2$RCl] with $LiAlH_4$ in THF	—	*d, e*
cis-[Ru(diphos)$_2$ArH]	diphos = $Me_2PCH_2CH_2$-PMe_2, Ar = phenyl, 2-naphthyl, anthryl, phenanthryl	*trans*-[Ru(diphos)$_2Cl_2$] with $Na^{\oplus}$ arene$^{\ominus}$ in THF	—	*e*
cis-[Ru(diphos)$_2H_2$]	diphos = $Me_2PCH_2CH_2$-PMe_2	*trans*-[Ru(diphos)$_2$HBr] with sodium naphthalenide in THF	—	*e*
cis-[Ru(diars)$_2$MeCl]	diars = *o*-$C_6H_4(AsMe_2)_2$	*cis*-[Ru(diars)$_2Cl_2$] with LiMe in ether–benzene	—	*d*
trans-[Ru(diphos)$_2$HX]	diphos = $R_2'PCH_2CH_2PR_2'$ (R′ = Me, Et or Ph), $Ph_2PCH_2PPh_2$, *o*-$C_6H_4(PEt_2)_2$	*cis*-[Ru(diphos)$_2X_2$] with $LiAlH_4$ in ether or THF with ethanol	Cl, Br, I, H, SCN, CN, NO_2	*f*
[Ru(MPh_3)$_4Cl_2$]	M = P, As, Sb	$RuCl_3 \cdot nH_2O$ in methanol with excess MPh_3; shaking	—	*g*
[Ru(MPh_3)$_3Cl_2$]	M = P, Sb	$RuCl_3 \cdot nH_2O$ in methanol with excess MPh_3; refluxing	—	*g*
[Ru(PR_2H)$_4Cl_2$]	R = Et, Ph	$RuCl_3 \cdot nH_2O$ in ethanol with PR_2H	—	*h*
[Ru(PR_3)$_3$(CO)X_2]	PR_3 = PEt_2Ph	(*i*) $RuCl_3$ under CO, shaken in 2-methoxyethanol (*ii*) [$Ru_2(PR_3)_6Cl_3$]Cl with KOH in boiling aqueous ethanol; X by metathesis in boiling 2-methoxyethanol (*iii*) $RuCl_3 \cdot nH_2O$ under ethylene in refluxing 2-methoxyethanol (*iv*) Ru(PR_3)$_3$(CO)HCl with HCl in methanol	Cl, Br, I	*i*
trans-[Ru(PR_3)$_2$(CO)$_2Cl_2$]	PR_3 = PEt_2Ph, PEt_3, PPh_3	Red solution obtained by passing CO through boiling ethanolic solutions or $RuCl_3 \cdot nH_2O$; *trans*-[Ru(PPh_3)$_4Cl_2$] with CO in warm acetone	—	*g, i*
cis-[Ru(PR_3)$_2$(CO)$_2Cl_2$]	PR_3 = PEt_2Ph, PEt_3	(*i*) $RuCl_3 \cdot nH_2O$ in boiling 2-methoxyethanol with PR_3 (*ii*) [$Ru_2(PR_3)_6Cl_3$]Cl in ethanol under CO pressure (*iii*) [$Ru_2(PR_3)_6Cl_3$]Cl boiled with KOH in allyl alcohol	—	*i*
trans-[Ru(diphos)$_2$HL]-[BPh_4]	diphos = $Et_2PCH_2CH_2PEt_2$, L = CO, *t*-BuNC, $P(OMe)_3$, N_2, PhCN, *p*-$MeOC_6H_4NC$	*trans*-[Ru(diphos)$_2$HCl] with $NaBPh_4$ and L in acetone or chloroform	—	*j*

Complex	Substituents on phosphine, arsine, stibine or complex	Synthetic method	X	Ref.
$[Ru(PPh_3)_3HCl]$	—	$[Ru(PPh_3)_3Cl_2]$ with hydrogen in benzene	—	k
$[Ru(PR_3)_3(CO)HX]$	$PR_3 = PEt_2$, Ph, $P(n\text{-}Pr)_2Ph$	$[Ru_2(PR_3)_6Cl_3]Cl$ in boiling ethanolic KOH; X by metathesis in diethylaminoethanol	Cl, Br, I	i, l

[a] This Table contains only compounds derived from Ru(III) halides. Other phosphine complexes may be obtained from $Ru_3(CO)_{12}$, $[Ru(CO)_4X_2]$, or $[Ru(CO)_3X_2]_2$.
[b] *J. Chatt, R. G. Hayter*, J. Chem. Soc., p. 896 (1961).
[c] *R. S. Nyholm, G. J. Sutton*, J. Chem. Soc., 567 (1965).
[d] *J. Chatt, R. G. Hayter*, J. Chem. Soc., p. 6017 (1963).
[e] *J. Chatt, J. M. Davidson*, J. Chem. Soc., p. 843 (1965).
[f] *J. Chatt, R. G. Hayter*, J. Chem. Soc., p. 2605 (1961).
[g] *T. A. Stephenson, G. Wilkinson*, J. Inorg. Nucl. Chem. *28*, 945 (1966).
[h] *R. G. Hayter*, Inorg. Chem. *3*, 301 (1964).
[i] *J. Chatt, B. L. Shaw, A. E. Field*, J. Chem. Soc., p. 3466 (1964).
[j] *G. M. Bancroft, M. J. Mays, B. E. Prater, F. P. Stefanini*, J. Chem. Soc. A, p. 2146 (1970).
[k] *R. S. Hallman, B. R. McGarvey, G. Wilkinson*, J. Chem. Soc. A, p. 3143 (1968); *R. A. Schunn, E. R. Wonchoba* Inorg. Synth., *13*, 131 (1972).
[l] *M. S. Lupin, B. L. Shaw*, J. Chem. Soc. A, p. 741 (1968).

$RuCl_3$ with the appropriate isonitriles in alcoholic solutions.[40,41]

Chelating diolefins, e.g., cycloocta-1,5-diene and *norbornadiene*, react with $RuCl_3$ in ethanol giving the polymeric, orange $[Ru(diolefin)Cl_2]_n$.[42] These polymers may be broken up with bases, *e.g.*, *p*-toluidine or phosphines (L), giving $[Ru(diolefin)L_2Cl_2]$.

Treatment of $[Ru(diolefin)Cl_2]_n$ with allylmagnesium bromide[43] gave $[Ru(\pi\text{-allyl})_2(diolefin)]$. Ruthenium trichloride reacted with $SnCl_2$ giving $[Ru(SnCl_3)_2Cl_2]^{2\ominus}$, isolated as $Me_4N^{\oplus}$ or $Ph_3HP^{\oplus}$ salts.[44]

An important aspect of the chemistry of ruthenium(II) complexes containing arylphosphine or phosphite ligands is the ability with which these ligands will undergo hydrogen transfer[45], usually with elimination of hydrogen or HX from the metal complex and formation of a bond between the metal and the *ortho* carbon atom in the phenyl group of the ligand. Thus, $[Ru(P(OPh)_3)_4HCl]$ readily undergoes H_2 elimination at ambient temperatures giving $\{Ru[P(OC_6H_4)(OC_6H_5)_2][P(OPh)_3]_3Cl\}$; recombination with hydrogen occurs under moderate pressures. Other complexes which behave in this way are $[Ru(PPh_3)_3HCl]$ and $[Ru(PPh_3)_3(N_2)H_2]$.

Carbonyl compounds of ruthenium, particularly of Ru(II), are an important aspect of the chemistry of this element. A more detailed treatment of this topic will be found in Chapter 29, but some indication of the variety of compounds and their synthesis is given in Scheme 3 (p. 290).

21.5 Ruthenium(III)

21.5.1 Halides and Halo Complexes

RuF_3 was obtained[46] by reaction of an excess of iodine with $[RuF_5]_4$ at 150°. The *trichloride* exists as water-insoluble and water-soluble forms; the former (which is found also as two forms) can be prepared[47] by heating spongy ruthenium at 370° in a 1:3 mixture of CO and chlorine, and the latter by prolonged reaction of HCl on RuO_4.[48] Water-soluble $RuCl_3$ initially appears to contain no ionic $Cl^{\ominus}$ when dissolved in water, but hydrolysis occurs slowly on standing, and both monomeric and polymeric aquated species are formed.[49] When isolated from water,

[40] *L. Malatesta, G. Padoa*, Rend. 1st, Lombardo Sci., Lett. *A*, 91, 277 (1957);
[41] *L. Malatesta, G. Padoa, A. Sonz*, Gazz. Chim. Ital. *85*, 1111 (1955).
[42] *E. W. Abel, M. A. Bennett, G. Wilkinson*, J. Chem. Soc. 3178 (1959).
[43] *J. Powell, B. L. Shaw*, J. Chem. Soc. *A*, 159 (1968).
[44] *J. F. Young, R. D. Gillard, G. Wilkinson*, J. Chem. Soc. 5176 (1964).
[45] *G. W. Parshall, W. H. Knoth, R. A. Schunn*, J. Amer. Chem. Soc. *91*, 4990 (1969).
[46] *E. E. Aynsley, R. D. Peacock, P. L. Robinson*, Chem. & Ind. (London) 1002 (1952).
[47] *K. R. Hyde, E. W. Hooper, J. Waters, J. M. Fletcher*, J. Less-Common Metals *8*, 428 (1965).
[48] *G. Brauer*, Handbook of Preparative Chemistry, p. 1597, Academic Press, New York 1597 (1965).
[49] *H. H. Cady, R. E. Connick*, J. Amer. Chem. Soc. *80*, 2646 (1958);
R. E. Connick, D. A. Fine, J. Amer. Chem. Soc. *83*, 3414 (1961);
R. E. Connick, D. A. Fine, J. Amer. Chem. Soc. *82*, 4187 (1960).

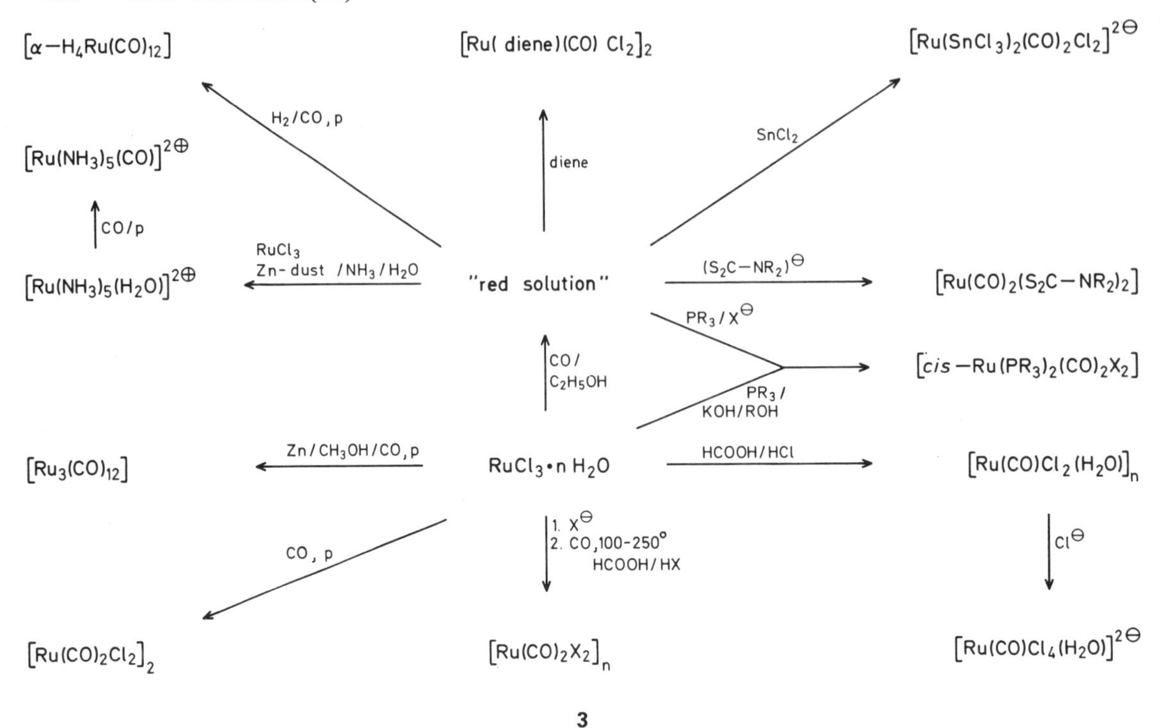

3

the trichloride has the approximate composition $RuCl_3 \cdot 3H_2O$ and is the starting material for the preparation of a large number of Ru(III), Ru(II), and Ru(0) complexes (Scheme 4, and *vide infra*).

The complex $[RuCl_6]^{3\ominus}$ is obtained[50] by reaction of HCl with $[RuCl_5(H_2O)]^{2\ominus}$, and $K_2[RuCl_5]$ is obtained[51] as brown crystals either by heating $K_2[RuCl_5(H_2O)]$ at 250° or by heating $K_2[RuCl_6]$ at 600° in a stream of dry HCl. The tetraalkylammonium salts, $[R_4N]_2[RuCl_5]$ (R = Me, Et, etc.), were prepared[52] from solutions of $RuCl_3$ in HCl when treated with a large excess of R_4NCl.

In aqueous solutions of ruthenium(III) chloride, a number of aquo species are present, *e.g.*, $[RuCl_n(H_2O)_{6-n}]^z$ ($n \geqslant 1$), and these have been separated by ion-exchange techniques.[49] The complex $K_2[RuCl_5(H_2O)]$ has been obtained[49,51] as red crystals by ethanolic reduction of solutions of RuO_4 in HCl in the presence of KCl. Reduction of $[NH_4]_2[RuO_4]$ by $SnCl_2$ in HCl afforded $[NH_4][RuCl_4(H_2O)_2]$.[53]

21.5.2 Aquo Species and Complexes Containing O- and S-Donor Atoms

The existence of $[Ru(H_2O)_6]^{3\oplus}$ is not confirmed, although there is polarographic evidence to suggest that it can be formed by reduction of Ru(IV) species in perchloric acid solutions or by the oxidation of $[Ru(H_2O)_6]^{2\oplus}$.[13,54]

Red $[Ru(acac)_3]$ (acacH = acetylacetone) was prepared[55] by the reaction of $RuCl_3$ with acetylacetone; it sublimes as a red vapor. Related complexes containing benzoylacetone and dibenzoylmethane have been prepared similarly.[56] The sulfur analog, $[Ru(SacSac)_3]$ (SacSac = dithioacetylacetone), was prepared by treating $RuCl_3$ in ethanol with acetylacetone, dry HCl, and H_2S at 0°.[57]

Reaction of RuO_4 in CCl_4 with glacial acetic acid and acetic anhydride afforded the blue $[Ru_3O(OCOMe)_6(H_2O)_3][OCOMe]$,[58] which is apparently similar to the basic acetate cations

[50] *R. Charronat*, Ann. Chim. (Paris) *16*, 179, 188, 235 (1931).

[51] *J. L. Howe, L. P. Haynes*, J. Amer. Chem. Soc. *47*, 2920 (1925).

[52] *A. Gutbier, F. Krauss*, J. Prakt. Chem. *91*, 103 (1915).

[53] *M. Buividaite*, Z. Anorg. Allgem. Chemie *222*, 279 (1935).

[54] *L. W. Niedrach, A. D. Tevebaugh*, J. Amer. Chem. Soc. *73*, 2835 (1951).

[55] *G. A. Barbieri*, Atti Accad. Naz. Lincei, Classe Sci. Fis., Mat. Natur., Rend. *23*, 336 (1914);

[56] *L. Wolf, E. Butter, H. Weinelt*, Z. Anorg. Allgem. Chemie *306*, 87 (1960).

[57] *G. A. Heath, R. L. Martin*, Australian J. Chem. *23*, 174 (1970).

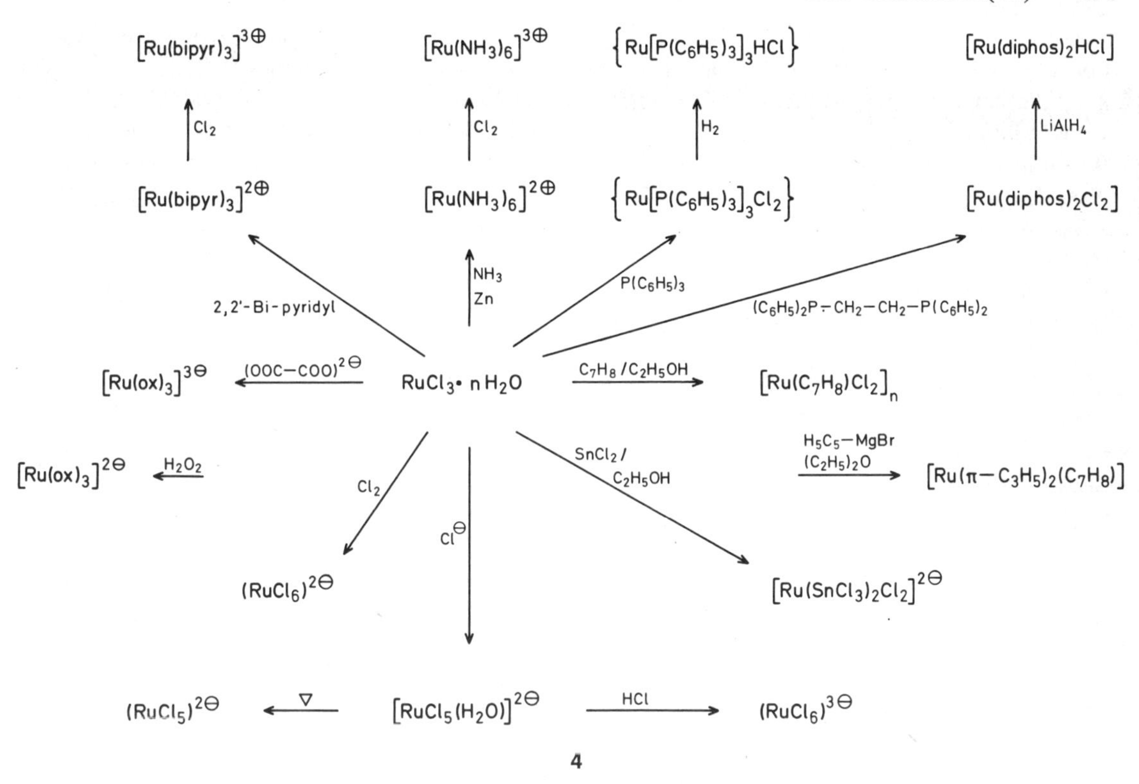

4

$[MO_3(OCOMe)_6(H_2O)_3]^+$ (M = Fe or Cr).[59] Reaction of $RuCl_3$ with carboxylic acids, RCO_2H (R = Me, Et or *n*-Pr), afforded[60] $[Ru_2(OCOR)_4]Cl$, which are 1:1 electrolytes in aqueous solution; the complexes are formally described as containing Ru(II) and Ru(III). Treatment of $[RuCl_6]^{3\ominus}$ with $\{R_2NCS_2\}_2$ (R = Me, Et, etc.), afforded $[Ru(S_2CNR_2)_3]$.[61]

21.5.3 Ammine Complexes and Species Containing N-Donor Atoms

The colorless hexammine, $[Ru(NH_3)_6]^{3\oplus}$, as the $Cl^\ominus$ salt, was obtained from $[Ru(NH_3)_6]^{2\oplus}$ by oxidation with chlorine.[16,62] The salts of $[Ru(NH_3)_6]^{3\oplus}$ are stable to acids, but turn yellow in alkalis, and evolve ammonia, and on subsequent addition of acid give bright blue solutions. These reactions are not observed with the pentammines, $[Ru(NH_3)_5(H_2O)]^{3\oplus}$ and $[Ru(NH_3)_5X]^{2\oplus}$ (X = halide). The chloropentammine[16,62,63] was obtained as yellow needles by the prolonged action of chlorine on $[Ru(NH_3)_6]Cl_2$ or by treatment of $[Ru(NH_3)_6]^{3\oplus}$ with HCl. Reaction of the Ru(II) hexammine, $[Ru(NH_4)_6]Cl_2$, with HCl gave the yellow $[Ru(NH_3)_5(H_2O)]Cl_3$.[16,62]

Oxidation of $[Ru(NH_3)_6]Cl_3$ in air gave[64] an intensely red species ("ruthenium red") whose structure is presently unknown. It is believed to have the empirical formula $[Ru_3O_2(NH_3)_{14}]^{6\oplus}$ and to contain a linear Ru–O–Ru–O–Ru arrangement. Oxidation of "ruthenium red" is believed to give the brown $[Ru_3O_2(NH_3)_{14}]^{7\oplus}$. When treated with ethylenediamine, $RuCl_3$ gave $[Ru(en)_2(OH)Cl]Cl$, which reacted with HCl giving $[Ru(en)_2Cl_2]Cl$.[65] Oxidation of $[Ru(en)_3]^{2\oplus}$ afforded $[Ru(en)_3]^{3\oplus}$ which, as the iodide salt, has been resolved using optically active

[58] *F. S. Martin*, J. Chem. Soc. 2682 (1952); *A. Spencer, G. Wilkinson*, J. Chem. Soc. Dalton 1570 (1972).

[59] *B. N. Figgis, G. B. Robertson*, Nature *205*, 694 (1965).

[60] *T. A. Stephenson, G. Wilkinson*, J. Inorg. Nucl. Chem. *28*, 2285 (1966).

[61] *L. Malatesta*, Gazz. Chim. Ital. *68*, 195 (1938).

[62] *F. M. Lever, A. R. Powell*, Chem. Soc. Spec. Publ. No. 13, 135 (1959).

[63] *K. Gleu, K. Rehm*, Z. Anorg. Allgem. Chemie *227*, 237 (1936); *K. Gleu, K. Rehm*, Z. Anorg. Allgem. Chemie *235*, 350 (1968).

[64] *J. M. Fletcher, B. F. Greenfield, C. J. Hardy, D. Scargill, J. L. Woodhead*, J. Chem. Soc. (London) 2000 (1961); *W. P. Griffith*, J. Chem. Soc. A, 2270 (1969).

[65] *G. T. Morgan, F. H. Burstall*, J. Chem. Soc. *41* (1936).

$[Rh(ox)_3]^{3\ominus}$ (ox = oxalate ion).[66] Reduction of optically pure $[Ru(en)_3]^{3\oplus}$ with zinc amalgam gave the corresponding ruthenium(II) complex without loss of optical configuration.

Reaction of $RuCl_3$ with pyridine (pyr) or bipyridyl in the appropriate aqueous acid afforded[19,67] $[Ru(pyr)_4Cl_2]X \cdot nH_2O$ or $[Ru(bipyr)_2Cl_2]X$ (X = Cl or ClO_4). Oxidation of $[Ru(bipyr)_3]Cl_2$ with ceric ion afforded $[Ru(bipyr)_3]^{3\oplus}$, which is blue, but has not been isolated. The corresponding red *o*-phenanthroline complex, $[Ru(phen)_3]^{3\oplus}$, was produced by chlorine oxidation of $[Ru(phen)_3]Cl_2$.[19,20]

21.5.4 Nitrosyl Complexes

Ruthenium forms more nitrosyl complexes than any other metal, and the Ru(NO) group is extremely stable. Treatment of almost any ruthenium compound with nitric acid, nitrate or nitrite ion, or nitrosonium salts will result in the formation of ruthenium nitrosyl species.

The aquated species, $Ru(NO)X_3 \cdot nH_2O$ (X = Cl or Br) was prepared[68] by evaporating a solution of RuO_4 in HX and HNO_3, and the iodide (X = I) was obtained from $Ru(NO)Cl_3 \cdot nH_2O$ by treatment with HI. The salts $K_2[Ru(NO)X_5]$ (X = Cl, Br or I) were obtained[68] by treatment of $Ru(NO)X_3 \cdot nH_2O$ with KX. Hydrolysis of $[Ru(NO)Cl_5]^{2\ominus}$ gave[69,70] $[Ru(NO)Cl_4(H_2O)]^{\ominus}$ and $[Ru(NO)Cl_4(OH)]^{2\ominus}$. Reaction of $K_2[Ru(NO)Cl_5]$ with KNCS gave[71] $K_2[Ru(NO)(NCS)_5]$ and oxidation of $K_4[Ru(CN)_6]$ with HNO_3 afforded[72] the brown $K_2[Ru(NO)(CN)_5]$. Treatment of $RuCl_3 \cdot nH_2O$ with $NaNO_2$ in water gave the orange $Na_2[Ru(NO)(NO_2)_4(OH)]$.[73]

Treatment of $Ru(NO)Cl_3 \cdot nH_2O$ with phosphines, arsines, and stibines (LR_3) or *o*-phenylenebis(dimethylarsine),[73a] bipyridyl, *etc.*(L–L), gave $[Ru(NO)(LR_3)_2Cl_3]$ or $[Ru(NO)(L\text{–}L)Cl_3]$.[74] When $[Ru(CO)_2(PPh_3)_2Cl_2]$ was treated with sodium nitrite, $[Ru(CO)_2(PPh_3)_2(ONO)_2]$ was formed.[75] This decomposed with loss of CO_2 into $[Ru(NO)_2(PPh_3)_2]$, which could also be obtained by action of NO on $[Ru(PPh_3)_3H_2]$.[76] This dinitrosyl reacted readily with oxygen giving[75] $[Ru(NO)(PPh_3)_2(O_2)(NO_3)]$ and with halogens (X_2) giving initially $[Ru(NO)(PPh_3)_2X]$ and then $[Ru(NO)(PPh_3)_2X_3]$. With oxygen in the presence of acids, $[Ru(NO)_2(PPh_3)_2]$ afforded salts of $[Ru(NO)_2(PPh_3)_2(OH)]^{\oplus}$. Reaction of $[Ru(CO)(PPh_3)_3HCl]$ with *N*-methyl-*N*-nitroso-*p*-toluenesulphonamide [*p*-$MeC_6H_4SO_2N(NO)Me$] gave[77] $[Ru(NO)(CO(PPh_3)_2Cl]$, which underwent oxidation-addition reactions with halogen giving $[Ru(NO)(PPh_3)_2ClX_2]$. Reduction of $[Ru(NO)(PPh_3)_2Cl_3]$ with zinc dust in boiling benzene afforded the emerald green $[Ru(NO)(PPh_3)_2Cl]$,[78] which, as expected, readily underwent oxidation-addition reactions, for example, with methyl iodide giving $[Ru(NO)(PPh_3)_2MeICl]$. Treatment of $[Ru(NO)(PPh_3)_2Cl]$ with $NOPF_6$ gave[79] $[Ru(NO)_2(PPh_3)_2Cl][PF_6]$, a compound containing one linear and one bent NO group.

Reaction of the polymeric $[Ru(CO)_2X_2]_n$ with NO at 230° gave[80] the polymeric $[Ru(NO)X_2]_n$, which on treatment with L (L = pyridine, ½-bipyridyl, or AsR_3) afforded $[Ru(NO)L_2X_2]_2$.[81] Nitric oxide reacted with $[Ru(S_2CEt_2)_3]$ giving[61] $[Ru(NO)(S_2CNEt_2)_3]$ in which one sulfur ligand is monodentate.

21.5.5 Phosphine, Arsine, and Stibine Complexes

Reaction of RuX_3 (X = Cl or Br) in ethanol (preferably acidified) with LR_3 (L = P or As)

---

[66] *H. Elsbernd, J. K. Beattie*, Inorg. Chem. *8*, 893 (1969).

[67] *J. Lewis, F. E. Mabbs, R. A. Walton*, J. Chem. Soc., *A*, 1366 (1967).

[68] *A. Joly* in *H. Remy*, Encyclopedie Chimique (Paris) *17*, 156 (1900); *A. Joly*, Compt. Rend. *108*, 854 (1889); *111*, 964 (1890).

[69] *M. N. Sinitsyn, O. E. Zvyagintsev*, Russian J. Inorg. Chem. *10*, 1397 (1965);

[70] *E. E. Mercer, W. M. Campbell, R. M. Wallace*, Inorg. Chem. *3*, 1018 (1964).

[71] *O. W. Haworth, R. E. Richards, L. M. Venanzi*, J. Chem. Soc. 3335 (1964).

[72] *W. Manchot, J. Düssing*, Ber. *63*, 1226 (1930).

[73] *A. Joly, M. Vezes*, Compt. Rend. *109*, 668 (1889); *J. M. Fletcher, I. L. Jenkins, F. M. Lever, F. S. Martin, A. R. Powell, R. Todd*, J. Inorg. Nucl. Chem. *1*, 378 (1955).

[73a] *J. Chatt, B. L. Shaw*, J. Chem. Soc. *A*, 1811 (1966);

[74] *M. B. Fairy, R. J. Irving*, J. Chem. Soc. *A*, 475 (1966).

[75] *K. R. Grundy, K. R. Laing, W. R. Roper*, Chem. Commun. 1500 (1970).

[76] *T. I. Eliades, R. O. Harris, M. C. Zia*, Chem. Commun, 1709 (1970).

[77] *K. R. Laing, W. R. Roper*, J. Chem. Soc. *A*, 2149 (1970).

[78] *M. H. B. Stiddard, R. E. Townsend*, Chem. Commun. 1372 (1969).

[79] *C. G. Pierpoint, D. G. Van Derveer, W. Durland, R. Eisenberg*, J. Amer. Chem. Soc. *92*, 4760 (1970).

[80] *W. Manchot, H. Schmid*, Z. Anorg. Allgem. Chemie *216*, 99 (1933).

[81] *R. J. Irving, P. G. Laye*, J. Chem. Soc. *A*, 161 (1966).

gave $[Ru(LR_3)_3X_3]$.[82] However, in the absence of acid but in methanol, ethanol, or acetone solution, the species $[Ru(LR_3)(S)X_3]$ (S = MeOH, EtOH, or Me_2CO) were obtained.[32] Reaction of these solvated species with HX afforded[83] the anionic complexes $[Ru(LR_3)_2X_4]^{\ominus}$, which were isolated as the tetraalkylammonium or tetraphenylarsonium salts. Similar anionic species were formed when the red solution, produced by treating $RuCl_3 \cdot 3H_2O$ dissolved in ethanol with CO, reacted with PR_3. Reaction of $[Ru(diars)_2X_2]$ with chlorine and $X^{\ominus}$ gave $[Ru(diars)_2X_2]X$; no hydrido species have yet been described.

21.5.6 Cyano and Stannous Complexes

$Ru(CN)_3$ was formed[84] as a green powder when chlorine was passed into solutions containing $K_4[Ru(CN)_6]$. Salts of $[Ru(CN)_6]^{3\ominus}$ have not yet been isolated, although there is polarographic evidence for such ions.[85] Salts of $[Ru(SnCl_3)_2Cl_2]^{\ominus}$ were precipitated[44] from solutions containing $RuCl_3$ and $SnCl_2$.

21.6 Ruthenium(IV)

21.6.1 Halides and Halo Complexes

Reaction of $[RuF_5]_4$ with iodine and IF_7 afforded RuF_4.[86] There is some doubt, however, about the existence of $RuCl_4$, believed to have been formed by the chlorination at 750° of $RuCl_3$.[87,88] The golden-yellow $K_2[RuF_6]$ was obtained[89] by the reaction of $[RuF_6]^{\ominus}$ with water; the red-brown $K_2[RuCl_6]$ was made either by passing chlorine through solutions containing $K_2[RuCl_5(H_2O)]$ or by fusing ruthenium metal with $KClO_3$.[48,50,90,91] On heating salts of the *hexachlororuthenate(IV)* ion, chlorine was driven off, and $[RuCl_5]^{2\ominus}$ was formed. Bromination[91] of $K_2[RuBr_5(H_2O)]$ afforded $K_2[RuBr_6]$ which forms deep blue aqueous solutions. Reaction of RuO_4 with HCl and KCl gave $K_4[Ru_2OCl_{10}]$[48,91] and the bromide[92] was prepared similarly; these compounds contain a linear Ru–O–Ru bond system.

Addition of increasing amounts of chloride ion to aquated Ru(IV) species in solution gave several complexes, among them the yellow $[Ru(H_2O)(OH)_2Cl_3]^{\ominus}$ and $[Ru(OH)_2Cl_4]^{2\ominus}$[93,94] and the violet $[Ru(H_2O)_2(OH)_2Cl_2]$ (there is some doubt about the nature of these species, and it has been suggested that the last is $[Ru_3O_2Cl_6(H_2O)_6]$ containing a linear Ru–O–Ru–O–Ru system as in "ruthenium red").

21.6.2 Oxides, Aquo Species, and Complexes Containing O- and S-Donor Atoms

Ruthenium dioxide, RuO_2, was prepared[90,95] by heating ruthenium metal or $RuCl_3$ in a stream of oxygen at 1000° or by reduction of Ru(VI) complexes. It is unaffected by acids, and, when anhydrous, forms blue crystals. The monohydrate, $RuO_2 \cdot H_2O$, could be prepared[96] by heating RuO_4 in a stream of hydrogen. "Ruthenium hydroxide", $RuO(OH)_2$, is said to be formed[97] by the reduction of RuO_4 or ruthenates, $[RuO_4]^{2\ominus}$, in alkaline solution. Ruthenites, $MRuO_3$ (M = Ca, Sr, or Ba), could be obtained[98] by heating ruthenium metal with the appropriate metal carbonate or peroxide.

Ru(IV) perchlorate solutions could be obtained[54,88,99] by the reduction of RuO_4 in per-

---

[82] *J. Chatt, G. J. Leigh, D. M. P. Mingos, R. J. Paske*, J. Chem. Soc. *A*, 2636 (1968).

[83] *T. A. Stephenson*, J. Chem. Soc. *A*, 889 (1970).

[84] *F. Krauss, G. Schrader*, Z. Anorg. Allgem. Chemie *173*, 63 (1928).

[85] *D. de Ford, A. W. Davidson*, J. Amer. Chem. Soc. *73*, 1469 (1951).

[86] *J. H. Holloway, R. D. Peacock*, J. Chem. Soc. (London) 3892 (1963).

[87] *N. I. Kolbin, A. N. Ryabov, V. M. Samoilov*, Russian J. Inorg. Chem. *8*, 805 (1963);
S. A. Shchukarev, N. I. Kolbin, A. N. Ryabov, Russian J. Inorg. Chem. *4*, 763 (1959).

[88] *O. Ruff, E. Vidic*, Z. Anorg. Allgem. Chemie *136*, 49 (1924).

[89] *M. A. Hepworth, R. D. Peacock, P. L. Robinson*, J. Chem. Soc. (London) 1197 (1954).

[90] *S. Aoyama*, Z. Anorg. Allgem. Chemie *138*, 249 (1924).

[91] *N. K. Pshenitsyn, N. A. Ezerskaya*, Russian J. Inorg. Chem. *6*, 312 (1961);
J. L. Howe, J. Amer. Chem. Soc. *32*, 775 (1901).

[92] *D. Hewkin, W. P. Griffith*, J. Chem. Soc. *A*, 472 (1966).

[93] *V. I. Shenskaya, A. A. Birynkov*, C.A. *59*, 13489h (1963).

[94] *P. Wehner, J. C. Hindman*, J. Phys. Chem. *56*, 10 (1952).

[95] *H. Remy, M. Kohn*, Z. Anorg. Allgem. Chemie *137*, 365 (1924).

[96] *J. L. Woodhead, J. M. Fletcher*, J. Less-Common Metals *4*, 460 (1962).

[97] *R. Charronat*, Ann. Chim. (Paris) *16*, 40, 68 (1931).

[98] *D. D. Khanolkar*, Curr. Sci. *30*, 52 (1961).

[99] *P. Wehner, J. C. Hindman*, J. Amer. Chem. Soc. *72*, 3911 (1950);
F. P. Gortsema, J. W. Cobble, J. Amer. Chem. Soc. *83*, 4317 (1961).

chloric acid with H_2O_2. Several hydrolyzed species are apparently present in solution, and ion-exchange, membrane, and electrochemical studies[99,100,101] indicate that the major species present is a tetramer with a tetrapositive charge. Other species present may include $[RuO(H_2O)_n]^{2\oplus}$. By allowing solutions of $[RuCl_5(H_2O)]^{2\ominus}$ to age, $[Ru(H_2O)_6]^{4\oplus}$ has apparently been prepared.[102] The oxalate, $[Ru(ox)_3]^{2\ominus}$, has been synthesized[103] by the hydrogen peroxide oxidation of $K_3[Ru(ox)_3]$.
Reaction of ruthenium metal with sulfur, selenium, or tellurium at high temperatures afforded[104] RuS_2, $RuSe_2$, or $RuTe_2$, respectively. The polysulfides, RuS_3 and RuS_6, were made[105] by reaction of $K_2[RuCl_6]$ with H_2S.

21.6.3 Ammino and Nitrogen-Containing Complexes

When $RuCl_3$ was oxidized by peroxide in the presence of pyridinium chloride, yellow $[Ru(pyr)_2Cl_4]$ was formed.[50] Nitric acid oxidation of $[BipyrH][Ru^{III}(bipyr)Cl_4]$ and its *o*-phenanthroline analog gave[106] $[Ru(bipyr)Cl_4]$ and $[Ru(phen)Cl_4]$, respectively.
Reduction of K_2RuO_4 with aqueous ammonia or $K_2[Ru(NO)Cl_5]$ with formaldehyde in base gave[107] $[Ru_2N(OH)_5 \cdot nH_2O]$. Reaction of this compound with HX (X=Cl or Br) afforded $[Ru_2NX_8(H_2O)]^{3\ominus}$; the chloride could be obtained independently by reduction of the pentachloronitrosyl Ru(III) species with $SnCl_2$ in HCl. Reaction of these octahalides with ammonia afforded $[Ru_2N(NH_3)_8X_2]^{3\oplus}$ and $[Ru_2N(NH_3)_6(H_2O)Cl_3]^{2\oplus}$. Treatment of the bromide with $AgNO_3$ in HNO_3 gave $[Ru_2N(NH_3)_8(NO_3)_2]^{3\oplus}$. All species are believed to contain linear Ru–N–Ru bonds.

21.6.4 Allylic and Olefinic Species

Ethanolic solutions of $RuCl_3$ reacted with isoprene giving a brown polymer, $[Ru(C_{10}H_{16})Cl_2]_n$.[108] This compound reacted in dichloromethane with CO giving $[Ru(C_{10}H_{16})(CO)Cl_2]$ and with pyridine giving $[Ru(C_{10}H_{16})(pyr)Cl_2]$. Buta-1,3-diene and $RuCl_3$ reacted together in 2-methoxyethanol forming[109] dodeca-2, 6, 10-triene-1, 12-diylchlororuthenate-(IV), $[Ru(C_{12}H_{18})Cl_2]$ (Fig. 1).

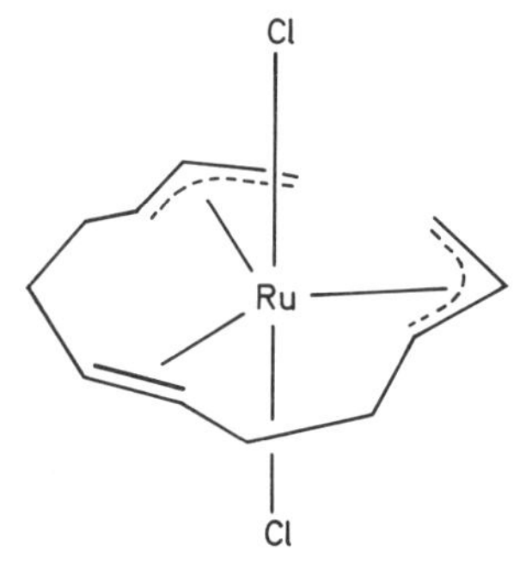

Fig. 1. The structure of $C_{12}H_{18}RuCl_2$ (H omitted for clarity).

21.7 Ruthenium(V)

21.7.1 Fluoride and Fluoro-Complexes

Tetrameric *ruthenium pentafluoride*, $[RuF_5]_4$, has been made by reaction of the metal with BrF_3.[89] The initial product was $[BrF_2^{\oplus}][RuF_6^{\ominus}]$, which decomposed *in vacuo* into $[RuF_5]_4$. It forms dark green crystals, which melt (85°–86°) to give a dark green liquid, which, in turn, boils (227°) to give a colorless vapor. Reaction of BrF_3 and bromine on $RuBr_3$ in the presence of KBr gave[89] $K[RuF_6]$, which may range in color from pink to blue depending on the conditions of synthesis. Salts such as $[BrF_2][RuF_6]$, $[BrF_4][RuF_6]$, and $[SeF_3][RuF_6]$ were obtained[110]

---

[100] *R. M. Wallace, R. C. Propst*, J. Amer. Chem. Soc. *91*, 3779 (1969).
[101] *H. Koch, H. Bruchertseifer*, Radiochem. Acta *4*, 82 (1965);
V. M. Vdovenko, L. N. Lazarev, Y. S. Kharitonov, Radiokhimiya 7, 232 (1965).
[102] *S. K. Shukla*, J. Chromatogr. *8*, 86 (1962).
[103] *R. Charronat*, Ann. Chim. (Paris) *16*, 123, 168, 188 (1931).
[104] *F. W. de Jong, A. Hoog*, Rec. Trav. Chim. Pays-Bas *46*, 173 (1927);
L. Thomassen, Z. Physik. Chem. (Frankfurt), Ser. *B2*, 349 (1929);
L. Wöhler, K. Ewald, H. G. Krall, Ber. *66*, 1638 (1933).
[105] *U. Anthony, A. Lucchesi*, Gazz. Chim. Ital. *30*, 11, 71, 539 (1900);
F. M. Jaeger, J. H. de Boer, Proc. Accad. Amst. *23*, 98 (1921).
[106] *F. P. Dwyer, H. A. Goodwin, E. C. Gyarfas*, Australian J. Chem. *16*, 42, (1963).
[107] *M. J. Cleare, W. P. Griffith*, J. Chem. Soc. *A*, 1117 (1970).
[108] *L. Porri, M. C. Gallazzi, A. Colombo, G. Allegra*, Tetrahedron Lett. 4187 (1965).
[109] *J. K. Nicholson, B. L. Shaw*, J. Chem. Soc. *A*, 807 (1966);
J. E. Lydon, J. K. Nicholson, B. L. Shaw, M. R. Truter, Proc. Chem. Soc. 421 (1964).
[110] *M. A. Hepworth, R. D. Peacock, P. L. Robinson*, Chem. & Ind. (London) 1516 (1955).

by addition of BrF_3, BrF_5, or SeF_4 to $[RuF_5]_4$. Xenon is reported to react with ruthenium pentafluoride to give $Xe[RuF_6]_n$.[111]

21.7.2 Oxides

Oxides and oxy complexes of ruthenium(V) are not well characterized, except for $Ba_3Ru_2MgO_9$, which was prepared by heating together an appropriate mixture of barium carbonate, magnesium oxide, and ruthenium metal at 1200°.[112]

21.8 Ruthenium(VI)

21.8.1 Halides and Halo Complexes

Direct fluorination of ruthenium metal gave[113] the dark brown RuF_6 (m.p. 54°), which is more reactive than its osmium analog. The oxytetrafluoride, $RuOF_4$, which is pale green (m.p. 115°), was prepared[114] by reacting the metal with BrF_3 and bromine. The compound claimed[46, 89] to be $K_2[RuF_8]$, which was obtained by the direct fluorination of $K_2[RuCl_6]$ at 200°, has been shown to be a mixed salt, KHF_2 and $K[Ru^VF_6]$. When "$RuCl_4$" was treated with HCl in the presence of CsCl, deep red $Cs_2[RuO_2Cl_4]$ was formed.[115]

21.8.2 Oxides and Aquo Species

The trioxide, RuO_3, has been detected in the vapor phase only.[116] Fusion of ruthenium metal with KNO_3 and KOH gave[117] the ruthenate, K_2RuO_4, as black crystals with a metallic green lustre. Aqueous solutions of ruthenate ions are orange, and unstable with respect to disproportionation to perruthenates and RuO_2.[118]

21.9 Ruthenium(VII)

When aqueous solutions of K_2RuO_4 were treated with chlorine until they became green,[48, 117, 119] $KRuO_4$ was formed. The perruthenate could also be obtained by fusing ruthenium metal with KOH and KNO_3 and it decomposes at 440° into K_2RuO_4 and RuO_2. It is unstable in aqueous solutions, decomposing into $[RuO_4]^{2\ominus}$ with the liberation of oxygen.

21.10 Ruthenium(VIII)

Ruthenium tetroxide, RuO_4, was easily prepared[48, 88, 118] by fusing ruthenium metal with KOH in the presence of an oxidizing agent and oxidizing the product, K_2RuO_4, with chlorine or permanganate ion in acid. Ruthenium tetroxide may also be prepared by the oxidation of RuO_2 with sodium metaperiodate in aqueous solution in the presence of CCl_4; RuO_4 is extracted into the chlorinated hydrocarbon. The tetroxide is yellow and is sparingly soluble in water.

RuO_4 is less stable chemically than osmium tetroxide, and is a much more powerful oxidizing agent. It forms weak complexes of unknown structure with PF_3, NH_3[120] and pyridine.[121] It is reduced by most acids and alkalis, being converted in the latter into the ruthenate(VI) ion. It is soluble, and apparently does not decompose, in bromine and liquid sulfur dioxide. It is the source of a great many compounds of Ru(IV), Ru(III), and Ru(II) (see Scheme 5).

Unlike OsO_4, ruthenium tetroxide has been little used in organic chemistry. However, it is known that the compound oxidizes[122] aldehydes to acids, alcohols to aldehydes or ketones, ethers to esters, olefins to aldehydes, and amides to imides. The reaction with olefins is different to that observed with OsO_4 in that the latter effects

---

[111] *N. Bartlett, N. K. Jha, H. H. Hyman*, in Noble Gas Compounds, Chicago 1963.

[112] *P. C. Donohue, L. Katz, R. Ward*, Inorg. Chem. *5*, 335, 339 (1966).

[113] *H. H. Claasen, H. Selig, J. G. Malm, C. L. Chernick, B. Weinstock*, J. Amer. Chem. Soc. *83*, 2390 (1961).

[114] *J. H. Holloway, R. D. Peacock*, J. Chem. Soc. (London) 527 (1963).

[115] *J. L. Howe*, J. Amer. Chem. Soc. *32*, 779 (1901).

[116] *W. E. Bell, M. Tagami*, J. Phys. Chem. *67*, 2432 (1963);
H. Schafer, A. Tebben, W. Gerhardt, Z. Anorg. Allgem. Chemie *321*, 41 (1963).

[117] *A. Gutbier, F. Falco, H. Zwicker*, Angew. Chem. *22*, 490 (1909);
F. Krauss, Z. Anorg. Allgem. Chemie *132*, 301 (1924).

[118] *G. Nowogrocki, G. Tridot*, Bull. Chim. Soc. (France) 688 (1965).

[119] *H. Debray, H. Joly*, Compt. Rend. *106*, 331, 1499 (1888).

[120] *M. L. Hair, P. L. Robinson*, J. Chem. Soc. (London) 2775 (1960); 106 (1958).

[121] *Y. Koda*, Inorg. Chem. *2*, 1306 (1963).

[122] *L. M. Berkowitz, P. N. Rylander*, J. Amer. Chem. Soc. *80* 6682 (1958);
P. N. Rylander, Engelhard Tech. Bull. *9*, 135 (1969).

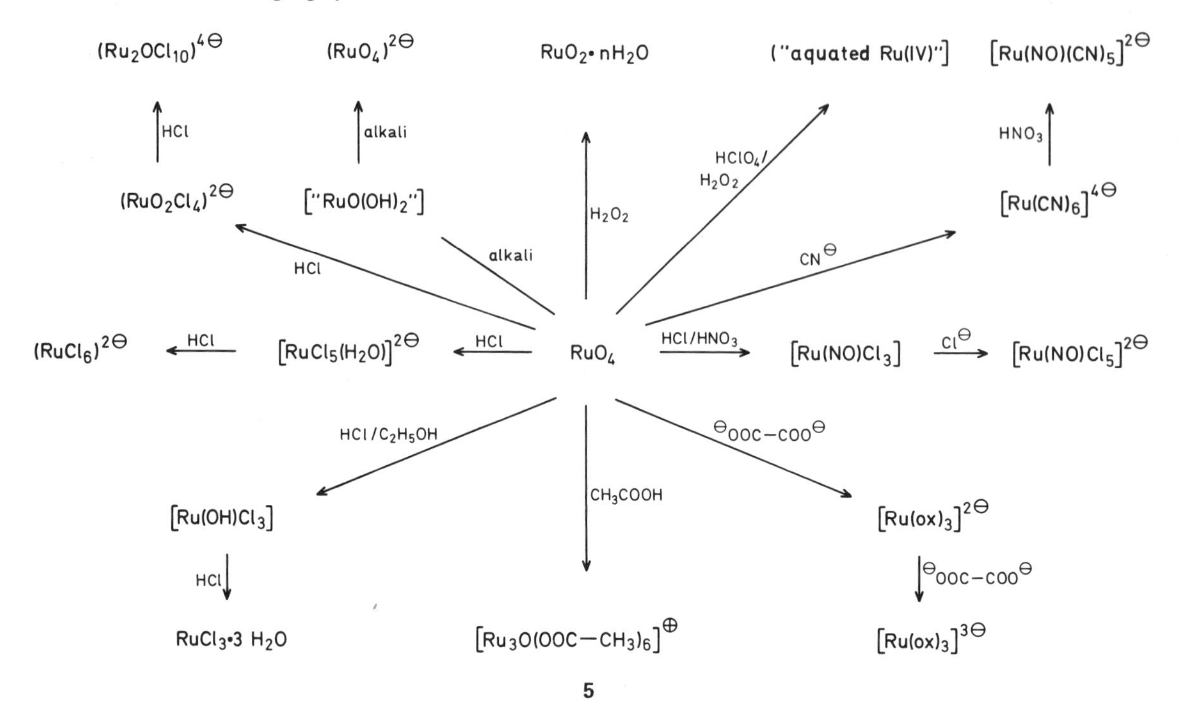

5

cis-hydroxylation of the double bond, whereas RuO_4 effects hydroxylation with C–C bond cleavage. Since RuO_4 reacts violently with almost all solvents, only CCl_4 or $CHCl_3$ can be used effectively[124] for these reactions, and convenient methods for the preparation of RuO_4/CCl_4 solutions have been described.
Ruthenium tetroxide is somewhat less toxic than osmium tetroxide, but being volatile, must be regarded as a dangerous poison.

[123] *C. Djerassi*, *R. R. Engle*, J. Amer. Chem. Soc. *75*, 3838 (1953).

21.11 Bibliography

Gmelin's Handbuch der Anorganischen Chemie, 8 Auflage, Vol. 64. Verlag Chemie, Berlin, 1939.

P. Pascal, Nouveau Traité de Chemie Minérale, Masson, Paris, 1958.

W. P. Griffith, The Chemistry of the Rarer Platinum Metals (Os, Ru, Ir and Rh). Wiley (Interscience), New York, 1967.

22 Osmium

B. J. McCormick
Department of Chemistry, West Virginia University,
Morgantown, West Virginia 26506, U.S.A.

Deviations from the general organizational scheme will be found in Section 22.5.1 where "osmyl" ($OsO_2^{2\oplus}$) and "nitridoosmate(VI)" ($OsN^{3\oplus}$) derivatives are discussed. All compounds involving $SCN^{\ominus}$ as a ligand are regarded as isothiocyanate derivatives, and nitrogen oxide, NO, is assumed to function as a three-electron donor.

22.1 Purification and Recovery of Elemental Osmium; Derivatives of Osmium(0) and Osmium(I)

Osmium is conveniently separated from impurities by converting it to the volatile tetroxide. This can be done by burning the metal in oxygen or by fusing it in an oxidizing alkali flux, such as a mixture of potassium hydroxide and potassium nitrate, followed by dissolution in water and distillation from nitric acid.[1,2] Osmium tetroxide may also be obtained from osmium-containing residues by roasting the residues in a stream of air or oxygen. Once formed, the tetroxide may be reduced directly to the metal or converted to a wide variety of osmium compounds.[3]

Osmium(0) compounds that do not contain carbonyl ligands are very rare. *Pentakis(trifluorophosphine)osmium(0)* has been produced at 250° and 500 atm in very low yield[4] in accordance with Eqn (1).

$$OsCl_3 + 3Cu + 5PF_3 \longrightarrow Os(PF_3)_5 + 3CuCl \quad (1)$$

If even small amounts of hydrogen-containing substances such as water are present, hydride complexes are formed in preference to $Os(PF_3)_5$. Only two osmium(I) compounds are sufficiently well known to be mentioned. *Osmium(I) iodide* has been prepared as a gray amorphous material by heating osmium tetroxide, hydroiodic acid, and ethanol for an extended period of time in a carbon dioxide atmosphere.[5] *Hexaammineosmium(I) bromide* has been synthesized by reducing hexaammineosmium(III) bromide with two molar equivalents of potassium in liquid ammonia.[6]

[1] *P. Pascal*, Nouveau Traité de Chimie Minérale, Vol. XI, Masson et Cie, Paris 1932.

[2] *G. Brauer*, Handbook of Preparative Inorganic Chemistry, Vol. II, 2nd Edition, Academic Press, New York 1963.

[3] *W. P. Griffith*, The Chemistry of the Rarer Platinum Metals, Chap. 3, Interscience Publishers, New York 1967.

[4] *Th. Kruck, A. Prasch*, Z. Anorg. Allgem. Chemie *356*, 118 (1968).

[5] *J. E. Fergusson, B. H. Robinson, W. R. Roper*, J. Chem. Soc. (London) 2113 (1962).

22.2 Derivatives of Osmium(II)

The chemistry of osmium(II) is probably more extensive than that for any other oxidation state, and complexes with a variety of donor atoms are known; however, in most cases at least one of the ligands is a good π-acceptor, such as 2,2′-bipyridyl or triphenylphosphine. A common method of preparation is the reduction of Os(III) or Os(IV) halides or halo complexes in the presence of ligands. It has also become increasingly clear that many other Os(III) and Os(IV) complexes may be reduced to the corresponding Os(II) compounds with relatively mild reducing agents. Complexes involving donors from Groups VII, VI, V, and IV are known, but those from Groups VII and V are the most important.

22.2.1 Hydride and Halide Derivatives

Many of the *hydride derivatives of osmium(II)* involve chelating ditertiary phosphine ligands. Colorless or yellow complexes of the general type *trans*-[OsHX(diphosphine)$_2$] can be prepared[7,8] by treating *cis*-[OsX$_2$(diphosphine)$_2$] with lithium aluminum hydride, *e.g.*:

$X = Cl, I, CH_3, C_2H_5$

diphosphine $= R_2P{-}CH_2{-}CH_2{-}PR_2$

$(R = CH_3, C_2H_5, C_6H_5)$

$(C_6H_5)_2P{-}CH_2{-}P(C_6H_5)_2$

o-$C_6H_4[P(C_2H_5)_2]_2$

trans-Dihydridobis(1,2-bis-diphenylphosphinomethane)osmium(II) has been made from the corresponding *cis*-dichloride and a large excess of lithium aluminum hydride.[9] Other dihydrides include those of the type *cis*-[OsH$_2$L$_4$] (L = $P(CH_3)_2(C_6H_5)$, $P(C_2H_5)_2(C_6H_5)$, $P(CH_3)(C_6H_5)_2$, $P(C_2H_5)(C_6H_5)_2$, $As(C_2H_5)_2(C_6H_5)$, or $As(C_2H_5)(C_6H_5)_2$), which can be obtained by treating [OsH$_4$L$_3$] with excess L in refluxing toluene,[10] and *cis-dihydridotetrakis(trifluorophosphine)osmium(II)* which has been synthesized by reacting

[6] *G. W. Watt, E. M. Potrafke, D. S. Klett*, Inorg. Chem. *2*, 868 (1963).

[7] *J. Chatt, R. G. Hayter*, J. Chem. Soc. (London) 2605 (1961).

[8] *J. Chatt, R. G. Hayter*, J. Chem. Soc. (London) 6017 (1963).

[9] *S. D. Ibekwe, U. A. Raeburn*, J. Organometal. Chem. *19*, 447 (1969).

[10] *B. Bell, J. Chatt, G. J. Leigh*, J. Chem. Soc. *D*, 576 (1970).

osmium(III) chloride with trifluorophosphine, hydrogen, and copper at elevated temperatures and pressures.[11] Several cationic hydrides have been isolated from acetone as tetraphenylborates, according to Eqn (2),

$$OsHCl[(C_2H_5)_2P(CH_2)_2P(C_2H_5)_2]_2 + NaB(C_6H_5)_4 + L \longrightarrow OsHL[(C_2H_5)_2P(CH_2)_2P(C_2H_5)_2]_2B(C_6H_5)_4 + NaCl \quad (2)$$

where L is p-$CH_3OC_6H_4NC$, $(CH_3)_3CNC$, $P(OCH_3)_3$, $P(OC_6H_5)_3$, or C_6H_5CN.[12]

Osmium halide compounds: the only simple halide known is *osmium(II) iodide*, which has been obtained as a black solid by treating osmium tetroxide with hydroiodic acid and ethanol under carefully controlled conditions.[5] On the other hand an extensive series of mixed-ligand osmium(II) compounds is known in which at least one of the ligands is a halide. An indication of the significance of some of these compounds is found in the report[13] that $[OsX_3(PR_3)_3]$ and $[OsX_3(AsR_3)_3]$ (X=Cl, Br; R=CH_3, C_2H_5, C_6H_5, n-C_3H_7, n-C_4H_9) in tetrahydrofuran under 100 atm of nitrogen are reduced by zinc amalgam to the complexes *mer*-$[OsX_2(N_2)(PR_3)_3]$ and *mer*-$[OsX_2(N_2)(AsR_3)_3]$, respectively. These compounds appear to be among the more well characterized complexes involving molecular nitrogen as a ligand.

A large number of compounds is known in which osmium(II) is bonded to an amine as well as to a halide. Hydrated species of the type *cis*-$[OsX_2(amine)_2]$, H_2O (X=Cl, Br; amine=2,2′-bipyridyl, 1,10-phenanthroline) have been obtained from the reduction of the corresponding osmium(III) halides with sodium dithionite.[14] One of the coordinated halides was found to be labile in refluxing 1,2-dihydroxyethane and was easily displaced by pyridine to give *cis*-[OsX(pyridine)(amine)$_2$]X′.H_2O (X′=Cl, I; n=1–3). It was also reported[14] that the passage of ammonia through a refluxing 1,2-dihydroxyethane solution of *cis*-dichlorobis(2,2′-bipyridyl)osmium(II) provided *chloroamminebis(2,2′-bipyridyl)osmium(II)*, which was isolated as the red iodide or chloride salt. The utility of refluxing 1,2-dihydroxyethane as a reaction medium for carrying out substitution reactions is further illustrated by the reactions of tetrachloro(2,2′-bipyridyl)osmium(IV)[15] and trichloro(2,2′,2″-terpyridyl)osmium(III)[16] with pyridine in this solvent to give *chlorotris(pyridine)(2,2′-bipyridyl)osmium(II)* and *chlorobis(pyridine)(2,2′,2″-terpyridyl)osmium(II)* respectively, which were isolated as *chloride*, *iodide*, or *perchlorate* salts. In these reactions the solvent acts as a reducing agent. Similar compounds with bromide or iodide rather than chloride in the coordination sphere also were reported. Complexes of the type [OsX(2,2′-bipyridyl)(2,2′,2″-terpyridyl)]X, nH_2O (X=Cl, Br, I; n=0–2) were obtained by similar procedures.[16] Some use of related solvents has been made. Thus potassium hexachloroosmate(IV) and ammonium hexabromoosmate(IV) give *dichlorotetrakis(pyridine)osmium(II)* and *dibromotetrakis(pyridine)osmium(II)*, respectively, when refluxed with pyridine in 1,2,3-trihydroxypropane.[15]

Several *osmium(II) nitrosyl complexes* are known, all of which contain a coordinated halide. The most simple of these compounds are of the form $K_2[Os(NO)X_5]$ (X=Cl, Br, I) and are obtained by treating potassium hydroxotetra(nitro)nitrosylosmate(II) with the appropriate halogen acid.[17] The remaining nitrosyl compounds have the general formula $OsX_3(NO)[Y(C_6H_5)_3]_2$ (X=Cl, Br, I; Y=P, As, Sb). The triphenylphosphine and triphenylarsine derivatives were prepared by reacting hexachloro-, hexabromo-, or hexaiodoosmate(IV) salts with nitrogen oxide and the phosphine or arsine,[18] but the triphenylstibine complexes were obtained by treating $OsX_3[Sb(C_6H_5)_3]_2$ with nitrogen oxide.[18,19] The novel use of N-methyl-N-nitrosotoluene-p-sulfonamide as a convenient source of nitrosyl ion in the preparation of compounds of this type has also been mentioned.[20]

---

[11] *Th. Kruck, A. Prasch,* Z. Anorg. Allgem. Chemie *371*, 1 (1969).
[12] *G. M. Bancroft, M. J. Mays, B. E. Prater, F. P. Stefanini,* J. Chem. Soc. *A*, 2146 (1970).
[13] *J. Chatt, G. J. Leigh, R. L. Richards,* J. Chem. Soc. *A*, 2243 (1970).
[14] *D. A. Buckingham, F. P. Dwyer, H. A. Goodwin, A. M. Sargeson,* Australian J. Chem. *17*, 325 (1964).
[15] *D. A. Buckingham, F. P. Dwyer, H. A. Goodwin, A. M. Sargeson,* Australian J. Chem. *17*, 315 (1964).
[16] *D. A. Buckingham, F. P. Dwyer, A. M. Sargeson,* Australian J. Chem. *17*, 622 (1964).
[17] *L. Wintrebert,* Ann. Chim. (Paris) *28* (7) 15 (1903).
[18] *A. Araneo, C. Bianchi,* Gazz. Chim. Ital. *97*, 885 (1967).
[19] *A. Araneo, V. Valenti, F. Cariati,* J. Inorg. Nucl. Chem. *32*, 1877 (1970).
[20] *J. J. Levison, S. D. Robinson,* J. Chem. Soc. A, 2947 (1970).

A variety of osmium(II) complexes are known in which halide ions and phosphines, arsines or stibines, as well as other molecules in some cases, simultaneously function as ligands. In this group are complexes having the formula $OsX_2(NR^1_3)(PR^2_3)_3$. Compounds of this type with $R^1=H$, CH_3, C_2H_5, and n-C_4H_9, $R^2=CH_3$, C_6H_5, and C_2H_5, and X=Cl or Br have been obtained from $[OsX_3(PR^2_3)_3]$ and the appropriate amine in ethanol, which functions both as the solvent and reducing agent.[21] Similar derivatives were also obtained with hydrazine and phenylhydrazine.[21] Monodentate phosphines provide dimeric chloro complexes of the form $[Os_2Cl_3(PR_3)_6]Cl$ (R= CH_3, C_2H_5, or C_6H_5) when treated with ammonium hexachloroosmate(IV) in the presence of ethanol[22] (see Eqn 3, Section 22.3.1). It is thought that the metal atoms in these complexes are bridged by the chloride ligands. The green complex obtained from the reaction of ammonium hexabromoosmate(IV) with triphenylphosphine which was formulated as $OsBr_2[P(C_6H_5)_3]_3$[23] may have the same type of dimeric structure. It is noteworthy that the phosphines in $OsBr_2[P(C_6H_5)_3]_3$ (dimer?) are readily replaced by triaryl phosphites to give complexes of the type $OsBr_2[P(OR)_3]_4$, with $R=C_6H_5$, p-C_6H_4Cl, and p-$C_6H_4(CH_3)$.[24] Tertiary arsines and stibines provide species analogous to those formed with triaryl phosphites; however, they are best obtained by reducing $[OsX_3(YR_3)_3]$ (X=Br, Cl; Y=As, Sb; $R=CH_3$, C_6H_5) with an alcohol,[23] hypophosphorous acid,[25] or sodium hydridoborate.[18] All of these *dihalo complexes* probably have a *trans* structure. Ditertiary phosphines provide similar compounds of the form $[OsCl_2(\text{ditertiary phosphine})_2]$, and complexes having both *cis* and *trans* structures can be synthesized.[22] *Trans* structures result when ammonium hexachloroosmate(IV) is reduced in the presence of the phosphine ligand. Easily oxidized compounds with *cis* structures arise when tri-μ-chlorohexakis(diethylphenylphosphine)diosmium(II) chloride dihydrate is treated with a ditertiary phosphine at 150–200° in the absence of solvent. Similar compounds, which are probably *trans*, are formed when an ammonium hexahaloosmate(IV) is treated with a ditertiary arsine in ethanol.[26]

Alkylation of *cis*- and *trans*-$[OsCl_2(\text{ditertiary phosphine})_2]$ has been accomplished by treatment with trimethyl- or triethylaluminum in the absence of solvent.[8] In no case were both chlorides replaced, but inversion of configuration was common.

22.2.2 Derivatives Containing Osmium(II)-Oxygen, -Nitrogen, -Phosphorus, -Arsenic, or -Carbon Bonds

The compound obtained from the action of potassium nitrite on potassium hexachloroosmate(IV), which was originally[17,27] thought to be potassium pentanitroosmate(III), is now formulated[28] as the mixed-ligand complex *potassium hydroxotetra(nitro)nitrosylosmate(II)* on the basis of results of infrared studies.[29,30] A few other complexes involving Group VI donors are known. Treatment of dichlorobis(2,2′-bipyridine)osmium(II) with sodium oxalate, or a basic solution of glycine or acetylacetone results in a substitution reaction in which the chlorides are displaced to give *oxalato*, *glycinato*, or *acetylacetonato osmium complexes*, respectively.[14]

Similar reactions were carried out with 1,10-phenanthroline complexes as starting materials and there seems to be little doubt that this type of reaction could be extended and used for the preparation of a wide variety of mixed-ligand species. The remaining osmium(II)-chalcogonides that should be mentioned are sulfito derivatives. Although several are known, they are not well enough characterized to be discussed here in detail; a discussion of these systems may be found elsewhere.[3,28]

Very stable octahedral complexes with aromatic bases are well known. The general preparative procedure that has found wide use is the treatment of a hexahaloosmate(IV) salt with the base at elevated temperature. In this way *tris(2,2′-bipyridyl)osmium(II)*,[31] *tris(1,10-phenanthro-*

[21] *J. Chatt, G. J. Leigh, R. J. Paske*, J. Chem. Soc. A, 854 (1969).

[22] *J. Chatt, R. G. Hayter*, J. Chem. Soc. (London) 896 (1961).

[23] *L. Vaska*, Chem. & Ind. (London) 1402 (1961).

[24] *J. J. Levison, S. D. Robinson*, J. Chem. Soc. A, 639 (1970).

[25] *F. P. Dwyer, R. S. Nyholm, B. T. Tyson*, J. Proc. Roy. Soc. N.S. Wales *81*, 272 (1947).

[26] *R. S. Nyholm, G. J. Sutton*, J. Chem. Soc. (London) 572 (1958).

[27] *L. Wintrebert*, Compt. Rend. *140*, 585 (1905).

[28] *W. P. Griffith*, Quart. Rev. Chem. Soc. *19*, 254 (1965).

[29] *P. Gans, A. Sabatini, L. Sacconi*, Inorg. Chem. *5*, 1877 (1966).

[30] *M. B. Fairey, R. J. Irving*, Spectrochim. Acta *22*, 359 (1966).

[31] *F. H. Burstall, F. P. Dwyer, E. C. Gyarfas*, J. Chem. Soc. (London) 953 (1950).

line)osmium(II),[32] and *bis(2,2′,2″-terpyridyl)osmium(II)*[33] ions have been prepared and isolated as a variety of salts. The enantiomorphous forms of the tris complexes have been isolated as perchlorates.[31,32] A rather extensive series of mixed-ligand octahedral complexes involving 2,2′-bipyridyl, 1,10-phenanthroline, and 2,2′,2″-terpyridyl is known (Table 1). The inner complex, *cis-dicyanobis(2,2′-bipyridyl)osmium(II)*, has been made by the reduction of ammonium hexachloroosmate(IV) with 2,2′-bipyridyl and sodium cyanide.[34,35]

Table 1. Mixed-Ligand Osmium Complex Ions with Aromatic Amines*

Complex ion†	Preparative method	Ref.
$[Ospy_2bipy_2]^{2\oplus}$	$[OsCl_2bipy_2]$ + py in 1,2-dihydroxyethane	a
$[Ospy_2phen_2]^{2\oplus}$	$[OsCl_2phen_2]$ + py in 1,2-dihydroxyethane	a
$[Ospy_2phenbipy]^{2\oplus}$	$[OsCl_2phenbipy]$ + py in 1,2-dihydroxyethane	a
$[Ospy_4bipy]^{2\oplus}$	$[OsCl_3pybipy]$ + py + NaH_2PO_2	b
$[Ospy_3terpy]^{2\oplus}$	$[OsterpyCl_3]$ + py in 1,2-dihydroxyethane	c
$[Ospybipyterpy]^{2\oplus}$	$[OsClbipyterpy]Cl$ + py	c
$[Osen_2bipy]^{2\oplus}$	$K[OsCl_4bipy]$ + en	c
$[Osenbipy_2]^{2\oplus}$	$[OsCl_2bipy_2]Cl$ + en	c
$[Osenphen_2]^{2\oplus}$	$[OsCl_2phen_2]Cl$ + en	c
$[Osbipyphen_2]^{2\oplus}$	$[OsCl_2phen_2]$ + bipy	c
$[Osbipy_2phen]^{2\oplus}$	$[OsCl_2bipy_2]$ + phen	c
$[Os(NH_3)_2bipy_2]^{2\oplus}$	$[OsCl_2bipy_2]$ + liquid NH_3; 120°	c
$[Os(NCS)bipyterpy]^{\oplus}$	$[OsClbipyterpy]Cl$ + KSCN in 1,2-dihydroxyethane	c
$[Os(NO_2)bipyterpy]^{\oplus}$	$[OsClbipyterpy]Cl$ + $NaNO_2$ in 1,2-dihydroxyethane	c
$[Os(CH_3CN)bipyterpy]^{2\oplus}$	$[OsClbipyterpy]Cl$ + CH_3CN reflux	c
$[Os(C_2H_5CN)bipyterpy]^{2\oplus}$	$[OsClbipyterpy]Cl$ + C_2H_5CN reflux	c
$[Os(py\text{-}R)bipyterpy]^{2\oplus}$	$[OsClbipyterpy]Cl$ + py-R; R = CH_3, C_2H_5, $i\text{-}C_4H_7$	c

* py = pyridine; bipy = 2,2′-bipyridyl; phen = 1,10-phenanthroline; terpy = 2,2′,2″-terpyridyl; en = ethylenediamine (1,2-diaminoethane).

† The crystalline complexes were isolated in general as chlorides, iodides, or perchlorates.

[a] *D. A. Buckingham, F. P. Dwyer, H. A. Goodwin, A. M. Sargeson*, Aust. J. Chem., *17*, 325 (1964).

[b] *D. A. Buckingham, F. P. Dwyer, H. A. Goodwin, A. M. Sargeson*, ibid. *17*, 315 (1964).

[c] *D. A. Buckingham, F. P. Dwyer, A. M. Sargeson*, ibid. *17*, 622 (1964).

A dinitrogen complex, *pentaamminedinitrogenosmium(II) chloride*, $[Os(NH_3)_5N_2]Cl_2$, has been synthesized by treating ammonium hexachloroosmate(IV) with hydrazine hydrate under reflux.[36] This compound can be diazotized to give *tetraamminebis(dinitrogen)osmium(II) chloride*, $[Os(NH_3)_4(N_2)_2]Cl_2$.[37]

Osmium–carbon complexes: the phosphine derivatives *cis- dimethylbis(1,2-bisdiphenylphosphinomethane)osmium(II)* and *cis-diphenylbis(1,2-bisdiphenylphosphinomethane) osmium (II)* are obtained when the corresponding *cis*-dichlorides are reacted with methyl and phenyllithium, respectively.[8] These are among the few compounds that contain an osmium-carbon σ-bond. Another example of osmium-carbon bonding in mixed ligand complexes is that provided by potassium pentacyanonitrosylosmate(II) dihydrate, which has been reported to form when potassium hexacyanoosmate(II) is treated with nitric acid.[38]

The remaining osmium(II) compounds are the products of isolated syntheses and as such represent unique species. The first compound in this category is *diisothiocyanatobis(o-phenylenebisdimethylarsine)osmium(II)*, which has been obtained from ammonium hexabromoosmate(IV), ammonium thiocyanate and the diarsine.[26] It is thought that the $NCS^{\ominus}$ groups are in *trans* positions. *Potassium hexacyanoosmate(II) trihydrate* is formed when excess potassium cyanide reacts with osmium tetroxide or potassium osmate, $K_2[OsO_2(OH)_4]$.[39] The corresponding acid, $H_4[Os(CN)_6]$, is obtained when the potassium salt is treated with hydrogen chloride in diethylether.[39–41]

22.3 Derivatives of Osmium(III)

Osmium(III) is not a well known oxidation state, partially because trivalent osmium complexes are easily

[32] *F. P. Dwyer, N. A. Gibson, E. C. Gyarfas*, J. Proc. Roy. Soc. N.S. Wales *84*, 68 (1950).

[33] *G. Morgan, F. H. Burstall*, J. Chem. Soc. (London) 1649 (1937).

[34] *A. A. Schilt*, J. Amer. Chem. Soc. *85*, 904 (1963).

[35] *A. A. Schilt*, Inorg. Chem. *3*, 1323 (1964).

[36] *A. D. Allen, J. R. Stevens*, Chem. Commun. 1147 (1967).

[37] *H. A. Scheidegger, J. N. Armor, H. Taube*, J. Amer. Chem. Soc. *90*, 3263 (1968).

[38] *E. J. Baran, A. Muller*, Z. Anorg. Allgem. Chemie *370*, 283 (1969).

[39] *F. Krauss, G. Schrader*, J. Prakt. Chem. *119*, 279 (1928).

[40] *D. F. Evans, D. Jones, G. Wilkinson*, J. Chem. Soc. (London) 3164 (1964).

[41] *A. D. Ginsberg, E. Koubek*, Inorg. Chem. *4*, 1186 (1965).

oxidized to the tetravalent state or, in the presence of strong π-acceptors, are readily reduced to the divalent state. Binary halides and halogeno complexes are known, along with a variety of mixed-ligand complexes involving halides. Group VI donors are of minor importance, although recent work suggests that it may be possible to prepare stable complexes with osmium-sulfur or -selenium bonds. The most important donors for this oxidation state are those of Group V. In view of some of the synthetic advances that have been made recently, it would appear that tertiary phosphine complexes will become well known. Compounds with Group IV donors are essentially unknown.

22.3.1 Halide Derivatives

Three binary osmium(III) halides have been synthesized. *Osmium(III) chloride* has been obtained from the thermal decomposition of osmium(IV) chloride in an atmosphere of chlorine.[42] A similar procedure has been used for the preparation of *osmium(III) bromide*.[43] *Osmium(III) iodide*, OsI_3, has been obtained by heating dihydronium hexaiodoosmate(IV) at 250° under a nitrogen atmosphere.[5] *Potassium hexachloro- and hexabromoosmate(III)* are both known. The former compound, K_3OsCl_6, can be made by heating a mixture of osmium and potassium chloride in a current of chlorine,[44] and the latter compound, K_3OsBr_6, has been made by the electrolytic reduction of potassium hexabromoosmate(IV) under an atmosphere of carbon dioxide and in the presence of excess bromide.[45]
Osmium(III) amine complexes: the reduction of potassium pentachloronitridoosmate(VI) with tin(II) chloride in hydrochloric acid produces *potassium pentachloroammineosmate(III)*, which was originally formulated as an amide.[46–48] Other halogeno-ammine complexes include *chloropentammineosmium(III)* and *bromopentammineosmium(III)*, which have been synthesized by ammoniating ammonium hexahalogenoosmate(IV) salts at elevated temperatures[49,50] and isolated as a variety of salts. These syntheses are difficult and very low yields are obtained. More recently it has been mentioned[36] that a much superior route to these compounds is the treatment of pentamminedinitrogenosmium(II) chloride with iodine and hydroiodic acid, to give *iodopentammineosmium(III) iodide*, from which the corresponding *chloride* and *bromide* can be obtained by treatment with the appropriate hydrohalic acid. The halogeno-amine complexes *dichlorobis(2,2′-bipyridyl)osmium(III)* chloride and *dichlorobis(1,10-phenanthroline)osmium(III) chloride* are well known. The best preparative method appears to be the reaction of potassium hexachloroosmate(IV) with the appropriate amine in refluxing *N,N*-dimethylformamide.[14]
Salts of *tetrachloro(1,10-phenanthroline)osmate(III)* and *tetrachloro(2,2′-bipyridyl)osmate(III)* have been obtained from the reduction of the corresponding osmium(IV) compounds with hypophosphorous acid.[15] Substitution reactions were carried out on these compounds in such a way that one or more chloride ligands was replaced with water, pyridine, acetylacetonate, or glycinate.[15] *Trichlorotris(pyridine)osmium(III)* also has been prepared.[15] Perchlorate salts of *chloro-* and *bromotris(pyridine)(2,2′-bipyridyl)*-osmium(III) are obtained upon oxidation of the corresponding osmium(II) compounds with mild oxidants.[15]
Group (V) osmium(III) complexes: a reaction scheme of great versatility has been used to prepare a collection of *halo-phosphine and -arsine osmium(III) complexes*.[51] When osmium tetroxide dissolved in ethanol containing a hydrohalic acid, HX, is treated with excess tertiary phosphine or arsine, L, a complex of the type $(LH)_2[OsX_6]$ is formed, which may undergo a series of stepwise reactions as indicated in Eqn (3). In general it is not possible to isolate all

$$(LH)_2[OsX_6] \longrightarrow [OsX_4L_2] \longrightarrow [OsX_3L_3] \longrightarrow [L_3OsX_3OsL_3]X \quad (3)$$

of the compounds from a particular ligand L; nevertheless, it is found that a wide variety of tertiary phosphines and arsines react in such a way as to provide easily isolated complexes of the type *mer*-$[OsX_3(YR_3)_3]$ (X = Cl, Br; Y = P, As; R = CH_3, C_2H_5, *n*-C_3H_7, *n*-C_4H_9, C_6H_5).

In some earlier work dealing with the preparation of tertiary arsine complexes[25] reactions like those shown in (3) were undoubtedly involved. Similarly, a series of

[42] *N. I. Kolbin, I. N. Semenov, Y. M. Shutov*, Russian J. Inorg. Chem. (Engl. Transl.) *8*, 1270 (1963).
[43] *S. A. Schukarev, N. I. Kolbin, I. N. Semenov*, Russian J. Inorg. Chem. (Engl. Transl.) *6*, 638 (1961).
[44] *C. Claus*, J. Prakt. Chem. *90*, 83 (1863).
[45] *W. R. Crowell, R. K. Brinton, R. F. Evenson*, J. Amer. Chem. Soc. *60*, 1105 (1938).
[46] *L. Brizard*, Compt. Rend. *123*, 182 (1896).
[47] *L. Brizard*, Ann. Chim. (Paris) *21* (7), 311 (1900).
[48] *W. P. Griffith*, J. Chem. Soc. A, 899 (1966).
[49] *F. P. Dwyer, J. W. Hogarth*, J. Proc. Roy. Soc. N.S. Wales *84*, 117 (1950).
[50] *G. W. Watt, L. Vaska*, J. Inorg. Nucl. Chem. *5*, 304 (1958).
[51] *J. Chatt, G. J. Leigh, D. M. P. Mingos, R. J. Paske*, J. Chem. Soc. A, 2636 (1968).

complexes of the form $[OsCl_3(P(CH_3)_2C_6H_5))_2L']$ where L′ is a tertiary phosphine or arsine, alkylphosphite, or alkyl phosphonite have been prepared from tetrachlorobis(dimethylphenylphosphine)osmium(IV).[52] It should also be noted that a number of complexes the type *fac*-$[OsCl_3L_3]$, where L is dimethylphenylphosphine, diethylphenylphosphine, or dimethylphenylarsine, have been synthesized by treating the corresponding tetrahydrido complexes with hydrogen chloride in methanol.[52]

There are other scattered reports on the synthesis of complexes of the type $[OsX_3L_3]$ by a variety of techniques. These compounds include *tribromotris(triphenylstibine)osmium(III)*,[23] *triiodotris(triphenylstibine)osmium(III)*,[23] *tribromotris(triphenylarsine)osmium(III)*,[23] *trichlorotris(trichlorophosphine)osmium(III)*, $[OsCl_3(PCl_3)_3]$,[53] and a compound in which one phosphine ligand is replaced by ammonia, *trichloroamminebis(triphenylphosphine)osmium(III)*, which has been prepared from oxotrichlorobis(triphenylphosphine)osmium(IV) and hydrazine dihydrochloride.[54]
Several complex ions of the form $[OsX_2$(*o*-phenylenebisdimethylarsine)$_2]^{\oplus}$ (X=Cl, Br, I), which probably are in the *trans*-configuration, have been prepared by oxidizing the corresponding osmium(II) compounds with halogens.[26]

22.3.2 Derivatives Containing Osmium(III)-Oxygen, -Sulfur, -Selenium, -Nitrogen, or Arsenic Bonds

Compounds containing osmium(III)-chalcogen bonds are not numerous. Dark red *tris(acetylacetonato)osmium(III)* is known[55] and acetylacetonate donors are found in several mixed-ligand complexes including *acetylacetonatobis(2,2′-bipyridyl)osmium(III) perchlorate* and *acetylacetonatobis(1,10-phenanthroline)osmium(III) perchlorate*, which were obtained from the oxidation of the corresponding osmium(II) compounds with iron(III) chloride,[14] and *bis(acetylacetonato)(2,2′-bipyridyl)osmium(III)* iodide hydrate.[15] Bis(glycinato)(2,2′-bipyridyl)osmium(III) iodide dihydrate also is known.[15] Osmium(III)-sulfur bonds are found in *tris(3-thiolo-1,3-diphenyl-2-propene-1-one)osmium(III)*, which was prepared from potassium hexachloroosmate(IV), sodium acetate, and the sulfur-containing ligand in ethanol,[56] and in *tris(4-thiolo-3-pentene-2-thione)osmium(III)*, which was prepared by reacting mixtures of osmium tetroxide, acetylacetone, and hydrogen chloride with hydrogen sulfide in ethanol.[57] The only example of selenium bonded to osmium is provided by *tris(N,N-diethyldiselenocarbamato)osmium(III)*, which is obtained from the reaction of diethylammonium *N,N*-diethyldiselenocarbamate with ammonium hexabromoosmate(IV).[58]
Several *osmium(III) amine complexes* are known; *hexaamineosmium(III) bromide* can be prepared in low yield by treating ammonium hexabromoosmate(IV) with liquid ammonia at 25°.[6,59] Chromatography on activated alumina is necessary for purification. Ethylenediamine reacts with hexabromoosmate(IV) ion to give *osmium(IV)-ethylenediamine complexes* in which one or more of the nitrogen atoms on the ligands is deprotonated.[60] One of these complexes has been protonated with sodium hydrogen sulfite to give tris(ethylenediamine)-osmium(III) iodide,[61] which is very easily oxidized by oxygen to osmium(IV) species. Complex ions of the form $[OsL_3]^{3\oplus}$, where L is 2,2′-bipyridyl or 1,10-phenanthroline, can be prepared by oxidizing the corresponding osmium(II) species with chlorine.[61,62] In both cases the enantiomorphous forms have been separated.
Only two isothiocyanate derivatives are known. *Tetra-*n*-butylammonium hexaisothiocyanatoosmate(III)* is readily obtained by refluxing potassium hexachloroosmate(IV) with potassium thiocyanate in water.[63] The remaining derivative, *bis(isothiocyanato)-bis(o-phenylenebisdimethylarsine)osmium(III) perchlorate* may be prepared in a somewhat similar fashion.[26]

---

[52] *P. G. Douglas, B. L. Shaw*, J. Chem. Soc. A, 334 (1970).
[53] *P. Machmer*, Inorg. Nucl. Chem. Lett. *4*, 91 (1968).
[54] *D. Bright, J. A. Ibers*, Inorg. Chem. *8*, 1078 (1969).
[55] *F. P. Dwyer, A. M. Sargeson*, J. Amer. Chem. Soc. *77*, 1285 (1955).
[56] *E. Uhlemann, Ph. Thomas*, Z. Naturforsch. B *23*, 275 (1968).
[57] *G. A. Heath, R. L. Martin*, Australian J. Chem. *23*, 1721 (1970).
[58] *B. Lorenz, R. Kirmse, E. Hoyer*, Z. Anorg. Allgem. Chemie *378*, 144 (1970).
[59] *F. P. Dwyer, J. W. Hogarth*, J. Proc. Roy. Soc. N.S. Wales *85*, 113 (1951).
[60] *F. P. Dwyer, J. W. Hogarth*, J. Amer. Chem. Soc. *77*, 6152 (1955).
[61] *F. P. Dwyer, E. C. Gyarfas*, J. Amer. Chem. Soc. *73*, 2322 (1951).
[62] *F. P. Dwyer, E. C. Gyarfas*, J. Amer. Chem. Soc. *74*, 4699 (1952).
[63] *H. H. Schmidtke, D. Garthoff*, Helv. Chim. Acta *50*, 1631 (1967).

22.4 Derivatives of Osmium(IV)

Osmium(IV) is the most stable oxidation state for this element, and it is found in connection with a diversity of donor groups. Halogen complexes are most common and important, as evidenced by the synthetic utility of hexahaloosmate(IV) salts. Group VI donors are of little importance. Derivatives of Group V elements are well known, but limited in number, and the only compounds of Group IV donors are derived from osmocene. There does not appear to be, as yet, a systematic approach to the synthesis of osmium(IV) compounds, and most of the references given are to isolated reports of specific syntheses.

22.4.1 Hydride, Halide, Chalcogen, and Nitrogen Derivatives

Hydride complexes of osmium(IV) are rare and only tetrahydrides of the type $[OsH_4(YR_3)_3]$ (Y=P, As; R=CH_3, C_2H_5, $n\text{-}C_4H_9$, C_6H_5) are known.[20,52]

Osmium(IV) chloride appears to exist in two forms. Direct chlorination of the metal at elevated temperatures[42] or of osmium tetroxide in carbon tetrachloride[64] at 470° gives rise to a black insoluble form, whereas treatment of the tetroxide with sulfinyl chloride produces a dark brown form that is somewhat soluble in organic solvents.[64,65] The two forms are polymorphous, but the details of the structures are not known. *Osmium(IV) bromide* has been synthesized by direct bromination of the metal[66] at 450–470°. The *hexafluoroosmate(IV) ion* is known as the potassium and cesium salts.[67] These compounds are prepared by the reaction of water with potassium or cesium hexafluoroosmate(V). The extremely useful synthetic starting materials *ammonium hexachloroosmate(IV)* and *ammonium hexabromoosmate(IV)* are quite easily prepared from osmium tetroxide.[68] The chloro-compound can be heated in an inert atmosphere to give the pure metal.[3] Other salts such as that of sodium and potassium are prepared similarly. Potassium and ammonium salts of *hexaiodoosmate(IV)* also are known,[5,17] as is the acid dihydronium hexaiodoosmate(IV), $(H_3O)_2[OsI_6]$.[5] Some mixed-halide species, *bromopentachloroosmate(IV)* and *trichlorotribromoosmate(IV)*, have been prepared as methylammonium salts.[69]

In its higher oxidation states osmium becomes distinctly class "a"[70] in character; consequently, very strong affinity is shown for nitride, oxide, and fluoride ions. Illustrative of this point in osmium(IV) chemistry is the compound *oxohexachlorodiosmium(IV)*, $[Os_2OCl_6]$, and the ion *oxodecachlorodiosmate(IV)*, $[Os_2OCl_{10}]^{4\ominus}$ which has been isolated in a hydrated form as an alkali metal or ammonium salt. The former compound can be prepared either from osmium tetroxide and concentrated hydrochloric acid under reflux[65] or by heating osmium(IV) chloride and oxygen in a temperature gradient.[71] The *oxodecachlorodiosmate(IV) ion*, prepared by heating osmium tetroxide, iron(II) sulfate, and ammonium chloride (or alkali metal chloride) in aqueous sulfuric-hydrochloric acid mixtures, was originally[69,72] formulated as $[OsCl_5(OH)]^{2\ominus}$; more recently the correct formulation has been established.[73,74] Treatment of osmium tetroxide in hydrochloric acid with gaseous hydrogen chloride gives *hydroxotrichloroosmium(IV)*, which probably is polymeric.[69] A similar bromide also has been made.[69]

Osmium(IV)-halide bonds also are found in several complexes in which there are bridging nitrido groups. When the product obtained from heating ammonium hexachloroosmate(IV) in chlorine at 400° is dissolved in hot, dilute hydrochloric acid and then treated with cesium chloride, cesium octachloro-μ-nitridodiaquodiosmate(IV), $Cs_3[Os_2NCl_8(H_2O)_2]$, is formed.[75] Treatment of the analogous potassium compound, which is similarly prepared, with aqueous ammonia at 100° gives *dichloro-μ-nitridooctaamminediosmium(IV) chloride*, $[Os_2N(NH_3)_8Cl_2]Cl_3$.[75] This same type of compound is obtained when hexahaloosmate(IV) salts are treated with

[64] *P. Machmer*, Chem. Commun. 610 (1967).
[65] *R. Colton, R. H. Farthing*, Australian J. Chem. *21*, 589 (1968).
[66] *I. N. Semenov, N. I. Kolbin*. Russian J. Inorg. Chem. (Engl. Transl.) *7*, 111 (1962).
[67] *M. A. Hepworth, P. L. Robinson, F. J. Westland*, J. Chem. Soc. (London) 4269 (1954).
[68] *F. P. Dwyer, J. W. Hogarth*, Inorg. Synth. *5*, 204, 206 (1957).
[69] *F. Krauss, D. Wilken*, Z. Anorg. Allgem. Chemie *137*, 349 (1924).
[70] *S. Ahrland, J. Chatt, N. R. Davies*, Quart. Rev. Chem. Soc. *12*, 265 (1958).
[71] *H. Schafer, K. H. Huneke*, J. Less-Common Metals *12*, 331 (1967).
[72] *F. P. Dwyer, J. W. Hogarth*, J. Proc. Roy. Soc. N.S. Wales *84*, 194 (1950).
[73] *R. R. Miano, C. S. Garner*, Inorg. Chem. *4*, 337 (1965).
[74] *B. Jezowska-Trzebiatowska, J. Hanuza, W. Wojciechowski*, J. Inorg. Nucl. Chem. *28*, 2701 (1966).
[75] *M. J. Cleare, W. P. Griffith*, J. Chem. Soc. A, 1117 (1970).

concentrated aqueous or liquid ammonia at 25–100°,[49,75,76] and similar compounds are known for which the halides are substituted for by other anions.[75]

The very few osmium(IV) complexes of amine, phosphine, or arsine ligands that have been reported also contain osmium(IV)-halogen bonds. *Tetrachloro(2,2'-bipyridyl)osmium(IV)* and *tetrachloro(1,10-phenanthroline)osmium-(IV)* have been obtained by pyrolysing 2,2'-bipyridylium and 1,10-phenanthrolinium hexachloroosmate(IV) salts.[15] A series of compounds having the general formula *trans*-$[OsX_4(YR_3)_2]$ (X=Cl, Br; Y=P, As; R=CH_3, C_2H_5, *n*-C_3H_7, *n*-C_4H_9, C_6H_5) has been synthesized either by oxidizing compounds of the type *mer*-$[OsX_3(YR_3)_3]$ with carbon tetrachloride or chlorine, or by refluxing osmium tetroxide in ethanol containing hydrogen chloride and the phosphine or arsine.[51,52] The corresponding purple triphenylarsine complex is said to be formed when ammonium hexabromoosmate(IV) is treated with triphenylarsine in methanol.[23]

Dihalobis(o-phenylenebisdimethylarsine)osmium(IV) perchlorates are obtainable when the analogous osmium(III) compounds are oxidized with concentrated nitric acid.[26]

Simple chalcogen derivatives are not common. *Osmium(IV) oxide* can be obtained in the anhydrous form by heating the metal and osmium tetroxide vapor together at 600° or in a hydrated form by the ethanol reduction of solutions of osmium tetroxide.[77,78] *Osmium(IV) sulfide*[79] results when osmium and elemental sulfur are heated at 700°.

Some *complexes of osmium(IV) and ethylenediamine* in which one or more of the ligand nitrogen atoms is deprotonated have been obtained from ammonium hexabromoosmate(IV) and anhydrous ethylenediamine.[60] The syntheses are difficult.

22.5 Derivatives of Osmium(V), Osmium(VI), Osmium(VII), and Osmium(VIII)

These four oxidation states are grouped together because class "a" character[70] is most pronounced in this group. In general the only donors of any consequence are fluoride, oxide, and nitride, although osmium(V) is known in one very unusual phosphine complex. Both osmium(V) and osmium(VII) are rare oxidation states and little is known of their chemistry. Osmium(VI) is well known, largely because of the stable *trans*-dioxoosmate-(VI) and nitridoosmate(VI) ions that are found in several complexes. Osmium(VIII) is at the same time both rare and of unquestioned importance because of the volatile tetroxide.

22.5.1 Fluoride, Oxide, and Nitride Derivatives

Osmium(V) fluoride has been obtained from osmium(VI) fluoride by treating it with hexacarbonyltungsten(0), reducing it with iodine, or by photolysis.[80] Several *hexafluoroosmate(V) salts* have been prepared by reacting bromine trifluoride with mixtures of osmium(IV) bromide and an alkali metal bromide.[80] Reaction of osmium(VI) fluoride with nitrogen oxide gives *hexafluoronitrosylosmium(V)*[81] and an interesting compound formulated as a *hexafluoroosmate(V)*, $(Xe_2F_3)[OsF_6]$, is reported to be isolable when osmium(V) fluoride reacts with xenon difluoride in bromine pentafluoride as solvent.[82]

As pointed out in Section 22.3.1 many tertiary phosphines undergo reaction sequence (3). Triphenylphosphine is anomolous, reacting with osmium tetroxide and hydrogen chloride or bromide in ethanol to give the extraordinary compound *oxotrichloro* (or *-bromo-*) *bis(triphenylphosphine)osmium(V)*, $OsOCl_3[P(C_6H_5)_3]_2$.[83]

The only binary hexahalide is the yellow osmium(VI) *hexafluoride*, which is obtained by heating the elements together.[84,85] *Oxotetrachloroosmium(VI)* can be obtained in a similar fashion.[86]

The osmium(VI) compounds remaining are either *trans-dioxoosmate(VI) ions* $OsO_2^{2\oplus}$, which are commonly referred to as *osmyl derivatives, or derivatives of nitridoosmate(VI)*, $OsN^{3\oplus}$. Many of the osmyl complexes are prepared either by

[76] *G. W. Watt, L. Vaska*, J. Inorg. Nucl. Chem. *6*, 246 (1958).
[77] *O. Ruff, H. Rathsburg*, Ber. *50*, 484 (1917).
[78] *N. Bartlett, N. K. Jha*, J. Chem. Soc. A, 536 (1968).
[79] *Sutarno, O. Knop, K. I. G. Reid*, Can. J. Chem. *45*, 1391 (1967).
[80] *G. B. Hargreaves, R. D. Peacock*, J. Chem. Soc. (London) 2618 (1960).
[81] *N. Bartlett, S. P. Beaton, N. J. Jha*, Chem. Commun. 168 (1966).
[82] *F. O. Sladky, P. A. Bulliner, N. Bartlett*, J. Chem. Soc. A, 2179 (1969).
[83] *J. Chatt, C. D. Falk, G. J. Leigh, R. J. Paske*, J. Chem. Soc. *A*, 2288 (1969).
[84] *B. Weinstock, J. G. Malm*, J. Amer. Chem. Soc. *80*, 4466 (1958).
[85] *G. B. Hargreaves, R. D. Peacock*, Proc. Chem. Soc. *85* (1959).
[86] *M. H. Hepworth, P. L. Robinson*, J. Inorg. Nucl. Chem. *4*, 24 (1957).

the action of an acid on osmium tetroxide or by reacting potassium *trans*-dioxotetrahydroxoosmate(VI) (potassium osmate) with an alkali metal salt. *Potassium osmate* itself is prepared by the ethanol reduction of osmium tetroxide in cold solutions of potassium hydroxide.[87,88] Some of the more well characterized osmyl complexes are given in Table 2. Nitrido derivatives, which are more common for osmium than for any other metal, are thought to involve a very strong osmium-nitrogen triple bond. Consequently the $OsN^{3\oplus}$ group is found as an identifiable entity in a number of complexes, which are given in Table 2.

Osmium(VII) is found only in conjunction with oxide or fluoride donors. *Osmium heptafluoride*, OsF_7, can be prepared from the union of the elements at 500–600° and 350–400 atm.[89] This yellow material is unstable at room temperature,

Table 2. "Osmyl" and Nitrido Derivatives of Osmium(VI)

Compound	Preparative method	Ref.
$K_2[OsO_2(OH)_4]$	$K_2[OsO_4(OH)_2](OsO_4 + KOH) + C_2H_5OH$	a, b
$K_2[OsO_2Cl_4]$	$K_2[OsO_2(OH)_4] + HCl$	c, d
$K_2[OsO_2(CN)_4]$	$OsO_4 + KCN$	e
$K_2[OsO_2(C_2O_4)_2]$	$OsO_4 + KOH + H_2C_2O_4$	c, d
$K_2[OsO_2(OH)_2(NO_2)_2]$	$OsO_4 + KNO_2$	b, c
$K_2[OsO_2(OH)_2C_2O_4]$	$OsO_4 + K_2C_2O_4$	b, c
$[OsO_2(NH_3)_4]Cl_2$	$K_2[OsO_2(OH)_4] + NH_4Cl$	c
$K_2[OsNCl_5]$	$K[OsO_3N] + KCl + HCl$	f
$K[OsN(H_2O)Br_4]$	$K[OsO_3N] + HBr$	g
$K[OsN(H_2O)(CN)_4]$, $2H_2O$	$K[OsO_3N] + HCN$	g
$K[OsN(H_2O)(OH)_2$-$(C_2O_4)]$, $2H_2O$	$K[OsO_3N] + K_2C_2O_4$	g
$Cs[OsN(H_2O)$-$(C_2O_4)_2]$, $4H_2O$	$K[OsO_3N] + H_2C_2O_4 + CsCl$	g
$Cs[OsN(H_2O)$-$(OH)_2F_2]$, $5H_2O$	$K[OsO_3N] + HF$ (aqueous)	g

[a] *O. Ruff, F. Bornemann*, Z. Anorg. Allg. Chem. *65*, 429 (1910).
[b] *W. P. Griffith*, J. Chem. Soc. *245* (1964).
[c] *L. Wintrebert*, Ann. Chim. Phys. (Paris) *28* (7) 15 (1903).
[d] *P. Pascal*, Nouveau Traite' de Chemie Minerale, Vol. II, Masson et. cie, Paris, 1932.
[e] *F. Krauss, G. Schrader*, J. Prakt. Chem *120*, 36 (1929).
[f] *A. F. Clifford, C. S. Kobayashi*, Inorganic Syntheses, *6*, 206 (1960).
[g] *W. P. Griffith*, J. Chem. Soc., 3694 (1965).

decomposing to give the hexafluoride and fluorine. *Oxopentafluoroosmium(VII)* is reported to form when osmium dioxide is fluorinated.[78]

Osmium(VIII) compounds are not numerous and in general are formed only under conditions that give rise to Os–O or Os–F bonds. *Trioxodifluoroosmium(VIII)* can be prepared by treating osmium tetroxide and potassium bromide with bromine trifluoride or by fluorinating osmium sponge in the presence of oxygen in an alumina apparatus.[86] Potassium, cesium, and silver salts of *trioxotrifluoroosmate(VIII)* can be prepared similarly.[86] The structures of these compounds are not known.

Perhaps the most important and common osmium compound is *osmium tetroxide* which can be prepared easily by burning the metal in oxygen.[2] This volatile compound is a very powerful oxidizing agent and it has been used as such for many years in organic syntheses. The tetroxide can be reduced to the element with hydrogen gas, thus providing a convenient purification method. Special precautions should be exercised in working with this material because of its high vapor pressure, strong oxidizing character, and toxicity. Further, it should be recognized that many osmium-containing compounds are readily converted to osmium tetroxide by oxidants such as nitric acid; consequently, all osmium compounds should be handled with care.

Treatment of osmium tetroxide with cold potassium hydroxide in slight excess gives unstable *potassium tetraoxodihydroxoosmate(VIII)*, $K_2[OsO_4(OH)_2]$, which is commonly referred to as *potassium perosmate*.[88,90] Other salts also are known. If ammonia is present when osmium tetroxide is treated with cold potassium hydroxide, *potassium nitridotrioxoosmate(VIII)* is formed.[91] This compound, which is frequently referred to as *potassium osmiamate*, is the common starting material for compounds containing the nitridoosmate(VI) ion.

22.6 Bibliography

Gmelin's Handbuch der Anorganischen Chemie, 8. Auflage, Bd. 66, Verlag Chemie Berlin 1939.
W. P. Griffith, Quart. Rev., Chem. Soc. *19*, 254 (1965).
W. P. Griffith, The Chemistry of the Rarer Platinum Metals, Chap. 3, Interscience Publishers, New York 1967.

[87] *O. Ruff, F. Bornemann*, Z. Anorg. Allgem. Chemie *65*, 429 (1910).
[88] *W. P. Griffith*, J. Chem. Soc. 245 (1964).
[89] *O. Glemser, H. W. Roesky, K.-H. Hellberg, H.-U. Werther*, Chem. Ber. *99*, 2652 (1966).
[90] *F. Krauss, D. Wilken*, Z. Anorg. Allgem. Chemie *145*, 151 (1925).
[91] *A. F. Clifford, C. S. Kobayaski*, Inorg. Synth. *6*, 204 (1960).

23 Cobalt

P. R. Mitchell
University Chemical Laboratory, University of Kent,
Canterbury, Kent, England

Present address: Institut für Anorganische Chemie, Universität Basel, CH–4056 Basel, Switzerland.

23.1 Introduction

Owing to the enormous literature on cobalt compounds, this chapter will emphasize the simplest methods of preparation of cobalt compounds frequently used as starting materials in research in cobalt chemistry, where these compounds are not commercially available at reasonable prices.

Although cobalt exhibits a wide range of oxidation states from -1 in the carbonyl anion to $+4$ in a complex fluoride (and possibly even $+5$ in a complex oxide), compounds of cobalt(II) and cobalt(III) predominate. Apart from a number of carbonyl derivatives, which are discussed in Chapter 29, there are relatively few examples of the lower oxidation states, and even fewer compounds of the higher oxidation states are known. As cobalt metal and most simple cobalt(II) salts are readily and cheaply available commercially in a reasonable pure state—the main impurities are usually iron and nickel salts—the preparation of these will not be discussed in any detail. Rather, the preparation of important compounds of the lower oxidation states of cobalt, Co^0 and Co^I, and of cobalt(III) complexes will be discussed. Even in this latter category no attempt will be made to refer to the preparation of all of the exceedingly numerous cobalt(III) compounds which have been reported, which would be impossible in the space available, but rather to refer to methods of general use in the synthesis of cobalt(III) complexes and to particular cases in which important cobalt(III) complexes can be easily prepared.

Two major compilations of cobalt chemistry, which are especially useful for the properties of cobalt compounds are in the tomes edited by Pascal[1,2] and Gmelin[3]: neither of these is particularly up to date, but they provide comprehensive compilations of the earlier work.

High purity cobalt metal is readily available commercially and is particularly useful as a standard for cobalt analyses [most cobalt(II) salts, although available chemically pure, are hydrated and therefore unsuitable for precise work], and as a starting material for synthesis of simple cobalt salts if these are required especially free from iron or nickel. Normal *commercial grades* of cobalt(II) salts for laboratory use may contain up to 0·1% iron and 0·5% nickel although analytical grades with 10–100 ppm of these metals are available. Methods for purifying cobalt from nickel and iron are summarized by Pascal,[4] one of the best of which, for use on a laboratory scale, is by preparation[5,6] and crystallization from concentrated hydrochloric acid of $[Co^{III}(NH_3)_5Cl]Cl_2$. This can then be reduced,[5] in a stream of hydrogen, to pure cobalt metal.

An alternative method for purification of cobalt, involving electrolytic separation, is given by Brauer,[7] who also describes the preparation of pure finely divided metallic cobalt by reduction of precipitated cobalt(II) oxalate with hydrogen at 500°. A pyrophoric form of cobalt metal can be prepared[8] by reduction of cobalt(III) oxide hydroxide, CoO(OH) (see Section 23.5.3) in hydrogen at 300°.

An excellent monograph[9] on cobalt which devotes considerable space to the occurrence, metallurgy, and properties of the metal and its alloys, has been produced by the Centre d'Information du Cobalt.

23.2 Binary Compounds*

The action of phenyl magnesium bromide on a suspension of $CoCl_2$ in ether under an atmosphere of hydrogen gives[10,11] a finely divided brown precipitate claimed to be CoH_2. The structure of this material is still unknown, but it certainly contains hydrogen in an active form, for it reacts[12] with carbon monoxide to give $HCo(CO)_4$ much more readily than does finely divided cobalt in the presence of hydrogen. It

[1] *J. Amiel, J. Besson,* Cobalt et Nickel, Vol. XVII, Part 2 in *P. Pascal,* Nouveau Traité de Chemie Minérale, Masson, Paris 1963.

[2] *J. Amiel, C. Duval, R. Duval, P. Job, A. Michel, P. Pascal,* Complexes du Fer, du Cobalt, et du Nickel, in *P. Pascal,* Nouveau Traité de Chemie Minérale, Masson, Paris 1959.

[3] *Gmelin,* Handbuch der Anorganischen Chemie Systemnummer 58, Kobalt, 8. Edition.

[4] *J. Amiel, T. Besson,* Cobalt et Nickel, Vol. XVII, p. 23, Part 2, in *P. Pascal,* Nouveau Traité de Chemie Minérale, Masson, Paris 1963.

[5] *S. P. L. Sörensen,* Z. Anorg. Allgem. Chemie *5,* 354 (1893);
S. P. L. Sörensen, J. Chem. Soc. Abstracts *66 ii,* 134 (1894).

[6] *H. Copaux,* Compt. Rend. *140,* 657 (1905);
H. Copaux, J. Chem. Soc. Abstracts *8 ii,* 254 (1905).

[7] *G. Brauer,* Handbook of Preparative Inorganic Chemistry, 2. Edition (English translation). Vol. 2, pp. 1513–1543, Academic Press, London 1965.

[8] *G. Brauer,* Handbook of Preparative Inorganic Chemistry, 2. Edition (English translation). Vol. 2, p. 1615, Academic Press, London 1965.

[9] Cobalt Monograph, Centre d'Information du Cobalt, Brussels, Belgium 1960.

[10] *T. Weichselfelder, B. Thiede,* Justus Liebigs Ann. Chem. 447, 64 (1926); C.A. *20,* 1363 (1926).

[11] *R. C. Ray, R. B. N. Sahai,* J. Indian Chem. Soc. 23, 61 (1946); C.A. *40,* 7034 (1946).

[12] *W. Hieber, H. Schulten, R. Marin,* Z. Anorg. Allgem. Chemie *240,* 261 (1939); C.A. *33,* 3285 (1939).

* Other than halides and oxides which are discussed in the section appropriate to their oxidation state.

has been suggested[13] that the hydrogen is chemisorbed and as more recent work[14] suggests that organic compounds are also present, Chatt and Shaw have suggested[15] that it is an organocobalt hydride similar to those known for chromium.

At least three sulfides of cobalt have been characterized: CoS, CoS_2, Co_3S_4, and perhaps also Co_9S_8, can be prepared[7,16] by heating the stoichiometric amount of finely divided cobalt and sulfur in a quartz tube at 650° for 2–3 days; CoS can also be prepared[17] as a black precipitate by the action of $S^{2\ominus}$ on aqueous $Co^{2\oplus}$, and dried by heating the precipitate to up to 540°. CoS, and CoS_2 which has the pyrite structure,[18] are active hydrogenation catalysts. The cobalt/selenium[1,19] and cobalt/tellurium[1,20] systems have also been studied in detail.

The only stoichiometric binary compound of cobalt and nitrogen is the explosive cobalt(II) azide,[21] $Co(N_3)_2$. Thermal decomposition of the polymeric *cobalt(III) amide*,[22] $Co(NH_2)_3$, (Section 23.5.4) at 50–70° in the absence of air evolves ammonia and a little nitrogen, to yield[23] CoN as a black pyrophoric powder which is usually[7] nitrogen deficient ($CoN_{0.9}$). Pyrolysis of $Co(NH_2)_3$ in a vacuum at 160° gives[23] $CoN_{0.4}$ which is further decomposed to metallic cobalt at 205°. "Co_2N" can also be prepared by the action of ammonia on pyrophoric cobalt freshly prepared by reduction of Co_3O_4 with hydrogen.[24] Lower cobalt nitrides, Co_4N and Co_3N, have also been reported.[1,24] Eight different binary cobalt phosphides have been described,[1] and three of these, CoP_3, CoP, and Co_2P are stated[7] to be readily prepared by heating together powdered cobalt and red phosphorus for 20 hr at 650°. The existence of cobalt arsenides and antimonides is discussed by Pascal.[1]

Of the various cobalt carbides reported[1] only Co_2C has an undisputed existence. It may be prepared[25,26] as a gray powder by the action of carbon monoxide on pyrophoric cobalt powder prepared *in situ* by reduction of cobalt(II) oxide with hydrogen. Of the five cobalt silicides observed[1] in the cobalt/silicon system the best characterized is Co_2Si, which can be prepared either by heating[27] a stoichiometric mixture of the elements or by passing[28] $SiCl_4$ vapor over cobalt at between 1200° and 1300°.

Co_3B, Co_2B, and CoB can be prepared[29] by heating the mixed elements; Co_3B_2, Co_3B_4, and $Co_{23}B_6$ have been reported as existing in ternary mixtures.[30,31] Co_2B, which is an effective heterogeneous catalyst in the hydrolytic decomposition of sodium hydridoborate[32] and in the decomposition of hydrogen peroxide[33] can easily be prepared[32,34] by mixing aqueous solutions of sodium hydridoborate and cobalt(II) chloride.

[13] *D. T. Hurd*, Chemistry of the Hydrides, p. 189, John Wiley & Sons, New York 1952.

[14] *B. Sarry*, Z. Anorg. Allgem. Chemie *286*, 211 (1956); C.A. *51*, 2439 (1957).

[15] *J. Chatt, B. L. Shaw*, Proc. XVIIth Intern. Cong. Pure Applied Chem. p. 147, 1959, Butterworth & Co., London 1961.

[16] *D. Lundqvist, A. Westgren*, Z. Anorg. Allgem. Chemie *239*, 85 (1938); C.A. *32*, 8872 (1938).

[17] *E. Dönges*, Z. Anorg. Allgem. Chemie *253*, 345 (1947); C.A. *43*, 2074 (1949).

[18] *W. F. de Jong, H. W. V. Willems*, Z. Anorg. Allgem. Chemie *160*, 185 (1927); C.A. *21*, 2231 (1927).

[19] *F. Bohm, F. Gronvold, H. Haraldsen, H. Prydz*, Acta Chem. Scand. *9*, 1510 (1955).

[20] *H. Haraldsen, F. Gronvold, T. Hurlen*, Z. Anorg. Allgem. Chemie *283*, 143 (1956); C.A. *50*, 10464 (1956).

[21] *F. P. Bowden, K. Singh*, Proc. Roy. Soc. London Ser. *A 227*, 22 (1954).

[22] *O. Schmitz-Dumont, J. Pilzecker, H. F. Piepenbrink*, Z. Anorg. Allgem. Chem. *248*, 175 (1941); C.A. *37*, 6205 (1943).

[23] *O. Schmitz-Dumont, N. Kron*, Angew. Chem. *67*, 231 (1955).

[24] *J. Clarke, K. H. Jack*, Chem. & Ind. (London) 1004 (1951).

[25] *H. A. Bahr, V. Jessen*, Ber. *63*, 2226 (1930); C.A. *25*, 22 (1931).

[26] *L. J. E. Hofer, W. C. Peebles*, J. Amer. Chem. Soc. *69*, 893 (1947).

[27] *E. Vigouroux*, Compt. Rend. *121*, 686 (1895); *E. Vigouroux*, J. Chem. Soc. Abstracts *70 ii*, 176 (1896).

[28] *E. Vigouroux*, Compt. Rend. *142*, 635 (1906); *E. Vigouroux*, J. Chem. Soc. Abstracts *90 ii*, 287 (1906).

[29] *J. D. Schoebel, H. H. Stadelmaier*, Z. Metallk. *57*, 323 (1966); C.A. *65*, 1890 (1966).

[30] *E. Ganglberger, H. Nowotny, F. Benesovsky*, Monatsh. Chem. *97*, 101 (1966); C.A. *65*, 3109 (1966).

[31] *F. N. Tavadze, I. A. Bairamashvili, D. V. Khantadze*, Dokl. Akad. Nauk SSSR *162*, 67 (1965); C.A. *63*, 5332 (1965).

[32] *T. Hirai, K. Takahashi*, Denki Kagaku, Oyobi Kogyo Butsuri Kagaku *35*, 886 (1967); C.A. *69*, 40773 (1968).

[33] *J. M. Pratt, G. Swinden*, Chem. Commun. 1321 (1969).

[34] *H. I. Schlesinger, H. C. Brown, A. E. Finholt, J. R. Gilbreath, H. R. Hoekstra, E. K. Hyde*, J. Amer. Chem. Soc. *75*, 215 (1953).

23.3 Low Oxidation States: Cobalt(—I), Cobalt(0), and Cobalt(I)

For many compounds of these oxidation states it is difficult to assign a meaningful oxidation state and the preparative method depends more on the type of ligands present than on the oxidation state. These oxidation states will therefore be discussed together.

The only simple salt of any of these oxidation states is *cobalt(I) iodide* which may be obtained[35] as a dark solid by decomposition of $[ICo(PF_3)_4]$ *in vacuo* at 0°. It is instantly decomposed by water to Co and CoI_2. A similar solid is obtained[36] by decomposition of $[ICo(CO)_4]$ at −20°. The *nitrosyl halides* $[Co(NO)_2X]_2$, X=Cl, Br may be prepared[8,37] by heating the anhydrous cobalt(II) halide with zinc at 70–80° in a stream of pure nitric oxide followed by sublimation at 115°. The *iodide* $[Co(NO)_2I]_2$ can be prepared by heating CoI_2 in a stream of nitric oxide,[8,37] and all three halides can be prepared in high yield by the action of nitric oxide on a methanolic solution of the cobalt(II) halide containing powdered copper.[38] Anionic derivatives,[38] $M^{\oplus}[CoX_2(NO)_2]^{\ominus}$, neutral nitrosyls containing pyridine,[38] ethylenediamine[38] and triphenylphosphine[39] *e.g.* $[CoX(L)(NO)_2]$, and a cationic triphenylphosphine nitrosyl,[39] $[Co(Ph_3P)_2(NO)](ClO_4)$, can be prepared from the nitrosyl halides in alcoholic solution. Hydridoborate reduction[39] of an ethanolic solution of $[Co(NO)_2Cl]_2$ and triphenylphosphine gives $[Co(Ph_3P)_3(NO)]$.

Reversible polarographic reduction of tris-phenanthroline[40] or tris-bipyridyl[41,42] complexes of cobalt(II) or cobalt(III) gives solutions containing cobalt(I); with $[Co(phen)_3]^{3\oplus}$ further reduction to cobalt(O) is possible.[40] In presence of excess ligand and in absence of air such solutions of cobalt(I) are quite stable. Dark brown crystals of $[Co^{I}(phen)_3](ClO_4)$ can be prepared[43] by reduction of $[Co^{III}(phen)_3](ClO_4)_3$ with sodium hydridoborate. Reduction of $[Co^{III}(bipy)_3](ClO_4)_3$ with sodium hydridoborate[41] or with sodium amalgam[44] gives the blue paramagnetic complex $[Co^{I}(bipy)_2](ClO_4)$; in presence of excess bipyridyl reduction with sodium amalgam[42] gives the tris complex $[Co^{I}(bipy)_3](ClO_4)$. Both $[Co(bipy)_3]^{\oplus}$ and $[Co(phen)_3]^{\oplus}$ react with tertiary phosphines in an atmosphere of hydrogen to give dihydrides, $[(LL)Co(R_3P)_2(H_2)](ClO_4)$.[45] Reduction of a solution of cobalt(II) chloride and bipyridyl in tetrahydrofuran with a solution of dilithiobipyridyl proceeds further to give[46] dark blue needles of $[Co^{0}(bipy)_3]$. Similarly, reduction[47] of cobalt phthalocyanine with lithium benzophenone in THF gives the intensely colored complexes $Li^{\oplus}[Co(phthalocyanine)]^{\ominus}$ and $Li^{\oplus}_2[Co(phthalocyanine)]^{2\ominus}$, formally containing cobalt(I) and cobalt(0) respectively. Reduction of cobalt(II)[48] or cobalt(III)[49] dimethylglyoxime complexes gives a solution of bis(dimethylglyoximato)cobalt(I), which is a useful intermediate in the synthesis of σ-bonded alkylcobalt derivatives.

Some of the reactions which can be used to prepare trifluorophosphine derivatives of cobalt(-I), cobalt(0) and cobalt(I) are outlined in Scheme 1. Unfortunately many of the reactions require high pressures (up to 400 atm.) of PF_3, especially in the direct syntheses from cobalt metal or halides. However, substitution of carbon monoxide by PF_3 will take place in $RCo(CO)_4$, (R=H[52,57], CF_3[52], C_2F_5[52], C_3F_7[52]) and in $Co(NO)(CO)_3$[58]

---

[35] *T. Kruck, W. Lang*, Z. Anorg. Allgem. Chemie *343*, 181 (1966).

[36] *B. L. Booth, R. N. Haszeldine, P. R. Mitchell*, J. Chem. Soc. A, 691 (1969).

[37] *W. Hieber, R. Marin*, Z. Anorg. Allgem. Chemie *240*, 241 (1939).

[38] *A. Sacco, M. Rossi, C. F. Nobile*, Ann. Chim. (Rom) *57*, 499 (1967); C.A. *67*, 60453 (1967).

[39] *T. Bianco, M. Rossi, L. Uva*, Inorg. Chim. Acta *3*, 443 (1969).

[40] *N. Maki, T. Hirano, S. Musha*, Bull. Chem. Soc., Japan *36*, 756 (1963).

[41] *A. A. Vlcek*, Nature *180*, 753 (1957).

[42] *G. M. Waind, B. Martin*, J. Inorg. Nucl. Chem. *8*, 551 (1958).

[43] *N. Maki, M. Yamagami, H. Itatani*, J. Amer. Chem. Soc. *86*, 514 (1964).

[44] *B. Martin, W. R. McWhinnie, G. M. Waind*, J. Inorg. Nucl. Chem. *23*, 207 (1961).

[45] *G. Mestroni, A. Camus, C. Cocevar*, J. Organometal. Chem. *29*, C17 (1971).

[46] *S. Herzog, R. Klausch, J. Lantos*, Z. Chem. *4*, 150 (1964).

[47] *R. Taube, M. Zach, K. A. Stauske, S. Heidrich*, Z. Chem. *3*, 392 (1963).

[48] *G. N. Schrauzer*, Inorg. Synth. *11*, 61 (1968).

[49] *G. N. Schrauzer, E. Deutsch*, J. Amer. Chem. Soc. *91*, 3341 (1969).

[50] *T. Kruck, W. Lang*, Agnew. Chem. Intern. Ed. Engl. *6*, 454.

[51] *P. L. Timms*, Chem. Commun. 1033 (1969).

[52] *C. A. Udovich, R. J. Clark*, Inorg. Chem. *8*, 938 (1969).

[53] *J. M. Campbell, F. G. A. Stone*, Angew. Chem. Intern. Ed. Engl. *8*, 140 (1969).

[54] *T. Kruck, W. Lang, A. Engelmann*, Angew. Chem. Intern. Ed. Engl. *4*, 148 (1965).

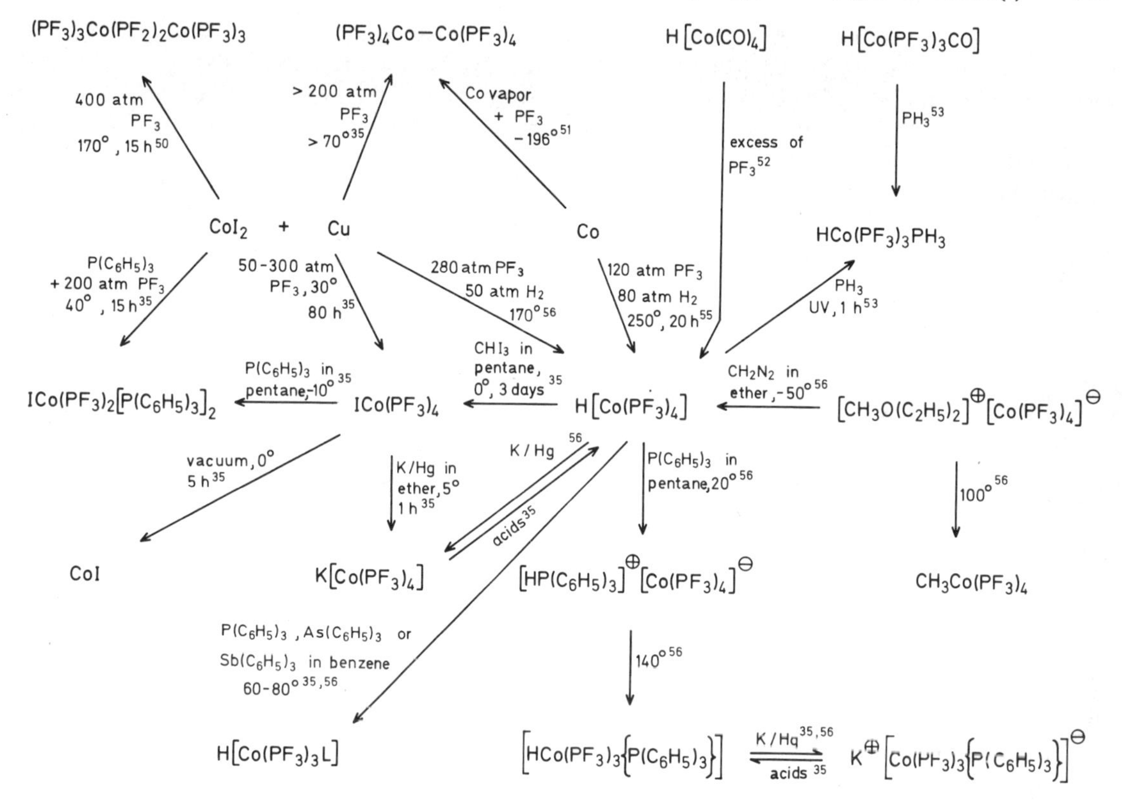

to give complete substitution if a large enough excess of PF_3 is used, preferably also with u.v. irradiation. $Co(NO)(PF_3)_3$ can also be prepared[59] by direct high pressure synthesis from $[Co(NO)_2Cl]_2$ and PF_3 in the presence of copper. The trifluorophosphine complexes of transition metals have been the subject of two reviews.[60, 61] Hydridoborate reduction of ethanolic solutions of CoX_2 and aryl phosphines gives[62] tetrahedral cobalt(I) complexes such as $[(Ph_3P)_3CoX]$. Some of the compounds originally claimed to be cobalt(0) derivatives of tertiary phosphines have been shown to be hydrido complexes. However, $[Co\{o\text{-}C_6H_4(PEt_2)_2\}_2]$ prepared[63] by heating finely divided cobalt powder and the diphosphine at 200° may be a genuine cobalt(0) derivative. A similar compound originally prepared[63] by reduction of $[CoBr_2(Ph_2PCH_2CH_2PPh_2)_2]$ with sodium hydridoborate has been shown[64] to be $[HCo(Ph_2PCH_2CH_2PPh_2)_2]$: this compound reacts readily with perchloric acid to give[64] a dihydride complex, $[CoH_2(Ph_2PCH_2CH_2PPh_2)_2](ClO_4)$. The cobalt(0) complex $[Co(Ph_2PCH_2CH_2PPh_2)_2]$ can in fact be prepared[65] by reduction of the cobalt(II) complex with potassium hydroxide in ethanol. A molecular nitrogen complex prepared[66] by reduction of a mixture of $[Co(acac)_3]$ and triphenylphosphine with diethyl-aluminum monoethoxide under nitrogen and formulated as $[(Ph_3P)_3Co(N_2)]$ is also a

---

[55] *T. Kruck, K. Baur, W. Lang*, Ber. *101*, 138 (1968).
[56] *T. Kruck, W. Lang, N. Derner, M. Stadler*, Chem. Ber. *101*, 3816 (1968).
[57] *B. J. Aylett, J. M. Campbell*, J. Chem. Soc. *A*, 1920 (1969).
[58] *R. J. Clark*, Inorg. Chem. *6*, 299 (1967).
[59] *T. Kruck, W. Lang*, Angew. Chem. Intern. Ed. Engl. *3*, 700 (1964).
[60] *T. Kruck*, Angew. Chem. Intern. Ed. Engl. *6*, 53 (1967).
[61] *J. F. Nixon*, Advan. Inorg. Chem. Radiochem. *13*, 363 (1970).
[62] *M. Aresta, M. Rossi, A. Sacco*, Inorg. Chim. Acta *3*, 227 (1969).
[63] *J. Chatt, F. A. Hart, D. T. Rosevear*, J. Chem. Soc. 5504 (1961).
[64] *A. Sacco, R. Ugo*, J. Chem. Soc. 3274 (1964).
[65] *A. Sacco, M. Rossi*, Chem. Commun. 602 (1965).
[66] *A. Yamamoto, S. Kitazume, L. S. Pu, S. Ikeda*, Chem. Commun. 79 (1967);
A. Misono, Y. Uchida, T. Saito, Bull. Chem. Soc., Japan *40*, 700 (1967).

hydrido complex[67] $[HCo(N_2)(Ph_3P)_3]$ although it has now been claimed[68] that $[(Ph_3P)Co(N_2)]$ can in fact be prepared by a slightly different method. $[HCo(N_2)(Ph_3P)_3]$ can also be prepared[69] by displacement of hydrogen from the trihydride:[70]

$$[CoCl_2\{(C_6H_5)_3P\}_2] + P(C_6H_5)_3 + BH_4^- \longrightarrow [CoH_3\{(C_6H_5)_3P\}_3] \underset{H_2}{\overset{N_2}{\rightleftharpoons}} [CoH(N_2)\{(C_6H_5)_3P\}_3]$$

The coordinated nitrogen in $[HCo(N_2)(Ph_3P)_3]$ can also be reversibly displaced[71] by nitriles:

$$\{CoH(N_2)[(C_6H_5)_3P]_3\} + RCN \rightleftharpoons \{CoH(RCN)[(C_6H_5)_3P]_3\} + N_2$$

Hydridoborate reduction of alcoholic solutions of cobalt(II) chloride and alkyl[72] or aryl[73] phosphites gives phosphite substituted hydrides such as $[HCo\{P(OEt)_3\}_4]$ and $[HCo\{P(OPh)_3\}_4]$. Sterically constrained phosphite ligands such as $P(OCH_2)_3CR$ ($R = CH_3$ and C_2H_5) cause[74,75] disproportionation of cobalt(II) in perchlorate solution to give a high yield of a mixture of colorless $[Co^{III}L_6](ClO_4)_3$ and yellow $[Co^{I}L_5](ClO_4)$ which can be purified by recrystallization from water and methanol respectively.

Complex cyanides of both cobalt(0)[76] and cobalt(I)[77] can be prepared by the reduction of $K_3[Co(CN)_6]$ with either an excess (to give "$K_4Co(CN)_4$") or with the stoichiometric amount (to give $K_3Co(CN)_4$) of potassium in liquid ammonia. The cobalt(0) complex is almost diamagnetic,[78] and therefore is presumably a dimer, $K_8[Co_2(CN)_8]$.

Many isonitrile complexes of low-valent cobalt are known: they have been reviewed.[79] They fall into two main categories:

Nitrosyl isonitrile complexes of the type $Co(NO)(CNR)_3$ can be prepared by reducing a mixture of $[Co(NO)(NH_3)_5]Cl_2$ and isonitriles with hydrazine[80] or by the reaction of hydroxylamine with alcoholic ammoniacal solutions of pentakis(isocyanide)cobalt(I) salts.[80,81]

Pentakis(isonitrile)cobalt(I) salts are readily prepared by mild reduction of tetrakis- or pentakis-(isonitrile)cobalt(II) salts with reducing agents such as hydrazine.[82,83] The pentakis salts are even reduced by simply boiling in ethanolic isonitrile solution. Reaction[84] of isonitriles with $[(Ph_3P)_2CoX_2]$ gives the cobalt(I) derivatives, $[(Ph_3P)_2(RCN)_3Co]X$ ($X = Cl$, I or ClO_4). $[Co(CNR)_5]^{\oplus}$ may also be prepared, as its $[Co(CO)_4]^{\ominus}$ salt by the action of isonitriles on dicobalt octacarbonyl.[85]

23.4 Cobalt(II)

This is the most readily available oxidation state of cobalt, although owing to the lability of its d^7 configuration it is impossible to synthesize as many complexes as with cobalt(III). In the absence of oxidizing and reducing agents cobalt(II) complexes are readily formed although with nitrogen donors even air is sufficient to cause oxidation to cobalt(III). Some phosphite ligands cause[74] disproportionation to cobalt(I) and cobalt(III) (see Section 23.3).

Tetrahedral and octahedral compounds of cobalt(II) are the most common, but an increasing number of

[67] *A. Misono*, Inorg. Synth. *12*, 12 (1970).
[68] *G. Speier, L. Markó*, Inorg. Chim. Acta *3*, 126 (1969).
[69] *A. Sacco, M. Rossi*, Chem. Commun. 316 (1967). *A. Sacco, M. Rossi*, Inorg. Chim. Acta *2*, 127 (1968).
[70] *A. Sacco, M. Rossi*, Inorg. Synth. *12*, 18 (1970).
[71] *A. Misono, Y. Uchida, M. Hidai, T. Kuse*, Chem. Commun. 208 (1969).
[72] *W. Kruse, R. H. Atalla*, Chem. Commun. *921* (1968).
[73] *J. J. Levison, S. D. Robinson*, Chem. Commun. 1405 (1968).
[74] *T. J. Huttemann, B. M. Foxman, C. R. Sperati, J. G. Verkade*, Inorg. Chem. *4*, 950 (1965).
[75] *J. G. Verkade, T. S. Piper*, Inorg. Chem. *2*, 944 (1963).
[76] *W. Hieber, C. Bartenstein*, Naturwissenschaften *39*, 300 (1952).
[77] *G. W. Watt, R. J. Thompson*, J. Inorg. Nucl. Chem. *9*, 311 (1959).
[78] *W. Hieber, C. Bartenstein*, Z. Anorg. Allgem. Chemie *276*, 12 (1954).
[79] *L. Malatesta*, Prog. Inorg. Chem. *1*, 283 (1959); *L. Malatesta, F. Bonati*, Isocyanide complexes of metals, Chapter 7, p. 131, Interscience Publishers, London 1969.
[80] *L. Malatesta, A. Sacco*, Z. Anorg. Allgem. Chemie *273*, 341 (1953).
[81] *L. Malatesta, A. Sacco*, Atti Accad. Naz. Lincei. Classe Sci. Fis. Mat. Natur. Rend. *13*, 264 (1952); C.A. *48*, 1189 (1954).
[82] *L. Malatesta, A. Sacco*, Z. Anorg. Allgem. Chemie *273*, 247 (1953).
[83] *A. Sacco, M. Freni*, Gazz. Chim. Ital. *89*, 1800 (1959); C.A. *55*, 4226 (1961).
[84] *A. Sacco, M. Freni*, Rend. ist. lombardo Sci. Pt. I., Classe Sci. mat. e nat. *94A*, 221 (1960); C.A. *55*, 15386 (1961).
[85] *W. Hieber, E. Böckly*, Z. Anorg. Allgem. Chemie *262*, 344 (1950); *A. Sacco*, Gazz. Chim. Ital. *83*, 632 (1953); C.A. *49*, 768 (1955).

five-coordinate complexes are being discovered, and recently a two-coordinate compound was reported.[86] A few cobalt(II) complexes have been proved by X-ray crystallography to be square planar.[87,155,159]

Many simple cobalt(II) salts are readily available commercially: indeed many of them have considerable industrial importance especially as catalysts, pigments, and paint driers. The industrial preparation and uses of simple cobalt salts have been reviewed.[88] In general cobalt(II) complexes are readily prepared, usually by direct addition of a cobalt(II) salt with the ligand in a suitable solvent; alcohol and acetone being especially useful as many cobalt(II) salts are soluble in them.

23.4.1 Cobalt(II) Halides and Halocomplexes

Anhydrous CoF_2 is most conveniently prepared[89,90] by passing of anhydrous hydrogen fluoride over $CoCl_2$ at 300°, or by dehydrating[91] $CoF_2 \cdot 4H_2O$ (which is itself prepared[91] by dissolving cobalt carbonate in hydrofluoric acid). An alternative method,[92] passing nitryl fluoride over cobalt metal, seems to have no advantages. The complex fluoroanions $[CoF_6]^{4\ominus}$, $[CoF_4]^{2\ominus}$, and $[CoF_3]^{\ominus}$ are all known: Ba_2CoF_6 is prepared[93] by heating together a 2:1 mixture of BaF_2 and CoF_2 at 1250°; alkali metal tetrafluorocobaltates are prepared[90] by fusing together 2:1 mixtures of MF and CoF_2 or of MHF_2 and CoF_2; and alkali metal trifluorocobaltates can be prepared either by fusion[90] of MF with CoF_2, from $CoBr_2$ and MF in methanol,[94] by evaporating[95] at 90° a mixed aqueous solution of $CoCl_2$ and KHF_2, or even by just heating[96] together saturated aqueous solutions of MF and CoF_2.

Dehydration of $CoCl_2 \cdot 6H_2O$ *in vacuo* or in a nitrogen stream gives an anhydrous salt which is contaminated with oxides. Pure $CoCl_2$ can be prepared by heating[97] the hexahydrate in a stream of HCl at 400° followed by sublimation in anhydrous HCl at 500°. Simpler methods are heating the $CoCl_2 \cdot 6H_2O$,[7,98] or other hydrated cobalt(II) salts,[99] including the acetate or carbonate, with thionyl chloride under reflux, or heating[7,100] $Co(CH_3COO)_2 \cdot 4H_2O$ with acetyl chloride in benzene solution. Cs_3CoCl_5, prepared by fusing[101] together CsCl and $CoCl_6$, or by evaporating their mixed aqueous solutions,[102] is not a salt of the $CoCl_5^{3\ominus}$ anion, but contains $CoCl_4^{2\ominus}$ and $Cl^{\ominus}$ ions.[103] Many salts of the blue tetrahedral $[CoCl_4]^{2\ominus}$ ion are known, and may be prepared by fusion of the appropriate chloride mixtures,[101] or by mixing solutions of the constituent chlorides in organic solvents such as ethanol[104] or nitromethane.[105] Although aqueous solutions can be used to prepare the cesium salt,[106] even in aqueous solutions saturated with HCl or LiCl conversion of the cobalt to $[CoCl_4]^{2\ominus}$ is incomplete,[107] and thus organic solvents are to be preferred. Hydration to the distorted octahedral $[CoCl_4H_2O)_2]^{2\ominus}$ readily occurs.[108] *Trichlorocobaltates(II), e.g.* $CsCoCl_3$, can also be prepared by fusion.[101]

Anhydrous *cobalt bromide* can be prepared by

---

[86] *D. C. Bradley, K. J. Fisher*, J. Amer. Chem. Soc. *93*, 2058 (1971).
[87] *R. P. Linstead, J. M. Robertson*, J. Chem. Soc. 1736 (1936).
[88] *E. De Bie, P. Doyen*, Cobalt *15*, 1 (1962).
[89] *C. Poulenc*, Ann. chim. Phys. *2*, 47 (1894); *G. Brauer*, Handbook of Preparative Inorganic Chemistry. 2. Edition (English translation). Vol. 1, p. 267, Academic Press, London 1965.
[90] *W. Rüdorff, G. Lincke, D. Babel*, Z. Anorg. Allgem. Chemie *320*, 150 (1963).
[91] *E. Birk, W. Biltz*, Z. Anorg. Allgem. Chemie *153*, 115 (1926); C.A. *20*, 2924 (1926).
[92] *E. E. Aynsley, G. Hetherington, P. L. Robinson*, J. Chem. Soc. 1119 (1954).
[93] *H. G. von Schnering*, Z. Anorg. Allgem. Chemie *353*, 1 (1967).
[94] *D. S. Crocket, H. M. Haendler*, J. Amer. Chem. Soc. *82*, 4158 (1960).
[95] *K. Hirakawa, K. Hirakawa, T. Hashimoto*, J. Phys. Soc., Japan *15*, 2063 (1960).
[96] *D. J. Machin, R. L. Martin, R. S. Nyholm*, J. Chem. Soc. 1490 (1963).
[97] *W. E. Hatfield, J. T. Yoke*, Inorg. Chem. *1*, 463 (1962).
[98] *A. R. Pray*, Inorg. Synth. *5*, 153 (1957).
[99] *D. Khristov, S. Karaivanov, V. Kolushki*, Godishnik, Sofiiskiya Univ., Fiz.-Mat. Fak. Khim. *55*, 49 (1960); C.A. *59*, 3513 (1963).
[100] *G. W. Watt, P. S. Gentile, E. P. Helvenston*, J. Amer. Chem. Soc. *77*, 2752 (1955).
[101] *H. J. Seifert*, Z. Anorg. Allgem. Chemie *307*, 137 (1961).
[102] *R. P. van Stapele, H. G. Beljers, P. F. Bongers, H. Zijlstra*, J. Chem. Phys. *44*, 3719 (1966).
[103] *B. N. Figgis, M. Gerloch, R. Mason*, Acta Cryst. *17*, 506 (1964).
[104] *N. S. Gill, R. S. Nyholm*, J. Chem. Soc. 3997 (1959); *N. S. Gill, F. B. Taylor*, Inorg. Synth. *9*, 136 (1967).
[105] *G. D. Stucky, J. B. Folkers, T. J. Kistenmacher*, Acta Cryst. *23*, 1064 (1967).
[106] *L. I. Katzin, E. Gebert*, J. Amer. Chem. Soc. *75*, 2830 (1953).
[107] *F. A. Cotton, D. M. L. Goodgame, M. Goodgame*, J. Amer. Chem. Soc. *83*, 4690 (1961).
[108] *N. Fogel, C. C. Lin, C. Ford, W. Grindstaff*, Inorg. Chem. *3*, 720 (1964).

dehydrating the hydrated salt in a stream of hydrogen bromide[97] or from cobalt acetate and acetyl bromide,[100] and can be purified[109] by sublimation *in vacuo* at 550°. A simpler method[110] is the dehydration of the hexahydrate over concentrated sulfuric acid at room temperature. Salts of the pentabromocobaltate(II)[102] and tetrabromocobaltate(II)[104,107] can be prepared by similar methods to those used for the chloro analogs.

Anhydrous *cobalt iodide* can likewise be prepared by dehydrating the hydrated salt in a stream of hydrogen iodide,[97] from cobalt and iodine[110] or hydrogen iodide[111] at red heat, or from Co_3O_4 and aluminum triiodide[112] at 230°, and can be purified by sublimation.[7,112] *Tetraiodocobaltate(II)* salts can be prepared[104,107] in a similar way to the chloro- and bromo-analogs.

A large number of *complexes* of the *cobalt(II) halides* with monodentate ligands have been prepared.[113] These are of three main types: CoX_2L_4, CoX_2L_2 and $[CoX_3L]^{\ominus}$. The *neutral adducts* of the types CoX_2L_4 and CoX_2L_2 may usually be prepared by the action of the ligand in organic solvents such as alcohol, chloroform or benzene on CoX_2, preferably as the anhydrous salt. The action of heat on CoX_2L_4 usually yields CoX_2L_2, which often exist as two isomeric forms, often with different colours, one tetrahedral, and the other a polymeric halide-bridged octahedral form. The ligand may be ammonia, pyridine,[113,114] substituted pyridines,[113,114] or nitriles.[83] Phosphines, phosphine oxides, arsines, and arsine oxides usually give only tetrahedral CoX_2L_2 complexes.[115] Although bidentate ligands such as bipyridyl[116] usually give analogous complexes, tri- and tetradentate ligands often react differently. Diethylenetriamine[117] gives a blue complex, Co(dien)-Cl_2, which is in fact[118] $[Co(dien)_2]^{2\oplus}[CoCl_4]^{2\ominus}$, whereas 1,1,7,7-tetraethyldiethylenetriamine, Et_4dien, gives a five-coordinate complex,[119,120] $[Co(Et_4dien)Cl_2]$, as does a tri-tertiary arsine.[121] Tetradentate amine, phosphine, and arsine ligands, such as $N(CH_2CH_2PPh_2)_3$, also give five-coordinate complexes.[122,123] Even the monodentate phosphine Ph_2PH gives,[124] with $CoBr_2$ in methanol, a complex $[CoBr_2(Ph_2PH)_3]$ which has been shown[125] to be five-coordinate in addition to $[CoBr(Ph_2PH)_3]Br$, tetrahedral, and $[CoBr(Ph_2PH)_4]_2Br_2$, bridged octahedral.

Anionic complexes of the type $[CoX_3L]^{\ominus}$ can be prepared by mixing alcoholic solutions of CoX_2, MX and the monodentate ligand.[113] Most of the salts prepared have been with tetraalkylammonium,[126] *N*-alkylpyridinium[126] and other organic cations, with ligands such as pyridine,[126] triphenylphosphine,[113] or benzimidazole.[127]

23.4.2 Cobalt(II) Compounds with Predominantly Group VI Donors

Most simple cobalt(II) salts of oxoacids and carboxylic acids are readily and cheaply available commercially, or can be readily prepared from the acid and cobalt carbonate. *Cobalt(II) carboxylates* can be readily dehydrated *in*

---

[109] *T. J. Wydeven, N. W. Gregory*, J. Phys. Chem. *68*, 3249 (1964).
[110] *G. L. Clark, H. K. Buckner*, J. Amer. Chem. Soc. *44*, 230 (1922).
[111] *W. Biltz, E. Birk*, Z. Anorg. Allgem. Chemie *127*, 34 (1923);
W. Biltz, E. Birk, J. Chem. Soc. Abstracts *124 ii*, 768 (1923).
[112] *M. Chaigneau, M. Chastagnier*, Bull. Soc. Chim. France 886 (1957).
[113] *R. Colton, J. H. Canterford*, Halides of the First Row Transition Metals, Chap. 7, Wiley Intersci. Publ., London 1969.
[114] *H. C. A. King*, J. Chem. Educ. *48*, 482 (1971).
[115] *G. Booth*, Advan. Inorg. Chem. Radiochem. *6*, 1 (1964).
[116] *R. J. H. Clark, C. S. Williams*, Spectrochim. Acta *23A*, 1055 (1967).
[117] *G. A. Barclay, A. K. Barnard*, J. Chem. Soc. 2540 (1958).
[118] *M. Ciampolini, G. P. Speroni*, Inorg. Chem. *5*, 45 (1966).
[119] *Z. Dori, H. B. Gray*, J. Amer. Chem. Soc. *88*, 1394 (1966).
[120] *Z. Dori, R. Eisenberg, H. B. Gray*, Inorg. Chem. *6*, 483 (1967).
[121] *G. A. Barclay, R. S. Nyholm*, Chem. & Ind. (London) 378 (1953).
[122] *L. Sacconi, I. Bertini*, J. Amer. Chem. Soc. *90*, 5443 (1968).
[123] *L. M. Venanzi*, Angew. Chem. Intern. Ed. Engl. *3*, 453 (1964);
J. G. Hartley, D. G. E. Kerfoot, L. M. Venanzi, Inorg. Chim. Acta *1*, 145 (1967).
[124] *K. Issleib, E. Wenschuh*, Z. Anorg. Allgem. Chemie *305*, 15 (1960).
[125] *J. A. Bertrand, D. L. Plymale*, Inorg. Chem. *5*, 879 (1966).
[126] *D. H. Brown, K. P. Forrest, R. H. Nuttall, D. W. A. Sharp*, J. Chem. Soc. A, 2146 (1968).
[127] *M. Goodgame, F. A. Cotton*, J. Amer. Chem. Soc. *84*, 1543 (1962).

vacuo,[128] or by heating,[129] to give anhydrous salts, which being soluble in organic solvents are useful intermediates: anhydrous cobalt(II) perchlorate, also a useful intermediate in the synthesis of cobalt(II) complexes, can be prepared[131] by stirring $Co(ClO_4)_2 \cdot 6H_2O$ with an excess of 2,2-dimethoxypropane at room temperature for several hours. Triethylorthoformate can also be used as a dehydrating agent.[130,132] Cobalt(II) sulfoxylate, $CoSO_2 \cdot xH_2O$, the only known salt of H_2SO_2, can be prepared[133] from sodium dithionate, cobalt(II) acetate, and sodium hydrogen carbonate.

The complex cation $[Co(OH_2)_6]^{2\oplus}$ is well known in simple cobalt(II) salts both in the solid and in aqueous solution; similar cationic complexes of other monodentate ligands, such as dimethyl sulfoxide, may also be readily prepared[134] from the ligand and simple cobalt(II) salts. A number of anionic *complexes of oxoanions* are known: the formally eight-coordinate[135] cobalt(II) complex $[Co(NO_3)_4]^{2\ominus}$ can be prepared[136] as its $[N(CH_3)_4]^{\oplus}$ salt from $[N(CH_3)_4]NO_3$ and $Co(NO_3)_2 \cdot 2H_2O$ in nitromethane; reaction of dinitrogen pentoxide with sodium or potassium hexanitrocobaltates(III) yields the salts $M_2[Co^{II}(NO_3)_4]$, rather than the expected cobalt(III) nitrato complex.[137] The tetrahedral[138] *carboxylate complex* $[Ph_4As]_2[Co(O_2CCF_3)_4]$ can be prepared from $[Ph_4As]Cl$, CF_3CO_2Ag and $CoCl_2$ in acetonitrile.[138] $Na_2[Co(CO_3)_2]$, prepared[139] from Na_2CO_3, $NaHCO_3$ and $Co(NO_3)_2$, and two complex oxalates, $(NH_4)_2[Co(C_2O_4)_2]$ and $(NH_4)_4[Co(C_2O_4)_3]$, (both prepared[140] by boiling together ammonium oxalate and cobalt(II) oxalate in aqueous solution) have been known for many years. The octahedral *dihydrate of bis-acetylacetonatocobalt(II)* $[Co(acac)_2H_2O)_2]$ can be prepared by several methods,[141,142] but is best prepared[143] by addition of a mixture of 2,4-pentanedione and sodium hydroxide solution to a solution of cobalt(II) chloride. It can be dehydrated[143] to the tetrameric[144] anhydrous complex, which reacts with bidentate nitrogenous ligands, such as 2-aminomethylpyridine, phenanthroline and bipyridyl, to give[143] high yields of complexes of the type $[Co(acac)_2(phen)]$: a dimeric hydrate $[Co(acac)_2(H_2O)]_2$ is also known.[145] Slow addition of a methanolic solution of cobalt(II) chloride to a methanolic solution of sodium hydroxide and 2,4-pentanedione gives[143] $Na[Co(acac)_3]$. Other ligands, such as β-keto amines[146] and picolinic acid *N*-oxide[147] give complexes of the type $[Co(LL)_2(H_2O)_2]$ by a similar method to that used for the acetylacetonate complexes (the handling techniques are, however, considerably different for β-keto-amines[146]). $[Co(hfac)_2(H_2O)_2]$ can be conveniently prepared[148] by heating 1,1,1,5,5,5-hexafluoropentanedione with dicobalt octacarbonyl under reflux.

Cobalt(II) *complexes with chelated S-donor ligands* are difficult to prepare as they are usually very readily oxidized to the cobalt(III) analogs. Aqueous solutions of $Co^{2\oplus}$ oxidize spontaneously[149] when mixed with sodium diethyldithiocarbamate, $Na_2S_2CN(C_2H_5)_2$, even with

---

128 *T. L. Davis, A. V. Logan*, J. Amer. Chem. Soc. *62*, 1276 (1940).
129 *W. P. Tappmeyer, A. W. Davidson*, Inorg. Chem. *2*, 823 (1963).
130 *R. W. Gray, M. B. Welch, R. O. Ragsdale*, Inorg. Chim. Acta *3*, 17 (1969).
131 *B. D. Catsikis, M. L. Good*, J. Inorg. Nucl. Chem. 1039, *36* (1974).
132 R. *Davis, J. E. Fergusson*, Inorg. Chim. Acta *4*, 23 (1970).
133 *D. T. Farrar, M. M. Jones*, Inorg. Synth. *9*, 116 (1967).
134 *D. W. Meek, D. K. Straub, R. S. Drago*, J. Amer. Chem. Soc. *82*, 6013 (1960).
135 *J. G. Bergman, F. A. Cotton*, Inorg. Chem. *5*, 1208 (1966).
136 *F. A. Cotton, T. G. Dunne*, J. Amer. Chem. Soc. *84*, 2013 (1962).
137 *R. J. Fereday, D. Sutton*, Chem. Commun. 510 (1966).
138 *J. G. Bergman, F. A. Cotton*, Inorg. Chem. *5*, 1420 (1966).
139 *M. P. Applebey, K. W. Lane*, J. Chem. Soc. *113*, 609 (1918).
140 *F. Ephraim*, Ber. *42*, 3850 (1909); *F. Ephraim*, J. Chem. Soc. Abstracts *96 i*, 876 (1909).
141 *F. Gach*, Monatsh. *21*, 98 (1900); *F. Gach*, J. Chem. Soc. Abstracts *78 i*, 276 (1900).
142 *R. G. Charles, M. A. Pawlikowsky*, J. Phys. Chem. *62*, 440 (1958).
143 *J. B. Ellern, R. O. Ragsdale*, Inorg. Synth. *11*, 82 (1968).
144 *F. A. Cotton, R. C. Elder*, Inorg. Chem. *4*, 1145 (1965).
145 *F. A. Cotton, R. C. Elder*, Inorg. Chem. *5*, 423 (1966).
146 *R. H. Holm, F. Röhrscheid, G. W. Everett*, Inorg. Synth. *11*, 72 (1968).
147 *A. B. P. Lever, J. Lewis, R. S. Nyholm*, J. Chem. Soc. 5262 (1962).
148 *M. Kilner, F. A. Hartman, A. Wojcicki*, Inorg. Chem. *6*, 406 (1967).
149 *J. P. Fackler, D. G. Holah*, Inorg. Nucl. Chem. Lett. *2*, 251 (1966).

rigorous exclusion of oxygen; cobalt(II) xanthates behave similarly.[150] Cobalt(II) complexes of dithiocarbamates[151] and xanthates[152] have been reported with larger alkyl groups, however. Although the reaction[153,154] of $NaS_2CN(CH_3)_2$ with $Co(NO_3)_2$ gives the cobalt(III) complex $[Co\{S_2CN(CH_3)_2\}_3]$, the diseleno analog[154] gives $[Co^{II}\{Se_2CN(CH_3)_2\}_2]$. Square planar[155] bis(dithioacetylacetonato)cobalt(II), $[Co(SacSac)_2]$ has been prepared[156] from a complicated reaction mixture containing acetylacetone, $Co^{2\oplus}$, and hydrogen sulfide in ethanolic hydrogen chloride: it is readily oxidized to $[Co^{III}(SacSac)_3]$.

Reaction of the anhydrous acid $HSPSF_2$ with metallic cobalt gives[157] $[Co(S_2PF_2)_2]$, but this is readily oxidized by air or by $(S_2PF_2)_2$ to $[Co^{III}(S_2PF_2)_3]$. The reaction[158] of sodium dithiomaleonitrile with cobalt(II) salts in aqueous solution gives square planar[159] $[Co\{S_2C_2(CN)_2\}_2]^{2\ominus}$ which is readily oxidized even by air to $[Co\{S_2C_2(CN)_2\}_2]^{\ominus}$. Bis(trifluoromethyl)-1,2-dithiolene displaces[160] carbon monoxide from $Co_2(CO)_8$ to give five-coordinate[161] $[Co\{S_2C_2(CF_3)_2\}_2]_2$ which can be reduced[160] to $[Co\{S_2C_2(CF_3)_2\}_2]^{\ominus}$ and $[Co\{S_2C_2(CF_3)_2\}_2]^{2\ominus}$. The preparations[162] and structures[163] of a number of related complexes, in which the oxidation state cannot be precisely defined, have been reviewed.

[150] *Z. A. Sheka, E. E. Kriss*, Zh. Neorg. Khim. *1*, 586 (1956); C.A. *51*, 118 (1957).

[151] *L. Compin*, Bull. Soc. Chim. France *27*, 464 (1920); *L. Compin*, J. Chem. Soc. Abstracts *118 i*, 536 (1920).

[152] *M. Delépine, L. Compin*, Bull. Soc. Chim. France *27*, 469 (1920); *M. Delépine, L. Compin*, J. Chem. Soc. Abstracts *118 i*, 522 (1920).

[153] *F. A. Cotton, J. A. McCleverty*, Inorg. Chem. *3*, 1398 (1964).

[154] *A. T. Pilipenko, N. V. Mel'nikova*, Russian J. Inorg. Chem. *14*, 236 (1969).

[155] *R. Beckett, B. F. Hoskins*, Chem. Commun. 909 (1967).

[156] *R. L. Martin, I. M. Stewart*, Nature *210*, 522 (1966).

[157] *F. N. Tebbe, H. W. Roesky, W. C. Rode, E. L. Muetterties*, J. Amer. Chem. Soc. *90*, 3578 (1968).

[158] *A. Davison, R. H. Holm*, Inorg. Synth. *10*, 8 (1967).

[159] *J. D. Forrester, A. Zalkin, D. H. Templeton*, Inorg. Chem. *3*, 1500 (1964).

[160] *A. Davison, N. Edelstein, R. H. Holm, A. H. Maki*, Inorg. Chem. *3*, 814 (1964).

[161] *J. H. Enemark, W. N. Lipscomb*, Inorg. Chem. *4*, 1729 (1965).

[162] *J. A. McCleverty*, Prog. Inorg. Chem. *10*, 49 (1968).

[163] *R. Eisenberg*, Prog. Inorg. Chem. *12*, 295 (1970).

Mercury tetrathiocyanatocobaltate(II), a convenient standard for the calibration of magnetic susceptibility determinations by the Gouy method, can be readily prepared by adding an aqueous solution containing $CoSO_4$ and NH_4SCN to hot aqueous $HgCl_2$.[164]

23.4.3 Cobalt(II) Complexes with Predominantly Group V Donors

A large number of cobalt(II) complexes with nitrogen donors have been prepared. Addition of the ligand to an aqueous or alcoholic solution of a cobalt(II) salt—a halide, or the nitrate or perchlorate is often used—will usually give the complex with the maximum number of ligand molecules. If the ligand is volatile, some dissociation of the complex may occur on heating to give complexes with smaller numbers of ligand molecules. Alternatively, reaction of the cobalt(II) salt with a deficiency of ligand will sometimes give lower complexes directly, although solvation may occur to complete the cobalt(II) coordination sphere. Thus aqueous solutions of $CoCl_2$, $Co(ClO_4)_2$, $Co(SCN)_2$ *etc.*, on boiling with an excess of terpyridine[165,167] give a bis-complex $[Co(terpy)_2]^{2\oplus}$, or with a deficiency[166] of terpyridine, $CoCl_2$ gives a mono-complex, $[Co(terpy)Cl_2]$, which can also be prepared[166] by heating $[Co(terpy)_2]Cl_2$ for 2 hr at 180° *in vacuo*. Many nitrogenous donors including amines,[168] diamines,[169] triamines,[169] pyridine,[168] nitriles,[168] phenanthroline,[170] formamide,[169] and ethanolamine[169] react with dicobalt octacarbonyl or dodecacobalt tetracarbonyl to give the *tetracarbonylcobaltate(−I) salts* of $[CoL_6]^{2\oplus}$, $[Co(LL)_3]^{2\oplus}$ or $[Co(LLL)_2]^{2\oplus}$ complexes.

Air oxidation of cobalt(II) complexes with ammonia and some other nitrogenous donors proceeds very readily: the preparation[7,171] of $[Co(NH_3)_6]Cl_2$ from $CoCl_2$ and concentrated

[164] *B. N. Figgis, R. S. Nyholm*, J. Chem. Soc. 4190 (1958).

[165] *J. S. Judge, W. A. Baker*, Inorg. Chim. Acta *1*, 68 (1967).

[166] *J. S. Judge, W. A. Baker*, Inorg. Chim. Acta *1*, 239 (1967).

[167] *G. Morgan, F. H. Burstall*, J. Chem. Soc. 1649 (1937).

[168] *W. Hieber, J. Sedlmeier*, Chem. Ber. *87*, 25 (1954).

[169] *W. Hieber, R. Wiesboeck*, Chem. Ber. *91*, 1146 (1958).

[170] *I. Wender, H. W. Sternberg, M. Orchin*, J. Amer. Chem. Soc. *74*, 1216 (1952).

[171] *G. B. Kauffman, N. Sugisaka*, Inorg. Synth. *9*, 157 (1967).

aqueous ethanolic ammonia solution must be carried out under strictly anaerobic conditions or oxidation to $[Co^{III}(NH_3)_6]Cl_3$ occurs: complexes with ethylenediamine and many other related ligands can be prepared in a similar manner.[3]

Octahedral low spin hexanitrocobaltates(II), *e.g.* $K_2Ca[Co(NO_2)_6]$, crystallize easily from aqueous solutions; the tetranitro complex $[Co(NO_2)_4]^{2\ominus}$ is formed if the nitrite ion concentration is relatively low.[172]

Phosphites cause disproportionation of cobalt(II) to pentakisphosphitocobalt(I) and hexakisphosphitocobalt(III) complexes.[74,75] A number of phosphine complexes of cobalt(II) are known: in most cases halide ligands are also present and the preparation is mentioned in Section 23.4.1. Nitric oxide reacts with solutions containing $Co^{2\oplus}$ (as the chloride, nitrate or perchlorate) and N- or P-donors to give complexes of the type $[L_4Co(NO)](ClO_4)_2$; with some ligands complexes of the type $[L_2Co(NO)_2](ClO_4)$ were obtained instead.[173] $Co^{II}As_6$ complexes can be prepared in a similar way with both bi- and tritertiary arsine ligands[121,174]: both $[Co(diars)_3](ClO_4)_2$[174] and $[Co(triars)_2](ClO_4)_2$[121] are readily oxidized by air to the cobalt(III) analogs.

23.4.4 Cobalt(II) Complexes with Predominantly Group IV Donors

The complicated chemistry of *cyanide complexes* of cobalt(II) and cobalt(III) has been reviewed.[175] Addition of potassium cyanide to a solution of $Co^{2\oplus}$ gives[176] a brownish precipitate of hydrated $Co(CN)_2$, which dissolves in an excess of potassium cyanide to give[176] a dark green paramagnetic solution containing predominantly $[Co(CN)_5]^{3\ominus}$: precipitation with ethanol[176] gives $K_6[Co_2(CN)_{10}]$ as a violet diamagnetic solid; early claims[177] of $K_4[Co(CN)_6]$ have not been substantiated.[176] $K_6[Co_2(CN)_{10}]$ dissolves in water to re-form dark green solutions, which deposit $Co_3[Co_2(CN)_{10}]$, as a green precipitate, if an excess of KCN is not present.[178] The solution of $[Co^{II}(CN)_5]^{3\ominus}$ is a very useful intermediate, being very readily oxidized[179] to $[Co^{III}(CN)_6]^{3\ominus}$, and reacting with hydrogen,[180] and to a small extent even with water[178] to give a hydrido species; it also reacts in the cold with sulfur dioxide,[180a] stannous chloride,[180a] tetrafluoroethylene,[181] and acetylene[182] to give $K_6[(NC)_5Co(SO_2)Co(CN)_5]$, $K_6[(NC)_5Co(SnCl_2)Co(CN)_5]$, $K_6[(NC)_5CoCF_2CF_2Co(CN)_5]$ and $K_6[(NC)_5CoCH{=}CHCo(CN)_5]$ respectively. The SO_2 insertion product is probably identical[183] with a compound claimed as a sulphito-bridged cobalt(III) derivative, $K_6[(NC)_5Co(SO_3)Co(CN)_5]$, over forty years ago.[184] Reaction with oxygen,[185] potassium ferrocyanide,[185] and alkyl halides[186,187] readily gives the cobalt(III) derivatives $K_6[(NC)_5CoOOCo(CN)_5]$, $Ba_3[(NC)_5Fe^{II}CNCo^{III}(CN)_5]$, and $K_3[(NC)_5CoR]$ respectively.

Cobalt(II) forms a wide range of *complexes with isonitriles*[79]: mixing ethanolic solutions of cobalt(II) halides and methyl isonitrile gives[83] both $[(CH_3NC)_4CoX_2]$ and $[(CH_3NC)_2CoX_2]$ depending on the conditions. Aqueous solutions of the CoL_2X_2 complexes contain[83] $[(CH_3NC)_4Co]^{2\oplus}[CoCl_4]^{2\ominus}$. Some of the CoL_4X_2 complexes, e.g. $(CH_3NC)_4CoI_2$, can exist as unstable diamagnetic and stable paramagnetic isomers of uncertain structure.[83] The stable form of $(CH_3NC)_4CoI_2$ reacts with aqueous $NaClO_4$ to give a compound said[83] to

---

172 *V. Cuttica, M. Paoletti*, Gazz. Chim. Ital. *52*, 279 (1922);
V. Cuttica, M. Paoletti, J. Chem. Soc. Abstracts *124 ii*, 76 (1923).

173 *T. B. Jackson, M. J. Baker, J. O. Edwards, D. Tutas*, Inorg. Chem. *5*, 2046 (1966).

174 *F. H. Burstall, R. S. Nyholm*, J. Chem. Soc. 3570 (1952)

175 *B. M. Chadwick, A. G. Sharpe*, Advan. Inorg. Chem. Radiochem. *8*, 83 (1966).

176 *A. W. Adamson*, J. Amer. Chem. Soc. *73*, 5710 (1951).

177 *W. Biltz, W. Eschweiler, A. Bodensiek*, Z. Anorg. Allgem. Chemie *170*, 161 (1928); C.A. *22*, 4017 (1928).

178 *W. P. Griffith, G. Wilkinson*, J. Chem. Soc. 2757 (1959).

179 *J. H. Bigelow*, Inorg. Synth. *2*, 225 (1946).

180 *N. K. King, M. E. Winfield*, J. Amer. Chem. Soc. *83*, 3366 (1960).

180a *A. A. Vlcek, F. Basolo*, Inorg. Chem. *5*, 156 (1966);

181 *M. J. Mays, G. Wilkinson*, J. Chem. Soc. 6629 (1965).

182 *W. P. Griffith, G. Wilkinson*, J. Chem. Soc. 1629 (1959).

183 *L. Cambi, E. Paglia*, Atti Accad. Naz. Lincei, Classe Sci. Fis. Mat. Natur. Rend. *24*, 378 (1958); C.A. *53*, 111 (1959).

184 *P. Ray*, Z. Anorg. Allgem. Chemie *208*, 392 (1932); C.A. *27*, 677 (1933).

185 *A. Haim, W. K. Wilmarth*, J. Amer. Chem. Soc. *83*, 509 (1961).

186 *J. Kwiatek, J. K. Seyler*, J. Organometal. Chem. *3*, 421 (1965).

187 *J. Halpern, J. P. Maher*, J. Amer. Chem. Soc. *86*, 2311 (1964).

be $[(CH_3NC)_4CoICo(CNCH_3)_4](ClO_4)_3$. $Co(ClO_4)_2$ with isonitrile gives $[(RNC)_5Co]^{2\oplus}(ClO_4^{\ominus})_2$, but there is only spectroscopic evidence for the existence of hexaisonitrile complexes.[83,188]

Several different structures have been found for the $[(RCN)_5Co]^{2\oplus}$ complexes: the very unstable blue paramagnetic $[(CH_3NC)_5Co]^{2\oplus}$,[83,189] which was presumed to be five-coordinate,[83] but which is probably hydrated,[189] $[(CH_3NC)_5Co(OH_2)]^{2\oplus}$, transforms to a red diamagnetic dimer,[190] $[(CH_3NC)_5Co–Co(CNCH_3)_5]^{4\oplus}$, but solutions of either compound are blue and paramagnetic (*cf* pentacyanocobalt(II) complexes). Solid $[(PhNC)_5Co](ClO_4)_3$ exists as either a paramagnetic blue hydrate, $[(PhNC)_5Co(OH_2)](ClO_4)_2$ or as a yellow paramagnetic anhydrous complex which is presumably a genuine five-coordinate complex.[189] Many of the cobalt(II) isonitrile complexes undergo reduction to Co(I) very readily, and, less readily, oxidation to cobalt(III).[79]

The related *hexaalkynyl cobalt(II) complexes* $Na_4[Co(C{\equiv}CCH_3)_6]$ and $K_4[Co(C{\equiv}CC_2H_5)_6]$ have been prepared[191] by the action of mono alkalimetal acetylides with $[Co(NH_3)_4(CNS)_2]$ in liquid ammonia. In solution in liquid ammonia they are oxidized by oxygen to the cobalt(III) derivative $M_3[Co(C{\equiv}CR)_6]$.

Although the action of Grignard reagents on cobalt(II) halides yields either the black "cobalt hydride" (see Section 23.2) or intractable black tars[192] of uncertain composition, if an *o*-disubstituted aryl Grignard reagent or aryllithium is reacted with phosphine complexes of cobalt(II) halides, square planar[193] alkyl derivatives, *e.g.* *trans*-$[Co(mesityl)_2(PEt_2Ph)_2]$, are obtained.[194] Cobalt(II) salts of the tricyanomethanide ion, e.g. $[Co\{C(CN)_3\}_2]$, and pyridine adducts, e.g. $[Co(py)_2\{C(CN)_3\}_2]$, can be prepared[195] by the action of potassium tricyanomethamide on $Co(NO_3)_2$ in alcohol: addition of pyridine gives an adduct $[Co(py)_2\{C(CN)_3\}_2]$. A number of phenanthroline and bipyridyl derivatives of cobalt(II) cyanides [and also of cobalt(III) cyanides] have been prepared.[196]

23.4.5 Cobalt(II) Complexes of Polydentate Ligands

When aqueous solutions of polydentate ligands are added to solutions of cobalt(II) salts, labile cobalt(II) complexes are formed: in a large number of cases the formation constants have been measured,[3] and they are often very high, but in comparatively few cases have solid complexes been isolated. $H_2[Co^{II}(edta)]\cdot 3H_2O$[197] and $[Co^{II}$(bis-salicylaldehyde-triethylenetetramine)$]$[198] are among the complexes prepared by this method. An alternative procedure (which in many cases has much to commend it in that extraneous ions are kept to a minimum, and thus the solution of the cobalt(II) complex—which is often very soluble—can be evaporated down to dryness in order to induce crystallization) is to dissolve $CoCO_3$ in an hot aqueous solution of the ligand.[199] Oxidation to cobalt(III) complex occurs with almost any oxidizing agent: in some cases even air causes initial formation of peroxy bridged cobalt(II) complexes,[200] and ultimate formation of cobalt(III) complexes.[200]

Of especial importance, because of their use as "models" for biological systems, are cobalt(II) complexes of planar tetradentate ligands. Cobalt(II) phthalocyanin is prepared[201] by heating cobalt metal with phthalodinitrile under reflux for 3 hr: it can be sublimed as dark blue needles. *N,N'*-disalicylalethylenediamine, H_2-salen, reacts with $CoCl_2$ under carefully controlled conditions to give[202] a monomeric[203]

---

[188] *N. Katoaka, H. Kon*, J. Amer. Chem. Soc. *90*, 2978 (1968).

[189] *J. M Pratt, P. R. Silverman*, Chem. Commun. 117 (1967);
J. M. Pratt, R. J. P. Williams, J. Chem. Soc. A, 1286 (1967).

[190] *F. A. Cotton, T. G. Dunne, J. S. Wood*, Inorg. Chem. *3*, 1495 (1964).

[191] *R. Nast, H. Levinsky*, Z. Anorg. Allgem. Chemie *282*, 210 (1955); C.A. *50*, 9200 (1956);
R. Nast, Angew. Chem. *72*, 26 (1960).

[192] *M. H. Abraham, M. J. Hogarth*, J. Organometal. Chem. *12*, 1 (1968).

[193] *P. G. Owston, J. M. Rowe*, J. Chem. Soc. 3411 (1963).

[194] *J. Chatt, B. L. Shaw*, J. Chem. Soc. 285 (1961).

[195] *H. Köhler*, Z. Anorg. Allgem. Chemie *331*, 237 (1964).

[196] *L. Cambi, E. Paglia*, J. Inorg. Nucl. Chem. *8*, 249 (1958).

[197] *I. A. W. Shimi, W. C. E. Higginson*, J. Chem. Soc. 260 (1958).

[198] *B. Das Sarma, J. C. Bailar*, J. Amer. Chem. Soc. *77*, 5476 (1955).

[199] *K. A. Fraser, H. A. Long, R. Candlin, M. M. Harding*, Chem. Commun. 344 (1965);
G. Lange, Diss. Bonn 1952; Gmelin System No. 58, B, Ergänzungsband Lieferung 1, p. 178, 179, 181.

[200] *R. D. Gillard, A. Spencer*, J. Chem. Soc. *A*, 2718 (1969).

[201] *P. A. Barrett, C. E. Dent, R. P. Linstead*, J. Chem. Soc. 1719 (1936).

[202] *H. Diehl, C. C. Hach*, Inorg. Synth. *3*, 196 (1950).

[203] *P. C. Hewlett, L. F. Larkworthy*, J. Chem. Soc. 882 (1965).

complex, [Co(salen)], which will reversibly absorb oxygen in the solid state,[204] and which absorbs oxygen when dissolved in DMF to give[205] a dimeric oxygen bridged[206] adduct. It also acts as a homogeneous oxidation catalyst.[207] Bis(acetylacetone)ethylenediamine, H_2BAE, also reacts[208,209] with $CoCl_2$ to give a monomeric complex $[Co(BAE)(H_2O)_2]$: the yield is low unless air is excluded during the preparation.[208] Recrystallization from benzene gives anhydrous[209] [Co(BAE)]. Under strictly anaerobic conditions the reaction of cobalt(II) acetate with dimethylglyoxime, Hdmg, similarly gives a high yield of $[Co(dmg)_2(H_2O)_2]$ as paramagnetic orange crystals, which can be dehydrated *in vacuo* at 80° to violet $[Co(dmg)_2]$ which absorbs oxygen avidly and reacts with various donors to give 1:1 and 1:2 adducts.[48] 1:1 Adducts, which can also be prepared[210] directly from cobalt(II) acetate, dimethylglyoxime (or diphenylglyoxime) and the stoichiometric amount of the donor (which can be pyridine, tri-*n*-butyl or triphenyl phosphine, triphenyl arsine or triphenyl stibine), also react with oxygen reversibly to give oxygen-bridged compounds, such as $[(py)(dmg)_2Co\text{–}O_2\text{–}Co(dmg)_2(py)]$, which can be isolated at −20°.[210] An alcoholic solution of the pyridine adduct $[(py)(dmg)_2Co\text{–}Co(dmg)_2(py)]$ acts as a homogeneous hydrogenation catalyst[211] and can be reduced[48] to a solution of bis(dimethylglyoxime)cobalt(I) complex which can be alkylated[48] to alkylcobalt(III) complexes (see 23.5). $[Co^{II}(BAE)]$[209] and [Co(salen)][212,213] can also be reduced to cobalt(I) derivatives, which can then be alkylated.

[204] *L. H. Vogt, H. M. Faigenbaum, S. E. Wiberley,* Chem. Rev. *63*, 269 (1963).

[205] *F. Calderazzo, C. Floriani, J. J. Salzmann,* Inorg. Nucl. Chem. Lett. *2*, 379 (1966).

[206] *M. Calligaris, G. Nardin, L. Randaccio,* Chem. Commun. 763 (1969).

[207] *L. H. Vogt, J. G. Wirth, H. L. Finkbeiner,* J. Org. Chem. *34*, 273 (1969).

[208] *G. T. Morgan, J. D. Main Smith,* J. Chem. Soc. 2030 (1925).

[209] *G. Costa, G. Mestroni, G. Tauzher, L. Stephani,* J. Organometal. Chem. *6*, 181 (1966).

[210] *G. N. Schrauzer, L. P. Lee,* J. Amer. Chem. Soc. *92*, 1551 (1970).

[211] *Y. Ohgo, S. Takeuchi, J. Yoshimura,* Bull. Chem. Soc., Japan *44*, 283 (1971).

[212] *F. Calderazzo, C. Floriani,* Chem. Commun. 139 (1967).

[213] *C. Floriani, M. Puppis, F. Calderazzo,* J. Organometal. Chem. *12*, 209 (1968).

23.5 Cobalt(III)

In aqueous solutions containing no other complexing agents $Co^{3\oplus}$ ions are sufficiently powerful oxidizing agents to oxidize water to oxygen. Poorer donors than water (*e.g.* the halogens) do not in general form cobalt(III) complexes, although both simple and complex cobalt(III) fluorides are known they are immediately reduced to cobalt(II) by water. Oxygen donors other than water likewise give very unstable cobalt(III) complexes unless the ligand is chelating, e.g. $CO_3^{2\ominus}$, $acac^{\ominus}$, which give fairly stable compounds although these can be readily reduced.

The Co(II)/Co(III) redox potential is drastically altered by N-donors:

$$[Co(H_2O)_6]^{3\oplus} + e \longrightarrow [Co(H_2O)_6]^{2\oplus} \qquad E^0 = +1{\cdot}84\ v$$

$$[Co(NH_3)_6]^{3\oplus} + e \longrightarrow [Co(NH_3)_6]^{2\oplus} \qquad E^0 = +0{\cdot}1\ v$$

and even atmospheric oxygen is adequate to oxidize cobalt(II) to cobalt(III) in their presence: with O-donors a more powerful oxidizing agent, e.g. H_2O_2, has to be used although in basic solution air is again adequate:

$$[CoO(OH)]_s + H_2O + e \longrightarrow [Co(OH)_2]_s + OH^{\ominus} \qquad E^0 = +0{\cdot}17\ v$$

The environment of cobalt(III) is almost exclusively regular octahedral: tetrahedral cobalt(III) is known[213a] only in a 12-heteropoly tungstate, and five-coordinate cobalt(III) of uncertain geometry has been reported[214] for the phosphine complexes of the type $[CoX_3(PR_3)_2]$. Octahedral cobalt(III) complexes are diamagnetic—the only exception being the complex fluoride $CoF_6^{3\ominus}$—and therefore fairly inert to substitution.

23.5.1 General Methods of Synthesis of Cobalt(III) Complexes

Many of the methods used today date back to the early workers, especially Werner: fortunately as a rule no difficulty is experienced in repeating Werner's syntheses. Valuable compilations of tested syntheses of cobalt(III) complexes are to be found in Inorganic Syntheses, and in Palmer.[215] The synthetic methods can be divided into four categories.

23.5.1.1 Syntheses Starting from Cobalt(II)

The air oxidation of a cobalt(II) solution con-

[213a] *L. C. W. Baker, V. E. Simmons,* J. Amer. Chem. Soc. *81*, 4744 (1959).

[214] *K. A. Jensen, B. Nygaard, C. T. Pedersen,* Acta Chem. Scand. *17*, 1126 (1963).

[215] *W. G. Palmer,* Experimental Inorganic Chemistry, pp. 529–553, Cambridge University Press 1965.

taining the appropriate ligands was frequently used by Werner, but it suffers from the disadvantage of requiring the passage of air for long periods—sometimes 24 hr or even longer. It is often useful[216] to add activated charcoal as a "catalyst". This may act by promoting electron exchange between cobalt(III) and the labile cobalt(II), accelerating attainment of the equilibrium composition. Thus aerial oxidation of a mixture of $CoCl_2$, NH_4Cl, and excess ammonia gives $[Co(NH_3)_5Cl]^{2\oplus}$, whereas if activated charcoal is added as well, a good yield of $[Co(NH_3)_6]Cl_3$ is obtained.[217] Ozonized oxygen (2% O_3 in O_2) is sometimes better than air: the synthesis of $[(NH_3)_5Co(\mu\text{-}O_2)Co(NH_3)_5](NO_3)_5$ carried out by Werner[218] using air works better with ozonized oxygen.[219] The same has also been found[220] for various other binuclear cobalt(III) ammines prepared by Werner.[221] When air or oxygen is used as the oxidant the required cobalt(III) complex is sometimes formed *via* well defined Co(II) and Co(III) intermediates which often have peroxo-bridges and can slow down formation of the required complex. For example, preparation of the bis(dipeptidato)cobalt(III) anion by air oxidation of aqueous $CoCl_2$ solutions containing the dipeptide proceeds[200] *via* a brown and a red oxygen-bridged dimeric bispeptide cobalt complex whereas when hydrogen peroxide is used, although an intermediate color change (possibly corresponding to one of these intermediates) is observed, the final cobalt(III) complex is rapidly formed.[222]

The main advantage of hydrogen peroxide over air or oxygen is however the speed of the reaction. Thus the oxidation step in the preparation[223] of $[Co(NH_3)_4CO_3]NO_3$ requires air oxidation for $2\frac{1}{2}$ hr: with hydrogen peroxide the oxidation is complete in 10 minutes. The addition of a small amount of charcoal as catalyst is useful when hydrogen peroxide is used as oxidant.[224] For ligands which are even only weakly acidic the mixture of an aqueous cobalt(II) salt and the ligand which is required in these oxidation procedures can often conveniently be a suspension of cobalt(II) carbonate in an aqueous solution of the ligand.

The commercial material varies considerably in its activity: if difficulty is experienced with this method, especially with very weakly acid ligands, then freshly precipitated cobalt(II) carbonate should be used instead.

When H_2O_2 is used as the oxidizing agent the cobalt carbonate readily dissolves in the weakly acidic ligand as oxidation of the small amount of cobalt(II) in solution takes place. This procedure works well for *tris(acetylacetonato)-cobalt(III)*[225] and for the *bis(dipeptidato)cobalt-(III)* anion[222] etc.

A wide range of other oxidizing agents may be used, but oxygen and hydrogen peroxide are of the widest utility as no contamination of the product with the reduced form of the oxidant can occur. Another slightly less convenient oxidant, which is, however, useful in the synthesis of oxalato complexes is lead dioxide which has been used in the preparation of $[Co(en)_2(ox)]Cl$[226] and of $K_3[Co(ox)_3]$.[215,227] Other oxidizing agents may destroy the free oxalate ion before complexation. Of the halogens chlorine[167] and bromine[228] have been used for the preparation of $[Co(terpy)_2]Cl_3$ and $[Co(py)_4Br_2]^{\oplus}(Br_3)^{\ominus}$ respectively. Sodium chlorite[229] is the preferred oxidizing agent for oxidizing $[Co^{II}(phen)_3]^{2\oplus}$ to $[Co^{III}(phen)_3]^{3\oplus}$, and can also be used in the preparation of many other complexes.[229]

216 *J. C. Bailar, J. B. Work*, J. Amer. Chem. Soc. *67*, 176 (1945).
217 *J. Bjerrum, J. P. McReynolds*, Inorg. Synth. *2*, 216 (1946).
218 *A. Werner, A. Mylius*, Z. Anorg. Allgem. Chemie *16*, 245 (1898);
A. Werner, A. Mylius, J. Chem. Soc. Abstracts *74 ii*, 334 (1898).
219 *J. S. Valentine, D. Valentine*, J. Amer. Chem. Soc. *93*, 1111 (1971).
220 *A. G. Sykes*, Trans. Faraday Soc. *58*, 543 (1962).
221 *A. Werner*, Justus Liebigs Ann. Chem. *375*, 1–144 (1910);
A. Werner, J. Chem. Soc. Abstracts *98 ii*, 857–871 (1910).
222 *R. D. Gillard, P. R. Mitchell*, unpublished observations.
223 *G. Schlessinger*, Inorg. Synth. *6*, 173 (1960).
224 *P. Spacu, C. Gheorghiu, M. Brezeanu, S. Popescu*, Analele Univ., C.I. Parhon, Bucuresti, Ser. stiint. nat. *19*, 43 (1958); C.A. *53*, 18722 (1959).
225 *B. E. Bryant, W. C. Fernelius*, Inorg. Synth. *5*, 188 (1957).
226 *F. P. Dwyer, I. K. Reid, F. L. Garvan*, J. Amer. Chem. Soc. *83*, 1285 (1961).
227 *J. C. Bailar, E. M. Jones*, Inorg. Synth. *1*, 35 (1939).
228 *A. V. Babaeva, I. B. Baranovskii*, Zhur. Neorg. Khim. *4*, 755 (1959); C.A. *54*, 7401 (1960).
229 *P. Spacu, C. Gheorghiu, M. Brezeanu, S. Popescu*, Rev. Chim. Acad. rép. populaire Roumaine *3*, 127 (1958); C.A. *53*, 14812 (1959).

The choice of oxidant can affect the final product. For example oxidation of cobalt(II) solutions containing potassium oxalate with two-electron oxidants such as hydrogen peroxide and hypochlorite ion gives[230] "Durrant's salt",

$$K_4\left[(C_2O_4)_2Co\overset{\displaystyle H\atop\displaystyle O}{\underset{\displaystyle O\atop\displaystyle H}{\diamond}}Co(C_2O_4)_2\right]\cdot 2\,H_2O$$

whereas the one-electron oxidant cerium(IV) gives instead $[Co(C_2O_4)_3]^{3\ominus}$. However, as mentioned above, the insoluble two-electron oxidant PbO_2 gives[215] $[Co(C_2O_4)_3]^{3\ominus}$.

One useful procedure, if contamination of the final product with unwanted ions is a problem, is to prepare cobalt(III) oxide hydroxide, CoO(OH), *in situ*, by oxidation of cobalt(II) hydroxide with hydrogen peroxide,[231] CoO(OH) will then dissolve in aqueous solutions of many acidic ligands, such as amino acids.[232] Alternatively CoO(OH) isolated by standard methods (see 23.5.3) can be used in a similar way.

23.5.1.2 Replacement Reactions of Cobalt(III)

As oxidation of cobalt(II) solutions containing ligands often results in mixtures of cobalt(III) complexes, especially if cobalt(III) complexes with several different ligands are required, it is often preferable to prepare such compounds by simple displacement reactions on preformed cobalt(III) complexes. Many of the methods used are fairly arbitrary, dating back to the original methods used by the early German workers, but there have been some attempts[233,234] to rationalize the syntheses of metal complexes on a mechanistic basis.

There are a number of cobalt(III) complexes containing only one fairly labile ligand. Examples of these are $[Co(CO_3)_3]^{3\ominus}$, $[Co(mal)_3]^{3\ominus}$, $[Co(SO_3)_3]^{3\ominus}$, $[Co(NO_2)_6]^{3\ominus}$ labile especially in weakly acidic solution, and $[Co(NH_3)_6]^{3\oplus}$, labile in weakly basic solution. $[Co(OH_2)_6]^{3\oplus}$ itself is too unstable to be of much use as an intermediate, although $[Co(urea)_6](ClO_4)_3$ can be used in nonaqueous solvents.[235] These complexes may be used either under conditions such that all the existing ligand molecules are lost—for example the displacement[236] of ammonia from $[Co(NH_3)_6]^{3\oplus}$ by the involatile hexadentate pentaethylenehexamine to give $[Co(penten)]^{3\oplus}$—or such that some are retained in a mixed complex. Thus $Na_3[Co(NO_2)_6]$ is widely used in the synthesis of nitrocomplexes such as $[Co(dien)(NO_2)_3]$[237] and $K[Co(aminoacidato)_2(NO_2)_2]$,[238] and there is a more restricted use of $Na_3[Co(SO_3)_3]$ in the preparation of sulfito complexes such as $Na[Co(SO_3)_2(NH_3)_2]$,[239] and of $[Co(mal)_3]^{3\ominus}$ in the preparation of malonato complexes like $Ba[Co(mal)_2(gly)]$.[240] There is clearly much unused scope for these last two intermediates.

A particularly convenient complex for this approach is the *tris-carbonatocobaltate(III) ion*, either as its potassium salt[241] which is very soluble and best prepared *in situ* as it is not easily isolated, or as its sodium salt[242] which is very easily isolated as an insoluble solid. Both of these salts have been used in the synthesis of many types of cobalt(III) complexes[241–243] such as $K[Co(NH_3)_2(CO_3)_2]$,[241] $[Co(benzoylacetonate)_3]$,[242] and $[Co(salicylate)_3]^{3\ominus}$.[242]

Complexes with four or five inert ligand molecules and two or one labile ligand molecules are readily available: it is thus possible to carry out systematic syntheses of many compounds. The most useful replacement reactions from the point of view of giving high yields of predictable products are the replacement of halo-

[230] *A. W. Adamson, H. Ogata, J. Grossman, R. Newbury*, J. Inorg. Nucl. Chem. *6*, 319 (1958).

[231] *S. Veil*, Compt. Rend. *182*, 1146 (1926); C.A. *20*, 2611 (1926).

[232] *H. Ley, H. Winkler*, Ber. *42*, 3894 (1909); *H. Ley, H. Winkler*, J. Chem. Soc. Abstracts *96 i*, 886 (1909).

[233] *F. Basolo*, Application of Reaction Mechanisms to the Syntheses of Metal Complexes, in *W. Schneider, G. Aneregg, R. Gut*, Essays in Co-ordination Chemistry, p. 201, Birkhäuser Verlag, Basel 1964.

[234] *J. L. Burmeister, F. Basolo*, The Application of Reaction Mechanisms to the Synthesis of Co-ordination Compounds, in *W. L. Jolly*, Preparative Inorganic Reactions, V, p. 1, Interscience, New York 1968.

[235] *L. E. Bennett*, Inorg. Chem. *9*, 1941 (1970).

[236] *F. P. Emmenegger, G. Schwarzenbach*, Helv. Chim. Acta *49*, 625 (1966).

[237] *P. H. Crayton*, Inorg. Synth. *7*, 207 (1963).

[238] *M. B. Celap, R. G. Denning, D. J. Radanovic, T. J. Janjic*, Inorg. Chim. Acta *5*, 9 (1971).

[239] *A. V. Babaeva, Y. Y. Kharitonov, I. B. Baranovskii*, Russian J. Inorg. Chem. *7*, 643 (1962).

[240] *K. Yamasaki, J. Hidaka, Y. Shimura*, Bull. Chem. Soc., Japan *42*, 119 (1969).

[241] *M. Mori, M. Shibata, E. Kyuno, T. Adachi*, Bull. Chem. Soc., Japan *29*, 883 (1956).

[242] *H. F. Bauer, W. C. Drinkard*, J. Amer. Chem. Soc. *82*, 5031 (1960).

[243] *K. V. Krishnamurty, G. M. Harris, V. S. Sastri*, Chem. Rev. *70*, 171 (1970).

gens, carbonate, and water. Thus the action of aqueous Na_2CO_3 on aqueous *cis*-$[Co(en)_2Cl_2]Cl$ on brief boiling[244] readily gives *cis*-$[Co(en)_2CO_3]^{\oplus}$. Other examples of replacement of halide ions are:

$$\alpha\text{-}[Co(trien)Cl_2]Cl + K_2C_2O_4 \longrightarrow \text{predominantly } \beta\text{-}[Co(trien)(C_2O_4)]^{\oplus} \quad \text{(Ref. 245)}$$

$$trans\text{-}[Co(en)_2Cl_2]^{\oplus} + NaSCN \xrightarrow{\text{charcoal}} cis\text{-}[Co(en)_2(NCS)_2]^{\oplus} \quad \text{(Ref. 246)}$$

Replacement of the carbonate ligand by other more strongly acidic ligands is also of wide utility:

$$[Co(en)_2CO_3]^{\oplus} + HOOC\text{—}CH_2\text{—}CH_2\text{—}COOH \longrightarrow [Co(en)_2(succinate)]^{\oplus} \quad \text{(Ref. 247)}$$

$$2[Co(en)_2CO_3]ClO_4 + H_4tart \longrightarrow [(en)_2Co(tart)Co(en)_2](ClO_4)_2 \quad \text{(Ref. 248)}$$

$$\alpha\text{-}[Co(trien)CO_3]^{\oplus} + H_2C_2O_4 \longrightarrow \alpha\text{-}[Co(trien)(ox)]^{\oplus} \quad \text{(Ref. 249)}$$

In the last case the acidic medium is said[249] to prevent isomerization of the α-$[Co(trien)(H_2O)_2]^{3\oplus}$ and α-$[Co(trien)(H_2O)(OH)]^{2\oplus}$ presumed to be formed as intermediates. Such aquo species are often formed from the chloro- or carbonato-complexes in a separate initial step, either by treatment of the chloro complex with base[250] or with silver oxide,[251] or by treatment of the carbonato complex with perchloric acid.[248] This step can often be observed by the color change and only when this is complete is the reactant ligand added.

Most syntheses of cobalt(III) complexes have been carried out in aqueous solution: however, some recent reports clearly show some advantages of the synthetic use of non-aqueous solvents. The synthesis of $[Co(NH_3)_5OCOR]^{\oplus}$ is normally readily achieved[252] from $[Co(NH_3)_5(CO_3)]^{\oplus}$ and RCO_2H in aqueous solutions if R = alkyl, but if the acid contains other reactive groups or is sterically hindered, either dimeric products may be obtained (e.g. terephthalic acid, see below), or no reaction may occur. However, the reaction[253] of acid anhydrides with a DMF solution of $[Co(NH_3)_5(OH_2)]^{3\oplus}$ in the presence of benzyldimethylamine readily gives carboxylato pentammine complexes even with R = $CH{=}CHCO_2H$, $CH{=}CH\text{—}CH{=}CH\text{—}CH_3$, and $C(CH_3)_3$. A monodentate ethylenediamine complex, *cis*-$[Co(en)_2(enH)Cl]Cl_3$, can be prepared[254] by the reaction of *trans*-$[Co(en)_2Cl_2]Cl$ with ethylenediamine in methanol: in water only $[Co(en)_3]Cl_3$ is obtained.

Trialkylphosphates have also been used as solvents for perchlorate salts of ammine and polyamine complexes; displacement of coordinated solvent by methoxide ion[255] and by amino acid esters[256] gives complexes which are immediately hydrolyzed by water. It has recently been reported[235] that $[Co\{OC(NH_2)_2\}_6](ClO_4)_3$, which is soluble in acetonitrile, is a convenient intermediate for the synthesis of cobalt(III) complexes under anhydrous conditions, but no examples of its use were mentioned.

23.5.1.3 Reactions of Ligands already Coordinated to Cobalt(III)

The reactions of ligands coordinated to transition metals in general have been reviewed.[257] Many interesting reactions are not applicable to cobalt(III) complexes as even mildly reducing conditions cause reduction to cobalt(II). However, there are a number of convenient syntheses using oxidizing conditions.

---

244 *A. Werner, McCutcheon*, Ber. *45*, 3281 (1912); *A. Werner, McCutcheon*, J. Chem. Soc. Abstracts *104 i*, 19 (1913).
245 *E. Kyuno, L. J. Boucher, J. C. Bailar*, J. Amer. Chem. Soc. *87*, 4458 (1965).
246 *R. Maskill*, Ph.D. Thesis, University of Kent at Canterbury, 1970; *R. D. Gillard, R. Maskill*, J. Chem. Soc. *A*. 2813 (1971).
247 *J. C. Duff*, J. Chem. Soc. *119*, 385 (1921).
248 *R. D. Gillard, M. G. Price*, J. Chem. Soc. *A*, 1813 (1969).
249 *G. R. Brubaker, D. P. Schaefer*, Inorg. Chem. *10*, 968 (1971).
250 *K. Garbett, R. D. Gillard*, J. Chem. Soc. *A*, 979 (1968).
251 *G. T. Morgan, J. D. Main Smith*, J. Chem. Soc. *125*, 1996 (1924).
252 *F. Basolo, J. G. Bergmann, R. G. Pearson*, J. Phys. Chem. *56*, 22 (1952).
253 *L. M. Jackman, R. M. Scott, R. H. Portman*, Chem. Commun. 1338 (1968).
254 *M. D. Alexander, C. A. Spillert*, Inorg. Chem. *9*, 2344 (1970).
255 *R. B. Jordan, A. M. Sargeson, H. Taube*, Inorg. Chem. *5*, 1091 (1966).
256 *D. A. Buckingham, L. G. Marzilli, A. M. Sargeson*, J. Amer. Chem. Soc. *89*, 2772 (1967).
257 *J. P. Collman*, Reactions of Ligands Coordinated with Transition Metals in *R. L. Carlin*, Transition Metal Chemistry, p. 1, Vol. II, Edward Arnold & Co., London 1966.

$$[Co(dien)(NO_2)_3] \xrightarrow{HNO_3} [Co(dien)(ONO_2)_3]$$ (Ref. 237)

$$[Co(en)_2(NCS)_2]^{\oplus} \xrightarrow{Cl_2} [Co(en)_2(NH_3)_2]^{3\oplus}$$ (Ref. 258)

$$[Co(NH_3)_5OOC-C_6H_4-CHO]^{2\oplus} \xrightarrow{Cl_2\ /\ H_2O} [Co(NH_3)_5OOC-C_6H_4-COOH]^{2\oplus}$$ (Ref. 259)

In this last case attempts at direct synthesis, *e.g.* from $[Co(NH_3)_5(OH_2)]^{3\oplus}$ and sodium terephthalate, yield only the dimeric complex

$$[(H_3N)_5Co-OOC-C_6H_4-COO-Co(NH_3)_5]^{4\oplus}$$

Complexes of ligands which are not readily available in an uncomplexed state may also be made by this method:

$$[Co(O-C(CH_3)=CH-C(CH_3)=O)_3] \xrightarrow{Cu(NO_3)_2} [Co(O-C(CH_3)=C(NO_2)-C(CH_3)=O)_3]$$ (Ref. 260)

$$[Co(cysteine)_3]^{3\ominus} \xrightarrow{H_2O_2} [Co(cysteinesulfinate)_3]^{3\ominus}$$ (Ref. 261)

23.5.1.4 Separation and Purification of Cobalt(III) Complexes

Some of the methods outlined above, especially those starting from cobalt(II) or CoO(OH) tend to give mixtures of products. Although the early workers usually relied on crystallization to separate the required complex, chromatographic methods are proving increasingly valuable. Uncharged complexes can often be separated on conventional adsorbants such as alumina (for tris α-aminoacid complexes[262]), or florisil (for tris β-diketone complexes[260]). As metal complexes are often fairly strongly absorbed on ion exchange resins, especially if the ionic change is high, ion exchange chromatography has been of limited use in the synthesis of pure cobalt(III) complexes. Elution with acids or alkali often causes some decomposition of the complex, and isolation of the pure complex from a solution containing a large amount of inorganic salts may be difficult. However, with singly charged ions, if the eluting salt is one that can be extracted with an organic solvent, useful separations can be achieved.[263] Sephadex ion exchange gels have proved very useful in the separation of cationic and anionic cobalt(III) complexes.[222]

One of the best absorbents for separating[200,264] cobalt complexes from each other and from excess ligand, is the dextran gel Sephadex-® G-10: it has proved useful,[246] for the separation of a wide variety of metal complexes, including both cationic, neutral and anionic complexes containing ligands such as ammonia, cyanide, thiocyanate, substituted phenanthrolines and bipyridyls, amino acids and peptides. For example,[265] oxidation with hydrogen peroxide of a mixture of $CoCl_2$, diethylenetriamine, and a dipeptide such as L-alanylglycine gives $[Co(dien)_2]^{3\oplus}$ and $[Co(L\text{-}ala\text{-}gly)_2]^{\ominus}$ in addition to the main product, $[Co(dien)(L\text{-}ala\text{-}gly)]^{\oplus}$; these three complexes are very easily separated on Sephadex. Sephadex G-10 is also very convenient for desalting the eluant from the ion exchange chromatography of cobalt(III) complexes.[222]

Paper chromatography, using aqueous ethylamine as solvent, is a useful method for purity checks on cobalt(III) complexes: complexes of different charge are separated particularly easily.[266]

23.5.2 Cobalt(III) Halides and Halo Complexes

Anhydrous *cobalt(III) fluoride*, which is very valuable as a fluorinating agent for organic compounds[267] can be prepared as a paramagnetic ($\mu = 2{\cdot}46$ B.M.) buff colored powder by the action of fluorine[267,268] on anhydrous

[258] *A. Werner, H. Müller, R. Klien, F. Bräunlich*, Z. Anorg. Allgem. Chemie *22*, 91 (1899); *A. Werner, H. Müller, R. Klien, F. Bräunlich*, J. Chem. Soc. Abstracts *78 i*, 86 (1900).

[259] *R. T. M. Fraser, H. Taube*, J. Amer. Chem. Soc. *82*, 4152 (1960).

[260] *J. P. Collman, R. L. Marshall, W. L. Young, S. D. Goldby*, Inorg. Chem. *1*, 704 (1962).

[261] *M. P. Schubert*, J. Amer. Chem. Soc. *55*, 3336 (1933).

[262] *R. G. Denning, T. S. Piper*, Inorg. Chem. *5*, 1056 (1966).

[263] *W. A. Freeman, C. F. Liu*, Inorg. Chem. *7*, 764 (1968);

[264] *Y. Wormser*, Compt. Rend. C, *263*, 805 (1966).

[265] *I. G. Browning, R. D. Gillard, J. Lyons, P. R. Mitchell, D. A. Phipps*, J. Chem. Soc. Dalton, 1815 (1972).

[266] *H. Yoneda*, Bull. Chem. Soc. Japan *40*, 2442 (1967).

[267] *M. Stacey, J. C. Tatlow*, Advan. Fluorine Chem. *1*, 166 (1960).

[268] *H. F. Priest*, Inorg. Synth. *3*, 175 (1950).

cobalt(II) chloride or cobalt(II) fluoride at 150–200°. Regeneration of spent CoF_3 in a fluorination reactor (a mixture of CoF_2 and CoF_3) can be achieved either with elemental fluorine,[267] or with chlorine trifluoride[269] or bromine trifluoride.[267] Hydrated cobalt(III) fluoride, $CoF_3 \cdot 3\frac{1}{2}H_2O$, which like the anhydrous material is decomposed by water, can be obtained by electrolytic oxidation[270] of a suspension of cobalt(II) fluoride in hydrofluoric acid. Magnetic measurements[271] over a wide temperature range (μ=4·47 B.M.) indicate that this compound contains octahedrally coordinated cobalt(III) as $[CoF_3(H_2O)_3] \cdot \frac{1}{2}H_2O$ rather than $[Co(H_2O)_6]^{3\oplus}[CoF_6]^{3\ominus} \cdot H_2O$.

Evidence for other cobalt(III) halides is poor. *Cobalt(III) chloride* cannot be prepared by dissolving the hydrated sesquioxide in hydrochloric acid as chlorine is immediately formed. It is said[272] however to be formed as a green solid by treating CoO(OH) with hydrogen chloride gas in ether at −5° in the dark. That the solid contains a cobalt(III) chloride of some sort is indicated by its reaction with ammonia to give $Co(NH_3)_6Cl_3$, although its green color suggests that it would be better formulated as an etherate, e.g. $[Co(OEt_2)_6]Cl_3$. There is no evidence for the existence of any simple cobalt(III) bromides or iodides.

Potassium hexafluorocobaltate(III), K_3CoF_6, can be obtained as a paramagnetic (μ=4·26 B.M.) blue complex by the reaction[273] of hydrated cobalt(III) fluoride with potassium fluoride in hydrofluoric acid, or by the fluorination[274] of potassium cobaltocyanide. A blue complex initially reported[275] as K_3CoF_7 prepared by the fluorination at 400° of a mixture of potassium fluoride and cobalt(II) fluoride has also been shown[274] to be K_3CoF_6. The chemistry of the fluorides and fluorocomplexes of Co(II), Co(III) and Co(IV) has been summarized by Sharpe.[276]

In view of the instability, or the uncertainty of even the existence, of cobalt(III) halides other than CoF_3, it is not surprising that no other Co(III) complexes containing only halogen donors have been characterized, although unstable solutions said to contain $[CoCl_6]^{3\ominus}$ have been prepared[277] by the action of hydrogen chloride in acetic acid on cobalt(III) acetate, $Na_3[Co(CO_3)_3]$, or $K_3[Co(C_2O_4)_3]$.

23.5.3 Cobalt(III) Complexes with Group VI Donors Only

Solutions containing the hexaquocobalt(III) ion can only be produced from cobalt(II) solutions by electrolytic oxidation[278] or by the action of the most powerful oxidizing agents such as fluorine,[279] ozone[280] or sodium bismuthate(V).[281] Such cobalt(III) solutions can be reasonably stable[278] if they are strongly acidic and kept at 0°. It is probable that genuine hexaquocobalt(III) can only be prepared[278,282] in aqueous solutions as the perchlorate salt, by electrolytic oxidation of cobalt(II) perchlorate.

Even in such solutions dimeric or polymeric species may predominate if the preparation is not carried out under the optimum conditions of Co(II) concentration, acidity, temperature, and electrode current density. Although strong solutions (up to 0.6 M) may be prepared, it is impossible to isolate solid cobalt(III) perchlorate.

Electrolytic oxidation of $CoSO_4$ in 4 M sulfuric acid at 0° gives[7,215,283] blue-green crystals of $Co_2(SO_4) \cdot 18H_2O$ which have been variously stated to be stable for several months[284] and to decompose within a few hours,[283] and which decompose rapidly in ice cold water, but dissolve in cold dilute sulfuric acid to give a green solution which is stable for several days. It has been suggested[285] that sulfate solutions of cobalt(III) prepared electrolytically contain sulfato-complexes, but electrolysis of a solution of cesium sulfate and cobalt(II) sulfate gives[286]

---

269 *E. G. Rochow, I. Kukin*, J. Amer. Chem. Soc. *74*, 1615 (1952).
270 *G. A. Barbieri, F. Calzolari*, Z. Anorg. Allgem. Chemie *170*, 109 (1928); C.A. *22*, 2120 (1928).
271 *H. C. Clark, B. Cox, A. G. Sharpe*, J. Chem. Soc. 4132 (1957).
272 *D. Hibbert, C. Duval*, Compt. Rend. *204*, 780 (1937); C.A. *31*, 2953 (1937).
273 *J. T. Grey*, J. Amer. Chem. Soc. *68*, 605 (1946).
274 *R. Hoppe*, Rec. Trav. Chim. Pays-Bas *75*, 569 (1956).
275 *W. Klemm, E. Huss*, Z. Anorg. Allgem. Chemie *258*, 221 (1949).
276 *A. G. Sharpe*, Advan. Fluorine Chem. *1*, 29 (1960).
277 *A. W. Chester, I. E. A. Heiba, R. M. Dessau, W. J. Koehl*, Inorg. Nucl. Chem. Lett. *5*, 277 (1969).
278 *G. Davies, B. Warnqvist*, Coord. Chem. Rev. *5*, 349 (1970).
279 *F. Fichter, H. Wolfmann*, Helv. Chim. Acta *9*, 1093 (1927); C.A. *21*, 2853 (1927).
280 *E. Brunner*, Helv. Chim. Acta *12*, 208, (1929); C.A. *23*, 2386 (1929).
281 *S. Kitashima*, Bull. Inst. Phys. Chem. Res., Japan *7*, 1035 (1928); C.A. *23*, 1074 (1929).
282 *B. Sramková, J. Zyka, J. Dolezal*, J. Elektroanal. Chem. *30*, 169 (1971).
283 *S. Swann, T. S. Xanthakos, R. Strehlow*, Inorg. Syn. *5*, 181 (1957).
284 *H. Marshall*, J. Chem. Soc. *59*, 760 (1891).
285 *L. H. Sutcliffe, J. R. Weber*, Trans. Faraday Soc. *57*, 91 (1961).

the alum $CsCo(SO_4)_2 \cdot 12H_2O$ as diamagnetic blue crystals, which if carefully washed undergo little decomposition in three months: as the alum is isomorphous with other alums it presumably contains $[Co^{III}(OH_2)_6]^{3\oplus}$ ions. *Cobalt(III) sulfate* can also be prepared[279] by the action of gaseous fluorine on an ice cold solution of $CoSO_4$ in sulfuric acid. Cobalt(III) selenate, $Co_2(SeO_4)_3 \cdot 18H_2O$, has been prepared[287] by electrolysis of a solution of cobalt(II) selenate at $-10°$.

Cobalt(III) nitrate may be prepared[288, 289] by the reaction of dinitrogen pentoxide with anhydrous cobalt(III) fluoride at temperatures between -78 and $+25°$. It is a dark green, weakly paramagnetic ($\mu < 1$ B.M.), solid which is very volatile even at room temperature, and soluble in non-oxidizable organic solvents (*e.g.* CCl_4) to give stable green solutions. It is rapidly decomposed by water giving cobalt(II) nitrate and oxygen, and reacts with ligands[289] such as pyridine to give adducts, e.g. $Co(NO_3)_3py_2$, which are probably polymeric.

Oxidation electrolytically[290, 291] or with chlorine[290] of solutions of cobalt(II) acetate in glacial acetic acid gives *cobalt(III) acetate* as a green solid. The structure is still uncertain.[292]

There is no evidence for the existence of cobalt(III) oxide, Co_2O_3, although a hydrated form, $Co_2O_3 \cdot H_2O$ or $CoO(OH)$, is well known. It may be prepared by oxidizing cobalt(II) oxide or hydroxide with a wide variety of oxidizing agents, *e.g.* air, peroxides, halogens, $MnO_4^{\ominus}$. It is convenient to prepare the cobalt(II) hydroxide *in situ*, and thus simultaneous addition of sodium hydroxide and either bromine[7] or hydrogen peroxide[293] to an aqueous solution of $Co^{2\oplus}$ readily gives a black precipitate which after drying has the composition $CoO(OH)$. This is also formed when many cobalt(III) complexes, especially ammine complexes, are decomposed by base: thus treatment of aqueous $[Co(NH_3)_6]Cl_3$ with aqueous alkali gives[294] the same black precipitate. This compound is especially useful as an intermediate in the synthesis of many cobalt(III) complexes especially of acidic ligands, such as oxalate, amino acids, peptides, which readily react with it to give complexes such as *tris(aminoacidate)cobalt(III)*[232] and the *bis(dipeptidato)cobalt(III) anion*[222] (see also Section 23.5.2).

Although complexes of cobalt(III) surrounded only by oxygen donors are well known, there are relatively few of these in comparison to the very large number of complexes of the hexamine type, due to the much lower stability of the CoO_6 complexes. The most important preparative methods of some of these compounds are summarized in Table 1.

Solutions containing Co^{III} surrounded only by oxygen-donors are not difficult to prepare: typical of these is the dark green *Field-Durrant*

Table 1. Synthesis of Cobalt(III) Complexes with Group VI Donors only

Complex	Reactants	Conditions	Remarks	Ref.
$Co^{III}O_6$				
$[Co(OH_2)_6](ClO_4)_3$	$Co(ClO_4)_2$ in $HClO_4$	electrolytic oxidation	as solution only	278, 282
$(SO_4)_{1.5}$	$CoSO_4$ in 8M H_2SO_4	electrolytic oxidation		7, 215, 283
$(SeO_4^{2\ominus})_{1.5}$	$CoSeO_4$	electrolytic oxidation		287
$Na_5[Co(H_2IO_6)_2(OH)_2]$	$CoCl_2 + NaIO_4 + NaOH + NaOCl$	precipitated with EtOH	polymerizes readily	295
$Na_3[Co(SO_3)_3]$	$Na_3[Co(NO_2)_6] + Na_2SO_3$	100°		296

continued

[286] *D. A. Johnson, A. G. Sharpe*, Inorg. Synth. *10*, 61 (1967); *D. A. Johnson, A. G. Sharpe*, J. Chem. Soc. *A* 798 (1966).
[287] *H. Copaux*, Ann. Chim. Phys. *6*, 508 (1905); *H. Copaux*, J. Chem. Soc. Abstracts *90 ii*, 91 (1906).
[288] *R. J. Fereday, N. Logan, D. Sutton*, Chem. Commun. 271 (1968).
[289] *R. J. Freeday, N. Logan, D. Sutton*, J. Chem. Soc. *A*, 2699 (1969).
[290] *H. Copaux*, Compt. Rend. *136*, 373 (1903); *H. Copaux*, J. Chem. Soc. Abstracts *84 i*, 309 (1903).
[291] *C. Schall, C. Thieme-Wiedtmarckter*, Z. Elektrochem. *35*, 337 (1929); C.A. *23*, 4412 (1929).
[292] *C. Oldham*, Prog. Inorg. Chem. *10*, 244 (1968).
[293] *A. Carnot*, Compt. Rend. *108*, 610 (1889); *A. Carnot*, J. Chem. Soc. Abstracts *56*, 678 (1889).
[294] *G. F. Hüttig, R. Kassler*, Z. Anorg. Allgem. Chemie *184*, 279 (1929); C.A. *24*, 799 (1930).
[295] *C. J. Nyman, T. A. Plane*, J. Amer. Chem. Soc. *83*, 2617 (1961).
[296] *G. Jantsch, K. Abresch*, Z. Anorg. Allgem. Chemie *179*, 345 (1929); C.A. *23*, 2898 (1929).

Table 1 continued

Complex	Reactants	Conditions	Remarks	Ref.
$[Co(NO_3)_3]$	$CoF_3 + N_2O_5$	$-78°$ to $+25°$	volatile dk. green solid	288, 289
Co(pyridine-2 carboxylate *N*-oxide)$_3 \cdot H_2O$	Co(acetate)$_2$ in CH_3CO_2H + ligand	oxidized with Cl_2		297
$[Co\{OC(NH_2)_2\}_6]^{3\oplus}$-$(ClO_4^{\ominus})_3$	$Co(ClO_4)_2$ in $HClO_4$ + urea	electrolytic oxidation		235
$K_3[Co(CO_3)_3] \cdot 3H_2O$	$CoCl_2 + KHCO_3 + H_2O_2$	0°, precipitation with EtOH in the dark		241
$Na_3[(Co(CO_3)_3] \cdot 3H_2O$	$Co(NO_3)_2 + H_2O_2 + NaHCO_3$	0°		242, 298
$[Co(NH_3)_6][Co(CO_3)_3]$	$[Co(NH_3)_4CO_3]_2(SO_4)_3$ + $KHCO_3 + K_2S_2O_8$	precipitated with $[Co(NH_3)_6](NO_3)_3$	anhydrous	299
$K_3[Co(ox)_3] \cdot 3\frac{1}{2}H_2O$	$CoCO_3 + H_2C_2O_4 + K_2C_2O_4$ + $PbO_2 + CH_3CO_2H$	precipitated with EtOH	unstable to light and heat	215, 227
$K_3[Co(mal)_3]$	$CoCO_3$ + KHmalonate + PbO_2	precipitated with EtOH		300
	K_2malonate + malonic acid + $CoCO_3 + H_2O_2$	precipitated with EtOH		301
$[Co(en)_3][Co(sal)_3]$	$Na_3[Co(CO_3)_3]$ + salicyclic acid	100°; precipitated with $[Co(en)_3](NO_3)_3$	15% yield	242
$Na_3[Co\{(+)\text{-}H_2tart\}_3]$	Ba_2H_2tart + $CoCl_2 + H_2O_2$	charcoal catalyst		302
$[Co(acac)_3]$	$CoCO_3$ + pentane-2,4-dione + H_2O_2			225
	$Na_3[Co(CO_3)_3]$ + pentane-2,4-dione	in acetone/water/nitric acid	95% yield	242
[Co(benzoylacetonate)$_3$]	$Na_3[Co(CO_3)_3]$ + 1-phenyl-buten-1,3-dione	in acetone/water/nitric acid	81% yield	242
[Co(2-acetyl-5-methoxy-phenolate)$_3$]	$[Co(NH_3)_5Cl]Cl_2$ + 2-hydroxy-4-methoxy-acetophenone	100°		303
[Co(hmc)$_3$]	$CoCl_2$ + 3-hydroxymethylene-2-oxo-1,7,7-trimethyl-bicyclo[2.2.1]heptane, sodium salt	air oxidation		304
	$Na_3[Co(CO_3)_3]$ + 3-hydroxy-methylene-2-oxo-1,7,7-trimethyl-bicyclo[2.2.1]-heptane in C_6H_6/H_2O	shake overnight		305
Ba[Co(gluconate)]	$Co(NO_3)_2$ + *d*-gluconic acid + NaOH	air oxidation; ppt. with barium salts		306
$H_5(NH_4)_3[Co(OH)$-$(MoO_4)_6]$	$Co(OOCCH_3)_2$ + $(NH_4)_2Mo_2O_7$ + aq. CH_3CO_2H	oxidize with $(NH_4)_2S_2O_8$ or H_2O_2		307, 308

[297] *A. B. P. Lever, J. Lewis, R. S. Nyholm*, J. Chem. Soc. 5262 (1962).

[298] *H. F. Bauer, W. C. Drinkard*, Inorg. Synth. *8*, 202 (1966).

[299] *T. P. McCutcheon, W. J. Schuele*, J. Amer. Chem. Soc. *75*, 1845 (1953).

[300] *W. Thomas*, J. Chem. Soc. *119*, 1140 (1921).

[301] *G. Lohmiller, W. W. Wendlant*, J. Inorg. Nucl. Chem. *32*, 2430 (1970).

[302] *F. Woldbye*, Optical Rotatory Dispersion of Transition Metal Complexes, Tech. U.S. Army Report, Contract No. DA-91-508-EUC-246, European Research Office, U.S. Army, Frankfurt a.M. 1959.

[303] *P. Pfeiffer, S. Golther, O. Agern*, Ber. *60*, 305 (1927); C.A. *21*, 1416 (1927).

[304] *I. Lifshitz*, Rec. Trav. Chim. Pays-Bas *69*, 1495 (1950); C.A. *45*, 2752 (1951).

[305] *J. H. Dunlop, R. D. Gillard, R. Ugo*, J. Chem. Soc. *A*, 1540 (1966).

[306] *W. Traube, F. Kuhbier, W. Schröder*, Ber. *69*, 2655 (1936);
B. Das Sarma, J. C. Bailar, J. Amer. Chem. Soc. *78*, 895 (1956).

[307] *C. Friedheim, F. Keller*, Ber. *39*, 4301 (1906);
C. Friedheim, F. Keller, J. Chem. Soc. Abstracts *92 ii*, 96 (1907).

[308] *L. C. W. Baker, B. Loev, T. P. McCutcheon*, J. Amer. Chem. Soc. *72*, 2374 (1950).

Table 1 continued

Complex	Reactants	Conditions	Remarks	Ref.
$Co^{III}O_3S_3$				
$Na_3[Co(SCH_2CO_2)_3]$	$Na_3[Co(CO_3)_3] + HSCH_2CO_2H$	90° for 30 min	85% yield	242
H_3[Co(cysteinate)$_3$]	KOH + cysteine · HCl + $CoSO_4$	air oxidation	red isomer 90% yield	309, 310
$Co^{III}S_6$				
$[Co\{S_2P(OC_2H_5)_2\}_3]$	$Co(ClO_4)_2 + H_2O_2$ $NH_4[S_2P(OC_2H_5)_2]$	in 90% EtOH		311
$[Co(S_2PF_2)_3]$	$[Co^{II}(S_2PF_2)_2] + (S_2PF_2)_2$			157
$[Co(S_2COR)_3$	$[Co(NH_3)_6](S_2COR)_3$	$R = C_2H_5$, 60° $R = CH_3$, 23°		312
	$CoCl_2 + ROCS_2^{\ominus}$		$R = C_2H_5$ and $CH(CH_3)_2$	313
	$K_3[Co(CO_3)_3] + C_2H_5OCSK$			314
$[Co\{S_2CN(CH_3)_2\}_3]$	$Co(NO_3)_2 + Na[S_2CN(CH_3)_2]$	in acetone or water		153, 154
$Na_3[Co(S_2C_2H_4)_3]$	$Na_3[Co(CO_3)_3] + HSCH_2CH_2SH$		black; 95% yield	242
KCa[Co(dithiooxalate)$_3$]	$[Co(NH_3)_6]Cl_3 + K_2S_2C_2O_2$	80°, 15 min		315
[Co(SacSac)$_3$]	Co^{II}(SacSac)$_2$ + acac + H_2S	in ethanolic HCl air oxidation		316

solution,[317] prepared by oxidation of a solution containing $CoCl_2$ and $KHCO_3$ with hydrogen peroxide; the solution contains the $[Co(CO_3)_3]^{3\ominus}$ ion which may be isolated with difficulty as its $K^{\oplus}$ salt,[241] or easily as its very insoluble $Na^{\oplus}$ salt.[242] (The sodium salt must be well dried before storage, or it decomposes slowly to a black solid.[298]) Either salt, or the Field-Durrant solution itself, is very useful in the preparation of a host of cobalt(III) complexes[241, 243]: they are especially useful for the preparation of other carbonato complexes. A green solid, supposedly cobalt(III) tris(carbonato)cobaltate(III), can be precipitated[318] from the Field-Durrant solution by glycerine. Similar dark green solutions can be prepared by hydrogen peroxide oxidation of cobalt(II) solutions containing organic dicarboxylic acids, such as malonic,[300,301] tartaric,[302] and succinic,[300] but only in the first two cases could solids be isolated. Lead dioxide oxidation of cobalt(II) oxalate in a solution containing potassium oxalate gives a good yield[227, 319, 320] of $K_3[Co(ox)_3] \cdot 3\frac{1}{2}H_2O$ as dark green crystals which are unstable both to light and heat.[320] If hydrogen peroxide is used as an oxidizing agent "Durrants salt", $K_4[(ox)_2Co(OH)_2Co(ox)_2] \cdot 3H_2O$, is obtained[215, 321] instead: this reacts with acetylacetone to give $Na_2[Co(ox)_2(acac)]$ and $Na[Co(acac)_2(ox)]$.[322]
The sodium salt $Na_4[(ox)_2Co(OH)_2Co(ox)_2]$, which is much more soluble[321] than the potas-

[309] *M. P. Schubert*, J. Amer. Chem. Soc. *53*, 3851 (1931).
[310] *R. G. Neville, G. Gorin*, J. Amer. Chem. Soc. *78*, 4893 (1956);
G. Gorin, J. E. Spessard, G. A. Wessler, J. P. Oliver, J. Amer. Chem. Soc. *81*, 3193 (1959).
[311] *C. K. Jorgensen*, Acta Chem. Scand. *16*, 2017 (1962).
[312] *G. W. Watt, B. J. McCormick*, J. Inorg. Nucl. Chem. *27*, 898 (1965).
[313] *E. E. Zaev, S. V. Larionov, Y. N. Molin*, Proc. Acad. Sci. USSR. Chem. *168*, 480 (1966).
[314] *F. Galsbol, C. E. Schäffer*, Inorg. Synth. *10*, 42 (1967).
[315] *F. P. Dwyer, A. M. Sargeson*, J. Amer. Chem. Soc. *81*, 2335 (1959).
[316] *G. A. Heath, R. L. Martin*, Chem. Commun. 951 (1969).
[317] *F. Field*, J. Chem. Soc. *14*, 48 (1862);
R. J. Durrant, J. Chem. Soc. Proc. *12*, 96 (1896).
[318] *C. Duval*, Compt. Rend. *191*, 615 (1930); C.A. *25*, 470 (1931).
[319] *F. M. Jaeger, W. Thomas*, Koninkl. Ned. Akad. Wetenschap. Proc. *21*, 693 (1919);
F. M. Jaeger, W. Thomas, J. Chem. Soc. Abstracts *116 i*, 252 (1919).
[320] *H. Copaux*, Compt. Rend. *134*, 1214 (1902);
H. Copaux, J. Chem. Soc. Abstracts *82 i*, 586 (1902).
[321] *A. M. Sargeson, I. K. Reid*, Inorg. Synth. *8*, 204 (1966).
[322] *F. P. Dwyer, I. K. Reid, A. M. Sargeson*, Australian J. Chem. *18*, 1919 (1965).

sium salt, is a useful intermediate in the synthesis of other oxalato-cobalt(III) complexes. Hydrogen peroxide oxidation[225] of a suspension of $CoCO_3$ in aqueous acetylacetone gives [Co-(acac)$_3$]. Other cobalt(III) tris (β-diketonate) complexes can be prepared[242, 305] from $Na_3Co(CO_3)_3$ or by electrophilic substitution of [Co(acac)$_3$].[257, 260, 323]

The preparations of several $Co^{III}O_3S_3$ and $Co^{III}S_6$ complexes are also listed in Table 1. The reaction of cysteine hydrochloride with aqueous cobalt(II) chloride in the presence of potassium hydroxide gives either a red complex H_3[Co(cysteinate)$_3$] or a green complex K_3[Co(cysteinate)$_3$] depending on the amount of base present.[309, 310] On the basis of the electronic spectra, the red complex is said to be a $Co^{III}O_3S_3$ complex,[310] and the green isomer the $Co^{III}N_3S_3$ isomer.

$Co^{III}S_6$ complexes can be prepared by most of the methods used for $Co^{III}O_6$ complexes, although the reaction of the thio-ligand with $Na_3[Co(CO_3)_3]$ is especially useful as $Co^{III}S_6$ complexes are much more thermally stable than many of the $Co^{III}O_6$ complexes, which are often auto-reduced to cobalt(II) especially in sunlight. Conversely, oxidation of $Co^{II}S_n$ complexes to $Co^{III}S_6$ complexes readily takes place in air or even spontaneously.[149–150] For example, reaction of a cobalt(II) salt with a dithiocarbamate salt[153, 154] or a xanthate salt[152] in air, or with a bis(dialkyldithiocarbamoyl) disulfide,[324, 325] gives the $Co^{III}S_6$ complex. The methyl- or ethyl-xanthate salt of $[Co(NH_3)_6]^{3\oplus}$ is very unstable, losing ammonia at room temperature or on slight warming or evacuation, to give the tris-xanthate complex, $[Co(S_2COR)_3]$.[312]

A number of other $Co^{III}S_6$ complexes have been mentioned but without full preparative details.[314] Some spectroscopic properties of some other $Co^{III}S_6$ complexes and of a $Co^{III}Se_6$ complex have been measured,[326] but no information on the synthesis was published.

23.5.4 Cobalt(III) Complexes with Group V Donors Only

The hexammine cobalt(III) salts have been known[327] since 1799 when the adduct $CoCl_3 \cdot 6NH_3$ was described. Air oxidation[217] of a $Co^{2\oplus}$ solution containing ammonium salts and excess ammonia only gives $[Co(NH_3)_6]^{3\oplus}$ salts in good yield in the presence of a catalyst such as activated charcoal: otherwise the reaction tends to stop at $[Co(NH_3)_5Cl]^{2\oplus}$.

Some of the more useful syntheses of some compounds of the $Co^{III}N_6$ type are given in Table 2.

Table 2. Syntheses of Co^{III} Complexes with Nitrogen Donors only

Complex	Reactants	Conditions	Remarks	Ref.
$[Co(NH_3)_6]X_3$	$CoX_2 + NH_4X + NH_3 +$ charcoal	oxidation by air or H_2O_2	X = Cl, Br, NO_3	217
$[Co(NH_2OH)_6]Cl_3$	$[Co(en)_2(NH_2OH)Cl]Cl_2 + NH_2OH \cdot HCl$	50–60° in alkaline solution		328
$[Co(en)_3]Cl_3$	$CoCl_2 + en \cdot HCl$	air oxidation	89%	329
	$Na_3[Co(CO_3)_3] + en \cdot 2HCl$		91%	242
$(+)$-$[(Co(en)_3]^{3\oplus}$	Cobalt tartrate + HCl + en	air oxidation	80% of (+)-isomer	330
cis-$[Co(en)_2(NH_3)_2]^{3\oplus}$	*cis*-$[Co(en)_2Cl_2] + NH_3$	−77 to −33		331
$[Co(\pm pn)_3]Cl_3$	$CoCl_2 + (\pm)$-pn + HCl	air oxidation: charcoal catalyst	separation of isomers	332

[323] *J. P. Collman, W. L. Young*, Inorg. Synth. *7*, 205 (1963).
[324] *I. V. Khodzhaeva, Y. V. Kissin*, Zh. Fiz. Khim. *37*, 791 (1963); C.A. *59*, 6224 (1963).
[325] *K. Lesz, T. Lipiec*, Roczniki Chem. *40*, 1117 (1966); C.A. *66*, 61360 (1967).
[326] *R. L. Martin, A. H. White*, Nature *223*, 394 (1969).
[327] *B. M. Tassaert*, Ann. Chim. *28*, *95* and *106* (1799).
[328] *A. Werner, E. Berl*, Ber. *38*, 893 (1905); *A. Werner, E. Berl*, J. Chem. Soc. Abstracts *88 ii*, 323 (1905).
[329] *J. B. Work*, Inorg. Synth. *2*, 221 (1946).
[330] *J. A. Broomhead, F. P. Dwyer, J. W. Hogarth*, Inorg. Synth. *6*, 186 (1960).
[331] *J. C. Bailar, J. H. Haslam, E. M. Jones*, J. Amer. Chem. Soc. *58*, 2226 (1936).
[332] *F. P. Dwyer, A. M. Sargeson, L. B. James*, J. Amer. Chem. Soc. *86*, 590 (1964).

Table 2 continued

Complex	Reactants	Conditions	Remarks	Ref.
$[Co(-pn)_3]Cl$ (tartrate)	$(-)$-pn + H_2SO_4 + $CoSO_4$ + silver tartrate	air oxidation: charcoal catalyst	Separation of geometric isomers[334]	333
$[Co(tn)_3]Cl_3$	$CoCl_2$ + tn + tnHCl	air oxidation: charcoal catalyst		335
$[Co(bn)_3]Br_3$	$[Co(NH_3)_5Cl]Cl_2$ + bn	120–140°	precipitated with HBr	336
$[Co\{(-)\text{-bn}\}_3]$	$CoCl_2$ + (–)-bn + (–)-bn·2HCl	air oxidation		302
$[Co(cptn)_3]Cl_3$	$CoCl_2$ + cptn + H_2O_2		CO_2 must be excluded	337
$[Co(chxn)_3]Cl_3$	*trans*-$[Co(chxn)_2Cl_2]^{\oplus}$ + chxn	3 hr at reflux temperature		338
$[Co(bipy)_3](ClO_4)_3$	bipy + $CoCl_2$ + H_2O_2	precipitated by $HClO_4$		174
$[Co(phen)_3](ClO_4)_3$	phen + $CoCl_2$ + $NaClO_2$	in ethanol: ppt. with $HClO_4$		229
$[Co(phenylbiguanido)_3](OH)_3$	phenylbiguanidinium chloride + $CoCl_2$ + NaOH	air oxidation		339
$[Co(dmg)_3]\cdot 2\frac{1}{2}H_2O$	dmgH + KOH + $Co(NO_3)_2$ + acetic acid	air oxidation		340
$[Co(dien)_2]Cl_3$	$[Co(NH_3)_5Cl]Cl_2$ + dien	70°		341
$[Co(terpy)_2]Cl_3$	$CoCl_2$ + terpy	Cl_2 oxidation	Co^{II} complex formed first	167
$K[Co(NHCONHCONH)_2]$	$[Co(NHCONHCONH)_3]^{3\ominus}$	Heating in aqueous solution		342
$[Co(penten)]I_3$	$[Co(NH_3)_6]I_3$ + penten	charcoal catalyst		236
$[Co(sen)]^{3\oplus}$	$Na_3[Co(CO_3)_3]$ + sen in HCl			343
$[Co(NH_3)_3(NO_2)_3]$	($CoCO_3$ + CH_3CO_2H) + ($NaNO_2$ + NH_4OH) + H_2O_2	charcoal catalyst; boil	53% yield	344
	$Co^{2\oplus}$ + NH_3 + NH_4Cl + $NaNO_2$	air oxidation	20% yield;	345
	$(CH_3CO_2)_2Co$ + $NaNO_2$ + NH_4OH	H_2O_2 oxidation		215
$K[Co(NH_3)_2(NO_2)_4]$	$CoCl_2$ + $NaNO_2$ + NH_4OH	air oxidation: precipitation with KCl	~50% yield	346
$K_2[Co(NH_3)(NO_2)_5]$	$K_2[Co(CO_3)(NO_2)_3(NH_3)]$ + $2NaNO_2$	evaporate aq. solution		347

continued

[333] *F. P. Dwyer, F. L. Garvan, A. Shulman*, J. Amer. Chem. Soc. *81*, 290 (1959).
[334] *T. E. MacDermott*, Inorg. Chim. Acta *2*, 81 (1968).
[335] *J. C. Bailar, J. B. Work*, J. Amer. Chem. Soc. *68*, 232 (1946).
[336] *J. Frejka, L. Zahlova*, Collect. Czech. Chem. Commun. *2*, 639 (1930); C.A. *25*, 1177 (1931).
[337] *F. M. Jaeger, H. B. Blumendal*, Z. Anorg. Allgem. Chemie *175*, 161 (1928); C.A. *23*, 573 (1929).
[338] *F. M. Jaeger, L. Bijkerk*, Koninkl. Ned. Akad. Wetenschap. Proc. *40*, 246 (1937); C.A. *32*, 448 (1938).
[339] *P. Ray*, Inorg. Synth. *6*, 71 (1960).
[340] *A. Nakahara, R. Tsuchida*, J. Amer. Chem. Soc. *76*, 3103 (1954).
[341] *J. Brigando*, Bull. Soc. Chim. France 211 (1957).
[342] *J. J. Bour, J. J. Steggerda*, Chem. Commun. 85 (1967).
[343] *R. W. Green, K. W. Catchpole, A. T. Phillip, F. Lions*, Inorg. Chem. *2*, 597 (1963).
[344] *G. Schlessinger*, Inorg. Synth. *6*, 189 (1960).
[345] *S. M. Jörgensen*, Z. Anorg. Allgem. Chemie *17*, 455 (1898);
S. M. Jörgensen, J. Chem. Soc. Abstracts *74 ii*, 592 (1898).
[346] *G. G. Schlessinger*, Inorg. Syn. *9*, 170 (1967).
[347] *M. Shibata, M. Mori, E. Kyuno*, Inorg. Chem. *3*, 1573 (1964).

Table 2 continued

Complex	Reactants	Conditions	Remarks	Ref.
$Na_3[Co(NO_2)_6]$	$Co(NO_3)_2 + NaNO_2 + CH_3CO_2H$	air oxidation		7, 215, 348
trans-$[Co(en)_2(NO_2)_2]NO_3$	$Co(NO_3)_2 + en \cdot HNO_3 + NaNO_2$	air oxidation	84% yield	349
cis-$[Co(en)_2(NO_2)_2]NO_3$	$K_3[Co(NO_2)_6] + en$	70°	15% yield	349
NO_2	*trans*-$[Co(en)_2Cl_2]Cl + NaNO_2$	−3°	60% yield: better than preceding method	350
cis-$[Co(phen)_2(NO_2)_2]NO_3$	$Co(NO_3)_2 + NaNO_2 + phen$			351
$[Co(dien)(NO_2)_3]$	$Co(NO_3)_2 + NaNO_2 + CH_3CO_2Na + dien$	air oxidation	78% yield	237
	$[Co(NH_3)_3(NO_2)_3] + dien$		80% yield	237
	$Na_3[Co(NO_2)_6] + dien$		59% yield	237
$[Co(trien)(NO_2)_2]Cl$	$CoCl_2 + NaNO_2 + trien \cdot HCl$	0°, vigorous aeration	70% yield	352

In general, oxidation with air or other oxidizing agents of a $Co^{2\oplus}$ solution containing the nitrogenous ligand gives the $Co^{III}N_6$ complex: a large excess of the ligand is often necessary or the *trans*-dichloro-N_4-cobalt(III) complex may be formed instead. With some ligands direct reaction does not proceed to completion and the $Co^{III}N_6$ complex has to be prepared by the action of ligand on the *trans*-dichloro-N_4-cobalt(III) complex, e.g. $[Co\{(-)\text{-chxn}\}_3]^{3\oplus}$.[338] The displacement of ammonia[236] from $[Co(NH_3)_6]^{3\oplus}$, and of carbon dioxide[242,343] from $[Co(CO_3)_3]^{3\ominus}$ in acidic solutions, provide other versatile methods.

Hexammine cobalt(III) and related complexes are of relatively little use as synthetic intermediates despite their easy synthesis as they are usually rather stable. However, several of these complexes have been of some importance in investigations[353,354] of the origins of the Cotton effect in octahedral cobalt(III) complexes.

The mixed $Co^{III}N_6$ complexes with both amine and nitro groups as ligands and also $Na_3[Co(NO_2)_6]$ are useful as synthetic intermediates. The systematic synthesis of all $[Co(NH_3)_n(NO_2)_{6-n}]^{3-n}$ $1<n<5$, has been described.[347] The very insoluble $[Co(NH_3)_3(NO_2)_3]$ can be prepared by at least five methods[355,356] all of which give a mixture[355,357] of 1,2,3- and 1,2,4-isomers together with ionic "isomers" such as $[Co(NH_3)_4(NO_2)_2]^{\oplus}[Co(NH_3)_2(NO_2)_4]^{\ominus}$. Jörgensen's method[345] gives a low yield (20%) of the purest (80%) material, while a more convenient method[344] gives a product which is stated[355] to be only 28% pure after recrystallization, but which an X-ray structural analysis[358] shows is the 1,2,4-isomer: Palmer's method[215] gives a very high yield (85%) of moderately pure (57%) material. *Cis*-$[Co(en)_2(NO_2)_2]^{\oplus}$ is best prepared[350] by the action of sodium nitrite on *trans*-$[Co(en)_2Cl_2]Cl$ at −3°: the earlier method,[349] the action of ethylenediamine on $K_3[Co(NO_2)_6]$, does not give consistent results. It is a useful synthetic intermediate, and when resolved, it is itself a useful resolving agent (see Section 23.5.7.3).

If nitric oxide is passed through an ammoniacal solution of cobalt(II) chloride, $[Co(NH_3)_5$-

[348] *M. Cunningham, F. M. Perkin*, J. Chem. Soc. *95*, 1562 (1909).

[349] *H. F. Holtzclaw, D. P. Sheetz, B. D. McCarthy*, Inorg. Syn. *4*, 176 (1953)

[350] *E. P. Harbulak, M. J. Albinak*, Inorg. Syn. *8*, 196 (1966).

[351] *A. V. Ablov*, Russian J. Inorg. Chem. *6*, 157 (1961); C.A. *56*, 3103 (1962).

[352] *A. M. Sargeson, G. H. Searle*, Inorg. Chem. *6*, 787 (1967).

[353] *S. F. Mason*, Pure Applied Chem. *24*, 335 (1970).

[354] *J. E. Sarneski, F. L. Urbach*, J. Amer. Chem. Soc. *93*, 884 (1971).

[355] *A. G. Maddock, A. B. J. B. Todesco*, J. Inorg. Nucl. Chem. *26*, 1535 (1964).

[356] *R. Duval*, Compt. Rend. *206*, 1652 (1938); C.A. *32*, 5720 (1938).

[357] *R. B. Hagel, L. F. Druding*, Inorg. Chem. *9*, 1496 (1970).

[358] *M. Laing, S. Baines, P. Sommerville*, Inorg. Chem. *10*, 1057 (1971).

NO]Cl$_2$ is formed as black crystals[359,360]: under different conditions of temperature and concentration a dimeric red complex of the same empirical formula is obtained.

Air oxidation of an aqueous solution of cobalt(II) bromide and *o*-aminobenzaldehyde causes self condensation of the aldehyde, around the cobalt as a template, to give two geometric isomers of a complex[361] of the type [Co(NNN)$_2$]$^{3\oplus}$ in addition to a *trans*-[Co(NNNN)Cl$_2$]$^{\oplus}$ complex,[362] where NNN and NNNN are ter- and quadri-dentate macrocyclic Schiff base ligands respectively.

Two Co^{III}N$_6$ complexes are known which contain the N-cyano ligand (*iso*-cyano): the reaction of sodium cyanide with [Co(CHCl$_2$CO$_2$)$_2$(trien)]ClO$_4$ or *trans*-[Co(CH$_3$CO$_2$)(trien)]ClO$_4$-HClO$_4$ gives *cis*-α- and *cis*-β-[Co(NC)$_2$(trien)$_2$]ClO$_4$ respectively[363]; other starting materials, e.g. [Co(CO$_3$)(trien)]ClO$_4$, gives the dicyano complex.

Attempts at preparation of Co^{III}P$_6$ complexes with monodentate phosphines lead to oxidation of the phosphine to a phosphine oxide.[364] However, the reaction[74,75] of phosphite ligands with cobalt(II) halides causes disproportionation, giving [Co^{III}\{(RO)$_3$P\}$_6$](ClO$_4$)$_3$ in addition to the cobalt(I) complex mentioned in 23.3. Co^{III}As$_6$ complexes with bi- and tri-dentate arsines are readily prepared by air oxidation of the cobalt(II) complexes.[174] The reaction[22] of KNH$_2$ in liquid ammonia with [Co(NH$_3$)$_6$](NO$_3$)$_3$ gives a polymeric cobalt(III) amide, Co(NH$_2$)$_3$. An analogous reaction[365] with KPH$_2$ gives Co(PH$_2$)$_3$, which reacts with an excess of KPH$_2$ to give K[Co$_2$(PH$_2$)$_7$].

23.5.5 Complexes with Combinations of N, O, S, and Halogen Donors

23.5.5.1 Co^{III}NO$_5$ Complexes

Relatively few Co^{III}NO$_5$ complexes are known: their stability is lower than that of cobalt(III) complexes containing more *N*-donors; with chelating ligands isolation is, however, possible. The reaction[240] of amino acids with a solution of [Co(mal)$_3$]$^{3\ominus}$ gives [Co(mal)$_2$(α)]$^{\ominus}$ complexes, and with a solution of Durrant's salt Na$_4$[(ox)$_2$Co(OH)$_2$Co(ox)$_2$], or a solution containing [Co(ox)$_2$(OH$_2$)$_2$]$^{\ominus}$, prepared *in situ*, deep blue [Co(ox)$_2$(α)]$^{\ominus}$ complexes[240] can be prepared: Ba[Co(ox)$_2$(gly)]$_2$ can also be prepared[323] by oxidation with hydrogen peroxide of a mixture of K$_2$C$_2$O$_4$, H$_2$C$_2$O$_4$, CoCO$_3$ and glycine, in aqueous solution. The displacement of nitro-groups in Na[Co(acac)$_2$(NO$_2$)$_2$] by α-amino acids provides a route[366] to the uncharged complexes [Co(acac)$_2$(α)] (Hα = glycine, alanine, phenylalanine etc.): in some cases small amounts of [Co(acac)(α)$_2$] were also formed.[366]

23.5.5.2 Co^{III}N$_2$O$_4$ Complexes

Complexes with two bidentate oxygen donors with either one bidentate or two monodentate nitrogen donors can readily be prepared. Oxidation[226] with lead dioxide of a cobalt(II) solution containing the appropriate amounts of the ligands yields Ca[Co(ox)$_2$(en)]$_2$ and K[Co(mal)$_2$(en)]. Addition[241] of ammonium carbonate to a solution containing K$_3$[Co(CO$_3$)$_3$] gives K[Co(NH$_3$)$_2$(CO$_3$)$_2$] as blue crystals claimed to be *cis*; heating the preparative mixture gives another isomer of K[Co(NH$_3$)$_2$(CO$_3$)$_2$] as violet crystals, claimed to be *trans*.

It is possible that a similar reaction on a solution of [Co(mal)$_3$]$^{3\ominus}$ might give [Co(mal)$_2$(L)$_2$]$^{\ominus}$ complexes, although the action of ethylenediamine on [Co(ox)$_3$]$^{3\ominus}$ gives only [Co(en)$_3$][Co(ox)$_3$].[226] Aqueous solutions of sodium acetylacetonate and sodium cobaltinitrite slowly deposit Na[Co(acac)$_2$NO$_2$)$_2$]·5H$_2$O on standing.[367,368] Addition of a monodentate N-donor ligand (N) to a solution of Na[Co(acac)$_2$(NO$_2$)$_2$] gives[368] an immediate precipitate of [Co(acac)$_2$(NO$_2$)(N)].

The reaction of glycine, L-alanine, or L-valine with a solution containing K$_3$[Co(CO$_3$)$_3$] gives[369] complexes of the type K[Co(α)$_2$(CO$_3$)].

23.5.5.3 Co^{III}O$_3$N$_3$, Co^{III}S$_3$N$_3$, and Co^{III}X$_3$N$_3$ Complexes

Tris α-amino acid complexes of cobalt can be pre-

[359] *O. Bostrup*, Inorg. Synth. *8*, 191 (1966).
[360] *J. L. Milward, W. Wardlaw, W. J. R. Way*, J. Chem. Soc. 233 (1938).
[361] *S. C. Cummings, D. H. Busch*, J. Amer. Chem. Soc. *92*, 1924 (1970).
[362] *S. C. Cummings, D. H. Busch*, Inorg. Chem. *10*, 1220 (1971).
[363] *K. Kuroda, P. S. Gentile*, Inorg. Nucl. Chem. Lett. *3*, 151 (1967).
[364] *G. Booth*, Advan. Inorg. Chem. Radiochem. *6*, 1 (1964).
[365] *O. Schmitz-Dumont, F. Nagel, W. Schaal*, Angew. Chem. *70*, 105 (1958).
[366] *S. H. Laurie*, Australian J. Chem. *21*, 679 (1968).
[367] *A. Rosenheim, A. Garfunkel*, Ber. *44*, 1865 (1911); *A. Rosenheim, A. Garfunkel*, J. Chem. Soc. Abstracts *100 i*, 619 (1911).
[368] *L. J. Boucher, J. C. Bailar*, J. Inorg. Nucl. Chem. *27*, 1093 (1965).
[369] *M. Shibata, H. Nishikawa, Y. Nishida*, Inorg. Chem. *7*, 9 (1968).

pared by at least five methods, but some of the methods give predominantly the α-isomer (also known as the *trans*, *mer* or 1,2,4-isomer) whereas others give the β-isomer (the *cis*, *fac* or 1,2,3-isomer), or a mixture of isomers. Both isomers exist as D- and L- form: if an optically active amino acid is used unequal amounts of the D–α and L–α or D–β and L–β isomers will be formed,[370] and the found individual isomers may be separated[371] in a favourable case, e.g., $[Co(ala)_3]$. If a racemic amino acid is used the possibilities for isomerism become very considerable[369, 371] and the complex mixtures obtained have never been studied in detail.

The main synthetic methods available are:

(a) The reaction of CoO(OH) with an amino acid[232] gives a mixture[371] of the α- and β-isomers in which the former predominates,[372] especially with optically active amino acids,[373] and from which the α-isomer can be readily separated.[372]

(b) The reaction of $[Co(NH_3)_6]Cl_6$ with an amino acid[372, 373] gives mainly the β-isomer; β-$[Co(gly)_3]$ and D-β-$[Co(L\text{-}ala)_3]$, and D-β-$[Co(L\text{-}leu)_3]$ are very insoluble in water and dilute acids, and can thus be readily separated.

(c) The reaction of $K_3Co(CO_3)_3$ with an excess of an amino acid[372–374] gives both isomers in relative amounts that depend on the amino acid. The isomer can be separated by chromatography on alumina.[374] If insufficient amino acid is used then the anionic bis(aminoacidato)(carbonato) complex can be isolated.[369]

(d) Oxidation of a solution of cobalt(II) aminoacidate with air[373] or with hydrogen peroxide.[224,370] With leucine this method gives[373] mainly the L-α and D-α isomers, whereas with valine all four isomers are formed.[373, 375]

(e) The reaction of $K^{\oplus}[Co(CO_3)(aminoacidate)_2]^{\ominus}$ with amino acids can be used to prepare complexes containing two different amino acids such as $[Co(gly)_2(L\text{-}val)]$, and $[Co(L\text{-}ala)_2(D\text{-}ala)]$.[369]

The evaporation of α-$[Co(gly)_3]$ with hydrochloric acid has been claimed to lead[376] to blue cationic complexes of protonated glycine, $[Co(Hgly)_n(gly)_{3-n}]Cl_n$, $n=1$, 2 or 3. However, recent work does not confirm this.[449]

Related to these tris(amino-acid) complexes is the tris(amino-alcohol) complex prepared[377] from $[Co(NH_3)_5(OH_2)](ClO_4)_3$ and L-(+)-alaninol, though this crystallizes as a partially protonated complex. A tris(*o*-amino-phenol) complex can be prepared[242] in 88% yield from $Na_3[Co(CO_3)_3]$ and the phenol heated under reflux in ethanol for an hour, and some other similar complexes have been prepared by the oxidation by hydrogen peroxide of an aqueous solution of cobalt(II) acetate and the ligand.[378, 379]

Relatively few other $Co^{III}O_3N_3$ and related complexes have been reported.[2, 3] Useful intermediates in the synthesis of triammine complexes are: $[Co(NH_3)_3(NO_2)_3]$ (see 23.5.4); $[Co(NH_3)_3(OH_2)_3](NO_3)_3$ which can be prepared by hydrolysis[380] of $[Co(NH_3)_3(NO_3)_3]$ in very dilute acetic acid, or by decomposition[344] of $[Co(NH_3)_3(NO_2)_3]$ with concentrated nitric acid; $[Co(NH_3)_3Cl_3]$ which can be prepared by heating[381] $[Co(NH_3)_3(OH_2)Cl_2]Cl$ at 115° for 40 hr or by the reaction[382] of $[(NH_3)_3Co(OH)_3Co(NH_3)_3]Cl_3$ with concentrated hydrochloric acid; and $[Co(NH_3)_3(OH_2)Cl_2]Cl$ which can be easily prepared[383] from the readily available[344] $[Co(NH_3)_3(NO_2)_3]$ by reaction with a mixture of H_2SO_4 and HCl in the presence of urea. Diethylenetriamine complexes can be prepared from $[Co(dien)(NO_2)_3]$ (see 23.5.4); $[Co(dien)(SCN)_2OH]$, which can be prepared by air

[370] *R. D. Gillard, N. C. Payne*, J. Chem. Soc. A, 1197 (1969).

[371] *J. H. Dunlop, R. D. Gillard*, J. Chem. Soc. 6531 (1965).

[372] *M. Mori, M. Shibata, E. Kyuno, M. Kanaya*, Bull. Chem. Soc., Japan *34*, 1837 (1961).

[373] *N. C. Payne*, Ph.D. Thesis, Sheffield 1967.

[374] *R. G. Denning, T. S. Piper*, Inorg. Chem. *5*, 1056 (1966).

[375] *R. D. Gillard, N. C. Payne, D. C. Phillips*, J. Chem. Soc. A, 973 (1968).

[376] *V. P. Ogoleva*, Trudy Dagestan. Sel'skokhoz Inst. *7*, 130 (1955); C.A. *52*, 6046 (1958); English translation, National Lending Library for Science and Technology, Great Britain, Translation No. RTS 6777, 1971.

[377] *J. Fujita, K. Ohashi, K. Saito*, Bull. Chem. Soc., Japan *40*, 2986 (1967).

[378] *A. Chakravorty, B. Behera, P. S. Zacharias*, Inorg. Chim. Acta *2*, 85 (1968).

[379] *K. C. Kalia, A. Chakravorty*, Inorg. Chim. Acta *2*, 154 (1968).

[380] *A. Werner, E. Bindschedler*, Ber. *39*, 2673 (1906); *A. Werner, E. Bindschedler*, J. Chem. Soc. Abstracts *90 ii*, 760 (1906).

[381] *M. Mori, M. Shibata, K. Hirota, K. Masuno, Y. Suzuki*, J. Chem. Soc., Japan, *79*, 1251 (1958); C.A. *53*, 12086 (1959).

[382] *E. Birk*, Z. Anorg. Allgem. Chemie *175*, 405 (1928); C.A. *23*, 1076 (1929).

[383] *G. Schlessinger*, Inorg. Synth. *6*, 180 (1960).

oxidation[237] of a neutral solution containing $Co(NO_3)_2$, KSCN and diethylenetriamine—if ethanol is used as a solvent [Co(dien)(SCN)$_3$] is formed instead[237]; and [Co(dien)Cl_3] or [Co(dien)(ONO_2)$_3$], which can be prepared[237] by the action of concentrated hydrochloric or nitric acid respectively on [Co(dien)(NO_2)$_3$]. If basic conditions are used for the preparation of other complexes from [Co(dien)Cl_3] care must be taken or disproportionation to $[Co(dien)_2]^{3\oplus}$ may occur.

A dark green tris(cysteinato)cobalt(III) complex $K_3[Co^{III}\{NH_2CH(CO_2)CH_2S\}_3]$ prepared[261] from cysteine hydrochloride, cobalt(II) chloride and potassium hydroxide, or from cysteine[310] and $[Co(NH_3)_6]Cl_3$ has been assigned[310] as the $Co^{III}S_3N_3$ isomer. It can be oxidized with hydrogen peroxide[261] to give a yellow tris(cysteinesulfinato) cobalt(III) complex. β-Mercaptoethylamine will displace edta from K[Co(edta)] to give a high yield of $[Co(SCH_2CH_2NH_2)_3]$ as blue crystals[384]; direct oxidation of cobalt(II) solutions containing β-mercaptoethylamine gives an intractable gelatinous solid.[384]

23.5.5.4 $Co^{III}N_4O_2$ and $Co^{III}N_4X_2$ and Related Complexes

Cis and *trans* $Co^{III}N_4O_2$ and $Co^{III}N_4Cl_2$ complexes are among the most useful intermediates in the synthesis of cobalt(III) complexes. The syntheses of some of the more useful of them are listed in Table 3. In general, the complexes

Table 3. Synthesis of Some Important $Co^{III}N_4O_2$ and $Co^{III}N_4X_2$ Intermediates

Complex	Synthesis	Conditions	Remarks	Ref.
cis-$[Co(NH_3)_4Cl_2]\frac{1}{2}S_2O_6$	$[Co(NH_3)_4(CO_3)]^{\oplus}$ + HCl in EtOH	0°		385
trans-$[Co(NH_3)_4Cl_2]HSO_4$	$[Co(NH_3)_4(CO_3)]^{\oplus}$ + HCl + conc. H_2SO_4			386
$[Co(NH_3)_4(CO_3)]NO_3$	$Co(NO_3)_3 + (NH_4)_2CO_3$ + conc. NH_4OH	oxidation with air or H_2O_2		223
trans-$[Co(py)_4Cl_2]Cl$	$CoCl_2$ + py/H_2O	Cl_2 oxidation		228, 387
trans-$[Co(py)_4Br_2]Br_3$	$[Co(py)_4Br_2] + Br_2$			228
trans-$[Co(en)_2Cl_2]Cl$	$CoCl_2$ + en	10 hours air oxidation (H_2O_2 oxidation is quicker and easier)		388 222
	$[Co(en)_2CO_3]Cl$ + HCl(excess)			389
cis-$[Co(en)_2Cl_2]Cl$	Evaporation of a neutral aqueous solution of the *trans*-isomer			388
	$[Co(en)_2CO_3]Cl$ + HCl	stoichiometric quantity of HCl		389
trans-$[Co(en)_2(OH_2)(OH)](Br)_2$	*cis*-$[Co(en)_2Cl_2]Cl$ + dilute ammonia soln.	precipitation with KBr	see also Ref. 390	385
$[Co(en)_2(CO_3)]Cl$	*cis*-$[Co(en)_2Cl_2]Cl$ + Na_2CO_3	heat in water bath		244
	$CoCl_2$ + HCl + en + Li_2CO_3	PbO_2 oxidation; CO_2 atmosphere		389
$[Co(en)_2(ox)]Cl$	$CoCl_2 + H_2C_2O_4$ + en	PbO_2 oxidation		226
	$[Co(en)_2Cl_2]Cl + K_2C_2O_4$			244
$[Co(tn)_2(ox)]Cl$	$(CH_3CO_2)_2Co$ + tn + $H_2C_2O_4$	PbO_2 oxidation		249

continued

[384] *D. H. Busch, D. C. Jicha*, Inorg. Chem. *1*, 884 (1962).

[385] *A. Werner*, Justus Liebigs Ann. Chem. *386*, 1–272 (1911);
A. Werner, J. Chem. Soc. Abstracts *102 i*, 74 (1912).

[386] *A. Werner, A. Klein*, Z. Anorg. Allgem. Chemie *14*, 29 (1897);
A. Werner, A. Klein, J. Chem. Soc. Abstracts *72 ii*, 264 (1897).

[387] *A. Werner, R. Feenstra*, Ber. *39*, 1538 (1906);
A. Werner, R. Feenstra, J. Chem. Soc. Abstracts *90 i*, 450 (1906).

[388] *J. C. Bailar*, Inorg. Synth. *2*, 222 (1946).

[389] *F. P. Dwyer, A. M. Sargeson, I. K. Reid*, J. Amer. Chem. Soc. *85*, 1215 (1963).

[390] *S. C. Chan, G. M. Harris*, Inorg. Chem. *10*, 1317 (1971).

Table 3 continued

Complex	Synthesis	Conditions	Remarks	Ref.
trans-$[Co(chxn)_2Cl_2]Cl$	$[Co(NH_3)_4(CO_3)]^{\oplus}$ + chxn·2HCl		*cis*-isomer also prepared	391
$[Co(bipy)_2Cl_2]Cl$	$CoCl_2$ + bipy	oxidation with H_2O_2[224,392] or Cl_2[393]		224, 392, 393
$[Co(bipy)_2(CO_3)]Cl$	$Na_2CO_3 + [Co(bipy)_2Cl_2]^{\oplus}$			392
$[Co(phen)_2(Cl_2)]Cl$	$CoCl_2$ + phen	oxidation with H_2O_2[224] or Cl_2[351]		224, 351
$[Co(phen)_2(CO_3)]Cl$	$[Co(phen)_2Cl_2]Cl + Na_2CO_3$			394

are synthesized by oxidation of a solution containing $Co^{2\oplus}$ and the ligands in the correct proportion, *e.g.* $[Co(en)_2(ox)]Cl$,[226] or by substitution of a bidentate O-donor ligand into another $Co^{III}N_4O_2$ or $Co^{III}N_4Cl_2$ complex. The reaction of $[Co(en)_2(CO_3)]^{\oplus}$ or $[Co(en)_2Cl_2]^{\oplus}$ with oxalic acid,[395] malonic acid[396,385] acetylacetone,[397] salicyclic acid,[398] and succinic acid and its derivatives[247] gives tris-chelated Co^{III}-N_4O_2 complexes. Reaction of a monocarboxylic acid with $[Co(NH_3)_4(CO_3)]^{\oplus}$ or $[Co(en)_2(CO_3)]^{\oplus}$ gives a *cis*-biscarboxylato complex initially: on heating the *trans*-isomer can also be isolated.[399] Hydrolysis[400] of *trans*-$[Co(en)_2Cl_2]Cl$ to *cis*-$[Co(en)_2Cl(OH_2)]SO_4$ followed by reaction of the *cis* salt with another ligand[400] can be used to prepare mixed complexes like *cis*-$[Co(en)_2BrCl]Br$.

In recent years the reaction of α-amino acids with $K_3[Co(NO_2)_6]$ has been used to prepare[238] several complexes of the type $K[Co(NO_2)_2(\alpha)_2]$ the isomerism of which has been studied in detail.[238] A simpler alternative route to the same complexes is by air oxidation of a solution containing potassium glycinate, potassium nitrite, and cobalt(II) acetate.[401] The reaction of amino acids with a solution of $[Co(en)_2(OH_2)_2](ClO_4)_3$ —itself prepared[402] from $[Co(en)_2CO_3](ClO_4)$ and perchloric acid—gives[403] *trans*-bis(ethylenediamine)bis(O-aminoacid)cobalt(III) complexes, whereas under different conditions bis(ethylenediamine)-(aminoacidato)cobalt(III) complexes are obtained (see 23.5.5.5).

23.5.5.5 $Co^{III}N_5O$ and $Co^{III}N_5X$ Complexes

A large number of pentammine cobalt(III) derivatives have been prepared for mechanistic studies, and although some of them, such as $[Co(NH_3)_5Br]Br_2$ and $[Co(NH_3)_5(OH_2)]Br_3$[404] may be prepared by direct oxidation of suitable cobalt(II) solutions the best method, applicable to almost all complexes of this type, is from $[Co(NH_3)_5(CO_3)]NO_3$. This is easily prepared[405] in large quantities by aerial oxidation of an ammoniacal solution containing $Co(NO_3)_2$ and $(NH_4)_2CO_3$, and it is readily converted[405] in acidic solution into a wide range of derivatives $[Co(NH_3)_5X]^{2\oplus}$, where X=F, I, $-NO_2$, $-ONO_2$ and $-OCOCH_3$. Many other carboxylate complexes can be prepared by a similar method.[252]

[391] *R. G. Asperger, C. F. Liu*, Inorg. Chem. *4*, 1492 (1965).

[392] *F. M. Jaeger, J. A. van Dijk*, Koninkl. Ned. Akad. Wetenschap. Proc. *39*, 164 (1936); C.A. *30*, 3350 (1936);
F. M. Jaeger, J. A. van Dijk, Z. Anorg. Allgem. Chemie *227*, 273 (1936); C.A. *30*, 7478 (1936).

[393] *A. A. Vlcek*, Inorg. Chem. *6*, 1425 (1967).

[394] *A. V. Ablov, D. M. Palade*, Russ. J. Inorg. Chem., *6*, 306 (1961); C.A. *57*, 5575 (1962).

[395] *A. G. Beaumont, R. D. Gillard*, J. Chem. Soc. *A* 1757 (1970).

[396] *H. Yoneda, Y. Morimoto*, Inorg. Chim. Acta *1*, 413 (1967);

[397] *J. K. Reid, A. M. Sargeson*, Inorg. Synth. *9*, 167 (1967).

[398] *K. Garbett, R. D. Gillard*, J. Chem. Soc. *A*, 979 (1968).

[399] *K. Kuroda, P. S. Gentile*, Bull. Chem. Soc., Japan *38*, 1362, 1368 and 2159 (1965).

[400] *J. K. Vaughn, R. D. Lindholm*, Inorg. Synth. *9*, 163 (1967).

[401] *M. B. Celap, T. J. Janjic, D. J. Radanovic*, Inorg. Syn. *9*, 173 (1967).

[402] *M. Linhard, G. Stirn*, Z. Anorg. Allgem. Chemie *268*, 105 (1952); C.A. *47*, 6344 (1953).

[403] *T. Yasui, J. Hidaka, Y. Shimura*, Bull. Chem. Soc., Japan *39*, 2417 (1966).

[404] *H. Diehl, H. Clark, H. H. Willard*, Inorg. Synth. *1*, 186 (1939).

[405] *F. Basolo, R. K. Murmann*, Inorg. Synth. *4*, 171 (1953).

Another useful intermediate is $[Co(NH_3)_5Cl]Cl_2$, which can be prepared[406] by direct oxidation of an ammoniacal solution of $CoCl_2$ with hydrogen peroxide, and can be converted into many derivatives including $[Co(NH_3)_5ONO]^{\oplus}$,[407] which readily isomerizes,[407] even in the solid state, to $[Co(NH_3)_5NO_2]^{2\oplus}$.

The reaction[408, 409] of $[Co(NH_3)_5OH_2](ClO_4)_3$ —which can be prepared by the action of $HClO_4$ on $[Co(NH_3)_5CO_3]^{\oplus}$—with amino acids leads to pentammine O-amino-acid complexes $[Co(NH_3)_5OCOCHRNH_3](ClO_4)_3$. If the preparative mixture is heated to 80° with charcoal, complexes of bidentate amino acids, $[Co(NH_3)_4(\alpha)]^{2\oplus}$, are obtained.[403] The reaction of *trans*-$[Co(en)_2Cl_2]Cl$ with the sodium salt of an amino acid[410–412] gives $[Co(en)_2(\alpha)]^{\oplus}$. A similar reaction[391] occurs with *cis*-[Co{(−)-chxn}$_2$Cl$_2$]Cl.

The preparation of *cis*-$[Co(en)_2(NH_3)Br]Br_2$ and *cis*- and *trans*-$[Co(en)_2(NH_3)(OH_2)]^{3\oplus}$ have been reported,[413] and these may also be used to make a wide range of $Co^{III}N_5O$ and $Co^{III}N_5X$ derivatives.

23.5.5.6 Cobalt(III) Complexes with Polydentate Ligands

A large number of cobalt(III) complexes with polydentate ligands have been prepared: in many cases polydentate ligands have been designed to alter the detailed stereochemistry of the complex with a view to interpreting the electronic transitions, especially those observed using the circular dichroism of optically active complexes.

The preparation of CoN_6 complexes of polydentate ligands is mentioned in 23.5.4, and the mono-diethylenetriamine intermediates in 23.5.5.3. Cobalt(III) complexes of amine or diamine polycarboxylic acids are readily prepared by dissolving cobalt(II) carbonate in an aqueous solution of the ligand and oxidizing with air or hydrogen peroxide: *e.g.* Ba[Co(edta)]$_2$[414] and complexes of other diaminetetracarboxylic acids.[222] If another ligand, *e.g.* ethylenediamine, is added to the preparative mixture with a tetradentate ligand, such as ethylenediamine diacetic acid and its homologs, then complexes such as $[Co(edda)(en)]ClO_4$ and Co[(edda)(L-ala)] can be prepared[415, 416] K[Co(edta)][417] and related complexes[418] can also be prepared in a very high yield by the oxidation with hydrogen peroxide of a solution containing a cobalt(II) salt, potassium acetate, and the ligand, either with[418] or without[417] charcoal as catalyst. $K[Co(nta)(OH)(H_2O)]$ can be prepared by a similar method.[419] Variants of this method are usually used for the more exotic ligands such as a hexadentate $N_2S_2O_2$-donor.[420] Bis(2,3-diaminopropionato)cobalt(III) bromide can be prepared by the action of the ligand on freshly prepared $Na_3[Co(CO_3)_3]$ in the presence of charcoal.[263]

Oxygenation of aqueous cobalt(II) solutions containing dipeptides[200] or tripeptides[421] leads ultimately to anionic bis(peptidato)cobalt(III) complexes. The very stable dipeptide complexes can also be prepared by a wide range of methods including oxidation with hydrogen peroxide of an aqueous suspension of cobalt(II) carbonate and the dipeptide,[222] and by the action of the dipeptide on cobalt(III) oxide hydroxide[222] or on $[Co(NH_3)_6]Cl_3$.[422]

A large number of both α- and β-triethylenetetramine complexes including [Co(trien)(aminoacidato)]$^{2\oplus}$ have been prepared usually from either $[Co(trien)(NO_2)_2]Cl$,[352] $[Co(trien)Cl_2]Cl$[352, 423] or $[Co(trien)(CO_3)]^{\oplus}$.[423] The dinitro-

---

[406] *G. G. Schlessinger*, Inorg. Synth. *9*, 160 (1967).
[407] *I. R. Beattie, D. P. N. Satchell*, Trans. Faraday Soc. *52*, 1590 (1956).
[408] *J. Fujita, T. Yasui, Y. Shimura*, Bull. Chem. Soc., Japan *38*, 654 (1965).
[409] *C. J. Hawkins, P. J. Lawson*, Inorg. Chem. *9*, 6 (1970).
[410] *J. Meisenheimer*, Justus Liebigs Ann. Chem. *438*, 217 (1924);
J. Meisenheimer, J. Chem. Soc. Abstracts *126 i*, 1035 (1924).
[411] *I. Lifschitz*, Rec. Trav. Chim. Pays-Bas *58*, 785 (1939); C.A. *33*, 8136 (1939).
[412] *C. T. Liu, B. E. Douglas*, Inorg. Chem. *3*, 1356 (1964).
[413] *M. L. Tobe, D. F. Martin*, Inorg. Synth. *8*, 198 (1966).
[414] *S. Kirschner*, Inorg. Synth. *5*, 186 (1957).
[415] *J. I. Legg, D. W. Cooke, B. E. Douglas*, Inorg. Chem. *6*, 700 (1967).
[416] *J. I. Legg, D. W. Cooke*, Inorg. Chem. *4*, 1576 (1965).
[417] *F. P. Dwyer, E. C. Gyarfas, D. P. Mellor*, J. Phys Chem. *59*, 296 (1955).
[418] *F. P. Dwyer, F. L. Garvan*, J. Amer. Chem. Soc. *81*, 2955 (1959).
[419] *M. Mori, M. Shibata, E. Kyuno, Y. Okubo*, Bull. Chem. Soc., Japan *31*, 940 (1958).
[420] *F. P. Dwyer, N. S. Gill, E. C. Gyarfas, F. Lions*, J. Amer. Chem. Soc. *74*, 4188 (1952).
[421] *R. D. Gillard, D. A. Phipps*, J. Chem. Soc. A, 1074 (1971).
[422] *E. D. McKenzie*, J. Chem. Soc. A, 1655 (1969).
[423] *I. G. Marzilli, D. A. Buckingham*, Inorg. Chem. *6*, 1042 (1967).

complex[352] and the dichloro-complex[424] are themselves prepared by air oxidation of a mixture of the appropriate cobalt(II) salt and triethylenetetramine. Five methods have been reported[425] for the synthesis of α- and β-isomers of $[Co(tetren)Cl]^{2\oplus}$. This can be prepared by oxidation of a $CoCl_2$ solution containing tetraethylenepentamine with hydrogen peroxide or sodium chlorite, or by the reaction of the ligand with $Na_3[Co(NO_2)_6]$, $[Co(NH_3)_3(NO_2)_3]$, or $Na_3[Co(CO_3)_3]$.

Interest in the reactions of bis(dimethylglyoximato)cobalt(III) complexes ("cobaloximes") and related cobalt(III) chelates has been considerable on account of the similarity[426] of the reactions of these complexes with the reactions of the various forms of coenzyme B_{12}. The chemistry of the alkyl cobaloximes has been reviewed by Schrauzer,[427] whose group has carried out much of the work on these compounds.

The air oxidation[428, 430] of an alcoholic solution of $CoCl_2$ and dimethylglyoxime forms $H[Co(dmg)_2Cl_2]$, incorrectly formulated as a cobalt(II) derivative[428] and later reformulated as a cobalt(III) compound.[430] Hydrogen peroxide oxidation gives a better yield,[431] and various other derivatives including $H[Co(dmg)_2ClBr]$, $[Co(dmg)_2Cl(H_2O)]$, can be prepared by a similar procedure,[48, 431, 432] or by substitution.[431] An alternative method[429, 433] of preparation of $[Co(dmg)_2(NH_3)_2]Cl$ is by the action of dimethylglyoxime on $[Co(NH_3)_6]^{3\oplus}$, $[Co(NH_3)_5(OH_2)]^{3\oplus}$, $[Co(NH_3)_5Cl]^{2\oplus}$, or $[Co(NH_3)_4(CO_3)]^{\oplus}$ in the presence of an excess of ammonium acetate. In the presence of acid, $[Co(NH_3)_5A]^{2\oplus}$ (A = Cl, Br, NO_2) with dialkylglyoximes gives[432] complexes of the type $[Co(dmg)_2(NH_3)A]$. Air oxidation of a methanolic solution containing $(CH_3CO_2)_2Co$, dimethylglyoxime, sodium cyanide and a pyridine gave cyanocobaloximes, but in rather low yield.[434]

Monoalkyl derivatives of cobaloximes, $[RCo(dmg)_2(L)]$ can be prepared by a variety of methods:

(a) Reduction of $[Co(dmg)_2(L)Cl]$ to solutions containing reactive bis(dimethylglyoximato)cobalt(I) species, followed by reaction of these with alkyl halides[434–436] or alkyl sulfates.[48]
(b) The action of Grignard reagents on $[Co(dmg)_2(L)Cl]$.[48]
(c) The reaction of $[Co(dmg)_2(H_2O)_2]$ or $[(py)Co(dmg)_2]_2$ with substituted olefins or with acetylenes in an atmosphere of hydrogen.[435, 436]
(d) The reaction[436] of $[(py)Co(dmg)_2]_2$ with a benzene solution of tri-*n*-butylborane followed by treatment with sodium hydroxide solution gives $[CH_3(CH_2)_3Co(dmg)(py)]$.

$[CH_3Co(dmg)_2(OH_2)]$ can be dehydrated[436] to the deep red, very insoluble $[CH_3Co(dmg)_2]$ in which the stereochemistry of the cobalt is unknown.

Reactions similar to the first three above can be carried out on $Na[Co(BAE)]$[437] or $[XCo(BAE)(L)]$[209, 437] and the first two reactions above also work for $Na[Co(salen)]$[213] and $[XCo(salen)(L)]$.[213, 438] Cobalt(III) derivatives of salen[438] and BAE,[209] e.g., $[Co(salen)L_2]^{\oplus}$ and $[Co(BAE)(Ph_3P)Br]$ can readily be prepared by air oxidation of solutions containing cobalt(II) halides, H_2salen, or H_2BAE and the appropriate ligand (ammonia, pyridine, *etc.*).

Modification of the macrocyclic ligand ring is also possible in some cases: the reaction of boron trifluoride etherate with $[RCo(dmg)_2(H_2O)]$ gives a macrocyclic complex in which the two dimethylglyoximato ligands are linked by two $>BF_2$ groups rather than by two hydrogen bonds:[436]

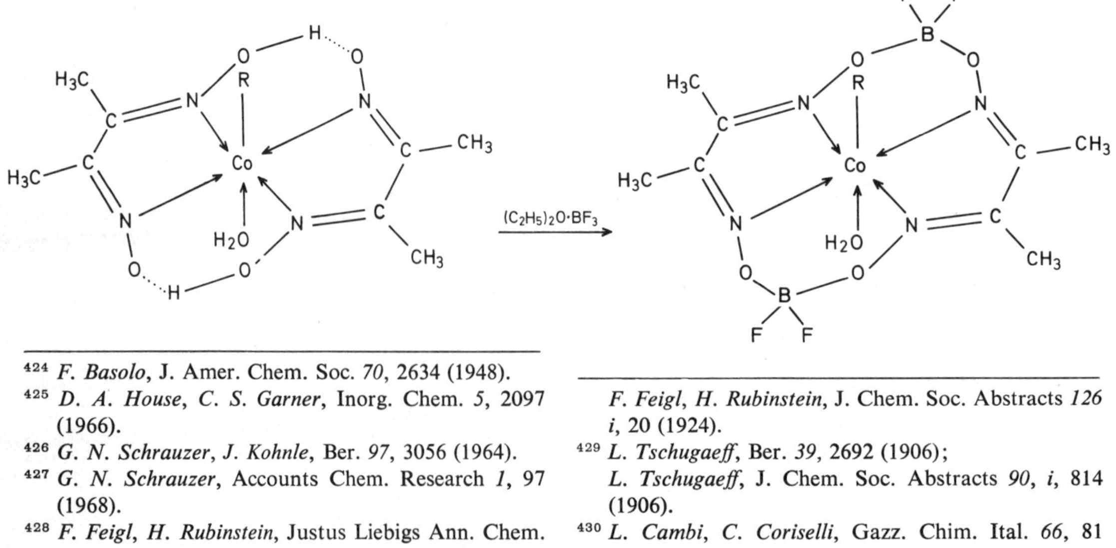

$\xrightarrow{(C_2H_5)_2O\cdot BF_3}$

[424] *F. Basolo*, J. Amer. Chem. Soc. *70*, 2634 (1948).
[425] *D. A. House, C. S. Garner*, Inorg. Chem. *5*, 2097 (1966).
[426] *G. N. Schrauzer, J. Kohnle*, Ber. *97*, 3056 (1964).
[427] *G. N. Schrauzer*, Accounts Chem. Research *1*, 97 (1968).
[428] *F. Feigl, H. Rubinstein*, Justus Liebigs Ann. Chem. *433*, 183 (1923);
F. Feigl, H. Rubinstein, J. Chem. Soc. Abstracts *126 i*, 20 (1924).
[429] *L. Tschugaeff*, Ber. *39*, 2692 (1906);
L. Tschugaeff, J. Chem. Soc. Abstracts *90, i*, 814 (1906).
[430] *L. Cambi, C. Coriselli*, Gazz. Chim. Ital. *66*, 81 (1936); C.A. *30*, 8158 (1936).

23.5.5.7 Binuclear Cobalt(III) Ammine and Related Complexes

The brown salts obtained[439] by air oxidation of ammoniacal cobalt(II) solutions were first studied in detail by Werner[221] who showed them to be dimeric peroxy-bridged cobalt(III) ammine complexes. Paramagnetic green salts can also be obtained, which Werner[221] formulated as dimeric peroxy-bridged cobalt(III)–cobalt(IV) complexes: owing to the similarity of the syntheses of the brown and green complexes the latter are also discussed in this section (see Section 6). In a lengthy paper Werner[221] summarizes the work of his group on the dimeric cobalt(III) ammine complexes including complexes containing various other bridging groups. More recently the peroxy-bridged cobalt(III) complexes have been reviewed,[440] and very recently the formation, structure and reactions of binuclear complexes of cobalt has been comprehensively reviewed.[441]

Binuclear ammine complexes of cobalt(III) can be prepared with one, two or three bridging groups.

(a) Singly-bridged complexes. The preparation of μ-peroxo- and μ-superoxo-bis{pentamminecobalt(III)} complexes are described in detail in a useful article in Inorganic Syntheses.[442] The peroxo-complex is prepared[442] by direct oxidation of an ammoniacal cobalt(II) solution. Further oxidation of the μ-peroxo complex with air,[221] or better, with ozonized oxygen[219, 220] or ammonium persulfate[442] gives the superoxo complex. μ-Amido-bis(pentamminecobalt(III)) may be prepared[442] via a series of reactions from $[(NH_3)_5CoO_2Co(NH_3)_5]^{4\oplus}$. The isoelectronic μ-hydroxy-bis[pentamminecobalt(III)] has recently been prepared[443] for the first time by the action of liquid ammonia on $[(NH_3)_3ClCo(\mu\text{-}NO_2,\mu\text{-}OH)CoCl(NH_3)_3]Cl_2$. The action of pure nitric oxide on an ammoniacal solution of cobalt(III) sulfate without cooling gives[444] the red isomer of $[Co(NH_3)_5(NO)](SO_4)$ which has been shown[445] by X-ray crystallography to be dimeric with a bridging hyponitrite ion.

(b) Doubly-bridged complexes. Full experimental details for the preparation of the μ-peroxo-μ-amido and the μ-superoxo-μ-amido-bis(tetramminecobalt(III)) complexes have been reported.[442] Displacement of the ammonia in $[(NH_3)_4Co(\mu\text{-}O_2,\mu\text{-}NH_2)Co(NH_3)_4]^{4\oplus}$ by ethylenediamine gives $[(en)_2Co(\mu\text{-}O_2,\mu\text{-}NH_2)Co(en)_2]^{4\oplus}$, which is reduced in basic solution to the analogous peroxo-complex.[442] Reactions of the peroxo- or superoxo-bridge of either the ammine or the ethylenediamine series lead to complexes with another bridging group such as sulfate,[221, 442] in addition to the amido bridges, and from these a whole range of complexes can be prepared.[221, 446] If *cis*-$[Co(NH_3)_4(H_2O)(OH)](SO_4)$ is heated to 100–110° the dihydroxy-bridged complex $[(NH_3)_4Co(OH)_2Co(NH_3)_4](SO_4)_2$ is formed.[451] Heating[447] *cis*-$[Co(en)_2(H_2O)(OH)]S_2O_6$ with acetic anhydride under reflux gives the ethylenediamine analog $[(en)_2Co(OH)_2Co(en)_2]^{4\oplus}$. Some dihydroxy-bridged compounds are formed even more readily: oxidation with hydrogen peroxide of a solution of $KHCO_3$, $CoCl_3$ and nitrilotriacetic acid gives[419] $[(nta)Co(OH)_2Co(nta)]^{2\ominus}$ in addition to the mononuclear $[Co(nta)(OH)(OH_2)]^{\ominus}$. Oxidation with

---

[431] *A. Ablov, N. M. Samus*, Russian J. Inorg. Chem. *5*, 410 (1960).

[432] *L. Tschugaeff*, Ber. *40*, 3498 (1907); *L. Tschugaeff*, J. Chem. Soc. Abstracts *92*, *i*, 904 (1907).

[433] *L. Tschugaeff*, Z. Anorg. Allgem. Chemie *46*, 144 (1905); *L. Tschugaeff*, J. Chem. Soc. Abstracts *88 i*, 743 (1905).

[434] *N. Yamazaki, Y. Hohokabe*, Bull. Chem. Soc., Japan *44*, 63 (1971).

[435] *G. N. Schrauzer, R. J. Windgassen*, J. Amer. Chem. Soc. *89*, 1999 (1967).

[436] *G. N. Schrauzer, R. J. Windgassen*, J. Amer. Chem. Soc. *88*, 3738 (1966).

[437] *G. Costa, G. Mestroni*, J. Organometal. Chem. *11*, 325 (1968).

[438] *G. Costa, G. Mestroni, L. Stephani*, J. Organometal. Chem. *7*, 493 (1967).

[439] *E. Frémy*, Justus Liebigs Ann. Chem. *83*, 227 (1852).

[440] *J. A. Connor, E. A. V. Ebsworth*, Advan. Inorg. Chem. Radiochem. *6*, 279 (1964).

[441] *A. G. Sykes, J. A. Weil*, Prog. Inorg. Chem. *13*, 1 (1970).

[442] *R. Davies, M. Mori, A. G. Sykes, J. A. Weil*, Inorg. Synth. *12*, 197 (1970).

[443] *H. Siebert, H. Feuerhake*, Chem. Ber. *102*, 2951 (1969).

[444] *W. P. Griffith, J. Lewis, G. Wilkinson*, J. Inorg. Nucl. Chem. *7*, 38 (1958).

[445] *B. F. Hoskins, F. D. Whillans, D. H. Dale, D. C. Hodgkin*, Chem. Commun. 69 (1969).

[446] *K. Garbett, R. D. Gillard*, J. Chem. Soc. A, 1725 (1968).

[447] *A. V. Ablov, N. I. Lobanov*, Zhur. Neorg. Khim *2*, 2570 (1957); C.A. *52*, 9838 (1958).

hydrogen peroxide of a suspension of $CoCO_3$ and an amino acid (glycine, L-alanine, or L-valine) gives[448,449] a low yield of the dimeric complexes $[(\alpha)_2Co(OH)_2Co(\alpha)_2]$, which can also be prepared[449] from $K[Co(\alpha)_2(CO_3)]$ by treatment first with acid and then alkali. The reaction of (+)-tartaric acid with $(\pm)$-$[Co(en)_2$-$(CO_3)](ClO_4)$ gives[248] two diastereoisomers of a tartrate bridged dimer $[(en)_2Co\{(+)$-tartrate$\}$-$Co(en_2)]^{\oplus}(ClO_4)_2$.

(c) Triply-bridged complexes. Hexammine-trioldicobalt salts, $[(NH_3)_3Co(OH)_3Co(NH_3)_3]^{3\oplus}$, can be prepared[450] by warming an aqueous solution of $[Co(NH_3)_3(OH_2)Br(Cl)]Br$ or by adding base to a solution of $[Co(NH_3)_3(OH_2)_2Cl]Cl_2$: one of the hydroxy bridges can be replaced by other groups such as nitro[443] or acetato.[221] $[(NH_3)_3Co\{(\mu\text{-}OH)_2, \mu\text{-}NH_2\}Co(NH_3)_3]I_3$ cannot be prepared from the trihydroxy-bridged complex, but is formed[221] by the addition of potassium iodide to $[(NH_3)_3(H_2O)Co(\mu\text{-}OH, \mu\text{-}NH_2)Co(OH_2)(NH_3)_3](NO_3)_4$.

23.5.5.8 Polynuclear Complexes

Compared with the large number of binuclear cobalt(III) complexes, few polynuclear cobalt(III) complexes have been well characterized. A trinuclear complex $[(H_2O)_2Co\{(HO)_2Co(en_2\}_2](SO_4)_2$, "pink sulfate", can be prepared[413,451] by allowing an aqueous solution containing cobalt(II) sulfate and ethylenediamine to stand in air for several days. Under similar conditions[452] cobalt(II) nitrate gives brown crystals of a tetranuclear complex $[Co\{(OH)_2Co(en)_2\}_3](NO_3)_6$, "hexol", which can in theory contain four pairs of enantiomers.[453] The problem of the separation of some of these isomers (and also resolution of the complex) has been discussed.[453,454] The preparation of the analogous dodecammine complex, $[Co\{(OH)_2Co(NH_3)_4\}_3](SO_4)_3$ is more difficult. It can be prepared by the reaction of aqueous ammonia with *cis*-$[Co(NH_3)_4Cl(H_2O)](SO_4)$,[215,455,456] or *cis*-$[Co(NH_3)_4(H_2O)_2]_2(SO_4)_3$,[455] or by the reaction[452] of ammonium sulfate with a solution of uncertain composition prepared by heating an aqueous solution of *trans*-$[Co(NH_3)_4Br_2]Br$.

The complex cation $[Co\{(OH)_2Co(NH_3)_4\}_3]^{6\oplus}$ is of considerable interest as the first non carbon-containing compound to be obtained optically active.[457]

The action of ethanolamine on $CoCl_2$ in the presence of pyridine gives[458] a trinuclear complex which can also be prepared[459] by the action of ethanolamine on $[Co(NH_3)_4(CO_3)]^{\oplus}$.

Although early workers claimed it to be a cobalt(II) complex[458,460] more recent work[459] reports that the complex is $[Co^{II}\{Co^{III}(OCH_2CH_2NH_2)_3\}_2]SO_4$; a crystal structure[461] shows two octahedral cobalt(III) atoms bridged by a central trigonal prismatic cobalt(II). The reaction of tris(2-aminoethanethiol)cobalt(III) with a hot aqueous solution of either $[Co(NH_3)_5Br]Br_2$ or $CoCl_2\cdot 6H_2O$ gives[384,462] a sulfur analog $[Co^{III}\{Co^{III}(SCH_2CH_2NH_2)_3\}_2]Br_3$ containing only cobalt(III), which can be resolved[463] with silver antimony (+)-tartrate.

23.5.5.9 Miscellaneous Mixed Donor Cobalt(III) Complexes

In addition to organo-cobalt(III) derivatives

[448] *D. C. Phillips*, Ph.D. Thesis, University of Kent at Canterbury, 1970.

[449] *R. D. Gillard, S. H. Laurie, D. C. Price, D. A. Phipps, C. F. Weick*, J. Chem. Soc. Dalton, 1385 (1974).

[450] *A. Werner, E. Bindschedler, A. Grün*, Ber. *40*, 4834 (1907);
A. Werner, E. Bindschedler, A. Grün, J. Chem. Soc. Abstracts *94 ii*, 43 (1908).

[451] *A. Werner, G. Jantsch*, Ber. *40*, 4426 (1907);
A. Werner, G. Jantsch, J. Chem. Soc. Abstracts *92 i*, 1012 (1907).

[452] *A. Werner, E. Berl, G. Jantsch, E. Zinggeler*, Ber. *40*, 2103 (1907);
A. Werner, E. Berl, G. Jantsch, E. Zinggeler, J. Chem. Soc. Abstracts *92 i*, 482 (1907).

[453] *H. A. Goodwin, E. C. Gyarfas, D. P. Mellor*, Australian J. Chem. *11*, 426 (1958).

[454] *R. D. Kern, R. A. D. Wentworth*, Inorg. Chem. *6*, 1018 (1967).

[455] *S. M. Jorgensen*, Z. Anorg. Allgem. Chemie *16*, 184 (1898);
S. M. Jorgensen, J. Chem. Soc. Abstracts *74 ii*, 226 (1898).

[456] *G. B. Kauffman, R. P. Pinnell*, Inorg. Synth. *6*, 176 (1960).

[457] *A. Werner*, Compt. Rend. *159*, 426 (1914);
A. Werner, J. Chem. Soc. Abstracts *106 ii*, 787 (1914).

[458] *W. Hieber, E. Levy*, Justus Liebigs Ann. Chem. *500*, 14 (1932); C.A. *27*, 1323 (1933).

[459] *V. V. Udovenko, A. N. Gerasenkova*, Russian J. Inorg. Chem. *11*, 1105 (1966).

[460] *H. Brintzinger, B. Hesse*, Z. Anorg. Allgem. Chemie *248*, 345 (1941); Chem. Abs. *36*, 4434 (1942).

[461] *J. A. Bertrand, J. A. Kelley, E. G. Vassian*, J. Amer. Chem. Soc. *91*, 2394 (1969).

[462] *D. H. Busch, J. A. Burke, D. C. Jicha, M. C. Thompson, M. L. Morris*, Advan. Chem. Ser. *37*, 125 (1963).

[463] *G. R. Brubaker, B. E. Douglas*, Inorg. Chem. *6*, 1562 (1967).

mentioned elsewhere (see 23.5.5.6 and 23.5.6), two other complexes deserve mention. The reaction[464] of diethylaluminum monoethoxide with $[Co(acac)_3]$ in the presence of the phosphine ligand $C(CH_2PPh_2)_4$ gives a hydrido diethyl complex $[(C_2H_5)_5HCo\{(Ph_2PCH_2)_3CCH_2PPh_2\}]$ in which the ligand is only tridentate. "Reduction" of a cobalt(II) chloride solution containing phenanthroline or bipyridyl with sodium hydridoborate in the presence of butadiene leads[465] to the formally cobalt(III) derivative $[Co(LL)(\pi\text{-methyl-allyl})_2]PF_6$.

$[CoI_3(triars)]$, prepared[121] by air oxidation of the cobalt(II) complex $[Co(I_2triars)]$ is of interest in that the cobalt is octahedral despite the size of the donor atoms. Air oxidation of a solution of CoX_2 in the presence of 1,2-bis(diethylphosphino)ethane (PP) gives the octahedral complexes $[Co(PP)_2X_2]X$.[466] Oxidation[467] of $[CoCl_2(PEt_3)_2]$ with nitrosyl halide gives the paramagnetic ($\mu = 3{\cdot}02$ B.M.) five-coordinate cobalt(III) complex $[CoCl_3(PEt_2)_2]$; some similar compounds have also been prepared.[467] The reaction of $[CoI_2\{(C_6H_{11})_3PO\}_2]$ with iodine gives the 1:1 electrolyte $[Co^{III}I_2\{(C_6H_{11})_3PO\}_2]I$ in which the cobalt(III) is said to be square planar.[468]

23.5.6 Cobalt(III) Complexes with Group IV Donors

Potassium hexacyanocobaltate(III) is readily prepared[7,179] as pale yellow crystals by dissolving cobalt(II) cyanide in a slight excess of cyanide, and heating the solution to boiling to oxidize Co(II) to Co(III). The strong *tribasic acid* $H_3[Co(CN)_6]$ is readily obtained by ion exchange,[175] and can be alkylated by heating with alcohols[469] (see below). The *pentacyano-complexes* $K_3[Co(CN)_5Br]$ and $K_3[Co(CN)_5I]$ can be prepared[470] by the action of the halogen on a solution of $K_3[Co^{II}(CN)_5]$. The *bromo-complex* can also be prepared[470] by the action of KCN on $[Co(NH_3)_5Br]Br_2$; the *thiocyanato-complex* can be prepared in a similar way.[471] The bromo-complex reacts[472] with hydride ion to give a solution containing $[Co(CN)_5H]^{3\ominus}$, which can also be prepared by the action of hydrogen[180] or hydridoborate[178,180] on $[Co(CN)_5]^{3\ominus}$ solutions. Salts of $[Co(CN)_5H]^{3\ominus}$ cannot be isolated, as attempts at crystallization yield only $K_6[Co_2(CN)_{10}]$ and hydrogen, but the solution is a useful intermediate, reacting with olefins to give alkylcobalt(III) complexes:

$$[Co(CN)_5H]^{3\ominus} + C_2F_4 \longrightarrow K_3[(NC)_5CoCF_2CF_2H] \quad \text{(Ref. 181)}$$

$$[Co(CN)_5H]^{3\ominus} + H_2C{=}CH{-}CH{=}CH_2 \longrightarrow K_3[(NC)_5Co(C_4H_7)] \quad \text{(Ref. 186)}$$

Many *alkyl pentacyanocobaltates(III)* can also be prepared[186,187] by the reaction of a wide range of alkyl halides with a solution of $K_3[Co(CN)_5]$. Ammine-,[183,473] aquo-,[185,474] and hydroperoxy-[475] pentacyanocobalt(III) complexes have been prepared.

Several other cobalt(III) cyanocomplexes can also be prepared by displacement of ammonia by aqueous potassium cyanide: indeed, in view of the large number of ammine complexes known, this method obviously has great scope. Thus alkaline KCN reacts with both the monomeric black[476] and the dimeric red[444] isomers of $[Co(NH_3)_5(NO)]Cl_2$ to give $K_3[Co(CN)_5(NO)]$, which can also be prepared[477] by the action of nitric oxide on a solution of $[Co(CN)_5]^{3\ominus}$. The dimeric compounds $[(NH_3)_5CoO_2Co(NH_3)_5](NO_3)_5$ and $[(NH_3)_4Co(\mu\text{-}NH_2,\mu\text{-}O_2)Co(NH_3)_4](NO_3)_4$ react with KCN to give $K_5[(CN)_5CoO_2Co(CN)_5]$ and $K_4[(CN)_4Co(\mu\text{-}NH_2,\mu\text{-}O_2)Co(CN)_4]$; in the latter case incomplete replacement, to give hexacyanodiammine and

[464] *J. Ellermann, W. H. Gruber*, Angew. Chem. Intern. Ed. Engl. *7*, 129 (1968).

[465] *G. Mestroni, A. Camus, E. Mestroni*, J. Organometal. Chem. *24*, 775 (1970).

[466] *C. E. Wymore, J. C. Bailar*, J. Inorg. Nucl. Chem. *14*, 42 (1960).

[467] *K. A. Jensen, B. Nygaard, C. T. Pedersen*, Acta Chem. Scand. *17*, 1126 (1963).

[468] *K. Issleib, B. Mitscherling*, Z. Anorg. Allgem. Chemie *304*, 73 (1960).

[469] *F. Hölzl, T. Meier-Mohar, F. Viditz*, Monatsh. *53–54*, 237 (1929); C.A. *24*, 1347 (1930).

[470] *A. W. Adamson*, J. Amer. Chem. Soc. *78*, 4260 (1956).

[471] *J. L. Burmeister*, Inorg. Chem. *3*, 919 (1964).

[472] *A. P. Ginsberg* in *R. L. Carlin*, Hydride Complexes of the Transition Metals, Transition Metal Chemistry, Vol. I, p. 112, Edward Arnold & Co., London 1965.

[473] *L. Cambi, E. Paglia*, Gazz. Chim. Ital. *88*, 691 (1958); C.A. *53*, 16793, (1959).

[474] *J. H. Bayston, R. N. Beale, N. K. King, M. E. Winfield*, Australian J. Chem. *16*, 954 (1963).

[475] *J. H. Bayston, M. E. Winfield*, J. Catalysis *3*, 123 (1964); C.A. *60*, 12699 (1964).

[476] *R. Nast, R. Thome*, Z. Anorg. Allgem. Chemie *309*, 283 (1961).

[477] *R. Nast, H. Rupert-Mesche, M. Helbig-Neubauer*, Z. Anorg. Allgem. Chemie *312*, 314 (1961).

heptacyanoammine complexes, also occurs.[478] $K_5[(CN)_5CoO_2Co(CN)_5]$ can also be prepared by bromine oxidation[185] of $K_6[(CN)_5CoO_2Co(CN)_5]$, which is itself prepared[185] by the action of oxygen on a solution of $[Co(CN)_5]^{3\ominus}$.

Complexes containing coordinated isocyanide are also known, see 23.5.4.

Cobalt(III) isonitrile complexes can be prepared either by alkylation of the $[Co(CN)_6]^{3\ominus}$ ion, or by oxidation of isonitrile derivatives of cobalt(I) or cobalt(II). $Ag_3Co(CN)_6$ can be alkylated using an excess of methyl iodide by warming at 45° for 8 days, to give two isomers of $[(CH_3NC)_3Co(CN)_3]$ (presumably 1,2,3- and 1,2,4-isomers) which can be separated by crystallization from alcohol.[479] A mono-isonitrile complex formulated as $H_2[(C_2H_5NC)Co(CN)_5]\cdot H_2O$ has been prepared by a similar method,[480] using the stoichiometric quantity of ethyl iodide in ethanol, and direct alkylation of $H_3[Co(CN)_6]$ with alcohols at 100° in a sealed tube gives mono- or bis-isonitrile complexes.[469] Oxidation[481] of $[(RNC)_4Co^{II}I_2]$ with hydrogen peroxide in presence of perchloric acid or of $[(RCN)_5Co^{I}]ClO_4$ with iodine gives $[(RCN)_4Co^{III}I_2]ClO_4$, whereas oxidation[481] of $[(RCN)_5Co^{II}](ClO_4)_2$ with iodine gives $[(RNC)_5Co^{III}I](ClO_4)_2$.

The *hexaalkynylcobaltate(III) complexes* $K_3[Co(C{\equiv}CH)_6]$ and $Na_3[Co(C{\equiv}CCH_3)_6]$ can be prepared[191] by air oxidation of a liquid ammonia solution of the analogous cobalt(II) compounds. The potassium salt is explosive.

23.5.7 Optically Active Cobalt(III) Complexes

It could be said that the two key factors in Werner's proof[482] of his "coordination theory" of the constitution of complexes of transition metals in general and of cobalt(III) in particular were his ability to predict the number of geometrical isomers that could be formed, and his resolution of those of his complexes which, according to his theory, were dissymmetric. This culminated in the first resolution[457] of a purely inorganic compound, $[Co\{(OH)_2Co(NH_3)_4\}_3]^{6\oplus}$.

The different methods of preparation of optically active complexes have been discussed in two reviews,[483, 484] and the optically active complexes which have been prepared have been listed:[484] the methods available for determining the absolute configurations of the resolved complexes have been reviewed.[485] There are four main methods of resolution available. Other methods are possible in principle[484] but are unlikely to be of any practical use.

23.5.7.1 Dissymmetric Synthesis

(a) 1st order. Preparation of the metal complex using only one enantiomer of an asymmetric ligand gives only or predominantly one enantiomer of the complex only if there is considerable stereoselectivity in the formation of the complex. Thus, preparation of $[Co(pdta)]^{\ominus}$ or $[Co(methylpenten)]^{3\oplus}$ from (−)-pdta[486] or (−)-methylpenten[487] gives $(+)_{546}$-$[Co\{(-)\text{-pdta}\}]^{\ominus}$ or $(-)$-$[Co\{(-)\text{-methylpenten}\}]^{3\oplus}$ as the only isomer. Preparation of $[Co(pn)_3]^{3\oplus}$ from (−)-pn gives predominantly[333] (85%) $(-)$-$[Co\{(-)\text{-}(pn)\}_3]^{3\oplus}$, but in this case the situation is complicated by the existence of 1,2,3 and 1,2,4 geometric isomers.[334] Some ligands, e.g. α-amino acids, have so little stereoselectivity[374] in the formation of their complexes that this method is usually inapplicable; however, the two products formed[412] in approximately equal amounts from $[Co(en)_2Cl_2]^{\oplus}$ and L-α-amino acids, $(+)$-$[Co(en)_2(L\text{-}\alpha)]^{2\oplus}$ and $(-)$-$[Co(en)_2(L\text{-}\alpha)]^{2\oplus}$ are diastereoisomers and in certain cases, *e.g.* with serine and threonine, can be separated[488] by methods in 23.5.7.3(b) below.

(b) 2nd order. The preparation of the metal complex, in the presence of a counter ion which preferentially precipitates one enantiomer of the complex, can give a high yield of this enantiomer, if a catalyst is present to interconvert

[478] *M. Mori, J. A. Weil, J. K. Kinnaird*, J. Phys. Chem. *71*, 103 (1967).

[479] *E. G. J. Hartley*, J. Chem. Soc. *105*, 521 (1914).

[480] *C. E. Bolser, L. B. Richardson*, J. Amer. Chem. Soc. *35*, 377 (1913).

[481] *A. Sacco*, Atti Accad. Naz. Lincei, Classe Sci. Fis. Mat. Natur, Rend. *15*, 82 (1953); C.A. *48*, 8109 (1954).

[482] *A. Werner*, Nobel Lecture «Uber die Konstitution und Konfiguration von Verbindungen höherer Ordnung», reprinted in Helvetica Chim. Acta, Alfred Werner Commemoration Volume, p. 24 (1966).

[483] *S. Kirschner*, Optically Active Coordination Compounds, in *W. L. Jolly*, Preparative Inorganic Reactions, Vol. I, Ch. II, p. 29, Interscience Publishers, New York 1964.

[484] *F. Woldbye*, Technique of Optical Rotatory Dispersion and circular, Dichroism, in *H. B. Jonassen, A. Weissberger*, Technique of Inorganic Chemistry, Vol. IV, p. 249, Interscience Publishers, New York 1965.

[485] *R. D. Gillard, P. R. Mitchell*, Struct. Bonding (Berlin) *7*, 46 (1970).

[486] *F. P. Dwyer, F. L. Garvan*, J. Amer. Chem. Soc. *83*, 2610 (1961).

[487] *J. R. Gollogly, C. J. Hawkins*, Australian J. Chem. *20*, 2395 (1967);
J. R. Gollogly, C. J. Hawkins, Chem. Commun. 873 (1966).

[488] *S. K. Hall, B. E. Douglas*, Inorg. Chem. *8*, 372 (1969).

the enantiomers. For example, air oxidation[330] of a complex mixture containing cobalt(II)-(+)-tartrate, ethylenediamine, concentrated hydrochloric acid, and charcoal, in 25% ethanol/water gives optically pure (+)-$[Co(en)_3]^{3\oplus}$ in 80% yield as its chloride (+)-tartrate salt. This method has so far found only limited application owing to the difficulty of finding suitable reaction conditions.

23.5.7.2 Preferential Destruction of One Isomer

Although photochemical decomposition of organic compounds using circularly polarized light gives slightly optically active products, there have been few reports[489] of a slight resolution of a cobalt complex (±)-$[Co(C_2O_4)_3]^{3\ominus}$, which is much more readily resolved by other methods.[490] Despite Pasteur's early success in resolving ammonium tartrate by bacterial destruction of the (+)-isomer, and the wide use of this method in the resolution of acylated racemic amino acids, attempts[491] to use it to resolve (±)-$[Co(en)_3]^{3\oplus}$ and other complexes were unsuccessful. However, recently (+)-α-tris(glycinato)cobalt(III) was prepared[492] by growth of *Enterobacter cloacae* in media containing the racemic complex, which is not easy to resolve by other methods: the optical purity of the bacterial product is much greater than that obtained[493] by chromatography on starch. (±)-$[Co(en)_2(phen)]^{3\oplus}$ has been prepared[494] with an optical purity greater than 80% by the growth of *Pseudomonas Stutzeri* on the racemic salt.

23.5.7.3 Resolution Procedures

(a) Spontaneous resolution. For most optically active compounds, including metal complexes, the racemate is less soluble than the optically active material. If the reverse is true at any temperature, then "spontaneous resolution" can occur,[484, 495] *i.e.* each individual crystal contains only one of the optical isomers, which can be separated by picking crystals and determining which isomer is present in each crystal either by the existence of hemihedral facets, or spectroscopically, now that the sensitivity of Optical Rotatory Dispersion and Circular Dichroism machines is such that the optical activity of a substance can be determined on < 1 mg of material. This method, originally used by Pasteur to resolve sodium ammonium tartrate, has not been used much for metal complexes as it is rather tedious, but it is quite convenient for the preparation of very small amounts of optically pure complex, if the total absence of other optically active species is desired.[222] It is sometimes possible to obtain a small sample which has a higher optical purity than the conventionally resolved material.[496] Complexes which have been resolved by this method include $K_3[Co(C_2O_4)_3]\cdot H_2O$,[319] $(NH_4)[Co(edta)]\cdot 2H_2O$,[497] and $[Co(en)_2(C_2O_4)]_2\cdot C_2O_4\cdot 8H_2O$[496]: attempts[222, 495] to repeat Werner's spontaneous resolution[498] of $K_3[Rh(C_2O_4)_3]\cdot H_2O$ at several temperatures, have not been successful.

It is occasionally possible to get a larger quantity of resolved material by spontaneous resolution if slow crystallization enables the first optically active crystal (which may be either enantiomer) to "seed" the solution such that a predominance of that enantiomer is obtained[499] —*e.g.* $[Co(en)_2(ox)]Cl$. Even with complexes that do not spontaneously resolve induced crystallization of one enantiomer by seeding with a single crystal of that enantiomer, or even of another optically active material, may be possible: Werner[500] used a seed crystal of

[489] *R. Tsuchida, A. Nakamura, M. Kobayashi*, J. Chem. Soc., Japan *56*, 1335 (1935); C.A. *30*, 963 (1936).

[490] *G. B. Kauffman, L. T. Takahashi, N. Sugisaka*, Inorg. Synth. *8*, 207 (1966);

[491] *M. J. S. Crespi*, Ph. D. Thesis, University of Illinois, 1960; Diss. Abstr. *20*, 4512 (1960).

[492] *R. D. Gillard, C. Thorpe*, Chem. Commun. 997 (1970).

[493] *H. Krebs, R. Rasche*, Z. Anorg. Allgem. Chemie *276*, 236 (1954); C.A. *49*, 15596 (1955).

[494] *L. S. Dollimore*, Ph.D. Thesis, University of Kent at Canterbury, 1973.

[495] *F. M. Jaeger*, Optical Activity and High Temperature Measurements, p. 118, McGraw Hill, New York 1930.

[496] *K. Yamasaki, H. Igarashi, Y. Yoshikawa, H. Kuroya*, Inorg. Nucl. Chem. Lett. *4*, 491 (1968).

[497] *H. A. Weakliem, J. L. Hoard*, J. Amer. Chem. Soc. *81*, 549 (1959).

[498] *A. Werner*, Ber. *47*, 1954 (1914);
A. Werner, J. Chem. Soc. Abstracts *106, i*, 922 (1914).

[499] *D. G. Brewer, K. T. Kan*, Canadian J. Chem. *49*, 965 (1971).

[500] *A. Werner, J. Bosshart*, Ber. *47*, 2171 (1914);
A. Werner, J. Bosshart, J. Chem. Soc. Abstracts *106 i*, 936 (1914).

(+)-$[Co(en)_2(C_2O_4)]Cl$ to resolve both (±)-$[Co(en)_2(C_2O_4)]Cl$ and (±)-$[Co(en)_2(NO_2)_2]Cl$.

(b) Diastereoisomer separation is undoubtedly the most useful method for the separation of ionic cobalt complexes. On addition of a suitable optically active counter ion it is usually possible to isolate the less soluble diastereoisomer without too much difficulty, and recrystallization usually gives material of high optical purity—due to the inertness of most cobalt(III) complexes, racemization on redissolving the complex, which is a problem with complexes of some other metals, is rarely a problem. Isolation of the more soluble diastereoisomer optically pure is usually difficult, however. It is obvious that at best only 50% of the complex should be obtained as the less soluble diastereoisomer. However, if the enantiomer left in solution can be readily racemized, higher yields are possible. Addition of $[Co^{II}(en)_3]^{2\oplus}$ to a solution containing (±)-$[Co(en)_3]^{3\oplus}$, chloride, and (+)-tartrate ions gives[501] at least 75% of (+)-$[Co(en)_3]^{3\oplus}$(Cl)-{(+)-tartrate}$^{2\ominus}$. An alternative approach if the enantiomer is racemized on heating, is to crystallize a crop of the less soluble diastereoisomer, concentrate the solution by boiling it vigorously, and cool to crystallize another crop of the less soluble diastereoisomer, etc. This has been used[222] to resolve $[Co(edta)]^{\ominus}$ using L-histidine.[502]

Many possible counter ions can be used: for cationic complexes (+)-tartrate,[501,503] antimony (+)-tartrate,[504,505] dibenzoyl-(+)-tartrate,[361] camphor 10-sulfonate,[504] and α-bromocamphor 10-sulfonate[504,506] have proved the most useful; for anionic complexes alkaloids including strychnine[507] and brucine, and basic amino acids such as L-histidine[502] have proved useful. In general naturally occurring optically active resolving agents and their derivatives are to be preferred, as they are commercially available optically pure cheaply, and do not racemize if the solution is heated during attempts at crystallization. However, the separation of the diastereoisomers by crystallization may be easier if an optically active metal complex is used. The use of metal complexes, such as $(-)_{546}$-$[Co(edta)]^{\ominus}$,[226] or (+)-$[Co(en)_2(NO_2)_2]^{\oplus}$ [508] suffers from the disadvantages that they have first to be prepared and resolved and that they may racemize: (−)-$[Co(ox)_3]^{3\ominus}$ does so readily even at room temperature,[490] and (−)-$[Co(edta)]^-$ racemizes on boiling:[417] however, (+)-$[Co(en)_3]^{3\oplus}$ readily prepared[330] by an asymmetric synthesis (see above), and stable to boiling[509] (in the absence of reducing agents or charcoal) has proved useful in the resolution of anionic complexes including $[Co(edta)]^{\ominus}$,[417] $[Co(C_2O_4)_3]^{3\ominus}$,[510] and $[Co(CO_3)_3]^{3\ominus}$.[222] The choice of a suitable counter ion is often a problem simply requiring trial and error: however, it is often convenient to choose a counter ion with the same size and charge as the ion to be resolved. Woldbye[484] tabulates many examples of resolutions by this method.

(c) Chromatographic separation of enantiomers is especially useful for uncharged complexes. A number of different absorbents for liquid–solid chromatography have been tried: powdered optically active quartz,[340,483,511] powdered optically active sodium chlorate,[483] starch,[493,512] (+)-lactose[513,514] tartaric acid absorbed on alumina,[515] cation exchange cellulose,[516] cation exchange Sephadex-®,[517] and an anion exchange resin in the (+)-tartrate form.[222] The last three absorbents can only be used to separate cationic complexes and the use of cation exchange cellulose or cation exchange Sephadex gives an optically active product dissolved in a strong salt solution from which it is virtually impossible to isolate the solid

---

[501] *D. H. Busch*, J. Amer. Chem. Soc. *77*, 2747 (1955).

[502] *R. D. Gillard, P. R. Mitchell, C. F. Weick*, J. Chem. Soc. Dalton, 1635 (1974).

[503] *F. McCullough, J. C. Bailar*, J. Amer. Chem. Soc. *78*, 714 (1956).

[504] *K. Garbett, R. D. Gillard*, J. Chem. Soc. *A*, 802 (1966).

[505] *F. P. Dwyer, F. L. Garvan*, Inorg. Synth. *6*, 195 (1960).

[506] *J. C. Bailar*, Inorg. Synth. *2*, 222 (1946).

[507] *D. H. Busch, J. C. Bailar*, J. Amer. Chem. Soc. *75*, 4574 (1953).

[508] *F. P. Dwyer, F. L. Garvan*, Inorg. Synth. *6*, 192 (1960).

[509] *B. E. Douglas*, J. Amer. Chem. Soc. *76*, 1020 (1954).

[510] *J. W. Vaughn, V. E. Magnuson, G. J. Seiler*, Inorg. Chem. *8*, 1201 (1969).

[511] *R. Tsuchida, M. Kobayashi, A. Nakamura*, J. Chem. Soc., Japan *56*, 1339 (1935); C.A. *30*, 926 (1936).

[512] *H. Krebs, J. Diewald, J. A. Wagner*, Angew. Chem. *67*, 705 (1955).

[513] *J. P. Collman, R. P. Blair, R. L. Marshall, L. Slade*, Inorg. Chem. *2*, 576 (1963).

[514] *T. Moeller, E. Gulyas*, J. Inorg. Nucl. Chem. *5*, 245 (1958).

[515] *T. S. Piper*, J. Amer. Chem. Soc. *83*, 3908 (1961).

[516] *J. I. Legg, B. E. Douglas*, Inorg. Chem. *7*, 1452 (1968).

[517] *Y. Yoshikawa, K. Yamasaki*, Inorg. Nucl. Chem. Lett. *6*, 523 (1970).

optically active complex: quartz and sodium chlorate columns are very hard to prepare requiring considerable care in picking the correct enantiomeric crystals. Starch has been successfully used to resolve[493] [Co(gly)$_3$], and (+)-lactose to resolve[513, 514] [Co(acac)$_3$]. The vapour phase chromatographic resolution[518] of [Cr(hfac)$_3$] on quartz has not yet been applied to the cobalt(III) analog, which is also very volatile. Counter current liquid–liquid extraction using diisoamyl-(+)-tartrate has been used to resolve[519] [Co(acac)$_3$].

23.5.7.4 Preparation from another Optically Active Complex

Any of the reactions mentioned in Section 23.5.1.3, such as the substitution reactions of [Co(acac)$_3$], may be used to prepare optically active complexes, and as in general no metal–ligand bonds are broken in these reactions, the absolute configuration of the cobalt in the reactant and product are the same. Substitutions at the cobalt in an optically active complex usually lead to an optically active product, although depending on the exact reaction conditions inversion or retention of configuration may occur:

e.g. (−)-[Co(en)$_2$Cl$_2$]$^{\oplus}$ + K$_2$CO$_3$ $\longrightarrow$ (+)-[Co(en)$_2$CO$_3$]$^{\oplus}$

(−)-[Co(en)$_2$Cl$_2$]$^{\oplus}$ + Ag$_2$CO$_3$ $\longrightarrow$ (−)-[Co(en)$_2$CO$_3$]$^{\oplus}$ (Ref. 520)

and (−)-[Co(en)$_2$Cl$_2$]$^{\oplus}$ $\xrightarrow{NH_3, -77°}$ (+)-[Co(en)$_2$(NH$_3$)$_2$]$^{3\oplus}$

$\xrightarrow{NH_3, 80}$ (−)-[Co(en)$_2$(NH$_3$)$_2$]$^{3\oplus}$ (Ref. 331)

Finally, there are a few cases in which oxidation of an optically active cobalt(II) complex is used to prepare the cobalt(III) complex: (+)-[Co(phen)$_3$]$^{3\oplus}$ [521] and (+)-[Co(bipy)$_3$]$^{3\oplus}$ [246] can be prepared by oxidation by chlorine of [Co^{II}(LL)$_3$]$^{2\oplus}$[(Sb$_2$(tartrate)$_2$]$^{2\ominus}$, itself precipitated by addition of sodium antimony (+)-tartrate to a solution of [Co^{II}(LL)$_3$]$^{2\oplus}$.[521]

[518] *R. E. Sievers, R. W. Moshier, M. L. Morris*, Inorg. Chem. *1*, 966 (1962).

[519] *N. S. Bowman, V. G'ceva, G. K. Schweitzer, I. R. Supernaw*, Inorg. Nucl. Chem. Lett. *2*, 351 (1966).

[520] *J. C. Bailar, R. W. Auten*, J. Amer. Chem. Soc. *56*, 774 (1934).

[521] *C. S. Lee, E. M. Gorton, H. M. Neumann, H. R. Hunt*, Inorg. Chem. *5*, 1397 (1966).

23.6 Cobalt(IV)

The series of dark green binuclear paramagnetic cobalt peroxo complexes first prepared by Werner[218, 221] were originally formulated as peroxo-bridged complexes containing both Co(III) and Co(IV). The E.S.R. spectrum[522] of these complexes typically shows a 15 line signal arising from coupling with two equivalent ^{59}Co nuclei indicating that the original formulation with non-equivalent cobalt atoms was incorrect. An alternative suggestion[442] is that two cobalt(III) atoms are bridged by a superoxide ion. As the preparations and reactions of these complexes are closely linked with those of binuclear cobalt(III) complexes with other bridging ligands, they are discussed in 23.5.5.7.

A dark green complex prepared[523] by reaction of 60% nitric acid with salicylatotetrammine-cobalt(III) was formulated[523] as 5-nitrosalicylatotetramminecobalt(IV) chloride on the basis of its paramagnetism. Further work[524] has suggested that it is a cobalt(III) complex containing the protonated *5-nitrososalicylato-tetramminecobalt(III) cation.*

An early report[275] of the isolation of blue K$_3$Co^{IV}F$_7$ has been shown[274] to be incorrect. Thus the only well characterized example of a cobalt(IV) complex is the hexafluorocobaltate(IV) anion which can be prepared[274] as its cesium salt by the fluorination of Cs$_2$CoCl$_4$ at 300°. Cs$_2$CoF$_6$ is a yellow paramagnetic solid (μ = 2·97 B.M.).

There are, however, some organo-cobalt derivatives originally reported[525] as RCoI$_3$ and R$_2$CoI$_2$ (R = α- and β-naphthyl), prepared by the reaction of ethereal solutions of anhydrous CoI$_2$ with the naphthyl magnesium iodides. If this formulation is correct these reactive greenish compounds must be considered formally to be Co(IV). However, as the very similar reaction[194] of CoX$_2$(PEt$_2$Ph)$_2$ with mesityl magnesium halides gives the Co(II) compounds [(mesityl)$_2$-Co(PEt$_2$Ph)$_2$] characterized[193] *inter alia* by

[522] *E. A. V. Ebsworth, J. A. Weil*, J. Phys. Chem. *63*, 1890 (1959).

[523] *Y. Yamamoko, K. Ito, H. Yoneda, M. Mori*, Bull. Chem. Soc., Japan *40*, 2580 (1967).

[524] *A. G. Beaumont, R. D. Gillard*, J. Chem. Soc. *A*, 2400 (1968).

[525] *D. A. E. Briggs, J. B. Polya*, J. Chem. Soc. 1615 (1951);
D. L. Ingles, J. B. Polya, Nature *164*, 447 (1949);
D. L. Ingles, J. B. Polya, J. Chem. Soc. 2280 (1949).

X-ray crystallography, it is more likely that $RCoI_3$ and R_2CoI_2 are in fact impure cobalt(II) compounds.

The cobalt(II) complexes of several chelating ligands (*e.g.* bis(salicylaldehyde)ethylenediamine) can act as reversible oxygen carrier both in the solid state and in solution. Although these oxygen adducts could be formulated as cobalt(IV) species they are discussed, with the parent compounds, in 23.4.5.

An oxide, CoO_2, prepared[526] as a brown precipitate by the addition of sodium hydroxide to a solution containing cobalt(II) chloride and hydrogen peroxide in ethanol at −25° is best considered as a cobalt(II) peroxide.[526]

23.7 Cobalt(V)

The sole example of cobalt(V) is provided by the *tetraoxocobaltate(V) anion*, $[CoO_4]^{3\ominus}$: the potassium salt can be prepared by heating a mixture of oxides of potassium and cobalt at 460–550° under oxygen at ½–1 atmosphere.[527,528] The product is isomorphous with the vanadate(V), manganate(V) and ferrate(V) ions, and the analysis and magnetic moment is consistent with the presence of Co(V).

[526] *V. A. Shcherbinin, G. A. Bogdanov*, Russian J. Inorg. Chem. *4*, 112 (1959); C.A. *53*, 12906 (1959).
[527] *C. Brendel, W. Klemm*, Z. Anorg. Allgem. Chemie *320*, 59 (1963); C.A. *58*, 9861 (1963).
[528] *R. Scholder*, Bull. Soc. Chim. France, 1112 (1965).

Appendix. Abbreviations used for ligands in the formulae of complexes

General			
R	alkyl, aryl	L	uncharged unidentate ligand
X	halide - $Cl^{\ominus}$, $Br^{\ominus}$, or $I^{\ominus}$	LL	uncharged bidentate ligand
A	anionic ligand	LLL	uncharged tridentate ligand
Specific			
en	diaminoethane	pn	1,2-diaminopropane
bn	(±)-2,3-diaminobutane	tn	1,3-diaminopropane
cptn	(±)-1,2-diaminocyclopentane	chxn	(±)-diaminocyclohexane
dien	diethylenetriamine	trien	triethylenetetramine
tetren	tetraethylenepentamine	penten	pentaethylenehexamine
phen	1,10-phenanthroline	bipy	2,2′-dipyridyl
terpy	2,2′,2″-terpyridine	py	pyridine
sen	1,1,1-tris(2′-aminoethylaminomethyl)-ethane		
methylpenten	N,N,N′,N′-tetrakis-(2′-aminoethyl)-1,2-diaminopropane		
diars	$o\text{-}C_6H_4\{(AsCH_3)_2\}_2$	triars	$CH_3As\{CH_2CH_2CH_2As(CH_3)_2\}_2$
H_2ox	oxalic acid	H_2mal	malonic acid
H_2sal	salicylic acid	H_4tart	tartaric acid
Hacac	acetylacetone	HSacSac	thioacetylthioacetone
Hhmc	hydroxymethylene-camphor	Hdmg	dimethylglyoxime
Hhfac	hexafluoroacetylacetone *i.e.* 1,1,1,5,5,5-hexafluoropentane-2,4-dione		
Hα	α-amino acid	Hgly	glycine
Hala	alanine	Hval	valine
H_3nta	nitrilotriacetic acid	H_4edta	ethylenediaminetetraacetic acid
H_4pdta	1, 2-propylenediaminetetraacetic acid		
H_2edda	ethylenediamine-N, N′-diacetic acid		
H_2salen	N,N′-bis(salycylidene)ethylenediamine		
H_2BAE	Bis(acetylacetone)ethylenediamine		

24 Rhodium

R. D. Gillard
Inorganic Chemistry Laboratory, University College, Cardiff, Wales

Many compounds of rhodium have been made.[1,2] However, they perhaps do not manifest the wide range of behavior found with such an element as iridium; this probably arises from two primary factors.
First, the common starting materials (commercially available) for work on rhodium are either the element itself or its soluble chloride (an ill defined substance), or chloro-complex ions of rhodium(III). Since the element is usually first treated with chlorine to bring it into reaction, starting materials for synthesis nearly always contain one or more bonds from chloride to rhodium. These are very stable, kinetically and thermodynamically, and so the vast bulk of known rhodium compounds still contain the $Rh{-}Cl_n$ moiety.
The other factor which underlies the limited numbers of distinct types of compound of rhodium is the dominance of two oxidation states, rhodium(III), particularly common with σ-bonding (and "innocent") ligands, and rhodium(I), found usually, though not always, in complexes with π-bonded (and "non-innocent") ligands.
With the recent continuing efforts in mechanistic approaches to synthesis, the accessibility of compounds of the lower oxidation states has increased.

24.1 Rhodium(−I) and Rhodium(0)

These oxidation states are known only among carbonyl derivatives. The well known *chlorocarbonyl bis-triphenylphosphine rhodium(I)* treated with sodium amalgam affords[3] *sodium carbonylbis-triphenylphosphinerhodate(−I)*, $Na[Rh(CO)_2(PPh_3)_2]$.
Rhodium(0) species, apart from the metal, include the *tetrameric tricarbonyl*, $[Rh(CO)_3]_4$, prepared[4] at 80° from carbon monoxide (200 atm) with rhodium trichloride and copper powder as chloride acceptor. When this is heated with excess of carbon monoxide at 150°, the very inert *hexameric carbonyl* $Rh_6(CO)_{16}$ is formed.

24.2 Rhodium(I)

This oxidation state has recently assumed a high level of importance through its coordinative unsaturation (its complexes are commonly tetragonal), which leads to useful additions of hydrogen, giving species which are catalytic in reducing organic molecules like olefins.

Few binary compounds of rhodium(I) have been described, and all seem to be ill-characterized, except perhaps the monochloride, RhCl, which is reportedly very stable, made[5] by reacting metallic rhodium with chlorine in a furnace at temperatures above 965°.
Syntheses of rhodium(I) complexes are highly empirical, in the sense that unexpected products are often obtained. These complexes have been made in a variety of ways, which fall into three primary routes. First, rhodium(III) species in the presence of ligands are treated with reducing agents (usually 2-electron reductants, like hydride or sources of it, *e.g.* ethanol). Secondly, the ligand itself (in excess) may serve as the reducing agent, i.e. reaction of a rhodium(III) species with ligand gives rise to a rhodium(I) product. Thirdly, since the ligands on rhodium(I) appear to be fairly labile, substitutions may be performed on easily available species containing rhodium(I). This third approach has been particularly useful for displacement on chlorodicarbonyl rhodium(I) dimer, $[Rh(CO)_2Cl]_2$, and on tetrapyridine rhodium(I) cation, $[Rh(py)_4]^{\oplus}$.
When *trans*-$[Rh(py)_4Cl_2]Cl$ in ethanol is treated at −80° batchwise with one mol of sodium borohydride (in diglyme), and the mixture allowed to warm to room temperature, the filtered solution contains[6] the $[Rh(py)_4]^{\oplus}$ ion. This may be precipitated with perchlorate. Its solution reacts with ethanolic HCl and $LiClO_4$, giving the colorless *trans*-$[Rh(py)_4HCl]ClO_4$, or with HBr, similarly, giving *trans*-$[Rh(py)_4HBr]ClO_4$. Similarly, dichlorine and diiodine form *trans*-$[Rhpy_4X_2]X$ or *trans*-$[py_4X_2]X_3$ (X = Cl or I). The freshly made brown solution of $[Rh(py)_4]Cl$ reacts with isonitriles, *e.g.* *p*-$CH_3C_6H_4NC$, giving $[Rh(p\text{-}CH_3C_6H_4NC)_4]Cl$, with triphenylphosphine giving the metastable orange isomer of $[Rh(PPh_3)_3Cl]$, and with bis[diphenylphosphine]ethane (diphos) affording $[Rh(diphos)_2]Cl$ (m.p. 282°).[7]
Quite apart from this displacement, numerous routes exist to the remarkable *Wilkinson's compound*, $[Rh(PPh_3)_3Cl]$, of which experimen-

[1] *W. P. Griffith*, The Chemistry of the Rarer Platinum Metals, John Wiley & Sons, New York; *Gmelin's* Handbuch der Anorganischen Chemie, 8. Aufl., Bd. 64, «Rhodium»: Verlag Chemie Weinheim 1938, Nachdruck 1955.
[2] *P. Pascal*, Nouveau Traité de Chimie Minérale, Vol. 19, Masson & Cie., Paris 1958.
[3] *J. P. Collman, F. D. Vastine, W. R. Roper*, J. Amer. Chem. Soc., *88*, 5035 (1966).
[4] *W. Hieber, H. Lagally*, Z. Anorg. Allgem. Chemie *251*, 96 (1943).
[5] *L. Wöhler, W. Müller*, Z. Anorg. Allgem. Chemie *149*, 132 (1925).
[6] *H. Shaw*, Ph.D. Thesis, University of Kent, Canterbury 1971.
[7] *A. Sacco, R. Ugo*, J. Chem. Soc. (London) 3274 (1964).

tal quantities are available commercially. The best procedure for its preparation is probably the reaction[8] of ethanolic rhodium trichloride and triphenylphosphine (which probably serves as the reducing agent). Other easily made complexes of rhodium(I) with Group V donors include *trans*-[Rh(PPh$_3$)(CO)$_2$Cl] prepared[9] in high yield through the reaction of an ethanolic solution of soluble rhodium trichloride and triphenylphosphine with an aqueous solution of formaldehyde; the corresponding [Rh(PPh$_3$)(CO)$_2$X] where X=Br, I, SCN, by treatment of the chloride complex with acetone solutions of LiBr, NaI, or KSCN; and the arsine analog [Rh(AsPh$_3$)(CO)$_2$Cl].

Among complexes of Group IV donors, [Rh$_2$(CO)$_4$Cl$_2$] is conveniently obtained[10] by reacting soluble rhodium trichloride as a solid with a stream of carbon monoxide at 100°. The product sublimes. It reacts with sodium cyclopentadienide giving[11] *π-cyclopentadienyl dicarbonylrhodium(I)*, an orange liquid, m.p. −11°, and, with β-diketones in basic media, gives[12] *β-diketonato-di-carbonylrhodium(I)*. [Rh(CNR)$_4$]$^{\oplus}$ salts may be made[13] simply by refluxing isonitriles with soluble rhodium trichloride. For the pure compounds, such as the chloride salts, the color of solutions shows a remarkable dependence on solvent, and solid solvates of different colors are known. These same compounds are readily made from rhodium dicarbonyl chloride dimer using an excess of isonitrile; using less isonitrile gives the intermediates [Rh(CO)(CNR)$_2$X]$_2$. When ethylene is bubbled into methanolic rhodium trichloride, some acetaldehyde is formed, and the chloro-bridged dirhodium(I,I) species [Rh(C$_2$H$_4$)Cl]$_2$ is produced.[14]

24.3 Rhodium(II)

The rarity and inaccessibility of the rhodium(II) state may be operational, rather than thermodynamic. Nevertheless, several complexes are now known, and are not particularly difficult to make. The bulk of currently known compounds are dimeric, and, in general, may be made by the route:

$$\mathrm{Rh^{III}} \xrightarrow{e\ominus} \mathrm{Rh^{II}} \xrightarrow{\text{fast}} \tfrac{1}{2}[\mathrm{Rh_2}^{\mathrm{II}}]$$

The 1-electron reduction may be achieved chemically (*e.g.* chromous ion, or, rather surprisingly, hydridoborate) or electrochemically (electrolysis of a halorhodium(III) species at a constant potential conveniently selected by inspection of the polarogram of the rhodium(III) substrate).

Among the complexes of Group VI ligands, the diamagnetic carboxylates of formula [(RCOO)$_2$RhL]$_2$ are noteworthy. Here, the carboxylic acid presumably serves as both reductant and ligand, since glacial acetic acid reacts with ammonium hexachlororhodate(III), or with hydrated rhodium(III) oxide to give,[15] respectively [Rh(CH$_3$COO)$_2$(H$_2$O)$_2$] or [Rh(CH$_3$COO)$_2$]$_2$. The latter compound forms many adducts of the type [Rh(CH$_3$COO)$_2$L$_2$] where L=pyridine, ammonia, or other simple donors. Similar series of compounds (made in similar ways) are known for formate and propionate as the carboxylate component.

The dimeric aquo-species [Rh(H$_2$O)$_5$]$_2$$^{4\oplus}$ is made[16] by the reaction in water of chloropenta-aquorhodium(III) cation with chromous solutions. An easier preparation of the free air-stable [Rh$_2$]$^{4\oplus}$(aqu) is achieved[17] by treating dimeric rhodium(II) acetate in water with aqueous tetrafluoroboric acid at 50°. A related species is [(H$_2$O)en$_2$Rh]$_2$$^{4\oplus}$, the product[18] of controlled potential electrolysis (at −1·0 V, using platinum electrodes) of a 10^{-3} M aqueous solution of *trans*-[Rh(en)$_2$Cl$_2$]Cl. The electrolysis is stopped when 1 Faraday per rhodium(III) ion has been provided. A related synthesis leads[19] to the dimer of rhodium(II), [Rh(DMG)$_2$(PPh$_3$)]$_2$, where DMG is the uninegative anion

[8] *J. A. Osborn, F. H. Jardine, J. F. Young, G. Wilkinson*, J. Chem. Soc. *A*, 19 771 (1966).
[9] *D. Evans, J. A. Osborn, G. Wilkinson*, Inorg. Synth. *11*, 99 (1968).
[10] *J. A. McCleverty, G. Wilkinson*, Inorg. Synth. *8*, 211 (1966).
[11] *E. O. Fischer, H. P. Fritz*, Angew. Chem. *73*, 353 (1961).
[12] *F. Bonati, G. Wilkinson*, J. Chem. Soc. (London) 3156 (1964).
[13] *L. Malatesta, L. Vallarino*, J. Chem. Soc. (London) 1867 (1956).
[14] *R. Cramer*, Inorg. Chem. *1*, 722 (1962).
[15] *S. A. Johnson, H. R. Hunt, H. M. Neumann*, Inorg. Chem. *2*, 960 (1963).
[16] *F. Maspero, H. Taube*, J. Amer. Chem. Soc. *90*, 7361 (1968).
[17] *P. Legzdins, G. L. Rempel, G. Wilkinson*, J. Chem. Soc. *D*, 825 (1969).
[18] *R. D. Gillard, B. T. Heaton, D. H. Vaughan*, J. Chem. Soc. *A*, 734 (1971).
[19] *K. G. Caulton, F. A. Cotton*, J. Amer. Chem. Soc. *91*, 6517 (1969).

of dimethylglyoxime. Reduction (with zinc amalgam) of the tetradentate Schiff base complex [Rh(salen)(py)Cl] affords[20] the rhodium–rhodium(II,II) bonded species, $[Rh(salen)(py)_2]$.

24.4 Rhodium(III)

The primary methods which have been used to synthesize rhodium(III) complexes may be subdivided.

(1) Reaction of "simple" soluble rhodium compounds (*e.g.* $RhCl_3$, or the nitrate) with ligands, usually in water. Such reactions are in general slow and may often be speeded up by,
(2) adding catalysts such as ethanol.
(3) Dissolution of an insoluble rhodium(III) compound, usually the oxide, in a solution of a protonic ligand.
(4) Substitution (sometimes catalyzed by 2-electron reductants) into a preformed rhodium(III) complex.

24.4.1 Starting Materials

The trifluoride (insoluble in water) is formed[21] when fluorine reacts at 380° with the tri-iodide (which may itself be precipitated as a brown-black powder from aqueous solutions of rhodium trichloride by potassium iodide).

The water soluble "*Rhodium trichloride*" is, despite its dominance as a synthetic agent, an ill-characterized material. Its properties, and in particular its reactivity depend very markedly on its provenance. Most samples of commercial material dissolve (apparently auto-catalytically) in water and in ethanol. It may be made[22] from the metal by:

(1) Chlorination for about half an hour of a mixture of sodium chloride and rhodium at 900°; the arrangement of the furnace and reagent is critical for the success of this step, the formation of $Na_3[RhCl_6]$.
(2) Precipitation from the aqueous solution of the hexachlororhodate(III) by potassium hydroxide of the hydrated sesquioxide, usually represented as $Rh_2O_3 \cdot 5H_2O$.
(3) Dissolution of the oxide in the minimum amount of concentrated hydrochloric acid.
(4) The most critical step of all: the evaporation of the solution obtained in (3) to dryness (usually done on a steam bath). The variability of the reactivity of the product seems to depend on its exact history at this stage.

Other "simple" compounds of rhodium(III) useful as starting materials in synthesis, which may be made by dissolving the oxide in acid and subsequent evaporation are $[Rh(H_2O)_6](ClO_4)_3$, lemon-yellow needles[23] from perchloric acid, the *tribromide* (again ill-characterized and variable in properties, but useful for making bromo-complexes) and the red *nitrate*, formulated $Rh(NO_3)_3 \cdot 2H_2O$, from nitric acid.

24.4.2 Group VI Donor Complexes

The orange *tris-acetylacetonatorhodium(III)* is obtained[24] when an aqueous solution of rhodium nitrate and acetylacetone is refluxed. The yield is not always high, and occasionally some rhodium metal may be deposited. The product may be recrystallized from benzene. Several other complexes of oxygen ligands, notably chelating carboxylato-complexes such as the species containing the ligands oxalato-, malonato-[25] and ethylenediaminetetra-acetato-[26] may be made by dissolving the hydrated rhodium(III) sesquioxide in a heated aqueous solution of the appropriate acid. In the case of the malonato complex, there is a suggestion that ethanol serves as a catalyst. The best preparation[27] of the well known *potassium tri-oxalatorhodate(III)* hydrate is through this oxide method. Several other routes have been mentioned, all of which have in common the use as starting material of either soluble rhodium trichloride or a chloro-complex of rhodium(III) (e.g. $[RhCl_5(H_2O)]^{2\ominus}$). The products here are liable to be contaminated with species containing Rh–Cl bonds. Further, there is a double salt of composition $\{K_3[Rh(C_2O_4)_3]\}_2 KCl \cdot 8H_2O$ which crystallizes readily in the presence of potassium and chloride ions. This is difficult to distinguish from the authentic trisoxalato salt since both form large deep-orange crystals. It is better to avoid the presence of chloride by using the route through the precipitated oxide.

[20] *R. J. Cozens, K. S. Murray, B. O. West*, J. Chem. Soc. *D*, 1262 (1970).
[21] *M. A. Hepworth, K. H. Jack, R. D. Peacock, G. J. Westland*, Acta Cryst. *10*, 63 (1957).
[22] *S. N. Anderson, F. Basolo*, Inorg. Synth. *7*, 214 (1963).
[23] *G. H. Ayres, J. S. Forester*, J. Inorg. Nucl. Chem. *3*, 365 (1957).
[24] *F. P. Dwyer, A. M. Sargeson*, J. Amer. Chem. Soc. *75*, 984 (1953).
[25] *F. M. Jaeger, W. Thomas*, Rec. Trav. Chim. Pays-Bas *38*, 300 (1919).
[26] *F. P. Dwyer, F. L. Garvan*, J. Amer. Chem. Soc. *82*, 4832 (1960).
[27] *A. Werner, J. Poupardin*, Ber. *47*, 1955 (1914).

A similar dissolution of an insoluble simple compound of rhodium, RhI_3, in amino acids such as L-alanine leads[28] to isomers of the tris-amino acidato species, $[Rh(L\text{-}ala)_3]$ etc. The best characterized of these is the colorless highly water insoluble β-isomer, which may be recrystallized by treating it with concentrated sulfuric acid, when it slowly dissolves, and carefully reprecipitating it from this solution by dilution with water.

Complexes of ligands which bond to rhodium(III) through sulphur may often be made directly from the ligand and chlororhodium(III) species. For example, an excess of thiourea (tu) when heated with the hexachlororhodate(III) ion in water gives[29] the complexes $[Rh(tu)_6]Cl_3$, $[Rh(tu)_5Cl]Cl_2$ and $[Rh(tu)_3Cl_3]$ (which isomer is not known). Similarly, dithiocarbamate complexes are readily made[30] from salts of the ligand and rhodium trichloride in acetone; the reaction is of unknown mechanism but sulfur-containing organic by-products are formed. Hexathiocyanatorhodate(III) is produced[31] on prolonged heating of aqueous chlororhodate-(III) species and potassium thiocyanate. It may be crystallized as the potassium salt.

24.4.3 Group V Donor Complexes

Complexes of nitrogenous ligands represent by far the largest number of derivatives of rhodium-(III). Methods of preparation have been fairly standard; for example, the uncatalyzed reaction of hexachlororhodate with ammonia slowly yields $[Rh(NH_3)_5Cl]Cl_2$. A better preparation[32] is to react rhodium trichloride in water with ammonium chloride and ammonium carbonate. To force the complete substitution by ammonia, it is necessary to treat preformed chloropentamminerhodium(III) with liquid ammonia, or with hot aqueous ammonia in a sealed tube.[33] Substitution of chloride in $[Rh(NH_3)_5Cl]^{2\oplus}$ by other ligands is not catalyzed[34] by 2-electron reductants but the aquation to $[Rh(NH_3)_5(H_2O)]^{3\oplus}$, although slow, is convenient, and subsequent anation by $X^{\ominus}$ to yield $[Rh(NH_3)_5X]^{2\oplus}$ is readily achieved in aqueous media by using excess of NaX or KX.

Other routes to $[Rh(NH_3)_5Cl]Cl_2$ include the rapid reaction (catalyzed[35] by glycerol or ethanol) of rhodium trichloride with ammonia, and the displacement of other nitrogenous ligands from rhodium(III) by ammonia, as in

$$trans\text{-}[Rh(py)_4Cl_2]Cl + 5NH_3 \longrightarrow [Rh(NH_3)_5Cl]Cl_2 + 4py \quad \text{(Ref. 36)}$$

$$1,2,4\text{-}[Rh(CH_3CN)_3Cl_3] + 5NH_3 \longrightarrow [Rh(NH_3)_5Cl]Cl_2 + 3CH_3CN \quad \text{(Ref. 6)}$$

In the ammine series, the ready preparation,[37] and remarkable stability as a solid, of the hydrido-species $[Rh(NH_3)_5H]SO_4$, is noteworthy. It is made as a colorless solid by treating chlororhodium(III) species and ammonia, in the presence of sulfate ions, with zinc powder. It provides a useful means of entering the tetrammine series, since with strong hydrochloric acid, the product is *trans*-dichlorotetramminerhodium(III) chloride. This last compound may also be obtained as a by-product of the synthesis[32] of $[Rh(NH_3)_5Cl]Cl_2$ but the yield is often very small indeed.

Ammonia may occasionally be displaced from rhodium(III) ammines by chelating ligands. Such reactions go with difficulty (for example, $[Rh(NH_3)_5Cl]Cl_2$ may be recrystallized unchanged from boiling pyridine) but among available examples are

$$[Rh(NH_3)_5Cl]Cl_2 + HDMG \xrightarrow[\text{tube}]{\text{sealed}} [Rh(DMG)_2(NH_3)_2]Cl \quad \text{(Ref. 38)}$$

$$[Rh(NH_3)_5Cl]Cl_2 + tpn + NaOH \longrightarrow [Rh(tpn)_2]Cl_3 \quad \text{(Ref. 39)}$$

(tpn = 1,2,3-tri-aminopropane)

and

---

[28] *J. H. Dunlop, R. D. Gillard*, J. Chem. Soc. (London) 6531 (1965).

[29] *W. Lebendenskii, E. S. Schapiro, I. P. Kasaltina*, Izvest, Inst. Izuch. Platin. Drug. Blag. Met *12*, 101 (1935).

[30] *F. A. Cotton, J. A. McCleverty*, Inorg. Chem. *4*, 1398 (1964).

[31] *O. W. Howarth, R. E. Richards, L. M. Venanzi*, J. Chem. Soc. (London) 3335 (1964).

[32] *S. A. Johnson, F. Basolo*, Inorg. Chem. *1*, 925 (1962).

[33] *S. M. Jorgensen*, J. Prakt. Chem. *44*, 48 (1891).

[34] *A. J. Poë, K. Shaw, M. J. Wendt*, Inorg. Chim. Acta *1*, 371 (1967).

[35] *R. D. Gillard, J. A. Osborn, G. Wilkinson*, J. Chem. Soc. 1951 (1965).

[36] *R. D. Gillard, E. D. McKenzie, M. D. Ross*, J. Inorg. Nucl. Chem. *28*, 1429 (1966).

[37] *K. Thomas, J. A. Osborn, A. R. Powell, G. Wilkinson*, J. Chem. Soc. *A*, 1801 (1968).

[38] *L. Tchugaev, W. Lebedenskii*, Z. Anorg. Allgem. Chemie *83*, 4 (1913).

[39] *F. G. Mann, W. J. Pope*, J. Chem. Soc. (London) 2675 (1926).

[39a] *F. Krauss, H. Umbach*, Z. Anorg. Allgem. Chemie *179*, 358 (1929).

$$[Rh(NH_3)_5Cl]Cl_2 + KCN \xrightarrow{\text{melt}} K_3[Rh(CN)_6] \quad \text{(Ref. 39a)}$$

Complexes of pyridine with rhodium(III) are well known and, in general, easily made. When warm aqueous rhodium chloride reacts with pyridine, in the absence of catalysts, the first product is impure *1,2,4-trichlorotripyridinerhodium(III)*. (This is better made by other routes, described later.) On prolonged heating, one more chloride is replaced by pyridine, giving *trans*-$[Rh(py)_4Cl_2]Cl \cdot 5H_2O$. This recrystallizes well from water, is soluble in a variety of solvents and, since the pyridine ligands are readily replaced, is a useful synthetic intermediate. This is particularly so because it can be made in a few minutes, using ethanol (or other two-electron reductants) as a catalyst. In general,[6] rhodium (chloride or bromide) (1 mol) in hot 30 to 50% aqueous ethanol is treated with a pyridine (4·1 mol) and the mixture boiled briefly. An initial red-brown precipitate redissolves, giving a yellow solution, which on cooling gives the product, e.g. *trans*-$[Rh(py)_4Cl_2]Cl \cdot 5H_2O$. The diiodo complex, which is rather less stable to loss of pyridine, may be made[6] at room temperature by bubbling hydrogen through a suspension of rhodium triiodide (0.1 g) in water (20 ml)/pyridine (40 ml)/ethanol (20 ml) giving a brown solution. This, after filtering and concentration *in vacuo* to *c.* 20 ml affords golden brown crystals (0·07 g) of *trans*-$[Rh(py)_4I_2]I \cdot 5H_2O$.

Pyridine may be displaced[36] from *trans*-$[Rh(py)_4X_2]^{\oplus}$ (X=Cl or Br) by a variety of reagents. This provides an easy route (by reaction in the warm with 1,2-diaminoethane (en) *added dropwise*) to *trans*-$[Rh(en)_2Cl_2]^{\oplus}$ and *trans*-$[Rh(en)_2Br_2]^{\oplus}$. Although these syntheses are convenient, they require constant attention.

On adding oxalate ion to a hot aqueous solution of $[Rh(py)_4Cl_2]Cl$, a precipitate of the oxalate salt forms, but on continued heating (severe bumping may occur at this stage) pyridine is released and steam-distilled away, the solid product being *1-chloro-2,4-oxalato-3,4,5-tripyridinerhodium(III)*. This dissolves in hot concentrated hydrochloric acid, the solution on cooling giving[40] large orange crystals of *1,2,4-trichlorotripyridinerhodium(III)*. This is the preferred method for this compound, giving it in a pure form, uncontaminated by the 1,2,3-isomer.

Acetonitrile is another nitrogenous ligand, useful, like pyridine, in synthetic rhodium chemistry, in the sense that it rapidly forms complexes (reaction 1) and is then easily displaced (reaction 2) by the ligands which are to be attached to rhodium(III) in the product.

$$RhCl_3 + CH_3CN \longrightarrow 1,2,4\text{-}[Rh(CH_3CN)_3Cl_3] \quad (1)$$

$$1,2,4\text{-}[Rh(CH_3CN)_3Cl_3] + L \longrightarrow 1,2,4\text{-}[RhL_3Cl_3] + 3CH_3CN \quad (2)$$

In boiling pyridine, the 1,2,4-complex of acetonitrile gives[6] 1,2,4-$[Rh(py)_3Cl_3]$ and similarly, *trans*-$[Rh(CH_3CN)_4Cl_2]ClO_4$ (made by adding silver perchlorate to the 1,2,4-trichloro-compound in acetonitrile solvent) gives *trans*-$[Rh(py)_4Cl_2]ClO_4$. With 2 mol of 1,10-phenanthroline, the tetrakis compound (in acetonitrile as solvent) reacts[6] to form *cis*-$[Rh(phen)_2Cl_2]ClO_4$, which gives crystals on addition of diethyl ether. Other phenanthroline complexes have also been made[40a] in a similar way. It is not always necessary to isolate the nitrile complex; rhodium trichloride may be dissolved in acetonitrile and the ligand (L) is added directly to this solution. Protonic ligands such as ammonia and aliphatic ammines may not be used in this method since they form amidines with the coordinated nitrile.

Complexes of diamines with rhodium(III) may be obtained by a variety of methods, some of which have been mentioned. $[Rh(en)_3]Cl_3$ precipitated[35] from an ethanolic solution of $RhCl_3$ heated with anhydrous ethylenediamine. The precipitate is gummy and deliquescent, but on dissolution in water, filtering and concentrating, gives colorless crystals of $[Rh(en)_3]Cl_3$ hydrate. A mixture of *trans*- and *cis*-isomers of $[Rh(en)_2Cl_2]^{\oplus}$ is made[22] by slowly and carefully neutralizing (KOH) an aqueous solution of chlororhodium(III) species and ethylenediamine (added with 2HCl). The *trans*-isomer crystallizes first; the yield of the *cis*-isomer is variable and often very small. There is currently no very satisfactory means of making *cis*-bisethylenediaminerhodium(III) species.

A similar preparation, using 1,4,7,10-tetra-azadecane (trien) gives[41] α-*cis*(1,2,3,6) [Rh(trien)Cl_2]Cl as the less soluble product, and β-*cis*-(1,2,3,4)-[Rh(trien)Cl_2]Cl may be obtained from the filtrate. *trans*-Dichloro-1,4,8,11-tetra-aza-

---

39a *A. W. Addison*, Ph.D. Thesis, University of Kent, Canterbury 1970.

40a *E. D. McKenzie, R. A. Plowman*, J. Inorg. Nucl. Chem. *32*, 199 (1970).

41 *P. M. Gidney*, Ph.D. Thesis, University of Kent, Canterbury 1971.

undecane rhodium(III) salts were easily made[42] from the solution obtained by reacting *trans*-[Rh(py)$_4$Cl$_2$]Cl with the ligand (tet).

Substitutions of chloride by other ligands may be catalyzed by 2-electron reductants. As examples of this, reactions (3)[34] and (4)[35] are interesting.

$$trans\text{-}[Rh(en)_2Cl_2]^{\oplus} \xrightarrow[\text{trace NaBH}_4]{\text{xsKBr}} trans\text{-}[Rh(en)_2Br_2]^{\oplus} \quad (3)$$

$$trans\text{-}[Rh(DMG)_2Cl_2]^{\ominus} + py \xrightarrow{H_3PO_2} trans\text{-}[Rh(DMG)_2(py)Cl] \quad (4)$$

Eqn (3) represents the easiest way of obtaining the dibromo product and (4) the only currently known way of obtaining that particular product. The species *trans*-[Rh(DMG)$_2$Cl$_2$]$^{\ominus}$ is itself very easily made[35] as its ammonium salt by warming aqueous ethanolic solutions of rhodium trichloride and dimethyglyoxime (DMG), in the presence of ammonium chloride, or as its pyridinium salt by displacing[36] pyridine from *trans*-[Rh(py)$_4$Cl$_2$]Cl with dimethylglyoxime.

Complexes of chelating *heterocyclic ligands* such as 2,2′,2″-terpyridyl, 2,2′-bipyridyl or 1,10-phenanthroline are readily made. By fusion of the ligand (270°) with rhodium trichloride, and subsequent work-up of the cooled melt with aqueous ethanol (a possible catalyst?) the compounds [Rh(bipy)$_3$]Cl$_3$[43] and [Rh(terpy)$_2$]Cl$_3$[44] may be obtained. Both salts should be colorless; a slight pink tinge often present is due to an impurity, which may be removed by careful recrystallization (charcoal helps).

A similar fusion of solid reagents {[NH$_4$]$_3$[RhCl$_6$] with KCN} is used[45] to make K$_3$[Rh(CN)$_6$]; the reaction is not simple, several uncharacterized colored intermediates forming, but the yield is good. The fusion of [Rh(NH$_3$)$_5$Cl]Cl$_2$ with KCN is supposedly simpler. The compound [Ph$_4$As]$_3$[Rh(CNO)$_6$], a non-explosive fulminato-complex, has been made[46] recently. The most accessible compounds of the bidentate heterocyclic ligands are those of the type *cis*-[Rh(phen)$_2$Cl$_2$]X. These are made[41,35] very easily by adding the ligand (0·3 g) to soluble rhodium trichloride (0·25 g) in hot aqueous ethanol (50% v/v, 15 ml). The brick-red *cis*-diiodo derivative, *cis*-[Rh(phen)$_2$I$_2$]I·2H$_2$O is made[40] by refluxing the complex chloride (0·1 g) in 1:1 aqueous ethanol (40 ml) with sodium iodide (1·0 g) for 30 min. The *trans*-bischelate complexes are currently unknown.

Some novel compounds, of the dirhodium (III,III) type have recently been made.[40,47] These contain a dioxo bridge, either as peroxo—e.g. in [Cl(L)$_4$RhO$_2$RhL$_4$Cl](BF$_4$)$_2$ or as superoxo—e.g. in [ClR$_4$RhO$_2$RhL$_4$Cl](BF$_4$)$_3$—and typical preparations involve treating a solution of *trans*-[RhL$_4$Cl$_2$]Cl in alkaline aqueous ethanol with 10% ozone in oxygen for about 2 hr, giving a deep blue solution which is chlorinated briefly and precipitated by strong tetrafluoroboric acid or perchloric acid. This superoxo-complex is converted to the yellow peroxo-bridged species by ferrous ion.

A variety of complexes with other Group V donors is known; many of the complexes with arsine and phosphine ligands also contain hydrido-ligands; for example, when trichlorotri-(methyl-diphenylarsine)rhodium(III) is treated with hypophosphorous acid, the product[48] is [RhHCl$_2$(AsMePh$_2$)$_3$]. The reaction of rhodium trichloride with 1,2-dimethylarsinobenzene (diars) gives[49] [Rh(diars)$_2$Cl$_2$]Cl.

24.5 Rhodium(IV), (V) and (VI)

Unlike iridium(IV) the oxidation state (IV) in rhodium is rare. As might be expected, it is most common among compounds of the highly electronegative elements.

Rhodium tetrafluoride, a violet solid, is made[50] by treating rhodium tribromide with liquid bromine trifluoride. Complex fluorides are made by exhaustive fluorination of rhodium(III), the fluorine serving as both oxidant and ligand. For example, fluorine reacts readily with the chloro-complexes of rhodium(III) (*e.g.* K$_2$[RhCl$_5$])[51] to give K$_2$[RhF$_6$].

---

[42] *B. Bosnich, R. D. Gillard, E. D. McKenzie, G. A. Webb*, J. Chem. Soc. (London) 1331 (1966).
[43] *B. Martin, G. M. Waind*, J. Chem. Soc. (London) 4284 (1958).
[44] *C. M. Harris, E. D. McKenzie*, J. Inorg. Nucl. Chem. *25*, 171 (1963).
[45] *W. Manchot, H. Schmid*, Ber. *64*, 2872 (1931).
[46] *W. Beck, P. Swoboda, K. Feldl, E. Schuierer*, Chem. Ber. *103*, 3591 (1970).
[47] *A. W. Addison, R. D. Gillard*, J. Chem. Soc. 2523 (1970).
[48] *F. P. Dwyer, R. S. Nyholm*, J. Proc. Roy. Soc., N.S. Wales *75*, 122 (1941);
J. Lewis, R. S. Nyholm, G. K. N. Reddy, Chem. & Ind. (London) 1396 (1960).
[49] *R. S. Nyholm*, J. Chem. Soc. (London) 857 (1950).
[50] *A. G. Sharpe*, J. Chem. Soc. (London) 3444 (1950).
[51] *E. Weise, W. Klemm*, Z. Anorg. Allgem. Chemie *272*, 211 (1953).

When the rich-strawberry-red ice-cold solution of hexachlororhodate(III) in the presence of cesium ions (10% cesium chloride may be used) is treated with a cold stream of chlorine, an immediate precipitation occurs[52] of cesium hexachlororhodate(IV). This green salt is a powerful oxidant. Attempts to recrystallize it in general decrease its purity.

Predictably, the currently known compounds of rhodium(V) and (VI) are fluoro-species. Deep-red *rhodium pentafluoride*, thought to be tetrameric $[RhF_5]_4$, requires the action[53] of elemental fluorine on rhodium metal (when it is obtained along with large amounts of the trifluoride and black hexafluoride) or (at 6 atm, 400°) on the preformed trifluoride. A solution of the pentafluoride in iodine pentafluoride treated with cesium fluoride precipitates[54] the dark red salt $Cs[RhF_6]$.

---

[52] *F. P. Dwyer, R. S. Nyholm, L. E. Rogers*, J. Proc. Roy. Soc. N.S. Wales *81*, 267 (1947).

[53] *C. L. Chernick, H. H. Classen, B. Weinstock*, J. Amer. Chem. Soc. *83*, 3165 (1961).

[54] *J. H. Holloway, P. R. Rao, N. Bartlett*, J. Chem. Soc. *D*, 306 (1965).

25 Iridium

B. T. Heaton
University Chemical Laboratory, University of Kent,
Canterbury, Kent, England

25.1 Iridium(−I)

Both $[Ir(NO)(PPh_3)_3]$ and $[Ir(NO)_2(PPh_3)_2]^{\oplus}$ ($Ph=C_6H_5$) formally contain iridium(−I), although it is sometimes difficult to assign a meaningful oxidation state to the metal in such cases.[1] X-ray studies[2] support an iridium(−I) formulation for $[Ir(NO)(PPh_3)_3]$, which can be prepared from readily available start-materials by the reactions given in Eqn (1).[3]

$$\{IrCl(CO)[P(C_6H_5)_3]_2\} \xrightarrow[P(C_6H_5)_3]{N_2H_4} \{IrH(CO)[P(C_6H_5)_3]_3\} \xrightarrow{4\text{—}CH_3\text{—}C_6H_4\text{—}SO_2\text{—}N(CH_3)(NO)} \{Ir(CO)(NO)[P(C_6H_5)_3]_2\} \xrightarrow{P(C_6H_5)_3} \{Ir(NO)[P(C_6H_5)_3]_3\} \quad (1)$$

However, the structure[4] of $[Ir(NO)_2(PPh_3)_2]ClO_4$, which is prepared by treating an ethanolic solution of $[IrH_2(PPh_3)_2]ClO_4$ with nitric oxide,[5] although somewhat favoring an iridium(−I), d^{10} state, is more tentative, since reaction (2) is atypical of such systems.

$$\{Ir(NO)_2[P(C_6H_5)_3]_2\}^{\oplus} + Hal^{\ominus} \longrightarrow \{IrHal(NO)_2[P(C_6H_5)_3]_2\} \quad (2)$$

The synthesis of $K[Ir(PF_3)_4]$ has been described,[6] but it involves $[HIr(PF_3)_4]$ which has only been available from high-pressure synthesis. Bennett *et al.*[7] have reduced $[IrCl(cyclooctene)_2]_n$ (see Section 25.3.3) with potassium amalgam under phosphorus trifluoride at atmospheric pressure to produce a high yield of the colorless crystalline salt $K[Ir(PF_3)_4]$, which is proving useful for the preparation of a whole range of new organometallic compounds,[7] *e.g.*,

$$K[Ir(PF_3)_4] + (C_6H_5)_3SnCl \longrightarrow \{[(C_6H_5)_3Sn]Ir(PF_3)_4\} + KCl \quad (3)$$

[1] *D. M. P. Mingos, W. T. Robinson, J. A. Ibers*, Inorg. Chem. *10*, 1043 (1971).
[2] *V. G. Albano, P. L. Bellon, M. Sansoni* in *D. M. P. Mingos, J. A. Ibers*, Inorg. Chem. *10*, 1035 (1971).
[3] *C. A. Reed, W. R. Roper*, J. Chem. Soc. *A*, 3054 (1970).
[4] *D. M. P. Mingos, J. A. Ibers*, Inorg. Chem. *9*, 1105 (1970).
[5] *L. Malatesta, M. Angoletta, G. Caglio*, Angew. Chem. Intern. Ed. Engl. *2*, 739 (1963).
[6] *T. Kruck, W. Lang*, Angew. Chem. Intern. Ed. Engl. *4*, 870 (1965).
[7] *M. A. Bennett, D. J. Patmore*, J. Chem. Soc. (D), 1510 (1969).

25.2 Iridium(0)

The iridium(0) oxidation state is confined to carbonyls except for the colorless crystalline complex $\{Ir[P(OPh)_3]_4\}_2$, which is prepared by heating $[IrH(CO)(PPh_3)_3]$ with triphenylphosphite.[8]

25.3 Iridium(I)

Many square-planar iridium(I) species are known, although an increasing number of five- and some seven-coordinate complexes are being prepared. One general feature of their reactivity is the ease of conversion from d^8 to octahedral d^6 complexes.

25.3.1 Hydrides

$[HIr(PF_3)_4]$ is a colorless liquid, prepared by reduction of iridium trichloride with phosphorus trifluoride under pressure,[6] or by addition of acid to $K[Ir(PF_3)_4]$[9] (see Section 25.1). The related complex $\{HIr[P(OPh)_3]_4\}$, a colorless crystalline solid, is formed on heating $[IrH_3(PPh_3)_3]$ with triphenylphosphite[10] at 100° for 1 hr. Further heating of $\{HIr[P(OPh)_3]_4\}$ in decalin results in cleavage of the carbon–hydrogen bonds and the formation of $\{IrH[P(OPh)_2(OC_6H_4)]_2[P(OPh)_3]\}$.[11]

25.3.2 Group V Donors

Trans-$[IrCl(N_2)(PPh_3)_2]$ is of interest not only as a complex containing molecular nitrogen, but as a valuable intermediate in the synthesis of other iridium(I) or (III) complexes. It is prepared in about 75% yield by reaction[12] of *p*-nitrobenzoylazide with a solution of *trans*-$[IrCl(CO)(PPh_3)_2]$ in chloroform containing a little ethanol at 0°. The resulting bright yellow nitrogen complex [$\nu(N{\equiv}N)$ 2105 cm^{-1} in chloroform solution] is moderately stable in air, but takes up oxygen rapidly in solution. Unsuccessful attempts have been made to isolate pure compounds containing other phosphines and the only other derivatives that could be isolated were $[IrX(N_2)(PPh_3)_2]$ ($X=Br, N_3$).[13]

[8] *S. D. Robinson*, J. Chem. Soc. (D), 521 (1968).
[9] *T. Kruck, W. Lang, N. Derner*, Chem. Ber. *101*, 3816 (1968).
[10] *G. Domenico*, Inorg. Nucl. Chem. Lett. *5*, 767 (1969).
[11] *E. W. Ainscough, S. D. Robinson*, J. Chem. Soc. (D), 863 (1970).
[12] *J. P. Collman, N. W. Hoffman, J. W. Hosking*, Inorg. Synth. *12*, 8 (1970).
[13] *J. Chatt, D. P. Melville, R. L. Richards*, J. Chem. Soc. (A), 2841 (1969).

The reactions of $[IrCl(N_2)(PPh_3)_2]$ may be classified into two types:

(a) Reaction with ligands ($L=PPh_3$, CO), which result in the elimination of nitrogen and formation of a four-coordinate iridium(I) species, $[IrCl(PPh_3)_2L]$.[14]
(b) Oxidation–addition reactions with substituted olefins or acetylenes to give either five-[15,16] or six-[17] coordinate iridium(III) compounds.

An unusual nitrosyl complex, $[Ir_2O(NO)_2(PPh_3)_2]$, obtained from the reaction of sodium nitrite with $[IrCl(CO)(PPh_3)_2]$, has been shown to have a μ-oxido group with near linear terminal nitrosyls.[18]
$[IrCl(PPh_3)_3]$, an orange crystalline solid, is prepared[19] by refluxing a ligroin solution of the cycloocta-1,5-diene complex, $[IrCl(1,5\text{-}COD)]_2$,[20] with an excess of triphenylphosphine. Although only slightly dissociated in solution, it undergoes ligand displacement [Eqn (4)],

$$\{IrCl[P(C_6H_5)_3]_3\} \xrightarrow{L} \{IrCl(L)[P(C_6H_5)_3]_2\} \quad (4)$$

($L=CO$ or PF_3) and the usual oxidation–addition reactions with H_2 and HCl.[19] Reaction (5) occurs readily on heating and testifies to the high affinity of iridium(III) for hydrogen.[19]

$$\{IrCl[P(C_6H_5)_3]_3\} \longrightarrow \text{(ortho-metallated } C_6H_4 \text{ ring bridging } (C_6H_5)_2P \text{ and Ir; Ir bearing H, Cl, } P(C_6H_5)_3, P(C_6H_5)_3) \quad (5)$$

$[Ir(diphos)_2]^{\oplus}$ (diphos $=Ph_2PCH_2CH_2PPh_2$) is prepared by the addition of diphos to $[IrCl(CO)(PPh_3)_2]$ followed by heating to 100°.[21,22] The bright orange diamagnetic solid reacts irreversibly with oxygen to give the cream-colored $[O_2Ir(diphos)_2]^{\oplus}$ which has a very long (1·625 Å) O–O bond length[23] and the coordinated oxygen has been likened to an excited state of oxygen.[24] The oxygen complex reacts further with sulfur dioxide to give the sulfatoiridium(III) complex, $[Ir(SO_4)(diphos)_2]^{\oplus}$.[25] Other iridium(III) complexes are formed by reaction with H_2,[21,22] HX,[21,22] $(EtO)_3SiH$,[26] allene,[27] and C_3S_2.[28]

25.3.3 Group IV Donors

The halogen-bridged iridium(I)–olefin complexes $[IrCl(1,5\text{-}COD)]_2$ and $[IrCl(C_8H_{14})_2]_2$ (1,5-COD = cycloocta-1,5-diene; C_8H_{14} = cyclooctene) have been much used for the preparation of other derivatives. $[IrCl(1,5\text{-}COD)]_2$ is prepared by refluxing an aqueous ethanolic solution of 1,5-COD with chloroiridic acid,[20] whereas it is necessary to reflux cyclo-octene in isopropanol (instead of water–ethanol) with chloroiridic acid to obtain $[IrCl(C_8H_{14})_2]_n$,[29] which is probably dimeric ($n=2$).[30] Some of the important reactions of the cycloocta-1,5-diene complex are shown in Scheme 1 (Refs 31–34).
Novel iridium(I)–ethylene complexes are obtained on reacting ethylene with the cyclooctene complex $[IrCl/C_8H_{14})_2]_2$. Thus, $[IrCl(C_2H_4)_4]$, the first formed product, is relatively unstable and at 30° gives $[IrCl(C_2H_4)_2]_2$.[30] This chloro-bridged dimer is not isomorphous with $[RhCl(C_2H_4)_2]_2$, which has been shown to have a bent structure,[35,36] but nevertheless undergoes the usual bridge-cleavage reactions [Eqn (6)].[37]

$$[IrCl(C_2H_4)_2]_2 \xrightarrow{L} [IrCl(C_2H_4)L_2] \quad (6)$$
$L=py$ or $P(C_6H_5)_3$

[14] *J. P. Collman, J. W. Kang*, J. Amer. Chem. Soc. *88*, 3459 (1966).
[15] *J. P. Collman, J. W. Kang*, J. Amer. Chem. Soc. *89*, 844 (1967).
[16] *B. Clarke, M. Green, F. G. A. Stone*, J. Chem. Soc. (A), 951 (1970).
[17] *J. P. Collman, K. Mitsuru, F. D. Vastine, J. Y. Sun, J. W. Kang*, J. Amer. Chem. Soc. *90*, 5430 (1968).
[18] *P. Karty, A. Walker, M. Mathew, G. J. Palenik*, J. Chem. Soc. (D), 1374 (1969).
[19] *M. A. Bennett, D. L. Milner*, Chem. Commun. 581 (1967).
[20] *G. Winkhaus, H. Singer*, Z. Naturforsch. *20b*, 602 (1965);
G. Winkhaus, H. Singer, Chem. Ber. *99*, 3610 (1966).
[21] *L. Vaska, D. L. Catone*, J. Amer. Chem. Soc. *88*, 5324 (1966).
[22] *A. Sacco, M. Rossi, C. F. Noble*, Chem. Commun. 589 (1966).
[23] *J. A. McGinnety, N. C. Payne, J. A. Ibers*, J. Amer. Chem. Soc. *91*, 6301 (1969).
[24] *R. Mason*, Nature *217*, 543 (1968).
[25] *J. J. Levinson, S. D. Robinson*, Inorg. Nucl. Chem. Lett. *4*, 407 (1968).
[26] *J. F. Harrod, C. A. Smith*, J. Amer. Chem. Soc. *92*, 2699 (1970).
[27] *J. A. Osborn*, J. Chem. Soc. (D), 1231 (1968).
[28] *A. P. Grinsberg, W. E. Silverthorn*, J. Chem. Soc. (D), 823 (1969).
[29] *B. L. Shaw, E. Singleton*, J. Chem. Soc. (A), 1683 (1967).
[30] *A. Van der Ent, T. C. Van Soest*, J. Chem. Soc. (D), 225 (1970).
[31] *J. R. Shapley, J. A. Osborn*, J. Amer. Chem. Soc. *92*, 6976 (1970).
[32] *J. R. Shapley, R. R. Schrock, J. A. Osborn*, J. Amer. Chem. Soc. *91*, 2816 (1969).
[33] *B. L. Shaw*, J. Chem. Soc. (D), 464 (1968).
[34] *G. R. Crooks, B. F. G. Johnston*, J. Chem. Soc. (A), 1662 (1970).

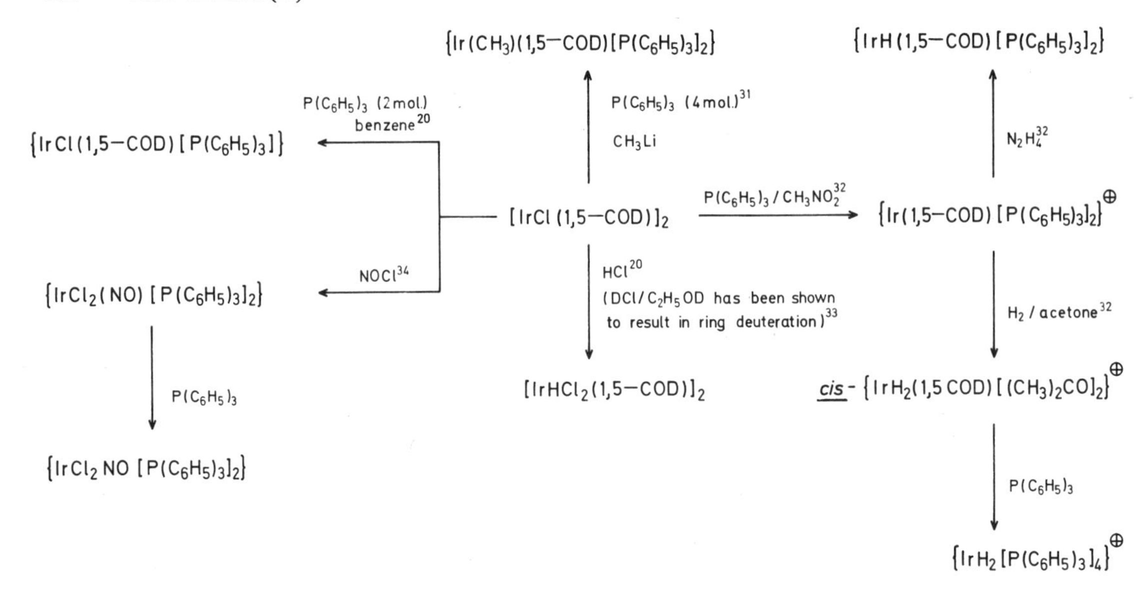

These monomeric complexes undergo rapid reactions with oxygen to give $[IrO_2Cl(C_2H_4)L_2]$ which can react further by losing ethylene to give $[IrO_2ClL_3]$.[37]

$[Ir(diene)_2]BPh_4$ has previously been described,[38] but Green *et al.* [39] have reported a more versatile synthesis which involves treatment of [Ir(diene)-acac][40] (acac = acetylacetonate) with $Ph_3C^{\oplus}BF_4^{\ominus}$ in the presence of excess diene (diene = 1,5-COD or norbornadiene). Dissociation of the purple–red solid in acetonitrile affords the yellow, crystalline, air-stable complex $[Ir(1,5\text{-}COD)(CH_3CN)_2]$-$BF_4$ which reacts with other ligands to give $[Ir(1,5\text{-}COD)L_2]BF_4$ ($L = PPh_3$, PPh_2Me, PPh_2-OMe; L_2 = diphos, bipy).[39] Reactions involving the coordinated diene have enabled a range of substituted diene–iridium(I) complexes to be prepared.[41]

The novel complex $[Ir(diene)BPh_4]$ has been prepared by addition of the diene (1,5-COD or 1,5-hexadiene) to an aqueous solution of iridium trichloride and sodium tetraphenylborate. It is a colorless crystalline solid and physical measurements suggest the structure[42]

$$\ominus B(C_6H_5)_3 \quad Ir^{\oplus} \quad L \quad L$$

L = 1,5-COD or 1,5-hexadiene

25.4 Iridium(II)

Iridium(II) is a very unusual oxidation state for iridium and the complexes $[Ir(diamine)Cl_2]$ (diamine = *N,N,N′,N′*= tetramethylethylenediamine or *N,N′*-dimethylpiperazine[43] which apparently contained iridium(II) have been shown to be complexes of platinum(II).[44] The ESR spectrum of $[Ir(cp)_2]$ (cp = cyclopentadiene) has been measured[45] at liquid nitrogen temperatures and confirms the presence of paramagnetic monomers; at room temperature diamagnetic dimers are formed.[46]

25.5 Iridium(III)

The most common oxidation state for iridium is iridium(III) and because of its inertness toward substitution reactions, forcing or catalytic conditions have often to be employed in synthetic work.

---

35 *D. M. Adams, P. J. Chandler*, J. Chem. Soc. (A), 588 (1969).

36 *K. Klandermann, L. F. Dahl* in *W. E. Oberhansli, L. F. Dahl*, J. Organometall. Chem. *3*, 43 (1965).

37 *H. Van Gaal, H. G. A. M. Cuppers, A. Van der Ent*, J. Chem. Soc. (D), 1964 (1970).

38 *J. F. Young, R. D. Gillard, G. Wilkinson*, J. Chem. Soc. (London) 5176 (1964).

39 *M. Green, T. A. Kuc, S. H. Taylor*, J. Chem. Soc. (D), 1553 (1970).

40 *B. F. G. Johnson, J. Lewis, D. A. White*, J. Amer. Chem. Soc. *91*, 5187 (1969).

41 *J. Lewis, A. W. Parkins*, J. Chem. Soc. (A), 953 (1969).

42 *R. R. Schrock, J. A. Osborn*, Inorg. Chem. *9*, 2339 (1970).

43 *F. G. Mann, H. R. Watson*, J. Chem. Soc. (London) 2772 (1958).

44 *D. W. Allen, F. G. Mann*, J. Chem. Soc. (A), 999 (1970).

45 *E. O. Fischer, H. Wawersik*, J. Organometall. Chem. *5*, 559 (1966).

46 *H. J. Keller, H. Wawersik*, J. Organometall. Chem. *8*, 185 (1967).

25.5.1 Hydrides

There are more hydrides containing iridium(III) than are to be found for any other element.

Monohydrides. The monohydrides are nearly all six-coordinate and of the types $[IrHX_2L_3]$ or *trans*-$[IrHX(MM)_2]^{\oplus}$ (X = halide; L = tertiary phosphine, arsine, or stibine; MM = tertiary phosphine or chelating phosphine), although the five-coordinate complex $[IrHCl_2(PPh_3)_2]$, has been prepared by the addition of hydrogen chloride (1 mole) to *trans*-$[IrCl(N_2)(PPh_3)_2]$.[47] $[IrHCl_2(PPh_3)_2]$ readily adds a further ligand, L′, to give $[IrHCl_2(PPh_3)_2L']$ (where L′ is *trans* to hydride and is MeOH, py, RCN, PhNC, or $AsPh_3$).[47] $[IrHX_2L_3]$ can exist in two isomeric forms (**1**) and (**2**). The yellow isomer (**1**) is formed by the reaction of *mer*-$[IrX_3L_3]$[48] with potassium hydroxide (1 mole) in ethanol.[49] Conversion to the colorless isomer (**2**) has been

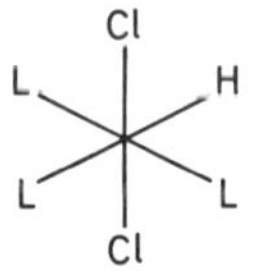

1 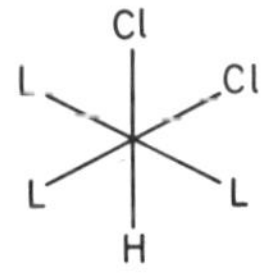 **2**

found to be accelerated by light.[50] Stereochemical assignments are based on dipole moments,[49] NMR,[51] and chemical studies.[52]

The interesting carboethoxyiridium(III) compound, $[IrH(COOEt)(Me_2PCH_2CH_2PMe_2)_2]^{\oplus}$ is formed when $[Ir(CO)(Me_2PCH_2CH_2PMe_2)_2]^{\oplus}$ is dissolved in ethanol.[53]

Complexes of the type $[IrHCl(UP)]BPh_4$[54] [UP = the quadridentate ligand, tris(*o*-diphenylphosphinephenyl)phosphine], and the unusual seven-coordinate species, $[IrHBr(PPh_3)(UP)]Br$[55] have been described.

Iridium(III) hydrides can be stabilized by ligands other than phosphines, arsines, and stibines and of note are the hydridotetrakis(piperidine)iridium(III) complex[56] and the complex $[IrHCl_2(DMSO)_3]$, which is formed on boiling $H[IrCl_4(DMSO)_2]$ in isopropanol.[57] $[IrHCl_2(DMSO)_3]$ has structure 1 (L = DMSO) and, in keeping with other X-ray structural determinations, the hydride is found to produce a significant *trans*-bond lengthening.[58]

The *cyanide hydride* $[IrH(CN)_5]^{3\ominus}$, believed to be present in a solution of $[Ir(CN)_6]^{3\ominus}$ containing excess hydridoborate ion,[59] has been isolated from the reaction of potassium cyanide in anhydrous methanol with $[IrCl(1,3\text{-cyclohexadiene})_2]_2$.[60] It is a colorless crystalline complex and the hydride resonance appears at $14{\cdot}37\tau$.[61]

There are a few monohydrides containing olefin ligands, e.g., $[IrH(PPh_3)_2(1,5\text{-COD})]$[62] and $[IrHCl_2(1,5\text{-COD})]$.[20, 63]

Dihydrides. Dihydrides of the type $[IrH_2XL_3]$ have configuration (**3**)

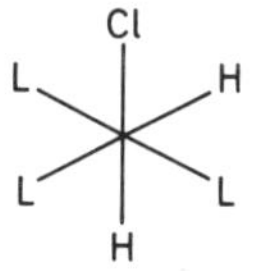

3

and can be prepared by further reduction of $[IrHX_2L_3]$.[49] $[IrH_2(diphos)_2]^{\oplus}$, prepared by addition of hydrogen to $[Ir(diphos)_2]^{\oplus}$, probably has a *cis* configuration.[21] Other *cis*-dihydrides of the type $[IrH_2(PPh_3)_2(MM)]$ (MM = β-diketonate[64] or *P,P*-dialkyldithiophosphinate[65]) have been described.

Trihydrides. Both *fac-* and *mer*-$[IrH_3L_3]$ have

---

[47] *D. M. Blake, M. Kubota,* J. Amer. Chem. Soc. *92,* 2587 (1970).

[48] *J. Chatt, A. E. Field, B. L. Shaw,* J. Chem. Soc. (London) 3371 (1963).

[49] *J. Chatt, R. S. Coffey, B. L. Shaw,* J. Chem. Soc. (London) 7391 (1965).

[50] *P. R. Brookes, B. L. Shaw,* J. Chem. Soc. (D), 918 (1968).

[51] *J. M. Jenkins, B. L. Shaw,* J. Chem. Soc. (A), 1407 (1966).

[52] *J. Powell, B. L. Shaw,* J. Chem. Soc. (A), 617 (1968).

[53] *S. D. Ibekwe, K. A. Taylor,* J. Chem. Soc. (A), 1 (1970).

[54] *J. W. Dawson, L. M. Venanzi,* J. Amer. Chem. Soc. *90,* 7229 (1968).

[55] *J. W. Dawson, D. G. E. Kerfoot, C. Preti, L. M. Venanzi,* J. Chem. Soc. (D), 1687 (1968).

[56] *E. R. Birnbaum,* J. Inorg. Nucl. Chem. *32,* 1046 (1970); *31,* 2626 (1969).

[57] *J. Trocha-Grimshaw, H. B. Henbest,* Chem. Commun. 544 (1967).

[58] *M. McPartlin, R. Mason,* J. Chem. Soc. (A), 2206 (1970).

[59] *M. L. H. Green,* Angew. Chem. *72,* 719 (1960).

[60] *K. Krogmann, W. Binder,* Angew. Chem. Intern. Ed. Engl. *6,* 881 (1961).

[61] *G. M. Whitesides, G. Maglio,* J. Amer. Chem. Soc. *91,* 4980 (1969).

[62] *H. Yamazaki, M. Takesada, N. Hagihara,* Bull. Chem. Soc. Japan *42,* 275 (1969).

[63] *S. D. Robinson, B. L. Shaw,* J. Chem. Soc. 4997 (1965).

[64] *A. Araneo,* J. Inorg. Nucl. Chem. *32,* 2925 (1970).

[65] *A. Araneo, F. Bonati, G. Minghetti,* Inorg. Chim. Acta *4,* 61 (1970).

been described[49] and their structure confirmed by physical measurements.[49,51] However, the action of lithium aluminum hydride on *mer*-$[IrCl_3(PEt_2Ph)_3]$ has been shown to give $[IrH_5(PEt_2Ph)_2]$[66] and not $[IrH_3(PEt_2Ph)_2]$ as previously formulated.[49] The addition of diethylphenylphosphine to $[IrH_5(PEt_2Ph)_2]$ gives >95% *mer*-$[IrH_3(PEt_2Ph)_3]$.[67] The phosphine *trans* to the hydride is labile and can be readily substituted by other uncharged ligands.[67] Araneo *et al.*[68] have also utilized this reaction by preparing a range of isonitrile complexes of the type $[IrH_3(AsPh_3)_2L]$ (L=isonitrile).

25.5.2 Group VII Donors

The *hexachloroiridate(III) ion*, $[IrCl_6]^{3\ominus}$ (sometimes referred to as hexachloroiridite), is readily prepared by reduction of hexachloroiridate(IV), $[IrCl_6]^{2\ominus}$, with either oxalate[69] or primary alcohols.

25.5.3 Group VI Donors

An improved preparation of $[Irox_3]^{3\ominus}$ (ox=oxalate) has been reported[70] and it has been suggested that the apparent water of crystallization in $K_3[Irox_3]\cdot 4\frac{1}{2}H_2O$ is possibly constitutive.[71]

The reaction of $[IrHCl_2(DMSO)_3]$ with diphenylacetylene[72] and benzylideneacetophenone[57] has been found to result in hydrogen addition to the carbon–carbon multiple bond and formation of iridium–carbon σ-bonded species, $[Ir(PhC{:}CHPh)Cl_2(DMSO)_3]$ and $[Ir(PhCH\cdot CH_2\cdot C{:}OPh)Cl_2(DMSO)_2]$, respectively.

25.5.4 Group V Donors

Numerous five- and six-coordinate nitrosyl complexes have been prepared by oxidation–addition reactions of $[Ir(NO)(PPh_3)_3]$[3]. The X-ray structures of $[IrCl_2(NO)(PPh_3)_2]$[73] and $[IrMeI(NO)(PPh_3)_2]$[1] have been determined and, besides showing $Ir\widehat{N}O$ to be 120° and 123°, respectively, have iridium–nitrogen bond lengths significantly longer than is found for species containing $Ir(I)/NO^{\oplus}$ and are therefore formulated as $Ir(III)/NO^{\ominus}$ compounds.

The previously described red form of $[IrCl_2phen_2]^{\oplus}$[74] has been shown to be phenH$[IrCl_4phen]$.[75] Physical measurements have shown that $[IrX_2(AA)_2]^{\oplus}$ (AA=bipy or phen) has the *cis*-configuration[76–78] and it seems probable that the previously described $[Irbipy_3]^{3\oplus}$[79] and $[Irphen_3]^{3\oplus}$[74] are *cis*-$[IrCl_2bipy_2]^{\oplus}$[76,77] and *cis*-$[IrCl_2phen_2]^{\oplus}$,[76] respectively. Both $[IrX(OH_2)phen_2]^{2\oplus}$[80] and $[Ir(OH_2)_2phen_2]^{3\oplus}$[81] are readily formed on irradiation of aqueous solutions of *cis*-$[IrX_2phen_2]^{\oplus}$.

A facile preparation of *trans*-$[IrCl_2py_4]^{\oplus}$ starting from $[IrCl_6]^{2\ominus}$[82] and a convenient synthesis of *cis*- and *trans*-$[IrX_2en_2]^{\oplus}$ using hypophosphite as catalyst[83,84] have been described. A photosynthetic route has been developed for the preparation of mixed complexes of the type *trans*-$[IrXYen_2]^{\oplus}$ (X=I, Br, ONO; Y=Cl or Br)[84] and their rates of substitution have been measured.[85]

mer-$[IrCl_3L_3]$ (L=tertiary phosphine or arsine), which is readily prepared,[48] is converted to the less readily accessible isomer by irradiation.[50]

25.5.5 Group IV Donors

The action of excess methylmagnesium chloride on *mer*-$[IrCl_3L_3]$ gives *mer*-$[IrMe_3L_3]$,[86,87] which can be converted to *fac*-$[IrMe_3L_3]$[87] and

---

[66] *B. E. Mann, C. Masters, B. L. Shaw*, J. Chem. Soc. (D), 703 (1970).

[67] *B. E. Mann, C. Masters, B. L. Shaw*, J. Chem. Soc. (D), 846 (1970).

[68] *A. Araneo, F. Bonati, G. Minghetti*, J. Organometall. Chem. *25*, 25c (1970).

[69] *G. B. Kauffman*, Inorg. Synth. *8*, 228 (1966).

[70] *D. Rose, C. J. N. Tyrell*, Englehard Ind. Tech. Bull. *8*, 99 (1967); C. A. 110885d (1967).

[71] *R. D. Gillard, S. H. Laurie, P. R. Mitchell*, J. Chem. Soc. (A), 3006 (1969).

[72] *J. Trocha-Grimshaw, H. B. Henbest*, J. Chem. Soc. (D), 757 (1968).

[73] *D. M. P. Mingos, J. A. Ibers*, Inorg. Chem. *10*, 1035 (1971).

[74] *B. Chiswell, S. E. Livingstone*, J. Inorg. Nucl. Chem. *26*, 47 (1964).

[75] *J. A. Broomhead, W. Grumley*, J. Inorg. Nucl. Chem. *29*, 2126 (1967).

[76] *R. D. Gillard, B. T. Heaton*, J. Chem. Soc. (A), 451 (1969).

[77] *R. E. Desimone, R. S. Drago*, Inorg. Chem. *8*, 2517 (1969).

[78] *P. Anderson, J. Josephson, G. Nord, E. C. Schaeffer, R. L. Tranter*, J. Chem. Soc. (D), 408 (1969).

[79] *B. Martin, G. M. Waind*, J. Chem. Soc. 4284 (1958).

[80] *J. A. Broomhead, W. A. Grumley*, J. Chem. Soc. (D), 1211 (1968).

[81] *L. H. Berka, G. E. Phillippon*, J. Inorg. Nucl. Chem. *32*, 3355 (1970).

[82] *R. D. Gillard, B. T. Heaton*, J. Chem. Soc. (D), 75 (1968).

[83] *R. A. Bauer, F. Basolo*, J. Chem. Soc. (D), 458 (1968).

[84] *R. A. Bauer, F. Basolo*, J. Amer Chem. Soc. *90*, 2437 (1968).

[85] *R. A. Bauer, F. Basolo*, Inorg. Chem. *8*, 2237 (1969).

[86] *J. Chatt, B. L. Shaw*, J. Chem. Soc. (A), 1836 (1966).

[87] *B. L. Shaw, A. C. Smithies*, J. Chem. Soc. (A), 1047 (1967).

has been used for the preparation of $[IrMe_2XL_3]$.[87]

Both σ- and π-allyl complexes can be obtained by the reaction of the allyl Grignard reagent with *mer*-$[IrCl_3L_3]$.[88] *Tris(allyl)iridium*, $[Ir(C_3H_5)_3]$,[89] has been obtained by the reaction of allylmagnesium chloride with $[Ir(acac)_3]$.[90]

The reaction of Dewar hexamethylbenzene or much better, a substituted ethylpentamethylcyclopentadiene (Me_5C_5CHXMe, X=OMe or Cl), with iridium trichloride produces the *pentamethylcyclopentadienyliridium chloride* dimer, $[IrCl_2(C_5Me_5)]_2$.[91] The usual bridge-cleavage reactions with phosphines[91] and the reactions with olefins have been described.[92,93] An interesting reaction is the oligomerization of the acetylene, (ROOCC≡CCOOR), by $[(C_5Me_5)Ir(OAc)_2]$.[94]

25.6 Iridium(IV)

25.6.1 Group VII Donors

Ammonium hexachloroiridate, $(NH_4)_2[IrCl_6]$, is the most readily prepared iridium salt. It can be precipitated pure, is not hygroscopic, and is prepared by passing chlorine over a mixture of iridium and sodium chloride at 625° followed by boiling with *aqua regia* and adding ammonium chloride.[95]

25.6.2 Group VI and V Donors

The best way of preparing complexes of the type $[IrCl_4L_2]$ is by oxidizing (with chlorine or nitric acid) the corresponding iridium(III) complex, $[IrCl_4L_2]^{\ominus}$. Thus, *trans*-$[IrCl_4L_2]$ (L= tertiary phosphine or arsine) has been prepared in high yield.[96]

25.7 Iridium(V)

The reaction of lithium aluminum hydride with *mer*-$[IrCl_3L_3]$ gives $[IrH_5L_2]$ ($L=PEt_2Ph$)[66] and it seems probable that the related complexes containing other phosphines,[49,97,98] which have been formulated $[IrH_3L_2]$, should be reformulated as pentahydrides, $[IrH_5L_2]$.

25.8 Literature

Literature coverage of iridium chemistry prior to the end of 1966 has been summarized by W. P. Griffith in "The Chemistry of the Rarer Platinum Metals", Chapter 5 (Wiley, New York, 1967). Therefore, this article deals mainly with the literature from the end of 1966 to the end of 1970 with a few later references. The reader is also referred to the earlier literature in "Handbuch der Anorganischen Chemie", 8 Auflage, Vol. 67 (Verlag Chemie, Berlin, 1939) and "Noveau Traité de Chimie Minerale", Vol. 18 (P. Pascal, ed. Masson, Paris, 1958).

[88] *J. Powell, B. L. Shaw*, J. Chem. Soc. (A), 780 (1968).
[89] *P. Chini, S. Martinengo*, Inorg. Chem. *6*, 837 (1967).
[90] *F. P. Dwyer, A. M. Sargeson*, J. Amer. Chem. Soc. *75*, 984 (1953).
[91] *J. W. Kang, K. Mosley, P. M. Maitlis*, J. Chem. Soc. (D), 1304 (1968).
[92] *B. L. Booth, R. N. Haszeldine, M. P. Hill*, J. Organometall. Chem. *16*, 491 (1969).
[93] *K. Mosley, P. M. Maitlis*, J. Chem. Soc. (A), 2884 (1970).
[94] *J. W. Kang, R. F. Childs, P. M. Maitlis*, J. Amer. Chem. Soc. *92*, 720 (1970).
[95] *G. B. Kauffman, L. A. Teter*, Inorg. Synth. *8*, 223 (1966).
[96] *J. Chatt, G. J. Leigh, D. M. P. Mingos, R. J. Paske*, J. Chem. Soc. (A), 2636 (1968).
[97] *L. Malatesta, G. Caglio, M. Angoletta*, J. Chem. Soc. (London) 6974 (1965).
[98] *M. Guitiniani, G. Dolcetti, M. Nicolini, U. Belluco*, J. Chem. Soc. (A), 1961 (1969).

26 Nickel

Kalojan R. Manolov
Department of Inorganic Chemistry,
Institute of Food Industry,
Plovdiv, Bulgaria

26.1 Purification of the Metal and Preparation of Powdered Nickel

The purity of commercial electrolytic nickel is about 99·95%. Further purification is accomplished electrolytically[1,2] and high-purity nickel has also been prepared by electrolysis[3] or by zone melting.[4,5]
Powdered nickel metal is an excellent catalyst for hydrogenation reactions. Two basic methods are used for preparing powdered nickel and two principal types of the powdered metal are known: Raney nickel and reduced nickel.

Raney nickel is prepared by dissolving nickel–aluminum alloy in ca. 30% sodium hydroxide at high temperature (not below 100°).[6,7] The fine powder contains various amounts of hydrogen and aluminum. It is pyrophoric and must be stored under water or alcohol but is stabilized by treatment with carbon dioxide.[8]
Reduced nickel is also pyrophoric. It is prepared by the reduction of nickel(II) oxide or hydroxide with hydrogen at 300–350°. The black powder can be used as a catalyst for reactions in liquid phase. Considerable work is being devoted to the preparation of nickel catalysts for gas phase reactions. Various oxides such as aluminum, thorium, and heavy metal oxides are used as supporting additives. The hydroxides of nickel and the supporting metals are either precipitated simultaneously, or the supporting oxide is impregnated with a solution of nickel salt.[9] The mixture is dried and then is reduced with hydrogen at appropriate temperatures.
If nickel compounds are reduced with hydrogen a fine nickel powder is obtained which can be used as catalyst. The catalytic power of the product depends on the conditions of the reduction (temperature and hydrogen pressure). Low temperatures are preferable. Reduction of nickel carbonate above 100° has been accomplished in a solution of ammonium carbonate and at 7 atm hydrogen pressure.[10] The reducing gas can be obtained by thermal decomposition of ammonia at 350–500°. The resultant hydrogen–nitrogen mixture reduces nickel oxide to produce a *powdered nickel* of good quality.[11]
The reduction of organic salts has been carried out at low temperatures and high pressure, *e.g.*, the oxalate was reduced at 282°,[12] the bis(cyclopentadienyl) derivative at 110–125° and high pressure in a solution of ethanol and cyclohexane.[13]
Other reducing agents have also been employed. Thus nickel(II) sulfate was reduced with sodium hypophosphite.[14] The mechanism of the reduction is presented by Eqns (1) and (2).[15]

$$H_2PO_2^{\ominus} + H_2O \longrightarrow HPO_3^{2\ominus} + 2H^{\oplus} + H^{\ominus} \quad (1)$$

$$Ni^{2\oplus} + 2H^{\ominus} \longrightarrow Ni + H_2 \quad (2)$$

Lithium hydride reduces nickel chloride in a melt of potassium chloride and lithium chloride at 360–700°.[16] The method of electrodeposition seems to be more convenient. The powder which is obtained by this latter procedure has a catalytic activity equal to that of Raney nickel.[17] Depending on the conditions of the electrolysis the metallic nickel deposits as a sponge[18] or as a fine crystalline powder.[11]

The reduction of fused salts systems yields a metal powder containing uniform polyhedra.[19] Particles with

---

[1] *H. N. Huntzicker, L. Kahlenberg*, Trans. Elektrochem. Soc. 349 (1933).
[2] *F. Ensslin*, Z. Erzbergbau Metallhütenw. *15* (8), 419 (1962).
[3] *J. G. Donaldson, K. K. Kershner, W. L. Balke*, U.S. Bur. Mines, Rept. Invest. No. 5993 (1962); C. A. *57*, 3188g (1962).
[4] *L. Kuchar, N. Deschamps, A. M. Wache, B. Dubois, O. Dimitrov*, Mem. Sci. Rev. Met. *64* (9), 785 (1967).
[5] *R. R. Soden, V. J. Albano*, J. Elektrochem. Soc. *115* (7), 766 (1968).
[6] *R. Cornubert, J. Phelisse*, Bull. Soc. Chim., France *19* (5), 399 (1952).
[7] *X. A. Dominguez, T. C. Lopez, R. Fanco*, J. Org. Chem. *26*, 1626 (1961).
[8] *T. Yamanaka, K. Taga, Y. Takagi*, Rikaraku Kenkyusho Hokoku *57*, 276 (1961).
[9] *L. Kh. Freidlin, N. V. Borunova, L. I. Gvinter, D. I. Lainer, N. M. Kagan*, Neftekhimiya *4* (4), 547 (1964).
[10] Fr. P. 1 436 201 (1966), Sheritt Gordon Mines, Erf.: *D. J. I. Evans, W. Kunda, H. A. Hancock, V. N. Mackiw;* C. A. *66*, 13627*x* (1967).
[11] *E. M. Minaev, O. K. Kolerov*, Tr. Kuibyshev. Aviats. Inst. *26*, 3 (1966).
[12] *A. A. Alchudzhan*, Izv. Akad. Nauk. Arm. SSR, Khim. Nauki *14* (2), 89 (1961).
[13] U.S. P. 2 999 075 (1955), Union Carbide Corp., Erf.: *R. L. Pruet;* C. A. *56*, 2930 i (1962).
[14] *P. Breteau*, Bull. Soc. Chim., France *9* (4), 518 (1911).
[15] *R. M. Lukes*, Plating *51*, 969 (1964).
[16] *A. Koutaissoff*, Promotionsarbeiten No. 3572,60 from J. Appl. Chem. (London) *16* (3), 204 (1966).
[17] *B. B. Bagalkote, J. G. Kane*, Indian J. Appl. Chem. *24*, 29 (1961).
[18] *V. Srb, R. Mraz*, Sbornik Vysoke Skoly Chem.-Technol. v Praze, Oddil Fak. Anorg. a Org. Technol. 505 (1958).
[19] *R. A. Lefever*, J. Elektrochem. Soc. *108*, 107 (1961).

globular surface were obtained by thermal decomposition of tetracarbonyl nickel in a stream of nascent nitrogen (from ammonia and oxygen).[20]

The *recovery of nickel catalysts* is described in numerous publications. The methods vary in minor details and are based in general on the electrolysis or the reduction with hydrogen. For example, the exhausted catalyst was cast into anodes and electrolyzed[21] in a solution of sodium chloride. Nickel hydroxide precipitates on the cathode and is subsequently reduced with hydrogen.
Nickel based catalysts can also be regenerated by heating in an inert atmosphere in order to remove the absorbed materials. Oxidation in air and at high temperature is followed by reduction with hydrogen.[22]
A *colloidal nickel* of very high catalytic activity has been prepared by the reduction of nickel salts with organometallic compounds. A convenient method involves the reduction of nickel dimethylglyoxime with diethylmagnesium in benzene under nitrogen atmosphere[23] or the reduction of nickel chloride with lithium or sodium naphthalene in tetrahydrofurane.[24]

26.2 Complexes of Nickel(0)

Zerovalent nickel can coordinate with donor molecules thus forming tetracoordinated nickel(0) complexes. Powdered nickel reacts directly with donors such as carbon monoxide (see Chapter 29), cyanide ion, aluminum chloride, and phosphine derivatives. On heating of a mixture of powdered nickel, mercury(II) chloride, and potassium cyanide *in vacuo* to 500° *potassium tetracyanonickelate(0)*, $K_4[Ni(CN)_4]$, is produced.[25] Raney nickel (suspended in benzene) coordinates with five mol at anhydrous aluminum chloride and forms the compound $Ni(C_6H_6)_2(AlCl_3)_5$.[26] Powdered nickel reacts with trifluorophosphine at 100° and 300 atm producing *tetrakis(trifluorophosphine)nickel(0)* in quantitative yield.[27] This reaction is catalyzed by iodine.[28] At mild conditions (60–80°) powdered nickel (from the thermal decomposition of oxalate) reacts with fluorophosphines (PF_3, CF_3PF_2, $(CF_3)_2PF$, CCl_3PF_2, $(CH_3)_2$, NPF_2, and $C_5H_{10}NPF_2$) by forming tetracoordinated complexes.[29]
A more convenient method for the preparation of nickel(0) complexes involves the replacement of carbon monoxide from nickel tetracarbonyl according to Eqn (3).

$$Ni(CO)_4 + 4L \longrightarrow NiL_4 + 4CO \quad (L = \text{donor molecule}) \qquad (3)$$

The reaction is carried out in organic solvents at suitable temperatures and high pressure. Nickel(0) compounds are readily oxidized by air, hence, working in inert atmosphere is necessary. The replacement of CO depends on the temperature. All four carbon monoxide molecules are displaced at temperatures of 80–100°. If the reaction is performed at lower temperature, the reaction leads to mixtures with $NiL_n(CO)_{4-n}$ ($n = 1$–4) (Tables 1 and 2). The ligand exchange rate depends on the solvent, the fastest rates were observed in solvents with electron-releasing groups.[30]
The stepwise replacement of the carbon monoxide presents a convenient method for the preparation of mixed complexes. For example, one or two molecules of carbon monoxide can be replaced at low temperatures by a ligand L, and the resultant product is treated at elevated temperatures with a ligand L′ to produce a complex of the type $NiL_2L'_2$. This procedure has been used for the preparation of *bis(triarylphosphine)-trichlorethylenenickel*(0),[31] and *bis(tributylphosphine)bis(carbondisulfide)nickel*(0).[32]
Bidentate and polydentate ligands can react with two or more molecules of tetracarbonylnickel by forming binuclear or polynuclear complexes.

[20] Neth. Appl. P. 6 510 895 (1966), International Nikkel Ltd., C. A. *65*, 1914f (1963).
[21] U.S. P. 3 019 181 Erf.: *M. H. Gwynn*; C. A. *56*, 9883f (1962).
[22] U.S. P. 2 868 631 (1961), Pure Oil. Co., Erf.: *N. L. Carr, V. Brozowski;* C. A. *55*, 10865b (1961).
[23] Ger. P. 1 063 385 (1959), Ethyl. Corp. Erf.: *T. H. Pearson;* C. A. 55, 12252e (1961).
[24] *G. Henrici-Olive, S. Olive*, Chimica *19* (1), 46 (1965).
[25] *S. Winbush, E. Griswold, J. Kleinberg*, J. Amer. Chem. Soc. *83*, 3197 (1961).
[26] Ger. P. 1 154 803 (1963), BASF, C.A. *60*, 13274b (1964).
[27] *Th. Kruck, K. Bauer, W. Lang*, Chem. Ber. *101* (1), 138 (1968).
[28] *Th. Kruck, K. Bauer*, Chem. Ber. *98* (9), 3070 (1965).
[29] *J. F. Nixon, M. D. Sexton*, Inorg. Nucl. Chem. Lett. *4* (5), 275 (1968).
[30] *R. J. Angelici, B. E. Leach*, J. Organometal. Chem. *11* (1), 203 (1968).
[31] Brit. P. 1 116 128 (1968), Intern. Nickel Ltd., C.A. *69*, 27529 (1968).
[32] U.S. P. 3472887 (1969), Intern. Nickel Co., Erf.: *J. T. Crriel, E. N. Hewitt;* C.A. *72*, 55664m (1970).

Table 1 Preparation of Nickel(0) Complexes [NiL_4] from $Ni(CO)_4$ according to Eqn (3)

Ligand L	Solvent	*t*°	press. [at]	Ref.
PF_3		70	100–200	*a*
PF_3		100	press. vessel	*b*
PX_3, $P(NCO)_3$, $P(NCS)_3$				*c, d*
CH_3PF_2				*e*
CF_3PF_2				*f*
$C_5H_{10}NPF_2$; $(CH_3)_2NPF_2$				*g*
$P(Oalkyl)Cl_2$				*h*
$P(Oalkyl)_2F$		high		*i*
$P(Oalkyl)_3$		high	reduced	*j*
R_2NPF_2, $(R_2N)_2PC_6H_5$	excess ligand	~20		*k*
substituted phosphines	benzene	20–70		*l*
bis-(phosphino)-*o*-carboranes	chloroform			*m*
F, F, F, F, $As(CH_3)_2$, $As(CH_3)_2$ (structure)				*n*
polycyclic phophite esters				*o*
H_3C, O, O, P, O (structure)				*p*
H_5C_6—CH=CH—CN	ether		reflux	*q*
alkylcyanamidecarbonyls	ethanol	55		*r*
maleic anhydride	benzene	−40		*s*
HN=CH—CH=NH				*t*
F_3C—C≡C—CF_3		50	sealed tube	*u*
O, O, P, O (structure)				*v*
F_3C, CF_3, S—S (structure)	hexane, benzene	~20°		*w*

[a] *A. Loutellier, M. Bigorgne*, Bull. Soc. Chim. France (11), 3198 (1965).
[b] *R. J. Clark, E. O. Brimm*, Inorg. Chem. *4* (5), 651 (1965).
[c] *D. H. Irvine, G. Wilkinson*, Science, N.Y. *113*, 742 (1951).
[d] *G. Wilkinson*, Z. Naturforsch. *9b*, 446 (1954).
[e] *F. Seel, K. Rudolph, R. Budenz*, Z. Anorg. Allg. Chemie *341* (3 4), 196 (1965).
[f] *A. B. Burg, G. B. Street*, Inorg. Chem. *5* (9), 1532 (1966).
[g] U.S. P. 3242171 (1966), DuPont, Erf.: *R. Schmutzler*; *C.A.* 64, 19683c.
[h] *P. Cassoux, J. F. Labarre*, C.R. Acad. Sci., Paris, Ser. C *264* (9), 736 (1967).
[i] *J. M. Savariault, P. Cassoux, J. F. Labarre*, C.R. Acad. Sci., Paris, Ser. C *269* (8), 496 (1969).
[j] *P. Cassoux, J. M. Savariault, J. F. Labarre*, Bull. Soc. Chim. France (3), 741 (1969).
[k] *R. Schmutzler*, Proc. Int. Conf. Coord. Chem., 8th, Vienna 188 (1964).
[l] *J. Chatt, F. A. Hart*, J. Chem. Soc. (London) 812 (1965).
[m] *F. Rohrscheid, R. H. Holm*, J. Organometal. Chem. *4* (4), 335 (1965).
[n] *W. R. Cullen, P. S. Dhaliwal, C. J. Stewart*, Inorg. Chem. *6* (12), 2256 (1967).
[o] *D. G. Hendricker*, U.S. At. Energy Comm. IS-T-31 (1965); C.A. *64*, 8233e.
[p] *J. G. Verkade, R. E. McCarley, D. G. Hendricker, R. B. King*, Inorg. Chem. *4* (2), 228 (1965).
[p] U.S. P. 3194848 (1965), National Distillers Chem. Corp., Erf.: *J. Feldman, B. A. Saffer, M. Thomas*; C.A. *63*, 9989d.
[r] *H. Bock, H. tom Dieck*, Chem. Ber. *99* (1), 213 (1966).
[s] *E. Weiss, K. Stark*, Z. Naturforsch. *20b* (5), 490 (1965).
[t] *H. Bock, H. tom Dieck*, Angew. Chem. *78* (10), 549 (1966).
[u] *R. B. King, M. J. Bruce, J. R. Phillips, F. G. A. Stone*, Inorg. Chem. *5* (4), 684 (1966).
[v] *D. G. Hendricker, R. E. McCarley, R. W. King, J. G. Verkade*, Inorg. Chem. *5* (4), 639 (1966).
[w] U.S. P. 3361777 (1968), DuPont, Erf.: *R. B. King*; C.A. *68*, 106510k.

Table 2 Preparation of Nickel(0) Complexes According to the Reaction: $Ni(CO)_4 + nL \longrightarrow NiL_n(CO)_{4-n} + nCO$

Ligand (L)	Conditions	*n*	Ref.
PH_3	*hv*	1	*a*
PF_3	100–200 atm., 0–70°	1, 2, 3, 4	*b*
$(tert\text{-}C_4H_9)_3P$	tetrahydrofuran	1	*c*
R_3P, R_3As, R_3Sb	below 40°, red. press.	3	*d*
R_3Sb, R_3Bi		1, 2	*e*
P, $(RO)_3As$, $(RO)_3Sb(RO)_3$	at 130°	3	*f*
$(CF_3)_2NP(CF_3)_2$	~20°	1	*g*
$(alkyl)_2NPF_2$		2, 4	*h*
$P[Z(CH_3)_3]_3$	(Z = Si, Ge, Sn, Pb) in tetrahydrofuran 20–100°	1	*i*
$(C_6H_5P)_4$	in ether at 15°	1	
triphenylphosphole		1	*k*
$C[CH_2PP(C_6H_5)_2]_4$	in benzene, ~20°	2	
$(Ph_2PC)_2B_{10}H_{10-n}X_n$ (n = 1–3)	in benzene, ~20°		*m*
RNC	low temp.	1	*n, o, p*
$LiSi(C_6H_5)_3$	pptn. by hexane in THF, above 0°	1	*q*
NO	in cyclopentadiene		*r*

[a] *I. H. Sabherwal, A. B. Burg*, Inorg. Nucl. Chem. Lett. *5* (4), 259 (1969).
[b] *A. Loutellier, M. Bigorgne*, Bull. Soc. Chim. France (11), 3198 (1965).
[c] *H. Schumann, O. Stelzer, U. Niederreuther*, J. Organometal. Chem. *16* (3), 64 (1969).
[d] Brit. P. 979553 (1965), Cities Service Research and Development Co.; C.A. *63*, 633.
[e] *D. Benlian, M. Bigorgne*, Bull. Soc. Chim. France 1583 (1963).
[f] Brit. P. 944574 (1963), Cities Serv. Res. Develop. Co.; C.A. *60*, 10570a (1964).
[g] *H. G. Ang*, J. Organometal. Chem. *19* 245 (1969).
[h] *R. Schmutzler*, Inorg. Chem. *3* (3), 415 (1964).
[i] *H. Schumann, O. Stelzer*, Angew. Chem. Intern. Ed. Engl. *6* (8), 701 (1967).
[j] *H. G. Ang. J. S. Shannon, B. O. West*, Chem. Commun. (1), 10 (1965).
[k] *A. R. Hands, A. J. H. Mercer*, J. Chem. Soc. (London) 6055 (1965).
[l] *J. Ellermann, K. Korn*, Angew. Chem. *78* (10), 547 (1966).
[m] *L. I. Zakharkin, G. G. Zhigareva*, Zh. Obshch. Khim. *37* (8), 1791 (1967).
[n] *W. Hieber*, Z. Naturforsch. *5b*, 129 (1950).
[o] *F. Klages, K. Mönkemeyer*, Chem. Ber. *83*, 501 (1950).
[p] *L. Malatesta, A. Sacco*, Atti Acad. Lincei *11*, 379 (1951).
[q] *Th. Kruck, E. Job, U. Klose*, Angev. Chem. Intern. Ed. Engl. *7* (5), 374 (1968).
[r] *R. D. Feltham, J. T. Carriel*, Inorg. Chem. *3* (1), 122 (1964).

Thus two molecules of methylbis[bis(trifluoromethyl)phosphine]amine, $(CF_3)_2PN(CH_3)P(CF_3)_2$, form the binuclear complex with two mols tetracarbonylnickel.[33]

```
      CF3      CO      CF3
        \      |      /
   F3C—P  —  Ni  —  P—CF3
       /       |       \
  H3C—N       CO        N—CH3
       \       |       /
   F3C—P  —  Ni  —  P—CF3
       /       |       \
     CF3      CO      CF3
```

Similarly, tetraphosphorus hexoxide, P_4O_6, can replace one molecule of carbon monoxide from each of four molecules of tetracarbonyl nickel producing the tetranuclear complex $P_4O_6[Ni(CO)_3]_4$.[34]

Similar displacement reactions can be performed with other starting materials which are more convenient to handle than the highly volatile and toxic tetracarbonylnickel.

(1) Replacement of the ligands of bis(cyclopentadienyl)nickel by four monodentate molecules. The reaction proceeds neat or in organic solvent[35] and leads to tetracoordinated nickel(0) complexes. This process avoids the difficulties to obtain pure products as encountered with displacement reactions utilizing nickel tetracarbonyl (which may produce mixtures of the type $NiL_n(CO)_{4-n}$). The interaction is performed in inert atmosphere at 20–120° (Table 3).

(2) Displacement of the ligands of tetrakis(trifluorophosphine)nickel(0) by other donor molecules. The reaction proceeds stepwise and leads to mixed complexes and mixtures with varying degree of ligand displacement. In benzene solution, one or two trifluorophosphine units can be displaced by triphenylphosphine to yield mixed complexes of the type $Ni(PF_3)_2(Ph_3P)_2$ and $Ni(PF_3)_3(Ph_3P)$.[36] Similarly, tetrakis(trifluorophosphine)nickel(0) can be reacted with triphenylarsine, triphenylstibine, pyridine, bipyridyl, and phenanthroline.[37] Complete displacement was achieved with dimethylamine,

[33] *A. B. Burg, R. A. Sinclair*, J. Amer. Chem. Soc. *88* (22), 5354 (1966).
[34] *J. G. Riess, J. R. Van Eazer*, J. Amer. Chem. Soc. *88* (10), 2166 (1966); *87* (23), 5506 (1965).
[35] *J. R. Olechowski, C. G. McAlister, R. F. Clark*, Inorg. Chem. *4* (2), 246 (1965).
[36] *Th. Kruck, K. Baur*, Chem. Ber. *98* (9), 3070 (1965).
[37] *Th. Kruck, K. Baur, K. Glinka, M. Stadler*, Z. Naturforsch. *23b* (9), 1147 (1968).

Table 3 Preparation of Nickel(0) Complexes of the Type NiL_4 Using Bis-[cyclopentadienyl]-nickel as Starting Material

Ligand (L)	Treatment	Ref.
PF_3	60° in sealed tube for 96 h	*a*
$PF_3, CF_3PF_2, (CF_3)_2PF, CCl_3PF_2, (C_2H_5)_2NPF_2, C_5H_{10}NPF_2$		*b*
$(C_6H_5O)_3P$	80° in cyclohexane, ppg. with methanol	*c*
2,2′-bipyridine,1 10-phenanthroline	60–120° cyclohexane	*d*
$(RO)_3P$	~20°, argon	*e*

[a] *J. F. Nixon*, Chem. Commun. (2), 34 (1966).
[b] *J. F. Nixon*, J. Chem. Soc. A (7), 1136 (1967).
[c] *J. R. Olechowski, C. G. McAlister, R. F. Clark*, Inorg. Syn. *9*, 181 (1967).
[d] *H. Behrens, K. Meyer*, Z. Naturforsch. *21b* (5), 489 (1966)
[e] U.S. P. 3152158 (1964), Cities Serv. Res. Develop. Co.; Erf.: *R. F. Clark*; C.A. *61*, 14230f (1964).

piperidine, and (dialkylamino)difluorophosphine.[38]

A displacement can also be achieved at the ligand molecule as is illustrated for the preparation of methoxophosphine complexes[39] which occurs according to Eqn (4).

$$N\cdot(PF_3)_4 + 12CH_3ONa \xrightarrow[\text{reflux}]{CH_3OH} Ni[P(OCH_3)_3]_4 + 12NaF \quad (4)$$

Reactions (5) and (6) depict similar processes [38]

$$Ni[PF_2N(CH_3)_2]_4 + 4HCl \longrightarrow Ni(PF_2Cl)_4 + 4(CH_3)_2NH \quad (5)$$

$$Ni(PF_2Cl)_4 + 4CH_3OH \longrightarrow Ni(PF_2OCH_2)_4 + 4HCl \quad (6)$$

Also, chlorophosphine complexes were converted into fluorophosphine complexes by fluorination with potassium thionylfluoride, KSO_2F.[40]

(3) Replacement of the ligands in bis(1,5-cyclooctadienyl)nickel(0). The reaction proceeds stepwise at low temperature and in inert atmosphere. Complexes of the type $(1,5\text{-}C_8H_{12})NiC_3F_6$ and $(R_3P)_2NiC_3F_6$ (R = phenyl, ethyl) have been prepared by this procedure[41] as well as complexes with hexafluoroacetone, bipyridine and phosphines.[42] Complete displacement was achieved by treating bis(1,5-cyclooctadienyl)nickel(0) with hexaphenylethane at temperatures from 20–70°.[43]

The π-bonded ethylene of bis(triphenylphosphine)(π-ethylene)nickel(0) is readily displaced by hexafluoroacetone.[44] The reaction with chloro- or bromotrifluoroethylene seems to produce a σ-bonded trifluoroethylene derivative, $(PPh_3)_2Ni(CF{=}CF_2)X$ (X = Cl, Br).[44]

Replacement reactions can also be performed by treating nickel(0) complexes with gaseous oxygen in ether solution at very low temperatures according to Eqn (7).[45,46]

$$[(C_6H_5)_3P]_4Ni + O_2 \longrightarrow \{[(C_6H_5)_3P]_2NiO_2\} + 2(C_6H_5)_3P \quad (7)$$

Isocyanide complexes, $Ni(RNC)_4$ (R = *t*-butyl, or cyclohexyl), behave similarly.[47]
Potassium tetracyanonickelate(0) reacts in liquid ammonia at room temperature according to Eqn (8).

$$K_4[Ni(CN)_4] + 4L \longrightarrow [NiL_4] + 4KCN \quad (8)$$

(L = $(C_6H_5)_3Z$, Z = P, As, or Sb; L = $Ph_2Z(CH_2)_2ZPh_2$, Z = P, or As, and 2L = *o*-phenanthroline or bipyridyl).[48]

Another route for the preparation of nickel(0) complexes involves the reduction of nickel(II) complexes.

[38] *Th. Kruck, M. Hoefler, H. Jung, H. Blume*, Angew. Chem., Intern. Ed. Engl. *8* (7), 522 (1969).
[39] *Th. Kruck, M. Hoefler*, Angew. Chem., Intern. Ed. Engl. *6* (6), 563 (1967).
[40] *F. Steel, K. Ballreich, R. Schmutzler*, Chem. Ber. *94*, 1173 (1961).
[41] *C. S. Cundy, M. Green, F. G. A. Stone*, J. Chem. Soc. A *1970* (10), 1647.
[42] *J. Brawning, C. S. Cundy, M. Green, F. G. A. Stone*, J. Chem. Soc. A (1), 20 (1969).
[43] *G. Wilkie, H. Schott*, Angew. Chem. *78* (11), 592 (1966).
[44] *J. A. Smith, M. Green, F. G. A. Stone*, J. Chem. Soc. A 1969, (19), 3019.
[45] *G. Wilke, H. Schott, P. Heimbach*, Angew. Chem. Intern. Ed. Engl. *6* (1), 92 (1967).
[46] Fr. P. 1542048 (1968), Studienges. Kohle CA *71*, 124657e (1969).
[47] *S. Otsuka, A. Nakamura, Y. Tatsuno*, Chem. Commun. (16), 836 (1967);
S. Otsuka, A. Nakamura, Y. Tatsuno, J. Amer. Chem. Soc. *91* (25), 6994 (1969).
[48] *H. Behrens, A. Mueller*, Z. Anorg. Allgem. Chemie *341* (3–4), 124 (1965).

Alkali metals reduce nickel(II) complexes to nickel(0) derivatives in liquid ammonia according to Eqn (9).

$$K_2[NiL_4] + 2K \longrightarrow K_4[NiL_4] \quad (9)$$

L=CN[49] or L=C:NH.[50,51]

A good reducing agent in aqueous alkaline solution at 90° is sodium pyrosulfite, $Na_2S_2O_5$. It appears that highly reactive colloidal nickel is formed in this reduction which then coordinates with four molecules of triethylphosphite.[52]

Complexes of zerovalent nickel with electron donors such as olefines, acetylenes, or diketones, were obtained by reducing nickel(II) bis(acetylacetonate) with diethylethoxyaluminum at 0–5° in benzene and in the presence of the ligand. This procedure has been used to obtain complexes with cyclooctatetraene,[53], butadiene,[54] and hexaphenylethane.[45] Bis(1,5-cyclooctadiene) nickel(0) was prepared from acetylacetonatonickel(II) and triethylaluminum at 5–10°.[54]

The reduction of nickel(II) bis(acetylacetonate) with ethylaluminum in toluene and in nitrogen atmosphere and in the presence of tris(cyclohexyl)phosphine (L) produced a nitrogen complex of the type $[NiL_2]N_2$.[55]

In organic solvents such as THF lithium benzophenone reduces the phthalocyanine-nickel(II) complex to the complex $Li_2[NiPc]6THF$ (Pc= phthalocyanine).[56,57]

Also, upon hydrogen reduction of nickel compounds in the presence of donor molecules nickel(0) complexes such as the complex with triethylphosphite[58] are obtained (Eqn (10)).

$$(CH_3COO)_2Ni + H_2 + 4P(OC_2H_5)_3 \longrightarrow Ni[P(OC_2H_5)_3]_4 + 2CH_3COOH \quad (10)$$

Trialkylphosphites act as reducing agents in the presence of alkylamines and reaction (11) proceeds in water, methanol, ethanol, or acetonitrile.[59]

$$NiX_2 + 5(RO)_3P + 2R_3N + H_2O \longrightarrow Ni[P(OR)_3]_4 + 2R_3NX + (RO)_3PO \quad (11)$$

$R = CH_3, C_2H_5$

Phosphite complexes of nickel(II) halides have been converted into the corresponding nickel(0) derivatives by reduction with potassium graphite, C_8K.[60]

Binuclear complexes such as $K_4[Ni_2(CN)_6]$ and $K_4[Ni_2(CN)_6(CO)_2]$ disproportionate in liquid ammonia solution and, in the presence of isocyanides, triphenylphosphine, triphenylarsine, or alkali acetylides, they form corresponding nickel(0) complexes.[61,62]

Dicoordinated nickel(0) complexes can coordinate with two additional donor molecules to form four-coordinated nickel(0) derivatives. This method provides for the preparation of mixed derivatives of the type NiL_2L_2' by merely refluxing NiL_2 with a ligand L′ (L=CH_2=CHCN or CH_2=CHCHO), L′=$(C_6H_5)_3Sb$ or $(CH_3O)_3P$),[63] or by treating NiL_2 with L′ at very low temperature[64] (L′=$(C_6H_5)_3P$).

26.3 Nickel(I) Derivatives

Nickel(I) compounds are generally prepared by the reduction of nickel(II) derivatives. Reducing agents are metals, amalgams, and virtually all common reductors. The reaction must be performed in presence of an excess of the appropriate divalent nickel compound; otherwise, zerovalent nickel compounds result.

Monovalent nickel derivatives are fairly unstable in air. Though nickel(I) coordination compounds are more stable, some simple inorganic compounds have also been prepared. Nickel(I) amidosulfonate

$$\left[H{-}N\begin{matrix}/SO_3Ni\\ \backslash SO_3Ni\end{matrix}\right]\cdot nH_2O$$

[49] *J. W. Eastes, W. M. Burgess*, J. Amer. Chem. Soc. *64*, 1187 (1942).
[50] *R. Nast*, Z. Naturforsch. *8b*, 381 (1953).
[51] *R. Nast, K. Vester*, Z. Anorg. Allgem. Chemie *279*, 146 (1955).
[52] U.S. P. 3 102 899 (1963); Shell Oil Co., Erf.: *L. G. Cannel;* CA *60*, 1362g (1964).
[53] Ger. P. 1 191 375 (1965), Studienges, Kohle, Erf.: *G. Wilke, E. W. Mueller, M. Kroener, P. Heimbach, H. Breil;* CA *63*, 7045b (1965).
[54] *T. Arakawa*, Kogyo Kagaku Zasshi *70* (10), 1738 (1967).
[55] *P. W. Jolly, K. Jonas*, Angew. Chem. Intern. Ed. Engl. *7* (9), 731 (1968).
[56] *R. Taube, M. Zach, K. A. Stauske, S. Heidrich*, Z. Chem. *3* (10), 392 (1963).
[57] *R. Taube*, Chem. Zvesty *19* (3), 215 (1965).
[58] U.S. P. 3 290 348 (1966), Shell Oil Co., Erf.: *R. D. Mullineaux;* CA *66*, 65080h (1967).
[59] *R. S. Vinal, L. T. Reynolds*, Inorg. Chem. *3* (7), 1062 (1964).
[60] *K. A. Jensen, B. Nygaard, G. Elisson, P. H. Nielsen*, Acta. Chem. Scand. *19* (3), 768 (1965).
[61] *R. Nast, H. Schulz*, Chem. Ber. *103* (3), 785 (1970).
[62] *R. Nast, H. D. Moerler*, Chem. Ber. *99* (21), 3787 (1966).
[63] U.S. P. 3 251 893 (1966), National Distill. Chem. Corp., Erf.: *J. Feldman, B. A. Saffer, M. Thomas;* CA *65*, 2148a (1966).
[64] *G. N. Schrauzer*, Chem. Ber. *94*, 642 (1961).

was prepared in aqueous solution from nickel halide, sodium hydrosulfite and sodium nitrite.[65] The hydroxide NiOH or the sulfide Ni_2S can be precipitated from such solutions.

Comparatively stable and most widely studied are the complexes of the monovalent nickel, in particular the cyano complexes. These are obtained by reduction of potassium tetracyanonickelate(II) with alkali metals in liquid ammonia[66] or by reduction of aqueous solutions of the same nickelate with potassium amalgam in a hydrogen atmosphere.[67] The potassium tricyanonickelate(I) $K_2[Ni(CN)_3]$, precipitates after dilution with alcohol. This compound was also prepared by electrolytic reduction,[68,69] by reduction with hypophosphite,[70] and from sodium tetrafulminatonickelate(II) and sodium amalgam[71] according to the following scheme (12)

$$Na_2[Ni(CNO)_4] \xrightarrow{Na-Hg} Na_4[Ni_2(CN)_6] + Na_2[Ni(CN)_4] \quad (12)$$

Potassium tricyanonickelate(I) is very unstable and exists as a dimer.[72,73] Its solution in liquid ammonia absorbs carbon monoxide at $-40°$ according to Eqn (13)

$$K_4[Ni_2(CN)_6] + 2CO \longrightarrow 2K_2[NiCO(CN)_3] \quad (13)$$

On heating of a mixture of metallic nickel, potassium cyanide, and potassium tetracyanonickelate(II) to 450–480° *in vacuo potassium tetracyanonickelate(I)* is obtained[74] according to Eqn (14).

$$K_2[Ni(CN)_4] + Ni + 4KCN \longrightarrow 2K_3[Ni(CN)_4] \quad (14)$$

The same compound was prepared by the reduction of potassium tetracyanonickelate(II), $K_2[Ni(CN)_4]$, with hydrogen in strong alkali solution.[75]

Potassium tricyanonickelate(I), when treated with nitrogen dioxide, converts into the nitrosylcomplex $K_2[Ni(NO)(CN)_3]$.[76,77] Nitrogen dioxide generally acts as a reducing agent and produces nickel(I) compounds in the presence of appropriate donor molecules. Thus the thiosulfatocomplex $K_3[Ni(NO)(S_2O_3)_2]2H_2O$ results from the reaction of an alcoholic solution of nickel(II) acetate and with nitrogen dioxide.[78,79] The same compound was prepared from nitroso compounds of the monovalent nickel and sodium thiosulfate.[16] On treatment of an alcoholic solution of nickel acetate and ethylmercaptane with nitrogen dioxide the corresponding nitrosomercaptane derivative, $[Ni(NO)SC_2H_5]$, is obtained.[78] The compound coordinates with one molecule of pyridine to yield $[Ni(NO)SC_2H_5]C_5H_5N$. Nitrosylphosphine complexes can be prepared from the corresponding nickel(II) phosphine complexes by treatment with nitrogen monoxide. By this method bis(tributylphosphine)nickel(II) chloride was converted to *bis(tributylphosphine)nitrosylnickel(I)* chloride.[81] Nitrosylphosphine complexes of the type $[(R_3P)_2Ni(X)NO]$, (R=butyl, or phenyl; X=Cl, Br, I, SCN, or NO_3), were prepared from the corresponding nickel(II) salt by reaction with nitrogen monoxide in the presence of the phosphine.[82]

Another route for the preparation of nickel(I) compounds involves the reduction of π-allyl halides in the presence of the ligands. For example, *tris(triphenylphosphine)nickel(I) halides*, $[Ni(PPh_3)_3X]$, (X=Cl, Br, I), were prepared from π-allyl halides in the presence of triphenylphosphine on treatment with norbornylene in benzene-ether solution.[83] The halocomplexes of

---

[65] *L. A. Tschugaeff, V. G. Ichlopine*, C.R. Acad. Sci., Paris *159*, 62 (1914).
[66] *J. W. Eastes, W. M. Burges*, J. Amer. Chem. Soc. *64*, 1187 (1942).
[67] *I. Bellucci, R. Corelli*, Atti. Acad. Naz. Lincei Classe Sci. Fis., Mat. Natur. Rend. *22*, (5), 485 (1913); *J. Bellucci, R. Corelli*, Z. Anorg. Allgem. Chemie *86*, 38 (1914).
[68] *G. Grube*, Z. Elektrochem. *32*, 561 (1926).
[69] *W. D. Traedwell, D. Huber*, Helv. Chim. Acta *26*, 10 (1943).
[70] *W. Manchot, H. Schmid*, Ber. *64B*, 2672 (1931).
[71] *W. Beck, F. Lux*, Chem. Ber. *95*, 1683 (1962).
[72] *R. Nast, H. Roos*, Z. Anorg. Allgem. Chemie *272*, 242 (1953).
[73] *R. Nast, T. von Krakkay*, Z. Anorg. Allgem. Chemie *272* (5–6), 233 (1953).
[74] *S. Winbush, E. Griswold, J. Kleinberg*, J. Amer. Chem. Soc. *83*, 3197 (1961).
[75] *R. Nast, T. von Krakkay*, Z. Naturforsch. *9b*, 798 (1954).
[76] *W. Mauchot*, Ber. *59*, 2445 (1926).
[77] *A. Job, A. Samuel*, C.R. Acad. Sci., Paris *177*, 188 (1923).
[78] *W. Manchot, G. Lehmann*, Justus Liebigs Ann. Chem. *470*, 255 (1929).
[79] *L. Cambi, A. Clerici*, Atti Acad. Naz. Lincei Classe Sci. Fis., Mat. Natur. Rend. *6* (6), 488 (1927).
[80] *W. Hieber, R. Nast*, Z. Anorg. Allgem. Chemie *244*, 23 (1940).
[81] *H. Brunner*, Angew. Chem. Intern. Ed. Engl. *6* (6), 566 (1967).
[82] *H. Brunner*, Chem. Ber. *101* (1), 143 (1968).
[83] *I. Porry, M. C. Gollazzi, G. Vitulli*, Chem. Commun. (5), 228 (1967).

nickel(I) with methylfumarate and methylmaleate have been prepared in similar fashion.[84] The same complexes were obtained from tetracarbonylnickel, allylbromide, and the appropriate ester by heating to 50° in benzene solution.[85]
The reduction of nickel(II) with strong reducing agents leads to nickel(0) compounds. If the latter are treated with hydrocarbon halides they convert into complex nickel(I) halides. Thus the reduction of nickel(II) bisacetylacetonate with organoaluminum derivatives in the presence of conjugated dienes initially produces nickel(0) but subsequent treatment with 1,1-diphenyl-1-bromoethane affords a *dienenickel(I) bromide*.[85]
Moderately strong reductors such as sodium tetrahydridoborate yield the corresponding nickel(I) complex directly. For example, bis-(diphenylphosphine)propanenickel(II) cyanide is reduced to *bis(diphenylphosphine)propanenickel-(I) cyanide*[86] upon interaction with $NaBH_4$.
Another convenient route to nickel(I) compounds involves the reduction of nickel(II) compounds with powdered nickel. This reaction is carried out with a mixture of nickel(II) halide, powdered nickel and donors such as triphenylphosphine in refluxing ether or tetrahydrofurane. The nitrosylphosphine complex was prepared by passing nitrogen oxide through this mixture.[87]
Some nickel(I) complexes can be prepared by ligand displacement. Thus cyanide complexes can be obtained from thiosulfato complexes according to Eqn (15).

$$K_3[Ni(NO)(S_2O_3)_2] + 3KNC \longrightarrow K_2[Ni(NO)(CN)_3] + 2K_2S_2O_3 \quad (15)$$

Mixed complexes are formed on treatment of a dimeric species with the appropriate ligand. For example, dimeric triphenylphosphinenitrosylnickel(I) iodide, $[(Ph_3P)Ni(NO)I]_2$, when treated in benzene with tributylphosphine, produces the mixed complex $[(Bu_3P)(Ph_3P)Ni(NO)I]$.[88] On treatment of bis(triphenylphosphine)dicarbonylnickel(0) with nitrogen oxide and subsequent hydrolysis the hydroxocomplex $[(Ph_3P)_2NiNO(OH)]$[89] is obtained. The analogous nitrosobromocomplex, $[(Ph_3P)_2Ni(NO)Br]$, was prepared from bis(triphenylphosphine)nickel(II) bromide and anhydrous sodium nitrite by refluxing in organic solvents in the presence of an excess of the ligand. Both the iodide and the chloride were prepared similarly.[90]

26.4 Nickel(II) Compounds

26.4.1 Hydrides and Hydridocomplexes

Metallic nickel absorbs large amounts of hydrogen to form nonstoichiometric hydrides. A defined *nickel(II) hydride* was prepared by hydrogenation of nickel(II) chloride in the presence of phenylmagnesium bromide[91,92,93] according to Eqn (16).

$$NiCl_2 + 2C_6H_5MgBr + 2H_2 \longrightarrow NiH_2 + 2C_6H_6 + 2MgBrCl \quad (16)$$

The same hydride is obtained on treatment of powdered nickel with hydrogen.[94] Electolytically hydrogenated nickel foils contain a nickel hydride though the latter is not stable. Neutron scattering measurements, however, indicated the presence of Ni and NiH.[95]
A good method for the preparation of NiH is the electrolytical hydrogenation (in the presence of thiourea) of a thin layer of nickel deposited on aluminum plates.[96]
If complexes of zerovalent nickel are treated in nonaqueous solution with strong acids the corresponding *hydridocomplexes* are formed; they are air and water sensitive. The reaction proceeds in an inert atmosphere. For example, bis(tricyclohexylphosphine)nickel(0) reacts with hydrogen chloride, anhydrous acetic acid, phenole, or pyrrole, to give a hydridocomplex of the type $[L_2Ni(H)R]$ (L = ligand, R = Cl, CH_3COO,

---

[84] *M. Dubini, F. Montino*, Chim. Ind. (Milan) *49* (12), 1283 (1967).
[85] Fr. P. 1 567 630 (1969), Mitsui Petrochem. Ind. Ltd., CA *72*, 111618p (1970).
[86] *B. Corain, M. Bressan, P. Rigo, A. Turco*, Chem. Commun. (9), 509 (1968).
[87] U.S. P. 3 481 710 (1969), Phillips Petroleum Co., Erf : *M. L. Perry;* CA *72*, 45601z (1970).
[88] *H. Brunner*, Z. Naturforsch. *24b* (3), 275 (1969).
[89] *R. D. Feltham*, Inorg. Chem. *3* (1), 119 (1964).
[90] *R. D. Feltham*, Inorg. Chem. *3* (1), 116 (1964).
[91] *W. Schlenk, Th. Weichselfelder*, Ber. *56*, 2230 (1923).
[92] *P. Job, R. Reich*, C.R. Acad. Sci. Paris *177*, 1439 (1923).
[93] *B. Sarry*, Naturwissenschaften *41*, 115 (1954); *B. Sarry*, Z. Anorg. Allgem. Chemie *280*, 65 (1955).
[94] *C. Vandael*, Ind. Chim. Belge *17*, 581 (1952).
[95] *J. W. Cable, E. O. Wollan, W. C. Koehler*, J. Phys. (Paris) *25* (5), 460 (1964).
[96] *B. Baranowski*, Roczniki Chem. *38* (6), 1019 (1964).

C_6H_5O).[97] Similarly,[98] complexes containing L = $Ph_2P(CH_2)_2PPh_2$ (Ph = phenyl) and X = $AlCl_4$, BF_4 or HCl_2) have been prepared. Some tetracoordinated nickel(0) complexes react analogously with acids below −50°.[99] Hydridocomplexes of the type $[(ZR_3)_4Ni(H)X]$, Z = P, As or Sb; R = OC_2H_5, $OC_6H_4CH_3$, or OC_6H_4-OCH_3; X = CF_3COO, CN, or $Zn(CN)Cl_2$, were obtained by this procedure.

These complexes can be used as starting material for the preparation of other hydridocomplexes by replacement of the anion X. The displacement can be achieved by the action of an organometallic compound[97] according to the following Eqn (17),

$$[(L)_2Ni(H)OOCCH_3] + C_6H_5Li \longrightarrow [(L)_2Ni(H)C_6H_5] + CH_3COOLi \quad (17)$$

or by treating a solution of the hydrido chloride complex in acetone–ethanol with the appropriate sodium salt[100] according to Eqn (18).

$$[(PR_3)_2Ni(H)Cl] + NaBH_4 \longrightarrow [(PR_3)_2Ni(H)(BH_4)] + NaCl \quad (18)$$

Another complex containing hydrogen was prepared by the reduction of nickel bisacetylacetonate with triisobutylaluminum in the presence of triphenylphosphine at −70° and under nitrogen. The complex is of the type $[(H)Ni(N_2)(PEt_3)_2]$ and is supposedly a nickel(0) derivative.[101]

The reduction of nickel(II) complexes with sodium tetrahydridoborate yields a hydride complex[102] according to Eqn (19).

$$[NiCl_2L_2] + NaBH_4 \longrightarrow [Ni(H)ClL_2] + NaCl + BH_3 \quad (19)$$

(L = tricyclohexylphosphine).

26.4.2 Halides and Halocomplexes

Nickel halides (hydrates) are commercial products. However, many syntheses of nickel compounds require anhydrous halides which can be prepared by special methods.

Anhydrous nickel fluoride is prepared by the thermal decomposition of ammonium tetrafluoronickelate in an atmosphere of hydrogen fluoride or by treating nickel(II) chloride or oxide with fluorinating agents such as fluorine,[103] chlorine trifluoride ClF_3,[104,105] or sulfur tetrafluoride.[106]

Anhydrous chloride can be obtained in good yield from nickel carbonate or organic salts by refluxing with acetyl chloride or prolonged storing with the same reagent at room temperature.[107] Both the chloride and the *bromide* can be prepared from nickel acetate and the appropriate acetyl halide in anhydrous acetic acid. The compounds crystallize as solvates of the type $2NiX_2\frac{1}{2}CH_3COOH$. Pure *nickel iodide* was obtained in the presence of iodide ions.[108] Also refluxing of nickel salts with organic acids in thionylchloride leads to the anhydrous chloride.[109]

The bromide and the iodide can be prepared by heating powdered nickel in bromine or iodine vapor. The products are purified by sublimation.

Hydroxihalides of the type $Ni_2(OH)_3X$ (X = Cl, Br, I) are formed from a mixture of the hydroxide and the corresponding halide at 200–300°. High pressure favors the reaction.[110]

Anhydrous nickel halides form complexes of the type $MNiX_3$ or M_2NiX_4 in the presence of excess alkali metal halide. *Fluoronickelates* have been prepared by fusion of a mixture of nickel fluoride with potassium fluoride[111] or ammonium fluoride,[112] by treating a methanol solution of nickel(II) bromide with alkali metal fluoride or ammonium fluoride,[113] or by reacting a mix-

[97] *K. Jonas, G. Wilke*, Angew. Chem. Intern. Ed. Engl. *8* (7), 519 (1969).
[98] *R. A. Schunn*, Inorg. Chem. *9* (2), 394 (1970).
[99] Ger. P., Offen. 1 808 434 (1969), DuPont, Erf.: *W. C. Drinkard jr., R. V. Lindsey jr.*, CA *71*, 70093x (1969).
[100] *M. L. H. Green, H. Munakata, T. Saito*, J. Chem. Soc. D 21, 1287 (1969).
[101] *S. C. Strivastava, M. Bigorgne*, J. Organometal. Chem. *18* (2), 30 (1969).
[102] *M. L. H. Green, T. Saito*, Chem. Commun. 5, 208 (1969).
[103] *O. Ruff, E. Ascher*, Z. Angew. Chem. *41*, 737 (1928).
[104] *E. G. Rochow, I. Kukin*, J. Amer. Chem. Soc. *74*, 1615 (1952).
[105] *R. L. Farrar jr., H. A. Smith*, J. Amer. Chem. Soc. *77*, 4502 (1955).
[106] U.S. P. 2952514 (1960); DuPont, Erf.: *W. C. Smith*, CA *55*, 3939c (1961).
[107] *D. Khristov*, Compt. Rend. Acad. Bulgare Sci. *16* (2), 177 (1963), *16* (7), 713 (1963).
[108] *H. D. Hardt, H. Pohlmann*, Z. Anorg. Allgem. Chemie *343* (1–2), 87 (1966).
[109] *D. Khristov, St. Karaivanov, V. Kolushki*, Godishnik Sofiiskiya Univ., Khim. Fac. *55*, 49 (1960).
[110] *H. R. Oswald, W. Feitknecht*, Helv. Chim. Acta *47* (1), 272 (1964).
[111] *C. Poulenc*, C.R. Acad. Sci., Paris *114*, 1426 (1892).
[112] *D. S. Crocket, R. A. Grossman*, Inorg. Chem. *3* (5), 644 (1964).
[113] *D. S. Crocket, H. M. Haendler*, J. Amer. Chem. Soc. *82*, 4158 (1960).

ture of nickel(II) chloride and alkali metal chloride with hydrogen fluoride.[114]

Tetrachloronickelates and *tetrabromonickelates* are obtained from mixtures containing stoichiometric amounts of the metal halides.[114]

26.4.3 Compounds with the Elements of the Oxygen Group bonded to Nickel

Binary compounds of nickel(II) with elements of the VIth group with definite constitution are formed with oxygen and sulfur only. Nickel oxide and sulfide are commercial products.

Special methods are suggested for the preparation of compounds with unusual composition. Thus *trinickel disulfide*, Ni_3S_2, is obtained by the reduction of nickel sulfate with hydrogen at 300°.[115] The same compound can be prepared from powdered nickel and sulfur dioxide at 460°[116] according to Eqn (20).

$$7Ni + 2SO_2 \longrightarrow 4NiO + Ni_3S_2 \quad (20)$$

Treatment of powdered nickel with hydrogen sulfide also produces trinickel disulfide.[117] The compound was likewise prepared from nickel sulfate by treatment with hydrogen sulfide and nitrogen at 500°.[118] Various mixed sulfides can be prepared by thermal treatment of the components in an evacuated tube. Thus powdered nickel, rhenium, and sulfur produce a mixed nickel-rhenium sulfide on treatment at 900°.[119]

The *selenide* was obtained by precipitation with hydrogen selenide or by treating nickel chloride with selenium vapor.[120] Also, reduction of nickel selenite in alkali solution leads to the selenide.[121]

Various *tellurides* such as NiTe, Ni_2Te_3, and $NiTe_2$ are obtained on heating of a mixture of the elements.[122] The telluride $NiTe_2$ was prepared by electrolysis from a solution of nickel cyanide and sodium tellurite.[123]; the telluride $NiTe_{1.85}$ was prepared hydrothermally from tellurium, nickel oxide, and sodium oxalate or formate in aqueous suspension at 265–343° and 50–150 atm.[124]

26.4.4 Binary Compounds with the Elements of the Fifth Group and Phosphine Complexes

Binary compounds of nickel with the elements of the nitrogen group are non-stoichiometric solid solutions, particularly the nitrides and phosphides. The *arsenides* and the *antimonides* have intermetallic character. Phase analysis of mixtures of nickel and Vth group elements indicated that various compounds are formed in the melt. Some of these compounds can be prepared by special procedures. For example, powdered nickel absorbs nitrogen at 500° to yield *trinickel nitride*, Ni_3N.[125] The same compound was prepared from nickel sponge and ammonia.[126] Powdered nickel reacts with ammonia to give a mixture of Ni_3N and Ni_4N.[127] This reaction can also be performed with a mixture of ammonia and hydrogen.[128] Heating of nickel oxide and nickel cyanide in an arc and in nitrogen atmosphere gives the nitride Ni_3N_2.[129]

The phase diagram of the system nickel–phosphorus indicates the existence of various *phosphide* phases such as Ni_3P, Ni_7P_3, Ni_2P, etc. These materials can be prepared by fused salt electrolysis[130,131] (such as electrolysis of a melt of alkali metal phosphates containing nickel

114 *P. Allamagny*, Bull. Soc. Chim. France 1099 (1960).

115 *G. Panetier, J. Abbeg*, Acta Chim. Acad. Sci., Hung. *30*, 127 (1962);
G. Panetier, J. Abbeg, Bull. Soc. Chim., France 186 (1961).

116 *G. Panetier, L. Davignon*, Bull. Soc. Chim. France (9), 2304 (1964).

117 *J. C. Colson*, C.R. Acad. Sci. Paris *259* (19), 3261 (1964).

118 *D. Delafosse, P. Barret*, C.R. Acad. Sci., Paris *251*, 2964 (1960).

119 *K. Koertes*, Rec. Trav. Chim. Pays-Bas *82* (11), 1099 (1963).

120 *H. Fonzes-Diacon*, C.R. Acad. Sci., Paris *131*, 556 (1900).

121 *V. M. Shul'man, V. L. Varand*, Izv. Akad. Nauk. SSSR Ser. Khim. (5), 943 (1966).

122 *S. Tengner*, Z. Anorg. Allgem. Chemie *239*, 126 (1938).

123 U.S. P. 3 419 484 (1968), Chrysler Corp., Erf.: *F. E. Ammerman, D. J. Schindehette;* CA *70*, 63590h (1969).

124 *L. Cambi, M. Elli*, Chim. Ind. (Milan) *50* (8), 869 (1968).

125 *G. T. Beilby, G. G. Henderson*, J. Chem. Soc. *79*, 1251 (1901).

126 *G. Rienaecker, K. H. Kohl*, Z. Anorg. Allgem. Chemie *333* (4–6), 291 (1964).

127 *R. Bernier*, Ann. Chim. *6*, 104 (1951).

128 *H. H. Stadelmair, T. S. Yun*, Z. Metallk. *52*, 477 (1961).

129 *A. C. Vournazos*, C.R. Acad. Sci., Paris *168*, 889 (1919).

130 *S. S. Hsu, P. N. Yocum, T. C. C. Cheng, K. B. Oldham, C. E. Meyers, K. Gingerich, C. H. Travaglini, J. C. Bailar, H. A. Laitinen, S. Swann*, U.S. Dept. Comm., Office Tech. Serv., P. B. Rept. 147079 (1961).

131 *F. Bertaut, P. Blum*, C.R. Acad. Sci., Paris *232*, 1566 (1951).

oxide[132,133] or alkali metal phosphates and nickel phosphate).[134] Powdered nickel and red phosphorus interact directly above 1000° to form a mixture of phosphides.[135,136] *Pure nickel diphosphide*, NiP_2, was obtained when this mixture was heated to 1100–1400° under a pressure of 65 kilobars.[137]

Other methods for preparing phosphides are the reaction of nickel with phosphorus halide[135] and the reduction of nickel phosphate with hydrogen.[138] Pure Ni_2P was obtained by the reduction of nickel diphosphate, $Ni_2P_2O_7$, with hydrogen at 450°.[139]

Nickel arsenides and antimonides are natural products; they can be synthesized from the elements.[140,141,142,143] Pure $NiAs_2$ is prepared from the powdered elements at 1400° and 60 kilobars[144] as well as at 1100–1400° and 65 kilobars.[137] Other methods for preparing nickel arsenides and antimonides involve the reduction of a solution containing stoichiometric amounts of the corresponding salts with hydrazine or with hypophosphorous acid.[145]

Phosphidocomplexes of the type $Ni(PH_2)_2$, xNH_3, and $K[Ni(PH_2)_3]$ have been prepared from tetraminonickel(II) thiocyanate, $[Ni(NH_3)_4](SCN)_2$, on reaction with potassium phosphide in liquid ammonia at −78°.[146] *Bis(diphenylphosphido)bis(diphenylphosphine)nickel(II)* was obtained from the reaction of nickel bromide with diphenylphosphine[147] according to Eqn (21).

$$NiBr_2 + 4(C_6H_5)_2PH \longrightarrow [(C_6H_5)_2P]_2Ni[HP(C_6H_5)_2]_2 + 2HBr \quad (21)$$

A variety of phosphine complexes were obtained by simple treatment of nickel(II) salts with the ligand in solution according to reaction (22)

$$NiX_2 + nL \longrightarrow NiL_nX_2 \quad (22)$$

where L is the ligand (phosphine, arsine, or stibine) and n is usually 2 (Table 4).

Phosphine, arsine, and stibine complexes of zerovalent nickel: see 26.2.

26.4.5 Compounds with the Elements of Groups IV and III

Binary compounds of nickel with the elements of groups IV and III are alloys. Even such elements as carbon, silicon and boron form nonstoichiometric compounds with nickel at high temperatures.

Relatively pure *nickel carbide*, Ni_3C, was prepared by thermal decomposition of the acetate.[148] Mixed carbides were obtained by heating a mixture of powdered nickel and the carbide of the appropriate metal. This procedure has been applied for the preparation of nickel–uranium carbide, $NiUC_2$.[149]

Phases such as NiSi, $NiSi_2$, Ni_2Si, Ni_3Si, Ni_3Si_7, and Ni_2Si_3 were formed by inductive heating of a mixture of the elements. The products were purified by zone refining.[150]

The *borides* consist of phases such as Ni_3B, Ni_2B, NiB, and Ni_4B_3 and were prepared analogously.[151] Ni_2B was precipitated from aqueous solution[152] according to Eqn (23).

$$2NiCl_2 + 4NaBH_4 + 9H_2O \longrightarrow Ni_2B + 12{\cdot}5H_2 + 3H_3BO_3 + 4NaCl \quad (23)$$

The product was contaminated with small amounts of other phases.

26.4.6 Nickel(II) Salts

The preparation of most nickel salts follows the conventional methods, *i.e.*, dissolving of the powdered metal, nickel carbonate, or hydroxide, in the appropriate acid. In this manner the following compounds are prepared:

Nickel perchlorate, $Ni(ClO_4)_2$,[153] its solvate with

---

132 *Fr. Weibke, G. Schrag*, Z. Elektrochemie *47*, 222 (1941).
133 *J. L. Andrieux*, Rev. Metall. *45*, 49 (1948); *J. L. Andrieux*, J. Four. Electr. *57*, 54 (1948).
134 *R. D. Blaugher, J. K. Huln, P. N. Yocum*, J. Phys. Chem. Solids *26* (12), 2037 (1965).
135 *Ganger*, Ann. Chim. Phys. *14* (7), 49 (1898).
136 *E. Larsson*, Arkiv Kemi *23* (32), 335 (1965).
137 *P. C. Donohue, T. A. Bither, H. S. Young*, Inorg. Chem. *7* (5), 998 (1968).
138 *H. Struve*, J. Prakt. Chem. *79*, 339 (1860).
139 *U. Hutter*, Ann. Chim. *8*, 450 (1953).
140 *J. Bolfa, R. Pastant*, Compt. Rend. Congr. Soc. Savantes, Dijon 309, (1959); (publ. 1960).
141 *R. J. Holmes*, Science *96*, 90 (1942).
142 *R. Pastant*, C.R. Congr. Nat. Soc. Savantes, Sect. Sci. *90* (2), 89 (1966).
143 *H. J. Goldschmidt, W. M. Ham*, J. Iron Steel Inst. (London), *202*, 347 (1964).
144 *R. A. Munson*, Inorg. Chem. *7* (2), 389 (1968).
145 *S. M. Kulifay*, J. Amer. Chem. Soc. *83*, 4916 (1962).
146 *O. S. DuMont, G. Uecker, W. Schaal*, Z. Anorg. Allgem. Chemie *370*, 67 (1969).
147 *K. Issleib, E. Wenschuh*, Z. Anorg. Allgem. Chemie *305*, 15 (1960).
148 *J. Leicester, M. Redman*, J. Appl. Chem. (London) *12*, 357 (1962).
149 *F. Anselin, D. Calais, G. Dean, A. Gaeynest*, C.R. Acad. Sci., Paris *257* (25), 3916 (1963).
150 *R. M. Ware*, Ultrapurif. Semicond. Water Proc. Conf., Boston, Mass. 192 (1961), (Publ. 1962).
151 *S. Rundquist*, Acta Chem. Scand. *13*, 1193 (1959).
152 *T. Hirai, K. Takahashi*, Denki Kagaku Oyobi Kogyo Butsuri Kagaku *35* (12), 886 (1967).
153 *H. Goldblum, F. Terlikovski*, Bull. Soc. Chim. France *11*, 103, 146 (1912).

Table 4 Preparation of Phosphine and Arsine Complexes According to Eqn (22) (p. 371)

Ligand (L)	Solvent	Conditions	X	*n*	Ref.
PH_3			C_5H_5	2	*a*
$(C_2H_5)_2PH$	ethanol, benzene		Br	4	*b*
$(CH_3)_3P$, $(CH_3)_2PCF_3$	evacd. sealed tube, ~20°		Cl, Br, I, NO_3	2	*c*
tertiary phosphines	from tetrafulminatonickel			2	*d*
$(aryl)_3P$	org. solvents	heating	CN	2	*e, f*
$arylPH_2$	ethanol	N_2-atm.	I	2	*g*
$C_6H_{11}PH_2$	benzene	reflux	Br	2	*h*
$(C_6H_5)_4P^{\oplus}$	toluene, benzene alcohols		Br	2	*i*
$(C_4H_9)_3PO$	$HCCO(C_2H_5)_3$	50°	ClO_4	4	*j*
$[(C_6H_5)_2P]_2$-*o*-carborane	$H_3C—COOC_2H_5$	reflux	Cl	2	*k*
$(aryl_3P$, $(alkyl)_3As$		ethanol		2	*l*
diarsine, triarsine	ethanol-ether		ClO_4	2	*m*

[a] *F. Klanberg, E. L. Muetterties*, J. Amer. Chem. Soc. *90* (12), 3296 (1968).
[b] *K. Issleib, G. Döll*, Z. Anorg. Allg. Chemie *305*, 1 (1960).
[c] *M. A. A. Beg, H. A. Clark*, Can. J. Chem. *39* (3), 595 (1961).
[d] *W. Beck, E. Schnierer*, Chem. Ber. *98* (1), 298 (1965).
[e] Ger. P. 1163814 (1964), BASF, Erf.: *G. N. Schrauzer*; C.A. *60*, 14541a (1964).
[f] *G. N. Schrauzer, P. Glockner*, Chem. Ber. *97* (9), 2451 (1964).
[g] *L. H. Pignolet, W. DeW. Horrocks, Jr.*, Chem. Commun. (17), 1012 (1968).
[h] *K. Issleib, H. R. Roloff*, Z. Anorg. Allgem. Chemie *324* (5–6), 250 (1963).
[i] *K. Issleib, G. Schwager*, Z. Anorg. Allgem. Chemie *310*, 43 (1961).
[j] *N. M. Karayannis, C. M. Mikulski, L. L. Pytlewski, M. M. Labes*, Inorg. Nucl. Chem. Lett. *5* (11), 897 (1969).
[k] *H. D. Smith, Jr.*, J. Amer. Chem. Soc. *87* (8), 1817 (1965).
[l] *K. A. Jensen*, Z. Anorg. Allgem. Chemie *229*, 265 (1936).
[m] *B. Bosnich, R. S. Nyholm, P. J. Pakling, M. L. Tobe*, J. Amer. Chem. Soc. *90* (17), 4741 (1968).

dioxane $Ni(ClO_4)_2 \cdot 6H_2O \cdot 5C_4H_8O_2$,[154] the *sulfite* $NiSO_3$,[155] *selenite* $NiSeO_3 \cdot 2H_2O$,[156] *selenate* $NiSeO_4 \cdot 6H_2O$,[157] *azide* $Ni(N_3)_2 \cdot H_2O$,[158] the *arsenates*,[159] the *tetrafluoroborate* $Ni(BF_4)_2 \cdot 6H_2O$,[160,161,162] *hexafluorosilicate* $NiSiF_6 \cdot 6H_2O$,[161,163] *tetrafluoroberillate* $Ni(BeF_4)_2$,[161] *pentafluoroaquoaluminate* $Ni(AlF_5H_2L)$[161] and its *hexahydrate*,[164] the *pentafluoroaquoferrate-(III)* Ni (FeF_5H_2O)[161] and its *hexahydrate*,[164] the *pentafluorogallate* $Ni(GaF_5)$[161] and its *heptahydrate*,[165] *hexafluorostanate* $Ni(SnF_6)$,[161] *hexafluorogermanate* $Ni(GeF_6)$,[161] *tetrafluorooxovanadate* $NiVOF_4$, $7H_2O$,[166] *sulfamate* $Ni(NH_2SO_3)_2$,[167,168,169] *hydroxyazide* Ni(OH)-

[154] *G. Vicentini, M. Perrier, E. Giesbrecht*, Chem. Ber. *94*, 1153 (1961).
[155] *J. S. Muspratt*, Justus Liebigs Ann. Chem. *50*, 259 (1844).
C. Rammelsberg, Justus Liebigs Ann. Chem. *67*, 391 (1846).
[156] *L. F. Nilson*, Bull. Soc. Chim. France *23* (2), 353 (1875).
[157] *J. Ferguson*, J. Chem. Soc. *127*, 2096 (1925).
[158] *L. Wöhler, F. Martin*, Z. Angew. Chem. *30*, 33 (1918).
[159] *B. C. Messance, A. M. Woerth*, Bull. Soc. Chim., France 574 (1962).
[160] *H. Funk, F. Binder*, Z. Anorg. Allgem. Chemie *155*, 327 (1926).
[161] *I. V. Tananaev, K. A. Avduevskaya*, Zh. Neorgan. Khim. *5*, 63 (1960).
[162] *A. Crouzet, S. Aleonard*, Bull. Soc. Fr. Mineral. Crystallogr. *92* (4), 388 (1969).
[163] *Topsoe, Christiansen*, Ann. Chim. Phys. *1* (5), 27 (1874).
[164] *R. F. Weinland, O. Köppen*, Z. Anorg. Allgem. Chemie *22*, 271 (1900).
[165] *E. Eineke*, Die Chemie, *55*, 40 (1942);
W. Pugh, J. Chem. Soc. 1937 (1959).
[166] *A. Piccini, G. Giorgis*, Gazz. Chim. Ital. *22*, 55 (1892).
[167] *B. F. Oberhauser, C. Urbina*, C.A. *41*, 1944 (1947).
[168] Neth. Appl. P. 6 515 184 (1966), Sherrit Gordon Mines Ltd., CA *65*, 14902d (1963).
[169] U.S. P. 3 321 273 (1967), Tenneco Chemicals Inc., Erf.: *A. Fischer;* CA *67*, 45673w (1967).

N_3,[170] the *chlorouranates* $Ni(UO_2Cl_4)\cdot 2H_2O$, and $Ni(UCl_6)$. $2H_2O$,[171] *tallium disulfitonickelate* $Tl_2[Ni(SO_3)_2]$,[172] *dibutylphosphate* Ni-$(Bu_2PO_2)_2$,[173] and the *diphosphatogermanate* Ni-$[GePO_4(HPO_2)_2]\cdot xH_2O$.[174]

Another conventional method involves heating of nickel oxide with a metal oxide. This method is widely utilized for the preparation of salts of the spinel type. The following salts were prepared by this procedure:

The *ferrate(III)* (*ferrite*) $NiFe_2O_4$[175] (900°), *rhodite* $NiRh_2O_4$ (1150°, oxygen atm),[176] *manganite* $NiMn_2O_4$ (760°),[177] *selenite* $NiSeO_3$,[178] *chromite* (400–500°),[179] (1350°, two days),[180] and $NiCrO_3$ (from NiO and CrO_2 at 1200° and 60 kilobars),[181] *molybdate* $NiMoO_4$ (1000–1500°),[182] *tungstate* $NiWO_4$ (1000–1500°),[182] (800–1000°),[183] Ni_3WO_6 (800–1000°),[184] and the *lithium nickel vanadate* (500–580°)[185] which was obtained according to Eqn (24).

$$6NiO + 3V_2O_5 + 2Li_2CO_3 + 2LiOH \longrightarrow 6LiNiVO_4 + 2CO_2 + H_2O \quad (24)$$

A modification of this latter method involves heating a mixture of the hydroxides or of the salts. Thus the *ferrite* $NiFe_2O_4$ was obtained by heating the hydroxides (400°),[186] the oxalates (500°),[187] nickel nitrate and hydrated iron oxide (480–1200°),[188] indignation of the complex $Ni_3Fe_6(OOCCH_3)_{17}O_3OH$, 12 pyridine,[189] or by oxidation of the metals at high temperature.[190]

The following compounds were prepared from the hydroxides; *metatitanate* $NiTiO_3$ (600°),[191] and the *rare earth nickelates* Pr_2NiO_4 and Nd_2NiO_4 (1200°).[192]

Various *phosphates* were prepared by fusion of alkali metal phosphates with nickel oxide, phosphate or carbonate: $Na_2NiP_2O_7$ (from NiO and $Na_3P_3O_9$),[193] $Ni(PO_3)_2$, $Ni_2P_2O_7$, and $Ni_3(PO_4)_2$ (900°),[194] polyphosphate (from NiO and $NaPO_3$ at 900°),[195] $[K_4Ni(PO_3)_6]_n$ (from KH_2PO_4, $(NH_4)_2HPO_4$, and $NiCO_3$ at 800°),[196] and $[Li_4Ni(PO_3)_6]_n$ (fusion of the phosphates).[197]

The *manganate(IV)* $NiMnO_3$ has been prepared from the oxalates at 600–780°.[198]

Heating of a mixture of nickel, sulfur, and phosphorus in a sealed tube produces *nickel thiopyrophosphate*, $Ni_2P_2S_7$.[199] The *thiophosphate*, $Ni_3(PS_4)_2$,[200] was prepared according to reaction (25).

$$3NiCl_2 + 2P_2S_5 \longrightarrow Ni_3(PS_4)_2 + 2PSCl_3 \quad (25)$$

The *thiomolybdate*, $NiMoS_4$, was obtained from powdered nickel and molybdenum disulfide under argon at high pressure at 1100°.[201]

---

[170] *Th. Curtius, J. Rissom*, J. Prakt. Chem. *58*, 261 (1898).
[171] *A. C. Vicente*, Acta Salamanticensia, Ser. Cienc. *6* (2), *43* (1963).
[172] *G. Canneri*, Gazz. Chim. Ital. *53*, 182 (1923).
[173] *V. A. Mikhailov, E. F. Grigor'eva*, Zh. Neorg. Khim. *9* (4), 867 (1964).
[174] *K. A. Avduevskaya, I. V. Tananaev, V. S. Mironova*, Izv. Akad. Nauk. SSSR, Neorgan. Mater. *1* (4), 554 (1965).
[175] *P. Kleinert*, Monatsber. Deut. Akad. Wiss. Berlin, *5*, 99 (1963).
[176] *S. Horiuchi, Sh. Miyahara*, J. Phys. Soc. Jap. *19* (3), 423 (1964).
[177] *G. Villers, R. Buhl*, C.R. Acad. Sci., Paris *260* (12) (Groupe 8), 3406 (1965).
[178] *R. L. Espil*, C.R. Acad. Sci., Paris *152*, 378 (1911).
[179] *H. Charosset, P. Turlier, Y. Trambouze*, J. Chim. Phys. *61* (9), 1257 (1964).
[180] *Y. Kino, Sh. Miyahara*, J. Phys. Soc., Jap. *20* (8), 1522 (1965).
[181] *B. L. Chamberland, W. H. Cloud*, J. Appl. Phys. *40* (1), 434 (1969).
[182] *P. S. Mamykin, N. A. Batrakov*, Tr. Ural. Politechn. Inst. *150*, 101 (1966).
[183] *L. G. Van Uitert, J. J. Rubin, W. A. Bonner*, J. Amer. Ceram. Soc. *46* (10), 512 (1963).
[184] *G. N. Belyaev, V. S. Filip'ev, E. G. Fesenko*, Zh. Strukt. Khim. *4* (5), 719 (1963).
[185] *J. C. Bernier, P. Poix, A. Michel*, Bull. Soc. Chim., France 1661 (1963).
[186] *A. Krause, A. Binkowna*, Monatsh. Chem. *95* (1), 1 (1963).
[187] *P. Kleinert, A. Funke*, Z. Chem. *1*, 155 (1961).
[188] *W. M. Keely, H. W. Maynor*, J. Chem. Eng. Data *9* (2), 170 (1964).
[189] *D. G. Wickham, E. R. Whipple, E. G. Larson*, J. Inorg. Nucl. Chem. *14*, 217 (1960).
[190] *E. Banks, N. H. Riederman, H. W. Schleuning, L. M. Silber*, J. Appl. Phys. *32* (3), 44 (1961).
[191] *Y. Saikali, J. M. Paris*, C.R. Acad. Sci., Paris, Ser. C *265* (19), 1041 (1967).
[192] *B. Willer, M. Daire*, C.R. Acad. Sci., Paris, Ser. C *267* (22), 1482 (1968).
[193] *R. Klement, E. Petz*, Monatsh. Chem. *95* (4–5), 1403 (1964).
[194] *J. F. Sarver*, Trans. Brit. Ceram. Soc. *65* (4), 191 (1966).
[195] *Yu. K. Delimarskii, V. N. Andreeva, T. N. Kapustova*, Izv. Akad. Nauk. SSR, Neorgan. Mater. *1* (1), 150 (1965).
[196] *R. C. Mehrotra, P. C. Vyas*, Indian J. Chem. *6* (4), 204 (1968).
[197] *R. C. Mehrotra, C. K. Oza*, Indian. J. Chem. *6* (3), 158 (1968).
[198] Fr. P. 1 421 055 (1965), Ampex Corp., Erf.: *D. G. Wickham;* CA *65*, 13007d.
[199] *L. Ferrand*, Ann. Chim. Phys. *17* (7), 415 (1899).
[200] *E. Glatzel*, Z. Anorg. Allgem. Chemie *4*, 186 (1893).
[201] *R. Chevrel, M. Sergent, J. Prigent*, C.R. Acad. Sci., Paris, Ser. C *267* (18), 1135 (1968).

Some salts were prepared by high temperature halogenation. For example, *nickel hexafluoraplumbate*, $NiPbF_6$, was prepared by fluorination of nickel oxide and ammonium hexachloroplumbate at 500°,[202] and the chlorosulfonate resulted from nickel chloride on treatment with a mixture of chlorine and sulfur trioxide.[203]

The thermal decomposition of nickel salts may lead to other salts. For example, *nickel metastanate*, $NiSnO_3$, was prepared from nickel hexahydrostanate.[204]

Another convenient route for the preparation of nickel salts is the anion exchange of nickel sulfate and alkali earth salts of the appropriate acid. The insoluble alkali earth metal sulfate is filtered off and the solution is evaporated. The reaction of nickel chloride or nitrate with alkali metal salts in alcohol yields an insoluble alkali metal chloride or nitrate and the alcohol soluble nickel salt. The basic procedure is useful for the preparation of the following salts:

(a) From nickel sulfate and alkali earth metal salt in aqueous solution, according to Eqn (26).

$$NiSO_4 + BaA \longrightarrow NiA + BaSO_4 \qquad (26)$$

$Ni(ClO_2)_2$,[205] $Ni(ClO_3)_2$,[206] $Ni(BrO_3)_2$; $6H_2O$,[207] *thiocyanate*,[208] NiS_2O_6,[209] *trithionate* NiS_3O_6; $6H_2O$,[210] *tetrathionate* NiS_4O_6 and *pentathionate* NiS_5O_6,[211] *thiosulfate*,[212,213] and *hypophosphate* $Ni(H_2PO_2)_2$; $6H_2O$.[208]

(b) From nickel nitrate and sodium salts in alcohol solution according to Eqn (27).

$$Ni(NO_3)_2 + 2NaX \longrightarrow NiX_2 + 2NaNO_3 \qquad (27)$$

(X = F, Cl, Br, I, NO_2, SCN, $(C_6H_5)_4B$),[214] (X = $N(CN)_2$, $C(CN)_3$),[215] and (X = NCSe).[216]

(c) From nickel halide and the appropriate silver salt the *monofluorophosphate*, $NiPO_3F \cdot H_2O$, was obtained.[217]

Double salts have been obtained by evaporation from solutions: $Ni[BiI_4]_2$,[218] $Ni[Bi(SCN)_4]_2$,[219] $[Ni(NH_3)_6](BH_4)_2$,[220,221] $M_4Ni(NO_2)_6$ (M = alkali metal),[222] $NiUO_2F_4$, $4H_2O$,[223] $NiUF_{10}$, $8H_2O$,[224] NiU_2F_{12}, $2H_2O$,[225] $NiHfF_6$, $6H_2O$, and Ni_2HfF_8, $12H_2O$,[226], $Cs_2Ni(B_{10}H_{10}CH)_2$.[227]

If anion exchange is not possible the nickel salt must be extracted and the extract is subsequently evaporated. The *thioselenophosphate*, $Ni[(C_6H_5O)_2PSeS]_2$, was extracted with carbon tetrachloride and precipitated with petrol ether.[228]

Nickel salts of carboxylic acids (C_3–C_{17}) are prepared by heating nickel acetate and the acid at 118° at normal or reduced pressure in solvents of higher b.p. than the b.p. of acetic acid.[229]

A special method for the preparation of nickel salts is the interaction of tetracarbonyl nickel

---

202 *R. Homann, R. Hoppe*, Z. Anorg. Allgem. Chemie *368*, 271 (1969).
203 *G. P. Luchinskii*, Zh. Obshch. Khim. *8*, 1864 (1938).
204 *Th. Dupuis, V. Lorenzelli*, C.R. Acad. Sci., Paris *259* (25), 4585 (1964).
205 *G. R. Levi*, Atti Acad. Naz. Lincei, Classe Sci. Fis., Mat. Natur., Rend. *32* (5), 165 (1923).
206 *A. Wachter*, J. Prakt. Chem. *33*, 321 (1844).
207 *M. Marbach*, Ann. Phys. *94*, 412 (1855).
208 *H. Grossmann*, Ber. *37*, 559 (1904).
209 *Baker*, Chem. News *36*, 203 (1877).
210 *A. Meuwsen, G. Heinze*, Z. Anorg. Allgem. Chemie *269*, 86 (1952).
211 *O. v. Deines, E. Christoph*, Z. Anorg. Allgem. Chemie *213*, 209 (1933).
212 *Letts*, J. Chem. Soc. *23*, 424 (1870).
213 *A. A. Blesa*, An. Soc. Esp. Fis. Qium. *43*, 1135 (1947).
214 *D. M. L. Goodgame, L. M. Venanzi*, J. Chem. Soc. 616 (1963).
215 *H. Koehler*, Z. Anorg. Allgem. Chemie *331* (5–6), 237 (1964).
216 *V. V. Skopenko, G. V. Tsintsadze*, Zh. Neorg. Khim. *9* (11, 2675 (1964).
217 *E. B. Singh, P. C. Sinha*, J. Indian Chem. Soc. *41* (6), 407, 411 (1964).
218 *J. Dick, A. Maurer*, Rev. Roumaine Chim. *10* (7), 633 (1965).
219 *A. Cyganski*, Zeszyty Nauk. Politech. Lodz. Chem. *14*, 41 (1964).
220 *V. I. Mikheeva, N. N. Mal'tseva, Z. K. Sterlyadkina*, Zh. Neorg. Khim. *10* (10), 2380 (1965).
221 Ger. P. 1 070 148 Farbenfabriken Bayer (1959). Erf.: *E. Zirngiebel, A. Buerger;* CA *55*, 18039e (1961).
222 *P. Pascal*, Nouveau Traite de Chim. Mineral *17*, 800 (1963).
223 *R. L. Davidovich, Yu. A. Buslaev, L. M. Murzakhanova*, Izv. Acad. Nauk. SSSR, Ser. Khim. 3, 687 (1968).
224 *F. Montoloy, S. Maraval, M. Capestan*, C.R. Acad. Sci. Paris, Ser. C *266* (11), 787 (1968).
225 *F. Montoloy, P. Plurien*, C.R. Acad. Sci., Paris, Ser. C. *267* (17), 1036 (1968).
226 *R. L. Davidovich, Yu. A. Buslaev, T. F. Levchishina*, Izv. Acad. Nauk. SSSR, Ser. Khim. *3*, 688 (1968).
227 *W. H. Knoth, J. L. Little, L. J. Todd*, Inorg. Synth. *11*, 41 (1968).
228 *N. I. Zemlyanskii, N. M. Chernaya*, Ukr. Khim. Zh. *33* (2), 182 (1967).
229 *N. G. Kostyuk, A. A. Levchenko*, Zh. Prikl. Khim. *39* (7), 1654 (1966).

with haloderivatives[230] according to the following Eqn (28).

$$Ni(CO)_4 + CF_3SO_2Cl \longrightarrow CF_3SO_2NiCl + 4CO \quad (28)$$

Insoluble salts have been prepared by precipitation processes: *hexacyanomanganate* $Ni_3[Mn(CN)_6]_2$,[231] the *selenite* $NiSeO_3$ and *tellurite* $NiTeO_3$,[232] *tellurite dihydrate* $NiTeO_3 \cdot 2H_2O$,[233] *tellurite* $NiTeO_4$,[234] *orthotellurate* $Na_2Ni_2TeO_6$ (hydrothermally at 200° and 10–15 atm under oxygen),[235] the *molybdate* $NiMoO_4$, H_2O,[236] *molybdomaleate* $Ni(MoO_3C_4H_4O_5) \cdot 4H_2O$,[237] *vanadatotellurate* $3NiO \cdot 2TeO_3 \cdot 3V_2O_5 \cdot 8H_2O$,[238] the *amide* $Ni(NH_2)_2$ (from $Ni(SCH)_2$ and $NaNH_2$ in liquid ammonia),[239] the *tripolyphosphate* $NiNa_3P_3O_{10}$, $12H_2O$,[240] *dithiophosphoamidates* $Ni[(R'O)(R''NH)P(S)S]_2$ (R′ and R″ = alkyl),[241] *diphenylthioselenophosphate*,[242] *niobate* $4K_2O \cdot 2NiO \cdot Nb_2O_5 \cdot 5H_2O$,[243] the *orthosilicate* $NiSiO_4$ (from tetraethylsilicate and nickel nitrate),[244] the *hexahydroxystanate* $Ni[Sn(OH)_6]$,[245,248] the *germanate* $NiGeO_3 \cdot yH_2O$,[246] the *borane* $[(CH_3)_4N]_2[Ni(B_{10}H_{12})_2]$,[247] the *tetrabromocadmate* $Ni[CdBr_4]$,[248] and the *uranocarbonate* $Ni_2[UO_2(CO_3)_3] \cdot 6H_2O$.[249]

26.4.7 Alcoholates and Phenolates

The alcoholates and the phenolates of nickel are sensitive to moisture. The *methylate* was prepared in anhydrous methanol[250] according to Eqn (29).

$$NiCl_2 + 2CH_3ON_a \longrightarrow Ni(OCH_3)_2 + 2NaCl \quad (29)$$

The reaction seems to lead to non-stoichiometric products.[251] Lithium methylate was suggested as more convenient reagent providing for defined products.[252]

Various *phenolates* can be prepared in similar manner by treating phenoles with sodium ethylates in the presence of nickel chloride in an anhydrous solvent[253] according to Eqn (30).

$$2PhOH + 2C_2H_5ONa + NiCl_2 \longrightarrow (PhO)_2Ni + 2C_2H_5OH + 2NaCl \quad (30)$$

(PhOH = ring substituted phenol).

Nickel(0) complexes such as ethylenebis(triphenylphosphine)nickel(0) react with alkyl- or arylperoxides by forming the corresponding alkolate or phenolate[254] as depicted in Eqn (31).

$$(C_2H_4)Ni(PPh_3)_2 + R_2O_2 \longrightarrow Ni(OR)_2 + C_2H_4 + 2PPh_3 \quad (31)$$

(R = alkyl or aryl).

Another method for the preparation of phenolates requires treatment of the phenol with sodium ethyleneglycolate to yield the phenolate. The latter is reacted at 80° with nickel chloride and on dilution with ethyleneglycol the corresponding nickel phenolate precipitates.[255] An

[230] *E. Lindner, H. Weber, G. Vitzthum*, J. Organometal. Chem. *13* (2), 431 (1968).

[231] *A. Ferrari, E. Morisi, M. E. Tani*, Gazz. Chim. Ital. *93* (11), 1455 (1963).

[232] *G. Gattow, O. J. Lieder*, Naturwissenschaften *50* (21), 662 (1963).

[233] *V. Lenher, E. Wolesensky*, J. Amer. Chem. Soc. *35*, 718 (1913).

[234] *F. Fouasson*, Bull. Soc. Chim. France *5* (5), 1380 (1938).

[235] *S. M. Boudin*, Avtoklavn. Metody Pererabotki Miner. Syr'ya, Acad. Nauk. SSSR, Lol'sk. Filial 10 (1964).

[236] *H. Pezerat*, C.R. Acad. Sci. Paris *261* (25), 5190 (1965).

[237] *S. Prasad, L. P. Pandey*, J. Indian Chem. Soc. *42* (11), 783 (1965).

[238] *S. Prasad, K. Ch. Pathak*, Indian J. Chem. *5* (1), 14 (1967).

[239] *P. Pascal*, Nouveau Traite Chim. Mineral *17*, 797 (1963).

[240] *L. I. Prodan, M. F. Yarmolenka*, Vesti Akad. Nauk. Belerusk. SSR, Ser. Fiz. Tekhn. Nauk 3, 63 (1961).

[241] Fr. P. 1 377 118 (1964), Farbenfabriken Bayer, Erf.: *H. Malz;* CA *62*, 9011e (1965).

[242] *L. A. Il'ina, N. I. Zemlyanskii, S. V. Larinov, N. M. Chernaya*, Izv. Akad. Nauk. SSSR, Ser. Khim. 1, 198 (1969).

[243] *A. V. Lapiskii, V. I. Bezrukov, L. G. Vlasov*, Zh. Neorg. Khim. *11* (2), 312 (1966).

[244] *N. A. Toropov, S. A. Babayan*, Zh. Neorg. Khim. *11* (1), 28 (1966).

[245] *T. Dupuis, C. Duval, J. Lecomte*, C.R. Acad. Sci., Paris *257* (21), 3080 (1963).

[246] *J. P. Labbe*, Ann. Chim. (Paris) *10* (7–8), 317 (1965).

[247] *A. R. Siedle, T. A. Hill*, J. Inorg. Nucl. Chem. *31* (12), 3874 (1969).

[248] *M. M. M. Sanchez*, Acta Salamanticensia, Ser. Cienc. *6* (2), 61 (1963).

[249] *J. Jindra, S. Skramovsky*, Collect. Czech. Chem. Commun. *31* (7), 2639 (1966).

[250] *M. Nehme*, Chim. Mod. *5*, 231 (1960).

[251] *W. L. German, T. W. Brandon*, J. Chem. Soc. 526 (1942).

[252] *R. W. Adams, E. Bishop, R. L. Martin, G. Winter*, Australian J. Chem. *19* (2), 207 (1966).

[253] Belg. P. 626 767 (1963), Hercules Powder Co., Erf.: *M. L. Soeder;* CA *60*, 9202d (1964).

[254] *H. Schott, G. Wilke*, Angew. Chem. Intern. Ed. Engl. *8* (11), 877 (1969).

[255] Japan P. 21820 (1967), Mitsubishi Rayon Co., Erf.: *K. Senda, A. Ichikawa, M. Sasaki, K. Hirose;* CA *69*, 18841j (1968).

analogous procedure has been used for the preparation of *thiophenolates* such as *nickel bis-(pentafluorothiophenolate)*.[256] Thioalcoholates and thiophenolates were also prepared from alkyl- or aryldisulfides R_2S_2 in analogy to Eqn (31).

Some thioalcoholates are stable in aqueous solution; they can be obtained from the mercaptane, RSH (R=alkyl, C_2–C_8), sodium hydroxide and nickel(II) salt.[257] Thiophenols[258] and alkali phenolates[259] react similarly.

A mixed *methylate-phenolate*[260] and a complex of *nickel ethylate*[261] have also been prepared.

26.4.8 Chelates

Methods for the preparation of chelates of nickel(II) are similar to the methods for the preparation of salts. Many chelates are insoluble and precipitate from solution. The precipitation can proceed in aqueous solution or in organic solvents. The formed hydrogen ion is removed by base addition (alkali metal hydroxides, carbonate, ammonia, etc.) and the formation of chelates is favored at elevated temperatures. In this manner numerous chelates with *organic acids*,[262-265] *phenols*,[266,267] *oximes*,[268-270] substituted *ketones*,[271-273] and substituted *mercaptanes*[274-276] have been prepared. In alcoholic solution the neutralization of the hydrogen ion can be achieved with alkali metal alcoholate rather than the hydroxide.[277-279]

A modification of the precipitation method is a titration procedure in which the chelate precipitates on proper adjustment of the pH.[280,281] In many cases the precipitation occurs quantitatively thus providing for gravimetric determination of nickel. Besides the well known *dimethyldioxime* various other *oximes*,[282-286] *substituted acids*,[287] *mercaptanes*[288] and other *thioderivatives*[289,290] have been suggested for that purpose. Soluble chelates can be isolated in solid state by extraction and subsequent evaporation or by diluent precipitation. Numerous chelates are extractable with carbon tetrachloride, chloroform or benzene. For example, the extraction procedure has been used for the preparation of

---

256 *C. R. Lucas, M. E. Peach*, Inorg. Nucl. Chem. Lett. *5* (2), 73 (1969).
257 *C. J. Swan, D. L. Trimm.* Chem. Ind. (London) *32*, 1363 (1967).
258 *L. F. Larkworthy, J. M. Murphy, D. J. Phillips*, Inorg. Chem. *7* (7), 1436 (1968).
259 *A. Davison, R. H. Holm*, Inorg. Synth. *10*, 8 (1967).
260 Fr. P. 1 565 805 (1969), Intern. Nickel Ltd., Erf.: *D. P. Jordan;* CA *72*, 90056m (1970).
261 *J. Ellermann, W. H. Gruber*, Z. Naturforsch. B *23* (10), 1307 (1968).
262 Japan P. 23 930 (1967), Yoshitomi Pharmaceutical Industries, Erf.: *G. Hasegawa;* CA *68*, 97199m.
263 *B. R. Rios, A. M. Perez*, An. Quim. *64* (1), 47 (1968).
264 *V. Ya. Temkina, N. M. Dyatlova, B. V. Zhadanov, M. N. Rusina*, Zh. Neorg. Khim. *13* (6), 1570 (1968).
265 *P. Lumme*, Suomen Kemistilehti *32B*, 253 (1959).
266 *V. Carassiti, A. Seminara*, Ann. Chim. (Rome) *54* (7), 969 (1964).
267 *H. ElKhadem, W. M. Orabi*, Z. Anorg. Allgem. Chemie *365* (5–6), 315 (1969).
268 *M. Bartusek, A. Okac*, Collect. Czech. Chem. Commun. *26*, 883 (1961).
269 *J. S. Dave, A. M. Talati*, J. Indian. Chem. Soc. *37*, 578 (1960).
270 *M. Kuras, V. Stuzka, F. Kasparek, J. Mollin, J. Slonka*, Collect. Czech. Chem. Commun. *26*, 315 (1961).
271 U.S. P. 3 361 709 (1968), Esso Res. Eng., Erf.: *D. E. Brown, D. E. Nicholson, R. I. McDougall;* CA *68*, 50638p (1968).
272 *M. W. Blackmore, R. W. Cattrall, R. J. Magee*, Inorg. Nucl. Chem. Lett. *4* (5), 305 (1968).
273 *E. Uhlemann*, J. Prakt. Chem. *21* (5–6), 277 (1963).
274 *E. Wenschuh*, Z. Chem. *6* (6), 226 (1966).
275 *E. J. Olszewski, M. J. Albinak*, J. Inorg. Nucl. Chem. *27* (6), 1431 (1965).
276 *E. Hoyer, B. Lorenz*, Z. Chem. *8* (1), 28 (1968).
277 *L. Wolf, E. Jaeger*, Z. Chem. *5* (10), 392 (1965).
278 U.S. P. 3 296 191 (1967), FMC Corp., Erf.: *H. T. Smallwood, S. Tryon;* CA *66*, 38501q (1967).
279 *G. W. Everet, R. H. Holm*, Proc. Chem. Soc. 238 (1964).
280 *N. N. Ghosh, A. Bhattacharyya*, J. Indian Chem. Soc. *4* (4), 311 (1964).
281 *V. P. Kurbatov, O. A. Osipov, K. N. Kovalenko, T. M. Yuzbasheva*, Zh. Neorg. Khim. *12* (6), 1557 (1967).
282 *J. S. Dave, A. R. Patel*, J. Sci. Ind. Research. (India) *20B*, 81 (1961).
283 *N. K. Mathur, C. K. Narani*, Talanta *11* (3), 647 (1964).
284 *P. D. Jones, E. J. Newman jr.*, Analyst *90* (1067), 112 (1965).
285 *B. S. K. Rao, O. E. Hileman jr.*, Talanta *14* (3), 299 (1967).
286 *H. Singh, K. Ch. Sharma*, Indian, J. Appl. Chem. *28* (2), 43 (1965).
287 *J. Weger, T. Gancarczyk*, Fresenius' Z. Anal. Chem. *235* (5), 418 (1968).
288 *J. A. W. Dalziel, D. Kealey*, Analyst *88* (1059), 411 (1964).
289 *L. Kekedy, F. Makkay*, Studia Univ. Babes-Bolyai, Ser. Chem. *1*, 135 (1962).
290 *N. K. Dutt, B. K. Bhattacharjee*, Indian J. Appl. Chem. *27* (5–6), 195 (1964).

nickel chelates with substituted 8-quinolinoles,[291,292] dioximes,[293] Schiff bases,[294] *o*-substituted mercaptopurine[295] and others.

Strongly acidic chelating agents react with powdered nickel under evolution of hydrogen. Thus acetylacetone dissolves nickel powder (in ethanol) and yields *bis(acetylacetonato)nickel-(II)*.[296] Difluorodithiophosphoric acid, $HSPSF_2$, reacts similarly.[297]

Nickel salts of unstable or volatile acids can be treated with more stable or non-volatile chelating acids. The reaction can be performed directly with or without an organic diluent. Starting materials for this type of reaction are:

(a) nickel carbonate (for the preparation of chelates with m-*phenylenediamine tetraacetic acid*,[298] *acetylacetone*,[299] *dithiocarbazide*,[300] *alkylsulfonylmethanic acid*,[301] etc.);
(b) nickel acetate (by heating the chelating compound with the nickel salt on an oil bath at 130° until the odor of acetic acid disappears[302]);
(c) nickel acetylacetonate (for the preparation of chelates of carboxilic acids such as *cinnamic acid*[303]).

If *nickel bis(acetylacetonate)* is treated in methanol with a base, the *methoxichelate*[304] results according to Eqn (32).

$$NiA_2 + 2CH_3OH \longrightarrow NiA(OCH_3)CH_3OH + AH \quad (AH = \text{acetylacetone}) \qquad (32)$$

The chelates $Ni(HN_2S_2)_2$, $Ni(NS_3)_2$, and $Ni(HN_2S_2)(NS_3)$ have been separated by chromatography.[305]

Some chelates, especially some thioderivatives, have been prepared *in situ* with the chelating agent. Typical examples are the nickel chelates with *α,β-dithioketones* which were prepared in toluene at 120° from arylacetylene and nickel polysulfide,[306] and from *β*-diketones in the presence of nickel salt with hydrogen sulfide and chloride.[307] The chelates with *diselenoacetylacetone*,[308] *thioacetylacetone*,[309] macrocycles (from aromatic amines and arylsubstituted diamines,[310] or from *β*-oxaldimines and *1,3-diaminopropane*[311]) have been prepared in similar fashion.

The nickel chelate with nitrogen sulfide, $Ni(HN_2S_2)_2$ has been refluxed in *p*-dioxane with phenylisothiocyanate[312] to give the chelate.

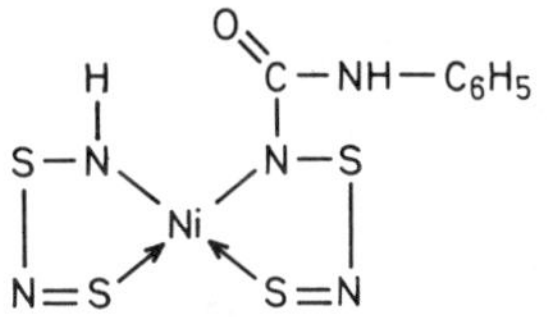

Complexes containing mobile hydrogen readily convert into chelates on treatment with base according to Eqn (33)

$$[Ni(LH)_2]X_2 + 2KOH \longrightarrow NiL_2 + 2H_2O + 2KX \qquad (33)$$

(X is an anion such as ClO_4.)

The deprotonation can be carried out by alkali

---

[291] *F. Ulmand, K. U. Meckenstack*, Z. Anal. Chem. *177*, 244 (1960).
[292] *K. Motojima, H. Hashitani*, Bunseki Kagaku *9*, 151 (1960).
[293] *A. H. I. Ben-Bassat, I. Binenboym*, Chemist Analyst *52* (4), 103 (1963).
[294] *O. Wawschinek, E. Weiss*, Mikrochim. Icnoanallyt Acta 5, 690 (1964).
[295] *A. Hampton, J. C. Fratantoni*, J. Med. Chem. *9*, 976 (1966).
[296] *O. Yoda, K. Suzuki, M. Shinohara, M. Nagakubo*, Nippon Kagaku Zasshi *87* (8), 890 (1966).
[297] *F. N. Tebbe, H. W. Roesky, W. C. Rode, E. L. Muetterties*, J. Amer. Chem. Soc. *90* (13), 3578 (1968).
[298] *E. Uhlig, D. Herrmann*, Z. Chem. *4* (11), 436 (1964).
[299] *V. A. Dontsova, N. F. Barabanshchikova*, Zh. Prikl. Khim. *40* (6), 1396 (1967).
[300] Brit. P. 963 924 (1964), Takeda Chem. Ind., Erf.: *R. Hatta, I. Sumina, J. Kinugawa, B. Tamura, H. Yamamoto, Sh. Suzuki;* CA *61*, 9987g (1964).
[301] *W. Beck, O. Johansen, W. P. Fehlhammer*, Z. Anorg. Allgem. Chemie *361* (3–4), 147 (1968).
[302] *K. Kitano, T. Mizuno, M. Yamaguchi*, Mem. Fac. Sci., Kyushu Univ. Ser. C *5* (4), 125 (1964).
[303] *E. M. Brainina, M. Kh. Minacheva, R. Kh. Freidlina, A. N. Nesmeyanov*, Dokl. Akad. Nauk. SSSR *138*, 598 (1961).
[304] *J. A. Bertrand, D. Caine*, J. Amer. Chem. Soc. *86* (11), 2298 (1964).
[305] *J. Weiss, U. Thewalt*, Z. Anorg. Allgem. Chemie *364* (5–6), 234 (1966).
[306] *G. N. Schrauzer, V. Mayweg*, Z. Naturforsch. *19b* (3), 192 (1964).
[307] *A. Ouchi, M. Nakatani, Y. Takahashi*, Bull. Chem. Soc., Japan *41* (9), 2044 (1968).
[308] *G. A. Heath, I. M. Stewart, R. L. Martin*, Inorg. Nucl. Chem. Lett. *5* (3), 169 (1969).
[309] *C. G. Barraclough, R. L. Martin, I. M. Stgewart*, Australian J. Chem. *22* (5), 891 (1969).
[310] *M. Green, P. A. Tasker*, Chem. Commun. *9*, 518 (1968).
[311] *E. Jeager*, Z. Chem. *8* (10), 392 (1968).
[312] *J. Weiss, U. Thewalt*, Z. Anorg. Allgem. Chemie *343* (5–6), 274 (1966).

hydroxide,[313–316] potassium amide,[317] or sodium tetrahydridoborate.[318]

26.4.9 Complexes

Many *nickel(II) complexes* are formed by coordination of donor molecules with a nickel salt according to Eqn (34).

$$NiX_2 + nL \longrightarrow NiL_nX_2 \quad (34)$$

(L = donor molecule; X = monovalent anion).

Often the complexes are less soluble than the starting materials and precipitate readily or crystallize on standing.

Many complexes containing nitrogen as donor atom[319–326] are prepared in aqueous solution. Others are prepared from nickel salts and the appropriate ligand in organic solvents such as alcohols, acetone, or tetrahydrofurane; alcohols such as methanol,[327–331], ethanol,[332–335] butanol,[336] and amylalcohol[337] are among the more common solvents. Many *aminocomplexes* crystallize after dilution of their aqueous solution with alcohol (C_1–C_3).[338, 339] In other instances the solvent must be evaporated until crystallization begins[340, 341]; this is particularly true for perchlorates.[342] Some ligands coordinate only in refluxing solution.[343–345]

Many *anhydrous nickel salts* (halides, nitrates, thiocyanates, etc.) are dissolved in a ligand solvent and formation occurs on heating; the desired compounds precipitate on cooling (Table 5).

Moisture-sensitive complexes can be prepared in anhydrous organic solvents in the presence of dehydrating agents such as 2,2-dimethoxypropane,[346, 347] or triethoxymethane.[348]

Some complexes are stable in solid state only and are usually prepared by interaction of stoichiometric quantities of the reactants.[349] The reaction proceeds at room temperature (nickel sulfate and thiourea 1:4[350]), on moderate heating (nickel chloride and 2-chloro-2-butene at 60°,[351] nickel bromide, aluminum bromide and hexamethylbenzene at 145°),[352] or, in some cases,

313 *M. A. Robinson, S. I. Trotz, T. J. Hurley*, Inorg. Chem. *6* (2), 392 (1967).
314 *W. Seidel*, Z. Chem. *7* (12), 462 (1967).
315 *R. L. Harris, A. W. Johnson, I. T. Kay*, Chem. Commun. *15*, 355 (1965).
316 *T. J. Hurley, M. A. Robinson*, Inorg. Chem. *7* (1), 33 (1968).
317 *I. W. Watt, J. F. Knifton*, Inorg. Chem. *6* (5), 1010 (1967).
318 *Y. Takahashi, M. Nakatani, A. Ouchi*, Bull. Chem. Soc., Japan *42* (1), 274 (1969).
319 *H. Grossmann, F. Hünseler*, Z. Anorg. Allgem. Chemie *46*, 392 (1905).
320 *N. I. Lobanov*, Zh. Neorg. Khim. *12* (10), 2582 (1967).
321 *A. Ravi, J. Gopalakrishnan, C. C. Patel*, Indian J. Chem. *5* (8), 356 (1967).
322 *L. P. Eddy, W. W. Levenhagen, Sh, K. McEven*, Inorg. Synth. *11*, 89 (1968).
323 *S. N. Avakyan, K. Voskanyan*, Zh. Obshch. Khim. *39* (5), 1098 (1969).
324 *T. A. Donavan, J. C. Bailar*, J. Inorg. Nucl. Chem. *26* (7), 1283 (1964).
325 *Sh. Utsuno, K. Sone*, Bull. Chem. Soc., Japan *37* (7), 1038 (1964).
326 *A. K. Ghosh, S. Chatterjee*, J. Inorg. Nucl. Chem. *26* (8), 1459 (1964).
327 *W. Hieber, E. Levi*, Z. Anorg. Allgem. Chemie *219*, 225 (1934).
328 *H. Brintzinger, B. Hesse*, Z. Anorg. Allgem. Chemie *252*, 293 (1944).
329 *F. L. Urbach, J. E. Sarneski, L. J. Turner, D. H. Busch*, Inorg. Chem. 1 *7* (10), 2169 (1968).
330 *J. A. Costamagna*, J. Inorg. Nucl. Chem. *30* (19), 2547 (1968).
331 *A. V. Ablov, N. V. Gerbeleu, N. Ya. Negryaste*, Zh. Neorg. Khim. *15* (1), 119 (1970).
332 *E. Lippmann, G. Vortmann*, Ber. *12*, 79 (1879).
333 *G. Spacu, C. C. Macarovici*, Bull. Soc. Stiinte Cluj *7*, 227 (1933).
334 *E. F. King, M. L. Good*, Inorg. Nucl. Chem. Lett. *5* (8), 631 (1969).
335 *S. N. Poddar*, Sci. Cult (Calcutta) *35* (1), 28 (1969).
336 *L. Sacconi, I. Bertini*, Inorg. Nucl. Chem. Lett. *2* (1), 29 (1966).
337 *P. Souchay*, Bull. Soc. Chim., France *7*, 797 (1940).
338 *J. Bradley*, School Sci. Rev. *47* (161), 65 (1965).
339 *A. V. Butcher, D. J. Phillips, J. P. Redfern*, J. Inorg. Nucl. Chem. *28* (11), 2765 (1966).
340 *A. Takeuchi, K. Sato, K. Sone, S. Yamada, K. Yamasaki*, Inorg. Chim. Acta *1* (3), 399 (1967).
341 *H. Greossmann, B. Schück*, Z. Anorg. Allgem. Chemie *50*, 1 (1906).
342 *J. L. Karn, D. H. Busch*, Nature *211* (5045), 160 (1966).
343 *E. A. Rick, R. L. Pruett*, Chem. Commun. *19*, 697 (1966).
344 Fr. P. 1 554 114 (1969), Union Carbide Corp. CA *72*, 78541h (1970).
345 *H. Buerger, U. Wannagatt*, Monatsh. Chem. *95* (4–5), 1099 (1964).
346 *K. M. Nykerk, D. P. Eyman, R. L. Smith*, Inorg. Chem. *6* (12), 2262 (1967).
347 *A. V. Butcher, D. J. Phillips, J. P. Redfern*, J. Inorg. Nucl. Chem. *30* (1), 325 (1968).
348 *P. W. N. M. van Leeuwen, W. L. Groeneveld*, Rec. Trav. Chim. Pays-Bas *86* (7), 721 (1967).
349 *D. Negoiu, V. Croitoru*, Analele Univ. Bucuresti, Ser. Stiini. Nat. *12* (41), 25 (1963).
350 *M. R. Udupa, G. Aravamudan*, Curr. Sci. *38* (1), 14 (1969).
351 *S. N. Avakyan, R. A. Karapetyan*, Arm. Khim. Zh. *19* (7), 490 (1966).
352 *G. H. Lindner, E. O. Fischer*, J. Organometal. Chem. *12* (2), 18 (1968).

Table 5 Formation of Complexes with the Solvent as a Ligand

Solvent = Ligand (L)	Treatment	Complex	X	Ref.
pyridine	heating	NiL_4X_2, NiL_6X_2	F. Cl, Br, I	*a*
quinoline	refluxing	NiL_2X_2	F, Cl, Br, I	*b*
acetonitrile	recrystalln.	NiL_6X_2, NiL_2X_2	ClO_4, Cl, Br	*c, d*
1,4-dioxane	recrystalln.	NiL_2X_2		*e*
dimethylsulfoxide	80°	NiL_6X_2	NO_3, Br, I, SCN	*f*
tetrahydroxy-thiophenoxide, on P_2O_5		NiL_6X_2	Cl, Br, I, SCN	*g*
acetylene (liquid)	high press., 25°, N_2	NiL_2X_2 NiL_3X_2	Br	*h*

[a] *I. I. Kalinichenko, A. A. Knyazeva*, Dokl. Akad. Nauk SSSR *169* (4), 887 (1966).
[b] *D. H. Brown, R. N. Nuttall, D. W. A. Sharp*, J. Inorg. Nucl. Chem. *26*, 1151 (1964).
[c] *P. W. N. M. van Leeuwen, W. L. Groeneveld*, Inorg. Nucl. Chem. Lett. *3* (4), 145 (1967).
[d] *B. J. Hathaway, D. G. Holah*, J. Chem. Soc. (London) 2400 (1964).
[e] *P. Pascal*, Nouveau Traite de Chim. Miner. *18*, 776 (1959).
[f] *F. A. Cotton, R. Francis*, J. Inorg. Nucl. Chem. *17*, 62 (1961).
[g] *R. Francis, F. A. Cotton*, J. Chem. Soc. (London) 2078 (1961).
[h] U.S. P. 3474120 (1969), Air Reduction Co., Erf.: *R. J. Tedeschi, G. L. Moore*; C.A. *71*, 125176j (1969).

exothermally (nickel chloride and 1-dimethylamino-5-methyl-2,3,4-hexatriene).[353]
Solid state reactions can also be used to convert hydrated nickel salts into solvates of acetic acid[354] according to Eqn (35)

$$NiX_2 \cdot 6H_2O + 6(CH_3CO)_2O \longrightarrow NiX_2(CH_3COOH)_6 + 6CH_3COOH \quad (35)$$

($X = BF_4$, ClO_4, NO_3).
The anions are replaceable. Thus a complex of the type $NiL_n(OH)_2$ can be transformed into a complex NiL_nX_2 by reaction with an ammonium salt, NH_4X.[355,356] In solution, replacement of the anions proceeds in the presence of a large excess of the replacing anion. Thus coordinated nickel chloride can be converted into the *bromide, iodide, cyanide*, or *thiocyanate*.[357] Sulfates can be prepared by anion exchange according to Eqn (36).[358]

$$NiL_n(SCN)_2 + Ag_2SO_4 \longrightarrow NiL_nSO_4 + 2AgSCN \quad (36)$$

Ligand displacement reactions are also used frequently. Thus the ammonia molecules of aminonickel halides can be replaced by amines in acetone solution and in the presence of triphenyltine bromide or triphenyllead bromide as activator.[359]
Various modifications of the cited general methods have been described. For example, alkylamines were distilled on a nickel halide and the mixture was allowed to stand for several days.[360] Nickel acetate and sodium nitrite have been reacted similarly with an excess of ethylenediamine in order to prepare *trisethylenediamine-nickel nitrite*.[361]
Some *nitrile*[362] or *trimethylphosphine*[363] *complexes* have been prepared by heating the ligand and nickel salt (halide, cyanide, thiocyanate, or nitrate) in a sealed tube.
The nitrogen atom of the donor is frequently bridged by hydrogen bonding. In this case no complex formation occurs in acid or neutral solution though coordination is readily accomplished in alkaline medium. Consequently, many complexes have been prepared by adding alkali

[353] *S. N. Avakyan, A. V. Mushegyan*, Zh. Obshch. Khim. *36* (3), 563 (1966).
[354] *P. W. N. M. van Leeuwen, W. L. Groeneveld*, Rec. Trav. Chim. Pays-Bas *87* (1), 86 (1968).
[355] *C. Gheorghiu, A. Nicolaescu*, An. Univ. Bucuresti, Stiint. Natur. Chim. *17* (1), 103 (1968).
[356] *S. P. Ghosh, A. K. Banerjee*, J. Indian Chem. Soc. *41* (4), 275 (1964).
[357] *W. Hieber, I. Bauer, G. Neumair*, Z. Anorg. Allgem. Chemie *335* (5–6), 250 (1965).
[358] *F. G. Mann*, J. Chem. Soc. *131*, 2904 (1927).
[359] *W. Jehn*, Z. Chem. *4* (8), 307 (1964).
[360] *E. Uhlig, K. Staiger*, Z. Anorg. Allgem. Chemie *336* (3–4), 179 (1965).
[361] *Yu-Pin Wang, Yu-Yun Tai, Ching-Illin*, Hua Hsueh Hsueh Pao *31* (4), 343 (1965).
[362] *A. V. Babaeva, Kh, U. Ikramov*, Zh. Neorg. Khim. *9* (3), 591 (1964), ibid. 596.
[363] *K. A. Jensen, O. Dahl*, Acta Chim. Scand. *22* (3), 1044 (1968).

metal hydroxide[364] or ammonia[365] to a solution of nickel salt and ligand.

Another method for the preparation of complexes involves treating a nickel salt with gaseous ligand. Complexes with ammonia,[366,367] or carbon dioxide,[368,370,371] were prepared in this manner.

Insoluble nickel salts such as the metaborate, $Ni(BO_2)_2$, chromate, $NiCrO_4$, or oxalate, NiC_2O_4, have been suspended in acetone in the presence of amines to form complexes of the type NiL_nX_2.[369]

Zerovalent nickel complexes can also be used as starting materials for the preparation of nickel(II) complexes. For example, two molecules of carbon monoxide or phosphine are readily lost on treatment with gaseous hydrogen chloride[370] or cyanogen.[371] Organic haloderivatives react with tetracarbonylnickel(0)[372] according to Eqn (37).

$$Ni(CO)_4 + LCl_2(\text{or } 2LCl) \longrightarrow LNiCl_2(\text{or } L_2NiCl_2) + 4CO \quad (37)$$

Complexes containing *macrocycles* as *ligands* are usually prepared *in situ* with the ligand.[373] Another route to nickel complexes involves conversion of a chelate by acetylation according to Eqn (38),[374]

$$2NiL_2 + 2AcCl + 2HCl \longrightarrow Ni(HL)_2Cl_2 + NiCl_2 + 2LAc \quad (38)$$

or by additional coordination of ligands as shown in Eqn (39).

$$NiL_2 + nZ \longrightarrow NiZ_nL_2 \quad (39)$$

Refluxing favors the addition of donor molecules to a chelate. Various chelates coordinate further with amines,[375,376] pyridine,[377,378] tin tetrachloride[379,380] and others.

Complexes with coordination number 2 or 4 have been prepared by thermal decomposition of hexacoordinated complexes. The decomposition proceeds *in vacuo* at moderate temperatures.[381,382]

Finally, ligand changes such as the oxidation of phosphine complexes to the corresponding phosphine oxide complexes should be mentioned.[383] (See also: Nickel complexes with Schiff bases, 26.4.11.)

26.4.10 Clathrates

Some crystalline nickel complexes can loosely bind organic molecules according to scheme (40).

$$NiL_nX_2 + mZ \longrightarrow NiL_nX_2 \cdot mZ \quad (40)$$

The resultant "*clathrates*" are similar in nature to the solvates. The organic molecule Z is mobile and is readily removed by moderate heating. Clathrates are prepared in the following manner.

(a) Suspension of a nickel complex in a solution of the compound to be clathrated.[384]
(b) Precipitation of a nickel complex in a solution containing the compound to be clathrated.[385–388]

---

[364] *R. Nasanen, P. Merilainen, P. Lappi*, Suomen Kemistilehti *37B* (11), 204 (1964).
[365] *P. Ray, S. K. Shiddhanta*, J. Indian Chem. Soc. *20*, 200 (1943);
P. Ray, C. Goutam, J. Indian Chem. Soc. *27*, 411 (1950).
[366] *C. Rammelsberg*, Ann. Phys. *48*, 119 (1839); *55*, 243 (1842).
[367] *E. Jona, T. Sramko, J. Gazo*, Chem. Zvesti *22* (9), 648 (1968).
[368] *A. Braibanti, G. Bigliardi, A. M. M. Lanfredi, A. Tiripicchio*, Nature *221* (5054), 1174 (1966).
[369] *G. Narain*, Z. Anorg. Allgem. Chemie *347* (3–4), 215 (1966); *G. Narain*, Curr. Sci. *35* (8), 202 (1966).
[370] *M. J. Hudson, R. S. Nyholm, M. H. B. Stiddard*, J. Chem. Soc. A 1, 40 (1968).
[371] *B. J. Argento, P. Fitton, J. E. McKeon, E. A. Rick*, J. Chem. Soc. D. *23*, 1427 (1969).
[372] *R. Criegee, G. Schroeder*, Justus Liebigs Ann. Chem. *623*, 1 (1959).
[373] *L. F. Lindoy, D. H. Busch*, J. Amer. Chem. Soc. *91* (17), 4690 (1969).
[374] *R. A. Krause, D. C. Jicha, D. H. Busch*, J. Amer. Chem. Soc. *83*, 528 (1961).
[375] Japan P. 17 563 (1965), Asahi Chem. Ind., Erf.: *S. Senoh, Sh. Ouchi, M. Kurihara;* CA *63*, 15896c (1965).
[376] Brit. P. 943 081 (1963), Amer. Cyanamide Co., CA *60*, 10261h (1964).
[377] *R. W. Kluiber, W. DeW. Horrocks jr.*, Inorg. Chem. *6* (1), 166 (1967).
[378] *J. Csaszar, J. Szeghalmi*, Acta Univ. Szeged, Acta Phys. Chem. *12* (3–4), 177 (1966).
[379] *V. T. Panyushkin, V. P. Kurbatov, A. D. Garnovskii, O. A. Osipov, V. I. Minkin, K. N. Kovalenko*, Zh. Neorg. Khim. *12* (3), 819 (1967).
[380] *D. Coucouvanis*, J. Amer. Chem. Soc. *92* (3), 707 (1970).
[381] *F. A. Cotton, R. Francis*, J. Inorg. Nucl. Chem. *17*, 62 (1961).
[382] *R. Duval, C. Duval*, Anal. Chim. Acta *5*, 71 (1951).
[383] *W. E. Daniels*, Inorg. Chem. *3* (12), 1800 (1964).
[384] *E. Boyl*, Chemiker-Ztg. *87* (24), 883 (1963).
[385] Belg. P. 59 1872 (1960), Labofina Soc. Anon. CA *55*, 19167a, c (1961).
[386] Belg. P. 59 5746 (1961), Labofina Soc. Anon. CA *55*, 14379a, c (1961).
[387] *G. Gawalek, H. Trautmann, H. Guenther*, Chem. Tech. (Leipzig), *16* (7), 409 (1964).
[388] U.S. P. 3 161 694 (1964), Union Oil Co. of California, Erf.: *W. D. Schaeffer;* CA *62*, 16116e (1965).

(c) Treatment of a suspension of a double salt with a ligand and the compound to be clathrated[389] according to Eqn (41).

$$CuNi(CN)_4 + 2NH_3 + C_6H_6 \longrightarrow [Cu(NH_3)_2Ni(CN)_4 \cdot C_6H_6] \quad (41)$$

Clathrates of the general constitution, $[M(NH_3)_2Ni(CN)_4 \cdot 2Z]$, (M = Fe, Co, Cu, Cd, Mn, Zn; Z = C_6H_6, $C_6H_5NH_2$) have been prepared from aqueous solutions of potassium tetracyannonickelate, the chlorides of the bivalent metals, and ammonia in the presence of benzene or aniline.[390]

Good clathrating agents are the aminonickel(II) cyanides and thiocyanates. They are obtained from nickel chloride, the amine, and potassium thiocyanate or cyanide[385–388, 391] in the presence of the compounds to be clathrated, or from nickel thiocyanate, the ligand and the clathrated molecule.[392]

The clathration proceeds specifically and provides a good method for the separation of *o*-, *m*- and *p*-isomers of benzene, toluene, and other aromatic compounds.[392]

26.4.11 Complexes and Chelates with Schiff Bases

Schiff bases are condensation products obtained from amines and aldehydes or ketones. Depending on the substituents Schiff bases form either chelates or complexes. The methods for the preparation of these compounds do not differ from the general methods for the preparation of chelates and complexes. The most useful are the following four methods.

(1) The Schiff base is treated with a nickel salt in an appropriate solvent. The treatment may include refluxing, alkalizing, or dehydration.[393–397]

(2) The Schiff base is synthesized in the presence of nickel salt. For example, the appropriate aldehyde or ketone is added to a solution of amine and nickel salt and the mixture is heated.[398, 399]

(3) Aminonickel complexes are treated with ketones[400–402] or aldehydes[403, 404] in solution.

(4) A solution of an aldehydatonickel or ketonickel complex is reacted with a solution of the amine.[405–407]

26.4.12 Polymers

The formation of complexes or chelates with bidentate or polydentate organic molecules leads to polymers of two different types.

(1) The monomers condense to macromolecular species which are then capable of forming complexes or chelates. Polymers of this type can be prepared by heating a solution or a mixture of the organic polymer and the nickel salt.[408–411]

A modification of this method involves the polymerization of the monomers in the presence of nickel salts. This procedure has been used for the preparation of polymer chelates with *pyromellitonitrile*,[412] *ethylenebis(dithiocarbamate)*, *hexamethylenebis(dithiocarbamate)*[413] and others.

---

[389] *R. Bauer, G. Schwarzenbach*, Helv. Chim. Acta *43*, 842 (1960).

[390] *T. Nakano, T. Miyoshi, T. Iwamoto, Y. Sasaki*, Bull. Chem. Soc. Japan *40* (5), 1297 (1967).

[391] *V. M. Bhatnagar*, Z. Anorg. Allgem. Chemie *350*, (3–4), 214 (1967).

[392] *E. M. Terent'eva, N. N. Kaptsov, G. M. Mamedaliev, T. N. Buturlova*, Neftekhimiya *8* (4), 633 (1968).

[393] *A. S. Kudryavtsev, I. A. Savich*, Zh. Obshch. Khim. *33* (11), 3763 (1963).

[394] *D. A. House, N. F. Curtis*, J. Amer. Chem. Soc. *86* (2), 223 (1964).

[395] *H. Jadamus, Q. Fernando, H. Freiser*, Inorg. Chem. *3* (6), 928 (1964).

[396] *M. Augustin, H. J. Kerrinnes, W. Langenbeck*, J. Prakt. Chem. *26* (3–4), 130 (1964).

[397] *L. Sacconi*, Inorg. Chem. *7* (5), 1034 (1968).

[398] *A. P. Terent'ev, G. V. Panova, E. G. Ruchadze*, Zh. Obshch. Khim. *34* (9), 3013 (1964).

[399] *M. S. Elder, G. A. Melson, D. H. Busch*, Inorg. Chem. *5* (1), 74 (1966).

[400] *N. F. Curtis, D. A. House*, J. Chem. Soc. 5502 (1965).

[401] *D. E. Goldbery*, J. Chem. Soc. A *11*, 2671 (1968).

[402] *M. J. Lacey*, Australian J. Chem. *23* (4), 841 (1970).

[403] *Th. E. MacDermott, D. H. Busch*, J. Amer. Chem. Soc. *89* (23), 5780 (1967).

[404] *D. A. House, N. F. Curtis*, J. Amer. Chem. Soc. *86* (7), 1331 (1964).

[405] *E. Uhlig, B. Machelett*, Z. Chem. *9* (4), 155 (1969).

[406] *A. Takeuchi, Sh. Yamada*, Bull. Chem. Soc., Japan *42* (10), 3046 (1969).

[407] *L. T. Taylor, W. M. Coleman*, J. Amer. Chem. Soc. *92* (5), 1449 (1970).

[408] *R. H. Horrocks, E. C. Winslow*, J. Polym. Sci. A *1* (12), 3655 (1963).

[409] *R. Liepins, G. S. P. Verma, C. Walker*, Macromolecules *2* (4), 419 (1969).

[410] *V. M. Bondarenko, A. F. Nikolaev*, Vysokomolekul. Soedin. *7* (12), 2105 (1965).

[411] *S. S. Ivanov, L. P. Gavryuchenkova, M. M. Koton*, Vysokomolekul. Soedin. *8* (3), 470 (1966).

[412] *Y. Nose, N. Sera, M. Hatano, Sh. Kambara*, Kogyo Kagaku Zasshi *67* (10), 1600 (1964).

[413] *L. C. Thompson, R. O. Moyer*, J. Inorg. Nucl. Chem. *27* (10), 225 (1965).

(2) The polymer represents organic molecules bridged by nickel atoms. Polymers of this type have been prepared from *5-amino-8-hydroxyquinoline* and *terephtalaldehyde*.[414] Its structure is illustrated by the formula below.

Polymers with *hydroxyanthraquinones* have been prepared in similar fashion.[415]

Products with similar structure can be prepared by heating a nickel chelate with monomer. For example, a mixture of nickel bisacetylacetonate and phyenylphosphoric acid was heated to produce a polychelate.[416]

26.4.13 Organonickel Compounds

Organonickel compounds can have one or two organic radicals σ-bonded to a nickel atom. They are prepared from anhydrous nickel(II) halide (the bromide is preferable) and a Grignard reagent according to reaction (42).

$$NiBr_2 + 2RMgBr \longrightarrow NiR_2 + 2MgBr_2 \quad R = \text{alkyl aryl} \qquad (42)$$

The reaction proceeds at low temperature in anhydrous solvents and in inert atmosphere (Table 6).

Table 6 Synthesis of Organonickel Compounds according to Scheme (42)

R	Solvent	Temp. °C	Ref.
$CH_2{=}CHCH_2$—	ether	−10	*a, b*
2,4,6-$(CH_3)_3C_6H_2$—	tetrahydrofurane	−20	*c*
C_6H_5—	ether		*d*
crotyl	ether	−10	*e*

[a] *G. Wilke, B. Bogdanovic*, Angew. Chem. *73*, 756 (1961).
[b] Belg. P. 631172 (1963), Studiengeselschaft Kohle, Erf.: *G. Wilkinson*; C.A. *61*, 690d (1964).
[c] *M. Tsutsui, H. H. Zeiss*, J. Amer. Chem. Soc. *82*, 6255 (1960).
[d] *M. Ryang, K. Yoshida, H. Yokoo, Sh. Tsutsumi*, Bull. Chem. Soc. Japan *38* (4), 636 (1960).
[e] *B. D. Babitskii, B. A. Dolgoplosk, V. A. Kormer, M. I Lobach, E. I. Tinyakova, V. A. Yakovlev*, Dokl. Akad. Nauk. SSSR *161* (3), 583 (1965).

Complex nickel halides react analogously by forming complexes of the corresponding organonickel compound according to the following scheme (43).

$$NiL_2Cl_2 + 2RMgCl \longrightarrow L_2NiR_2 + 2MgCl_2 \qquad (43)$$

L = bipyridyl or *t*-phosphine ; R = ClH_5

Reactions (42) and (43) may lead to the formation of organonickel halides of the type NiRX or L_nNiRX,[417, 418, 419] *i.e.*, only one halogen atom of the nickel halide may be replaced by the organic radical. Nickel organohalides, NiRX, can also be prepared from zerovalent nickel complexes and organohalides according to reactions (44) and (45) (Table 7).

$$Ni(CO)_4 + RX \longrightarrow RNiX + 4CO \qquad (44)$$

$$2Ni(CO)_4 + 2RX \longrightarrow [RNiX]_2 + 8CO \qquad (45)$$

Hydrocarbon complexes of nickel(0) have also been used as starting materials rather than tetracarbonylnickel(0) (Ref. *b* in Table 7). *Nickel(II) bisacetylacetonate* has been reduced by diethylethoxyaluminum and an organohalide was added to interact with the resultant zerovalent nickel complex according to (44) or (45) (Ref. *c*, Table 2). Interaction of butadiene with a solution of bis(cyclooctadienyl)nickel at −20° probably leads to zerovalent nickel since the addition of triarylmethyl bromide results in the formation of triarylmethylnickel bromide.[420]

Organonickel halides have been prepared from bis(organo)nickel by reaction with equimolar amounts of hydrogen chloride. Hence, bis-(π-allyl)nickel reacts with hydrogen chloride at −80° in ether solution to give π-allylnickel chloride.[421]

Mixed organonickel compounds have been prepared from organonickel halide and Grignard reagent or organolithium compound; according to Eqns (46) and (47) (Table 8).

$$L_nRNiX + R'MgX \longrightarrow L_nRNiR' + MgX_2 \qquad (46)$$

$$L_nRNiX + R'Li \longrightarrow L_nRNiR' + LiX \qquad (47)$$

[414] *F. Horowitz, M. Tryon, R. G. Christensen, T. P. Perros*, J. Appl. Polym. Sci. *9* (7), 2321 (1965).
[415] U.S. P. 312 236 (1965), Monsanto Res. Corp., Erf.: *J. J. O'Connell;* CA *63*, 8517d (1965).
[416] *V. V. Korshak, S. P. Krukovskii, J. H. Wang*, Vysokomolekul. Seedin. Ser. B *9* (8), 583 (1967).
[417] *J. R. Phillips, D. T. Rosevear, F. G. A. Stone*, J. Organometal. Chem. *2* (6), 455 (1964).
[418] *K. P. MacKinnon, B. O. West*, Australian J. Chem. *21* (1), 2801 (1968).
[419] *A. J. Rest, D. T. Rosevear, F. G. A. Stone*, J. Chem. Soc. A. *1*, 66 (1967).
[420] Japan P. 30 876 (1969), Mitsui Petrochem. Ind., Erf.: *T. Arakawa;* CA *72*, 67115w (1970).
[421] Neth. Appl. P. 6 409 180 (1965), Studienges. Kohle, CA *63*, 11617b (1961).

Table 7 Synthesis of Organo-nickel(II)-halides from Zerovalent Nickel Complexes

$Ni(CO)_4 + RH \longrightarrow RNiX + 4CO$

$L_nNi(CO)_{4-n} + RX \longrightarrow L_nRNiX(CO)_{2-n} + 2CO$

$2Ni(CO)_4 + 2RH \longrightarrow [RNiX]_2 + 8CO$

Ni(0)complex	Organohaloderivative	Conditions	Organo-nickel	Ref.
$Ni(CO)_4$	$CH_2{=}CHCH_2X$ (X = Br, I)	−78°	$[CH_2{=}CH{=}CH_2NiX]_2$	*a, b*
$Ni(CO)_4$	1,2-I_2-C_6H_4	70° cyclohexane	$(NiC_7H_4I_2O_2)_2$	*k*
$(RH)_4Ni(0)$	cyclo(C_3–C_{12})enylX		RNiX	*c*
Ni(II)-pentern-2,4-dionat	$(C_6H_5)_3CCl$	reductn. $(C_2H_5)_2Al{-}OC_2H_5$ ether −15°	RNiCl	*d*
$Ni(CO)_3[P(C_6H_5)_3]$	$CH_2{=}CHCH_2Br$	C_6H_6, 0–30°	$(C_3H_5)NiBr[P(C_6H_5)_3]CO$	*e, f, g*
$[P(C_6H_5)_3]_2Ni(C_2H_4)$	$CF_2{=}CFX$ (X = Cl, Br)		$[P(C_6H_5)_3]_2Ni(CF{=}CF_2)X$	*h*
Ni(0) complex	RX (R = aryl, vinyl)		RNiX	*i, j*
$Ni(C_8H_{12})_2$	$(C_6H_5)_3CCl$	toluene	$(C_6H_5)_3CNiCl$	*l*

[a] *E. O. Fischer, G. Buorger*, Chem. Ber. *94*, 2409 (1961).
[b] Ger. P. 1249866 (1967), Studienges. Kohle, Erf.: *G. Wilke*; C.A. *68*, 49790a (1968).
[c] Neth. Appl. P. 6409178 (1965), Studienges. Kohle; C.A. *63*, 5276e (1965).
[d] Japan P. 30977 (1969), Mitsui Petrochem. Ind., Erf.: *T. Arakawa*; C.A. *72*, 67116x (1970).
[e] Ital. P. 791655 (1967), Montecatini Edison; C.A. *71*, 91660v (1969).
[f] *F. Guerrieri, G. P. Chiusoli*, Chem. Commun. (15), 781 (1967).
[g] *F. Guerrieri, G. P. Chiusoli*, J. Organometal. Chem. *15* (1), 209 (1968).
[h] *J. A. Smith, M. Green, F. G. A. Stone*, J. Chem. Soc. A (19), 3019 (1969).
[i] *D. R. Fahey*, J. Amer. Chem. Soc. *92*, 402 (1970).
[j] *M. D. Rausch, F. E. Tibbets*, Inorg. Chem. *9* (3), 512 (1970).
[k] *E. W. Gowling, S. F. A. Kettle, G. M. Sharples*, Chem. Commun. (1), 21 (1968).
[l] *T. Arakawa*, Kogyo Kagaku Zasshi, *70* (10), 1738 (1967).

Table 8 Preparation of Mixed Organonickel Compounds According to Reactions (46) and (47)

R	R′	Solvent	L	*n*	X	Ref.
$CH_2{=}CHCH_2$—	CH_3—	ether, −78°	—	0	Br	*a*
C_5H_5—	C_6Z_5 (Z = H, F, Cl)		$(C_6H_5)_3P$	1	Cl	*b, c*
C_5H_5—	CH_3—, C_6H_5—, C_6X_5—		$R_2''R'''P$	1	Cl	*d*

[a] *B. Bogdanovic, H. Boennemann, G. Wilke*, Angew. Chem. *78* (11), 591 (1966).
[b] *M. R. Churchill, T. A. O'Brien, M. D. Rausch, Y. F. Chang*, Chem. Commun. (19), 992 (1967).
[c] *P. M. Treichel, R. L. Shubkin*, Inorg. Chim. Acta *2* (4), 485 (1968).
[d] *M. D. Rausch, Y. F. Chang, H. B. Gordon*, Inorg. Chem. *8* (6), 1355 (1969).

Zerovalent nickel complexes react with aryllithium ArLi and form *aroylnickel carbonyl* $Li^{\oplus}[ArCONi(CI)_3]^{\ominus}$.[422]

Another route for the synthesis of organonickel compounds involves the interaction of a nickel chelate with organoaluminum derivatives at low temperature. Thus nickel bisacetylacetonate was reacted with triethylaluminum or diethylethoxyaluminum in the presence of α,α′-dipyridyl at −20° in ether to give *dipyridyldiethylnickel*.[423]

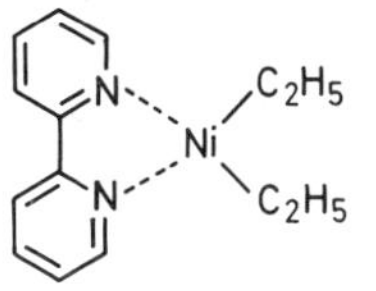

[422] *M. Ryang, S. K. Myeong, Y. Sawa, Sh. Tsutsumi*, J. Organometal. Chem. *5* (4), 305 (1966).

[423] *A. Yamamoto, K. Morifuji, S. Ikeda, T. Saito, Y. Uchida, A. Misono*, J. Amer. Chem. Soc. *87* (20), 4652 (1965); *88* (22), 5198 (1966).

Dimeric π-allylnickel bromide has been prepared according to Eqn (48).[424]

$$2(C_3H_5)_2Ni + 2C_3H_5Br \longrightarrow [C_3H_5NiBr]_2 + 2(C_3H_5)_2 \quad (48)$$

The reaction can be reversed. The distillation (*in vacuo*) of the dimer yields (π-allyl)nickel which was trapped.[425] A similar conversion was achieved when dimeric π-allylnickel chloride was shaken in water and toluene under argon atmosphere; bis(π-allyl)nickel and nickel dichloride were isolated.[426] The decomposition of dimeric organonickel halides proceeds readily in ether on treatment with ammonia or water[427] according to Eqn (49).

$$[RNiBr]_2 + 6NH_3 \longrightarrow R_2Ni + [Ni(NH_3)_6]Br_2 \quad (49)$$

(R = allyl, crotyl, 1-vinylallyl, or methyl).

[424] *V. A. Vashkevich, A. M. Lazutkin, B. I. Mitsner*, Zh. Obshch. Khim. *37* (8), 1926 (1967).

[425] *E. J. Corey, L. S. Hegedus, M. F. Semmelhack*, J. Amer. Chem. Soc. *90* (9), 2417 (1968).

[426] *B. A. Dolgoplosk*, Dokl. Akad. Nauk. SSSR *183* (5), 1083 (1968).

[427] Fr. P. 1 543 303 (1968), Studienges. Kohle; CA *72*, 43881s (1970).

The coordination power of nickel is not saturated in normal organonickel compounds and the nickel atom can coordinate with donors according to Eqn (50).

$$RNiX + nL \longrightarrow RNi(L)_nX \quad (50)$$

The reaction proceeds at room temperature and in solvents such as ether, benzene, or ethanol (Table 9).

Nickel acetylides were obtained[428,429] according to reaction (51).

$$[Ni(NH_3)_6](SCN)_2 + 4K{-}C{\equiv}C{-}R \ NH \longrightarrow K_2[Ni(C{\equiv}C{-}R)_4]\cdot 2NH_3 + 4NH_3 + 2KSCN \quad (51)$$

Ammonia is readily lost *in vacuo* and $K_2[Ni(C{\equiv}C{-}R)_4]$ is formed.

Another method for the preparation of nickel-alkyne compounds is illustrated by the interaction between organonickel halide, alkyne, and diethylamine according to scheme (52).

$$RNi[P(C_6H_5)_3]Cl + HC{\equiv}C{-}R^1 + (C_2H_5)_2NH \longrightarrow \{RNi[P(C_6H_5)_3](C{\equiv}C{-}R^1)\} + (C_2H_5)_2NH_2Cl \quad (52)$$

[428] *R. Nast*, Z. Naturforsch. *8B*, 381 (1953).

[429] *R. Nast, K. Vester*, Z. Anorg. Allgem. Chemie *279*, 146 (1955).

Table 9 Coordination Compounds with Organonickel

$$RNiX + uL \longrightarrow [RNiL_n]X$$

R	L	X	*n*	Ref.
$H_2C{=}CH{-}CH_2{-}$	$S{=}C(NR'R'')_2$	Cl, Br, I	1	*a*
	thiourea	halides	2	*b*
$H_2C{=}H{-}CH_2{-}$, PR_3	CO	Cl, Br, I	1	*c*
$CH_2{=}C(CH_3){-}CH_2{-}$	$F_2C{=}CF_2$	$H_2C{=}C(CH_3){-}CH_2{-}$	1	*d*
	$(C_6H_5)_2P{-}CH_2{-}CH_2{-}P(C_6H_5)_2$	Br	1	*e*
	$O{=}C_6Cl_4{=}O$			
$H_3C{-}CH{=}CH{-}CH_2{-}$	$H_2C{=}C(CH_3){-}CH_2{-}$	Cl	1	*f*
$C_5H_5{-}$	$(C_6H_5)_3P$	Br	1	*g*
	$(CF_3)_2PH$	C_5H_5	2	*h*
	C_2H_2; $H_5C_6{-}C{\equiv}CH$	C_5H_5	1	*i*

[a] Ital. P. 807705 (1968), Montecatini Edison S.p.A; C.A. *71*, 81542u (1969).

[b] *F. Guerrieri*, Chem. Commun. (16), 983 (1968).

[c] Ref. *f*, Table 2, p. 364.

[d] *J. Browning, D. J. Cook, C. S. Cundy, M. Green, F. G. A. Stone*, Chem. Commun. (16), 929 (1968).

[e] *M. R. Churchill, T. A. O'Brien*, Chem. Commun. (5), 246 (1968).

[f] *V. I. Skoblikova, Z. D. Stepanova, B. D. Babitskii, V. A. Kormer*, Zh. Obshch. Khim. *39* (1), 219 (1969).

[g] U.S. P. 3476769 (1969), Phillips Petroleum; C.A. *72*, 43880 (1970).

[h] *R. C. Dobbie, M. Green, F. G. A. Stone*, J. Chem. Soc. A (12), 1881 (1969).

[i] Brit. P. 966860 (1964), Ethyl. Corp., Erf.: *M. Duberk*; C.A. *61*, 16098b.

This method was utilized for the reaction of *cyclopentadienyltriphenylphosphinenickel chloride* with *o*- or *p-phenylenacetylide*.[430]
Isocyanides interact with organonickel compounds in an insertion reaction[431] according to scheme (53).

$$(C_2H_5)NiLR^1 + 2R^2{-}NC \longrightarrow (C_5H_5)Ni(R^2{-}NC)(\overset{R^1}{\overset{|}{C}}{=}N{-}R^2) + L \quad (53)$$

(Z = donor mol).

26.5 Nickel(III) Compounds

Simple nickel(III) derivatives are unstable. They act as oxidizing agents and the metal is reduced to nickel(II). Only the *oxide* Ni_2O_3 and the *hydroxide* $Ni(OH)_3$ are relatively stable.
The oxide has been prepared by calcination of either the nitrate[432] or a mixture of nickel chloride and potassium chlorate[433]; this oxide has been considered as nickel(II) nickelate(IV) $NiO \cdot NiO_2$.[434] The thermal decomposition of nickel nitrate hexahydrate leads to nickel(III) or nickel(IV) oxides.[435,436]
The hydroxide $Ni(OH)_3$ appears as black precipitate on reaction of nickel(II) hydroxide with hypochlorite, hypobromite, or persulfate at 70°.[437]
Anodic oxidation of an ethanol solution of nickel(II) chloride in the presence of chlorine and hydrogen chloride produces a change of color indicating the formation of $Ni^{3\oplus}$. However, $NiCl_3$ could not be isolated.[438] If the oxidation is performed in the presence of donors comparatively stable nickel(III) complexes are obtained. The following oxidizing agents have been used: alkalimetal hypochlorite, hypobromite, or persulfate, gaseous oxygen, air, and others.

Alkalimetal *hexafluoronickelates(III)* have been prepared by fluorination of a mixture of stoichiometric amounts of alkali metal chloride and nickel sulfate at 300°. A mixture of stoichiometric amounts of hexaminonickel(II) chloride and sodium carbonate has been converted to sodium hexafluoronickelate(III), $Na_3[NiF_6]$, by heating to 450° in a stream of fluorine.[439]
Air oxidation of a solution of nickel(II) salts in the presence of donors has been described. Some nickel(III) complexes which were prepared by oxidation of nickel(II) compounds with various oxidating agents[440,441,442] are listed in Table 10. These complexes are of the type $Ni(L)_nX_3$ (L = ligand, X = Cl, Br, I, ClO_4, *etc.*) (Ref. *e*, *i*, *j*); $Ni(L)_nX_2Z$ (X and Z monovalent anions) (Ref. *f*);

$M[Ni(L)_2]$ ($M = [(C_3H_7)_4N]^{\oplus}$ or $[(C_4H_9)_4N]^{\oplus}$, L = bivalent ligand) (Ref. *d*, *c*)

The oxidation of nickel(II) nitrate dihydrate with dinitrogen pentoxide N_2O_5 (Ref. *l*) leads to the complex $(NO_2)^{\oplus}[Ni'''(NO_3)_4]^{\ominus}$.

The reaction of some nickel(II) complexes with bromine or iodine gives compounds which are considered to be molecular bromine or iodine complexes. Both the iodo and the bromo complexes (which were prepared from bis(diphenylglyoximato)nickel(II) and iodine or bromine) are of the type $Ni(L)_2X$ (X = Br or I). They are nickel(II) derivatives.[443]

Nickel(III) complexes with long-chain cyclic tetramines have polymeric structure.[444]
Some special oxidating agents have been used. The complex bis(dithiocarbamate)nickel(II) was extracted with benzene and was oxidized with bis(diethylthiocarbamoyl)disulfide to give *nickel(III) diethyldithiocarbamate*.[445]
Another method for the preparation of nickel(III) complexes utilizes nickel(IV) complexes as starting material. Nickel(IV) complexes are unstable in aqueous solution and convert successively into nickel(III) and nickel(II) compounds. If an alcoholic solution of a nickel(IV)

---

[430] *P. J. Kim, H. Masai, K. Sonogashira, N. Hagihara*, Inorg. Nucl. Chem. Lett. *6* (2), 181 (1970).
[431] *Y. Yamamoto, H. Yamazaki, N. Hagihara*, J. Organometal. Chem. *18* (1), 189 61969;
Y. Yamamoto, H. Yamazaki, N. Hagihara, Mem. Inst. Sci. Ind. Res., Osaka, Univ. *26*, 167 (1969).
[432] *Wächter*, J. Prakt. Chem. *30*, 321 (1848).
[433] *Delffs*, Ber. *12*, 2182 (1879).
[434] *G. L. Clark, W. C. Asbury, R. M. Wick*, J. Amer. Chem. Soc. *47*, 2661 (1925).
[435] *D. P. Bogatskii*, Zh. Obshch. Khim. *7*, 1397 (1937).
[436] *G. I. Chufarov, M. G. Zhuravleva, E. P. Tat'evskaya*, Dokl. Acad. Nauk. SSSR *73*, 1209 (1950).
[437] *J. Labat*, Ann. Chim. (Paris) *9* (7–8), 399 (1964).
[438] *C. Schall*, Z. Elektrochem. *38*, 27 (1932).
[439] *H. Henkel, R. Hoppe*, Z. Anorg. Allgem. Chemie *364* (5–6), 253 (1969).
[440] *A. K. Babko*, Zh. Anal. Khim. *3*, 284 (1948).
[441] *K. B. Yatsimirskii, Z. M. Grafova*, Zh. Obshch. Khim. *23*, 935 (1953).
[442] *L. S. Nadezhina, P. N. Kovalenko*, Zh. Obshch. Khim. *24*, 1734 (1954).
[443] *A. S. Foust, R. H. Soderberg*, J. Amer. Chem. Soc. *89* (21), 5507 (1967).
[444] *N. F. Curtis, D. F. Cook*, Chem. Commun. *18*, 962 (1967).
[445] *P. M. Solozhenkin, N. L. Kopist*, Dokl. Akad. Nauk. Tadzh. SSR *12* (5), 30 (1969).

Table 10 Nickel(III) Complexes by Oxidation of Nickel(II) Compounds $\{[NiL_nX_3], [NiL_nX_2Y], M[NiL_2]\}$

Ligand	Oxidant	Conditions	Ref.
o-$C_6H_4(As(CH_3)_2)_2$	air	acid. soln. reflux	*a*
(3)-1,2-dicarbolide ion $B_9C_2H_{11}^{2\ominus}$	air	40% NaOH	*b*
$F_4C_6(SH)_2$	air	CH_3OH	*c*
KSC_6H_4SeK	air	C_2H_5OH	*d*
$H_2N—CH_2—CH_2—NH_2$	chlorine	CH_3OH	*e*
$H_2N—CH_2—CH_2—CH_2—NH_2$	bromine	C_2H_5OH	*f*
$CH_3As(As_2C_5H_{11})_2$	bromine	$CHCl_3$	*g*
$H_5C_6-CH_2-C(=NOH)-CH=NOH$	Cl_2, Br_2, I_2		*h*
$Cl-C_6H_4-NH-C(NH_2)=N-C(NH_2)=N-CH(CH_3)_2$ (paludrine)	$NaClO_2$	org. solvents	*i*
substituted biguanide	$NaClO_2$	ethanol	*j*
$NN^{\ominus}$	NO_2	CCL_4	*k*
$NO_3^{\ominus}$	N_2O_5		*l*
biuret	$K_2S_2O_8$	5*N* KOH	*m*
$CN^{\ominus}$	$NH_2OH \cdot HCl$	KOH, N_2	*n*

[a] *R. S. Nyholm*, Nature *165*, 154 (1950); *R. S. Nyholm*, J. Chem. Soc. (London), 2061 (1950).
[b] *L. F. Warren, Jr.*, *M. F. Hawthorne*, J. Amer. Chem. Soc. *89* (2), 470 (1967).
[c] *A. Callaghan*, *A. J. Layton*, *R. S. Nyholm*, J. Chem. Soc. D. (18), 399 (1969).
[d] *C. G. Pierpont*, *B. J. Corden*, *R. Eisenberg*, J. Chem. Soc. D. (8), 401 (1969).
[e] *A. V. Babaeva*, *I. B. Baranovskii*, *G. G. Afanas'eva*, Zh. Neorg. Khim. *10* (5), 1268 (1965).
[f] *A. V. Babaeva*, *V. I. Belova*, *Ya. K. Syrkin*, *G. G. Afanas'eva*, Zh. Neorg. Khim. *13* (5), 1261 (1968).
[g] *G. A. Barclay*, *R. S. Nyholm*, Chem. & Ind. (London) 378 (1953).
[h] *L. E. Edelman*, J. Amer. Chem. Soc. *72*, 5765 (1950).
[i] *P. Spacu*, *C. Gheorghiu*, *A. Nicolaescu*, Inorg. Chim. Acta *2* (4), 413 (1968).
[j] *C. Gheorghiu*, *A. Nicolaescu*, Z. Chem. *8* (5), 184 (1968).
[k] *A. Job*, *A. Samuel*, C.R. Acad. Sci. Paris *177*, 188 (1923).
[l] *C. C. Addison*, *B. G. Ward*, Chem. Commun. (22), 819 (1966).
[m] *J. J. Bour*, *J. J. Steggerda*, Chem. Commun. (2), 85 (1967).
[n] *L. Malatesta*, *R. Pizzotti*, Gazz. Chim. Ital. *72*, 174 (1942).

compound is added to an aqueous solution of donor molecules a precipitate of the corresponding nickel(III) compound can be obtained. For example, an alcohol solution of tetrakis(benzamidoximato)nickel(IV) was added to an aqueous solution of serine to produce *tris-(serinato)nickel(III)*.[446]

26.6 Nickel(IV) Derivatives

Simple nickel(IV) compounds are unstable and are exceedingly rare. They act as very strong oxidants but in the presence of donor molecules stable nickel(IV) complexes can be formed by oxidation of nickel(II) derivatives.

Nickel dioxide, NiO_2 (known as nickel peroxide as well), has been prepared by oxidation of nickel(II) hydroxide with alkalimetal hypochlorite,[447,448] hypobromite,[449] or sodium peroxodisulfate.[449] It appears that nitric acid

[446] *K. Manolov*, *B. Angelov*, Monatsh. Chem. *102*, 763 (1971).

[447] Brit. P. 975 390 (1964), Shionogi, CA *62*, 5192e (1965).
[448] Fr. P. 1381867 (1964), Farbwerke Hoechst, CA *62*, 9017a (1965).
[449] *J. Labat*, Ann. Chim. (Paris) *9* (7–8), 399 (1964).

oxidizes nickel nitrate.[450] An alcoholic solution of nickel(II) chloride has been reacted with 30% hydrogen peroxide to yield nickel dioxide.[451] The compound has been considered either as peroxide, Ni(O–O), or as dioxide, O=Ni=O [452]

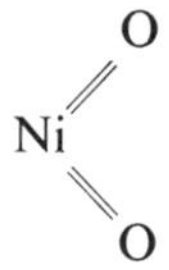

Potassium nickel(IV) orthoperiodate, $KNiIO_6$, has been prepared by treating nickel sulfate and potassium orthoperiodate with alkalimetal peroxodisulfate.[453] The same salt was obtained by the oxidation of nickel(II) salts with potassium hypochlorite in alkaline solution and in the presence of potassium orthoperiodate.[454] *Potassium nickel(IV) orthotellurate*, K_2NiTeO_6, has been prepared in similar manner.[454] *Nickel(IV) niobate*, $Na_{12}NiNb_{12}O_{38}$ 48–50H_2O, has been obtained by oxidation of nickel(II) salts with an excess of sodium hypobromite in the presence of sodium niobate, $Na_7HNb_6O_{19} \cdot 15H_2O$.[455]

Sodium hexafluoronickelate(IV), Na_2NiF_6, has been prepared from sodium fluoride and nickel(II) fluoride by treatment with fluorine at 350° and 350 atm.[456] A high pressure fluorination of sodium nickelate(IV), Na_2NiO_3 (from sodium oxide and nickel dioxide), also yields sodium hexafluoronickelate(IV).[457]

Nickel(IV) complexes with organic ligands are prepared by oxidation of nickel(II) chelates or complexes with various oxidants. Nickel(II) bis(dimethylglyoximate) has been oxidized to *nickel(IV) glyoximate* with lead dioxide,[458] halogens, air or persulfate.[459] This oxidation was questioned,[460] but later the alkali metal nickel (IV) *tris(dimethylglyoximates)*, $K_2[Ni(dmgl)_3] \cdot 6H_2O$ and $Na_2[Ni(dmgl)_3] \cdot 5H_2O$, were isolated in crystalline state from the oxidation of nickel(II) bis(dimethylglyoximate) with iodine in alkaline solution.[461,462] Hexaminonickel(II) ion, $[Ni(NH_3)_6]^{2\oplus}$, has been oxidized in ammonia solution and in the presence of *o*-phenylenediamine or 1-methyl-3,4-diaminobenzene to yield the corresponding *arylaminonickel-(IV)* complex.[463] Bis(diarsino)dichloronickel(IV) diperchlorate, $[Ni(diars)_2Cl_2](ClO_4)_2$, has been obtained from the corresponding complex of the bivalent nickel.[464]

Air oxidation is accomplished only in strongly alkaline solution. For example, nickel(II) chloride was oxidized in sodium hydroxide solution at 40° and in the presence of cesium carborane, $CsB_{10}H_{12}CH$, to give *cesium nickel(IV) biscarborane*, $Cs[Ni(B_{10}H_{10}CH)_2]$.[465] An alkaline suspension of bis(*o*-aminothiophenolato)nickel(II) was reacted with air to yield *bis(o-aminothiophenolato)oxonickel(IV)*, $[(C_6H_4(NH_2)S)_2NiO]_2$.[466] Oxidation with alkalimetal polysulfide in alkaline solution has been used for the preparation of the complex $Na_2[S_2Ni(SSCNHC_6H_5)_2]$.[466]

The oxidation with halogens is accomplished in carbon tetrachloride suspension. Thus, bis(thioacetylacetone)nickel(II) chloride has been oxidized with chlorine to yield *bis(thioacetylacetone)nickel(IV) chloride*, $[Ni(thac)_2]Cl_4$,[467] and bis(propylenediamine)nickel(II) bromide was oxidized with bromine to yield *bis(propylenediamine)nickel(IV) bromide* $[Ni(pn)_2]Br_4$.[468]

The reaction of nickel(II) complexes with iodine yields molecular complexes containing the ion $I_3^{\ominus}$ rather than nickel(IV). Bis(ethylenediamine)nickel(II) iodide, $[Ni(en)_2]I_2$, does not form a nickel(IV) complex on reaction with iodine but a molecular nickel(II) complex with iodine.[469] Also, the complex obtained (by

[450] *W. Vubel*, Chemiker-Ztg. *46*, 978 (1922).
[451] *G. Pellini, D. Meneghini*, Z. Anorg. Allgem. Chemie *60*, 178 (1908).
[452] *C. Tubandt, W. Riedel*, Ber. *44*, 2565 (1911).
[453] *P. Ray, B. Sarma*, Nature *157*, 627 (1946); *P. Ray, B. Sarma*, J. Indian Chem. Soc. *25*, 205 (1948).
[454] *Y. Yoshino, T. Takeuchi, H. Kinoshita*, Nippon Kagaku Zasshi *86* (10), 978 (1965).
[455] *C. M. Flynn jr., G. D. Stucky*, Inorg. Chem. *8* (2), 332 (1969).
[456] *R. Bougon*, C.R. Acad. Sci., Paris, Ser. C *267* (11), 681 (1968).
[457] *H. Henkel, R. Hoppe, G. C. Allen*, J. Inorg. Nucl. Chem. *31* (12), 3855 (1969).
[458] *F. Feigl*, Ber. *57*, 758 (1924).
[459] *M. A. P. Rollet*, C.R. Acad. Sci., Paris *183*, 212 (1926).
[460] *A. Okac, M. Polster*, Collect. Czech. Chem. Commun. *13*, 561 (1948).
[461] *A. Okac*, Theory Struct. Complex, Compds. p. 167, Paper Symp. Wroclaw, Poland 1962 (Publ. 1964).
[462] *M. Simek*, Collect. Czech. Chem. Commun. *27*, 220 (1962).
[463] *F. Feigl, M. Furth*, Monatsh. Chem. *48*, 445 (1927).
[464] *R. S. Nyholm*, J. Chem. Soc. 2061 (1950); 2602 (1951).
[465] *D. H. Hyatt, J. L. Little, J. T. Moran, F. R. Scholer, L. F. Todd*, J. Amer. Chem. Soc. *89* (13), 3342 (1967).
[466] *W. Hieber, R. Brück*, Naturwissenschaften *36*, 312 (1949).
[467] *A. Furuhashi, K. Watanuki, A. Ouchi*, Bull. Chem. Soc. Japan *42* (1), 260 (1969).
[468] *A. V. Babaeva, V. I. Belova, Ya. K. Syrkin, G. G. Afanas'eva*, Zh. Neorg. Khim. *13* (5), 1261 (1968).
[469] *I. B. Baranovskii, V. I. Belova*, Zh. Neorg. Khim. *10* (1), 306 (1965).

extraction) from a melt of phenol, diphosphorus trisulfide, tetraethylammonium chloride, and nickel chloride, was shown to be a nickel(II) chelate[470] and not a nickel(IV) complex as was previously stated.[471]

Iron(III) chloride oxidizes some nickel(III) compounds to the corresponding nickel(IV) complexes. Thus, the ion $[Ni(B_9C_2H_{11})_2]^{\ominus}$ containing tervalent nickel was oxidized to *nickel(IV) bis(dicarbolide)* $[Ni(B_9C_2H_{11})_2]$[472] in aqueous solution with stoichiometric amounts of iron(III) chloride.[472]

---

[470] *J. P. Fackler jr., D. Coucouvanis,* J. Amer. Chem. Soc. *89* (7), 1745 (1967).
[471] *W. Hieber, R. Brück,* Z. Anorg. Allgem. Chemie *269*, 28 (1952).
[472] *L. F. Warren jr., M. F. Hawthorne,* J. Amer. Chem. Soc. *89* (2), (1967).

26.7 Bibliography

R. Cotton, J. H. Canterford, Halides of the First Row Transition Metals. Wiley Interscience Publ. New York (1969).

J. Hanotier, Ind. Chem. Belge *31* (1), 19 (1966).

J. H. Harwood, Chem. Process. Eng. *48* (6), 100 (1967).

L. Malatesta, R. Ugo, S. Cencini, Advan. Chem. Ser. No. *62*, 318 (1967).

J. Powel, Organometal. Chem. Rev., Sect. B *5* (3), 507 (1969).

P. Ray, Chem. Revs. *61*, 313 (1961).

M. Rushdy, D. Sandulescu, M. Popescu, V. Vintu, Rev. Chim. Acad. Rep. Populaire Roumaine, *7* (2), 1227 (1962).

G. N. Schrauzer, Advan. Organometal. Chem. *2*, 1 (1964).

B. West, New Pathways Inorg. Chem. 303 (1968).

G. Wilke, Pure Appl. Chem. *17* (1–2), 179 (1968).

G. Wilke et al, Agew. Chem. Intern. Ed. Engl. *5* (2), 151 (1966).

27 Palladium

D. M. Adams and R. D. W. Kemmitt
Department of Chemistry, University of Leicester,
Leicester LE1 7RH, England

The chemistry of palladium is predominantly that of the +2 oxidation state. In contrast to platinum the +4 state is of little importance, the compounds tending to be thermally unstable.

27.1 Purification and Recovery

Residues containing palladium are first ignited to remove organic material; the metal is then extracted with hot aqua regia. The nitric acid present is removed by evaporating the solution to dryness, adding hydrochloric acid, and then heating the solution at 85 to 90°. Palladium is precipitated as $PdCl_2(NH_3)_2$ by addition of excess ammonia followed by hydrochloric acid. The salt is purified and ignited in a stream of hydrogen to give palladium sponge.[1]

Colloidal palladium is prepared by adding $PdCl_2$ to a solution of sodium protalbinate in water containing a slight excess of NaOH. Reduction is effected with $N_2H_4 \cdot H_2O$ and the black solution is purified by dialyzation. Concentration and drying *in vacuo* gives shiny black platelets which are "soluble" in water. One volume of the colloidal solution can absorb about 3000 volumes of hydrogen.[2]

Reduction of an aqueous solution of a Pd(II) salt with sodium formate produces palladium black.[2]

27.2 Palladium(0)

$K_4[Pd(CN)_4]$ is prepared by treatment of $K_2[Pd(CN)_4]$ with potassium in liquid ammonia.[3] It is yellow-white, thermally unstable but can be kept for at least 12 hr at −33° in liquid ammonia.

Di-isonitrile compounds $Pd(RNC)_2$, (R=Ph, p-MeC_6H_4, p-$MeO \cdot C_6H_4$) are made from $PdI_2(RNC)_2$ and RNC by treatment with alcoholic KOH. The compounds are insoluble in most solvents and cannot be recrystallized from those in which they do dissolve. With phosphites further compounds $Pd(RNC)\{(RO)_3P\}_3$ and $\{(RO)_3P\}_4Pd$ are formed.[4,5]

Treatment of methanolic solutions of palladous chloride or sulfate with NO yields $Pd(NO)_2Cl_2$ or $Pd(NO)_2SO_4$ as unstable solids which evolve NO in moist air.[6] $[Pd(NO)Cl]_n$ is formed when palladous chloride is reacted with NO in moist air.[7,8]

$Pd(PF_3)_4$ is formed by interaction of $PdCl_2$ and PF_3 at 100° in a bomb using copper as halogen acceptor.[9,10] Many *Pd(0) complexes* containing tertiary phosphines, arsines, stibines or phosphites have been made, and by a variety of methods. They are generally thermally stable but are oxidized by air, slowly when solid, rapidly in solution. Although they may be made by reaction of donor ligand with $Pd(NO_3)_2$, PdO or hydrazine reduction of complexes $Pd^{II}Cl_2L_2$[5] two more recent methods are generally preferred for convenience.

(i) Reduction of cationic complexes $[PdL_4]^{2\oplus}$ with aqueous sodium hydridoborate or sodium naphthalide[11]; isolation of the intermediate is unnecessary as in the scheme[12]

$$K_2PdCl_4 + 4L + NaBH_4 \xrightarrow[\text{alcohol}]{\text{Aqueous}} PdL_4$$

(ii) Displacement reaction of ligand with, e.g. PR_3 with (π-allyl)(π-cyclopentadienyl)Pd,[13] or $Pd(acac)(C_8H_2acac)$.[14] Reduction of anhydrous $PdCl_2$ in absolute ethanol has recently been used in preparation of $Pd(PhP(OR)_3)_4$.[15]

All zerovalent compounds appear to dissociate

---

[1] *T. J. Walsh, E. A. Hausman* in *I. M. Kolthoff, P. J. Elving*, Treatise on Analytical Chemistry, Part II, Vol. VIII, p. 379. Interscience Publishers, New York, London 1963.

[2] *G. Brauer*, Handbook of Preparative Inorganic Chemistry, Vol. II, Second Edition, Academic Press, New York 1965.

[3] *J. J. Burbage, W. C. Fernelius*, J. Amer. Chem. Soc. 65, 1484 (1943).

[4] *L. Malatesta*, J. Chem. Soc. (London) 3924 (1955).

[5] *L. Malatesta, M. Angoletta*, J. Chem. Soc. (London) 1186 (1957).

[6] *W. Manchot, H. Waldmüller*, Ber. *59*, 2363 (1926).

[7] *J. Smidt, R. Jira*, Chem. Ber. *93*, 162 (1960).

[8] *W. P. Griffith, J. Lewis, G. Wilkinson*, J. Chem. Soc. (London) 1775 (1959).

[9] *G. F. Svatos, E. E. Flagg*, Inorg. Chem. *4*, 422 (1965).

[10] *Th. Kruck, K. Baur*, Angew. Chem. Intern. Ed. Engl. *4*, 521 (1965).

[11] *J. Chatt, F. A. Hart, H. R. Watson*, J. Chem. Soc. (London) 2537 (1962).

[12] *D. T. Rosevear, F. G. A. Stone*, J. Chem. Soc. (A), 164 (1968).

[13] *A. J. Mukhedkar, M. Green, F. G. A. Stone*, J. Chem. Soc. (A), 3023 (1969).

[14] *B. F. G. Johnson, T. Keating, J. Lewis, M. S. Subramanian, D. A. White*, J. Chem. Soc. (A) 1793 (1969).

[15] *A. A. Orio, B. B. Chastain, H. B. Gray*, Inorg. Chim. Acta *3*, 8 (1969).

in solution giving mixtures in which coordinatively unsaturated species are present:

$$Pd(PR_3)_4 \rightleftharpoons Pd(PR_3)_3 \rightleftharpoons Pd(PR_3)_2$$

The monomeric dicoordinated species has no been isolated but a yellow solid $[Pd(PPh_3)_2]_n$ has been obtained by bubbling vinyl chloride through a solution of $Pd(PPh_3)_4$.[16] Complexes of the tertiary phosphine $Me \cdot C(CH_2 \cdot PPh_2)_3$ may be made in two isomeric forms, α and β, which differ in having two or three of the phosphorus atoms bonded to the metal. The β isomer is formed by method (i) above; when recrystallized from benzene it yields the α-form.[11]

27.3 Palladium(II)

In this section coordination complexes of donor ligands are treated together as methods of preparation are usually little affected by a simple change of ligand. Thus, for example, complexes of sulfur donor ligands appear in Section 27.3.7 whilst the simpler compounds of sulfur are in Section 27.3.3. Ligands such as dithiolate which are formally anionic are included in Section 27.3.3.

27.3.1 Palladium Hydrides

In marked contrast to platinum and despite the complexity of the Pd/H_2 system (discussed fully in reference 17) the Pd–H bond in complexes is unstable and few *complex hydrides* are known. $[PdHCl_2(CO)]^{\ominus}$ may be isolated as salts of large cations (e.g. $Ph_4As^{\oplus}$) from alcoholic solutions of $PdCl_2$ through which CO has been passed. The only other palladium hydride complex obtained pure is $PdHCl(PEt_3)_2$ from reaction of *trans*-$[PdCl_2(PEt_3)_2]$ with Me_3GeH. It decomposes in 5 min at 20° in benzene solution.[18, 19]

27.3.2 Halides and Halocomplexes

Although PdF_2 may be prepared by fluorination of the dichloride at *ca* 500° the product is contaminated with palladium.[20] The most convenient routes to the pure product are (i) thermal decomposition according to the following scheme,[21]

$$PdGeF_6 \xrightarrow[\text{vacuum}]{350°} PdF_2 + GeF_4,$$

and (ii) reaction of PdS_2 with SF_4 in a bomb at 150–300°.[22]

$PdCl_2$ has three polymorphs, the form obtained being dependent upon the method of preparation. Transition temperatures are at 401° and 504°.[23] Powdered palladium begins to react with chlorine at 260°.[24] The reaction is said to be "extremely violent" at 300°,[25] but others say that the reaction is most vigorous at or above 525°.[24] Sublimation of product may be prevented by carrying out the reaction in oxygen. Palladium also reacts with HCl but only in the presence of oxygen.[24]

Preparation at *ca* 300° yields the stable low-temperature form as a red powder, Pd_6Cl_{12}.[26] This is converted above 504° to the well-known chain $(PdCl_2)_n$ form[27] which is metastable, reverting to the low-temperature polymorph over a period of months.[23] The red form is very hygroscopic, dissolving in water from which crystals of red-brown $PdCl_2 \cdot 2H_2O$ can be recovered. $PdCl_2$ may also be prepared by a wet method directly from the elements.[28]

The only viable preparation reported for $PdBr_2$ is dissolution of the metal in nitric acid containing bromine. It is not known whether this compound is polymorphic. The maroon solid reacts according to the following scheme[21] to yield the starting material for one route to PdF_2.

$$PdBr_2 + GeO_2 + 2BrF_3 \longrightarrow PdGeF_6 + O_2 + 2Br_2$$

Indeed a series of complex fluorides $Pd^{II}[MF_6]$ have been made by reactions such as the above and variants of it.

---

[16] *P. Fitton, J. E. McKeon*, Chem. Commun. 4 (1968).
[17] *K. M. Mackay*, Hydrogen Compounds of the Metallic Elements, Spon, London 1966.
[18] *E. H. Brooks, F. Glockling*, Chem. Commun. 510 (1965).
[19] *E. H. Brooks, F. Glockling*, J. Chem. Soc. (A), 1030 (1967).
[20] *O. Ruff, E. Ascher*, Z. Anorg. Allgem. Chemie *183*, 204 (1929).
[21] *N. Bartlett, P. R. Rao*, Proc. Chem. Soc. 393 (1964).
[22] *J. H. Canterford, R. Colton*, Halides of the Second and Third Row Transition Metals, Wiley, London 1968.
[23] *J. R. Soulen, P. Saumagne*, J. Phys. Chem. *69*, 3669 (1965).
[24] *Y. I. Ivashentsev, R. I. Timonova*, Zh. Neorg. Khim. *12*, 592 (1967).
[25] *F. Keiser, J. Breed*, Amer. Chem. J. *16*, 20 (1894).
[26] *H. Schäfer, U. Wiese, K. Rinke, K. Brendel*, Angew. Chem. Intern. Ed. Engl. *6*, 253 (1967).
[27] *A. F. Wells*, Z. Kristallogr. *100*, 189 (1938).
[28] British Patent, 879,074 (1961); C.A. *56*, 8298c (1962).

PdI_2 is made by addition of KI to a solution of $PdCl_2$ in hydrochloric acid.[29] Like $PdCl_2$ it is tri-morphic.[30]

Complex Halides. $CsPdF_3$, a pink-brown solid, is the product of refluxing an equimolar mixture of CsF and PdF_2 in SeF_4.[31]

Tetrachloropalladates(II), $M_2^I[PdCl_4]$, are made by evaporating a hydrochloric acid solution of $PdCl_2$ and appropriate halide M^IX.[22] H_2-$[PdCl_4]$ is present in solutions prepared by combination of palladium sponge and chlorine in methanol; addition of KCl forms $K_2[PdCl_4]$, and other salts are made similarly.[28] A wet method, suitable for regeneration of spent reaction mixture residues is reported.[32] Tetrachloropalladates may also be obtained by reduction of the corresponding hexachloropalladates-(IV).[33–35] A detailed procedure for the preparation of $[Pd(NH_3)_4][PdCl_4]$, the palladium analog of Magnus' green salt, is available.[36]

Tetrabromopalladates(II), $M_2^I[PdBr_4]$, are made from $PdBr_2$ as for the chlorides, by halogen exchange with $[PdCl_4]^{2\ominus}$, or by reduction of hexabromopalladate(IV) with the calculated amount of oxalic acid.[37]

PdI_2 is slightly soluble in excess of aqueous KI to give a red solution from which a solid, said to be $K_2[PdI_4]$, may be obtained by evaporation.[38, 39]

The planar *μμ-dihalogenotetrahalogenodipalladates-(II)*, $[Pd_2X_6]^{2\ominus}$, (X = Cl, Br, I), are obtained by interaction of aqueous solutions of $[PdX_4]^{2\ominus}$ with the desired tetra-alkyl or -aryl ammonium or arsonium salt.[40, 41]

[29] *S. A. Shchukarev, T. A. Tolmacheva, Y. L. Pazukhina*, Russian J. Inorg. Chem. *9*, 1354 (1964).
[30] *G. Thiele, K. Brodersen, E. Kruse, B. Holle*, Naturwissenschaften *54*, 615 (1967).
[31] *N. Bartlett, J. W. Quail*, J. Chem. Soc. (London) 3728 (1961).
[32] U.S. Patent, 3,210,152; C.A. *63*, 15895a (1965).
[33] *R. Hoppe, W. Klemm*, Z. Anorg. Allgem. Chemie *268*, 364 (1952).
[34] *F. Puche*, Compt. Rend. *200*, 1206 (1935).
[35] British Patent 879,674 (1959);
[36] *G. B. Kauffman, J. Hwa-Sam Tsai*, Inorg. Synth. *8*, 234 (1970).
[37] *E. Büllmann, A. C. Anderson*, Ber. *36*, 1565 (1903).
[38] *S. Lassaigne*, J. Chim. Medic. *9*, 447 (1833).
[39] *D. M. Adams, D. M. Morris*, J. Chem. Soc. (A), 756 (1969).
[40] *C. M. Harris, S. E. Livingstone, N. C. Stephenson*, J. Chem. Soc. (London) 3697 (1958).
[41] *D. M. Adams, P. J. Chandler, R. G. Churchill*, J. Chem. Soc. (A), 1272 (1967).

27.3.3 Oxygen, Sulfur, and Other Group VI Derivatives of Palladium(II)

Oxygen Compounds. Black anhydrous *PdO* is soluble in concentrated HBr but insoluble in all other mineral acids including aqua regia. For this reason it is not of great value as a starting material for palladium chemistry although its importance as a catalyst in organic chemistry is considerable.

PdO is made by heating finely divided palladium in oxygen,[42] more readily by interaction of $PdCl_2$ with $NaNO_3$ at 600°,[43, 44] $PdCl_2$ with oxygen at 780° [24] or thermal decomposition of $Pd(NO_3)_2$ between 200° and 400°.[45] An electrolytic method of preparation from a palladium anode is recorded which yields the oxide in nearly 100% yield in a highly active catalytic form.[46]

$PdO \cdot nH_2O$ is obtained as a yellow gelatinous precipitate upon addition of dilute alkali to solutions of palladous salts (*e.g.* the nitrate), and by boiling a solution of $K_2[PdCl_4]$ in a large excess of caustic soda.[2] It is acid soluble. Upon heating it progressively loses water, changing color and eventually forming anhydrous PdO with much decomposition to the elements. Palladium is almost unique in second-row chemistry in forming a simple aquo-ion, $[Pd(H_2O)_4]^{2\oplus}$, present in solutions of $PdO \cdot$-nH_2O in dilute nitric, sulfuric or perchloric acids.[47] It also forms several salts.

The reactions: $Pd + HNO_3/H_2SO_4$, $Pd + KHSO_4$, hydrated $Pd(NO_3)_2 + H_2SO_4$ yield red-brown, very deliquescent $PdSO_4 \cdot 2H_2O$. On gentle heating $PdSO_4 \cdot H_2O$ (said to be olive green) is formed, whilst hydrolysis gives a complicated basic sulfate.[47a, 48] From a solution of the metal in selenic acid red-brown $PdSeO_4$ may be isolated.[49]

[42] *L. Wöhler*, Z. Elektrochem. *11*, 836 (1905).
[43] *R. L. Shriner, R. Adams*, J. Amer. Chem. Soc. *46*, 1683 (1924).
[44] *G. Brauer*, Handbook of Preparative Inorganic Chemistry, Vol. II, 2nd Ed., Academic Press, London 1965.
[45] U.S. Patent, 2,450,176 (1948).
[46] C.A. *64*, 13767g (1966).
[47] *S. E. Livingstone*, J. Chem. Soc. (London) 5091 (1957).
[47a] *M. Kane*, Proc. Roy. Soc. *132*, 275 (1842).
[48] *M. Fischer*, Pogg. Ann. *9*, 256 (1827);
M. Fischer, Pogg. Ann. *10*, 607 (1827);
M. Fischer, Pogg. Ann. *12*, 504 (1828);
M. Fischer, Pogg. Ann. *18*, 256 (1830);
M. Fischer, Pogg. Ann. *71*, 431 (1847);
M. Fischer, Pogg. Ann. *74*, 123 (1848).
[49] *S. Hradecki*, Monatsch. Chem. *36*, 289 (1915).

Yellow-brown, highly deliquescent hydrated $Pd(NO_3)_2$ is obtained by dissolution of palladium in nitric acid of density 1·35–1·40.[50] Treatment of this salt with liquid N_2O_5 yields a volatile anhydrous salt $[Pd(NO_3)_2]_n$.[51] $K_2[Pd(NO_3)_4]$ results from interaction of $K_2[Pd(NO_2)_4]$ with concentrated nitric acid.[52]
Brown $[Pd(OCOCH_3)_2]_3$ is prepared by boiling a solution of palladium in glacial acetic acid containing some nitric acid. Other carboxylates are obtained similarly, although the fluorocarboxylates are monomeric in solution. These compounds react with donor ligands such as NR_3, PR_3, AsR_3 to form yellow complexes *trans*-$[PdL_2(OCOR)_2]$.[53]
O_2 Compounds. Addition of $Pd(PPh_3)_4$ in carbon dioxide-free CH_2Cl_2 to diethyl ether through which oxygen is bubbled gives green crystals of $PdO_2(PPh_3)_2$.[54] Passage of oxygen and carbon dioxide through $Pd(PPh_3)_4$ in benzene gives the peroxycarbonate complex, $PdOCO_3(PPh_3)_2$ which also results from the action of CO_2 on $PdO_2(PPh_3)_2$.[55] $PdO_2\{(CH_3)_3CNC\}_2$ results from the action of oxygen on $Pd\{CNC(CH_3)_3\}_2$.[56]
Chelating Oxygen Bonded Complexes. The addition of KHC_2O_4 to $PdCl_2(NH_3)_2$ in water gives golden yellow needles of $K_2Pd(C_2O_4)_2 \cdot 4H_2O$. The salt, $Na_2Pd(C_2O_4)_2 \cdot 2H_2O$ is prepared similarly.[57] Treatment of palladium(II) nitrite with one equivalent of $K_2C_2O_4$ gives the mixed salt $K_2[Pd(C_2O_4)(NO_2)_2]$.[57a]
Other oxygen-bonded complexes of palladium are also very easily prepared. Thus addition of acetylacetone, neutralized with KOH, to $PdCl_2$ in water gives $Pd(acac)_2$.[58] Addition of salicylaldehyde to a buffered (pH 5–6) solution of $PdCl_2$ gives yellow-orange $Pd(salicylaldehyde)_2$. Treatment of the crude product with an alkylamine in refluxing chloroform gives *N*-alkyl derivatives.[59]
Sulfur Compounds. Many binary palladium chalcogenides are known but they are rather inert chemically. For the Pd–S system the phases Pd_4S, $Pd_{14}S_5$, $Pd_{11}S_5$, PdS and PdS_2 are known, and all can be made from the elements under specified conditions. The simpler members can also be obtained by wet methods. Thus, PdS is conveniently prepared either by heating $PdCl_2$ with sulfur, or by treating Pd(II) solutions with H_2S.[60] Dark-brown PdSe is made similarly using H_2Se.[61]
PdS_2 is soluble in aqua regia but not in other mineral acids. Preparation is from $PdCl_2$ and excess of sulfur at 400°–500°, or from a Pd(IV) complex halide (*e.g.* K_2PdCl_6) and sulfur.[62]
$Pd(SCN)_2$ is obtained as a brick-red precipitate upon addition of KSCN to a solution of $K_2[PdCl_4]$.[63] $Pd(SeCN)_2$ is made similarly.

Sulfur-bonded Complexes. Palladium(II) has a stronger affinity for a sulfur donor than an oxygen donor and thus many complexes are known. Thiourea, dimethylsulfoxide and the numerous dialkylsulfide complexes are dealt with under Section 27.3.7.

$K_2[Pd(SCN)_4]$ forms ruby-red needles upon addition of KSCN to a solution of palladium chloride.[64] $[Pd(SeCN)_4]^{2\ominus}$ salts have been made in acetone solution.[65] In substituted complexes with these ligands linkage isomerism is known. *Trans*-$[Pd(SCN)_2(Ph_3As)_2]$[66] isomerizes to the isothiocyanato form when heated to 150°, as do many other similar complexes.[67] In the $[PdXL_3]^{\oplus}$ series the bulkier ligands (*e.g.* Et_4dien) allow both Pd–S(Se) and Pd–N isomers to be formed but with (dien) only the S(Se)-

[50] *N. Schneider*, J. Amer. Chem. Soc. *40*, 582 (1918).
[51] *B. O. Field, C. J. Hardy*, J. Chem. Soc. (London) 4428 (1964).
[52] *R. Eskenazi, R. Levitus, J. Raskovan*, Chem. Ind. (London) 1327 (1962).
[53] *T. A. Stephenson, S. M. Morehouse, A. R. Powell, J. P. Heffer, G. Wilkinson*, J. Chem. Soc. (London) 3632 (1965).
[54] *C. J. Nyman, C. E. Wymore, G. Wilkinson*, J. Chem. Soc. (A), 561 (1968).
[55] *P. J. Hayward, D. M. Blake, C. Wilkinson, C. J. Nyman*, J. Amer. Chem. Soc. *92*, 5873 (1970).
[56] *S. Otsuka, A. Nakamura, Y. Tatsuno*, J. Amer. Chem. Soc. *91*, 6994 (1969).
[57] *G. Landesen*, Z. Anorg. Allgem. Chemie *154*, 429 (1926).
[57a] *C. Duval* in *P. Pascal*, Nouveau Traité de Chimie Minérale, Vol. XIX, p. 577, Masson et Cie, Paris 1958.
[58] *A. A. Grinberg, L. K. Simonova*, Zh. Prikl. Khim. *26*, 880 (1953).
[59] *L. Sacconi, M. Ciampolini, F. Maggio, G. Del Re*, J. Amer. Chem. Soc. *82*, 815 (1960).
[60] *L. Wöhler, K. Ewald, H. G. Krall*, Ber. *66*, 1638 (1933).
[61] *L. Moser, K. Atynski*, Monatsh. Chem. *45*, 235 (1925).
[62] *W. Biltz, J. Larr*, Z. Anorg. Allgem. Chemie *228*, 257 (1936).
[63] *L. Harbeck, D. Lunge*, Z. Anorg. Allgem. Chemie *16*, 50 (1898).
[64] *S. Ivanoff*, Chem. Ztg. *47*, 209 (1924).
[65] *D. Forster, D. M. L. Goodgame*, Inorg. Chem. *4*, 1712 (1965).
[66] *J. L. Burmeister, F. Basolo*, Inorg. Synth. *12*, 218 (1970).
[67] *F. Basolo, J. L. Burmeister, A. J. Poe*, J. Amer. Chem. Soc. *85*, 1700 (1963).

bonded type is known.[68–70] Both isomers exist for Pd(bipyridyl)$(CNS)_2$ but only the S-bonded form in the phenanthroline complex.[71]

$Pd(CS_2)(PPh_3)_2$ results from the reaction of CS_2 with $Pd(PPh_3)_4$.[72] A number of complexes containing either a terminal or bridging *monoalkylthio group*, *RS*$^{\ominus}$, exist and have been prepared by a variety of methods. The reaction of C_2H_5SH with $[PdCl_2\{P(n\text{-}C_3H_7)_3\}]_2$ in acetone gives

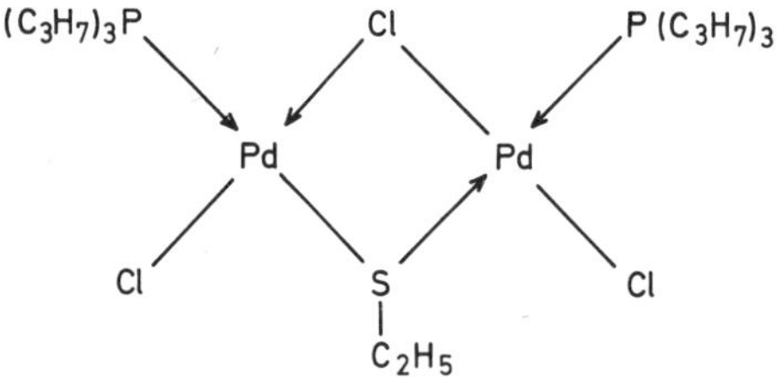

$(C_3H_7)_3P$ Cl $P(C_3H_7)_3$
Pd Pd
Cl S Cl
C_2H_5

and the C_2H_5 dimer containing two bridging $-SC_2H_5$ groups, $[PdClSC_2H_5P(C_3H_7)_3]_2$, can be isolated from the mother liquor.[73] $(C_6H_5)_2S_2$ reacts with Na_2PdCl_4 in methanol to give the polymer $[Pd(SC_6H_5)Cl]_n$[74] which contains alternating halogen and sulfur bridges. The halogen bridges are cleaved by neutral ligands L (L= pyridine, triphenylphosphine, triphenylarsine) to give S-bridged complexes $[Pd(SC_6H_5)ClL]_2$. The reactions of thiols with Na_2PdCl_4 gives the polymers $[Pd(SR)_2]_n$ which in the case of the higher aliphatic derivatives are probably hexamers.[75,76] $PdCl_2(PPh_3)_2$ reacts with $NaSC_2H_5$ to give $Pd(SEt)_2(PPh_3)_2$ which contains terminal SEt groups.[76] Similarly the reaction of $NaSC_6Cl_5$ and $PdCl_2$ in aqueous solution followed by addition of Ph_4AsCl gives the salt $[Ph_4As]_2$-$[Pd(SC_6Cl_5)_4]$.[77]

$PdCl_2$ in water reacts with potassium dithiooxalate to give $K_2Pd(S_2C_2O_2)_2$.[78] At 150° $[(C_6H_5)_2I]_2[Pd(S_2C_2O_2)_2]$ decomposes to give palladium metal and $[Pd(SPh)_2]_n$.[79] The green maleonitriledithiolate complex, $[(C_4H_9)_4N]_2Pd$-$[S_2C_2(CN)_2]_2$, is readily prepared from sodium maleonitriledithiolate and $PdCl_2$ in aqueous methanol followed by addition of $(C_4H_9)_4$-NBr.[80] This type of complex readily undergoes one electron transfer reactions, and addition of iodine or $NiS_4C_4(CF_3)_4$ to the tetraalkylammonium salt gives black crystals of $[R_4N]$-$[PdS_4C_4(CN)_4]$.[81,82] The addition of bis-(trifluoromethyl)-1,2-dithietene to $Pd[S_2C_2(CF_3)_2]$-$(PPh_3)_2$ followed by addition of *N*-methylquinolinium iodide gives red-brown $[C_9H_{10}N]$-$[PdS_4C_4(CF_3)_4]$.[81] A related neutral complex, bis-(dithiobenzil)palladium, forms 1:1 adducts with butadiene and other olefins.[83]

Addition of the salts $KS_2CN(CH_3)_2$ and KS_2-COMe to either $PdCl_2$ or Na_2PdCl_2 in water gives precipitates of the complexes $Pd(S_2CN$-$Me_2)_2$[84] and $Pd(S_2COMe)_2$[85] respectively and other dithiocarbamates and xanthates may be similarly prepared. Phosphines cause C–O bond cleavage in the xanthates to give for example $Pd(S_2CO)[PCH_3(C_6H_5)_2]_2$.[86] $NaS_2CC_6H_5$ and aqueous K_2PdCl_4 give $Pd(S_2CC_6H_5)_2$[87] and with K_2CS_3, $Pd(S_2CS)_2{}^{2\ominus}$ is formed which can be precipitated as $[(C_6H_5)_4As]_2Pd(S_2CS)_2$.[88] Passage of dry HCl gas followed by H_2S through a solution of acetylacetone in ethanol at −78° fol-

---

[68] *J. L. Burmeister, V. J. DeStefano*, Inorg. Chem. *8*, 1546 (1969).
[69] *J. L. Burmeister, H. J. Gysling, J. C. Lim*, J. Amer. Chem. Soc. *91*, 44 (1969).
[70] *J. V. Kingston, G. R. Scollary*, Chem. Commun. 455 (1969).
[71] *W. R. McWhinnie, J. D. Miller*, Adv. in Inorg. Radiochem. *12*, 135 (1969).
[72] *T. Kashiwagi, N. Yasuoka, T. Ueki, N. Kasai, M. Kakudo, S. Takahashi, N. Hagihara*, Bull. Chem. Soc., Japan *41*, 296 (1968).
[73] *J. Chatt, F. A. Hart*, J. Chem. Soc. (London) 1953 (1953).
[74] *T. Boschi, D. Crociani, L. Toniolo, U. Belluco*, Inorg. Chem. *9*, 533 (1970).
[75] *F. G. Mann, D. Purdie*, J. Chem. Soc. (London) 1549 (1935).
[76] *R. G. Hayter, F. S. Humiec*, J. Inorg. Nucl. Chem. *26*, 807 (1964).
[77] *C. R. Lucas, M. E. Peach*, Inorg. Nucl. Chem. Lett. 5, 73 (1969).
[78] *C. S. Robinson, H. O. Jones*, J. Chem. Soc. *101*, 62 (1912).
[79] *G. E. Hunter, R. A. Krause*, Inorg. Chem. *9*, 537 (1970).
[80] *E. Billig, R. Williams, I. Bernal, J. H. Waters, H. B. Gray*, Inorg. Chem. *3*, 663 (1964).
[81] *A. Davison, N. Edelstein, R. H. Holm, A. H. Maki*, Inorg. Chem. *2*, 1227 (1963);
A. Davison, N. Edelstein, R. H. Holm, A. H. Maki, Inorg. Chem. *3*, 814 (1964).
[82] *A. Davison, R. H. Holm*, Inorganic Synthesis, Vol. X, 9, McGraw-Hill Book Co., New York 1967.
[83] *G. N. Schrauzer, V. P. Mayweg*, J. Amer. Chem. Soc. *87*, 1483 (1965).
[84] *L. Malatesta*, Gazz. Chim. Ital. *68*, 195 (1938).
[85] *G. W. Watt, B. J. McCormick*, J. Inorg. Nucl. Chem. *27*, 898 (1965).
[86] *J. P. Fackler, W. C. Seidel*, Inorg. Chem. *8*, 1631 (1969).
[87] *C. Furlani, M. L. Luciani*, Inorg. Chem. *7*, 1586 (1968).
[88] *J. P. Fackler, D. Coucouvanis*, J. Amer. Chem. Soc. *88*, 3913 (1966).

lowed by addition of $PdCl_2$ gives bright red crystals of $Pd(SC(CH_3)CHC(CH_3)S)_2$.[89]

Sulfites and Sulfur Dioxide Complexes. Slow addition of Ag_2SO_3 to $PdCl_2$ in water gives hydrated palladium(II) sulfite which was formulated as $Pd(SO_3)(H_2O)_3$[90] but has more recently been given the formula $Pd(SO_3)(H_2O)_2$.[91] The compound reacts with 1,10-phenanthroline to give $Pd(SO_3)$(phen) whilst with stoichiometric quantities of ammonia stepwise replacement of water results to give $[Pd(SO_3)(H_2O)_2(NH_3)]$, $[Pd(SO_3)(H_2O)(NH_3)_2]$ and $[PdSO_3(NH_3)_3]$.[90] $Na_2[Pd(SO_3)_2]\cdot H_2O$ results from the reaction of stoichiometric quantities of sodium sulfite and $Pd(SO_3)(H_2O)_2$ in water.[90] In warm water this dissolves to give $[Pd(SO_3)(H_2O)_2]$.[2-91] $K_2[Pd(SO_3)_2]$ results from the reaction of potassium disulfite and $PdCl_2$ in water.[90] It is less soluble than the sodium salt but on prolonged boiling solutions containing $[Pd(SO_3)_2(H_2O)]^{2\ominus}$ and $[PdSO_3(H_2O)_2]$ are formed.[91] $Na_6[Pd(SO_3)_4]\cdot 2H_2O$ separates as a yellow solid when sodium sulfite is added to $PdCl_2$ in water containing hydrochloric acid.[91] $Pd(NH_3)_4Cl_2$ reacts with excess of sodium sulfite to produce $Na_2[Pd(SO_3)_2(NH_3)_2]\cdot 6H_2O$.[90] Passage of sulfur dioxide through a benzene solution of $Pd(PPh_3)_4$ gives $[Pd(PPh_3)_4(SO_2)]_n$.[72]

Thiosulfates. $[Pd(NH_3)_4]Cl_2$ reacts with $Na_2S_2O_3\cdot 5H_2O$ in water to give a precipitate of yellow, $Na_2[Pd(S_2O_3)_2]\cdot H_2O$. A soluble form results when acetone is added to the above complex in aqueous ammonia.[92] The complexes $K_2Pden(S_2O_3)_2$ and $PdS_2O_3\cdot 1\frac{1}{2}en$ (en = ethylenediamine) are known.[92, 93]

27.3.4 Nitrogen Derivatives

Reaction of NaN_3 with solutions of Pd(II) in HCl acetone leads to formation of $[Pd(N_3)_4]^{2\ominus}$ and *μμ-diazidotetraazido palladium(II)*, $[Pd_2(N_3)_6]^{2\ominus}$. The latter may be isolated as the $[Fe(o\text{-phen})_3]^{2\oplus}$ salt and the former with $Ph_4As^{\oplus}$. Although these compounds are not explosive intermediate products are.[94-96] Tetraazidopalladium(II) reacts with Ph_3P to give *trans*-$[Pd(N_3)_2(Ph_3P)_2]$; a series $Pd(N_3)_2L_2$ exists where L = monodentate or L_2 = bidentate donor ligand, the members of which may also be made directly by the following scheme:[96a]

$$Na_2PdCl_4 + L \longrightarrow [PdCl_2L_2] \xrightarrow[NaN]{DMF} Pd(N_3)_2L_2.$$

Reaction of *trans*-$[Pd(N_3)_2(Ph_3P)_2]$ with CO gives $Pd(NCO)_2(Ph_3P)_2$, whilst with nitriles an interesting addition occurs with formation of a cyclic system[97, 98]

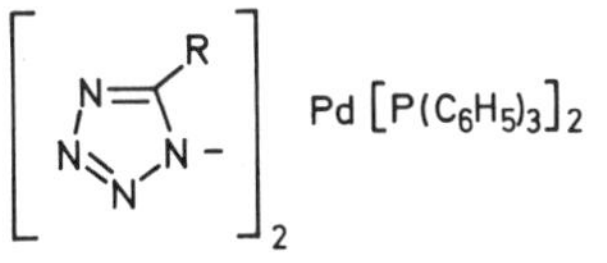

$$\left[\begin{array}{c} \text{R} \\ \text{N=C} \\ \text{N}\diagdown\text{N}\text{–} \\ \text{N} \end{array}\right]_2 Pd[P(C_6H_5)_3]_2$$

$[Pd(NCO_4)_2]^{2\ominus}$, formed by interaction of $[PdCl_4]^{2\ominus}$ with AgNCO,[99] also reacts with tertiary phosphines yielding *trans*-$[Pd(NCO)_2(PR_3)_2]$.[100] The same products result from interaction of $PdCl_2L_2$ and AgNCO.[101] The complexes $[Pd(NCO)(dien)]^{\oplus}$ and $[Pd(NCO)(Et_4\text{-}dien)]^{\oplus}$ are also Pd–N bonded as is $Pd(bipy)(NCO)_2$.[71] The cyanide complexes $PdX_2(RCN)_2$ are made by treatment of aqueous $[MX_4]^{2\ominus}$ (X = Cl, Br, I) solutions with RCN.[102]

Complexes with the thiocyanate ion can exhibit linkage isomerism. They are treated in Section 27.3.3.

$K_2[Pd(NO_2)_4]$, known as a dihydrate and anhydrous, is made from interaction of KNO_2 and Pd(II) solutions.[57a] When treated with HCl $K_2[PdCl_2(NO_2)_2]$ is formed.

---

[89] *O. Siimann, J. Fresco,* Inorg. Chem. *8,* 1846 (1969).

[90] *G. A. Earwicker,* J. Chem. Soc. (London) 2620 (1960).

[91] *R. Eskenazi, J. Raskovan, R. Levitus,* J. Inorg. Nucl. Chem. *27,* 371 (1965).

[92] *R. Eskenazi, R. Levitus,* J. Inorg. Nucl. Chem. *31,* 2195 (1969).

[93] *D. I. Riabchicov, I. Issakova,* Dokl. Akad. Nauk. SSSR *41,* 161 (1943).

[94] *F. G. Sherif, K. F. Michail,* J. Inorg. Nucl. Chem. *25,* 999 (1963).

[95] *W. Beck, K. Feldl, E. Schwierer,* Angew. Chem. Intern. Ed. Engl. *4,* 439 (1965).

[96] *W. Beck, W. P. Fehlhammer, P. Pöllmann, E. Schwierer, K. Feldl,* Chem. Ber. *100,* 2335 (1967).

[96a] *K. Bowman, Z. Dori,* Inorg. Chem. *9,* 395 (1970).

[97] *W. Beck, W. P. Fehlhammer, H. Bock, M. Bauder,* Chem. Ber. *102,* 3637 (1969); *W. P. Fehlhammer, W. Beck, P. Pöllmann,* Chem. Ber. *102,* 3903 (1969).

[98] *W. Beck, W. P. Fehlhammer, P. Pöllmann, H. Schächl,* Chem. Ber. *102,* 1976 (1969).

[99] *D. Forster, D. M. L. Goodgame,* J. Chem. Soc. (London) 1286 (1965).

[100] *W. Beck, E. Schwierer,* Chem. Ber. *98,* 298 (1965).

[101] *A. H. Norbury, A. I. P. Sinha,* J. Chem. Soc. (A), 1598 (1968).

[102] *R. A. Walton,* Spectrochim. Acta *21,* 1795 (1965).

27.3.5 Complexes with Palladium–Carbon Bonds

$Pd(CN)_2$ is made by addition of KCN to a slightly acid solution of either $PdCl_2$ or $(NH_4)_2[PdCl_4]$ in water.[103] $K_2[Pd(CN)_4]$ is made by dissolving $Pd(CN)_2$ in excess of KCN.[103] It crystallizes from water as the trihydrate, when dried at 100° forms a monohydrate, and becomes anhydrous at 200°. The anhydrous acid $H_2[Pd(CN)_4]$ is obtained by HCl/ether extraction of cyanide solutions.[104] $Na_2[Pd(CNO)_4]\cdot 5H_2O$ results from interaction of palladous nitrate with sodium fulminate.[105] Treatment with various tertiary phosphine ligands yields compounds of the type $Pd(CNO)_2(PR_3)_2$.[106]

Organo-Palladium Compounds: Alkyl, Aryl and Acyl Complexes. Mono-alkyl compounds of the type *trans*-$PdXR(L)_2$ are obtained from the interaction of alkylmagnesium halides or alkyllithium reagents with the complexes PdX_2L_2, (X=Cl or Br; L=tertiary phosphine or arsine).[107] Analogous reactions with arylmagnesium halides yield the corresponding aryl compounds.[107] Dialkyl- and diaryl-palladium complexes of the general formula PdR_2L_2 (R=alkyl or aryl; L=tertiary phosphine, arsine and amine, or dialkylsulfide) are prepared by the reaction of the dihalopalladium complexes with excess of organolithium compounds in ether at low temperatures.[107] Metathetical reactions of the halo (methyl) palladium complexes with alkali metal cyanides and thiocyanates give the corresponding cyano and thiocyanato derivatives.[107] Acetyl complexes, $PdX(COMe)(PEt_3)_2$ can be prepared by the action of carbon monoxide on the complexes *trans*-$PdXMe(PEt_3)_2$ at atmospheric pressure and room temperature.[108]

Heptafluoropropyl iodide reacts with $Pd(CH_3)_2$(bipyridyl) to give the perfluoroalkyl complexes $Pd(CH_3)(C_3F_7)$(bipyridyl) and $Pd(C_3F_7)_2$(bipyridyl).[109] Perfluoroalkyl complexes, $PdIR_fL_2$, ($R_f=CF_3$, C_2F_5, C_3F_7; L=PPh_3, $PMePh_2$; $L_2=Ph_2PCH_2CH_2PPh_2$) also result from the addition of a perfluoroalkyl iodide to the complexes PdL_4.[110,111] CF_3COCl reacts with $Pd(PMePh_2)_4$ to give the acyl complex *trans*-$PdCl(COC_3F_7)(PMePh_2)_2$.[111] The action of LiC_6F_5 upon *trans*-$PdCl_2(PEt_3)_2$ gives *trans*-$PdCl(C_6F_5)(PEt_3)_2$ and the *cis* and *trans* isomers of $Pd(C_6F_5)_2(PEt_3)_2$.[112] *Trans*-$PdCl(C_6F_5)(PMePh_2)_2$ also results from the action of C_6F_5COCl upon $Pd(PMePh_2)_4$. Complexes containing the ligand $CCl(CF_3)_2$ *i.e.* $Pd\{CCl(CF_3)_2\}_2(PhCN)_2$ and $[PdCl\{CCl(CF_3)_2\}(PhCN)]_2$ can be prepared from $PdCl_2(PhCN)_2$ and $(CF_3)_2CN_2$.[113]

Certain halogen-bridged dimeric complexes of palladium(II) containing palladium–carbon bonds are obtained by the reactions of azobenzene,[114] 2-phenylpyridine,[115] benzo[*h*]quinoline[116] and 8-methylquinoline(II)[114] with $PdCl_4^{2\ominus}$.

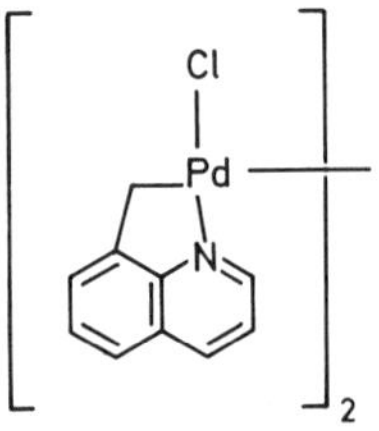

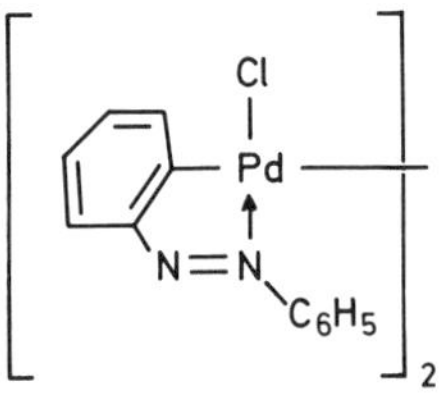

Vinyl and Acetylide Complexes. The perfluorovinyl complex, *trans*-$Pd(CF{=}CF_2)_2(PEt_3)_2$ can be prepared by the reaction of $LiCF{=}CF_2$ with *trans*-$PdCl_2(PEt_3)_2$.[117] Addition of $CF_2{=}CFBr$ or $CF_2{=}CFCl$ to PdL_4, (L=PPh_3, $PMePh_2$), gives the vinyl complexes $PdX(CF{=}CF_2)L_2$ (X=Br, L=PPh_3; X=Cl or Br, L=PPh_3, $PMePh_2$).[111] Similar products, *e.g.*, $PdCl(CCl{=}CCl_2)(PPh_3)_2$, also result from reactions with chloro-olefins.[118]

Acetylide complexes of the type $K_2[Pd(C{\equiv}CR)_2]$, (R=H, Me or Ph) are produced by reacting

[103] *J. H. Bigelow*, Inorg. Synth. *2*, 245 (1946).
[104] *D. F. Evans, D. Jones, G. Wilkinson*, J. Chem. Soc. (London) 3164 (1964).
[105] *W. Wöhler, A. Berthmann*, Ber. *62*, 2748 (1929).
[106] *W. Beck, E. Schwierer*, Chem. Ber. *98*, 298 (1965).
[107] *G. Calvin, G. E. Coates*, J. Chem. Soc. (London) 2009 (1960).
[108] *G. Booth, J. Chatt*, Proc. Chem. Soc. 67 (1961).
[109] *P. M. Maitlis, F. G. A. Stone*, Chem. & Ind. (London) 1865 (1962).
[110] *D. T. Rosevear, F. G. A. Stone*, J. Chem. Soc. (A), 164 (1968).
[111] *A. J. Mukhedkar, M. Green, G. F. A. Stone*, J. Chem. Soc. (A), 3023 (1969).
[112] *F. J. Hopton, A. J. Rest, D. T. Rosevear, F. G. A. Stone*, J. Chem. Soc. (A), 1326 (1966).
[113] *J. Ashley-Smith, J. Clemens, M. Green, F. G. A. Stone*, J. Organometal. Chem. *17*, P 23 (1969).
[114] *A. Kasahara*, Bull. Chem. Soc. Japan *41*, 1272 (1968).
[115] *G. E. Hartwell, R. V. Lawrence, M. J. Smas*, Chem. Commun. 912 (1970).
[116] *S. Trofimenko*, J. Amer. Chem. Soc. *93*, 1808 (1971).
[117] *A. J. Rest, D. T. Rosevear, F. G. A. Stone*, J. Chem. Soc. (A), 66 (1967).
[118] *P. Fitton, J. E. McKeon*, Chem. Commun. *4* (1968).

palladium(II) cyanide and $K_2[Pd(CN)_4]$ with potassium acetylides in liquid ammonia followed by reduction with potassium in liquid ammonia.[119] Barium salts, $Ba[Pd(C{\equiv}CR)_2(CN)_2]\cdot(NH_3)_3$ can be obtained by metathesis using barium thiocyanate. The acetylide complexes containing the unsubstituted ligand $C{\equiv}CH^{\ominus}$ tend to be the least stable, but all the complexes decompose in air.[119] More stable complexes, e.g. $Pd(C{\equiv}CR)_2(PEt_3)_2$, ($R{=}CF_3$ or C_6H_5), can be obtained from $PdX_2(PEt_3)_2$ ($X{=}Cl$ or Br) and the Grignard reagent $RC{\equiv}CMgX$.[120,121]

Olefin Compounds. Palladium forms a series of formally zerovalent and divalent olefin complexes. They are less stable than their platinum analogs and consequently fewer compounds have been prepared.

Zerovalent olefin complexes of the type $Pd(olefin)(PPh_3)_2$ are stabilized by electron withdrawing substituents on the olefins and are prepared by the addition of the olefin to $Pd(PPh_3)_4$. This method gives complexes of the type $Pd(olefin)(PPh_3)_2$, (olefin = maleic anhydride, ethylfumarate, tetracyanoethylene and fumaronitrile),[118,122] which are air stable and soluble in chloroform. A palladium(0) olefin complex, Pd(cyclopentadiene)(cyclohexa-1,3-diene) also results from the action of $K^{\oplus}C_5H_5^{\ominus}$ on $PdCl_2(1,3\text{-}C_6H_8)$.[123] $Pd(C_5H_6)(1,3\text{-}C_6H_8)$ forms red, air-stable crystals which are soluble in most organic solvents.

Treatment of Na_2PdCl_4 in methanol with dibenzylideneacetone in the presence of sodium acetate gives a quantitative yield of the brown crystalline complex $Pd(PhCH{=}CHCOCH{=}CHPh)_2$.[124]

Olefins react with $PdCl_2(PhCN)_2$ in benzene or in 50% acetic acid to form the divalent complexes $Pd_2Cl_4(olefin)_2$,[125] Successful preparations of this type of complex by the direct action of the olefin with $PdCl_2$ suspended in an inert solvent[126] or liquid olefin[127] have also been achieved.

Cis-olefins appear to be more reactive towards $PdCl_2$ than their *trans*-isomers.[127] The olefin complexes are stable for a few hours in air but are more stable under nitrogen at 0°. They readily react with moisture to give palladium metal; thus the complex $[PdCl_2(C_2H_4)]_2$ decomposes rapidly in water to give acetaldehyde and is only stable under an atmosphere of ethylene. The π-complexes of straight-chain olefins dissolve in chlorinated solvents in which they slowly dissociate with deposition of $PdCl_2$. The solubility in aliphatic hydrocarbons increases with the increase of molecular weight of the olefins, but the π-complexes from butene isomers are practically insoluble.[127] Olefin complexes are probably formed initially when oct-1-ene and dec-1-ene react with $PdCl_2$ but these complexes rapidly lose hydrogen chloride to give a π-allyl complex. The cyclo-octene complex is very stable and is not converted to a π-allyl complex on heating.

$[PdCl_2(C_2H_4)]_2$ reacts with pyridine-*N*-oxide to give one of the few monomeric palladium(II) olefin complexes, *trans*-$PdCl_2(C_2H_4)(C_5H_5NO)$.[128] Other monomeric olefin complexes are only known in solution.[129]

Chelate complexes of the type PdX_2(diene), ($X{=}Cl$ or Br) are readily formed by the dienes such as hexa-1,5-diene,[130] dicyclopentadiene,[131] bicyclo[2.2.1]heptadiene,[132] cyclo-octa-1,5-diene[131] and cyclo-octatetraene[133] usually by the action of the diene on $PdCl_2(PhCN)_2$ in benzene or chloroform, Na_2PdCl_4 or $PdCl_2$ in polar solvents. Metathesis with lithium bromide gives the corresponding dibromo derivatives.[132] The cyclo-octa-1,5-diene complex also results from the reaction of PdX_2 with either cyclo-octa-1,3-diene or 4-vinylcyclohexene. Rearrangement of *cis,trans*-cyclodeca-1,5-diene, upon reaction with $PdCl_2(PhCN)_2$ in benzene, also occurs to give $PdCl_2$(*cis*-1,2-divinylcyclohexane).[134] However, the isomeric *cis,cis*-cyclodeca-1,6-diene does not rearrange but produces the

---

119 *R. Nast, W. Hoerl*, Chem. Ber. *95*, 1470 (1962).

120 *G. Calvin, G. E. Coates*, Chem. & Ind. (London) 160 (1958).

121 *M. I. Bruce, D. A. Harbourne, F. Waugh, F. G. A. Stone*, J. Chem. Soc. (A), 356 (1968).

122 *G. L. McClure, W. H. Baddley*, J. Organometal. Chem. *27*, 155 (1971).

123 *E. O. Fischer, H. Werner*, Chem. Ber. *93*, 2075 (1960).

124 *Y. Takahashi, Ts. Ito, S. Sakai, Y. Ishii*, Chem. Commun. 1065 (1970).

125 *M. S. Kharasch, R. C. Seyler, F. R. Mayo*, J. Amer. Chem. Soc. *60*, 882 (1938).

126 *W. M. MacNevin, S. A. Giddings*, Chem. & Ind. (London) 1191 (1960).

127 *G. F. Pregaglia, M. Donati, F. Conti*, Chem. & Ind. (London) 1923 (1966).

128 *W. H. Clement*, J. Organometal. Chem. *10*, 19 (1967).

129 *F. R. Hartley*, Chem. Rev., *69*, 799 (1969).

130 *K. A. Jensen*, Acta Chem. Scand. *7*, 866 (1953).

131 *J. Chatt, L. M. Vallarino, L. M. Venanzi*, J. Chem. Soc. (London) 2496 (1957).

132 *E. W. Abel, M. A. Bennett, G. Wilkinson*, J. Chem. Soc. (London) 3178 (1959).

133 *H. P. Fritz, H. Keller*, Chem. Ber. *95*, 158 (1962).

134 *J. C. Trebellas, J. R. Olechowski, H. B. Jonassen*, J. Organometal. Chem. *6*, 412 (1966).

expected complex.[134] The addition of Dewar hexamethylbenzene, DHMB (bicyclo[2.2.0]hexa-2,5-diene) to $PdCl_2(PhCN)_2$ in dichloromethane or $Na_2PdCl_4 \cdot 4H_2O$ in methanol gives $PdCl_2$-(DHMB).[135]

In chloroform solution at 34° $PdCl_2$(DHMB) decomposes almost completely into $PdCl_2$ and hexamethylbenzene in about 20 minutes. The dibromide decomposes more slowly. Treatment of $PdCl_2$(DHMB) with triphenylphosphine gives $PdCl_2(PPh_3)_2$ and HMDB. When a methanolic of $PdCl_2$(HMDB) is treated with $NaOCH_3$ in methanol the halogen bridged allyl complex $Pd_2Cl_2(C_{12}H_{17})_2$ is obtained.[135]

When $PdCl_2(C_8H_{12})$, [C_8H_{12}=cyclo-octa-1,5-diene] is suspended in boiling methanol and treated with sodium carbonate the air-stable halogen bridged dimer, $Pd_2Cl_2(C_8H_{12}\ OMe)_2$ is readily formed.[136] Similar reactions occur when $PdCl_2(C_8H_{12})$ and other diene complexes of palladium are treated with the conjugate bases of ketoesters,[137] malonic esters,[137] carboxylic acids[138] and amines.[139] The halogen bridges may be cleaved by amines[136] and treatment of the Pd(acac)-(C_8H_{12}acac) [acac=$CH_3COCHCOCH_3$] with $P(C_6H_5)_3$, $As(C_6H_5)_3$ or $Sb(C_6H_5)_3$ gives the zerovalent complexes, $Pd(MPh_3)_4$, (M=P, As or Sb).[14]

At −40° buta-1,3-diene displaces pentene from $[PdCl_2(pentene)]_2$ to give $[PdCl_2$(buta-1,3-diene)$]_2$. However, above −20° this complex is unstable and at room temperature isomerizes[140] to the π-allyl compound, $\{PdCl(CH_2CH\text{-}CHCH_2Cl)\}_2$, which is the product obtained at room temperature.[141]

Complexes of Olefins which Contain Additional Donor Groups. A number of olefin complexes have been prepared with olefins which contain an additional donor atom. Thus the ligands AA and AP react with a chloroform solution of $PdCl_2(PhCN)_2$ to give air stable yellow complexes $PdCl_2AA$ and $PdCl_2AP$ in which both the arsenic or phosphorus donor atom and the double bond are coordinated. Both these chelate complexes react with KNCS to give complexes of the type $Pd(NCS)_2(ligand)_2$ in which the double bonds are uncoordinated.[142,143] These complexes are probably formed by a disproportionation reaction.

$$2PdCl_2\ (ligand) + 6NCS^{\ominus} \longrightarrow Pd(NCS)_2(ligand)_2 + Pd(SCN)_4^{2\ominus} + 4Cl^{\ominus}$$

The reaction of sodium iodide with $PdCl_2AA$ gives the black dimer $(PdI_2AA)_2$ in which the double bond is not coordinated. However, $PdCl_2AP$ and sodium iodide gives the red-black PdI_2AP which is probably polymeric and in this complex the double bonds are coordinated.[142,143] In contrast to these studies CH_2=$CH(CH_2)_3MR_2$, (M=P or As; R=CH_3 or C_6H_5) form halogen bridged dimers, $[PdCl_2(ligand)]_2$ in which only the phosphine or arsine ligand is coordinated.[142,143] The addition of but-3-enylbutyl sulfide, BBS to $Na_2PdCl_4 \cdot 4H_2O$ in 1:1 ethanol water acidified with 2·5MHCl gives a chelate complex, $PdCl_2BBS$. The sulfur ligand is readily displaced by *p*-toluidine and triphenylphosphine to give $PdCl_2(p\text{-toluidine})_2$ and $PdCl_2(PPh_3)_2$.[144]

Acetylene Compounds. The palladium(II) acetylene complexes are more complex and less well understood than the platinum(II) complexes mainly because acetylenes are very easily polymerized by palladium(II) compounds.[129,145,146] Thus diphenylacetylene and $PdCl_2(PhCN)_2$ in benzene on gentle warming gives mainly hexaphenylbenzene.[145–147] However, in ethanol-chloroform the orange red crystalline complex (I) readily separates.[145–147] The addition of $(CH_3)_2(OH)CC{\equiv}CC(OH)(CH_3)_2$ to either an acidified aqueous solution of $PdCl_2$ or K_2PdCl_4 gives a fine crystalline green powder stable in air which appears to be a bis-acetylene complex, $PdCl(acetylene)_2$. Bromo, iodo and thiocyanato derivatives are known.[148]

[135] *B. L. Shaw, G. Shaw,* J. Chem. Soc. 602 (1969).
[136] *J. Chatt, L. M. Vallarino, L. M. Venanzi,* J. Chem. Soc. (London) 3413 (1957).
[137] *J. Tsuji, M. Takahashi,* J. Amer. Chem. Soc. *87*, 3275 (1965).
[138] *C. B. Anderson, B. J. Burreson,* Chem. & Ind. (London) 620 (1967).
[139] *G. Riaro, A. Dehenzi, R. Palumbo,* Chem. Commun. 1150 (1967).
[140] *S. D. Robinson, B. L. Shaw,* J. Chem. Soc. (London) 4806 (1963).
[141] *M. Donati, F. Conti,* Tetrahedron Lett. 1219 (1966).
[142] *M. A. Bennett, W. R. Kneen, R. S. Nyholm,* Inorg. Chem. 7, 556 (1968).
[143] *M. A. Bennett, H. W. Kouwenhoven, J. Lewis, R. S. Nyholm,* J. Chem. Soc. (London) 4570 (1964).
[144] *D. C. Goodall,* J. Chem. Soc. (A), 887 (1968).
[145] *H. Dietl, H. Reinheimer, J. Moffat, P. M. Maitlis,* J. Amer. Chem. Soc. *92*, 2276 (1970).
[146] *H. Reinheimer, J. Moffat, P. M. Maitlis,* J. Amer. Chem. Soc. *92*, 2285 (1970).
[147] *A. T. Blomquist, P. M. Maitlis,* J. Amer. Chem. Soc. *84*, 2329 (1962);
P. M. Maitlis, D. Pollock, M. L. Games, W. J. Pryde, Can. J. Chem. *43*, 470 (1965).
[148] *A. V. Babaeva, T. I. Beresneva,* Zh. Neorg. Khim. *11*, 1966 (1966).

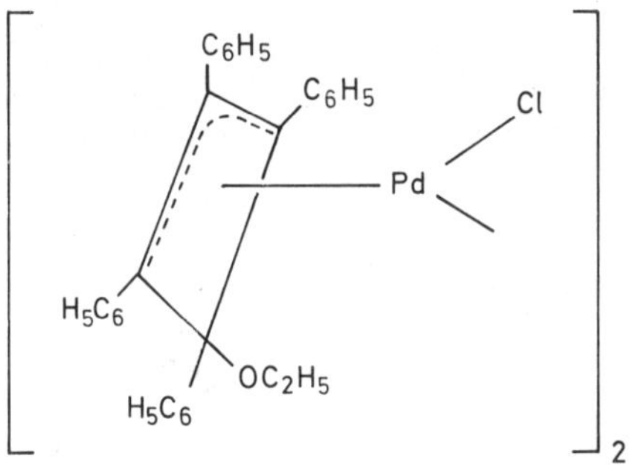

I

In contrast to the studies on palladium(II) well defined, airstable palladium(0) acetylene complexes of the type Pd(acetylene)$(PPh_3)_2$ are readily formed. Thus $CF_3C{\equiv}CCF_3$ and MeOCOC≡CCOOMe react with $Pd(PPh_3)_4$ to give Pd(acetylene)$(PPh_3)_2$.[149] Pd(fumaronitrile)-$(PPh_3)_2$ with dicyanoacetylene gives Pd-(NCC≡CCN)$(PPh_3)_2$.[122]

π-Allyl Compounds. $[PdCl(\pi\text{-}C_3H_5)]_2$ is probably best obtained by passage of carbon monoxide through an aqueous methanolic solution of sodium chloropalladite containing allyl chloride.[150] Addition of allyl chloride to a mixture of anhydrous tin(II) chloride, palladium(II) chloride and lithium chloride in methanol also gives a good yield.[151] Other reducing agents, *e.g.*, iron, zinc or copper, may be used in place of CO/H_2O or $SnCl_2$ but the yields are lower.[152] $[PdCl(\pi\text{-}C_3H_5)]_2$ is a yellow crystalline compound, stable in air, and readily soluble in polar solvents and dilute aqueous acids and alkalis. It is decomposed by boiling water to yield propane and acrolein.[153] Treatment with $Na^{\oplus}C_5H_5^{\ominus}$ in tetrahydrofuran-benzene gives $Pd(\pi\text{-}C_3H_5)(\pi\text{-}C_5H_5)$[154]; with thallous acetylacetonate it gives Pd(acac)-$(\pi\text{-}C_3H_5)$.[140] The chloride bridge is also readily cleaved by donor ligands.[155,156] With ethylenediamine or bipyridyl, ionic complexes are formed, $[Pd(\pi\text{-}C_3H_5)en]^{\oplus}Cl^{\ominus}$ and $[Pd(\pi\text{-}C_3H_5)\text{-}bipy]^{\oplus}Cl^{\ominus}$. With the stoichiometric amount of bipyridyl $[Pd(\pi\text{-}C_3H_5)(bipy)][Pd(\pi\text{-}C_3H_5)Cl_2]$ results.[157] Anionic π-allyl compounds, *e.g.*, $[Ph_4P][PdCl_2(\pi\text{-}C_3H_5)]$ are also readily extracted from concentrated aqueous solutions of $[PdCl(\pi\text{-}C_3H_5)]_2$ and potassium chloride.[158] Metathesis with sodium iodide or potassium thiocyanate in acetone gives $[PdI(\pi\text{-}C_3H_5)]_2$ and $[PdSCN(\pi\text{-}C_3H_5)]_2$ respectively.[155,156] Substituted π-allyl palladium complexes may also be prepared in a similar manner to that described for $[PdCl(\pi\text{-}C_3H_5)]_2$[156] and other methods which have been used successfully include the reaction of palladium(II) halides with substituted allyl alcohols[159] at 50°, reactions of palladium(II) halides with allyl halides[160] or reactions of branched chain alkenes[161–163] in 50% acetic acid at 50–100° with various palladium compounds, *e.g.*, $PdCl_2$, Na_2PdCl_4 and $PdCl_2(PhCN)_2$. In general the complexes contain the bulkiest substituent in the *syn*-position.[164]

Passage of butadiene through a benzene solution of $PdCl_2(PhCN)_2$ gives $[PdCl\{CH_2CHCHCH_2\text{-}X\}]_2$ which reacts with methanol to give the less stable methoxy derivative (X=MeO).[140] This complex is best prepared by passage of butadiene into a suspension of Na_2PdCl_4 in methanol; other 1,4-dienes behave similarly.[140]

When allene is bubbled into a solution of $PdCl_2\text{-}(PhCN)_2$ in benzene a rapid reaction occurs and the compound $[PdCl(CH_2CClCCH_2)]_2$ is formed. However, when solid $PdCl_2(PhCN)_2$ is added to benzene, through which allene is bubbling, a different complex, **(II)** results.

A further product, **(III)**, results when allene is passed through a solution of $PdCl_2(PhCN)_2$ in benzonitrile.[165]

π-Allyl compounds with substituted allyl ligands are oxidized in acid solutions in the

[149] *E. O. Greaves, C. J. L. Lock, P. M. Maitlis*, Can. J. Chem. *46*, 3879 (1968).

[150] *W. T. Dent, R. Long, A. J. Wilkinson*, J. Chem. Soc. (London) 1585 (1964);
J. K. Nicholson, J. Powell, B. L. Shaw, Chem. Commun. 174 (1966).

[151] *M. Sakabibava, Y. Takahashi, S. Sakai, Y. Ishii*, Chem. Commun. 396 (1969).

[152] *J. H. Lukas, J. E. Blom*, J. Organometal. Chem. *26*, C25 (1971).

[153] *R. Hüttel, J. Kratzer, M. Bechter*, Chem. Ber. *94*, 766 (1961).

[154] *B. L. Shaw*, Proc. Chem. Soc. 247 (1960).

[155] *J. Powell, B. L. Shaw*, J. Chem. Soc. 1839 (1967).

[156] *S. D. Robinson, B. L. Shaw*, J. Chem. Soc. (London) 5002 (1964).

[157] *G. Paiaro, A. Musco*, Tetrahedron Lett. 1583 (1965).

[158] *R. J. Goodfellow, L. M. Venanzi*, J. Chem. Soc. (A), 784 (1966).

[159] *J. Smidt, W. Hafner*, Angew. Chem. *71*, 284 (1959).

[160] *J. C. W. Chien, H. C. Dehm*, Chem. & Ind. (London) 745 (1961).

[161] *R. Hüttel, H. Christ*, Chem. Ber. *96*, 3101 (1963).

[162] *R. Hüttel, J. Kratzer*, Chem. Ber. *94*, 766 (1961).

[163] *R. Hüttel, H. Christ*, Chem. Ber. *97*, 1439 (1964).

[164] *H. C. Volger*, Rec. Trav. Chim. Pays-Bas *88*, 225 (1969);
J. Lukas, S. Coren, J. E. Blom, Chem. Commun. 1303 (1969).

[165] *R. G. Schultz*, Tetrahedron *20*, 2809 (1964).

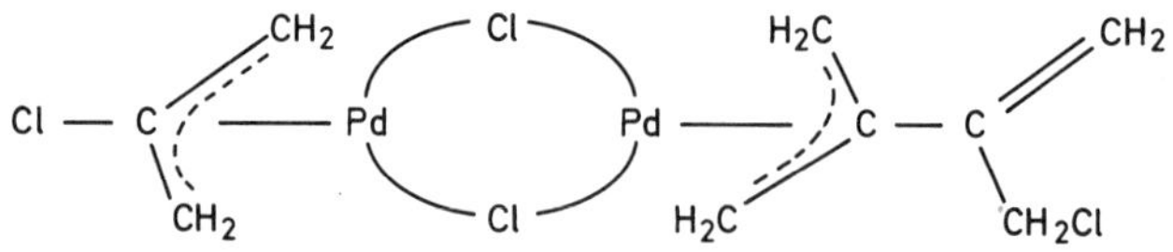

II

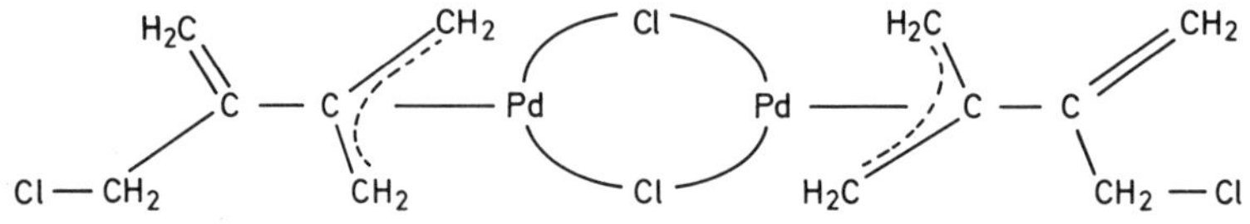

III

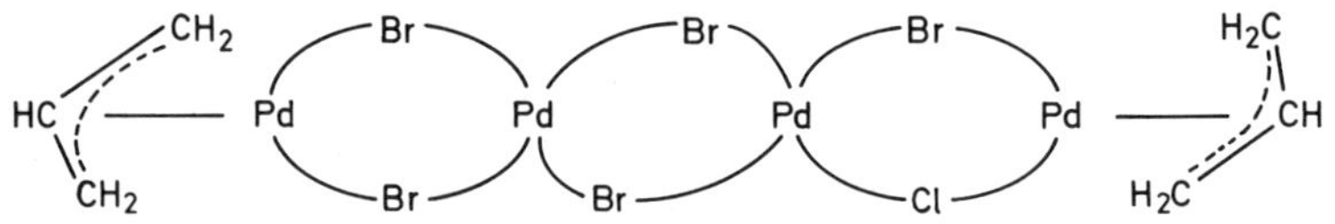

IV

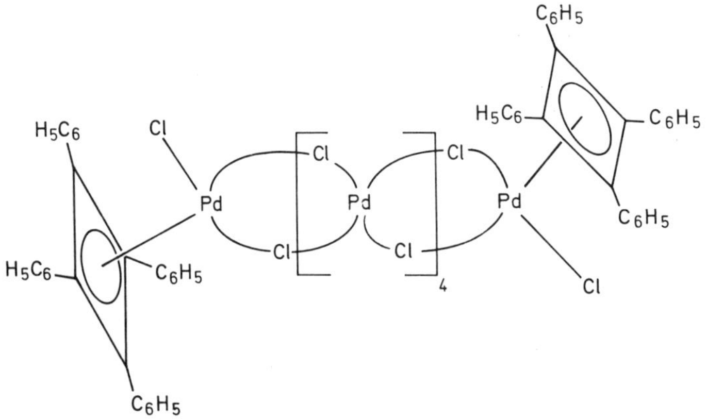

V

presence of manganese dioxide, chromic acid or palladium(II) chloride to give αβ-unsaturated carbonyl compounds.[163] Carbonylation of $[PdCl(\pi\text{-}C_3H_5)]_2$ yields but-3-enoyl chloride. In methanol solutions methyl vinylacetate is formed.[166]

When allyl alcohol reacts with $PdBr_2$ in hydrobromic acid $[PdBr(\pi\text{-}C_3H_5)]_2$ results but on increasing the concentration of $PdBr_2$, the compound (**IV**) is formed.[167]

Bis-(π-allyl)palladium and bis-(π-methylallyl-palladium) are obtained by the reaction of anhydrous $PdCl_2$ with allyl- and methylallyl-magnesium chloride, respectively, in ethereal solution at −40°. $Pd(\pi\text{-}C_3H_5)_2$ is a yellow crystalline substance, is readily soluble in organic solvents and does not react with alcohol or water. Hydrogen chloride or $PdCl_2$ reac with $Pd(\pi\text{-}C_3H_5)_2$ in ether at −80° to form $[PdCl(\pi\text{-}C_3H_5)]_2$.[168]

Cyclopropenyl and Cyclobutadiene Compounds. In addition to cyclopentadienyl compounds which are covered elsewhere cyclopropenyl and cyclobutadiene complexes of palladium exist.

The reduction of $PdCl_2$ with ethylene in aqueous acetonitrile in the presence of triphenylcyclopropenyl chloride gives the orange red solid, $\{PdCl(\pi\text{-}C_3Ph_3)\}_2$.[169]

Cyclobutadiene complexes of varying stoichiometry can result from the action of diphenylacetylene on $PdCl_2(PhCN)_2$.[147] With a 1:1 ratio and slow addition of diphenylacetylene $[(Ph_4C_4)(PdCl_2)_3]_2$, (**V**) is formed. A 2:1 ratio gives $[(Ph_4C_4)(PdCl_2)_{2,5}]_2$. The addition of

[166] *J. Tsuji, J. Kiji, M. Morikawa,* Tetrahedron Lett. 1811 (1963).

[167] *I. I. Moiseev, M. N. Vargaftik, Ya. K. Syrkin,* Izv. Akad. Nauk SSSR Ser. Khim. 775 (1964).

[168] *M. L. Lobach, B. D. Babitskii, V. A. Kormer,* Uspekhi Khimii 36, 1158 (1967).

[169] *I. I. Moiseev, M. N. Vargaftik, Ya. K. Syrkin,* Izv. Akad. Nauk SSSR Ser. Khim. 775 (1964).

Na_2PdCl_4 in water to an ethanolic solution of diphenylacetylene gives (**I**). This complex reacts with hydrogen chloride to provide a good yield of $[(Ph_4C_4)PdCl_2]_2$.[147]

Carbene Compounds. Palladium metal reacts with one equivalent 1,1-dichloro-2,3-diphenylcyclopropene in benzene to give the halogen bridged dimer, bis-[dichloro(2,3-diphenylcyclopropenylidene)palladium(II)], **VI**.[170]

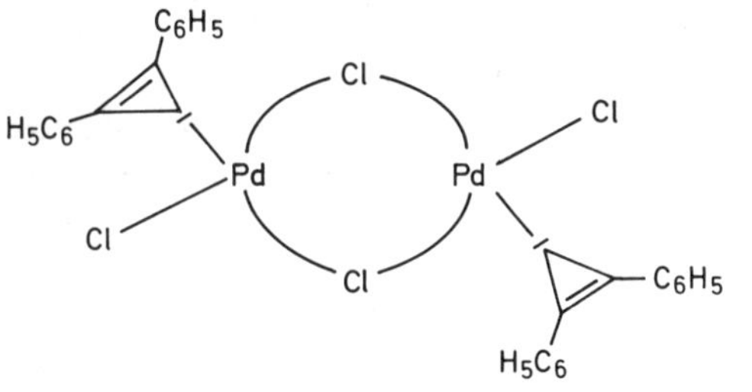

VI

Carbene complexes, $PdCl_2L\{CY(C_6H_5NH)\}$, ($Y=OCH_3$, $CH_3C_6H_4NH$), also result when the complexes $PdCl_2(L)(C_6H_5NC)$, ($L=C_6H_5$-NC or $P(C_6H_5)_3$) are treated with either methanol or *p*-toluidine.[171] The carbene complex $PdCl_2\{C(OCH_3)NHC_6H_5\}(PPh_3)$ reacts with alcoholic KOH to give the palladium–carbon bonded compound, **VII**.

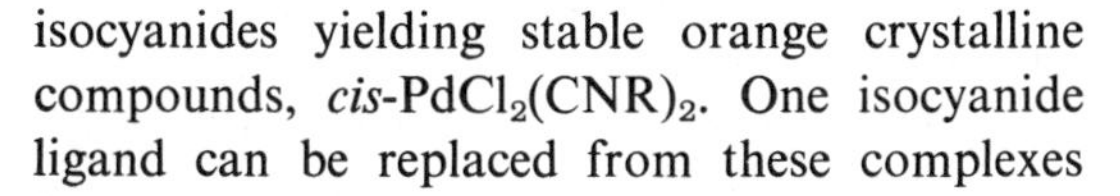

VII

which with hydrochloric acid regenerates the original complex.[172]

Isocyanide Compounds. Palladium(II) halides[173] or $PdCl_2(CH_3CN)_2$[172] react readily with aryl isocyanides yielding stable orange crystalline compounds, *cis*-$PdCl_2(CNR)_2$. One isocyanide ligand can be replaced from these complexes by triphenylphosphine or triphenylarsine to give complexes of the type *cis*-$PdCl_2(CNR)(L)$.[172] These latter complexes react with methanol or *p*-toluidine to give carbene complexes.[171]

The formally zerovalent compounds $Pd(CNR)_4$ and $Pd(CNR)_2$ have been discussed in Section 27.2.

27.3.6 Metal–Metal Bonded Compounds

The addition of $SnCl_2$ to solutions of $PdCl_2$ in a mixture of 2M-HCl and methanol followed by addition of Ph_4AsCl gives the palladium–tin bonded compounds, $[Ph_4As]_2[Pd(SnCl_3)_3Cl]$ and $[Ph_4As]_2[Pd(SnCl_3)_4]_2$.[174,175]

A variety of compounds containing Pd–$SnCl_3$ bonds can be prepared by the addition of $SnCl_2$ to such complexes as $PdCl_2(PPh_3)_2$,[176] $PdCl_2(CNR)_2$[177,178] or $PdCl(\pi\text{-allyl})(PPh_3)$.[176,179] Such reactions lead to the stable complexes $PdCl(SnCl_3)(PPh_3)_2$, $Pd(SnCl_3)_2(RNC)_2$ and $PdSnCl_3(\pi\text{-allyl})PPh_3$. Reactions with $GeCl_2$ ead to analogous complexes $Pd(GeCl_3)_2(RNC)_2$.[177]

Metathetical reactions of the type

$$trans\text{-}PdCl_2Py_2 + 2Na[Mo(\pi\text{-}C_5H_5)(CO)_3] \xrightarrow[-2\,NaCl]{} trans\text{-}Pd\{Mo(\pi\text{-}C_5H_5)(CO)_3\}Py_2$$

$$trans\text{-}PdCl_2[P(C_2H_5)_3]_2 + 2LiGe(C_6H_5)_3 \xrightarrow[-2\,LiCl]{} Pd[Ge(C_6H_5)_3]_2[P(C_2H_5)_3]_2$$

also lead to palladium–metal bonded compounds. *Trans*-$Pd\{Mo(\pi\text{-}C_5H_5(CO)_3\}Py_2$ decomposes slowly in air and rapidly in solution.[180] $Pd(GePh_3)_2(PEt_3)_2$ is stable in the solid state to 97° but decomposes in toluene solution even at −20°.[181] The anionic complex, $K_2[Pd(CN)_2(GePh_3)_2]$, is obtained from KCN and $Pd(GePh_3)_2(PEt_3)_2$.[181] A lead compound $Pd(PbPh_3)_2(PEt_3)_2$ is also known.[182]

The action of $Fe(CO)_4(PPh_2H)$ upon $[PdCl(\pi\text{-}C_3H_5)]_2$ gives the complex (**VIII**).[183]

---

170 *K. Öfele*, J. Organometal. Chem. 22, C9 (1970).
171 *B. Crociani, T. Boschi, U. Belluco*, Inorg. Chem. *9*, 2021 (1970).
172 *B. Crociani, T. Boschi*, J. Organometal. Chem. *24*, C1 (1970).
173 *F. Bonati, L. Malatesta*, Isocyanide Complexes of Metals, Wiley Intersci. Publ., New York 1969.
174 *M. A. Khattak, R. J. Magee*, Chem. Commun. 400 (1965).
175 *G. E. Batley, J. C. Bailar jr.*, Inorg. Nucl. Chem. Lett. *4*, 577 (1968).
176 *M. Sakakibara, Y. Takahashi, S. Sakai, Y. Ishii*, Inorg. Nucl. Chem. Lett. *5*, 427 (1969); *M. Sakakibara, Y. Takahashi, S. Sakai, Y. Ishii*, J. Organometal. Chem. *27*, 139 (1971).
177 *B. Crociani, T. Boschi, M. Nicolini*, Inorg. Chim. Acta *4*, 577 (1970).
178 *F. Bonati, T. Boschi, B. Crociani, G. Minghetti*, J. Organometal. Chem. *25*, 255 (1970).
179 *J. N. Crosby, R. D. W. Kemmitt*, J. Organometal. Chem. *26*, 277 (1971).
180 *P. Braunstein, J. Dehand*, J. Organometal. Chem. *24*, 497 (1970).
181 *E. H. Brooks, F. Glockling*, J. Chem. Soc. (A), 1241 (1966).
182 *G. Carturan, G. Deganello, T. Boschi, U. Belluco*, J. Chem. Soc. (A), 1143 (1969).
183 *B. C. Benson, R. Jackson, K. K. Joshi, D. T. Thompson*, Chem. Commun. 1506 (1968).

$(C_6H_5)_2P$ — Cl — $Fe(CO)_4$
Pd Pd
$(CO)_4Fe$ — Cl — $P(C_6H_5)_2$

VIII

The reaction of $AlCl_3$, Al and $PdCl_2$ in boiling benzene produces two unique crystalline complexes, **(IX)** and **(X)**.[184]

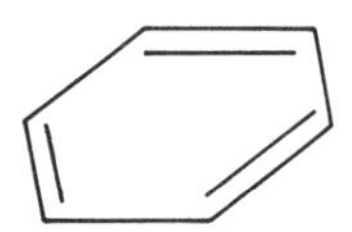

Cl_3AlCl —— Pd —————— Pd —— $ClAlCl_3$

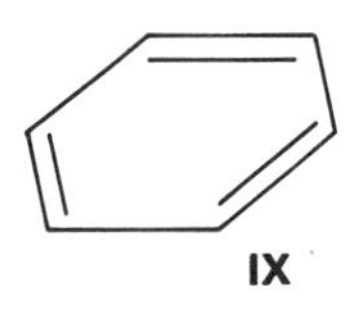

IX

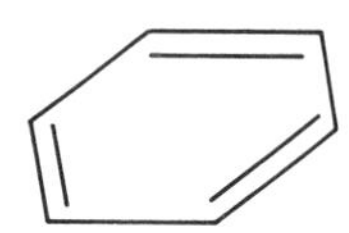

Cl_6Al_2Cl —— Pd —————— Pd —— $ClAl_2Cl_6$

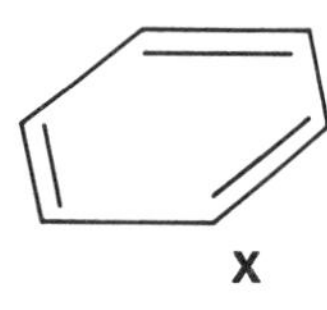

X

27.3.7 Complexes of Palladium(II) with Donor Ligands

$[PdL_4]^{2\oplus}$ *Series and Related Complexes.* Members of this series are prepared by two broad approaches. (a) Combination of the components as in the synthesis of $[Pd(NH_3)_4]^{2\oplus}$ from Pd(II) salts and excess of ammonia; (b) from PdX_2L_2.[57a] By simple variations of procedure compounds $[PdL_4]X_2$ with X=F, Cl, Br, I, OH, SO_3, CO_3, NO_3, SO_4, ClO_4, *etc.* can be made,[57a] as well as those containing complex ions such as $[Pd(NO_2)_4]^{2\ominus}$ or $[MX_6]^{2\ominus}$.[185] For example,

$$[Pd(NH_3)_4]Cl_2 + 2AgClO_4 \xrightarrow[-2\,AgCl]{} [Pd(NH_3)_4](ClO_4)_2 \xrightarrow{1.\ HClO_4\ \ 2.\ NaSCN} [Pd(NH_3)_4][Pd(SCN)_4]^{57a,186,187}$$

$$trans\text{-}[Pd(NH_3)_2F_2] \xrightarrow{1.\ AgF\ \ 2.\ NH_3} [Pd(NH_3)_4]F_2{}^{57a}$$

It has recently been shown that the reaction of palladous nitrate with ammonia yields not $[Pd(NH_3)_4](NO_3)_2$ but $[Pd(NH_3)_3(NO_2)]_2$-$[Pd(NH_3)_4](NO_3)_4$.[188]

A large number of similar complexes have been made where L is monodentate (*e.g.* NH_2OH, py, etc.), [57a] bi-,[189,190] tri-,[191,192] or tetradentate[191–193] with appropriate mono-dentate donor ligands completing the coordination arrangement.[194] There are also many complexes of polydentate ligands with two or more different donor atoms, 2,[195] 3 or 4[196] of which may coordinate, *e.g.*, *bis*[phenyl(*o*-thio-methylphenyl)arsino]-ethane and -propane.[197] Complexes of *o*-phenylene bis(dimethylarsine) and 8-dimethylarsinoquinoline (8-dmaq), $[Pd(diars)_2X_2]$ and $[Pd(8\text{-}dmaq)_2X_2]$ appear to be six-coordinate in solution.[198,199] Doubt has recently been expressed concerning the ready preparation of $[Pd(phen)_2]^{2\oplus}$.[200]

$[PdL_3X]^{\ominus}$ *Series.* The compound formed by reaction of cold $NaNO_2$ solution with *trans*-$[PdCl_2(NH_3)_2]$ in aqueous ammonia is $[Pd(NH_3)_3(NO_2)]Cl$ and not $[Pd(NH_3)_2(NO_2)_2]$.[201]

---

[184] *G. Allegra, G. Casagrande, A. Immirzi, L. Porri, G. Vitulli,* J. Amer. Chem. Soc. *92*, 289 (1970).

[185] *P. Deville, M. Debray,* Compt. Rend. *86*, 926 (1878).

[186] *J. S. Coe, M. D. Hussain, A. A. Malik,* Inorg. Chim. Acta *2*, 65 (1968).

[187] *J. S. Coe, J. R. Lyons,* Inorg. Chem. *9*, 1775 (1970).

[188] *F. P. Boer, V. B. Carter, J. W. Turley,* Inorg. Chem. *10*, 651 (1971).

[189] *H. C. Freeman, J. F. Geldard, F. Lions, M. R. Snow,* Proc. Chem. Soc. 258 (1964).

[190] *J. Chatt, F. G. Mann,* J. Chem. Soc. 1622 (1939).

[191] *B. B. Smith, D. T. Sawyer,* Inorg. Chem. *8*, 1154 (1969).

[192] *F. G. Mann, H. R. Watson,* J. Chem. Soc. 2772 (1958).

[193] *D. St. C. Black, E. Markham,* Rev. Pure Appl. Chem. *15*, 109 (1965).

[194] *L. Rasmussen, C. K. Jorgensen,* Inorg. Chim. Acta *3*, 543 (1969).

[195] *P. Ray,* Chem. Rev. *61*, 313 (1961).

[196] *H. A. O. Hill, K. A. Raspin,* J. Chem. Soc. (A), 619 (1969).

[197] *R. L. Dutta, D. W. Meek, D. H. Busch,* Inorg. Chem. *9*, 1215 (1970).

[198] *G. A. Barclay, R. S. Nyholm, R. V. Parish,* J. Chem. Soc. (London) 4433 (1961).

[199] *G. A. Barclay, M. A. Collard, C. M. Harris, J. V. Kingston,* J. Chem. Soc. (A), 830 (1969).

[200] *L. Rasmussen, C. K. Jorgensen,* Inorg. Chim. Acta *3*, 547 (1969).

[201] *J. S. Coe, R. Hulme, A. A. Malik,* J. Chem. Soc. (London) 138 (1964).

Heating (XI) in cyclohexane yields [Pd(en)-$(PBu_3)Cl]Cl$.[202]

$$\left[\begin{array}{c} (C_4H_9)_3P \rightarrow Pd(\mu\text{-}Cl)_2Pd \leftarrow (en) \\ (en) \rightarrow \quad\quad\quad\quad \leftarrow P(C_4H_9)_3 \end{array} \right] Cl_2$$

XI

$[Pd(Me_2As(CH_2)_3As(Me)(CH_2)_3AsMe_2)Cl]^{\oplus}$[202] and $[Pd(H_4EDTA)Cl]^{\oplus}$[191] are made by simple procedures from the components.

$[PdL_2X_2]$ Series. These compounds are generally made by treating a Pd(II) solution with the appropriate ligand (NH_3, py, SR_2, PR_3, etc.). The procedures are simple and straightforward and are typified by the following route to *trans*-$[PdCl_2(NH_3)_2]$ which yields a particularly pure product.[36]

$$Pd + 2HNO_3 + 4HCl \xrightarrow[-2H_2O \atop -NO_2]{} H_2[PdCl_4] \xrightarrow[-2HCl]{+NaCl}$$

$$Na_2[PdCl_4] \xrightarrow[-2NaCl]{+4NH_3} [Pd(NH_3)_4]Cl_2 \xrightarrow{+2HCl} trans\text{-}[PdCl_2(NH_3)_2]$$

Although several methods of preparation of the *cis*-isomers are known not all give pure products and, frequently, isomerization to the more stable *trans* form occurs, apparently even in the solid state. A reliable method has recently been developed based upon reaction of *cis*-$[Pd(NH_3)_2(H_2O)_2]^{2\oplus}$.[187] In perchloric acid solutions aquation of $[Pd(NH_3)_4]^{2\oplus}$ only takes place as far as *cis*-$[Pd(NH_3)_2(H_2O)_2]^{2\oplus}$: this cation may then react with appropriate salts yielding pure *cis*-$[PdX_2(NH_3)_2]$ complexes. The first *cis*-di-iodo complex of palladium was made by this route, which seems capable of more general use. *Cis*-$[PdCl_2(SbPr_3)_2]$ is obtained upon evaporation of petroleum solution, being much less soluble than the *trans*-isomer which predominates in solution.[203]

Complexes of ligands of heavier donor atoms such as SR_2, SeR_2, TeR_2, PR_3, AsR_3, SbR_3 are usually made by reaction of the ligand with $[PdX_4]^{2\ominus}$,[47] or from suspensions of $PdCl_2$ in ethanol or benzene.[204] For ligands which do not react with $PdCl_2$ suspensions or which undergo solvolysis, a metathetical reaction with *trans*-$[PdCl_2(PhCN)_2]$ is convenient.[205] It seems that the configuration of the product of metathesis depends upon the substituting ligand, being *cis* when L is $P(OMe)_3$, or the solvolytically unstable caged ligands $P(CH_2O)_3CCH_3$ and $P(NCH_3CH_2)_3CCH_3$.

Some unusual complexes of a phosphine bearing two phosphorus atoms (one of them quaternary) have been made by interaction of the ligand, 3-[(diphenylphosphino)methyl]-3-methyl-1,1-diphenylphosphetanium ion, with Na_2PdCl_4. The quarternary phosphorus atom is not involved in coordination.[206]

Reaction of $PdX_2(PR_3)_2$ with further ligands L (L$=PEt_2H$, PEtPhH) yields $PdX_2(PR_3)L$ which may be converted to phosphide-bridged *trans*-$[PdX(PR_2)L]_2$ on treatment with base. Interaction of chelate ligands with these bridged species yields $[Pd_2(PR_3)_2(chelate)_2]X_2$.[207]

Procedures analogous to those described above for monodentate ligands may be employed in formation of complexes of bidentate ligands.[208–211] The potentially tridentate $(o\text{-}Me_2NC_6H_4)_2PPh$ forms PdX_2L, (X = halogen), in which coordination is *via* phosphorus and *one* nitrogen.[212] The diazapropellane complex made by reaction of ligand with $[PdCl_4]^{2-}$ is said to be formed only with palladium, possibly due to steric limitations.[213] Complexes of 1,2-bis(dimethylarsino)-*o*-carboranes have been made by reaction of it with the benzonitrile complex $PdCl_2(C_7H_5N)_2$.[214] $PdX_2(L_2)$ complexes, (L_2 = bidentate ligand), may also be formed from uncharged starting materials as, for example, by reaction of bis(dimethylglyoximato)- and bis(2-pyridinaldoxinato)-palladium(II) with acetyl chloride.[215]

---

202 *F. G. Mann, D. Purdie*, J. Chem. Soc. (London) 1549 (1935); 873 (1936).
203 *J. Chatt, R. G. Wilkins*, J. Chem. Soc. (London) 70 (1953).
204 *G. Booth*, Adv. Inorg. Radiochem. *6*, 1 (1964).
205 *J. M. Jenkins, J. G. Verkade*, Inorg. Synth. *11*, 108 (1968).
206 *O. Berglund, D. W. Meek*, Inorg. Chem. *8*, 2602 (1969).
207 *R. G. Hayter, F. S. Humiec*, Inorg. Chem. *2*, 306 (1963).
208 *H. D. G. Drew, F. W. Pinkard, G. H. Preson, W. Wardlow*, J. Chem. Soc. (London) 1895 (1932).
209 *R. V. Rund*, Inorg. Chem. *9*, 1211 (1970).
210 *K. A. Hoffmann, G. Bugge*, Ber. *40*, 1772 (1907).
211 *D. W. Allen, I. T. Mollar, F. G. Mann, R. M. Canadine, J. Walker*, J. Chem. Soc. (A), 1097 (1969).
212 *H. P. Fritz, J. R. Gordon, K. E. Schwarzhans, L. M. Venanzi*, J. Chem. Soc. (London) 5210 (1965).
213 *M. Korat, G. Schmuckler, D. Ginsberg*, J. Chem. Soc. (A), 1784 (1970).
214 *H. D. Smith*, Inorg. Chem. *8*, 696 (1969).
215 *R. A. Krause, D. C. Jicha, D. H. Busch*, J. Amer. Chem. Soc. *83*, 528 (1961).

Finally, compounds *cis*-$[Pd(PPh_3)_2(tetrazolato)_2]$ have been synthesized by oxidative addition of tetrazoles to $Pd(PPh_3)_4$.[216] A similar reaction with imides yields such compounds as $Pd(PPh_3)_2(succinimide)_2$.[217]

$[PdLX_3]^{\ominus}$ Series. $[Pd(NH_3)Cl_3]^{\ominus}$ has been claimed (as the $[Pd(NH_3)_4]^{2\oplus}$ salt) as a by-product in the reaction $Pd(NH_3)_2Cl_2$ with $[Pd(NH_3)_4]Cl_2$.[57a] Better established products are obtained by reaction of $[PdCl_4]^{2\ominus}$ with alanine,[57a] *N*-methylquinuclidinium iodide $(C_8H_{16}NI)$,[218] and a quarternary phosphine.[206]

$[PdX_2L]_2$ Series. The members of this series are halogen-bridged. Heating a mixture of palladium sponge and PCl_5 yields $[PdCl_2(PCl_3)_2]$ which can be extracted into benzene and re-crystallized. Corresponding esters with $L = P(OMe)_3$, $P(OEt)_3$ are known.[57a] Hydrated Na_2PdCl_4 reacts with $PdCl_2(MR_2)_2$, (M = S, Se, Te) to give $Pd_2Cl_4(MR_2)_2$.[219] Alternatively refluxing or boiling solutions of PdX_2L_2 yields the dimers $[PdX_2L]_2$ in which X = halogen, CNS, NO_2.[202, 220, 221] Sometimes the bridged dimer is the only product isolated from reaction of ligand with $[PdCl_4]^{2\ominus}$.[222] The bridges are readily cleaved by amines yielding monomers *trans*-$[PdCl_2(Amine)L]$.[223]

Five-Coordinate Species. $[Pd(diars)_2X]^{\oplus}$ is known in nitromethane solution.[224] $[Pd(8\text{-}dmaq)_2X]ClO_4$ is made by addition of an ethanolic solution of Ph_4AsCl to a similar solution of $[Pd(8\text{-}dmaq)_2](ClO_4)_2 \cdot H_2O$.[199] $[Pd(qas)X]^{\oplus}$, (X = Cl, Br), a trigonal-bipyramidal complex is also known,[225] (qas = tris(*o*-diphenylarsinophenyl)arsine).

27.4 Palladium(III)

The existence of palladium in the +3 oxidation state is improbable. Pd_2O_3 has been claimed.[226] It is described as a chocolate-brown powder, made by anodic oxidation of solutions of palladous nitrate. It is likely to be a mixed-valence compound containing Pd(II) and Pd(IV).

A compound of formula PdF_3 is well established but is correctly formulated as $Pd^{II}[Pd^{IV}F_6]$. It may be obtained by (*i*) fluorination of the metal at 500°[227] (*ii*) fluorination of $PdBr_2$ at 400°,[228] (*iii*) by use of BrF_3 in the scheme.[20, 229, 230]

$$2PdBr_2 + 4BrF_3 \longrightarrow 2[PdF_3, BrF_3] \xrightarrow{180^\circ} Pd[PdF_6] + 2BrF_3$$

27.5 Palladium(IV)

In marked contrast to the chemistry of Pt(IV), Pd(IV) compounds are few in number and generally of rather low stability. Almost all complexes are of the types $M_2^I[PdX_6]$ and PdX_4L_2.

PdF_4, a pink diamagnetic solid violently hydrolyzed by water has been made from $Pd[PdF_6]$ (Section 27.4) and fluorine at 100 lb/in² and 350°.[20] There are no other palladium(IV) binary halides.

Alkali metal *salts of $[PdF_6]^{2\ominus}$* have been made[31] using the reaction sequence:

Other routes are from $M_2^I[PdCl_4]$ or $M_2^I[PdCl_6]$ by dissolution in BrF_3[229] or by reaction with fluorine.[33, 230] These compounds are yellow, diamagnetic and sensitive to moisture.

Hexachloropalladates(IV) are readily prepared by dissolving the metal in aqua regia, or in concentrated HCl saturated with chlorine, and removed from solution by addition of alkali metal chloride.[44, 231, 232] The *hexabromo-salts* are most simply made by treating an HBr solution of the corresponding tetrabromide, $M_2^I[PdBr_4]$, with bromine.[232] These complexes lose halogen upon gentle heating, yielding $M_2^I[PdX_4]$ salts.

Substituted Pd(IV) halocomplexes are of very low stability. Treatment of a suspension of

---

[216] *J. H. Nelson, D. L. Schmitt, R. A. Henry, D. W. Moore, H. B. Jonassen,* Inorg. Chem. *9*, 2678 (1970).
[217] *D. M. Roundhill,* Inorg. Chem. *9*, 254 (1970).
[218] *A. K. Banerjee, L. M. Vallarino, J. V. Quagliano,* Co-ord. Chem. Rev. *1*, 239 (1966).
[219] *J. Chatt, L. M. Venanzi,* J. Chem. Soc. (London) 2351 (1957).
[220] *F. G. Mann, A. F. Wells,* J. Chem. Soc. (London) 702 (1938).
[221] *D. M. Adams, P. J. Chandler,* J. Chem. Soc. (A), 588 (1969).
[222] *K. C. Kalia, A. Chakravorty,* Inorg. Chem. *8*, 2586 (1969).
[223] *J. Chatt, L. M. Venanzi,* J. Chem. Soc. (London) 2787 (1955).
[224] *C. M. Harris, R. S. Nyholm,* J. Chem. Soc. (London) 4375 (1956).
[225] *C. A. Savage, L. M. Venanzi,* J. Chem. Soc. (London) 1584 (1962).
[226] *L. Wöhler, F. Martin,* Z. Anorg. Allgem. Chemie *182*, 159 (1929).
[227] *O. Ruff, E. Ascher,* Z. Anorg. Allgem. Chemie *183*, 204 (1929).
[228] *M. A. Hepworth, K. H. Jack, R. D. Peacock, G. J. Westland,* Acta Cryst. *10*, 63 (1957).
[229] *N. Bartlett, J. W. Quail,* J. Chem. Soc. 3728 (1961).
[230] *N. Bartlett,* in *W. L. Jolly,* Preparative Inorganic Reactions, Vol. II, p. 301.
[231] *F. Puche,* Ann. Chim. (Paris) *9*, 233 (1938).
[232] *N. V. Sidgwick,* Chemical Elements and their Compounds, Oxford 1950.

$$KBr \xrightarrow{BrF_3} K[BrF_4] \xrightarrow{SeF_4} K[SeF_5]$$
$$PdBr_2 \xrightarrow{BrF_3} 3PdF_3 \cdot BrF_3 \xrightarrow{SeF_4} PdF_4 \cdot 2SeF_4$$
$$\xrightarrow{\text{reflux in } SeF_4} K_2[PdF_6]$$

$PdCl_2Py_2$ in chloroform with chlorine yields deep orange $PdCl_4py_2$ which loses chlorine rapidly on standing in moist air: it is also formed when *trans*-$[PdCl_2py_2]$ is treated with NOCl.[233] $PdCl_4(NH_3)_2[Pd(diars)_2X_2]^{2\oplus}$ ($X=Cl$, Br) and $PdCl_4(en)$ are made similarly.[57d, 224] $PdCl_2(NO_3)_2(NH_3)_2$ is a little more stable. A complex $\{[Pd(NH_3)_4]Cl_2[Pd(NH_3)_4Br_2]Br_2\}$ has been described.[234]

$K_2[Pd(CN)_6]$ has been prepared from KCN and K_2PdCl_6 in the presence of $K_2S_2O_8$ to prevent reduction.[235]

$PdO_2 \cdot nH_2O$ has been claimed but it is of low stability and doubtful structure. It is described as a dark-red powder which loses oxygen even at 0°.[236, 237]

---

[233] *R. Eskenazi, J. Raskovan, R. Levitus*, Anales. Asoc. Quim. Arg. *51*, 306 (1963); C.A. *63*, 12655f (1965); *58*, 2119b (1963).

[234] *A. V. Babaeva, E. Ya. Khananova*, Dokl. Akad. Nauk SSSR *159*, 1209 (1964).

[235] *H. Siebert, A. Siebert*, Angew. Chem. Intern. Ed. Engl. *8*, 600 (1969).

[236] *Berzelius*, Pogg. Ann. *13*, 454 (1828).

[237] *Bellucci*, Atti. Acad. Naz. Lincei. Classe Sci. Fis., Mat. Natur. Rend. *13*, 306 (1904).

28 Platinum

Flavio Bonati
Istituto di Chimica Generale ed Inorganica,
20133 Milan, Italy

28.1 Platinum Purification and Recovery

Sufficiently pure platinum is commercially available. The element is often recovered in the laboratory, which means that a purification is needed. This purification is generally carried out by crystallization of a suitable salt, *e.g.* potassium tetrachloroplatinate(IV).

Details about the separation of platinum from other elements, the preparation of platinum sponge by thermal decomposition of $(NH_4)_2PtCl_6$, of platinum black by formate reduction of $PtCl_6^{2\ominus}$, of platinized asbestos, as well as the handling of platinum equipment and platinum electroplating are generally described in standard textbooks of preparative chemistry.[1]

28.2 Platinum(0)

With few exceptions, all platinum(0) compounds contain tertiary phosphines or arsines as ligands.
The rather unstable *ammonia* and *ethylenediamine* (en) *derivatives*, $Pt(NH_3)_5$ and $Pt(en)_2$, were obtained by reduction of $[Pt(NH_3)_4]Br_2$[2] or of $[Pt(en)_2]I_2$[3] with potassium in liquid ammonia. *Nitrosyl derivatives* include the unstable, paramagnetic $(Ph_3P)_2Pt(NO)CF_3$,[4] containing formally Pt(−I), from $(Ph_3P)_4Pt$ and *trifluoromethylnitrosyl*, $[(NO(Ph_3P)_2Pt]_2$, obtained from the same compound and nitric oxide, and $(NO)(Ph_3P)_2Pt–HgCl$, prepared by reaction of the dimer with mercury(II) halide. In addition, very volatile π-$C_5H_5Pt(NO)$ was isolated[5] from the treatment of the reaction mixture of NO and $[Pt(CO)Cl_2]_2$ with sodium cyclopentadienyl.
Brown, probably tetrameric, *mono(triphenylphosphine)platinum(0)* was obtained by allowing $(Ph_3P)_4Pt$ to react in air with 1,5-cyclooctadiene,[6] while trimeric red $([Ph_3P)_2Pt]_3$ was isolated after melting $(Ph_3P)_2Pt$ under nitrogen.[6] The related compound $(Ph_2PCH_2CH_2PPh_2)Pt$ was obtained[7] by action of the ligand on $[Pt(CO)_2]_n$. *Monomeric bis(triphenylphosphine)platinum(0)* is obtained[8] in many ways, *e.g.* by displacement of ethylene, by means of a nitrogen stream on a suspension of $(Ph_3P)_2Pt(C_2H_4)$, by oxidation according to the following equation

$$(Ph_3P)_3Pt + O_2 \longrightarrow (Ph_3P)_2Pt + Ph_3PO$$

by reduction of *cis*-$(Ph_3P)_2PtCl_2$ with hydrazine[9] or of $(Ph_3P)_2PtO_2$ with sodium tetrahydridoborate,[10] or by removal of HCN from *trans*-$(Ph_3P)_2Pt(CN)$. It is best prepared[8] from *trans*-$(Ph_3P)_2PtHCl$ and *n*-butyllithium at −10°; it is generally used without isolation. *Dimeric*, bright yellow, *bis(triphenylphosphine)platinum(0)* is obtained[11] by a photochemical decomposition of an oxalate complex:

$$2(Ph_3P)_2Pt(C_2O_4) \longrightarrow (Ph_3P)_2Pt{-}Pt(PPh_3)_2 + 4CO_2$$

Yellow *tris(triphenylphosphine)platinum(0)*, m.p. 205°,[12] can be obtained by hydrazine reduction[13] of *cis*-$(Ph_3P)_2PtCl_2$ or boiling[14] $(Ph_3P)_4Pt$ in ethanol; colorless $(Ph_3P)_4Pt$, m.p. 196°,[12] is obtained by reduction of *cis*-$(Ph_3P)_2PtCl_2$ with hydrazine hydrate,[15] by irradiation of $(Ph_3P)_2Pt(C_2O_4)$, or by sodium tetrahydridoborate reduction[15] of potassium tetrachloroplatinate(IV), always in the presence of excess ligand; but in the best preparation the tetrachloroplatinate(II) is reduced[14] by the ligand in warm ethanolic potassium hydroxide. The related $(Ph_2PCH_2CH_2PPh_2)Pt$,[16] $(Ph_2PMe)_4Pt$,[12] $(Ph_2PCH_2Ph_2)_3Pt$,[12] $(Ph_2PC{\equiv}CPh)_4Pt$[17] and $[(Ph_3P)_4(Ph_2PC{\equiv}CPPh_2)_2Pt_2]$

[1] *G. Brauer*, Handbook of Preparative Inorganic Chemistry, Vol. II, p. 1560, New York 1965.
[2] *G. W. Watt, M. T. Walling jr., P. J. Mansfield*, J. Amer. Chem. Soc. *75*, 6175 (1966).
[3] *G. W. Watt, R. E. McCarley, J. W. Dawes*, J. Amer. Chem. Soc. *79*, 1163 (1957).
[4] *M. Green, R. B. L. Osborne, A. J. Rest, F. G. A. Stone*, J. Chem. Soc. (A), 2525 (1968).
[5] *E. O. Fischer, H. Schuster-Woldan*, Z. Naturforsch. *19b*, 766 (1964).
[6] *R. D. Gillard, R. Ugo, F. Cariati, S. Cenini, F. Bonati*, Chem. Commun. 869 (1966).
[7] *G. Booth, J. Chatt*, J. Chem. Soc. (A), 2131 (1969).
[8] *R. Ugo, G. LaMonica, F. Cariati, S. Cenini, F. Conti*, Inorg. Chim. Acta *4*, 390 (1970).
[9] *G. D. Dobinson, R. Mason, G. B. Robertson, R. Ugo, F. Conti, S. Cenini, D. Morelli, F. Bonati*, Chem. Commun. 739 (1967).
[10] *C. D. Cook, G. S. Jauhal*, J. Amer. Chem. Soc. *90*, 1464 (1968).
[11] *D. M. Blake, C. J. Nyman*, J. Amer. Chem. Soc. *92*, 5359 (1970).
[12] *P. Chini, G. Longoni*, J. Chem. Soc. (A), 1542 (1970).
[13] *C. D. Cook, G. S. Jauhal*, Can. J. Chem. *45*, 301 (1967).
[14] *R. Ugo, F. Cariati, G. LaMonica*, Inorg. Synth. *11*, 105 (1968).
[15] *L. Malatesta, C. Cariello*, J. Chem. Soc. (London) 2323 (1958).
[16] *D. T. Rosevear, F. G. A. Stone*, J. Chem. Soc. (A), 164 (1968).
[17] *K. S. Wheelock, J. H. Nelson, H. B. Jonassen*, Inorg. Chim. Acta *4*, 399 (1970).

are obtained similarly, while red-violet $(Ph_3P)_4Pt_3$ was obtained from *cis*-$(Ph_3P)_2PtCl_2$, hydrazine hydrate and potassium hydroxide.

Phosphorus trifluoride reacts[18] with platinum-(II) or -(IV) chloride (100°, 100 atm) and excess copper powder to yield $(F_3P)_4Pt$, b.p. 86°/730; this compound reacts with tertiary phosphines to yield $(F_3P)_{4-n}(R_3P)_nPt$ (n=1 or 2, R=Ph; n=4, R=OPh)[18] and with isocyanides to yield $(F_3P)_3(RNC)Pt$.[18,19] *Trialkyl and triaryl phosphite complexes*, $[P(OR)_3]_4Pt$, are prepared by reduction of a tetrachloroplatinate(II) with the ligand in ethanolic potassium hydroxide.[15,20]

One of the ligands present in the $(R_3E)_3Pt$ compounds (E=P, As) can be substituted by an unsaturated hydrocarbon (L), such as olefin,[21] acetylene[21] or allene[22,23] to yield $(R_3E)_2P{\cdot}L$. The ligand L includes hydroxyacetylenes,[24,25] hexafluorobut-2-yne,[26] methyl butynoate,[26] cyclooctyne,[27] fluoroolefins,[3,28,29] fumaronitrile, tetracyanoethylene,[30] acrylonitrile,[31] 1,1-dichloro-2,2-dicyanoethylene,[32] methyl fumarate.[31] These compounds are obtained by reaction of $(R_3E)_3Pt$ with the appropriate ligand L,[15,20,28,31] by reduction of *cis*-$(Ph_3P)_2PtCl_2$ or of *cis*-$[(MeO)_3P]_2PtCl_2$ with hydrazine hydrate[33,34] or of $(Ph_3P)_2PtO_2$ with sodium tetrahydridoborate[34] in the presence of L, by displacement of ethylene,[21] *trans*-stilbene[3,28,32] or phenylacetylene[30] by another ligand. Special cases include $(Ph_3P)_2Pt(CF_3C{\equiv}CCF_3)$, from[35] *cis*-$(Ph_3P)_2Pt(OCOCF_3)_2$ and the ligand, and $(Ph_3P)_2PtL$, from[21,30] *cis*-$(Ph_3P)_2PtHCl$ and the ligand L, fumaronitrile or tetracyanoethylene.

Reduction of tetrakis(alkynyl)platinate(II) with potassium in liquid ammonia affords $K_2[Pt(C{\equiv}CR)_2]$ derivatives;[36] *dicyanoplatinate-(0)* compounds are prepared similarly.[37] *Platinum(0) acetylene complexes*, Pt(α-hydroxyacetylene)$_2$, are prepared[38] by addition of the appropriate hydrocarbon to an aqueous-methanolic solution of tetrachloroplatinate(II).

Bis(1,5-cyclooctadiene)platinum(0) was obtained by ultraviolet irradiation[38] of (1,5-cyclooctadiene)bis(isopropyl)platinum(II).

An extensive number of platinum(0) carbonyl derivatives is known (see Chapter 29).

28.3 Platinum(I)

Although a few platinum(I) derivatives are mentioned, e.g. a chloride[39] and a tricyanoplatinate(I),[40] their identity ought yet to be confirmed.

28.4 Platinum(II)

28.4.1 Hydrides

Although platinum absorbs hydrogen, no simple binary hydride is known. Platinum(II) hydrides are known where four ligands, only one of which is a hydride group, surround the metal

---

[18] *T. Kruck, K. Baur*, Z. Anorg. Allgem. Chemie *364*, 193 (1969);
J. F. Nixon, D. M. Sexton, J. Chem. Soc. (A), 321 (1970).
[19] *R. D. Johnston, F. Basolo, R. G. Pearson*, Inorg. Chem. *10*, 247 (1971).
[20] *M. Meier, F. Basolo, R. G. Pearson*, Inorg. Chem. *8*, 795 (1969).
[21] *U. Belluco, B. Crociani, R. Pietropaolo, P. Uguagliati*, Inorg. Chim. Acta Rev. *3*, 19 (1969).
[22] *J. A. Osborn*, Chem. Commun. 1231 (1968).
[23] *S. Otsuka, K. Nakamura, K. Tani*, J. Organometal. Chem. *14*, P30 (1968).
[24] *J. H. Nelson, H. B. Jonassen, D. M. Roundhill*, Inorg. Chem. *8*, 2591 (1969);
J. H. Nelson, J. J. R. Reed, H. B. Jonassen, J. Organometal. Chem. *29*, 164 (1971).
[25] *D. M. Roundhill, H. B. Jonassen*, Chem. Commun. 1233 (1968).
[26] *E. O. Greaves, C. J. L. Lock, P. Maitlis*, Can. J. Chem. *46*, 3879 (1968).
[27] *T. L. Gilchrist, F. J. Graveling, C. W. Rees*, Chem. Commun. 821 (1968).
[28] *W. J. Bland, R. D. W. Kemmitt, G. W. Littlecott*, J. Chem. Soc. (A), 2062 (1969).
[29] *D. J. Cook, M. Green, N. Mayne, F. G. A. Stone*, J. Chem. Soc. (A), 1771 (1968).
[30] *W. H. Baddley, L. M. Venanzi*, Inorg. Chem. *5*, 33 (1966);
P. Uguagliati, W. H. Baddley, J. Amer. Chem. Soc. *90*, 5446 (1968).
[31] *S. Cenini, R. Ugo, F. Bonati, G. LaMonica*, Inorg. Nucl. Chem. Lett. *3*, 191 (1967).
[32] *A. McAdam, J. N. Francis, J. A. Ibers*, J. Organometal. Chem. *29*, 1499 (1971).
[33] *J. Chatt, B. L. Shaw, A. A. Williams*, J. Chem. Soc. (London) 3269 (1962).
[34] *C. D. Cook, G. S. Jauhal*, Inorg. Nucl. Chem. Lett. *3*, 31 (1967).
[35] *D. M. Barlex, R. D. W. Kemmitt, G. W. Littlecott*, Chem. Commun. 613 (1969).
[36] *R. Nast, W. D. Heinz*, Chem. Ber. *95*, 1478 (1962).
[37] *F. D. Rochon, T. Theophanides*, Can. J. Chem. *46*, 2973 (1968).
[38] *J. Mueller, P. Goeser*, Angew. Chem. Intern. Ed. Engl. *6*, 364 (1967).
[39] *L. Woehler, S. Streicher*, Ber. *46*, 1591 (1913).
[40] *L. Manchot, G. Lehmann*, Ber. *63*, 2775 (1930).

atom in a square planar arrangement. The reported dihydride, $(Ph_3P)_2PtH_2$ has been reformulated[41] as $(Ph_3P)_2Pt(CO_3)$. All the platinum hydrides belong either to the type $(R_3E)_2PtHX$ (E=P, As; X=halide, etc.) or $[(ligand)_3PtH]^{\oplus}$ or to some pentacoordinated species containing one or more $SnCl_3^{\ominus}$ groups. The pentacoordinated $(Et_3P)_2(\pi\text{-}C_2F_4)PtHCl$ has been reformulated[42] as $[(Et_3P)_2Pt(CO)Cl]^{\oplus}$-$BF_4^{\ominus}$.

Trans-$(R_3E)_2PtHX$ is prepared[43] from $(R_3E)_2$-PtX_2 (either isomer) and alcoholic potassium hydroxide, or 90% aqueous hydrazine hydrate, or lithium alanate in tetrahydrofuran, or gaseous hydrogen (95°, 50 atm). The alleged *cis*-isomer is only a different crystallographic form of the *trans*-isomer.[44] The compound can be obtained in many other ways: by oxidative hydrolysis[45] according to

$$[trans\text{-}(R_3P)Pt(CO)Cl]BF_4 + H_2O \rightleftharpoons CO_2 + trans\text{-}[(R_3P)_2PtHCl]$$

by oxidative addition[46] of hydrochloric acid or hydrogen chloride to $(Ph_3P)_nPt$ (n=3, 4), by hydrogenation[47] of *trans*-$(Et_3P)_2Pt(COMe)I$ (150°, 200 atm), by action[48] of cyclohexylmagnesium bromide and subsequent hydrolysis to $(R_3P)_2PtBr_2$, by olefin loss[43] from an alkyl compound, e.g.

$$trans\text{-}[(C_6H_5)_3P]_2Pt(C_2H_5)Cl \longrightarrow trans\text{-}[(C_6H_5)_3P]_2PtHCl + C_2H_4$$

by action[49] of hydrogen chloride on *trans*-$(Et_3P)_2Pt(Cl)(GePh_3)$, by room temperature hydrogenolysis[50] of *trans*-$(R_3P)_2Pt(Cl)$-(EMe_3), or by hydrolysis[50] of the latter compound according to the reaction

$$trans\text{-}[(C_2H_5)_3P]_2PtCl[E(CH_3)_3] + H_2O \longrightarrow trans\text{-}[(C_2H_5)_3P]_2PtHCl + \tfrac{1}{2}[(CH_3)_3E]_2O$$

(E = Si, Ge).

Other *trans*-$(R_3E)_2PtHX$ compounds (E=P, As) are readily prepared by ligand exchange from the chloride, X being another halide, a pseudohalide or a carboxylate,[51] or by oxidative addition of HX to $(R_3P)_nPt$ (n=3, 4), especially when the X group has a strong coordinating power.[46] If this condition is not fulfilled, as is the case with $Cl^{\ominus}$, it is possible to isolate $[(R_3P)_3PtH]^{\oplus}X^{\ominus}$ intermediates, accessible also by other routes.[52,53] If diphenylphosphine is used, a bridged compound[54] is obtained, $[(Et_3P)_2Pt_2H_2(\mu\text{-}PPh_2)]Cl_2$. Oxidative addition is possible also with succinimide, phthalimide and saccharin or tetrazole to yield the corresponding $(R_3P)_2PtH$(imide) or $(R_3P)_2PtH$(tetrazolato) compounds.[55]

Other compounds containing one group VI, V or IV ligand in addition to tertiary phosphines or arsines are prepared according to the reactions:

$$(R_3P)_2PtHX + L \longrightarrow [(R_3P)_2PtLH]^{\oplus}X^{\ominus} \quad \text{(Refs 52, 56)}$$

$$(R_3P)_2PtHX + L' + NaClO_4 \longrightarrow [(R_3P)_2PtL'H]^{\oplus}ClO_4^{\ominus} + NaCl \quad \text{(Ref. 57)}$$

where L is an amine, thiourea, Ph_3P, Ph_3As, and L′ an isocyanide, tertiary phosphine, phosphite, pyridine or carbon monoxide. Derivatives containing both tertiary phosphines and group IV substituents are prepared[43] by alkylation of *trans*-$(R_3P)_2PtHX$ with a Grignard or organolithium reagent, by oxidative addition of certain fluorine substituted triarylsilanes to $(Ph_3P)_4Pt$ affording $(Ph_3P)_2Pt(SiR_3)H$, by hydrogenolysis of a platinum–germanium,

[41] *F. Cariati, R. Mason, G. B. Robertson, R. Ugo*, Chem. Commun. 408 (1967);
C. J. Nyman, C. E. Wymore, G. Wilkinson, Chem. Commun. 407 (1967).

[42] *H. C. Clark, P. W. R. Corfield, K. R. Dixon, J. A. Ibers*, J. Amer. Chem. Soc. *89*, 3360 (1967).

[43] *J. Chatt, B. L. Shaw*, J. Chem. Soc. (London) 5075 (1962).

[44] *A. F. Kemmitt, F. Glockling*, J. Chem. Soc. (A), 2163 (1969).

[45] *H. C. Clark, K. R. Dixon, W. J. Jacobs*, J. Amer. Chem. Soc. *91*, 1346 (1969).

[46] *F. Cariati, R. Ugo, F. Bonati*, Inorg. Chem. *5*, 1128 (1966).

[47] *G. Booth, J. Chatt*, J. Chem. Soc. (A), 634 (1966).

[48] *R. J. Cross, F. Glockling*, J. Organometal. Chem. *3*, 253 (1965).

[49] *F. Glockling, K. A. Hooton*, J. Chem. Soc. (A), 826 (1968).

[50] *F. Glockling, K. H. Hooton*, J. Chem. Soc. (A), 1066 (1967).

[51] *R. R. Dean, J. C. Green*, J. Chem. Soc. (A), 3047 (1968);
P. W. Atkins, J. C. Green, M. L. H. Green, J. Chem. Soc. (A), 2275 (1968).

[52] *L. Toniolo, M. Giustiniani, U. Belluco*, J. Chem. Soc. (A), 2666 (1969).

[53] *H. Clark, K. R. Dixon*, J. Amer. Chem. Soc. *91*, 596 (1969).

[54] *J. Chatt, J. M. Davidson*, J. Chem. Soc. (London) 2433 (1964).

[55] *D. M. Roundhill*, Chem. Commun. 567 (1970);
J. H. Nelson, D. L. Schmitt, R. A. Henry, D. W. Moore, H. B. Jonassen, Inorg. Chem. *9*, 2678 (1970).

[56] *M. Giustiniani, G. Dolcetti, U. Belluco*, J. Chem. Soc. (A), 2047 (1969).

[57] *M. J. Church, M. J. Mays*, J. Chem. Soc. (A), 3074 (1968).

-silicium, or -lead bond, e.g.[58]

$$\{[(C_2H_5)_3P]_2Pt[Ge(C_6H_5)_3][Ge(CH_3)_3]\} + H_2 \rightleftharpoons (CH_3)_3GeH + \{[(C_2H_5)_3P]_2PtH[Ge(C_6H_5)_3]\}$$

Compounds with Pt–Sn bonds are prepared by addition of tin(II) chloride to *trans*-$(R_3P)_2$-PtHCl, yielding[59] *trans*-$(R_3P)_2Pt(H)(SnCl_3)$, or by reaction[60] of $(Et_3P)_2PtCl_2$ with tin(II) chloride, hydrogen and $Me_4N^{\oplus}Cl^{\ominus}$ to yield $[Me_4N]^{\oplus}[(Et_3P)_2PtH(SnCl_3)_2]^{\ominus}$: obtained[60] also from $Me_4N^{\oplus}SnCl_3^{\ominus}$ and $(Et_3P)_2PtH(SnCl_3)$.
Hydride derivatives include tetramethylammonium hydridopentakis(trichlorogermyl)platinate(II), from potassium tetrachloroplatinate-(II) and trichlorogermane in hydrochloric acid,[61] and the homolog tin derivative, obtained under hydrogen pressure (30°, 500 atm) from $[Pt(SnCl_3)_5]^{3\ominus}$.

28.4.2 Halides and Halocomplexes

The reported PtF_2 could not be confirmed.[62] The *chloride*, Pt_6Cl_{12}, can be prepared[63] heating platinum sponge at 500° in a chlorine stream, or, better, by thermal decomposition of $PtCl_4$ or $H_2PtCl_6 \cdot 6H_2O$; the latter method is used also to prepare the bromide, while the iodide is made from the chloride and potassium iodide.[64]
Solutions of *hydrogen tetrachloroplatinate* can be obtained by hydrazine reduction of the corresponding platinum(IV) compound in acid solution; no corresponding bromide or iodide is known. Very many stable salts,[65] e.g. K_2PtX_4,[66] can be prepared by reduction[67] from the corresponding platinum(IV) salts; with certain cations ($R_4N^{\oplus}$, $R_4As^{\oplus}$) binuclear derivatives are obtained,[68] such as $Pt_2X_6^{2\ominus}$ (X=Br, I). Derivatives are also known where the cation is a complex platinum(II) species, e.g. $[(NH_3)_4Pt]$-$[PtCl_4]$ (*Magnus' salt*), from hydrogen tetrachloroplatinate(II) and ammonia.[69]
The $[PtX_3(ligand)]^{\ominus}$ compounds are prepared by interaction of a ligand, such as ammonia or pyridine,[70] with $Pt_2X_6^{2\ominus}$ (X=Br, I) or K_2Pt-Cl_4;[71] the reaction of ethylene with K_2PtCl_4 is catalyzed by tin(II) chloride.[72,73] The ethylene molecule can be easily displaced by other olefins[74] or by acetylenes.[75] The cation can be either inorganic or a quaternary ammonium, phosphonium or arsonium group where one of the alkyl substituents carries an olefinic function.[76]
Dimeric, *halogen bridged complexes*, Pt_2X_4-$(ligand)_2$, are prepared by addition[77] of platinum(II) halide to $(ligand)_2PtX_2$ (ligand is PR_3 or AsR_3) or by action[78] of hydrochloric acid on an olefinic complex according to the reaction

$$2K[PtCl_3(olefin)] \longrightarrow 2KCl + Pt_2Cl_4(olefin)_2$$

Many complexes of the type $(ligand)_2PtX_2$, either *cis* or *trans*, are known.[65,79] They can be prepared[65,79] by one of the following routes:

---

[58] *G. Deganello, G. Carturan, U. Belluco*, J. Chem. Soc. (A), 2873 (1968);
J. Chatt, C. Eaborn, P. N. Kapoor, J. Organometal. Chem. *13*, P21 (1968).
[59] *R. V. Lindsey, G. W. Parshall, U. G. Soltberg*, J. Amer. Chem. Soc. *87*, 658 (1965).
[60] *R. D. Cramer, R. V. Lindsey, C. T. Previtt, U. G. Stolberg*, J. Amer. Chem. Soc. *87*, 658 (1965).
[61] *J. K. Wittle, G. Urry*, Inorg. Chem. *7*, 560 (1968).
[62] *N. Bartlett, D. H. Lohmann*, J. Chem. Soc. (London) 619 (1964).
[63] *W. E. Cooley, D. H. Busch*, Inorg. Synth. *5*, 208 (1957).
[64] *L. Woehler, F. Mueller*, Z. Anorg. Allgem. Chemie *149*, 377 (1925).
[65] *P. Pascal*, Nouveau Traité de Chimie Minérale, Vol. XIX, p. 780, Paris 1958.
[66] *S. E. Livingstone*, Synth. Inorg. Metal-Org. Chem. *1*, 1 (1971).
[67] *R. N. Keller*, Inorg. Synth. *2*, 247 (1946).
[68] *C. M. Harris, S. E. Livingstone, I. H. Reece*, J. Chem. Soc. (London) 3697 (1958).
[69] *G. Brauer*, Handbook of Preparative Inorganic Chemistry II, Academic Press, New York 1965.
[70] *S. E. Livingstone, A. Whitley*, Australian J. Chem. *15*, 175 (1962).
[71] *G. Pajaro, A. Panunzi*, Tetrahedron Lett. 441 (1965).
[72] *R. D. Cramer, E. L. Jenner, R. V. Lindsey, U. G. Stollberg*, J. Amer. Chem. Soc. *85*, 1391 (1963).
[73] *F. R. Hartley*, Organometal. Chem. Rev. *A6*, 119 (1970).
[74] *T. Kinugasa, M. Nakamura, H. Yamada, A. Saika*, Inorg. Chem. *7*, 2649 (1968).
[75] *P. D. Kaplan, M. Orchin*, Inorg. Chem. *6* (1967);
J. Chatt, R. Guy, L. A. Duncanson, J. Chem. Soc. 5287 (1961);
J. Chatt, R. Guy, L. A. Duncanson, D. T. Thompson, J. Chem. Soc. (London) 5170 (1963).
[76] *R. G. Denning, F. R. Hartley, L. M. Venanzi*, J. Chem. Soc. (A), 324 (1967);
D. V. Claridge, L. M. Venanzi, J. Chem. Soc. 3419 (1963);
R. G. Denning, L. M. Venanzi, J. Chem. Soc. (London) 3241 (1963).
[77] *J. R. Goodfellow, L. M. Venanzi*, J. Chem. Soc. 7533 (1967).
[78] *J. Chatt, M. L. Searle*, Inorg. Synth. *5*, 211 (1957).
[79] *G. Brauer*, Handbook of Preparative Inorganic Chemistry, II, p. 1577, Academic Press, New York 1965.

reaction of the ligand with a tetrahaloplatinate(II), generally K_2PtCl_4 in aqueous solution; by addition of the ligand to a platinum(II) halide; bridge-splitting of a binuclear, halide bridged complex; displacement of an olefin or another suitable ligand from the coordination sphere; or rearrangement of a $[(ligand)_4Pt]^{2\oplus}[PtX_4]^{2\ominus}$ salt, generally on heating.

The ligands may be the same or of different nature. They belong to group VI (amine oxides,[80] cacodyl oxide,[81] diphenylcyclopropenone,[82] thioethers,[83,84] dithioethers, thiourea and related Se and Te homologs,[83] dialkylsulfoxide,[83] trialkyl-phosphine and -arsine sulfides and selenides[85]), to group V (ammonia,[86,87] amines,[88] hydroxylamines,[89] hydrazines,[90] nitriles,[91] R_3E compounds[79,92–94] (E=P, As, Sb), phosphites,[94,95] phosphorus(III) halides,[95] $(R_2N)_2PF$ or $(R_2N)PF_2$[96]) and to group IV (carbon monoxide (see Chapter 29), isocyanides,[97] olefins and acetylenes,[98,99] $SnCl_3^{\ominus}$ [100] and $GeCl_3^{\ominus}$ [101]).

Compounds of the type *cis*-(ligand)(carbene)-$PtCl_2$, where carbene means $(R–NH)_2C$: or (RNH)(RO)C: and ligand may be an isocyanide or a phosphine, are prepared[102,103] by addition of an amine or an alcohol to (RNC)-(ligand)$PtCl_2$. Addition of an amine, or of an alkoxide, can also occur at the coordinated olefinic double bond and the reaction, e.g.[104]

Pt, Cl, Cl + $H_2N–CH(CH_3)–C_6H_5$ ⟶ $Pt^{\ominus}$, Cl, Cl, CH_2, $\overset{\oplus}{N}H_2–CH(CH_3)–C_6H_5$

can be stereoselective.

Although in the $(ligand)_2PtX_2$ complexes the halide groups are generally Cl, Br or I, fluorides are known: they are obtained[105] by interaction of $(R_3P)_4Pt$ or of $(RO)_3P_4Pt$ with hydrogen fluoride, which on acting with $(R_3P)_2PtX_2$ or $(R_3P)_2$-PtHX affords[106] the mixed halides, $(R_3P)_2PtXF$. Compounds with only one halogen include fluorides[106] and belong either to the type $[(ligand)_3PtX]^{\oplus}$ or $(ligand)_2PtXY$, where Y is another mononegative group. Both isomers of $(Ph_3P)_2PtCl[(O)S(O)CH{=}CHCH_3]$ are obtained[107] by insertion of sulfur dioxide into

80 *A. R. Brause, M. Rycheck, M. Orchin,* J. Amer. Chem. Soc. *89*, 6500 (1967).
81 *K. A. Jensen, E. Fredericksen,* Z. Anorg. Allgem. Chemie *230*, 34 (1936).
82 *W. L. Fichtmann, P. Schmidt, M. Orchin,* J. Organometal. Chem. *12*, 249 (1968).
83 *S. E. Livingstone,* Quart. Rev. *19*, 386 (1965).
84 *B. E. Aires, J. E. Ferguson, D. T. Howarth, J. M. Miller,* J. Chem. Soc. (A), 1944 (1971).
85 *P. Nicpon, D. W. Meak,* Chem. Commun. 398 (1966).
86 *P. Pascal,* Nouveau Traité de Chimie Minérale, Vol. XIX, p. 852, Paris 1958.
87 *R. N. Keller,* Inorg. Synth. *2*, 243 (1966).
88 *M. Orchin, P. J. Schmidt,* Inorg. Chim. Acta Rev. *2*, (1968).
89 *R. Uhlenhuth,* Justus Liebigs Ann. Chem. *311*, 120 (1900);
L. A. Tschugaev, H. Tscherniaev, Compt. Rend. *161*, 367 (1915);
L. A. Tschugaev, H. Tscherniaev, J. Chem. Soc. *113*, 884 (1918).
90 *F. Para, N. Klynchnikov,* Russian J. Inorg. Chem. *14*, 122 (1969).
91 *R. A. Walton,* Quart. Rev. *19*, 126 (1965);
R. A. Walton, Can. J. Chem. *44*, 1480 (1966).
92 *G. B. Kauffmann,* Inorg. Synth. *7*, 249 (1963).
93 *G. Booth,* Adv. Inorg. Chem. Radiochem. *6*, 1 (1964).
94 *L. N. Essen, T. N. Bukhitiyarova,* Russian J. Inorg. Chem. *14*, 242 (1969);
L. N. Essen, T. N. Bukhitiyarova, Russian J. Inorg. Chem. *14*, 543 (1969).
95 *A. Rosenheim, W. Loewenstamm,* Z. Anorg. Allgem. Chemie *37*, 394 (1903);
J. Chatt, A. A. Williams, J. Chem. Soc. (London) *1*, 3061 (1959).
96 *J. F. Nixon, D. M. Sexton,* Chem. Commun. 827 (1969).
97 *L. Malatesta, F. Bonati,* Isocyanide Complexes of Metals, p. 167, London 1969;
F. Bonati, G. Minghetti, J. Organometal. Chem. *24*, 251 (1970).
98 *U. Belluco, B. Crociani, R. Pietropaolo, P. Uguagliati,* Inorg. Chim. Acta Rev. *3*, 19 (1969).
99 *F. R. Hartley,* Chem. Rev. *69*, 799 (1969).
100 *D. F. Young, R. D. Gillard, G. Wilkinson,* J. Chem. Soc. (London) 5176 (1964);
D. F. Young, Adv. Inorg. Chem. Radiochem. *11*, 119 (1968).
101 *J. K. Wittle, C. Urry,* Inorg. Chem. *7*, 560 (1968).
102 *F. Bonati, G. Minghetti, T. Boschi, B. Crociani,* J. Organometal. Chem. *25*, 255 (1970).
103 *E. M. Baddley, J. Chatt, R. L. Richards, G. A. Sim,* Chem. Commun. 1322 (1969).
104 *A. DeRenzi, R. Palumbo, G. Pajaro,* J. Amer. Chem. Soc. *93*, 880 (1971).
105 *J. McAvoy, K. C. Moss, D. W. A. Sharp,* J. Chem. Soc. (London) 1376 (1965).
106 *R. D. W. Kemmitt, R. D. Peacock, J. Stocks,* J. Chem. Soc. (A), 866 (1971).
107 *H. C. Volger, K. Vrieze,* J. Organometal. Chem. *6*, 297 (1966);
H. C. Volger, K. Vrieze, J. Organometal. Chem. *9*, 527 (1969);
H. C. Volger, K. Vrieze, J. Organometal. Chem. *13*, 495 (1968).

$[(\pi\text{-}C_3H_5)Pt(PPh_3)_2]^{\oplus}Cl^{\ominus}$; carbon monoxide insertion into the corresponding molecule affords[108] *trans*-$(R_3P)_2Pt(COR)Cl$; reaction of *trans*-$(R_3P)_2PtHX$ with alkyl azide affords an amido complex, *trans*-$(R_3P)_2Pt(NHR)X$.[109] Group IV substituents include also alkyl, aryl, perchloro- or perfluoroaryl groups, which can be introduced by controlled alkylation of the dihalide,[110] or by cleavage of one of the two Pt–C bonds present in the dialkyls with hydrogen halide,[111] or by addition of R_FI compounds to zerovalent platinum complexes, such as $(Ph_3P)_3Pt$.[112] Compounds containing unsaturated groups, e.g. –CR=CR′R″,[113] –CH=C=CRR′,[53] or –C≡CR,[54] can be obtained by interaction of the required halosubstituted or reactive and unsaturated hydrocarbon with zerovalent phosphine complexes, sometimes without isolation of the intermediate $(R_3P)_2Pt$(olefin) compound.[116] The ligand Y include the –C(O)OR group, which is formed[117] by reaction of an alcohol with the carbonyl group of $[PtX(CO)(R_3'P)_2]^{\oplus}$, or it can be the addition product of an alkoxide to a coordinated olefinic group, such as in *triphenylphosphine(8-methoxycyclo-oct-4-enyl)-chloroplatinum(II)*,[118] obtained by a bridge splitting reaction with Ph_3P on the corresponding binuclear complex. The group Y can be also a R_3E group[119] (R=alkyl or halogen; E=Ge, Sn, Pb); these compounds are obtained by hydrogen cleavage[120] of compounds like *trans*-$(R_3P)_2Pt(GePh_3)_3$, by action of alkali metal[121,122] or mercury[123,124] derivatives of the $R_3E^{\ominus}$ group on platinum(II) complex halides, or by reaction between *trans*-$(R_3P)_2PtHX$ and HEX_3.[124,125]

Cationic $[(ligand)_3PtX]^{\oplus}$ derivatives can be obtained by reaction of various $(ligand)_2PtX_2$ complexes with the same or another ligand in the presence of a poor nucleophile, such as perchlorate[126] or tetrafluoroborate.[127]

28.4.3 Oxygen and Sulfur Derivatives

The action of oxygen under pressure[128,129] affords the oxide, PtO, while melting platinum sponge with sulfur,[130] selenium[131] or tellerium[131] the sulfide, PtS, the selenide and the telluride are produced. The hydroxide is obtained[132] by action of alkali on a tetrachloroplatinate(II). Compounds of the type $[Pt(NH_3)_2X_2]$ react with water, the halides and nitrites slowly, the sulfates and nitrates completely, to give[133] aquo complexes, $[Pt(NH_3)_2(H_2O)]X_2$, Sulfur complexes, $[(R_2S)_4Pt][PtCl_4]$, are prepared[134] from the

[108] *G. Booth, J. Chatt*, J. Chem. Soc. (A), 634 (1966).
[109] *W. Beck, M. Bauder*, Chem. Ber. *103*, 583 (1970).
[110] *D. T. Rosevear, E. G. A. Stone*, J. Chem. Soc. (London) 5275 (1965);
C. H. Bamford, G. C. Eastmond, K. Hargreaves, Trans. Faraday Soc. *64*, 175 (1968).
[111] *J. D. Ruddick, B. L. Shaw*, J. Chem. Soc. (A), 2801 (1969).
[112] *D. T. Rosevear, F. G. A. Stone*, J. Chem. Soc. (A), 164 (1968).
[113] *W. J. Bland, R. D. Kemmitt*, J. Chem. Soc. (A), 1278 (1968);
W. J. Bland, R. D. Kemmitt, J. Organometal. Chem. *14*, 201 (1968).
[114] *J. P. Collmann, R. Caruse, J. Kang*, Inorg. Chem. *8*, 2574 (1969).
[115] *C. D. Cook, G. S. Jauhal*, Can. J. Chem. *45*, 301 (1967).
[116] *M. Green, R. B. L. Osborne, A. J. Rest, F. G. A. Stone*, J. Chem. Soc. (A), 2525 (1968).
[117] *H. C. Clark, K. R. Dixon, W. J. Jacobs*, J. Amer. Chem. Soc. *91*, 1346 (1969).
[118] *B. Crociani, P. Uguagliati, T. Boschi, U. Belluco*, J. Chem. Soc. (A), 2869 (1960).
[119] *F. Glockling*, Quart. Rev. *20*, 45 (1966);
R. J. Cross, Organometal. Chem. Rev. *2*, 97 (1967).
[120] *F. Glockling, K. Hooton*, J. Chem. Soc. (A), 826 (1968).
[121] *E. H. Brooks, R. J. Cross, F. Glockling*, Inorg. Chim. Acta *2*, 17 (1968).
[122] *G. Deganello, G. Carturan, U. Belluco*, J. Chem. Soc. (A), 2873 (1968).
[123] *F. Glockling, K. Hooton*, J. Chem. Soc. (A), 1066 (1967).
[124] *A. F. Klemmitt, F. Glockling*, J. Chem. Soc. (A), 1164 (1971).
[125] *J. E. Bentham, S. Craddock, E. A. V. Ebsworth*, J. Chem. Soc. (A), 587 (1961).
[126] *H. C. Clark, K. R. Dixon, W. J. Jacobs*, J. Amer. Chem. Soc. *90*, 2259 (1968);
H. C. Clark, K. R. Dixon, W. J. Jacobs, J. Amer. Chem. Soc. *91*, 596 (1969).
[127] *M. J. Church, M. J. Mays*, J. Chem. Soc. (A), 3074 (1968).
[128] *A. Woehler, W. Frey*, Z. Elektrochem. *15*, 129 (1909).
[129] *P. Laffitte, P. Grandadam*, Compt. Rend. *200*, 456 (1935).
[130] *W. Biltz, R. Jura*, Z. Anorg. Allgem. Chemie *190*, 161 (1930).
[131] *F. Roessler*, Z. Anorg. Allgem. Chemie *15*, 406 (1897);
L. Thomassen, Z. Physik. Chem. (Leipzig) *2*, 349 (1929).
[132] *L. Woehler, F. Martin*, Z. Elektrochem. *15*, 791 (1909).
[133] *K. A. Jensen*, Z. Anorg. Allgem. Chemie *242*, 87 (1939).
[134] *L. A. Tschugaev, W. Subbotin*, Chem. Ber. *43*, 1200 (1910).

ligand and K_2PtCl_4, while $K_2Pt(NO_2)_4$ yields $(R_2S)_2Pt(NO_2)_2$.

Anionic, O- or S-bonded derivatives of platinum(II) are prepared by exchange from $PtCl_4^{2\ominus}$, or from $PtCl_2$, and the required anion, like thiosulfate,[135] oxalate,[136] 1,1-dicyanoethylene-2,2-dithiolate,[137] sulfite,[138] to yield complexes of the type $[Pt(anion)_2]^{2\ominus}$. Neutral inner complexes are obtained similarly, e.g. *cis*- and *trans*-$[Pt(NH_2–CH_2–(O)CO–)_2$,[139,140] Pt(β-diketonato)$_2$,[141,142] or $Pt(O_2C–CH_2SEt)_2$;[143] while dithio-α-diketone complexes, *e.g.* bis(dithiobenzyl)platinum(II) are prepared from benzoin, P_4S_{10} and potassium tetrachloroplatinate(II).[144] This complex can be alkylated to yield *[1,2-(benzylmercapto)-1,2-phenylethylene](1,2-diphenylethylene-1,2-dithiolato)platinum(II)*.[145]

β-Diketones, and especially acetylacetone, react with K_2PtCl_4 affording an interesting series[142] of derivatives, such as $K[(C_5H_7O_2)Pt(CH\{COCH_3\}_2)_2]$. Pentan-2,4-dionato derivatives containing other ligands include

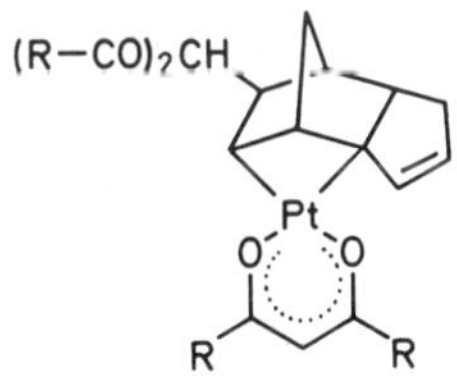

prepared[146] from dichloro(endo-dicyclopentadiene)platinum(II), β-diketone and sodium carbonate, and some allyl derivatives, such as[147] *bis-(pentan-2,4-dionato)bis(μ-allyl)diplatinum-(II)*, from diallylplatinum(II) and cyclopentadienylthallium, and *(pentan-2,4-dionato)-(2-methylallyl)platinum(II)*.

A number of complexes with tertiary phosphines or arsines as ligands can be prepared by halide exchange, *e.g.*[148] $(Ph_3P)_2Pt(CO_3)$ by the use of silver carbonate. The latter complex is then used[148] to prepare other salts, *e.g.* acetate, oxalate, benzoate,[149] by reaction with the required acid. The same carbonate and the arsine homolog can also be obtained from $(Ph_3P)_4Pt$, oxygen and carbon dioxide.[148,149] A carboxylate, $(Ph_3P)_2Pt(OCOCF_3)_2$ is prepared from trifluoacetic acid and $(Ph_3P)_2Pt(PhC{\equiv}CPh)$, while other zerovalent compounds, such as $(R_3P)_2Pt(CF_3C{\equiv}CCF_3)$ or $(R_3P)_2Pt(CF_2{=}CF_2)$ react to produce[150] $(R_3P)_2Pt(OCOCF_3)[C(CF_3{=}CHCF_3]$ and $(R_3P)_2Pt(OCOCF_3)(CF_2–CF_2H)$. Reaction of a carboxylate in alcohol with carbon monoxide affords[150] $(Ph_3P)_2Pt(OCOCF_3)(CO_2R')$.

Tris- or tetrakis-(triphenylphosphine)platinum(0) and some of the homologs react to a whole series of complexes (Table 1).

Arylplatinum(II) derivatives of the type $(R_3P)_2Pt(C_6H_5)(SR)$ are prepared[151] from the corresponding chloride and sodium mercaptide. Finally, some allegedly hexa- or pentacoordinate platinum(II) complexes have been reformulated as square planar complexes, e.g. as $[(R_3P)_2Pt(R_2PS_2)](R_2^1PS_2)$ or $[R_3P(R_2^1PS_2)Pt(R_2PS_2)]$ respectively.[152]

28.4.4 Complexes with Group V Donor Ligands

The reduction of potassium hexachloroplatinate(IV) according to the reaction

$$K_2PtCl_6 + 6NaNO_2 \longrightarrow K_2[Pt(NO_2)_4] + 2NO_2 + 6NaCl$$

affords tetranitroplatinate(II).[153,154] The latter

135 *D. J. Rjabschikov*, C.R. Acad. Sci. USSR *41*, 208 (1943).
136 *C. Vèzes*, Bull. Soc. Chim. France [3] *19*, 875 (1898).
137 *B. G. Werden, E. Billig, H. B. Gray*, Inorg. Chem. *5*, 78 (1966).
138 *P. Bergsöe*, Z. Anorg. Allgem. Chemie *19*, 318 (1898).
139 *H. Ley, K. Ficken*, Ber. *45*, 377 (1912).
140 *A. A. Gruenberg, B. W. Ptilyn*, J. Prakt. Chem. *136*, II, 143 (1933).
141 *H. Werner*, Ber. *34*, 2584 (1901).
142 *D. Gibson*, Coord. Chem. Rev. *4*, 225 (1969).
143 *L. Ramberg*, Ber. *46*, 1696 (1913).
144 *G. N. Schrauzer, V. P. Mayweg*, J. Amer. Chem. Soc. *87*, 1483 (1965).
145 *G. N. Schrauzer, H. N. Rabinowitz*, J. Amer. Chem. Soc. *90*, 4297 (1968).
146 *J. K. Stille, D. B. Fox*, Inorg. Nucl. Chem. Lett. *5*, 157 (1969).
147 *W. S. McDonald, B. E. Mann, G. Roper, B. L. Shaw, L. Shaw*, Chem. Commun. 1254 (1969).
148 *C. J. Nyman, C. E. Wymore, G. Wilkinson*, J. Chem. Soc. (A), 561 (1968).
149 *D. M. Blake, C. J. Nyman*, J. Amer. Chem. Soc. *92*, 5359 (1970).
150 *D. Barlex, B. Kemmitt, G. Littlecott*, Chem. Commun. 613 (1969).
151 *D. T. Rosevear, F. G. A. Stone*, J. Chem. Soc. (London) 5275 (1965).
152 *P. Hayward, D. Blake, G. Wilkinson, C. Nyman*, J. Amer. Chem. Soc. *92*, 5873 (1970).
153 *A. F. Nilson*, Ber. *9*, 1722 (1876); *A. F. Nilson*, Ber. *10*, 934 (1877).
154 *I. I. Chernyaev, L. A. Nazarova, A. S. Mironova*, Russian J. Inorg. Chem. *6*, 1238 (1961).

Table 1 Bis-[triphenylphosphine]-platinum(II)-complexes from tris- or tetrakis-[triphenylphosphine]-platinum(0)

Reaction with	Obtained complex	Ref.
o-$C_6H_4(COO^{\ominus})(N_2^{\oplus})$	L_2Pt–O–CO–C_6H_4–N=N (chelate) $\xrightarrow[-N_2]{\nabla}$ L_2Pt–O–CO–C_6H_4	a
H_2C=C–O–C(=O) (diketene)	H_3C–C(–O–)=CH–C(=O)–PtL_2	b
$H_3F-CO-CF_3$	L_2Pt–O–$C(CF_3)(F_3C)$	c
CS_2	$L_2Pt(S_2C)$	d, e
benzothiadiazole-S,S-dioxide ($C_6H_4N_2SO_2$)	L_2Pt–O–$S(O_2)$–C_6H_4–N=N; $\xrightarrow{\nabla/C_2H_5OH}$ L_2Pt–O–S(O)–C_6H_4–N=N; $\xrightarrow{\nabla}$ L_2Pt–O–S(O)–C_6H_4	a
$(F_3C)_2C$=C(S–CH$_2$–S)–C$(CF_3)_2$; $(F_3C)_2C$=C(S)–S–C(=C$(CF_3)_2$)–S	$L_2Pt(S_2C{=}C(CF_3)_2)$	f
O_2	$[(C_6H_5)_3P]_2PtO_2$ or $[(C_6H_5)_3P]_2PtO_2 \cdot$ solv. solv. = C_6H_6,[g] $C_6H_5CH_3$,[h] $CHCl_3$[i]	

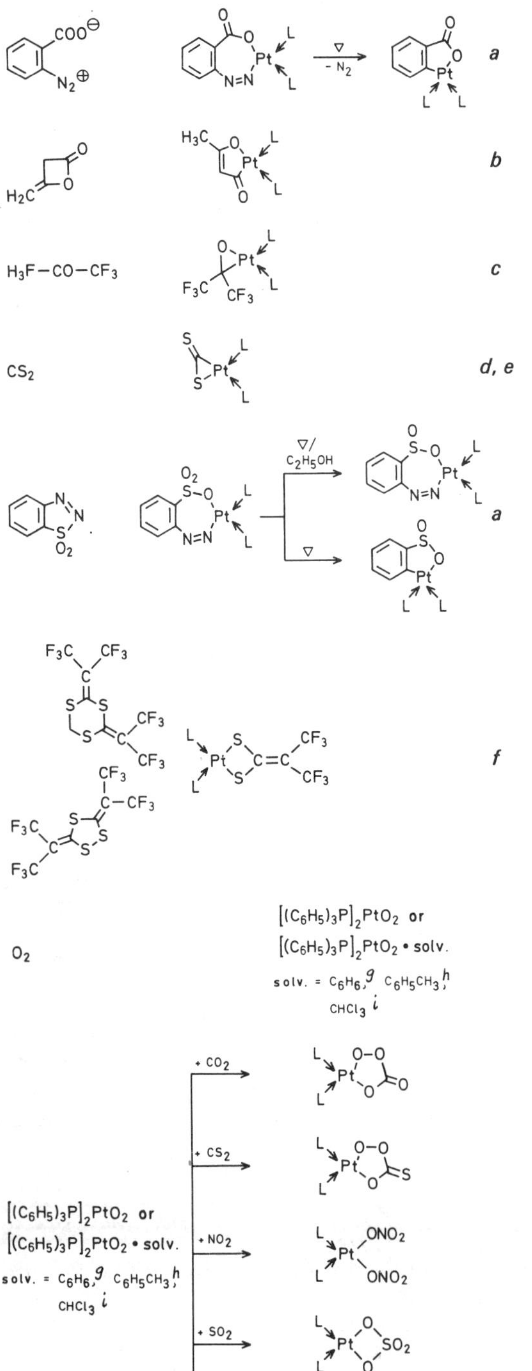

$[(C_6H_5)_3P]_2PtO_2$ or $[(C_6H_5)_3P]_2PtO_2 \cdot$ solv. (solv. = C_6H_6,[g] $C_6H_5CH_3$,[h] $CHCl_3$[i])

- $+ CO_2 \rightarrow$ L_2Pt(O–O–C(=O)–O)
- $+ CS_2 \rightarrow$ L_2Pt(O–O–C(=S)–O)
- $+ NO_2 \rightarrow$ $L_2Pt(ONO_2)_2$
- $+ SO_2 \rightarrow$ L_2Pt(O–SO_2–O)
- $+ R_2CO$, R–CHO $\rightarrow$ L_2Pt(O–C(R)((H)R)–O–O)

[a] *C. D. Cook, G. S. Jauhal*, J. Amer. Chem. Soc. *90*, 1464 (1968).

[b] *T. Kobayashi, Y. Takahashi, S. Sakai, Y. Ishii*, Chem. Commun. 1373 (1968).

[c] *B. Clarke, M. Green, R. B. L. Osborn, F. G. A. Stone*, J. Chem. Soc. (A), 167 (1968).

[d] *M. Baird, G. Hartwell jr., R. Mason, A. I. Rae, G. Wilkinson*, Chem. Commun. 92 (1967).

[e] *T. Kashiwagi, N. Yasuoka, T. Ueki, N. Kasai, M. Kakudo, S. Takahashi, N. Hagihara*, Bull. Chem. Soc., Japan *41*, 296 (1968).

[f] *M. Green, R. B. L. Osborn, F. G. A. Stone*, J. Chem. Soc. (A), 944 (1970).

[g] *C. D. Cook, P. T. Cheng, S. C. Nyburg*, J. Amer. Chem. Soc. *91*, 2123 (1969).

[h] *T. Kashiwagi, N. Yasuoka, N. Kosai, M. Kakudo, S. Takahashi, N. Hagihara*, Chem. Commun. 743 (1969).

[i] *J. Powell*, Organometal. Chem. *B6*, 803 (1970).

reacts[155,156] with concentrated ammonia to yield *cis*-$(NH_3)_2Pt(NO_2)_2$ and subsequently $[(NH_3)_3PtNO_2]NO_2$. Ammonia, hydroxylamine, hydrazine, or amine derivatives of the type $[Pt(ligand)_4](NO_2)_2$ or $[Pt(ligand)_4](NO_3)_2$ are prepared[157] from the chloride and the silver salt of the anion. On heating $[Pt(NH_3)_4](NO_2)_2$-*cis*-$(NH_3)_2Pt(NO_2)_2$ is formed.[158] *Cis*-$(NH_3)_2Pt(CNS)_2$ is obtained[159] by reaction of the stoichiometric quantity of ammonia with $K_2[Pt(CNS)_4]$ while the trans isomer is generated[159] by action of KSCN on *trans*-$(NH_3)_2PtCl_2$. *Tetraazidoplatinato(II)* complexes are obtained from tetrachloroplatinate(II) and excess sodium azide.[160]

The polymeric form of platinum(II) acetylacetonate reacts with pyridine to give[161] *cis*-$(pyridine)_2Pt[CH(COCH_3)_2]_2$. *Phthalocyanine complexes of platinum(II)* are prepared[162] heating together platinum(II) chloride and phthalonitrile.

Reaction of $[(C_2H_8N_2)_2Pt]I_2$ ($C_2H_8N_2$=ethylenediamine) with potassium in liquid ammonia

[155] *L. A. Tschugaev, S. S. Kiltinovic*, J. Chem. Soc. Abstracts *109*, 1286 (1916).

[156] *G. Brauer*, Handbook of Preparative Inorganic Chemistry, II, p. 1580, Academic Press, New York 1965.

[157] *P. Pascal*, Nouveau Traité de Chimie Minérale, Vol. XIX, p. 905, Paris 1958;
T. G. Appleton, J. R. Hall, Inorg. Chem. *9*, 1802 (1970).

[158] *L. A. Tschugaev, S. Krassikov*, Z. Anorg. Allgem. Chemie *131*, 299 (1923).

[159] *A. A. Gruenberg*, Z. Anorg. Allgem. Chemie *157*, 299 (1926).

[160] *W. Beck, W. P. Fehlhammer, P. Poellmann, E. Schuierer, K. Feldl*, Chem. Ber. *100*, 2335 (1967).

[161] *D. Gibson, J. Lewis, C. Oldham*, J. Chem. Soc. (A), 72 (1967).

[162] *P. A. Barrett, C. E. Dent, R. P. Linstead*, J. Chem. Soc. 1719 (1936).

yields $K[Pt(C_2H_7N_2)(C_2H_6N_2)]$, while $Pt(C_2H_7N_2)_2$ is formed[163,164] from the zerovalent complex, $Pt(C_2H_8N_2)_2$ with loss of hydrogen; the reaction does not occur with NH_3 derivatives.

Cis-L_2PtR_2 compounds (L = pyridine, PR_3, AsR_3, SbR_3) are prepared[165] by displacement of cyclo-octatetraene from $C_8H_8PtR_2$ or $C_8H_8Pt_2R_4$. When R constitutes a perfluoroaryl group an organolithium reagent can be reacted with the corresponding dichloride;[166] organolithium or organomagnesium bromide are used[166] with alkyl, aryl, fluoro- or fluorochloro-alkenyl groups, like $-CF{=}CF_2$ or $CF_2{=}CFCl$. *Cis*- and *trans*-$(R_3P)_2Pt(C{\equiv}CR)_2$ compounds are obtained[167,168] from the chloro-compounds and sodium acetylide in liquid ammonia or from an acetylenic alcohol and a base. Interaction of KOH with aromatic isocyanide and *cis*-$(R_3P)_2PtCl_2$ in methanol affords[169]

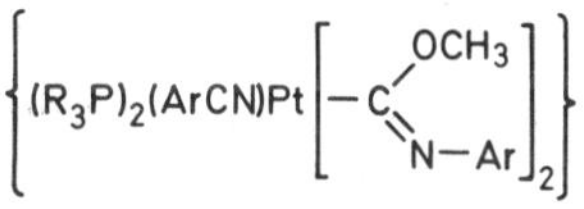

$$\left\{(R_3P)_2(ArCN)Pt\left[-C{\genfrac{}{}{0pt}{}{OCH_3}{\text{=}N-Ar}}\right]_2\right\}$$

Ionic organoplatinum compounds, e.g. $[(Ph_2PCH_2CH_2PPh_2)(PR_3)PtR']^{\oplus}$, are prepared from the ligand and $(R_3P)_2PtR'Cl$.[170] $(R_3P)_2Pt(CN)_2$ is obtained by exchange from the chloride, or from $(Ph_3P)_4Pt$ and cyanogen or tetracyanoethylene.[171] Oxidative addition on the zerovalent platinum compound yields A with trifluoroacetonitrile,[172] B with diethyl azodicarboxylate,[173] C with 4-phenyl-1,2,4-triazoline-3,5-dione,[173] while D[172] is obtained by olefin displacement by trifluoroacetonitrile from $(Ph_3P)_2Pt$(*trans*-stilbene) and E by reaction[174] of $(Ph_3P)_2PtO_2$ with nitric oxide.

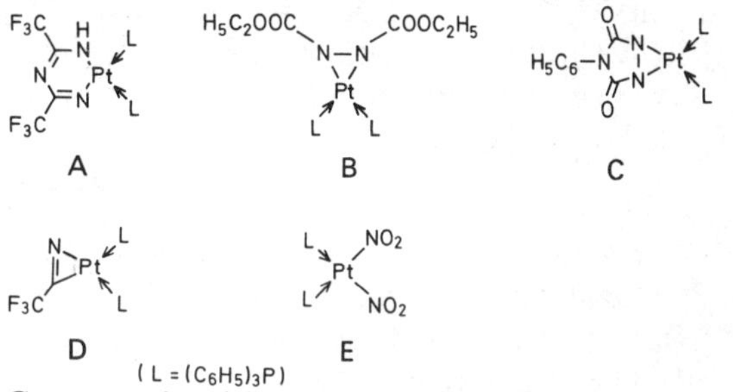

A B C

D E

(L = $(C_6H_5)_3P$)

Compounds with Pt–Ge and with Pt–Sn bonds are known.[175,176,177] The first type is obtained[175] by reaction of $(R_3P)_2PtCl_2$ and Ph_3GeLi; *trans*-$(R_3E)_2Pt(PbPh_3)_2$ (E = P, As) is prepared[178] similarly. The other type is prepared by insertion of tin(II) chloride into a Pt–Cl bond, to yield *cis*-$(R_3P)_2Pt(SnCl_2)(p\text{-}FC_6H_4)$[176] or $(Ph_3P)_2Pt(SnCl_3)_2$.[177]

28.4.5 Complexes with Platinum-Group IV Bonds

Complexes of platinum(II) containing bonds with group IV elements include compounds with Pt–C and with Pt–Ge or Pt–Sn bonds.

Ionic compounds with four equivalent Pt–C bonds can belong to the $[PtL_4]^{2\oplus}$ or to the $[PtX_4]^{2\ominus}$ type. Compounds of the first type are prepared[179,180] by reaction of an isocyanide with K_2PtCl_4 and the resulting $[(RNC)_4Pt][PtCl_4]$ can be converted, by exchange reaction, into a picrate or other salts. The anionic derivatives are prepared[181] by reaction of platinum(II) chloride with potassium cyanide or by fusing platinum with potassium hexacyanoferrate(II). The resultant $K_2[Pt(CN)_4]$ is subsequently

[163] *G. W. Watt, J. W. Dawes*, J. Amer. Chem. Soc. *81*, 8 (1959).
[164] *G. W. Watt, M. T. Walling jr., P. J. Mayfield*, J. Amer. Chem. Soc. *75*, 6175 (1953).
[165] *C. R. Kistner, J. D. Blackman, W. C. Harris*, Inorg. Chem. *8*, 2165 (1969).
[166] *J. D. Ruddick, B. L. Shaw*, J. Chem. Soc. (A), 2801 (1969);
A. J. Rest, D. T. Rosevear, F. G. A. Stone, J. Chem. Soc. (A), 66 (1967);
G. W. Parshall, J. Amer. Chem. Soc. *88*, 704 (1966).
[167] *A. Furlani, P. Bicev, M. V. Russo, P. Carusi*, J. Organometal. Chem. *29*, 321 (1971).
[168] *I. Collamati, A. Furlani*, J. Organometal. Chem. *17*, 457 (1969).
[169] *G. Minghetti, F. Bonati*, Atti Accad. Naz. Lincei, Classe Sci. Fis. Mat. Natur. Rend. [VIII] *49*, 287 (1970).
[170] *K. A. Hooton*, J. Chem. Soc. (A), 1896 (1970).
[171] *W. H. Baddley, L. M. Venanzi*, Inorg. Chem. *5*, 33 (1966).
[172] *W. J. Bland, R. D. W. Kemmitt, I. W. Nowell, D. R. Russell*, Chem. Commun. 1065 (1968).
[173] *M. Green, R. B. L. Osborne, F. G. A. Stone*, J. Chem. Soc. (A), 3083 (1968).
[174] *J. P. Collmann, M. Kubota, J. W. Hosking*, J. Amer. Chem. Soc. *89*, 4809 (1967).
[175] *E. H. Brooks, R. J. Cross, F. Glockling*, Inorg. Chim. Acta *2*, 17 (1968).
[176] *R. V. Lindsay, C. W. Parshall, U. G. Stolberg*, J. Amer. Chem. Soc. *87*, 659 (1965).
[177] *R. D. Cramer, E. L. Jenner, R. V. Lindsey, U. G. Stolberg*, J. Amer. Chem. Soc. *85*, 1691 (1963).
[178] *G. Deganello, G. Carturan, U. Belluco*, J. Chem. Soc. (A), 2873 (1968).
[179] *F. Bonati, G. Minghetti*, J. Organometal. Chem. *24*, 251 (1970).
[180] *L. Malatesta, F. Bonati*, Isocyanide Complexes of Metals, p. 168, London 1969.
[181] *L. Chugaev, P. Teearu*, Ber. *47*, 568 (1914).

transformed into the free acid, $H_2Pt(CN)_4 \cdot 5H_2O$, or into other salts.[182,183] The tetrakis(thiocyanato)- or -(fulminato)platinate derivatives are prepared by reaction of platinum(II) chloride and excess alkali salt of the required anion;[184,185] the acids are unstable. The $[(RC{\equiv}C)_4Pt]^{2\ominus}$ compounds are prepared[186] according to the reaction (in liquid ammonia)

$$K_2[Pt(SCN)_4] + 4KC{\equiv}C{-}R \longrightarrow K_2[Pt(C{\equiv}C{-}R)_4] + 4KSCN$$

A covalent compound with four equivalent Pt–C bonds is the volatile *bis(π-allyl)platinum(II)*, prepared[187] from the chloride and ethereal allylmagnesium chloride at low temperatures. Compounds with nonequivalent Pt–C bonds include $(RNC)_2Pt(CN)_2$, obtained[188] directly from $K_2[Pt(CN)_4]$ and isocyanide or by reaction with methyl iodide according to the following equation

$$2CH_3I + K_2[Pt(CN)_4] \longrightarrow (CH_3NC)_2Pt(CN)_2 + 2KI$$

and some (diolefin)PtR_2 compounds, prepared[189–192] by reaction of the corresponding halide and Grignard reagents under strict conditions.

Compounds containing bonds between platinum and heavier group IV elements include $[(CH_3)_4N][Pt(CN)_2(GePh_3)_2]$, from the reaction[193] of $(R_3P)_2Pt(GePh_3)_2$ with potassium cyanide, and the anion $[Pt(SnCl_3)_5]^{3\ominus}$, obtained[194,195] from tin(II) chloride and tetrachloroplatinate(II). An acetone solution of tin(II) chloride reacts[194,195] with platinum(II) chloride or with $[Cl_2Pt(SnCl_3)_2]^{2\ominus}$ yielding $[Et_4N]_4[Pt_3Sn_8Cl_{20}]$ which reacts with 1,5-cyclooctadione to afford $(C_8H_{12})_3Pt_3Sn_2Cl_6$; both compounds contain a Pt_3Sn_2 cluster.

182 *G. Brauer*, Handbook of Preparative Inorganic Chemistry, II, 1576, Academic Press, New York 1965.
183 *H. Baumhauer*, Z. Krist. *43*, 356 (1907).
184 *G. Buckton*, J. Chem. Soc. *7*, 32 (1855); *G. Buckton*, Justus Liebigs Ann. Chem. *92*, 86 (1954).
185 *M. Beck, E. Schuierer*, Chem. Ber. *98*, 298 (1965); *L. Woehler, H. Berthmann*, Ber. *62*, 2748 (1929).
186 *R. Nast, W.-D. Heinz*, Chem. Ber. *95*, 1478 (1972).
187 *G. Wilke, H. Bogdanovic, P. Hardt, P. Heimbach, W. Keim, M. Kroner, W. Oberkirch, T. Tanaka, E. Steinbruecke, D. Walter, H. Zimmermann*, Angew Chem. *78*, 157 (1966); *J. K. Becconsall, B. E. Job, S. O'Brien*, J. Chem. Soc. (A), 423 (1967); *W. S. McDonald, B. E. Mann, G. Raper, B. L. Shaw, G. Shaw*, Chem. Commun. 1254 (1969).
188 *L. Chugaev, B. Teearu*, Ber. *47*, 2643 (1914).
189 *J. Mueller, P. Goeser*, Angew. Chem. Intern. Ed. Engl. *6*, 364 (1967).
190 *H. P. Fritz, D. Sellmann*, Spectrochim. Acta *23A*, 1991 (1967); *H. P. Fritz, D. Sellmann*, Z. Naturforsch. *22b*, 20 (1967).
191 *C. R. Kistner, J. H. Hutchinson, J. R. Doyle, J. C. Storlie*, Inorg. Chem. *2*, 1255 (1963).
192 *J. R. Doyle, J. H. Hutchinson, N. C. Baezinger, L. W. Tresselt*, J. Amer. Chem. Soc. *83*, 2768 (1961).
193 *E. H. Brooks, R. J. Cross, F. Glockling*, Inorg. Chim. Acta *2*, 17 (1968).
194 *R. V. Lindsey, G. W. Parshall, U. G. Stollberg*, Inorg. Chem. *5*, 109 (1966).
195 *R. D. Cramer, E. L. Jenner, R. V. Lindsey, U. G. Stolberg*, J. Amer. Chem. Soc. *85*, 1691 (1963).

28.5 Platinum(III)

Platinum(III) compounds should be paramagnetic, unless a metal–metal bond (or other causes) eliminate the effect of the unpaired electron. However, only one paramagnetic compound, green $(NH_3)_2PtI(SCN)_2$, can be obtained,[196] by reaction of $(NH_3)_2Pt(SCN)_2$, either isomer, with iodine.

Owing to diamagnetism or to chemical evidence, all the other compounds which formally should be considered as platinum(III) derivatives have to be regarded, at least in principle, as mixed Pt(II)–Pt(IV) compounds: *e.g.*[197] $(EtNH_2)_3PtCl_3 \cdot 2H_2O$ is $[(EtNH_2)_4Pt^{IV}Cl_2][(EtNH_2)_4Pt^{II}]Cl_4 \cdot 4H_2O$.

The *hydroxide*, $Pt_2O_3 \cdot xH_2O$ is obtained from the trichloride and potassium hydroxide.[198] The halides, PtX_3 (X = Cl, Br, I), can be prepared by heating[198–200] another halide, either of platinum(II) or (IV) in a rather limited temperature range in the presence of the halogen. Dark green Cs_2PtCl_5 is obtained on careful chlorination[201] in the cold of Cs_2PtCl_4 or from cesium chloride and $PtCl_3$. A cyanocomplex, $KPt(CN)_4$ is produced[202] similarly and the free acid is generated by hydrogen peroxide oxidation of hydrogen tetracyanoplatinate(II).

Amine complexes such as the red $(H_2NCH_2CH_2NH_2)PtCl_3$ can be obtained by careful oxidation[203] of a solution of a corresponding platinum(II) complex, with

196 *G. S. Muraveiskaya, G. M. Lanin, V. F. Sorokina*, Russian J. Inorg. Chem. *13*, 771 (1968).
197 *T. D. Ryan, R. E. Rundle*, J. Amer. Chem. Soc. *83*, 2814 (1961); *B. M. Craven, D. Hall*, Acta Cryst. *14*, 475 (1961).
198 *L. Woehler, F. Martin*, Ber. *42*, 3958 (1909).
199 *L. Woehler, F. Mueller*, Z. Anorg. Allgem. Chemie *149*, 377 (1925).
200 *J. Krustinsons*, Z. Elektrochem. *44*, 537 (1938).
201 *L. Tschugaev, J. Tschernaiev*, Z. Anorg. Allgem. Chemie *182*, 159 (1929).
202 *L. A. Levy*, J. Chem. Soc. *101*, 1081 (1912).
203 *H. D. K. Drew, F. W. Pinkard, W. Wardlaw, E. G. Cox*, J. Chem. Soc. (London) 1013 (1932).

air[204,205] in acidic solution or with persulfate;[201] alternatively they can be obtained by grinding or heating together the required platinum(II) and platinum(IV) complexes.

28.6 Platinum(IV)

28.6.1 Hydrides

Although the action of HX on a platinum(II) compound should give a hydrido derivative of platinum(IV) (oxidative addition), the equilibrium is generally unfavorable to the isolation of the compound or the latter is unstable. The reported $[(Ph_2PCH_2CH_2PPh_2)PtGeCl_3(PEt_3)(H)Cl]Cl$ has been reformulated[206] as $[(Ph_2PCH_2CH_2PPh_2)PtGeCl_3(PEt_3)]^{\oplus}[HCl_2]^{\ominus}$. The other hydrides are prepared by reaction[207] of 1-ethynylcyclohexanol with $(Ph_3P)_4Pt$, *trans*-$H_2Pt(PPh_3)(C_2R)$ being the reaction product, or by addition[208] of tetracyanoethylene to $(Ph_3P)_2PtH(CN)$, but not to other analog platinum(II) derivatives.

28.6.2 Halides and Halocomplexes

All tetrahalides and hexahaloplatinates(IV) are known. The *fluoride*, PtF_4, brown solid, is obtained[209] by thermal decomposition as depicted in the following equation,

$$[ClF_2]^{\oplus}[PtF_6]^{\ominus} \longrightarrow PtF_4 + ClF_3 + \tfrac{1}{2}F_2$$

or, rather impure, by reduction of PtF_6 with a large excess of tetrafluorohydrazine. The *chloride* can be obtained by chlorination[210] of the metal at 250–300°, from platinum and $AsCl_3$ or $SeCl_4$,[211] or, better, by dehydration[212] of $H_2PtCl_6 \cdot 6H_2O$ in a stream of chlorine at 300°. The *bromide*[213,214] is obtained from the metal, bromine and hydrobromic acid (sealed tube, 180°); the *iodide*[215] from $H_2PtCl_6 \cdot 6H_2O$ and hydriodic acid.

Many 1:1 or 1:2 adducts of PtF_4 with other fluorides are known, *e.g.*

$$BrF_3 + [O_2]^{\oplus}[PtF_6]^{\ominus} \longrightarrow PtF_4 \cdot BrF_3$$

$$BrF_3 + PtF_4 \longrightarrow PtF_4 \cdot 2BrF_3 \qquad \text{(Ref. 216)}$$

The *hexahaloplatinates* are obtained by action[217,218] of halide ion on the tetrahalide, from $PtCl_6^{2\ominus}$ and potassium iodide,[219,220] by reacting[221] platinum with bromine and hydrobromic acid or with aqua regia and subsequent treatment with hydrochloric acid,[222] by action[223] of boiling BrF_6 on chloro- or bromoplatinate, $PtX_6^{2\ominus}$, or by hydrolysis[224] of $[O_2]^{\oplus}[PtF_6]^{\ominus}$. Mixed derivatives, *e.g.* $K_2PtCl_3F_3$, can be obtained from the chlorocomplexes and BrF_3.[225] Complex acids and their salts of the type $H_2[PtCl_5(OH)]$ can be obtained[226,227] by melting H_2PtCl_6 with potassium hydroxide under reduced pressure, or by dissolving platinum(IV) chloride in water.

An extensive number of *haloderivatives* of platinum(IV) with group VI and group V donor ligands are known, the types varying from $[PtX(\text{ligand})_5]^{3\oplus}$ to $[Pt(\text{ligand})X_5]^{\ominus}$. The ligands include amines,[228] phosphines,[229] ars-

---

[204] *H. Reihlen, E. Flohr*, Ber. *67*, 2010 (1934).
[205] *H. D. K. Drew, H. J. Tress*, J. Chem. Soc. (London) 1244 (1935).
[206] *K. A. Hooton*, J. Chem. Soc. (A), 680 (1969).
[207] *J. H. Nelson, H. B. Jonassen, D. Roundhill*, Inorg. Chem. *8*, 2591 (1969).
[208] *P. Uguagliati, W. H. Baddley*, J. Amer. Chem. Soc. *90*, 5446 (1968).
[209] *F. P. Cotsema, R. H. Toeniskoetter*, Inorg. Chem. *5*, 1925 (1966).
[210] *L. Woehler, S. Streicher*, Ber. *46*, 920 (1913).
[211] *A. Gutbier, F. Heinrich*, Z. Anorg. Allgem. Chemie *81*, 378 (1913).
[212] *A. Gutbier, F. Heinrich*, Inorg. Synth. *2*, 247 (1946).
[213] *V. Meyer, H. Zublin*, Ber. *13*, 404 (1880).
[214] *L. Woehler, H. Mueller*, Z. Anorg. Allgem. Chemie *149*, 378 (1925).
[215] *H. Topsoe*, Overs. Danske Selsk. Forh. 77 (1869).
[216] *N. Bartlett, D. H. Lohmann*, J. Chem. Soc. (London) 5253 (1952);
N. Bartlett, D. H. Lohmann, J. Chem. Soc. (London) 619 (1964).
[217] *A. Gutbier, F. Bauriedel*, Ber. *41*, 4243 (1908).
[218] *A. Gutbier, F. Krauss, L. v. Mueller*, Sitzb. Phys.-Med. Sozi. Erlangen *45*, 25 (1914).
[219] *R. L. Datta*, J. Chem. Soc. Abstracts *103*, 426 (1913).
[220] *R. L. Datta, T. Gosh*, J. Amer. Chem. Soc. *36*, 1017 (1914).
[221] *E. Biilmann, A. C. Anderson*, Ber. *36*, 1365 (1903).
[222] *G. Brauer*, Handbook of Preparative Inorganic Chemistry, II, p. 1569, New York 1965.
[223] *G. Sharpe*, J. Chem. Soc. (London) 197 (1953).
[224] *N. Bartlett, D. H. Lohmann*, Proc. Chem. Soc. 115 (1962).
[225] *D. H. Brown, K. R. Dixon, D. W. A. Sharpe*, J. Chem. Soc. (A), 1244 (1966).
[226] *A. Miolati, I. Bellucci*, Atti Accad. Naz. Lincei, Classe Sci. Fis. Mat. Natur. Rend. [5] *9*, II, 51 (1900).
[227] *A. Miolati*, Z. Anorg. Allgem. Chemie *33*, 251 (1903).
[228] *I. I. Chernayev, N. N. Zheligovskaja, L. T'i-k'eng, D. V. Kurganovic*, Russian J. Inorg. Chem. *9*, 312 (1964);
I. I. Chernayev, A. V. Babkov, N. N. Zheligovskaia, Russian J. Inorg. Chem. *9*, 319 (1964).
[229] *W. J. Bland, R. D. V. Kemmitt*, J. Chem. Soc. (A), 2062 (1969).

ines[230,231] and the group X can be either identical or different; in the latter case X may include $OH^{\ominus}$, $CN^{\ominus}$, $NO_3^{\ominus}$, $NO_2^{\ominus}$. Although some compounds can be readily obtained from the corresponding platinum(II) complexes by oxidation, their preparation is not amenable to a simple generalization: an extensive survey is available.[232]

One obtains for example by chlorination from

$$[(NH_3)_3PtCl_3] \xrightarrow{Cl_2} [(NH_3)_2Pt(NCl_2)Cl_3]^{233}$$

$$cis\text{- and } trans\text{-}[(NH_3)_2(Py)_2PtCl_2]Cl_2 \xrightarrow{Cl_2} [(Py)_2Pt(NCl_2)_2] + [(NH_3)(Py)_2Pt(NCl_2)Cl_2]Cl_2{}^{233}$$

$$[(R_3P)_2Pt(C{\equiv}C{-}C_6H_5)_2] \xrightarrow{Cl_2} [(R_3P)_2Pt({-}CCl{=}CCl{-}C_6H_5)_2Cl_2]^{234}$$

$$\xrightarrow{CH_3OH,\ reflux} [(R_3P)_2Pt({-}CCl{=}CCl{-}C_6H_5)Cl_3]$$

$$[Pt(CN)_4]^{2\ominus} \xrightarrow{Cl_2} [Pt(CN)_4Cl_2]^{2\ominus\,235}$$

One or more halogen groups of PtX_4 can be replaced by an organic group. The reaction of $H_2PtCl_6 \cdot 6H_2O$ and cyclopropene in acetic acetic/acetic anhydride or the addition of the hydrocarbon to $[(C_2H_4)PtCl_2]_2$ affords[236] $[(CH_2)_3PtCl_2]_4$. Reaction of potassium hexachloroplatinate(IV) with CH_3MgI affords[237] $[(CH_3)_3PtI]_4$, while the corresponding chloride can be obtained[238] by action of a dialkylmercury on $PtCl_4$. Reaction of 1,2-bis(diphenylethynyl)benzene with $PtCl_4$ yields[239] the *cis*-1:1 adduct.

Many platinum(IV) mixed alkyl-halo-derivatives with tertiary phosphines[240,241] or arsines[241,242] as ligands (L) can be prepared from *cis*-$L_2Pt(CH_3)_2$ by oxidative addition of halogens, methyl halide, acetyl chloride, methylsulfonyl chloride, or perfluoroalkyl iodide.[241]

28.6.3 Oxygen and Sulfur Derivatives

Although hydrated platinum(IV) oxide is the starting material for the so-called Adam's catalyst which is employed in the hydrogenation of organic compounds, the compound cannot be prepared in pure state by either melting[243,244] ammonium hexachloroplatinate(IV) and sodium nitrate (the product being approximately $PtO_2 \cdot H_2O$) or heating[245] platinum sponge with oxygen under pressure. *Platinum(IV) sulfide*,[246] *selenide*,[247] and *telluride*[248] are prepared from the elements at high temperature or, in the case of $PtS_2 \cdot 5H_2O$,[249] by addition of sodium sulfide or of hydrogen sulfide[250] to a solution of hexachloroplatinate(IV).

If platinum(IV) chloride is boiled with excess alkali hydroxide and then either acetic acid[251] or ethanol[252] is added, $H_2[Pt(OH)_6]$ or the alkaline derivative, respectively, precipitate. Complex sulfur derivatives, *e.g.* $[NH_4]_2[Pt(S_5)_3] \cdot 2H_2O$, are obtained[253,254] from platinum(IV) chloride, ammonium chloride and sulfur; thiocyanate[255]

---

[230] *M. A. Bennett, J. Chatt, G. L. Erskine, J. Lewis, R. F. Long, R. S. Nyholm*, J. Chem. Soc. (A), 501 (1967).
[231] *M. A. Bennett, G. J. Erskine, R. S. Nyholm*, J. Chem. Soc. (A), 1260 (1967).
[232] *P. Pascal*, Nouveau Traité de Chimie Minérale, p. 920, Paris 1958.
[233] *Y. N. Kukuskin*, Russian J. Inorg. Chem. *4*, 1131 (1959).
[234] *R. Ettorre*, J. Organometal. Chem. *19*, 247 (1969).
[235] *A. Miolati, L. Bellucci*, Gazz. Chim. Ital. *30*, II, 588 (1901).
[236] *S. E. Binns, R. H. Cragg, R. D. Gillard, B. T. Heaton, M. F. Pilbrow*, J. Chem. Soc. (A), 1227 (1969).
[237] *D. E. Clegg, J. R. Hall*, Spectrochim. Acta *21*, 357 (1965).
[238] *S. F. A. Kettle*, J. Chem. Soc. 5737 (1965).
[239] *E. Mueller, K. Munk, P. Ziemek, M. Sauerbier*, Justus Liebigs Ann. Chem. *40*, 713 (1968).
[240] *J. D. Ruddick, B. L. Shaw*, J. Chem. Soc. (A), 2801 (1969).
[241] *H. C. Clark, J. D. Ruddick*, Inorg. Chem. *9*, 2556 (1970).
[242] *J. D. Ruddick, B. L. Shaw*, J. Chem. Soc. (A), 2964 (1969).
[243] *A. Vogel*, A Textbook of Practical Organic Chemistry, p. 470, London 1957.
[244] *W. F. Bruce*, J. Amer. Chem. Soc. *48*, 687 (1936).
[245] *P. Laffitte, P. Grandadam*, Compt. Rend. *200*, 456 (1935).
[246] *W. Biltz, R. Juza*, Z. Anorg. Allgem. Chemie *190*, 161 (1930).
[247] *L. Moser, K. Atynsky*, Monatsh. Chem. *45*, 235 (1925).
[248] *C. Roessler*, Z. Anorg. Allgem. Chemie *15*, 405 (1897).
[249] *I. K. Taimni, G. S. Salaria*, Chim. Acta *11*, 329 (1954).
[250] *G. Brauer*, Handbook of Preparative Inorganic Chemistry, II, p. 1576, New York 1965.
[251] *L. Woehler*, Z. Anorg. Allgem. Chemie *40*, 423 (1904).
[252] *G. Brauer*, Handbook of Preparative Inorganic Chemistry, II, p. 1575, New York 1965.
[253] *P. E. Jones, L. Katz*, Chem. Commun. 842 (1967).
[254] *K. A. Hoffmann, F. Heschtlen*, Ber. *36*, 3090 (1903).
[255] *G. B. Buckton*, Justus Liebigs Ann. Chem. *22*, 284 (1954).

and selenocyanate[256] complexes are prepared from hexachloroplatinate(IV). The thiocyanato complex, $[Pt(SCN)_6]^{2\ominus}$, reacts[257] with donor molecules, *e.g.* ethylendiamine, to afford $[(C_2H_8N_2)_2Pt(SCN)_2](SCN)_2$. Derivatives containing organic groups include a series of salts of the $(CH_3)_3Pt(IV)$ moiety. They are prepared from the hydroxide, $[(CH_3)_3PtOH]_4$, which is formed by action[258] of moist silver oxide on the iodide, or from the halides and β-diketones,[259] 8-hydroxyquinoline,[260] salicylaldehyde or their thallium(I) salts, giving compounds with remarkable and unexpected structures;[261] from these compounds their adducts with nitrogen ligands are easily prepared.[260,261] *Trimethylplatinum(IV) derivatives of carboxylic acids* and of oxyacids can be prepared in similar fashion.[262]

28.6.4 Amine and Nitrogen Complexes

Derivatives of the cations $[(NH_3)_6Pt]^{4\oplus}$, $[(\text{ethylendiamine})_3Pt]^{4\oplus}$ are obtained[263] by action of liquid ammonia on ammonium hexachloroplatinate(IV) or from[264] ethylenediamine and platinum(IV) chloride; the latter compound could be resolved in optical antimers with optically active tartaric acid.[264]

Reaction[265] of hexamminoplatinum(IV) nitrate with potassium amide in liquid ammonia affords $K_2[Pt(NH_2)_6]$ or $[(NH_3)_3Pt(NH_2)_3]NO_3$, depending on the conditions used. Hexanitroplatinate(IV), $[Pt(NO_3)_6]^{2\ominus}$, is prepared[266] by oxidation of tetranitroplatinate(II), $[Pt(NO_2)_4]^{2\ominus}$; *hexaazidoplatinate(IV)* is prepared by exchange from the chlorocomplex and sodium azide.[267]

Trimethylplatinum(IV) halide can add[268] an amine ligand, affording derivatives of the $[(CH_3)_3Pt(\text{amine})_3]^{\oplus}$ cation. A similar addition takes place on $[(CH_3)_3Pt(C_5H_7O_2)]_2$ yielding hexacoordinated $(CH_3)_3Pt(CH[COCH_3]_2)$(bipyridyl), $C_5H_7O_2$ being pentane-2,4-dionate.

Tertiary phosphine or arsine complexes of tetramethylplatinum(IV), *cis*-$(R_3E)_2Pt(CH_3)_4$, are obtained by action of methyllithium on $(ER_3)_2PtBr(CH_3)_3$ or $(ER_3)_2PtCl_2(CH_3)_2$.[269]

28.6.5 Complexes with Platinum-Group(IV) Bonds

Platinum silicide, PtSi, can be obtained[270,271] by combining the elements at red heat.

Although a tetramethylplatinum(IV) complex, $[Pt(CH_3)_4]_4$, had been reported, and though its phosphine and arsine derivatives are known (see 28.6.4), several attempts to duplicate the preparation have been unsuccessful.[272] Similarly, the so-called trimethylplatinum, $[Pt(CH_3)_3]_2$ or $[Pt(CH_3)_3]_{12}$, was prepared by action of potassium on trimethyliodoplatinum(IV), but the data could not be verified.[272]

28.7 Platinum(V)

By interaction of the elements at 350° dark-red $[PtF_5]_4$ (together with PtF_6) is formed.[273] *Pentafluoroplatinum(V)* is obtained also by dissociation of PtF_6, by reaction of platinum(II) chloride with fluorine at 350°, or by reduction[274] with tetrafluorohydrazine according to the reaction:

$$2PtF_6 + N_2F_4 \longrightarrow 2PtF_5 + NF_3$$

---

[256] *G. Spacu, V. Armeanu*, Bull. Soc. Stiinte Cluj *7*, 610 (1934).
[257] *H. Grossmann, B. Schneck*, Ber. *39*, 1900 (1906).
[258] *W. J. Pope, S. S. Peachey*, J. Chem. Soc. Abstracts *95*, 571 (1909).
[259] *G. R. Hoff, C. H. Brubaker*, Inorg. Chem. *7*, 1655 (1968).
[260] *K. Kite, M. R. Truter*, J. Chem. Soc. (A), 207 (1966).
[261] *R. Mason, G. B. Robertson, P. Pauling*, J. Chem. Soc. (A), 485 (1969);
R. N. Hargreaves, M. R. Truter, J. Chem. Soc. (A), 2282 (1969);
D. Gibson, Coord. Chem. Rev. *4*, 225 (1969).
[262] *K. Kite, J. A. S. Smith, E. J. Wilkins*, J. Chem. Soc. (A), 1744 (1966).
[263] *L. Tschugaev*, Z. Anorg. Allgem. Chemie *137*, 1 (1924).
[264] *A. Werner*, Vierteljaresschr. naturf. Ges. Zürich *62*, 553 (1917); Gmelins Handbuch der anorganischen chemie, Platin, Teil D, pp. 470–474, Weinheim, 1957.
[265] *L. Heck*, Z. Naturforsch. *25b*, 428 (1970).
[266] *M. J. Nolan, D. W. James*, Australian J. Chem. *23*, 1043 (1970).
[267] *W. Beck, W. P. Fehlhammer, P. Poellmann, E. Schuierer, K. Feldl*, Chem. Ber. *100*, 2335 (1967).
[268] *D. E. Clegg, J. R. Hall*, Australian J. Chem. *20*, 2025 (1967).
[269] *J. D. Ruddick, B. L. Shaw*, Chem. Commun. 1135 (1967).
[270] *P. Lebeau, A. Novitzky*, Compt. Rend. *145*, 241 (1907).
[271] *E. Vigoroux*, Compt. Rend. *145*, 376 (1907).
[272] *M. N. Hoechstetter, C. H. Brubaker*, Inorg. Chem. *8*, 400 (1969).
[273] *H. Bartlett, D. H. Lohmann*, Proc. Chem. Soc. 14 (1960);
N. Bartlett, D. H. Lohmann, J. Chem. Soc. (London) 619 (1964).
[274] *F. P. Gortsema, R. H. Toeniskoetter*, Inorg. Chem. *5*, 1925 (1966).

A number of yellow or orange hexafluoroplatinate(V) derivatives are known. They are accessible through the following reactions:

$NO + PtF_6 \longrightarrow [NO]^{\oplus}[PtF_6]^{\ominus}$ (Refs. 275, 276)
$NO_xF + PtF_6 \longrightarrow [NO_x]^{\oplus}[PtF_6]^{\ominus} + \frac{1}{2}F_2$ $(x = 1, 2)$ (Ref. 275)
$Xe + PtF_6 \longrightarrow [Xe]^{2\oplus}[PtF_6]^{2\ominus}$ (Refs. 277, 278)
$O_2 + PtF_6 \longrightarrow [O_2]^{\oplus}[PtF_6]^{\ominus}$ (Ref. 279)

The last salt can be obtained also by fluorination of platinum in a glass or silica apparatus.[273]

The potassium salt is obtained from the dioxygenyl salt and potassium iodide in IF_7 solvent. Other salts are also obtained, e.g.:

$XeF_2 + PtF_5 \xrightarrow{BrF_5} [Xe_2F_3][PtF_6] + [XeF][PtF_6]$[278]
$ClF_3 + PtF_6 \longrightarrow [Cl_2F][PtF_6]$
$Xe + PtF_5 + F_2 \xrightarrow{200°/5\ at.} [XeF_5][PtF_6]$[280]

28.8 Platinum(VI)

Dark-red *hexafluoroplatinum(VI)*, m.p. 61°, b.p. 69°, is obtained[281,282] on burning platinum wire in a fluorine atmosphere (300–400 Torr). When platinum and fluorine react at 350°, *oxotetrafluoroplatinum(VI)* is formed,[283] in addition to hexafluoroplatinum(VI) and tetrafluoroplatinum(IV).

By anodic oxidation of an alkaline solution of platinum(IV) hydroxide the compound $K_3[PtO_{10}]$ is formed, which, after washing with dilute acetic acid, leaves a brownish-red solid, assumed[284] to be impure PtO_3.

Other compounds claimed[285] to be derivatives of platinum(VI) include platinum triselenide, $PtSe_3$, obtained[286] by treating hydrogen hexachloroplatinato(IV) and selenium dioxide with an alkaline solution of formaldehyde, metallic PtP_2 and $PtAs_2$, prepared from the elements at high temperature.[287]

28.9 Platinum(VIII)

The material reported to be $Pt(CO)F_8$ has properties[288] identical to those of *cis*-$(CO)_2PtBr_2$. The evidence supporting the existence of PtO_4 is rather scarce.[289]

28.10 Bibliography

Gmelins Handbuch der Anorganischen Chemie, System-Nummer 68, Teil C, Lieferung 1, Verlag Chemie, Berlin 1939.

Gmelins Handbuch der Anorganischen Chemie, System-Nummer 68, Teil C, Lieferung 2, 3, Verlag Chemie, Berlin 1940.

Gmelins Handbuch der Anorganischen Chemie, System-Nummer 68, Teil D, Verlag Chemie, Weinheim/Bergstr. 1957.

P. Pascal, Nouveau Traité de Chimie Minérale, Tome 14, p. 656, Masson et Cie, Paris 1958.

G. Brauer, Handbook of Preparative Inorganic Chemistry, p. 1560, Academic Press, New York 1965.

F. R. Hartley, Organometal. Chem. *A6*, 119 (1970).

F. R. Hartley, The Chemistry of Platinum and Palladium, Applied Science Publ. Ltd., London 1973.

U. Belluco, Organometallic and Coordination Chemistry of Platinum, Academic Press, London 1973.

---

[275] *F. P. Gortsema, R. H. Toeniskoetter*, Inorg. Chem. *5*, 1217 (1966).

[276] *N. Bartlett, S. P. Beaton*, Chem. Commun. 167 (1967).

[277] *N. Bartlett*, Proc. Chem. Soc. 218 (1962).

[278] *F. O. Sladky, P. A. Bulliner, N. Bartlett*, J. Chem. Soc. *A*, 2179 (1969).

[279] *N. Bartlett, D. H. Lohmann*, Proc. Chem. Soc. 115 (1962);
N. Bartlett, D. H. Lohmann, J. Chem. Soc. 5253 (1962).

[280] *N. Bartlett, F. Einstein, D. F. Stewart, J. Trotter*, Chem. Commun. 550 (1966).

[281] *F. P. Gortsema, R. H. Toeniskoetter*, Inorg. Chem. *5*, 1217 (1966).

[282] *B. Weinstock, J. C. Malm, E. E. Weaver*, J. Amer. Chem. Soc. *83*, 4310 (1961).

[283] *N. Bartlett*, Proc. Chem. Soc. 14 (1960).

[284] *L. Woehler, F. Martin*, Ber. *42*, 3326 (1909);
L. Woehler, F. Martin, Z. Elektrochem. *15*, 792 (1909).

[285] *N. V. Sidgwick*, The Chemical Elements and their Compounds, p. 1625, Oxford 1950.

[286] *A. Minozzi*, Atti Accad. Naz. Lincei, Classe Sci. Fis. Mat. Natur. Rend. [5] *18*, *11*, 150 (1909).

[287] *L. Woehler*, Z. Anorg. Allgem. Chemie *186*, 324 (1930).

[288] *R. D. W. Kemmitt, R. D. Peacock, I. L. Wilson*, Chem. Commun. 772 (1968).

[289] *A. Schneider, U. Esch*, Z. Elektrochem. *49*, 55 (1943).

29 Transition Metal Carbonyls

R. B. King
Department of Chemistry, University of Georgia,
Athens, Georgia 30601, U.S.A.

This chapter summarizes the most useful procedures for the preparation of important metal carbonyl derivatives. The compounds discussed are arranged according to the position of the central metal atom in the periodic table. Metals in the same column of the periodic table are discussed together. This chapter starts with titanium, the metal most to the left in the periodic table that forms carbonyl derivatives. In the case of each column of the periodic table, the pure metal carbonyls are discussed first followed by metal carbonyl anions, metal carbonyl hydrides, metal carbonyl halides, sulfur derivatives of metal carbonyls, metal carbonyl nitrosyls, and phosphorus derivatives of metal carbonyls. Finally metal carbonyl derivatives also containing other carbon groups are discussed such as alkylmetal carbonyls, olefin–metal carbonyls, arene–metal carbonyls, and finally cyclopentadienylmetal carbonyl derivatives.

Special precautions are often necessary when handling metal carbonyls. Most metal carbonyls are toxic and thus should be handled in an efficient hood. In some cases, such as $Ni(CO)_4$, this toxicity hazard is quite considerable. Furthermore, many metal carbonyl derivatives are very air-sensitive and thus must be handled in an inert atmosphere. Normally pure nitrogen is sufficiently inert for protection of metal carbonyls from atmospheric oxidation although in exceptional cases argon should be used.

29.1 Bis(cyclopentadienyl)dicarbonyltitanium

The only known carbonyl derivative of the titanium, zirconium, hafnium triad is the titanium compound *bis(cyclopentadienyl)dicarbonyltitanium*, $(C_5H_5)_2Ti(CO)_2$, **1**, a red very air-sensitive solid. This titanium compound can be prepared by carbonylation of the commercially available $(C_5H_5)_2TiCl_2$ at 100°/110 atm in the presence of a strong reducing agent such as butyllithium or preferably sodium cyclopentadienide.[1]

1

[1] *J. G. Murray*, J. Amer. Chem. Soc. *83*, 1287 (1961).

29.2 The Carbonyl Derivatives of Vanadium, Niobium, and Tantalum

29.2.1 Hexacarbonyl Derivatives

The $V(CO)_6^{\ominus}$ anion is conveniently prepared by the carbonylation of a mixture of anhydrous vanadium(III) chloride, sodium metal, and diglyme at 160°/200 atmospheres according to the following equation:[2,3]

$$VCl_3 + 4Na + 2H_3C{-}O{-}CH_2{-}CH_2{-}O{-}CH_2{-}CH_2{-}O{-}CH_3\text{(diglyme)} + 6CO \longrightarrow [Na(diglyme)_2][V(CO)_6] + 3NaCl$$

The product $[Na(diglyme)_2][V(CO)_6]$, a yellow crystalline solid, can be handled in the air for short periods of time but should be stored for longer periods in the dark and in an inert atmosphere. Reaction of $[Na(diglyme)_2][V(CO)_6]$ with aqueous hydrochloric acid followed by extraction with diethyl ether gives a yellow-orange ethereal solution which evolves hydrogen upon evaporation to give the neutral $V(CO)_6$,[4] a blue-green pyrophoric solid which can be purified by sublimation at 50°/15 mm. The neutral $V(CO)_6$ can also be prepared in one step from vanadium(III) chloride by carbonylation at 135°/200 atm in the presence of magnesium as a reducing agent and pyridine as a solvent followed by acidification and ether extraction.[5]

The *niobium* and *tantalum hexacarbonyl anions* of the type $M(CO)_6^{\ominus}$ (M = Nb or Ta) can be isolated as their bis(diglyme)sodium salts by carbonylation of the metal pentahalides at 105°/270 atm, in the presence of sodium metal as a reducing agent, diglyme as a solvent, and an iron compound ($FeCl_3$ or $Fe(CO)_5$) as a catalyst according to the following equations:[2,6]

$$MCl_5 + 6Na + 2\text{diglyme} + 6CO \longrightarrow [Na(diglyme)_2][M(CO)_6] + 5NaCl$$

These compounds are yellow solids which oxidize in air much more rapidly than the vanadium compound. The $M(CO)_6^{\ominus}$ anions cannot be converted to neutral $M(CO)_6$ derivatives in the case of niobium and tantalum. However, reaction of the hexacarbonyltantalate anion with alkylmercuric halides gives red volatile neutral *alkylmercury tantalum hexacarbonyl* derivatives according to the following equation:[7]

$$[Na(diglyme)_2][Ta(CO)_6] + RHgCl \longrightarrow RHgTa(CO)_6 + NaCl + 2\text{diglyme}$$

29.2.2 Substitution Products of Hexacarbonylvanadium

Reaction of $V(CO)_6$ with triphenylphosphine in hexane at 25° results in the replacement of two carbonyl groups to give *trans*-$[(C_6H_5)_3P]_2$-$V(CO)_4$, **2**.[8] Other types of Lewis bases generally react with $V(CO)_6$ to give vanadium(II) derivatives of the *hexacarbonylvanadate anion* of the type $[V(base)_n][V(CO)_6]_2$.[9]

The triphenylphosphine derivative *trans*-$[(C_6H_5)_3P]_2V(CO)_4$, **2**, can be used to prepare other triphenylphosphine derivatives of vanadium carbonyl.[8] Thus the reduction of **2** with sodium amalgam results in disproportionation to give the sodium salt $Na[(C_6H_5)_3PV(CO)_5]$. This is the most useful method for preparing Lewis base substitution products of the hexacarbonylvanadate anion, since this anion does not react with Lewis bases upon heating to give substitution products. Reaction of *trans*-$[(C_6H_5)_3P]_2V(CO)_4$, **2**, with nitric oxide results in the replacement of one triphenylphosphine ligand to give the nitrosyl derivative $(C_6H_5)_3PV(CO)_4NO$. A similar reaction of $V(CO)_6$ with nitric oxide at −78° gives a red solution containing $V(CO)_5NO$, but this compound decomposes upon warming to room temperature.

$(H_5C_6)_3P \rightarrow V(CO)_4 \leftarrow P(C_6H_5)_3$

2

29.2.3 Arene and Olefin Derivatives of Vanadium Carbonyl

Hexacarbonylvanadium (Section 2.1) reacts with aromatic hydrocarbons, including benzene and its methylated derivatives, upon warming to 35–50° to give 5 to 30% yields of the red salts

---

[2] *R. P. M. Werner, H. E. Podall*, Chem. & Ind. (London) 144 (1961).

[3] *R. B. King*, Organometal. Synth. *1*, 82 (1965).

[4] *F. Calderazzo, R. Ercoli*, Chim. Ind. (Milan) *44*, 990 (1962).

[5] *R. Ercoli, F. Calderazzo, A. Alberola*, J. Amer. Chem. Soc. *82*, 2966 (1960).

[6] *R. P. M. Werner, A. H. Filbey, S. A. Manastyrskyj*, Inorg. Chem. *3*, 298 (1964).

[7] *K. A. Keblys, M. Dubeck*, Inorg. Chem. *3*, 1646 (1964).

[8] *R. P. M. Werner*, Z. Naturforsch. *16b*, 499 (1961).

[9] *W. Hieber, J. Peterhans, E. Winter*, Chem. Ber. *94*, 2572 (1961); *W. Hieber, E. Winter, E. Schubert*, Chem. Ber. *95*, 3070 (1962).

$[(arene)V(CO)_4][V(CO)_6]$, **3**, according to the following equation:[10]

$$2V(CO)_6 + \text{arene} \longrightarrow [(\text{arene})V(CO)_4][V(CO)_6] + 2CO$$

Other salts (*e.g.*, $PF_6^{\ominus}$ and $B(C_6H_5)_4^{\ominus}$) of the $[(arene)V(CO)_4]^{\oplus}$ cations can be obtained by appropriate metathesis reactions. Furthermore, hydride addition to the $[(arene)V(CO)_4]^{\oplus}$ cations by means of sodium *hydroborate* in tetrahydrofuran solution gives *π-cyclohexadienylvanadium tetracarbonyl* derivatives.[11] For example, the benezene derivative $[C_6H_6V(CO)_4]^{\oplus}$ reacts with sodium borohydride in tetrahydrofuran solution to give the brown, volatile, unsubstituted π-cyclohexadienyl derivative $C_6H_7V(CO)_4$, **4**.

3

4

5

The reaction between cycloheptatriene and $V(CO)_6$ in boiling hexane gives two products of interest. The first product, reported[12] to be formed only when traces of air are present, is the non-ionic π-cycloheptatrienyl derivative $C_7H_7V(CO)_3$, **5**, a green volatile diamagnetic solid which is soluble in hexane.[13] The second product[14] is the hexacarbonylvanadate salt $[C_7H_7VC_7H_8][V(CO)_6]$, **6**, a paramagnetic brown solid which is insoluble in hydrocarbon solvents.

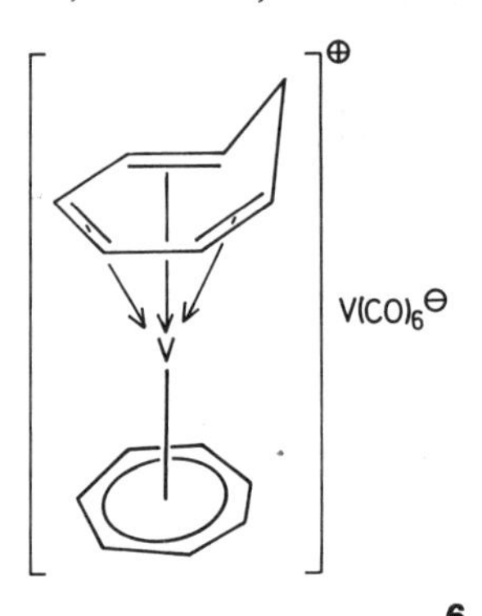

6

29.2.4 Cyclopentadienylmetal Carbonyl Derivatives

All three metals form compounds of the type $C_5H_5M(CO)_4$ (**7**, M = V, Nb, and Ta). These are orange volatile solids which apparently increase in air-stability upon going from vanadium through niobium to tantalum. The vanadium compound $C_5H_5V(CO)_4$ (**7**, M = V) can be prepared by carbonylation of $(C_5H_5)_2V$ at 120°/200 atm.[15] Addition of hydrogen to the carbon monoxide in this carbonylation reaction appears to improve the yield and simplify the product isolation. The handling of the rather air-sensitive $(C_5H_5)_2V$ in this preparation of $C_5H_5V(CO)_4$ can be avoided by using a tetrahydrofuran solution containing VCl_3 or VCl_4 and excess sodium cyclopentadienide instead of pure $(C_5H_5)_2V$ for the carbonylation reaction.[16] The niobium and tantalum derivatives $C_5H_5M(CO)_4$ (**7**, M = Nb or Ta) can similarly be prepared by carbonylation of a tetrahydrofuran solution containing the metal pentachloride and excess sodium cyclopentadienide.[17] However, in the case of tantalum derivative $C_5H_5Ta(CO)_4$ the best reported yield of product prepared by this method is only 8%.

The *cyclopentadienylmetal tetracarbonyls*, $C_5H_5M(CO)_4$ (**7**, M = V, Nb, and Ta) can also be obtained from the bis(diglyme)sodium salts of the hexacarbonylmetallates $[Na(diglyme)_2][M(CO)_6]$ (M = V, Nb, and Ta) (Section 29.2.1) by reaction with cyclopentadienylmercuric chloride in tetrahydrofuran solution at room temperature according to the following equation:[6]

$$[Na(diglyme)_2][M(CO)_6] + C_5H_5HgCl \longrightarrow C_5H_5M(CO)_4 + Hg + 2CO + NaCl + 2\,\text{diglyme}$$

---

[10] *F. Calderazzo*, Inorg. Chem. *3*, 1207 (1964).
[11] *F. Calderazzo*, Inorg. Chem. *5*, 429 (1966).
[12] *G. M. Whitesides, H. L. Mitchell*, J. Amer. Chem. Soc. *91*, 2245 (1969).
[13] *R. P. M. Werner, S. A. Manastyrskyj*, J. Amer. Chem. Soc. *83*, 2023 (1961).
[14] *F. Calderazzo, P. L. Calvi*, Chim. Ind. (Milan) *44*, 1217 (1962).
[15] *E. O. Fischer, S. Vigoureux*, Chem. Ber. *91*, 2205 (1958).
[16] *R. B. King*, Organometal. Synth. *1*, 105 (1965).
[17] *K. N. Anisimov, N. E. Kolobova, A. A. Pasynski*, Izv. Akad. Nauk. SSSR. Ser. Khim., 2238 (1969).

This reaction proceeds readily at room temperature. The yields of $C_5H_5M(CO)_4$ from $[Na(diglyme)_2]$-$[M(CO)_6]$ are moderate to high (40–80%), but the preparations of the hexacarbonylmetallate starting materials from the commercially available metal halides (Section 29.2.1) may be difficult, particularly in the case of niobium.[6] The cyclopentadienylmercuric chloride reagent required for this reaction is prepared simply by mixing mercuric chloride and sodium cyclopentadienide in a 1:1 mole ratio in tetrahydrofuran at room temperature.

The vanadium compound $C_5H_5V(CO)_4$ (**7**, M=V) can be converted to other cyclopentadienylvanadium carbonyl derivatives. Reduction of $C_5H_5V(CO)_4$ with sodium metal, either dissolved in liquid ammonia or as 1% sodium amalgam in tetrahydrofuran, gives the yellow sodium salt $Na_2[C_5H_5V(CO)_3]$ which can be converted to other salts of the $[C_5H_5V(CO)_3]^{2\ominus}$ anion by metathesis.[15] Acidification of an aqueous solution of the $[C_5H_5V(CO)_3]^{2\ominus}$ anion gives the green *bimetallic complex* $(C_5H_5)_2V_2$-$(CO)_5$, **8**.[18] This latter complex is very reactive. Thus, it is cleaved by aliphatic tertiary phosphines at room temperature according to the following equation:[18]

$$(C_5H_5)_2V_2(CO)_5 + 3R_3P \longrightarrow C_5H_5V(CO)_3PR_3 + C_5H_5V(CO)_2(PR_3)_2$$

However, the more weakly basic triphenylphosphine does not break the vanadium–vanadium bond in $(C_5H_5)_2V_2(CO)_5$, **8**; instead, the green bimetallic derivative $(C_5H_5)_2V_2(CO)_4$-$P(C_6H_5)_3$ is formed. The monometallic tertiary phosphine derivatives of the types C_5H_5V-$(CO)_3PR_3$ and sometimes $C_5H_5V(CO)_2(PR_3)_2$ can also be prepared by direct reactions of $C_5H_5V(CO)_4$ with the tertiary phosphine, either with ultraviolet irradiation or at elevated temperatures.

The carbonyl groups in $C_5H_5M(CO)_4$ (**7**, M=V or Nb) can be replaced by olefins and acetylenes. Most such reactions are performed using ultraviolet irradiation in a saturated hydrocarbon solvent. Thus ultraviolet irradiation of C_5H_5V-$(CO)_4$ with conjugated diolefins such as butadiene and 1,3-cyclohexadiene gives the red complexes $C_5H_5V(CO)_2$(diene).[19] A similar ultraviolet irradiation of $C_5H_5V(CO)_4$ with alkynes of the type RC≡CR′ gives green complexes of the type $C_5H_5V(CO)_2(RC_2R')$.[20] Several complexes have been obtained by u.v. irradiation of $C_5H_5Nb(CO)_4$ with diphenylacetylene including the monocarbonyl $C_5H_5Nb(CO)$-$(C_6H_5C_2C_6H_5)_2$.[21]

Another type of cyclopentadienylvanadium carbonyl derivative is the hexacarbonylvanadate salt $[(C_5H_5)_2V(CO)_2][V(CO)_6]$ which is prepared by reaction of $(C_5H_5)_2V$ with $V(CO)_6$ in the presence of carbon monoxide at atmospheric pressure according to the following equation:[22]

$$(C_5H_5)_2V + 2CO + V(CO)_6 \longrightarrow [(C_5H_5)_2V(CO)_2][V(CO)_6]$$

Other salts of the $[(C_5H_5)_2V(CO)_2]^{\oplus}$ cation (**9**), such as the tetraphenylborate salt, can be prepared by metathesis.

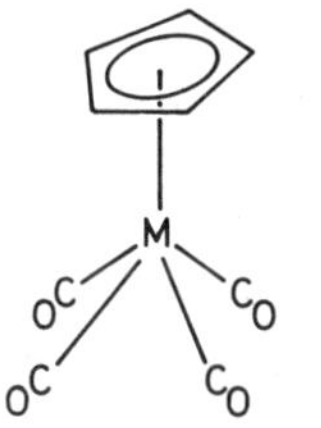

7 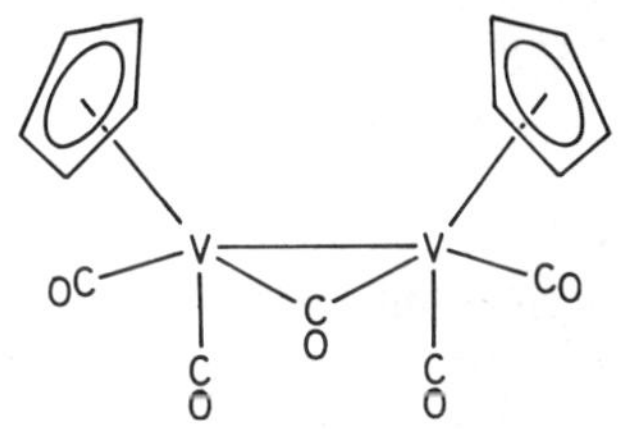 **8**

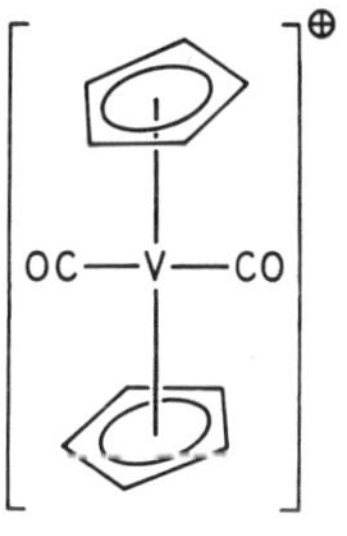

9

29.3 The Carbonyl Derivatives of Chromium, Molybdenum, and Tungsten

29.3.1 Hexacarbonyls

The metal hexacarbonyls $M(CO)_6$ (M=Cr, Mo, and W) are colorless air-stable solids which possess sufficient volatility and stability to air and water that they evaporate upon standing in air, sublime easily at room temperature, and can

[18] *E. O. Fischer, R. J. J. Schneider*, Angew. Chem. Intern. Ed., Engl. *6*, 569 (1967).

[19] *E. O. Fischer, H. P. Kögler, P. Kuzel*, Chem. Ber. *93*, 3006 (1960).

[20] *R. Tsumura, N. Hagihara*, Bull. Chem. Soc. Japan *38*, 1901 (1965).

[21] *A. N. Nesmeyanov, K. N. Anisimov, N. E. Kolobova, A. A. Pasynski*, Izv. Akad. Nauk. SSSR, Ser. Khim., 2814 (1968).

[22] *F. Calderazzo, S. Bacciarelli*, Inorg. Chem. *2*, 721 (1963).

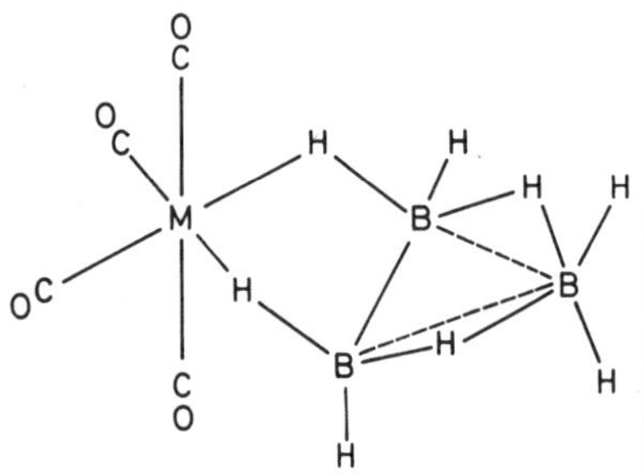

10 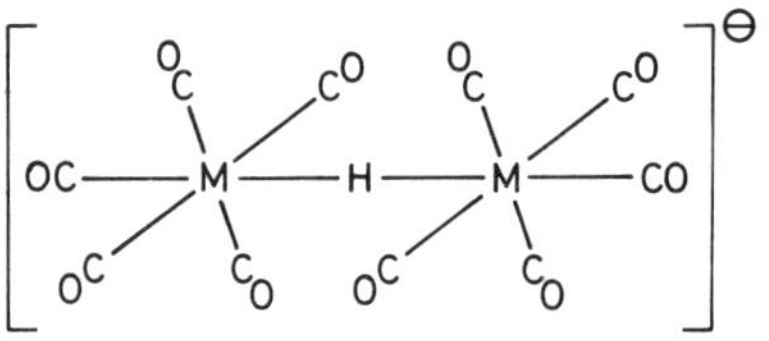 11 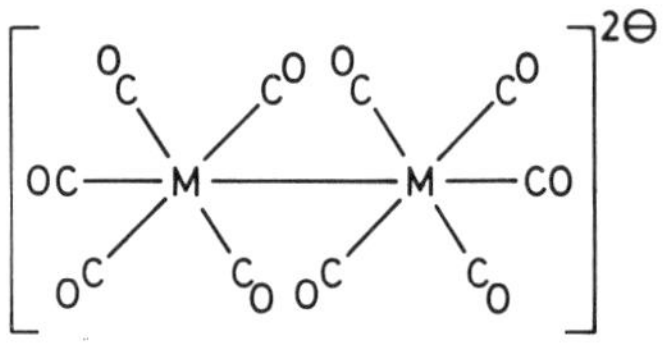 12

be purified by steam distillation. Hexacarbonylchromium can be prepared efficiently by either of the following two methods.

(1) Reduction of a pyridine solution of chromium(III) acetylacetonate with magnesium metal in the presence of carbon monoxide under pressure (140°/270 atm.) with a small amount of iodine and/or $Co_2(CO)_8$ added as a catalyst.[23]

(2) Carbonylation of a mixture of anhydrous chromium(III) chloride, anhydrous aluminum chloride, aluminum powder, and benzene at 140°/300 atm.[24]

The hexacarbonyls of molybdenum and tungsten, $M(CO)_6$ (M = Mo and W) are best prepared by reaction of a diethyl ether solution of the metal halide ($MoCl_5$ or WCl_6) with zinc at room temperature in the presence of carbon monoxide under pressure.[25] Best results are obtained in this preparation if the zinc and metal halide solution do not come into contact until the carbon monoxide pressure is reached.

The use of gaseous carbon monoxide as a source of carbonyl groups in the preparations of the hexacarbonyls of molybdenum and tungsten can be avoided by preparing these carbonyls by reaction of the halides ($MoCl_5$ or WCl_6) with pentacarbonyliron, $Fe(CO)_5$, in the presence of hydrogen under pressure.[26] In many places $Fe(CO)_5$ is an inexpensive commercial product.

In all of these preparations of the metal hexacarbonyls, the product is best isolated by hydrolysis of the reaction mixture followed by steam distillation of the metal hexacarbonyl. Further purification, if necessary, can be accomplished by sublimation.

29.3.2 Metal Carbonyl Anions and Metal Carbonyl Hydrides

Reactions of the metal hexacarbonyls $M(CO)_6$ (M = Cr, Mo, and W) with sodium hydroborate in boiling tetrahydrofuran gives the *anions* $[HM_2(CO)_{10}]^{\ominus}$ (**10**), which can be isolated as their tetraethylammonium[27] or bis(triphenylphosphine)iminium[28] salts. Ultraviolet irradiation of the metal hexacarbonyls $M(CO)_6$ (M = Cr, Mo, and W), with a dilute sodium amalgam in tetrahydrofuran solution gives the $[M_2(CO)_{10}]^{2\ominus}$ anions (**11**, M = Cr, Mo, and W) which can likewise be isolated as their tetraethylammonium[27] or bis(triphenylphosphine)iminium[28] salts. Reactions of the metal hexacarbonyls $M(CO)_6$ (M = Cr, Mo, and W) with the octahydrotriborate anion, $B_3H_8^{\ominus}$, gives the yellow anions $[M(CO)_4B_3H_8]^{\ominus}$ (**12**, M = Cr, Mo, and W), which can be isolated as their tetraethylammonium salts.[29]

29.3.3 Metal Carbonyl Halides

The metal carbonyls $M(CO)_6$ (M = Cr, Mo, and W) react with halide ions (except fluoride) with loss of one carbonyl group to form the *halopentacarbonylmetallates* according to the following equation:[30, 31]

$$M(CO)_6 + M'X \xrightarrow[65°-120°]{\text{ethers}} M'[M(CO)_5X] + CO$$

---

[23] *G. Natta, R. Ercoli, F. Calderazzo, A. Rabizzoni*, J. Amer. Chem. Soc. *79*, 3611 (1957);
R. Ercoli, F. Calderazzo, G. Bernardi, Gazz. Chim. Ital. *89*, 809 (1959).

[24] *E. O. Fischer, W. Hafner, K. Öfele*, Chem. Ber. *92*, 3050 (1959).

[25] *K. A. Kocheskov, A. N. Nesmeyanov, M. M. Nadj, I. M. Rossinskaya*, C.R. Acad. Sci. URSS *26*, 54 (1940).

[26] *A. N. Nesmeyanov, E. P. Mikheev, K. N. Anisimov, V. L. Volkov, Z. P. Valueva*, Zh. Neorg. Khim. *4*, 403, 249 (1959); C.A. *53*, 21327h, 12907c (1959).

[27] *R. G. Hayter*, J. Amer. Chem. Soc. *88*, 4376 (1966).

[28] *J. K. Ruff*, Inorg. Chem. *6*, 2080 (1967).

[29] *F. Klanberg, L. J. Guggenberger*, Chem. Comm., 1293 (1967).

[30] *E. O. Fischer, K. Öfele*, Chem. Ber. *93*, 1156 (1960).

[31] *E. W. Abel, I. S. Butler, J. G. Reid*, J. Chem. Soc. (London) 2068 (1963).

13

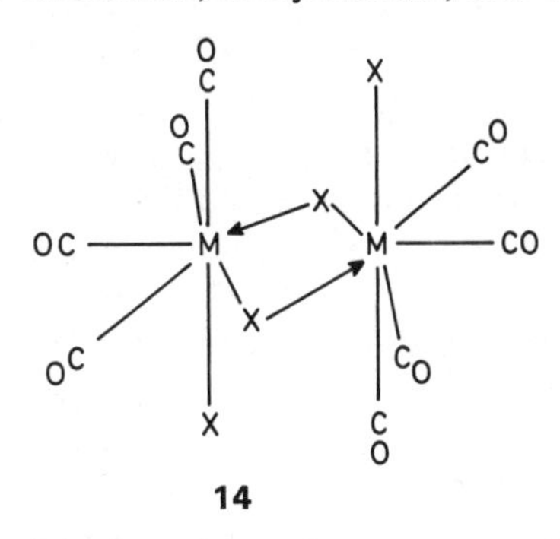

14

M = Cr, Mo, or W; M′ = *N*-methylpyridinium, tetraalkylammonium, or bis(triphenylphosphine)-iminium; ether solvents: tetrahydrofuran, dioxane, or diglyme; X = Cl, Br, I, or NCS.

A non-hydroxylic but coordinating solvent appears to be necessary for this reaction. Alkali-metal halides also react with the metal hexacarbonyls in this manner, but pure alkali metal halopentacarbonylmetallate salts cannot be isolated.

The halopentacarbonylmetallates undergo some *oxidation* and *halogenation reactions* of interest. Thus mild oxidation of the iodopentacarbonylchromate anion, $[Cr(CO)_5I]^{\ominus}$, with $Fe^{3\oplus}$ or $I_3^{\ominus}$ gives the neutral volatile thermally unstable deep blue $Cr(CO)_5I$ according to the following equation:[32]

$$Cr(CO)_5I^{\ominus} + Fe^{3\oplus} \longrightarrow Cr(CO)_5I + Fe^{2\oplus}$$

The molybdenum and tungsten anions of the type $[M(CO)_5X]^{\ominus}$ (M = Mo and W; X = Cl, Br, and I) do not undergo a similar oxidation to a neutral $M(CO)_5X$ complex. Instead they react with the free halogens, X_2^1, to form a heptacoordinate *metal(II)-tetracarbonyltrihalide complex* according to the following equation:[33, 34]

$$M(CO)_5X^{\ominus} + X_2^1 \longrightarrow [M(CO)_4X(X_2^1)]^{\ominus} + CO$$

If X and X^1 are different, mixed halogen metal carbonyl anions are obtained. The smaller chromium atom does not appear to form analogous heptacoordinate complexes.

The halopentacarbonylmetallates of molybdenum and tungsten react with allyl halides according to the following equation:[35]

$$2M(CO)_5X^{\ominus} + 3C_3H_5X^1 \longrightarrow [M_2X_3^1(C_3H_5)_2(CO)_4]^{\ominus} + X^{\ominus} + C_3H_5X + 6CO$$

The *binuclear anions* appear to have structure **13** with three bridging halides and one π-allyl group bonded to each metal atom. Chromium does not appear to form similar compounds.

Some *neutral metal carbonyl halides* of molybdenum and tungsten are also known. The metal hexacarbonyls, $M(CO)_6$ (M = Mo or W), react with chlorine or bromine at −78° to form the yellow carbonyl halides $[M(CO)_4X_2]_2$ according to the following equation:[36]

$$2M(CO)_6 + 2X_2 \longrightarrow [M(CO)_4X_2]_2 + 4CO$$

where M = Mo or W and X = Cl or Br. These binuclear complexes appear to have structure **14** with heptacoordinate metal atoms and two halogen bridges. Chromium does not seem to form similar compounds. The compounds $[M(CO)_4X_2]_2$ (**14**) react at room temperature with ligands such as triphenylphosphine and pyridine with rupture of the halogen bridges to form the deep yellow monometallic heptacoordinate complexes of the type $L_2M(CO)_3X_2$.

29.3.4 Substituted Octahedral Metal Carbonyl Derivatives

Numerous compounds can be obtained in which one or more carbonyl groups of the metal hexacarbonyls are replaced by other Lewis base ligands including tertiary phosphines,[37] tertiary arsines,[37] tertiary stibines,[37] amines,[38] pyridines,[39] nitriles,[40] isocyanides,[41] *etc.* Many such compounds, particularly those of the type

[32] *H. Behrens, H. Zizlsperger*, Z. Naturforsch. *16b*, 349 (1961).

[33] *R. B. King*, Inorg. Chem. *3*, 1039 (1964).

[34] *M. C. Ganorkhar, M. H. B. Stiddard*, J. Chem. Soc. (London), 3494 (1965).

[35] *H. D. Murdoch*, J. Organometal. Chem. *4*, 119 (1965).

[36] *R. Colton, I. B. Tomkins*, Australian J. Chem. *19*, 1143, 1519 (1966).

[37] *T. A. Magee, C. N. Matthews, T. S. Wang, J. H. Wotiz*, J. Amer. Chem. Soc. *83*, 3200 (1961).

[38] *C. S. Kraihanzel, F. A. Cotton*, Inorg. Chem. *2*, 533 (1963).

[39] *W. Hieber, F. Mühlbauer*, Z. Anorg. Allgem. Chemie *221*, 337 (1935).

[40] *B. L. Ross, J. G. Grasselli, W. M. Ritchey, H. D. Kaesz*, Inorg. Chem. *2*, 1023 (1963).

[41] *F. A. Cotton, F. Zingales*, J. Amer. Chem. Soc. *83*, 351 (1961).

$LM(CO)_5$, can be prepared by heating the metal hexacarbonyl with the ligand in an inert solvent or in a sealed tube. In some cases (*e.g.* acetonitrile[40] and pyridine[39]) excess ligand can serve as the reaction solvent. Most reactions of the metal hexacarbonyls $M(CO)_6$ (M=Cr, Mo, and W) with Lewis base ligands are carried out in the temperature range 100–150°, but in special cases they can be carried out at temperatures as low as 50° or as high as 250°. In many instances u.v. irradiation[42] can be used to effect reaction between the metal hexacarbonyl and a Lewis base ligand; this technique is particularly useful for the preparation of thermally unstable complexes.

Certain substitution products of the octahedral metal carbonyls are more reactive than the parent metal carbonyls towards Lewis base ligands. This enables the reaction between the metal carbonyl derivative and the Lewis base to be carried out under milder conditions. This is of value when thermally unstable products are synthesized. Thus the halopentacarbonylmetallates react with many Lewis bases (e.g. amines, hydrazines, and isocyanides) at or near room temperature according to the following equation:[43]

$$M(CO)_5X^{\ominus} + L \longrightarrow LM(CO)_5 + X^{\ominus}$$

Sometimes the ligands L will react further with the $LM(CO)_5$ products to replace additional carbonyl groups forming $L_2M(CO)_4$ or $L_3M(CO)_3$ compounds. Similarly, the octahydrotriboratotetracarbonylmetallate anions, $[M(CO)_4B_3H_8]^{\ominus}$ (Section 29.3.2) react with Lewis bases (e.g. PH_3, $(C_6H_5)_3P$, and certain isocyanides) according to the following equation:[44]

$$[M(CO)_4B_3H_8]^{\ominus} + 2L \longrightarrow cis\text{-}L_2M(CO)_4 + B_3H_8^{\ominus}$$

This latter reaction is stereospecific.

Olefin–metal carbonyls of chromium, molybdenum, and tungsten (Section 19.3.6) are often useful reagents for the preparation of substituted octahedral metal carbonyl derivatives by reactions with Lewis base ligands. Under sufficiently mild conditions most such reactions are stereospecific, since the Lewis base ligands must replace the coordinated olefin ligands without affecting the carbonyl groups. Thus the norbornadiene-metal tetracarbonyls, $C_7H_8M(CO)_4$ (M=Cr, Mo, and W) normally react with Lewis base ligands (*e.g.*, tertiary phosphines and isocyanides) to form *cis*-$L_2M(CO)_4$ derivatives (**15**) with complete displacement of the coordinated norbornadiene according to the following equation:[45]

$$C_7H_8M(CO)_4 + 2L \longrightarrow cis\text{-}L_2M(CO)_4 + C_7H_8$$

Similarly the cycloheptatriene–metal tricarbonyls, $C_7H_8M(CO)_3$ (M=Cr, Mo, and W) normally react with Lewis base ligands to form *fac.*-$L_3M(CO)_3$ derivatives (**16**) with complete displacement of the coordinated cycloheptatriene according to the following equation:[46]

$$C_7H_8M(CO)_3 + 3L \longrightarrow fac\text{-}L_3M(CO)_3 + C_7H_8$$

Arene–metal tricarbonyls (Section 29.3.5) react similarly.[47] In a few particularly favorable cases the bis(1,3-cyclohexadiene)-metal dicarbonyls, $(C_6H_8)_2M(CO)_2$ (M=Mo and W) react analogously with tertiary phosphines, particularly chelating ligands such as *cis*-$(C_6H_5)_2PCH{=}CHP(C_6H_5)_2$, with displacement of the coordinated 1,3-cyclohexadiene ligands but not the carbonyl ligands to give compounds of the type *cis*-$L_4M(CO)_2$.[48]

15 **16**

29.3.5 Arene-metal Tricarbonyl Derivatives

The metal hexacarbonyls $M(CO)_6$ (M=Cr, Mo, and W) react with arenes upon heating to give (arene)$M(CO)_3$ derivatives. Thus heating $Cr(CO)_6$ with various arenes at temperatures from 110° to 220° results in the formation of

[42] *W. Strohmeier*, Angew. Chem. Intern. Ed. Engl. *3*, 730 (1964).

[43] *H. D. Murdoch, R. Henzi*, J. Organometal. Chem. *5*, 166, 463 (1966).

[44] *F. Klanberg, E. L. Muetterties*, J. Amer. Chem. Soc. *90*, 3296 (1968).

[45] *J. M. Jenkins, J. G. Verkade*, Inorg. Chem. *6*, 2250 (1967).

[46] *E. W. Abel, M. A. Bennett, G. Wilkinson*, J. Chem. Soc. (London) 2323 (1959).

[47] *G. W. A. Fowles, D. K. Jenkins*, Inorg. Chem. *3*, 257 (1964).

[48] *R. B. King, L. W. Houk, P. N. Kapoor*, Inorg. Chem. *8*, 1792 (1969).

arene–chromium tricarbonyl derivatives according to the following equation.[49,50,51]

$$Cr(CO)_6 + \text{arene} \longrightarrow (\text{arene})Cr(CO)_3 + 3CO$$

Numerous arenes form *arene–chromium tricarbonyl* derivatives upon such heating with $Cr(CO)_6$. Benzene, toluene, all three xylenes, mesitylene, durene, hexamethylbenzene, aniline, dimethylaniline, anisole, phenol, and chlorobenzene all form yellow arene–chromium tricarbonyl derivatives upon heating with $Cr(CO)_6$. Benzaldehyde, benzoic acid, nitrobenzene, hexachlorobenzene, and hexafluorobenzene do not form arene–chromium tricarbonyl derivatives upon heating with $Cr(CO)_6$. However, benzaldehydetricarbonylchromium can be prepared by acid cleavage of the chromium tricarbonyl complex of benzaldehyde-diethylacetal $[C_6H_5CH(OC_2H_5)_2]Cr(CO)_3$, which can be obtained by heating the free acetal with $Cr(CO)_6$. Similarly, the benzoic acid–chromium tricarbonyl complex $(C_6H_5COOH)Cr(CO)_3$ can be prepared by alkaline hydrolysis of methylbenzoate–chromium tricarbonyl $(C_6H_5CO_2CH_3)Cr(CO)_3$ followed by acidification. The methylbenzoate–chromium tricarbonyl can be prepared by heating $Cr(CO)_6$ with methylbenzoate. The polycyclic aromatic hycrocarbons naphthalene (orange), phenanthrene (orange), anthracene (purple), and pyrene (red) form chromium tricarbonyl complexes of the indicated colors.[52]

Some *heterocyclic* analogs of the *arene–chromium tricarbonyls* can also be prepared. Thiophene reacts with $Cr(CO)_6$ to form the orange complex $C_4H_4SCr(CO)_3$ in very low yield (3%).[53] A greatly improved yield of the thiophene complex $C_4H_4SCr(CO)_3$ can be obtained by reaction of thiophene with a mixture of (α-picoline)$_3Cr(CO)_3$ or (pyridine)$_3Cr(CO)_3$ and boron trifluoride diethyl etherate at room temperature.[54] This latter method also can be used to make chromium tricarbonyl complexes of selenophene derivatives.

Molybdenum and *tungsten* also form *arene–metal tricarbonyl derivatives*, but only with a much more limited series of arenes, mainly benzene and its methylated derivatives. The acetonitrile derivative $(CH_3CN)_3W(CO)_3$ is a useful intermediate for the preparation of arene–tungsten tricarbonyls, since it reacts with arenes under much milder conditions than $W(CO)_6$.[55] This acetonitrile intermediate $(CH_3CN)_3W(CO)_3$ is readily prepared by heating $W(CO)_6$ in acetonitrile solution for five days.

29.3.6 Olefin and Acetylene Metal Carbonyl Derivatives

The following types of complexes can be prepared by reactions of carbonyl derivatives of chromium, molybdenum, and tungsten with various olefinic derivatives:[55,56]

(a) Triolefins: (triene)$M(CO)_3$: triene = cycloheptatriene, 1,3,5-cyclooctatriene, 5,6-dimethylenebicycloheptene-2, and cyclooctatetraene (only three of the four double bonds of the latter are complexed with the metal atom).
(b) Conjugated diolefins: (diene)$_2M(CO)_2$: diene = 1,3-cyclohexadiene, butadiene, and tetraphenylcyclopentadienone.
(c) Nonconjugated diolefins: (diene)$M(CO)_4$: diene = 1,5-cyclooctadiene, norbornadiene, and substituted bicyclo-2,2,2-octatrienes (two of the three double bonds are complexed with the metal atom).
(d) Unsaturated ketones: (enone)$_3$M: enone = methyl vinyl ketone.

Metal carbonyls that form the above derivatives upon reaction with appropriate olefins include the hexacarbonyls $M(CO)_6$ (M = Cr, Mo, and W), the diglyme complex $(CH_3OCH_2CH_2OCH_2CH_2OCH_3)Mo(CO)_3$,[57,58] the acetonitrile complexes $(CH_3CN)_3M(CO)_3$ (M = Cr, Mo,[59] and W[55]) and the ammine $(NH_3)_3Cr(CO)_3$. The

---

[49] *E. O. Fischer, K. Öfele, H. Essler, W. Fröhlich, J. P. Mortensen, W. Semmlinger*, Z. Naturforsch. *13b*, 458 (1958);
E. O. Fischer, K. Öfele, H. Essler, W. Fröhlich, J. P. Mortensen, W. Semmlinger, Chem. Ber. *91*, 2763 (1958).
[50] *B. Nichols, M. C. Whiting*, J. Chem. Soc. (London) 551 (1959).
[51] *G. Natta, R. Ercoli, F. Calderazzo, S. Santambrogio*, Chim. Ind. (Milan) *40*, 1003 (1958).
[52] *R. B. King, F. G. A. Stone*, J. Amer. Chem. Soc. *82*, 4557 (1960);
E. O. Fischer, N. Kriebitzsch, R. D. Fischer, Chem. Ber. *92*, 3214 (1959).
[53] *E. O. Fischer, K. Öfele*, Chem. Ber. *91*, 2395 (1958).
[54] *K. Öfele*, Chem. Ber. *99*, 1732 (1966).
[55] *R. B. King, A. Fronzaglia*, Inorg. Chem. *5*, 1837 (1966).
[56] *R. B. King*, Organometal. Synth. *1*, 122 (1965).
[57] *S. Winstein, H. D. Kaesz, C. G. Kreiter, E. C. Friedrich*, J. Amer. Chem. Soc. *87*, 3267 (1965).
[58] *H. D. Kaesz, S. Winstein, C. G. Kreiter*, J. Amer. Chem. Soc. *88*, 1319 (1966).
[59] *R. B. King*, J. Organometal Chem. *8*, 139 (1967).

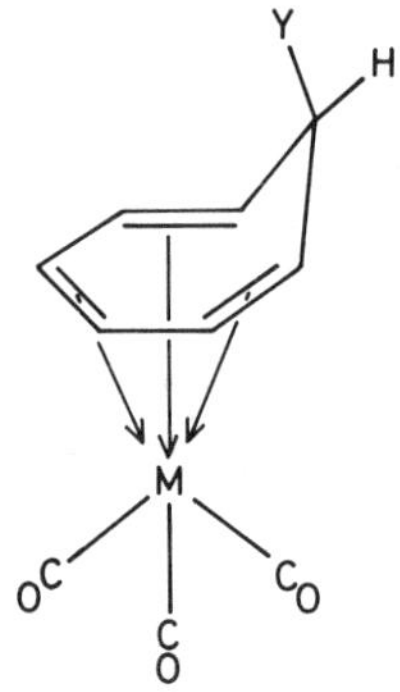

diglyme, acetonitrile, and ammonia complexes require milder reaction conditions (30° to 80°) for reactions with the olefinic derivatives than the unsubstituted metal hexacarbonyls (100° to 175°) and therefore are used for preparation of the less thermally stable complexes such as those of cyclooctatetraene. The reactivity of $W(CO)_6$ towards most diolefins and triolefins is so low that it is normally better to use the acetonitrile complex $(CH_3CN)_3W(CO)_3$ for the preparation of tungsten carbonyl complexes of diolefins and triolefins.[55]

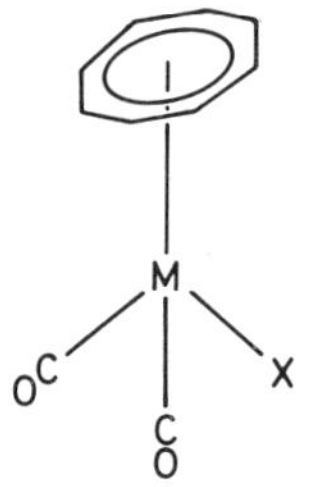

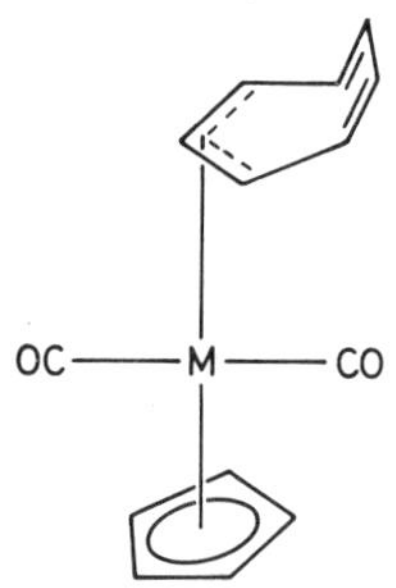

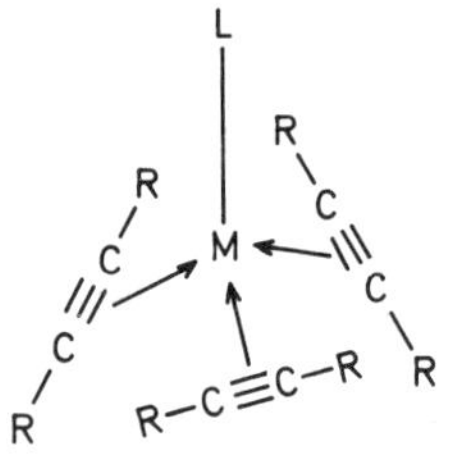

The metal tricarbonyl complexes of cycloheptatriene of formula $C_7H_8M(CO)_3$ (**17**: Y=H; M=Cr, Mo, and W) can be converted to some interesting π-cycloheptatrienyl (C_7H_7) derivatives. Reactions of the $C_7H_8M(CO)_3$ derivatives (**17**, Y=H; M=Cr, Mo, and W) with triphenylmethyl salts results in hydride abstraction to give salts of the $[C_7H_7M(CO)_3]^{\oplus}$ cations according to the following equation:[60]

$$C_7H_8M(CO)_3 + [(C_6H_5)_3C][PF_6] \longrightarrow [C_7H_7M(CO)_3][PF_6] + (C_6H_5)_3CH$$

The $[C_7H_7Cr(CO)_3]^{\oplus}$ cation reacts with some nucleophiles $Y^{\ominus}$ to give *exo*-substituted cycloheptatrienetricarbonylchromium derivatives $C_7H_7YCr(CO)_3$ (**17**, M=Cr).[61] However, with other nucleophiles (e.g. cyclopentadienide) the $[C_7H_7Cr(CO)_3]^{\oplus}$ cation either undergoes ring contraction to give the benzene derivative $C_6H_6Cr(CO)_3$[62] or coupling to form the ditropyl (bicycloheptatrienyl) derivative $(OC)_3CrC_7H_7$–$C_7H_7Cr(CO)_3$.[63] The $[C_7H_7M(CO)_3]^{\oplus}$ cations (M=Mo or W but not Cr) react with halide ions in acetone solution with carbon monoxide evolution to give the green halides $C_7H_7M(CO)_2X$ (**18**, M=Mo or W; X=Cl, Br, or I).[64] The iodides $C_7H_7M(CO)_2I$ (M=Mo or W) are useful intermediates for the preparation of other π-cycloheptatrienyl derivatives by reactions with strongly nucleophilic reagents. Thus reactions of the iodides $C_7H_7M(CO)_2I$ (**18**, M=Mo or W; X=I) with a tetrahydrofuran solution of sodium cyclopentadienide at room temperature gives the orange allylic π-cycloheptatrienyl derivative $C_5H_5M(CO)_2C_7H_7$ (**19**, M=Mo or W).[64] The iodide $C_7H_7Mo(CO)_2I$ reacts with $NaMn(CO)_5$ in tetrahydrofuran to give green $(OC)_5Mn$–$Mo(CO)_2C_7H_7$ which has a manganese–molybdenum bond holding the two parts of the molecule together. Reaction of $C_7H_7Mo(CO)_2I$ with pentafluorophenyllithium gives the σ-pentafluorophenyl derivative $C_6F_5Mo(CO)_2C_7H_7$.[65]

Reactions between alkynes and carbonyl derivatives of molybdenum and tungsten give products (**20**, L=CO or CH_3CN) in which three alkyne ligands are bonded to a single metal atom. Treatment of the acetonitrile complex $(CH_3CN)_3W(CO)_3$ with the alkynes diphenylacetylene, methylphenyiacetylene, and hexyne-2(diethylacetylene) results in the complete elimination of the coordinated acetonitrile to give pale yellow compounds of the type $(RC{\equiv}CR')_3WCO$ (**20**, M=W; L=CO).[66] However, $(CH_3CN)_3M(CO)_3$ (M=Mo or W) reacts with hexafluorobutyne-2 to give the white very stable volatile

[60] *H. J. Dauben jr., L. R. Honnen*, J. Amer. Chem. Soc. *80*, 5570 (1958).

[61] *J. D. Munro, P. L. Pauson*, J. Chem. Soc. (London) 3475 (1961).

[62] *J. D. Munro, P. L. Pauson*, J. Chem. Soc. (London) 3479 (1961).

[63] *J. D. Munro, P. L. Pauson*, J. Chem. Soc. (London) 3484 (1961).

[64] *R. B. King, M. B. Bisnette*, Inorg. Chem. *3*, 785 (1964).

[65] *M. D. Rausch, A. K. Ignatowicz, M. R. Churchill, T. A. O'Brien*, J. Amer. Chem. Soc. *90*, 3242 (1968).

[66] *D. P. Tate, J. M. Augl, W. M. Ritchey, B. L. Ross, J. G. Grasselli*, J. Amer. Chem. Soc. *86*, 3261 (1964).

complexes $(CF_3C_2CF_3)_3MNCCH_3$ (**20**, M = Mo or W; L = CH_3CN).[55]

The octahedral metal carbonyls of chromium, molybdenum, and tungsten also form carbene complexes. Reactions of the metal hexacarbonyls $M(CO)_6$ (M = Cr, Mo and W) with organolithium compounds RLi (R = CH_3 or C_6H_5) result in addition of the organolithium compound to one of the carbonyl groups to give the $[RCOM(CO)_5]^{\ominus}$ (M = Cr, Mo, and W) anions according to the following equation:[67]

$$M(CO)_6 + RLi \longrightarrow Li[(R{-}CO)M(CO)_5]$$

This anion can be isolated as its tetramethylammonium salt. Alkylation of this anion with a very strong alkylating agent such as the trialkyloxonium tetrafluoroborates give the neutral *alkoxyalkylcarbenemetal pentacarbonyl* complexes $(OC)_5MC(OR)R'$ (**21**, M = Cr, Mo, and W; R = CH_3 or C_6H_5; R′ = CH_3 or C_2H_5) according to the following equation:[67, 68]

$$[(R{-}CO)M(CO)_5]^{\ominus} + R_3{}^1O^{\oplus} \longrightarrow \{(OC)_5M[C(OR{-}R^1)]\} + R_2{}^1O$$

Other carbene–metal pentacarbonyl complexes can be prepared from the $(OC)_5M[C(OR)R^1]$ complexes (**21**) by "transesterification" reactions where the alcohol ROH is eliminated by reaction with an active hydrogen compound according to the following general equation:

$$(OC)_5M[C(OR)R^1] + Y{-}H \longrightarrow (OC)_5M[C(Y){-}R^1] + ROH$$

However, such reactions only take place when the Y group has at least one lone electron pair; otherwise the carbene complex $(OC)_5CYR^1$ is probably not stable. Thus, thiophenol[69] reacts with $(OC)_5MC(OR)R'$ derivatives (**21**) with elimination of ROH to give the carbene complexes $(OC)_5MC(EC_6H_5)R'$ (**22**, E = S). However, selenophenol reacts with $(OC)_5MC(OR)R'$ derivatives (**21**) without elimination of ROH to give the adducts $(OC)_5M \leftarrow Se(C_6H_5)[CH(OR){-}R']$ with transfer of the coordinated carbene from the transition metal to the selenium.[70]

Many amines[71] RR″NH react with $(OC)_5MC(OR)R'$ derivatives (**21**) to give aminocarbene complexes $(OC)_5MC(NRR'')R'$.

21 **22**

29.3.7.1 Cyclopentadienylmetal Carbonyls

Preparation of the green *cyclopentadienylchromium carbonyl* $[C_5H_5Cr(CO)_3]_2$ is relatively difficult but can be accomplished by either of the following two methods:

(1) Oxidation of the hydride $HCr(CO)_3C_5H_5$ (Section 29.3.7.2) under carefully controlled conditions according to the following equation:[71]

$$4HCr(CO)_3C_5H_5 + O_2 \longrightarrow 2[C_5H_5Cr(CO)_3]_2 + 2H_2O$$

This reaction is best done in a solvent such as pentane in which the $HCr(CO)_3C_5H_5$ is soluble but the $[C_5H_5Cr(CO)_3]_2$ is insoluble. Thus the $[C_5H_5Cr(CO)_3]_2$ precipitates out as it is formed providing some protection from its destruction by excess oxygen.

(2) Reaction of the anion $[C_5H_5Cr(CO)_3]^{\ominus}$ (Section 29.3.7.2) with an appropriate oxidizing agent such as tropylium bromide according to the following equation:[72]

$$2NaCr(CO)_3C_5H_5 + 2C_7H_7Br \longrightarrow [C_5H_5Cr(CO)_3]_2 + C_7H_7{-}C_7H_7 + 2NaBr$$

This second method avoids the use of the very air-sensitive $HCr(CO)_3C_5H_5$.

The molybdenum and tungsten compounds of the type $[C_5H_5M(CO)_3]_2$ (M = Mo and W) can be prepared by air oxidation of the corresponding hydrides $HM(CO)_3C_5H_5$ (Section 29.3.7.2).[73]

---

[67] *E. O. Fischer, A. Maasböl,* Chem. Ber. *100,* 2445 (1967).
[68] *E. O. Fischer, A. Maasböl,* J. Organometal. Chem. *12,* P15 (1968).
[69] *U. Klabunde, E. O. Fischer,* J. Amer. Chem. Soc. *89,* 7141 (1967).
[70] *E. O. Fischer, V. Kiener,* Angew. Chem. Intern. Ed. Engl. *6,* 961 (1967).
[71] *J. A. Connor, E. O. Fischer,* Chem. Commun. 1042 (1967);
E. Moser, E. O. Fischer, J. Organometal. Chem. *16,* 275 (1969).
[72] *R. B. King, F. G. A. Stone,* Inorg. Synth. *7,* 104 (1963).
[73] *E. O. Fischer, W. Häfner, H. O. Stahl,* Z. Anorg. Allgem. Chemie *282,* 47 (1955).

In these preparations, it is not necessary to isolate the hydrides $HM(CO)_3C_5H_5$. Instead a tetrahydrofuran solution of $NaMo(CO)_3C_5H_5$ can be prepared from sodium cyclopentadienide and the metal hexacarbonyl (Section 29.3.7.2), acidified with acetic acid, and air then bubbled through the reaction mixture to effect the oxidation. The molybdenum compound $[C_5H_5Mo(CO)_3]_2$ can also be prepared by heating $Mo(CO)_6$ with excess dicyclopentadiene at about 150°.[74] Pentamethylcyclopentadiene reacts with $Mo(CO)_6$ under similar conditions to form a red compound $[(CH_3)_5C_5Mo(CO)_2]_2$ rather than a strictly analogous $[(CH_3)_5C_5Mo(CO)_3]_2$.[75]

29.3.7.2 Cyclopentadienylmetal Carbonyl Anions and Hydrides

The *cyclopentadienylmetal tricarbonyl anions* $[C_5H_5M(CO)_3]^{\ominus}$ (M = Cr, Mo, and W) can be prepared in solution as the sodium salts $NaM(CO)_3C_5H_5$ by reactions of the metal hexacarbonyls with sodium cyclopentadienide according to the following equation:[76]

$$M(CO)_6 + NaC_5H_5 \longrightarrow NaM(CO)_3C_5H_5 + 3CO$$

These reactions are best performed in ethereal solvents ranging from boiling tetrahydrofuran (~65°) for conversion of $Mo(CO)_6$ to $NaMo(CO)_3C_5H_5$ to boiling diglyme (diethylene glycol dimethyl ether, b.p. ~160°) for conversion of $Cr(CO)_6$ to $NaCr(CO)_3C_5H_5$. Warm dimethylformamide can also be used as a solvent for this reaction.[77] If one wishes to avoid the use of sodium cyclopentadienide, the sodium salts $NaM(CO)_3C_5H_5$ (M = Cr, Mo, and W) can also be made by reduction of tetrahydrofuran solutions of the bimetallic derivatives $[C_5H_5M(CO)_3]_2$ (M = Cr, Mo, and W) with a dilute sodium amalgam,[78] but this method has the disadvantage of requiring the prior preparation of the $[C_5H_5M(CO)_3]_2$ derivatives from the metal hexacarbonyls (Section 29.3.7.1) and also may introduce mercury derivatives of the type $Hg[M(CO)_3C_5H_5]_2$, which are formed in side-reactions,[79] as possible contaminants in the reaction mixtures.

The sodium salts $NaM(CO)_3C_5H_5$ are generally not isolated but reacted further in the ethereal solvents in which they are prepared.[80] In cases where a solid derivative of the $[C_5H_5M(CO)_3]^{\ominus}$ anion is needed, the tetraethylammonium salts $[(C_2H_5)_4N][C_5H_5M(CO)_3]$ are more conveniently isolated.[81] The tetraethylammonium salts can be obtained as crystalline solids by evaporating the solvent from the preparation of the $NaM(CO)_3C_5H_5$ derivative and treating a filtered ethanolic extract of the residue with the tetraethylammonium halide.

Acidification of the $NaM(CO)_3C_5H_5$ (M = Cr, Mo, and W) salts with a non-oxidizing acid such as acetic or phosphoric acids gives the yellowish *hydrides* $HM(CO)_3C_5H_5$ (M = Cr, Mo, and W). The molybdenum and chromium derivatives $HM(CO)_3C_5H_5$ (M = Cr and Mo) oxidize almost immediately upon exposure to air and thus must be handled in an inert atmosphere.

Both of these derivatives are also somewhat thermally unstable, since they decompose at their melting points (~60°) with hydrogen evolution. The tungsten derivative $HW(CO)_3C_5H_5$ is considerably more stable since it can be handled for brief periods in the air and can be heated above its melting point; however, $HW(CO)_3C_5H_5$ readily turns red upon exposure to light.

Alternative preparations of the $HM(CO)_3C_5H_5$ (M = Cr, Mo, and W) derivatives are also available. Thus the chromium derivative $HCr(CO)_3C_5H_5$ can be prepared by reaction of $(C_5H_5)_2Cr$ with a mixture of carbon monoxide and hydrogen under pressure.[82] The tungsten derivative $HW(CO)_3C_5H_5$ can be prepared by reaction of $(CH_3CN)_3W(CO)_3$ with monomeric cyclopentadiene[55,83] in an inert solvent such as tetrahydrofuran or hexane.

29.3.7.3 Cyclopentadienylmetal Carbonyl Halides

The red *cyclopentadienylmetal carbonyl chlorides* $C_5H_5M(CO)_3Cl$ (M = Mo and W) are best prepared by reactions of the corresponding hydrides $HM(CO)_3C_5H_5$ (M = Mo and W) with carbon tetrachloride at 0° according to the following equation:[76]

$$HM(CO)_3C_5H_5 + CCl_4 \longrightarrow C_5H_5M(CO)_3Cl + CHCl_3$$

The *cyclopentadienylmetal carbonyl bromides* $C_5H_5M(CO)_3Br$ (M = Mo and W) are best prepared by reactions of the corresponding hydrides $HM(CO)_3C_5H_5$ (M = Mo and W) with *N*-bromosuccinimide according to the following equation:

$$HM(CO)_3C_5H_5 + BrNC_4H_4O_2 \longrightarrow C_5H_5M(CO)_3Br + HNC_4H_4O_2$$

---

[74] *R. B. King*, Organometal. Synth. *1*, 109 (1965).
[75] *R. B. King, M. B. Bisnette*, J. Organometal. Chem. *8*, 287 (1967).
[76] *T. S. Piper, G. Wilkinson*, J. Inorg. Nucl. Chem. *3*, 104 (1956).
[77] *E. O. Fischer*, Inorg. Synth. *7*, 136 (1963).
[78] *R. G. Hayter*, Inorg. Chem. *2*, 1031 (1963).
[79] *R. B. King*, J. Inorg. Nucl. Chem. *25*, 1296 (1963).
[80] *R. B. King*, Advan. Organometal. Chem. *2*, 157 (1964).
[81] *R. B. King, M. B. Bisnette, A. Fronzaglia*, J. Organometal. Chem. *5*, 341 (1966).
[82] *E. O. Fischer, W. Häfner*, Z. Naturforsch. *10b*, 140 (1954).
[83] *S. A. Keppie, M. F. Lappert*, J. Organometal. Chem. *19*, P5 (1969).

The required $HM(CO)_3C_5H_5$ starting materials for these reactions need not be isolated but instead can be used as generated in tetrahydrofuran solution by acidification with acetic acid of the solution obtained by boiling $M(CO)_6$ (M = Mo and W) with a slight excess of sodium cyclopentadienide in tetrahydrofuran.

The red–black *iodides* $C_5H_5M(CO)_3I$ (M = Mo or W) are best prepared by reactions of iodine with either $HM(CO)_3C_5H_5$ or $[C_6H_5M(CO)_3]_2$ according to one of the following equations:[76,84]

$$[C_5H_5M(CO)_3]_2 + I_2 \longrightarrow 2C_5H_5M(CO)_3I$$

$$HM(CO)_3C_5H_5 + I_2 \longrightarrow C_5H_5M(CO)_3I + HI$$

Reaction of the molybdenum derivative $[C_5H_5Mo(CO)_3]_2$ with excess iodine or bromine in benzene solution results in the loss of some carbon monoxide to give the brown *trihalides* $C_5H_5Mo(CO)_2X_3$ (X = Br or I) according to the following equation:[85]

$$[C_5H_5Mo(CO)_3]_2 + 3X_2 \longrightarrow 2C_5H_5M(CO)_2X_3 + 2CO$$

29.3.7.4 Substituted Cyclopentadienylmetal Carbonyl Derivatives

The molybdenum compound $[C_5H_5Mo(CO)_3]_2$ reacts with triphenylphosphine to form a monosubstituted derivative $(C_5H_5)_2Mo_2(CO)_5P(C_6H_5)_3$.[86] Ultraviolet irradiation of $[C_5H_5Mo(CO)_3]_2$ with triphenylphosphite gives the red nearly insoluble disubstituted derivative $[C_5H_5Mo(CO)_2P(OC_6H_5)_3]_2$;[87] the molybdenum–molybdenum bond in this latter compound is cleaved by sodium amalgam in tetrahydrofuran solution to give the sodium salt $Na[C_5H_5Mo(CO)_2P(OC_6H_5)_3]$. This sodium salt undergoes reactions similar to those of the unsubstituted sodium salt $NaMo(CO)_3C_5H_5$ (Sections 29.3.7.2 and 29.3.7.6). The hydride $HMo(CO)_3C_5H_5$ (Section 29.3.7.2) reacts with dimethyldisulfide at room temperature to give the *bimetallic derivative* $[C_5H_5Mo(CO)_2SCH_3]_2$.[88]

The $C_5H_5M(CO)_3X$ (M = Mo or W; X = Cl, Br, or I) halides react with various Lewis base ligands (*e.g.*, amines, tertiary phosphines, tertiary arsines, isocyanides, *etc.*) in one of the following two ways:[89]

(1) Replacement of carbonyl groups with the ligand introduced forming a non-ionic derivative according to the following equation:

$$C_5H_5M(CO)_3X + L \longrightarrow C_5H_5M(CO)_2LX + CO$$

This type of reaction occurs particularly readily with phosphorus ligands such as $[(CH_3)_2N]_3P$ and $(C_6H_5)_3P$.

(2) Replacement of the halogen with the ligand introduced without carbon monoxide evolution forming an ionic derivative according to the following equation:

$$C_5H_5M(CO)_3X + L \longrightarrow [C_5H_5M(CO)_3L]X$$

This type of reaction is favored if a Lewis acid catalyst, such as anhydrous aluminum chloride, is added to the reaction mixture. The complex cations formed in this reaction can often be conveniently isolated as hexafluorophosphates by hydrolysis of the reaction mixture followed by addition of ammonium hexafluorophosphate.

The yellow *cyclopentadienylmetal tetracarbonyl cations* $[C_5H_5M(CO)_4]^{\oplus}$ (M = Mo or W) can be prepared by reaction of a mixture of $C_5H_5M(CO)_3Cl$ and anhydrous aluminum chloride with carbon monoxide.[90] Similarly, the red *benzenecyclopentadienylmetal carbonyl cations* $[C_5H_5M(CO)C_6H_6]^{\oplus}$ can be synthesized by addition of $C_5H_5M(CO)_3Cl$ to boiling benzene containing aluminum chloride.[91] The cations $[C_5H_5M(CO)_4]^{\oplus}$ and $[C_5H_5M(CO)C_6H_6]^{\oplus}$ are both conveniently isolated as hexafluorophosphate salts.

29.3.7.5 Nitrosyl and Arylazo Derivatives of Cyclopentadienylmetal Carbonyls

The chromium compound $C_5H_5Cr(CO)_2NO$ can be prepared by treatment of $[C_5H_5Cr(CO)_3]_2$ (Section 29.3.7.1) with nitric oxide in benzene solution at room temperature.[92] An analogous

---

[84] *E. W. Abel, A. Singh, G. Wilkinson*, J. Chem. Soc. (London) 1321 (1960).

[85] *R. J. Haines, R. S. Nyholm, M. H. B. Stiddard*, J. Chem. Soc. *A*, 1606 (1966);
M. L. H. Green, W. E. Lindsell, J. Chem. Soc. *A*, 686 (1967).

[86] *K. W. Barnett, P. M. Treichel*, Inorg. Chem. *6*, 294 (1967).

[87] *R. J. Haines, R. S. Nyholm, M. H. B. Stiddard*, J. Chem. Soc. *A*, 43 (1968);
R. B. King, K. H. Pannell, Inorg. Chem. *7*, 2356 (1968).

[88] *P. M. Treichel, J. H. Morris, F. G. A. Stone*, J. Chem. Soc. 720 (1963).

[89] *P. M. Treichel, K. W. Barnett, R. L. Shubkin*, J. Organometal. Chem. *7*, 449 (1967).

[90] *E. O. Fischer, K. Fichtel, K. Öfele*, Chem. Ber. *95*, 3172 (1962).

[91] *E. O. Fischer, F. J. Kohl*, Z. Naturforsch. *18b*, 504 (1963).

[92] *E. O. Fischer, O. Beckert, W. Hafner, H. O. Stahl*, Z. Naturforsch. *10b*, 598 (1955).

preparation does not work for the molybdenum and tungsten compounds $C_5H_5M(CO)_2NO$ (M = Mo or W). However, the molybdenum and tungsten compounds can be prepared by reaction of the sodium salt $NaM(CO)_3C_5H_5$ (M = Mo or W) or the hydride $HM(CO)_3C_5H_5$ with the nitrosating reagent Diazald® (*N*-methyl-*N*-nitroso-*p*-toluenesulfonamide).[76,93] If Diazald® is not available, trifluoroacetyl nitrite (nitrosyl trifluoroacetate, $CF_3C(O)ONO$) or sodium nitrite in the presence of acetic acid can be used to nitrosate $NaM(CO)_3C_5H_5$ to give $C_5H_5M(CO)_2NO$, but with some loss in yield.

The *cation* $[C_5H_5Cr(NO)_2CO]^{\oplus}$ can be obtained by reaction of $C_5H_5Cr(NO)_2Cl$ with carbon monoxide under pressure in the presence of anhydrous aluminum chloride.[94] It is conveniently isolated as its green–brown hexafluorophosphate salt.

Arylazo compounds of cyclopentadienylmolybdenum carbonyl of the type $RN_2Mo(CO)_2C_5H_5$ (**23**, R = phenyl, *p*-tolyl, *p*-nitrophenyl, *p*-anisyl, etc.) can be prepared by reactions of the aryldiazonium salts $RN_2^{\oplus}BF_4^{\ominus}$ with $NaMo(CO)_3C_5H_5$ in tetrahydrofuran at −78°.[95] Similar compounds are obtained by reactions of the hydrides $HM(CO)_3C_5H_5$ (Section 29.3.7.2), M = Mo and W) with arylhydrazines at elevated temperatures.[96]

23

29.3.7.6 Cyclopentadienylmetal Carbonyl Derivatives Containing Additional Metal–Carbon or Metal–Metal Bonds

The anions $[C_5H_5M(CO)_3]^{\ominus}$ (M = Mo or W; occasionally also Cr) react with a variety of inorganic and organic halides to give non-ionic $RM(CO)_3C_5H_5$ derivatives (**24**) according to the following equation:[76]

$$[C_5H_5M(CO)_3]^{\ominus} + RX \longrightarrow RM(CO)_3C_5H_5 + X^{\ominus}$$

In most cases the sodium salts of the anions $[C_5H_5M(CO)_3]^{\ominus}$ as obtained by heating the metal hexacarbonyl with sodium cyclopentadienide in an ethereal solvent (Section 29.3.7.2) are used for syntheses of this type. This synthetic technique may be used for the preparation of numerous compounds with molybdenum or tungsten σ-bonded to carbon, germanium, tin, lead, mercury, gold, and other transition metals. Halides reacting according to the above equation include methyl iodide, ethyl bromide, allyl chloride, benzyl chloride, chloromethyl methylsulfide, heptafluorobutyryl chloride, 1,3-dibromopropane (one bromine only), heptafluorobutyryl chloride, trimethyltin chloride, triphenyllead chloride, $(C_6H_5)_3PAuCl$, and $C_5H_5Fe(CO)_2I$. The $RM(CO)_3C_5H_5$ compounds (**24**) where R is bonded to the molybdenum or tungsten atom through carbon, germanium, tin, or lead are pale yellow solids which can be handled in the air for at least brief periods and which can be purified by vacuum sublimation. The tungsten derivatives are more stable than the molybdenum derivatives, both to air oxidation and to thermal decomposition. The chromium derivatives $RCr(CO)_3C_5H_5$ are too unstable for isolation in most cases.

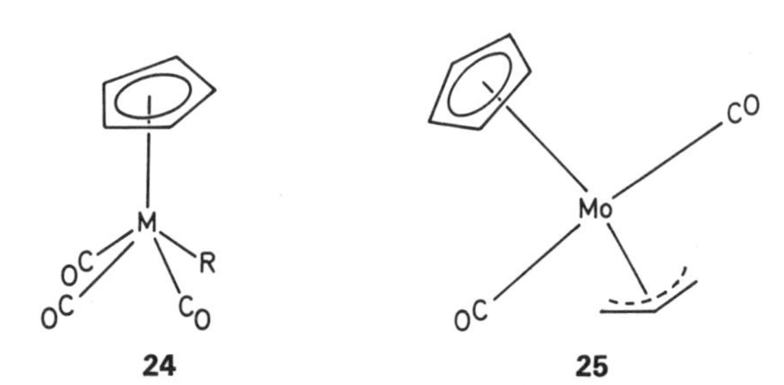

24 **25**

Perfluoroalkyl derivatives of the type $R_fM(CO)_3C_5H_5$ (**24**, R = perfluoroalkyl = R_f) cannot be prepared by reaction of the anions $[C_5H_5M(CO)_3]^{\ominus}$ with perfluoroalkyl halides. The molybdenum derivatives $R_fMo(CO)_3C_5H_5$ (**24**, R = CF_3 or C_3F_7) can be prepared by thermal decarbonylation at ~120° of the corresponding acyl derivatives $R_fCOMo(CO)_3C_5H_5$.[97]

Some compounds of the type $RM(CO)_3C_5H_5$ form interesting derivatives when treated with tertiary phosphines and similar ligands. Such reactions proceed particularly readily in acetonitrile solution.[98] Thus

[93] *R. B. King*, Inorg. Chem. *6*, 30 (1967).
[94] *E. O. Fischer, P. Kuzel*, Z. Anorg. Allgem. Chemie *317*, 226 (1962).
[95] *R. B. King, M. B. Bisnette*, Inorg. Chem. *5*, 300 (1966).
[96] *M. L. H. Green, T. R. Sanders, R. N. Whiteley*, Z. Naturforsch. *23b*, 106 (1968).
[97] *R. B. King, M. B. Bisnette*, J. Organometal. Chem. *2*, 15 (1964).
[98] *P. J. Craig, M. Green*, J. Chem. Soc. *A*, 1978 (1968).

reactions of $CH_3Mo(CO)_3C_5H_5$ with tertiary phosphines in acetonitrile solution give the substituted acetylmolybdenum derivatives $CH_3COMo(CO)_2(PR_3)$-(C_5H_5). The corresponding ethyl and benzylmolybdenum derivatives react similarly to form propionyl and phenylacetyl derivatives, respectively. However, the trifluoromethyl derivative $CF_3Mo(CO)_3C_5H_5$ (24: $R=CF_3$) does not form acyl derivatives when reacted similarly with tertiary phosphines.[99] Instead, photochemical reactions of $CF_3Mo(CO)_3C_5H_5$ with tertiary phosphines result only in replacement of carbonyl groups to give derivatives of the type $CF_3Mo(CO)_2$-$(PR_3)(C_5H_5)$.

Some $RM(CO)_3C_5H_5$ derivatives (**24**) can be converted to interesting *cyclopentadienylmolybdenum dicarbonyl derivatives.* Thus photolysis of the σ-allyl derivative $C_3H_5M(CO)_3C_5H_5$ (**24**, M=Mo or W; $R=CH_2{=}CHCH_2$) results in the elimination of one equivalent of carbon monoxide to give the corresponding yellow π-allyl derivative $C_3H_5M(CO)_2C_5H_5$ (**25**, M=Mo or W).[100,101] However, pyrolysis of C_3H_5Mo $(CO)_3C_5H_5$ (**24**, M=Mo; $R=CH_2{=}CHCH_2$) results only in complete elimination of the allyl group giving the coupling product $[C_5H_5$-$Mo(CO)_3]_2$. A similar photolysis or pyrolysis of the σ-benzyl derivative $C_6H_5CH_2Mo(CO)_3$-C_5H_5 (**24**, M=Mo; $R=C_6H_5CH_2$) gives a low yield of the red π-benzyl derivative $C_6H_5CH_2$-$Mo(CO)_2C_5H_5$ (**26**).[102] Photolysis or preferably

26

pyrolysis of the sulfur compound CH_3SCH_2M-$(CO)_3C_5H_5$ (**24**, M=Mo or W; $R=CH_3SCH_2$) results in the elimination of one equivalent of carbon monoxide to give the dicarbonyl CH_3-$SCH_2M(CO)_2C_5H_5$ of structure **27** (M=Mo or W).[103]

27

The anion $[C_5H_5Mo(CO)_3]^{\ominus}$ does *not* react with chlorosilanes of the type R_3SiCl to give R_3SiMo-$(CO)_3C_5H_5$ derivatives. However, some R_3SiMo-$(CO)_3C_5H_5$ derivatives (*e.g.*, $R=CH_3$) have been prepared according to the following equation:[104]

$$HMo(CO)_3C_5H_5 + R_3SiNR'_2 \longrightarrow R_3SiMo(CO)_3C_5H_5 + R'_2NH$$

29.4 Carbonyl Derivatives of Manganese, Technetium, and Rhenium

29.4.1 The Metal Carbonyls

Decacarbonyldimanganese, $Mn_2(CO)_{10}$, is a yellow crystalline solid which is sufficiently water-stable and volatile to be purified by steam distillation. It can be prepared by any of the following three methods:

(1) Reaction of the commercially available $CH_3C_5H_4Mn(CO)_3$ with sodium metal in diglyme solution in the presence of carbon monoxide. This reaction can be run at atmospheric pressure[105] but better yields are obtained when the reaction is run at elevated pressures.[106]
(2) Reaction of anhydrous manganese(II) acetate in diisopropyl ether with excess triethylaluminum or triisobutylaluminum in the presence of carbon monoxide under pressure.[107]
(3) Reaction of anhydrous manganese(II) chloride with sodium benzophenone ketyl in

[99] *R. B. King, R. N. Kapoor, K. H. Pannell*, J. Organometal. Chem. *20*, 187 (1969).
[100] *M. Cousins, M. L. H. Green*, J. Chem. Soc. 889 (1963).
[101] *M. L. H. Green, A. N. Stear*, J. Organometal. Chem. *1*, 230 (1964).
[102] *R. B. King, A. Fronzaglia*, J. Amer. Chem. Soc. *88*, 709 (1966).
[103] *R. B. King, M. B. Bisnette*, Inorg. Chem. *4*, 486 (1965).
[104] *D. J. Cardin, M. F. Lappert*, Chem. Commun. 506 (1966).
[105] *R. B. King, J. C. Stokes, T. F. Korenowski*, J. Organometal. Chem. *11*, 641 (1968).
[106] *H. E. Podall, A. P. Giraitis*, J. Org. Chem. *26*, 2587 (1961).
[107] *H. E. Podall, J. H. Dunn, H. Shapiro*, J. Amer. Chem. Soc. *82*, 1325 (1960);
F. Calderazzo, Inorg. Chem. *4*, 293 (1965).

tetrahydrofuran solution followed by treatment of the resulting manganese ketyl solution with carbon monoxide under pressure.[108]

In the cases of the first and third method listed above hydrolysis with sulfuric or phosphoric acid is necessary to liberate the product. The first method listed above which uses $CH_3C_5H_4Mn(CO)_3$ is the preferred method when this manganese starting material is readily available and reasonably inexpensive. If $CH_3C_5H_4$-$Mn(CO)_3$ is not readily available, then the second method listed above, which uses aluminum alkyls, is preferred. If aluminum alkyls are neither readily available nor safely handled in the available facilities, then the third method using sodium benzophenone ketyl is the preferred method.

The *dimetal decarbonyls of technetium*[109] and *rhenium*[110] are most conveniently prepared by reactions of the metal oxides (*e.g.*, TcO_2 or Re_2O_7) or their oxoacid salts such as $KReO_4$ with carbon monoxide at elevated temperatures and pressures according to the following equations:

$$2TcO_2 + 14CO \longrightarrow Tc_2(CO)_{10} + 4CO_2$$

$$Re_2O_7 + 17CO \longrightarrow Re_2(CO)_{10} + 7CO_2$$

Conversion of oxoacid salts such as $KReO_4$ to the corresponding dimetal decacarbonyls requires higher temperatures and pressures than conversion of the corresponding metal oxides to the corresponding dimetal decacarbonyls. An alternative preparation of $Re_2(CO)_{10}$ utilizes the reaction of rhenium pentachloride with sodium metal in diglyme solution in the presence of carbon monoxide under pressure.[111]

29.4.2 Metal Carbonyl Anions and Metal Carbonyl Hydrides

The dimetal decacarbonyls $M_2(CO)_{10}$ (M = Mn and Re) react with dilute sodium amalgam in tetrahydrofuran solution to give the sodium *pentacarbonylmetallates* according to the following equation:[112]

$$M_2(CO)_{10} + 2Na \longrightarrow 2NaM(CO)_5$$
$$(M = Mn \text{ or } Re).$$

The resulting tetrahydrofuran solutions of the sodium salts $NaM(CO)_5$ can be used for many preparative purposes.

Reactions of the sodium salts $NaM(CO)_5$ (M = Mn or Re) with a strong non-volatile non-oxidizing acid such as phosphoric acid gives the volatile liquid *hydrides* $HM(CO)_5$ (M = Mn or Re). The $HM(CO)_5$ (M = Mn or Re) derivatives are weak acids, stable thermally to about 100°, but oxidized rapidly by air to form the corresponding $M_2(CO)_{10}$ derivative. The volatility of the $HM(CO)_5$ (M = Mn or Re) derivatives is sufficient for them to be handled by high-vacuum techniques.

Some *trimetallic carbonyl hydrides* of manganese, technetium, and rhenium are known. Reaction of $Mn_2(CO)_{10}$ with excess concentrated aqueous potassium hydroxide at 60° gives a dark green solution which upon acidification with concentrated phosphoric acid gives dark red $[HMn(CO)_4]_3$ (**28**, M = Mn) which can be purified by vacuum sublimation.[113] The technetium analog $[HTc(CO)_4]_3$ (**28**, M = Tc) is the major product obtained when $Tc_2(CO)_{10}$ is reduced with sodium amalgam in tetrahydrofuran and the resulting reaction mixture acidified with phosphoric acid; only traces of $HTc(CO)_5$ are obtained from this reaction.[114] The rhenium compound $[HRe(CO)_4]_3$ (**28**, M = Re) is obtained by reaction of $Re_2(CO)_{10}$ with sodium hydroborate in boiling tetrahydrofuran followed by acidification of the reaction mixture with phosphoric acid.[115]

A similar reaction of $Mn_2(CO)_{10}$ with sodium hydroborate in boiling tetrahydrofuran gives the red, volatile boron–manganese derivative $HB_2H_6Mn_3(CO)_{10}$ (**29**).[116]

Another type of trimetallic rhenium carbonyl hydride is the compound $HRe_3(CO)_{14}$ (**30**, M = M′ = Re) which can be prepared by reaction of $Re_2(CO)_{10}$ with sodium hydroborate in boiling tetrahydrofuran followed by addition of a preformed tetrahydrofuran solution of $NaRe(CO)_5$ (see above) prior to acidification with phosphoric acid.[117] Modifications of this synthetic procedure lead to *mixed metal carbonyl hydrides* of the

---

[108] *R. D. Closson, L. R. Buzbee, G. C. Ecke*, J. Amer. Chem. Soc. *80*, 6167 (1958).
[109] *J. C. Hileman, D. K. Huggins, H. D. Kaesz*, Inorg. Chem. *1*, 933 (1962).
[110] *W. Hieber, H. Fuchs*, Z. Anorg. Allgem. Chemie *248*, 256 (1941).
[111] *A. Davison, J. A. McCleverty, G. Wilkinson*, J. Chem. Soc. 1133 (1963).
[112] *W. Hieber, G. Wagner*, Z. Naturforsch. *12b*, 478 (1957); *13b*, 339 (1958).
[113] *B. F. G. Johnson, R. D. Johnston, J. Lewis, B. H. Robinson*, J. Organometal. Chem. *10*, 105 (1967).
[114] *H. D. Kaesz, D. K. Huggins*, Can. J. Chem. *41*, 1250 (1963).
[115] *D. K. Huggins, W. Fellmann, J. M. Smith, H. D. Kaesz*, J. Amer. Soc. *86*, 4841 (1964).
[116] *H. D. Kaesz, W. Fellmann, G. R. Wilkes, L. F. Dahl*, J. Amer. Chem. Soc. *87*, 2753 (1965).
[117] *W. Fellmann, H. D. Kaesz*, Inorg. Nucl. Chem. Lett. *2*, 63 (1966).

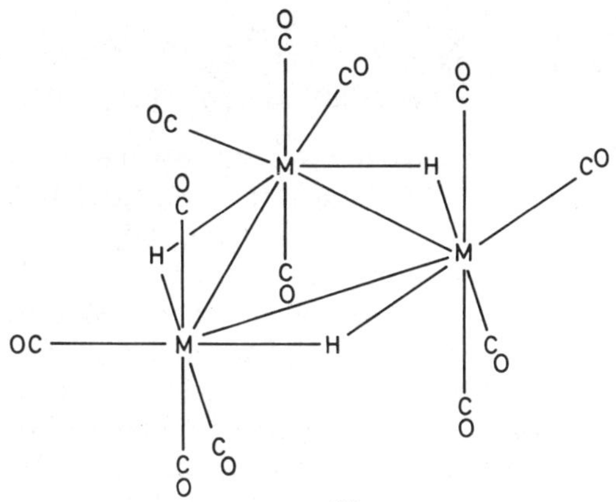

28

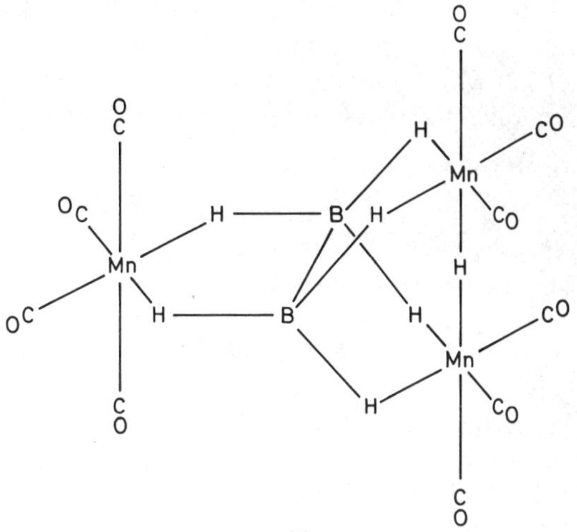

29

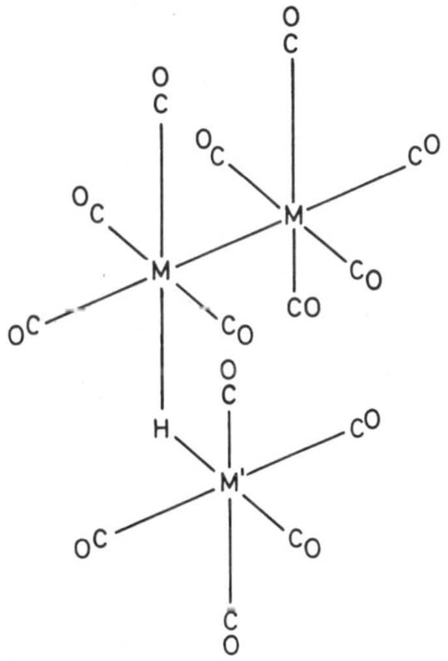

30

type $HM_2M'(CO)_{14}$. Thus the reaction of $Re_2(CO)_{10}$ with sodium hydroborate followed by addition of a preformed tetrahydrofuran solution of $NaMn(CO)_5$ gives a mixture which reacts with phosphoric acid to liberate the mixed trimetallic carbonyl hydride $HRe_2Mn(CO)_{14}$ (**30**, M=Re; M'=Mn).

The reaction of $Re_2(CO)_{10}$ with sodium hydroborate in boiling tetrahydrofuran without subsequent acidification gives a complex mixture of polymetallic rhenium carbonyl anions from which $[H_2Re_3(CO)_{12}]^{\ominus}$, $[H_6Re_4(CO)_{12}]^{2\ominus}$, and $[Re_4(CO)_{16}]^{2\ominus}$ can be isolated as tetraalkylammonium or tetraarylphosphonium salts.[118]

[118] *M. R. Churchill, R. Bau*, Inorg. Chem. *7*, 2606 (1968);
M. R. Churchill, P. H. Bird, H. D. Kaesz, R. Bau, B. Fontal, J. Amer. Chem. Soc. *90*, 7135 (1968);
H. D. Kaesz, B. Fontal, R. Bau, S. W. Kirtley, M. R. Churchill, J. Amer. Chem. Soc. *91*, 1021 (1969).

29.4.3 Metal Carbonyl Halides

Manganese, technetium, and rhenium all form monometallic carbonyl halides of the type $M(CO)_5X$ (M=Mn, Tc, or Re; X=Cl, Br, or I, but not F). The *chlorides* and *bromides* can be prepared by cleaving the metal–metal bonds in the dimetal decacarbonyl with the free halogens according to the following equation:[119]

$$M_2(CO)_{10} + X_2 \rightarrow 2M(CO)_5X$$
(M = Mn, Tc, or Re; CX = Cl or Br)

This reaction occurs readily with chlorine or bromine at room temperature in an unreactive solvent such as carbon tetrachloride or dichloromethane. However, iodine requires heating to react with the dimetal decacarbonyls. The *iodides*

[119] *E. W. Abel, G. Wilkinson*, J. Chem. Soc. (London) 1501 (1959).

$M(CO)_5I$ are therefore better prepared by reactions of the sodium salts $NaM(CO)_5$ (Section 29.4.2) with iodine in tetrahydrofuran solution according to the following equation:

$$NaM(CO)_5 + I_2 \longrightarrow M(CO)_5I + NaI$$

The technetium and rhenium derivatives of the type $M(CO)_5X$ (M = Tc or Re; X = Cl, Br, or I) can also be prepared by carbonylation of a mixture of the appropriate dipotassium hexahalometallates(IV), K_2MX_6, with excess copper powder at elevated temperatures and pressures.[120] This latter method avoids the need to prepare the dimetal decacarbonyls $M_2(CO)_{10}$ (M = Tc or Re) in an intermediate step.

Bimetallic manganese, technetium, and rhenium *carbonyl halides* of the type $[M(CO)_4X]_2$ (**31**, M = Mn, Tc, or Re; X = Cl, Br, or I)

31

can be prepared by pyrolysis of the corresponding monometallic derivatives $M(CO)_5X$ at temperatures up to ~100° according to the following equation:[119]

$$2M(CO)_5X \longrightarrow [M(CO)_4X]_2 + 2CO$$

29.4.4 Chalcogen Derivatives of Manganese, Technetium, and Rhenium Carbonyls

The bimetallic derivatives $[RSMn(CO)_4]_2$ (**32**, R = CH_3, C_2H_5, etc.; M = Mn)

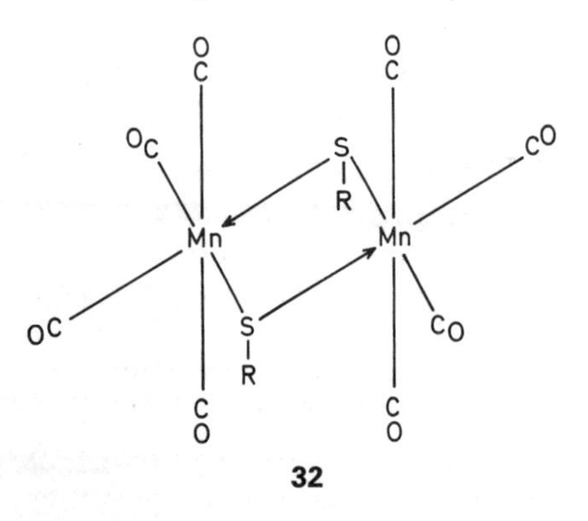

32

are prepared by reaction of $HMn(CO)_5$ (Section 29.4.2) with dialkyl disulfides according to the following equation:[88]

$$2HMn(CO)_5 + 2RSSR \longrightarrow [RSMn(CO)_4]_2 + 2CO + 2RSH$$

Rhenium carbonyl mercaptides of the types $RSRe(CO)_5$, $[RSRe(CO)_4]_2$, and $[RSRe(CO)_3]_n$ are prepared by reaction of $HRe(CO)_5$ or $Re(CO)_5Cl$ with thiols under various conditions.[121] Reaction of $[Mn(CO)_4Br]_2$ (Section 4.3) with disodium maleonitriledithiolate, $Na_2S_2C_2(CN)_2$, in methanol solution gives the brick red anion $[Mn(CO)_4S_2C_2(CN)_2]^{\ominus}$ (**33**), which may be isolated as its methyltriphenylphosphonium salt.[122] A somewhat similar reaction of $Mn(CO)_5Br$ with disodium ethylenedithiolate, $Na_2S_2C_2H_2$, gives a neutral red bimetallic derivative $H_2C_2S_2Mn_2(CO)_6$ (**34**).[123]

Many σ-alkyl manganese carbonyl derivatives $RMn(CO)_5$ (R = methyl, benzyl, *etc.*) (Section 29.4.8) react with liquid SO_2 at its boiling point (−10°) or with gaseous SO_2 in an inert solvent to give *manganese pentacarbonyl S-alkylsulfinates*, $RSO_2Mn(CO)_5$ (**35**), according to the following equation:[124]

$$RMn(CO)_5 + SO_2 \longrightarrow RSO_2Mn(CO)_5$$

Pentacarbonylmanganese derivatives, $RMn(CO)_5$, with very electronegative R groups (*e.g.* $CF_3Mn(CO)_5$) do not undergo this reaction because of the strength of the manganese–carbon σ-bond.

29.4.5 Manganese Carbonyl Nitrosyls

The two known manganese carbonyl nitrosyls are red $Mn(CO)_4NO$, m.p. 0°, and green $MnCO(NO)_3$, m.p. 27°. Both of these compounds are very volatile and can be handled in a conventional vacuum system. The red $Mn(CO)_4NO$ can be obtained by reaction of $HMn(CO)_5$ (Section 29.4.2) with Diazald (*N*-methyl-*N*-nitroso-*p*-toluenesulfonamide) in diethyl ether solution at room temperature.[125] A frequently

---

[120] *W. Hieber, R. Schuh, H. Fuchs,* Z. Anorg. Allgem. Chemie *248*, 243 (1941).

[121] *A. G. Osborne, F. G. A. Stone,* J. Chem. Soc. *A*, 1143 (1966).

[122] *J. Locke, J. A. McCleverty,* Chem. Commun. *102* (1965).

[123] *R. B. King, C. A. Eggers,* Inorg. Chem. *7*, 1214 (1968).

[124] *F. A. Hartman, A. Wojcicki,* Inorg. Chem. *7*, 1504 (1968).

[125] *P. M. Treichel, E. Pitcher, R. B. King, F. G. A. Stone,* J. Amer. Chem. Soc. *83*, 2593 (1961).

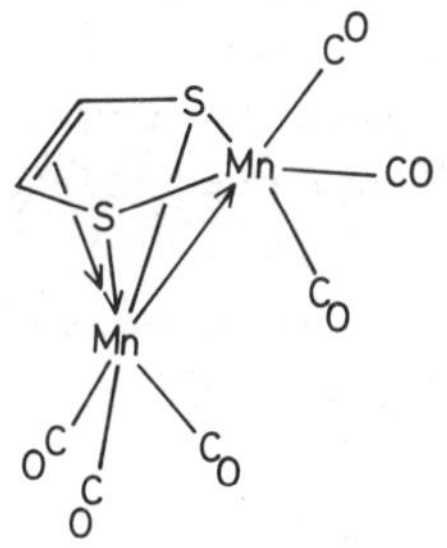

33 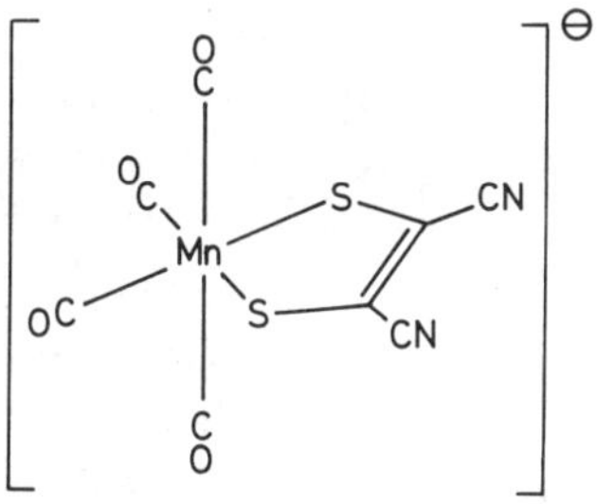 34 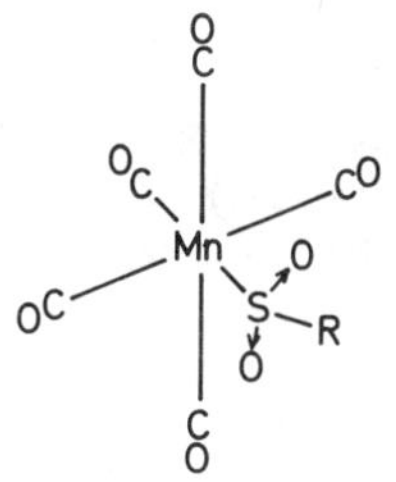 35

preferred preparation of $Mn(CO)_4NO$ utilizes the reaction of the triphenylphosphine complex $[(C_6H_5)_3PMn(CO)_4]_2$ (Section 29.4.6) with nitric oxide in tetrahydronaphthalene at 95° according to the following equation:[126]

$$[(C_6H_5)_3PMn(CO)_4]_2 + 2NO \longrightarrow Mn(CO)_4NO + (C_6H_5)_3PMn(CO)_3NO + CO + (C_6H_5)_3P$$

The green $MnCO(NO)_3$ can be prepared, but in very low yield (~3%) by reaction of $Mn(CO)_5I$ (Section 29.4.3) with nitric oxide at 90–100°.[127] A much more efficient method (but possibly a longer one when only very small quantities are needed) utilizes the reaction between $Mn(CO)_4$-NO and nitric oxide in xylene solution at 90° according to the following equation:[126]

$$Mn(CO)_4NO + 2NO \longrightarrow MnCO(NO)_3 + 3CO$$

29.4.6 Carbonyl Derivatives Containing Organophosphorus and Similar Ligands

Decacarbonyldimanganese, $Mn_2(CO)_{10}$, reacts with tertiary phosphines and similar ligands at temperatures above 100° to give bimetallic substitution products of the type $[R_3PMn(CO)_4]_2$; the triphenylphosphine derivative has been studied in the greatest detail.[128] The bimetallic derivatives $[R_3PMn(CO)_4]_2$ react with dilute sodium amalgam in tetrahydrofuran solution to give the sodium salts $Na[R_3PMn(CO)_4]$ ($R = C_6H_5$, OC_6H_5, $N(CH_3)_2$, *etc.*) which undergo many of the same reactions as the unsubstituted $NaMn(CO)_5$ (Section 29.4.2) including acidification to give the *substituted manganese carbonyl hydrides* $HMn(CO)_4PR_3$.[128] Ultraviolet irradiation of $Mn_2(CO)_{10}$ with tertiary phosphines and similar ligands also may give products of the type $R_3PMn_2(CO)_9$.[129]

The halides $M(CO)_5X$ (M = Mn, Tc, or Re; X = Cl, Br, or I) react readily with many Lewis base ligands such as tertiary phosphines, tertiary arsines, tertiary stibines, amines, and isocyanides upon mild heating in inert solvents. In most cases the monosubstituted products $LM(CO)_4X$ and/or the disubstituted products *cis*- and *trans*-$L_2M(CO)_3X$ are obtained.[130] However, in the case of phenyl isocyanide all possible substitution products of the type $(RNC)_nMn(CO)_{5-n}X$ can be obtained by varying the reaction conditions and the solvent.[131] A preparative method which often gives exclusively monosubstituted derivatives of the type $LM(CO)_4X$ utilizes the cleavage of the halogen bridges in the $[M(CO)_4X]_2$ derivatives with stoichiometric amounts of the ligand according to the following equation:

$$[M(CO)_4X]_2 + 2L \longrightarrow 2LM(CO)_4X$$

Sufficiently mild reaction conditions are used so that no carbonyl substitution takes place.

Tertiary phosphine and similar *derivatives* of the manganese carbonyl nitrosyls can be prepared. Nitric oxide treatment of the $[R_3PMn(CO)_4]_2$ derivatives gives the red compounds $R_3PMn(CO)_3NO$.[126, 132] The compound $MnCO(NO)_3$ reacts readily with tertiary phosphines at room temperature to give the green derivatives $R_3PMn(NO)_3$.[133]

Some dialkylphosphido and dialkylarsenido derivatives of manganese carbonyl can be pre-

---

[126] *H. Wawersik, F. Basolo*, Inorg. Chem. *6*, 1066 (1967).

[127] *C. G. Barraclough, J. Lewis*, J. Chem. Soc. (London) 4842 (1960).

[128] *W. Hieber, G. Faulhaber, F. Theubert*, Z. Anorg. Allgem. Chemie *314*, 125 (1962).

[129] *J. Lewis, R. S. Nyholm, A. G. Osborne, S. S. Sandhu, M. H. B. Stiddard*, Chem. & Ind. (London) 1398 (1963).

[130] *R. J. Angelici, F. Basolo, A. J. Poe*, J. Amer. Chem. Soc. *85*, 2215 (1963).

[131] *K. K. Joshi, P. L. Pauson, W. H. Stubbs*, J. Organometal. Chem. *1*, 51 (1962).

[132] *H. Wawersik, F. Basolo*, J. Amer. Chem. Soc. *89*, 4626 (1967).

[133] *R. G. Hayter*, J. Amer. Chem. Soc. *86*, 823 (1964).

pared by reactions of either of the following types:[133]

(1) Reaction of $Mn_2(CO)_{10}$ with a R_2EER_2 derivative (E=P or As) at about 100° according to the following equation:

$$Mn_2(CO)_{10} + R_2EER_2 \longrightarrow \quad + 2CO$$

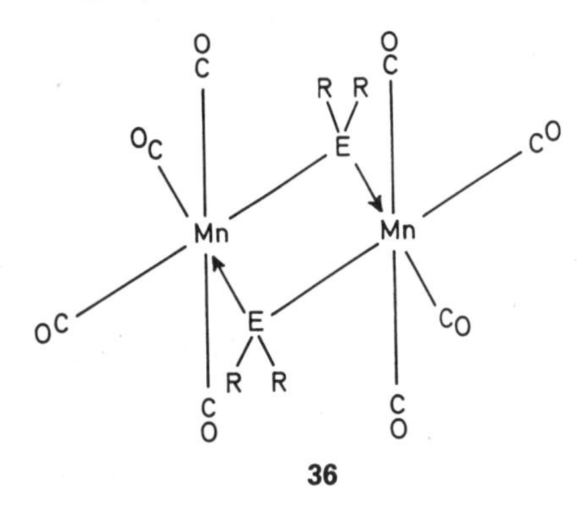

36

(2) Reaction of $NaMn(CO)_5$ with R_2ECl (R=CH_3, C_6H_5, etc., E=P or As) in tetrahydrofuran according to the following equation:

$$2NaMn(CO)_5 + 2R_2ECl \longrightarrow [R_2EMn(CO)_4]_2 + 2NaCl + 2CO$$

Sometimes heating is necessary to effect the final stage of this reaction. If the mole ratio $NaMn(CO)_5/R_2ECl$ is 2:1, a bridging hydride derivative of the type $R_2EMn_2(CO)_8H$ (**37**) may be obtained.

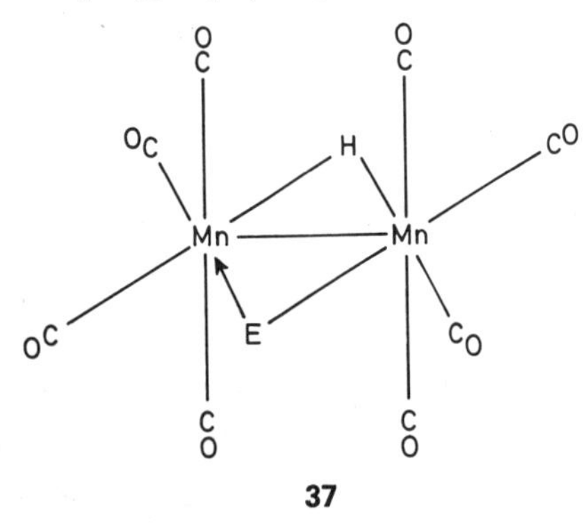

37

29.4.7 Metal Carbonyl Derivatives containing Additional Metal–Carbon or Metal–Metal Bonds

The sodium salts $NaM(CO)_5$ (M=Mn and Re) (Section 29.4.2) react with numerous organic and organometallic halides to form non-ionic $RM(CO)_5$ compounds according to the following equation:[134]

$$NaM(CO)_5 + RX \longrightarrow RM(CO)_5 + NaX$$

Only alkyl halides with the following two characteristics react as above with $NaM(CO)_5$ derivatives to give good yields of the corresponding $RM(CO)_5$ derivatives:

(1) Sufficient reactivity of the carbon–halogen bond. Thus methyl and benzyl halides react with $NaMn(CO)_5$ to give reasonable yields of the corresponding $RMn(CO)_5$ derivatives whereas phenyl and vinyl halides are too unreactive to form the corresponding $RM(CO)_5$ derivative when treated with $NaMn(CO)_5$.

(2) Absence of hydrogen atoms bonded to a *saturated* (sp^3 hybridized) carbon atom bonded to the carbon atom which is bonded to the halogen atom. Presence of such a hydrogen atom appears to lead to side reactions including formation of $HM(CO)_5$. Thus methyl iodide forms stable $CH_3Mn(CO)_5$ when reacted with $NaMn(CO)_5$ whereas ethyl iodide forms very unstable $C_2H_5Mn(CO)_5$ when reacted with $NaMn(CO)_5$.[135] Higher alkyl halides such as isopropyl and *tert*-butyl iodides do not appear to form stable $RMn(CO)_5$ derivatives when reacted with $NaMn(CO)_5$.

Numerous acyl chlorides react with the sodium salts $NaM(CO)_5$ (M=Mn or Re) to form the corresponding $RCOM(CO)_5$ derivatives (**38**, R=CH_3, C_6H_5, CF_3, C_2F_5, C_3F_7, etc.).[134] The following equilibrium exists between corresponding alkylmetal derivatives $RM(CO)_5$ (M=Mn or Re) and acylmetal derivatives $RCOM(CO)_5$:

$$RCOM(CO)_5 \rightleftharpoons RM(CO)_5 + CO$$

As a consequence of this equilibrium many $RM(CO)_5$ derivatives react with carbon monoxide particularly under pressure to form the corresponding $RCOM(CO)_5$ derivatives. However, this reaction is only rarely preparatively useful since most $RCOM(CO)_5$ derivatives are more easily prepared from $NaM(CO)_5$ and the corresponding acyl chloride. The reverse reaction where acylmetal derivatives $RCOM(CO)_5$ are decarbonylated upon heating (generally at about 80° in the case of manganese and at about 120° in the case of rhenium) at give the corresponding $RM(CO)_5$ derivatives is much more useful preparatively since many $RM(CO)_5$ derivatives cannot be prepared from the sodium

[134] *R. D. Closson, J. Kozikowski, T. H. Coffield*, J. Org. Chem. *22*, 598 (1957).

[135] *M. L. H. Green, P. L. I. Nagy*, J. Organometal. Chem. *1*, 58 (1962).

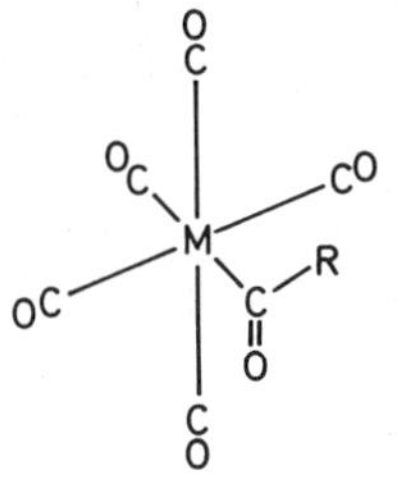

38 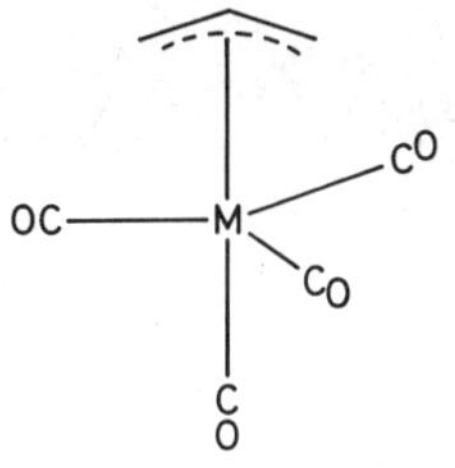 **39** 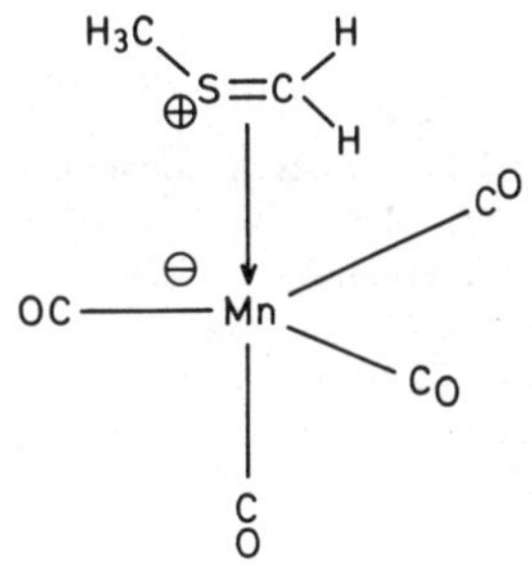 **40**

salts $NaM(CO)_5$ and the corresponding halides RX. Thus *aryl derivatives* such as $C_6H_5Mn(CO)_5$ and *perfluoroalkyl derivatives* of the type $R_fM(CO)_5$ ($R_f = CF_3$, C_2F_5, C_3F_7, *etc.*; M = Mn or Re) are best prepared by the following two-step sequence involving preparation and decarbonylation of the intermediate acylmetal pentacarbonyl:

$$NaM(CO)_5 + RCOCl \longrightarrow RCOM(CO)_5 + NaCl$$

$$RCOM(CO)_5 \longrightarrow RM(CO)_5 + CO$$

Another preparative method which is useful for the preparation of $RM(CO)_5$ derivatives with heavily fluorinated R groups generally with a CF_2 group directly bonded to the transition metal consists of the addition of the hydrides $HM(CO)_5$ (Section 29.4.2) to highly fluorinated olefins, *e.g.*[136]

$$HMn(CO)_5 + CF_2{=}CF_2 \longrightarrow HCF_2CF_2Mn(CO)_5$$

Hydrometallation reactions of this type often proceed at room temperature and atmospheric pressure.

Pyrolysis of other types of $RM(CO)_5$ derivatives can lead to interesting *organomanganese pentacarbonyls.* Thus pyrolysis of the π-allylmanganese derivative $CH_2{=}CHCH_2Mn(CO)_5$ at about 80° results in the loss of one carbonyl group to give the π-allylmanganese derivative $C_3H_5Mn(CO)_4$ (**39**, M = Mn) according to the following equation:[137,138]

$$CH_2{=}CHCH_2Mn(CO)_5 \longrightarrow C_3H_5Mn(CO)_4 + CO$$

Similar *π-allylic metal tetracarbonyl derivatives* of manganese and rhenium can be prepared by either of the following two methods:

(1) Addition of $HMn(CO)_5$ to a conjugated diene.[138] Thus the reaction of $HMn(CO)_5$ with butadiene at ~95° gives the π-crotyl derivative π-$C_4H_7Mn(CO)_4$.

(2) Reaction of the metal pentacarbonyl halide with an allyltin compound according to the following equation:[139]

$$M(CO)_5Br + (CH_3)_3SnCH_2CH{=}CH_2 \longrightarrow C_3H_5M(CO)_4 + (CH_3)_3SnBr + CO$$

39, M = Mn or Re

This method is the only one which has been suitable for the preparation of π-allylic rhenium tetracarbonyl derivatives up to the present time.

Reaction of $NaMn(CO)_5$ with $ClCH_2SCH_3$ gives first the pentacarbonyl $CH_3SCH_2Mn(CO)_5$ which then undergoes decarbonylation upon heating to give the yellow liquid tetracarbonyl $CH_3SCH_2Mn(CO)_4$ (**40**).

The sodium salts $NaM(CO)_5$ (M = Mn or Re) can be used to prepare metal carbonyl derivatives with unusual metal–metal bonds. Thus reactions of the sodium salts $NaM(CO)_5$ (M = Mn or Re) with organotin halides of the type R_3SnCl (R = methyl, phenyl, etc.) gives stable compounds of the type $R_3SnM(CO)_5$ (M = Mn or Re) with tin–manganese or tin–rhenium bonds:[140,141] The metal–metal bonds in these compounds are so strong that halogens (chlorine or bromine) do not cleave the metal–metal bonds in $(C_6H_5)_3SnMn(CO)_5$ but instead remove the phenyl groups to give $X_3SnMn(CO)_5$ derivatives (X = Cl or Br) according to the following equation:[140]

---

[136] *P. M. Treichel, E. Pitcher, F. G. A. Stone,* Inorg. Chem. *1,* 511 (1963).

[137] *H. D. Kaesz, R. B. King, F. G. A. Stone,* Z. Naturforsch. *15b,* 682 (1960).

[138] *W. R. McClellan, H. H. Hoehn, H. N. Cripps, E. L. Muetterties, B. W. Howk,* J. Amer. Chem. Soc. *83,* 1601 (1961).

[139] *E. W. Abel, S. M. Moorhouse,* J. Chem. Soc. Dalton 1706 (1973).

[140] *R. D. Gorsich,* J. Amer. Chem. Soc. *84,* 2486 (1962).

[141] *J. A. J. Thompson, W. A. G. Graham,* Inorg. Chem. *6* 1365 (1967).

[142] *C. E. Coffey, J. Lewis, R. S. Nyholm,* J. Chem. Soc., 1741 (1964);
A. S. Kasenally, J. Lewis, A. R. Manning, J. R. Miller, R. S. Nyholm, M. H. B. Stiddard, J. Chem. Soc., 3407 (1965).

$$2(C_6H_5)_3SnMn(CO)_5 + 2X_2 \longrightarrow 2X_3SnMn(CO)_5 + 6C_6H_5X$$

This reaction provides a convenient synthesis of the $X_3SnMn(CO)_5$ derivatives. Reaction of $NaMn(CO)_5$ with the gold compound $(C_6H_5)_3PAuCl$ gives the product $(C_6H_5)_3PAuMn(CO)_5$ with a manganese–gold bond.[142] Similarly, the reaction of $NaMn(CO)_5$ with $C_5H_5Fe(CO)_2I$ (Section 29.5.8.3) gives the product $C_5H_5Fe(CO)_2Mn(CO)_5$ of structure **41** with a manganese–iron bond.

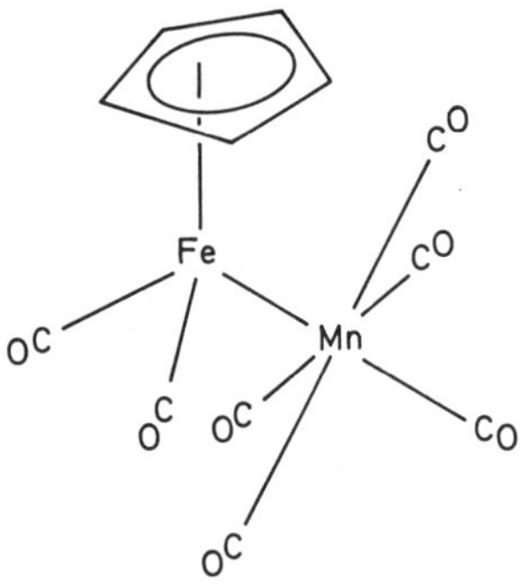

41

29.4.8 Arene and Olefin Derivatives

The *benzenetricarbonylmanganese cation* $[C_6H_6Mn(CO)_3]^{\oplus}$ is obtained by treatment of $Mn(CO)_5X$ (X=Cl or Br) with aluminum chloride in boiling benzene.[143] This cation is conveniently isolated as the hexafluorophosphate salt or, less desirably because of the possible explosion hazard, as the perchlorate salt. Similar $[(arene)M(CO)_3]^{\oplus}$ cations can be obtained from alkylated benzenes.

Solid salts of the $[C_6H_6Mn(CO)_3]^{\oplus}$ cation react with lithium aluminum hydride in tetrahydrofuran solution to give the yellow sublimable *π-cyclohexadienyl complex* $C_6H_7Mn(CO)_3$ (**42**, R=H) as the major product.[143] In addition a low yield of the diene-bonded π-cyclohexadiene complex $C_6H_8Mn(CO)_3H$ (**43**)[144] can also be isolated from this reaction mixture. Unfortunately, no better way for preparing $C_6H_8Mn(CO)_3H$ (**43**) is currently known. The π-cyclohexadienyl complex $C_6H_7Mn(CO)_3$ (**42**, R=H) can also be prepared by heating $Mn_2(CO)_{10}$ with 1,3-cyclohexadiene in a boiling saturated hydrocarbon solvent at about 130°.[143] This latter synthesis of **42** (R=H) involves fewer steps but gives a much lower yield and requires a more difficult product purification. Reactions of the cation $[C_6H_6Mn(CO)_3]^{\oplus}$ with alkyllithium compounds

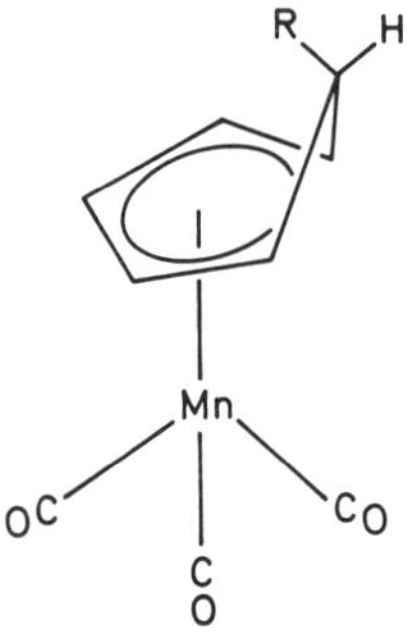

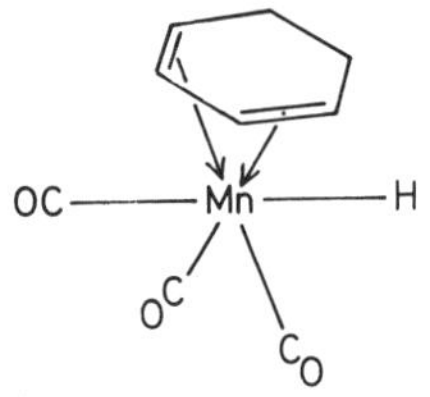

42 **43**

gives *exo*-substituted π-cyclohexadienylmanganese derivatives of the type $RC_6H_6Mn(CO)_3$ (**42**).

29.4.9 Cyclopentadienylmetal Carbonyl Derivatives

The manganese compound $C_5H_5Mn(CO)_3$ (**44**, X=H) can be prepared by reaction of $(C_5H_5)_2Mn$ with carbon monoxide preferably under pressure.[145] A more convenient preparation of $C_5H_5Mn(CO)_3$, which gives yields up to 80%, utilizes the reaction between the pyridine complex $(C_5H_5N)_2MnCl_2$, magnesium metal, cyclopentadiene, and carbon monoxide under pressure in dimethylformamide solution in the presence of hydrogen.[146] The technetium and rhenium compounds $C_5H_5M(CO)_3$ (M=Tc or Re) are best prepared by reaction of the corresponding metal pentacarbonyl chloride $M(CO)_5Cl$ with sodium cyclopentadienide in boiling tetrahydrofuran solution.[147]

The π-cyclopentadienyl ring in $C_5H_5Mn(CO)_3$ (**44**, X=H) can undergo substitution with electrophilic reagents in a manner similar to benzene and ferrocene. Thus, the treatment of $C_5H_5Mn(CO)_3$ with acyl chlorides in the presence of aluminum chloride in carbon disulfide or dichloromethane solution results in the formation of the acyl derivatives $RCOC_5H_4Mn(CO)_3$ (**44**, X=RCO) according to the following equation:[147]

$$C_5H_5Mn(CO)_3 + RCOCl \longrightarrow RCOC_5H_4Mn(CO)_3 + HCl$$

where R=CH_3, C_6H_5, etc.

---

[143] *G. Winkhaus, L. Pratt, G. Wilkinson*, J. Chem. Soc. (London) 3807 (1961).

[144] *G. Wilkinson*, Z. Anorg. Allgem. Chemie *319*, 404, (1963).

[145] *E. O. Fischer, R. Jira*, Z. Naturforsch. *9b*, 618 (1954);
T. S. Piper, F. A. Cotton, G. Wilkinson, J. Inorg. Nucl. Chem. *1*, 165 (1965).

[146] *J. F. Cordes, D. Neubauer*, Z. Naturforsch. *17b*, 791 (1962).

[147] *E. O. Fischer, W. Fellmann*, J. Organometal. Chem. *1*, 191 (1963).

These ketones are useful intermediates for the preparation of other substituted cyclopentadienylmanganese tricarbonyls by conventional organic reactions. For example, the compounds $RCOC_5H_4Mn(CO)_3$ (**44**, X=RCO) react with alkylmagnesium halides R′MgX in diethyl ether solution to give, after hydrolysis, the alcohols $RR'C(OH)C_5H_4Mn(CO)_3$ (**44**, X=RR′C-(OH)). These alcohols can be easily dehydrated to olefins when the R or R′ groups have a hydrogen on the α-carbon atom. The sulfonic acid $(OC)_3MnC_5H_4$-SO_3H (**44**, X=SO_3H) may be obtained by sulfonation of $C_5H_5Mn(CO)_3$ with sulfuric acid in acetic anhydride; this sulfonic acid is conveniently isolated as its *p*-toluidinium salt.[148] The chloromercury derivative $ClHgC_5H_4Mn(CO)_3$ (**44**, X=ClHg) may be obtained by treatment of the sulfonic acid with mercuric chloride.

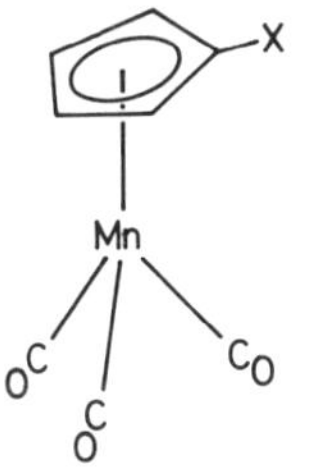

X
Mn
OC CO
CO

44

Another useful cyclopentadienylmanganese tricarbonyl derivative is the amine $H_2NC_5H_4Mn$-$(CO)_3$ (**44**, X=NH_2) which may be obtained by treatment of the acid chloride $ClCOC_5H_4Mn$-$(CO)_3$ (**44**, X=COCl) with sodium azide in acetone followed by pyrolysis in boiling benzene and base hydrolysis of the resulting isocyanate $OCNC_5H_4Mn(CO)_3$ (**44**, X=NCO).[149] The amine $H_2NC_5H_4Mn(CO)_3$ (**44**, X=NH_2) can be diazotized with a mixture of hydrogen chloride and isoamyl nitrite in isopropanol to give a yellow diazonium ion $[(OC)_3MnC_5H_4N_2]^{\oplus}$. Pyrolysis of the tetrafluoroborate of this diazonium salt gives the fluoro-derivative FC_5H_4Mn-$(CO)_3$ (**44**, X=F). Similar pyrolysis of the complex halides $[N_2C_5H_4Mn(CO)_3][HgX_3]$ (X=Cl or Br) gives the halides $XC_5H_4Mn(CO)_3$ (**44**, X=Cl or Br). The iodide $IC_5H_4Mn(CO)_3$ can be obtained by treatment of the diazonium salt with aqueous potassium iodide.

The carbonyl groups in $C_5H_5Mn(CO)_3$ can be partially replaced with other ligands by ultraviolet irradiation with the ligand in an inert solvent (generally a saturated hydrocarbon). The following equations exemplify photochemical reactions of this general type:[19, 150]

$$C_5H_5Mn(CO)_3 + L \xrightarrow{h\nu} C_5H_5Mn(CO)_2L + CO$$

L=amines, tertiary phosphines, tertiary arsines, sulfoxides, olefins, and isocyanides

$$C_5H_5Mn(CO)_3 + \text{diene} \xrightarrow{h\nu} C_5H_5Mn(CO)(\text{diene}) + 2CO$$

diene=butadiene, *etc.*[19]

In all of these reactions u.v. irradiation is necessary for the reaction to take place. Mere heating without ultraviolet irradiation is insufficient to effect these reactions.

A particularly important *cyclopentadienylmanganese carbonyl intermediate* is the cation $[C_5H_5Mn(CO)_2NO]^{\oplus}$ which may be prepared by reaction of $C_5H_5Mn(CO)_3$ with sodium nitrite and concentrated aqueous hydrochloric acid in boiling ethanol.[151] This cation is conveniently isolated and handled as the yellow hexafluorophosphate salt $[C_5H_5Mn(CO)_2NO][PF_6]$ which is readily obtained by adding ammonium hexafluorophosphate to the reaction mixture. The analogous *rhenium* compound $[C_5H_5Re(CO)_2$-$NO][PF_6]$ cannot be prepared analogously but can be obtained by reaction of $C_5H_5Re(CO)_3$ with nitrosylsulfuric acid ($NO^{\oplus}HSO_4^{\ominus}$) in an inert solvent such as dichloromethane followed by addition of aqueous ammonium hexafluorophosphate.[152]

The salt $[C_5H_5Mn(CO)_2NO][PF_6]$ is a useful intermediate for the preparation of other cyclopentadienylmanganese carbonyl nitrosyl derivatives. Thus it reacts with numerous tertiary phosphines and similar ligands upon mild heating in a polar organic solvent such as methanol or acetone in the absence of ultraviolet irradiation. Substitution products of the types $[C_5H_5Mn(CO)(NO)L][PF_6]$ and $[C_5H_5Mn(NO)L_2]$-$[PF_6]$ are readily obtained from such reactions.[153,154] For example, triphenylphosphine forms the compound $[C_5H_5Mn(CO)(NO)P(C_6H_5)_3][PF_6]$. Reaction of $[C_5H_5Mn(CO)_2NO][PF_6]$ with methanolic sodium methoxide gives the relatively unstable red crystalline methoxycarbonyl derivative $C_5H_5Mn(CO)(NO)$-(CO_2CH_3) (**45**, X=CO_2CH_3) which reacts with methylmagnesium bromide to give, after hydrolysis, the

[148] *M. Cais, J. Kozikowski*, J. Amer. Chem. Soc. *82*, 5667 (1960).
[149] *M. Cais, N. Narkis*, J. Organometal. Chem. *3*, 168, 269 (1965).
[150] *E. O. Fischer, M. Herberhold*, Experientia Suppl. *9*, 259 (1964).
[151] *R. B. King, M. B. Bisnette*, Inorg. Chem. *3*, 791 (1964).
[152] *E. O. Fischer, H. Strametz*, Z. Naturforsch. *23b*, 278 (1968).
[153] *R. B. King, A. Efraty*, Inorg. Chem. *8*, 2374 (1969).
[154] *H. Brunner, H. D. Schindler*, J. Organometal. Chem. *19*, 135 (1969).

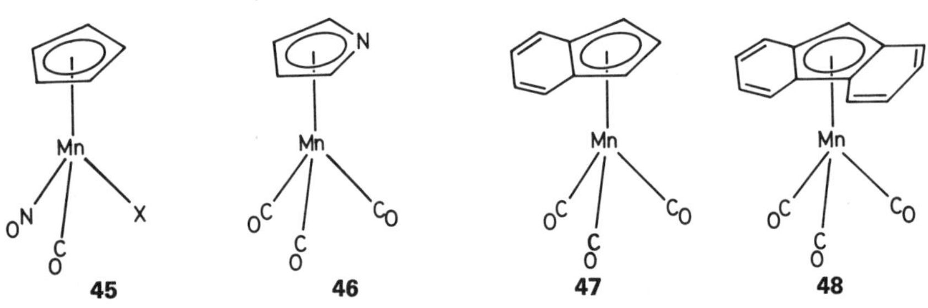

orange-brown unstable liquid acetyl derivative $C_5H_5Mn(CO)(NO)(COCH_3)$ (**45**, X=$COCH_3$).[155] Reduction of $[C_5H_5Mn(CO)_2NO][PF_6]$ with aqueous sodium hydroborate gives the red-purple $[C_5H_5Mn(CO)(NO)]_2$.[153]

Manganese carbonyl forms some π-pyrrolyl derivatives which may be regarded as aza-analog of the π-cyclopentadienyl derivatives discussed above. The yellow-orange *π-pyrrolyltricarbonylmanganese*, $C_4H_4NMn(CO)_3$ (**46**) may be obtained either by reaction of $Mn_2(CO)_{10}$ with pyrrole in boiling octane[156,157] or by reaction of $Mn(CO)_5Br$ with potassium pyrrolide (KC_4H_4N) in boiling tetrahydrofuran.[156] Similar compounds are obtained from alkylpyrroles.[156]

Some benzo-derivatives of cyclopentadienyltricarbonylmanganese are also known. Reaction of $Mn(CO)_5Br$ with sodium indenide in boiling tetrahydrofuran gives the yellow indenyl derivative $C_9H_7Mn(CO)_3$ (**47**).[158] Similarly the reaction of $Mn(CO)_5Br$ with sodium fluorenide gives the yellow-orange fluorenyl derivative $C_{13}H_9Mn(CO)_3$ (**48**).[158]

29.5 Carbonyl Derivatives of Iron, Ruthenium, and Osmium

29.5.1 The Metal Carbonyls

Iron forms the three carbonyls $Fe(CO)_5$, $Fe_2(CO)_9$, and $Fe_3(CO)_{12}$. *Pentacarbonyliron*, $Fe(CO)_5$, is prepared by reaction of elemental iron with carbon monoxide at elevated temperatures and pressures.[159] It is rarely prepared in the laboratory since it is generally an inexpensive commercial product. *Enneacarbonyldiiron*, $Fe_2(CO)_9$ (**49**),

49

may be prepared by u.v. irradiation of $Fe(CO)_5$ in glacial acetic acid solution at or below room temperature.[160,161] It is an orange insoluble solid which precipitates from the reaction mixture in the pure state as it is formed. Dodecacarbonyltriiron, $Fe_3(CO)_{12}$ (**50**)

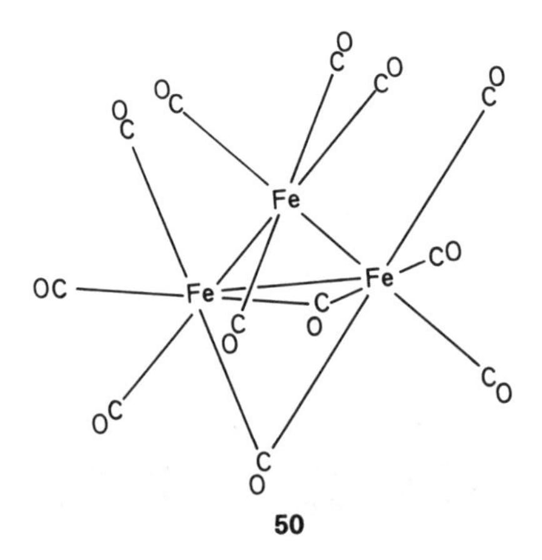

may be prepared by either of the following two methods:

(1) Reaction of $Fe(CO)_5$ with aqueous methanolic sodium hydroxide gives a solution containing the $HFe(CO)_4^{\ominus}$ anion. Oxidation of this anion with a mild oxidizing agent such as

[155] *R. B. King, M. B. Bisnette, A. Fronzaglia*, J. Organometal. Chem. *5*, 341 (1966).
[156] *K. K. Joshi, P. L. Pauson, A. R. Qazi, W. H. Stubbs*, J. Organometal. Chem. *1*, 471 (1964).
[157] *R. B. King, A. Efraty*, J. Organometal. Chem. *20*, 264 (1969).
[158] *R. B. King, A. Efraty*, J. Organometal. Chem. *23*, 527 (1970).
[159] *A. Mittasch*, Angew. Chem. *30*, 827 (1928).
[160] *E. Speyer, H. Wolf*, Ber. *60*, 1424 (1927).
[161] *E. H. Braye, W. Hübel*, Inorg. Synth. *8*, 178 (1966).

freshly prepared manganese dioxide in the presence of an ammonium chloride buffer gives a dark red solution, probably containing the $HFe_3(CO)_{11}^{\ominus}$ anion. Acidification of this solution with sulfuric acid gives a black precipitate of $Fe_3(CO)_{12}$.[162, 163]

(2) Reaction of $Fe(CO)_5$ with triethylamine in aqueous solution at 80° gives the reddish black salt $[(C_2H_5)_3NH][HFe_3(CO)_{11}]$. Acidification of this salt with methanolic sulfuric acid gives a black precipitate of $Fe_3(CO)_{12}$.[164]

Ruthenium forms the two carbonyls $Ru(CO)_5$ and $Ru_3(CO)_{12}$. *Pentacarbonylruthenium*, $Ru(CO)_5$, may be obtained in low yield by reaction of metallic ruthenium with carbon monoxide at elevated temperatures and pressures.[165] However, better yields of $Ru(CO)_5$ are obtained from the reaction of ruthenium(III) acetylacetonate with a mixture of carbon monoxide and hydrogen at 180°/200 atm. preferably in the presence of an inert solvent such as heptane.[166] Pentacarbonylruthenium, $Ru(CO)_5$, is a volatile liquid which rapidly forms $Ru_3(CO)_{12}$ upon exposure to light. *Dodecacarbonylruthenium*, $Ru_3(CO)_{12}$ (**51**, M = Ru),

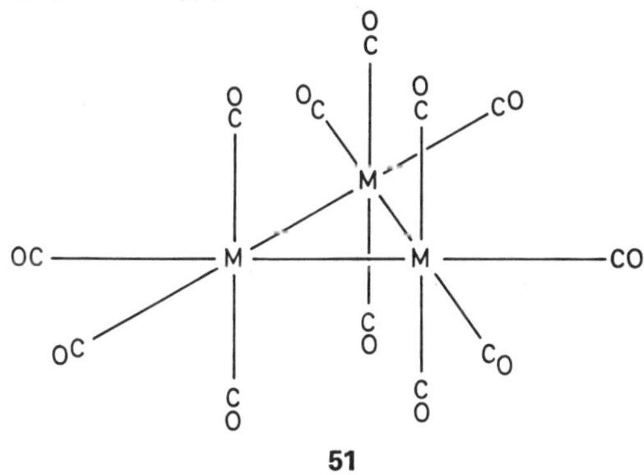

may be obtained by reaction of a methanolic solution of hydrated ruthenium(III) chloride with zinc at 60° in the presence of carbon monoxide under some pressure (~10 atm.).[167] The use of carbon monoxide under pressure can be avoided in the conversion of hydrated ruthenium(III) chloride to $Ru_3(CO)_{12}$ if the following three-step procedure is used:[168]

$$RuCl_3 \cdot 3H_2O + 1{,}3\text{-}C_6H_8 \xrightarrow{C_2H_5OH} [C_6H_6RuCl_2]_n + ?$$

$$2/n[C_6H_6RuCl_2]_n + 6CO \xrightarrow{CH_3OH} [Ru(CO)_3Cl_2]_2 + 2C_6H_6$$

$$3[Ru(CO)_3Cl_2]_2 + 6Zn + 6CO \xrightarrow{CH_3OH} 2Ru_3(CO)_{12} + 6ZnCl_2$$

All three of these steps are relatively easy ones and both carbonylation steps can be performed at atmospheric pressure.

Osmium forms the three carbonyls $Os(CO)_5$, $Os_2(CO)_9$, and $Os_3(CO)_{12}$. *Pentacarbonylosmium*, $Os(CO)_5$, is obtained from the reaction of osmium tetroxide, OsO_4, with carbon monoxide at 160°/100 to 200 atm either in the absence of a solvent or in a saturated hydrocarbon solvent such as heptane.[165] Like $Ru(CO)_5$, $Os(CO)_5$ is unstable at room temperature and decomposes in light at or above room temperature to give $Os_3(CO)_{12}$. Ultraviolet irradiation of $Os(CO)_5$ at −40° gives $Os_2(CO)_9$ (**52**), an unstable yellow-orange solid which decomposes even below room temperature.[169] Reaction of OsO_4 with carbon monoxide at elevated temperatures in xylene[170] or preferably methanol[171] solution gives the yellow crystalline $Os_3(CO)_{12}$ (**51**, M = Os).

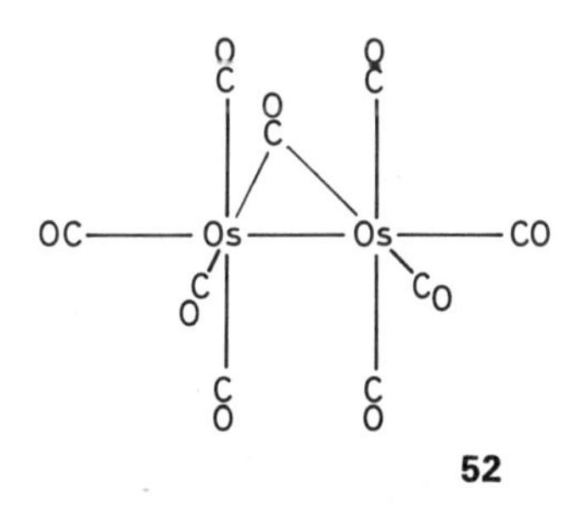

29.5.2 Metal Carbonyl Anions and Metal Carbonyl Hydrides

Several iron carbonyl anions can be prepared by treatment of various iron carbonyls with bases

[162] *W. Hieber*, Z. Anorg. Allgem. Chemie *204*, 165 (1932).
[163] *R. B. King, F. G. A. Stone*, Inorg. Synth. *7*, 193 (1963).
[164] *W. McFarlane, G. Wilkinson*, Inorg. Synth. *8*, 181 (1966).
[165] *W. Manchot, W. J. Manchot*, Z. Anorg. Allgem. Chemie *226*, 385 (1936).
[166] *F. Calderazzo, F. L'Eplattenier*, Inorg. Chem. *6*, 1220 (1967).
[167] *M. I. Bruce, F. G. A. Stone*, Chem. Commun. 684 (1966); J. Chem. Soc. *A*, 1238 (1967).
[168] *R. B. King, P. N. Kapoor*, Inorg. Chem. *11*, 336 (1972).
[169] *J. R. Moss, W. A. G. Graham*, Chem. Commun. 835 (1970).
[170] *C. W. Bradford, R. S. Nyholm*, Chem. Commun. 384 (1967).
[171] *B. F. G. Johnson, J. Lewis, P. A. Kilty*, Chem. Commun. 180 (1968).

or reducing agents.[172] The colorless mono*metallic anion* $HFe(CO)_4^{\ominus}$ can be prepared by reaction of $Fe(CO)_5$ with aqueous or alcoholic alkali. The *dinegative anion* $Fe(CO)_4^{2\ominus}$ can be prepared by reduction of $Fe(CO)_5$ with alkali metals in liquid ammonia or by reduction of $Fe_3(CO)_{12}$ with alkali metals in tetrahydrofuran. The dark red–orange dinegative anion $Fe_2(CO)_8^{2\ominus}$ can be prepared either by ultraviolet irradiation of $Fe(CO)_5$ with dilute sodium amalgam in tetrahydrofuran solution[173] or by reaction of $Fe_2(CO)_9$ with aqueous alkali. The dark red–violet $HFe_3(CO)_{11}^{\ominus}$ can be prepared by reaction of $Fe_3(CO)_{12}$ with aqueous alkali or (as the triethylammonium salt $[(C_2H_5)_3NH][HFe_3(CO)_{11}]$) by reaction of $Fe(CO)_5$ with aqueous triethylamine at 80°. A red–black tetrametallic anion $Fe_4(CO)_{13}^{2\ominus}$ can be obtained by reaction of $Fe(CO)_5$ with piperidine or a similar amine at 85°.

The only properly characterized *neutral iron carbonyl hydride* is $H_2Fe(CO)_4$, which is liberated upon acidification of $HFe(CO)_4^{\ominus}$ solutions with a strong mineral acid.[174] The compound $H_2Fe(CO)_4$ is a gas which is unstable at room temperature except when mixed with excess carbon monoxide.

Tetrametallic *ruthenium carbonyl hydrides* $H_4Ru_4(CO)_{12}$ and $H_2Ru_4(CO)_{13}$ can be prepared from $Ru_3(CO)_{12}$ by the following methods:[175]

(1) Reaction of $Ru_3(CO)_{12}$ with sodium amalgam or sodium hydroborate in tetrahydrofuran solution or with methanolic potassium hydroxide followed by acidification with phosphoric or sulfuric acid.

(2) Pyrolysis of $Ru_3(CO)_{12}$ in an oxygenated solvent such as dibutyl ether.[176]

Osmium forms a monometallic carbonyl hydride $H_2Os(CO)_4$ which is a colorless liquid stable at room temperature even in the presence of air.[177] This hydride is best prepared by reaction of OsO_4 with a mixture of carbon monoxide and hydrogen at 160°/180 atm either in heptane solution or in the absence of solvent. The *tetrametallic osmium carbonyl hydride* $H_4Os_4(CO)_{12}$ can be prepared by heating $H_2Os(CO)_4$ in boiling heptane.[178] Reaction of $Os_3(CO)_{12}$ with methanolic potassium hydroxide or sodium hydroborate in tetrahydrofuran solution followed by acidification with phosphoric acid gives dark red $H_2Os_3(CO)_{10}$ in 30% yield in addition to orange $H_2Os_3(CO)_{12}$ and yellow $H_4Os_4(CO)_{12}$ as minor products.[179]

29.5.3 Metal Carbonyl Halides

Iron tetracarbonyl dihalides of the type $Fe(CO)_4X_2$ (X=Cl, Br, or I) can be prepared by reactions of $Fe(CO)_5$ with the free halogens X_2 in an inert solvent at or below room temperature according to the following equation:[180]

$$Fe(CO)_5 + X_2 \longrightarrow Fe(CO)_4X_2 + CO$$

Mild heating of the $Fe(CO)_4X_2$ halides above room temperature results in the complete loss of carbon monoxide to give the corresponding iron(II) halides with their thermal stability increasing in the following sequence: $Fe(CO)_4Cl_2$ (least stable) < $Fe(CO)_4Br_2$ < $Fe(CO)_4I_2$ (most stable).

Reaction of $Fe_3(CO)_{12}$ with a deficiency of iodine in boiling tetrahydrofuran gives the *bimetallic derivative* $[Fe(CO)_4I]_2$ in low yield (~2%) according to the following equation:[181]

$$2Fe_3(CO)_{12} + 3I_2 \longrightarrow 3[Fe(CO)_4I]_2$$

Reaction of $Fe(CO)_5$ with the perfluoroalkyl iodides, R_fI ($R_f=CF_3$, C_2F_5, $n\text{-}C_3F_7$, $(CF_3)_2CF$, and $n\text{-}C_7F_{15}$) at ~70° results in the formation of the red volatile $R_fFe(CO)_4I$ derivatives according to the following equation:[182]

$$Fe(CO)_5 + R_fI \longrightarrow R_fFe(CO)_4I + CO$$

Heating certain $R_fFe(CO)_4I$ derivatives (R_f= C_2F_5, $n\text{-}C_3F_7$, *etc.*) results in decarbonylation to give the bimetallic derivatives $[R_fFe(CO)_3I]_2$.[182]

---

[172] *W. Hieber, W. Beck, G. Braun*, Angew. Chem. *22*, 795 (1960).
[173] *J. K. Ruff*, Inorg. Chem. *7*, 1818 (1968).
[174] *W. Hieber, H. Vetter*, Z. Anorg. Allgem. Chemie *212* 145 (1933);
H. W. Sternberg, R. Markby, I. Wender, J. Amer. Chem. Soc. *79*, 6116 (1957).
[175] *B. F. G. Johnson, R. D. Johnston, J. Lewis, B. H. Robinson, G. Wilkinson*, J. Chem. Soc. *A*, 2856 (1968).
[176] *B. F. G. Johnson, J. Lewis, I. G. Williams*, J. Chem. Soc. *A*, 901 (1970).
[177] *F. L'Eplattenier, F. Calderazzo*, Inorg. Chem. *6*, 1092 (1967).
[178] *J. R. Moss, W. A. G. Graham*, J. Organometal. Chem. *23*, C47 (1970).
[179] *B. F. G. Johnson, J. Lewis, P. A. Kilty*, J. Chem. Soc. *A*, 2859 (1968).
[180] *W. Hieber, G. Bader*, Ber. *61*, 1717 (1928).
[181] *F. A. Cotton, B. F. G. Johnson*, Inorg. Chem. *6*, 2113 (1967).
[182] *R. B. King, S. L. Stafford, P. M. Treichel, F. G. A. Stone*, J. Amer. Chem. Soc. *83*, 3604 (1961).

Ruthenium forms several types of *carbonyl halides*. The most readily accessible are the compounds $[Ru(CO)_2X_2]_n$ (X=Cl or Br) which can be obtained by heating hydrated ruthenium(III) *chloride* in a mixture of concentrated aqueous formic and hydrohalic acids followed by complete removal of volatile materials.[183] The corresponding *iodide* $[Ru(CO)_2I_2]_n$ can be prepared by carbonylation of anhydrous ruthenium(III) iodide at atmospheric pressure and elevated temperatures.[184] The ruthenium carbonyl halides $[Ru(CO)_2X_2]_n$ (X=Cl, Br, or I) are solids which are non-volatile and insoluble in nonreactive, noncoordinating solvents. They presumably have a polymeric structure with ruthenium atoms linked with halogen bridges. The *bimetallic ruthenium carbonyl chloride* $[Ru(CO)_3Cl_2]_2$ can be prepared either by carbonylation of ruthenium(III) chloride in methanol solution at elevated pressures or by carbonylation of the diolefin complexes $[(diene)RuCl_2]_n$ at atmospheric pressure.[167,168] High pressure carbonylation of ruthenium iodides gives the *monometallic* $Ru(CO)_4I_2$;[185] the chloride and bromide analogs can be prepared but are unstable at room temperature with respect to formation of the corresponding bimetallic $[Ru(CO)_3X_2]_2$ derivatives.

Osmium also forms several types of *metal carbonyl halides*. The monometallic derivatives $Os(CO)_4X_2$ (X=Cl or Br) can be prepared by reaction of $H_2Os(CO)_4$ (Section 29.5.2) with the carbon tetrahalides, CX_4 (X=Cl or Br) at room temperature.[177] The *bimetallic* derivatives $[Os(CO)_3X_2]_2$ (X=Cl or Br) are obtained by carbonylation of the corresponding osmium halides at 270°.[186] Reactions of $Os_3(CO)_{12}$ at room temperature with chlorine or bromine in benzene or dichloromethane solution gives the yellow *trimetallic* carbonyl halides $Os_3(CO)_{12}X_2$.[179,187] Prolonged heating of these $Os_3(CO)_{12}X_2$ derivatives in boiling benzene results in decarbonylation to give the corresponding $Os_3(CO)_{10}X_2$ derivatives.[188]

29.5.4 Chalcogen Derivatives of Iron, Ruthenium and Osmium Carbonyls

Several types of organosulfur derivatives of iron carbonyls have been prepared. Red crystalline air-stable derivatives of the type $[RSFe(CO)_3]_2$ (**53**, R=CH_3, C_2H_5, $(CH_3)_2CH$, C_6H_5, etc.) can be prepared by reaction of $Fe_3(CO)_{12}$ (Section 29.5.1) with the corresponding thiol (RSH) or disulfide (RSSR) in boiling benzene.[189] The methyl derivative $[CH_3SFe(CO)_3]_2$ (**53**, R=CH_3) can be separated into two stereoisomers by chromatography on alumina in pentane solution.[190] An alternative preparation of the methylthio derivative $[CH_3SFe(CO)_3]_2$ (**53**, R=CH_3)

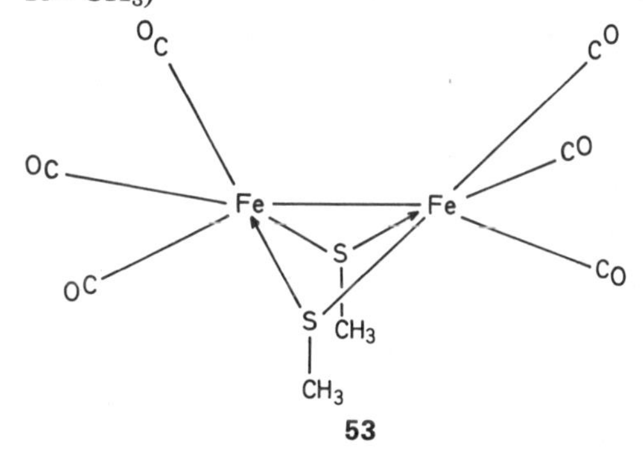

53

utilizes the reaction of dimethyl disulphide with $Fe(CO)_5$ in the presence of carbon monoxide under pressure.[191] If high pressure equipment is available, this latter method is the best way to prepare $[CH_3SFe(CO)_3]_2$ (**53**, R=CH_3). Reaction of $Fe(CO)_5$ with excess dimethyldisulfide in a closed system in the absence of added carbon monoxide gives the red insoluble polymeric $[(CH_3S)_2Fe(CO)_2]_n$.

Iron carbonyl sulfides without organic groups attached to the sulfur are also known. The purple volatile $Fe_3(CO)_9S_2$ can be prepared by reaction of $Fe_3(CO)_{12}$ with cyclohexene episul-

---

[183] *R. Colton, R. H. Farthing*, Australian J. Chem. *20*, 1283 (1967);
B. F. G. Johnson, R. D. Johnston, J. Lewis, J. Chem. Soc. *A*, 792 (1969);
M. J. Cleare, W. P. Griffith, J. Chem. Soc. *A*, 372 (1969).

[184] *W. Manchot, J. König*, Ber. *57*, 2130 (1924).

[185] *E. R. Corey, M. V. Evans, L. F. Dahl*, J. Inorg. Nucl. Chem. *24*, 926 (1962).

[186] *W. Manchot, J. König*, Ber. *58*, 229 (1925).

[187] *J. P. Candlin, J. Cooper*, J. Organometal. Chem. *15*, 230 (1968).

[188] *A. J. Deeming, B. F. G. Johnson, J. Lewis*, J. Organometal. Chem. *17*, P40 (1969).

[189] *W. Hieber, P. Spacu*, Z. Anorg. Allgem. Chemie *233*, 353 (1937);
W. Hieber, C. Scharfenberg, Ber. *73*, 1012 (1940);
W. Hieber, W. Beck, Z. Anorg. Allgem. Chemie *305*, 265 (1960);
S. F. A. Kettle, L. E. Orgel, J. Chem. Soc. (London) 3890 (1960).

[190] *R. B. King*, J. Amer. Chem. Soc. *84*, 2460 (1962).

[191] *R. B. King, M. B. Bisnette*, Inorg. Chem. *4*, 1663 (1965).

fide.[192] The red volatile $Fe_2(CO)_6S_2$, which has structure **54** with a sulfur–sulfur bond, can be prepared by reaction of the $HFe(CO)_4^{\ominus}$ anion (Section 29.5.2) with aqueous sodium polysulfide followed by acidification.[193]

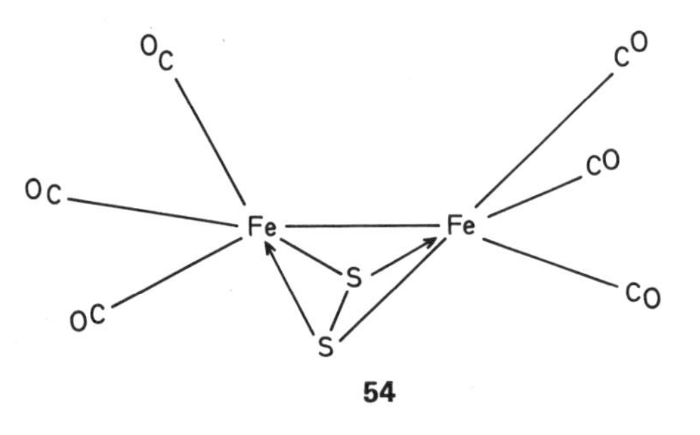

54

Several types of organosulfur derivatives of ruthenium carbonyl are known. Reactions of thiols with $Ru_3(CO)_{12}$ under mild conditions[194] form orange trimetallic derivatives $HRu_3(CO)_{10}$-SR (**55**, M=Ru; R=C_2H_5, n-C_4H_9, and C_6H_5) in which the ruthenium triangle of $Ru_3(CO)_{12}$ (**51**, M=Ru) is retained.

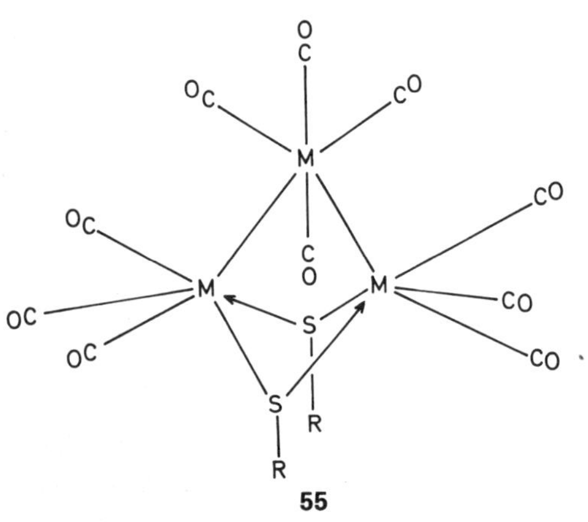

55

Reactions of thiols with $Ru_3(CO)_{12}$ under more vigorous conditions or reactions of disulfides with $Ru_3(CO)_{12}$ result in rupture of the ruthenium triangle to give derivatives of the type $[RSRu(CO)_3]_2$, $[RSRu(CO)_3]_n$, and $[(RS)_2Ru(CO)_2]_n$ which are mainly similar to known organosulfur derivatives of iron carbonyls.[195]
Reactions of $Os_3(CO)_{12}$ with the thiols RSH (R=C_6H_5, C_2H_5, and n-C_4H_9) give the trimetallic derivatives $HOs_3(CO)_{10}SR$ (**55**, M=Os; R=C_6H_5, C_2H_5, and n-C_4H_9) analogous to the trimetallic ruthenium derivatives described above.[194]

29.5.5 Iron Carbonyl Nitrosyl Derivatives

Of particular preparative importance in the area of iron carbonyl nitrosyl derivatives is the $[Fe(CO)_3NO]^{\ominus}$ anion which can be obtained as the dark yellow potassium salt by reaction of $Fe(CO)_5$ with potassium nitrite in methanol solution at 35° according to the following equation:[196]

$$Fe(CO)_5 + KNO_2 \longrightarrow K[Fe(CO)_3NO] + CO + CO_2$$

Reaction of this potassium salt with mercury(II) cyanide in aqueous solution gives a red precipitate of the non-ionic water-insoluble mercury derivative $Hg[Fe(CO)_3NO]_2$.[197] Reaction of $K[Fe(CO)_3NO]$ with allyl chloride gives the red, volatile, liquid π-allyl derivative $C_3H_5Fe(CO)_2$-NO (**56**),[198] which can also be prepared by reaction of $C_3H_5Fe(CO)_3$ (Section 29.5.7.7) with nitric oxide.[199]
Another *iron carbonyl nitrosyl* derivative of major significance is the red, volatile, air-sensitive $Fe(CO)_2(NO)_2$ which can be prepared by any of the following methods:

(1) Nitrosation of $HFe(CO)_4^{\ominus}$ or $[Fe(CO)_3$-$NO]^{\ominus}$ with sodium nitrite and a weak acid such as acetic acid or carbon dioxide.[196]
(2) Reaction of $Fe_3(CO)_{12}$ (Section 29.5.1) with nitric oxide at 95°.[200]
(3) Pyrolysis of $Hg[Fe(CO)_3NO]_2$[196]

The preparative method of choice depends upon the availability of the required starting materials.

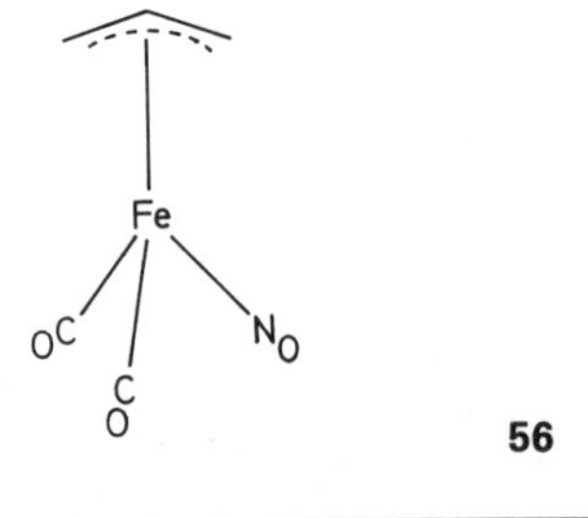

56

[192] *R. B. King*, Inorg. Chem. *2*, 326 (1963).
[193] *W. Hieber, J. Gruber*, Z. Anorg. Allgem. Chemie *296*, 91 (1958).
[194] *G. R. Crooks, B. F. G. Johnson, J. Lewis, I. G. Williams*, J. Chem. Soc. *A*, 797 (1969).
[195] *G. Cetini, O. Gambino, E. Sappa, M. Valle*, J. Organometal. Chem. *15* P4 (1968).
[196] *W. Hieber, H. Beutner*, Z. Anorg. Allgem. Chemie *320*, 101 (1963).
[197] *R. B. King*, Inorg. Chem. *2*, 1275 (1963).
[198] *R. Bruce, F. M. Chaudhary, G. R. Knox, P. L. Pauson*, Z. Naturforsch. *20b*, 73 (1965).
[199] *H. D. Murdoch*, Z. Naturforsch. *20b*, 179 (1965).
[200] *J. S. Anderson*, Z. Anorg. Allgem. Chemie *208*, 238 (1932);
R. L. Mond, A. E. Wallis, J. Chem. Soc. *121*, 32 (1922).

29.5.6 Carbonyl Derivatives Containing Organophosphorus and Similar Ligands

Reactions of $Fe(CO)_5$ with tertiary phosphines may give either the monosubstituted derivatives $R_3PFe(CO)_4$ or the disubstituted derivatives *trans*-$(R_3P)_2Fe(CO)_3$ depending upon the reaction conditions.[201] The monosubstituted derivatives $R_3PFe(CO)_4$ can be made with less risk of contamination with disubstituted derivatives *trans*-$(R_3P)_2Fe(CO)_3$ by the reaction of $Fe_2(CO)_9$ with the tertiary phosphine according to the following equation:[202]

$$Fe_2(CO)_9 + R_3P \longrightarrow R_3PFe(CO)_4 + Fe(CO)_5$$

The disubstituted derivatives $(R_3P)_2Fe(CO)_3$ can be prepared by displacement of a coordinated diolefin from a (diene)$Fe(CO)_3$ complex (Section 29.5.7.2); diene = butadiene, two carbon–carbon double bonds of cyclooctatetraene, *etc.*) with the tertiary phosphine according to the following equation:[203]

$$(\text{diene})Fe(CO)_3 + 2R_3P \longrightarrow (R_3P)_2Fe(CO)_3 + \text{diene}$$

The $(R_3P)_2Fe(CO)_3$ derivative formed in such reactions is normally the *trans*-isomer (**57**) except

57

when the two coordinating phosphorus atoms come from a chelating polytertiary phosphine. Reactions of $Fe_3(CO)_{12}$ with tertiary phosphines may give the monometallic derivatives $R_3PFe(CO)_4$ and *trans*-$(R_3P)_2Fe(CO)_3$.[204] However, if the reactions of $Fe_3(CO)_{12}$ with certain tertiary phosphines are carried under sufficiently mild conditions for a sufficiently short time, green-black to black trimetallic derivatives of the type $(R_3P)_nFe_3(CO)_{12-n}$ (n=1, 2, or 3) may be obtained. Examples of such compounds include the triphenylphosphine derivative $(C_6H_5)_3PFe_3(CO)_{11}$[205] and the phenyldimethylphosphine derivative $[C_6H_5P(CH_3)_2]_3Fe_3(CO)_9$.[206] Many of these reactions of iron carbonyls with tertiary phosphines occur similarly with related ligands such as tertiary arsines, tertiary stibines, and isocyanides.

Most reactions of $Ru_3(CO)_{12}$ and $Os_3(CO)_{12}$ with tertiary phosphines and related ligands result in retention of the metal triangle. Dark violet to orange trimetallic ruthenium derivatives of the type $[LRu(CO)_3]_3$

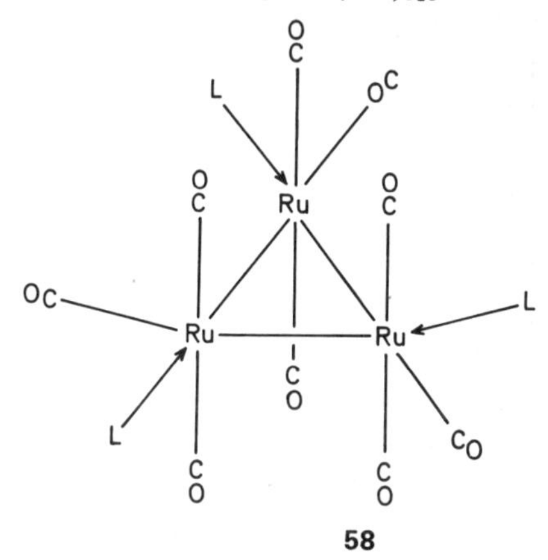

58

(**58**, L = $(C_6H_5)_3P$, $(n\text{-}C_4H_9)_3P$, $(C_2H_5)_3P$, $(C_6H_5O)_3P$, $(C_6H_5)_3As$, $(C_6H_5)_3Sb$, etc.) may be obtained by heating $Ru_3(CO)_{12}$ with the ligand in methanol or acetone solution at 55 to 60°.[207,208] The ruthenium triangle in the triphenylphosphine derivative $[(C_6H_5)_3PRu(CO)_3]_3$ is cleaved by carbon monoxide under pressure (160 atmospheres) at 150° to give pale yellow monometallic $(C_6H_5)_3PRu(CO)_4$ and with triphenylphosphine in methyl isobutyl ketone at 125° for 4 hr to give *trans*-$[(C_6H_5)_3P]_2Ru(CO)_3$.[209] The compound *trans*-$[(C_6H_5)_3P]_2Ru(CO)_3$ can also be obtained by the reductive carbonylation of hydrated ruthenium(III) chloride in the presence of triphenylphosphine and zinc.[210]

The metal carbonyl halides $Fe(CO)_4X_2$ (X = Cl, Br, or I) and $[Ru(CO)_2Cl_2]_n$ react readily with

---

[201] *F. A. Cotton, R. V. Parish*, J. Chem. Soc. 1440 (1960).

[202] *R. B. King, T. F. Korenowski*, Inorg. Chem. *10*, 1188 (1971).

[203] *A. Reckziegel, M. Bigorgne*, J. Organometal. Chem. *3*, 341 (1965).

[204] *A. F. Clifford, A. K. Mukherjee*, Inorg. Chem. *2*, 151 (1963).

[205] *R. J. Angelici, E. E. Siefert*, Inorg. Chem. *5*, 1457 (1966).

[206] *W. S. McDonald, J. R. Moss, G. Raper, B. L. Shaw, R. Greatrex, N. N. Greenwood*, Chem. Commun. 1295 (1969).

[207] *F. Piacenti, M. Bianchi, E. Benedetti, G. Sbrana*, J. Inorg. Nucl. Chem. *29*, 1389 (1967).

[208] *J. P. Candlin, A. C. Shortland*, J. Organometal. Chem. *16*, 289 (1969).

[209] *F. Piacenti, M. Bianchi, E. Benedetti, G. Braca*, Inorg. Chem. *7*, 1815 (1968).

[210] *J. P. Collman, W. R. Roper*, J. Amer. Chem. Soc. *87*, 4008 (1965).

Lewis bases to form substitution products particularly $L_2M(CO)_2X_2$.[211] Reactions of $Fe(CO)_2(NO)_2$ with tertiary phosphines and similar ligands give red derivatives of the types $LFeCO(NO)_2$ and $L_2Fe(NO)_2$.[212, 213]
Iron carbonyls react with the compounds R_2PPR_2 under mild conditions to form $(OC)_4FePR_2PR_2Fe(CO)_4$ (**59**) derivatives where the phosphorus–phosphorus bond is maintained.[214]

59

Iron carbonyls react with the compounds R_2PPR_2 under more vigorous conditions to give derivatives of the type $[R_2PFe(CO)_3]_2$ (**60**)

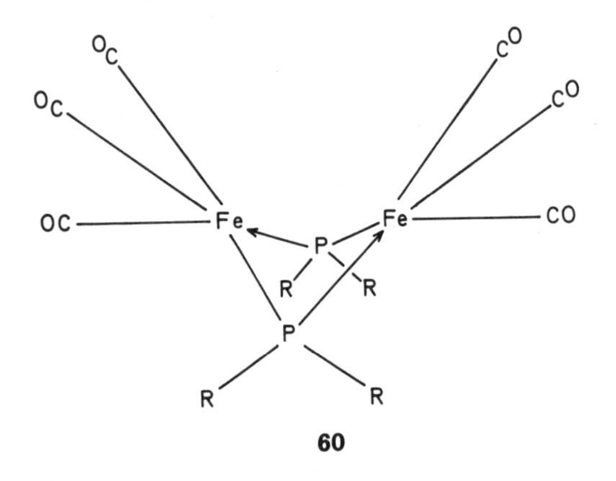

60

in which the phosphorus–phosphorus bond has been broken, but in which an iron–iron bond has been formed. The iron–iron bond in $[R_2PFe(CO)_3]_2$ (**60**) may be cleaved with halogens or similar reagents.

[211] *R. C. Taylor, W. D. Horrocks jr.*, Inorg. Chem. *3*, 584 (1964).
[212] *L. Malatesta, A. Araneo*, J. Chem. Soc. (London) 3803 (1957).
[213] *D. W. McBride, S. L. Stafford, F. G. A. Stone*, Inorg. Chem. *1*, 386 (1962).
[214] *R. G. Hayter*, Inorg. Chem. *3*, 711 (1964).

29.5.7 Olefin Derivatives of the Metal Carbonyls

29.5.7.1 Monoolefin Derivatives of Iron Carbonyls

Monoolefin derivatives of iron carbonyl of the type (olefin)$Fe(CO)_4$ can be prepared by reactions of $Fe_2(CO)_9$ (Section 29.5.1) with the monoolefin in a hydrocarbon solvent at 25 to 50° according to the following equation:[215]

$$Fe_2(CO)_9 + \text{olefin} \longrightarrow (\text{olefin})Fe(CO)_4 + Fe(CO)_5$$

Monoolefins forming (olefin)$Fe(CO)_4$ derivatives in the above reactions include particularly ones with electronegative substituents such as maleic anhydride, maleimide, fumaric acid, maleic acid, acrylic acid, methyl acrylate, acrylamide, acrolein, acrylonitrile, and cinnamaldehyde. The yellow volatile liquid ethylene complex $C_2H_4Fe(CO)_4$ can be prepared by reaction of $Fe_2(CO)_9$ with ethylene under pressure;[216] careful fractional vacuum distillation is necessary to separate the $C_2H_4Fe(CO)_4$ from the $Fe(CO)_5$ concurrently produced.

29.5.7.2 Diolefin Derivatives of Iron Carbonyls

Most conjugated diolefins form iron carbonyl complexes of the type (diene)$Fe(CO)_3$. The butadiene complex, $C_4H_6Fe(CO)_3$ (**61**), a yellow liquid, m.p. 19°, b.p. 47–49°/0·1 mm, can be prepared by reaction of $Fe(CO)_5$ with butadiene in a sealed stainless steel cylinder at ~140°.[217] In other cases, $Fe(CO)_5$ at ~130° or $Fe_3(CO)_{12}$ at ~80° may be used to convert conjugated diolefins into their corresponding iron tricarbonyl complexes. Conjugated diolefins besides butadiene known to form diene-iron tricarbonyl complexes include isoprene,[218] piperylene,[218] various derivatives of sorbic acid ($CH_3CH{=}CH{-}CH{=}CHCOOH$),[218] 1,3-cyclohexadiene,[217] octafluoro-1,3-cyclohexadiene,[219] and 1,3-cycloheptadiene.[220] Reaction of cycloheptatriene with $Fe(CO)_5$ gives both the corresponding C_7H_8Fe-

[215] *E. Weiss, K. Stark, J. E. Lancaster, H. D. Murdoch*, Helv. Chim. Acta *46*, 288 (1963).
[216] *H. D. Murdoch, E. Weiss*, Helv. Chim. Acta *46*, 1588 (1963).
[217] *B. F. Hallam, P. L. Pauson*, J.Chem. Soc. (London) 642 (1958).
[218] *R. B. King, T. A. Manuel, F. G. A. Stone*, J. Inorg. Nucl. Chem. *16*, 233 (1961).
[219] *H. H. Hoehn, L. Pratt, K. F. Watterson, G. Wilkinson*, J. Chem. Soc. (London) 2738 (1961).
[220] *R. Burton, L. Pratt, G. Wilkinson*, J. Chem. Soc. (London) 594 (1961);
H. J. Dauben jr., D. J. Bertelli, J. Amer. Chem. Soc. *83*, 497 (1961).

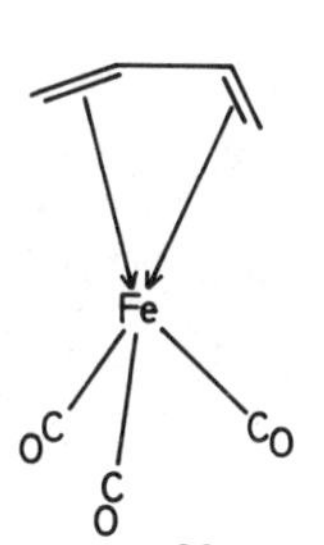

61 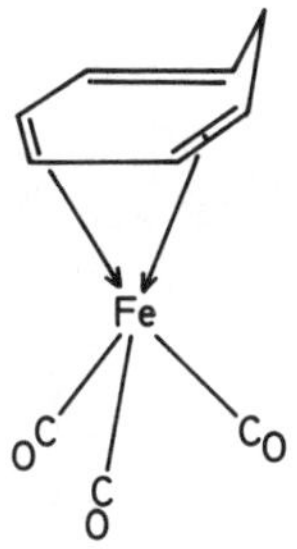 62 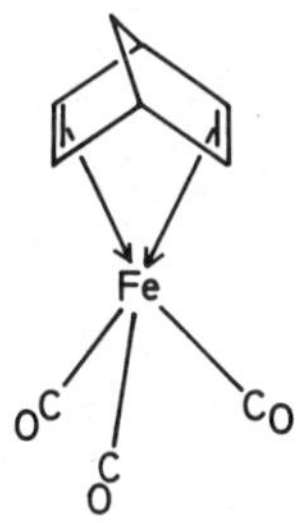 63 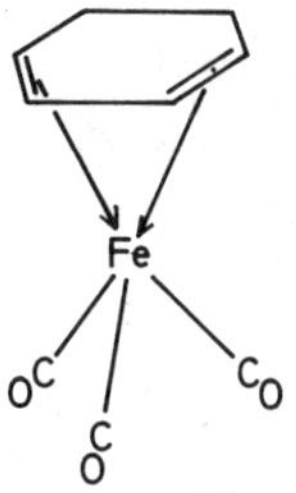 64

$(CO)_3$ complex (**62**) where two of the three double bonds of the cycloheptatriene are bonded to the iron atom and the 1,3-cycloheptadiene complex $C_7H_{10}Fe(CO)_3$ identical to that obtained from $Fe(CO)_5$ and 1,3-cycloheptadiene.[220] The non-conjugated diolefins which form stable (diene)$Fe(CO)_3$ complexes are mostly bicyclic ones such as bicyclo[2,2,1] heptadiene (norbornadiene) which reacts with $Fe(CO)_5$ at 100° to give the yellow-orange liquid complex norbornadienetricarbonyliron, $C_7H_8Fe(CO)_3$ (**63**).[221] Similar diene–iron tricarbonyl complexes can only rarely be prepared from other types of non-conjugated diolefins, since they readily undergo isomerization to conjugated diolefins when heated with iron carbonyls. Indeed, reactions of iron carbonyls with non-conjugated diolefins are often better methods for preparing (diene)$Fe(CO)_3$ complexes of the conjugated diolefins formed upon isomerization since many non-conjugated diolefins have a lower tendency to form oligomers and polymers upon heating than the corresponding isomeric conjugated diolefins. Thus the preferred method for preparing 1,3-cyclohexadienetricarbonyliron, C_6H_8-$Fe(CO)_3$ (**64**) utilizes the reaction of $Fe(CO)_5$ with 1,4-cyclohexadiene rather than 1,3-cyclohexadiene in a sealed stainless steel cylinder at 150°[222]; this method avoids contamination of the products with dimers, etc., of 1,3-cyclohexadiene.

29.5.7.3 Iron Carbonyl Derivatives of Cyclobutadiene Derivatives

Many cyclobutadiene derivatives form iron tricarbonyl complexes similar to the (diene)Fe-$(CO)_3$ complexes (Section 29.5.7.2). However, special techniques are necessary for the syntheses of iron tricarbonyl derivatives of cyclobutadienes, since cyclobutadienes are not normally obtainable in the free state for use as starting materials.

The following synthetic techniques are the most useful for preparing iron tricarbonyl complexes of cyclobutadiene and its substitution products: (*1*) *Reactions of 3,4-dichlorocyclobutenes with* $Fe_2(CO)_9$: The unsubstituted cyclobutadienetricarbonyliron, $C_4H_4Fe(CO)_3$ (**65**, X=H) can be prepared by reaction of 3,4-dichlorocyclobutenes with $Fe_2(CO)_9$ according to the following equation:[223]

$$C_4H_4Cl_2 + Fe_2(CO)_9 \longrightarrow C_4H_4Fe(CO)_3 + FeCl_2 + 6CO$$

The required 3,4-dichlorocyclobutene can be prepared from cyclooctatetraene by a three-step synthesis.[224]

(*2*) *Reactions of cis-3,4-carbonyldioxycyclobutenes with iron carbonyls:* Photolysis of vinylene carbonate with acetylenes forms the *cis*-3,4-carbonyldioxycyclobutenes (**66**).[225] These compounds react with $Na_2Fe(CO)_4$ (Section 5.2) or, less desirably, $Fe_2(CO)_9$ (Section 29.5.1) to give cyclobutadiene-iron tricarbonyl derivatives. This method is particularly useful for preparing 1,2-disubstituted cyclobutadieneiron tricarbonyl derivatives. It can also be used for the preparation of certain monosubstituted cyclobutadiene-iron tricarbonyl derivatives and also provides one way of avoiding the need for cyclooctatetraene as a starting material for the preparation of unsubstituted $C_4H_4Fe(CO)_3$ (**65**, X=H).

(*3*) *Photochemical Reaction of* $Fe(CO)_5$ *with α-pyrone:* u.v. irradiation of α-pyrone (**67**) with $Fe(CO)_5$ results in the elimination of carbon

[221] *R. Pettit*, J. Amer. Chem. Soc. *81*, 1266 (1959).

[222] *R. B. King*, Organometal. Synth. *1*, 129 (1965).

[223] *G. F. Emerson, L. Watts, R. Pettit*, J. Amer. Chem. Soc. *87*, 131 (1965); *L. A. Paquette, L. D. Wise*, J. Amer. Chem. Soc. *89*, 6659 (1967).

[224] *M. Avram, I. Dinulescu, M. Elian, M. Farcasiu, E. Marica, G. Mateescu, C. D. Nenitzescu*, Chem. Ber. *97*, 372 (1964).

[225] *R. H. Grubbs*, J. Amer. Chem. Soc. *92*, 6693 (1970).

dioxide to give the unsubstituted cyclobutadiene-tricarbonyliron (**65**, X=H) according to the following equation:[226]

$$C_5H_4O_2(\mathbf{67}) + Fe(CO)_5 \longrightarrow C_4H_4Fe(CO)_3(\mathbf{65}, X=H) + 2CO + CO_2$$

(*4*) *Reactions of* $Fe(CO)_5$ *with Alkynes* (*see Section 29.5.7.8*): certain alkynes, such as diphenylacetylene, react with $Fe(CO)_5$ under relatively vigorous conditions to give substituted cyclobutadiene-iron tricarbonyls, $R_4C_4Fe(CO)_3$, in addition to other products.[227] This method is occasionally useful for preparing completely substituted cyclobutadiene-iron tricarbonyl derivatives.

Certain substituted cyclobutadiene-iron tricarbonyl derivatives can be prepared from $C_4H_4Fe(CO)_3$ (**65**, X=H) by conventional types of organic reactions. The cyclobutadiene ring in $C_4H_4Fe(CO)_3$ (**65**, X=H) is susceptible to electrophilic substitution reactions. Its reactivity towards electrophilic substitution reactions is greater than that of $C_5H_5Mn(CO)_3$ (Section 29.4.9).[228] Thus the acetyl derivative $CH_3COC_4H_3Fe(CO)_3$ (**65**, X=$COCH_3$) can be prepared by Friedel–Crafts acylation of $C_4H_4Fe(CO)_3$ (**65**, X=H) with acetyl chloride in the presence of aluminum chloride. The aldehyde $HCOC_4H_3Fe(CO)_3$ (**65**, X=CHO) can be prepared by reaction of $C_4H_4Fe(CO)_3$ (**65**, X=H) with a mixture of *N*-methylformanilide and phosphorus oxychloride. Ketones and aldehydes of these types can be converted to various alcohols by conventional organic methods using hydridic reducing agents such as $NaBH_4$ or alkylmagnesium compounds. The chloromethyl derivative $ClCH_2C_4H_3Fe(CO)_3$ (**65**, X=CH_2Cl) may be obtained by reaction between $C_4H_4Fe(CO)_3$, formaldehyde, and hydrochloric acid. Similarly, the dimethylaminomethyl derivative $(CH_3)_2NCH_2C_4H_3Fe(CO)_3$ (**65**: X=$(CH_3)_2NCH_2$) may be obtained by reaction between $C_4H_4Fe(CO)_3$, formaldehyde, and dimethylamine. The chloromercuri derivative $ClHgC_4H_3Fe(CO)_3$ (**65**, X=ClHg) can be prepared by mercuration of $C_4H_4Fe(CO)_3$ (**65**, X=H) with a mixture of mercuric acetate and sodium chloride.

[226] *M. Rosenblum, C. Gatsonis*, J. Amer. Chem. Soc. *89*, 5074 (1967)

[227] *F. L. Bowden, A. B. P. Lever*, Organometal. Chem. Rev. *3*, 227 (1968).

[228] *L. Watts, J. D. Fitzpatrick, R. Pettit*, J. Amer. Chem. Soc. *87*, 3253 (1965).

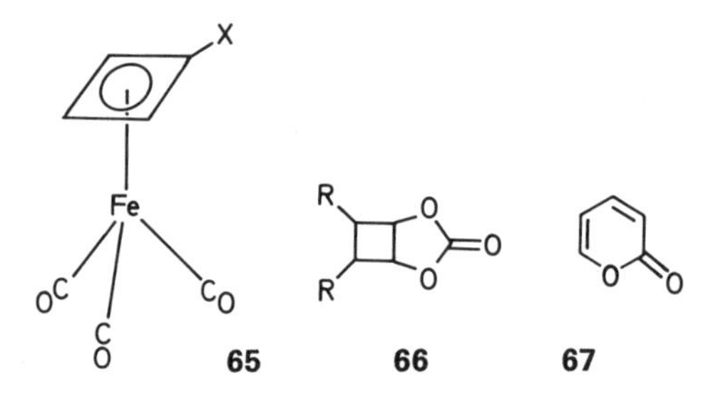

29.5.7.4 Cyclohexadienyl and Cycloheptadienyl Derivatives of Iron Carbonyl

Iron tricarbonyl complexes of the type $[(dienyl)Fe(CO)_3]^{\oplus}$ can be prepared either by abstraction of hydride from a $(diene)Fe(CO)_3$ complex with a CH_2 group adjacent to the complexed diolefin or by addition of a proton to an uncomplexed carbon–carbon double bond in a $(diene)Fe(CO)_3$ complex. Both methods are illustrated by the following two alternative preparations of the cycloheptadienyltricarbonyliron cation, $[C_7H_9Fe(CO)_3]^{\oplus}$ (**68**):

(1) Protonation of the cycloheptatriene complex $C_7H_8Fe(CO)_3$ (**62**, Section 29.5.7.2) according to the following equation:[220]

$$C_7H_8Fe(CO)_3 + HBF_4 \longrightarrow [C_7H_9Fe(CO)_3][BF_4]$$

Reactions of this type are generally performed in acetic anhydride or propionic anhydride. Protonation can also be effected in some cases by reaction of the $(diene)Fe(CO)_3$ with gaseous hydrogen halide in an inert solvent.

(2) Abstraction of hydride from the 1,3-cycloheptadiene complex $C_7H_9Fe(CO)_3$ according to the following equation:[220]

$$C_7H_{10}Fe(CO)_3 + [(C_6H_5)_3C][BF_4] \longrightarrow [C_7H_9Fe(CO)_3][BF_4] + (C_6H_5)_3CH$$

Reactions of this type are generally performed in dichloromethane or a similar solvent in which the reactants are soluble but the $[C_7H_9Fe(CO)_3][BF_4]$ product is insoluble. In most cases the $[(dienyl)Fe(CO)_3]^{\oplus}$ cations are most conveniently isolated as their tetrafluoroborate or hexafluorophosphate salts. Other $[(dienyl)Fe(CO)_3]^{\oplus}$ cations of some importance include the cyclohexadienyltricarbonyl cation, $[C_6H_7Fe(CO)_3]^{\oplus}$ (**69**) prepared by halide abstraction from 1,3-cyclohexadienetricarbonyliron, $C_6H_8Fe(CO)_3$[229] and the cation $[C_8H_9Fe(CO)_3]^{\oplus}$ prepared by protonation of cyclooctatetrae-

[229] *E. O. Fischer, R. D. Fischer*, Angew. Chem. *72*, 919 (1960).

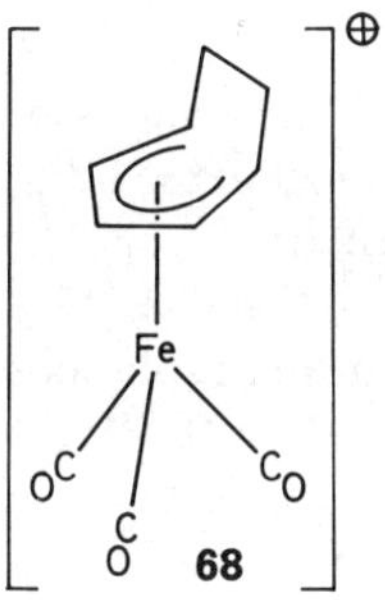

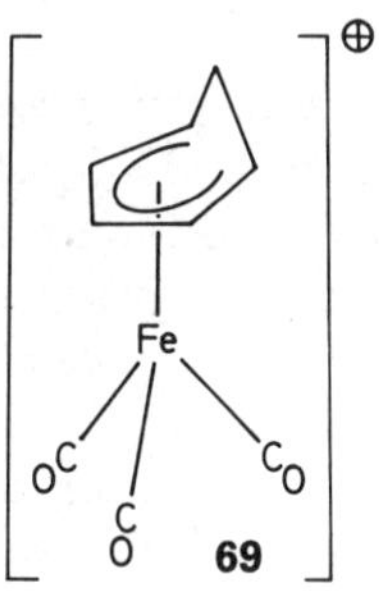

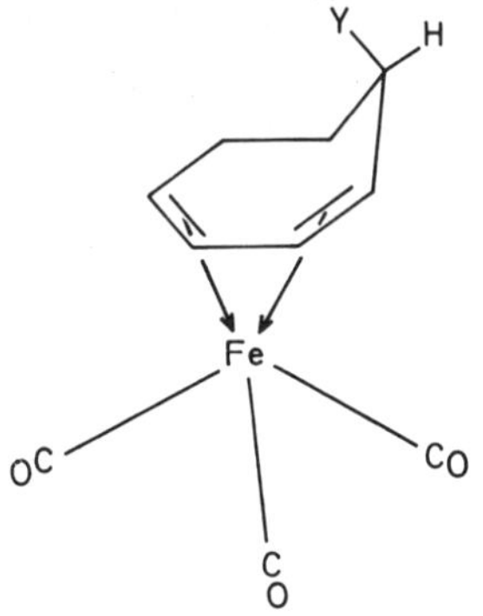

netricarbonyliron, $C_8H_8Fe(CO)_3$ (Section 29.5.7.5).[230]

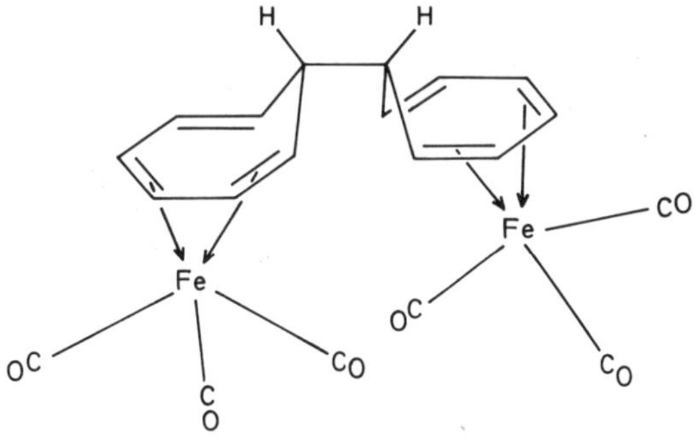

The $[(dienyl)Fe(CO)_3]^{\oplus}$ cations are relatively reactive and can thus be converted to other dienyliron carbonyl derivatives. Some of these reactions are conveniently illustrated by the chemistry of the cycloheptadienyl derivatives. The maroon iodide $C_7H_9Fe(CO)_2I$ can conveniently be prepared by reaction of $[C_7H_9Fe(CO)_3][BF_4]$ with potassium iodide in acetone solution.[220] This iodide reacts with a stoichiometric amount of sodium amalgam in tetrahydrofuran to give the bimetallic $[C_7H_9Fe(CO)_2]_2$ and with potassium cyanide in acetone to give the orange cyano derivative $C_7H_9Fe(CO)_2CN$. The cation $[C_7H_9Fe(CO)_3]^{\oplus}$ reacts with numerous nucleophiles (e.g. cyanide, alkoxides, and the diethyl malonate anion) to give ring-substituted 1,3-cycloheptadiene-iron tricarbonyl derivatives of the type $C_7H_8YFe(CO)_3$ (**70**, X=H; Y=CN, OR, $CH(CO_2C_2H_5)_2$; R=CH_3 or C_2H_5). Reaction of the cation $[C_7H_9Fe(CO)_3]^{\oplus}$ with zinc dust in tetrahydrofuran results in coupling to give the bimetallic complex $(OC)_3FeC_7H_9$–$C_7H_9Fe(CO)_3$ (**71**).

[230] *G. N. Schrauzer*, J. Amer. Chem. Soc. *83*, 2966 (1961);
A. Davison, W. McFarlane, L. Pratt, G. Wilkinson, J. Chem. Soc. (London) 4821 (1962);
M. Brookhart, E. R. Davis, J. Amer. Chem. Soc. *92*, 7622 (1970).

29.5.7.5 Eight-membered Carbocyclic Ring Derivatives of Iron, Ruthenium, and Osmium Carbonyls

Several interesting iron carbonyl complexes can be prepared from iron carbonyls and cyclic polyolefins with eight-membered rings such as 1,3,5-cyclooctatriene and cyclooctatetraene. The yellow-orange liquid bicyclo[4,2,0] octadienetricarbonyliron,[231] $C_8H_{10}Fe(CO)_3$ (**72**) can be obtained by reaction of 1,3,5-cyclooctatriene with $Fe(CO)_5$ at ~130°. The compounds $C_8H_{10}Fe(CO)_3$, a metal complex of 1,3,5-cyclooctatriene of apparent structure **73**, and $C_8H_{10}Fe_2(CO)_6$ (**74**) can be obtained if the reaction between 1,3,5-cyclooctatriene and iron carbonyls is carried out under milder conditions

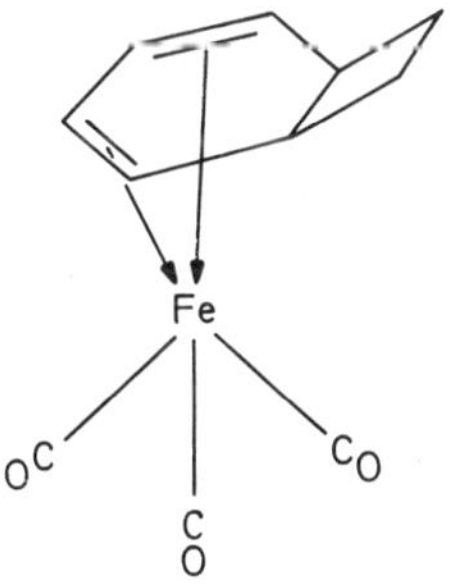

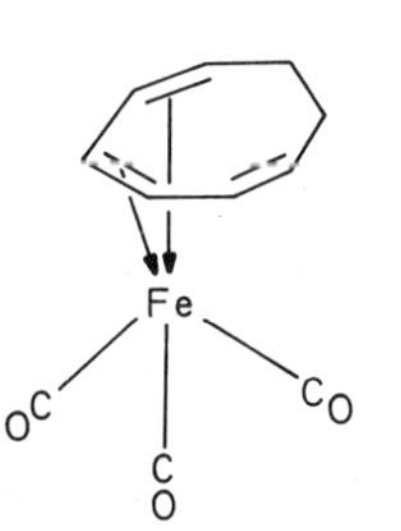

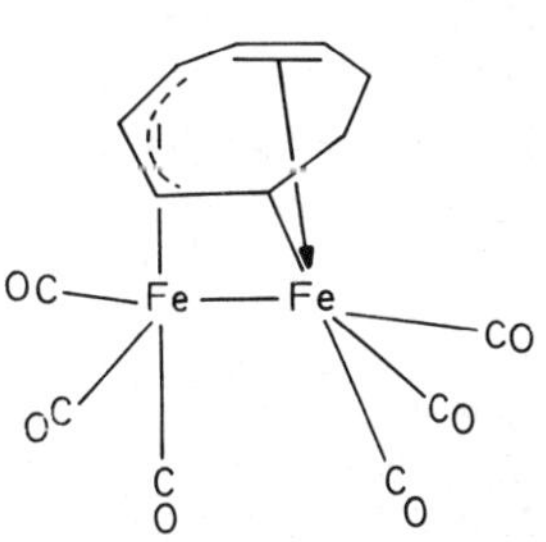

[231] *T. A. Manuel, F. G. A. Stone*, J. Amer. Chem. Soc. *82*, 366 (1960).

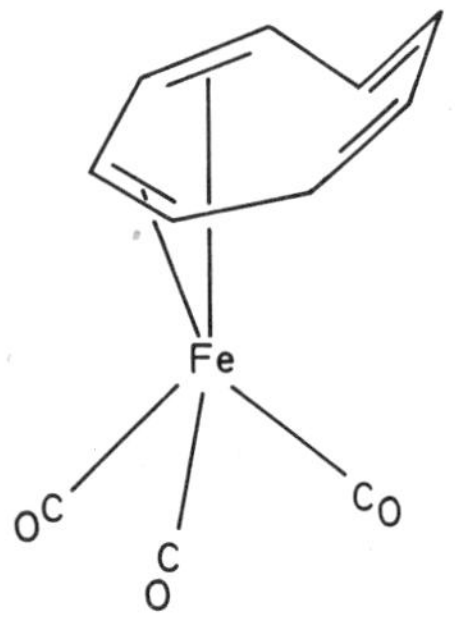

(~80°) using the more reactive $Fe_3(CO)_{12}$.[232] Other cyclic triolefins (e.g. cycloheptatriene, tropone, etc.) react with $Fe_2(CO)_9$ under mild conditions to form (triene)$Fe_2(CO)_6$ derivatives similar to (**74**).[233]

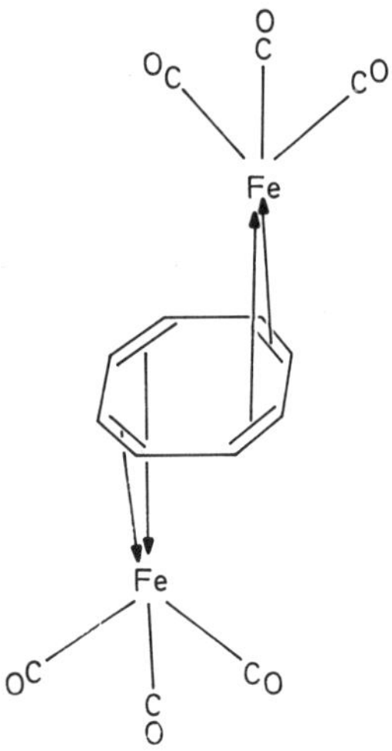

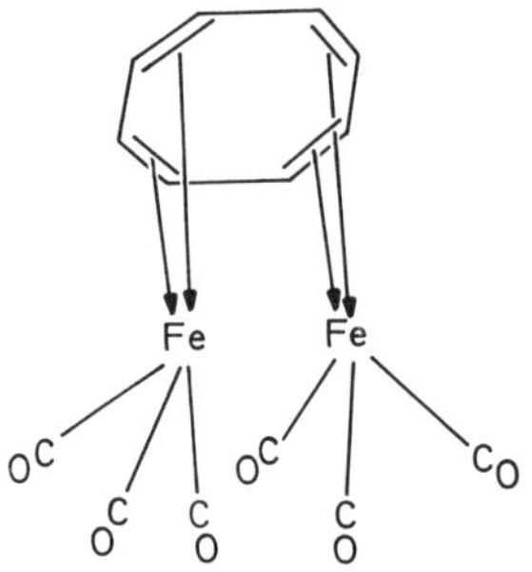

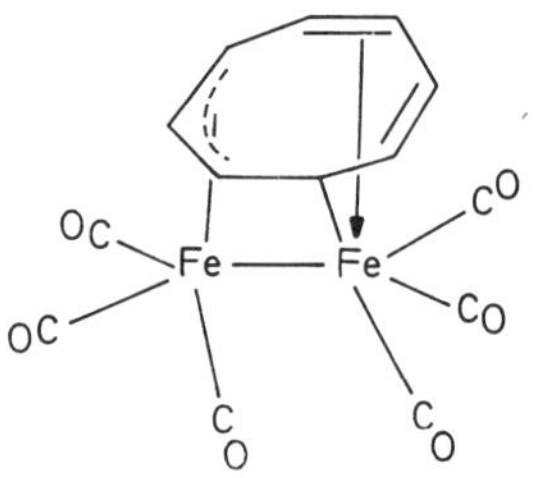

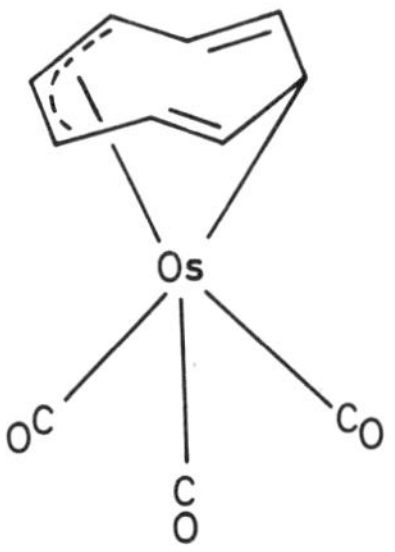

Several *iron carbonyl complexes* containing *cyclooctatetraene* can be prepared. The most readily available of these is red $C_8H_8Fe(CO)_3$ (**75**) which is the major iron carbonyl derivative formed when $Fe(CO)_5$ is reacted with cyclooctatetraene at temperatures above 100°.[231] Three isomeric $C_8H_8Fe_2(CO)_6$ (**76**, **77**, and **78**) complexes can also be obtained from cyclooctatetraene and iron carbonyls under the proper conditions.[231, 234] The *trans*-isomer **76** is a by-product from the preparation of $C_8H_8Fe(CO)_3$ (**75**) using the reaction of $Fe(CO)_5$ with cyclooctatetraene at elevated temperatures. The *cis*-isomer **77** and the metal–metal bonded isomer **78** require very mild conditions for their formation; reaction of $Fe_2(CO)_9$ with cyclooctatetraene at room temperature gives the best results.[234] The black diiron pentacarbonyl derivative $C_8H_8Fe_2(CO)_5$ (**79**) can be obtained by heating solutions of either $C_8H_8Fe_2(CO)_6$ isomer **77** or **78**.[234]

Cyclooctatetraene also forms complexes with *ruthenium* and *osmium carbonyls*. The compounds $C_8H_8Ru(CO)_3$ (**75**, M=Ru), $C_8H_8Ru_2(CO)_6$ (**77**, M=Ru), and $C_8H_8Ru_2(CO)_5$ (**79**, M=Ru) analogous to known cyclooctatetraene-iron carbonyl complexes can be obtained from the reaction of cyclooctatetraene with $Ru_3(CO)_{12}$ (Section 29.5.1) in boiling heptane.[235] The red air-stable crystalline trimetallic derivative $(C_8H_8)_2Ru_3(CO)_4$ of structure **80** not analogous to any of the known cyclooctatetraene-iron carbonyls can be prepared in relatively good yield by the reaction of $Ru_3(CO)_{12}$ with cyclooctatetraene in boiling octane.[235]

---

[232] *T. A. Manuel, F. G. A. Stone*, J. Amer. Chem. Soc. *82*, 6240 (1960);
R. B. King, Inorg. Chem. *2*, 807 (1963);
F. A. Cotton, W. T. Edwards, J. Amer. Chem. Soc. *91*, 843 (1969).

[233] *G. F. Emerson, J. E. Mahler, R. Pettit, R. Collins*, J. Amer. Chem. Soc. *86*, 3590 (1964).

[234] *C. E. Keller, G. F. Emerson, R. Pettit*, J. Amer. Chem. Soc. *87*, 1388 (1965).

[235] *M. I. Bruce, M. Cooke, M. Green*, J. Organometal. Chem. *13*, 227 (1968);
F. A. Cotton, A. Davison, T. J. Marks, A. Musco, J. Amer. Chem. Soc. *91*, 6598 (1969).

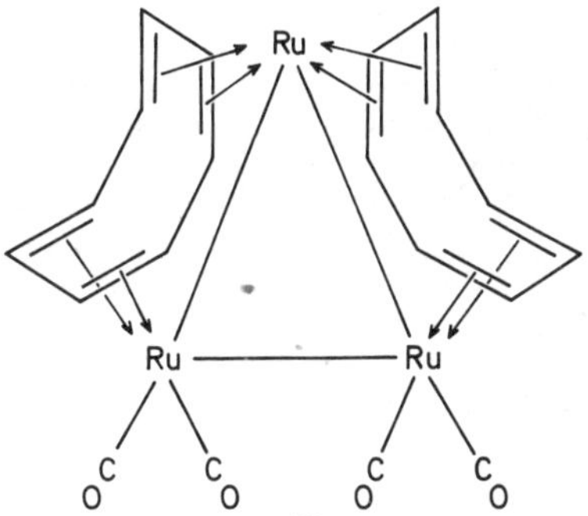

Ru
Ru Ru
C O C O C O C O

80

Ultraviolet irradiation of cyclooctatetraene with $Os_3(CO)_{12}$ gives a $C_8H_8Os(CO)_3$ of structure **81** which is not analogous to that of $C_8H_8Fe(CO)_3$ (**75**, M = Fe). However, $C_8H_8Os(CO)_3$ (**81**) rearranges to an isomeric $C_8H_8Os(CO)_3$ upon heating which has a structure (**75**, M = Os) analogous to that of $C_8H_8Fe(CO)_3$.[236]

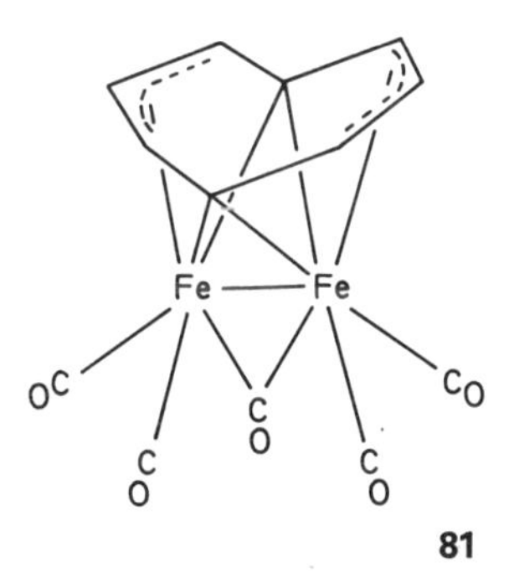

81

29.5.7.6 Azulene Derivatives of Iron and Ruthenium Carbonyls

Azulene and its alkyl derivatives, such as 4,6,8-trimethylazulene and guaiazulene, form unusual metal complexes with iron and ruthenium carbonyls. Dark red $C_{10}H_8Fe_2(CO)_5$ (**82**) is the major product obtained from the reaction of azulene with $Fe(CO)_5$.[237] Minor products from this azulene-$Fe(CO)_5$ reaction are $(C_{10}H_8)_2Fe_4(CO)_{10}$[238] and $[C_{10}H_8Fe(CO)_2]_2$.[239] The red tetrametallic complex $(CH_3)_3C_{10}H_5Ru_4(CO)_9$ is the major product obtained from the reaction of 4,6,8-trimethylazulene with $Ru_3(CO)_{12}$.[240]

---

[236] *M. I. Bruce, M. Cooke, M. Green*, Angew. Chem. Intern. Ed. Engl. *7*, 639 (1968).
[237] *R. Burton, L. Pratt, G. Wilkinson*, J. Chem. Soc. (London) 4290 (1960);
M. R. Churchill, Chem. Commun. 450 (1966).
[238] *M. R. Churchill, P. H. Bird*, Inorg. Chem. *8*, 1941 (1969).
[239] *R. B. King*, J. Amer. Chem. Soc. *88*, 2075 (1966).
[240] *M. R. Churchill, K. Gold, P. H. Bird*, Inorg. Chem. *8*, 1956 (1969).

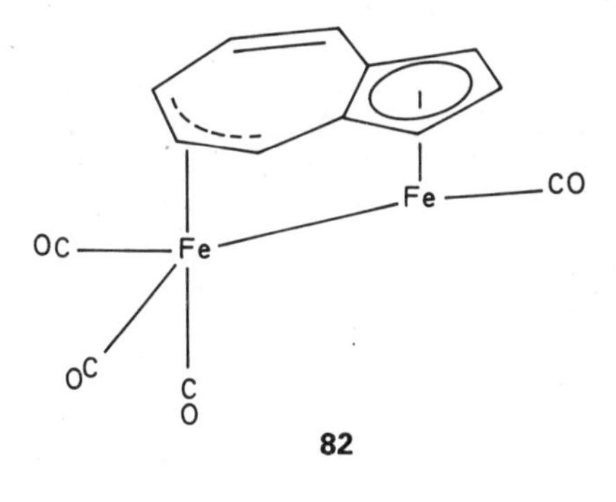

82

29.5.7.7 Allyl Derivatives of Iron and Ruthenium Carbonyls

Numerous allylic halides react with $Fe(CO)_5$ or $Fe_2(CO)_9$[241] under mild conditions to give *π-allylic iron tricarbonyl halides* as exemplified by the following equation for the preparation of the unsubstituted π-allyltricarbonyliodoiron, $C_3H_5Fe(CO)_3I$, from allyl iodide and $Fe(CO)_5$.[242]

$$Fe(CO)_5 + C_3H_5X \longrightarrow C_3H_5Fe(CO)_3X + 2CO$$

The reactions between the iron carbonyls and the allyl halides must be carried out at temperatures below 50°. If this temperature is exceeded, further reaction of the allyl halides with the iron carbonyls will occur to form ultimately ferrous halides.

The π-allyliron tricarbonyl halides $C_3H_5Fe(CO)_3X$ are yellow (X = Cl) to brown (X = I) solids subliming readily at 40°/0·1 mm. Reaction of the halides $C_3H_5Fe(CO)_3X$ with silver salts results in removal of the halogen to give salts of the $[C_3H_5Fe(CO)_3]^{\oplus}$ cation.[243] Chromatography of the iodide $C_3H_5Fe(CO)_3I$ on alumina in benzene solution removes the iodine atom giving $[C_3H_5Fe(CO)_3]_2$ as a red pyrophoric solid subliming at 25°/1 mm.[244]

The *π-allylruthenium tricarbonyl halides*, $C_3H_5Ru(CO)_3X$ (X = Cl, Br, or I) can be prepared by reaction of $Ru_3(CO)_{12}$ with allyl halides at 60–70° in isooctane solution.[245]

---

[241] *R. A. Plowman, F. G. A. Stone*, Z. Naturforsch. *17b*, 575 (1962);
H. D. Murdoch, E. Weiss, Helv. Chim. Acta *45*, 1927 (1962);
R. F. Heck, C. R. Boss, J. Amer. Chem. Soc. *86*, 2580 (1964).
[242] *R. B. King*, Organometal. Synth. *1*, 176 (1965).
[243] *G. F. Emerson, R. Pettit*, J. Amer. Chem. Soc. *84*, 4591 (1962).
[244] *H. D. Murdoch, E. A. C. Lucken*, Helv. Chim. Acta *47*, 1517 (1964).
[245] *G. Sbrana, G. Braca, F. Piacenti, P. Pino*, J. Organometal. Chem. *13*, 240 (1968).

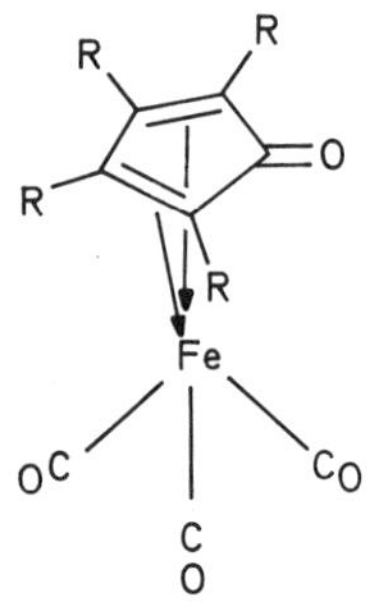

83

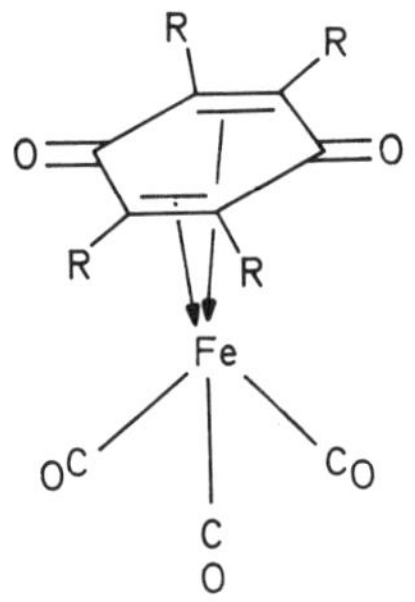

84

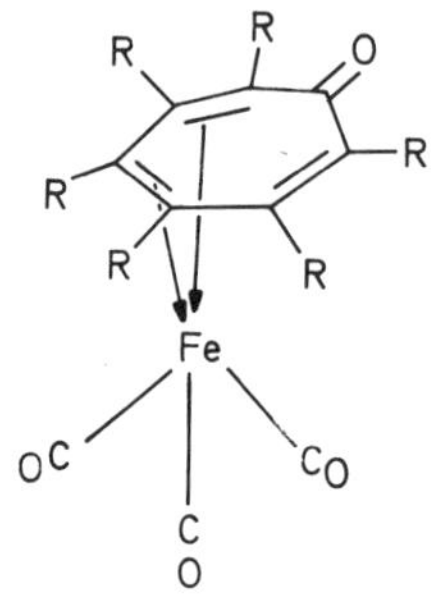

85

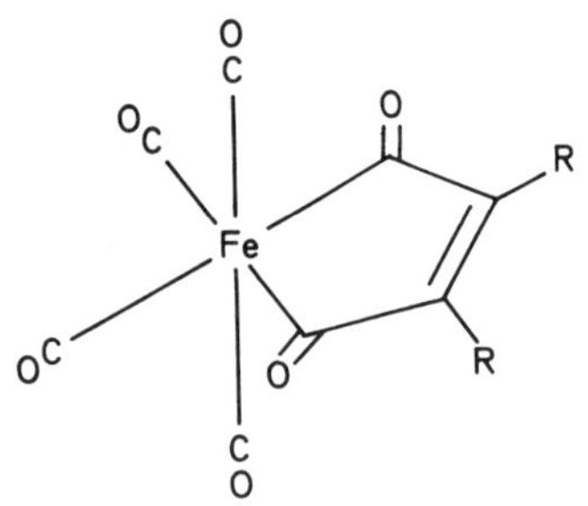

86

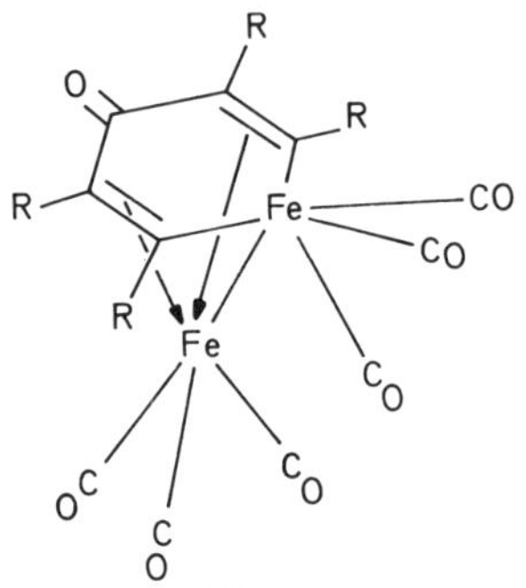

87

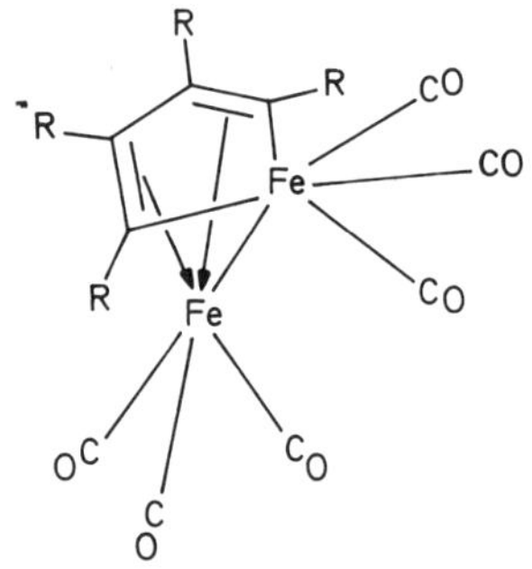

88

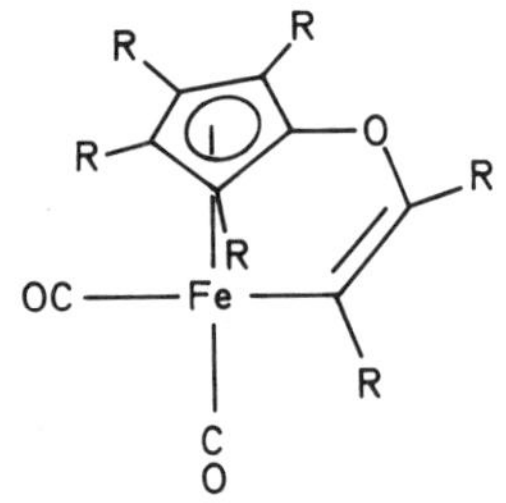

89

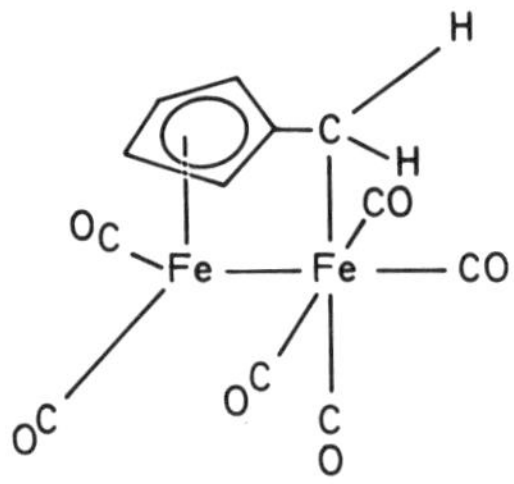

90

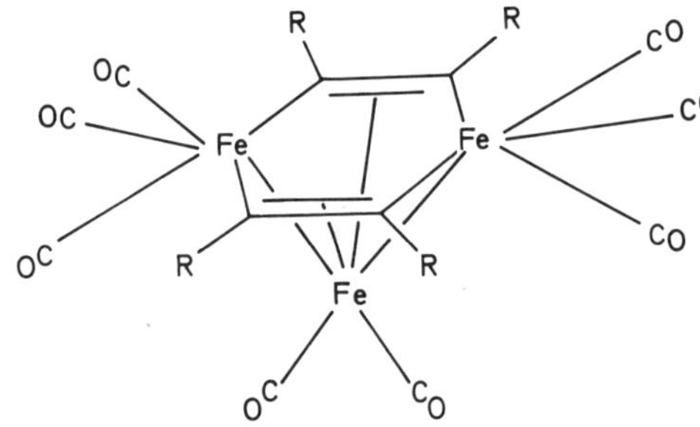

91

29.5.7.8 Complexes from Acetylenes and Iron and Ruthenium Carbonyls

Numerous unusual iron carbonyl complexes can be prepared by reactions of various acetylenes with iron carbonyls under a variety of conditions.[227] The preparation of substituted cyclobutadiene-iron tricarbonyls, $R_4C_4Fe(CO)_3$, from $Fe(CO)_5$ and disubstituted acetylenes $RC{\equiv}CR$ has already been discussed (Section 29.5.7.3). Other types of monometallic iron carbonyl complexes obtained from acetylenes and iron carbonyls include iron tricarbonyl π-complexes of substituted cyclopentadienones (**83**), quinones (**84**), and tropones (**85**).

In these complexes two or three acetylene units have joined together with or without the incorporation of carbon monoxide to form a ring system. Under fairly mild conditions an iron tetracarbonyl complex (**86**) with a σ-bonded maleyl group can be obtained. Binuclear complexes (*e.g.* **87** and **88**) with one iron tricarbonyl group in a heterocyclic ring and an iron–iron dative bond are also obtained.

Substituted π-cyclopentadienyl derivatives such as **89** and **90** can also be obtained from certain reactions between acetylenes and iron carbonyls; however, certain electronegative substituents such as alkoxycarbonyl appear to be necessary for the formation of complexes of the type **90**. More complex products obtained from reactions of acetylenes and iron carbonyls include the trimetallic derivative **91** and the iron carbonyl carbide $Fe_5(CO)_{15}C$.[246]

The ruthenium carbonyl $Ru_3(CO)_{12}$ reacts with alkynes such as $C_6H_5C{\equiv}CC_6H_5$ to give products

---

[246] *E. H. Braye, L. F. Dahl, W. Hübel, D. L. Wampler,* J. Amer. Chem. Soc. *84*, 4663 (1962).

similar to the iron carbonyl derivatives discussed above including **83** and **87**. In addition, under mild conditions (~70°) the reaction of $Ru_3(CO)_{12}$ with diphenylacetylene gives the trimetallic derivative $(C_6H_5C_2C_6H_5)_3Ru_3(CO)_9$ in which the ruthenium triangle of $Ru_3(CO)_{12}$ is maintained.[247]

29.5.8 Cyclopentadienylmetal Carbonyl Derivatives

29.5.8.1 Cyclopentadienylmetal Carbonyls

Iron, ruthenium, and osmium each form cyclopentadienylmetal carbonyl derivatives of the type $[C_5H_5M(CO)_2]_2$ (M = Fe, Ru, or Os). The red-violet *iron* derivative $[C_5H_5Fe(CO)_2]_2$ can be prepared either by heating $Fe(CO)_5$ with monomeric cyclopentadiene in an autoclave[248] at ~135° or more conveniently by heating $Fe(CO)_5$ with excess dicyclopentadiene at ~130°.[249] The orange *ruthenium* derivative $[C_5H_5Ru(CO)_2]_2$ is prepared by reaction of $[Ru(CO)_3Cl_2]_2$ or $[Ru(CO)_2Cl_2]_n$ (Section 29.5.3) with sodium cyclopentadienide in boiling tetrahydrofuran.[250] The yellow osmium *derivative* $[C_5H_5Os(CO)_2]_2$ is considerably more difficult to prepare but can be obtained in modest yield (~20%) by reaction of $[Os(CO)_3Cl_2]_2$ (Section 29.5.3) with sodium cyclopentadienide suspended in benzene at ~220°.[251]

Iron and ruthenium also form *tetrametallic cyclopentadienylmetal carbonyls* of the type $[C_5H_5MCO]_4$. The dark green air-stable $[C_5H_5FeCO]_4$ is prepared by heating $[C_5H_5Fe(CO)_2]_2$ in boiling xylene for several days.[252] The purple ruthenium analog $[C_5H_5RuCO]_4$ is prepared similarly from $[C_5H_5Ru(CO)_2]_2$.[253]

29.5.8.2 Cyclopentadienylmetal Carbonyl Anions and Hydrides

The anions $[C_5H_5M(CO)_2]^{\ominus}$ (M = Fe[254] and Ru[255]) can be prepared in solution as their sodium salts by reduction of the bimetallic derivatives $[C_5H_5M(CO)_2]_2$ (M = Fe or Ru, Section 29.5.8.1) with sodium amalgam in tetrahydrofuran solution according to the following equation:

$$[C_5H_5M(CO)_2]_2 + 2Na \longrightarrow 2NaM(CO)_2C_5H_5$$

Magnesium–iron derivatives of similar chemical reactivity as the $[C_5H_5Fe(CO)_2]^{\ominus}$ anion can be prepared according to either of the following equations:[256]

$$C_5H_5Fe(CO)_2Cl + Mg \longrightarrow C_5H_5Fe(CO)_2MgCl$$

$$[C_5H_5Fe(CO)_2]_2 + 1{,}2\text{-}C_2H_4Br_2 + 2Mg \longrightarrow 2C_5H_5Fe(CO)_2MgBr + CH_2{=}CH_2$$

Both of these reactions can be carried out in tetrahydrofuran solution. The use of the magnesium derivatives $C_5H_5Fe(CO)_2MgX$ rather than the sodium salt $NaFe(CO)_2C_5H_5$ for preparative purposes eliminates the need to use relatively large amounts of mercury and the resulting risk of having $Hg[Fe(CO)_2C_5H_5]_2$ as a by-product in the reaction mixture.[79]

The *hydrides* $HM(CO)_2C_5H_5$ (M = Fe or Ru) can be prepared by either of the following two methods:

(1) Reduction of $C_5H_5M(CO)_2Cl$ (M = Fe[257] or Ru[255]) with sodium borohydride in tetrahydrofuran solution.

(2) Reaction of $NaFe(CO)_2C_5H_5$ with *tert*-butyl chloride according to the following equation:[135]

$$NaFe(CO)_2C_5H_5 + (CH_3)_3CCl \longrightarrow HFe(CO)_2C_5H_5 + (CH_3)_2C{=}CH_2 + NaCl$$

These hydrides $HM(CO)_2C_5H_5$ (M = Fe or Ru) are unstable liquids which oxidize rapidly in air and which decompose rapidly at or above room temperature to give hydrogen and $[C_5H_5M(CO)_2]_2$.

29.5.8.3 Cyclopentadienylmetal Carbonyl Halides

The cyclopentadienylmetal carbonyl halides $C_5H_5M(CO)_2X$ (M = Fe or Ru; X = Cl, Br, or I)

---

[247] *C. T. Sears jr., F. G. A. Stone*, J. Organometal. Chem *11*, 644 (1968); *G. Cetini, O. Gambino, E. Sappa, M. Valle*, J. Organometal. Chem. *17*, 437 (1969); *O. Gambino, G. Cetini, E. Sappa, M. Valle*, J. Organometal. Chem. *20*, 195 (1969).

[248] *B. F. Hallam, P. L. Pauson*, J. Chem. Soc. (London) 3030 (1956)

[249] *T. S. Piper, F. A. Cotton, G. Wilkinson*, J. Inorg. Nucl. Chem. *1*, 165 (1955); *R. B. King, F. G. A. Stone*, Inorg. Synth. *7*, 110 (1963); *R. B. King*, Organometal. Synth. *1*, 114 (1965).

[250] *E. O. Fischer, A. Vogler*, Z. Naturforsch. *17b*, 421 (1962).

[251] *E. O. Fischer, A. Vogler, K. Noack*, J. Organometal. Chem. *7*, 135 (1967).

[252] *R. B. King*, Inorg. Chem. *5*, 227 (1966).

[253] *T. Blackmore, J. D. Cotton, M. I. Bruce, F. G. A. Stone*, J. Chem. Soc. *A*, 2931 (1968).

[254] *E. O. Fischer, R. Böttcher*, Z. Naturforsch. *10b*, 600 (1955).

[255] *A. Davison, J. A. McCleverty, G. Wilkinson*, J. Chem. Soc. (London) 1133 (1963).

[256] *J. M. Burlitch, S. W. Ulmer*, J. Organometal. Chem. *19*, P21 (1969).

[257] *M. L. H. Green, C. N. Street, G. Wilkinson*, Z. Naturforsch. *14b*, 738 (1959).

can be prepared from the corresponding $[C_5H_5M(CO)_2]_2$ (M = Fe or Ru) derivatives by either of the following two methods:

(1) Reaction of the $[C_5H_5M(CO)_2]_2$ derivative with the free halogen X_2 according to the following equation:

$$[C_5H_5M(CO)_2]_2 + X_2 \longrightarrow 2C_5H_5M(CO)_2X$$

(M = Ru, X = Br[250]; M = Fe, X = Br;[249] M = Fe, X = I[258]).

This reaction is best carried out in dichloromethane or chloroform.

(2) Air oxidation of $[C_5H_5Fe(CO)_2]_2$ in the presence of the corresponding hydrohalic acid according to the following equation:[259]

$$2[C_5H_5Fe(CO)_2]_2 + 4HX + O_2 \longrightarrow 4C_5H_5Fe(CO)_2X + 2H_2O$$

This air oxidation may be carried out in ethanol solution.

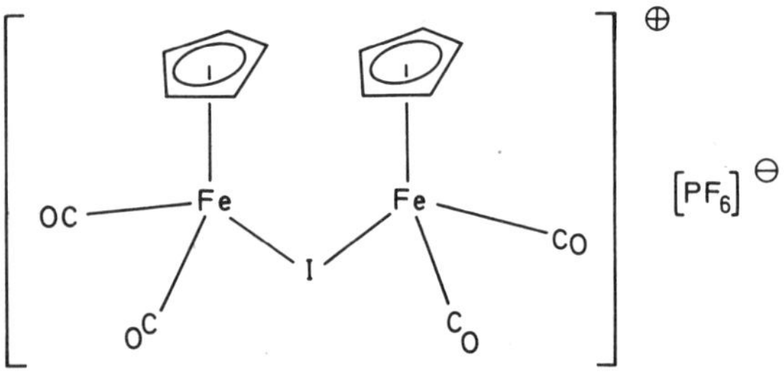

92

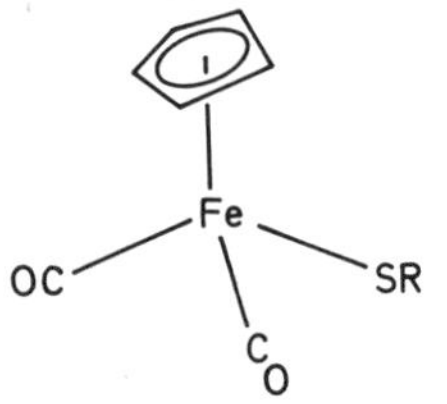

Another cyclopentadienyliron carbonyl iodide derivative of interest is the substituted iodonium ion $[C_5H_5Fe(CO)_2IFe(CO)_2C_5H_5]^{\oplus}$. **92** which may be isolated as its red hexafluorophosphate salt by reaction of $C_5H_5Fe(CO)_2I$ with anhydrous aluminium chloride in liquid sulfur dioxide solution followed by evaporation of the sulfur dioxide, hydrolysis, and treatment with ammonium hexafluorophosphate.[260]

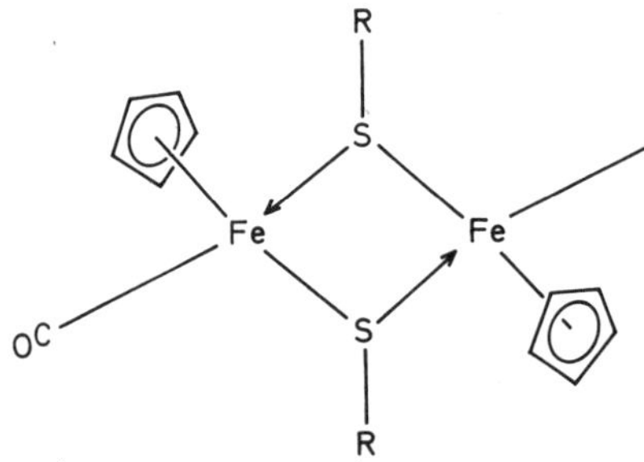

29.5.8.4 Organosulfur Derivatives of the Cyclopentadienyliron Carbonyls

Cyclopentadienyliron carbonyls form monometallic *alkylthio derivatives* of the general formula $C_5H_5Fe(CO)_2SR$ (**93**) and bimetallic alkylthio derivatives of the general formula $[C_5H_5FeCOSR]_2$ (**94**).

Thermal reaction of $[C_5H_5Fe(CO)_2]_2$ with a dialkyl disulfide such as CH_3SSCH_3 occurs at about ~100° to give bimetallic derivatives $[C_5H_5FeCOSR]_2$ (**94**).[261] However, the photochemical reaction of $[C_5H_5Fe(CO)_2]_2$ with dimethyl disulfide at room temperature gives the monometallic derivative $C_5H_5Fe(CO)_2SCH_3$.[261] The fluorinated analog, $C_5H_5Fe(CO)_2SCF_3$ (**93**, R = CF_3), can be prepared by reaction of the silver derivative CF_3SAg with $C_5H_5Fe(CO)_2I$.[262] The intense blue-green radical cation $[C_5H_5FeCOSCH_3]_2^{\oplus}$ can be obtained as its hexafluoroantimoniate salt by oxidation of $[C_5H_5FeCOSCH_3]_2$ (**94**, R = CH_3) with $AgSbF_6$ according to the following equation:[263]

$$[C_5H_5FeCOSCH_3]_2 + AgSbF_6 \longrightarrow [C_5H_5FeCOSCH_3]_2[SbF_6] + Ag$$

S-Alkylsulfinato derivatives of cyclopentadienyliron dicarbonyl (**95**) can be prepared by SO_2 insertion into the iron–carbon σ-bond in many $RFe(CO)_2C_5H_5$ derivatives (Section 29.5.8.6) according to the following equation:[264]

93

94

$$RFe(CO)_2C_5H_5 + SO_2 \longrightarrow RSO_2Fe(CO)_2C_5H_5$$

Many such reactions can be effected by simply dissolving the $RFe(CO)_2C_5H_5$ derivative in liquid sulfur dioxide and allowing the sulfur dioxide to evaporate. Alkyliron derivatives with a CH_2 group bonded to the

---

[258] *T. S. Piper, G. Wilkinson,* J. Inorg. Nucl. Chem. *2*, 38 (1956);
R. B. King, Organometal. Synth. *1*, 175 (1965).

[259] *T. S. Piper, F. A. Cotton, G. Wilkinson,* J. Inorg. Nucl. Chem. *1*, 165 (1955).

[260] *E. O. Fischer, E. Moser,* Z. Naturforsch. *20b*, 184 (1965).

[261] *R. B. King, M. B. Bisnette,* Inorg. Chem. *4*, 482 (1965).

[262] *R. B. King, N. Welcman,* Inorg. Chem. *8*, 2540 (1969).

[263] *R. B. King, M. B. Bisnette,* Inorg. Chem. *6*, 469 (1967).

[264] *J. P. Bibler, A. Wojcicki,* J. Amer. Chem. Soc. *88*, 4862 (1966).

iron atom undergo this sulfur dioxide insertion reaction more readily than aryliron derivatives. However, perfluoroalkyliron and perfluoroaryliron derivatives such as $CF_3Fe(CO)_2C_5H_5$ and $C_6H_5Fe(CO)_2C_5H_5$ do not undergo at all this sulfur dioxide insertion reaction.

The iron thiocarbonyl cation, $[C_5H_5Fe(CO)_2CS]^{\oplus}$ (**96**) can be prepared by the following synthetic procedure:[265]

$$NaFe(CO)_2C_5H_5 + CS_2 \longrightarrow Na[C_5H_5Fe(CO)_2CS_2]$$

$$Na[C_5H_5Fe(CO)_2CS_2] + CH_3I \longrightarrow C_5H_5Fe(CO)_2C(S)SCH_3 + NaI$$

$$C_5H_5Fe(CO)_2C(S)SCH_3 + HCl \longrightarrow [C_5H_5Fe(CO)_2CS]^{\oplus} + Cl^{\ominus} + CH_3SH$$

The $[C_5H_5Fe(CO)_2CS]^{\oplus}$ cation is conveniently isolated as its tetraphenylborate.

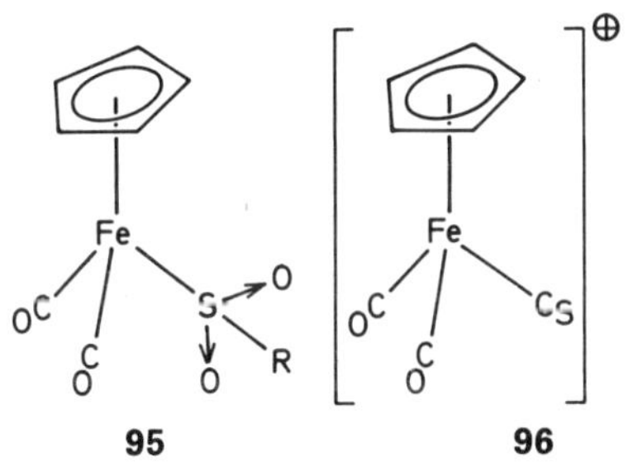

95 **96**

29.5.8.5 Organophosphorus Derivatives of the Cyclopentadienyliron Carbonyls

The bimetallic derivative $[C_5H_5Fe(CO)_2]_2$ reacts with numerous tricovalent phosphorus derivatives R_3P in boiling benzene to form the monosubstituted derivatives $(C_5H_5)_2Fe_2(CO)_3PR_3$ (**97**, $R = C_2H_5$, n-C_3H_7, and n-C_4H_9 [green]; $R = OCH_3$, OC_2H_5, O-i-C_3H_7, O-n-C_4H_9, and OC_6H_5 [red]).[266]

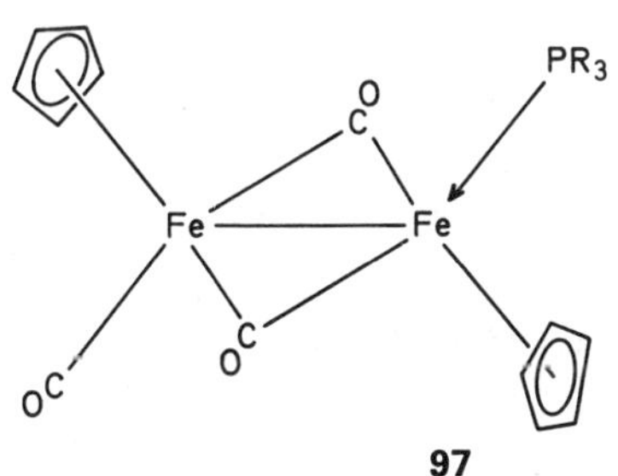

97

The bimetallic $[C_5H_5Fe(CO)_2]_2$ reacts with chelating ditertiary phosphines (*e.g.*, $(C_6H_5)_2PCH_2CH_2P(C_6H_5)_2$] in boiling toluene to give green derivatives of the type $(C_5H_5)_2Fe_2(CO)_2$(diphos).[267] However, the very reactive $(CH_3)_2PCH_2CH_2P(CH_3)_2$ reacts with $[C_5H_5Fe(CO)_2]_2$ at room temperature to give the orange pyrophoric salt $[C_5H_5Fe(CO)(CH_3)_2PCH_2CH_2P(CH_3)_2][C_5H_5Fe(CO)_2]$.[268]

Reactions of the halides $C_5H_5Fe(CO)_2X$ with tertiary phosphines may give either the nonionic derivatives $C_5H_5Fe(CO)(PR_3)X$ or the ionic derivatives $[C_5H_5Fe(CO)_2PR_3]^{\oplus}$ depending upon the reaction conditions.[89] Addition of anhydrous aluminum chloride to the reaction mixture favors formation of the ionic derivatives, which are conveniently isolated as their hexafluorophosphate salts. The iodonium ion $[C_5H_5Fe(CO)_2IFe(CO)_2C_5H_5]^{\oplus}$ (**91**) reacts with tertiary phosphines (and other similar Lewis base ligands) under very mild conditions according to the following equation:[260]

$$[C_5H_5Fe(CO)_2IFe(CO)_2C_5H_5]^{\oplus} + R_3P \longrightarrow [C_5H_5Fe(CO)_2PR_3]^{\oplus} + C_5H_5Fe(CO)_2I$$

Some alkyl derivatives of cyclopentadienyliron dicarbonyl such as the methyl derivative $CH_3Fe(CO)_2C_5H_5$ react with many tertiary phosphines in boiling tetrahydrofuran or acetonitrile without carbon monoxide evolution according to the following equation:[269]

$$RFe(CO)_2C_5H_5 + R_3P \longrightarrow RCOFe(CO)(PR_3)(C_5H_5)$$

In this reaction the alkyl group becomes an acyl group. Tetraalkylbiphosphines, R_2PPR_2, react with $[C_5H_5Fe(CO)_2]_2$ in boiling toluene to form the brown bimetallic derivatives $[C_5H_5Fe(CO)PR_2]_2$ (**98**).[270] Similar organoarsenic compounds can also be prepared.

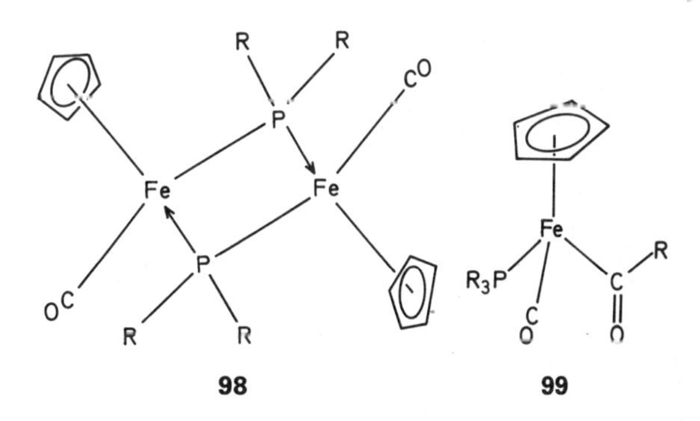

98 **99**

[265] *L. Busetto, U. Belluco, R. J. Angelici,* J. Organometal. Chem. *18*, 213 (1969).

[266] *R. J. Haines, A. L. du Preez,* Inorg. Chem. *8*, 1459 (1969).

[267] *R. J. Haines, A. L. du Preez,* J. Organometal. Chem. *21*, 181 (1970).

[268] *R. B. King, K. H. Pannell, C. A. Eggers, L. W. Houk,* Inorg. Chem. *7*, 2353 (1968).

[269] *J. P. Bibler, A. Wojcicki,* Inorg. Chem. *5*, 889 (1966).

[270] *R. G. Hayter,* J. Amer. Chem. Soc. *85*, 3120 (1963).

29.5.8.6 Cyclopentadienylmetal Carbonyl Derivatives of Iron and Ruthenium with Additional Metal–Carbon or Metal–Metal Bonds

Many $RFe(CO)_2C_5H_5$ derivatives can be prepared by either of the following two methods:

(1) Reactions of $NaFe(CO)_2C_5H_5$ (Section 29.5.8.2) with alkyl halides according to the following general equation:

$$NaFe(CO)_2C_5H_5 + RX \longrightarrow RFe(CO)_2C_5H_5 + NaX$$

A very large variety of alkyl halides undergo this reaction including methyl iodide, ethyl iodide, isopropyl iodide, benzyl chloride, chloromethyl methylsulfide, and 2-chloroethyl methylsulfide. Even the unreactive iodobenzene[76] and vinyl bromide[271] give small yields (~5%) of the corresponding $RFe(CO)_2C_5H_5$ derivatives when reacted with $NaFe(CO)_2C_5H_5$. Numerous acyl chlorides, RCOCl, react with $NaFe(CO)_2C_5H_5$ to give the corresponding $RCOFe(CO)_2C_5H_5$ derivatives.[272] Polyfluoroaromatic compounds react with $NaFe(CO)_2C_5H_5$ to form polyfluoroaryl derivatives of the $C_5H_5Fe(CO)_2$ system as exemplified by the following reaction of $NaFe(CO)_2C_5H_5$ with hexafluoro benzene to give the pentafluorophenyl derivative $C_6F_5Fe(CO)_2C_5H_5$:[273]

$$NaFe(CO)_2C_5H_5 + C_6F_6 \longrightarrow C_6F_5Fe(CO)_2C_5H_5 + NaF$$

Highly fluorinated olefins also react with $NaFe(CO)_2C_5H_5$ to form polyfluoroalkenyl derivatives of the $C_5H_5Fe(CO)_2$ system as exemplified by the following reaction of $NaFe(CO)_2C_5H_5$ with tetrafluoroethylene to give the trifluorovinyl derivative $CF_2{=}CFFe(CO)_2C_5H_5$:[274]

$$NaFe(CO)_2C_5H_5 + CF_2{=}CF_2 \longrightarrow CF_2{=}CFFe(CO)_2C_5H_5 + NaF$$

Ruthenium analogs of many of the above compounds can be prepared by using $NaRu(CO)_2C_5H_5$ instead of $NaFe(CO)_2C_5H_5$.

(2) Reactions of the halides $C_5H_5Fe(CO)_2X$ (Section 29.5.8.3; X=Cl, Br, or preferably I) with organometallic derivatives of reactive metals (i.e. Mg, Li, Na, etc.) according to the following general equations:[76]

$$C_5H_5Fe(CO)_2X + RM \longrightarrow RFe(CO)_2C_5H_5 + MX$$

The yields of $RFe(CO)_2C_5H_5$ derivatives obtained when this reaction is used are generally lower than those of $RFe(CO)_2C_5H_5$ derivatives obtained from $NaFe(CO)_2C_5H_5$ and RX. However, this synthesis of $RFe(CO)_2C_5H_5$ derivatives from $C_5H_5Fe(CO)_2X$ is useful when the halide RX, which would be required for the synthesis of the $RFe(CO)_2C_5H_5$ derivative by reaction with $NaFe(CO)_2C_5H_5$, is not readily obtained. Thus the π-cyclopentadienyl-σ-cyclopentadienyl derivative $C_5H_5Fe(CO)_2C_5H_5$ (**100**) is best prepared from $C_5H_5Fe(CO)_2I$ and sodium cyclopentadienide according to the following equation:[76]

$$C_5H_5Fe(CO)_2I + NaC_5H_5 \longrightarrow C_5H_5Fe(CO)_2C_5H_5 + NaI$$

Some other synthetic methods for the preparation of $RFe(CO)_2C_5H_5$ derivatives are occasionally useful. Thus the perfluoroalkyl derivatives ($R_fFe(CO)_2C_5H_5$ $R_f = CF_3$, C_2F_5, C_3F_7, etc.) can be prepared by either of the following two methods:

(1) Reactions of the $R_fFe(CO)_4I$ (Section 29.5.3) derivatives with thallium(I) cyclopentadienide in tetrahydrofuran solution at room temperature according to the following equation:[275]

$$R_fFe(CO)_4I + TlC_5H_5 \longrightarrow R_fFe(CO)_2C_5H_5 + TlI + 2CO$$

This is the preferred method in cases where the required $R_fFe(CO)_4I$ starting material (Section 29.5.3) is readily obtained in good yield from $Fe(CO)_5$ and the perfluoroalkyliodide (*e.g.*, $R_f = C_2F_5$ or C_3F_7).

(2) Photochemical decarbonylation of the $R_fCOFe(CO)_2C_5H_5$ derivatives.[276] This is the presently preferred preparation of the trifluoromethyl derivative $CF_3Fe(CO)_2C_5H_5$.

The photochemical decarbonylation of the benzoyl derivative $C_6H_5COFe(CO)_2C_5H_5$ provides a useful synthesis of the phenyl derivative $C_6H_5Fe(CO)_2C_5H_5$.[276] Still another alternative method for the synthesis of $C_6H_5Fe(CO)_2C_5H_5$ utilizes the reaction between $NaFe(CO)_2C_5H_5$ and diphenyliodonium iodide.[277]

[271] *M. L. H. Green, M. Ishaq, T. Mole*, Z. Naturforsch. *20b*, 598 (1965).
[272] *R. B. King*, J. Amer. Chem. Soc. *85*, 1918 (1963).
[273] *R. B. King, M. B. Bisnette*, J. Organometal. Chem. *2*, 38 (1964).
[274] *P. W. Jolly, M. I. Bruce, F. G. A. Stone*, J. Chem. Soc. (London) 5830 (1965).
[275] *R. B. King, R. N. Kapoor, L. W. Houk*, J. Inorg. Nucl. Chem. *31*, 2179 (1969).
[276] *R. B. King, M. B. Bisnette*, J. Organometal. Chem. *2*, 15 (1964).
[277] *A. N. Nesmeyanov, Yu. A. Chapovskii, B. V. Lokshia, I. V. Polovyanyuk, L. G. Makarova*, Dokl. Akad. Nauk. SSSR, *166*, 1125 (1966).

Some cyclopentadienyliron dicarbonyl derivatives with metal–metal bonds are known. Thus the reaction of $NaFe(CO)_2C_5H_5$ with the halides R_3ECl (E = Si, Ge, Sn, or Pb) gives stable orange products of the type $R_3EFe(CO)_2C_5H_5$.[140] Reaction of $[C_5H_5Fe(CO)_2]_2$ with stannous chloride gives the orange iron–tin derivative $[C_5H_5Fe(CO)_2]_2SnCl_2$ **(101)**.[278] The chlorine atoms attached to the tin in $[C_5H_5Fe(CO)_2]_2SnCl_2$ **(101)** can readily be replaced with other groups.[279]

100 **101**

29.5.8.7 Olefin Complexes of Cyclopentadienyliron Carbonyls

The yellow ethylene complex $[C_5H_5Fe(CO)_2C_2H_4]^{\oplus}$ **(102)** can be prepared by either of the following two methods:

(1) Heating a mixture of $C_5H_5Fe(CO)_2Cl$ and aluminum chloride in the presence of ethylene under pressure followed by hydrolysis and addition of ammonium hexafluorophosphate to give $[C_5H_5Fe(CO)_2C_2H_4][PF_6]$.[280]

(2) Abstraction of hydride from $C_2H_5Fe(CO)_2C_5H_5$ with triphenylmethyl tetrafluoroborate according to the following equation:[134]

$$C_2H_5Fe(CO)_2C_5H_5 + [(C_6H_5)_3C][BF_4] \longrightarrow [C_5H_5Fe(CO)_2C_2H_4][BF_4] + (C_6H_5)_3CH$$

Analogous methods can be used to prepare other complexes of the type $[C_5H_5Fe(CO)_2(olefin)]^{\oplus}$.

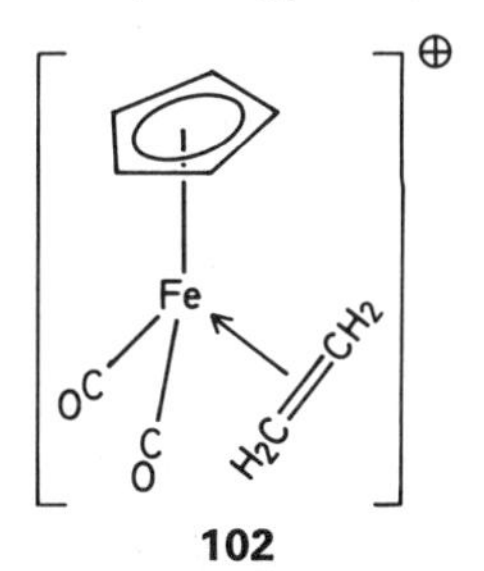

102

[278] *F. Bonati, G. Wilkinson*, J. Chem. Soc. 179 (1964).

[279] *A. N. Nesmeyanov, K. N. Anisimov, N. E. Kolobova, V. V. Skripkin*, Izv. Akad. Nauk. SSSR, Ser. Khim. 1292 (1966).

[280] *E. O. Fischer, K. Fichtel*, Chem. Ber. *94*, 1200 (1961).

29.6 Carbonyl Derivatives of Cobalt, Rhodium, and Iridium

29.6.1 The Metal Carbonyls

Octacarbonyldicobalt, $Co_2(CO)_8$, is an orange crystalline solid which oxidizes in air over a period of a few minutes and which decomposes upon warming above room temperature. Although $Co_2(CO)_8$ can be prepared by reaction of finely divided pyrophoric cobalt metal with carbon monoxide at elevated temperatures and pressures,[281] the following two preparative methods are generally more convenient in actual practice:

(1) Reaction of a cobalt(II) salt of a weak acid such as cobalt(II) carbonate[282] or cobalt(II) acetate[283] with a mixture of carbon monoxide and hydrogen at elevated temperatures and pressures according to the following equations:

$$2CoX_2 + 8CO + 2H_2 \longrightarrow Co_2(CO)_8 + 4HX$$

(2) Reaction of an aqueous solution of cobalt(II) chloride with an alkaline solution of potassium cyanide in the presence of carbon monoxide at atmospheric pressure followed by acidification with concentrated hydrochloric acid and decomposition of the evolved $HCo(CO)_4$ (Section 29.6.2) at about $-20°$.[284]

If good high pressure equipment is available, the first method listed above for the preparation of $Co_2(CO)_8$ is more convenient. If good high pressure equipment is not available, then the second method is satisfactory.

Dodecacarbonyltetracobalt, $Co_4(CO)_{12}$, a black crystalline solid, can be prepared by either of the following two methods:

(1) Decarbonylation of $Co_2(CO)_8$ by heating a toluene solution to 85–95°.[285]

(2) Reaction of a heptane solution of $Co_2(CO)_8$ and cobalt(II) 2-ethylhexanoate in an auto-

[281] *L. Mond, H. Hirtz, M. D. Cowap*, J. Chem. Soc. *97*, 798 (1910).

[282] *I. Wender, H. W. Sternberg, S. Metlin, M. Orchin*, Inorg. Synth. *5*, 190 (1957).

[283] *P. Szabó, L. Markó, G. Bor*, Chem. Tech. (Berlin) *13*, 549 (1961).

[284] *R. J. Clark, S. E. Whiddon, R. E. Serfass*, J. Organometal. Chem. *11*, 637 (1968).

[285] *P. Chini, V. Albano, S. Martinengo*, J. Organometal. Chem. *16*, 471 (1969).

clave with hydrogen at 30°/60 atmospheres.[286]

The first method is more convenient but the second method makes more efficient use of the complexed carbon monoxide in the $Co_2(CO)_8$ starting material.

Hexadecacarbonylhexacobalt, $Co_6(CO)_{16}$, can be prepared by oxidation of the $Co_6(CO)_{15}^{2\ominus}$ anion (Section 29.6.2) with either ferric chloride or mercuric chloride in the presence of sodium chloride.[287]

Rhodium forms the two *carbonyls* $Rh_4(CO)_{12}$ (red) and $Rh_6(CO)_{16}$ (black). A compound $Rh_2(CO)_8$ analogous to $Co_2(CO)_8$ does not appear to exist. The most convenient and efficient preparation of $Rh_4(CO)_{12}$ utilizes the carbonylation of $[Rh(CO)_2Cl]_2$ (Section 29.6.3) at atmospheric pressure in hexane solution in the presence of sodium bicarbonate as an insoluble mild base.[288] The following two preparations of $Rh_6(CO)_{16}$ appear to be the most convenient and efficient:

(1) Carbonylation of $[Rh(CO)_2Cl]_2$ (Section 29.6.3) at room temperature and atmospheric pressure in aqueous methanol in the presence of a lithium acetate buffer.[288]
(2) Further carbonylation of $[Rh(CO)_2Cl]_2$ by treatment with excess $Fe(CO)_5$ in boiling methanol.[289]

Alternatively $K_3RhCl_6 \cdot H_2O$ can be carbonylated at atmospheric pressure in aqueous solution in the presence of copper followed by further carbonylation of the resulting solution of $[Rh(CO)_2Cl_2]^{\ominus}$ in the presence of a sodium citrate buffer.[290]

Iridium, like rhodium, forms the two *carbonyls* $Ir_4(CO)_{12}$ (yellow) and $Ir_6(CO)_{16}$ (red). *Dodecacarbonyltetrairidium*, $Ir_4(CO)_{12}$, is best prepared by carbonylation of hydrated iridium trichloride at 60/50° atmospheres in methanol solution.[291] *Hexadecacarbonylhexairidium*, $Ir_6(CO)_{16}$, can be prepared by carbonylation of the tetraethylammonium salt $[(C_2H_5)_4N],[Ir_6(CO)_{15}]$ in the presence of acetic acid.[292]

29.6.2 Metal Carbonyl Anions and Metal Carbonyl Hydrides

The anion $Co(CO)_4^{\ominus}$ can be generated from an aqueous ammoniacal solution of cobalt(II) by treatment with carbon monoxide at atmospheric pressure in the presence of an appropriate reducing agent such as potassium cyanide or sodium dithionite.[284, 293] If $Co_2(CO)_8$ is available, it can be converted to the $Co(CO)_4^{\ominus}$ anion by reaction with sodium amalgam or sodium hydroxide in tetrahydrofuran solution.[294] The $Co(CO)_4^{\ominus}$ anion is most often employed in the solution in which it is prepared. If a solid $Co(CO)_4^{\ominus}$ salt is desired, the tetraethylammonium bis(triphenylphosphine)iminium salt can be isolated and used. Alternatively, reaction of $Co_2(CO)_8$ with tertiary phosphines or isocyanides may give solid $Co(CO)_4^{\ominus}$ salts of complex cobalt(I) cations according to the following equations:[295, 296]

$$Co_2(CO)_8 + 2R_3P \longrightarrow [(R_3P)_2Co(CO)_3][Co(CO)_4] + CO$$

$$Co_2(CO)_8 + 5RNC \longrightarrow [(RNC)_5Co][Co(CO)_4] + 4CO$$

Acidification of solutions of $Co(CO)_4^{\ominus}$ with a strong mineral acid such as hydrochloric acid or sulfuric acid liberates $HCo(CO)_4$, a gas which decomposes rapidly at room temperature but which can be handled at room temperature in the gas phase in a stream of carbon monoxide.[297]

Several *polymetallic cobalt carbonyl anions* can also be prepared. Reduction of $Co_2(CO)_8$ with alkali metals in diethyl ether solution gives the red $Co_3(CO)_{10}^{\ominus}$ anion.[298] Dissolution of $Co_2(CO)_8$ in ethanol gives a red solution of $[Co(C_2H_5OH)_n][Co(CO)_4]_2$ which upon pyrolysis at 60° for 8 hr gives a deep green solution from which dark green $K_2Co_6(CO)_{15}$ can be obtained by evapora-

[286] *R. Ercoli, P. Chini, M. Massi-Mauri*, Chim. Ind. (Milan) *41*, 132 (1959);
R. B. King, Organometal. Synth. *1*, 103 (1965).
[287] *P. Chini*, Chcm. Commun. 440 (1967);
P. Chini, Inorg. Chem. *8*, 1206 (1969).
[288] *P. Chini, S. Martinengo*, Inorg. Chim. Acta *3*, 315 (1969).
[289] *S. Martinengo, P. Chini, G. Giordano*, J. Organometal. Chem. *27*, 389 (1971).
[290] *B. L. Booth, M. J. Else, R. Fields, H. Goldwhite, R. N. Haszeldine*, J. Organometal. Chem. *14*, 417 (1968).
[291] *S. H. H. Chaston, F. G. A. Stone*, Chem. Commun. 964 (1967).
[292] *L. Malatesta, G. Caglio, M. Angoletta*, Chem. Commun. 532 (1970).
[293] *W. Hieber, E. O. Fischer, E. Böckly*, Z. Anorg. Allgem. Chemie *269*, 308 (1952).
[294] *W. F. Edgell, J. Lyford, IV*, Inorg. Chem. *9*, 1932 (1970).
[295] *W. Hieber, W. Freyer*, Chem. Ber. *93*, 462 (1960).
[296] *W. Hieber, J. Sedlmeier*, Chem. Ber. *87*, 789 (1954).
[297] *H. W. Sternberg, I. Wender, R. A. Friedel, M. Orchin*, J. Amer. Chem. Soc. *75*, 2717 (1953).
[298] *S. A. Fieldhouse, B. H. Freeland, C. D. M. Mann, R. J. O'Brien*, Chem. Commun. 181 (1970).

tion followed by treatment with a concentrated aqueous solution of potassium bromide. The potassium salt $K_2Co_6(CO)_{15}$ can be converted to the corresponding cesium and tetraalkylammonium salts by metathesis.[299] Reduction of $Co_4(CO)_{12}$ (Section 29.6.1) with alkali metals (lithium, sodium, and potassium) in tetrahydrofuran solution gives first the green $Co_6(CO)_{15}^{2\ominus}$ and then the dark red $Co_6(CO)_{14}^{4\ominus}$; the potassium and tetraethylammonium salts of the latter anion can be isolated.[285, 300]

Several *rhodium carbonyl anions* can be prepared. Reaction of metallic sodium with a tetrahydrofuran solution of $[Rh(CO)_2Cl]_2$ (Section 29.6.3) in an atmosphere of carbon monoxide until the solution becomes colorless gives the monometallic $Rh(CO)_4^{\ominus}$ anion which can be isolated as its tetramethylammonium salt.[301] Reaction of $[Rh(CO)_2Cl]_2$ with carbon monoxide in tetrahydrofuran or methanol in the presence of an acetate buffer gives the violet dodecametallic $Rh_{12}(CO)_{30}^{2\ominus}$ anion which can be isolated as either a tetraalkylammonium salt or an alkali-metal salt.[302] Other more difficultly preparable rhodium carbonyl anions include the yellow trimetallic[303] $Rh_3(CO)_{10}^{\ominus}$ and the dark green heptametallic[303] $Rh_7(CO)_{16}^{3\ominus}$.

Several *iridium carbonyl anions* can also be prepared. Reaction of $Ir_4(CO)_{12}$ with potassium carbonate in methanol in the presence of carbon monoxide gives the yellow anion $HIr_4(CO)_{11}^{\ominus}$ isolated as its benzyltrimethylammonium or tetraethylammonium salt.[292, 304] If the reaction of $Ir_4(CO)_{12}$ with potassium carbonate in methanol is carried out in the absence of added carbon monoxide, the brown anion $Ir_8(CO)_{20}^{4\ominus}$ is obtained; this anion can be isolated as its benzyltrimethylammonium salt. Reaction of $Ir_4(CO)_{12}$ with sodium metal in tetrahydrofuran first gives the tetrametallic anion $HIr_4(CO)_{11}^{\ominus}$ and then the dark brown hexametallic anion $Ir_6(CO)_{15}^{2\ominus}$ which can be isolated as its tetraethylammonium salt.[292]

[299] *P. Chini, V. Albano*, J. Organometal. Chem. *15*, 433 (1968).
[300] *P. Chini*, Chem. Commun. 440 (1967).
[301] *P. Chini, S. Martinengo*, Inorg. Chim. Acta *3*, 21 (1969).
[302] *P. Chini, S. Martinengo*, Inorg. Chim. Acta *3*, 299 (1969).
[303] *P. Chini, S. Martinengo*, Chem. Commun. 1092 (1969).
[304] *L. Malatesta, G. Caglio*, Chem. Commun. 420 (1967).

29.6.3 Metal Carbonyl Halides of Rhodium and Iridium

Cobalt does not form any stable unsubstituted metal carbonyl halides. *Rhodium* forms the compound $[Rh(CO)_2Cl]_2$ which is an important starting material for many rhodium carbonyl syntheses. The most convenient preparation of $[Rh(CO)_2Cl]_2$ utilizes the reaction of hydrated rhodium trichloride with carbon monoxide at atmospheric pressure in a special apparatus which enables the red $[Rh(CO)_2Cl]_2$ to sublime out of the hot reaction zone as it is formed.[305] The *iridium* compound $Ir(CO)_3Cl$ can be prepared by carbonylation of iridium trichloride at elevated temperatures at atmospheric pressure. Best results are obtained if the iridium trichloride is adsorbed on silica gel.[306]

29.6.4 Cobalt Carbonyl Nitrosyl Derivatives

Nitrosyltricarbonylcobalt $Co(CO)_3NO$, is a dark-red liquid, m.p. $-1°$, b.p. 50°/760 mm. This compound may be prepared by either of the following two methods:

(1) Nitrosation of $Co(CO)_4^{\ominus}$ (Section 29.6.2) in aqueous solution according to the following equation:[307]

$$Co(CO)_4^{\ominus} + HNO_2 \longrightarrow Co(CO)_3NO + CO + OH^{\ominus}$$

The nitrous acid required to effect this reaction may be generated from nitrite and a weak acid (acetic acid or carbon dioxide) or from nitric oxide and water.

(2) Reaction of $Co_2(CO)_8$ with nitric oxide according to the following equation:[308]

$$Co_2(CO)_8 + 2NO \longrightarrow 2Co(CO)_3NO + 2CO$$

This second method is the simplest if $Co_2(CO)_8$ is readily available.

29.6.5 Carbonyl Derivatives Containing Organophosphorus and Similar Ligands

Octacarbonyldicobalt, $Co_2(CO)_8$, reacts with tricovalent phosphorus derivatives of the type R_3P ($R = C_6H_5$, OC_6H_5, etc.) to form substitution products of the type $[R_3PCo(CO)_3]_2$.[309]

[305] *J. A. McCleverty, G. Wilkinson*, Inorg. Synth. *8*, 211 (1966).
[306] *E. O. Fischer, K. S. Benner*, Z. Naturforsch. *17b*, 774 (1962).
[307] *F. Seel*, Z. Anorg. Allgem. Chemie *269*, 40 (1952).
[308] *R. L. Mond, A. E. Wallis*, J. Chem. Soc. *121*, 34 (1922).
[309] *W. Hieber, W. Freyer*, Chem. Ber. *91*, 1230 (1958); *W. Hieber, W. Freyer*, Chem. Ber. *93*, 462 (1960).

Reactions of these $[R_3PCo(CO)_3]_2$ derivatives with sodium amalgam in tetrahydrofuran solution give the sodium salts $Na[Co(CO)_3PR_3]$.[295] These sodium salts undergo many reactions similar to $NaCo(CO)_4$ (Section 29.6.2) including acidification with hydrochloric or phosphoric acid to give the *hydrides* $HCo(CO)_3PR_3$ which are somewhat more stable than the unsubstituted $HCo(CO)_4$ but which still decompose around room temperature.[310] Reactions of the sodium salts $Na[Co(CO)_3PR_3]$ with positive halogen sources (*e.g.*, perfluoroalkyl iodides and *N*-bromosuccinimide) give substituted cobalt carbonyl halides of the general formula $R_3PCo(CO)_3X$ (X = Br or I).[311]

Nitrosyltricarbonylcobalt, $Co(CO)_3NO$, reacts with tricovalent phosphorus derivatives and similar ligands under relatively mild conditions to form the substitution products $LCo(CO)_2NO$ and $L_2Co(CO)(NO)$.[212]

The procedures useful for the preparation of tertiary phosphine derivatives of *rhodium* and *iridium* carbonyl halides are conveniently illustrated by the triphenylphosphine derivatives, which are particularly important in homogenous catalysis. The rhodium derivative *trans*-$[(C_6H_5)_3P]_2Rh(CO)Cl$ (**103**: M = Rh) can be prepared either by reaction of $[Rh(CO)_2Cl]_2$ with triphenylphosphine at room temperature in an inert solvent[312] or by heating hydrated rhodium trichloride with triphenylphosphine in ethanol solution followed by addition of a carbonylating reagent such as aqueous formaldehyde.[313] The analogous iridium compound *trans*-$[(C_6H_5)_3P]_2Ir(CO)Cl$ (**103**: M = Ir); "Vaska's Catalyst") is conveniently prepared by heating a mixture of hydrated iridium trichloride and excess triphenylphosphine in a good carbonylating solvent such as 2-methoxyethanol or preferably dimethylformamide.[314] Reactions of the compounds *trans*-$[(C_6H_5)_3P]_2M(CO)Cl$ (**103**: M = Rh or Ir) with excess triphenylphosphine and hydrazine or sodium hydroborate in ethanol solution gives the hydrides $[(C_6H_5)_3P]_3M(CO)H$ (M = Rh or Ir).[315]

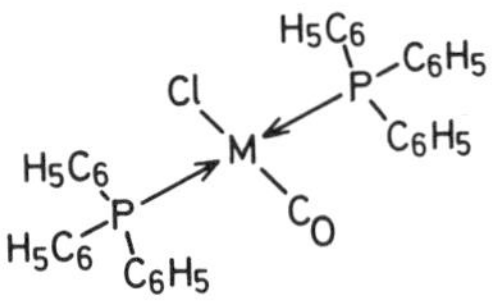

103

29.6.6 Cobalt Carbonyl Derivatives with Additional Metal–Carbon or Metal–Metal Bonds

Several $RCo(CO)_4$ derivatives can be prepared by reactions of the $Co(CO)_4^{\ominus}$ anion. The light yellow very unstable *σ-methyl* derivative $CH_3Co(CO)_4$, dec. > −35°, can be prepared by reaction of methyl iodide with the $Co(CO)_4^{\ominus}$ anion.[316] Stable *perfluoroalkylcobalt* derivatives of the type $R_fCo(CO)_4$ can be prepared by reactions of $NaCo(CO)_4$ with perfluoroacyl chlorides according to the following equation:[317]

$$NaCo(CO)_4 + R_fCOCl \longrightarrow R_fCo(CO)_4 + CO + NaCl$$

These $R_fCo(CO)_4$ derivatives are stable liquids which can be purified by distillation. A silyl derivative $H_3SiCo(CO)_4$, b.p. 112°, can be prepared by the reaction between $NaCo(CO)_4$ and SiH_3I.[318] Similarly, several derivatives of the type $R_3SnCo(CO)_4$ ($R = CH_3$, C_6H_5, *etc.*) with cobalt–tin bonds can be prepared by reaction of $NaCo(CO)_4$ with R_3SnCl derivatives.[319] Silylcobalt tetracarbonyls can also be prepared by reaction of $Co_2(CO)_8$ with trialkylsilanes, R_3SiH, according to the following equations:[320]

$$Co_2(CO)_8 + 2R_3SiH \longrightarrow 2R_3SiCo(CO)_4 + H_2$$

Compounds of the type $R_3ECo(CO)_4$ (R = alkyl or aryl; E = Si, Ge, Sn, but not C) are stable at room temperature.

The *π-allyl* derivative, $C_3H_5Co(CO)_3$ (**104**), an air-sensitive distillable red-yellow liquid, can be prepared by reaction of $NaCo(CO)_4$ with allyl chloride in diethyl ether.[321] Substituted π-allyl

[310] *W. Hieber, E. Lindner*, Z. Naturforsch. *16b*, 137 (1961);
W. Hieber, E. Lindner, Chem. Ber. *94*, 1417 (1961).
[311] *W. Hieber, E. Lindner*, Chem. Ber. *95*, 273 (1962).
[312] *J. A. McCleverty, G. Wilkinson*, Inorg. Synth. *8*, 214 (1966).
[313] *D. Evans, T. A. Osborn, G. Wilkinson*, Inorg. Synth. *11*, 99 (1968).
[314] *J. P. Collman, J. W. Kang*, J. Amer. Chem. Soc. *89*, 944 (1967).
[315] *S. S. Bath, L. Vaska*, J. Amer. Chem. Soc. *85*, 3500 (1963).
[316] *W. Hieber, O. Vohler, G. Braun*, Z. Naturforsch. *13b*, 192 (1958).
[317] *W. R. McClellan*, J. Amer. Chem. Soc. *83*, 1598 (1961).
[318] *B. T. Aylett, T. M. Campbell*, J. Chem. Soc. *A*, 1910 (1969).
[319] *S. Breitschaft, F. Basolo*, J. Amer. Chem. Soc. *88*, 2702 (1966).
[320] *A. J. Chalk, J. F. Harrod*, J. Amer. Chem. Soc. *89*, 1640 (1967).
[321] *R. F. Heck, D. S. Breslow*, J. Amer. Chem. Soc. *82*, 750 (1960).

derivatives can be prepared by addition of $HCo(CO)_4$ (Section 29.6.2) to 1,3-dienes. Thus the reaction of $HCo(CO)_4$ with butadiene gives the π-1-methyllallylcobalt derivative π-$CH_3C_3H_4Co(CO)_3$.[322] An unusual π-allylic cobalt tricarbonyl derivative is the deep red air-sensitive liquid π-cycloheptatrienyl derivative $C_7H_7Co(CO)_3$ (**105**) which can be prepared by ultraviolet irradiation of a mixture of $Co_2(CO)_8$ and cycloheptatriene.[64]

104 **105**

Among the most stable cobalt carbonyls are the purple volatile *methinyltricobalt enneacarbonyls* $YCCo_3(CO)_9$ (**106**). Most of these $YCCo_3(CO)_9$ derivatives can be prepared by reaction of $Co_2(CO)_8$ with the YCX_3 trihalides according to the following equation:[323]

$$9Co_2(CO)_8 + 4YCX_3 \longrightarrow 4YCCo_3(CO)_9 + 36CO + 6CoX_2$$

($Y = H$, F, Cl, Br, I, CH_3, C_6H_5, CF_3, CH_3OCO, etc.; $X = Cl$, Br, or I).

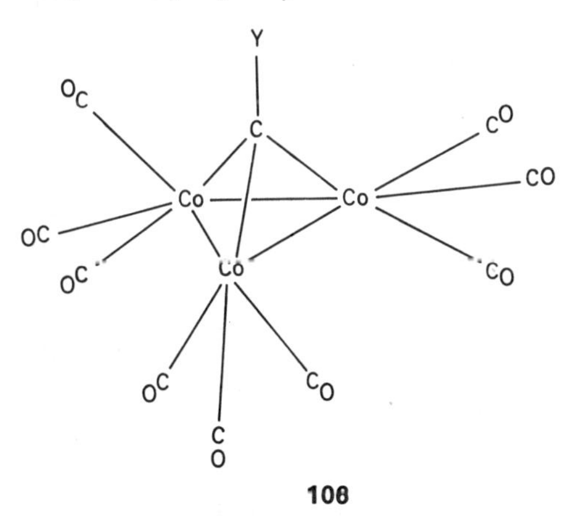

106

If Y is a halogen in the above reaction, X cannot be lighter than Y. In addition the $YCCo_3(CO)_9$ compounds of the type $RCH_2CCo_3(CO)_9$ (**106**, $Y = RCH_2$) can be prepared by reactions of the alkyne-dicobalt hexacarbonyls $(RC_2H)Co_2(CO)_6$ (Section 29.6.7) with strong mineral acids.[324]

Several methods have been devised for the interconversion of $YCCo_3(CO)_9$ (**106**) derivatives. The hydrogen derivative $HCCo_3(CO)_9$ (**106**: $Y = H$) can be alkylated with dialkylmercury derivatives according to the following equation:[325]

$$HCCo_3(CO)_9 + R_2Hg \longrightarrow RCCo_3(CO)_9 + RH + Hg$$

where $R = C_6H_5$, C_6F_5, $C_6H_5CH_2$, $C_5H_5FeC_5H_4$, p-XC_6H_4 ($X = Cl$, F, CH_3, and CH_3O), α-$C_{10}H_7$, n-C_5H_{11}, and $CH_3OCH_2CH_2$. The yields in this reaction are highest if the reaction is carried out in the presence of a carbon monoxide atmosphere. Another method for the preparation of $YCCo_3(CO)_9$ (**106**) derivatives where Y is an aryl group utilizes the Friedel–Crafts reaction of $ClCCo_3(CO)_9$ with aromatic hydrocarbons according to the following equation:[326]

$$ClCCo_3(CO)_9 + RH \longrightarrow RCCo_3(CO)_9 + HCl$$

where $R = C_6H_5$, p-XC_6H_4 ($X = Cl$, Br, or C_6H_5), $C_5H_5FeC_5H_4$, *etc.* The phenyl derivative $C_6H_5CCo_3(CO)_9$ (**106**: $Y = C_6H_5$) can also be acylated under Friedel–Crafts conditions ($AlCl_3$)[327] to give compounds such as the acetyl derivative p-$CH_3COC_6H_4CCo_3(CO)_9$ (**106**: $Y = CH_3COC_6H_4$).

29.6.7 Compounds Obtained by Interactions of Alkynes with Cobalt Carbonyls

Several interesting types of compounds can be prepared by reactions of various alkynes with cobalt carbonyls. The red *alkyne-dicobalt hexacarbonyls* $(Y_2C_2)Co_2(CO)_6$

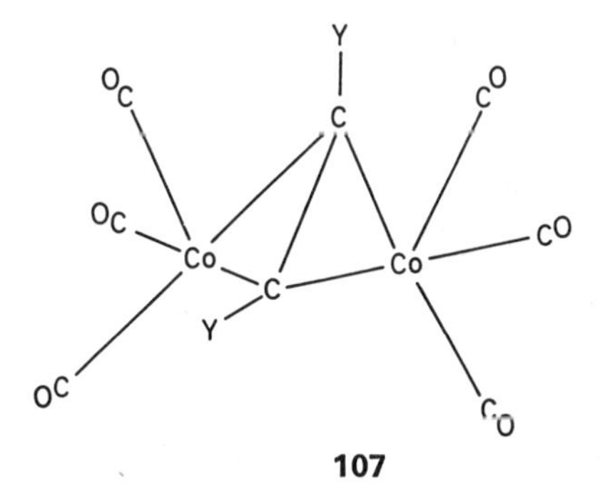

107

[322] *H. B. Jonassen, R. I. Stearns, J. Kentämaa, D. W. Moore, A. G. Whittaker*, J. Amer. Chem. Soc. *80*, 2586 (1958).

[323] *W. T. Dent, L. A. Duncanson, R. G. Guy, H. W. B. Reed, B. L. Shaw*, Proc. Chem. Soc. 169 (1961); *G. Bor, L. Markó, B. Markó*, Chem. Ber. *95*, 333 (1962); *R. Ercoli, S. Santambrogio, G. Tettamanti Casagrande*, Chim. Ind. (Milan) *44*, 1344 (1962).

[324] *R. Markby, I. Wender, R. A. Friedel, F. A. Cotton, H. W. Sternberg*, J. Amer. Chem. Soc. *80*, 6529 (1958).

[325] *D. Seyferth, J. E. Hallgren, R. J. Spohn*, J. Organometal. Chem. *23*, C55 (1970).

[326] *R. Dolby, B. H. Robinson*, Chem. Commun. 1058 (1970).

[327] *D. Seyferth, A. T. Wehman*, J. Amer. Chem. Soc. *92*, 5520 (1970).

(**107**, Y=H, CH_3, C_6H_5, CF_3, CO_2CH_3, CN, *etc.*) are prepared by the facile reactions of the alkynes YC≡CY with $Co_2(CO)_8$ at room temperature in inert hydrocarbon solvents.[328] Reactions of some of the alkyne-dicobalt hexacarbonyls $(Y_2C_2)Co_2(CO)_6$ (**107**, Y=H, CH_3, C_6H_5, *etc.*) with excess carbon monoxide at 70–75°/210 atm. gives the red complexes $[Y_2C_2(CO)_2]Co_2(CO)_7$ (**108**).[329]

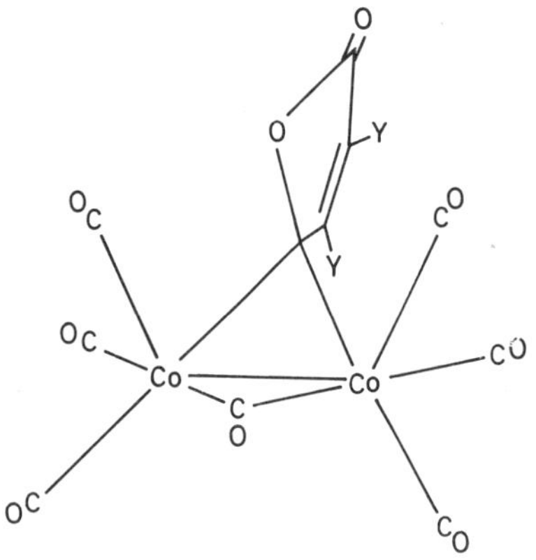

108

Reaction of the alkyne-dicobalt hexacarbonyls $(Y_2C_2)Co_2(CO)_6$ (**107**, Y=H, CH_3, $(CH_3)_3C$, C_6H_5, *etc.*) with an excess of a hindered alkyne Y′C≡CY′ (Y′=*tert*-butyl *etc.*) gives purple $Y_2Y'_4C_6Co_2(CO)_4$ derivatives of structure **109**.[330]

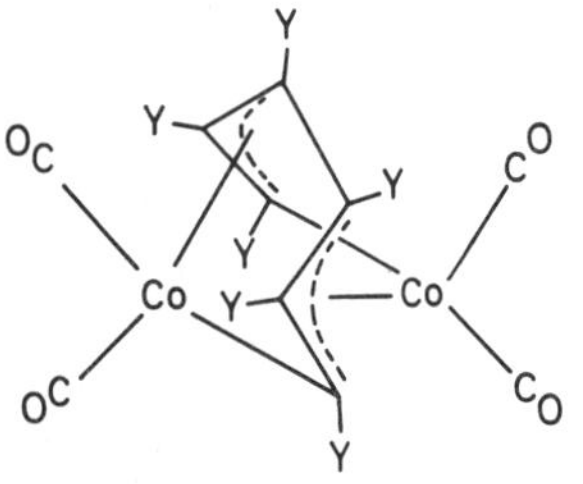

109

The dark-blue complexes $Y_2C_2Co_4(CO)_{10}$ may be prepared by reaction of $Co_4(CO)_{12}$ with the alkynes YC≡CY.[331]

29.6.8 Cyclopentadienylmetal Carbonyls

Cobalt, rhodium, and irridium each form the volatile liquid cyclopentadienylmetal dicarbonyls $C_5H_5M(CO)_2$. The dark red *cobalt* derivative $C_5H_5Co(CO)_2$ may be prepared by either of the following two methods:

(1) *Carbonylation of $(C_5H_5)_2Co$:* The reaction between $(C_5H_5)_2Co$ and carbon monoxide can be carried out at atmospheric pressure in triglyme solution at 180–210°.[332] However, the low yields (~12%) make this method suitable only when neither $Co_2(CO)_8$ nor a high pressure autoclave is available. The reaction between $(C_5H_5)_2Co$ and carbon monoxide in benzene solution at 130°/60 atmospheres gives a higher yield of $C_5H_5Co(CO)_2$.[333]

(2) *Reaction of $Co_2(CO)_8$ with Cyclopentadiene:* Reaction of $Co_2(CO)_8$ with freshly distilled monomeric cyclopentadiene in boiling dichloromethane gives a good yield of $C_5H_5Co(CO)_2$.[334]

The orange liquid *rhodium* derivative $C_5H_5Rh(CO)_2$ can be prepared by reaction of $[Rh(CO)_2Cl]_2$ (Section 29.6.3) with sodium cyclopentadienide[335] or preferably thallium cyclopentadienide.[336] The yellow liquid *iridium* derivative $C_5H_5Ir(CO)_2$ can be prepared by reaction of $Ir(CO)_3Cl$ (Section 29.6.3) with sodium cyclopentadienide at room temperature. The crystalline pentamethylcyclopentadienylmetal dicarbonyls $(CH_3)_5C_5M(CO)_2$ (M=Rh or Ir) can be obtained by reductive carbonylation of the halides $[(CH_3)_5C_5MCl_2]_2$ (M=Rh or Ir) using either $Fe_2(CO)_9$ in boiling benzene (for M=Ir) or carbon monoxide at atmospheric pressure in the presence of metallic zinc (for M=Rh).[337] *Polymetallic cyclopentadienylmetal carbonyls* of *cobalt* and *rhodium* can be prepared by decarbonylation of the monometallic $C_5H_5M(CO)_2$ derivatives. The black trimetallic $[C_5H_5CoCO]_3$ is obtained by u.v. irradiation of a hexane solution of $C_5H_5Co(CO)_2$.[252] Red crystalline $(C_5H_5)_2Rh_2(CO)_3$ is obtained when C_5H_5Rh-

[328] *H. Greenfield, H. W. Sternberg, R. A. Friedel, J. H. Wotiz, R. Markby, I. Wender,* J. Amer. Chem. Soc. *78*, 120 (1956).
[329] *H. W. Sternberg, J. G. Shukys, C. D. Donne, R. Markby, R. A. Friedel, I. Wender,* J. Amer. Chem. Soc. *81*, 2339 (1959).
[330] *U. Krüerke, C. Hoogzand, W. Hübel,* Chem. Ber. *94*, 2817 (1961).
[331] *U. Krüerke, W. Hübel,* Chem. Ber. *94*, 2829 (1961).
[332] *R. B. King,* J. Amer. Chem. Soc. *84*, 2460 (1962).
[333] *R. B. King, P. M. Treichel, F. G. A. Stone,* J. Amer. Chem. Soc. *83*, 3600 (1961); *R. B. King,* Organometal. Synth. *1*, 115 (1965).
[334] *R. B. King,* Organometal. Synth. *1*, 118 (1965).
[335] *E. O. Fischer, K. Bittler,* Z. Naturforsch. *16b*, 225 (1961).
[336] *T. Knight, M. J. Mays,* J. Chem. Soc. *A*, 654 (1970).
[337] *J. W. Kang, K. Moseley, P. M. Maitlis,* J. Amer. Chem. Soc. *91*, 5970 (1969); *J. W. Kang, P. M. Maitlis,* J. Organometal. Chem. *26*, 393 (1971).

$(CO)_2$ is allowed to stand for several days at room temperature while exposed to air and light.[338] Ultraviolet irradiation of $C_5H_5Rh(CO)_2$ gives two isomers of black $[C_5H_5RhCO]_3$.[338]

Other cyclopentadienylmetal carbonyl derivatives of cobalt, rhodium, and iridium can be prepared by oxidative addition reactions of the $C_5H_5M(CO)_2$ derivatives. Reaction of the $C_5H_5M(CO)_2$ derivatives (M = Co and Rh) with iodine in diethyl ether solution proceeds according to the following equation to give the black $C_5H_5M(CO)I_2$ derivatives:[339]

$$C_5H_5M(CO)_2 + I_2 \longrightarrow C_5H_5M(CO)I_2 + CO$$

Similar reactions of the $C_5H_5M(CO)_2$ derivatives with perfluoroalkyl iodides, R_fI ($R_f = CF_3$, C_2F_5, n-C_3F_7, $(CF_3)_2CF$, *etc.*) in warm benzene gives the perfluoroalkyl derivatives $C_5H_5M(CO)R_fI$ (**110**: M = Co(black) or Rh (red)).[333] Reaction of $C_5H_5Co(CO)_2$ with allyl halides gives the yellow salts $[C_5H_5Co(CO)(\pi\text{-}C_3H_5)]X$ (**111**) in addition to occasional small quantities of the carbonyl-free derivatives $C_5H_5Co(\pi\text{-}C_3H_5)X$.[340] Reactions of the $C_5H_5M(CO)_2$ (M = Co, Rh, and Ir) derivatives with tertiary phosphines gives monocarbonyls of the type $C_5H_5M(CO)PR_3$.[341]

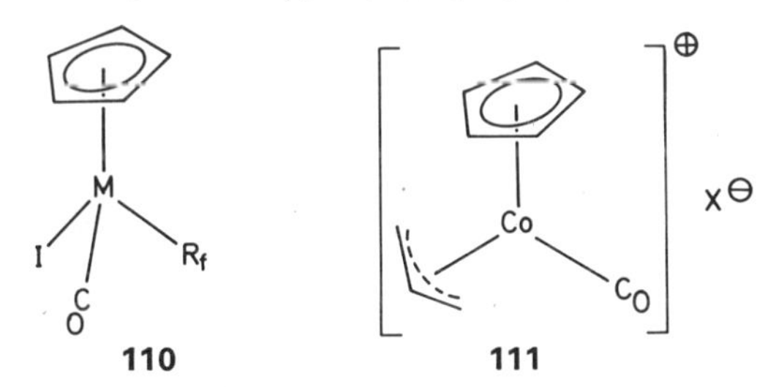

29.7 Carbonyl Derivatives of Nickel, Palladium, and Platinum

29.7.1 Nickel Carbonyl Derivatives not Containing Cyclopentadienyl Groups

Tetracarbonylnickel, $Ni(CO)_4$, is a colorless, extremely volatile (b.p. 43°/760 mm) liquid which decomposes to metallic nickel and carbon monoxide upon heating above room temperature or upon transferring in a vacuum system at room temperature. The toxicity of $Ni(CO)_4$ is comparable to that of the very toxic HCN making hazardous the handling of $Ni(CO)_4$. Tetracarbonylnickel, $Ni(CO)_4$, is frequently inexpensively commercially available in metal cylinders. It can be prepared by either of the following two methods:

(1) Reaction of finely divided metallic nickel with carbon monoxide even at atmospheric pressure.[342]
(2) Reduction of nickel salts with dithionite ion in aqueous ammonia solution in the presence of carbon monoxide at atmospheric pressure.[293]

Reactions of $Ni(CO)_4$ can be used to prepare other nickel carbonyl derivatives. Treatment of $Ni(CO)_4$ with tricovalent phosphorus derivatives under mild conditions results in the replacement of carbonyl groups to give products of the type $(R_3P)_nNi(CO)_{4-n}$. For example, triphenylphosphine reacts with $Ni(CO)_4$ in boiling diethyl ether to give the disubstituted derivative $[(C_6H_5)_3P]_2Ni(CO)_2$.[343] Reaction of $Ni(CO)_4$ with tri(2-cyanoethyl)phosphine gives a small yield of the orange tetrametallic derivative $[(NCCH_2CH_2)_3P]_4Ni_4(CO)_6$.[344] Another type of tetrametallic derivative $[(CF_3)_2C_2]_3Ni_4(CO)_3$ is obtained from the reaction of $Ni(CO)_4$ with hexafluorobutyne-2 at 50°.[345]

29.7.2 Cyclopentadienylnickel Carbonyl Derivatives

Red-violet crystalline $[C_5H_5NiCO]_2$ can be prepared by heating $(C_5H_5)_2Ni$ with excess $Ni(CO)_4$ in benzene solution at 80° according to the following equation:[346, 347]

$$(C_5H_5)_2Ni + Ni(CO)_4 \longrightarrow [C_5H_5NiCO]_2 + 2CO$$

At higher temperatures $(C_5H_5)_2Ni$ and $Ni(CO)_4$ form the green-brown trimetallic $(C_5H_5)_3Ni_3(CO)_2$.[346]

Several other cyclopentadienylnickel carbonyl derivatives can be obtained from $[C_5H_5NiCO]_2$. Reaction of $[C_5H_5NiCO]_2$ with iodine at low temperatures gives black $C_5H_5Ni(CO)I$ which decomposes at 0°.[346] A similar reaction of $[C_5H_5NiCO]_2$ with perfluoroalkyl iodides, R_fI ($R_f = CF_3$, C_2F_5, and C_3F_7) gives the dark red

[338] *O. S. Mills, E. F. Paulus*, Chem. Commun. 815 (1966);
O. S. Mills, E. F. Paulus, J. Organometal. Chem. *10*, 331 (1967);
E. F. Paulus, E. O. Fischer, H. P. Fritz, H. Schuster-Woldan, J. Organometal. Chem. *10*, P3 (1967).
[339] *R. B. King*, Inorg. Chem. *5*, 82 (1966).
[340] *R. F. Heck*, J. Org. Chem. *28*, 604 (1963).
[341] *H. G. Schuster-Woldan, F. Basolo*, J. Amer. Chem. Soc. *88*, 1657 (1966).
[342] *W. R. Gilliland, A. A. Blanchard*, Inorg. Synth. *2*, 234 (1946).
[343] *J. D. Rose, F. S. Statham*, J. Chem. Soc. 69 (1950); *R. B. King*, Organometal. Synth. *1*, 181 (1965).
[344] *M. J. Bennett, F. A. Cotton, B. H. C. Winquist*, J. Amer. Chem. Soc. *89*, 5366 (1967).
[345] *R. B. King, M. I. Bruce, J. R. Phillips, F. G. A. Stone*, Inorg. Chem. *5*, 684 (1966).
[346] *E. O. Fischer, C. Palm*, Chem. Ber. *91*, 1725 (1958).
[347] *J. F. Tilney-Bassett*, J. Chem. Soc. (London) 577 (1961);
R. B. King, Organometal. Synth. *1*, 119 (1965).

volatile $R_fNi(CO)C_5H_5$ derivatives.[348] Reaction of $[C_5H_5NiCO]_2$ with stannous chloride in boiling tetrahydrofuran gives green $C_5H_5Ni(CO)SnCl_2Ni(CO)C_5H_5$.[349]

29.7.3 Palladium and Platinum Carbonyl Derivatives

Palladium and *platinum* both form relatively unstable metal *carbonyl chlorides* of the general formula $M(CO)Cl_2$ (M = Pd or Pt). The palladium compound $Pd(CO)Cl_2$ is obtained by carbonylation of anhydrous $PdCl_2$ at room temperature and atmospheric pressure in the presence of methanol vapor.[350] Under other conditions another palladium carbonyl chloride, $[Pd(CO)Cl]_n$, may be obtained.[351] The platinum compound $Pt(CO)Cl_2$ may be prepared by reaction of platinum metal with a mixture of chlorine and carbon monoxide at 250°.[352] Another product of this reaction is the monometallic $Pt(CO)_2Cl_2$. The compounds $[M(CO)Cl_2]_2$ react with chloride ion to give the *complex anions* $[M(CO)Cl_3]^{\ominus}$, and with selected Lewis base ligands (tertiary phosphines, pyridine oxide, *etc.*) to form the monometallic derivatives $LPt(CO)Cl_2$. However, compounds of the type $LPt(CO)Cl_2$ are more conveniently prepared by atmospheric pressure carbonylation of $[LPtCl_2]_2$ derivatives.[353]

Several halide-free carbonyl derivatives of palladium and platinum are known. Reaction of palladium(II) acetylacetonate with triethylaluminum in the presence of both triphenylphosphine and carbon monoxide in toluene solution give the cream yellow monometallic derivative $[(C_6H_5)_3P]_3PdCO$. This compound is thermally unstable and loses triphenylphosphine at room temperature to give the orange-yellow $[(C_6H_5)_3P]_4Pd_3(CO)_3$.[354] Similar red trimetallic platinum derivatives of the type $(R_3P)_4Pt_3(CO)_3$ can be prepared by either of the following two methods:[355]

(1) Carbonylation of K_2PtCl_4 in the presence of the tertiary phosphine, carbon monoxide, and hydrazine.
(2) Reaction of the polymeric platinum carbonyl $[Pt(CO)_2]_n$ with the tertiary phosphine. The polymeric carbonyl $[Pt(CO)_2]_n$ can be prepared either by carbonylation of K_2PtCl_4 in the presence of hydrazine or by hydrolysis of $Pt(CO)_2Cl_2$ in an atmosphere of carbon monoxide.

The cyclopentadienylplatinum carbonyl $[C_5H_5PtCO]_2$ can be prepared in low (~5%) yield by reaction of sodium cyclopentadienide with the crude $Pt(CO)_2Cl_2$ obtained by carbonylation of $PtCl_2$.[356] Reaction of $[C_5H_5PtCO]_2$ with iodine gives the monometallic iodide $C_5H_5Pt(CO)I$.

29.8 Bibliography

R. B. King, Organometallic Syntheses: Volume I: Transition Metal Compounds, Academic Press, New York 1965.

R. B. King, Transition Metal Organometallic Chemistry: An Introduction, Academic Press, New York 1969.

J. C. Hileman, Metal Carbonyls, Preparative Inorg. Reactions *1*, 77 (1964).

R. L. Pruett, Cyclopentadienyl and Arene Metal Carbonyls, Preparative Inorg. Reactions *2*, 187 (1965).

R. B. King, Reactions of Alkali-Metal Derivatives of Metal Carbonyls and Related Compounds, Advan. Organometal. Chem. *2*, 157 (1964).

T. A. Manuel, Lewis-Base Metal Carbonyl Complexes, Advan. Organometal. Chem. *3*, 181 (1965).

G. R. Dobson, I. W. Stolz, R. K. Sheline, Substitution Products of the Group VIB Metal Carbonyls, Advan. Inorg. Chem. Radiochem. *8*, 1 (1966).

R. Pettit, G. F. Emerson, Diene-iron Carbonyl Complexes and Related Species, Advan. Organometal. Chem. *1*, 1 (1964).

[348] *D. W. McBride, E. Dudek, F. G. A. Stone*, J. Chem. Soc. (London) 1725 (1964).
[349] *D. J. Patmore, W. A. G. Graham*, Inorg. Chem. *3*, 1405 (1966).
[350] *W. Manchot, J. König*, Ber. *59*, 883 (1926).
[351] *W. Schnabel, E. Kober*, J. Organometal. Chem. *19*, 455 (1969).
[352] *F. Canziani, P. Chini, A. Quarta, A. Di Martino*, J. Organometal. Chem. *26*, 285 (1971).
[353] *J. Chatt, N. P. Johnson, B. L. Shaw*, J. Chem. Soc. (London) 1663 (1969);
A. C. Smithies, M. Rycheck, M. Orchin, J. Organometal. Chem. *12*, 199 (1968).
[354] *A. Misono, Y. Uchida, M. Hidai, K. Kudo*, J. Organometal. Chem. *20*, P7 (1969).
[355] *G. Booth, J. Chatt*, J. Chem. Soc. *A*, 2131 (1969);
J. Chatt, P. Chini, J. Chem. Soc. *A*, 1538 (1970).
[356] *E. O. Fischer, H. Schuster-Woldan, K. Bittler*, Z. Naturforsch. *18b*, 429 (1963).

30 Ferrocenes

K. Schlögl and H. Falk
Institute of Organic Chemistry, University of Vienna,
Vienna, Austria

30.1 General Aspects

The discovery[1] of *bis-(π-cyclopentadienyl-)iron* (ferrocene) in 1951* has caused the development of a new era of organometallic chemistry and hardly any other compound has attracted the attention of various fields of chemistry to a similar extent (*cf.* Section 30.4).

30.1.1 Structure and Properties of Ferrocene

X-ray analysis has revealed that crystalline ferrocene exists in the "antiprismatic" conformation.[2] In ferrocene derivatives deviations from the fully staggered conformation may occur, primarily, if both rings—coplanar in ferrocene itself—are linked by bridging groups.

(Cf. amongst others: 1.1′-dibenzoylferrocene[3]; [2]-ferrocenophane[4]; [3]ferrocenophane and 1.2-(α-oxotetramethylene)ferrocene[5]; 1.1′-dimethylferrocene-β-carboxylic acid[6]; biferrocenyl.[7])

In the vapor phase an equilibrium exists between prismatic and antiprismatic (eclipsed and staggered) conformation with a rotational barrier of about 0·9 kcal/mol.[8]

Ferrocene crystallizes (*e.g.*, from dichloromethane–hexane or from ether) in monoclinic orange crystals[9] (m.p. 173°–174°) and sublimes above 100°. It is thermally quite stable decomposing at temperatures above 470°;[10] the density at 25° is 1·49 g/cm³. Ferrocene is diamagnetic[11] and has no dipole moment.[12] *Cf.* Section 30.3.3 for spectroscopic data.

The chemical properties of ferrocene are determined primarily by two factors:

(1) The redox-equilibrium ferrocene⇌ferricenium-ion + $1e^{\ominus}$ with an oxidation potential of −0·367 V (in acetic acid–perchloric acid),[13] and
(2) its "aromatic" behavior which is demonstrated by the high reactivity towards electrophile substitution.[14] This fact provides the basis for the synthesis of a large number of more or less easily accessible ferrocene derivatives.

Several publications deal with the bonding and the electronic structure of ferrocene (and its derivatives).[15,16] The use of ferrocene as an additive for oils and fuels may be mentioned.[17]

30.1.2 Preparation of Ferrocene

Most methods for the synthesis of ferrocene are based on the reaction of iron(II) salts with the cyclopentadienyl anion ($C_5H_5^{\ominus}$).

The reaction of $FeCl_2 \cdot 4H_2O$ with potassium hydroxide and cyclopentadiene in dimethylsulfoxide–dimethoxyethane appears to be the best route for the preparation of ferrocene;[18] the reaction of cyclopentadienyl sodium (obtained from cyclopentadiene and sodium in xylene) with iron(III) chloride and iron in tetrahydrofuran gives similar results[19] but is much more laborious. The "diethyl amine method" is less advantageous,[19] although an improvement

[1] *T. J. Kealy, P. L. Pauson*, Nature *168*, 1039 (1951).
[2] *J. D. Dunitz, L. E. Orgel, A. Rich*, Acta Cryst. *9*, 373 (1956).
[3] *Y. T. Struchkov, T. L. Khotsyanova*, Kristallografiya *2*, 382 (1957); C.A. *52*, 3457 (1958).
[4] *M. B. Laing, K. N. Trueblood*, Acta Cryst. *19*, 373 (1965).
[5] *N. D. Jones, R. E. Marsh, J. H. Richards*, Acta Cryst. *19*, 330 (1965);
E. B. Fleischer, S. Hawkinson, Acta Cryst. *22*, 376 (1967).
[6] *O. L. Carter, A. T. McPhail, G. A. Sim*, J. Chem. Soc. *A*, 365 (1967).
[7] *A. C. MacDonald, J. Trotter*, Acta Cryst. *17*, 872 (1964).
[8] *R. K. Bohn, A. Haaland*, J. Organometal. Chem. *5*, 470 (1966);
A. Haaland, J. E. Nilsson, Chem. Commun. 88 (1968).
[9] *J. D. Dunitz, L. Orgel*, Nature *171*, 121 (1953).
[10] *E. O. Fischer, H. Grubert*, Chem. Ber. *92*, 2302 (1959).
[11] *G. Wilkinson, M. Rosenblum, M. C. Whiting, R. B. Woodward*, J. Amer. Chem. Soc. *74*, 2125 (1952).
[12] *E. Weiss*, Z. Anorg. Allgem. Chemie *287*, 236 (1956).

* Here and in the following the quotations of reviews (as far as general aspects of ferrocene chemistry are concerned) have been omitted, since they are summarized in Section 30.4 (Bibliography).

[13] *J. G. Mason, M. Rosenblum*, J. Amer. Chem. Soc. *82*, 4206 (1960);
T. Kuwana, D. E. Bublitz, G. Hoh, J. Amer. Chem. Soc. *82*, 5811 (1960);
J. Komenda, J. Tirouflet, Compt. Rend. *254*, 3093 (1962).
[14] *R. B. Woodward, M. Rosenblum, M. C. Whiting*, J. Amer. Chem. Soc. *74*, 3458 (1952).
[15] *D. R. Scott, R. S. Becker*, J. Chem. Phys. *35*, 516 (1961).
[16] *A. T. Armstrong, D. G. Carroll, S. P. McGlynn*, J. Chem. Phys. *47*, 1104 (1967);
H. P. Fritz, J. H. Keller, K. E. Schwarzhans, J. Organometal. Chem. *13*, 505 (1968);
Y. S. Sohn, D. N. Hendrickson, H. B. Gray, J. Amer. Chem. Soc. *93*, 3603 (1971);
D. Nielson, D. Boone, H. Eyring, J. Phys. Chem. *76*, 511 (1972);
S. Evans, M. L. H. Green, B. Jewitt, A. F. Orchard, C. F. Pygall, Faraday Transact. II, 1847 (1972).
[17] *E. G. Nottes, J. F. Cordes*, Erdöl und Kohle *18*, 885 (1965); C.A. *64*, 7937 (1966).
[18] *W. L. Jolly*, Inorg. Synth. *11*, 120 (1968).
[19] *G. Wilkinson*, Org. Synth. Coll. Vol. IV, 473 (1963).

of it supposedly increases the yields to 70–90%.[20] A thorough study on the ferrocene synthesis (employing sodium ethoxide as base) has been published;[21] yields up to 95% are recorded and the process may be performed continuously.

Less important synthetic procedures are: the direct interaction of cyclopentadiene with iron at 300°,[22] the reaction of cyclopentadienyl magnesium bromide with iron(III) chloride in benzene–ether (first synthesis of ferrocene!)[1] or of cyclopentadiene with iron pentacarbonyl at elevated temperatures.[23]

30.1.3 Isomerism and Nomenclature of Substituted Ferrocenes

Due to the particular molecular geometry, special problems of isomerism are encountered in dealing with ferrocene derivatives. Besides positional and geometrical isomerism suitably substituted (chiral) ferrocenes may also exhibit optical isomerism.

There are three possible positional-isomeric disubstituted ferrocenes: two *homo*anular* (α and β, or 1,2 and 1,3, resp.) and one *hetero*anular (1,1′) product. Due to the virtually free rotation about the ring-iron-axis (*cf.* Section 30.1.1) no 1,1′-disubstituted isomers can be isolated.[24]

On disubstitution with two different groups (R,R′) both *homo*anular isomers become chiral (point group C_1), whereas the *hetero*anular product is "symmetric" (point group C_s) (*cf.* Section 30.2.5) due to the unrestricted (average) rotation.

With an increasing number of substituents the complexity of isomerism increases (*cf.* Table 1). For the drawing of ferrocene stereoformulas—especially with respect to isomerism and stereochemical problems—the Newman type projection formulas as shown in **1** or **2** have been very useful.

[20] *R. Gelius, W. Uhlmann, W. Sperling*, Z. Chem. *6*, 227 (1966).

[21] *J. F. Cordes*, Z. Naturforsch. *21b*, 746 (1966).

[22] *S. A. Miller, J. A. Tebboth, J. F. Tremaine*, J. Chem. Soc. (London) 632 (1952); *R. Riemschneider, D. Helm*, Z. Naturforsch. *16b*, 234 (1961).

[23] *G. Wilkinson, P. L. Pauson, F. A. Cotton*, J. Amer. Chem. Soc. *76*, 1970 (1954); *B. F. Hallam, P. L. Pauson*, J. Chem. Soc. (London) 3030 (1956).

[24] *M. Rosenblum, R. B. Woodward*, J. Amer. Chem. Soc. *80*, 5443 (1958).

* *anular* is derived from the latin "*anulus*" and should therefore—in contrast to the prevailing usage (as well as *anulenes*)—be spelled with *one* n.

Table 1 Positional Isomers of Ferrocenes

Substitution		Number of possible isomers		
Ring 1	Ring 2	Total	Chiral	Achiral
R, R		3	0	3
R, R′		3	2	1
R, R, R		4	0	4
R, R, R′		8	3	5
R, R′, R″		13	13	0
R, R′	R, R′	6	4	2
R, R′, R″	R, R′, R″	13	8	5

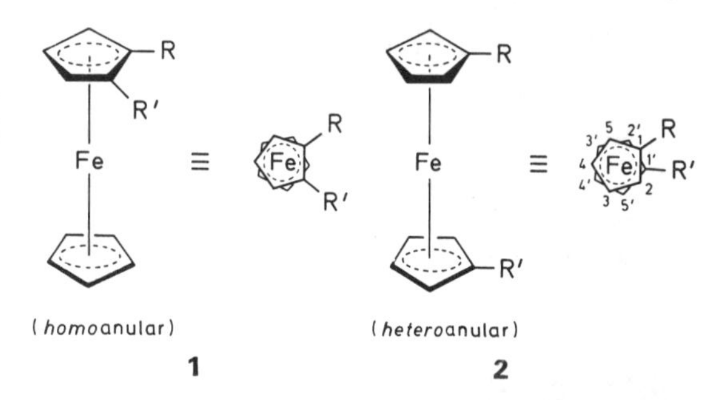

(homoanular) **1** (heteroanular) **2**

The trivial name ferrocene[14] for bis-(π-cyclopentadienyl)-iron (ferrocenyl, *Fc*, is used for the group C_5H_5–Fe–C_5H_4–) has been accepted by IUPAC[25] and is also used in Chemical Abstracts. Both rings are numbered clockwise from 1–5 and 1′–5′, resp., with the primed numbers used for the un- or "less"-substituted ring. It should be noted that *homo*anularly, unsymmetrically substituted ferrocenes are chiral; consequently, positional isomers become enantiomers (*cf.* also Section 30.2.5). Hence, utilization of a name such as 1-ethyl-2-methylferrocene or 1-methylferrocene-3-carboxylic acid implies that the absolute configuration of the compound in question is known. (*Cf.* Ref. 26 for the configurational nomenclature of optically active ferrocenes.) Therefore, for racemates or optically active ferrocenes of unknown configuration the designation α (for 1,2-) or β (for 1,3-) is preferable: α-ethyl-methylferrocene, methylferrocene-β-carboxylic acid.

Although not provided for in the (tentative) IUPAC rules,[25] the prefix "ferroceno" has been accepted for *homo*anularly bridged ferrocenes, and for *hetero*anularly bridged compounds the ferrocenophane nomenclature has been

[25] IUPAC, Information Bulletin No. 35, 31 (1969).

[26] *K. Schlögl* in *N. L. Allinger, E. L. Eliel*, Topics in Stereochem. Bd. I, p. 39; Interscience Publishers New York 1967.

Table 2 Important Monosubstituted Ferrocenes (*Fc*–R)

Compound No.	R	Crystallized from	M.-P. [°C]	B.-P.	Torr	Molecular formula	Molecular weight
3	$-CH_2N(CH_3)_2$	—	—	91	0·5 (n_D=1·5893)	$C_{13}H_{17}FeN$	243·1
	$-CH_2\overset{\oplus}{N}(CH_3)_3J^{\ominus}$	—	~200(d.)	—		$C_{14}H_{20}FeIN$	385·0
4	–CHO	Hexane oder Ethanol–Water	130–132	70	0·01	$C_{11}H_{10}FeO$	214·0
5	$-COCH_3$	Hexane	85	90	0·01	$C_{12}H_{12}FeO$	228·0
6	–COOH	Benzene–Hexane	210 (d.)	—		$C_{11}H_{10}FeO_2$	230·0
7	$-B(OH)_2$	—	~140	—		$C_{10}H_{11}BFeO_2$	229·8
8	–Li	—	—	—		$C_{10}H_9FeLi$	192·0

adopted:[27, 28, 28a] 1,1′-trimethylene-ferrocene = [3]-ferrocenophane.

30.2 Syntheses of Ferrocene Derivatives

30.2.1 Monosubstituted Ferrocenes

30.2.1.1 Important Starting Materials

Several monosubstituted ferrocenes (*Fc*–R) are of considerable interest as starting materials for the synthesis of numerous ferrocene derivatives; hence, they are treated separately. The unique utility of these starting materials resides in the reactivities of their functional groups; they are primarily the Mannich-base **3** (and its methiodide), the carboxaldehyde **4**, the acetylderivative **5**, the carboxylic acid **6**, the ferrocenylboronic acid **7**, and ferrocenyllithium **8**. Some data of these compounds are summarized in Table 2, the important methods for their preparations are discussed in Sections 30.2.1.1.1 through 30.2.1.1.6. For subsequent and transformation products *cf.* Section 30.2.1.2.

30.2.1.1.1 *N,N*-Dimethylaminomethylferrocene

Mannich reaction of ferrocene with *N,N,N′,N′*-tetramethyl methylenediamine in acetic acid in the presence of phosphoric acid yields the desired product **3** in 80% yield.[29] **3** is usually pure enough for further use. Its methoiodide, a very valuable starting material for the preparation of many ferrocene derivatives (*cf.* Section 30.2.1.2), is easily accessible by reaction of **3** with methyl iodide in methanol.[29]

30.2.1.1.2 Ferrocenecarboxaldehyde

By far the best method for the synthesis of the title compound is the Vilsmeier formylation of ferrocene with *N*-methyl formanilide and $POCl_3$.[30, 31] The purification of **4** is achieved *via* its bisulfite addition product; yields up to 80% may be obtained.[31]

Further methods for preparing **4** are the Friedel-Crafts alkylation of ferrocene with dichloro methylether (and subsequent hydrolysis; yield ca 50%)[32] or the nearly quantitative oxidation of hydroxymethylferrocene ($Fc-CH_2-OH$) with manganese dioxide.[33] Lower yields (23 and 54% resp.) are obtained if methylferrocene[34] or the Mannich base **3**[35] are oxidized with MnO_2. The Sommelet reaction of the methiodide of **3** affords **4** in 37% yield.[36]

30.2.1.1.3 Acetylferrocene

The synthesis of acetylferrocene is an excellent

[27] *B. H. Smith*, Bridged Aromatic Compounds, Academic Press, New York 1964;
W. E. Watts, Organometal. Chem. Rev. *2*, 231 (1967).

[28] *H. Falk, O. Hofer, K. Schlögl*, Monatsh. Chem. *100*, 624 (1969).

[28a] *F. Vögtle, P. Neumann*, Tetrahedron *26*, 5847 (1970).

[29] *D. Lednicer, C. R. Hauser*, Org. Synth. *40*, 31 (1960).

[30] *K. Schlögl*, Monatsh. Chem. *88*, 601 (1957).

[31] *M. Rosenblum, A. K. Banerjee, N. Danieli, R. W. Fish, V. Schlatter*, J. Amer. Chem. Soc. *85*, 316 (1963).

[32] *P. L. Pauson, W. E. Watts*, J. Chem. Soc. (London) 3880 (1962).

[33] *J. K. Lindsay, C. R. Hauser*, J. Org. Chem. *22*, 355 (1957).

[34] *K. L. Rinehart jr., A. F. Ellis, C. J. Michejda, P. A. Kittle*, J. Amer. Chem. Soc. *82*, 4112 (1960).

[35] *K. Schlögl, M. Walser*, Tetrahedron Lett. 5885 (1968).

[36] *G. D. Broadhead, J. M. Osgerby, P. L. Pauson*, J. Chem. Soc. 650 (1958).

example for the Friedel-Crafts acylation of ferrocene. Usually, methylene chloride is employed as solvent, though the acylation reagent as well as the Lewis acid may be varied.

Amongst the various methods described for the acetylation of ferrocene, such as employing acetic anhydride and hydrofluoric acid (yield 80%),[37] phosphoric acid (yield 85%)[38] or boron trifluoride (yield 73% and 80%, resp.)[36,39] the synthesis of **5** is accomplished best by the reaction of ferrocene with acetic anhydride and $BF_3 \cdot Et_2O$ in methylene chloride;[40] under these conditions a very pure product is obtained (yield 90%).

The use of acetyl chloride and aluminum chloride cannot be recommended, since even a slight excess of the catalyst causes formation of appreciable amounts of the disubstituted product (1,1′-diacetylferrocene, *cf.* Section 30.2.2.2.1).[41]

30.2.1.1.4 Ferrocenecarboxylic acid

Reaction of ferrocene with *N,N*-diphenylcarbamylchloride-aluminum chloride and subsequent hydrolysis of the amide **9** affords the acid **6** in 70% yield.[42] This reaction seems to be the best method for the synthesis of **6**. The reaction of ferrocene with methyl chlorothioformate is also basically a Friedel-Crafts reaction; the product **10** can be hydrolyzed to give **6** with 70% yield.[43]

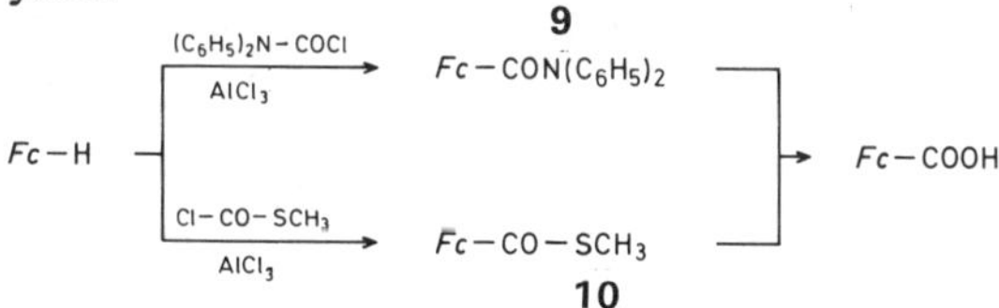

Recently, **6** was prepared with 91% yield by cleavage of (2-chlorobenzoyl) ferrocene with K-t. butoxide.[43a] This method is also useful for the synthesis of substituted ferrocene mono- and dicarboxylic acids such as **87**.[43b]

Satisfactory results are also obtained with the oxidation of acetyl-ferrocene (**5**) with hypochlorite[44] or iodine in pyridine;[37] the direct oxidation of ferrocenecarboxaldehyde (**4**) with silver oxide seems to be restricted to small scale preparations.[45]

Carboxylation of ferrocenyllithium (**8**) proceeds smoothly to give **6**; since, however, one usually deals with a mixture of the mono- and di-lithio derivatives (*cf.* Section 30.2.1.1.6), ferrocene-1,1′-dicarboxylic acid (Section 30.2.1.1.6) is practically always present as by-product.[46,47] The chloride of **6**, *Fc*–COCl, is of particular interest; it is obtained by treatment of **6** with phosphorus tri- or penta-chloride in benzene[30,48] (preferably with PCl_3 in the presence of catalytic amounts of pyridine). After purification by recrystallization from pentane[48] or distillation[49] the chloride is a crystalline material (m.p. 50°). Esters,[30,46] amides and the hydrazide[30] are readily accessible by standard procedures originating from either the acid **6** or its chloride.

30.2.1.1.5 Ferrocenylboronic acid

For the preparation of **7**, ferrocenyllithium (**8**, *cf.* 30.2.1.1.6) is treated with tris-(*n*-butoxy)-borane. Lithiation affords a mixture of mono- and dilithio ferrocene resulting in a mixture of **7** and ferrocene-1,1′-bis(boronic) acid; they can be separated either by careful acidification of a solution of the sodium salts (preferred precipitation of the bisacid occurs)[50] or, better, by extraction of the mixture with ether, in which solvent the bisacid (18% yield) is less soluble than **7** (44%).[51]

30.2.1.1.6 Ferrocenyllithium

If an excess of *n*-butyllithium is employed in the lithiation of ferrocene (as is normally done with

[37] *V. Weinmayr*, J. Amer. Chem. Soc. *77*, 3009 (1955).
[38] *Cf. R. Reimschneider, D. Helm*, Justus Liebigs Ann. Chem. *646*, 10 (1962).
[39] *C. R. Hauser, J. K. Lindsay*, J. Org. Chem. *22*, 482 (1957).
[40] *Y. Nagai, J. Hooz, R. A. Benkeser*, Bull. Chem. Soc. Japan *37*, 53 (1964).
[41] *M. Rosenblum, J. O. Santer, W. G. Howells*, J. Amer. Chem. Soc. *85*, 1450 (1963).
[42] *W. F. Little, R. Eisenthal*, J. Amer. Chem. Soc. *82*, 1577 (1960).
[43] *D. E. Bublitz, G. H. Harris*, J. Organometal. Chem. *4*, 404 (1965).
[43a] *E. R. Biehl, P. C. Reeves*, Synthesis *1973*, 360.
[43b] *V. Rapić, K. Schlögl, B. Steinitz*, J. Organometal. Chem. *94*, 87 (1975).
[44] *V. Weinmayr*, J. Amer. Chem. Soc. *77*, 3012 (1955).
[45] *K. Schlögl, M. Walser*, Monatsh. Chem. *100*, 840 (1969).
[46] *R. A. Benkeser, D. Goggin, G. Schroll*, J. Amer. Chem. Soc. *76*, 4025 (1954).
[47] *J. Tirouflet, E. Laviron, R. Dabard, J. Komenda*, Bull. Soc. Chim. France 857 (1963).
[48] *H. H. Lau, H. Hart*, J. Org. Chem. *24*, 280 (1959).
[49] *K. Schlögl, H. Seiler*, Naturwissenschaften *45*, 337 (1958).
[50] *A. N. Nesmeyanov, V. A. Sazonova, V. N. Drozd*, Dokl. Akad. Nauk. SSSR *126*, 1004 (1959); C.A. *54*, 6673 (1960).
[51] *H. Shechter, F. Helling*, J. Org. Chem. *26*, 1034 (1961).

aromatic compounds), mixtures of mono-(**8**) and dilithioferrocene are obtained.[52] A selective monolithiation can be achieved with equimolar amounts of butyllithium and ferrocene.[52]

Further possibilities for the preparation of pure **8** are the metal exchange of *Fc*–HgCl[53] or *Fc*–Hg–*Fc*[54] (*cf.* Section 30.2.1.2.9) with *n*-butyllithium (yields 60 and 80%, resp.) or the reaction of butyllithium with *Fc*–I or *Fc*–Br (*cf.* Section 30.2.1.2.5) which quantitatively yields **8**. Lithiation of *Fc*-Cl occurs on the carbonatom adjacent to chlorine.[55] **8** can be isolated (red crystals); it is, however, contaminated with dilithioferrocene.[56]

30.2.1.2 Transformation Products of Monosubstituted Ferrocenes (described in Section 30.2.1.1)

Ferrocene derivatives discussed in the following sections are "organic" compounds with respect to their properties and, in particular, the methods of synthesis and reactions employed. Consequently, they are arranged (in Sections 30.2.1.2.1 through 30.2.1.2.9) according to the classical principles of organic chemistry.

30.2.1.2.1 Alkylferrocenes

The direct Friedel–Crafts alkylation of ferrocene affords primarily mixtures of mono- and poly-alkylferrocenes[57] which can be separated—if at all—only with difficulty (*cf.* Section 30.2.2.1). Hence, this procedure is not recommended for the synthesis of pure products. However, reduction of the easily accessible acyl and aroyl ferrocenes, resp. (Sections 30.2.1.1.2 through 30.2.1.1.4) by the Clemmensen method, with lithium aluminum hydride–aluminum chloride or catalytically (H_2/Pt) offers no problems.

The Clemmensen reduction usually proceeds smoothly[58] and was successfully utilized for the reduction of ferrocenoylcarboxylic acids [Fc–CO$(CH_2)_n$COOH] to give ω-ferrocenylcarboxylic acids [Fc–$(CH_2)_{n+1}$COOH].[59] Such ketoacids can also be reduced with H_2/Pt in acetic acid or ethanol[49,60] The Clemmensen reduction of aroylferrocenes may lead to the formation of dimeric byproducts.[61]

Constitutions and stereochemistry of the nine products formed from benzoylferrocene were established.[61a]

Reduction with the mixed hydride $LiAlH_4$–$AlCl_3$ (in ether or tetrahydrofurane) gives excellent results not only with acylferrocenes, but also with ferrocenyl ethers and carbinols (Section 30.2.1.2.4), *i.e.* compounds which contain the structural unit *Fc*–C–O–. In all cases alkylferrocenes or aralkylferrocenes are obtained in almost quantitative yield.[62,63] Ferrocene-carboxaldehyde (**4**), the carboxylic acid **6** (or its esters) are reduced to methylferrocene (*Fc*–CH_3, **11**). **11** is also accessible from the methiodide of the Mannich-base **3** by reduction with sodium amalgam[64] or with sodium in liquid ammonia.[65]

Reaction of this methiodide with Grignard compounds affords alkylferrocenes of the general formula *Fc*–CH_2–R.[66]

For the preparation of branched alkylferrocenes, such as *iso*propylferrocene (**13**), the sequence shown on p. 475 was successfully employed (the overall yield of **13** being 71%).[67]

Alkylferrocenes may also be obtained by catalytic hydrogenation of the alkenyl ferrocenes **15** (Section 30.2.1.2.2) which are easily accessible by dehydration of ferrocenylcarbinols **14**.

Diferrocenylmethane (*Fc*–CH_2–*Fc*) can be

---

[52] *S. I. Goldberg, L. H. Keith*, T. S. Prokopov, J. Org. Chem. *28*, 850 (1963).

[53] *D. Seyferth, H. P. Hoffman, R. Burton, J. F. Helling*, Inorg. Chem. *1*, 227 (1962).

[54] *M. D. Rausch*, Inorg. Chem. *1*, 414 (1962).

[55] *F. L. Hedberg, H. Rosenberg*, Tetrahedron Lett. 4011 (1969).

[56] *K. Schlögl, M. Fried*, Monatsh. Chem. *94*, 537 (1963).

[57] *T. Leigh*, J. Chem. Soc. (London) 3294 (1964).

[58] *K. L. Rinehart jr., R. J. Curby jr., D. H. Gustafson, K. G. Harrison, R. E. Bozak, D. E. Bublitz*, J. Amer. Chem. Soc. *84*, 3263 (1962).

[59] *G. Haller, K. Schlögl*, Monatsh. Chem. *98*, 2044 (1967).

[60] *K. L. Rinehart jr., R. J. Curby jr., P. E. Sokol*, J. Amer. Chem. Soc. *79*, 3420 (1967).

[61] *M. D. Rausch, D. L. Adams*, J. Org. Chem. *32*, 4144 (1967);
H. Patin, Compt. Rend. *270*, 243 (1970).

[61a] *S. I. Goldberg, W. D. Bailey, M. L. McGregor*, J. Org. Chem. *36*, 761 (1971).

[62] *K. Schlögl, A. Mohar, M. Peterlik*, Monatsh. Chem. *92*, 921 (1961).

[63] *R. A. Benkeser, J. L. Bach*, J. Amer. Chem. Soc. *86*, 890 (1964).

[64] *A. N. Nesmeyanov, E. G. Perevalova, L. S. Shilovtseva, V. A. Beinoravichute*, Dokl. Akad. Nauk. SSSR *121*, 117 (1968); C.A. *53*, 323 (1959).

[65] *D. W. Slocum, W. E. Jones*, J. Organometal. Chem. *15*, 262 (1968).

[66] *A. N. Nesmeyanov, E. G. Perevalova, L. S. Shilovtseva*, Izv. Akad. Nauk. SSSR 1982 (1961); C.A. *56*, 10185 (1962).

[67] *K. Schlögl, M. Fried*, Monatsh. Chem. *95*, 558 (1964).

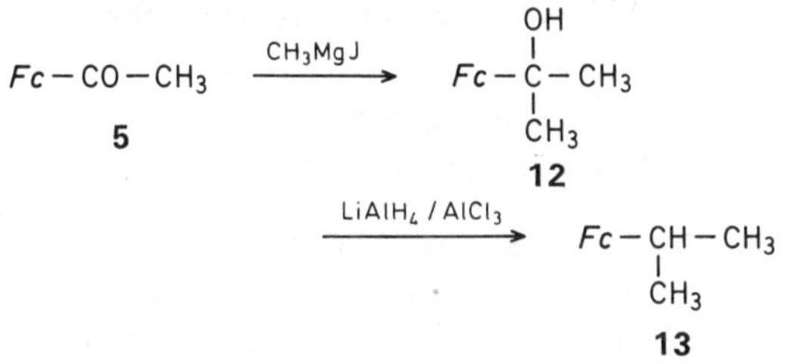

$$Fc-CO-CH_3 \xrightarrow{CH_3MgJ} Fc-C(OH)(CH_3)-CH_3$$

5 → 12

$$\xrightarrow{LiAlH_4/AlCl_3} Fc-CH(CH_3)-CH_3$$

13

prepared from ferrocene and paraformaldehyde; triferrocenylmethane is obtained by the reaction of ferrocenyllithium (**8**) with diferrocenylketone (*Fc*–CO–*Fc*) and subsequent reduction of the carbinol Fc_3C–OH.[32]

30.2.1.2.2 Alkenyl and Alkynylferrocenes

Vinylferrocene derivatives **15** are conveniently prepared by dehydration of the corresponding ferrocenyl-alkyl-carbinols **14** (Section 30.2.1.2.4) with acid alumina or potassium hydrogen sulfate.

$$Fc-CH(OH)-CH(R)-R^1 \xrightarrow[-H_2O]{} Fc-CH=C(R)(R^1)$$

14 → 15

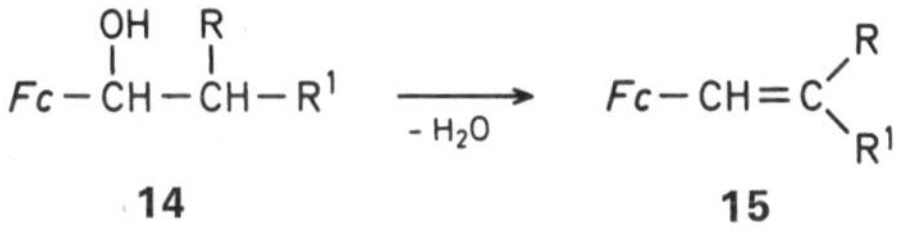

The parent vinylferrocene (**15**, R=R′=H) is accessible from **14** (R=R′=H) with alumina in benzene (50% yield);[68–70] the required carbinol is readily obtained from acetylferrocene (**5**). The dehydration of **14** with potassium hydrogen sulfate in refluxing benzene also proceeds very smoothly. After 30 min reaction time vinylferrocene is obtained in *ca.* 90% yield.

Copper sulfate has also been suggested (yield 97%)[71] for the dehydration of **14**.

The Wittig reaction of acetylferrocenes with appropriate phosphoranes has been employed for the synthesis of alkenylferrocenes[72] and several ferrocenylpolyenes were prepared[73] by this method and by the Knoevenagel condensation. Similarly the condensation of *Fc*–CHO (**4**) with C–H–acidic compounds yields products of the general structure **15**: malonic and hippuric acid,[30] acetophenone,[74] substituted acetophenones, and certain acetylated heterocyclic compounds[75] provided the expected products in high yields.

The preparation of ferrocenylacetylenes is shown in the following scheme:

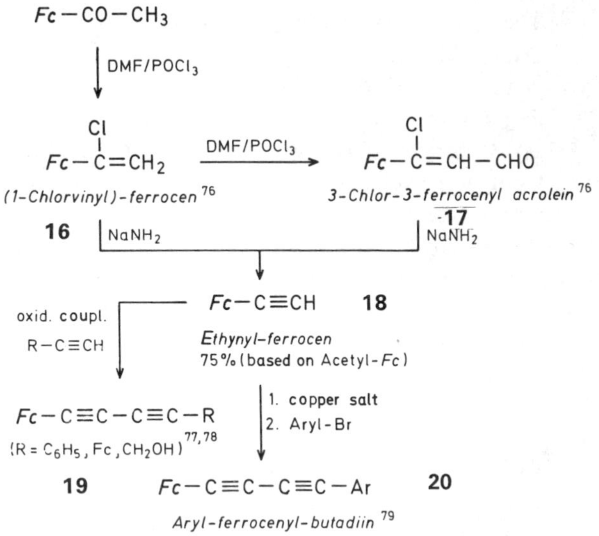

$$Fc-CO-CH_3$$

↓ DMF/$POCl_3$

$$Fc-C(Cl)=CH_2 \xrightarrow{DMF/POCl_3} Fc-C(Cl)=CH-CHO$$

(1-Chlorvinyl)-ferrocen[76] **16** — *3-Chlor-3-ferrocenyl acrolein*[76] **17**

↓ $NaNH_2$ (from 16 and 17)

$$Fc-C\equiv CH \quad \mathbf{18}$$

Ethynyl-ferrocen 75% (based on Acetyl-*Fc*)

oxid. coupl. $R-C\equiv CH$ ↓

$$Fc-C\equiv C-C\equiv C-R^{77,78}$$

($R = C_6H_5$, *Fc*, CH_2OH) **19**

↓ 1. copper salt 2. Aryl-Br

$$Fc-C\equiv C-C\equiv C-Ar \quad \mathbf{20}$$

Aryl-ferrocenyl-butadiin[79]

Cf. below[80] for the synthesis of ferrocenyl poly-ynes.

Base-catalyzed rearrangement of propargyl ferrocenes (**21**, R=H or C_6H_5) affords mixtures of allenes (**22**) and isomeric acetylenes **23**.[81]

$$\underset{\mathbf{21}}{Fc-CH_2-C\equiv C-R} \rightleftharpoons \underset{\mathbf{22}}{Fc-CH=C=CH-R} \rightleftharpoons \underset{\mathbf{23}}{Fc-C\equiv C-CH_2-R}$$

Ferrocenyl-1,2- and 1,3-dienes are also obtained from dibromoferrocenyl cyclopropanes and *n*-butyllithium.[81a]

---

[68] *K. Schlögl, A. Mohar*, Naturwissenschaften *48*, 376 (1961).
[69] *K. Schlögl, A. Mohar*, Monatsh. Chem. *92*, 219 (1961).
[70] *M. D. Rausch, A. Siegel*, J. Organometal. Chem. *11*, 317 (1968).
[71] *I. Pascal, W. Borecki*, U.S. Pat. 3132165 (1964); C.A. *61*, 4396 (1964).
[72] *P. L. Pauson, W. E. Watts*, J. Chem. Soc. (London) 2990 (1963).
[73] *K. Schlögl, H. Egger*, Justus Liebigs Ann. Chem. *676*, 76 (1964).
[74] *C. R. Hauser, J. K. Lindsay*, J. Org. Chem. *22*, 906 (1957).
[75] *J. Boichard, J. P. Monin, J. Tirouflet*, Bull. Soc. Chim. France 851 (1963).
[76] *K. Schlögl, W. Steyrer*, Monatsh. Chem. *96*, 1520 (1965);
M. Rosenblum, N. Brawn, J. Papenmeier, M. Applebaum, J. Organometal. Chem. *6*, 173 (1966).
[77] *K. Schlögl, H. Egger*, Monatsh. Chem. *94*, 376 (1963).
[78] *H. Egger, K. Schlögl*, Monatsh. Chem. *95*, 1750 (1964).
[79] *M. D. Rausch, A. Siegel, P. Klemann*, J. Org. Chem. *31*, 2703 (1966).
[80] *K. Schlögl, W. Steyrer*, J. Organometal. Chem. *6*, 399 (1966).
[81] *K. Schlögl, A. Mohar*, Monatsh. Chem. *93*, 861 (1962).
[81a] *W. M. Horspool, R. G. Sutherland, B. J. Thomson*, J. Chem. Soc. (London) (*C*) 1554 (1971).

30.2.1.2.3 Arylferrocenes

The most convenient method for the arylation of ferrocene is the Gomberg-Bachmann-reaction, *i.e.* the reaction with an aryldiazonium salt in acetic acid or acetone[82,83] which results in yields of about 30–40%. The mechanism of this reaction was investigated.[84]
The Ullmann-reaction of haloferrocenes (*cf.* Section 30.2.1.2.5) leads to biferrocenyls in high yield (90%), whereas the mixed Ullmann coupling for the synthesis of arylferrocenes is less satisfactory.[85,86] The synthesis of arylferrocenes from aryl cyclopentadienyls is of no practical importance.[87]
Biferrocenyls can also be prepared by coupling of ferrocenyllithium (**8**) with cobalt bromide[88] or chloride,[89] or, preferably, by treating ferrocenylboronic acid(s) (*cf.* Section 30.2.1.1.5) with copper(II) acetate.[90]
For the synthesis of α,α'-disubstituted biferrocenyls coupling of α-substituted lithoferrocenes with $CoCl_2$ can be recommended.[90a]
Pyridyl and quinolyl ferrocenes can be obtained from ferrocenyllithium (**8**) and the corresponding heterocyclic compounds;[56] the cyclization of ferrocenylbutadiynes (**19**, Section 30.2.1.2.2) with $S^{2\ominus}$ has been successfully performed[78,91] for the preparation of 2-thienylferrocenes (**24**).

$$Fc-C{\equiv}C-C{\equiv}C-R \xrightarrow{H_2S/RO^{\ominus}} \text{Fc-(2-thienyl)-R}$$

19 → **24**

Cyclo-trimerization of ferrocenyl acetylene (**18**) and its derivatives under various conditions leads to 1,2,4-tri-ferrocenylbenzenes;[92] the symmetrical product (1,3,5-tri-ferrocenyl benzene) is obtained by acid-catalyzed trimerization of acetylferrocene (**5**).[92]

30.2.1.2.4 Ferrocenes with Functional Groups in the Side-chain

A variety of monosubstituted ferrocenes containing functional groups in the side-chain can be prepared by conventional organic reactions though special requirements of ferrocene chemistry have to be considered. In many instances the compounds described in Section 30.2.1.1 are utilized as starting materials.
For cyclization reactions leading to defined products (Sections 30.2.2.2.3 and 30.2.2.3.2), for chemical methods for the elucidation of structures (Section 30.3.1) or in context with stereochemical studies (Section 30.2.5), it is frequently important—starting from easily accessible ferrocenes, such as the Mannich base **3**, the carboxaldehyde **4** or the carboxylic acid **6**—to lengthen the side-chain of a particular ferrocene derivative by one or two carbon atoms in an unambiguous manner.
Some important ferrocene derivatives and methods for side-chain extension are briefly discussed in this section. The arrangement follows the principle of organic chemistry: alcohols, ethers, oxocompounds, carboxylic acids, amines, amides, nitrils, amino acids and heterocyclic compounds.
Ferrocenyl carbinols can be obtained as shown in **25** to **29**
The properties of the carbinols having an OH-group at the α-carbon atom of the side-chain are primarily determined by the stability of ferrocenyl carbonium ions $Fc{-}C^{\oplus}$,[94] which are readily formed from the former.
The reduction with lithium aluminum hydride-aluminum chloride (or with Pd/H_2) leading to alkylferrocenes was mentioned in Section 30.2.1.2.1 and the smooth dehydration to

---

[82] *G. D. Broadhead, P. L. Pauson*, J. Chem. Soc. (London) 367 (1955).
[83] *M. Rosenblum*, J. Amer. Chem. Soc. *81*, 4530 (1959).
[84] *M. Rosenblum, W. G. Howells, A. K. Banerjee, C. Bennett*, J. Amer. Chem. Soc. *84*, 2726 (1962);
A. L. J. Beckwith, R. J. Leydon, Tetrahedron *20*, 791 (1964).
[85] *M. Rausch*, J. Org. Chem. *26*, 1802 (1961).
[86] *S. I. Goldberg, R. L. Matteson*, J. Org. Chem. *29*, 323 (1964).
[87] *P. L. Pauson*, J. Amer. Chem. Soc. *76*, 2187 (1954).
[88] *I. J. Spilners, J. P. Pellegrini jr.*, J. Org. Chem. *30*, 3800 (1965).
[89] *A. Watanabe, I. Motoyama, K. Hata*, Bull. Chem. Soc., Japan *39*, 790 (1966).
[90] *G. Marr, R. E. Moore, B. W. Rockett*, Tetrahedron *25*, 3477 (1969).
[90a] *D. J. Booth, G. Marr, B. W. Rockett*, J. Organometal. Chem. *32*, 227 (1971).
[91] *H. Falk, H. Lehner, K. Schlögl*, Monatsh. Chem. *101*, 967 (1970).
[92] *K. Schlögl, H. Soukup*, Tetrahedron Lett. 1181 (1967); Monatsh. Chem. *99*, 927 (1968).
[93] *H. Egger, K. Schlögl*, J. Organometal. Chem. *2*, 398 (1964).
[94] *M. Cais*, Organometal. Chem. Rev. *1*, 435 (1966);
J. Feinberg, M. Rosenblum, J. Amer. Chem. Soc. *91*, 4324 (1969);
M. J. Nugent, R. E. Carter, J. H. Richards, J. Amer. Chem. Soc. *91*, 6145 (1969);
S. Lupan, M. Kapon, M. Cais, F. H. Herbstein, Angew. Chem. *84*, 1104 (1972).
T. D. Turbitt, W. E. Watts, J. Chem. Soc. (Perkin II), 177, 185, 189, 195 (1974).

$$Fc{-}CHO \text{ or } Fc{-}CO{-}R \xrightarrow[\sim 100\%]{LiAlH_4{}^{30},\ NaBH_4{}^{36}} Fc{-}CH_2{-}OH\ (\mathbf{25}) \text{ or } Fc{-}CH(R){-}OH\ (\mathbf{26})$$

$$\xrightarrow{Na{-}C{\equiv}CH\ ^{69}} Fc{-}CH(OH){-}C{\equiv}CH\ (\mathbf{27}) \text{ or } Fc{-}C(R)(OH){-}C{\equiv}CH\ (\mathbf{28})$$

$$\xrightarrow{R^1{-}MgX\ ^{93}} Fc{-}CH(OH){-}R^1 \text{ or } Fc{-}C(R)(OH){-}R^1\ (\mathbf{29})$$

$$Fc{-}CH_2{-}\overset{\oplus}{N}(CH_3)_3J^{\ominus} \xrightarrow{OH^{\ominus}\ ^{33}} Fc{-}CH_2{-}OH$$

alkenylferrocenes was noted in Section 30.2.1.2.2. Nucleophilic attack of ROH affords ethers of the general structure **30**.[95] Ferrocenylmethyl ethers (**31**) can be obtained from the methiodide of **3** by reaction with RO⊖[96] and diferrocenyl dimethylether (**32**) is obtained by treatment of **25** with acid alumina.[69]

The rearrangement of unsaturated ferrocenyl carbinols usually provides the expected products: *e.g.* **27** is rearranged by acids to ferrocenylacroleine (**33**).[81]

$$Fc{-}\overset{|}{\underset{|}{C}}{-}OR\ (\mathbf{30}) \qquad Fc{-}CH_2OR\ (\mathbf{31})$$

$$Fc{-}CH_2OCH_2{-}Fc\ (\mathbf{32}) \qquad Fc{-}CH{=}CH{-}CHO\ (\mathbf{33})$$

Carbinols, such as **26** or **29** can be converted with malonic esters to the corresponding substituted esters (*cf.* also **41**).[97] Acyl- and aroylferrocenes are obtained by Friedel-Crafts reactions of ferrocene with carboxylic acid chlorides or anhydrides (*cf.* Section 30.2.1.1.3 for acetyl ferrocene). For the preparation of benzoyl, α-thenoyl or ferrocenoyl ferrocene (diferrocenyl ketone, *Fc*–CO–*Fc*) *cf.*[69,98,99]

Acylferrocenes, such as acetylferrocene (**5**) can also be prepared from the carboxylic acid **6** and its chloride, resp., (Section 30.2.1.1.4) by reaction with alkylcadmium compounds[100] (chain lengthening). The Mannich reaction of **5** gives **34** which, on the basis of its reactive end group, may (by nucleophilic substitution of the corresponding ammonium salts) be converted into other derivatives.[59,101]

Ferrocenoyl acetaldehyde (**35**), which can be obtained from **5** and ethylformate,[102] is a useful starting material for several syntheses (*vide infra*).

$$Fc{-}COCH_2CH_2N(CH_3)_2\ (\mathbf{34})$$

$$Fc{-}COCH_2CHO \rightleftharpoons Fc{-}COCH{=}CHOH\ (\mathbf{35})$$

ω-Ferrocenylcarboxylic acids (**37**) are prepared by reduction of the corresponding ketoacids **36** (*cf.* Section 30.2.1.2.1).[49,59,60] The acetic acid (**37**, $n=0$) may be obtained from its nitril **38**, which is accessible by nucleophilic substitution of the methoiodide of **3** with CN⊖ in 96% yield.[103] Chain lengthening by two carbon atoms is accomplished by a Knoevenagel condensation with the aldehyde **4** and subsequent hydrogenation of the ferrocenylacrylic acid (**39**) to yield the propionic acid **37** ($n=1$).[30] **37** can also be

$$Fc{-}CO(CH_2)_nCOOH\ (\mathbf{36}) \qquad Fc{-}(CH_2)_{n+1}COOH\ (\mathbf{37})$$

---

[95] *E. G. Perevalova, A. Y. Ustynyuk, A. N. Nesmeyanov*, Izv. Akad. Nauk. SSSR 1972 (1963); C.A. *60*, 6865 (1964).

[96] *A. N. Nesmeyanov, E. G. Perevalova, A. Y. Ustynyuk, L. S. Shilovtseva*, Izv. Akad. Nauk. SSSR 554 (1960); *A. N. Nesmeyanov, E. G. Perevalova, A. Y. Ustynyuk, L. S. Shilovtseva*, Dokl. Akad. Nauk. SSSR *133*, 1105 (1960); C.A. *54*, 22540 and 24616 (1960).

[97] *C. Moise, J. Tirouflet*, Bull. Soc. Chim., France 2656 (1970).

[98] *M. D. Rausch, M. Vogel, H. Rosenberg*, J. Org. Chem. *22*, 903 (1957).

[99] *K. Schlögl, H. Pelousek*, Justus Liebigs Ann. Chem. *651*, 1 (1962).

[100] *H. Falk, K. Schlögl*, Tetrahedron *22*, 3047 (1966).

[101] *G. Tainturier, J. Tirouflet*, Bull. Soc. Chim., France 600 (1966).

[102] *K. Schlögl, H. Egger*, Monatsh. Chem. *94*, 1054 (1963).

[103] *D. Lednicer, J. K. Lindsay, C. R. Hauser*, J. Org. Chem. *23*, 653 (1958); *D. Lednicer, J. K. Lindsay, C. R. Hauser*, Org. Synth. *40*, 45 (1960).

Fc—CH_2—C≡N **38** *Fc*—CH=CH—COOH **39**

Fc—$CH_2CH(COOH)_2$ **40**

obtained by a malonic ester synthesis from the methoiodide of **3** and subsequent decarboxylation of ferrocenylmethylmalonic acid (**40**).[104] β-Ferrocenylpropionic acids of the general structure **42** (R,R^1 or R^2=H, CH_3 or C_6H_5) can also be prepared by reaction of the carbinols **29** with malonic ester(s) and subsequent hydrolysis and decarboxylation of the esters **41**.[97]

Fc—C(R)(OH)—R^1 (**29**) ⟶ *Fc*—C(R)(R^1)—C(R^2)($COOC_2H_5$)—$COOC_2H_5$ (**41**) ⟶ *Fc*—C(R)(R^1)—C(R^2)(H)—COOH (**42**)

R, R^1, R^2 = H, CH_3, C_6H_5

Aminomethylferrocene (**43**) is obtained by lithium aluminum hydride reduction of ferrocenecarboxamide[30] (*Fc*–$CONH_2$, Section 30.2.1.1.4) or azidomethylferrocene (*Fc*–CH_2N_3) which in turn is prepared from the methoiodide of **3** and $N_3^{\ominus}$.[105]
Some primary and secondary ferrocenylamines **44** and **45** were obtained by reduction of acylferrocene oximes with $LiAlH_4$.[106] α-Ferrocenylethylamine (**44**, R=CH_3), however, is better prepared by catalytic hydrogenation of the oxime of **5** or by reductive amination of **5**.[67] The degree of the rearrangement to **45** depends on the substituent R. For R=C_6H_5 *ca.* 60% of the secondary amine **45** (R=C_6H_5) are obtained; aminoferrocene (*cf.* Section 30.2.1.2.6) can be prepared[106] from the latter by oxidation with manganese dioxide and hydrolysis of the resultant Schiff-base (*Fc*—N=CH—C_6H_5).

Fc—CH_2NH_2 **43** *Fc*—CH(NH_2)—R **44** *Fc*—NH—CH_2—R **45**

Amides of the cited ferrocenylcarboxylic acids are prepared in normal fashion. Several *N*-substituted amides of **6** were obtained by acylation of ferrocene with isocyanates and aluminum chloride.[107]
Cyanoferrocene (**46**) is prepared by dehydration of ferrocenealdoxime, preferably with *N,N*-dicyclohexyl carbodiimide (78% yield), but acetic anhydride or phosphorous pentachloride also have been used.[36] **46** is also formed in 84% yield if bromoferrocene (*cf.* Section 20.2.1.2.5) is reacted with copper(I) cyanide[108] or by direct "cyanation" of ferrocene with hydrocyanic acid in the presence of iron(III) chloride in tetrahydrofurane (86% yield).[109] Hydrolysis of **46** gives **6**.[110]

Fc—CH=NOH ⟶ *Fc*—C≡N (**46**) ⟶ *Fc*—COOH (**6**)

In order to synthesize ferrocenylalanine (**47**), formaminomalonic ester was treated either with the methiodide of **3**[111] or with chloromethylferrocene (*Fc*–CH_2Cl)[30] and the substituted malonic ester was hydrolyzed and decarboxylated.[30,111]
Some ferrocenyl substituted heterocyclic compounds are described.[102,112] For example, 3-ferrocenylpyrazole (**48**) and 4-ferrocenyl-2-pyrimidone (**49**) were obtained from ferrocenoyl acetaldehyde (**35**) and hydrazine or urea respectively.[102] (*Cf.* also Ref. 112a.)

Fc—CH_2—CH(NH_2)—COOH **47** **48** **49**

30.2.1.2.5 Ferrocenyl Halogen Compounds

Due to the high sensitivity of ferrocene towards oxidizing agents direct halogenation of the molecule is impossible; hence, exchange reactions have to be employed for the synthesis of haloferrocenes. For this purpose either *Fc*–HgCl (**64**, Section 30.2.1.2.9) or ferrocenylboronic

[104] *C. R. Hauser, J. K. Lindsay*, J. Org. Chem. *22*, 1246 (1957).
[105] *D. E. Bublitz*, J. Organometal. Chem. *23*, 225 (1970).
[106] *K. Schlögl, H. Mechtler*, Monatsh. Chem. *97*, 150 (1966).
[107] *M. D. Rausch, P. Shaw, D. Mayo, A. M. Lovelace*, J. Org. Chem. *23*, 505 (1958).
[108] *A. N. Nesmeyanov, V. A. Sazonova, V. N. Drozd*, Chem. Ber. *93*, 2717 (1960).
[109] *A. N. Nesmeyanov, E. G. Perevalova, L. P. Jurjeva*, Chem. Ber. *93*, 2729 (1960).
[110] *A. N. Nesmeyanov, E. G. Perevalova, L. P. Jurjeva, K. I. Grandberg*, Izv. Akad. Nauk. SSSR 1377 (1963); C.A. *59*, 15310 (1963).
[111] *J. M. Osgerby, P. L. Pauson*, J. Chem. Soc. 646 (1958).
[112] *F. D. Popp, E. B. Moynahan*, J. Heterocycl. Chem. *7*, 351, 739 (1970).
[112a] *F. D. Popp, E. B. Moynahan*, Advan. Heterocycl. Chem. *13*, 1 (1971).

$$Fc{-}N{=}N{-}Fc \; (\mathbf{56}) \xrightarrow{} Fc{-}NH_2 \; (\mathbf{54})$$

$$Fc{-}NO_2 \; (\mathbf{50}) \xrightarrow{\text{red.}} Fc{-}NH_2 \; (\mathbf{54}) \xleftarrow{LiAlH_4} Fc{-}N_3 \; (\mathbf{55})$$

$$Fc{-}NH_2 \; (\mathbf{54}) \xrightarrow{COCl_2} Fc{-}N{=}C{=}O \; (\mathbf{52})$$

$$Fc{-}CON_3 \; (\mathbf{51}) \longrightarrow Fc{-}N{=}C{=}O \; (\mathbf{52}) \xrightarrow{C_6H_5CH_2OH} Fc{-}NHCOOCH_2C_6H_5 \; (\mathbf{53})$$

acid (**7**, Section 30.2.1.1.5) are used as starting materials: **64** is reacted with *N*-iodo or *N*-bromo succinimide to yield iodo or bromoferrocene (yields: 85 and 58%, resp., m.p. 46° and 31°, resp.);[113] reaction of **7** with copper(II) bromide or chloride affords the desired compounds in ca. 80% yield (m.p. of *Fc*–Cl: 59°).[108] Fluoroferrocene (m.p. 116°–118°) was obtained in 10% yield from ferrocenyllithium (**8**) and perchlorylfluoride.[113a]

30.2.1.2.6 Ferrocenyl Nitrogen Compounds

In contrast to nitrobenzene nitroferrocene (**50**) cannot be obtained by a "classical" nitration and is therefore not easily accessible (oxidation potential of ferrocene Section 30.1.1). It may be prepared in low yield by reaction of ferrocenyllithium (**8**) either with dinitrogen tetroxide (2% yield)[114] or with propylnitrate (no yields reported)[115] in ether at −70°. Use of ethyl nitrate increases the yield up to 17%.[116] Two different melting points have been reported for **50**: 96°–97°[114] and 130°.[115,116]

The rather unstable aminoferrocene (**54**) is obtained by Curtius degradation of the azide **51**; the latter is readily obtained from ferrocenoylchloride (Section 30.2.1.1.4).[49,117] When **51** is heated in benzyl alcohol the urethane **53** is produced (the intermediate isocyanate **52** can be isolated[49]); **53** is then subjected to alkaline hydrolysis[117] or hydrogenolytic cleavage[117,118] (the overall yield based on **6** is *ca.* 35%).

Reaction of **54** with phosgene leads to **52** (80% yield); various ferrocenyl ureas can be prepared[49] from the latter.

Some less important methods for the preparation of **54** are the reaction of haloferrocenes with copper phthalimide and subsequent hydrazinolysis of the resultant *N*-ferrocenyl phthalimide,[109] or a direct reaction of ferrocenecarboxylic acid (**6**) with sodium azide in polyphosphoric acid.[119]

A classical diazotation of **54** is not possible; however, reaction of **54** with isoamyl nitrite in benzene gives a 1% yield of phenylferrocene.[119] **54** can be acylated and alkylated in the usual manner.

In order to synthesize azidoferrocene (**55**) either ferrocenyllithium (**8**) is treated with *p*-tolylazide or (preferably) bromoferrocene (Section 30.2.1.2.5) is reacted with sodium azide in dimethylformamide (yields are 28 and 98%, resp.).[120] Reduction of **55** with lithium aluminiumhydride gives **54**.[120]

Reaction of **8** with dinitrogen monoxide affords azoferrocene (**56**) (30% yield), which can be hydrogenated to give aminoferrocene (**54**);[121] phenylazoferrocene (Fc–N=N–C_6H_5) was prepared originating from aminocyclopentadiene.[122]

30.2.1.2.7 Ferrocenyl Oxygen Compounds

Reaction of ferrocenylboronic acid (**7**, Section 30.2.1.1.5) with copper(II) acetate affords acetoxyferrocene (**57**) which can be hydrolyzed to give the rather air-sensitive hydroxyferrocene (**58**) in nearly 90% yield.[108] If **57** is treated with methylsulfate in concentrated potassium hydroxide solution, the (stable) methylether **59** is obtained.[108]

---

[113] *R. W. Fish, M. Rosenblum*, J. Org. Chem. *30*, 1253 (1965).

[113a] *F. L. Hedberg, H. Rosenberg*, J. Organometal. Chem. *28*, C-14 (1971).

[114] *J. F. Helling, H. Shechter*, Chem. & Ind. 1157 (1959).

[115] *H. Grubert, K. L. Rinehart jr.*, Tetrahedron Lett. *12*, 16 (1959).

[116] *D. W. Slocum, T. R. Engelmann, C. Ernst, C. A. Jennings, W. Jones, B. Koonsvitsky, J. Lewis, P. Shenkin*, J. Chem. Ed. *46*, 144 (1969).

[117] *F. S. Arimoto, A. C. Haven jr.*, J. Amer. Chem. Soc. *77*, 6295 (1955).

[118] *E. M. Acton, R. M. Silverstein*, J. Org. Chem. *24*, 1487 (1959).

[119] *H. Lehner, K. Schlögl*, Monatsh. Chem. *101*, 895 (1970).

[120] *A. N. Nesmeyanov, V. N. Drozd, V. A. Sazonova*, Dokl. Akad. Nauk. SSSR *150*, 321 (1963); C.A. *59*, 5196 (1963).

[121] *A. N. Nesmeyanov, E. G. Perevalova, T. V. Nikitina*, Tetrahedron Lett. 1 (1960).

[122] *G. R. Knox, P. L. Pauson*, J. Chem. Soc. (London) 4615 (1961).

$$Fc\text{—}B(OH)_2 \ (\mathbf{7}) \xrightarrow{Cu(OAc)_2} Fc\text{—}OCOCH_3 \ (\mathbf{57}) \longrightarrow Fc\text{—}OH \ (\mathbf{58})$$
$$Fc\text{—}OCOCH_3 \ (\mathbf{57}) \downarrow Fc\text{—}OCH_3 \ (\mathbf{59})$$

30.2.1.2.8 Ferrocenyl Sulfur Compounds

Ferrocenesulfonic acid (**60**) is readily obtained by sulfonation of ferrocene with 0·25 mol of concentrated sulfuric acid in acetic anhydride (yield: 25% as ammonium salt),[37] with the SO_3-dioxane complex in dichloroethane (yield 62%),[123] or with chlorosulfonic acid in acetic anhydride (yield 66%).[124] The modest yields of isolated product are due to the high solubility of **60**. However, **60** can be isolated as the sparingly soluble *p*-toluidine salt; yield of the salt is about 90%.[125] For the preparation of the ferrocene-sulfonyl chloride (*Fc*-SO_2Cl, m.p. 100°) which is a useful intermediate for further synthesis, the toluidine salt can be treated with phosphorus pentachloride.[125]
Ferrocenylthiol (**61**) can be prepared by the reduction of the sulfonic acid **60** with lithium aluminum hydride and the sulfinic acid **62** is obtained by hydrogenation of ferrocenesulfochloride.[124] Thioethers, such as **63**, are obtained in good yield by alkylation of **61** (e.g. with methylsulfate) and aryl ferrocenyl sulfides (*Ar*–S–*Fc*) have been prepared from iodoferrocene (Section 30.2.1.2.5) and sodium arylmercaptides.[126]

$$Fc\text{—}SO_2H \ (\mathbf{62}) \longleftarrow Fc\text{—}SO_3H \ (\mathbf{60}) \xrightarrow{LiAlH_4} Fc\text{—}SH \ (\mathbf{61}) \longrightarrow Fc\text{—}SCH_3 \ (\mathbf{63})$$

30.2.1.2.9 Ferrocenyl Metal and Metalloid Compounds

For the important lithium derivative **8** see Section 30.2.1.1.6. The introduction of sodium *via* phenyl[127] or *n*-amyl sodium[53] affords mixtures of mono- and di-metallated products and offers no advantage over the lithiation (Section 30.2.1.1.6). Grignard compounds are of no significance in the ferrocene field. They may be prepared, however, by reaction of haloferrocenes (Section 30.2.1.2.5) with magnesium in the presence of methyl iodide or dibromoethane in tetrahydrofurane.[51] The reactivity of the halogenides is in the expected order I > Br > Cl.
Chloromercurioferrocene (*Fc*–HgCl, **64**) is conveniently obtained from ferrocene and mercury-(II) acetate in ether–methanol and by subsequent reaction with lithium chloride to give a mixture of the mono and bis-substituted ferrocene. **64** is isolated by extraction with *n*-butanol (yield 50%)[128] or methylene chloride (yield 73%).[113] Reduction of **64** with sodium yields differrocenyl mercury (*Fc*–Hg–*Fc*, **65**) in 70% yield.
Some organo-silicon derivatives of ferrocene were obtained from ferrocenyllithium (**8**) and the corresponding silanes.[46,129] The *Fc*–Si–bond is readily cleaved by acid reagents.[129] Disilylated ferrocenes are prepared by a "total synthesis", namely by reaction of the corresponding cyclopentadienyl silicon compounds with iron(II) chloride[130] (*cf.* Section 30.1.2).

The preparation of ferrocene germanium compounds has been reported;[53] ferrocenyl phosphorus and arsenic compounds were obtained by reaction of ferrocene with suitable phosphines,[131] phosphorus halides[132] and $AsCl_3$,[133] resp., in the presence of aluminum chloride.

30.2.2 Disubstituted Ferrocenes

As noted above electrophilic substitution of ferrocene may result in di- (or poly-) substituted products (*cf.* the Friedel-Crafts alkylation, Section 30.2.1.2.1); thus, due to the deactivating effect of the acyl groups acylation produces almost exclusively *hetero*anular diacyl derivatives. On electrophilic substitution of alkylferrocenes, however, a mixture of the three possible isomers is formed (Section 30.1.3) and no orientation effect of substituents already present is observed; electronic and steric effects seem to govern the substitution to a similar

---

[123] *A. N. Nesmeyanov, E. G. Perevalova, S. S. Churanov,* Dokl. Akad. Nauk. SSSR *114*, 335 (1957); C.A. *52*, 368 (1959).
[124] *G. R. Knox, P. L. Pauson,* J. Chem. Soc. (London) 692 (1958).
[125] *H. Falk, Ch. Krasa, K. Schlögl,* Monatsh. Chem. *100*, 1552 (1969).
[126] *M. D. Rausch,* J. Org. Chem. *26*, 3579 (1961).
[127] *S. I. Goldberg, D. W. Mayo, M. Vogel, H. Rosenberg, M. D. Rausch,* J. Org. Chem. *24*, 824 (1959).
[128] *M. D. Rausch, M. Vogel, H. Rosenberg,* J. Org. Chem. *22*, 900 (1957).
[129] *G. Marr, D. E. Webster,* J. Organometal. Chem. *2*, 99 (1964).
[130] *P. T. Kan, C. T. Lenk, R. L. Schaaf,* J. Org. Chem. *26*, 4038 (1961).
[131] *G. P. Sollott, H. E. Mertwoy, S. Portnoy, J. L. Snead,* J. Org. Chem. *28*, 1090 (1963).
[132] *G. P. Sollott, E. Howard jr.,* J. Org. Chem. *27*, 4034 (1962).
[133] *G. P. Sollott, W. R. Peterson jr.,* J. Org. Chem. *30*, 389 (1965).

extent. The situation is further complicated by the fact that *two* rings can be substituted.

Whereas Friedel-Crafts acetylation of toluene leads almost exclusively to *p*-methylacetophenone (only traces of the *o*-product are formed: 0·6%),[134] the corresponding reaction of methylferrocene (**11**) gives a mixture of 30% α-, 31% β-, and 49% of the *hetero*anular isomer (1-acetyl-1′-methylferrocene).[135] For a reasonable comparison of the results of such substitutions of ferrocene and its derivatives the concept of site reactivities, *i.e.*, the relative reactivities of ring positions (sites) (α, β, and *hetero*) has been shown to be quite useful.[24,83]

In the case of monosubstituted ferrocenes one can refer to a position of the unsubstituted ring (*i.e.*, 1′)[136] or, in the case of 1.1′-disubstituted derivatives to an α-position.[24,83,136] In both cases reference to the β-position is possible;[137] this has the advantage, that mono and disubstituted ferrocenes can be compared as to their reactivities.

In all cases the reactivities of the reference positions are arbitrarily set as 1.0; the yields of the isomers in question are divided by the number of free ring positions: for α and β by two, for 1′ (unsubstituted ring) by five.

The results of some typical electrophilic substitutions of ferrocenes are summarized in Table 3.

The small orientation effect of substituents R can be clearly recognized. Only bulky groups such as *tert*.-butyl impair substitution in the α-position by steric effect as noted above, deactivating groups (*e.g.*, acetyl) promote *hetero*-anular substitution. For the preparation of α-diacetylferrocene *cf.* Ref. 24.

For these reasons pure isomers are not readily obtained by substitution reactions. Consequently, the resultant isomeric mixtures have to be separated (*cf.* Section 30.2.2.1) or such reactions must be utilized in which either by cyclizations, "total syntheses" or by utilization of neighboring effects (*cf.* Sections 30.2.2.2 and 30.2.2.3) pure isomers are formed.

Table 3 Relative Site Reactivities of the Ring Positions for Electrophilic Substitution of Ferrocenes [24, 83, 119, 136, 137, 138, 138a]

R	R′	(Group introduced)	Relative site reactivities based on β=1·0: α	1′
CH_3	H	$COCH_3$	0·75	0·63
CH_3	H	CHO	0·32	0·37
CH_3	H	$CH_2N(CH_3)_2$	0·58	0·15
CH_3	H	COC_6H_5	0·46	0·49
CH_3	CH_3	$COCH_3$	0·59	—
C_2H_5	H	$COCH_3$	0·33	0·24
iso-Propyl	H	$COCH_3$	0·25	0·42
tert.-Butyl	H	$COCH_3$	0·08	0·30
OCH_3	H	$CH_2N(CH_3)_2$ [a]	1·60	0·43
SCH_3	H	$CH_2N(CH_3)_2$	1·45	0·60
C_6H_5	H	$COCH_3$	1·64	2·13
C_6H_5	H	CHO	0·43	0·57
$COCH_3$	H	$COCH_3$ [b]	0	40

[a] Revised values.[138a]

[b] In addition, *ca.* 1% of the α-isomer is formed;[24] however, since the β-isomer can hardly be separated from 1.1′-diacetylferrocene (*cf.* Sections 30.2.2.1 and 30.3.2) exact figures cannot be presented.

30.2.2.1 Separation of Isomeric Mixtures

For the separation of positional isomers chromatographic methods have been applied primarily. Attempts to separate isomers by crystallization have been unsuccessful (*cf.* however, Ref. 43b). Due to the possible formation of many isomers (*cf.* Table 1), chromatography too is only feasible in the case of di- (or simple tri-) substituted ferrocenes.

Thin-layer chromotagraphy (*TLC*) provides not only for an elegant qualitative analysis of isomeric mixtures,[139,140] but often has been successfully applied to difficult separation problems on a preparative scale (up to a gram scale). Kieselgel-G is used as adsorbent with hexane, benzene, or benzene ether mixtures as eluants.[139,140] Migration rates of isomeric acyl-

[134] *H. C. Brown, G. Marino*, J. Amer. Chem. Soc. *81*, 5611 (1959).

[135] *R. A. Benkeser, Y. Nagai, J. Hooz*, J. Amer. Chem. Soc. *86*, 3742 (1964).

[136] *K. Schlögl, H. Falk, G. Haller*, Monatsh. Chem. *98*, 82 (1967).

[137] *G. R. Knox, I. G. Morrison, P. L. Pauson, M. A. Sandhu, W. E. Watts*, J. Chem. Soc. (London) (*C*) 1853 (1967).

[138] *J. H. Richards, T. J. Curphy*, Chem. & Ind. (London) 1456 (1956).

[138a] *D. W. Slocum, C. R. Ernst*, Tetrahedron Lett. 5217 (1972).

[139] *K. Schlögl, H. Pelousek, A. Mohar*, Monatsh. Chem. *92*, 533 (1961).

[140] *R. E. Bozak, J. K. Fukuda*, J. Chromatography *26*, 501 (1967).

alkyl-ferrocenes are fairly characteristic for the type of substitution (α, β, or *hetero*) thus providing initial information about the structure of a given isomer in simple manner (*cf.* Section 30.3.2).

Similar results are obtained utilizing column chromatography (mostly on alumina) which provides for separation on larger scale. According to our experience a combination of column and thin-layer chromatography (preliminary purification and subsequent separation!) is particularly effective.

The following examples illustrate this point:

Acetyl methylferrocenes,[141] acetyl ethylferrocenes,[24] *iso*propylferrocenecarboxamides,[67] acetyl-1.1′-dimethyl and di*iso*propylferrocenes,[142] methyl methylferrocenecarboxylates,[143] 1.1′-dimethylferrocene-*N*,*N*′-diphenylcarboxamides[144] and 1.1′-dimethyl ferrocenecarboxylic acid *p*-bromophenacylesters.[139]

Since ferrocene derivatives are colored, no reagent is necessary for their detection on *TLC*-plates or columns. Alkylferrocenes and related compounds (such as ethers or carbinols) are yellow, acylferrocenes orange, diacylferrocenes or α,β-unsaturated acylferrocenes are red (with extensive conjugation violet). Traces as minute as 2 μg of alkyl and alkenylferrocenes may be detected on a *TLC*-plate by treatment with bromine vapor since the resultant ferricenium salts are blue-green.[139] Spots of acylferrocenes often fade during this treatment.

TLC offers the advantage that reaction processes can be monitored readily and optimal conditions for many ferrocene syntheses can be rapidly elucidated.[145]

In a few cases gas–liquid chromatography was utilized in separation problems; presently, *VPC* is restricted to the more volatile compounds such as triethylsiyl alkylferrocenes or acetyl alkylferrocenes.[63,135,146]

Counter-current distribution is rather laborious and offers no advantages as compared to the outlined chromatographic methods.[147]

High pressure liquid–liquid chromatography (LLC) can be employed successfully for many separation problems in ferrocene (and metallocene) chemistry.[147a]

30.2.2.2 *Hetero*anularly Disubstituted Ferrocenes

30.2.2.2.1 Symmetrical Derivatives

As noted above (Sections 30.2.1 and 30.2.2) electrophilic substitution of monosubstituted ferrocenes *Fc*–R, in which R is a deactivating substituent (such as acyl) yields *hetero*anular 1.1′-disubstituted products almost exclusively. Therefore (Friedel-Crafts) acylation of ferrocene with an excess of reagent (at least two moles) leads to the expected 1.1-diacylferrocenes in excellent yield. Separation from the monoacylferrocenes (and thus a purification) is readily accomplished by chromatographic methods (*cf.* Section 30.2.2.1) on the basis of the different adsorbabilities.

Thus many (symmetrically) disubstituted ferrocenes are easily accessible as transformation products of the diacylferrocenes.

Though here too (as was the case for the monosubstituted products (*cf.* 30.2.1.1)) only a few compounds are important starting materials (*e.g.*, diacetylferrocene **79**, the di-carboxylic acid **87** or the bisboronic acid **103**), the 1.1′-disubstituted ferrocenes are systematically arranged according to the principle used in Sections 30.2.1.2.1 through 30.2.1.2.9.

1.1′-Dialkyl (and aralkyl, resp.) ferrocenes are conveniently prepared by reduction of di-acyl (aroyl) ferrocenes with lithium aluminum hydride-aluminum chloride (or by the Clemmensen method).[62] For example, 1.1′-dimethylferrocene (**66**) is readily obtained from the dimethylester **90**,[141] diethylferrocene (**67**) from diacetylferrocene (**79**).[62]

66 is also obtained from the bis-Mannich base **91** by reduction of the methiodide[148] (*vide infra*). For the preparation of di-benzyl or α-thenyl ferrocene (**68** and **69**, resp.) from the corresponding diaroylferrocenes (**81** and **82**, resp.).[98,99]

---

[141] *E. A. Hill, J. H. Richards*, J. Amer. Chem. Soc. *83*, 4216 (1961).

[142] *K. L. Rinehart jr., K. L. Motz, S. Moon*, J. Amer. Chem. Soc. *79*, 2749 (1957).

[143] *H. Falk, G. Haller, K. Schlögl*, Monatsh. Chem. *98*, 592 (1967).

[144] *L. Westman, K. L. Rinehart jr.*, Acta Chem. Scand. *16*, 1199 (1962).

[145] *J. E. Herz*, J. Chem. Educ. *43*, 599 (1966).

[146] *Y. Nagai, J. Hooz, R. A. Benkeser*, Bull. Chem. Soc., Japan *36*, 482 (1963), *37*, 53 (1964).

[147] *K. Bauer, H. Falk, K. Schlögl*, Monatsh. Chem. *99*, 2186 (1968).

[147a] *R. Eberhardt, H. Lehner, K. Schlögl*, Monatsh. Chem. *104*, 1409 (1973); *H. Falk, H. Lehner, K. Schlögl*, J. Organometal. Chem. *55*, 191 (1973).

[148] *P. L. Pauson, M. A. Sandhu, W. E. Watts*, J. Chem. Soc. (London) (*C*) 251 (1966).

	R		R
66	CH_3	**82**	OC–(2-thienyl)
67	C_2H_5	**83**	$CO(CH_2)_nCOOH$
68	$CH_2C_6H_5$	**84**	$(CH_2)_{n+1}COOH$
69	H_2C–(2-thienyl)	**85**	$COCH{=}CH{-}R$
70	$CH{=}CH_2$	**86**	$COCH_2COR$
71	$CH(CH_3)_2$	**87**	COOH
72	$C(C_6H_5)_3$	**88**	COCl
73	C_6H_5	**89**	CON_3
74	CH_2OH	**90**	$COOCH_3$
75	$-CH(OH)-CH_3$	**91**	$CH_2N(CH_3)_2$
76	$-C(R)(OH)-CH_3$	**92**	$C{\equiv}N$
77	CH_2OCH_3	**93**	Cl
78	CHO	**94**	Br
79	$COCH_3$	**95**	I
80	CO—R	**96**	NH_2
81	COC_6H_5	**97**	$N{=}NC_6H_5$
		98	N_3
		99	$N{=}C{=}O$
		100	SO_3H
		101	Li
		102	HgCl
		103	$B(OH)_3$

The quite unstable 1.1′-divinylferrocene (**70**) was prepared in 70% yield by dehydration of the bis-carbinol **75** which in turn is prepared from diacetylferrocene (**79**).[69]

Di-*iso*propylferrocene (**71**) is obtained similar to various other *hetero*anularly disubstituted ferrocenes by a "total synthesis" through the corresponding fulvenes (*vide infra*).[149] This route can also be used for the preparation of some di-aralkyl ferrocenes, such as 1.1′-bis-tritylferrocene (**72**).[150]

1.1′-Diphenylferrocene (**73**) is prepared from the reaction of phenylferrocene with phenyldiazonium chloride in 50% yield[82] (*cf.* also Section 30.2.1.2.3).

Bis-hydroxymethyl ferrocene (**74**) and sec. alcohols (*e.g.*, **75**) are obtained by lithium aluminum hydride reduction of the dimethylester **90**[151] or of diacetylferrocene **79**[152] in *ca.* 90% yield and *tert.* carbinols **76** by reaction of **79** with Grignard compounds[67,153] Ethers (e.g. 1.1′-bismethoxymethyl ferrocene, **77**) are accessible from the bis-Mannich-base **91**.[148]

The dialdehyde **78** was prepared by MnO_2 oxidation of **74**.[154]

1.1′-Diacetylferrocene (**79**) is probably the most important diacylferrocene. It is obtained by acylation of ferrocene with acetyl chloride or acetic anhydride in the presence of aluminum chloride or boron trifluoride-etherate in dichloromethane or carbon disulfide in 70–90% yield.[14,24,36,153,155,156]

Besides **79** many other 1.1′-diacylferrocenes **80** (R=C_8 to C_{14}),[155] diaroylferrocenes (such as dibenzoyl[98] and di-α-thenoyl ferrocene,[99] **81** and **82**, resp.) and the bis-ketoacids **83** have been prepared. The latter were obtained from **79** by condensation with diethyl carbonate (83, $n=1$, as diethylester),[157] by Friedel-Crafts reaction of ferrocene with succinicanhydride-aluminum chloride ($n=2$)[152] or with the corresponding monoester chlorides ($n=2,3$).[157] The ketoacids **83** can be catalytically hydrogenated to give the ferrocene-1.1′-biscarboxylic acids (**84**) and the latter may be subjected to an acyloin cyclization.[157]

Further conversion products of diacetylferrocene (**79**) are chalcones of the general formula **85** which are obtained by an aldol condensation with the corresponding aldehydes,[158] and the bis-dicarbonylcompounds **86** accessible by condensation with esters or ketones (employing potassium amide in liquid ammonia).[157,159]

The most important transformation product of **79**, however, is ferrocene-1.1′-dicarboxylic acid **87** which is prepared by oxidation with hypoiodite[14] or hypochlorite[160] in *ca.* 80% yield. **87** can be utilized as starting material for the preparation of many 1.1′-disubstituted ferrocenes.

---

149 *G. R. Knox, P. L. Pauson,* J. Chem. Soc. (London) 4610 (1961).

150 *R. C. Koestler, W. F. Little,* Chem. & Ind. (London) 1589 (1958).

151 *M. Okawara, Y. Takemoto, H. Kitaoka, E. Haruki, E. Imoto,* Kogyo Kagaku Zasshi *65*, 685 (1962); C.A. *58*, 577 (1963).

152 *P. J. Graham, R. V. Lindsay, G. W. Parshall, M. L. Peterson, G. M. Whitman,* J. Amer. Chem. Soc. *79*, 3416 (1957).

153 *R. Riemschneider, D. Helm,* Chem. Ber. *89*, 155 (1956).

154 *J. M. Osgerby, P. L. Pauson,* J. Chem. Soc. (London) 4604 (1961).

155 *M. Vogel, M. D. Rausch, H. Rosenberg,* J. Org. Chem. *22*, 1016 (1957).

156 *M. D. Rausch, E. O. Fischer, H. Grubert,* J. Amer. Chem. Soc. *82*, 76 (1960).

157 *K. Schlögl, H. Seiler,* Monatsh. Chem. *91*, 79 (1960).

158 *T. A. Mashburn jr., C. E. Cain, C. R. Hauser,* J. Org. Chem. *25*, 1982 (1960).

159 *C. R. Hauser, C. E. Cain,* J. Org. Chem. *23*, 1142 (1958);
C. E. Cain, T. A. Mashburn jr., C. R. Hauser, J. Org. Chem. *26*, 1030 (1961).

160 *F. W. Knobloch, W. H. Rauscher,* J. Polymer. Sci. *54*, 651 (1961).

87 can also be obtained by lithiation of ferrocene with an excess of butyllithium and subsequent carboxylation. The mixture contains *ca.* 30% of **87**;[161] the yield of 1.1'-dilithioferrocene (**101**) and therefore of **87** can be increased to 94% by employing butyllithium in the presence of tetramethyl ethylenediamine in hexane.[162] Consequently, this latter procedure is (together with a new one[43b]) the best method for preparing **87**. The dichloride **88** is obtained from the acid with phosphorus[163]—or oxalyl chloride[160] in *ca.* 80% yield. Reaction of **88** with sodium azide affords the diazide **89**, which on heating in toluene gives the bis-isocyanate **99**.[161] Esterification of **87** to the dimethylester **90** proceeds smoothly;[46,164] reduction of **90** leads to **74**[151] or **66**[141] depending on the conditions.

The above mentioned bis-Mannich-base **91** is obtained from ferrocene and a three molar excess of *N,N,N',N'*-tetramethyl diaminomethane (in analogy to the mono-product **3**, cf. Section 30.2.1.1.1) though only in low yield;[148] the expected conversion products, such as **66**, **77** or bridged ethers can be prepared[143] from **91**. A better access to **91** can also be obtained by the "total synthesis"[165] (*vide infra*).

1.1'-Dicyanoferrocene (**92**) is synthesized by dehydration of the bis-aldoxime of the above mentioned dialdehyde **78**.[154]

1.1'-Dichloro and -dibromo ferrocene, resp. (**93**, **94**), are synthesized in good yield from ferrocene-1.1'-bis-boronic acid (**103**, Section 30.2.1.1.5) and copper(II) chloride and bromide, resp.[50] Diiodoferrocene (**95**) is better prepared from the bis-mercury-compounds **102** (*cf.* Section 30.2.1.2.9) and iodine.[166]

1.1'-Bis-phenylazoferrocene (**97**) was obtained from the reaction of diazocyclopentadiene, phenyllithium and iron(II) chloride (80% yield).[167] Catalytical hydrogenation of **97** with platinum-hydrogen in acetic acid yields 1.1'-diaminoferrocene (**96**),[167,168] which is also obtained from diazidoferrocene (**98**) by reduction with lithium aluminum hydride (62% yield).[120] **98** was prepared from dibromoferrocene **94** and sodium azide in the presence of copper(I) bromide.[120] The synthesis of the di-isocyanate **99** from the di-azide **89** has been cited above.[161]

Ferrocene-1.1'-disulfonic acid (**100**) is obtained if an excess of sulfonating agent is employed.[37,124] In analogy to the mono-sulfonic acid **60** (*cf.* Section 30.2.1.2.8), several further transformation products can be made originating from **100**.

Ferrocene-bis-metal and -metalloid compounds have been mentioned above, *e.g.*, the important di-lithio derivative **101** (in connection with the dicarboxylic acid **87**) and the bis-mercurio compound **102** (*cf.* the preparation of di-iodo-ferrocene **95**).

Symmetrical *hetero*anularly disubstituted ferrocenes can be synthesized either by substitution of monosubstituted ferrocene or by the often mentioned "total synthesis" originating from substituted cyclopentadienes and iron(II) compounds (*cf.* also Section 30.1.2).

For the preparation of the required cyclopentadienyl anions suitably substituted fulvenes are useful, which then can be reduced with lithium alkyls (aryls) or with lithium aluminum hydride. The fulvenes can be made from cyclopentadiene and carbonyl compounds; the choice of the substituents R, however, is rather limited, since they have to be inert towards the cited reducing agents.

By this route 1.1'-bis-diphenylmethyl-ferrocene[150] (**72**)[150] was prepared from 6.6-diphenylfulvene, phenyllithium and iron(II) chloride, and from 6-dimethylaminoethyl fulvene, lithium aluminum hydride and ferrous chloride the already mentioned bis-Mannich-base **91** (yield 59%)[165] and from 6.6-dimethylfulvene the di-*iso*propyl ferrocene (**71**).[149] The synthesis of bis-phenylazoferrocene (**97**) was mentioned previously.[167] The influence of substituents on this ferrocene synthesis *via* fulvenes has been investigated.[169, 169a]

30.2.2.2.2 Unsymmetrical Derivatives

Strongly deactivating substituents (*e.g.*, acyl) direct further electrophilic attack on a mono-

[161] *R. L. Schaaf, C. T. Lenk*, J. Chem. Eng. Data *9*, 103 (1964).

[162] *M. D. Rausch, D. J. Ciappanelli*, J. Organometal. Chem. *10*, 127 (1967).

[163] *S. I. Goldberg*, J. Org. Chem. *25*, 482 (1960).

[164] *R. L. Schaaf*, J. Org. Chem. *27*, 107 (1962).

[165] *G. R. Knox, J. D. Munro, P. L. Pauson, G. H. Smith, W. E. Watts*, J. Chem. Soc. (London) 4619 (1961).

[166] *A. N. Nesmeyanov, E. G. Perevalova, O. A. Nesmeyanova*, Dokl. Akad. Nauk. SSSR *100*, 1099 (1955); C.A. *50*, 2558 (1956).

[167] *G. R. Knox*, Proc. Chem. Soc. *56* (1959).

[168] *G. R. Knox, P. L. Pauson*, J. Chem. Soc. (London) 4615 (1961).

[169] *P. L. Pauson, M. A. Sandhu, W. E. Watts*, J. Chem. Soc. (London) (*C*), 860 (1968).

[169a] *U. Müller-Westerhoff*, Tetrahedron Lett. 4639 (1972).

substituted ferrocene into the unsubstituted ring. This principle was used for the syntheses of many symmetrical *hetero*anularly disubstituted ferrocenes (**2**, R=R′; *cf.* Section 30.2.2.2.1); it can, however, also be utilized for the preparation of unsymmetrically substituted products (**2**, R≠R′). If the first substituent R is an activating one (*e.g.*, alkyl), (frequently laborious) chromatographic separation of the isomers may be necessary (*cf.* Section 30.2.2.1).

From unsymmetrically (or occasionally also from symmetrically) 1.1′-disubstituted ferrocenes many disubstituted ferrocenes of the structural type **2** (R≠R′) may be obtained by appropriate subsequent reactions.

This possibility is illustrated below for some typical examples; within the groups of "basic compounds" (acetyl ferrocenes, carboxylic acids and esters, resp., halo- and nitroferrocenes, resp.; R = $COCH_3$, COOH and $COOCH_3$, Hal and NO_2, resp.) the derivatives are arranged with respect to R′ according to the general scheme (as outlined in Sections 30.2.1.2. and 30.2.2.2.1).

The acetylation of ferrocenecarboxamides, particularly that of the *N,N*-diphenyl derivative **9** (Section 30.2.1.1.4) giving a yield of *ca.* 70%,[42] offers access to compounds of type **2**. Hydrolysis of the acetyl derivative of **9** produces acetylferrocenecarboxylic acid (**113**) in 80% yield; however, attempts to hydrolyze 1′-acetylferrocenecarboxamide were unsuccessful. Similarly, benzoylation of **9** and subsequent hydrolysis affords benzoylferrocenecarboxylic acid (**114**).[42] The acid **113** can also be obtained (though only in moderate yield) by acetylation of methyl ferrocenecarboxylate and subsequent saponification.[170]

113 is a useful starting material for further transformations: reduction with lithium aluminum hydride affords the bis-alcohol **122**, which, on oxidation with MnO_2, gives acetyl ferrocenecarboxaldehyde **104** (yield 64%); **104** can be converted into the acrylic acid derivative **106** in a Knoevenagel condensation.[171]

Reaction of 1.1′-diacetylferrocene (**79**) with *o*-, *m*-, or *p*-nitrobenzaldehyde gives chalcones of the general structure **105**.[172]

[170] *A. N. Nesmeyanov, O. A. Reutov*, Dokl. Akad. Nauk. SSSR *115*, 518 (1957); C.A. *52*, 5393 (1958).

[171] *K. Schlögl, M. Peterlik, H. Seiler*, Monatsh. Chem. *93*, 1309 (1962).

[172] *M. Furdik, S. Toma, J. Suchy, P. Elecko*, Chem. Zvesti *15*, 45 (1961); C.A. *55*, 18692 (1961).

	R	R′
104	$COCH_3$	CHO
105		COCH=CH—R
106		CH=CHCOOH
107		C≡N
108		Cl
109		Br
110		OH
111		SO_3H
112	COOH	CH_3
113		$COCH_3$
114		COC_6H_5
115		Br
116		SO_3H
117	$COOCH_3$	CH_3
118		CH_2OH
119		CHO
120		COOH
121		COCl
122	CH_2OH	$CHOHCH_3$
123	Cl	I
124		NH_2
125		HgCl
126		$B(OH)_2$
127	Br	C_2H_5
128		C≡N
129		I
130		HgCl
131		$B(OH)_2$
132	NO_2	C≡N

Acetylation of chloro- or bromoferrocene (Section 30.2.1.2.5) with acetyl chloride–aluminum chloride in dichloromethane at 0° provides the corresponding acetyl derivatives **108** and **109**, resp. in *ca.* 75% yield.[173]

1′-Bromo-acetylferrocene (**109**) undergoes an iodoform reaction (with iodine in pyridine) to form the carboxylic acid **115**,[174] Clemmensen reduction of **109** produces bromoethylferrocene (**127**)[175] and reaction of **109** with copper(II) acetate yields 1′-hydroxy acetylferrocene (**110**).[175]

Sulfonation of acetylferrocene (**5**) or of ferrocenecarboxylic acid (**6**) with the SO_3-dioxane complex in dichloromethane affords the corresponding sulfonic acids **111** and **116**, resp. in *ca.* 50% yield.[176, 177]

[173] *D. W. Hall, J. H. Richards*, J. Org. Chem. *28*, 1549 (1963).

[174] *A. N. Nesmeyanov, V. A. Sazonova, V. N. Drozd*, Izv. Akad. Nauk. SSSR 45 (1962); C.A. *57*, 865 (1962).

[175] *A. N. Nesmeyanov, V. A. Sazonova, V. N. Drozd*, Dokl. Akad. Nauk. SSSR *137*, 102 (1961); C.A. *55*, 21081 (1961).

[176] *A. N. Nesmeyanov, B. N. Strunin*, Dokl. Akad. Nauk. SSSR *137*, 106 (1961); C.A. *55*, 19885 (1961).

[177] *A. N. Nesmeyanov, O. A. Reutov*, Izv. Akad. Nauk. SSSR 926 (1959); C.A. *54*, 469 (1960).

The formation of isomers causes difficulties when preparing pure 1-acyl-1′-alkylferrocenes by Friedel-Crafts-acylation of alkylferrocenes. For the synthesis of the latter compounds the method *via* symmetrical 1.1′-disubstitution products (as described in Section 30.2.2.2.1) is preferred.

For example, 1′-methylferrocenecarboxylic acid (**112**) has been prepared from the dicarboxylic acid **87** *via* the monoester **120**; the latter is obtained from **87** and methanol-sulfuric acid in 52% yield;[178] the chloride **121** was reduced with lithium aluminum tris-(butoxy)hydride to give the aldehyde **119** (46% yield), which can be reduced by the Clemmensen method and the resultant methyl 1′-methylferrocenecarboxylate (**117**) is hydrolyzed to **112** (92% yield).

112 is better prepared by reduction of the dimethylester **90** with 0·5 moles of lithium aluminum hydride, oxidation of the resultant mixture with manganese dioxide (**118**→**119**) and subsequent Clemmensen reduction. Separation of the reaction mixture (**90**, **117**, and 1.1′-dimethylferrocene, **66**) on a silica column with benzene gives the desired methylester **117** in a overall yield of 37% (based on unrecovered bisester **90**).[28]

The best synthesis for the acid **112** and its ester **117** seems to be *hetero*anular formylation of methyl ferrocenecarboxylate with dichloromethyl ether and aluminum chloride and subsequent Clemmensen reduction of the aldehyde **119** (yield *ca.* 90%). *Cf.* Ref. 179 for the corresponding preparation of 1.1′-dimethylferrocene-β-carboxylic acid.

In anology to the transformation products of the monocarboxylic acid **6** (Section 30.2.1.2) the carboxy group of **112** may be converted to produce other compounds such as the acetyl or ethynyl derivative.[178]

Mixed dihaloferrocenes are prepared by interaction of 1′-chloromercurio- chloro- or bromoferrocene (**125**, **130**) with iodine: chloro- and bromo-iodoferrocene, resp. (**123**, **129**), are obtained in good yield. The required mercury derivatives **125** and **130** are readily obtained from ferrocene-1,1′-bisboronic acid (**103**, Section 30.2.1.1.5 and 30.2.2.2.1) by reaction of the latter with copper(II) chloride or bromide to yield the haloferrocenylboronic acids (**126** and **131**, resp., yields *ca.* 50%) which are reacted with mercury(II) chloride in aqueous acetone.[180]

1-Chloroferrocene-1′-boronic acid (**126**) reacts with copper phthalimide and is subsequently hydrazinolyzed to give a 48% yield of 1-chloro-1′-aminoferrocene (**124**).[181]

A direct "cyanation" of acetyl-, bromo-, or nitroferrocene with hydrogen cyanide and iron(III) chloride in tetrahydrofurane affords the desired nitrils **107**, **128**, and **132** in 18, 78, and 3%[182] yield, respectively.

Some chloro-, bromo-, and hydroxy-ferrocenes (**108**, **109**, **115**, **127**, and **110**) were cited above.

The Mannich-reaction of monosubstituted ferrocenes yields mixtures of isomers and therefore, has no preparative value.[143, 183, 184]

30.2.2.2.3 *Hetero*anularly Bridged Materials (Ferrocenophanes)

Ferrocenophanes represent special cases of *hetero*-anular disubstitution of ferrocenes in which both rings are linked from position 1 to 1′ by a two- to *n*-membered ring (bridge). Such compounds **133** (*cf.* Section 30.1.3 for their nomenclature) can be obtained by a "total synthesis" originating from corresponding α,ω-bis-cyclopentadienyl

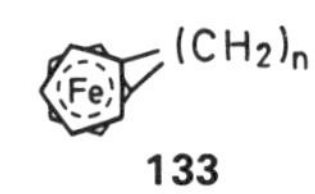

133

alkanes and iron salts, by cyclization of monosubstituted ferrocenes (*e.g.*, by Friedel-Crafts cyclization of ferrocenylcarboxylic acids **37**) or by condensation, dehydration, or acyloin cyclization of symmetrically or unsymmetrically 1,1′-disubstituted ferrocenes.

The first procedure is particularly useful for the synthesis of compounds with a two-carbon bridge, *i.e.*, [2]ferrocenophanes, though in some cases [3] to [5]ferrocenophanes were obtained. For example, 6-substituted fulvenes react with

---

[178] *K. Schlögl, H. Soukup,* Monatsh. Chem. *99,* 927 (1968).

[179] *H. Falk, K. Schlögl,* Monatsh. Chem. *102,* 33 (1971).

[180] *A. N. Nesmeyanov, V. A. Sazonova, V. N. Drozd, L. A. Nikonova,* Dokl. Akad. Nauk. SSSR *131,* 1088 (1960); C.A. *54,* 21025 (1960).

[181] *A. N. Nesmeyanov, V. A. Sazonova, V. I. Romanenko,* Dokl. Akad. Nauk. SSSR *157,* 922 (1964); C.A. *61,* 13343 (1964).

[182] *A. N. Nesmeyanov, E. G. Perevalova, L. P. Jurjeva, K. I. Grandberg,* Izv. Akad. Nauk. SSSR 1772 (1962); C.A. *58,* 7971 (1963).

[183] *A. N. Nesmeyanov, E. G. Perevalova, L. S. Shilovtseva,* Izv. Akad. Nauk. SSSR 1767 (1962); C.A. *58,* 7972 (1963).

[184] *S. McVey, I. G. Morrison, P. L. Pauson,* J. Chem. Soc. (London) (*C*), 1847 (1967).

sodium and iron(II) chloride in ether to yield the corresponding (in the bridge substituted) [2]-ferrocenophanes.

The high yields (up to 70%) reported in the literature[185] could not be reproduced. We obtained only yields of *ca.* 3% which is in the same order of magnitude as the yields obtained in the synthesis of [2] to [5]ferrocenophanes according to Refs 186 and 187; in this latter preparation α,ω-dibromoalkanes are reacted with sodium cyclopentadienate and the mixture is subsequently treated with iron(II) or (III) chloride (yields range from 0·002 to 2%).

For a "total synthesis" of bridged ferrocene containing silicon in the bridge see below.[188]

For [2]ferrocenophanes the "total synthesis" is the only feasible method of preparation (see Ref. 4 for their structure). (**135** which is accessible by photo-cyclization of ferrocenesulfonyl azide in 14% yield[188a] has *no* carbon-bridge!) For the higher homologs ([3]ferrocenophane and higher), however, the other two synthetic routes mentioned (*vide supra*) are suited much better. For a "total synthesis" of [2]ferrocenophane and its tetramethyl derivative see Ref. 189.

(Friedel-Crafts)-cyclization of ω-ferrocenylcarboxylic acids **37** (cf. Section 30.2.1.2.4) is unsuccessful in the case of ferrocenylacetic acid but for ferrocenylpropionic acid and its derivatives it proceeds (with only a few exceptions, such as 2,2-dimethyl-3-ferrocenylpropionic acid)[31] *via hetero*anular cyclization[58,190,191] and formation of [3]ferrocenophanes (*e.g.*, **134** from **37**, $n=1$). **134** can be conveniently prepared by reaction of ferrocene with acryloylchloride and $AlCl_3$ at $-78°$.[191a]

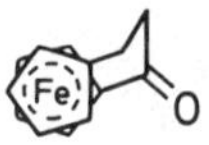

134 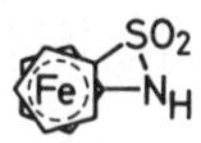 **135**

From γ-ferrocenylbutyric acids (as well as from the just mentioned ferrocenyldimethylpropionic acid) one obtains, however, *homo*anularly bridged ferrocenes (*cf.* Section 30.2.2.3.2); higher homologs afford (by *inter*molecular acylation) polymeric products. Thus the stereoselectivity of the cyclization of ferrocenylcarboxylic acids (for which trifluoroacetic anhydride usually is employed—polyphosphoric acid gives less satisfactory results) provides a chance to prepare either [3]ferrocenophanes or anellated ferrocenes (ferroceno compounds).

Catalytic reduction of [3]ferrocenophanone (**134**) ($Pt–H_2$)[157] or reduction with lithium aluminum hydride–aluminum chloride yields [3]ferrocenophane (**133**, $n=3$). Treatment of the *p*-tosyl-hydrazone of **134** with sodium hydride or methoxide provide the corresponding *endo*cyclic unsaturated "hydrocarbon"[192] and by ring enlargement (of **134**) with diazomethane isomeric [4]ferrocenophanones can be prepared.[31]

As a rule, [4]ferrocenophanes are difficult to prepare, but [5]ferrocenophanes can be obtained by base-catalyzed condensation of 1,1′-diacetyl or -diacylferrocenes (**79** and **80**, resp.) with aldehydes; products of the general formula **136** are formed in this manner,[158,193] their structures have been elucidated.[172,194]

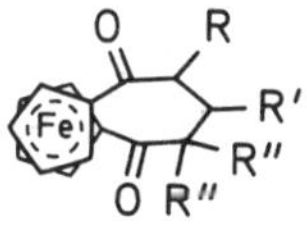

136

[*n*]Ferrocenophanes ($n=3$ to 10) can be obtained by suitable cyclization reactions from 1,1′-disubstituted, symmetrical or unsymmetrical ferrocenes (*cf.* Sections 30.2.2.2.1 and 30.2.2.2.2). Thus, Dieckmann-condensation of dimethyl ferrocene-1,1′-bis-acetate with triphenylmethyl sodium yields the bridged β-ketoester **137**,

---

185 *R. L. Pruett, E. L. Morehouse*, U.S. Pat. 3063974 (1962); C.A. *58*, 11404 (1963);
N. M. Sweeney, U.S. Pat. 3035075 (1962); C.A. *58*, 10242 (1963).

186 *A. Lüttringhaus, W. Kullick*, Makromol. Chem. *44–46*, 669 (1961).

187 *H. L. Lentzner, W. E. Watts*, Chem. Commun. 26 (1970).

188 *R. L. Schaaf, P. T. Kan, C. T. Lenk*, J. Org. Chem. *26*, 1790 (1961).

188a *R. A. Abramovitch, C. I. Azogu, R. G. Sutherland*, Chem. *Commun.* 1439 (1969).

189 *H. L. Lentzner, W. E. Watts*, Tetrahedron *27*, 4343 (1971).

190 *B. Gautheron, J. Tirouflet*, Compt. Rend. *258*, 6443 (1964);
J. W. Huffman, R. L. Asbury, J. Org. Chem. *30*, 3941 (1965);
A. Dormond, J. P. Ravoux, J. Decombe, Bull. Soc. Chim. France 1152 (1966).

191 *K. L. Rinehart jr., R. J. Curby jr.*, J. Amer. Chem. Soc. *79*, 3290 (1957).

191a *T. D. Turbitt, W. E. Watts*, J. Organometal. Chem. *46*, 109 (1972).

192 *N. M. Applebaum, R. W. Fish, M. Rosenblum*, J. Org. Chem. *29*, 2452 (1964).

193 *T. H. Barr, W. E. Watts*, Tetrahedron *24*, 3219 (1968).

194 *M. Furdik, S. Toma, J. Suchy*, Chem. Zvesti *15*, 789 (1961); *16*, 449 (1962); C.A. *58*, 11398 (1963).

which can be converted to the [3]ferrocenophanone **138** (isomeric with **134**).[195]

137 **138**

Derivatives of higher [*n*]ferrocenophanes (n = 4, 6, 8, and 9) were prepared by acyloin condensation of the corresponding dimethyl ferrocene-1,1′-bis-carboxylates (*cf.* **84**, Section 30.2.2.2.1) the yield ranging from 20 (n=4) to 75% (n=9); however the acyloins **139** could not be reduced.[157]

139

[3]Ferrocenophanes with a heteroatom in the bridge, *i.e.*, oxa-, aza- or thia-ferrocenophanes, also can be prepared by cyclization of 1,1′-disubstituted ferrocenes: dehydration of carbinols such as **74** or **75** (Section 30.2.2.2.1) with acid alumina or potassium hydrogen sulfate affords cyclic ethers **140** (R=H,[196] CH_3[171,197] or C_6H_5[69]). The dimethyl and diphenyl derivatives could be separated into the racemate and the meso compound.[148,198]

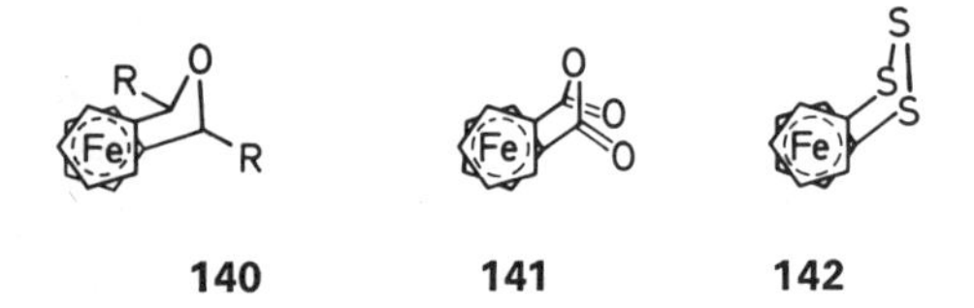

140 **141** **142**

Reaction of the dichloride **88** of ferrocene-1,1′-dicarboxylic acid with water yields the cyclic anhydride **141**,[199] and 1,1′-dilithio ferrocene (**101**) reacts with sulfur to yield trithia[3]ferrocenophane (**142**).[200]

For the synthesis of bridged siloxanyl ferrocenes *cf.* Ref. 188.

[*m,n*]Ferrocenophanes **143** are interesting bridged ferrocenes; in the most simple case ($m=n=0$) one ferrocene molecule is bridged *hetero*anularly by a second ferrocenylene moiety, or else by a 1,1′-bisalkylene or -bis-ethynylidene ferrocene.

Whereas [4,4]ferrocenophane (**143**, $m=n=4$) has been prepared in traces from 1,4-biscyclopentadienylbutane (disodium salt) and iron(III) chloride,[186,201] a dimethyl derivative of [1,1]ferrocenophane was first obtained in 1966 (from a bis-fulvene and iron(II) chloride[202]). [0,0]Ferrocenophane (1,1′-biferrocenylene) and [2,2]ferrocenophane-1,13-diyne (**144**) were prepared in low yield by coupling of diiodoferrocene (**95**) with copper bronze[203] and of the copper salt of 1′-iodoferrocenylacetylene, resp.[204]

[*m,n*] Ferrocenophanes with "mixed valence states" were studied in some detail.[204]

143 **144**

30.2.2.3 *Homo*anularly Disubstituted Ferrocenes

Besides the separation of isomeric mixtures (such as those formed on electrophilic substitution of *e.g.*, alkylferrocenes) as described in Section 30.2.2.1, selective substitution reactions can be employed for the preparation of pure α-(*i.e, homo*anularly) disubstituted ferrocenes (**1**), *i.e.*, α-lithiation of appropriate ferrocenes or *homo*anular cyclizations. For a review on 1,2-disubstituted ferrocene see Ref. 204b.

30.2.2.3.1 Selective α-Lithiation

Heteroatoms containing a free electron pair (such as nitrogen or oxygen) and being located in a side-chain in α-or β-position to the ferrocenyl direct the lithiation (*e.g.*, with butyl

[195] *W. Mock, J. H. Richards*, J. Org. Chem. *27*, 4050 (1962);
A. Sonoda, I. Moritani, J. Organometal. Chem. *26*, 133 (1971).

[196] *K. L. Rinehart jr., A. K. Frerichs, P. A. Kittle, L. F. Westman, D. H. Gustafson, R. L. Pruett, J. E. McMahon*, J. Amer. Chem. Soc. *82*, 4111 (1960).

[197] *E. C. Winslow, E. W. Brewster*, J. Org. Chem. *26*, 2982 (1961).

[198] *K. Yamakawa, M. Hisatome*, Tetrahedron Lett. 2827 (1967).

[199] *A. N. Nesmeyanov, O. A. Reutov*, Dokl. Akad. Nauk. SSSR *120*, 1287 (1958); C.A. *53*, 1292 (1959).

[200] *A. Davison, J. C. Smart*, J. Organometal. Chem. *19*, P7 (1969).

[201] *A. Lüttringhaus, W. Kullick*, Angew. Chem. *70*, 438 (1958).

[202] *W. E. Watts*, J. Amer. Chem. Soc. *88*, 855 (1966);
W. E. Watts, J. Organometal. Chem. *10*, 191 (1967).

[203] *F. L. Hedberg, H. Rosenberg*, J. Amer. Chem. Soc. *91*, 1258 (1969). See also: *U. T. Mueller-Westerhoff, P. Eilbracht*, J. Amer. Chem. Soc. *94*, 9272 (1972).

[204] *M. Rosenblum, N. M. Brawn, D. Ciappenelli, J. Tancrede*, J. Organometal. Chem. *24*, 469 (1970).

[204a] *D. O. Cowan, C. LeVenda, J. Park, F. Kaufman*, Acc. Chem. Res. *6*, 1 (1973).

[204b] *J. H. Peet, B. W. Rockett*, Rev. Pure Appl. Chem. *22*, 145 (1972).

Table 4 Reactions and Subsequent Products of (Dimethylamino-methyl) -α-lithio ferrocene (145)

Reaction of 145 with	Primary products (146) R′	R (1)	R′	Ref.
$(n\text{-}C_4H_9\text{–}O)_3B$	$B(OH)_2$	$CH_2N(CH_3)_2$	Cl, Br, J	206
	"	CH_2OH	Cl, Br, J	206
	"	$CH_2NHC_6H_5$	Cl, Br, J	206
	"	$CH_2N(CH_3)_2$	$-(CH_2NMe_2)$*Fc*	207
HCHO	CH_2OH	CHO	CHO	208
CO_2	COOH	CH_3	COOH	209
$(C_6H_5)_2CO$	$C(OH)(C_6H_5)_2$	CH_2OH	$C(OH)(C_6H_5)_2$	210
	"	CH_2CN	"	210
	"	CH_2COOH	"	210

lithium, *cf.* Section 30.2.1.1.6) almost completely into the neighboring ring position, *i.e.* α to the first substituent; this fact has been substantiated by the structures of the products of reactions of many lithio ferrocenes. Probably a complex formation between lithium and the heteroatom is responsible for this effect; if different acidities of various ring positions were caused by electronic effects, mixtures of isomers would result, which, however, were never observed.[116]

Besides compounds of the type *Fc*–CH_2CH_2–N<, *Fc*–CH_2O–R[205,205a] or *Fc*–Cl[205a] the easily accessible *N,N*-dimethylaminomethylferrocene (**3**, Section 30.2.1.1.1) is suitable for this purpose; treatment of **3** with *n*-butyllithium privides the α-lithio derivative **145** in a smooth reaction.

The preparative value of this reaction resides in the great variability of substituents which results from both the substitution of lithium (to give **146**) and from an exchange of the group $-N(CH_3)_2$ (to give **1**). (See also Section 30.2.1.2.) The yields are greater than 50%; therefore, this reaction is the method of choice for preparing α-disubstituted ferrocenes **1**.

CH_2 CH_3 N Fe Li CH_3 **145** ⟶ $CH_2-N(CH_3)_2$ Fe R′ **146** ⟶ R Fe R′ **1**

Some examples of such reactions of **145** are summarized in Table 4.

Other amines which may be utilized in a α-lithiation are 2-ferrocenylpyridine[211] or 1-methyl-2-ferrocenyl piperidine[212] which in analogy to 1-(dimethylamino)-1-ferrocenylethane[213] permits (if optically active) asymmetric syntheses by stereoselective introduction of the substituent R′ (*via* lithium); consequently, this method provides an elegant access to optically active, α-disubstituted ferrocenes (*cf.* Section 30.2.5).

In the reaction of α-chloromethylferrocene with *n*-butyllithium "ferrocin" (dehydroferrocene) could be detected as an intermediate.[213a]

A further access to 1,2-disubstituted ferrocenes

---

205 *D. W. Slocum, T. R. Engelmann, C. A. Jennings,* Australian J. Chem. *21*, 2319 (1968).
D. W. Slocum, B. P. Koonsvitsky, Chem. Commun. 846 (1969).

205a *D. W. Slocum, B. P. Koonsvitsky, C. R. Ernst,* J. Organometal. Chem. *38*, 125 (1972).

206 *G. Marr, R. E. Moore, B. W. Rockett,* J. Chem. Soc. (London) (C), 24 (1968).

207 *G. Marr, R. E. Moore, B. W. Rockett,* Tetrahedron *25*, 3477 (1969);
K. Schlögl, M. Walser, Monatsh. Chem. *100*, 1515 (1969).

208 *J. Tirouflet, C. Moise,* Compt. Rend. *262*, 1890 (1966).

209 *H. Falk, K. Schlögl, W. Steyrer,* Monatsh. Chem. *97*, 1029 (1966).

210 *D. W. Slocum, B. W. Rockett, C. R. Hauser,* J. Amer. Chem. Soc. *87*, 1241 (1965).

211 *D. W. Booth, B. W. Rockett,* Tetrahedron Lett. 1438 (1967).

212 *T. Aratani, T. Gonda, H. Nozaki,* Tetrahedron Lett. 2265 (1969);
T. Aratani, T. Gonda, H. Nozaki, Tetrahedron *26*, 5453 (1970).

213 *D. Marquarding, H. Klusacek, G. Gokel, P. Hoffmann, I. Ugi,* Angew. Chem. *82*, 361 (1970);
D. Marquarding, H. Klusacek, G. Gokel, P. Hoffmann, I. Ugi, J. Amer. Chem. Soc. *92*, 5389 (1970).

213a *J. W. Huffmann, J. F. Cope,* J. Org. Chem. *36*, 4068 (1971).

is based on the mercuration of *Fc*-I (Section 30.2.1.2.5, see also 30.2.1.2.9). The intermediate α-isomer is isolated as bis(2-iodoferrocenyl)-mercury and can be transformed into 1,2-diiodoferrocene or ferrocene-1,2-dicarboxylic anhydride.[213b] For the preparation of substituted ferrocene carboxylic acids see Ref. 43b.

30.2.2.3.2 *Homo*anularly Bridged Derivatives

In analogy to the formation of ferrocenophanes (Section 30.2.2.2.3) syntheses of *homo*anularly bridged ferrocenes can be accomplished primarily by three routes: the cyclization either of suitable monosubstituted or of symmetrically and unsymmetrically α-disubstituted ferrocenes.

Ring closure of γ-ferrocenylbutyric acids (**147**) (conveniently obtained by acylation of ferrocene with succinic anhydride(s) and subsequent reduction of the ketoacids formed (*cf.* Section 30.2.1.2.4)) with polyphosphoric acid or (preferably) trifluoroacetic anhydride affords cyclic ketones [ferroceno cyclohexenones or (α-oxotetramethylene) ferrocenes] of the structural type **148** in good to excellent yield.[58,141,191,214] A Friedel-Crafts-cyclization of the acid chlorides is also possible.[215]

$$Fc-CH_2-\underset{|}{\overset{R}{C}}H-\underset{|}{\overset{R^1}{C}}H-COOH \xrightarrow[-H_2O]{} \mathbf{148}$$

147 **148**

Alkyl or aryl derivatives of **148** (*e.g.*, R or R′ = methyl or phenyl) are of some stereochemical interest, since stereoselective ring closures of the corresponding ferrocenyl butyric acids **147** are possible and the substituents may occupy preferred positions (*exo* or *endo* with respect to the *Fc*-moiety in **148**).[216,217]

Ketones **148** can be subjected to several subsequent reactions (mostly in analogy to acylferrocenes, Section 30.2.1.2); reduction of C=O to >CHOH or the reaction with organometallic compounds poses similar stereochemical problems (*exo*- or *endo*-OH).[215,218]

Oxidation with manganese dioxide of **148** gives benzoquinone derivatives,[34,219] which yield the corresponding ferroceno cyclohexenediones by catalytic hydrogenation.

Friedel-Crafts-cyclizations can also lead to ferroceno cyclopentenones; for example, reaction of the chloride of *o*-ferrocenylbenzoic acid with aluminum chloride yields ferroceno[*b*]-indenone (**149**).[220]

149

Steric hindrance may direct cyclizations of suitably substituted ferrocenylpropionic acids to *homo*anular rather than *hetero*anular bridging, resulting in the formation of ferroceno cyclopentenones (*cf.* also Sections 30.2.2.2.3 and 30.2.4).

Bischler-Napieralski-cyclization of *N*-formyl-2-ferrocenylethylamine produces **150**, which can be reduced with lithium aluminum hydride to tetrahydroferroceno[*c*]pyridine (**151**),[221] however, **151** is better prepared by reaction of 2-ferrocenylethylamine with formic acid and formaldehyde.[215]

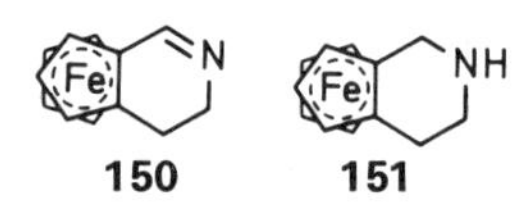

150 **151**

The "total synthesis" of α-disubstituted ferrocenes is unsatisfactory since a cross-reaction

---

[213b] *P. V. Roling, M. D. Rausch*, J. Org. Chem. *39*, 1420 (1974).

[214] *J. W. Huffmann, D. J. Rabb*, J. Org. Chem. *26*, 3588 (1961).

[215] *K. Schlögl, M. Fried, H. Falk*, Monatsh. Chem. *95*, 576 (1964).

[216] *R. Dabard, B. Gautheron*, Bull. Soc. Chim. France 667 (1963); *J. Tirouflet, B. Gautheron, R. Dabard*, Bull. Soc. Chim. France 96 (1965); *B. Gautheron, J. Tirouflet*, Compt. Rend. *265*, 273 (1967).

[217] *H. Falk, K. Schlögl*, Monatsh. Chem. *96*, 1065 (1965).

[218] *D. S. Trifan, R. Bacskai*, Tetrahedron Lett. *13*, 1 (1960); *J. Boichard, J. Tirouflet*, Compt. Rend. *253*, 1337 (1961); *E. A. Hill, J. H. Richards*, J. Amer. Chem. Soc. *83*, 4216 (1961); *K. Schlögl, M. Fried*, Tetrahedron Lett. 1473 (1963).

[219] *H. Egger, H. Falk*, Tetrahedron Lett. 437 (1966); *H. Egger, H. Falk*, Monatsh. Chem. *97*, 1590 (1966).

[220] *E. Bublitz, W. E. McEwen, J. Kleinberg*, J. Amer. Chem. Soc. *84*, 1845 (1962); *M. Cais, A. Modiano, A. Raveh*, J. Amer. Chem. Soc. *87*, 5607 (1965).

[221] *J. M. Osgerby, P. L. Pauson*, J. Chem. Soc. (London) 4600 (1961); *D. Lednicer, C. R. Hauser*, J. Org. Chem. *24*, 43 (1959).

takes place and mixtures are obtained. Benzoferrocene was prepared in only 9% yield from the reaction of sodium salts of cyclopentadiene and indene with iron(II) chloride.[222]
For the synthesis of 1,3-diferrocenylferrocene from 1,4-diferrocenyl cyclopentadiene *cf.* Ref. 222a.
Condensation of ferrocene-1,2-dicarboxaldehyde (Table 4, Section 30.2.2.3.1) with ketones yields ferroceno-4,5-tropones (**152**).[208]

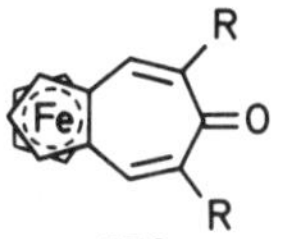

152

30.2.3 Trisubstituted Ferrocenes

The principal routes for the preparation of trisubstituted ferrocenes are summarized in the following scheme where A and D stand for activating and deactivating substituents, resp. (such as alkyl and acyl); E indicates substituents which can be introduced in a S_E-reaction, such as acylation, alkylation or lithiation.

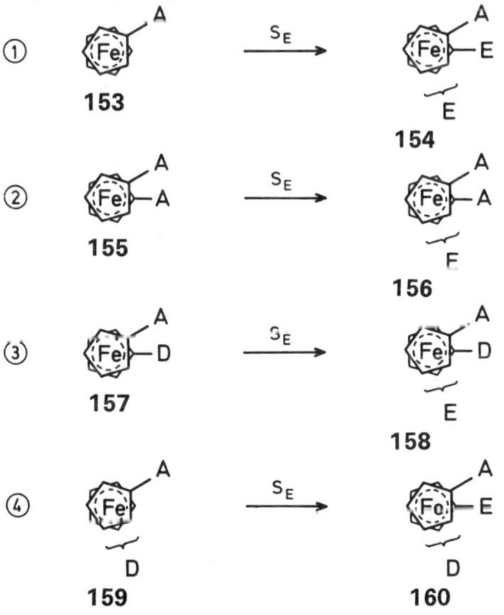

① **153** $\xrightarrow{S_E}$ **154**

② **155** $\xrightarrow{S_E}$ **156**

③ **157** $\xrightarrow{S_E}$ **158**

④ **159** $\xrightarrow{S_E}$ **160**

If starting materials (**155**) with two different (activating) substituents (A and A′) are employed (in a reaction such as ②) the number of isomers increases. The isomeric mixtures (**154**, **156**, and **158**) obtained in the reactions ①, ②, and ③ usually can be separated smoothly by chromatographic methods (Section 30.2.2.1) since only *two* isomers are formed. Many further ferrocene derivatives are accessible from the primary substitution products by appropriate reactions (*cf.* Section 30.2.1.2).
Some typical examples illustrate the variety of the reactions leading to trisubstituted ferrocenes.
Friedel-Crafts-acetylation of compounds **153** (A = benzyl or 2-thienyl) affords β-acetyl derivatives **154** (E = $COCH_3$) which are additionally acetylated in the benzene or thiophene ring (in *p*- and 5-position, resp.).[99]
Metallation of **153** (A = C_2H_5) with *n*-amylsodium and subsequent carboxylation and esterification[63] yields α- and β-bis-carbomethoxy derivatives **154** (E = $COOCH_3$).
Products of the general structure **155** are very useful for the synthesis of trisubstituted ferrocenes. Acetylation of **155** (A = CH_3) yields a mixture of the α- and β-isomer (**156**, E = $COCH_3$).[142] Acetylation, succinoylation or Vilsmeier-formylation of **155** (A = C_2H_5) provide the expected products **156** (E = $COCH_3$, $COCH_2CH_2COOH$ or CHO);[223,224] the β-isomers predominate in the latter reaction (as well as in the case of dimethylferrocene)[143] due to the steric bulk of the Vilsmeier-complex. Acetylation of diphenylferrocene (**155**, A = phenyl) affords a mixture of acetyl derivatives (**156**, E = $COCH_3$).[83]
Aminomethylation of di methyl, -methoxy, or -methylthio ferrocene (**155**, A = CH_3, OCH_3, or SCH_3) yields the expected mixtures of isomeric Mannich-bases [**156**, E = $CH_2N(CH_3)_2$].[143,184]
If dimethylferrocene (**155**, A = CH_3) is lithiated with *n*-butyllithium and subsequently carboxylated, a mixture of isomeric carboxylic acids (**156**, E = COOH) is formed;[140,144] the latter are better prepared by acylation with *N,N*-diphenylcarbamyl chloride and hydrolysis of the diphenylamides [**156**, A = CH_3, E = $CON(C_6H_5)_2$][143] which can be separated easily.[43b] These isomeric-dimethylcarboxylic acids can be reduced to give trimethylferrocenes.[143]
[3]Ferrocenophanes (**133**, Section 30.2.2.2.3) are special representatives of **155**, where A,A = $-CH_2CH_2CH_2-$. In the Vilsmeier formylation of **133** the β-isomer **156** (E = CHO) predominates,[171]

---

[222] *R. B. King, M. B. Bisnette,* Angew. Chem. *75*, 642 (1963);
R. B. King, M. B. Bisnette, Inorg. Chem. *3*, 796 (1964).
[222a] *E. W. Neuse, R. K. Crossland,* J. Organometal. Chem. *43*, 385 (1972).

[223] *J. Tirouflet, J. P. Monin, G. Tainturier, R. Dabard,* Compt. Rend. *256*, 433 (1963);
G. Tainturier, J. Tirouflet, Compt. Rend. *258*, 5666 (1964).
[224] *D. E. Bublitz,* Can. J. Chem. *42*, 2381 (1964).

but Friedel-Crafts-acetylation leads to a mixture of the α- and β-isomers ($E=COCH_3$).[141,225,226,227] Reaction of [3]ferrocenophane with diphenylcarbamylchloride–aluminum chloride and hydrolysis of the (easily separable) diphenylamides [**156**, $E=CON(C_6H_5)_2$] offers a convenient method for preparing the [3]ferrocenophanecarboxylic acids (**156**, A,A= $-CH_2CH_2CH_2-$, E=COOH).[28] See Ref. 43b. Cyclization of 1-methylferrocene-1′-propionic acid (**155**, $A=CH_3$, $A'=CH_2CH_2COOH$) to yield the isomeric methyl[3]ferrocenophanones is a special case.[28]

On the basis of the cited examples for the substitution of compounds of type **155** it can be concluded, that substitution rules and site reactivities similar to those discussed for monosubstituted ferrocenes in Section 30.2.2 (*cf.* Table 3) are valid here.

The acetylation of **148** (R=R′=H, Section 30.2.2.3.2) to 1′-acetyl-1,2-(α-oxotetramethylene)-ferrocene[228] may be considered as an example for the substitution type **159**→**160**.

30.2.4 Tetra- and Polysubstituted Ferrocenes

The synthetic principles presented in the previous chapters can be applied to the synthesis of higher substituted ferrocenes.

The most important reaction sequence is: acylation, reduction of –COR to produce an "activating" substituent ($-CH_2R$), repeated acylation (if necessary with subsequent separation of isomers), repeated reduction and so on, until the desired number of substituents has been introduced. This sequence can be carried to decasubstitution.[229]

As mentioned in Section 30.1.3, many positional isomers as well as diastereomers (meso and racemic forms) may be encountered with polysubstituted ferrocenes (*cf.* Table 1).

The following tetrasubstituted ferrocenes are of some interest:

Bis-cyclization of ferrocene-1,1′-bispropionic acid (**84**, $n=1$) affords a mixture of the two isomeric [3][3]ferrocenophanediones **161** and **162**,[171] but ring closure of [3]ferrocenophane-α-propionic acid with trifluoroacetic anhydride yields the bis-*hetero*anular compound **163** and the *homo*anularly bridged derivative **164**.[226]

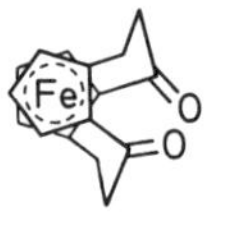

161 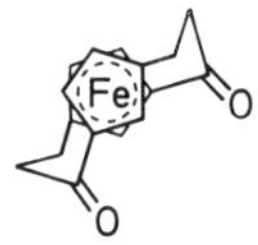 **162** 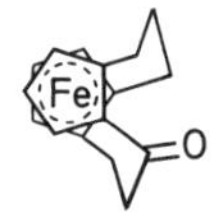 **163** 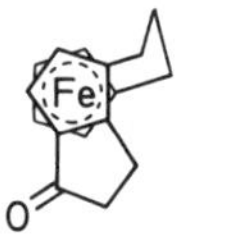 **164**

Some diacyl [3]ferrocenophanes have been reported.[229a]

Cyclization of the homolog ferrocene-1,1′-bisbutyric acid (**84**, $n=2$) provides a mixture of both bis-*homo*anular ketones.[223] Meso form and racemate (**165** and **166**, resp.) were separated by chromatography (m.p. 170° and 160°, resp.); the configurations were assigned by resolution of **166**.[230]

("*meso*") **165** 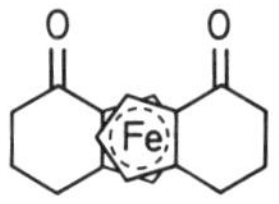 (one enantiomer) **166**

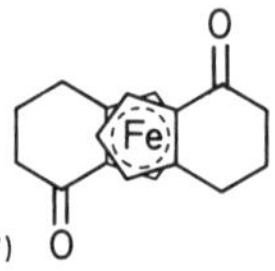

In analogy to the stereoselective α-lithiation of the Mannich-base **3** (*cf.* Section 30.2.2.3.1) the lithiation of the bis-Mannich-base (**91**, Section 30.2.2.2.1) proceeds selectively in both α-ring positions; appropriate reactions yield α,α'-tetrasubstituted ferrocenes of the structural type **167** (as mixture of meso- and racemic forms).[231]

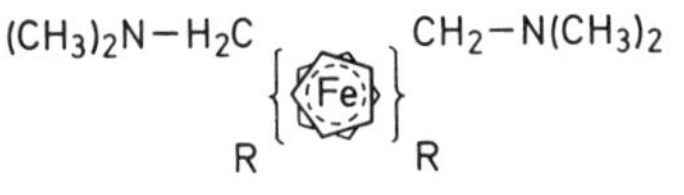

$(CH_3)_2N-H_2C$ … $CH_2-N(CH_3)_2$, R, R

167

Finally, symmetrical α,α'-tetrasubstituted products **168**, such as 1,1′,2,2′-tetrabenzyl ferrocene (**168**, R=benzyl) can be prepared by a "total synthesis" originating from corresponding cyclopentadiene derivatives.[232]

---

[225] *D. W. Hall, J. H. Richards*, J. Org. Chem. *28*, 1549 (1963).

[226] *K. L. Rinehart jr., D. E. Bublitz, D. H. Gustafson*, J. Amer. Chem. Soc. *85*, 970 (1963).

[227] *T. H. Barr, E. S. Bolton, H. L. Lentzner, W. E. Watts*, Tetrahedron *25*, 5253 (1969).

[228] *G. Haller, K. Schlögl*, Monatsh. Chem. *98*, 603 (1967).

[229] *K. Schlögl, M. Peterlik*, Monatsh. Chem. *93*, 1328 (1967).

[229a] *J. A. Winstead, R. R. McGuire, R. E. Cochoy, A. D. Brown, G. J. Gauthier*, J. Org. Chem. *37*, 2055 (1972).

[230] *H. Falk, K. Schlögl*, Monatsh. Chem. *96*, 266 (1965).

[231] *E. S. Bolton, P. L. Pauson, M. A. Sandhu, W. E. Watts*, J. Chem. Soc. (London) (*C*), 2260 (1969).

[232] *W. F. Little, R. C. Koestler*, J. Org. Chem. *26*, 3245 (1961).

168

Pentasubstituted ferrocenes are formed primarily on electrophilic substitution of the [3][3]ferrocenophane system thus establishing the constitutions of the products.[226]
For example, Friedel-Crafts acylation of [3][3]-(1,3)ferrocenophane with *N,N*-diphenylcarbamyl chloride and aluminum chloride and subsequent hydrolysis of the resultant amide provides the carboxylic acid **169**, from which the usual derivatives and subsequent products, such as acetyl or vinyl [3][3]ferrocenophane can be prepared.[28]
Appropriate synthetic and cyclization procedures (as discussed previously) provide a route for the preparation of up to [3][3][3]ferrocenophanes (*i.e.*, threefold bridged ferrocenes) and their substitution products, such as the fourfold bridged (i.e. octasubstituted) ferrocene **170**.[226, 229]

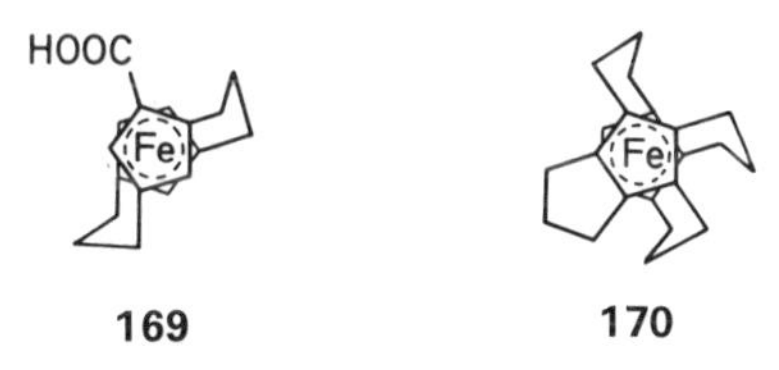

169 **170**

Amongst the decasubstituted ferrocenes both decamethylated and ethylated ferrocenes have been prepared: the first one was obtained from the reaction of pentamethylcyclopentadienate with iron(II) chloride[233] and the latter compound was prepared by repeated acetylation and reduction of ferrocene.[229]
Polychlorinated ferrocenes including the decachloro ferrocene were obtained by repeated lithiation of 1,1′-dichloroferrocene (**93**) with butyllithium and subsequent exchange reaction with hexachloroethane; the lithiation invariably takes place adjacent to Cl (at the α-position, *cf.* Section 30.2.1.1.6).[234] Dechachloroferrocene has a remarkably high thermal stability.[234]

30.2.5 Optically Active (Chiral) Ferrocenes

For ferrocene derivatives the necessary chirality for the occurrence of enantiomers (and hence for the optical activity) can be attained twofold:

(1) Ferrocenyl itself can be a ligand of a chiral compound
(2) *Homo*anular ferrocene derivatives with at least two different substituents (*cf.* **1**, Section 30.1.3) belong to the point group C_1; consequently, they are chiral and can be resolved into enantiomers. This "ferrocene chirality" was investigated in some detail.[26, 235]

Many representatives of the first group have been prepared, in which ferrocenyl is attached either to the centre of chirality or separated from it by one or more atoms. This group includes (amongst others) α-ferrocenylethylamine (**171**),[67, 236, 237] α-phenyl-β-ferrocenylpropionic acid (**172**),[236] α-phenyl-γ-ferrocenylbutyric acid (**173**),[217] and *N*-ferrocenylsulfonyl-α-aminoacids **174**.[125]

Fc–CH(NH_2)–CH_3 **171**

Fc–CH_2–CH(C_6H_5)–COOH **172**

Fc–CH_2–CH_2–CH(C_6H_5)–COOH **173**

Fc–SO_2–NH–CH(R)–COOH **174**

The resolution of such compounds can be achieved by conventional methods, such as recrystallization of diastereomeric salts *e.g.*, those derived from α-phenethylamine or tartaric acid.
The second group of chiral ferrocenes is larger, since numerous di- and polysubstituted ferrocenes are possible which fulfil the criteria mentioned above. The rotational barrier about the molecular axis is rather low (*cf.* Section 30.1.1); therefore, unsymmetrical *hetero*anular disubstitution (*cf.* Section 30.2.2.2.2) is not sufficient for chirality, since chiral conformers are readily converted into each other (*i.e.* racemized) by rotation around the ring-iron-axis.
Systemic variation of the substituents of a few optically active key compounds (*cf.* the Sections 30.2.1.2 and 30.2.2) led to some 250 optically active (ferrocenechiral) ferrocenes, the relative configurations of which were established by chemical transformations. The absolute con-

---

[233] *R. B. King, M. B. Bisnette*, J. Organometal. Chem. *8*, 287 (1967).
[234] *F. L. Hedberg, H. Rosenberg*, J. Amer. Chem. Soc. *92*, 3239 (1970).
F. L. Hedberg, H. Rosenberg, J. Amer. Chem. Soc. *95*, 870 (1973).
[235] *K. Schlögl*, Pure Appl. Chem. *23*, 413 (1970).
[236] *H. Falk, Ch. Krasa, K. Schlögl*, Monatsh. Chem. *100*, 254 (1969).
[237] *D. Marquarding, P. Hoffman, H. Heitzer, I. Ugi*, J. Amer. Chem. Soc. *92*, 1969 (1970).

figurations of all compounds were initially elucidated by assignment of the configurations to some basic compounds with chemical methods[217,230] (*cf.* Volume I, Chapter 3.8); and later were confirmed by anomalous X-ray diffraction[6,179] (*cf.* Volume I, Chapter 7.2).

The following key compounds may be illustrative:

(+)(1*S*)* - 1,2 - (α-Oxotetramethylene)ferrocene (**175**, cf. Section 30.2.2.3.2 for the synthesis of the racemic compound), was obtained by resolution with (−)-menthydrazide[215] and (+)-(1*S*)-methylferrocene-α-carboxylic acid (**176**)[100] with (−)-α-phenethylamine (for the preparation of racem. **176** *cf.* Section 30.2.2.3.1, Table 4).

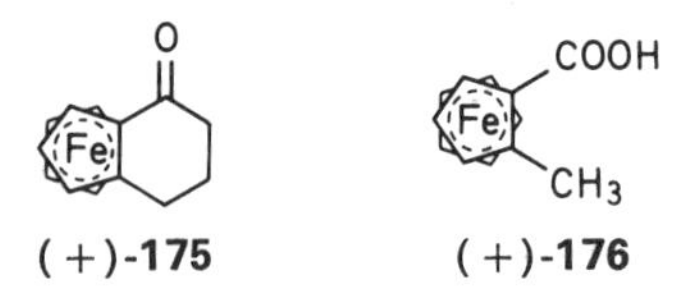

(+)-**175** (+)-**176**

The optical purity of **176** was determined by the method of isotope dilution; thereby the maximum rotations of all compounds correlated with this acid are known.[179,238]

A further route to optically active ferrocenes is provided by the highly stereoselective α-lithiation (*cf.* Section 30.2.2.3.1) of compounds belonging to the first group (centrochiral ferrocenes). Lithiation of (+)-(*R*)-1-(dimethylamino)-1-ferrocenylethane (dimethylderivative of **171**) with butyllithium gives the α-lithioderivative in an optical yield of 96%; various (active) α-disubstituted ferrocenes are accessible from the latter.[212,213]

Although ferrocenes usually exhibit a considerable chemical stability, a surprisingly smooth racemization of optically active ferrocenes was observed in some cases.[239]

The theoretical aspects of the ferrocene chirality including optical rotatory dispersion and circular dichroism (*cf.* Vol. I, Chapter 5.7) have been investigated in some detail; a discussion of the results would stand beyond the scope of this chapter; *cf.* below[26,91,119,235,236,240,241] for details.

30.2.6 Ligand Exchange Reactions

Reductive cleavage of ferrocene can be performed with lithium or sodium in diethylamine and with calcium in liquid ammonia.[242]

Ferrocene and its alkyl-derivatives undergo ligand exchange under the influence of Lewis acids, as shown by equilibration-experiments with α-diethyl-ferrocene (**1**, $R = R' = C_2H_5$). If this compound is treated with aluminum chloride in dichloromethane, a mixture of starting material, α,α′-tetraethylferrocene (**168**, $R = C_2H_5$) and ferrocene results.[224]

Such ligand exchange reactions are of some preparative interest for exchanging cyclopentadienyl and benzene; a variety of cations of the general formula **177** has been prepared by this procedure.[243]

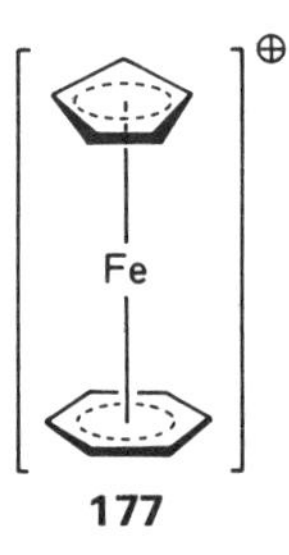

177

Quite obviously, an intramolecular ligand exchange takes place during the racemization of optically active ferrocenes (Section 30.2.5).[239]

If a =CH– group of the five-membered ring in ferrocene is replaced by the isoelectronic =N– group, azaferrocene (**178**) is obtained. The latter **178** and its alkyl derivatives (such as 2-methyl and 2.5-dimethylazaferrocene) are prepared by reaction of the potassium salts of pyrrol or methylpyrrols with $C_5H_5Fe(CO)_2$ (*cf.* Ref. 244 for its synthesis) in benzene in *ca.* 20%

---

[238] *P. Reich-Rohrwig, K. Schlögl*, Monatsh. Chem. *99*, 1752 (1968).

[239] *D. W. Slocum, S. P. Tucker, T. R. Engelmann*, Tetrahedron Lett. 621 (1970);
H. Falk, H. Lehner, J. Paul, U. Wagner, J. Organometal. Chem. *28*, 115 (1971).

* For the configurational (*R*)(*S*)-nomenclature of optically active ferrocenes (and metallocenes) *cf.* Ref. 26.

[240] *H. Falk, O. Hofer*, Monatsh. Chem. *100*, 1499, 1507 (1969).

[241] *H. Falk, H. Lehner*, Tetrahedron *27*, 2279 (1971).

[242] *D. S. Trifan, L. Nicholas*, J. Amer. Chem. Soc. *79*, 2746 (1957).

[243] *A. N. Nesmeyanov, N. A. Volkenau, I. N. Bolesova*, Tetrahedron Lett. 1725 (1963);
A. N. Nesmeyanov, N. A. Volkenau, I. N. Bolesova, Dokl. Akad. Nauk. SSSR *166*, 607 (1966); C.A. *64*, 17635 (1966);
A. N. Nesmeyanov, Pure Appl. Chem. *17*, 211 (1968).

[244] *R. B. King, F. G. A. Stone*, Inorg. Synth. *8*, 110 (1963).

yield.[245] The yield increases (36% for **178**) on use of dioxane as solvent.[246]

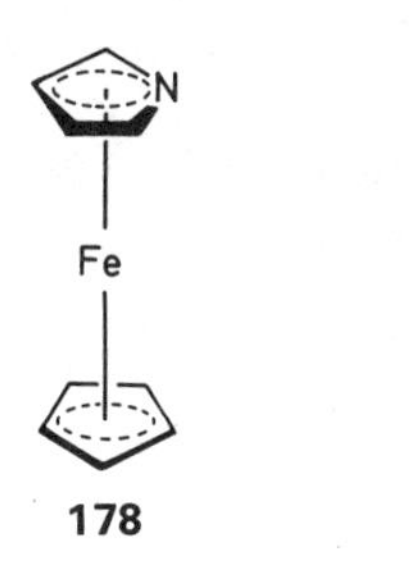

178

30.2.7 Ferrocene Polymers

Polymeric ferrocenes[247] can be divided in two groups: in one group (**179**) ferrocenyl moieties are attached as side-chains to a polymeric backbone, and in the other group (**180** and **181**, resp.) ferrocene is part of the main chain (backbone).

179 **180** **181**

Polymers of type **179** are obtained by radical polymerization of vinylferrocene[117, 248] (Section 30.2.1.2.2) or *trans*-cinnamoyl-ferrocene[249] or by copolymerization with styrene and methacrylates (vinylferrocene)[117, 248] and styrene, acrylonitril, acrylates, butadiene, and isoprene, resp. (cinnamoylferrocene).[249] Azo-bis-isobutyronitril is mainly used as initiator, since peroxides can cause oxidative destruction.

Poly(ferrocenyl)ethylenes (potential semiconductors) (**182**) are prepared by heating acetylferrocene (**5**) with zinc chloride.[250] Polycondensation of **5** with urea in the presence of zinc chloride affords polymers of the general structure **183**.[251]

The ferrocenyl group can also be introduced into

182 **183** **184**

preformed polymers: for example, reaction of diazotized poly(*p*-aminostyrene) with ferrocene (*cf.* the arylation of ferrocene, Section 30.2.1.2.3) gives polyferrocenyl polystyrene.[252] Such resins are of some interest as "redox-exchangers". Natural polymers, such as proteins, can be "labelled" at the amino group sites by reaction with ferrocene derivatives (such as ferrocenyl imidoester **184**).[253, 254]

Polymers of the general type **180** and **181**, resp., represent the majority of the known polymeric ferrocenes. Only a few examples may be mentioned:

In the most simple case –X– stands for a single bond; poly-ferrocenylens of this type can be prepared by direct (radical) polymerization of ferrocene (e.g., with *tert*-butylperoxide).[255] Ferrocenyllithium (**8**) or haloferrocenes (*cf.* Sections 30.2.1.1.6 and 30.2.1.2.5) can be polymerized with cobalt(II) chloride or under the conditions of the Ullmann reactions; in both reactions only oligomers are obtained.[88]

Polymers of type **181** with –X–=–CH_2– or –CH(R)– are prepared by polycondensation of monomers of the general structure **26** (*cf.* Section 30.2.1.2.4) with acid catalysts,[256] from dimethylaminomethylferrocene (**3**) and zinc chloride[257] or by condensation of ferrocene with aldehydes.[258]

Many examples of polymers are known, which contain carbonyl, nitrogen, silicon, or phosphorous in the –X– group [*e.g.*, polyurethanes, prepared from ferrocene-1.1′-diisocyanate (**99**, Section 30.2.2.2.1) and glycols[259]].

[245] *K. K. Joshi, P. L. Pauson, A. R. Quazi, W. H. Stubbs*, J. Organometal. Chem. *1*, 471 (1964).
[246] *K. Bauer, H. Falk, K. Schlögl*, Angew. Chem. *81*, 150 (1969).
[247] *H. J. Lorkowski*, Fortschr. Chem. Forsch. *9*, 207 (1967); *E. W. Neuse*, Advan. Macromol. Chem. 1 (1968).
[248] *Y. H. Chen, M. Fernandez-Refojo, H. G. Cassidy*, J. Polym. Sci. *40*, 433 (1959).
[249] *L. E. Coleman jr., M. D. Rausch*, J. Polym. Sci. *28*, 207 (1958).
[250] *Y. Paushkin, L. S. Polak, T. P. Vishnyakova, I. I. Patalakh, F. F. Machus, T. A. Sokolinskaya*, J. Polym. Sci. *2 C*, 1481 (1964).
[251] *I. A. Golubeva, T. P. Vishnyakova*, Plast. Massy 10 (1965); C.A. *64*, 128202 (1966).
[252] *B. Sansoni, O. Sigmund*, Angew. Chem. *73*, 299 (1961).
[253] *H. Falk, M. Peterlik, K. Schlögl*, Monatsh. Chem. *100*, 787 (1969).
[254] *H. Franz*, Z. Chem. *7*, 235 (1967).
[255] *A. N. Nesmeyanov, V. V. Korshak, V. V. Voevodskii, N. S. Kochetskova, S. L. Sosin, R. B. Materiokova, T. N. Bolotnikova, V. M. Chibrikin, N. M. Bazhin*, Dokl. Akad. Nauk. SSSR *137*, 1370 (1961); C.A. *55*, 21081 (1961).
[256] *E. W. Neuse, D. S. Trifan*, J. Amer. Chem. Soc. *85*, 1952 (1963).
[257] *E. W. Neuse*, U.S. Pat. 3238185 (1966); C.A. *64*, 16012 (1966).
[258] *E. W. Neuse, E. Quo*, Bull. Chem. Soc. Japan *39*, 1508 (1966).
[259] *P. Petrovich, H. Valot*, Compt. Rend. *263*, 214 (1966).

For additional condensation polymers (employing the dichloride **88**, Section 30.2.2.2.1) *cf.* Ref. 160.

30.3 Identification and Structural Determination of Ferrocenes

30.3.1 Analysis

The analysis of ferrocene and its derivatives is based on the application of classical methods. In addition some special procedures for the analysis of iron in ferrocenes have been developed.

For example, the iron(III) ion resulting from the oxidative degradation of ferrocene can be determined quantitatively by the usual methods[260] (*cf.* also Vol. VI). Ferrocene derivatives can be precipitated from mixtures with hexacyanoferrate(II)[261] and $Fe^{3\oplus}$ can also be determined by titration with standardized ferrocene solutions.[262]

We have employed the gravimetric determination of iron as iron(III) oxide by treatment of ferrocene derivatives with concentrated nitric acid in a platinum crucible. Satisfactory results were obtained.

Molecular weight determination of alkylferrocenes can be accomplished by potentiometric titration with dichromate; for the analysis of acylferrocenes these have to be reduced quantitatively[263] (*cf.* also Ref. 264).

A chemical method has been reported for the identification of α-, β-, or *hetero*anularly (1.1′) disubstituted ferrocenes.[265] It is based on the reaction sequence: succinoylation, reduction of the ketoacid to ferrocenylbutyric acid, cyclization of the latter and reduction of the obtained ketone (**148**) to the "hydrocarbon" (1.2-tetramethylene ferrocene) which then can be compared with the corresponding compound obtained from the isomeric ferrocene derivative. This method is obviously quite laborious and cannot compete with spectroscopic methods (especially *NMR*, Section 30.3.3.2).

30.3.2 Chromatographic Behavior of Isomeric Ferrocenes

Chromatographic methods are especially advantageous in the ferrocene field for two reasons: one, due to their inherent color no special reactions generating a colored species are required for the detection (*cf.* also Section 30.2.2.1). Secondly, their chromatographic behavior permits valuable conclusions with respect to the structure including isomeric mixtures.

Ferrocenecarboxylic acids (**6**, **37**, **84**) and ferrocenyl aminoacids (*e.g.*, **47**) can be smoothly separated by paper chromatography;[67,157] homologous ferrocene derivatives can be clearly distinguished on thin layer chromatograms by their (different) R_f-values,[24,56,81,139] and a useful relationship exists between adsorbability and structure of isomeric acyl alkylferrocens which is valuable for an initial rapid structural determination (also of mixtures).

The three possible isomers (*cf.* Section 30.1.3) exhibit increasing adsorbabilities in the order: $\alpha(1.2) <$ *hetero*$(1.1') < \beta(1.3)$;[24] consequently, in all cases the α-isomer is eluted first from the column and has the highest R_f-value (on *TLC*). This rule does not hold only for alumina and silica (column and *TLC*) as adsorbents but also for gas–liquid-chromatography[40,63,135,146] and the paper chromatography of alkylferrocenecarboxylic acids.[67] For application of high pressure liquid–liquid chromatography (LLC) to separation problems in the ferrocene (and metallocene) field see Ref. 147a.

30.3.3 Spectroscopy

30.3.3.1 Infrared and Electron Absorption Spectroscopy

Typical i.r. and electron absorption bands of ferrocene are listed in Table 5.[15,266]

The i.r.-bands at 1100 and 1000 cm^{-1} are of particular importance for the solution of structural problems: these bands are observed only if at least one ring is unsubstituted (*i.e.*, in ferrocene and its mono- and *homo*anularly polysubstitution-products) but are lacking in *hetero*anularly di- (or poly) substituted ferrocenes.[156]

---

[260] *A. N. Belder, E. J. Bourne, J. B. Pridham*, Chem. & Ind. 996 (1959);
S. I. Goldberg, Anal. Chem. *31*, 486 (1959);
H. Rosenberg, C. Riber, Microchem. J. *6*, 103 (1962).

[261] *J. D. Behun*, Talanta *9*, 83 (1962).

[262] *L. Wolf, H. Franz, H. Hennig*, Z. Chem. *1*, 27, 220 (1960).

[263] *M. Peterlik, K. Schlögl*, Z. Anal. Chem. *195*, 113 (1963).

[264] *D. M. Knight, R. C. Schlitt*, Anal. Chem. *37*, 470 (1965).

[265] *G. Tainturier, J. Tirouflet*, Bull. Soc. Chim. France 595, 600 (1966).

[266] *E. R. Lippincott, R. D. Nelson*, Spectrochim. Acta *10*, 307 (1958).

Table 5 Infrared- and Ultraviolet-Bands of Ferrocene

i.r. (Nujol) [cm^{-1}]	Assignment	u.v. (Ethanol) [nm]	[ϵ]
811	C–H bend (in plane)	528	7·5
1002	C–H bend (perpendicular to plane)	440	96
1108	Antisymmetric ring pulsation	324	55
1411	Antisymmetric C–C stretch	260	~2200 (shoulder)
3085	C–H stretch	230	~4600 (shoulder)

In the electron absorption spectrum a broad low-intensity band around 440 nm is characteristic; it is responsible for the color of ferrocene. Its location remains constant for alkylferrocenes, whereas a bathochromic shift is observed and its intensity increases in such derivatives where *Fc* is conjugated with C=O or other chromophors. This feature was illustrated by systematic studies of u.v.-absorption of acylferrocenes,[24, 267] ferrocenyl polyenes[268] and polyynes[80] or ferrocenyl thiophenes.[78]

30.3.3.2 Magnetic Resonance Spectroscopy

^{1}H–*NMR* spectroscopy (*cf.* Vol. I, Chapter 5.5.2) is without doubt the most powerful tool for the elucidation of structures in the ferrocene field. It facilitates the identification of isomeric di- and polysubstituted derivatives as well as the identification of isomeric mixtures. For *NMR*-studies on deuterated ferrocenes *cf.* Ref. 269.
Ferrocene exhibits a singlet at 4·04 δ (in CCl_4, 60 MHz, *TMS* as internal standard) in its *NMR*-spectrum. Due to the virtually unrestricted rotation of the rings (*cf.* Section 30.1.1), a superposition of an A_2B_2- or A_2X_2-system (according to the substituted ring) onto the singlet of the ring-protons of the unsubstituted ("*hetero*") ring is observed in monosubstituted ferrocenes. Which of both systems is present, depends on the substituent R: alkylferrocenes give rise to an A_2B_2-system,[40, 270] but in the case of acylferrocenes an A_2X_2-system is more likely,[270a] the "*hetero*-ring" signal is only slightly shifted (*ca.* −0·05 ppm) as compared with ferrocene.
The elucidation of the structure of isomeric disubstitution products can be based either on the ring-proton signals of ferrocene[40, 271] or, in the case of acyl alkylferrocenes, on the chemical shifts of the alkyl protons.[136]
For many functional groups R increment-systems have been reported for the positions of the ring proton signals.[40, 271, 272] On that basis the chemical shifts of the α- and β-ring protons (relative to R) can be estimated and the structures of the corresponding isomers (α, β, or 1.1′) are deduced from the relative intensities of these signals.
For ferrocenophanes (*cf.* Section 30.2.2.2.3) *NMR*-spectroscopy has been very valuable not only for the elucidation of structural problems,[226, 272a] but also for the conformational analysis.[31, 194, 226, 273] In the case of *homo*anular di- or higher substituted ferrocenes the signals of the "*hetero* ring" protons (5H) are influenced in a characteristic manner by the stereochemistry of certain substitutes.[274]
In some cases—*e.g.*, for the dimethylamino-methyl haloferrocenes—a geminal coupling was

[267] *K. L. Rinehart jr., K. L. Motz, S. Moon*, J. Amer. Chem. Soc. *79*, 2749 (1957);
R. T. Lundquist, M. Cais, J. Org. Chem. *27*, 1167 (1962).

[268] *K. Schlögl, H. Egger*, Justus Liebigs Ann. Chem. *676*, 88 (1964).

[269] *M. D. Rausch, A. Siegel*, J. Organometal. Chem. *17*, 117 (1969).
D. W. Slocum, C. R. Ernst, Organometal. Chem. Rev. *6*, 337 (1970).

[270] *D. W. Slocum, W. E. Jones, C. R. Ernst*, J. Org. Chem. *37*, 4278 (1972).

[270a] *M. D. Rausch, V. Mark*, J. Org. Chem. *28*, 3225 (1963).

[271] *J. Tirouflet, G. Tainturier, H. Singer*, Bull. Soc. Chim. France 2565 (1966).

[272] *G. G. Dvoryantseva, S. L. Portnova, Y. N. Sheinker, L. P. Yurjeva, A. N. Nesmeyanov*, Dokl. Akad. Nauk. SSSR *169*, 1083 (1966); C.A. *66*, 15276 (1967).

[272a] *G. J. Gauthier, J. A. Winstead, A. D. Brown, Jr.*, Tetradron Lett. 1593 (1970).

[273] *H. Falk, O. Hofer*, Monatsh. Chem. *101*, 477 (1970).

[274] *P. Reich-Rohrwig, K. Schlögl*, Monatsh. Chem. *99*, 2175 (1968).

observed in structures which contained diastereotopic protons.[275]
Lanthanide "shift reagents" such as Eu (dpm)$_3$ were successfully applied to analytical and stereochemical problems in the ferrocene field.[275a]
For ^{13}C and ^{1}H-NMR spectra of monosubstituted ferrocenes with a chiral centre in the substituent see Ref. 275b.

30.3.3.3 Mass Spectroscopy

In addition to exact molecular weight determination of ferrocenes by means of mass spectrometry, their fragmentation pattern is of interest; from the latter important conclusions regarding their structure can be drawn and in some cases even their stereochemistry can be elucidated.
At low ionization voltage (*e.g.*, 8 eV, *cf.* Vol. I, Chapter 6.1 for the technique and fundamentals of mass spectrometry) usually only molecular ions are formed;[276] consequently, it is advantageous to work at higher voltage (70 eV). For ferrocene the molecular ion ($m/e = 186$) and the fragments $C_5H_5Fe^{\oplus}$ ($m/e = 121$) and $Fe^{\oplus}$ (56) are typical.[277]
Acylferrocenes having a carbonyl group attached to the ring (*Fc*–COR, R=OH, CH_3, OCH_3, $NHCH_3$ *etc.*) exhibit a characteristic transfer of R to the iron:[278] $Fc\text{-}COR^{\oplus} \rightarrow C_5H_5FeR^{\oplus} \rightarrow FeR^{\oplus}$. The behavior of primary secondary, and tertiary ferrocenyl alcohols (*cf.* Section 30.2.1.2.4) is also of some interest.[279] Key fragments in the spectra of these compounds are: ($M-H_2O^{\oplus}$), $C_5H_5Fe^{\oplus}$ and $C_5H_5FeOH^{\oplus}$ ($m/e = 138$). The ion 121 ($C_5H_5Fe^{\oplus}$) frequently occurs in ferrocenes having an unsubstituted ring; its absence is, however, not conclusive for *hetero*anular (di)substitution.
In ferrocenylcarbinols the ratio of the intensities of the peaks 138 and 121 ($C_5H_5FeOH^{\oplus}/C_5H_5Fe^{\oplus}$) depends on the steric position of the OH-group; this is of some interest for anellated systems (*e.g.*, reduction products of **148**).[219]
Besides the intensive molecular ion only a slight and noncharacteristic fragmentation is observed in the mass spectra of ferrocenophanes (*cf.* Section 30.2.2.2.3).[193]

[275] *G. R. Knox, P. L. Pauson, G. V. D. Tiers*, Chem. & Ind. (London) 1046 (1959);
P. Smith, J. J. McLeskey, D. W. Slocum, J. Org. Chem. *30*, 4356 (1965);
D. W. Slocum, T. R. Engelmann, J. Org. Chem. *34*, 4101 (1969).
[275a] *J. Paul, K. Schlögl, W. Silhan*, Monatsh. Chem. *103*, 243 (1972).
[275b] *A. N. Nesmeyanov*, J. Org. Chem. *69*, 429 (1974).
[276] *L. Friedmann, I. P. Irsa, G. Wilkinson*, J. Amer. Chem. Soc. *77*, 3689 (1955).
[277] *F. W. McLafferty*, Anal. Chem. *28*, 306 (1956).
[278] *A. Mandelbaum, M. Cais*, Tetrahedron Lett. 3847 (1964).
[279] *H. Egger*, Monatsh. Chem. *97*, 602 (1966).

30.4 Bibliography

30.4.1 Important General Information on Ferrocenes

M. Dub, Organometallic Compounds (Methods of Synthesis, Physical Constants and Chemical Reactions), 2. Ed. (Literature 1937–1964); Vol. I, p. 220, Springer Verlag, Berlin 1966. First Supplement (Literature 1965–1968) p. 389 (1975).

Organometallic Chem. Reviews, Section B—Annual Surveys, from *1* (1965).

30.4.2 Reviews and Monographs

K. Plesske, Ringsubstitutionen und Folgereaktionen an Aromaten-Metall-π-Komplexen, I und II., Angew. Chem. *74*, 301, 347 (1962).

W. F. Little, Metallocenes, Survey of Progress in Chem. *1*, 133 (1963).

M. Rosenblum, Chemistry of the Iron Group Metallocenes, Bd. I; John Wiley & Sons, New York 1965.

M. Cais, Problem of Metal Participation in the Properties of Metallocenyl Carbonium Ions, Organometal. Chem. Rev. *1*, 435 (1966).

K. Schlögl, Stereochemie von Metallocenen, Fortschr. Chem. Forsch. *6*, 479 (1966).

W. E. Watts, Ferrocenophane Systems, Organometal. Chem. Rev. *2*, 231 (1967).

K. Schlögl, Stereochemistry of Metallocenes in *N. L. Allinger, E. L. Eliel*, Topics in Stereochemistry, Bd. I, p. 39, Interscience Publishers, New York 1967.

H. J. Lorkowski, Ferrocen als Grundbaustein der makromolekularen Chemie, Fortschr. Chem. Forsch. *9*, 207 (1967).

G. E. Coates, M. L. H. Green, K. Wade, Organometallic Compounds, Band II, Methuen & Co., London 1968.

A. N. Nesmeyanov, Développements Récents de la Chimie Organique du Ferrocène, Pure Appl. Chem. *17*, 211 (1968).

E. Neuse, Ferrocene Polymers, Advan. Macromol. Chem. 1 (1968).

D. E. Bublitz, K. L. Rinehart jr., The Synthesis of Substituted Ferrocenes and Other π-Cyclopentadienyl-Transition Metal Compounds, Org. Reactions *17*, 1 (1969).

A. N. Nesmeyanov, Kimia Ferrocena, Akademia Nauk. SSSR, Moskau 1969.

D. W. Slocum, T. R. Engelmann, C. Ernst, C. A. Jennings, W. Jones, B. Koonsvitsky, J. Lewis, P. Shenkin, Metalation of Metallocenes, J. Chem. Educ. *46*, 144 (1969).

D. W. Slocum, C. E. Ernst, Metallocene Homoanular Electronic Effects, Organometal. Chem. Rev. (A) *6*, 337 (1970).

K. Schlögl, Configurational and Conformational Studies in the Metallocene Field, Pure Appl. Chem. *23*, 413 (1970).

M. Cais, M. S. Lupin, Mass Spectra of Metallocenes and Related Compounds, Advan. Organometal. Chem. *8* (1970).

F. D. Popp, E. B. Moynahan, Heterocyclic Ferrocenes, Advan. Heterocycl. Chem. *13*, 1 (1971).

J. H. Peet, B. W. Rockett, 1.2-Disubstituted Ferrocenes, Rev. Pure Appl. Chem. *22*, 145 (1972).

J. C. Johnson, Metallocene Technology (Collection of US Patents); Noyes Data Corporation, Park Ridge und London 1973.

31 Sandwich Compounds

Part A: Metallocenes

Harold Rosenberg
Air Force Materials Laboratory,
Wright-Patterson Air Force Base, Ohio 45433, U.S.A.

Those transition metal complexes in which two (or more) cyclopentadienyl ligands are π-bonded to the metal atom in a "sandwich" structure are commonly known as metallocenes. In such sandwich complexes the metal atom is situated equidistant from the five equivalent carbon atoms in each ring and the compounds exhibit a high degree of aromaticity. The original and most well-known of the metallocenes, whose discovery in 1951 was responsible for the extraordinary growth of transition metal complex chemistry, is di-π-cyclopentadienyliron, or ferrocene. The interest in, and investigations of, ferrocene, have been so extensive that the preparation of ferrocene and its numerous ring-substituted derivatives constitutes the subject matter of Chapter 30 of this volume.

Part A of the present chapter deals with procedures for the synthesis of metallocenes or di-π-cyclopentadienyl sandwich complexes and their derivatives of all other transition metals other than iron. The preparation of those cyclopentadienylmetal compounds in which the bonding between the cyclopentadienyl ring(s) and transition metal is of the ionic or covalent (2-electron) type is not covered in this chapter but may be found in other chapters of this volume under the appropriate metal heading.

Sandwich compounds of the transition elements not containing the cyclopentadienyl group are combined in part B of this chapter. However, carbonyl complexes have been omitted since they are discussed in Chapter 29 of this volume.

31.1 Derivatives of the Inner Transition Elements

31.1.1 Poly-π-cyclopentadienylmetal Compounds of Actinides

While scandium, yttrium and most of the lanthanide elements combine with cyclopentadienyl radicals to yield derivatives which are essentially ionic in nature, members of the actinide or 5-*f* series are known to form π-bonded cyclopentadienylmetal complexes. The latter are thus the only cyclopentadienylmetal derivatives of Group III of the Periodic Table which may be considered members of the metallocene class.

Both tri- and tetra-π-cyclopentadienyl compounds have been prepared from the actinide elements. Although the synthesis of *tri-π-cyclopentadienyluranium(III)* (**1**, M=U) by the reaction of uranium(III) chloride and sodium cyclopentadienide has been reported, the red, air-sensitive and thermally unstable complex was obtained in less than 5% yield and was incompletely characterize.[1] However, *tri-π-cyclopentadienlymetal* complexes of *plutonium* and *americium* (**1**, M=Pu or Am) are obtained in 50–60% yield by treatment of the metal trichloride with molten beryllium cyclopentadienide at temperatures of 65–70°, according to the following equation:[2,3]

$$2MCl_3 + 3Be(C_5H_5)_2 \longrightarrow 2M(\pi\text{-}C_5H_5)_3 + 3BeCl_2$$

Although both have similar structures and are hydrolyzed by water, the flesh-colored americium complex, unlike its moss-green pyrophoric plutonium analog, decomposes only slowly in air and does not melt below 330°, indicating a resemblance to the cyclopentadienylmetal complexes of the lanthanides.

Tetra-π-cyclopentadienylmetal complexes are formed by thorium, protactinium, uranium and

[1] *L. T. Reynolds, G. Wilkinson*, J. Inorg. Nucl. Chem. *2*, 246 (1956).

[2] *F. Baumgärtner, E. O. Fischer, B. Kanellakopulos*, Angew. Chem. *88*, 866 (1965);
F. Baumgärtner, Engl. Angew. Chem. Intern. Ed. Engl. *4*, 878 (1965).

[3] *F. Baumgärtner, E. O. Fischer, B. Kanellakopulos, P. Lauberau*, Angew. Chem. Intern. Ed. Engl. *78*, 112 (1966);
F. Baumgärtner, E. O. Fischer, B. Kanellakopulos, P. Lauberau, Angew. Chem. Intern. Ed. Engl. *5*, 134 (1966).

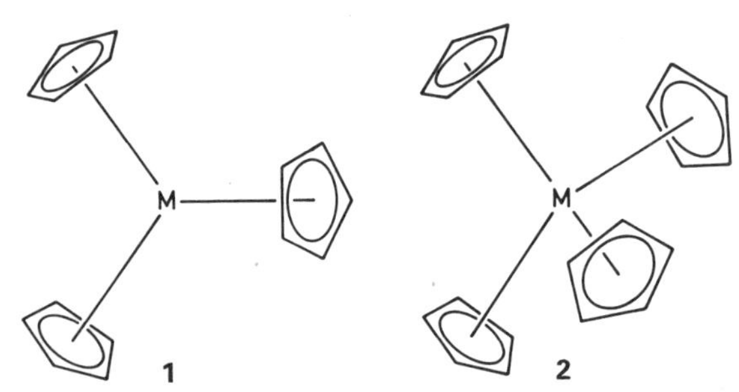

neptunium (**2**, M=Th, Pa, U or Np). The thorium and neptunium derivatives (**2**, M=Th or Np) are obtained by reaction of the anhydrous metal tetrahalide with an excess of freshly prepared potassium cyclopentadienide in diethyl ether or suspended in benzene.[4,5] In addition, *tetra-π-cyclopentadienylthorium(IV)*, as well as its uranium analog (**2**, M=U), may be prepared by heating magnesium cyclopentadienide and the appropriate metal tetrafluoride in the absence of any solvent at 200–230° for 0.5–2 hr.[6] Although *tetra-π-cyclopentadienyluranium(IV)* is also formed by the refluxing of uranium(IV) chloride with either sodium or potassium cyclopentadienide in diethyl ether/benzene or tetrahydrafaran, the yields are low (6–20%).[7,8] The orange-yellow *tetra-π-cyclopentadienylprotactinium(IV)* (**2**, M=Pa) is isolated in 54% yield by treatment of protactinium(V) chloride with beryllium cyclopentadienide in the melt.[9]

31.1.2 Tri-π-cyclopentadienylmetal Halides and Alkoxides of Thorium, Uranium, and Neptunium

Thorium, uranium and neptunium form tri-π-cyclopentadienylmetal chlorides of the type $(\pi\text{-}C_5H_5)_3MCl$ (**3**, M=Th, U or Np, X=Cl). The air-sensitive *π-cyclopentadienylmetal chlorides* of *thorium and uranium* are prepared by the reaction of the metal tetrachlorides with potassium[10] or sodium cyclopentadienides[1] in either diethyl ether or tetrahydrofuran in accordance with the following equation:

$$3MC_5H_5 + M'Cl_4 \longrightarrow (\pi\text{-}C_5H_5)_3M'Cl + MCl$$
$$(M = K \text{ or } Na;\ M' = Th \text{ or } U)$$

An alternate method for the synthesis of the tri-π-cyclopentadienyluranium(IV) halide in 90% yield involves the reaction of uranium(IV) chloride with thallium cyclopentadienide in either tetrahydrofuran or in a benzene suspension.[7] The dark brown, air-sensitive *chloro-(tri-π-cyclopentadienyl)neptunium(IV)* is formed in a melt reaction of neptunium(IV) chloride and beryllium cyclopentadienide at 65–70°.[11]
Chlorotri-π-cyclopentadienyluranium(IV) (**3**, M=U; X=Cl), is a useful intermediate for the preparation of other tri-π-cyclopentadienyluranium derivatives. Thus it reacts with sodium tetrahydridoborate in tetrahydrofuran at room temperature to form the *tetrahydridoborate* derivative (**3**, M=U; X=BH_4) in near quantitative yield.[7] *Salts* of the *tri-π-cyclopentadienyluranium cation* (**4**, X=Reineckate, silicotungstate, hexachloroplatinate or triiodide) are formed when aqueous solutions of the chloride are treated with the appropriate anion under a nitrogen atmosphere.[7]
Tri-π-cyclopentadienylmetal alkoxides of *thorium* and *uranium* (**3**, M=Th or U; X=CH_3O or $n\text{-}C_4H_8O$) can be prepared in modest yields by the addition of sodium cyclopentadienide to the reaction mixture (presumably containing the trichlorometal alkoxide) obtained by refluxing the metal tetrachloride and freshly-prepared sodium methoxide or *n*-butoxide in 1,2-dimethoxyethane.[10]

31.2 Compounds of Titanium, Zirconium, and Hafnium

31.2.1 Di-π-cyclopentadienylmetals

Of the Group IVB metals, titanium and zirconium form the parent, unsubstituted metallocenes, *di-π-cyclopentadienyltitanium* and *-zirconium(II)*. The former is most conveniently prepared in high yield by the reduction of dichlorodi-π-cyclopentadienyltitanium(IV) with

---

[4] *E. O. Fischer, A. Treiber*, Z. Naturforsch. *17b*, 276 (1962).
[5] *F. Baumgärtner, E. O. Fischer, B. Kanellakopulos, P. Lauberau*, Angew. Chem. *80*, 661 (1968); *F. Baumgärtner, E. O. Fischer, B. Kanellakopulos, P. Lauberau*, Angew. Chem. Intern. Ed. *7*, 634 (1968).
[6] *A. F. Reid, P. C. Wailes*, Inorg. Chem. *5*, 123 (1966).
[7] *M. L. Anderson, L. R. Crisler*, J. Organometal. Chem. *17*, 345 (1969).
[8] *E. O. Fischer, Y. Hristidu*, Z. Naturforsch. *17b*, 275 (1962).
[9] *F. Baumgärtner, E. O. Fischer, B. Kanellakopulos, P. Lauberau*, Angew. Chem. *81*, 182 (1969); *F. Baumgärtner, E. O. Fischer, B. Kanellakopulos, P. Lauberau*, Angew. Chem. Intern. Ed. Engl. *8*, 202 (1969).
[10] *G. L. Ter Haar, M. Dubeck*, Inorg. Chem. *3*, 1648 (1964).
[11] *E. O. Fischer, B. Kanellakopulos, P. Lauberau*, J. Organometal. Chem. *5*, 583 (1966).

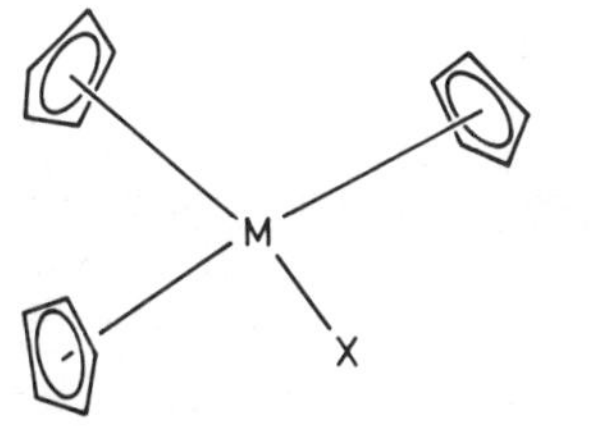

3

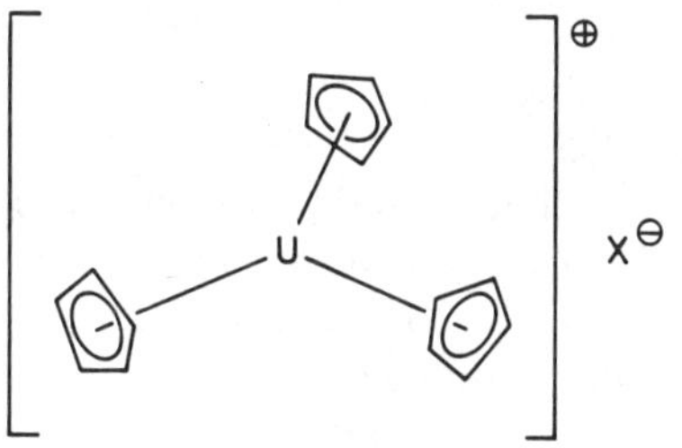

4

sodium naphthalide in tetrahydrofuran solution,[12] sodium amalgam in aromatic hydrocarbons,[13] or sodium sand in toluene.[14]

The dark-green crystalline, oxygen-sensitive solid, which decomposes without melting above 200°, has been found to be a dimer in solution but may exist as a monomer at low temperatures.

Its purple-black, insoluble and pyrophoric zirconium analog is formed similarly through the reduction of dichlori-di-π-cyclopentadienylzirconium(IV) by sodium naphthalenide in tetrahydrofuran.[15] A hetero-annularly-disubstituted metallocene, *bis(tert-butyl-π-cyclopentadienyl)zirconium(IV)*, can be obtained in 76% yield by treating 1 molar equivalent of anhydrous zirconium(IV) chloride with 2 molar equivalents of thallium *tert*-butylcyclopentadienide in benzene.[16]

31.2.2 Di-π-cyclopentadienylmetal Hydrides and Hydridoborates

The bridged hydride(**5**), *bis(di-π-cyclopentadienyl-μ-hydrido)titanium(III)*, is prepared by reduction of di-π-cyclopentadienyldimethyltitanium(IV) with hydrogen at 0°.[17]

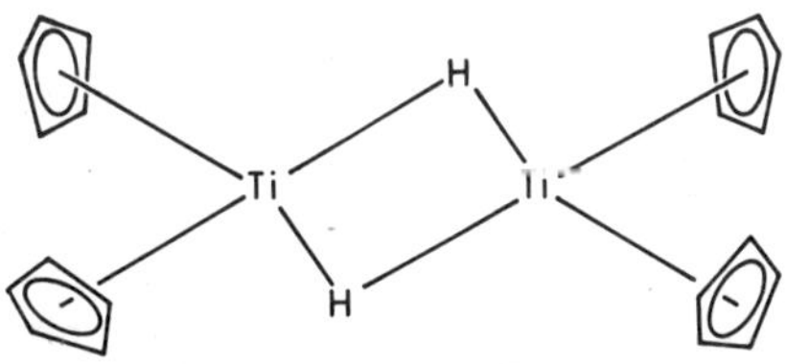

5

[12] *G. W. Watt, L. J. Baye, F. O. Drummond*, J. Amer. Chem. Soc. *88*, 1138 (1966).

[13] *K. Shikata*, Kogyo Kagaku Zasshi *68*, 1248 (1965).

[14] *J. J. Salzmann, P. Mosimann*, Helv. Chim. Acta *50*, 1831 (1967).

[15] *G. W. Watt, F. O. Drummond*, J. Amer. Chem. Soc. *88*, 5926 (1966).

[16] *A. N. Nesmeyanov, R. B. Materikova, E. M. Brainina, N. S. Kochetkova*, Izv. Akad. Nauk. SSSR, Ser. Khim. 1323 (1969); engl.: 1220.

[17] *J. E. Bercaw, H. H. Brintzinger*, J. Amer. Chem. Soc. *91*, 7301 (1969).

The *monohydrides* (**6**, X=Cl or CH_3) are obtained by lithium aluminum hydride reduction of dichlorodi-π-cyclopentadienylzirconium(IV) or chlorodi-π-cyclopentadienylmethylzirconium(IV) in tetrahydrofuran.[18]

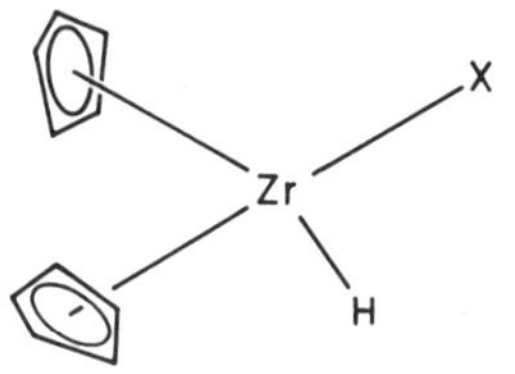

6

Although the simple dihydrido derivative (**6**, X=H), *di-π-cyclopentadienyldihydridozirconium(IV)*, is reportedly formed in the reduction of oxybis[chlorodi-π-cyclopentadienylzirconium(IV)] by lithium aluminum hydride in tetrahydrofuran, it may not be monomeric.[18] The *dihydride* (**7**), prepared in 78% yield by stirring a benzene solution of di-π-cyclopentadienylbis(tetrahydridoborato)zirconium(IV) with trimethylamine, exists as a polymer.[19]

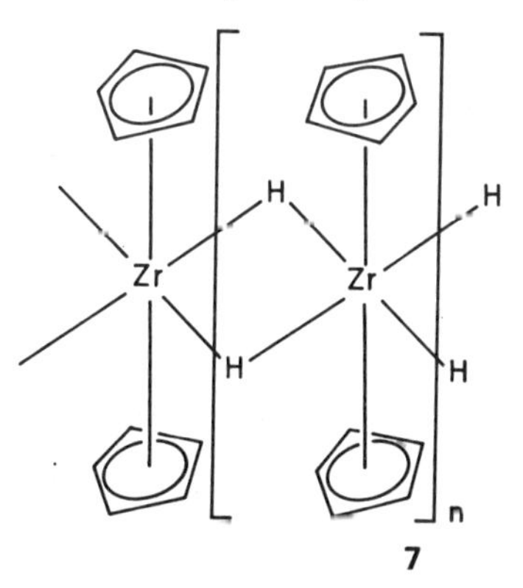

7

Both mono- and di-tetrahydridoborate derivatives are formed by the di-π-cyclopentadienylmetal compounds of the Group IVB elements. The extremely air-sensitive, black-violet *di-π-*

[18] *B. Kautzner, P. C. Wailes, H. Weigold*, Chem. Commun. 1105 (1969).

[19] *B. D. James, R. K. Nanda, M. G. H. Wallbridge*, Inorg. Chem. *6*, 1979 (1967).

cyclopentadienyl(tetrahydridoborato)titanium(III) is prepared by the reaction of equimolar quantities of dichloro-di-π-cyclopentadienyl-titanium(IV) with lithium tetrahydridoborate in diethyl ether.[20] When either dichlorodi-π-cyclopentadienylzirconium or -hafnium(IV) is treated with an excess of lithium tetrahydridoborate, the corresponding *bis(tetrahydridoborato) derivatives* (**8**, M=Zr or Hf), with bridged hydrogen–boron bonding to the transition metal are formed.[21,22]

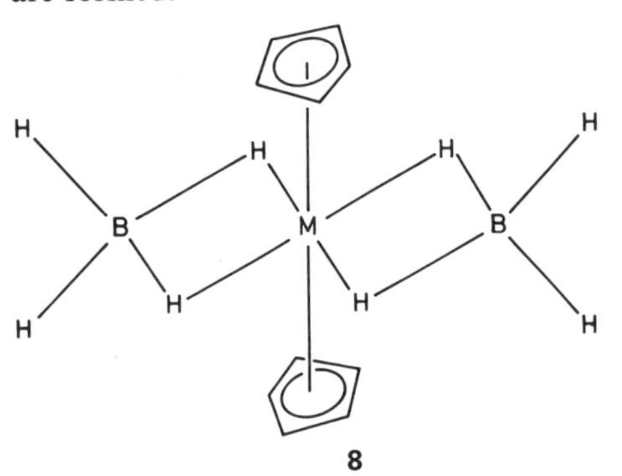

8

The rather oxidatively and hydrolytically stable *di-π-cyclopentadienylbis(tetrahydridoborato)zirconium(IV)* (**8**, M=Zr) may be used, in turn as a convenient starting material for the preparation of other hydride species. By treating **8** (M=Zr) with an equimolar quantity of trimethylamine in benzene at 0°, *di-π-cyclopentadienylhydrido (tetrahydridoborato) zirconium (IV)* is obtained, while use of 2 molar equivalents of the amine with **8** leads to the formation of the dihydride (**6**, X=H) as indicated above.[19] A unique hydridotriborate (**9**) is formed when dichlorodi-π-cyclopentadienyltitanium(IV) is reacted with cesium octahydridotriborate.[23]

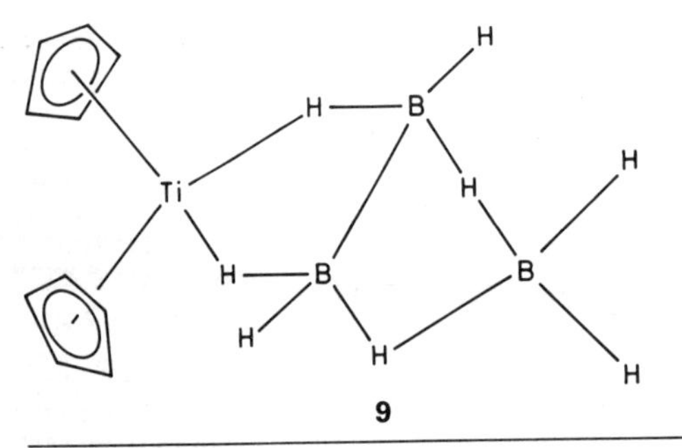

9

²⁰ *H. Nöth, R. Hartwimmer*, Chem. Ber. *93*, 2238 (1960).

²¹ *R. K. Nanda, M. G. H. Wallbridge*, Inorg. Chem. *3*, 1798 (1964).

²² *N. Davies, B. D. James, M. G. H. Wallbridge*, J. Chem. Soc. *A*, 2601 (1969).

²³ *F. Klanberg, E. L. Muetterties, L. J. Guggenberger*, Inorg. Chem. 7, 2272 (1968).

31.2.3 Halides

31.2.3.1 Di-π-cyclopentadienylmetal Monohalides

The green *chlorodi-π-cyclopentadienyltitanium(III)* (**10**, X=Cl) exists as a dimer and is prepared from the commercially available dichlorodi-π-cyclopentadienyltitanium(IV) by reduction with zinc in tetrahydrofuran[24] or by heating with chlorodimethylaluminum at 100°.[25] It can also be obtained directly from titanium(III) chloride by reaction with sodium cyclopentadienide in tetrahydrofuran.[26] The analogous *bromide*, *iodide*, and *tetrafluoroborate* (the latter with metal–fluorine bridged-bonding to the boron atom) (**10**, X=Cl, Br or BF_4) are formed in good yields by the reaction of di-π-cyclopentadienyl(tetrahydridoborato)titanium(III) with the corresponding hydrogen halides in diethyl ether or benzene, or with boron trihalides under a nitrogen atmosphere.[27] In a similar manner, *halodi-π-idenyltitanium(III)* compounds (**11**, X=Cl, Br or I), fused ring analogs of the di-π-cyclopentadienylmetal halides, are prepared by reduction of the corresponding dihalodi-π-indenyl derivatives with triethylaluminum in petroleum ether at −10°, followed by treatment with trimethylamine.[28]

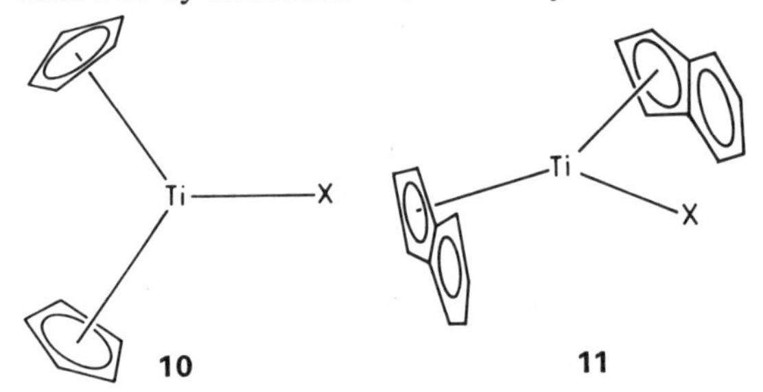

10 **11**

Mixed monohalide-alkoxy, -aryloxy and related oxygen-containing *derivatives* of *di-π-cyclopentadienyltitanium, zirconium*, and *hafnium* are generally prepared from the reaction of a dihalodi-π-cyclopentadienylmetal compound with an equimolar quantity or slight excess of a hydroxy derivative or its sodium salt. While *bromodi-π-cyclopentadienylhydroxytitanium (IV)*

²⁴ *J. Birmingham, A. Fischer, G. Wilkinson*, Naturwissenschaften *42*, 96 (1955).

²⁵ *K. Clauss, H. Bestian*, Justus Liebigs Ann. Chem. *654*, 8 (1962).

²⁶ *G. Natta, G. Dallasta, G. Mazzanti, U. Giannini, S. Cesca*, Angew. Chem. *71*, 205 (1959).

²⁷ *H. Nöth, R. Hartwimmer*, Chem. Ber. *93*, 2246 (1960).

²⁸ *W. Marconi, M. L. Santostasi, M. De Malde*, Chim. Ind. (Milan) *44*, 229 (1962); C.A. *57*, 4689 (1962).

is obtained by dissolving dibromodi-π-cyclopentadienyltitanium(IV) in water and saturating the solution with hydrogen bromide,[29] *chlorodi-π-cyclopentadienylethoyx-*, *isopropoxy-*, *allyloxy-*, and *phenoxy-titanium*[30] or *-zirconium(IV)*[31] are obtained by treating the corresponding dichloro compound with an excess of the appropriate alcohol in a benzene solution containing triethylamine as an acid acceptor. On stirring dibromo- or dichlorodi-π-cyclopentadienylhafnium(IV) with 8-hydroxyquinoline in benzene at 20° for 1 hr, the corresponding *monohalo-8-hydroxyquinolato derivatives* (**12**, X=Cl or Br) are formed in 66–68% yield.[32]

Chlorotrialkylsiloxy derivatives (**13**, R=CH_3 or C_6H_5; R′=CH_3, n-C_4H_9, or C_6H_5) result when dichlorodi-π-cyclopentadienyltitanium(IV) is reacted with the appropriate silanol in the presence of triethylamine[33,34] or with sodium silanolate in toluene,[33,35] according to the following equation:

$$(\pi\text{-}C_5H_5)_2TiCl_2 + R^1R_2SiOH(Na) \xrightarrow[-HCl(NaCl)]{} (\pi\text{-}C_5H_5)_2Ti(Cl)(OSiR_2R^1)$$

Binuclear tetravalent *dihalometalloxanes* of *titanium*, *zirconium*, and *hafnium* (**14**, M=Tl, Zr or Hf; X=Cl or Br) are prepared from the corresponding dihalodi-π-cyclopentadienylmetal compounds by either oxidation or hydrolysis.

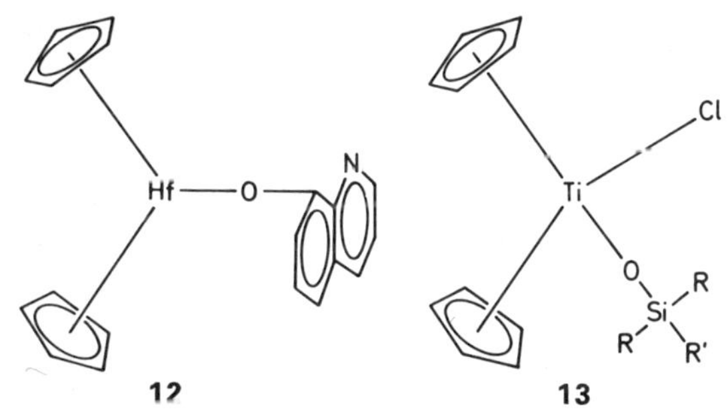

12 **13**

Thus, the dichlorotitanoxane (**14**, M=Ti; X=Cl) is formed in 60% yield when dichlorodi-π-cyclopentadienyltitanium(IV) is reduced in acetone with zinc dust and the reduction product is air-oxidized.[36] While the zirconium analog can be obtained similarly upon oxidation of an amine-reduction product of the dichloro precursor[37,38] the zirconoxane as well as the dichlorohafnoxane (**14**, M=Zr or Hf, X=Cl) are best prepared by treating the di-π-cyclopentadienylmetal dichloride with ethanol, triethylamine and water, presumably through hydrolysis of a monochloroethoxy intermediate.[39,40] A dibromozirconoxane (**14**, M=Zr; X=Br) results from hydrolysis of dibromodi-π-cyclopentadienylzirconium(IV).[41] When the dimeric chlorodi-π-cyclopentadienyltitanium(III) is oxidized by reaction with an alkyl or aryl disulfide in benzene solution, a chlorodi-π-cyclopentadienylalkyl- (or aryl) thiotitanium(IV) is formed,[42] while treatment with alkyl or aryl azides leads to *substituted dichlorotitanazanes* of structure **15** (R=C_2H_5, C_6H_5, p-CH_3-C_6H_4, p-ClC_6H_4 and p-$O_2NC_6H_4$).[43]

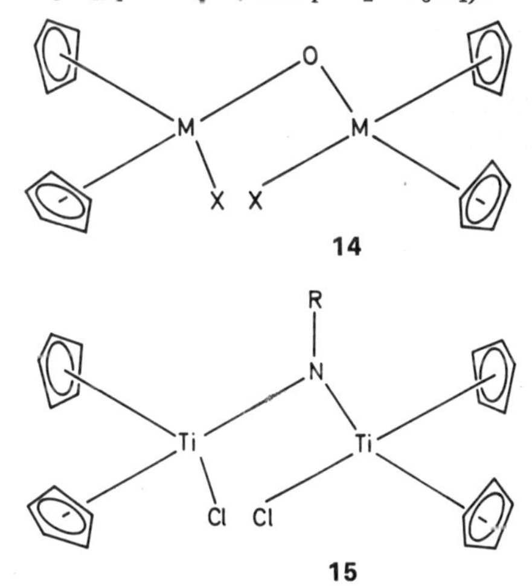

14

15

Monochlorodi-π-cyclopentadienyl-alkyl and *-ary derivatives of titanium and zirconium* are

[29] *G. Wilkinson, J. M. Birmingham*, J. Amer. Chem. Soc. *76*, 4281 (1954).

[30] *A. N. Nesmeyanov, O. V. Nogina, A. M. Berlin, A. S. Girchovich, G. V. Shatalov*, Izv. Akad. Nauk SSSR, Otdel. Khim. Nauk. 2146 (1961); engl.: 2008.

[31] *E. M. Brainina, R. Kh. Freidlina, A. N. Nesmeyanov*, Dokl. Akad. Nauk. SSSR *154*, 1113 (1964); engl.: 143.

[32] *E. M. Brainina, M. Kh. Minacheva, B. V. Lokshin*, Izv. Akad. Nauk. SSSR, Ser. Khim. 817 (1968); engl.: 784.

[33] *J. G. Noltes, G. J. M. van der Kerk*, Rec. Trav. Chim. Pays-Bas *81*, 39 (1962).

[34] *T. Takiguchi, H. Suzuki*, Bull. Chem. Soc., Japan *41*, 2810 (1968).

[35] *J. S. Skelcey*, Diss. Abstracts *22*, 4177 (1962).

[36] *S. A. Giddings*, Inorg. Chem. *3*, 61 (1964).

[37] *E. Samuel, R. Setton*, Compt. Rend. *254*, 308 (1962).

[38] *A. F. Reid et al.*, Australian J. Chem. *18*, 173 (1965).

[39] *E. Samuel, R. Setton*, Compt. Rend. *256*, 443 (1963).

[40] *M. Kh. Minacheva, E. M. Brainina, R. Kh. Freidlina*, Dokl. Akad. Nauk. SSSR *173*, 581 (1967); engl.: 282.

[41] *E. M. Brainina, G. G. Dvoryantseva, R. Kh. Freidlina*, Dokl. Akad. Nauk. SSSR *156*, 1375 (1964); engl.: 633.

[42] *R. S. P. Coutts, J. R. Surtees, J. M. Svan, P. C. Wailes*, Australian J. Chem. *19*, 1377 (1966).

[43] *R. S. P. Coutts, J. R. Surtees*, Australian J. Chem. *19* 387 (1966).

prepared by reaction of dichlorodi-π-cyclopentadienyltitanium or -zirconium(IV) with an approximately equimolar quantity of an alkyl (or aryl) Grignard or lithium reagent. Thus, while the methyl- and ethyl-substituted chlorotitanium and zirconium compounds are obtained from the corresponding Grignard reagents,[44,45,46] *n*-propyl- and pentafluorophenyl-substituted chlorotitanium derivatives are formed in reactions with the appropriate lithium compounds in diethyl ether.[47,48] In contrast, a single cyclopentadienyl-substituted monohalide, the yellow, extremely hygroscopic chlorotris(methyl-π-cyclopentadienyl)zirconium(IV) is obtained by treating sodium methylcyclopentadienide with zirconium(III) chloride in toluene at room temperature.[49]

By procedures analogous to that used for the alkyl and aryl derivatives, unique Group IVA metal-substituted *monochlorodi-π-cyclopentadienyl titanium* or *zirconium* compounds (**16**, M=Ti or Zr; M′=Si, Ge or Sn), with structures containing metal-to-metal bonds, may be prepared from the corresponding dichlorodi-π-cyclopentadienylmetal precursor. *Chlorodi-π-cyclopentadienyl)triphenylsilyl)zirconium(IV)* (**16** M=Zr; M′=S) is formed by the reaction of the dihalide with triphenylsilyllithium in tetrahydrofuran at −50°,[50] while triphenylgermyl- and triphenylstannyl-titanium analogs

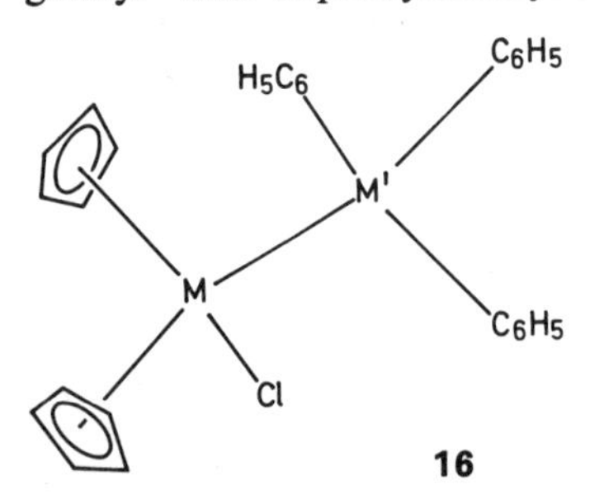

16

x = Cl, Br,
y = Cl, Br, C_6H_5

17

(**16**, M=Ti; M′=Ge or Sn) are obtained from triphenylgermyllithium and triphenylstannylsodium, respectively.[51]

Related *chloroborano compounds* (**17**, X=Cl or Br; Y=Cl, Br or C_6H_5) are prepared in 75–80% yields when di-π-cyclopentadienyltitanium(II) dimer is reacted with halodiphenylboranes or boron trihalides.[52]

31.2.3.2 Di-π-cyclopentadienylmetal Dihalides

As the primary starting materials used for the preparation of a large number of di-π-cyclopentadienylmetal derivatives of Group IVB elements, the di-π-cyclopentadienylmetal dihalides are compounds of ever increasing importance in the development of the chemistry and utility of di-π-cyclopentadienyl-titanium, -zirconium and, to a lesser extent, -hafnium complexes.

Di-π-cyclopentadienyldifluorotitanium(IV) is best obtained by a halogen-exchange reaction of chlorodi-π-cyclopentadienyltitanium(III) and sodium fluoride in aqueous medium at room temperature.[53] However, such a reaction fails when employed for the preparation of the corresponding difluorides of zirconium and hafnium.[54] The latter compounds, however, can be synthesized in good yield by the ligand exchange of boron trifluoride-diethyl etherate and di-π-cyclopentadienylbis(dimethylamido)-zirconium or -hafnium(IV), derived from the reaction of the corresponding dichloride and dimethylamine.[53]

The commercially-available, dichlorodi-π-cyclopentadienyltitanium(IV), a dark red, high-melting crystalline solid, has been of considerable interest as a catalyst for olefin polymerizations, in addition to serving as the precursor for a wide variety of π-cyclopentadienyltitanium

[44] *W. P. Long*, J. Amer. Chem. Soc. *81*, 5312 (1959).
[45] *W. P. Long, D. S. Breslow*, J. Amer. Chem. Soc. *82*, 1953 (1960).
[46] *H. Sinn, G. Opperman*, Angew. Chem. *78*, 986 (1966);
H. Sinn, G. Opperman, Angew. Chem. Intern. Ed. Engl. *5*, 962 (1966).
[47] Ger. Pat. 1 037 446 (1958), Farbwerke Hoechst; Erf.: *H. Bestian, K. Clauss*; C.A. *54*, 18546 (1960).
[48] *M. A. Chaudhari, P. M. Treichel, F. G. A. Stone*, J. Organometal. Chem. *2*, 206 (1964).
[49] *E. M. Brainina, E. J. Mortikova, L. A. Petrashkovich, R. Kh. Freidlina*, Dokl. Akad. Nauk SSSR *169*, 335 (1966); engl.: 681.
[50] *D. J. Cardin, S. A. Keppie, B. M. Kingston, M. F. Lappert*, Chem. Commun. 1035 (1967).
[51] *R. S. P. Coutts, P. C. Wailes*, Chem. Commun. 260 (1968).
[52] *G. Schmid et al.*, Angew. Chem. *79*, 683 (1967);
G. Schmid, Angew. Chem. Intern. Ed. Engl. *6*, 696 (1967).
[53] *P. M. Druce, B. M. Kingston, M. E. Lappert, T. R. Spalding, R. C. Srivastava*, J. Chem. Soc. *A*, 2106 (1969).
[54] *E. Samuel*, Bull. Soc. Chim. France 3548 (1966).

complexes. Although a number of methods are available for its synthesis in the laboratory, it is most readily prepared by the inverse addition of a suspension of sodium cyclopentadienide in 1,2-dimethoxyethane to anhydrous titanium-(IV) chloride in carefully-dried benzene.[29,55] An interesting alternate method of preparation involves the reaction of the metal tetrachloride with the relatively stable and storable thallium cyclopentadienide in benzene, although the yield (61%) of the dichloride is somewhat lower than that obtained by use of the sodium salt.[56]

The commercially-available *zirconium*, as well as *hafnium analogs*, are prepared, in a similar manner to the titanium dichloride, from sodium cyclopentadienide and the corresponding metal tetrachlorides.[55] By a slight modification, involving addition of tetrahydrofuran to dissolve the sodium cyclopentadienide-1,2-dimethoxyethane suspension, the *dichlorodi-π-cyclopentadienylhafnium(IV)* may be obtained in 81% yield.[52]

In the case of the other dihalides, a simple route to *dibromo-* and *diiododi-π-cyclopentadienyl-titanium, -zirconium* and *-hafnium* is available by a ligand exchange of the dichloride with boron(III) bromide or iodide in accordance with the following equation:[52]

$$3(\pi\text{-}C_5H_5)_2MCl_2 + 2BX_3 \longrightarrow 3(\pi\text{-}C_5H_5)_2MX_2 + 2BCl_3$$

(M = Ti, Zr or Hg; X = Br or I)

The reaction, which is run in dichloromethane at room temperature, is essentially quantitative after a few minutes. In addition, a mixed dihalo derivative, bromochloro-di-π-cyclopentadienyltitanium(IV), may be prepared by treating anhydrous titanium tetrachloride with a 4-fold excess of a 1:1 mixture of cyclopentadienylmagnesium bromide and cyclopentadienylmagnesium chloride in ether–benzene solution at −70°.[57]

Through utilization of one of several reactions a number of *monoalkyl* or *heteroannularly substituted dialkyl derivatives* of *di-π-cyclopentadienyl dichlorides* of *titanium* and *zirconium* can be synthesized. By treating π-cyclopentadienyltrichlorotitanium(IV) with an equimolar quantity of sodium or lithium alkylcyclopentadienide[58] in tetrahydrofuran, monoalkyl- and alkaryl-substituted di-π-cyclopentadienyltitanium dichlorides [**18**, M=Ti; R=CH_3, $(CH_3)_2CH$, $(C_6H_5)_2CH$, $n\text{-}C_4H_9(CH_3)_2C$, and $C_6H_5(CH_3)_2C$; R^1=H] are formed.

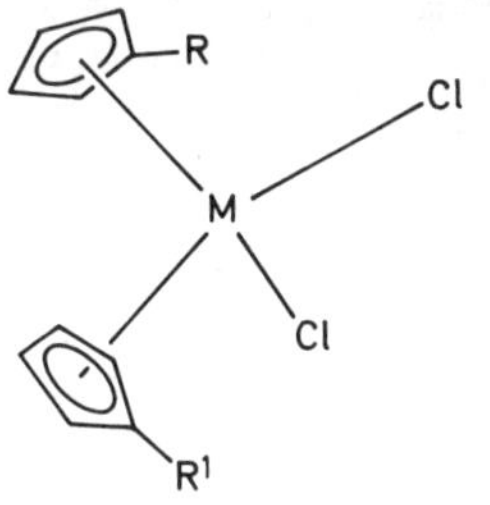

R Cl M Cl R^1

R^1 = H

R = CH_3, $CH(CH_3)_2$, $CH(C_6H_5)_2$, $-C(CH_3)_2-C_4H_9$, $-CH(CH_3)-C_6H_5$

18

The lithium isopropyl-, (1,1-diphenylmethyl-), (1,1-dimethylpentyl-), and (1,1-dimethylbenzyl)-cyclopentadienides are obtained by reduction of the appropriately substituted fulvene with lithium aluminum hydride or *n*-butyl-(or phenyl)-lithium in diethyl ether or diethyl ether/benzene. In a somewhat converse manner, the homoannularly substituted polyalkyl dichloride, *dichloro(π-cyclopentadienyl)(pentamethyl-π-cyclopentadienyltitanium(IV)* is formed in the reaction of sodium cyclopentadienide in tetrahydrofuran with pentamethyl-π-cyclopentadienyltrichlorotitanium(IV), derived by treating lithium pentamethylcyclopentadienide with titanium(IV) chloride.[59] A *mono-ethyl-substituted dihalide* (**18**, M=Ti; R=C_2H_5; R^1=H) may be prepared by the reaction of dichlorodi-π-cyclopentadienyltitanium(IV) with triethylamine.[60]

Similar to their monosubstituted analogs, *dialkyl-, dialkayl-* and *diallyldi-π-cyclopentadienyl-dichlorides* are formed by interaction of 2 molar equivalents of the appropriately-substituted sodium or lithium cyclopentadienide in tetrahydrofuran or 1,2-dimethoxyethane. Heteroannular 1,1′-dimethyl-[61] and -diallyl-[62] derivatives (**18**, M=Ti or Zr; R and R^1=CH_3

[55] *R. B. King*, Organometallic Syntheses, Vol. I, p. 75, Academic Press, New York 1965.

[56] *C. C. Hunt, J. R. Doyle*, Inorg. Nucl. Chem. Lett, *2*, 283 (1966).

[57] U.S. Pat. 2 983 741 (1961); Union Carbide Corp. Invent.: *J. C. Brantley;* C.A. *55*, 22339 (1961).

[58] *M. F. Sullivan, W. F. Little jr.*, J. Organometal. Chem. *8*, 277 (1967).

[59] *A. N. Nesmeyanov, E. J. Fedine, O. V. Nogina, N. S. Kochetkova, V. A. Dubovitsky, P. V. Petrovsky*, Tetrahedron, Supplement 8, Part II, 389 (1966).

[60] *A. N. Nesmeyanov, O. V. Nogina, T. P. Surikova*, Izv. Akad. Nauk. SSSR, Otd. Khim. Nauk, 1314 (1962); engl.: 1236.

[61] *L. T. Reynolds, G. Wilkinson*, J. Inorg. Nucl. Chem. *9*, 86 (1958).

[62] *H. W. Post, W. T. Schwartz*, U.S. Dept. Com., O.T.S. Report No. AD 255,545 (1961).

or $CH_2{=}CHCH_2$) are obtained from the sodium salt and the metal tetrachloride, while the corresponding 1,1′-bis(isopropyl)-, -(1,1-diphenylmethyl), -(1,1-dimethylpentyl), and -(1,1-dimethylbenzyl) compounds[58] [**18**, M=Zr; $R = R^1 = (CH_3)_3CH$, $(C_6H_5)_2CH$, $n\text{-}C_4H_9(CH_3)_2C$, and $C_6H_5(CH_3)_2C$] are obtained from the lithium salt. A disubstituted *tert*-butyl derivative [**18**, M=Zr; $R = R^1 = (CH_3)_3C$] is formed by treatment of thallium *tert*-butylcyclopentadienide [obtained from thallium(I) hydroxide and *tert*-butylcyclopentadiene] with anhydrous zirconium(IV) chloride in benzene.[16] Similarly, the heteroannularly tetrasubstituted dihalide, *dichlorobis(1-methyl-3-phenyl-π-cyclopentadienyl) titanium(IV)*, can be prepared by treating the lithium salt, derived from reacting 1-methyl-3-phenycyclopentadiene and *n*-butyl-lithium in diethyl ether/hexane, with titanium tetrachloride.[63]

Analogous to the preparation of the dialkyl-substituted dichlorides, the heteroannularly-substituted *dichlorobis(phenylazo-π-cyclopentadienyl)titanium(IV)* (**19**) and dichlorobis-[trisubstituted silyl (or germyl)]-π-cyclopentadienyltitanium(IV) derivatives (**20**, M=Si or Ge; $R = CH_3$ or C_6H_5; $R^1 = C_2H_5O$ or C_6H_5) are obtained by the reaction of two equivalents of lithium phenylazocyclopentadienide[64] or an appropriately substituted silyl (or germyl) cyclopentadienide[65,66] with titanium(IV) chloride in an ether solvent.

19

$R = CH_3, C_6H_5$, $R^1 = OC_2H_5, OC_6H_5$ **20**

Fused ring analogs of the di-π-cyclopentadienyl dihalides of titanium and zirconium are formed by procedures identical to those used for the synthesis of the cyclopentadienyl compounds. *Dichlorobis(π-indenyl)*-(or *π-fluorenyl*)-*titanium* and *-zirconium(IV)* (**21** and **22**, M=Zr; X=Cl) may be prepared by treatment of sodium indenide or fluorenide with the appropriate anhydrous metal tetrachloride in tetrahydrofuran.[54,67] The indenide derivatives (**21**, M=Ti and Zr; X=Cl) can be hydrogenated over platinium to the corresponding *dichlorobis-(4,5,6,7-tetrahydrido-π-idenyl)titanium* and *-zirconium(IV)*. Dichlorodi-π-indenyltitanium-(IV) (**21**, M=Ti; X=Cl) is also formed by the reaction of lithium indenide and titanium(IV) chloride in diethyl ether at 0°.[28] If the metal tetrachloride is replaced by titanium(IV) bromide or iodide, the corresponding dibromide and diiodide (**21**, M=Ti; X=Br or I) are obtained.[28] A mixed ligand dihalide, dichloro-π-cyclopentadienyl-π-indenyltitanium(IV), results when π-cyclopentadienyltrichlorotitanium(IV) is treated with sodium indenide.[68]

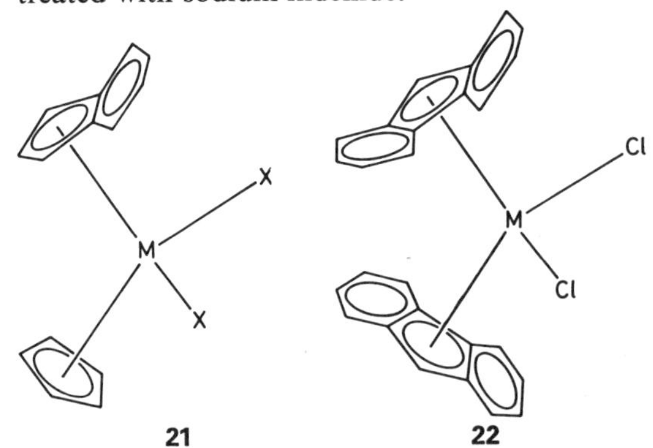

Dichlorodi-π-cyclopentadienyltitanium(IV) forms complexes when heated in aliphatic or aromatic hydrocarbons with aluminum halides[44,69] or alkylaluminum compounds[69-71] of the type RR′R″Al (R,R′ and $R'' = CH_3$, C_2H_5, $(CH_3)CHCH_2$ or Cl), some of which (complexes) are of importance as catalysts for the polymerization of ethylene, α-olefins or vinyl ethers. In addition to forming stable tetraammoniate and tetrakis(methylamine) complexes upon reaction with liquid ammonia or methylamine at temperatures below 25°,[72] the

63 *R. F. Heck*, J. Org. Chem. *30*, 2205 (1965).

64 *G. R. Knox, J. K. Munro, P. L. Pauson, G. H. Smith, W. E. Watts*, J. Chem. Soc. 4619 (1961).

65 *G. V. Drozdov, V. A. Bartashev, T. P. Maksimova, N. V. Kozlova*, Zh. Obshch. Khim. *37*, 2558 (1967); engl.: 2434.

66 *D. Seyferth, H. P. Hofmann, R. Burton, J. F. Helling*, Inorg. Chem. *1*, 227 (1962).

67 *E. Samuel, R. Setton*, J. Organometal. Chem. *4*, 156 (1965).

68 *E. Samuel*, J. Organometal. Chem. *19*, 265 (1969).

69 *G. Natta, P. Pino, G. Mazzanti, U. Giannini*, J. Inorg. Nucl. Chem. *8*, 612 (1958).

70 *A. H. Maki, E. W. Randall*, J, Amer. Chem. Soc. *82*, 4109 (1960).

71 *A. K. Zefirova, A. E. Shilov*, Dokl. Akad. Nauk. SSSR *136*, 599 (1961); engl.: 77.

72 *A. Anagnostopoulos, D. Nicholls*, J. Inorg. Nucl. Chem. *27*, 339 (1965).

dichloride when treated with zinc in toluene forms a green complex with bridged structure **23**,[73] similar to that proposed for those complexes derived from a number of the aluminum halides and alkyls.

23

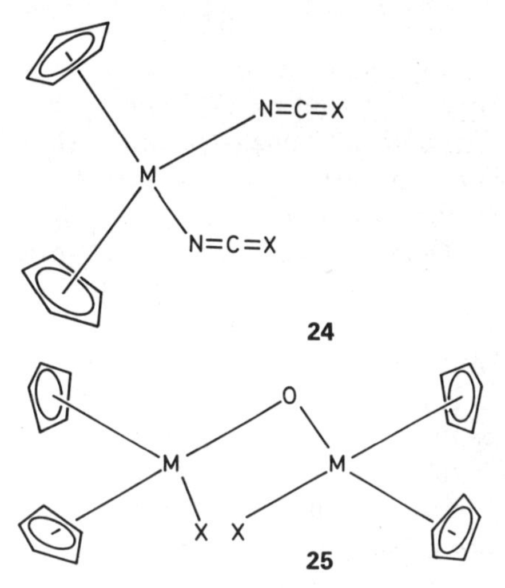

31.2.4 Di-π-cyclopentadienylmetal Pseudohalides

A single pseudohalide of trivalent di-π-cyclopentadienylmetals of Group IVB, the pale green, extremely air-sensitive *di-π-cyclopentadienylisocyanatotitanium(III)*, is formed by adding an air-free aqueous solution of sodium cyanate to chlorodi-π-cyclopentadienyltitanium(III) in air-free water.[74] Metal–oxygen-bonded *dicyanato derivatives of di-π-cyclopentadienyltitanium, -zirconium* and *-hafnium*, [$(\pi\text{-}C_5H_5)_2M(OCN)_2$, where M=Ti, Zr or Hf], are prepared from the corresponding dichlorides by reaction with potassium or silver cyanate.[75,76] In a like manner, *diisothiocyanato complexes* (**24**, M=Ti, Zr or Hf; X=NCS) are obtained from the appropriate dichlorodi-π-cyclopentadienylmetal(IV) derivative by treatment with potassium thiocyanate in acetone.[71,72,73] An analogous *isoselenocyanate complex* (**24**, M=Ti; X=NCSe) is formed in 69% yield upon reaction of the dichloride with potassium selenocyanate in acetone.[71] *Metalloxane pseudohalides* (**25**, M=Ti or Zr; X=NCS or OCN) are prepared by treating di-π-cyclopentadienyldiisothiocyanatotitanium(IV) (incorrectly believed to be the dithiocyanato complex) with *n*-butylamine in acetone.[77] or by heating dichlorodi-π-cyclopentadienylzirconium(IV) with silver cyanate in benzene.[78]

Although not a pseudohalide, orange crystalline *di-π-cyclopentadienylbis(perchlorato)titanium(IV)* which explodes on heating, is formed, similar to the pseudohalides, from the dichloro compound when it is reacted with silver perchlorate in tetrahydrofuran.[79]

31.2.5 Oxygen-containing Derivatives of Di-π-cyclopentadienyl-titanium and -zirconium

In addition to forming the orange crystalline hydroxide, *bromodi-π-cyclopentadienylhydroxytitanium(IV)* (**26**, M=Ti; R=Br), whose preparation is described in Section 31.2.3.1, the di-π-cyclopentadienylhydroxytitanium(III) cation [resulting from hydrolysis of dibromodi-π-cyclopentadienyltitanium(IV)] reacts with a saturated solution of potassium picrate to form a hydroxypicrate [**26**, M=Ti; $R=(O_2N)_3C_6H_2O$].[29] Pentafluorophenyl hydroxides (**26**, M=Ti or Zr; $R=C_6F_5$) of di-π-cyclopentadienyltitanium and -zirconium are obtained upon base hydrolysis of chlorodi-π-cyclopentadienyl(pentafluorophenyl)titanium(IV)[48] or aqueous hydrolysis of di-π-cyclopentadienylbis(pentafluorophenyl)zirconium(IV).[80]

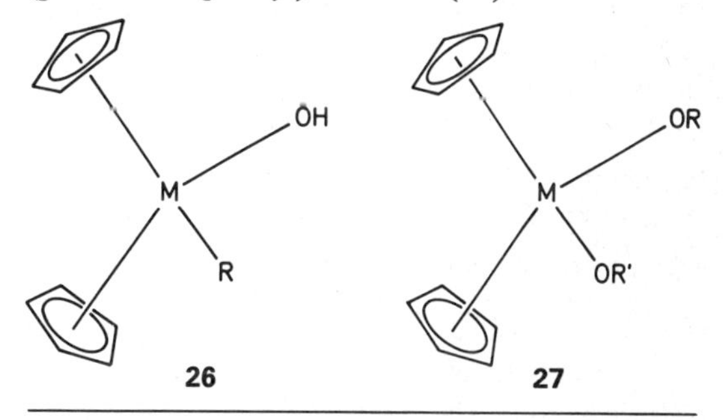

[73] *J. J. Salzmann*, Helv. Chem. Acta *51*, 526 (1968).
[74] *R. Coutts, P. C. Wailes*, Inorg. Nucl. Chem. Lett. *3*, 1 (1967).
[75] *J. L. Burmeister, E. A. Deardorff, C. E. Van Dyke*, Inorg. Chem. *8*, 170 (1969).
[76] *J. L. Burmeister, E. A. Deardorff, A. Jensen, V. H. Christiansen*, Inorg. Chem. *9*, 58 (1970).
[77] *S. A. Giddings*, Inorg. Chem. *6*, 849 (1967).
[78] *R. S. P. Coutts, P. C. Wailes*, Australian J. Chem. *19*, 2069 (1966).
[79] *G. Doyle, R. S. Tobias*, Inorg. Chem. *6*, 1111 (1967).
[80] *M. A. Choudhari, F. G. A. Stone*, J. Chem. Soc. *A*, 838 (1966).

Similar to its hydroxy analog, *di-π-cyclopentadienylethoxy(pentafluorophenyl)titanium-(IV)* is formed by refluxing chlorodi-π-cyclopentadienyl(pentafluorophenyl)titanium(IV) with sodium ethoxide under nitrogen.[48] A mixed *alkoxymethacrylate* [**27**, M=Ti; R=C_2H_5; R′=$COC(CH_3){=}CH_2$] can be obtained in quantitative yield by heating chlorodi-π-cyclopentadienylethoxytitanium(IV) with potassium methacrylate at 60–70° in chloroform.[30] A *dialkoxide* (**27**, M=Ti; R=R′=CF_3CH_2) results when dichlorodi-π-cyclopentadienyltitanium(IV) is treated with 2 equivalents of sodium trifluoroethanolate in absolute toluene.[81] *D-iπ-cyclopentadienyldiphenoxytitanium(IV)* (**27**, M=Ti; R=R′=C_6H_5) or substituted bis-(phenoxides) [**27**, M=Ti; R and R′=*o*-(or *p*-)*Cl*-C_6H_4), *o*-(or *p*-)$CH_3C_6H_4$) *o*-$O_2NC_6H_4$, 3,5-$(CH_3)_2C_6H_3$, and *p*-$C_6H_5C_6H_4$] are formed in 65–95% yield when dichlorodi-π-cyclopentadienyltitanium(IV) is treated with phenol[82] or a suitably-substituted phenol[83] in the presence of sodium amide. The related diphenoxy (**27**, M=Zr; R and R′=C_6H_5) and substituted-*diphenoxyzirconium compounds* can be prepared similarly by reacting dibromodi-π-cyclopentadienylzirconium(IV) with the desired phenol in the presence of an amine.[84, 85] A monopicrate, *di-π-cyclopentadienyl(2,4,6-trinitrophenoxy)titanium(III)* is obtained when an aqueous solution of dibromodi-π-cyclopentadienyltitanium(IV) is converted to the perchlorate, which is passed through a Jones reductor and its eluate run into a 2,4,6-trinitrophenol (picric acid) solution under nitrogen.[29] The corresponding dipicrate [27, M=Ti; R=R′=2,4,6-$(O_2N)_3C_6H_3$] can be prepared simply by adding a saturated solution of 2,4,6-trinitrophenol to a saturated aqueous solution of the dibromide and recrystallizing the product from dilute 2,4,6-trinitrophenol.[29] *Bridged phenoxy compounds*, **28** and **29**, are formed when dichlorodi-π-cyclopentadienyltitanium(IV) is treated with 1,2-dihydroxybenzene (pyrocatechol) or 2,2′-dihydroxybiphenyl in the presence of sodium amide.[83]

28 **29**

A *disiloxy derivative* [**27**, M=Ti; R and R′=$Si(C_6H_5)_3$] is best prepared by reaction of dichlorodi-π-cyclopentadienyltitanium(IV) with 2 equivalents of sodium triphenylsilanolate in refluxing toluene.[33, 86] The bis(triphenylsiloxy) compound, as well as its methyldiphenyl- and dimethylphenyl-siloxy analogs [**20**, M=Ti; R and R′=$Si(C_6H_5)_2CH_3$ or $Si(CH_3)_2C_6H_5$] can also be obtained conveniently by treatment of the dichloro compound with the appropriate silanol in the presence of triethylamine.[34]

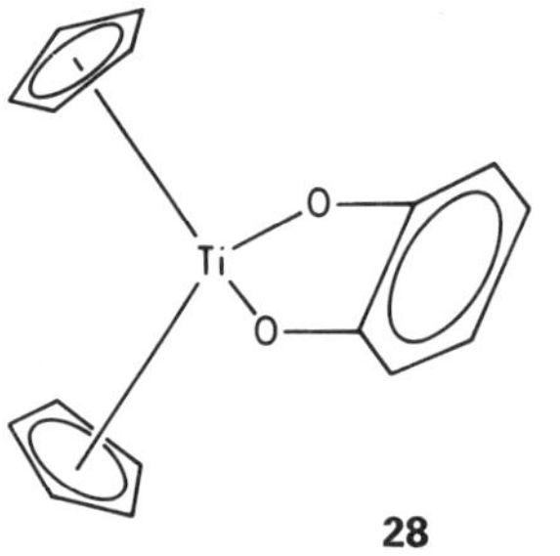

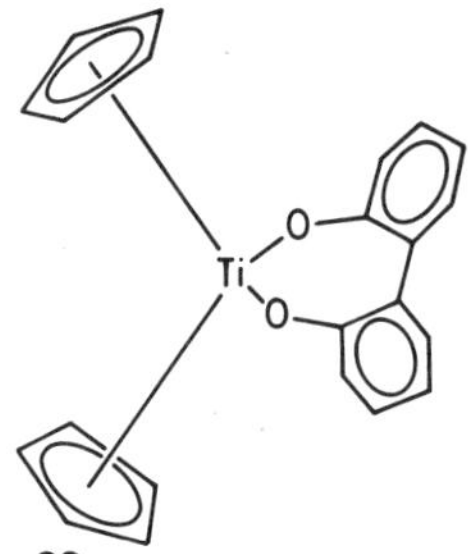

When acetylacetone is added to dimeric chlorodi-π-cyclopentadienyltitanium(III) in the form of its cation, derived from the reduction of dichlorodi-π-cyclopentadienyltitanium(IV) with zinc dust in air-free water, the monomeric and paramagnetic *acetylacetonato complex* (**30**) is precipitated.[87]

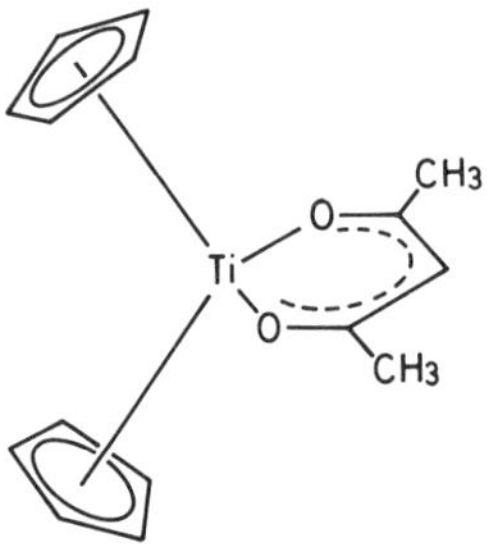

30

If sulfate or carbonate ion is added, instead of the diketone, to the aqueous solution containing the blue di-π-cyclopentadienyltianium-(III) cation, complexes with titanium-oxygen bridged structures, such as that shown for the pale green sulfate complex (**31**), are formed.[88]

In a similar manner, by treatment of the dimeric chlorodi-π-cyclopentadienyltitanium(III) with the sodium salts of organic acids, such as formic, acetic, decanoic, octadecanoic and benzoic, complexes of type **32** (R=H, CH_3, *n*-C_9H_{19},

---

[81] *G. V. Drozdov, A. L. Klebanskii, V. A. Bartashev*, Zh. Obshch. Khim. *33*, 2422 (1963); engl.: 2362.
[82] *K. Andrä*, Z. Chem. *7*, 318 (1967).
[83] *K. Andrä*, J. Organometal. Chem. *11*, 567 (1968).
[84] *K. Andrä, E. Hille*, Z. Chem. *65* (1968).
[85] *K. Andrä, E. Hille*, Z. Naturforsch. *246*, 169 (1969).
[86] U.S. Pat. 3 030 394 (1962), Amer. Cynamide Co., Invent.: *S. A. Giddings;* C.A. *57*, 3589 (1962).
[87] *R. B. Coutts, P. C. Wailes*, Australian J. Chem. *22*, 1549 (1969).
[88] *R. B. Coutts, P. C. Wailes*, Australian J. Chem. *21*, 1181 (1968).

31

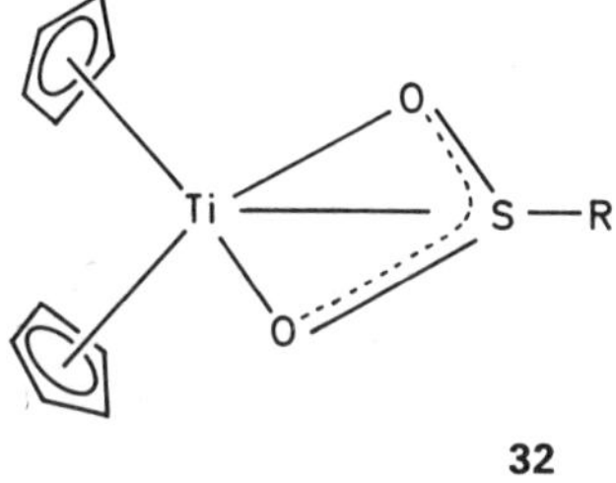

32

n-$C_{17}H_{35}$, and C_6H_5) **32** can be obtained.[89] A number of *diacyloxy* or *dicarboxylato derivatives* of *di-π-cyclopentadienyl-titanium* or *-zirconium(IV)* (**27**, M=Ti or Zr; R and R′= alkyl-CO or aryl-CO) and their π-indenyl analogs are prepared from metathetical reactions of the corresponding dichlorides and metal salt of a carboxylic acid in an organic solvent. Reactants for the preferred synthesis and properties of various dicarboxylato compounds of the latter type are summarized in Table 1.

Table 1 Bis-[π-cyclopentadienyl]-dicarboxylato-titanium(IV) and -zirconium(IV) from Bis-[π-cyclopentadienyl]-dichlorotitanium(IV) and -zirconium(IV) and Salts of Carboxylic Acids

R^1–COOM′		Reaction solvent	Complex-metal	Yield [%]	Lit. ref. no.
R′	M′				
CH_3	K	C_6H_6	Ti	93	30
$C_{17}H_{35}$	Na	THF	Ti		90
$ClCH_2$	Na	THF	Ti		90
Cl_2CH	Na	THF	Ti		90
Cl_3C	Na	THF	Ti	80	90
F_3C	Ag	C_6H_6	Ti	100	91
	Ag	CH_2Cl_2	Zr	70	92
F_3C–CF_2	Ag	CH_2Cl_2	Ti	82	92
F_3C–CF_2–CF_2	Ag	CH_2Cl_2	Ti	00	92
			Zr	82	92
$HOCH_2$	Na	CH_3COOH	Ti		90
$HSCH_2$	Na	CH_3COOH	Ti		90
$H_2C{=}C(CH_3)$	K	$CHCl_3$	Ti	90	30
C_6H_5	Ag	C_6H_6	Ti	40	93

31.2.6 Sulfur- and Selenium-containing Species

Dichlorodi-π-cyclopentadienyltitanium(IV) reacts with a 1:1 aqueous ethanol solution of ammonium(V)-pentasulfide in acetone,[94] or with an aqueous solution of sodium(V)-pentaselenide in acetone[95] to form the dark red-to-violet *cyclic pentachalcogenides* (**33**, X=S or Se) in high yields.

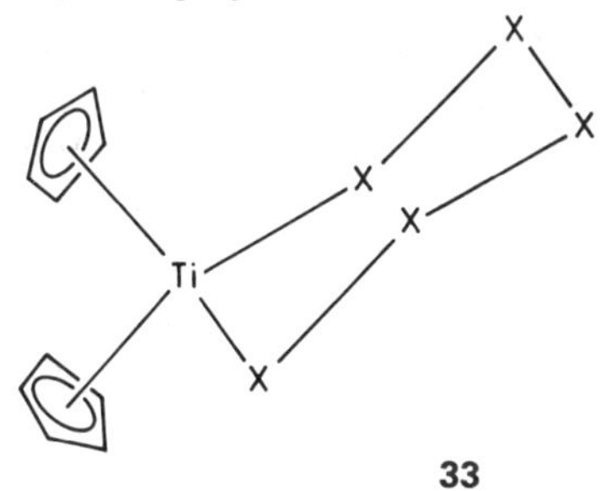

33

Although the preparation of the corresponding cyclic tri- and tetra-sulfido derivatives from the dichloride and disodium tri- or tetrasulfide is reported,[96] when the syntheses were repeated, only the penta-sulfide (**33**, X=S) could be isolated.[94]

Dimercapto and *bis[alkyl-(or aryl)thio] derivatives* of *di-π-cyclopentadienyl-titanium* or *-zirconium(IV)* (**34**, M=Ti or Zr; R=H, C_2H_5, n-C_3H_7, $C_6H_5CH_2CH_2$, $C_6H_5CH_2$, C_6H_5, p-ClC_6H_4, and p-$CH_3C_6H_4$) are most conveniently prepared in yields of 72–100% by treating the appropriate dichlorodi-π-cyclopentadienylmetal compound with two equivalents of either hydrogen sulfide,[97] an alkylthiol[98] or a thiophenol[98, 99] in either diethyl ether or benzene and in the presence of triethylamine as a proton acceptor. The deep red, air-stable *di-π-cyclopentadienylbis(methylthio)titanium(IV)*[77] (**34**, M=Ti, X=S; R=CH_3), as well as the bis-(ethyl-, benzyl-, and phenythio) compounds,[42]

[89] *R. B. Coutts, P. C. Wailes*, Australian J. Chem. *20*, 1579 (1967).

[90] Brit. Pat. 858 930 (1961); Nat. Lead Co.; C.A. *55*, 16565 (1961).

[91] *G. V. Drozdov, A. L. Klebanskii, V. A. Bartashov*, Zh. Obshch. Khim. *32*, 2390 (1962); engl.: 2359.

[92] *R. B. King, R. N. Kapoor*, J. Organometal. Chem. *15*, 457 (1968).

[93] *G. A. Razuvaev, V. N. Latyaeva, L. I. Vyshinskaya*, Dokl. Akad. Nauk. SSSR *138*, 1126 (1961); engl.: 592.

[94] *H. Köpf, B. Block*, Chem. Ber. *102*, 1540 (1969).

[95] *H. Köpf, B. Block, M. Schmidt*, Chem. Ber. *101*, 272 (1968).

[96] *R. Ralea, C. Ungurenasu, S. Cihodara*, Rev. Roumaine-Chim. *12*, 861 (1966).

[97] *H. Köpf, M. Schmidt*, Angew. Chem. *77*, 965 (1965); *H. Köpf, M. Schmidt*, Angew. Chem. Intern. Ed. Engl. *4*, 953 (1965).

[98] *H. Köpf, M. Schmidt*, Z. Anorg. Allgem. Chemie *340*, 139 (1965).

[99] *H. Köpf*, J. Organometal. Chem. *14*, 353 (1968).

can be obtained by an alternate procedure involving reaction of the dichloride with sodium alkyl- (or aryl)thiolates in toluene or benzene.

34

Several related *disilenyl derivatives* (**34**, M=Ti or Zr; X=Se; R=C_2H_5 or C_6H_5) are prepared, similar to their thio analogs, by the reaction of dichlorodi-π-cyclopentadienyl-titanium or -zirconium(IV) with ethane-[100] or benzeneselenol.[99] By analogous reactions, *cis-1,2-enedithiolates*, or 5-membered ring *S*-coordinated chelate complexes of di-π-cyclopentadienyl-titanium and -zirconium(IV) (**35** and **36**) are formed. Thus, when the appropriate dichloride is reacted with disodium 1,2-ethylenedithiolate,[101,102] disodium 1,2-maleonitriledithiolate,[103,104] or disodium 1-phenyl-1,2-ethylenedithiolate,[102] in tetrahydrofuran or chloroform, *cis*-1,2-ethylenedithiolates of type **35** (M=Ti or Zr; R and R'=H, CN or C_6H_5) can be prepared. If the dichlorides are treated with a benzene solution of benzene-1,2-dithiol or toluene-3,4-dithiol[99,103] in the presence of triethylamine, *1,2-phenylenedithiolates* of structure **36** (M=Ti or Zr; R=H or CH_3) can be obtained. In contrast to the corresponding titanium compounds, the zirconium complexes of the latter type are extremely sensitive to air and moisture.

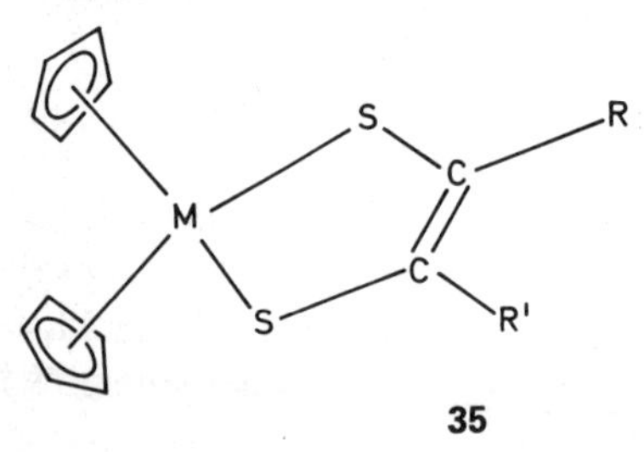

35

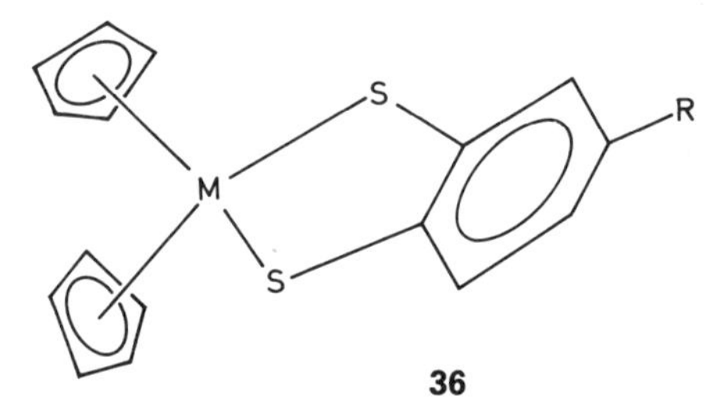

36

31.2.7 Nitrogen- and Phosphorus-containing Derivatives

Cyclopentadiene reacts with tris(dimethylamino)titanium(III) to form the brown 4-membered *alkylamido-bridged titanazane* complex **37** (M=Ti; X=N; R=CH_3).

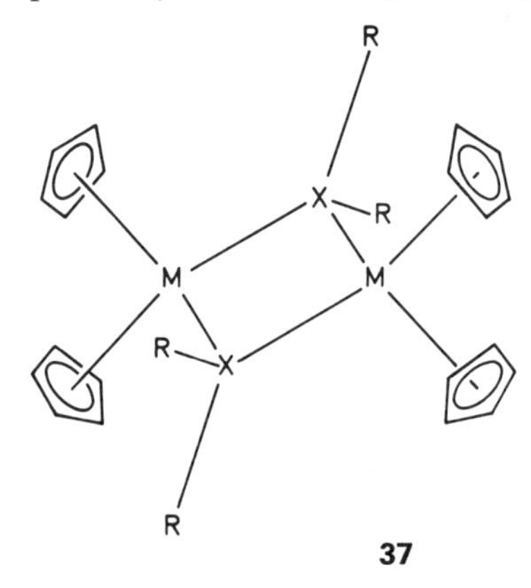

37

While *bis(dimethylamido)di-π-cyclopentadienyl-titanium(IV)* is obtained only in low yield by treatment dichlorodi-π-cyclopentadienyltitanium(IV) with lithium diethylamide,[105] highly colored bis(diphenylamido), -pyrollato, -indolato and -carbazolato derivatives of di-π-cyclopentadienyl-titanium or -zirconium are conveniently prepared from the dihalide (dichloride for titanium and dibromide for zirconium complexes) and the potassium or lithium salt of the diaryl or heterocyclic amine in tetrahydrofuran,[106] according to the following equation:

$$(\pi\text{-}C_5H_5)_2MX_2 + 2M'NR_2 \longrightarrow (\pi\text{-}C_5H_5)_2M(NR_2)_2 + 2M'X$$

[M = Ti or Zr; X = Cl or Br; $NR_2 = N(C_6H_5)_2$, NC_4H_4, NC_8H_6 or $NC_{12}H_8$]

However, although the zirconium dihalide reacts to form the bis(diphenylamido) complex in

[100] *H. Köpf, B. Block, M. Schmidt*, Z. Naturforsch *22b*, 1077 (1967).
[101] *R. B. King, C. A. Eggers*, Inorg. Chem. *7*, 340 (1968).
[102] *H. Köpf*, Z. Naturforsch. *23b*, 1531 (1968).
[103] *H. Köpf, M. Schmidt*, J. Organometal. Chem. *4*, 426 (1965).
[104] *J. Locke, J. A. McCleverty*, Inorg. Chem. *5*, 1157 (1966);
E. C. Alyea, D. C. Bradley, M. F. Rappert, A. R. Sauger, Chem. Commun. 1064 (1969).
[105] *G. Chandra, M. F. Lappert*, Inorg. Nucl. Chem. Lett. *1*, 83 (1965).
[106] *K. Isslieb, G. Bätz*, Z. Anorg. Allgem. Chemie *369*, 83 (1969).

56% yield, reaction of dichloro-π-cyclopentadienyltitanium(IV) with potassium diphenylamide leads to reduction of the latter to the dimeric *chlorodi-π-cyclopentadienyltitanium-(III)*.[106] Substitution of potassium phthalimide for the metal amides in the reaction with the dihalides results in the formation of imido complexes (**38**, M=Ti or Zr).[106] *Di-π-cyclopentadienyl-dinitrato-zirconium* or *-hafnium(IV)* can be obtained by stirring the appropriate tetracyclopentadienylmetal complex with 56% nitric acid in 1,2-dichloroethane at −35° and allowing the reaction mixture to come to room temperature over a 2-hr period.[40,107]

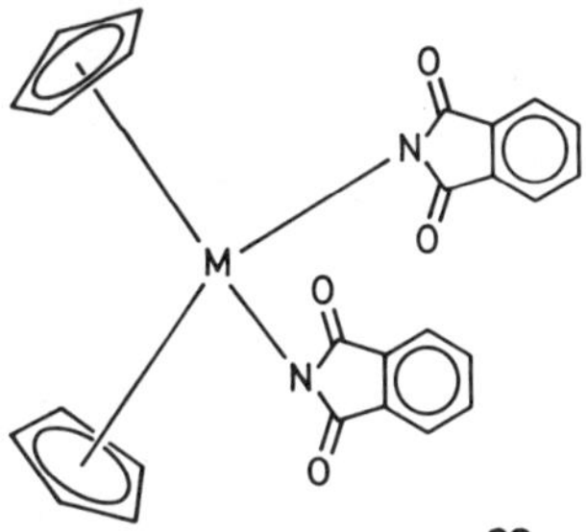

38

Dichlorodi-π-cyclopentadienyl-titanium and -zirconium(IV), when treated with lithium dialkylphosphides, form dark violet or brown, moisture-sensitive dimeric 4-membered ring complexes (**37**, M=Ti or Zr; X=P; R=C_2H_5 or n-C_4H_9), similar in structure to the titanazanes.[108] However, the reaction fails to yield the cycloaliphatic- or aryl-substituted analogs when carried out with the corresponding dicyclohexyl- or diphenyl-metal phosphides.

An interesting phosphorus-containing metal complex is the spirohetorocyclic derivative (**39**) which is obtained when dichlorodi-π-cyclopentadienylzirconium(IV) in tetrahydrofuran is reacted with tetrakis[(sodium phenylphosphino)-methyl]methane derived from the reaction of tetrakis [(diphenylphosphino) methyl] methane and metallic sodium in liquid ammonia at −33°.[109]

39

31.2.8 Di-π-cyclopentadienylmetal Alkyl, Alkynyl, and Aryl Derivatives of Titanium and Zirconium

Dichlorido-π-cyclopentadienyl- and dichlorobis-(methyl-π-cyclopentadienyl)titanium(IV) react with methyllithium in an ether solvent to readily yield the stable dimethyl compounds (**40**, M=Ti; R=H or CH_3; R′=CH_3).[25,47] When the former dichloride is treated with benzylmagnesium chloride in diethyl ether at −10°, a 65% yield of the dibenzyl derivative (**40**, M=Ti; R=H; R′=$C_6H_5CH_2$) is obtained.[110] In a similar manner, reaction of the dichloride or its zirconium analog with sodium 1-phenylethyne in benzene-diethyl ether, gives the corresponding *bis(phenylethynyl)* com-

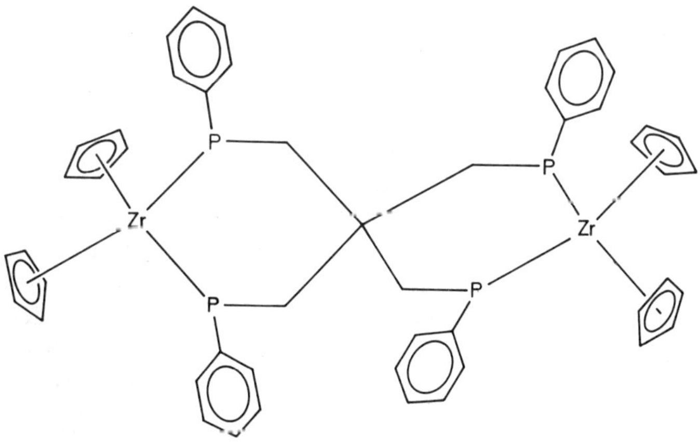

[107] *E. M. Brainina, M. Kh. Minacheva, R. Kh. Freidlina*, Izv. Akad. Nauk SSSR, Ser. Khim. 1877 (1965); engl.: 1839.

[108] *K. Isslieb, H. Hackert*, Z. Naturforsch. *21b*, 519 (1966).

[109] *J. Ellerman, F. Poersch*, Angew. Chem. *79*, 323 (1967);
J. Ellerman, F. Poersch, Angew. Chem. Intern. Ed. *6*, 355 (1967).

[110] *G. A. Razuvaev, V. N. Latyaeva, L. I. Vyshinskaya*, Dokl. Akad. Nauk. SSSR *189*, 103 (1969); engl.: 884.

pounds (**40**, M=Ti or Zr; R=H; R′=C_6H_5-C≡C) in good yield.[111,112]

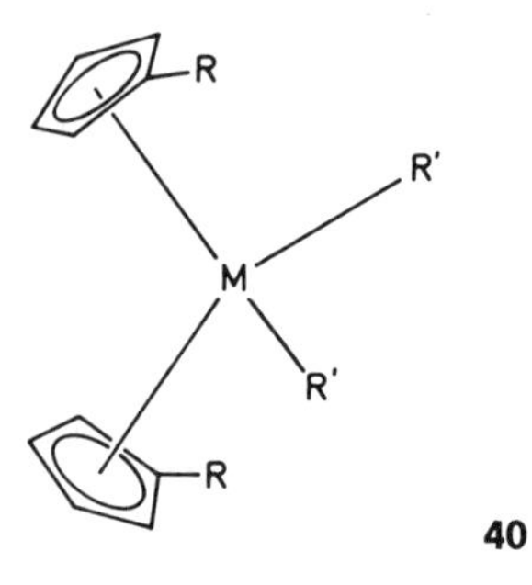

40

In addition to the monosubstituted di-π-cyclopentadienylphenyl- and pentafluorophenyl-titanium(III) formed by the reaction of the dimeric chlorodi-π-cyclopentadienyltitanium(III) and an appropriate aryl organometallic,[112] a number of stable *diaryl derivatives* of *di-π-cyclo pentadienyl-titanium* and *-zirconium* can be prepared from the dichlorides and an excess of an aryllithium under nitrogen in diethyl ether in accordance with the following equation:[48,80,113,114,115]

$$(\pi\text{-}C_5H_5)_2MCl_2 + 2RLi \longrightarrow (\pi\text{-}C_5H_5)_2MR_2 + 2LiCl$$

M = Ti, Zr
R = C_6H_5, 3-and 4-CH_3–C_6H_4, C_6F_5, 4-$(CH_3)_2N$–C_6H_4

A unique *heterocyclic derivative* (**41**) is formed when dichlorodi-π-cyclopentadienyltitanium(IV) is allowed to interact with 2,2′-dilithiooctafluorobiphenyl.[116]

41

[111] *H. Köpf, M. Schmidt*, J. Organometal. Chem. *10*, 383 (1967).

[112] *H. J. de Liefde Meyer, F. Jellinek*, Inorg. Chim. Acta *4*, 651 (1970).

[113] *L. Summers, R. Uloth*, J. Amer. Chem. Soc. *76*, 2278 (1954).

[114] *L. Summers, R. Uloth, A. Holmes*, J. Amer. Chem. Soc. *77*, 3604 (1955).

[115] *H. C. Beachall, S. A. Butler*, Inorg. Chem. *4*, 1133 (1965).

[116] *S. C. Cohen, A. G. Massey*, J. Organometal. Chem. *10*, 471 (1967).

Reaction of the corresponding zirconium dichloride with phenyl- or *p*-totyl-lithium in diethyl ether under argon, followed by hydrolysis in an aqueous dichloromethane solution, leads to aryl-substituted zirconoxanes of type **42** (R=C_6H_5 or *p*-$CH_3C_6H_4$) in yields of 44–53%.[41]

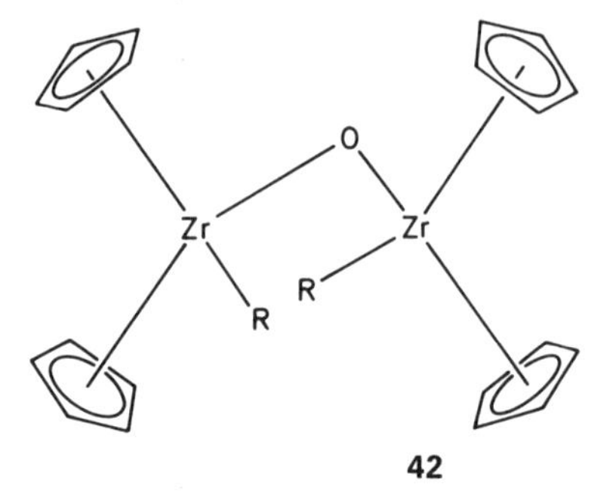

42

31.2.9 Olefin, Acetylene, and Bimetallic Complexes

The deep-green, air-sensitive and paramagnetic, symmetrical *tricyclopentadienyltitanium(III)* (**43** M=,Ti) formed by the reaction of dichlorodi-π-cyclopentadienyltitanium(IV) with a large excess of sodium cyclopentadienide in which the latter acts as a reducing agent,[117] is most likely composed of a single normally σ-bonded and two centrally σ- or π- bonded cyclopentadienyl rings.[118] The compound can also be prepared in somewhat higher (70%) yield by treating the dimeric chlorodi-π-cyclopentadienyltitanium(III) with an equimolar amount of sodium cyclopentadienide in tetrahydrofuran.[119] In the case of the *tetracyclopentadienyl derivatives of titanium, zirconium* and *hafnium*, three different types of structure are involved. The violet-black, thermally-stable, tetracyclopentadienyltitanium complex, obtained by heating 1 molar equivalent of dichlorodi-π-cyclopentadienyltitanium(IV) with 2 molar equivalents of sodium cyclopentadienide in tetrahydrofuran at 0°,[118] has been shown to consist of two cyclopentadienyl rings π-bonded and two σ-bonded to the metal atom.[120] In contrast, the structure of the corresponding moisture-sensitive zirconium derivative (**44**), prepared in 54% yield by the addition of a benzene solution of

[117] *E. O. Fischer, A. Löchner*, Z. Naturforsch. *15b*, 266 (1960).

[118] *F. W. Siegert, H. J. de Liefde Meyer*, J. Organometal. Chem. *20*, 141 (1969).

[119] *A. J. Canty, R. S. P. Coutts, P. C. Wailes*, Australian J. Chem. *21*, 807 (1968).

[120] *J. L. Caulderon, E. M. Brainina, N. G. Bokii, Yu. T. Struchkov*, J. Amer. Chem. Soc. *92*, 3801 (1970).

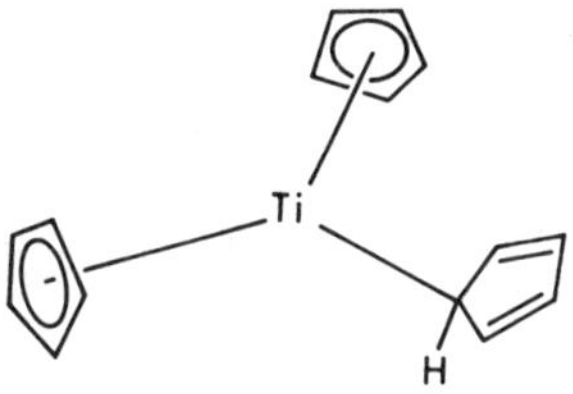

43 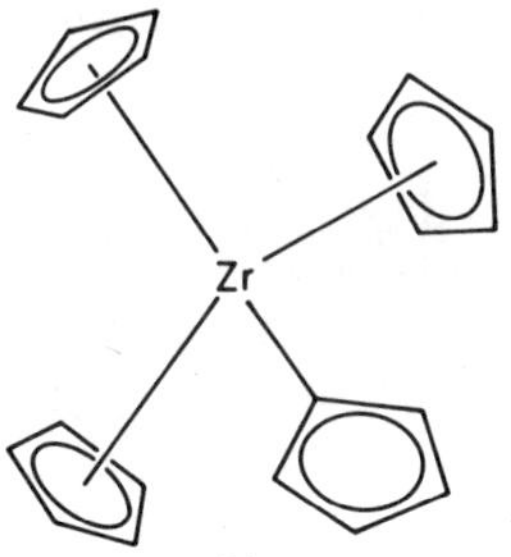 **44** 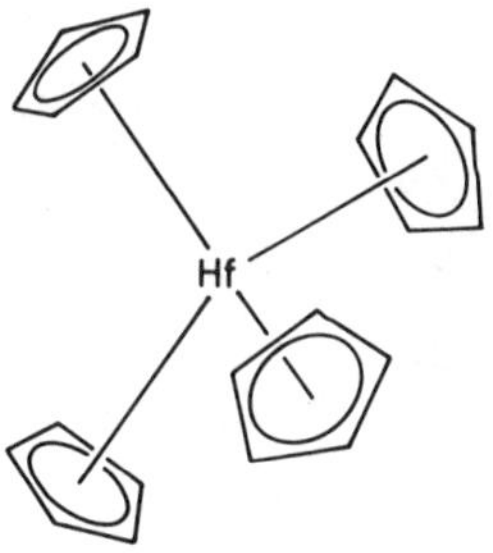 **45**

zirconium(IV) chloride to excess sodium cyclopentadienide under an argon atmosphere,[107] has been found by X-ray crystallographic measurements to be composed of a single σ-bonded and three π-bonded cyclopentadienyl rings. In still further contrast, the hafnium analog, similarly derived by treatment of a suspension of sodium cyclopentadienide in toluene with a benzene solution of hafnium(IV) chloride under argon,[40] has been shown by X-ray study to possess a symmetrical structure (**45**), consisting of four completely equivalent cyclopentadienyl rings, presumably all π-bonded to the central metal atom.[121]

The stable but extremely air-sensitive, violet to black monomeric and paramagnetic *π-allyldi-π-cyclopentadienyltitanium(III)* and its 1-(or 2-)π-methylallyl, as well as the 1-(or 2-)π-dimethylallyl homologs are obtained in 64–80% yields by the room temperature addition of a tetrahydrofuran solution of 2 molar equivalents of the appropriate allylmagnesium chloride (or bromide in diethyl ether) to a tetrahydrofuran solution of 1 molar equivalent of dichlorodi-π-cyclopentadienyltitanium(IV) according to the following equation:[122]

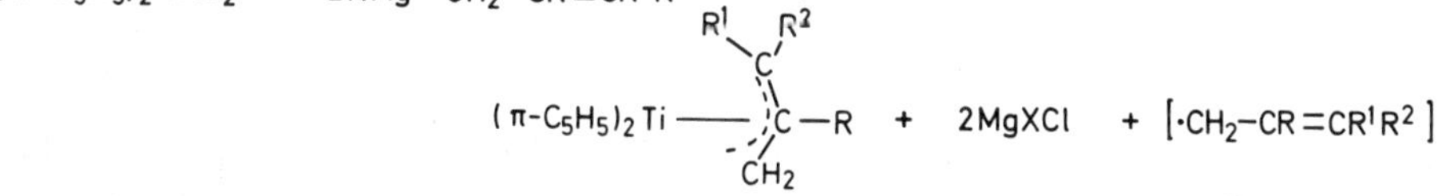

$$(\pi\text{-}C_5H_5)_2TiCl_2 + 2XMg{-}CH_2{-}CR{=}CR^1R^2 \longrightarrow (\pi\text{-}C_5H_5)_2Ti{-}[\pi\text{-}CH_2{-}C(R){-}CR^1R^2] + 2MgXCl + [\cdot CH_2{-}CR{=}CR^1R^2]$$

An alternative, and sometimes more convenient, method for the preparation of the π-bonded methylallyl and dimethylallyl complexes involves the reaction of two equivalents of isopropylmagnesium bromide in diethyl ether with a mixture of one equivalent of dichlorodi-π-cyclopentadiene and two equivalents of a suitable diene at 0°.[123] The yields of the π-allyldi-π-cyclopentadienyltitanium(III) complexes from 1,3-dienes are generally 70–80%, but are lower for 1,4- and 1,5-dienes, while allenes fail to give isolable complexes. In both methods the tetravalent (π-C_5H_5) titanium dichloride is reduced to chlorodi-π-cyclopentadienyltitanium(III) which, if used as the initial reactant with the Grignard reagent, can also yield the π-allyl complex.[113] However, in contrast, when 1 molar equivalent of the zirconium analogue, dichlorodi-π-cyclopentadienylzirconium(IV), is treated with 2 molar equivalents of allyl- or 2-methylallylmagnesium chloride in tetrahydrofuran at 0°, the corresponding cream-colored, air-sensitive bis(allyl) complexes (**46**), ($R{=}H$ or CH_3) are formed, in which one allyl (or methylallyl) group is π-bonded and one is σ-bonded to the metal atom.[124]

Dicarbonyldi-π-cyclopentadienyltitanium(IV) reacts, when heated with phenylacetylene in hexane or 1,2-diphenylacetylene in benzene, to yield the novel green, crystalline, air-stable heterocyclopentadiene-type complexes with structure **47** ($M{=}Ti$; $R{=}H$ or C_6H_5; $R'{=}C_6H_5$).[125]

The complex from 1,2-diphenylacetylene can also be prepared in the same relatively low (15%) yield by treating a large excess of sodium cyclopentadienide in tetrahydrofuran with a mixture of diphenylacetylene and titanium(IV) chloride at room temperature under an argon atmos-

[121] *V. I. Kulishov, E. M. Brainina, N. G. Bokii, Yu. T. Struchkov*, Chem. Commun. 475 (1970).
[122] *H. A. Martin, F. Jellinek*, J. Organometal. Chem. *8*, 115 (1967).
[123] *H. A. Martin, F. Jellinek*, J. Organometal. Chem. *12*, 149 (1968).
[124] *H. A. Martin, P. J. Lemaire, F. Jellinek*, J. Organometal. Chem. *14*, 149 (1968).
[125] *K. Sonogashira, N. Hagihara*, Bull. Chem. Soc., Japan *39*, 1178 (1966).

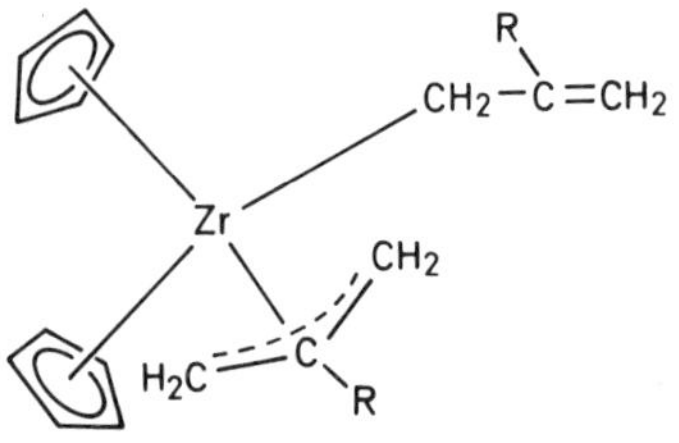

R = H, CH_3

46

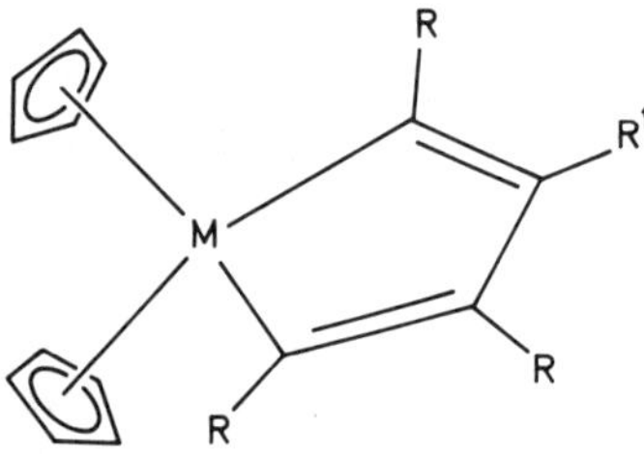

R = H, R' = C_6H_5
R = R' = C_6H_5
M = Ti, Zr

47

48

49

phere.[126] An analogous zirconium complex (**47**, M = Zr, R and R' = C_6H_5) is obtained in 53% yield by refluxing a diethyl ether solution of dichlorodi-π-cyclopentadienylzirconium(IV) and 1,4-dilithiobutadiene, the latter derived by shaking an ethereal solution of 1,2-diphenylacetylene with excess lithium shavings.[127] An interesting *heterocyclic indene-like complex* (**48**) is formed by the reaction of 1,2-diphenylacetylene and di-π-cyclopentadienyldiphenyltitanium(IV) in benzene.[128]

Chloro- and dichlorodi-π-cyclopentadienyltitanium form complexes with aluminum halides and alkyls, as indicated in Section 31.2.3.1, a number of which are presumed to have structures involving metal-to-metal bonding. The red bimetallic complex dimer (**49**), formed by heating chlorodi-π-cyclopentadienyltitanium(III) with excess triethylaluminum in benzene at 80°,[129] contains both titanium–titanium and titanium–aluminum bonds, as shown by X-ray crystallographic data.[130]

Other *bimetallic complexes* of type **50** (M = Si or Sn) can be prepared by the reaction of dichlorodi-π-cyclopentadienyltitanium(IV) with triphenylsilylpotassium in 1,2-dimethoxyethane[131] or triphenylstannylsodium in tetrahydrofuran.[51]

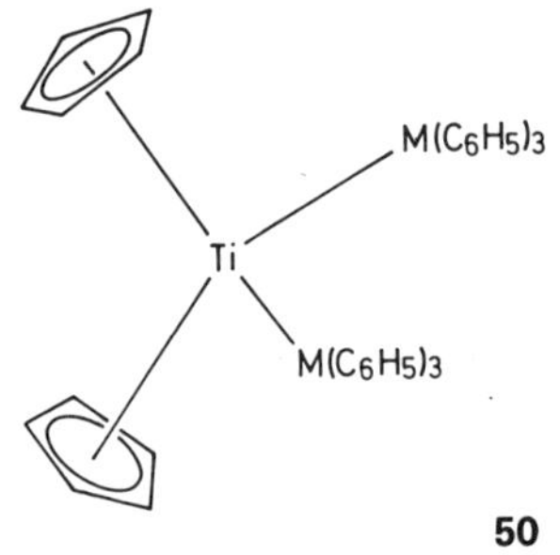

50

31.3 Metallocenes of Vanadium, Niobium, and Tantalum

31.3.1 Di-π-cyclopentadienylmetal Compounds and Hydrides

Although a report presenting data on di-π-cyclopentadienyltantalum(II), without any evidence for its synthesis, has been published,[132]

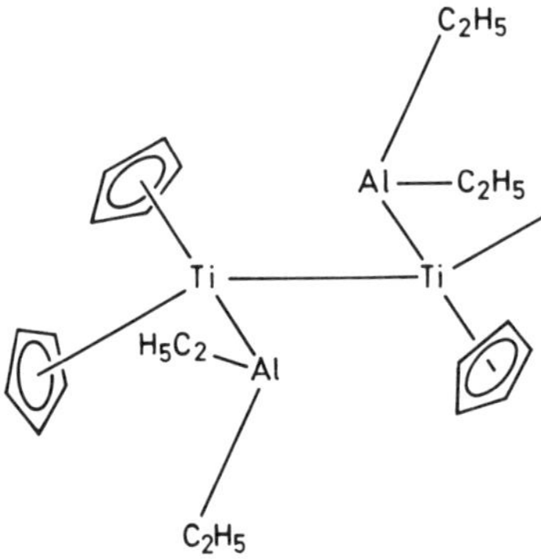

[126] *M. E. Volpin, V. A. Dubovitsky, O. V. Nogina, D. N. Kursanov*, Dokl. Akad. Nauk. SSSR, *151*, 1100 (1963); engl.: 623.

[127] *E. H. Braye, W. Hübel, I. Caplier*, J. Amer. Chem. Soc. *83*, 4406 (1961).

[128] *H. Masai, K. Sonogashira, N. Hagihara*, Bull. Chem. Soc., Japan *41*, 750 (1968).

[129] *G. Natta, G. Mazzanti, P. Corradini, U. Giannini, S. Cesca*, Atti. Accad. Naz. Lincei, Rend. Classe Sci. Fiz., Mat. Natur. *26*, 150 (1959); C.A. *54*, 17357 (1960).

[130] *P. Corradini, A. Sirign*, Inorg. Chem. *6*, 601 (1967).

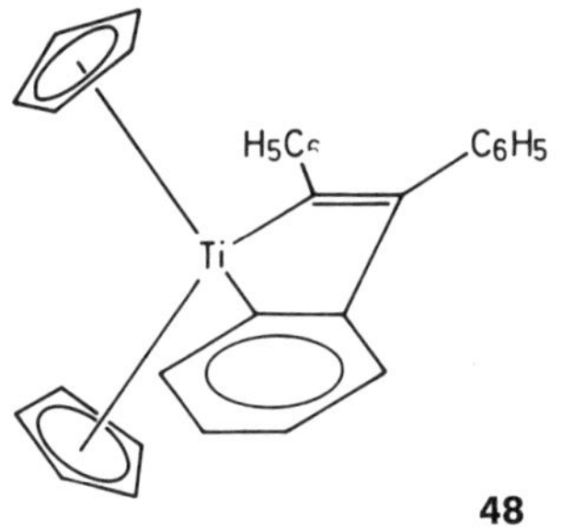

[131] *E. Hengge, H. Zimmerman*, Angew. Chem. *80*, 153 (1968);
E. Hengge, H. Zimmerman, Angew. Chem. Intern. Ed. Engl. *7*, 142 (1968).

[132] *S. S. Batsanov*, Izv. Sibirsk Otd. Akad. Nauk. SSSR, Ser. Khim. Nauk. 110 (1962); C.A. *58*, 4039 (1963).

of the Group V transition metals well-defined preparative methods exist only for the parent metallocenes derived from vanadium and niobium. The purple, air-sensitive *di-π-cyclopentadienylvanadium(II)* is most conveniently and efficiently prepared in *ca.* 50% yield by the room-temperature addition of vanadium(III) chloride to sodium cyclopentadienide in completely moisture- and peroxide-free tetrahydrofuran under a nitrogen atmosphere, according to the following equation:[133]

$$3NaC_5H_5 + VCl_3 \longrightarrow (\pi\text{-}C_5H_5)_2V + 3NaCl + [C_5H_5]$$

Care must be taken to exclude oxygen from all reactants during the reaction and purification steps, as well as to avoid exposing the product to air since the metallocene oxidizes rapidly with deflagration. A recent modification of this procedure involves the use of sodium hydride, instead of sodium metal, for improving the synthesis of the sodium cyclopentadienide used in the preparation of the vanadium complex.[134]

In a related manner, *di-π-cyclopentadienylniobium(II)* is obtained by the reduction of niobium(V) chloride with excess sodium cyclopentadienide in an inert solvent.[135]

A heteroannularly-disubstituted metallocene, *bis(methyl-π-cyclopentadienyl)vanadium(II)* can be prepared similar to its unsubstituted analog by treating sodium methylcyclopentadienide with the vanadium trihalide in tetrahydrofuran at 0°. It is obtained as a purple oil which also ignites violently on exposure to air.[134]

While tantalum may not form a stable unsubstituted di-π-cyclopentadienylmetal compound, it does form a stable hydride, *di-π-cyclopentadienyltrihydridotantalum(V)*. The white, crystalline trihydride is prepared in good yield by heating a mixture of sodium cyclopentadienide and tantalum(V) chloride in tetrahydrofuran and in the presence of sodium tetrahydridoborate.[136]

31.3.2 Di-π-cyclopentadienylmetal Halides

Monohalides of di-π-cyclopentadienyl-vanadium and -niobium are obtained from the parent or substituted metallocenes by treatment with organic or organometallic halides, hydrogen halides or halogens. Although it can be prepared directly from vanadium(IV) chloride, the blue, crystalline *chlorodi-π-cyclopentadienylvanadium(III)* is most conveniently formed by treatment of di-π-cyclopentadienylvanadium(II) with chloromethane in petroleum ether[134] or by a redistribution reaction of equimolar quantities of the metallocene and dichlorodi-π-cyclopentadienylvanadium(IV) in the absence of oxygen.[138] However, while *bromodi-π-cyclopentadienylvanadium(III)* is formed, similar to the chloride, in 85% yield by the reaction of di-π-cyclopentadienylvanadium(II) and bromoethane in petroleum ether,[137] the corresponding iodide is prepared by treating the metallocene with iodine.[137] In contrast, *chlorodi-π-cyclopentadienylniobium(III)* is obtained by a cleavage reaction of allyldi-π-cyclopentadienylniobium(III) and hydrogen chloride.[139] An interesting niobium monohalide complex, **51**, is formed in the autoclave reaction of sodium cyclopentadienide and niobium(V) chloride under carbon monoxide and hydrogen pressure in the presence of lithium tetrahydridoborate.[140]

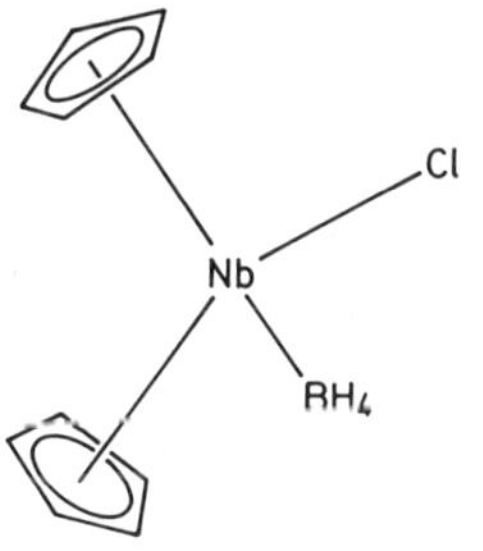

Cl
Nb
BH_4

51

The green, thermally-stable dihalide, *dichlorodi-π-cyclopentadienylvanadium(IV)*, is best obtained by a procedure analogous to that used for its titanium analog, as described in Section 31.2.3.2, involving the addition of a suspension of sodium cyclopentadienide in 1,2-dimethoxyethane to a benzene solution of anhydrous vanadium(IV) chloride under nitrogen.[74] This addition sequence is essential in order to prevent reduction of the tetravalent

[133] *R. B. King*, Organometallic Syntheses, Vol. I, p. 64, Academic Press, New York 1965.

[134] *M. F. Rettig, R. S. Drago*, J. Amer. Chem. Soc. *91*, 1361 (1969).

[135] U.S. Pat. 2 921 948 (1960), Union Carbide Corp., Invent.: *J. C. Brantley;* C.A. *54*, 11048 (1960).

[136] *M. L. H. Green, J. A. McCleverty, L. Pratt, G. Wilkinson*, J. Chem. Soc. 4854 (1961).

[137] *H. J. de Liefde Meyer, M. J. Janssen, G. J. van der Kerk*, Rec. Trav. Chim. Pays-Bas *80*, 831 (1961).

[138] *H. J. de Liefde Meyer, M. J. Janssen, G. J. van der Kerk*, Chem. & Ind. (London) 119 (1960).

[139] *F. W. Siegert, H. J. de Liefde Meyer*, Abstr. Papers, 4th Intern. Conf. Organometal. Chem. Paper J13, Bristol 1969.

[140] *R. B. King*, Z. Naturforsch. *18b*, 157 (1963).

vanadium complex to di-π-cyclopentadienylvanadium(II) by excess sodium cyclopentadienide. The corresponding brown-black niobium dihalide is obtained, together with *di-π-cyclopentadienyltrichloroniobium(V)* as a by-product in a similar manner from niobium(V) chloride.[135] An alternate two-step procedure, which circumvents the necessity for separation of halides and provides the dichloride in near-quantitative yield, involves the treatment of tetracyclopentadienylniobium(IV) [obtained from the reaction of 4 equivalents of sodium cyclopentadienide and niobium(V) chloride] with an excess of hydrogen chloride in diethyl ether.[141]

Dibromodi-π-cyclopentadienylvanadium(IV) is formed from the reaction of cyclopentadienylmagnesium bromide with vanadium(IV) chloride, followed by treatment with hydrobromic acid.[29] A mixed halide, *bromochlorodi-π-cyclopentadienylvanadium(IV)*, is prepared similarly by the treatment of vanadyl(III) chloride in benzene with cyclopentadienylmagnesium bromide in diethyl ether at 0° under nitrogen.[142]

Di-π-cyclopentadienyldiiodoniobium(IV) is obtained when dichlorodi-π-cyclopentadienylhydroxyniobium(V) (**52**) is reacted with ammonium iodide in the presence of benzylthiol, as a reducing agent, in accordance with the following equation:[143]

$$2(\pi\text{-}C_5H_5)_2Nb(OH)I_2 + 2NH_4I + 2C_6H_5SH \longrightarrow 2(\pi\text{-}C_5H_5)_2NbI_2 + C_6H_5SSC_6H_5 + 4Cl^{\ominus} + 2H_2O$$

The yellow *hydroxydichloride*,[135] as well as the corresponding *dibromide*,[29] are formed by hydrolysis of the appropriate di-π-cyclopentadienyltrihaloniobium(V), **53** (M = Nb; X = Cl or Br).

52 **53**

In addition to the pentavalent niobium trichloride derivative (**53**, M = Nb; X = Cl) obtained as a by-product from the reaction of sodium cyclopentadienide and niobium(V) chloride,[135] the corresponding *tantalum trihalide* (**53**, M = Ta; X = Cl) can be prepared by heating calcium cyclopentadienide, derived from the reaction of calcium hydride and cyclopentadiene, with tantalum(V) chloride in a sealed tube for 6 hr.[144] The related *tribromides* (**53**, M = Nb or Ta; X = Br) are formed in 62–70% yield by treating the appropriate metal pentachloride with sodium cyclopentadienide in tetrahydrofuran or 1,2-dimethoxyethane.[29] Similar to the formation of its *diiodo* analog, di-π-cyclopentadienyltriiodoniobium(V) (**53**, M = Nb; X = I) is prepared by the reaction of the dichlorohydroxy derivative with three equivalents of ammonium iodide in the presence of benzylthiol.[143]

31.3.3 Di-π-cyclopentadienylmetal Pseudohalides and Chalcogenides of Vanadium and Niobium

Green to brown-colored solid *dipseudohalides* of *di-π-cyclopentadienylvanadium*, including diazido, cyano, isocyanato, isothiocyanato and isoselenocyanato derivatives, are formed in excellent yields when dichlorodi-π-cyclopentadienylvanadium(IV) is treated with the appropriate anion in an aqueous, acetone or acetone-diethyl ether medium, according to the following equation:[145]

$$(\pi\text{-}C_5H_5)_2VCl_2 + 2MX \longrightarrow (\pi\text{-}C_5H_5)_2VX_2 + 2MCl$$
$$(M = Na \text{ or } K;\ X = N_3,\ CN,\ OCN,\ NCS \text{ or } NCSe)$$

The dipseudohalide obtained from this reaction using potassium cyanate is shown by spectral studies to be the nitrogen-bonded diisocyanato complex.[76] Furthermore, it should be noted that both the diazido and diselenocyanato complexes explode violently when heated or ignited, and extreme caution should be used in their preparation and handling.

In addition to a single dicarboxylate derivative, dibenzoatodi-π-cyclopentadienylvanadium(IV), prepared in moderately good yield by the reaction of equimolar quantities of di-π-cyclopentadienylvanadium(II) and dibenzoyl peroxide in diethyl ether,[146] an oxide, *chlorodi-π-cyclo-*

[141] *F. W. Siegert, H. J. de Liefde Meyer*, Rec. Trav. Chim. Pays-Bas *87*, 1445 (1968).
[142] U.S. Pat. 2 882 288 (1959), Union Carbide Corp., Invent.: *J. C. Brantley, E. L. Morehouse;* C.A. *54*, 3453 (1960).
[143] *P. M. Treichel, G. P. Werber*, J. Organometal. Chem. *12*, 479 (1968).
[144] U.S. Pat. 3 030 393 (1962); Commonwealth Engineering Co., Invent.: *J. Bulloff;* C.A. *57*, 7312 (1962).
[145] *G. Doyle, R. S. Tobias*, Inorg. Chem. *7*, 2479 (1968).
[146] *G. A. Razuvaev, V. N. Latyaeva, A. N. Lineva*, Dokl. Akad. Nauk. SSSR *187*, 340 (1969); engl.: 545.

pentadienyloxoniobium(V), **54**, can be obtained in 15% yield by treating either chlorodi-π-cyclopentadienyl-dithio- or -diiodo-niobium(V) with iodomethane in dichloromethane in an unexplained oxidation reaction.[147]

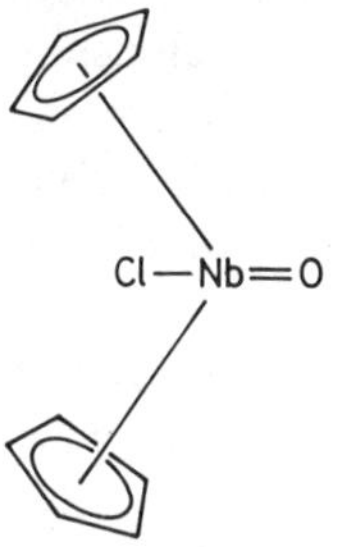

54

Red to brown-colored monomeric halide and pseudohalide-substituted cyclic dithio derivatives of di-π-cyclopentadienylniobium having structure **55** (X=Cl, Br, I or SCN) are formed by the reaction of the hydroxydichloride **52** with hydrogen sulfide and either the appropriate potassium salt or, in the case of the iodo compound (**55**, X=I), ammonium iodide in methanol solution.[147]

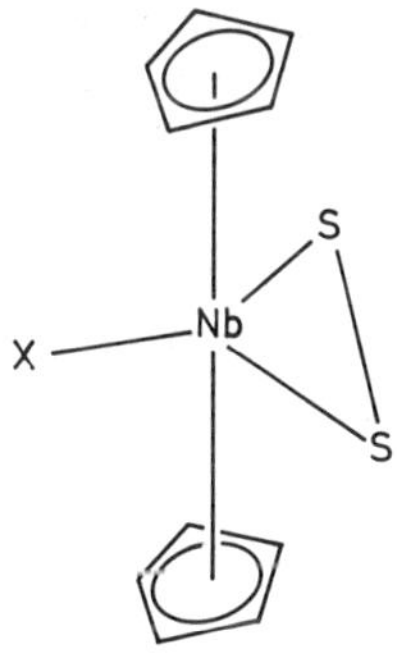

55

When oxygen-containing bidentate ligands, such as acetylacetone, benzoylacetone, dibenzoylmethane or hexafluoroacetylacetone, as well as tropolone, neat or in tetrahydrofuran, are added to an aqueous solution of di-π-cyclopentadienyldiperchloratovanadium(IV) [prepared by adding silver perchlorate to a suspension of dichlorodi-π-cyclopentadienylvanadium(IV) in water], highly explosive *cationic chelate perchlorates* with structure **56** (R=CH_3, C_6H_5 or CF_3; X=ClO_4) are formed.[145] Furthermore, a thermally stable trifluoromethylsulfonate salt (**56**, R=CH_3; X=F_3CSO_3) of the acetylacetonato complex is obtained in the same manner as the perchlorate, but using silver trifluoromethylsulphonate in place of silver perchlorate. Other *di-π-cyclopentadienyl-2,4-pentadionatovanadium(IV)* salts [**56**, R=CH_3; X=Cl_3CSO_3, BF_4, $(C_6H_5)_4B$, PF_6, I, NCS or NCSe] are readily prepared by adding a solution of the appropriate anion, in the form of its sodium or potassium salt, to a solution of the chloride chelate salt derived by the addition of an excess of acetylacetone to an aqueous suspension of dichlorodi-π-cyclopentadienylvanadium(IV).[145]

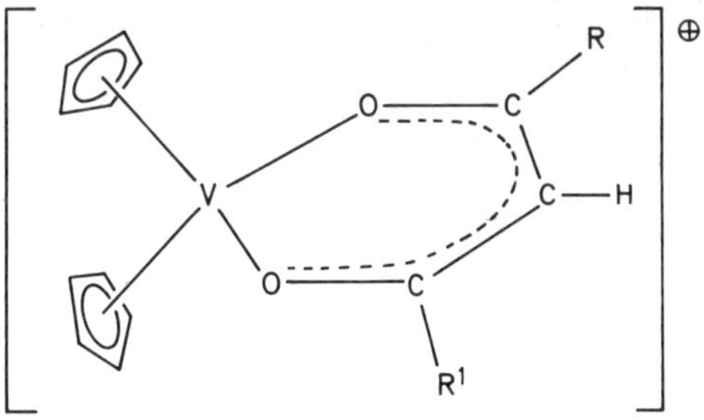

56

31.3.4 Metal-alkyl and -aryl Derivatives of Di-π-cyclopentadienyl-vanadium and -niobium

Monoalkyl complexes of di-π-cyclopentadienylvanadium (**57**, R=CH_3, C_2H_5 or $C_6H_5CH_2$) are formed in fair to good yields by the low-temperature reaction of chlorodi-π-cyclopentadienylvanadium(III) with the appropriate Grignard reagent in diethyl ether and/or 1,2-dimethoxyethane.[137] The corresponding monoaryl derivatives [**57**, R=C_6H_5, *p*-$CH_3C_6H_4$, *p*-$(CH_3)_2NC_6H_4$] are best obtained from the monochloride and an aryllithium in diethyl ether/1,2-dimethoxyethane or tetrahydrofuran.[138, 148, 149]

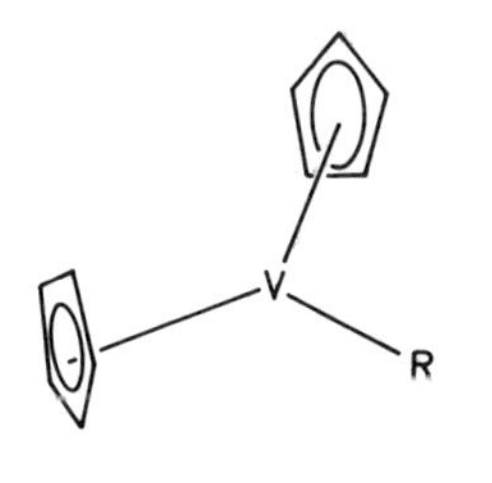

57

An alternate method for preparing the monomethyl as well as the phenyl and pentafluoro-

[147] *P. M. Treichel, G. P. Werber*, J. Amer. Chem. Soc. *90*, 1753 (1968).

[148] *H. J. de Liefde Meyer, M. J. Janssen, G. J. van der Kerk*, Studies in the Organic Chemistry of Vanadium, Institute for Org. Chem. TNO, Utrecht 1963.

[149] *F. W. Siegert, H. J. de Liefde Meyer*, J. Organometal. Chem. *15*, 131 (1968).

phenyl derivatives involves the reduction of dichlorodi - π - cyclopentadienylvanadium (IV) when treated with an alkyl- or aryllithium[150] or phenylsodium[141] according to the following equation:

$$(\pi\text{-}C_5H_5)_2VCl_2 + 2RM \longrightarrow (\pi\text{-}C_5H_5)_2VR + 2MCl + [Re]$$

$(R = CH_3,\ C_6H_5 \text{ or } C_6F_5;\ M + Li \text{ or } Na)$

Unlike its vanadium analog which, as indicated above, is reduced to trivalent vanadium when treated with an excess of organometallic, dichlorodi-π-cyclopentadienylniobium(IV) reacts with two equivalents of phenylsodium[141] or phenyllithium[151] in diethyl ether to yield the diphenyl derivative **58**.

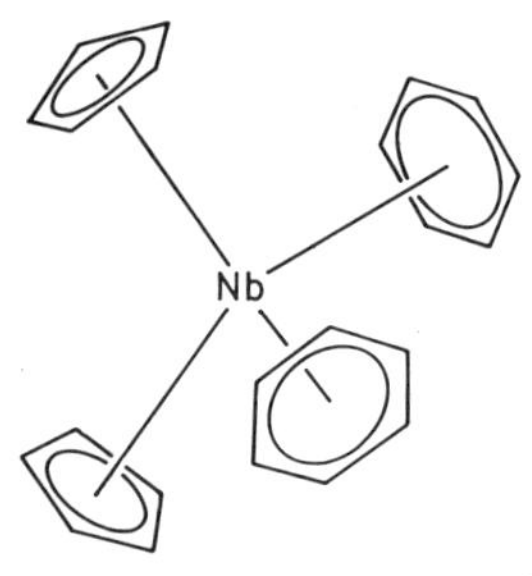

58

However, the tetravalent vanadium dichloride compound, similar to dichlorodi-π-cyclopentadienyltitanium(IV), does interact with an equimolar quantity of 2,2′-dilithiooctafluorobiphenyl to form the thermally stable biphenylene complex **59**.[112]

59

31.3.5 Cyclopentadienyl, Olefin and Acetylene Complexes of the Metallocenes

Similar to its behavior with alkyl and aryl organometallics, dichlorodi-π-cyclopentadienylvanadium(IV) is reduced when treated with excess of sodium cyclopentadienide to yield the rather unstable black σ-cyclopentadienyldi-π-cyclopentadienylvanadium(III);[118] the latter is best prepared, however, by the reaction of chlorodi-π-cyclopentadienylvanadium(III) with an equimolar quantity of the sodium salt.[149] On the other hand and as a further example of the close resemblance of di-π-cyclopentadienylniobium derivatives to their titanium, rather than vanadium analogs, the blue-violet, oxygen-sensitive *tetracyclopentadienylniobium(IV)*, **60** (M = Nb), can be formed in 56% yield by adding solid niobium(IV) chloride to a suspension of sodium cyclopentadienide in benzene.[141]

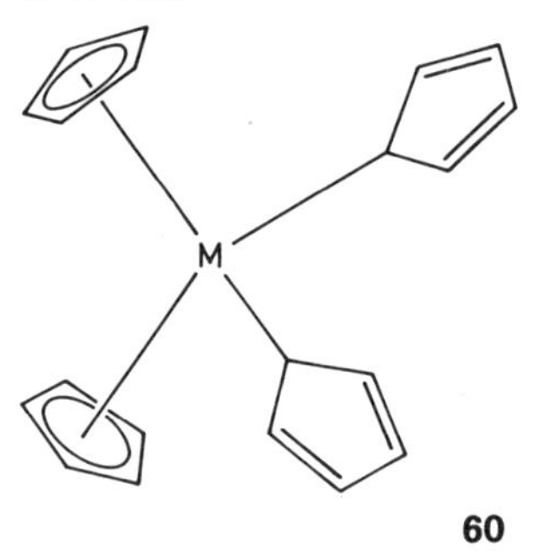

60

Both this niobium complex, as well as its red-violet tantalum analog (**60**, M = Ta), obtained similarly from the tantalum pentachloride using either diethyl ether or benzene as solvent,[152] are presumed from infrared spectral data to contain two π-bonded and two σ-bonded cyclopentadienyl rings.[153]

Chlorodi-π-cyclopentadienylvanadium(III) reacts at low temperatures with allyl, methylallyl and 2-butenyl Grignard reagents to form black, thermally unstable and oxygen-sensitive, paramagnetic complexes in which the allylic substituents are all σ-bonded to the metal atom.[149]

$$(\pi\text{-}C_5H_5)_2VCl + R^1\text{—}CH{=}\underset{}{\overset{R}{\overset{|}{C}}}\text{—}CH_2\text{—}MgX \xrightarrow[-MgCl_2]{} (\pi\text{-}C_5H_5)_2V\text{—}CH_2\text{—}\overset{R}{\overset{|}{C}}{=}CH\text{—}R^1$$

$(R = H \text{ or } CH_3;\ R' = H \text{ or } CH_3;\ X = Cl \text{ or } Br)$

Unlike dichlorodi-π-cyclopentadienylvanadium(IV) which is reduced to the σ-bonded *allyldi-π-cyclopentadienylvanadium(III)* when treated with excess allylmagnesium halide,[150] the corresponding niobium dichloride is also reduced but is converted to a π-bonded monoallyl complex.[130, 151]

[150] *H. J. de Liefde Meyer, F. Jellinek*, Abstr. Papers, 3rd Intern. Symp. Organometal. Chem. p. 250, München 1967.

[151] *F. W. Siegert, H. J. de Liefde Meyer*, J. Organometal. Chem. *23*, 177 (1970).

[152] *E. O. Fischer, A. Treiber*, Chem. Ber. *94*, 2193 (1961).

[153] *D. A. Bochvar, N. P. Gambaryan, E. M. Brainina, R. Kh. Friedlina*, Dokl. Akad. Nauk. SSSR *183*, 1324 (1968); engl.: 1104.

Both trivalent mono- and tetravalent *di-ethynyl* σ-bonded *complexes* of *di-π-cyclopentadienylvanadium* can be prepared by the reaction of sodium phenylacetylide with chlorodi-π-cyclopentadienylvanadium(III) or dichlorodi-π-cyclopentadienylvanadium(IV).[154] In contrast, di-π-cyclopentadienylvanadium(II) adds reversibly to acetylene derivatives to form the tetravalent complexes with structure **62** [R and R′=C_6H_5, C_6F_5, p-$(CH_3)_3CC_6H_4$, CF_3 or CH_3COO; R=C_6H_5, R′=p-$CH_3C_6H_4$; R=CH_3, R′=C_6H_5; R=CH_3; R′=$(C_2H_5)_2N$].[112, 155]

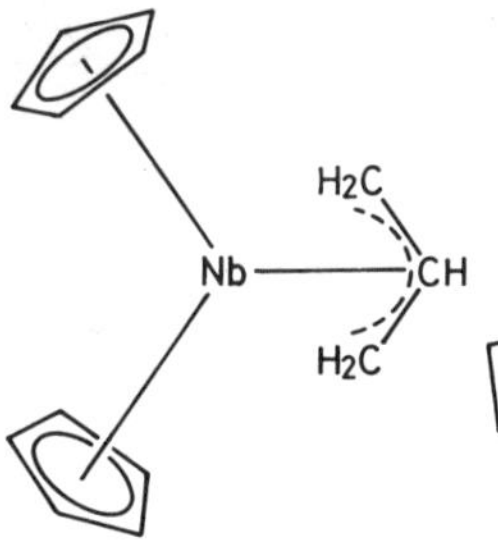

61

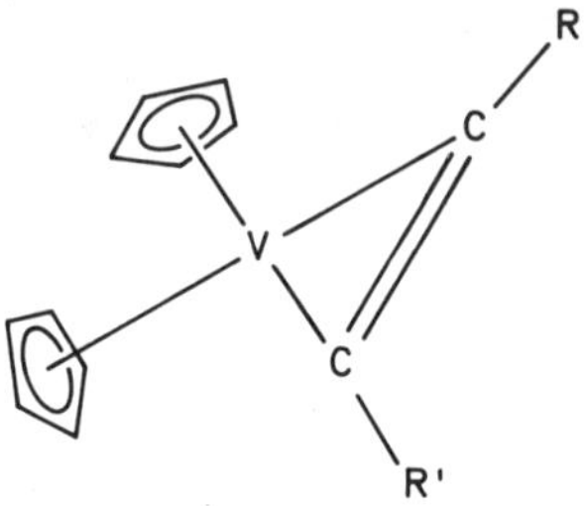

62

31.4 Metallocene Derivatives of Chromium, Molybdenum, and Tungsten

31.4.1 Di-π-cyclopentadienylmetal Compounds and Cationic Salts of Chromium and Molybdenum

Although a number of methods are available for the preparation of *di-π-cyclopentadienylchromium(II)*, the preferred procedure involves the addition of chromium(III) chloride to a solution of sodium cyclopentadienide in tetrahydrofuran.[156] The dark red crystalline metallocene oxidizes rapidly in air and precautions must be taken in its synthesis and handling to exclude oxygen.

An alternate procedure, which gives a high yield and, for certain purposes, may be more convenient, utilizes the corresponding thallium cyclopentadienide for reaction with chromium(II) chloride in tetrahydrofuran.[157] A yellow molybdenum analog, *di-π-cyclopentadienylmolybdenum(^{99}Mo)(II)*, is obtained in low yield by treating a tetrahydrofuran solution of sodium cyclopentadienide with molybdenum(^{99}Mo) pentachloride in the presence of sodium tetrahydridoborate at −5°.[158]

A heteroannularly-disubstituted mettallocene, *bis(methyl-π-cyclo-pentadienyl) chromium(II)* can be synthesized, similar to its unsubstituted analog, from sodium methylcyclopentadienide and the chromium trihalide in tetrahydrofuran. When 1,2-diphenylacetylene is heated with either molybdenum hexacarbonyl or with 1,2-dimethoxyethane molybdenum tricarbonyl in benzene, the novel, thermally stable, red perarylated compound, *bis(pentaphenyl-π-cyclopentadienyl)molybdenum(II)*, **63**, is formed in very low yield.[159]

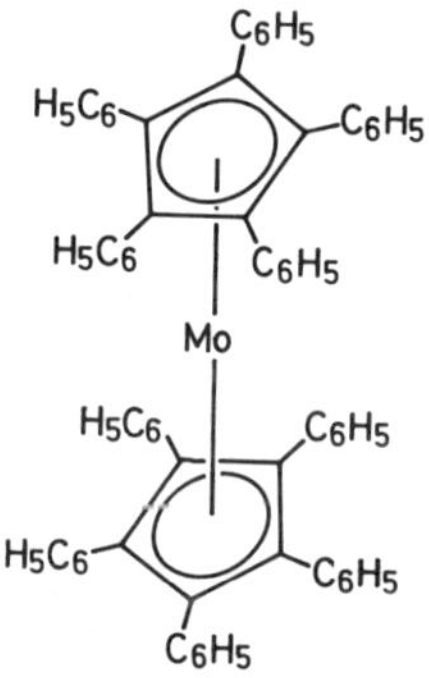

63

Complex salts of the *cationic di-π-cyclopentadienylchromium(III)*, [**64**, X=$C_5H_5CrCl_3$ or $C_5H_5Cr(CO)_3$] are prepared by reaction of di-π-cyclopentadienylchromium(II) with tetrachloromethane in tetrahydrofuran,[160] or with carbon monoxide at 100–110° and under pressure of 100 atmospheres.[161] If the cyclopenta-

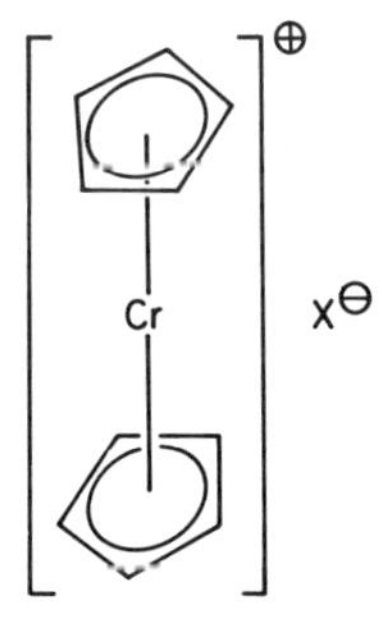

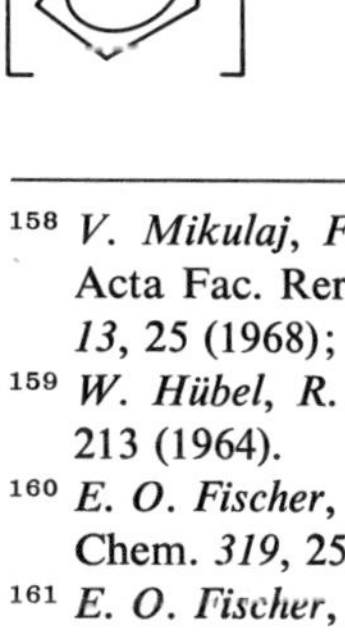

64

[154] *J. H. Teuben, H. J. de Liefde Meyer*, J. Organometal. Chem. *17*, 87 (1969).

[155] *R. Tsumara, N. Hagihara*, Bull. Chem. Soc., Japan *38*, 861 (1965).

[156] *R. B. King*, Organometallic Syntheses, Vol. I, p. 66, Academic Press, New York 1965.

[157] Fr. Pat. 1 343 770 (1963); Studiengesellschaft Kohle m.b.H; Erf.: *G. O. Schenk, E. K. von Gustorf;* C.A. *60*, 8062 (1964).

[158] *V. Mikulaj, F. Macasek, R. Kopunek, P. Matton*, Acta Fac. Rerum Natur. Univ. Comenianae Chim. *13*, 25 (1968); C.A. *71*, 50169y (1969).

[159] *W. Hübel, R. Merenyi*, J. Organometal. Chem. *2*, 213 (1964).

[160] *E. O. Fischer, K. Ulm, P. Kuzel*, Z. Anorg. Allgem. Chem. *319*, 253 (1963).

[161] *E. O. Fischer, W. Hafner*, Z. Naturforsch. *10b*, 140 (1955).

dienyltrichlorochromate or iodide salt (**64**, $X = C_5H_5CrCl_3$ or I) is treated with aqueous sodium tetraphenylborate, the corresponding tetraphenylborate [**64**, $X = B(C_6H_5)_4$] can be obtained.[162]

31.4.2 Di-π-cyclopentadienylmetal Hydrides of Molybdenum and Tungsten

Pentafluorophenyl-substituted di-π-cyclopentadienyl-molybdenum and -tungsten *monohydrides*, **65** (M = Mo or W), are the products formed in the reaction of dichlorodi-π-cyclopentadienyl-molybdenum or -tungsten(IV) with phenyllithium in diethyl ether as a result of reduction of the halides.[163]

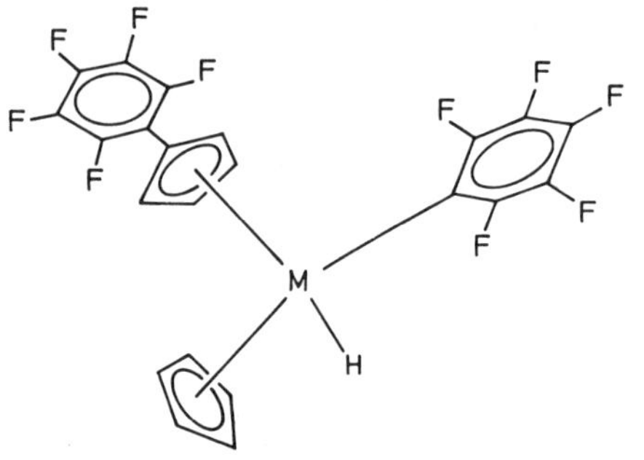

65

The yellow *dihydrides* of *di-π-cyclopentadienyl-molybdenum* or *-tungsten*, **66** (M = Mo, W; R = H), are conveniently prepared in yields of 50–65% by the reaction of excess sodium cyclopentadienide with the appropriate metal pentachloride in tetrahydrofuran and in the presence of sodium tetrahydridoborate.[136]

The melting point (163–165°) indicated for the dihydridotungsten derivative in the published procedure is undoubtedly incorrect and 193–195° is most likely the valid melting range.[164] This dihydride behaves as a base and also undergoes ring metallation with *n*-butyllithium in diethyl ether to yield a buff-yellow crystalline, spontaneously air-flammable dilithio derivative (**66**, M = W; R = Li) which on deuterolysis with deuterium oxide forms the corresponding dideuteride (**66**, M = W; R = D).[165]

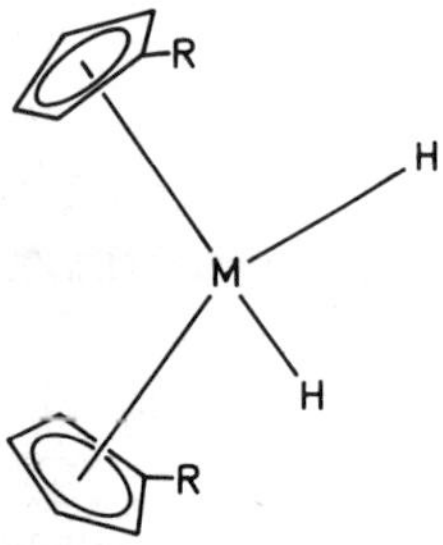

66

Substituted dihydrides containing metal to metal bonds, **67** [M = Mo or W; $M'X_3 = Al(CH_3)_3$ or BF_3] can be obtained by treating the appropriate dihydride (**66**) with trimethylaluminum[166] or boron trifluoride in benzene or toluene.[167]

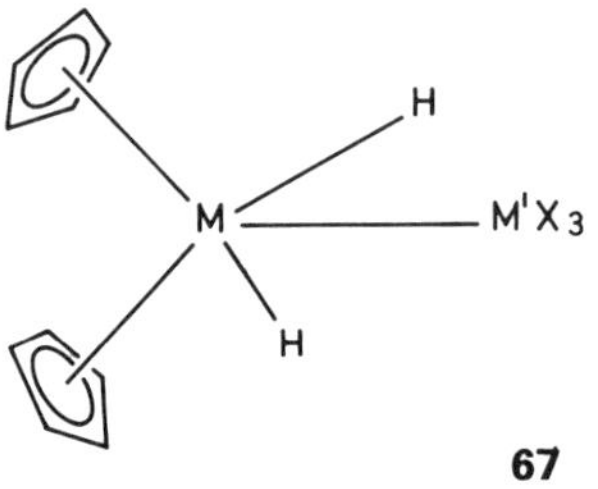

67

31.4.3 Di-π-cyclopentadienylmetal Halides

Monohalides of *di-π-cyclopentadienylchromium-(III)* are formed directly from the metal halide or from the parent metallocene. Thus, the dark blue, high melting bromodi-π-cyclopentadienylchromium(III) (**68**, M = Cr; X = Br; R = H) can be prepared by treating chromium-(III) bromide with lithium cyclopentadienide in diethyl ether,[168] while its yellow iodo analog (**68**, M = Cr; X = I; R = H) is obtained from the reaction of di-π-cyclopentadienylchromium(II) with iodine under nitrogen[169] or with allyl iodide in benzene.[162] A trivalent *molybdenum*

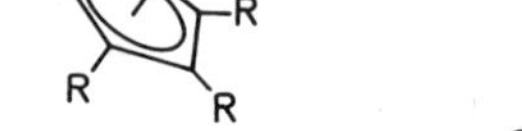

68

---

[162] *M. L. H. Green, W. E. Lindsell*, J. Chem. Soc. *A*, 2215 (1969).

[163] *E. O. Fischer, K. Ulm*, Chem. Ber. *95*, 692 (1962).

[164] *D. F. Shriver*, J. Amer. Chem. Soc. *85*, 3509 (1963).

[165] *R. L. Cooper, M. L. H. Green, J. T. Moelwyn-Hughes*, J. Organometal. Chem. *3*, 261 (1965).

[166] *H. Brunner, P. C. Wailes, H. D. Kaesz*, Inorg. Nucl. Chem. Lett. *1*, 125 (1965).

[167] *J. M. Johnson, D. F. Shriver*, J. Amer. Chem. Soc. *88*, 301 (1966).

[168] U.S. Pat. 2 864 843 (1958), Ethyl Corp., Invent.: *E. G. DeWitt, H. Shapiro, J. E. Brown;* C.A. *53*, 5284h (1959).

[169] *E. O. Fischer, H. P. Kögler*, Angew. Chem. *68*, 462 (1956).

bromide (**68**, M=Mo; X=Br; R=C_6H_5) is synthesized in 60% yield by treating bis(pentaphenyl-π-cyclopentadienyl)molybdenum(II) with bromine in dichloromethane.[159] Pentafluorophenyl metal- and ring-substituted monochlorides, **69** (M=Mo or W), are prepared by treating the corresponding monohydrides (**65**, M=Mo or W) with carbon tetrachloride.[163]

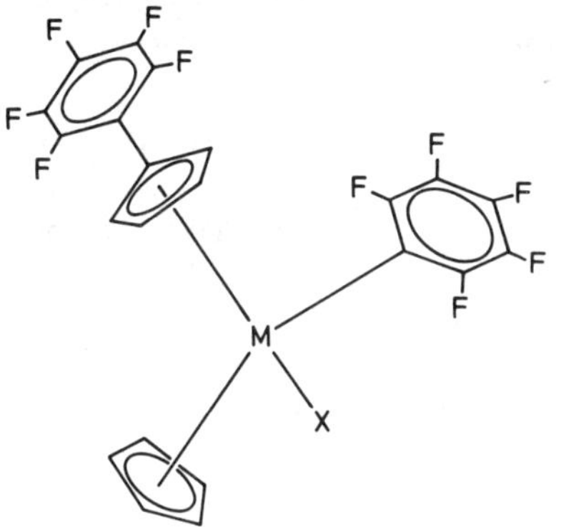

69

Green *dichlorides* and *dibromides* of *di-π-cyclopentadienyl-molybdenum* and *-tungsten* (**70**, M=Mo or W; X=Cl or Br) are best prepared by treating the appropriate dihydride (**66**, M=Mo or W) with trichloro- or tribromo-methane.[170] The corresponding *diiodides* (**70**, M=Mo or W; X=I) can be obtained similarly by reacting the dihydrides with iodomethane or iodine in diethyl ether.[170]

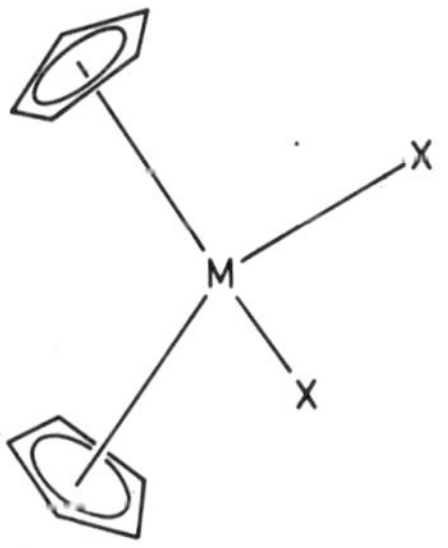

70

When two equivalents of ether cyclopentadiene[171] or sodium cyclopentadienide[172] are heated with one equivalent of tungsten(VI) oxide tetrachloride in tetrahydrofuran, *dichlorodi-π-cyclopentadienyloxotungsten(VI)*, **71**, is formed in high yield. The corresponding *di-π-indenyl dichloride* is prepared analogously in *ca.* 80% yield by reacting indene with the tungsten halide.[171]

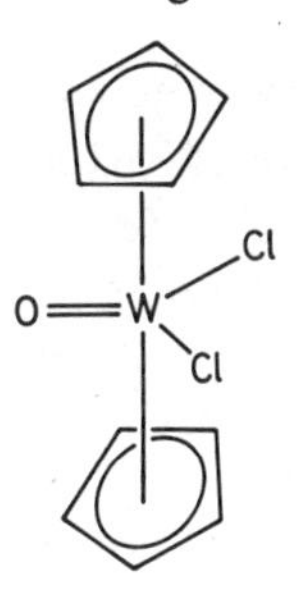

71

Brown *salts* of cationic *dihalodi-π-cyclopentadienyl-molybdenum* and *-tungsten(V)*, **72** [M=Mo or W; X=Cl or Br; Y=PF_6, $PtCl_6$, $SiW_{12}O_{40}(H_2O)_x$ or $Cr(NCS)_4(NH_3)_2$], are formed when the dichloro- (or dibromo-) di-π-cyclopentadienylmetal(IV) derivative is oxidized with 8 M nitric acid or concentrated hydrochloric acid in air, and the resulting solution treated with ammonium hexafluorophosphate, chloroplatinic acid, tungstosilicic acid or Reinecke acid.[170, 173] Corresponding purple to maroon-colored *hydrogen halide salts* (**72**, M=Mo or W; X=Cl or Br; Y=HX_2) are obtained by oxidation of a solution of the appropriate dihydride derivative in chloroform with chlorine or bromine.[170]

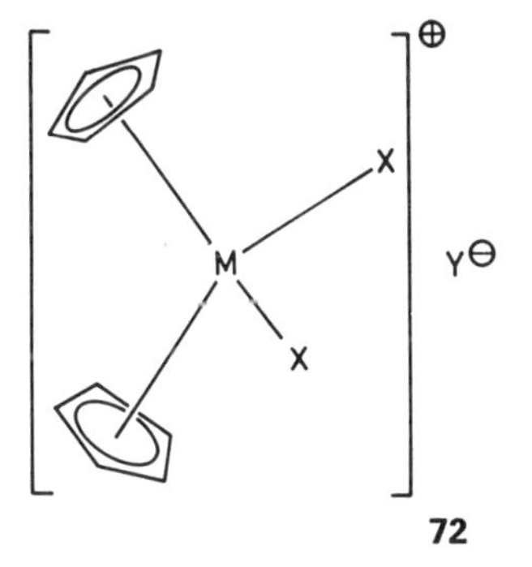

72

31.4.4 Pseudohalide and Chalcogen-containing Derivatives of Di-π-cyclopentadienyl-molybdenum and -tungsten

Dichlorodi-π-cyclopentadienyl-molybdenum or -tungsten(IV) undergoes displacement reactions when treated with two equivalents of lithium azide in acetone–methanol or with aqueous potassium thiocyanate to yield the purple, explosive diazido or purple *diisothiocyanato complexes* in good yields, according to the following equation:[174]

---

170 *R. L. Cooper, M. L. H. Green*, J. Chem. Soc. *A*, 1155 (1967).

171 *S. P. Anand, R. K. Multani, B. D. Jain*, Curr. Sci. *37*, 487 (1968); C.A. *70*, 87931w (1970).

172 *S. P. Anand, R. K. Multani, B. D. Jain*, J. Organometal. Chem. *17*, 423 (1969).

173 *F. A. Cotton, G. Wilkinson*, Z. Naturforsch. *9b*, 417 (1954).

174 *M. L. H. Green, W. E. Lindsell*, J. Chem. Soc. *A*, 2150 (1969).

$$(\pi\text{-}C_5H_5)_2MCl_2 + 2M'X \longrightarrow (\pi\text{-}C_5H_5)_2MY_2 + 2M'Cl$$
$$(M = MO \text{ or } W;\ M' = Li \text{ or } K;\ X = N_3 \text{ or } SCN)$$

A similar reaction of dichlorodi-π-cyclopentadienyltungsten(IV) with sodium cyanide in acetone–methanol or acetone–ethanol results in the formation of the orange *alkoxycyano complexes* **73** (M = Mo or W; R = CH_3 or C_2H_5).[174] Treatment of the dichloride with aqueous sodium cyanate produces red-brown *carbamates*, **74** (M = Mo or W), (presumably arising from hydrolysis of the isocyanato complexes initially formed), in addition to a deep red *diisocyanato monohydrate complex* formed in the case of the tungsten derivative.[174]

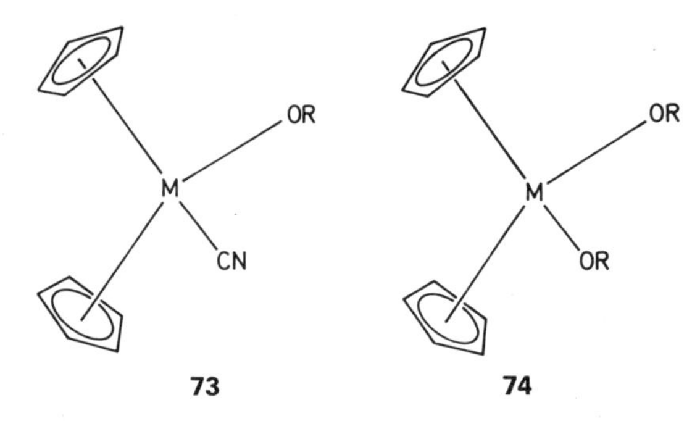

73 **74**

Analogous to the formation of the pseudohalide derivatives, the dichlorides of di-π-cyclopentadienyl-molybdenum and -tungsten(IV) react with sodium carboxylates in acetone–ethanol to give green or purple *dicarboxylato derivatives* (**74**, M = Mo or W; R = CF_3CO or C_6H_5CO) in yields of *ca.* 80%.[175] Related brown or reddish-brown *bis(carboxylato)di-π-cyclopentadienyloxotungsten(VI)* complexes (**75**, R = CH_3, C_2H_5, n-C_3H_7 or C_6H_5) are obtained in high yields by treating dichlorodi-π-cyclopentadienyloxotungsten(VI) with the sodium salt of the desired acid in benzene, toluene or tetrahydrofuran.[176]

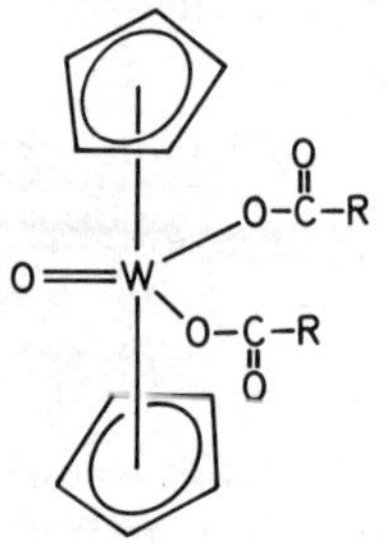

75

In a somewhat similar manner, *1,2-dioxophenylene derivatives* (**76**, M = Mo or W; X and X′ = 0) or an analogous 9,10-dioxophenanthrenylene complex, **77**,[177] can be prepared by reaction of the appropriate dichloride with 1,2-benzenediol (catechol)[178] or 9,10-phenanthrenediol[177] in the presence of aqueous or ethanolic sodium hydroxide. A *mixed amido-metal-oxo chelate* (**76**, X = NH; X′ = 0) is obtained when 1-amino-2-hydroxybenzene is substituted for the aromatic diol in the reaction with dichlorodi-π-cyclopentadienylmolybdenum.[177] If the dichloride is treated with acetylacetone and aqueous sodium hydroxide, the cationic *acetylacetonato chloride* complex (**78**) is formed in 70% yield.[177]

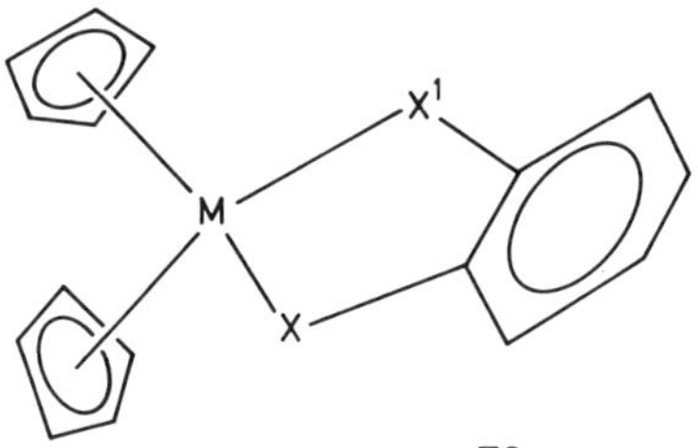

76

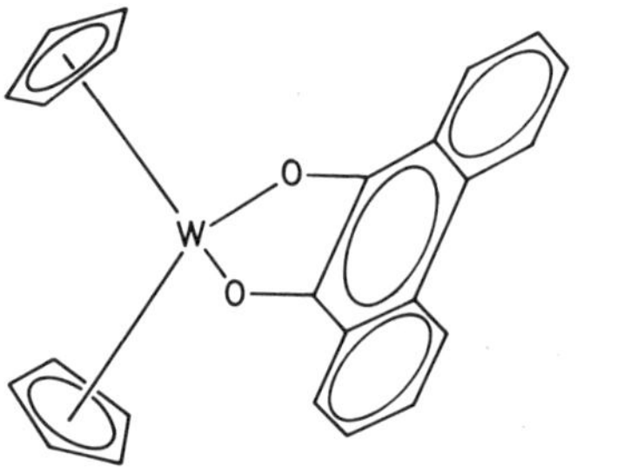

77

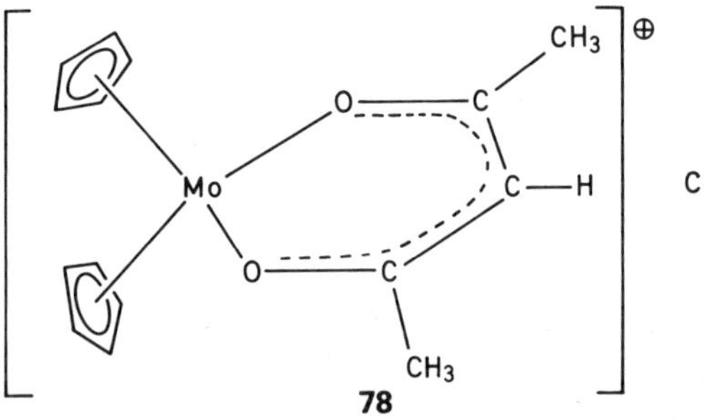

78

Treatment of dichlorodi-π-cyclopentadienylmolybdenum(IV) with ammonium pentasulfide gives the dark red, air-stable cyclic *tetrachalcogenide* (**79**).[179]

[175] *M. G. Harris, M. L. H. Green, W. E. Lindsell*, J. Chem. Soc. *A*, 1453 (1969).
[176] *S. P. Anand, R. K. Multani, B. D. Jain*, J. Organometal. Chem. *19*, 387 (1969).
[177] *E. Gore, M. L. H. Green, M. G. Harris, W. E. Lindsell, H. Shaw*, J. Chem. Soc. *A*, 1981 (1969).
[178] *M. L. H. Green, W. E. Lindsell*, J. Chem. Soc. *A*, 1455 (1967).
[179] *H. Köpf*, Angew. Chem. *81*, 332 (1969); *H. Köpf*, Angew. Chem. Intern. Ed. Engl. *8*, 375 (1969).

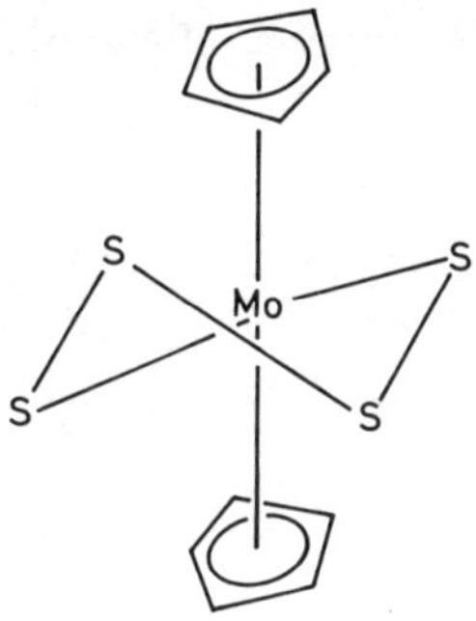

79

Dimercapto and *bis[alkyl- (or -aryl)thio] derivatives* of *di-π-cyclopentadienyl-molybdenum* and *-tungsten(IV)* (**80**, M=Mo or W; R=H; CH_3, C_2H_5 or C_6H_5) are conveniently prepared by heating the appropriate dichlorodi-π-cyclopentadienylmetal compound with two equivalents of either hydrogen sulfide,[178] an alkylthiol,[175] or a thiophenol[178] in the presence of either sodium or sodium hydroxide in ethanol. If an alkanedithiol or a sodium mercaptocarboxylate and sodium hydroxide is added to a suspension of an equimolar quantity of the dichloride in ethanol, red or brown cyclic dithiolato and thiocarboxylato complexes are formed in 70–75% yield:[175]

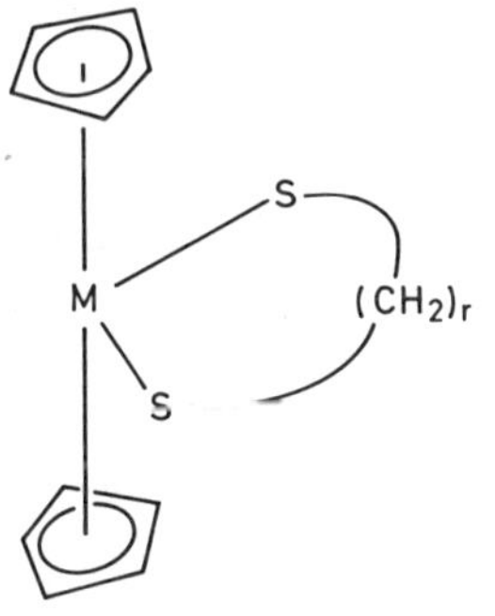

a 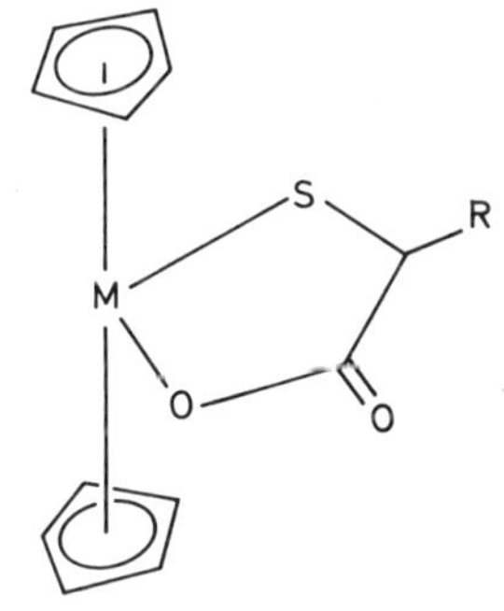 *b*

By analogous reactions, *cis-1,2-enedithiolates* of *di-π-cyclopentadienyl-molybdenum* and *-tungsten(IV)* (**81**, M=Mo or W; R=H or CN) are obtained when the appropriate dichloride is treated with disodium 1,2-ethylenedithiolate[102] or disodium 1,2-maleonitriledithiolate[178] in an aqueous or ethanolic medium.

If the dichlorides are reacted with benzene-1,2-dithiol in the presence of triethylamine in aqueous benzene,[180] or with toluene-3,4-dithiol in 2 M sodium hydroxide,[178] bright colored (orange to purple) *1,2-phenylenedithiolates* of

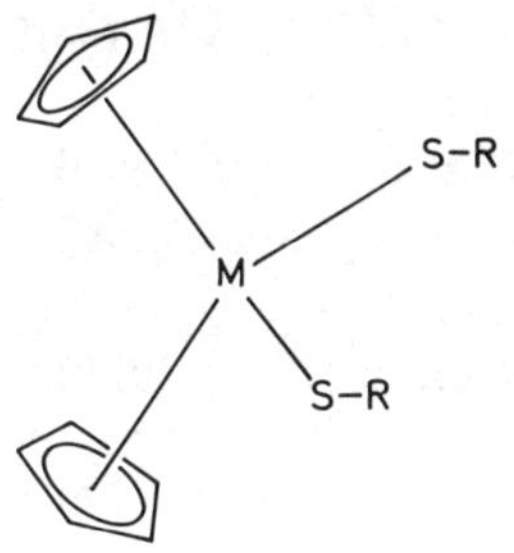

80

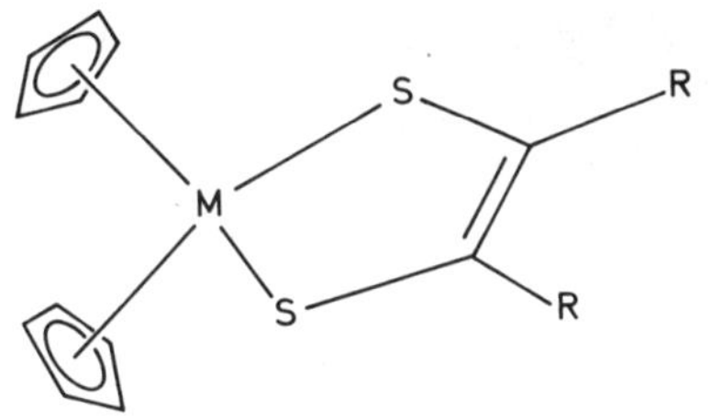

81

structure **82** (M=Mo or W; X=S; R=H or CH_3) can be prepared. Substitution of an arylene-1,2-dithiol by 2-amino-benzenethiol in ethanol in the reaction results in the formation of the *amidothiolate complexes* **82** (M=Mo or W; X=NH).[177] Interesting *bimetallic thiolato complexes* (**83**, M=Mo or W; X=Cl, Br or PF_6) can be synthesized by the reaction of the bis(methylthiolato) derivatives of di-π-cyclopentadienyl-molybdenum and tungsten(IV) (**80**, M=Mo or W; R=CH_3) with platinum(II) chloride, bromide, or hexafluorophosphate in ethanol or tetrahydrofuran.[181]

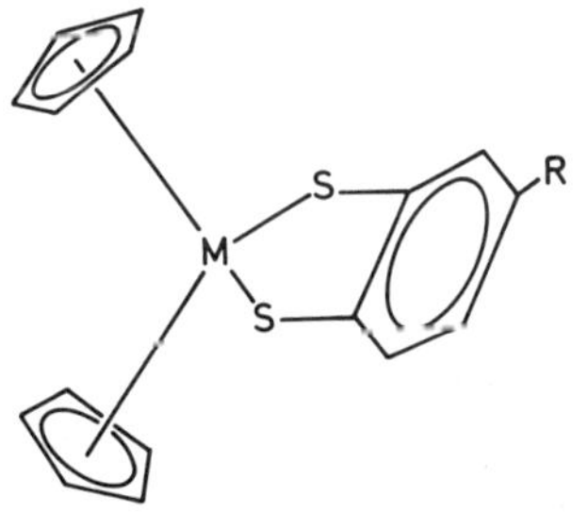

82

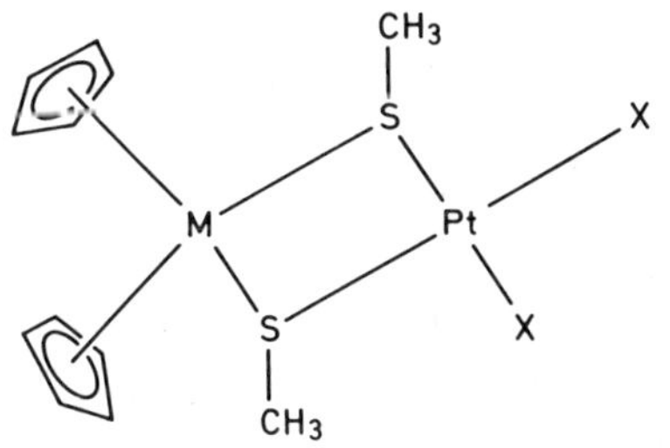

83

[180] *A. Kutoglu, H. Köpf,* J. Organometal. Chem. *25,* 455 (1970).

[181] *A. Dias, M. Green,* Chem. Commun. 962 (1969).

31.4.5 Di-π-cyclopentadienylmetal Derivatives of Molybdenum and Tungsten with Additional Metal–Carbon or with Metal–Nitrogen Bonds

Dichlorodi-π-cyclopentadienyloxotungsten(VI) reacts with alkyl or aryl Grignard or lithium reagents in ether solvents, as well as with presumably α,α-dichlorobenzyllithium in dimethylaniline, to form yellow- to brown-colored bis-alkyl, -aryl and -benzoyl-substituted derivatives of di-π-cyclopentadienyloxotungsten(VI), **84** (R = CH_3, C_2H_5, $C_6H_5CH_2$, C_6H_5 or C_6H_5-CO).[176]

O═W(R)(R)

84

The thermally stable but air-sensitive acetylene complex, **85** (X–X = C═C), but not its tungsten analog, can be prepared by treating the di-π-cyclopentadienyldihydridomolybdenum with 1,2-diphenylacetylene in boiling toluene or tetrahydrofuran.[182] When either the molybdenum or tungsten dihydride is added to other alkyne derivatives, such as hexafluorobut-2-yne or 1,2-bis(methoxycarbonyl)ethyne in tetrahydrofuran at 0°, insertion of the metal–hydrogen bond in the dihydride occurs with formation of *cis* (**86**, M = Mo or W; X–X = C═C; R = CF_3 or CH_3OOC) or *trans* adducts.[182] The reaction of di-π-cyclopentadienyldihydridomolybdenum-(IV) with azobenzene in refluxing toluene or tetrahydrofuran, or with 1,2-bis(methoxycarbonyl)azine in ethanol yields products (**85**, X–X = N═N; R = C_2H_5OOC) with structures similar to those obtained from the alkynes.[182] No comparable compounds, however, are obtained from reactions of the tungsten dihydride with the azo derivatives.

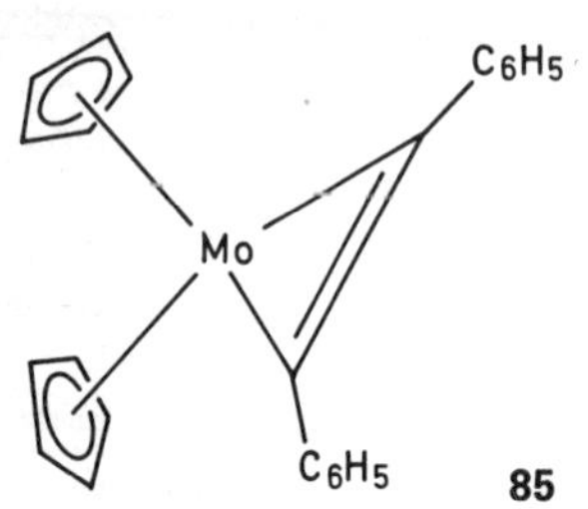

85

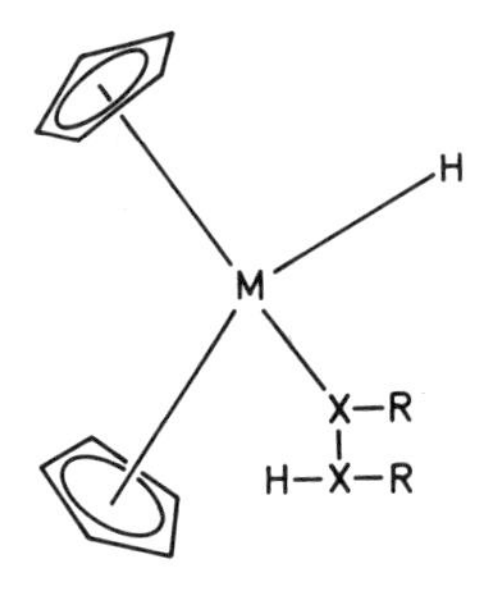

86

[182] *S. Otsuka, A. Nakamura, H. Minamida,* Chem. Commun. 1148 (1969).

31.5 Metallocenes of Manganese, Technetium, and Rhenium

31.5.1 Di-π-cyclopentadienylmetals and Hydrides

Unlike its neighboring transition elements, such as vanadium, iron, cobalt and nickel, which form neutral covalent π-bonded dicyclopentadienyl derivatives, manganese forms a sandwich complex with cyclopentadienyl ligands which clearly has an ionic structure and thus is not a metallocene.[183] *Manganese(II) cyclopentadienide*, the only complex of the *d*-block transition metals for which a primarily ionic ring-metal bond is proposed, is best prepared by refluxing sodium cyclopentadienide with highly reactive manganese(II) bromide (prepared *in situ* from manganese and bromine) in 1,2-dimethoxyethane under nitrogen.[184] It is obtained as dark brown paramagnetic crystals which are extremely sensitive to air and are hydrolyzed instantly by water.

In contrast to manganese, the other transition metals of this group, technetium and rhenium, do not form stable neutral di-π-cyclopentadienyl derivatives but, instead, yield *di-π-cyclopentadienylmetal monohydrides*. The golden yellow, thermally stable but very air-sensitive di-π-cyclopentadienylhydridotechnetium(III) is obtained in only 14% yield by treating a cold tetrahydrofuran suspension of technetium(IV) chloride with sodium cyclopentadienide and then heating the mixture for 10 hours at 50° in the presence of sodium tetrahydridoborate,[185] according to the following equation:

$$6NaC_5H_5 + 2TcCl_4 + 2NaBH_4 \longrightarrow 2(\pi\text{-}C_5H_5)_2TcH + 8NaCl + B_2H_6$$

[183] *G. Wilkinson, F. A. Cotton, J. M. Birmingham,* J. Inorg. Nucl. Chem. *2*, 95 (1956).

[184] *R. B. King,* Organometallic Synthesis, Vol. I, p. 67, Academic Press, New York 1965.

[185] *E. O. Fischer, M. Schmidt,* Angew. Chem. *79*, 99 (1967);
E. O. Fischer, M. Schmidt, Angew. Chem. Intern. Ed. Engl. *6*, 93 (1967).

The analogous and better-known lemon-yellow rhenium monohydride is formed in somewhat better (35–40%) yield by a similar reaction with rhenium(IV) chloride but involving a shorter (5-hour) heating period.[186] If the rhenium hydride, in turn, is dissolved in a solution of deuterium chloride in deuterium oxide and the salt neutralized with a deuterium oxide solution of sodium deuteroxide, the corresponding metal deuteride can be prepared.[186]

The neutral hydride, *di-π-cyclopentadienylhydridorhenium(III)*, unlike other metal hydride derivatives, is a base of similar strength to ammonia and is capable of undergoing ring-substitution reactions. Thus, when the hydride reacts with *n*-butyllithium in diethyl ether, a white solid dilithio derivative (**87**, R=Li) is formed which turns purple on contact with the air and which, on deuterolysis with deuterium oxide, yields the heteroannular dideuteride (**87**, R=D).[165] If an ethereal solution of the dilithio derivative is treated with mercury(II) chloride in tetrahydrofuran, the yellow, air-stable *bis(chloromercuri) derivative* (**87**, R=HgCl) is obtained.[165] The Friedel-Crafts reaction of the rhenium hydride with benzylmethyl ether in the presence of tin(IV) chloride results in the formation of a heteroannularly-substituted *dibenzyl compound* (**87**, R=$C_6H_5CH_2$).[187]

Re——H

87

As a base di-π-cyclopentadienylhydridorhenium(III) reacts with Lewis acids, such as boron trifluoride or -trichloride, but not with diborane(6), to form stable yellow and orange adducts with structures involving metal–metal donor–acceptor bonding, presumably of type **88** (M=B; X=F or Cl).[167] Similar *stable adducts* are prepared by treating the rhenium hydride with trimethylborane, -aluminum or -gallium and, although their exact structures are not known, they can probably be represented by **88** (M=B, Al or Ga; X=CH_3).[166]

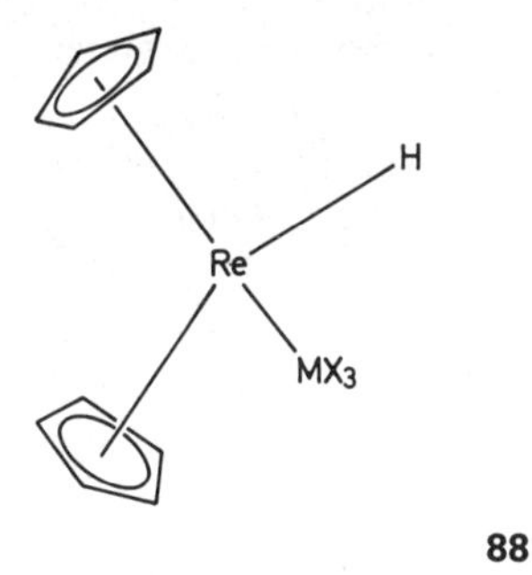

88

31.5.2 Acetylene Complexes and Cationic Salts from Di-π-cyclopentadienylmetal Hydrides of Technetium and Rhenium

Di-π-cyclopentadienylhydridorhenium(III) undergoes a facile addition, involving insertion of the metal–hydrogen bond, when treated with dimethyl acetylenedicarboxylate in tetrahydrofuran to form a *cis* adduct, the dark red, oxidatively unstable substituted maleate, **94** (R=CH_3OOC; R′=H), in good yield.[188]

The maleate may be converted nearly quantitatively into its geometric fumarate isomer, **89** (R=H; R′=CH_3OOC) by heating with freshly reduced platinum in benzene. However, in contrast to the maleate, the orange-brown substituted acrylate, **89** (R and R′=H), which is obtained as a *cis* adduct from the reaction of the rhenium hydride and methyl propiolate, cannot be isomerized over platinum.[188] Although the di-π-cyclopentadienylrhenium hydride readily forms adducts with acetylenic esters, it is unreactive towards acetylene and hex-3-yne at room temperature, as well as towards the olefinic esters, dimethyl maleate and dimethyl fumarate.

Re——C(R)=C(R′)OOCCH3

89

When di-π-cyclopentadienylhydridotechnetium is added to 2 M hydrochloric acid, it is transformed into the cationic chloride **90** (M=Tc; R=H; X=Cl), from whose solution the cation may be precipitated by treatment with hexafluorophosphate anion.[185] The analogous cationic salts, **90** [M=Re; R=H; X=Cl, Br, PF_6 and $Cr(NCS)_4(NH_3)_2$] are obtained when the corre-

[186] *M. L. H. Green, L. Pratt, G. Wilkinson*, J. Chem. Soc. 3916 (1958).

[187] U.S. Pat. 2 916 503 (1959), Ethyl Corp., Invent.: *J. Kozikowski;* C.A. *54*, 5693 (1960).

[188] *M. Dubeck, R. A. Schell*, Inorg. Chem. *3*, 1757 (1964).

sponding rhenium hydride is treated in a related manner with hydrogen halides or the appropriate anion.[185,186] A ring-substituted yellow, air-stable cationic salt whose structure is probably that of **90** (M=Re) can be prepared by dissolving the bis(chloromercuri) derivative of type **87** (R=HgCl) in concentrated hydrochloric acid.[165] Although neutral halides of di-π-cyclopentadienylrhenium(III) are unknown, maroon and green *halide salts* of *dihalodi-π-cyclopentadienylrhenium(V)* (**91**, X and Y=Cl, Br or I) are formed when a chloroform solution of di-π-cyclopentadienylhydridorhenium(III) is treated with chlorine, bromine or iodine.[170] The red-brown hexafluorophosphate **91** (X=Cl; Y=PF_6) may be obtained by dissolving the corresponding chloride in concentrated hydrochloric acid and adding aqueous ammonium hexafluorophosphate.[170]

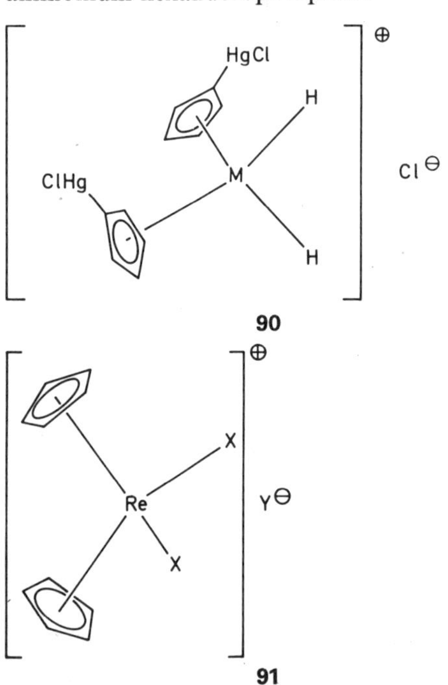

31.6 Ruthenium and Osmium Derivatives

31.6.1 Di-π-cyclopentadienylmetals and Cationic Salts

While a number of methods are available for the preparation of *di-π-cyclopentadienylruthenium(II)*, the oxidatively and very thermally-stable metallocene is best prepared in yields up to 65% by a modification[189] of the well-established procedure[190] in which excess sodium cyclopentadienide in 1,2-dimethoxyethane is heated with anhydrous ruthenium(III) chloride and ruthenium metal, according to the following equation:

$$6NaC_5H_5 + 2RuCl_3 + 2Ru \longrightarrow 3(\pi\text{-}C_5H_5)_2Ru + 6NaCl$$

An alternate synthesis, which leads to a purer product, although in somewhat lower yield, and which may have certain other advantages in the laboratory, involves the reaction of ruthenium(III) chloride with thallium(I) cyclopentadienide in tetrahydrofuran.[191]

The light yellow crystalline metallocene which is the most thermally stable (to almost 600°) of all metallocenes, serves as the starting material, similar to its congenor, ferrocene, for the preparation of a large number of substituted di-π-cyclopentadienylruthenium(II) derivatives.

The corresponding white, high melting and thermally stable osmium compound, *di-π-cyclopentadienylosmium(II)*, is obtained in relatively low yields of up to 25% by the reaction (similar to that used for its ruthenium analog) of excess sodium cyclopentadienide and osmium(IV) chloride in 1,2-dimethoxyethane, in which the halide is initially reduced to osmium(II) chloride.[192]

Cationic red *polyhalide* (**92**, X=I_3 or I_5) and yellow *perchlorate* (**92**, X=ClO_4) *salts* of *di-π-cyclopentadienylruthenium(III)* can be prepared by treating the parent metallocene with iodine in either carbon tetrachloride[193] or benzene,[194] or with perchloric acid in ethanol.[193] Other metallocenium salts, including the 2,4-dinitrophenoxide, 2,4,6-trinitrophenoxide (picrate), 7-iodo-8-hydroxyquinoline-5-sulfonate, 6-chloro-5-nitrotoluene-3-sulfonate, and tungstosilicate, are obtained from the perchlorate (**92**, X=ClO_4) by reprecipitation with the appropriate anion.[193] A cationic di-π-cyclopentadienylhydroxyosmium(IV) salt (**93**) can be synthesized in high yield by oxidizing di-π-cyclopentadienylosmium(II) with either iron(III) chloride or hydrated

---

[189] *O. Hofer, K. Schlögl,* J. Organometal. Chem. *13,* 443 (1968).

[190] *D. E. Bublitz, W. E. McEwen, J. Kleinberg,* Org. Synth. *41,* 96 (1961).

[191] *H. Rosenberg, R. A. Ference,* Abstr. Papers, 160th INOR Nat. Mtg., Amer. Chem. Soc. Chicago 1970.

[192] *E. O. Fischer, H. Grubert,* Chem. Ber. *92,* 2302 (1959).

[193] *G. Wilkinson,* J. Amer. Chem. Soc. *74,* 6146 (1952).

[194] *A. V. Savitskii, Ya. K. Syrkin,* Trudy po Khirn. i Khirn. Tekhnol. *4,* 165 (1961); C.A. *55,* 27246 (1961).

ammonium iron(III) sulfate and precipitating the cation with ammonium hexafluorophosphate.[192]

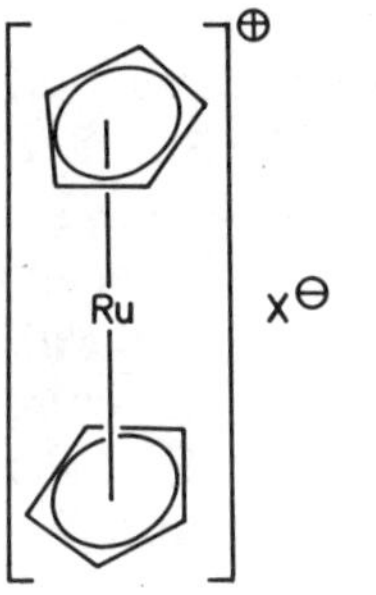

92

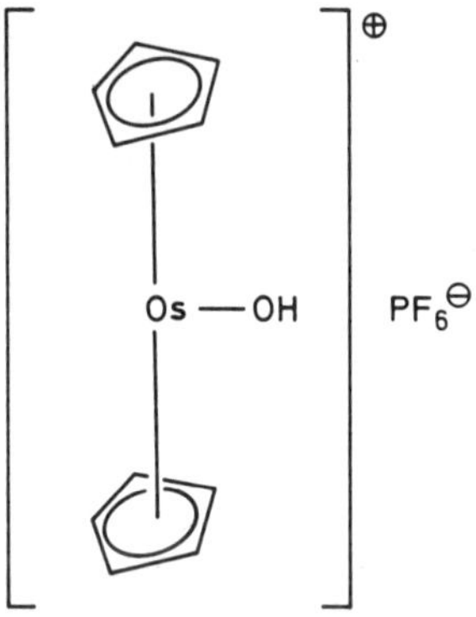

93

31.6.2 Hydrocarbon Derivatives of Di-π-cyclopentadienyl-ruthenium and -osmium

π-Cyclopentadienyl(methyl-π-cyclopentadienyl)ruthenium(II) can be prepared in over 90% yield by reduction of π-cyclopentadienyl-[(hydroxymethyl)-π-cyclopentadienyl]ruthenium(II) with lithium tetrahydroaluminate and aluminum(III) chloride in diethyl ether at room temperature.[189] When the reaction is extended to the monoacetyl or benzoyl derivatives of di-π-cyclopentadienyl-ruthenium and -osmium-(II), reduction of the carbonyl group is readily effected with formation of the corresponding monoethyl and -benzyl derivatives in equally high yields.[195]

Homoannularly-*substituted dialkyl compounds*, π-cyclopentadienyl[bis-1,2-(and 1,3-)dimethyl-π-cyclopentadienyl]ruthenium(II), are formed when isomeric 1-methoxy-carbonyl-2 (or 3)-methyl derivatives (resulting from the lithiation and carboxymethylation of π-cyclopentadienyl-(methyl-π-cyclopentadienyl)ruthenium(II)) are reduced in diethyl ether with the same reducing agents and separated by preparative thin-layer chromatography.[189] Although the corresponding *heteroannular* dimethyl compound (**94**, $R=CH_3$) can also be obtained from the above reaction mixture,[189] it is more directly and conveniently synthesized in higher purity by the addition of anhydrous ruthenium(III) chloride at $-78°$ to a tetrahydrofuran solution of lithium methylcyclopentadienide, derived from the reaction of *n*-butyllithium and freshly distilled methylcyclopentadiene.[196] An analogous colorless cotton-like dibenzyl derivative (**99**, $R=C_6H_5CH_2$) is formed in good yield by the reduction of the corresponding dibenzoyl compound with a nickel–zinc catalyst in a boiling aqueous dioxane medium.[196]

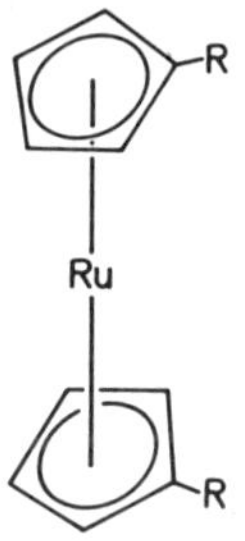

94

Monovinyl derivatives of *di-π-cyclopentadienyl-ruthernium* and *-oxmium(II)* can be conveniently prepared in high purity and in yields of 70–90% by the aluminum oxide-catalyzed dehydration of the corresponding 1-hydroxyethyl compounds by means of sublimation pyrolysis for 6–8 hours at 175–180° under reduced pressure (12–16 mm).[197] A *heteroannular divinyl compound*, yellow *bis(vinyl-π-cyclopentadienyl)ruthenium-(II)* (**94**, $R=CH_2{=}CH$) is obtained similarly in 65% yield from bis(α-hydroxyethyl-π-cyclopentadienyl)ruthenium(II).[197]

Both the mono- and di-vinyl ruthenium derivatives, unlike their air-sensitive ferrocene analogs (*cf.* Chapter 30), are quite stable in the crystalline state in air, but π-cyclopentadienyl(vinyl-π-cyclopentadienyl)-osmium(II) darkens after standing for several days in the open. A halovinyl compound, *π-cyclopentadienyl-(1-chlorovinyl-π-cyclopentadienyl)ruthenium(II)*, can be synthesized simply by stirring the corresponding monoacetyl compound with the Vilsmeier complex [phosphorus trichloride and *N,N*-dimethylformamide] for a very brief time at 20°.[189]

The monosubstituted alkyne, *π-cyclopentadienyl(ethynyl-π-cyclopentadienyl)ruthenium(II)*,

---

[195] *V. Mark, M. D. Rausch,* Inorg. Chem. *3*, 1067 (1964).

[196] *H. Rosenberg, R. A. Ference,* Air Force Materials Lab. W-PAFB, Ohio unpubl.

[197] *M. D. Rausch, A. Siegel,* J. Organometal. Chem. *11*, 317 (1968).

is readily formed when [(1-chloro-2-formylvinyl)-π-cyclopentadienyl]-π-cyclopentadienylruthenium(II) (obtained by chloroformylation of the corresponding monoacetyl derivative by 1–2 hr stirring with the Vilsmeier complex at room temperature) is treated with sodium amide in liquid ammonia[189] or, somewhat more conveniently, is refluxed with aqueous sodium hydroxide in dioxane solution.[198] Oxidative coupling of the ethynyl derivative can be effected by heating with vacuum-dried copper(II) acetate in a pyridine/methanol/diethyl ether solution to yield the 1,4-disubstituted buta-1,3-diyne (**95**) in 75% yield.[189]

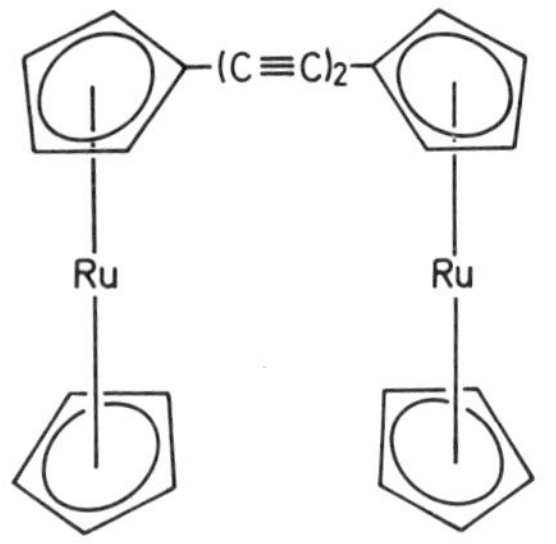

95

If an ethanolic solution of the ethynylmetallocene is treated with an ammoniacal solution of copper(I) iodide and the result copper acetylide derivative refluxed under nitrogen with iodoferrocene in pyridine, a good yield of the orange, high-melting mixed dimetallocenylacetylene, **96**, can be obtained.[198]

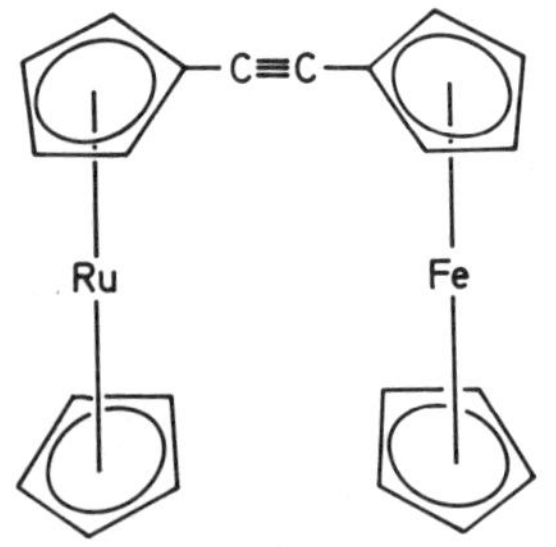

96

31.6.3 Halides and Pseudohalides

Mono-, di-, and tri-chloro derivatives (**97**, X=Cl; X′, X″ and X‴=H; or X and X′=Cl; X″ and X‴=H; or X, X′ and X″=Cl, X‴=H) of di-π-cyclopentadienylruthenium(II) can be obtained as a mixture, readily separable by preparative thin-layer chromatography, from the reaction of either *p*-toluene-sulfonyl chloride or hexachloroethane at −70° with the mixed mono-, di- and tri-lithio compounds resulting from treating di-π-cyclopentadienylruthenium(II) in dry hexane with *n*-butyllithium and *N,N,N′,N′*-tetramethylethylenediamine under nitrogen.[191] The corresponding mono-, di- and tri-bromo and -iodo analogs, as well as a tetrabromo derivative (**97**, X,X′,X″ and X‴=Br), separably by preparative gas chromatography, are prepared similarly by treatment with the novel halogenating agents, 1,2-dibromo-(or diiodo)tetrafluoroethane, developed for the synthesis of other ring-substituted halometallocenes.[199]

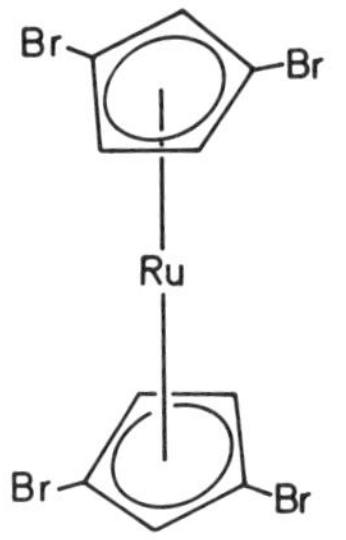

97

The di-, tri- and tetra-halo derivatives were shown by proton nuclear magnetic resonance spectroscopy in all cases to be the 1,1′-, 1,1′,3-, and 1,1′,3,3′-isomers, respectively. The mono-bromo and -iodo as well as dibromo compounds can be interconverted to other halides by metal–halogen exchange reactions involving addition of *n*-butyllithium in diethyl ether at −70° to the halo derivative, followed by treatment with the appropriate halogenated ethane.[191]

An interesting polyhalogenated compound, *bis-(1,2,3,4,5-pentachloro-π-cyclopentadienyl)ruthenium(II)*, **98** can be formed directly from di-π-cyclopentadienylruthenium(II) by six repetitive lithiation (with *n*-butyllithium in hexane) and halogenation (with hexachloroethane) steps, involving metal-halogen exchange reactions,

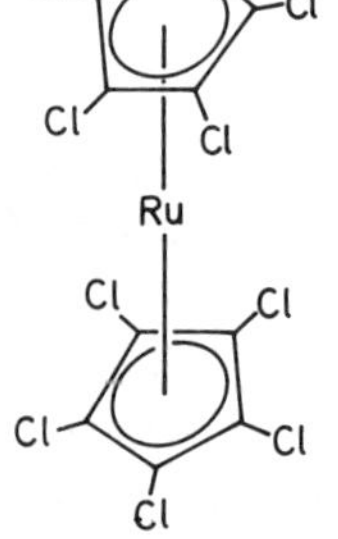

98

[198] *M. D. Rausch, A. Siegel,* J. Org. Chem. *34,* 1974 (1969).

[199] *R. F. Kovar, M. D. Rausch, H. Rosenberg,* Organometal. Chem. Synth. *1,* 173 (1971).

without isolation of the various polychloro intermediates.[200]

This novel light yellow decasubstituted metallocene exhibits unusually high thermal stability (>375°) in contrast to its ferrocene analog (270°)[201] and similar to the unsubstituted metallocene. In addition, it is extraordinarily resistant to attack by strong oxidants such as concentrated nitric acid, from whose carbon tetrachloride solution it is recovered unchanged after stirring for 3 days. The perchloro compound can be metallated quantitatively with *n*-butyllithium in tetrahydrofuran-hexane at −70° to yield the heteroannular dilithio derivative which, in turn, can be readily converted into other difunctional perchlorinated species, such as the 1,1′-diiodooctachloro compound, by iodination, carbonation and hydration.[200]

By cyanation of di-π-cyclopentadienylruthenium(II), or in reality its cationic salt, with hydrogen cyanide in tetrahydrofuran in the presence of iron(III) chloride, the monocyano derivative, *(cyano-π-cyclopentadienyl)-π-cyclopentadienylruthenium(II)*, can be prepared although in only 5% yield.[202] In contrast, yellow *α-azidobenzyl compounds*, **99** (M=Ru or Os) are formed in yields of *ca.* 90% by treating π-cyclopentadienyl-[(α-hydroxybenzyl)-π-cyclopentadienyl]-ruthenium or -osmium(II) (obtained from the lithium tetrahydridoaluminate reduction of the corresponding monobenzoyl derivatives) with a solution of hydrogen azide in benzene at room temperature over a three day period, using formic acid as a catalyst.[203]

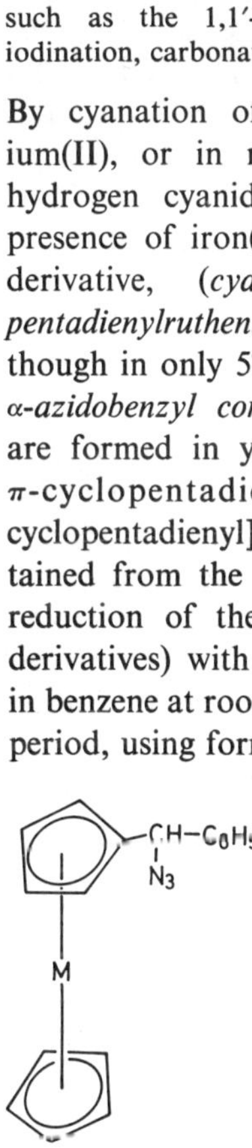

99

31.6.4 Hydroxy and Carbonyl Derivatives

The monocarbinol, *π-cyclopentadienyl[(hydroxymethyl)-π-cyclopentadienyl]ruthenium(II)*, **100** (M=Ru; R=H), m.p. 79–81°, may be synthesized in over 80% yield by heating π-cyclopentadienyl[(dimethylamino)methyl-π-cyclopentadienyl]ruthenium(II) with iodomethane in the presence of aqueous sodium hydroxide.[189] However, the carbinol may be contaminated with traces of *oxybis[(methylene-π-cyclopentadienyl)-π-cyclopentadienylruthenium(II)]* since the same compound, but with a melting point of 88·5–89·5°, can be prepared, in somewhat lower yield, but in high purity, by lithium tetrahydridoaluminate reduction of the mono(methoxycarbonyl) compound in diethyl ether.[196] The analogous heteroannularly-substituted dicarbinol, *bis[(1-hydroxymethyl)-π-cyclopentadienyl]ruthenium(II)*, is obtained in a similar manner in 50% yield by reduction of the corresponding bis(methoxycarbonyl) compound.[196]

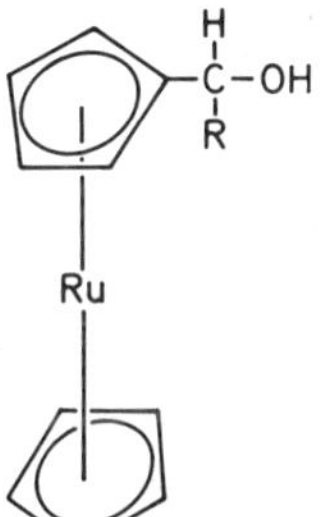

100

Hydroxyethyl and *hydroxymethylphenyl derivatives* (**100**, M=Ru or Os; R=CH_3 or $CH_2C_6H_5$) of both di-π-cyclopentadienyl-ruthenium and -osmium(II) are formed in excellent yields when the appropriate monoacetyl or monobenzoyl compound is reduced with sodium tetrahydridoborate in aqueous methanol[204] or with lithium tetrahydridoaluminate in anhydrous diethyl ether.[203] If the ruthenium compound, **100** (M=Ru; R=C_6H_5), is heated with copper powder at 210° for 10 min., the metallocenylmethyl(phenyl) ether, **101**, is produced in *ca.* 40% yield.[203]

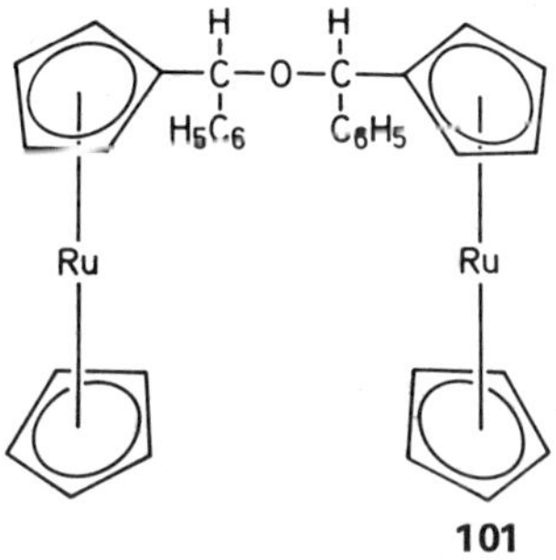

101

[200] *H. Rosenberg, F. L. Hedberg*, Abstr. Papers, 5th Intern. Conf. Organometal. Chem. Paper *89*, p. 236, Moscow 1971.

[201] *F. L. Hedberg, H. Rosenberg*, J. Amer. Chem. Soc. *92*, 3239 (1970).

[202] *A. N. Nesmeyanov, A. A. Lubovich, L. P. Yureva, S. P. Gubini, E. G. Perelova*, Izv. Akad. Nauk. SSSR, Ser. Khim. 935 (1967); engl.: 906.

[203] *D. E. Bublitz, W. E. McEwen, J. Kleinberg*, J. Amer. Chem. Soc. *84*, 1845 (1962).

[204] *G. R. Buell, W. E. McEwen, J. Kleinberg*, J. Amer. Chem. Soc. *84*, 40 (1962).

By oxidation of the monohydroxy compound, **100** (M = Ru; R = H), with activated manganese(II) oxide in chloroform, the monoaldehyde *π-cyclopentadienyl(formyl)π-cyclopentadienyl)ruthenium(II)* can be prepared in very high yield.[189] A bright yellow crystalline metallocene-substituted β-chloroacrylaldehyde, **102**, is formed in high purity and in yields of up to 90% when a mixture of the monoacetyl derivative and Vilsmeier complex [phosphoryl(III)chloride and *N,N*-dimethylformamide] is stirred for 1–2 hr at 25°.[198]

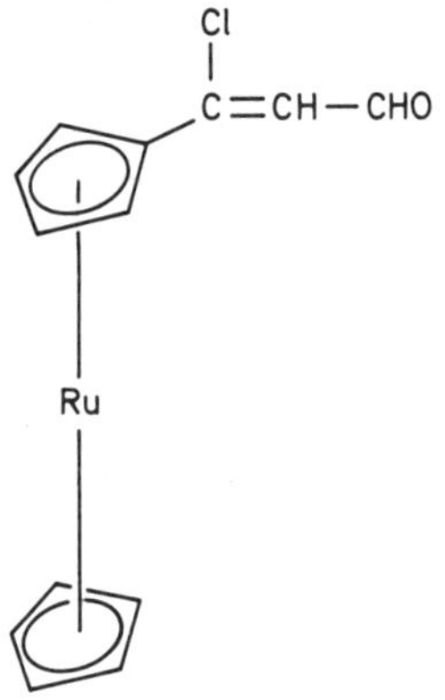

102

Di-π-cyclopentadienylruthenium(II) can be readily acetylated, in a Friedel-Crafts reaction, by treatment with a dichloromethane solution of an equimolar quantity of acetyl chloride in the presence of anhydrous aluminum chloride to furnish the yellow *monoacetyl derivative*, **103** (M = Ru; R = CH_3; R′ = CH_3), in 45% yield.[205] By an identical reaction but utilizing three equivalents of the acetyl halide, a 22% yield of the heteroannular diacetyl compound (**103**, M = Ru; R = CH_3; R′ = CH_3CO) can be obtained.[205] The light yellow *monoacetyl osmium* analog, **103** (M = Os; R = CH_3; R′ = H), can be prepared in 85% yield by acetylation of the less reactive osmium-derived metallocene with a large excess of acetic anhydride in the presence of phosphoric acid,[206] while the corresponding heteroannular diacetyl derivative can only be isolated as a by-product of the reaction in less than 1% yield.[207]

Monobenzoyl derivatives (**103**, M = Ru and Os;

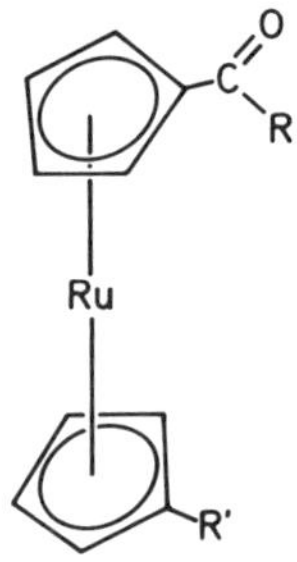

103

R = C_6H_5) can be synthesized in yields of 60–65% by a similar Friedel-Crafts reaction involving the refluxing of a mixture of the appropriate metallocene with a benzoyl chloride–aluminum(III) chloride complex in dichloromethane for 20 hours.[205] Both a *2-methoxy-* and *2-hydroxybenzoyl compound* (**103**, M = Ru; R = o-CH_3O-C_6H_4CO and o-HOC_6H_4CO; R′ = H) are prepared in a like manner by reaction of the metallocene with o-methoxybenzoyl chloride in presence of varying amounts of the aluminum halide catalyst.[208] From the benzoylation reaction of di-π-cyclopentadienylruthenium(II) both a homoannular 1,2- and heteroannular *1,1′-dibenzoyl derivative* (**103**, M = Ru; R = C_6H_5; R′ = C_6H_5CO), the latter in yields of 19–24%, can be obtained in addition to the monobenzoyl compound. The interesting orange-red mixed ketone, **104**, is readily formed by an extension of the aroylation procedure in which a solution of di-π-cyclopentadienylruthenium(II) and chlorocarbonylferrocene is added to a mixture of aluminum(III) chloride in dichloromethane.[205]

104

31.6.5 Carboxy, Alkoxycarbonyl, and Carbamoyl Derivatives

The monocarboxylic acid, *π-cyclopentadienyl-(carboxy-π-cyclopentadienyl)ruthenium(II)* (**105**, X = O; R and R′ = H) is prepared, in combina-

[205] *M. D. Rausch, E. O. Fischer, H. Grubert*, J. Amer. Chem. Soc. *82*, 76 (1960).

[206] *E. A. Hill, J. H. Richards*, J. Amer. Chem. Soc. *83*, 3225 (1963).

[207] *M. D. Rausch, V. Mark*, J. Org. Chem. *28*, 3225 (1963).

[208] *C. Kashima, R. Kobayashi, N. Sugiyama*, Nippon Kagaku Zasshi *90*, 1053 (1969); C.A. *72*, 31989v (1970).

tion with the diacid, by treating di-π-cyclopentadienylruthenium(II) with n-butyllithium in diethyl ether under nitrogen, followed by carbonation and hydrolysis of the monolithio derivative formed *in situ*.[205] The corresponding heteroannularly-substituted dicarboxy compound (**105**, X=O; R=H; R′=HOOC) is obtained in 86% yield, with 1% of the monoacid as by-product, by a similar procedure but carrying out the lithiation in tetrahydrofuran-diethyl ether solution.[205]

Homoannular 1,2- and 1,3-, as well as heteroannular (**105**, X=O; R=H; R′=CH_3), methyl-substituted monocarboxylic acids are formed and chromatographically separated upon saponification of a mixture of the three monoesters derived from lithiation of π-cyclopentadienyl-(methyl-π-cyclopentadienyl)ruthenium(II) and carbonation of the mixed lithio compounds, followed by esterification.[189]

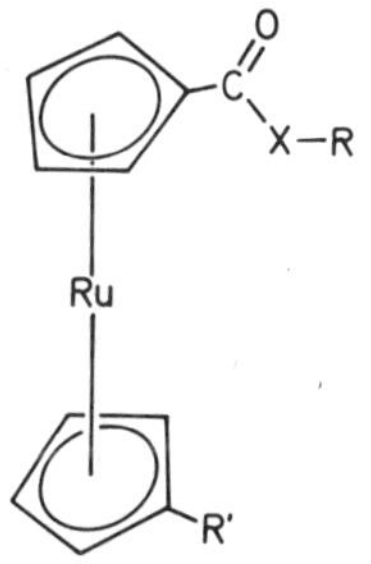

105

Metallocene-substituted alkanoic acids are prepared by the aluminum chloride-catalyzed acylation of di-π-cyclopentadienylruthenium(II) in dichloromethane to give a 67% yield of the γ-keto derivative (**106**, R=CO) which, on Clemmenson reduction with zinc amalgam and hydrochloric acid in methanol–benzene, followed by saponification of the resulting ester, yields the γ-substituted butyric acid (**106**, R=CH_2).[189]

When heteroannular bis(carboxy-π-cyclopentadienyl)ruthenium(II) is heated with succinic anhydride in dry dichloromethane, the corresponding yellow bis(chloroformyl) compound can be obtained, although in only relatively low yield.[196] Treatment of the monoacid with methanol, in the presence of a trace amount of sulfuric acid, yields the monoester (**105**, X=O; R=CH_3; R′=H).[205] Similar treatment of the mixed acids (derived from lithiation of di-π-cyclopentadienylruthenium(II) and carbonation) results in a good yield of the heteroannular diester, in addition to a smaller amount of the monoester.[196] As previously indicated, a mixture of 1- and 2-homoannular, as well as 1′-heteroannular, methyl-substituted monomethyl esters, separable by preparative thin-layer chromatography in the ratio of 10:20:70%, can be prepared by esterification (with diazomethane) of the mixed acids obtained from the lithiation and carbonation of the monomethyl metallocene.[189]

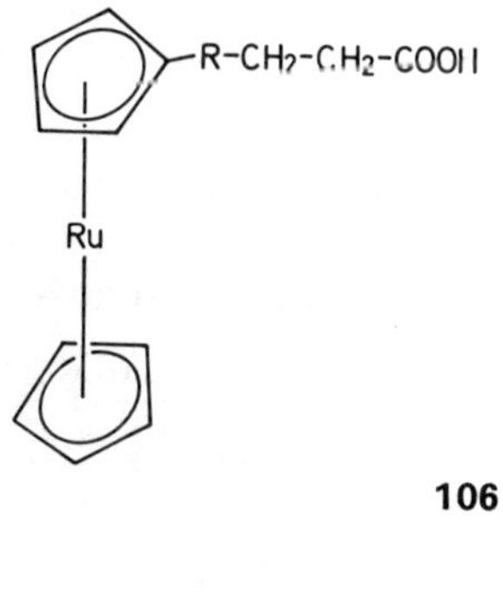

106

Acetoxy derivatives of the *1-hydroxyethylmetallocenes* (**107**, M=Ru or Os) are formed in yields of 75–90% by reaction of the appropriate 1-hydroxyethyl compound with acetic anhydride in pyridine.[204, 206]

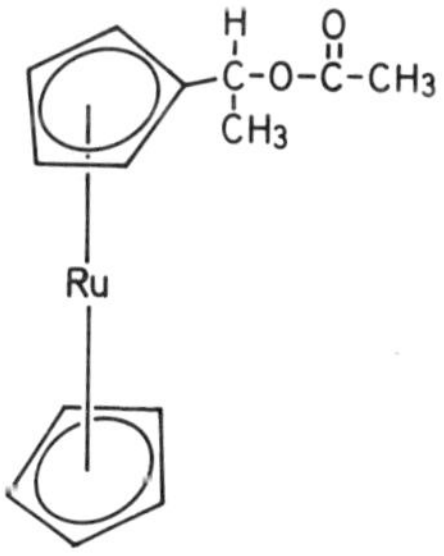

107

An alkylthio ester (**105**, X=S; R=CH_3; R′=H) can be obtained by the aluminum chloride catalyzed Friedel-Crafts reaction of di-π-cyclopentadienylruthenium(II) with o-methylchlorothioformate in dichloromethane.[209]

Similar to the synthesis of the methylthio ester, the N-phenylcarbamoyl derivative (**105**, X=NH; R=C_6H_5; R′=H) is prepared by refluxing a mixture of excess phenylisocyanate and di-π-cyclopentadienylruthenium(II) in dichloromethane, in the presence of aluminum chloride.[205] Related mixed homoannular, as well as heteroannular [**105**, X=N; R=$(C_6H_5)_2$; R′=H] methyl-substituted N,N-diphenylcarbamoyl compounds are obtained in an analogous manner from the monomethyl metallocene by an aluminum chloride catalyzed reaction with

[209] *D. E. Bublitz, G. H. Harris*, J. Organometal. Chem. *4*, 404 (1965).

N,N-diphenylcarbamoyl chloride in 1,2-dichloroethane.[189]

Although not a carbamoyl derivative, the light yellow, low-melting methyleneamino compound, *π-cyclopentadienyl[(N.N-dimethylamino)methyl-π-cyclopentadienyl]-ruthenium(II)* can be formed, similar to carbamoyl-derivatives, by the Friedel–Crafts alkylation of di-π-cyclopentadienylruthenium(II) with bis(dimethylamino)methane in acetic acid–phosphoric acid at 120°.[189]

31.6.6 Dinuclear Derivatives, Fused Ring Analogs and Bimetallic Derivatives of Di-π-cyclopentadienylruthenium

When π-cyclopentadienyl(iodo-π-cyclopentadienyl)ruthenium(II) is heated under nitrogen with copper bronze at 120°–140° for 16 hours, the light yellow dinuclear compound, *bicyclopentadienylbis[cyclopentadienylruthenium (II)]* (**108**, $R=C_5H_5RuC_5H_4$; R′=H) is formed, although in relatively low yields.[191]

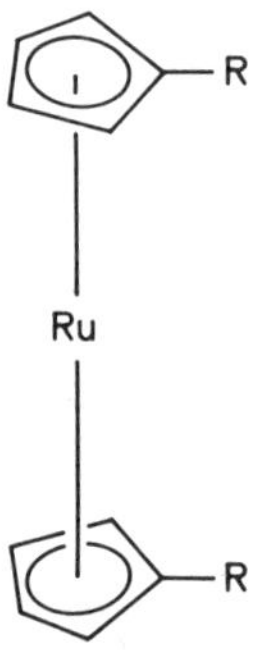

R
Ru
R

108

Similarly, by heating the corresponding heteroannular diiodo derivative with copper bronze in diethylbenzene at 170°–180;, the interesting and novel, extremely high melting and thermally stable, metallocenophane, *bis[bicyclodienyleneruthenium(II)]*, **109**, can be obtained in a comparable yield.[191]

Nonsymmetrical *derivatives* of *di-π-cyclopentadienylruthenium(II)* with a single fused ring

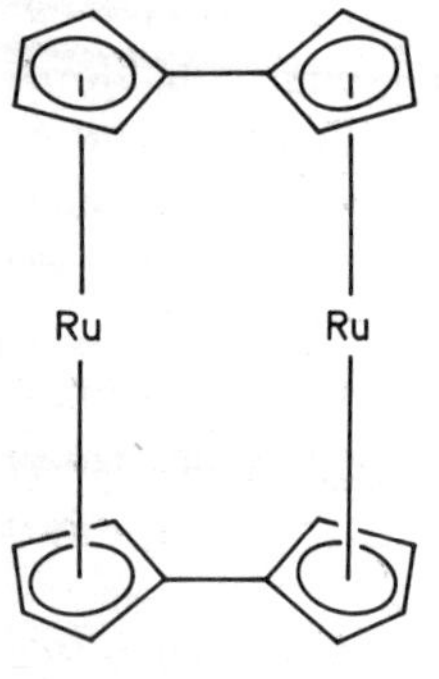

Ru Ru

109

ligand may be formed by cyclization reactions of carboxy- and azidomethyl- (or carbonyl)-substituted precursors. Thus, *π-cyclopentadienyl-(4,5,6,7-tetrahydro-4-oxoindenyl)ruthenium(II)* can be prepared in high yield by the thermal dehydration and ring closure of [(3-carboxypropyl)-π-cyclopentadienyl]-π-cyclopentadienylruthenium(II) (**106**, $R=CH_2$), in the presence of polyphosphoric acid.[189] The tricyclic fused indanone derivative *π-cyclopentadienyl(dihydro-8-oxocyclopent[a]indenyl)ruthenium(II)* is obtained as a by-product from the reaction of [(a-azidobenzyl)-π-cyclopentadienyl]-π-cyclopentadienylruthenium(II) with concentrated sulfuric acid in chloroform or in acetic acid.[203] The orange, crystalline symmetrical fused ring analog, *di-π-indenylruthenium(II)* (**110**), ehose two ligands are fully eclipsed in the solid state, can be synthesized by the addition of a suspension of specially-dried ruthenium(III) chloride and ruthenium in tetrahydrofuran to a tetrahydrofuran solution of sodium indenide.[210] The corresponding air-stable bis(4,5,6,7-tetrahydro-π-indenyl) derivative is formed by the catalytic hydrogenation of **110** in 95% ethanol over platinum(IV) oxide at 25°.[210]

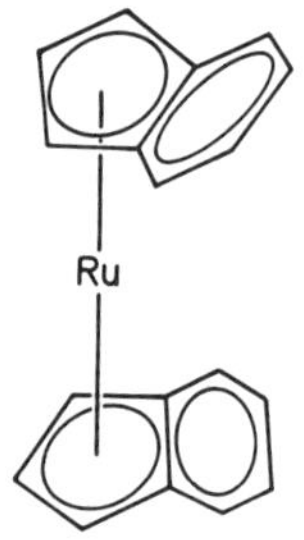

Ru

110

White or light-yellow, low melting monosubstituted and heteroannularly-disubstituted Group IV *organometallic derivatives* [**108**, $R=(CH_3)_3Si$, $(CH_3)_3Ge$ or $(CH_3)_3Sn$, R′=H; and R and $R'=(CH_3)_3Si$, $(CH_3)_3Ge$ or $(CH_3)_3Sn$] of *di-π-cyclopentadienylruthenium(II)* may be prepared by treatment of a mixture of the mono- and di-lithio compounds (derived from lithiation of the metallocene in diethyl ether) with chlorotrimethylsilane, bromotrimethylgermane or chlorotrimethylstannane at −70° and separation of the organometallated species by dry-column or preparative thin-layer chromatography.[191]

---

[210] *J. H. Osiecki, C. J. Hoffman, D. P. Hollis*, J. Organometal. Chem. *3*, 107 (1965).

31.7 Cobalt, Rhodium and Iridium, Metallocenes

31.7.1 Di-π-cyclopentadienylmetals

Cobalt, rhodium and iridium form di-π-cyclopentadienylmetal compounds which are relatively unstable but which are readily oxidized to stable cations. While commercially available as a benzene solution, *di-π-cyclopentadienylcobalt(II)* can be prepared in yields of up to 90% by the addition of pure blue anhydrous cobalt(II) chloride to a solution of sodium cyclopentadienide in tetrahydrofuran and refluxing of the mixture for 2 hr.[211] An alternate high-yield synthesis of the compound which may have certain advantages in the laboratory involves the heating of anhydrous cobalt(II) chloride with thallium cyclopentadienide in benzene under a nitrogen atmosphere.[56]

The very air-sensitive, purple-black crystalline and paramagnetic metallocene may be stable for very brief periods in air but is best handled in the absence of oxygen.

The brown-black paramagnetic analog, *di-π-cyclopentadienylrhodium(II)*, monomeric at or below −196°, is isolated in 8% yield as a dimer from the molten sodium metal reduction at 120° *in vacuo* of di-π-cyclopentadienylrhodium-(III)(1+)hexafluorophosphate(1−), **111** (M=Rh). The salt is synthesized by treatment of an aqueous solution of the product from the reaction of cyclopentadienylmagnesium bromide and rhodium(III) chloride with aqueous ammonium hexafluorophosphate.[212] A low yield of the corresponding pale yellow iridium compound with near-identical properties is obtained by a similar reduction of the analogous iridium-containing hexafluorophosphate salt (**111**, M=Ir) which, in turn, is derived by heating a mixture of iridium(III) chloride and cyclopentadienylmagnesium bromide in tetrahydrofuran in an autoclave at 200°–220° for 5 hr.[212] The reduction of the hexafluorophosphate in both cases must be carried on a small scale in order to avoid explosions.

A heteroannularly disubstituted derivative, the extremely air-sensitive brown-purple *bis(methyl π-cyclopentadienyl)cobalt(II)*, **112**, can be prepared by the addition of a tetrahydrofuran solution of sodium methylcyclopentadienide to anhydrous cobalt(II) chloride.[133]

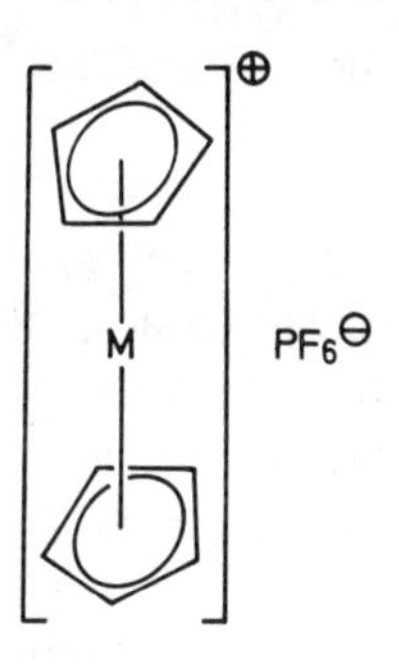

111

112

If the hexa-amine salt which is obtained from the reaction of potassium indenide and dithiocyanatotetra-aminecobalt(II) in liquid ammonia carried out according to the following equation:

$$2KC_9H_7 + Co(NH_3)_4(SCN)_2 + 2NH_3 \longrightarrow (\pi\text{-}C_9H_7)_2Co(NH_3)_6 + 2KSCN$$

is decomposed *in vacuo*, the shiny black crystalline, paramagnetic fused ring analog of di-π-cyclopentadienylcobalt(II), *di-π-idenylcobalt-(II)* (**113**), is formed.[213]

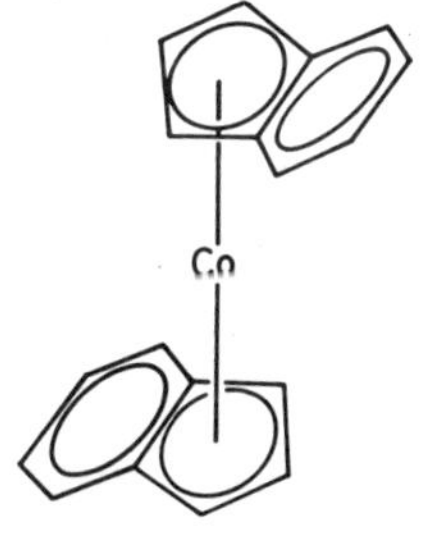

113

31.7.2 Salts of Cationic Unsubstituted Metallocenes

As indicated in 31.7.1, di-π-cyclopentadienylcobalt, -rhodium and -iridium(II) form stable cations which can be utilized in solution or isolated as quite stable salts. The cations can

[211] *R. B. King*, Organometallic Syntheses, Vol. I, p. 70. Academic Press, New York, 1965.

[212] *E. O. Fischer, H. Wawersik*, J. Organometal. Chem. *5*, 559 (1966).

[213] *E. O. Fischer, D. Seus, R. Jira*, Z. Naturforsch. *8b*, 692 (1953).

be prepared by oxidation of the metallocenes with halogens, hydrogen halides, carbon tetrachloride or hydrogen peroxide, as well as by direct synthesis, in certain cases, from the reaction of the respective metal halides or acetylacetonates with alkali or alkaline earth metal cyclopentadienides. A summary of the more important salts derived from the unsubstituted *cationic di-π-cyclopentadienylmetal compounds* of *cobalt*, *rhodium*, and *iridium*, and common anions, together with reactants from the preferred method of synthesis and some physical properties, is contained in Table 2. A similar summary of salts obtained from the cationic metallocenes and metal-containing or organic anions is presented in Table 3.

31.7.3 Salts of Cationic Ring-substituted Di-π-cyclopentadienylmetal Compounds of Cobalt and Rhodium

Cationic salts of monosubstituted and heteroannularly-disubstituted di-π-cyclopentadienylcobalt and -rhodium(III), similar to their unsubstituted analogs, are generally prepared by reactions of substituted cyclopentadienides or cyclopentadienones with cobalt or rhodium halides, as well as by substitution of the appropriate metallocene or metallocenyl cation, fol-

Table 2 Cationic Salts of Unsubstituted Di-π-cyclopentadienyl-cobalt, -rhodium and -iridium(III)

$[(\pi\text{-}C_5H_5)_2M]X$		Reactants		Physical data		Ref. no.
M	X	Cationic	Anionic	Color	F [°C]	
Co	Cl	$(\pi\text{-}C_5H_5)_2Co$	CCl_4			214
Rh	Cl	$\pi\text{-}C_5H_5RhC_5H_5$	2N HCl			215
Co	Br	$C_5H_5MgBr + CoBr_2$	H_2O_2–H_2O	yellow		216
Co	Br_3	$C_5H_5Tl + CoBr_2$	$Br_2 + HBr$	yellow		217
Rh	Br_3	$C_5H_5MgBr + RhCl_3$	$Br_2 + KBr$	yellow	182–5 d.	218
Ir	Br_3	$C_5H_5MgBr + IrCl_3$	$Br_2 + KBr$	yellow	>185 d.	212
Co	I_3	$[(\pi\text{-}C_5H_5)_2Co]^{\oplus}$	KI_3	brown		219
Rh	I_3	$C_5H_5MgBr + Rh(C_5H_7O_2)_3$	$I_2 + KI$			220
Ir	I_2	$C_5H_5MgBr + Ir(C_5H_7O_2)_3$	$I_2 + KI$			220
Co	SCN	$(\pi\text{-}C_5H_5)_2Co$	$(SCN)_2$	yellow		221
Co	ClO_4	$[(\pi\text{-}C_5H_5)_2Co]^{\oplus}$	$ClO_4^{\ominus}$			216
Co	OH	$[(\pi\text{-}C_5H_5)_2Co]Cl$	Ag_2O			219
Rh	OH	$[(\pi\text{-}C_5H_5)_2Rh]Cl$	Ag_2O			220
Ir	OH	$[(\pi\text{-}C_5H_5)_2Ir]Cl$	Ag_2O			220
Co	SO_4	$[(\pi\text{-}C_5H_5)_2Co]^{\oplus}$	$SO_4^{2\ominus}$			216
Co	NO_3	$[(\pi\text{-}C_5H_5)_2Co]^{\oplus}$	$NO_3^{\ominus}$	yellow		216
Co	PF_6	$[(\pi\text{-}C_5H_5)_2Co]^{\oplus}$	$PF_6^{\ominus}$	yellow		222
Rh	PF_6	$[(\pi\text{-}C_5H_5)_2Rh]^{\oplus}$	$PF_6^{\ominus}$		>300	212
Ir	PF_6	$[(\pi\text{-}C_5H_5)_2Ir]^{\oplus}$	$PF_6^{\ominus}$		>300	212

[214] *S. Katz, J. F. Weiher, A. F. Voigt*, J. Amer. Chem. Soc. *80*, 6459 (1958).
[215] *E. O. Fischer, U. Zahn*, Chem. Ber. *92*, 1624 (1959).
[216] *E. O. Fischer, R. Jira*, Z. Naturforsch. *8b*, 1 (1953).
[217] *A. N. Nesmeyanov, R. B. Materikova, N. S. Kochetkova*, Izv. Akad. Nauk. SSSR, Otd. Khim. Nauk, 1334 (1963); engl.: 1211.
[218] *R. J. Angelici, E. O. Fischer*, J. Amer. Chem. Soc. *85*, 3733 (1963).
[219] *G. Wilkinson*, J. Amer. Chem. Soc. *74*, 6148 (1952).
[220] *F. A. Cotton, R. O. Whipple, G. Wilkinson*, J. Amer. Chem. Soc. *75*, 3586 (1953).
[221] *M. F. A. Dove, D. R. Sowerby*, Z. Naturforsch. *20b*, 394 (1965).
[222] *C. Funlani, I. Collamati*, Chem. Ber. *95*, 2928 (1962).

Table 3 Salts of Unsubstituted Di-π-cyclopentadienyl-cobalt, -rhodium and -iridium(II) and Metal-containing or Organic Anions

$[(\pi\text{-}C_5H_5)_2M]X$		Reactants		Color	Ref. no.
M	X	Cationic	Anionic		
Co	$B(C_6H_5)_4$	$(\pi\text{-}C_5H_5)_2Co + H_2O_2\text{–}H_2O$	$NaB(C_6H_5)_4$	yellow	223
Rh	$B(C_6H_5)_4$	$\pi\text{-}C_5H_5RhH_5C_5 + H_2O_2\text{–}HCl$	$NaB(C_6H_5)_4$		215
Ir	$B(C_6H_5)_4$	$\pi\text{-}C_5H_5IrC_5H_5 + H_2O_2\text{–}HCl$	$NaB(C_6H_5)_4$		215
Co	$CoCl_4$	C_5H_5Na	xs $CoCl_2$	green	222
Co	$CoBr_4$	$C_5H_5MgBr + CoBr_2$	dil. HBr	green	224
Co	$Co(CO)_4$	$[(\pi\text{-}C_5H_5)_2Co]Br$	$NaCo(CO)_4$	red	216
Co	$Cr(NCS)_4(NH_3)_2$	$[(\pi\text{-}C_5H_5)_2Co]^{\oplus}$	$[Cr(NCS)_4(NH_3)_2]^{\ominus}$		216
Rh	$Cr(NCS)_4(NH_3)_2$	$[(\pi\text{-}C_5H_5)_2Rh]^{\oplus}$	$[Cr(NCS)_4(NH_3)_2]^{\ominus}$		220
Ir	$Cr(NCS)_4(NH_3)_2$	$[(\pi\text{-}C_5H_5)_2Ir]^{\oplus}$	$[Cr(NCS)_4(NH_3)_2]^{\ominus}$		220
Co	$C_5H_5Cr(CO)_3$	$[(\pi\text{-}C_5H_5)_2Co]^{\oplus}$	$C_5H_5Cr(CO)_3Na$		161
Co	$HFe(CO)_4$	$[(\pi\text{-}C_5H_5)_2Co]^{\oplus}$	$HFe(CO)_4^{\ominus}$		216
Co	$Fe_2(CO)_8$	$[(\pi\text{-}C_5H_5)_2Co]^{\oplus}$	$Fe_2(Co)_8^{2\ominus}$		225
Co	$Fe_4(CO)_{13}$	$(\pi\text{-}C_5H_5)_2Co$	$Fe_4(CO)_{13}^{2\ominus}$	black	226
Co	$Mn(CO)_5$	$[(\pi\text{-}C_5H_5)_2Co]OH$	$KMn(CO)_5$	violet	227
Co	$PtCl_6$	$[(\pi\text{-}C_5H_5)_2Co]Cl$	H_2PtCl_6		219
Co	$SiW_{12}O_{40}$	$[(\pi\text{-}C_5H_5)_2Co]Cl$	$H_4SiW_{12}O_{40}$	yellow	219
Rh	$SiW_{12}O_{40}$	$[(\pi\text{-}C_5H_5)_2Rh]^{\oplus}$	$H_4SiW_{12}O_{40}$		220
Ir	$SiW_{12}O_{40}$	$[(\pi\text{-}C_5H_5)_2Ir]^{\oplus}$	$H_4SiW_{12}O_{40}$		220
Co	$C_5H_5W(CO)_3$	$[(\pi\text{-}C_5H_5)_2Co]^{\oplus}$	$C_5H_5W(CO)_3Li$		228
Co	$(O_2N)_2C_6H_3O$	$[(\pi\text{-}C_5H_5)_2Co]^{\oplus}$	$(O_2N)_2C_6H_3ONa$	yellow	219
Co	$(O_2N)_3C_6H_2O$	$[(\pi\text{-}C_5H_5)_2Co]^{\oplus}$	$(O_2N)_3C_6H_4OH$	orange	219
Rh	$[(O_2N)_3C_6H_2N]_2$	$[(\pi\text{-}C_5H_5)_2Rh]^{\oplus}$	$[(O_2N)_3C_6H_2]_2NH$	scarlet	220
Ir	$[(O_2N)_3C_6H_2]_2N$	$[(\pi\text{-}C_5H_5)_2Ir]$	$[(O_2N)_3C_6H_2]_2NH$	scarlet	220
Co	$Cl_4C_6O_2$	$[(\pi\text{-}C_5H_5)_2Co]I_3$	$Cl_4H_4C_6O_2$	brown	229
Co	$[(CN)_2C]_2C_6H_4$	$[(\pi\text{-}C_5H_5)_2Co]Cl$	$(CN)_2{=}C(CN)_2$	black	230

lowed by oxidation and/or precipitation with the desired anion.

The *tribromide* and *Reineckate salts* of the cationic heteroannular dimethyl compound, **114** [R and R′ = CH_3; X = Br_3 or $Cr(NCS)_4(NH_3)_2$], are formed by precipitation of the cation, resulting from oxidation of the product of the reaction of sodium methylcyclopentadienide and cobalt(II) chloride, with bromine water or Reinecke's salt.[231]

[223] *E. O. Fischer, R. Jira,* Z. Naturforsch. *8b*, 327 (1953).

[224] U.S. Pat. 2 988 563 (1961), Union Carbide Corp., Invent.: *J. C. Brantley;* C.A. *55*, 24789 (1961).

[225] *W. Hieber, R. Werner,* Chem. Ber. *90*, 1116 (1957).

[226] *G. A. Razuvaev et al.,* Zh. Obshch. Khim. *37*, 672 (1967); engl.: 630.

[227] *W. Hieber, W. Schropp jr.,* Chem. Ber. *93*, 455 (1960).

[228] *E. O. Fischer, W. Hafner, H. O. Stahl,* Z. Anorg. Allgem. Chem. *282*, 47 (1955).

[229] U.S. Pat. 3 379 740 (1968), American Cyanamide Co., Invent.: *Y. Matsunga;* C.A. *69*, 44025R (1968).

[230] *L. R. Melby, R. J. Harder, W. R. Hertler, W. Mahler, R. E. Benson, W. E. Mochel,* J. Amer. Chem. Soc. *84*, 3374 (1962).

[231] *M. L. H. Green, L. Pratt, G. Wilkinson,* J. Chem. Soc. 3753 (1959).

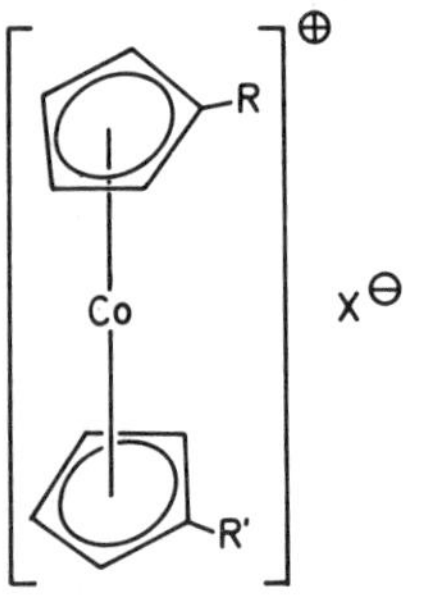

114

In a similar manner, but utilizing a thallium alkyl-substituted cyclopentadienide, the tribromides of the corresponding heteroannular diethyl, di-*t*-butyl and bis(1,3-di-*t*-butyl) derivatives can be obtained in yields of 63–75%.[232] The tribromide salt of the diethyl as well as of the homologous di-*n*-propyl derivative (**114**, R and R'$=C_2H_5$ or *n*-C_3H_7; X$=Br_3$) can be used, in turn, to synthesize the corresponding *hexachloroplatinates*[232,233] or a *trichloromercurate*, by conversion with silver(I) oxide and hydrochloric acid to the soluble chloride, followed by precipitation with hexachloroplatinic acid or mercury(II) chloride. A rather unique homoannular polyalkyl-substituted cationic salt, π-cyclopentadienyl(pentamethyl-π-cyclopentadienylrhodium(III)(1+) *hexafluorophosphate(1−)* (**115**) is formed in 76% yield when the product of the reaction of μ-dichloro-dichlorobis(pentamethyl-π-cyclopentadienul)dirhodium(II) and cyclopentadiene in ethanol in the presence of sodium carbonate is treated with hexafluorophosphate anion.[234]

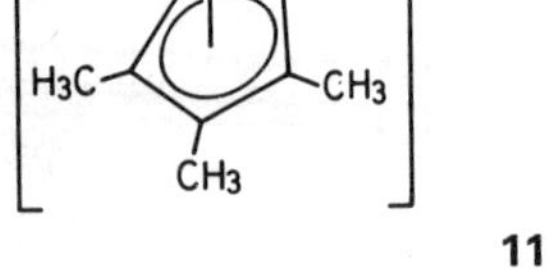

115

[232] *A. N. Nesmeyanov, R. B. Materikova, N. S. Kochetkova, E. V. Leonova*, Dokl. Akad. Nauk. SSSR *177*, 131 (1967); engl.: 994.

[233] *A. N. Nesmeyanov, R. B. Materikova, N. S. Kochetkova, L. A. Tsurgozen*, Dokl. Akad. Nauk SSSR *160*, 137 (1965); engl.: 39.

[234] *K. Moseley, P. M. Maitlis*, Chem. Commun. 616 (1969).

Tetraphenylborate and reineckate salts of the heteroannularly-disubstituted bis(dimethylamino)methyl as well as the diisopropenyl cations together with *tribromide, perchlorate* and *picrate* salts of the latter [**114**, R and R'$=(CH_3)_2NCH_2$ or $CH_2=C(CH_3)$; X$=B(C_6H_5)_4$ or $Cr(NCS)_4(NH_3)_2$, Br_2, ClO_4 or $(O_2N)_3$-C_6H_3O], are prepared by precipitation of the oxidized metallocenes resulting from reactions of cobalt(II) chloride and substituted cyclopentadienyl anions derivable from substituted fulvenes.[64] *Hexafluorophosphate* and tetraphenylborate salts of a monophenyl-substituted derivative of di-π-cyclopentadienylcobalt(III), **114** [R$=C_6H_5$; R'$=$H; X$=PF_6$ or $B(C_6H_5)_4$], may be obtained by treating di-π-cyclopentadienylcobalt(II) with phenylmagnesium bromide in tetrahydrofuran and precipitating with the desired anion after hydrolysis of the reaction product.[222] In contrast, the hexafluorophosphate of the corresponding rhodium analog is formed by the reaction of 2*N* hydrochloric acid and π-cyclopentadienyl-(1-*exo*-phenylcyclopentadiene)rhodium(III) derived from treatment of the zinc reduction product of di-π-cyclopentadienylrhodium(III)(1+)tribromide(1−) with phenyllithium in diethyl ether), followed by precipitation of the cation with ammonium hexafluorophosphate.[218]

A heteroannular tetrasubstituted dialkyl-dihydroxy cationic derivative, [*bis(1-hydroxy-3-methyl)-π-cyclopentadienyl*]*cobalt(III)*, can be synthesized in the form of its reineckate by treating bis(benzyloxy-π-cyclopentadienyl)-cobalt(II) [obtained by adding cobalt(II) chloride to the reaction product of 3-methyl-2-cyclopenten-1-one and sodium amide in liquid ammonia and shaking with benzoyl chloride) with hydrogen peroxide and precipitating with Reinecke's salt.[235] The yellow high-melting bromide, tetrafluoroborate and perchlorate of the homoannular hydroxytetraphenyl-substituted cation (**116**, R$=$H; X$=$Br; BF_4 or ClO_4) are formed in good yields by shaking a chloroform solution of π-cyclopentadienyl(1,2,3,4-tetraphenylcyclopena-1,3-dien-5-one) cobalt (II) resulting from the reaction of π-cyclopentadienyldicarbonylcobalt and 1,2,3,4-tetraphenylcyclopenta-1,3-dien-5-one, with a concentrated aqueous solution of the appropriate anion.[236]

[235] U.S. Pat. 2 849 470 (1958), du Pont de Nemours, Invent.: *R. E. Benson;* C.A. *53*, 406 (1953).

[236] *J. E. Sheats, M. D. Rausch*, J. Org. Chem. *35*, 3245 (1970).

The hexafluorophosphate and tetraphenylborate of the corresponding acetoxy-substituted cation [**116**, $R=CH_3CO$; $X=PF_6$ or $B(C_6H_5)_4$] are obtained in a similar manner after initially forming the (acetoxy) cation by treating π-cyclopentadienyl(1,2,3,4-tetraphenylcyclopenta-1,3-diene-5-one)cobalt(II) with acetyl chloride.[236]

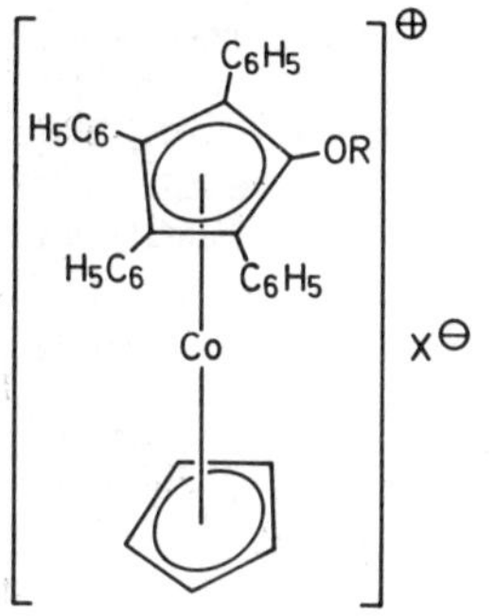

116

A heteroannular diacetyl-substituted cation can be prepared in the form of its reineckate or tetraphenylborate [**117**, $R=CH_3$; $R'=CH_3CO$; $X=Cr(NCS)_4(NH_3)_2$ or $B(C_6H_5)_4$] in yields of up to 65% by oxidizing the dialkenyl-substituted perchlorate [**114**, R and $R'=CH_2{=}CH(CH_3)_2$; $X=ClO_4$] with ozone and precipitating with the desired anion.[64] The hexafluorophosphate and tetraphenylborate salts of the *monocarboxy-*

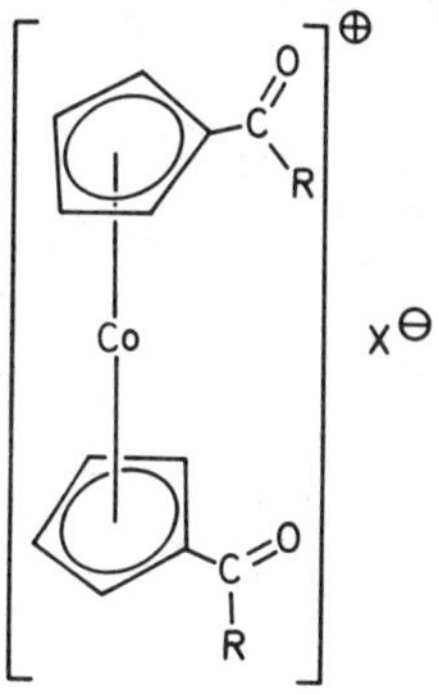

117

pentadienyl)cobalt(III)(1+)tribromide(1−) with either bromine or silver perchlorate.[234] Yellow hexafluorophosphate salts of the corresponding mono- (**117**, $R=Cl$; $R'=H$; $X=PF_6$)[236] and bis-(chlorocarbonyl)cationic derivatives (**117**, $R=Cl$; $R'=ClCO$; $X=PF_6$)[238] are readily synthesized by refluxing the hexafluorophosphate salts of the analogous carboxy-substituted cations with thionyl chloride. The chlorocarbonyl-substituted cationic hexafluorophosphates can be utilized, in turn, to form salts of the corresponding mono- [**117**, $R=CH_3O$; $R'-H$; $X=PF_6$ or $B(C_6H_5)_4$][236] and bis-(methoxycarbonyl)-containing compounds [**117**, $R=CH_3O$; $R'=CH_3OCO$; $X=B(C_6H_5)_4$][238] by refluxing the latter with methanol and precipitating with the sodium salt of the appropriate anion in methanol.

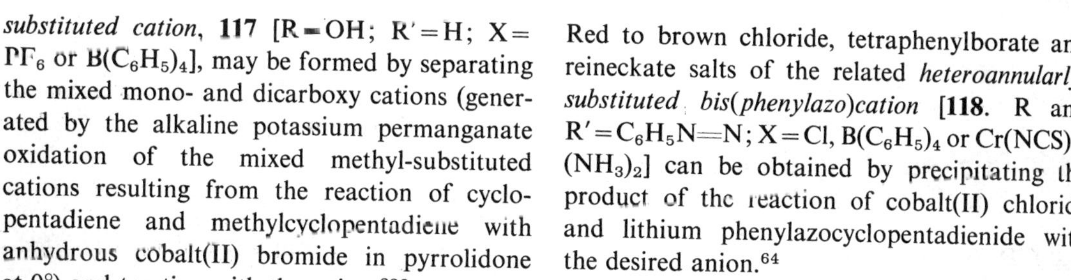

$[R{-}COOH][PF_6] \longrightarrow [R{-}COOCH_3][PF_6];[B(C_6H_5)_4]$ [231,237]

$\longrightarrow [R{-}COCl][PF_6] \longrightarrow [R{-}CO{-}N_3][PF_6]$ [236] $\longrightarrow [R{-}NH_2][PF_6];[B(C_6H_5)_4]$ [236]

(mono) $[R{-}N{=}N{-}C_6H_4{-}OH][B(C_6H_5)_4]$ [236] $[R{-}NO_2][PF_6];[B(C_6H_5)_4]$ [236]

substituted cation, **117** [$R=OH$; $R'=H$; $X=PF_6$ or $B(C_6H_5)_4$], may be formed by separating the mixed mono- and dicarboxy cations (generated by the alkaline potassium permanganate oxidation of the mixed methyl-substituted cations resulting from the reaction of cyclopentadiene and methylcyclopentadiene with anhydrous cobalt(II) bromide in pyrrolidone at 0°) and treating with the anion.[236]

The brown bromide and perchlorate of the corresponding heteroannularly-substituted dicarboxy cation (**117**, $R=OH$; $R'=COOH$; $X=Br$ or ClO_4) are best obtained in a similar manner by precipitating the alkaline potassium permanganate oxidation product of bis(methyl-π-cyclo-

Red to brown chloride, tetraphenylborate and reineckate salts of the related *heteroannularly-substituted bis(phenylazo)cation* [**118**. R and $R'=C_6H_5N{=}N$; $X=Cl$, $B(C_6H_5)_4$ or $Cr(NCS)_4(NH_3)_2$] can be obtained by precipitating the product of the reaction of cobalt(II) chloride and lithium phenylazocyclopentadienide with the desired anion.[64]

Similar to its di-π-cyclopentadienyl analog, di-π-indenylcobalt(II) forms cationic salts by

---

[237] *E. O. Fischer, G. E. Herberich*, Chem. Ber. *94*, 1517 (1961).

[238] *C. U. Pittman, O. E. Ayres, S. P. McManus, J. E. Sheats, C. E. Whitten*, Macromolecules *4*, 360 (1971).

118

direct synthesis from a cobalt halide and metal indenide, or by oxidation of the preformed fused-ring metallocene and precipitation with an appropriate anion. Thus, the tetracobaltate of di-π-indenylcobalt(III), **119**, is obtained by dilute hydrobromic acid hydrolysis of the product of the reaction of indenylmagnesium bromide and cobalt(II) chloride in diethyl ether–toluene.[224] This salt, in turn, can be converted into the corresponding tribromide, **119** ($X = Br_3$), by treatment with hydrogen peroxide in dilute hydrobromic acid.[224] Other salts [**119**, X=Br, I, I_3, ClO_4, NO_3, PF_6, $B(C_6H_5)_4$ and $Cr(NCS)_4(NH_3)_2$] are prepared by oxidation of di-π-indenylcobalt(II) with hydrogen peroxide or peroxodisulfuric acid and precipitation with the appropriate anion.[213]

119

31.8 Nickel, Palladium, and Platinum Derivatives

31.8.1 Di-π-cyclopentadienylmetals and Cationic Salts

Di-π-cyclopentadienylnickel(II) can conveniently be prepared in yields as high as 80% by the "amine" method in which freshly distilled cyclopentadiene is added to a solution of specially prepared anhydrous nickel(II) bromide in diethylamine under nitrogen and the mixture stirred overnight, according to the following reaction:[239]

$$NiBr_2 + 2C_5H_6 + 2(C_2H_5)_2NH \longrightarrow (\pi\text{-}C_5H_5)_2Ni + 2[(C_2H_5)_2NH]Br$$

In a modification of this procedure, nickel(II) chloride may be substituted for the metal bromide in the reaction, but the product yields are slightly lower.[240]

The dark green crystalline paramagnetic metallocene, which is commercially available as an 8–10% solution in toluene, oxidizes slowly in air over a period of days and must be stored for longer periods in the absence of air. Solutions of di-π-cyclopentadienylcobalt(II) are even more oxygen-sensitive and must be handled at all times in an inert atmosphere. While useful as a hydrogenation and polymerization catalyst, as well as a motor fuel additive, the metallocene is moderately toxic and care must be taken to avoid inhalation of vapors or contact with the skin and mucous membranes.

Although *di-π-cyclopentadienylpalladium(II)* has been described as a cherry red crystalline, low-melting[241] and thermally unstable compound,[242] no details or even method of preparation have been reported. In addition, the corresponding di-π-cyclopentadienylplatinum(II) could not be obtained presumably by the synthetic route used for its palladium analog.[242] However, when 1,3,5,7-cyclooctatetraene is reacted with platinum(II) iodide and a cyclopentadienylmagnesium halide, a cyclooctatetracene-bis[di-π-cyclopentadienylplatinum(II)] complex, presumably having structure **120**, is formed.[243]

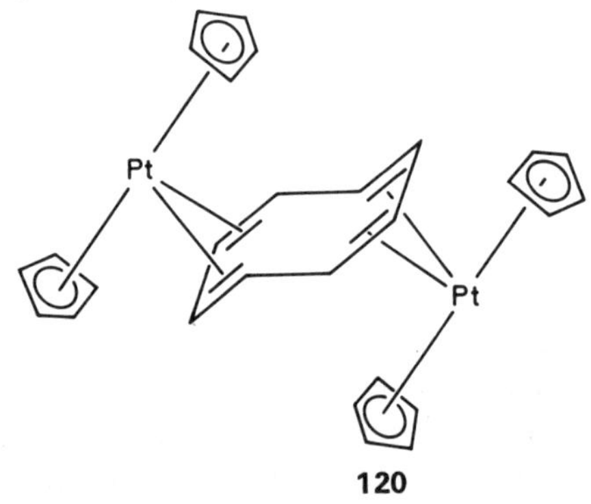

120

[239] *R. B. King*, Organometallic Syntheses, Vol. I, p. 71, Academic Press, New York 1965.

[240] *H. Watanabe, I. Motoyama, K. Hata*, Bull. Chem. Soc., Japan *38*, 853 (1965).

[241] *B. E. Douglas*, The Chemistry of the Coordination Compounds, *J. C. Bailar jr.*, p. 498, Reinhold Publ. Co., New York, 1956.

[242] *G. Wilkinson, F. A. Cotton*, Progr. Inorg. Chem. *1*, 12 (1959).

[243] *N. C. Baenziger, J. R. Doyle*, U.S. Dept. Comm., O.T.S., PB Rept. 156 075 (1961); C.A. *58*, 5241 (1963).

The nickel-containing metallocene is readily oxidized to the rather unstable yellow-orange cation, *di-π-cyclopentadienylnickel(III)*, which can be precipitated, like its cobalt analog, by a variety of anions to yield stable cationic salts. Thus, when the metallocene is treated with bromine or iodine in petroleum ether, the brown *tribromide* and *triiodide* salts (**121**, $X=Br_3$ or I_3) are obtained.[244]

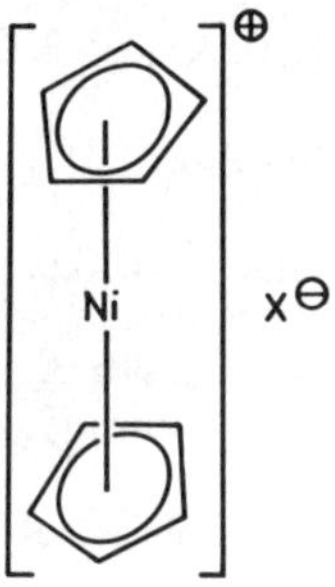

121

Tetraphenylborate and reineckate, as well as picrate and tungstosiicate salts[245] are prepared similarly by oxidation of the metallocene and precipitation with the desired anion. If di-π-cyclopentadienylnickel(II) is treated with two molar equivalents of tetrachloro-*p*-benzoquinone in benzene under an argon atmosphere, a dark brown, very thermally stable (>350°) charge transfer complex of structure **122** is formed.[246]

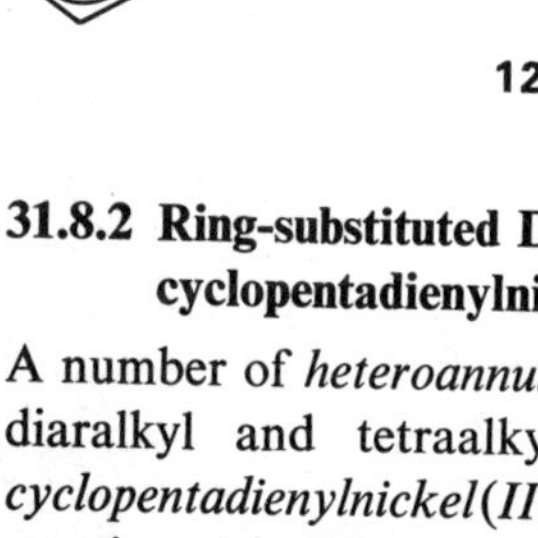

122

31.8.2 Ring-substituted Derivatives of Di-π-cyclopentadienylnickel

A number of *heteroannularly-substituted* dialkyl, diaralkyl and tetraalkyl *derivatives* of *di-π-cyclopentadienylnickel(II)* may be prepared by reactions involving a metal alkyl (or -aralkyl-) cyclopentadienide and a nickel salt. Thus, the low melting, dark green *bis(methyl)-π-cyclopentadienyl)nickel(III)* (**123**, $R=CH_3$) is obtained in high purity by refluxing a mixture of sodium methylcyclopentadienide and freshly-prepared nickel(II) bromide in tetrahydrofuran for 2·5 hr.[133] The extremely air-sensitive dialkylmetallocene, which oxidizes instantly on contact with atmospheric oxygen, may also be synthesized in high yield and with certain possible other advantages by the reaction of the less stable thallium methylcyclopentadienide and nickel(II) chloride monohydrate in diethyl ether.[157]

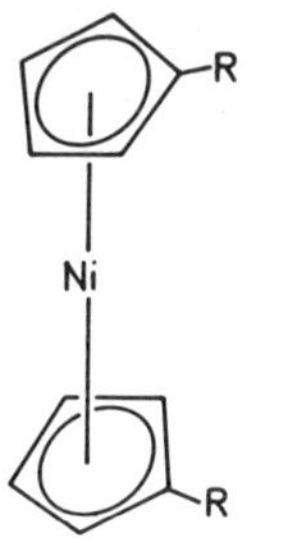

123

The heteroannular diethyl and di-*t*-butyl derivatives (**123**, $R=C_2H_5$ or t-C_4H_9), together with the tetrasubstituted bis(1,3-di-*t*-butyl) compound (**124**), are formed in yields of 70–75% by heating a tetrahydrofuran solution of the appropriate alkyl-substituted sodium cyclopentadienide with hexaaminenickel dichloride in the presence of 1-chloropentane.[247]

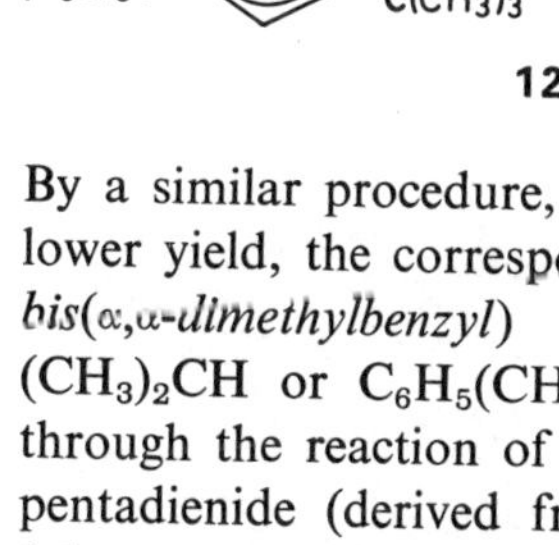

124

By a similar procedure, although in somewhat lower yield, the corresponding diisopropyl and *bis(α,α-dimethylbenzyl) derivatives* [**123**, R= $(CH_3)_2CH$ or $C_6H_5(CH_3)_2C$] may be formed through the reaction of the lithium alkylcyclopentadienide (derived from treating dimethylfulvene with an ethereal solution of lithium tetrahydridoaluminate or phenyllithium) and

---

[244] *E. O. Fischer, R. Jira*, Z. Naturforsch. *8b*, 217 (1953).

[245] *G. Wilkinson, P. L. Pauson, F. A. Cotton*, J. Amer. Chem. Soc. *76*, 1970 (1954).

[246] *J. C. Goan, E. Berg, H. E. Podall*, J. Org. Chem. *29*, 975 (1964).

[247] *A. N. Nesmeyanov, E. V. Leonova, N. S. Kochetkova, S. M. Butyugin, J. S. Meisner*, Izv. Akad. Nauk. SSSR, Ser. Khim. 106 (1971); engl.: 89.

dichlorotetrakis(pyridine)nickel(II) in tetrahydrofuran.[64]

Bis[(cyclopentadienyldimethylsily)-π-cyclopentadienyl]nickel(II) [**123**, R = $C_5H_5(CH_3)_2Si$], is prepared by treating the lithium silycyclopentadienide, resulting from the reaction of dicyclopentadienyldimethylsilane and *n*-butyllithium in benzene–petroleum ether with anhydrous nickel(II) chloride.[248] The deep red-brown *di-π-indenylnickel(II)* (**125**), which decomposes rapidly in air and especially when in solution, is obtained when nickel(II) acetylacetonate is added to indenylmagnesium bromide in diethyl ether–toluene and the mixture is stirred at room temperature under an atmosphere of highly-purified nitrogen.[249]

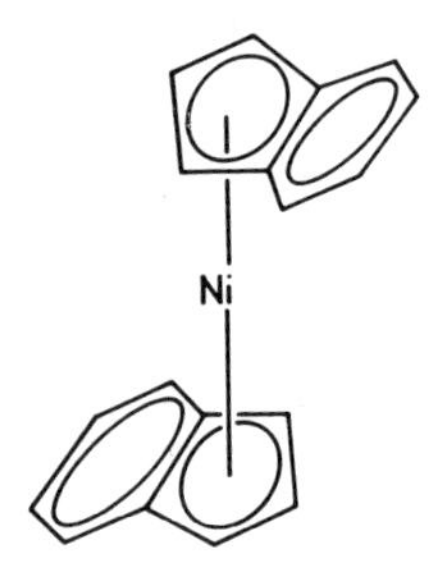

125

31.9 Bibliography

W. F. Little, Surv. Prog. Chem. *1*, 133 (1963).

J. M. Birmingham, Advan. Organometal. Chem. *2*, 365 (1964).

M. D. Rausch, J. M. Birmingham, Ann. N.Y. Acad. Sci. *125*, 57 (1965).

M. Dub, Organometallic Compounds, 2nd Ed., Vol. I, Springer Verlag, New York 1966.

M. L. H. Green, Organometallic Compounds, 3rd Ed., Vol. II, p. 90, Methuen Co., London 1968.

M. Tsutsui, M. N. Lery, A. Nakamura, M. Jchikawa, K. Mori, Introduction to Metal-π-Complex Chemistry, Plenum Press, New York 1970.

R. B. King, Transition-Metal Organometallic Chemistry, Academic Press, New York 1969.

A. P. Ginsberg in *R. L. Carlin*, Transition-Metal Chemistry, Vol. I, p. 140, Marcel Dekker, New York 1965.

---

248 U.S. Pat. 3 060 215 (1962), U.S. Air Force, Invent.: *H. Rosenberg, M. D. Rausch;* C.A. *58*, 6865 (1963).

249 *H. P. Fritz, F. H. Kohler, K. E. Schwarzhaus*, J. Organometal. Chem. *19*, 449 (1969).

31 Sandwich Compounds

Part B: Other Sandwich Complexes

John R. Wasson

Department of Chemistry, University of Kentucky,
Lexington, Kentucky 40506

This section of Chapter 31 is concerned with those metal sandwich compounds not containing the cyclopentadienyl group; mixed complexes containing cyclopentadienyl and other ligands are mentioned only peripherally. In contrast to the metallocenes discussed in the preceding section A of this chapter large numbers of well-characterized sandwich compounds with other ligands are not known. Consequently, it has been useful to discuss compounds according to types of ligands they contain rather than attempt a systematic treatise by central metal atoms. The classification of ligands employed here is essentially that used by Coates, Green, and Wade in their two volume text.[1] A sampling of representative structures of metal sandwich compounds is given in Table 1.

Table 1 Metal Sandwich Compounds

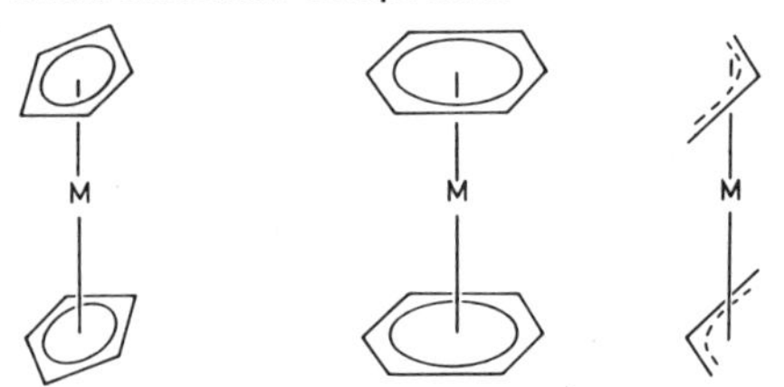

bis(π-cyclopentadienyl)- di-π-benzene- di-π-allyl-

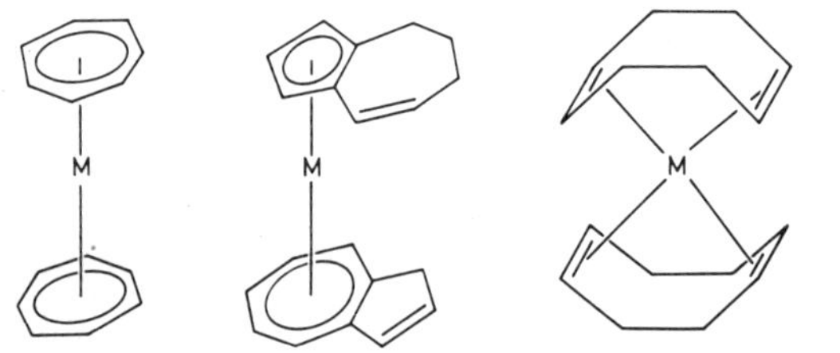

bis(π-tropylium)-[a] bis(π-azulene)- bis(π-cyclooctadiene)-

31.1 π-Enyl (π-Allyl) Complexes[1]

π-allyl(π-C_3H_5) complexes are best prepared[2] by treating metal halides with allyl Grignard reagents, *e.g.*,

$$NiCl_2 + 2C_3H_5MgBr \xrightarrow[-10^\circ]{\text{diethylether}} Ni(C_3H_5)_2 + MgBr_2 + MgCl_2$$

Mixed π-allyl compounds may also be prepared by this route.[3]

Allyl halides react with a variety of anionic metal complexes[4] to form *π-allyl compounds*—usually mixed complexes not of the type considered here. The sandwich structure has been established for *bis(2-methyl-π-allyl)nickel.*[5]

The π-ligand character of allylic donors have been reviewed.[6] Several reviews of complexes with allylic ligands[7] and a review concerned with the use of π-allylnickel intermediates in organic synthesis[8] have appeared.

31.2 Four- and Five-electron Ligands

A variety of olefins, *e.g.*, *butadiene*, *substituted butadienes*, *cyclobutadiene*, *cyclopentadiene*, *cyclohexa-1,3-diene*, *cyclohepta-1,3,5-triene*, *cyclopentadienones*, *cycloocta-1,3,4-triene*, *acetylenes*, *cinnamaldehyde*, *penta-1,3-diene*, and *cyclooctatetraene*, can act as four-electron ligands with transition metals. Large numbers of complexes with these ligands are known and they frequently exhibit intriguing properties, e.g., fluctional behavior which can be characterized by NMR studies.[9] Recently, *trialkylcyclopropenium compounds*[10] have been added to the preceding list of four-electron ligands. Tri-*t*-butylcyclopropenium fluoroborate[11] and sodium bromide in methanolic solution under nitrogen react[10] with excess nickel carbonyl to yield *tri-t-butylcyclopropenium nickel dicarbonyl bromide.* This product readily reacts with sodium salts of other π-ligands to give mixed sandwich com-

[1] *G. E. Coates, M. L. H. Green, K. Wade*, Organometallic Compounds, Methuen & Co., London, Vol. I, 1967; Vol. II, 1968.

[2] *G. Wilke*, Angew. Chem. Intern. Ed. Engl. *2*, 105 (1963);
G. Wilke, B. Bogdanovic, Angew. Chem. *73*, 756 (1961).

[3] *E. O. Fischer, G. Burger*, Z. Naturforsch. *16b*, 77 (1961);
W. R. McClellan, H. H. Hoehn, H. N. Cripps, J. Amer. Chem. Soc. *83*, 160 (1961).

[4] *H. D. Kaesz, R. B. King, F. G. A. Stone*, Z. Naturforsch. *15b*, 682 (1960);
M. Cousins, M. L. H. Green, J. Chem. Soc. (London) 889 (1963);
J. Kwatek, J. K. Seyler, J. Organometal. Chem. *3*, 421 (1965).

[5] *R. Uttech, H. Dietrich*, Z. Kristallogr. *122*, 60 (1965).

[6] *M. R. Churchill, R. Mason*, Advan. Organometal. Chem. *5*, 93 (1967).

[7] *M. L. H. Green, P. L. I. Nagy*, Advan. Organometal. Chem. *2*, 325 (1965);
R. G. Guy, B. L. Shaw, Adv. Inorg. Chem. Radiochem. *4*, 111 (1962);
E. O. Fischer, H. Werner, Angew. Chem. *2*, 147 (1962);
G. Wilke, Angew. Chem. Intern. Ed. Engl. *5*, 151 (1966).

[8] *P. Heimbach, P. W. Jolly G. Wilke*, Advan. Organometal. Chem. *8*, 29 (1970).

[9] *J. D. Warren, M. A. Busch, R. J. Clark*, Inorg. Chem. *11*, 452 (1972).

[10] *W. K. Olander, T. L. Brown*, J. Amer. Chem. Soc. *94*, 2139 (1972).

[11] *J. Ciabattoni, E. C. Nathan*, J. Amer. Chem. Soc. *91*, 4766 (1969).

pounds. Mixed sandwich compounds rather than unmixed compounds can be expected for four-electron π-ligands on the basis of the effective atomic number (EAN) rule which is a useful guideline for predicting the stoichiometries of transition metal π-complexes.[12] Unmixed complexes as anionic and neutral and cationic paramagnetic species can be expected for several of the transition metals but the synthetic possibilities remain to be explored. A review of cyclobutadiene metal complexes has appeared.[13]

Five-electron π-ligands include[1] five-, six-, and seven-membered hydrocarbon ligands and non-cyclic ligands. The π-cyclopentadienyl group is the most common ligand in this class and has already been discussed in the first section of this chapter. Metal π-complexes formed by seven-membered and eight-membered carbocyclic compounds have been the subject of reviews.[7,14]

A number of *π-cyclohexadienyl compounds* are known but only as mixed complexes. Reduction of the cation $[\pi\text{-}C_5H_5FeC_6H_6]^{\oplus}$ with sodium tetrahydridoborate yields the *cyclohexadienyl compound*,[15] $\pi\text{-}C_6H_7Fe\text{-}\pi\text{-}C_5H_5$. The cation, $[(\pi\text{-}C_6H_6)_2Re]^{\oplus}$, upon reduction[16] with sodium tetrahydridoborate, yields the mixed *arene-cyclohexadienyl* $\pi\text{-}C_6H_6Re\text{-}\pi\text{-}C_6H_7$, while the dibenzeneruthenium cation gives a mixture of products.

The majority of unsubstituted cyclohexadienyl complexes have an intense infrared band at about 2700 cm^{-1} which is assigned to the *exo* C–H_α stretching frequency. Apparently, a pure-cyclohexadienyl metal sandwich compound has not yet been reported.

31.3 Six-electron Ligands

Ligands in this class include, the *arenes*, *cyclooctatetraene*, *butatrienes*, *cycloocta-1,3,5-triene*, *bicyclo-4,3,0-nonatriene*, and *heterocycles* such as *thiophene* and *pyridine* among others. Of these ligands the arene sandwich compounds have been the most extensively examined.[17–19]

The history of arene π-complexes, traced by Zeiss,[17] is of interest since it demonstrates many of the synthetic problems encountered. The initial preparation of polyphenylchromium compounds by Hein[20] employed a synthetic route still of importance for the preparation of arene sandwich compounds. The basic procedure consists of reaction of a transition metal halide with an aryl Grignard reagent at low temperatures ($<0°$) with exclusion of oxygen and moisture. The experimental conditions (stoichiometry, temperature, and solvent) are critical. Usually several products can result, *e.g.*, in the preparation of dibenzenechromium the sigma-bonded arene-chromium species di(biphenyl)chromium and benzene-biphenylchromium are obtained in addition to the desired product. Cationic bis-arene species are also possible and the separation of the products of the reactions of metal halides with aryl Grignard reagents can be very difficult.

The *Fischer-Hafner synthesis*[21] (also called the aluminum method or the reducing Friedel–Crafts method) is the most general approach to the preparation of bis-π-arene complexes. The method involves the reduction of a metal salt with aluminum powder followed by addition of arene ligand to the metal. The reaction (1) is catalyzed by a large excess (seven-to-eight-fold) aluminum(III) chloride.

$$3CrCl_3 + 2Al + AlCl_3 + 6C_6H_6 \longrightarrow 3[(\pi\text{-}C_6H_6)_2Cr]AlCl_4 \quad (1)$$

Aluminum(III) bromide, but not the fluoride or iodide, can also be employed as a catalyst in this reaction. Triethylaluminum can also be substituted for the aluminum halide/aluminum mixture.

The bis-π-arene metal complex cation can be reduced to the neutral complex with dithionite ion (Eqn 2) or aluminum powder in aqueous

[12] *R. B. King*, Transition-Metal Organometallic Chemistry, Academic Press, New York 1969; *M. Tsutsui, M. N. Levy, A. Nakamura, M. Ichikawa, K. Mori*, Introduction to Metal-π-Complex Chemistry, Plenum Press, New York 1970.

[13] *P. M. Maitlis*, Advan. Organometal. Chem. *4*, 94 (1966).

[14] *M. A. Bennett*, Advan. Organometal. Chem. *4*, 353 (1966).

[15] *D. Jones, G. Wilkinson*, Chem. & Ind. (London), 1408 (1961).

[16] *D. Jones, G. Wilkinson*, J. Chem. Soc. (London) 2479 (1964).

[17] *H. Zeiss*, Organometallic Chemistry, p. 380, Reinhold Publishing Co., New York 1960.

[18] *E. O. Fischer, H. Werner*, Metal π-Complexes, Vol. I, Elsevier Publ. Co., Amsterdam 1966.

[19] *H. Zeiss, P. J. Wheatley, H. J. S. Winkler*, Benzenoid-Metal Complexes, Ronald Press Co., New York 1966.

[20] *F. Hein*, Ber. *52*, 195 (1919); *F. Hein*, Ber. *54*, 1905, 2708, 2727 (1921).

[21] *E. O. Fischer, W. Hafner*, Z. Naturforsch. *10b*, 665 (1955); *E. O. Fischer, W. Hafner*, Z. Anorg. Allgem. Chemie *286*, 146 (1956).

alkali.[22] Alternately, the neutral complex can be

$$2[(\pi\text{-}C_6H_6)_2Cr]^{\oplus} + S_2O_4^{2\ominus} + 4OH^{\ominus} \longrightarrow 2(\pi\text{-}C_6H_6)_2Cr + 2SO_3^{2\ominus} + 2H_2O \quad (2)$$

obtained by disproportionation of the bis-π-arene cation in aqueous alkali solution[23] (Eqn 3).

$$2[(\pi\text{-}C_6H_6)_2Cr]^{\oplus} \xrightarrow{H_2O} (\pi\text{-}C_6H_6)_2Cr + 2C_6H_6 + Cr^{2\oplus} \quad (3)$$

Usually substituted benzenes, *e.g.*, *mesitylene* and *hexamethylbenzene*, afford more readily prepared complexes than benzene itself. The experimental procedure has been described in detail by Fischer[22] and is generally applicable with minor modifications[17–19] for the preparation of transition metal arene complexes. It is noted that *bis-arene-chromium complexes* can equilibrate with each other and free arene ligands in the presence of aluminum(III) chloride or bromide.[24] A substantial number of bis-arene sandwich compounds have been obtained employing the Fischer–Hafner synthesis and these have been tabulated.[1,17,25]

Disubstituted acetylenes, *e.g.*, 2-butyne, are trimerized at low temperatures by organometallic compounds, *e.g.*, $Cr(C_6H_6)_3(THF)_3$ (THF = tetrahydrofuran), with the formation of bis-arene complexes,[17,26] Eqn (4).

Manganese, cobalt, and nickel organometallic compounds have been employed for the cyclic condensation of acetylenes.[27]

The *bis-benzenetechnetium cation* is prepared[28] by the normal Fischer–Hafner process as well as by the neutron bombardment[29] of bis-benzene-molybdenum, *i.e.*,

$$(\pi\text{-}C_6H_6)_2Mo \xrightarrow{n\gamma} (\pi\text{-}C_6H_6)_2Mo^{99} \xrightarrow{\beta^-} (\pi\text{-}C_6H_6)_2Tc^{99m} \longrightarrow (\pi\text{-}C_6H_6)_2Tc^{98}$$

It is possible to metalate dibenzenechromium with amylsodium.[30] The reaction does not proceed readily, a mixture of products is obtained and yields are usually low.

Treatment of metal carbonyls with arene yields metal carbonyl complexes.[31] Ultraviolet light irradiation[32] of the reaction mixture or the use of high boiling, coordinating solvents,[33] *e.g.*, diglyme, does not yield bis-arene complexes. It is noted, however, that extensive use of photochemical methods of synthesis has not been made. The review of the photochemistry of metal carbonyls, metallocenes and olefin complexes cited below[34] is particularly suggestive for future work.

Dibenzenechromium and, presumably, other bis-arene sandwich complexes can act as electron donors in the formation of various charge-transfer or molecular complexes with strong π-acids such as tetracyanoethylene (TCNE).[35] The molecular complexes are readily pre-

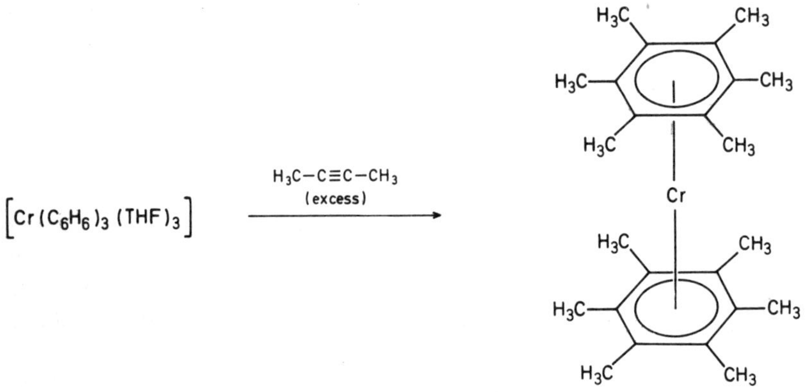

$$[Cr(C_6H_6)_3(THF)_3] \xrightarrow[\text{(excess)}]{H_3C-C\equiv C-CH_3} \text{bis(hexamethylbenzene)chromium} \quad (4)$$

[22] *E. O. Fischer*, Inorg. Synth. *6*, 132 (1960).

[23] *E. O. Fischer, J. Seeholzer*, Z. Anorg. Allgem. Chemie *312*, 244 (1961).

[24] *Yu. A. Sorokin, G. G. Petukhov*, Zh. Obshch. Khim. *35*, 2135 (1965);
F. Hein, K. Kartte, Z. Anorg. Allgem. Chemie *307*, 22 (1960);
E. O. Fischer, J. Seeholzer, Z. Anorg. Allgem. Chemie *312*, 244 (1961).

[25] *E. O. Fischer, H. P. Fritz*, Angew. Chem. *73*, 353 (1961).

[26] *W. Herwig, W. Metlesics, H. Zeiss*, J. Amer. Chem. Soc. *81*, 6203 (1959);
H. Zeiss, W. Herwig, J. Amer. Chem. Soc. *80*, 2913 (1958).

[27] *M. Tsutsui, H. Zeiss*, J. Amer. Chem. Soc. *81*, 6090 (1959);
M. Tsutsui, H. Zeiss, J. Amer. Chem. Soc. *82*, 6255 (1960);
M. Tsutsui, H. Zeiss, J. Amer. Chem. Soc. *83*, 825 (1961).

[28] *C. Palm, E. O. Fischer, F. Baumgärtner*, Tetrahedron Lett. *6*, 253 (1962).

[29] *F. Baumgartner, E. O. Fischer, U. Zahn*, Naturwissenschaften *48*, 478 (1961);
F. Baumgartner, E. O. Fischer, U. Zahn, Chem. Ber. *94*, 2198 (1961).

[30] *E. O. Fischer, H. Brunner*, Z. Naturforsch. *16b*, 406 (1961);

pared by mixing the constituents together in an appropriate solvent. Unfortunately, the formulation of the resulting compounds is not without difficulty since the reactions can give rise to conventional molecular complexes, salts containing cations of the sandwich compound and π-acid anions, or involve chemical addition of the π-acid to the arene. The semiconductor and magnetic properties of such molecular complexes are frequently very interesting. Dibenzenechromium forms a 1:1 complex with tetracyanoethylene which has been formulated[35] as a salt, $[\pi\text{-}C_6H_6)_2Cr]^{\oplus}TCNE^{\ominus}$.

Many other six-electron ligands, *e.g.*, *pyrrole*,[36] *pyridines*,[37] *thiophenes*,[38] and *borazines*,[39] among others, which could form sandwich compounds, are known to form *half-sandwich metal carbonyls* but pure sandwich compounds have apparently not yet been reported. Red-brown paramagnetic borabenzenecobalt complexes of

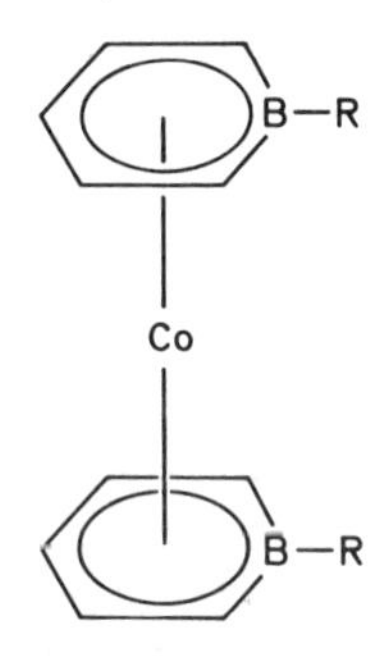

R = C_6H_5, CH_3, OCH_3

type I have been prepared[40] by the reaction of bis-(π-cyclopentadienyl)cobalt with organoboron dihalides and boron trihalides. Chromatography on alumina and crystallization from hexane at low temperatures affords purification of the products of the reaction. The use of an excess of boron tribromide with $(\pi\text{-}C_5H_5)_2Co$ yields:

$$5(\pi\text{-}C_5H_5)_2Co + 6BBr_3 \longrightarrow 4[(\pi\text{-}C_5H_5)_2Co]BBr_4 + (\pi\text{-}C_5H_5BBr]_2Co$$

Mass spectra data[40] suggests that the neutral C_5H_5BBr radical is a better electron acceptor in bonding to the metal than the C_5H_5 radical, in accord with theoretical considerations.[41]

Cycloheptatriene, C_7H_8, and the *tropylium ion*, $C_7H_7^{\oplus}$, behave formally as six-electron donors.[42] Syntheses of tropylium π-complexes start with cycloheptatriene. Attempts to prepare tropylium π-complexes from alkali salts of metal carbonylates, *e.g.*, $Na[Mn(CO)_5]$, and tropylium bromide yield *ditropyl*, $C_7H_7\text{–}C_7H_7$.[43] *Bis(π-cycloheptatriene)*- and *bis(π-tropylium)-metal complexes* analogous to dibenzenechromium have apparently not been obtained but a number of half-sandwich and mixed sandwich compounds are known.[42] The complex $(\pi\text{-}C_5H_5)V(CO)_4$ yields[44] purple paramagnetic $(\pi\text{-}C_5H_5)V(\pi\text{-}C_7H_7)$ whose sandwich structure has been confirmed.[45] Red-brown paramagnetic (π-tropylium)-(cycloheptatriene)vanadium cation, $[(\pi\text{-}C_7H_7)VC_7H_8]^{\oplus}$, is isolated as its $[V(CO)_6]^{\ominus}$ salt from the reaction[46] of cycloheptatriene with vanadium carbonyl. The complex $(\pi\text{-}C_7H_7)V(CO)_7$ is also formed in the reaction. The *tetraphenylborate salt* is prepared by anion exchange.

Azulene π-complexes offer possibilities for bonding to the metal with the seven-membered ring, the five-membered ring, or both. A number of metal carbonyl complexes with one metal bound

---

[30] *E. O. Fischer, H. Brunner*, Chem. Ber. *95*, 1999 (1962);
E. O. Fischer, H. Brunner, Chem. Ber. *98*, 175 (1965).
[31] *E. O. Fischer, K. Ofele*, Chem. Ber. *90*, 2532 (1957);
E. O. Fischer, K. Ofele, Chem. Ber. *91*, 2763 (1958);
G. Natta, R. Ercoli, F. Calderazzo, Chim. Ind. (Milan) *40*, 287 (1958);
D. A. Brown, J. R. Raju, J. Chem. Soc. *A*, 40 (1966).
[32] *W. Strohmeier*, Chem. Ber. *94*, 3337 (1961).
[33] *R. P. M. Werner, T. H. Coffield*, Chem. & Ind. (London) 936 (1960).
[34] *E. K. von Gustorf, F. W. Grevels*, Topics in Current Chem. *13*, 366 (1969).
[35] *J. W. Fitch, J. J. Lagowski*, Inorg. Chem. *4*, 864 (1965);
G. Huttner, E. O. Fischer, R. D. Fischer, J. Organometal. Chem. *6*, 288 (1966);
J. W. Fitch, J. J. Lagowski, J. Organometal. Chem. *5*, 480 (1966).
[36] *K. K, Joshi, P. L. Puason*, Proc. Chem. Soc. 326 (1962);
R. B. King, M. B. Bisnette, Inorg. Chem. *3*, 796 (1964).
[37] *E. O. Fischer, K. Ofele*, Z. Naturforsch. *14b*, 736 (1959);
E. O. Fischer, K. Ofele, Chem. Ber. *93*, 1156 (1960).
[38] *E. O. Fischer, K. Ofele*, Z. Naturforsch. *91*, 2395 (1958);
M. F. Bailey, L. F. Dahl, Inorg. Chem. *4*, 1306 (1965);
H. Singer, Z. Naturforsch. *21b*, 810 (1966).
[39] *R. Prinz, H. Werner*, Angew. Chem. *79*, 63 (1967);
H. Noth, W. Regnet, Z. Anorg. Allgem. Chemie *352*, 1 (1967).
[40] *G. E. Herberich, G. Greiss, H. F. Heil, J. Muller*, Chem. Commun. 1328 (1971).
[41] *P. J. Busse, K. Niedenzu, J. R. Wasson*, unpubl.
[42] *M. A. Bennett*, Advan. Organometal. Chem. *4*, 354 (1966).
[43] *E. W. Abel, M. A. Bennett, R. Burton, G. Wilkinson*, J. Chem. Soc. (London) 4559 (1958).
[44] *R. B. King, F. G, A. Stone*, J. Amer. Chem. Soc. *81*, 5263 (1959).
[45] *G. Engebretson, R. E. Rundle*, J. Amer. Chem. Soc. *85*, 481 (1963).
[46] *F. Calderazzo, P. L. Calvi*, Chem. Ind. (Milan) *44*, 1217 (1962).

to each ring are known. Mixed complexes containing the *azulenium cation*, $C_{10}H_9^{\oplus}$, have been prepared by the following reaction:[47]

$$(\pi\text{-}C_6H_6)Cr(\pi\text{-}C_5H_5) + C_{10}H_8 + (C_2H_5)_2O \cdot BF_3 \longrightarrow$$
$$\xrightarrow{+H_2O} (\pi\text{-}C_{10}H_9)Cr(\pi\text{-}C_5H_5)^{\oplus}$$

(isolated as $PF_6^{\ominus}$ salt) yellow, paramagnetic

$$\xrightarrow[KOH]{Na_2S_2O_4} (\pi\text{-}C_{10}H_9)Cr(\pi\text{-}C_5H_5)$$

45% yield, dark green, diamagnetic air sensitive

The preparations and NMR spectra of analogous complexes with 4,6,8-*trimethylazulene* have been described.[48] Green, diamagnetic bis(π-azulene) chromium results from the reaction:[49]

$$CrCl_3 + C_{10}H_8 + iso\text{-}C_3H_7MgBr \xrightarrow[2.\ CH_3OH]{1.\ h\nu} (\pi\text{-}C_{10}H_9)_2Cr$$

$(\pi\text{-}C_{10}H_9)_2Cr$ is readily hydrogenated to yield diamagnetic, blue $(\pi\text{-}C_{10}H_{12})_2Cr$ which, in solution, is air oxidized to the mono-cation which can be isolated as its hexafluorophosphate salt. *Bis(π-azulene)iron*, $(\pi\text{-}C_{10}H\ )_2Fe$, results upon substitution of iron(III) chloride for chromium(III) chloride in the above reaction.[49] A number of arene complexes are listed in Table 2.

31.4 Other Organic Ligands

The cycloheptatrienyl group, C_7H_7, derived from cycloheptatriene, C_7H_8, is the best known seven-electron ligand. Its complexes have been considered in the previous section since its distinction from the tropylium cation is not always unambiguous. Here we are primarily concerned with complexes of cyclooctatetraene and its derivatives.

Free cyclooctatetraene exists in a tub conformation essentially without conjugation between the double bonds. Cyclooctatetraene, COT, C_8H_8, is readily reduced by alkali metals or polarographically to produce the aromatic, planar cyclooctatetraenyl dianion, $C_8H_8^{2\ominus}$, which has 10 π-electrons. Complexes with this ligand have been reviewed[1,18] and a theoretical

[47] *E. O. Fischer, S. Breitschaft*, Chem. Ber. *96*, 2451 (1963).

[48] *H. P. Fritz, C. G. Kreiter*, J. Organometal. Chem. *1*, 70 (1963).

[49] *E. O. Fischer, J. Muller*, J. Organometal. Chem. *1*, 464 (1964).

Table 2 Transition Metal Arene Sandwich Compounds

Neutral Compounds	Ref.
Neutral Compounds	
$V(C_6H_6)_2$	50, 51, 52
$V[1,3,5\text{-}(CH_3)_3C_6H_3]_2$	52
$Cr(C_6H_6)_2$	50, 53–56
$Fe[C_6(CH_3)_6]_2$	57
$Co[C_6(CH_3)_6]_2$	58
$Ni[1,3,5\text{-}(CH_3)_3C_6H_3]_2$	59
$Mo(C_6H_6)_2$	60, 61
$W(C_6H_6)_2$	61
Cationic Complexes	
$V[1,3,5\text{-}(CH_3)_3C_6H_3]_2^{\oplus}$	52, 62
$Cr(C_6H_6)_2^{\oplus}$	53, 54, 55
$Mn[(CH_3)_6C_6]_2^{\oplus}$	63
$Fe[(CH_3)_6C_6]_2^{\oplus}$	57
$Fe(C_6H_6)_2^{\oplus}$	64
$Co[(CH_3)_6C_6]_2^{\oplus}$	63, 65
$Co[(CH_3)_6C_6]_2^{2\oplus}$	63
$Mo(C_6H_6)_2^{\oplus}$	60, 61
$^{99}Tc(C_6H_6)_2^{\oplus}$	66

[50] *E. O. Fischer, H. P. Kogler*, Chem. Ber. *90*, 250 (1957).

[51] *E. O. Fischer, A. Reckziegel*, Chem. Ber. *94*, 2204 (1961).

[52] *E. O. Fischer, G. Joos, W. Meer*, Z. Naturforsch. *13b*, 456 (1958).

[53] *E. O. Fischer, W. Hafner*, Z. Naturforsch. *10b*, 665 (1955).

[54] *E. O. Fischer, W. Hafner*, Z. Anorg. Allgem. Chemie *286*, 146 (1956).

[55] *H. H. Zeiss, W. Herwig*, J. Amer. Chem. Soc. *78*, 5959 (1956).

[56] *E. O. Fischer, J. Seeholzer*, Z. Anorg. Allgem. Chemie *312*, 244 (1961).

[57] *E. O. Fischer, F. Rohrscheid*, Z. Naturforsch. *17b*, 483 (1962).

[58] *E. O. Fischer, H. H. Lindner*, J. Organometal. Chem. *2*, 222 (1964).

[59] *M. Tsutsui, H. Zeiss*, J. Amer. Chem. Soc. *82*, 6255 (1960).

[60] *E. O. Fischer, H. O. Stahl*, Chem. Ber. *89*, 1805 (1956).

[61] *E. O. Fischer, F. Scherer, H. O. Stahl*, Chem. Ber. *93*, 2065 (1960).

[62] *F. Calderazzo*, Inorg. Chem. *3*, 810 (1964).

[63] *M. Tsutsui, H. Zeiss*, J. Amer. Chem. Soc. *83*, 5825 (1961).

[64] *M. Tsutsui, H. Zeiss*, Naturwissenschaften *44*, 420 (1957).

[65] *E. O. Fischer, H. H. Lindner*, J. Organometal. Chem. *1*, 307 (1964).

[66] *F. Baumgartner, E. O. Fischer, U. Zahn*, Chem. Ber. *94*, 2198 (1961).

Table 2 (contd.)

Neutral Compounds	Ref.
$Ru[1,3,5\text{-}(CH_3)_3C_6H_3)_2{}^{2\oplus}$	67
$Rh[(CH_3)_6C_6]_2{}^{\oplus}$	65
$Rh([CH_3)_6C_6]_2{}^{2\oplus}$	65
$Rh[1,3,5\text{-}(CH_3)_3C_6H_3]_2{}^{3\oplus}$	68
$W(C_6H_6)_2{}^{\oplus}$	61
$Re(C_6H_6)_2{}^{\oplus}$	69
$Re[1,3,5\text{-}(CH_3)_3C_6H_3]_2{}^{\oplus}$	69
$Ir[1,3,5\text{-}(CH_3)_3C_6H_3]_2{}^{3\oplus}$	68

description of the bonding presented.[70] Of particular interest is the green pyrophoric compound *bis(π-cyclooctatetraenyl)uranium*, "*uranocene*", prepared by the following reaction:[71]

$$UCl_4 + 2K_2C_8H_8 \xrightarrow[\text{dry, oxygen-free tetrahydrofuran}]{-30°} U(\pi\text{-}C_8H_8)_2 + 4KCl$$

After stirring overnight green crystals are isolated in 80% yield by addition of oxygen-free water and extracting the resulting precipitate with benzene. The compound is stable to water, aqueous sodium hydroxide and acetic acid. The sandwich structure has been confirmed.[72] *1,3,5,7-Tetramethylcyclooctatetraene* forms[73] similar complexes with *uranium(IV)* and *neptunium(IV)*. Organo-lanthanide and -actinide complexes have been the subject of a review.[74] Reaction of *nickel(II) acetylacetonate* and *1,5,9-cyclododecatriene* with *diethylethoxyaluminum* in diethyl ether yields[75] red, air-sensitive *cyclododecatriene-nickel*, $C_{12}H_{18}Ni$, which has a half-sandwich structure.[76] $C_{12}H_{18}Ni$ reacts[75] with

1,5-cyclooctadiene at room temperature to yield yellow *bis(1,5-cyclooctadiene)nickel*, $(C_8H_{12})_2Ni$, which reacts with cyclooctatetraene to give a brown insoluble polymer $[(C_8H_8)Ni]_n$ which presumably has the structure[75,77]

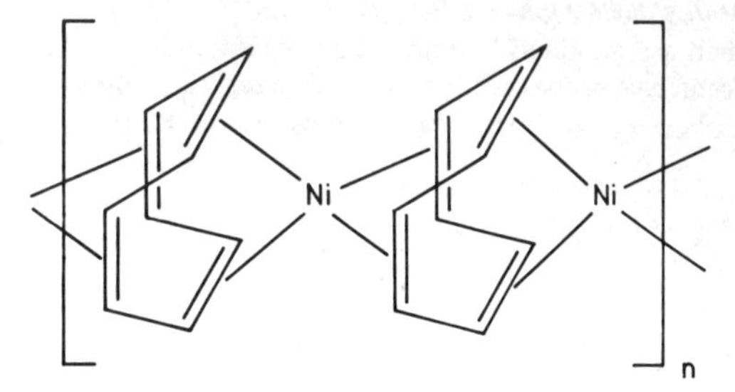

Many cyclooctatetraene complexes have been prepared but most of them are half-sandwich or mixed sandwich compounds, *e.g.*, $Ti(C_8H_8)(C_5H_5)$.[78] The vanadium complex, $V(C_8H_8)_2$, has been reasonably well-characterized.[79,80] *Duroquinone*, which is sometimes regarded as a diolefin, reacts,[75,81] with nickel tetracarbonyl to yield red *bis(duroquinone)nickel*, $[(CH_3)_4C_6O_2]_2Ni$. Mixed sandwich compounds, (olefin)-nickel(duroquinone), are obtained in high yields by direct reaction of nickel tetracarbonyl, duroquinone and excess olefin.[81,82]

The spectroscopic properties of sandwich compounds, as well as other organometallic compounds, are discussed in the two volumes edited by M. Tsutsui.[83] The

[67] *E. O. Fischer, R. Bottcher*, Z. Anorg. Allgem. Chemie *291*, 305 (1957).

[68] *E. O. Fischer, H. P. Fritz*, Angew. Chem. *73*, 353 (1961).

[69] *E. O. Fischer, A. Wirzmuller*, Chem. Ber. *90*, 1725 (1957).

[70] *R. D. Fischer*, Theor. Chim. Acta *1*, 418 (1963).

[71] *A. Streitwieser, jr., U. Muller-Westerhoff*, J. Amer. Chem. Soc. *90*, 7364 (1968);
D. G. Karraker, J. A. Stone, E. R. Jones jr., N. Edelstein, J. Amer. Chem. Soc. *92*, 4841 (1970).

[72] *A. Zalkin, K. N. Raymond*, J. Amer. Chem. Soc. *91*, 5667 (1969);
A. Avdeef, K. N. Raymond, Inorg. Chem. *11*, 1083 (1972).

[73] *A. Streitwieser, jr., D. Dempf, G. N. Lamar, D. G. Karraker, N. Edelstein*, J. Amer. Chem. Soc. *93*, 7343 (1971).

[74] *H. Gysling, M. Tsutsui*, Advan. Organometal. Chem. *9*, 361 (1970).

[75] *B. Bogdanovic, M. Kroner, G. Wilke*, Justus Liebigs Ann. Chem. *699*, 1 (1966).

[76] *G. N. Schrauzer, H. Thyret*, J. Amer. Chem. Soc. *82*, 6420 (1960).

[77] *G. Wilke*, Angew. Chem. *72*, 581 (1960);
G. Wilke, Angew. Chem. *75*, 10 (1963).

[78] *H. O. van Oven, H. J. de Liefde Meijer*, J. Organometal. Chem. *19*, 373 (1969).

[79] *H. Breil, G. Wilke*, Angew. Chem. Intern. Ed. Engl. *5*, 989 (1966).

[80] *J. L. Thomas, R. G. Hayes*, Inorg. Chem. *11*, 348 (1972).

[81] *G. N. Schrauzer, H. Thyret*, Z. Naturforsch. *16b*, 353 (1961);
G. N. Schrauzer, H. Thyret, Z. Naturforsch. *17b*, 73 (1962);
G. N. Schrauzer, Advan. Organometal. Chem. *2*, 3 (1964).

[82] *G. N. Schrauzer, H. Thyret*, Chem. Ber. *96*, 1755 (1963).

[83] *M. Tsutsui*, Ed., Characterization of Organometallic Compounds, Wiley-Intersci., New York, Part I, 1970; Part II, 1971.

mass spectra of metal sandwich compounds have been recently reviewed.[84]

31.5 Transition Metal Carborane Compounds

Within recent years metal carborane complexes with structures akin to those found for organometallic sandwich compounds have received a great deal of attention.[85,86] Much of this work has been due to M. F. Hawthorne and his colleagues.[85] Many half-sandwich and mixed sandwich complexes have been characterized.[86]

The *dicarbollide ions* 1,2-$B_9C_2H_{11}^{2\ominus}$ and 1, 7-$B_9C_2H_{11}^{2\ominus}$, which are obtained by removing a $BH^{2\oplus}$ unit from the parent carboranes with base,[87,88] have the following numbering system (hydrogen atoms are omitted):

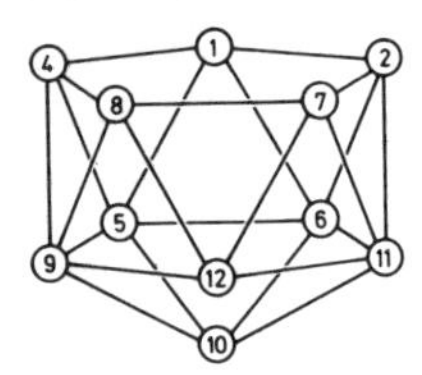

Bis(π-1,2-dicarbollyl)chromium complexes are produced[89] by treating chromium(III) chloride with the 1,2-$B_9C_2H_{11}^{2\ominus}$ anion. Cr(π-1,2-$B_9C_2H_{11})^{2\ominus}$ salts are stable in hot aqueous solutions but decomposed under basic conditions. An X-ray structural study of CsCr[π-1,2-$B_9H_9C_2(CH_3)_2]_2$ established a symmetrical sandwich structure for the compound.[89]

Photochemical reaction of *chromium, molybdenum,* and *tungsten hexacarbonyls* with 1,2-$B_9C_2H_{11}^{2\ominus}$ yields[90] compounds of the type (π-1,2-$B_9C_2H_{11}$)$M(CO)_3^{2\ominus}$ (where M=Cr, Mo, W) which can be isolated as *tetramethylammonium salts.* Using the conditions employed for the preparation of ferrocene the 1,2-dicarbollide ion reacts with anhydrous iron(II) chloride in tetrahydrofuran in the presence of air to yield the stable red Fe(1,2-$B_9C_2H_{11})_2^{\ominus}$ ion. This anion can be reduced with sodium amalgam in acetonitrile to yield the air-sensitive Fe(1,2-$B_9C_2H_{11})_2^{2\ominus}$ anion. The iron(III) complex can also be prepared from the reaction of iron(II) chloride and 1,2-$B_9C_2H_{11}^{2\ominus}$ in 40% aqueous sodium hydroxide.[91] Neutral, sublimable Fe-(π-C_5H_5)(π-1,2-$B_9C_2H_{11}$) is prepared[92] by reaction of an equimolar mixture of *cyclopentadienide 1,2-dicarbollide ion,* and *iron(II) chloride* followed by air oxidation. The sandwich structure of this mixed sandwich compound has been confirmed.[93] The reaction:

$$1{\cdot}5CoCl_2 + 2(1,2\text{-}B_9C_2H_{11})^{2\ominus} \longrightarrow Co(1,2\text{-}B_9C_2H_{11})_2^{\ominus} + 3Cl^{\ominus} + 0{\cdot}5Co$$

which is an example of an unusual disproportionation reaction, has been employed[90,94] to produce *bis(π-1,2-dicarbollyl)cobalt(III)* salts. Exhaustive bromination of Rb[Co(π-1,2-$B_9C_2H_{11})_2$] yields[95] a hexabromo derivative in which three hydrogen atoms of each dicarbollide ion are replaced by bromide in positions as distant as possible from the carbon atoms.

The Ni(π-1,2-$B_9C_2H_{11})_2^{\ominus}$ ion is formed[90,96] in the reaction of nickel(II) salts with 1,2-dicarbollide ion followed by air oxidation. Reduction of the ion with one equivalent of ferric ion yields sublimable Ni(π-1,2-$B_9C_2H_{11})_2$ while reductions with sodium amalgam yields air-sensitive, paramagnetic Ni(π-1,2-$B_9C_2H_{11})_2^{2\ominus}$. X-ray studies indicate that the nickel(III) species has a normal sandwich structure whereas the nickel(II) compound has a "slipped" structure.[97] Palladium acetylacetonate reacts[98] with excess 1,2-dicarbollide ion in cold glyme solution to produce air-sensitive, unstable Pd(π-1,2-$B_9C_2H_{11})_2^{2\ominus}$. The neutral species Pd(π-1,2-$B_9C_2H_{11})_2$ is produced upon oxidation of the palladium(II) species with iodine. Deep blue salts containing the Cu(π-1,2-$B_9C_2H_{11})_2^{2\ominus}$ anion are obtained from the reaction[90] of copper(II) sulfate with 1,2-dicarbollide anion generated

[84] *M. Cais, M. S. Lupin,* Advan. Organometal. Chem. *8,* 211 (1970).
[85] *M. F. Hawthorne,* Accounts Chem. Res. *1,* 281 (1968).
[86] *L. J. Todd,* Advan. Organometal. Chem. *8,* 87 (1970).
[87] *R. A. Wiesboeck, M. F. Hawthorne,* J. Amer. Chem. Soc. *86,* 1642 (1964);
P. M. Garrett, F. N. Tebbe, M. R. Hawthorne, J. Amer. Chem. Soc. *86,* 5016 (1964).
[88] *M. F. Hawthorne, D. C. Young, P. M. Garrett,* J. Amer. Chem. Soc. *90,* 862 (1968).
[89] *H. W. Ruhle, M. F. Hawthorne,* Inorg. Chem. *7,* 2279 (1968).
[90] *M. F. Hawthorne, D. C. Young, T. D. Andrews,* J. Amer. Chem. Soc. *90,* 879 (1968).
[91] *M. F. Hawthorne, T. D. Andrews, P. M. Garrett,* Inorg. Synth. *10,* 111 (1967).
[92] *M. F. Hawthorne, R. L. Pilling,* J. Amer. Chem. Soc. *87,* 3987 (1965).
[93] *A. Zalkin, D. H. Templeton, T. E. Hopkins,* J. Amer. Chem. Soc. *87,* 3988 (1965).
[94] *M. F. Hawthorne, T. D. Andrews,* Chem. Commun. 443 (1965).
[95] *B. G. Deboer, A. Zalkin, D. H. Templeton,* Inorg. Chem. *7,* 2288 (1968).
[96] *L. F. Warren, M. F. Hawthorne,* J. Amer. Chem. Soc. *89,* 470 (1967).
[97] *R. M. Wing,* J. Amer. Chem. Soc. *90,* 4828 (1968).
[98] *L. F. Warren, M. F. Hawthorne,* J. Amer. Chem. Soc. *90,* 4823 (1968).

in 40% aqueous sodium hydroxide. The blue copper(II) salts are air oxidized to red copper(III) salts. Both the copper(II) and copper(III) salts have the "slipped" structure[97,99] shown in Figure 1. The reaction[42] of anhydrous gold(III)

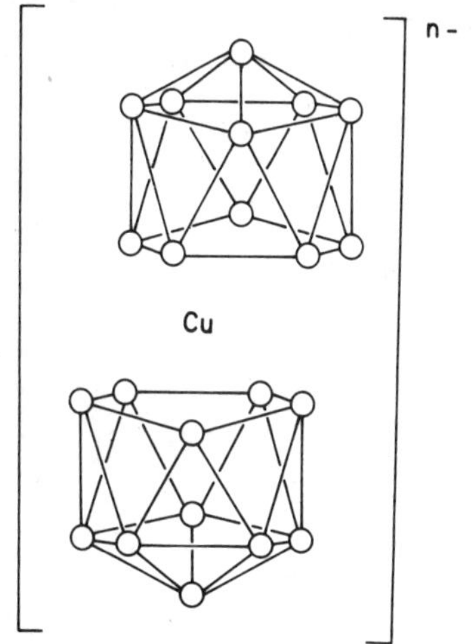

chloride with the 1,2-dicarbollide ion in dimethoxyethane followed by oxidation with acidic hydrogen peroxide yields red $Au(\pi\text{-}1,2\text{-}B_9C_2H_{11})^{\ominus}_2$ which can be reduced with sodium amalgam to deep blue-green $Au(\pi\text{-}1,2\ B_9C_2H_{11})_2^{2\ominus}$. The 1,7-dicarbollide anion undergoes reactions[90] similar to the 1,2-dicarbollide anion but the 1,2-dicarbollide complexes appear to be more stable.

The preparation of carbollide ligands $B_{10}H_{10}\text{-}CNH_3^{2\ominus}$ and $B_{10}H_{10}CH^{3\ominus}$ has been described in detail.[100] The preparations of the transition metal sandwich compounds follow procedures[101] very similar to those described above for 1,2-dicarbollide complexes. The molecules 1,2- and 1,7-$B_{10}H_{10}CHP$[102] lose a boron atom upon refluxing in excess piperidine for several hours to form piperidinium salts of the 1,2- and 1,7-$B_9H_{10}CHP^{\ominus}$ anions[103] which are isoelectronic with the $B_9C_2H_{11}^{2\ominus}$ anions discussed above. These anions react with transition metal halides to yield sandwich compounds. The cage molecules 1,2- and 1,7-$B_{10}H_{10}CHAs$ behave similarly to the phosphorus compounds.

The $B_7C_2H_{11}^{2\ominus}$ anion results[104] upon treatment of $B_7C_2H_{13}$ with two equivalents of sodium hydride in diethyl ether. Reaction of the anion with cobalt(II) chloride yielded a salt containing the $Co(B_7C_2H_9)_2^{\ominus}$ anion which undergoes thermal isomerization[105] at 315°.

In the preparation of $Co(\pi\text{-}1,2\text{-}B_9C_2H_{11})_2^{\ominus}$ using the 1,2-$B_9C_2H_{11}^{2\ominus}$ anion generated by aqueous base a bright red salt of composition $Cs_2\text{-}[(\pi\text{-}B_9C_2H_{11})\ Co\ (\pi\text{-}B_8C_2H_{10})\ Co\ (\pi\text{-}B_9C_2H_{11})]\cdot 2H_2O$ may be precipitated.[106] The difunctional $B_8C_2H_{10}^{4\ominus}$ anion has been given the trivial name (3,6)-1,2-dicarbacanastide.

---

[99] *R. M. Wing*, J. Amer. Chem. Soc. *89*, 5599 (1967).

[100] *W. H. Knoth, J. L. Little, J. R. Lawrence*, Inorg. Synth. *11*, 33 (1968);
D. E. Hyatt, D. A. Owen, L. J. Todd, Inorg. Chem. *5*, 1749 (1966);
W. H. Knoth, J. Amer. Chem. Soc. *89*, 1274 (1967).

[101] *W. H. Knoth, J. L. Little, L. J. Todd*, Inorg. Synth. *11*, 41 (1968);
W. H. Knoth, J. Amer. Chem. Soc. *89*, 3342 (1967);
D. E. Hyatt, J. L. Little, J. T. Moran, J. Amer. Chem. Soc. *89*, 3342 (1967).

[102] *J. L. Little, J. T. Moran, L. J. Todd*, J. Amer. Chem. Soc. *89*, 5495 (1967).

[103] *L. J. Todd, I. C. Paul, J. L. Little*, J. Amer. Chem. Soc. *90*, 4489 (1968).

[104] *M. F. Hawthorne, T. A. George*, J. Amer. Chem. Soc. *89*, 7114 (1967).

[105] *T. A. George, M. F. Hawthorne*, J. Amer. Chem. Soc. *90*, 1661 (1968).

[106] *J. N. Francis, M. F. Hawthorne*, J. Amer. Chem. Soc. *90*, 1663 (1968).

32 Heteropoly Compounds

George A. Tsigdinos
Research Laboratory, Climax Molybdenum Company,
Ann Arbor, Michigan 48105, U.S.A.

32.1 Introduction

The heteropoly compounds are a large, fundamental family of salts and free acids with each member containing a complex and high-molecular weight anion. These anions contain two to eighteen hexavalent molybdenum or tungsten or pentavalent vanadium or niobium atoms around one or more central atoms. In some cases, pentavalent vanadium or niobium replaces some of the molybdenum or tungsten atoms in the structure. They are all highly oxygenated. Examples are $[PMo_{12}O_{40}]^{3\ominus}$, $[NiW_6O_{24}H_6]^{4\ominus}$, $[As_2Mo_{18}O_{62}]^{6\ominus}$, $[TeMo_6O_{24}]^{6\ominus}$, $[MnNb_{12}O_{38}]^{12\ominus}$, and $[PMo_{10}V_2O_{40}]^{5\ominus}$, where $P^{5\oplus}$, $Ni^{2\oplus}$, $As^{5\oplus}$, $Te^{6\oplus}$, and $Mn^{4\oplus}$ are the central atoms. Approximately 36 different elements have been reported to function as central atoms in distinct heteropoly anions. Moreover, many of these elements can act as central atoms in more than one series of heteropoly anions. As early as 1826 Berzelius prepared and analyzed ammonium 12-molybdophosphate, the first heteropoly compound. Marignac prepared 12-tungstosilicic acid in 1862 and recognized such compounds as a distinct class rather than double salts.[1] The first systematic attempt to understand the structure of heteropoly compounds was made in 1908 by Miolati, who suggested a structure for these compounds based on the ionic theory and Werner's coordination theory. Prior to 1908 a large number of compounds had been described which were later proven to be heteropoly electrolytes. In that early literature these were reported by means of dualistic formulas, *e.g.*, $2Na_2O \cdot SiO_2 \cdot 12MoO_3 \cdot xH_2O$. Such formulas were merely a means of expressing elemental composition and conveyed no structural information in a modern sense. Miolati's theory was extensively developed and applied by Rosenheim and his coworkers.[1] Since then the properties and several structures of these compounds have been elucidated.

X-ray structure determinations have been made on many series of heteropoly compounds. Some typical structures are given in Figures 1–6. In these, the anion structure is represented by polyhedra that share corners, edges or faces with one another. Each Mo, W, V, or Nb is at the centre of an octahedron and each oxygen atom is located at each vertex. These octahedra are usually considerably distorted. The central atom may be located at the centre of an XO_4 tetrahedron which in turn is surrounded by MoO_6 or WO_6 octahedra which share corners and edges with one another to give the Keggin structure typified by the *12-molybdophosphate anion*, $[PMo_{12}O_{40}]^{3\ominus}$, shown in Figure 1. 12-heteropoly anions having this type of structure belong to Series A.[1] Unlike the 12-molybdophosphate

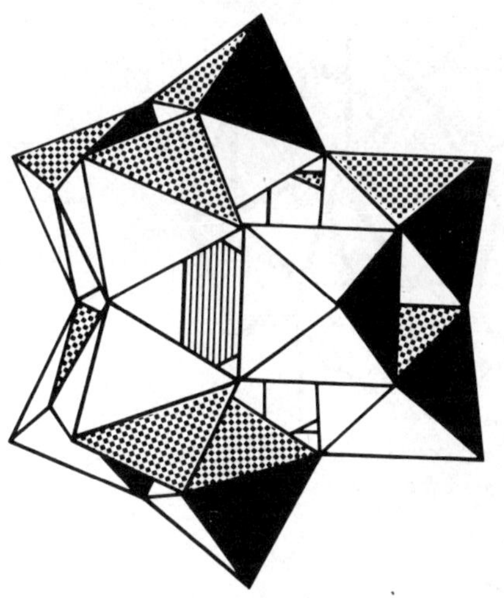

Fig. 1. Polyhedral diagram of the 12-molybdophosphate anion $[PMo_{12}O_{40}]^{3\ominus}$.

anion, the *12-molybdocerate(IV) anion* structure (Series B),[1] shown in Figure 2, consists of a 12-coordinate cerium atom surrounded by twelve MoO_6 octahedra, the added feature of this structure involves the first example in which MoO_6 octahedra share common faces. The *12-niobomanganate(IV) anion*, $[MnNb_{12}O_{38}]^{12\ominus}$, does not have the cage-like structure of the previously mentioned heteropoly anions but consists of an MnO_6 octahedron with two Nb_6O_{19} groups on its side as shown in Figure 3. The structure of *6-heteropoly anions*, shown in Figure 4 and typified by the *6-molybdotellurate anion*, $[TeMo_6O_{24}]^{6\ominus}$, consists of an annular arrangement of the six MoO_6 octahedra about the central TeO_6 octahedron. The *9-molybdomanganate(IV) anion*, $[MnMo_9O_{32}]^{6\ominus}$, has a structure, shown in Figure 5, which consists of a central MnO_6 octahedron surrounded by nine MoO_6 octahedra in such a way that the structure is asymmetric. The structure of the *18-molybdo-(or tungsto)diphosphate anions*, Figure 6, is related to the Keggin structure[1] and consists of two PO_4 tetrahedra surrounded by $18MoO_6$ or WO_6 octahedra. In the tungsten compound rotation of one of the half units by 60° gives rise to isomerism.[1]

In view of the complexity and confusion of the voluminous literature that has accumulated since Berzelius first observed compounds of this type, great caution should be exercised in its interpretations. For example, analyses reported in the older literature are often imprecise. The molecular weights of heteropoly compounds are so high that small analytical errors produce great errors in the formulas reported. Degradation in solution was often overlooked and much of the work was unwittingly performed on mixtures thus obtained.

[1] *G. A. Tsigdinos*, Heteropoly Compounds of Molybdenum and Tungsten: Their Structure and Properties, Climax Molybdenum Company Bulletin Cdb-12a (Revised), November 1969.

Fig. 2. Idealized sketch of the $[CeMo_{12}O_{42}]^{8\ominus}$ ion showing the linkage of the MoO_6 octahedra.

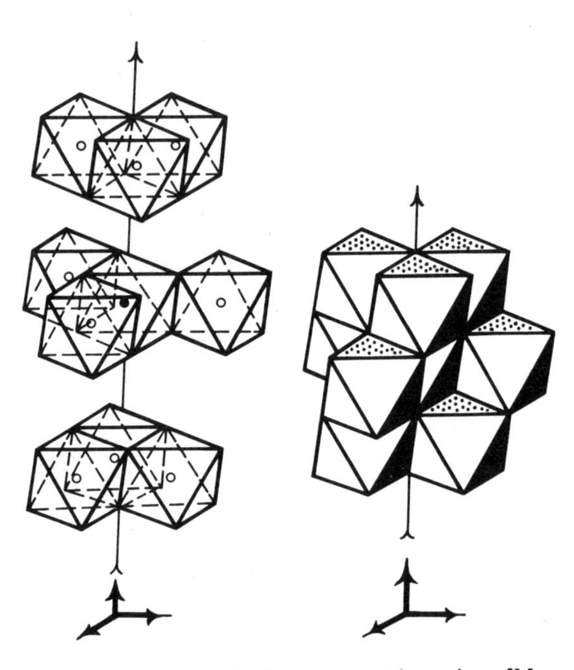

Fig. 5. Structure of the asymmetric anion $[MnMo_9O_{32}]^{6\ominus}$. Left: exploded view showing manganese atom (black circle) and molybdenum atoms (open circles). Right: view of complete ion.

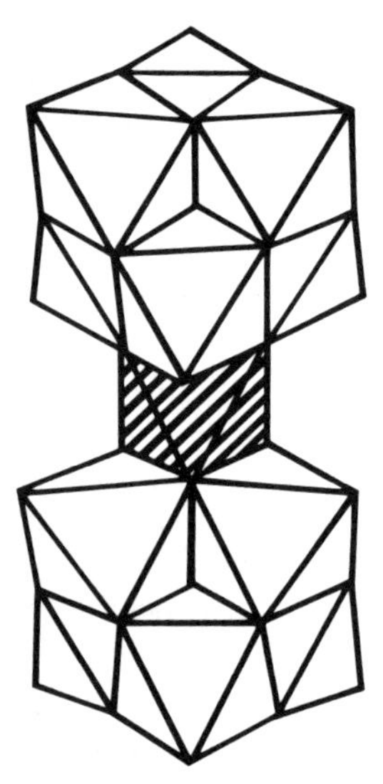

Fig. 3. Polyhedral diagram showing the structure of the 12-niobomanganate(IV) anion, $[MnNb_{12}O_{38}]^{12\ominus}$.

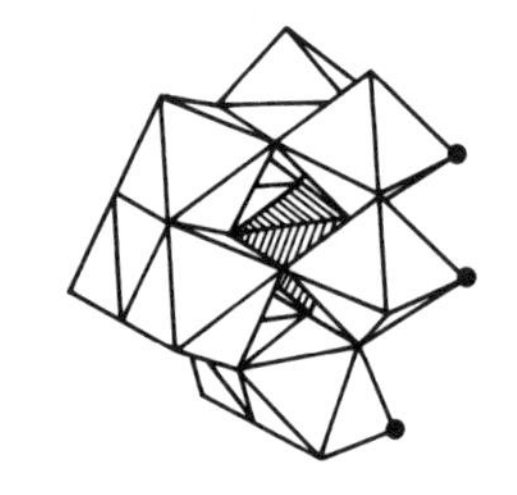

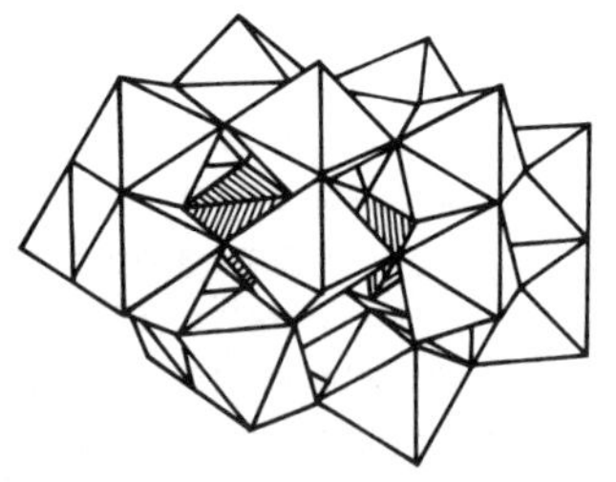

Fig. 6. Lower: complete structure of the dimeric anions $[P_2W_{18}O_{62}]^{6\ominus}$ and $[P_2Mo_{18}O_{62}]^{6\ominus}$. Upper: the half unit obtained by splitting $[P_2W_{18}O_{62}]^{6\ominus}$ structure. This half unit may be also obtained by removing WO_6 octahedra from the 12-heteropolytungstate structure (Figure 1).

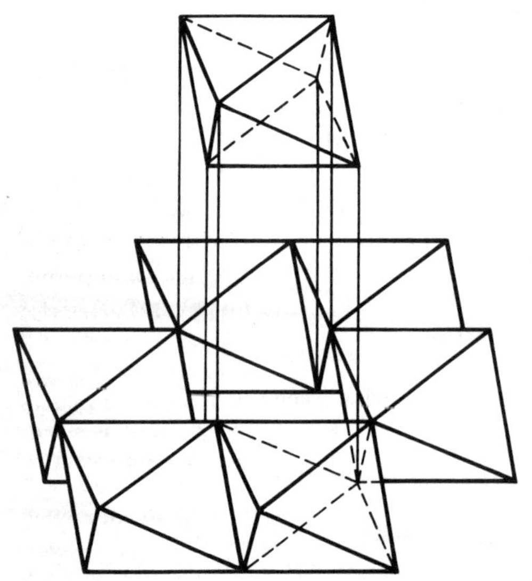

Fig. 4. Structure of the ion $[TeMo_6O_{24}]^{6\ominus}$ with the central TeO_6 octahedron elevated to show the annular arrangement of the six MoO_6 octahedra.

Unfortunately this trend continues to persist though to a much smaller extent in modern literature. It is the purpose of this chapter, to delineate the most reliable synthetic routes of heteropoly compounds. The structure, properties and nomenclature of these compounds[1] and the related isopoly compounds of molybdenum, tungsten, and vanadium[2] has been given in the review articles cited.

32.2 12-Heteropolyvanadophosphates

A large number of heteropoly vanadophosphates have been reported in the literature containing various ratios of vanadium to phosphorus but the composition of the crystalline salts obtained depends on the ratio of vanadate to phosphate present in solution, the hydrogen ion concentration, concentration of reactants and the type of cation used.[3] These species remain so far unidentified but many of these are undoubtedly mixtures rather than true compounds.

The *12-vanadophosphate anion* has been prepared from sodium metavanadate and phosphoric acid solutions acidified with acetic acid.[4, 5] The sodium, potassium, and ammonium salts of the 12-vanadophosphate anion have been isolated[4, 6] and carefully analyzed.[6] The results indicate[6] that the true formula of the vanado phosphate anion is $[PV_{12}O_{40}H_8]^{7\ominus}$ (see Figure 1).

32.3 Heteropolyvanadates of Manganese(IV) and Nickel(IV)

The preparation and properties of heteropoly vanadates with transition metals as central atoms have been recently reported for the first time.[7] The potassium, sodium, and ammonium salts of the *13-heteropolyvanadates* of *manganese(IV)* and *nickel(IV)* were prepared as red-orange crystalline solids.[7] Typical formulas of some of these salts are: $K_7[MnV_{13}O_{38}]\cdot18H_2O$, $Na_7[MnV_{13}O_{38}]\cdot24H_2O$, and $K_7[NiV_{13}O_{38}]\cdot16H_2O$. The preparative route involves the oxidation of Mn(II) ion by peroxydisulfate in the presence of isopolyvanadate ions.[7]

Two additional vanadomanganate(IV) heteropoly anions have been prepared by the reaction of manganese(II) and isopolyvanadate ions and peroxydisulfate.[8] Salts of the *11-vanadomanganate(IV)* anion, $K_5[MnV_{11}O_{32}]\cdot10H_2O$, $(NH_4)_{4.5}H_{0.5}[MnV_{11}O_{32}]\cdot12H_2O$ and $Cs_{4.5}H_{0.5}[MnV_{11}O_{32}]\cdot7H_2O$ were isolated as dark red crystalline solids which are moderately stable in solutions of pH 2–3. The *potassium* and *ammonium salts* of the *11-vanado-3-manganate(IV) anion*, namely, $K_3H[Mn_3V_{12}O_{39}]\cdot10H_2O$ and $(NH_4)_5[HMn_3V_{12}O_{39}]\cdot15H_2O$, were obtained as black crystalline solids by recrystallization of the corresponding salts of the 11-vanadomanganate(IV) anion in the pH range 2–3.

No evidence for heteropoly complex formation between vanadium(V) and the metals Cr(III), Co(II), Co(III), Fe(III), Ti(IV), Cu(II), Cu(III), Zn(II), Ag(II), Ag(III), B(III), Al(III), Sn(IV), and Pb(IV) has been reported.

32.4 Heteropolyniobates

Oxidation of solutions of manganese(II) and 6-niobate with peroxydisulfate gives the *12-niobomanganate(IV) anion.*[9] The salts $K_{10}H_2[MnNb_{12}O_{38}]\cdot21H_2O$ and $Na_{12}[MnNb_{12}O_{38}]\cdot50H_2O$ were prepared as bright orange solids. The corresponding *12-tantalomanganate(IV)* anion could not be prepared but a mixed sodium *12-niobotantalomanganate(IV)* was isolated as yellow needles of composition $Na_{12}[Mn(Nb, Ta)_{12}O_{38}]\cdot50H_2O$ where Nb:Ta = 1·36:1.[9] *Sodium 12-niobomanaganate(IV)* was also prepared by the reaction of the managanese(II)–EDTA complex with sodium 6-niobate and hydrogen peroxide.[10] The mixed salt $K_8Na_4[MnNb_{12}O_{38}]\cdot21H_2O$ was also reported.[10] Oxidation of nickel(II) salts with sodium hypobromite in the presence of sodium 6-niobate yielded *sodium 12-niobonickelate(IV)*, a dark maroon solid of composition $Na_{12}[NiNb_{12}O_{38}]\cdot48\text{–}50H_2O$.[10]

---

[2] *G. A. Tsigdinos, C. J. Hallada*, Isopoly Compounds of Molybdenum, Tungsten, and Vanadium, Climax Molybdenum Company Bulletin Cdb-14, February 1969.

[3] *G. Jander, K. F. Jahr*, Kolloid-Beihefte *41*, 297 (1934).

[4] *P. Souchay, S. Dubois*, Ann. Chim. (Paris), [12], *3*, 88 (1948).

[5] *A. Rosenheim, M. Pieck*, Z. Anorg. Allgem. Chemie *98*, 223 (1916).

[6] *M. T. Pope*, Heteropoly Compounds of VB Elements, U.S. Clearinghouse Fed. Sci. Tech. Inform. AD 1970 No. 708 503; C.A. *74*, 18859z (1971).

[7] *C. M. Flynn jr., M. T. Pope*, J. Amer. Chem. Soc. *92*, 85 (1970).

[8] *C. M. Flynn jr., M. T. Pope*, Inorg. Chem. *9*, 2009 (1970).

[9] *B. W. Dale, J. T. Buckley, M. T. Pope*, J. Chem. Soc. *A*, 301 (1969).

[10] *C. M. Flynn jr., G. D. Stucky*, Inorg. Chem. *8*, 332 (1969).

Sodium 6-niobo(ethylenediamine)cobaltate(III), $Na_5[Co(C_2H_8N_2)Nb_6O_{19}]\cdot 18H_2O$ is a greenish-blue solid prepared by boiling a solution of sodium 6-niobate with *trans*-$[Co(en)_2Cl_2]Cl$.[11] When $Cr(en)_3I_3\cdot H_2O$ is employed, *sodium 6-niobo(ethylenediamine)chromate(III)* results, $Na_5[Cr(C_2H_8N_2)Nb_6O_{19}]\cdot 18H_2O$, a purplish-pink powder.[11]

32.5 12-Heteropoly Anions (General Aspects)

The 12-heteropoly anions constitute the bulk of the heteropoly anions thus far isolated and studied. These heteropoly anions can be subdivided into two series.[1] The structure of Series A is built around a central XO_4 tetrahedron and the anion contains 40 oxygen atoms (see Figure 1). Series B is typified by the *12-molybdocerate(IV) anion*, $[CeMo_{12}O_{42}]^{8\ominus}$, in which there are 42 oxygen atoms present within the anion and in which the coordination about the cerium atom is twelve-fold as shown in Figure 2. There are no known 12-heteropolytungstates belonging to Series B. The *12-heteropolymolybdates* reported earlier to contain *manganese* and *boron* as central atoms probably do not exist.[12,13] The *12-heteropolymolybdates* and *12-heteropolytungstates* of Series A all have basicity ($8-n$) where n is the oxidation state of the central atom. Higher basicities that are reported in solution[14,15] probably reflect degradation of the heteropoly anion to simpler species. Both the *12-molybdophosphoric acid* and *12-tungstophosphoric acids*[14–19] have been reported to be heptabasic instead of tribasic. These results arose since both heteropoly anions are extensively hydrolyzed in water, especially in dilute solutions as determined by conductivity and pH measurements of compounds of this type.[20,21] Consequently, salts of this type of heteropoly acids often reported in the literature[18,19,21–23] containing more cations than required by the established basicity of the acid are to be considered impure. However, mixtures of water with oxygen-containing organic solvents impart hydrolytic stability to such *heteropoly anions*.[21]

In the Keggin structure of the 12-heteropolymolybdates and 12-heteropolytungstates, it is possible to substitute atoms such as vanadium or niobium or even other transition elements for molybdenum or tungsten. The tungsten heteropoly species will be described later. The most important molybdenum analogs of this type are presented below.

32.5.1 Molybdovanadophosphate Anions

The *11-molybdo-1-vanadophosphoric acid*, $H_4[PMo_{11}VO_{40}]\cdot 32H_2O$, can be prepared as an orange crystalline solid by ether extraction of acidified solutions of molybdate, metavanadate and phosphate solutions.[24] The acid is a strong electrolyte and has greater hydrolytic stability than 12-molybdophosphoric acid.[20,21] The salts $Na_4[PMo_{11}VO_{40}]\cdot 8H_2O$ and $(NH_4)_3H[PMo_{11}O_{40}]\cdot 7\frac{1}{2}H_2O$ have also been prepared,[24] the former by ion exchange and the latter by precipitation when ammonium chloride is added to solutions of the corresponding free acid.

The *10-molybdo-2-vanadophosphate anion*, $[PMo_{10}V_2O_{40}]^{5\ominus}$, can be prepared by acidification of molybdate, metavanadate, and phosphate solutions and subsequent extraction of the free acid with ether.[24] The orange solid thus obtained, $H_5[PMo_{10}V_2O_{40}]\cdot 34H_2O$, effloresces to a lower hydrate. It can also be prepared by refluxing sodium metavanadate, phosphoric acid and molybdenum trioxide, the free acid being obtained by ether extraction after acidification of the solution.[24] *The ammonium salt*, $(NH_4)_3H_2[PMo_{10}V_2O_{40}]\cdot 7\frac{1}{2}H_2O$, is obtained by addition of ammonium chloride to solutions of the corresponding acid.[24]

The *9-molybdo-3-vanadophosphoric acid*, $H_6[PMo_9V_3O_{40}]\cdot 34H_2O$, can be prepared in a manner similar to that used for the previous two acids although in a much lower yield.[24] The red crystalline acid effloresces slowly to give a lower hydrate. It is the least hydrolytically stable of the three acids in this series.[20]

---

[11] *C. M. Flynn jr., G. D. Stucky*, Inorg. Chem. *8*, 178 (1969).
[12] *P. Souchay, A. Tchakirian*, Ann. Chim. (Paris) *1*, 249 (1946).
[13] *W. C. Schumb, W. H. Hartford*, J. Amer. Chem. Soc. *56*, 2613 (1934).
[14] *E. Matijevic, M. Kerker*, J. Amer. Chem. Soc. *81*, 1307 (1959).
[15] *E. Matijevic, M. Kerker*, J. Amer. Chem. Soc. *62*, 1271 (1958).
[16] *J. B. Goehring, M. Kerker, E. Matijevic, S. Y. Tyree jr.*, J. Amer. Chem. Soc. *81*, 5280 (1959).
[17] *E. Matijevic, M. Kerker*, J. Amer. Chem. Soc. *81*, 5560 (1959).
[18] *W. P. Thistlethwaite*, J. Inorg. Nucl. Chem. *19*, 1581 (1967).
[19] *R. Ripan*, C.R. Acad. Sci., Paris *227*, 474 (1948).
[20] *C. J. Hallada, G. A. Tsigdinos, B. S. Hudson*, J. Phys. Chem. *72*, 4304 (1968).
[21] *G. A. Tsigdinos, C. J. Hallada*, Inorg. Chem. *9*, 2488 (1970).
[22] *Gmelins* Handbuch der Anorganischen Chemie, p. 324, System Nummer 54 (Tungsten), Verlag Chemie, Berlin 1933.
[23] *Gmelins* Handbuch der Anorganischen Chemie, p. 312, System Nummer 53 (Molybdenum), Verlag Chemie, Berlin 1935.
[24] *G. A. Tsigdinos, C. J. Hallada*, Inorg. Chem. *7*, 437 (1968).

Finally, the *potassium* and *ammonium salts* of the anion $[PMo_6V_6O_{38}]^{5\ominus}$ have been prepared but not characterized.[25] The anion may have the Keggin structure and could thus be formulated as $[PMo_6V_6O_{40}H_4]^{5\ominus}$.

32.5.2 The 10-Molybdo-2-niobophosphate Anion

Molybdoniobophosphoric acid has been used in the colorimetric determination of niobium, but the free acid cannot be isolated. Although in a brief report it has been claimed that the free acid has been made[26] it has been shown[27] that the free *10-molybdo-2-niobophosphoric acid* is unstable in aqueous media. However, the free acid can be extracted into butanol or butyl acetate from acidified solutions of molybdate, 6-niobate and phosphate. These yellow solutions are stable up to one hour. The *pyridinium or tetramethylammonium salts* of the *10-molybdo-2-niobophosphate anion*, $(C_5H_5NH)_4H[PMo_{10}Nb_2O_{40}]$ and $[(CH_3)_4N]_4[PMo_{10}Nb_2O_{40}]$, can be precipitated directly out of the organic phase by addition of the corresponding ammonium chlorides.[27]

32.5.3 The 11-Molybdo-1-nickelo(II)silicate Anion and Related Species

First row transition elements may replace molybdenum or tungsten in the Keggin structure but this substitution is far less common in heteropolymolybdates than in heteropolytungstates. *Potassium 11-molybdo-1-nickelosilicate*, $K_6[SiMo_{11}NiO_{40}H_2]\cdot 12H_2O$, can be prepared by treating a solution of 12-molybdosilicic acid with nickel sulfate and potassium acetate.[28] The salt can be recrystallized from water but is unstable towards acids and ion exchange resins in the hydrogen form by the loss of nickel.

Like the nickel analog, *potassium 11-molybdo-1-cobalt(II)-silicate*, $K_6[SiMo_{11}CoO_{40}H_2]\cdot 14H_2O$, can be prepared by treating a solution of 12-molybdosilicic acid with cobalt(II) acetate and potassium acetate.[29] The red-brown solid can be recrystallized from water; the *guanidinium* salt, $(CN_3H_6)_6[SiMo_{11}CoO_{40}H_2]\cdot 7H_2O$, can be prepared by metathesis. The heteropoly anion is extensively decomposed below pH 4·5 or above pH 6. Strong mineral acids and cation-exchange resins in the hydrogen form remove all the cobalt from the anion.[29]

Other 12-triheteropolyanions $[XMo_{11}ZO_{40}H_2]^{n\ominus}$ where X = P, Si, Ge, and Z = $Mn^{2\oplus}$, $Cu^{2\oplus}$, have been reported.[30] The anions could be isolated only as the sodium salts. Other salts were difficult to obtain pure, either because of their solubility or because the anion is unstable. The method of preparation involves partly degrading the $[XMo_{12}O_{40}]^{n\ominus}$ anion to an *11-molybdate heteropoly species* by adjusting the pH of the solution between 4·3–4·5 with potassium acetate or bicarbonate and then adding the desired divalent cation. The sodium salts of the *11-molybdomangano(II)phosphate*, *-silicate*, and *-germanate* were also prepared by similar procedures.[30] These anions are less stable than their tungsten analogs.

32.6 12-Heteropoly Anions: Series A

By far the most important compounds in this series are the *12-molybdophosphoric* and *12-molybdosilic acids* and their tungsten analog.

32.6.1 12-Molybdosilicic Acid and Salts

The 12-molybdosilicic acid, $H_4[SiMo_{12}O_{40}]$, exists in two forms in solution.[31] The α-form (the commonly known form) is produced when there are less than 1·5 equivalents of acid per mol of molybdate in solution during formation. With more than 2 equivalents of acid per mol of molybdate, the β-form is produced.[31] The β-form is more easily reduced than the α-form and changes spontaneously and irreversibly to the α-form over a period of several hours, and has not been isolated from aqueous solutions.

Alpha *12-molybdosilicic acid*, $H_4[SiMo_{12}O_{40}]\cdot xH_2O$, can be prepared as a yellow crystalline solid in quantitative yield by ether extraction of acidified solutions of sodium molybdate and sodium metasilicate.[31] Alternatively, quantitative

---

[25] *P. Souchay, P. Courtin*, C.R. Acad. Sci. Paris, Ser. C, *270*, 1714 (1970).

[26] *J. H. Kennedy*, Abstracts of Papers, p. 13N, American Chemical Society, 142nd Meeting, Atlantic City, September 1962.

[27] *G. A. Tsigdinos*, Climax Molybdenum Company, Unpubl.

[28] *S. A. Malik*, J. Inorg. Nucl. Chem. *32*, 2425 (1970).

[29] *T. J. R. Weakley, S. A. Malik*, J. Inorg. Nucl. Chem. *29*, 2935 (1967).

[30] *C. M. Tourne, G. F. Tourne, S. A. Malik, T. J. Weakley*, J. Inorg. Nucl. Chem. *32*, 3875 (1970).

[31] *J. D. H. Strickland*, J. Amer. Chem. Soc. *74*, 862, 868, 872 (1952).

preparation of this acid has been accomplished by ion exchange[27] from its sodium salt (see below). The solid acid undergoes slow decomposition during storage. Samples develop increasing amounts of insolubles during storage but the ratio of Mo/Si in the soluble portion of such mixture is 12/1.[27]

Beta 12-molybdosilicic acid has been isolated for the first time as a yellow crystalline solid by ion exchange of mixed solvent solutions of the corresponding sodium salt. It is photosensitive thereby easily attaining a green tint.[27] Salts of the β-12-molybdosilicate anion have not thus far been isolated. The sodium salt has been prepared only in solution.[27]

Ammonium 12-molybdosilicate, $(NH_4)_4[SiMo_{12}O_{40}]$, has been prepared directly by acidification of solutions of ammonium paramolybdate and sodium metasilicate.[23] It is insoluble in cold water but it has moderate solubility at higher temperatures.

Sodium 12-molybdosilicate, $Na_4[SiMo_{12}O_{40}]\cdot 14H_2O$, can be prepared directly and in quantitative yield from sodium molybdate, sodium metasilicate and molybdenum trioxide.[27] Concentrated solutions of the salt can be used to prepare the free acid by ion exchange as described above.[27] There is no need to isolate the salt for this purpose.

Although several salts of 12-molybdosilicic acid have been reported in the early literature,[23] considerable doubt exists about their purity. Contrary to claims,[23] $Ag_4[SiMo_{12}O_{40}]$ cannot be prepared by the addition of silver nitrate to solutions of tetrasodium 12-molybdosilicate but only to solutions of the corresponding free acid.[27] Alkaline earth and first row transition elements salts of 12-molybdosilicic acid are best prepared by the reaction of the metal carbonate with 12-molybdosilicic acid.[27,32,33] Several of these salts are highly soluble in water.

32.6.2 12-Molybdophosphoric Acid and Salts

Failure to recognize the hydrolytic instability of *12-molybdophosphoric acid*, $H_3[PMo_{12}O_{40}]$, has lead to the development of incorrect preparative procedures and misleading physical chemical results even in recent years. A commercial form of molybdophosphoric acid labelled as $P_2O_5\cdot 20MoO_3\cdot xH_2O$ is a mixture consisting primarily of 12-acid. No heteropoly species containing Mo/P ratio of 10/1 exists.[34] This composition was incorrectly assumed to be the 11-acid, $H_7[PMo_{11}O_{39}]\cdot xH_2O$, in physicochemical studies of this type of compound.[35] The preparation of the free acid and salts of the 12-molybdophosphate anion should therefore be undertaken with caution.

12-Molybdophosphoric acid can be prepared as a yellow crystalline solid by ether extraction of acidified solutions of sodium molybdate and phosphate.[36] The crude product thus obtained is recrystallized from water at room temperature to yield the desired acid, $H_3[PMo_{12}O_{40}]\cdot 29H_2O$, as large yellow crystals. These readily effloresce to a lower hydrate. Frequently, small amounts of a fine precipitate develop during the early stages of recrystallization due to the hydrolytic decomposition of the acid and should be removed by filtration. Alternately, 12-molybdophosphoric acid can be prepared by boiling molybdenum trioxide in phosphoric acid but the procedure as described in the literature[37,38] yields only a crude product.[27] The pure product is obtained only after recrystallization of the crude acid from water at room temperature.[27]

The *ammonium salt*, $(NH_4)_3[PMo_{12}O_{40}]$, is highly insoluble and can be prepared directly by the careful acidification of ammonium paramolybdate and phosphate with nitric acid.[39]

Soluble salts of the *12-molybdophosphate anion*, like the *sodium salt*, cannot be prepared by ion exchange because of degradation of the heteropoly anion by the resin. However, sodium 12-molybdophosphate, $Na_3[PMo_{12}O_{40}].xH_2O$, can be prepared by the addition of the stoichiometric quantity of sodium bicarbonate to a concentrated solution of 12-molybdophosphoric acid at room temperature and subsequent evaporation (at $\sim 25°$) till crystals are obtained. Solutions of this and other soluble salts of this acid have pH values between 1–2 due to the hydrolytic degradation of the 12-heteropoly anion. The reversibility of this degradation allows the formation of such salts.

Several alkaline earth and transition metal salts of 12-molybdophosphoric acid have been isolated.[32,33,40] These are usually very soluble in

---

[32] *A. Ferrari, L. Cavalca, M. Nardelli*, Gazz. Chim. Ital. *80*, 352 (1950).

[33] *M. Nardelli*, Ric. Sci. *22*, 443 (1952).

[34] *P. Souchay, J. Faucherre*, Bull. Soc. Chim. France *18*, 355 (1951).

[35] *M. C. Baker, P. A. Lyons, S. J. Singer*, J. Phys. Chem. *59*, 1074 (1955).

[36] *H. Wu*, J. Biol. Chem. *43*, 189 (1920).

[37] *A. Linz*, Ind. Eng. Chem. Anal. *15*, 459 (1942).

[38] *D. H. Killefer, A. Linz*, Molybdenum Compounds, p. 87, Interscience Publishers, New York 1952.

[39] *G. Brauer*, Handbook of Preparative Inorganic Chemistry, p. 1698, 2nd Ed., Vol. II, Academic Press, New York 1965. (English Translation.)

[40] *A. Ferrari, L. Cavalca, M. Nardelli*, Gazz. Chim. Ital. *79*, 61 (1949).

water and are best prepared by the addition of the stoichiometric amount of the metal carbonate to a solution of 12-molybdophosphoric acid. Salts of 12-molybdophosphoric acid that are reported to have basicities greater than 3 are either incorrectly formulated or are impure.[21]

32.6.3 12-Molybdogermanates

12-Molybdogermanates are less stable than 12-molybdosilicates. The free *12-molybdogermanic acid*, $H_4[GeMo_{12}O_{40}]\cdot 30H_2O$, has been isolated as a yellow crystalline solid[23] but further characterization of the solid is lacking. Solutions of this acid are stable, but upon neutralisation degradation begins before all four replaceable hydrogen ions are neutralized yielding the *11-molybdogermanate anion*, which is stable in the range of pH 3·5 to 4·2.[41] Salts of this acid reported in the older literature[23] with basicities higher than four are undoubtedly mixtures.

32.6.4 12-Molybdoarsenates

The *12-molybdoarsenate anion*, $[AsMo_{12}O_{40}]^{3\ominus}$, is hydrolyzed readily by water and does not exist in aqueous solutions. Mistaken claims for the presence of such an anion in water[42] were no doubt due to the more stable *18-molybdodiarsenate anion*, $[As_2Mo_{18}O_{62}]^{6\ominus}$. Only the very insoluble ammonium and potassium salts of the 12-molybdoarsenate anion can be prepared from aqueous solutions.[39] Mixed solvents, however, stabilize the 12-molybdoarsenate anion[43,44] but the free acid, $H_3[AsMo_{12}O_{40}]$, reportedly isolated from water dioxane mixtures,[10] was shown to be the salt $Na_2H[AsMo_{12}O_{40}]$ rather than the free acid.[27]

32.6.5 12-Tungstosilicic and 12 Tungstophosphoric Acids and Salts

The *hydrate* $H_4[SiW_{12}O_{40}]\cdot 7H_2O$ can be prepared by an ether route.[45,46] The acid has greater hydrolytic stability than the corresponding molybdenum analog. Salts of this acid are prepared in a manner similar to that for the analogous 12-molybdosilicic acid.[32] The acid is tetrabasic in aqueous solutions. However, the apparent basicity of solid salts is governed by the size of the cations. For example, the compound $K_4[SiW_{12}O_{40}]\cdot 18H_2O$ can be obtained with potassium ion, but cesium forms only the salt $Cs_3H[SiW_{12}O_{40}]$ regardless of the proportions of cesium salt and 12-tungstosilicic acid that are mixed.[47]

12-Tungstophosphoric acid can be prepared as an almost white crystalline solid by ether extraction of acidified solutions of tungstate and phosphate.[36,46] Like its molybdenum analog it is hydrolyzed in water.[1] Samples of the solid acid undergo deterioration upon storage. Salts can be prepared by procedures used for the molybdenum analogs.

32.6.6 12-Tungstogermanates

The *12-tungstogermanate anion*, $[GeW_{12}O_{40}]^{4\ominus}$, is formed by acidification of solutions of sodium germanate and sodium paratungstate. The free acid can be isolated from such solutions by ether extraction.[22,48,49] The colorless crystalline acid is tetrabasic[50] and has the composition H_4-$[GeW_{12}O_{40}]\cdot 30H_2O$. The *sodium salt*, Na_4-$[GeW_{12}O_{40}]\cdot 21H_2O$, has been obtained by neutralization of the acid with sodium carbonate.[48] The 12-tungstogermanate anion is stable below pH 3·6. Above this pH it degrades into the *11-tungstogermanate anion*, this conversion being complete at pH 4·2.[50] The latter is stable up to pH 7·3. The octabasic guanidinium 12-tungstogermanate reported[22,48] is thus not the pure compound.

32.6.7 12-Heteropolytungstates with Transition Metals as Peripheral Atoms

Several heteropolytungstate anions of the type $[XW_{11}ZO_4H_n]^{y\ominus}$ have been prepared and it was shown that in each of these the Z peripheral atom has taken the place of one tungsten atom within a WO_6 octahedron forming part of the Keggin structure of the 12-tungstophosphate anion (see

---

[41] *P. Souchay, A. Tchakirian*, Ann. Chim. (Paris) *1*, 249 (1946).

[42] *W. Kemula, S. Rosolowskii*, Roczniki Chem. *38*, 905 (1964).

[43] *P. Souchay, R. Contant*, C.R. Acad. Sci. Paris, Ser. C *265*, 723 (1967).

[44] *G. A. Tsigdinos*, Electrochemical Properties of Heteropoly Molybdates, Climax Molybdenum Company Bulletin Cdb-15, May 1971.

[45] *W. L. Jolly*, The Synthesis and Characterization of Inorganic Compounds, Prentice Hall, New Jersey 1970.

[46] *E. O. North*, Inorg. Synth. *1*, 129 (1939).

[47] *J. R. de A. Santos*, Proc. Roy. Soc. Ser. A *150*, 309 (1935).

[48] *A. Burkl*, Monatsh. Chem. *56*, 179 (1930).

[49] *R. Schwarz, H. Giese*, Ber. *63* B, 2428 (1930).

[50] *A. Tchakirian, P. Souchay*, Ann. Chim. (Paris) *1*, 232 (1946).

Figure 1).[51] In these anions Z=Co(II) and X=Si, Co(II), and Co(III) (formerly regarded as 12-tungstates).[52] An anion related to metatungstate, Z=Co(III), $X=H_2$, was also reported.[51] In addition, several other heteropolytungstates of the same general formula have been prepared.[29,30] These are:

with X=Si	Z=Co(III), Ni(II), Mn(II), Mn(III), Cu(II)
X=P	Z=Ni(II), Mn(II), Mn(III), Cu(II)
X=Ge	Z=Co(II), Co(III), Ni(II), Mn(II), Mn(III), Cu(II)
X=B or Zn	Z=Mn(II), Mn(III), Cu(II)

Like the corresponding heteropolymolybdates (Section 32.5.3) they are prepared by converting the 12-tungstate into the 11-tungstate with potassium carbonate and then adding the desired transition metal salt. Of these the Co(III) species are the most resistant to acid and base attack and solutions of the free acids have been obtained.[29] Stable free acids of the $[XW_{11}Mn^{3\oplus}O_{40}H_2]^{n\ominus}$ anions can be prepared.[30] All other anions have been isolated as various salts.

32.6.8 Other 12-Heteropolytungstates

Several transition metal atoms form 12-heteropolytungstates in which the metal is present as a tetrahedrally coordinated central atom with the Keggin structure. With cobalt, the *12-tungstocobaltate(II)* and *12-tungstocobaltate(III) anion* are known to exist.[52,53] For example, preparations for the blue-green crystalline compound $K_5H[Co^{2\oplus}W_{12}O_{40}]\cdot 15H_2O$ and the yellow solid $K_5[Co^{3\oplus}W_{12}O_{40}]\cdot 11H_2O$ have been described.[52,53] Stable solutions of the yellow *12-tungstocobaltic(III) acid* have been prepared by ion exchange.[53] *12-Heteropolytungstate anions* with Fe(III), Cr(III), Zn(II), Al(III), and Mn(IV) as central atoms have been prepared and isolated either as the salts or the free acids.[54–57] Some of these are not too stable and are therefore difficult to prepare. Efforts to reproduce the preparation of 12-tungstochromic(III) acid were not successful.[58] *12-Heteropolytungstates* in which the *tungsten* has been substituted in part by *vanadium* have been reported and although the free acids and salts of these have been isolated they have not been studied in detail.[59–61] These are, $H_5[PW_{10}V_2O_{40}]\cdot 32H_2O$, $H_5[AsW_{10}V_2O_{40}]\cdot 30H_2O$, and $H_6[SiW_{10}V_2O_{40}]\cdot 21H_2O$. *Tetraheteropoly acids* like $H_5[PMo_8W_2V_2O_{40}]$ have also been reported but their identity has not been definitely established.[62] A *tungstovanadoselinous acid* of composition $H_6[SeW_{10}V_2O_{40}]\cdot xH_2O$ has been isolated as a red glassy solid and partly characterised.[63] The acid is hexabasic.[63] The free *12-tungstotellurous acid*, $H_4[TeW_{12}O_{40}]\cdot xH_2O$, has been isolated as a yellow crystalline powder. The acid is tetrabasic and is prepared by ether extraction of acidified solutions of sodium tungstate and tellurous acid.[64] Finally, *12-tungstoboric acid* can be isolated, in small yields, as a colorless crystalline solid of composition $H_5[BW_{12}O_{40}]\cdot 65H_2O$.[39]

32.7 12-Heteropoly Anions: Series B

This series of *12-heteropoly compounds* is typified by the anion $[CeMo_{12}O_{42}]^{8\ominus}$, the structure of which has been determined in the salt $(NH_4)_6H_2[CeMo_{12}O_{42}]\cdot 12H_2O$ (see Figure 2).[65] The *12-molybdothorate(IV) anion* typified by the salt $(NH_4)_8[ThMo_{12}O_{42}]\cdot 8H_2O$ may be isostructural with the cerium complex. There are no known tungsten analogs in this series.

Ammonium 12-molybdocerate(IV), $(NH_4)_8[CeMo_{12}O_{42}]\cdot 8H_2O$, is prepared by the addition of ammonium hexanitratocerate(IV) to a boiling

---

[51] *L. C. W. Baker, V. S. Baker, K. Eriks, M. T. Pope, M. Shibata, O. W. Rollins, J. H. Fang, L. L. Koh*, J. Amer. Chem. Soc. *88*, 2329 (1966).

[52] *L. C. W. Baker, T. P. McCutcheon*, J. Amer. Chem. Soc. *78*, 4503 (1956).

[53] *V. E. Simmons*, Ph.D. Thesis, Boston University 1963.

[54] *D. H. Brown, J. A. Mair*, J. Chem. Soc. (London) 3946 (1962).

[55] *D. H. Brown*, J. Chem. Soc. (London) 3181 (1962).

[56] *D. H. Brown*, J. Chem. Soc. (London) 3281 (1962).

[57] *D. H. Brown* J. Chem. Soc. (London) 4408 (1962).

[58] *P. G. Rasmussen*, University of Michigan, Private Communication.

[59] *I. Krivy, J. Krtil*, Collect. Czech. Chem. Commun. *29*, 587 (1964).

[60] *A. I. Kokorin, N. A. Polotebnova*, J. Gen. Chem. (USSR) *27*, 339 (1957).

[61] *V. Kourzhim, A. K. Lavrukhina*, J. Anal. Chem. USSR *15*, 313 (1960).

[62] *A. I. Kokorin*, J. Gen. Chem. (USSR) *27*, 615 (1957).

[63] *N. A. Polotebnova*, Russian J. Inorg. Chem. *10*, 1498 (1965).

[64] *E. S. Ganelina*, Russian J. Inorg. Chem. *1*, 812 (1962).

[65] *D. D. Dexter, J. V. Silverton*, J. Amer. Chem. Soc. *90*, 3509 (1968).

solution of ammonium paramolybdate.[66,67] The yellow crystalline solid is sparingly soluble in water. It can be converted to the acid salt $(NH_4)_6H_2[CeMo_{12}O_{42}]\cdot 12H_2O$ by recrystallization from 2% H_2SO_4.[66] The free acid $H_8[CeMo_{12}O_{42}]$ can be prepared in solution by ion exchange[67,68] and isolated as a yellow glass-like mass by prolonged evaporation in the air.[68] The corresponding salts of the *12-molybdothorate anion* $(NH_4)_8[ThMo_{12}O_{42}]\cdot 8H_2O$ and $(NH_4)_6H_2[ThMo_{12}O_{42}]\cdot 11H_2O$ are prepared by similar routes[69] but the reported preparation of $Na_8[ThMo_{12}O_{42}]\cdot 15H_2O$[69] could not be reproduced.[27] Efforts to prepare and isolate the free 12-molybdothoric acid by ion exchange were not successful.[70]

32.8 6-Heteropoly Compounds of Molybdenum

The 6-heretopoly compounds of molybdenum have been studied extensively and are well understood. Typical examples are the *6-molybdotellurate anion*, $[TeMo_6O_{24}]^{6\ominus}$, and the *6-molybdochromate(III) anion*, $[CrMo_6O_{24}H_6]^{3\ominus}$, the structure of which as determined by X-ray analysis is shown in Figure 4. This class of anions is divided into two series.[1] Series B, typified by the 6-molybdochromate anion, is distinguished from Series A (6-molybdotellurate) in that three molecules of water of constitution are required so that the total number of oxygen atoms within the anion is twenty-four.

32.8.1 Series A: Central Atom: $Te^{6\oplus}$, $I^{7\oplus}$

The general formula $[X^{n\oplus}Mo_6O_{24}]^{(12-n)\ominus}$ describes the anions in this particular series where X is the central atom. Only *salts* of these heteropoly anions have been isolated, the acids being too unstable. These salts are white and soluble in water. The salts $(NH_4)_6[TeMo_6O_{24}]\cdot 7H_2O$ and $(NH_4)_6[TeMo_6O_{24}]\cdot Te(OH)_6\cdot 7H_2O$ are prepared by the reaction of telluric acid, molybdenum trioxide, and ammonium hydroxide.[71] The sodium salt, $Na_6[TeMo_6O_{24}]\cdot 22H_2O$, can be similarly prepared employing sodium hydroxide.[72] The *iodine analog*, $Na_5[IMo_6O_{24}]\cdot 15H_2O$, is prepared from sodium hydrogen periodate, molybdenum trioxide and sodium carbonate.[39] The yellow free *6-molybdoiodic acid*, prepared by ion exchange, evolves gaseous products upon isolation,[39] therefore, probably undergoes some decomposition.

32.8.2 Series B: Central Atom: $Al^{3\oplus}$, $Cr^{3\oplus}$, $Fe^{3\oplus}$, $Co^{3\oplus}$, $Ni^{2\oplus}$, $Ga^{3\oplus}$, $Rh^{3\oplus}$

The heteropolyanions in this series have the general formula $[X^{n\oplus}Mo_6O_{24}H_6]^{(6-n)\ominus}$ where X is the central atom. In general, they are prepared by mixing hot solutions of the paramolybdate anion and a simple salt of the central atom. An oxidizing agent such as hydrogen peroxide or persulfate must be added during the preparation of the Co(III) complex. Solutions of the free acids may be prepared from salts by ion exchange[73,74] but only the *cobalt* and *rhodium*[74] *anions* are not degraded by the ion exchange resin.

Ammonium 6-molybdoaluminate(III) has been prepared as a white crystalline solid, $(NH_4)_3[AlMo_6O_{24}H_6]\cdot 7H_2O$.[75] The corresponding *ammonium 6-molybdochromate(III)* readily forms as a pink crystalline hydrate, $(NH_4)_3[CrMo_6O_{24}H_6]\cdot 7H_2O$.[75] *Ammonium 6-molybdocobaltate(III)*, a green-blue solid of composition $(NH_4)_3[CoMo_6O_{24}H_6]\cdot 7H_2O$, is best prepared by oxidizing a solution of ammonium paramolybdate and cobaltous ion with ammonium persulfate.[76] When hydrogen peroxide is used as the oxidizing agent the dark *green ammonium 10-molybdodicobaltate(III)*, $(NH_4)_6[Co_2Mo_{10}O_{38}H_4]\cdot 8H_2O$, is also formed although in smaller quantities.[74,76] *Ammonium 6-molybdoferrate*, $(NH_4)_3[FeMo_6O_{24}H_6]\cdot 7H_2O$, can be obtained in very pure form only by careful control of the temperature.[73] The *rhodium complex*, $(NH_4)_3[RhMo_6O_{24}H_6]\cdot 7H_2O$, is obtained as a yellow crystalline solid from boiling solutions of

[66] *G. A. Barbieri*, Atti Accad. Naz. Lincei, Classe Sci. Fis. Mat. Natur., Rend. Ser. 5, *23*, Part I, 805 (1914).
[67] *L. C. W. Baker, G. A. Gallagher, T. P. McCutcheon*, J. Amer. Chem. Soc. *75*, 2493 (1953).
[68] *Z. F. Shakhova, S. A. Gavrilova*, J. Inorg. Chem. USSR *8*, 1370 (1958).
[69] *G. A. Barbieri*, Atti Accad. Naz. Lincei, Classe Sci. Fis. Mat. Natur., Rend. Ser. 5, *21*, Part I, 781 (1913).
[70] *Z. F. Shakhova, S. A. Gavrilova, V. F. Zakharova*, Russian J. Inorg. Chem. *7*, 904 (1962).
[71] *V. W. Meloche, W. Woodstock*, J. Amer. Chem. Soc. *51*, 171 (1929).
[72] *S. R. Wood, A. Carlson*, J. Amer. Chem. Soc. *61*, 1810 (1939).
[73] *L. C. W. Baker, G. F. Foster, W. Tan, F. Scholnick, T. P. McCutcheon*, J. Amer. Chem. Soc. *77*, 2136 (1955).
[74] *G. A. Tsigdinos*, Ph.D. Thesis, Boston University 1961.
[75] *R. D. Hall*, J. Amer. Chem. Soc. *29*, 692 (1907).
[76] *C. Friedheim, F. Keller*, Ber. *39*, 4302 (1906).

ammonium paramolybdate and sodium hexachlororhodate.[77] Solutions of *6-molybdorhodic acid* have been prepared by ion exchange and the acid was shown to be tribasic.[74] *Ammonium 6-molybdogallate*, $(NH_4)_3[GaMo_6O_{24}H_6]\cdot 7H_2O$, can be precipitated as a white crystalline solid by the addition of gallium sulfate to ammonium paramolybdate.[78,79] The potassium salt is prepared similarly.[79] Solutions of the 6-molybdogallic acid were prepared by ion exchange, the acid thus prepared was shown to be tribasic.[79] *Ammonium 6-molybdonickelate(II)* can be precipitated out of boiling solutions of ammonium paramolybdate and nickel nitrate as a light blue solid of composition $(NH_4)_4[NiMo_6O_{24}H_6]\cdot 5H_2O$.[80] When solutions of this salt are heated above 55° they turn green due to the hydrolytic degradation of the anion but the process is reversed upon cooling.[74] The potassium salt is prepared similarly employing potassium paramolybdate.[74] The free acid of the 6-heteropolynickelate anion is unstable in solution.[74] The ammonium salts of these heteropoly anions are not too soluble in water (~3 g salt per 100 ml water at 25°) but their more soluble salts can be prepared directly by the reaction of sodium paramolybdate solutions with the corresponding metal nitrates.[74]

32.8.3 Other 6-Heteropolymolybdate Anions

The *ammonium salts* of the *6-triheteropolynickelate(II) anions* corresponding to the general formula $(NH_4)_4[NiMo_{6-n}W_nO_{24}H_6]\cdot 5H_2O$ with $n=0$ to 6 were prepared from boiling solutions of molybdate, tungstate and nickel ions at pH = 6·7.[80] The free acids were also prepared by ion-exchange but these solutions decompose with time.[80] Other 6-heteropolymolybdates of general formula $[X^{n\oplus}Mo_6O_x]_m^{m(2x-36-n)\ominus}$ with central atoms such as $Cu^{2\oplus}$, $Mn^{2\oplus}$, $Co^{2\oplus}$ have been reported.[23] But their composition is indefinite due to their extensive hydrolytic degradation in solution.[74]

---

[77] *G. A. Barbieri*, Atti Accad. Naz. Lincei, Classe Sci. Fis. Mat. Natur., Rend. Ser. 5, *23*, Part I, 338 (1914).

[78] *B. N. Ivanov-Emin, Y. I. Rabovik*, Zh. Neorg. Khim. *3*, 2429 (1958).

[79] *O. W. Rollins, J. E. Earley*, J. Amer. Chem. Soc. *81*, 5571 (1959).

[80] *E. Matijevic, M. Kerker, H. Beyer, F. Thenbert*, Inorg. Chem. *2*, 581 (1963).

32.9 6-Heteropoly Compounds of Tungsten

The 6-heteropoly anions of tungsten are probably structurally related to those of molybdenum[1] and consequently they may have the general formula $[X^{n\oplus}W_6O_{24}H_6]^{(6-n)\ominus}$, however, only $Te^{6\oplus}$, $I^{7\oplus}$ (Series A) and $Ni^{2\oplus}$ and $Ga^{3\oplus}$ (Series B) have been established to act as central atoms. A *guanidinium salt* of the *6-tungstotellurate anion* $(CN_3H_6)_6[TeW_6O_{24}]\cdot 3H_2O$ has been prepared.[22,39] The preparation of the *sodium* and *potassium salts* of the *6-tungstoiodate(VII) anion*, $Na_5[IW_6O_{24}]\cdot 8H_2O$, and $K_5[IW_6O_{24}]\cdot 4H_2O$, has also been described[22] but further work on these compounds is lacking. The *ammonium, potassium*, and *sodium salts* of the *6-tungstonickelate(II) anion*, $Na_4[NiW_6O_{24}H_6]\cdot 16H_2O$, $K_4[NiW_6O_{24}H_6]\cdot 9H_2O$, and $(NH_4)_4[NiW_6O_{24}H_6]\cdot 6H_2O$, have been prepared from boiling solutions of nickel nitrate and paratungstate ion at pH 6·5.[81]

Careful control of reaction conditions are necessary to yield these blue solids in good yield. Description of these compounds in the early literature[22] as green indicates that the nickel atom was outside the sphere of the tungstate anion.[81] A *6-tungstogallate(III) anion* has been reported but details on its preparation are lacking.[82]

32.10 9-Heteropolymolybdates

The 9-heteropolymolybdates are of general formula $[X^{n\oplus}Mo_9O_{32}]^{(10-n)}$ and are built about a central XO_6 octahedron as shown in Figure 5. Tetravalent manganese and tetravalent nickel are the only known central atoms. There are no known corresponding tungstates in this series.[1]

The salt $(NH_4)_6[MnMo_9O_{32}]\cdot 6H_2O$ can be isolated as orange red crystals. It is prepared by the careful oxidation of manganous ion by persulfate in the presence of ammonium paramolybdate.[83] The anion is susceptible to hydrolytic degradation which may lead to the formation of impure material in spite of the excellent appearance of the crystals obtained.[74] Thus, compounds of this type containing various ratios of manganese to molybdenum reported in the early literature[23] have been shown to be mixtures.[74] The very soluble sodium salt of the

---

[81] *U. C. Agarwala*, Thesis, Boston University 1960.

[82] *O. W. Rollins*, J. Inorg. Nucl. Chem. *33*, 75 (1971).

[83] *L. C. W. Baker, T. J. R. Weakley*, J. Inorg. Nucl. Chem. *28*, 447 (1966).

9-molybdomanganate(IV) anion has been prepared in solution and the heteropoly anion thus obtained shown to be monomeric.[83] Unstable $H_6[MnMo_9O_{32}]$ was prepared in solution by ion exchange.[83]

32.11 Dimeric 9-Heteropoly Compounds (2:18 Series)

The *9-heteropolymolybdates* and *9-heteropolytungstates* are binuclear complexes containing two central XO_4 tetrahedra surrounded by $18MoO_6$ or $18WO_6$ octahedra as shown in Figure 6. They are of general formula $[X_2^{n\oplus}$-(Mo or W)$_{18}O_{62}]^{(16-2n)\ominus}$ where X is the central atom $P^{5\oplus}$ or $As^{5\oplus}$. The structure of these anions shown in Figure 6 has been characterized in the salt $K_6[P_2W_{18}O_{62}]\cdot 14H_2O$.[84] The free acids of the molybdenum anions and their salts are all bright yellow. The analogous tungstates are light yellow. The acids are 6-basic. *Guanidinium, cesium, silver*, and *thallium salts* of different basicities have sometimes been reported,[23] but these are almost certainly mixtures.[1] The preparation of a *9-tungstoberyllic(II) acid* has also been reported.[85] Although the free acid is difficult to isolate, the *guanidinium* and *lead salts* of this acid have been formulated as $3(CN_3H_6)_2O\cdot BeO\cdot 9WO_3\cdot 3H_2O$ and $3PbO\cdot BeO\cdot 9WO_3\cdot 35H_2O$.[85] The *9-tungstoberyllate(II) anion* may belong to the above mentioned series with the formula $[Be_2W_{18}O_{62}]^{12\ominus}$.

32.11.1 9-Heteropolymolybdates of $P^{5\oplus}$ and $As^{5\oplus}$

Ammonium 18-molybdodiphosphate, $(NH_4)_6$-$[P_2Mo_{18}O_{62}]\cdot 14H_2O$ can be prepared from highly acidified solutions of molybdate and phosphate by prolonged boiling.[36,38] The soluble ammonium salt can be subsequently isolated by the addition of ammonium chloride. It can be purified either by careful recrystallization or by purification from water-dioxane.[74,86,87] It can revert to the highly insoluble $(NH_4)_3$-$[PMo_{12}O_{40}]$ upon warming of its solutions. The free acid $H_6[P_2Mo_{18}O_{62}]\cdot 11H_2O$ can be prepared by ether extraction and subsequently isolated as orange crystals.[36,38] Solutions of this acid have also been prepared by ion exchange and the acid shown to be hexabasic.[74] The corresponding *18-molybdodiarsenic acid*, $H_6[As_2$-$Mo_{18}O_{62}]\cdot 25H_2O$, has been also isolated.[23,39] However, isolation of this acid may lead to some decomposition due to its greater hydrolytic instability than the phosphorus analog. Solutions of the 18-molybdodiarsenic acid were prepared by ion exchange.[74] The acid was shown to be hexabasic at 15° but at 25° it behaves as heptabasic due to some hydrolytic degradation.[74]

32.11.2 9-Heteropolytungstates of $P^{5\oplus}$ and $As^{5\oplus}$

The *18-tungstodiphosphate anion*, $[P_2W_{18}O_{62}]^{6\ominus}$ exists in two forms. The alpha or symmetrical form and the beta or unsymmetrical form. The latter is shown in Figure 6. The corresponding *heteropolymolybdates* exist only in the beta form. The ammonium salts of both forms have been isolated from acidified solutions of tungstate and phosphate upon prolonged boiling.[36] Crystals of the alpha form thus produced are hexagonal, less soluble and paler in color than those of the more soluble beta form which are triclinic.[36] The free acids of these forms have been prepared by ether extraction and isolated.[36] The ammonium salt of the *18-tungstodiarsenate anion* $(NH_4)_6[As_2W_{18}O_{62}]\cdot 14H_2O$ has been prepared as well as the free acid.[39] The preparation of the salt $(NH_4)_6[As_2W_{18}O_{62}]\cdot 3\frac{1}{2}C_4H_8O_2\cdot 17H_2O$ has also been described.[86] These compounds are prepared by procedures similar to those used for the preparation of the phosphorus analogs.

32.11.3 Heteropolytungstates (2:18 Series) with Transition Metals as Peripheral Atoms

Like the 12-heteropolytungstate anions which are capable of substituting one tungsten atom for a transition element (see Section 32.6.7), heteropolytungstates of the 2:18 series are capable of undergoing similar substitutions. The anions $[X_2W_{18}O_{62}]^{6\ominus}$ (X=P or As) have structures related to the 12-heteropolytungstates (Figure 1),[84] they undergo partial decomposition to X_2W_{17} heteropoly anions at pH ~7, and are therefore capable of substituting a transition element in the lost site.

New *17-tungstate heteropolyanions*, $[X_2W_{17}$-$ZO_{62}H_2]^{8\ominus}$ (X=P or As; Z=$Mn^{2\oplus}$, $Co^{2\oplus}$, $Ni^{2\oplus}$, or $Cu^{2\oplus}$), were prepared when the pH of a solution containing $[X_2W_{18}O_{62}]^{6\ominus}$ at 80° was raised to 6 with aqueous potassium acetate in the presence of the $Z^{2\oplus}$ cation. The element Z was

[84] *B. Dawson*, Acta Crystallogr. *6*, 113 (1953).
[85] *D. H. Brown*, J. Inorg. Nucl. Chem. *14*, 129 (1960).
[86] *M. T. Pope, E. Papaconstantinou*, Inorg. Chem. *6*, 1147 (1967).
[87] *E. Papaconstantinou, M. T. Pope*, Inorg. Chem. *6*, 1152 (1967).

always found in the anion. These anions were isolated as the potassium, rubidium, guanidinium, or dimethylammonium salts.[30,88,89] The anions containing manganese(II) can be oxidized either electrolytically or by persulfate to the Mn(III) species which remains within the anion.[89] Anions containing cobalt(III) are prepared similarly.[89] These new heteropoly anions are stable in aqueous solution and are decomposed slowly by base at 25°. A cation exchange resin in the hydrogen form completely removes the divalent Z atoms from the anions. However, the free acids of anions with Mn(III) and Co(III) can be prepared by ion exchange.[89] The acids are heptabasic.

32.12 Preparation of Reduced Species of Heteropoly Anions (Heteropolyblues)

The investigation of the oxidation-reduction behavior of heteropoly compounds in aqueous and nonaqueous solvents has received increasing attention since knowledge from such an investigation has lead to useful preparative electrochemical procedures for preparing reduced forms of *heteropoly anions* (heteropolyblues). Information of this type has been obtained polarographically.[44] In particular, cyclic voltammetry and alternating current polarography have been applied to elucidate the oxidation-reduction properties of heteropoly compounds.[44]

32.12.1 12-Heteropolymolybdates and -tungstates

Most of the earlier polarographic work on heteropoly compounds has been restricted to the reduction of the *12-molybdophosphate* and *12-molybdosilicate anions* in aqueous solution. These anions, and especially the phosphorus one, are particularly susceptible to hydrolytic degradation and it is doubtful that the reduced species obtained from this anion from aqueous solutions are pure. However, the two and four electron reduction products of 12-molybdosilicic acid, $H_6[SiMo_{12}O_{40}]\cdot 6H_2O$ and $H_8[SiMo_{12}O_{40}]\cdot 6H_2O$, have been prepared by controlled potential electrolysis as stable blue solids.[90] The polarographic behavior of 12-heteropolymolybdates in water-dioxane solutions has been examined in detail.[44] 12-heteropolytungstates are less powerful oxidizing agents than the corresponding molybdates and their reduction behavior has been studied in detail in solution.[91,92] No reduced heteropolytungstate species have been isolated in the solid form.

32.12.2 Dimeric 9-Heteropolymolybdates (2:18 Series)

The first definitive investigation of a *heteropoly* blue compound involved the reduction of ammonium 18-molybdodiphosphate, $(NH_4)_6$-$[P_2Mo_{18}O_{62}]$, in acidic solutions using various reducing agents.[36] The blue solids obtained were shown to be the two, four, and six-electron reduction products of the $[P_2Mo_{18}O_{62}]^{6\ominus}$ anion. These reduced species have also been obtained by electrolytic reduction and crystalline samples of the ammonium salts and solutions of the free acids of the three reduced 18-molybdodiphosphates have been characterized.[87] Unlike the oxidized forms, the reduced forms have considerable stability in alkaline solutions. Titration of the free acids of the blues prepared by ion exchange showed the expected basicities for $H_8[P_2Mo_{18}O_{62}]$ and $H_{10}[P_2Mo_{18}O_{62}]$, but the last two protons in $H_{12}[P_2Mo_{18}O_{62}]$ could not be replaced before the anion decomposed.[87] The reduction of the anions $[As_2Mo_{18}O_{62}]^{6\ominus}$, α-$[P_2W_{18}O_{62}]^{6\ominus}$, β-$[P_2W_{18}O_{62}]^{6\ominus}$, and $[As_2W_{18}O_{62}]^{6\ominus}$ has been carried out[86,87] but the isolation of reduced forms of these anions has not been reported.[86,87] The oxidation-reduction behavior of the 6-heteropolymolybdates (see Section 32.8) has also been reported[44] but it was shown that the reduced species of these anions are unstable thus precluding their isolation.

32.13 Miscellaneous Heteropoly Compounds

There are a number of heteropoly compounds that either do not belong to any of the above categories mentioned or at present have not been studied in detail. However, some of these have been sufficiently characterized to justify inclusion in the present context.

---

88 *S. A. Malik, T. J. R. Weakley*, Chem. Commun. 1094 (1967).

89 *S. A. Malik, T. J. R. Weakley*, J. Chem. Soc. *A*, 2647 (1968).

90 *J. Amiel, P. Rabette, D. Olivier*, C.R. Acad. Sci., Paris, Ser. C, *267*, 1703 (1968).

91 *M. T. Pope, G. M. Varga jr.*, Inorg. Chem. *5*, 1249 (1966).

92 *G. M. Varga jr., E. Papaconstantinou, M. T. Pope*, Inorg. Chem. *9*, 662 (1970).

32.13.1 The 10-molybdodicobaltate(III) Anion

The *ammonium salt* of this anion has been prepared entirely free from its related ammonium 6-molybdocobaltate(III) (Section 32.8.2) by the oxidation of a solution of cobaltous ion and paramolybdate by hydrogen peroxide in the presence of activated charcoal.[74, 93] The compound has the composition $(NH_4)_6[Co_2Mo_{10}O_{38}H_4]\cdot 5H_2O$. In the absence of charcoal, a mixture of this species and $(NH_4)_3[CoMo_6O_{24}H_6]\cdot 7H_2O$ (the latter in the greater quantity) results.[74, 76] Activated charcoal decomposes the 6-molybdocobaltate(III) anion but not the 10-molybdodicobaltate species.[74] The 10-molybdodicobaltate(III) anion because of its structure[94] can be resolved into optical antipodes. It is the first wholly inorganic species to have been thus resolved.[95] Both optically active isomers were isolated.[95]

32.13.2 Other Heteropoly Anions

Heteropolytungstates of lanthanide elements have been recently synthesized.[96] Salts of the anions $[L^{3\oplus}W_{10}O_{35}]^{7\ominus}$ where L=La, Ce, Pr, Nd, Ho, Er, Yb, and Y and of the anion $[Ce^{4\oplus}W_{10}O_{35}]^{6\ominus}$ have been prepared and analyzed. All the anions are stable in the pH range 5·5–8·5. Stable solutions of their free acids cannot be obtained.[96] The reported 8-tungstocerate(III or IV) anions[97–99] have been shown to be *10-tungstocerates*.[96] Lanthanide cations react rapidly with the heteropolyanions $[SiW_{11}O_{40}]^{8\ominus}$, $[PW_{11}O_{39}]^{7\ominus}$, and $[P_2W_{17}O_{61}]^{10\ominus}$ to give new heteropoly complexes several salts of which have been isolated.[96] Mixed heteropoly compounds containing both molybdenum and tungsten that correspond to the composition $H_3[PMo_nW_mO_{40}]$ and $H_4[SiMo_nW_mO_{40}]$, where $n+m=12$ have been reported.[100–102] These were isolated as crystalline materials, but further work would be required to establish their identity.

32.14 General Literature on Heteropoly Compounds

The footnotes throughout the chapter are references to literature dealing with aspects of heteropoly compounds, and the related area of isopoly compounds. Although most of this literature is recent, in several cases it is uncritically written both from the preparative and interpretative viewpoint. However, it contains valuable information and cannot be ignored.

A review of the heteropoly compounds of molybdenum, tungsten, and vanadium up to the end of 1968 (with 190 references) is given in Ref. 1. The isopoly compounds of molybdenum, tungsten, and vanadium are reviewed (up to the end of 1967) similarly in Ref. 2 (with 160 references). Experimental work dealing with the electrochemical behavior of heteropoly compounds which contain valuable references in this area is given in Refs 44, 86, 87, 91, 92, and 103. A detailed review containing 862 references on the properties in solution of 12-heteropolymolybdates with emphasis on the analytical aspects is also available.[104] The properties in solution with emphasis on the nature of the species and equilibria present in such media is given in Ref. 105. Several laboratory experiments describing the preparation of some heteropoly compounds designed for students are prescribed in Ref. 106. General information on heteropoly anions may be found in Ref. 107. Specific references to the preparation of heteropoly compounds are given in Refs 22, 23, and 39 but all these are not critical. Sound preparative procedures for preparing the 1:12 and 2:18 heteropolymolybdates and -tungstates are outlined in Ref. 36. An extensive review of the literature on heteropoly compounds may also be found in Ref. 108.

---

93 *G. A. Tsigdinos, M. T. Pope, L. C. W. Baker*, Abstracts of Papers Presented before the Division of Inorganic Chemistry, p. 48M, American Chemical Society National Meeting, Boston, Mass. 1959.

94 *H. T. Evans, J. S. Showell*, J. Amer. Chem. Soc. *91* 6881 (1969).

95 *T. Ama, J. Hidaka, Y. Shimura*, Bull. Chem. Soc., Japan *43*, 2654 (1970).

96 *R. D. Peacock, T. J. R. Weakley*, J. Chem. Soc. *A*, 1836 (1971).

97 *R. Ripan, I. Todorut*, Roczniki Chem. *38*, 1587 (1964).

98 *R. Ripan, I. Todorut*, Rev. Roumaine-Chem. *11*, 691 (1966).

99 *R. Ripan, I. Todorut*, Rev. Roumaine-Chim. *11*, 1279 (1966).

100 *A. I. Kokorin, N. A. Polotebnova*, Zh. Obshch. Khim. *26*, 3 (1956).

101 *N. A. Polotebnova, Y. L. Neimark*, Russian J. Inorg. Chem. *13*, 1583 (1968).

102 *B. E. Reznik, G. M. Ganzburg, G. V. Mal'tseva, V. P. Stotsenko*, Russian J. Inorg. Chem. *15*, 1444 (1970).

103 *R. Massart, G. Hervé*, Rev. Chim. (Bucharest) 5, 501 (1968).

104 *M. Jean*, Chim. Anal. (Paris) *44*, 195, 243 (1962).

105 *P. Souchay*, Polyanions et Polycations, p. 49, Cauthier-Villars Editeur, Paris 1963.

106 *G. B. Kauffman, P. F. Vartamian*, J. Chem. Educ. *47*, 213 (1970).

107 *P. G. Rasmussen*, J. Chem. Educ. *44*, 726 (1967).

108 *L. Malaprade* in *P. Pascal* (Editor), Nouveau Traité de Chimie Minérale, p. 903, Vol. 14, Masson et Cie, Paris 1959.

Index

Note
Problems arise in indexing chemical names in that the first letter of a name may be useless in locating the substance. Thus while a name such as *zinc chloride* would be sought under *Z*, a name such as *potassium chlorozincate(II)* would not be sought under *P*, but would be expected under *Zinc*, and more especially under *Zinc(II)*. To accommodate this, such compounds will be found indented under the parent metals. Similarly a name such as *Alkylzinc bromide* would not be sought under *A* but again under *Zinc*, and will be rendered as *Zinc bromide, alkyl-*.
Further names such as Dimethylgallium will be found as Gallium, dimethyl-.
The general order is: Element
Element compounds
Element prefixes
Element subcompounds
Thus one might get
Copper, metallocenes involving
production and purification,
Copper acetylide,
complexes with Group V elements,
Copper, dimethyl-,
pentabromodiphenyl-,